THE HANDBOOK OF Formulas and Tables for Signal Processing

The Electrical Engineering Handbook Series

Series Editor
Richard C. Dorf
University of California, Davis

Titles Included in the Series

The Avionics Hanbook, Cary R. Spitzer
The Biomedical Engineering Handbook, Joseph D. Bronzino
The Circuits and Filters Handbook, Wai-Kai Chen
The Communications Handbook, Jerry D. Gibson
The Control Handbook, William S. Levine
The Digital Signal Processing Handbook, Vijah K. Madisetti & Douglas Williams
The Electrical Engineering Handbook, Richard C. Dorf
The Electric Power Engineering Handbook, L.L. Grigsby
The Electronics Handbook, Jerry C. Whitaker
The Engineering Handbook, Richard C. Dorf
The Handbook of Formulas and Tables for Signal Processing, Alexander D. Poularikas
The Industrial Electronics Handbook, J. David Irwin
Measurements, Instrumentation, and Sensors Handbook, J. Michael Golio
The Mechanical Systems Design Handbook, Osita D.I. Nwokah
The Microwave Engineering Hanbook, J. Michael Golio
The Mobile Communications Handbook, Jerry D. Gibson
The Ocean Engineering Handbook, Ferial El-Hawary
The Technology Management Handbook, Richard C. Dorf
The Transforms and Applications Handbook, Alexander D. Poularikas
The VLSI Handbook, Wai-Kai Chen

THE HANDBOOK OF Formulas and Tables for Signal Processing

Alexander D. Poularikas
Department of Electrical and Computer Engineering
The University of Alabama in Huntsville

CRC Press
Taylor & Francis Group
Boca Raton London New York

CRC Press is an imprint of the
Taylor & Francis Group, an **informa** business

CRC Press
Taylor & Francis Group
6000 Broken Sound Parkway NW, Suite 300
Boca Raton, FL 33487-2742

First issued in paperback 2019

ISBN-13: 978-0-8493-8579-7 (hbk)
ISBN-13: 978-0-367-40031-6 (pbk)

Library of Congress Cataloging-in-Publication Data

Poularikas, Alexander D., 1933-
The handbook of formulas and tables for signal processing / by Alexander D. Poularikas
p. cm. — (Electrical engineering handbook series)
Includes index.
ISBN 0-8493-8579-2 (alk. paper)
1. Signal processing—Handbooks, manuals, etc. I. Title.
II. Series.
TK5102.9.P677 1998
621.382'2--dc21
98-10347
CIP

Library of Congress Card Number 98-10347

Visit the Taylor & Francis Web site at
http://www.taylorandfrancis.com

and the CRC Press Web site at
http://www.crcpress.com

About The Author

Alexander D. Poularikas is a professor in the Department of Electrical and Computer Engineering at the University of Alabama in Huntsville. He received a B.S. degree in Electrical Engineering in 1960, an M.S. degree in Physics in 1963, and a Ph.D. in 1965, all at the University of Arkansas, Fayetteville.

He has held positions as assistant, associate, and professor at the University of Rhode Island (1965–1983), professor and Chairman of the Engineering Department at the University of Denver (1983–1985), and professor (1985–) and Chairman (1985–1989) at the University of Alabama in Huntsville. Dr. Poularikas was a visiting scientist at MIT (1971–1972), and summer faculty fellow at NASA (1968, 1972), at Stanford University (1966), and at Underwater Systems Center (1971, 1973, 1974).

He has coauthored the books *Electromagnetics* (Marcel Dekker, 1997), *Electrical Engineering: Introduction and Concepts* (Matrix Publishers, 1982), *Workbook for Electrical Engineers* (Matrix Publishers, 1983), *Signals and Systems* (Brooks/Cole, 1985), *Elements of Signals and Systems* (PWS-KENT, 1987), and *Signals and Systems* (2nd edition) (PWS-KENT, 1992). He is Editor-in-Chief for the books *Transforms and Applications Handbook* (CRC Press, 1995) and *Handbook of Formulas and Tables for Signal Processing* (CRC Press, 1999).

Dr. Poularikas is a senior member of the IEEE, was a Fulbright scholar and was awarded the Outstanding Educator's Award by the IEEE Huntsville Section in 1990 and 1996. His main interest is in the area of signal processing.

PREFACE

The purpose of *The Handbook of Formulas and Tables for Signal Processing* is to include in a single volume the most important and most useful tables and formulas that are used by engineers and students involved in signal processing. This includes deterministic as well as statistical signal processing applications. The handbook contains a large number of standard mathematical tables, so it can also be used as a mathematical formulas handbook.

The handbook is organized into 45 chapters. Each contains tables, formulas, definitions, and other information needed for the topic at hand. Each chapter also contains numerous examples to explain how to use the tables and formulas. Some of the figures were created using MATLAB and MATHEMATICA.

The editor and CRC Press would be grateful if readers would send their opinions about the handbook, any error they may detect, suggestions for additional material for future editions, and suggestions for deleting material.

The handbook is testimony to the efforts of colleagues whose contributions were invaluable, Nora Konopka, Associate Editor at CRC Press, the commitment of the Editor-in-Chief of the series, Dr. Richard Dorf, and others. Special thanks go to Dr. Yunlong Sheng for contributing Chapter 42.

Alexander D. Poularikas
Huntsville, Alabama
July 1998

CONTENTS

1

Fourier Series

1.1 Definitions and Series Formulas

1.1.1 A function is ***periodic*** if $f(t) = f(t + nT)$, where n is an integer and T is the period of the function.

1.1.2 The function $f(t)$ is ***absolutely integrable*** if $\int_a^b |f(t)|\,dt < \infty$.

1.1.3 An ***infinite series*** of function

$$f_1(t) + f_2(t) + \cdots + f_k(t) + \cdots = \sum_{k=1}^{\infty} f_k(t)$$

converges at a given value of t if its ***partial sums***

$$s_n(t) = \sum_{k=1}^{\infty} f_k(t), \quad (n = 1,2,3,\cdots)$$

have a finite limit $s(t) = \lim_{n\to\infty} s_n(t)$.

1.1.4 The series in 1.1.3 is ***uniformly convergent*** in [a,b] if, for any positive number ε, there exists a number N such that the inequality $|s(t) - s_n(t)| \le \varepsilon$ holds for all $n \ge N$ and for all t in the interval [a,b].

1.1.5 ***Complex form*** of the series:

$$f(t) = \sum_{n=-\infty}^{\infty} \alpha_n e^{jn\omega_o t} = \sum_{n=-\infty}^{\infty} |\alpha_n| e^{j(n\omega_o t + \varphi_n)}, \quad t_o \le t \le t_o + T$$

$$\alpha_n = \frac{1}{T}\int_{t_o}^{t_o+T} f(t)e^{-jn\omega_o t}\,dt, \qquad \omega_o = \frac{2\pi}{T}, \quad T = \text{period}$$

$$\alpha_n = |\alpha_n| e^{j\varphi_n} = |\alpha_n|\cos\varphi_n + j|\alpha_n|\sin\varphi_n, \qquad \alpha_{-n} = \alpha_n^*, \; t_o = \text{any real value.}$$

1.1.6 ***Trigonometric form*** of the series

$$f(t) = \frac{A_o}{2} + \sum_{n=1}^{\infty}(A_n\cos n\omega_o t + B_n \sin n\omega_o t), \quad A_o = 2\alpha_o = \frac{2}{T}\int_{t_o}^{t_o+T} f(t)\,dt$$

$$A_n = (\alpha_n + \alpha_n^*) = \frac{2}{T}\int_{t_o}^{t_o+T} f(t)\cos n\omega_o t\,dt, \quad B_n = j(\alpha_n - \alpha_n^*) = \frac{2}{T}\int_{t_o}^{t_o+T} f(t)\sin n\omega_o t\,dt$$

$$f(t) = \frac{A_o}{2} + \sum_{n=1}^{\infty} C_n\cos(n\omega_o t + \varphi_n), \quad C_n = (A_n^2 + B_n^2)^{1/2}, \; \varphi_n = -\tan^{-1}(B_n / A_n)$$

1.1.7 ***Parseval's*** formula

$$\frac{1}{T}\int_{t_o}^{t_o+T} |f(t)|^2\,dt = \sum_{n=-\infty}^{\infty}|\alpha_n|^2 = \frac{A_o^2}{4} + \sum_{n=1}^{\infty}\left(\frac{A_n^2}{2} + \frac{B_n^2}{2}\right) = \frac{A_o^2}{4} + \sum_{n=1}^{\infty}\frac{C_n^2}{2}$$

1.1.8 ***Sum*** of cosines

$$\frac{1}{2} + \cos t + \cos 2t + \cdots + \cos nt = \frac{\sin(n+\frac{1}{2})t}{2\sin\frac{t}{2}}$$

1.1.9 ***Truncated*** Fourier series

$$f_N(t) = \frac{A_o}{2} + \sum_{n=1}^{N}(A_n\cos n\omega_o t + B_n\sin n\omega_o t)$$

$$= \frac{1}{T}\int_{-T/2}^{T/2} f(v)\frac{\sin\left[(2N+1)\omega_o \frac{t-v}{2}\right]}{\sin\left[\omega_o \frac{t-v}{2}\right]}\,dv$$

1.1.10 ***Sum*** and ***difference*** functions

$$p(t) = C_1 f(t) \pm C_2 h(t) = \sum_{n=-\infty}^{\infty}[C_1\beta_n \pm C_2\gamma_n]e^{jn\omega_o t} = \sum_{n=-\infty}^{\infty}\alpha_n\, e^{jn\omega_o t},$$

C_1 = constant, C_2 = constant, β_n = Fourier expansion coefficients of $f(t)$, γ_n = Fourier expansion coefficients of $h(t)$, $\alpha_n = C_1\beta_n \pm C_2\gamma_n$, $f(t)$ and $h(t)$ are periodic with same period.

1.1.11 ***Product*** of two functions

$$p(t) = f(t)h(t) = \sum_{n=-\infty}^{\infty}\sum_{m=-\infty}^{\infty}(\beta_{n-m}\gamma_m)e^{jn\omega_o t} = \sum_{n=-\infty}^{\infty}\alpha_n\, e^{jn\omega_o t}$$

$$\alpha_n = \frac{1}{T}\int_{-T/2}^{T/2} f(t)h(t)e^{-jn\omega_o t}\,dt = \sum_{m=-\infty}^{\infty}(\beta_{n-m}\gamma_m)$$

β_n = Fourier expansion coefficients of $f(t)$, γ_n = Fourier expansion coefficients of $h(t)$, $f(t)$ and $h(t)$ are periodic with same period.

1.1.12 ***Convolution*** of two functions

$$g(t) = \frac{1}{T}\int_{-T/2}^{T/2} f(\tau)h(t-\tau)\,d\tau = \sum_{n=-\infty}^{\infty}\alpha_n\, e^{jn\omega_o t} = \sum_{n=-\infty}^{\infty}\beta_n\gamma_n e^{jn\omega_o t}$$

$\alpha_n = \beta_n\gamma_n$. β_n = Fourier expansion coefficients of $f(t)$, γ_n = Fourier expansion coefficients of $h(t)$, $f(t)$, and $h(t)$ are periodic with same period.

1.1.13 If $H(\omega)$ (***transfer function***) is the Fourier transform of the ***impulse response*** $h(t)$ of a linear time invariant system (LTI), then its output due to a periodic input function $f(t)$ is

$$y(t) = \frac{A_o}{2}H(0) + \sum_{n=1}^{\infty}\left|H(n\omega_o)\right|\,[A_n\cos[n\omega_o t + \varphi(n\omega_o)] + B_n\sin[n\omega_o t + \varphi(n\omega_o)]]$$

$$H(n\omega_o) = H_r(n\omega_o) + jH_i(n\omega_o) = [H_r^2(n\omega_o) + H_i^2(n\omega_o)]^{1/2}e^{j\varphi(n\omega_o)}$$

$$\varphi(n\omega_o) = \tan^{-1}[H_i(n\omega_o)/H_r(n\omega_o)]$$

$H_r(\cdot)$ and $H_i(\cdot)$ are real functions.

1.1.14 Lanczos ***smoothing*** factor

$$f_N(t) = \frac{A_o}{2} + \sum_{n=1}^{N}\frac{\sin(n\pi/N)}{n\pi/N}\,[A_n\cos n\omega_o t + B_n\sin n\omega_o t]$$

where A_0, A_n, and B_n are the trigonometric expansion Fourier series coefficients (see 1.1.6).

1.1.15 Fejé ***smoothing*** series

$$f_N(t) = \frac{A_o}{2} + \sum_{n=1}^{N}\frac{N-n}{N}\,[A_n\cos n\omega_o t + B_n\sin n\omega_o t]$$

where A_0, A_n, and B_n are the trigonometric expansion Fourier series coefficients (see 1.1.6).

1.1.16 ***Transformation*** from 2ℓ to 2π

If the period is 2ℓ, then the Fourier series of f(t) is

$$f(t) = \frac{A_o}{2} + \sum_{k=1}^{N}\left[A_k\cos\frac{\pi k t}{\ell} + B_k\sin\frac{\pi k t}{\ell}\right]$$

If we set $\pi t/\ell = x$ or $t = x\ell/\pi$, we obtain the equivalent series

$$\varphi(x) = f\left(\frac{x\ell}{\pi}\right) = \frac{a_o}{2} + \sum_{k=1}^{\infty} [a_k \cos kx + b_k \sin kx]$$

The above means: If $f(t)$ has period 2ℓ, then $\varphi(x) = f(x\ell/\pi)$ has a period 2π.

$$a_k = \frac{1}{\pi}\int_{-\pi}^{\pi} f(x)\cos kx\,dx \qquad k = 0,1,2,\cdots$$

$$b_k = \frac{1}{\pi}\int_{-\pi}^{\pi} f(x)\sin kx\,dx \qquad k = 1,2,\cdots$$

1.1.17 Table of Fourier Series Expansions

1. $f(t) = \frac{1}{\pi}\sum_{n=1,3,5\cdots} \frac{1}{n}\sin\frac{n\pi t}{L}$

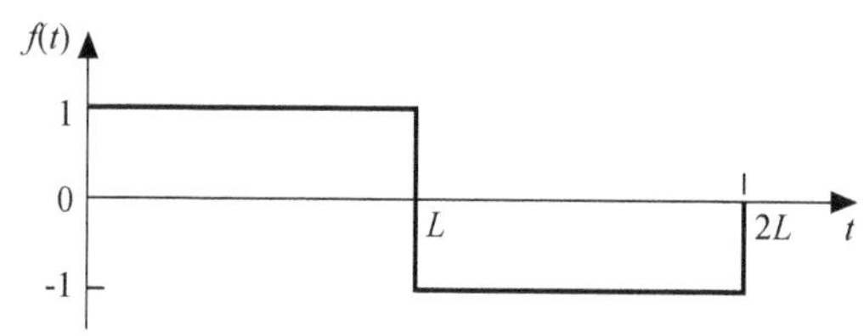

2. $f(t) = \frac{2}{\pi}\sum_{n=1}^{\infty} \frac{(-1)^n}{n}\left(\cos\frac{n\pi c}{L} - 1\right)\sin\frac{n\pi t}{L}$

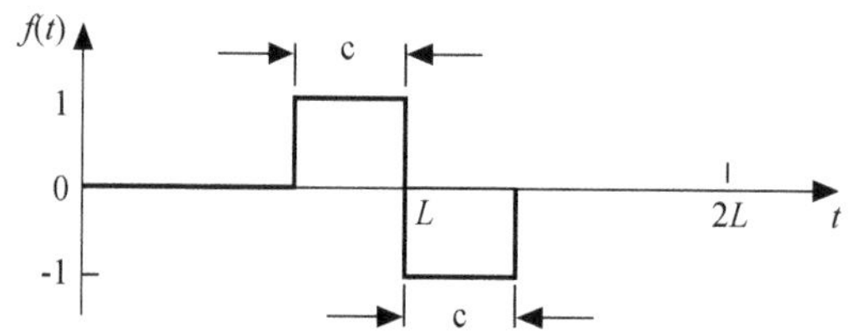

3. $f(t) = \frac{c}{L} + \frac{2}{\pi}\sum_{n=1}^{\infty} \frac{(-1)^n}{n}\sin\frac{n\pi c}{L}\cos\frac{n\pi t}{L}$

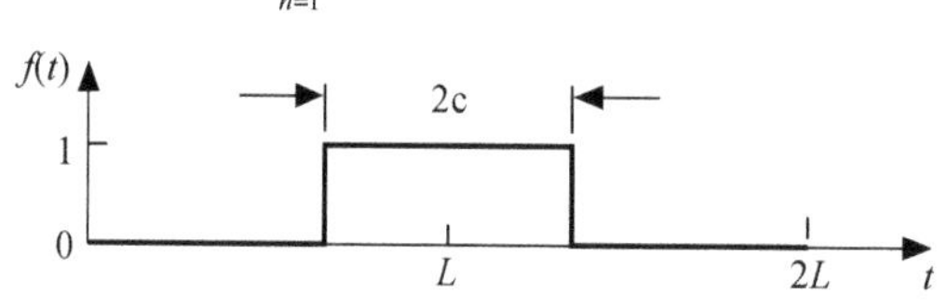

4. $f(t) = \frac{2}{L}\sum_{n=1}^{\infty} \sin\frac{n\pi}{2}\,\frac{\sin\frac{1}{2}n\pi c/L}{\frac{1}{2}n\pi c/L}\sin\frac{n\pi t}{L}$

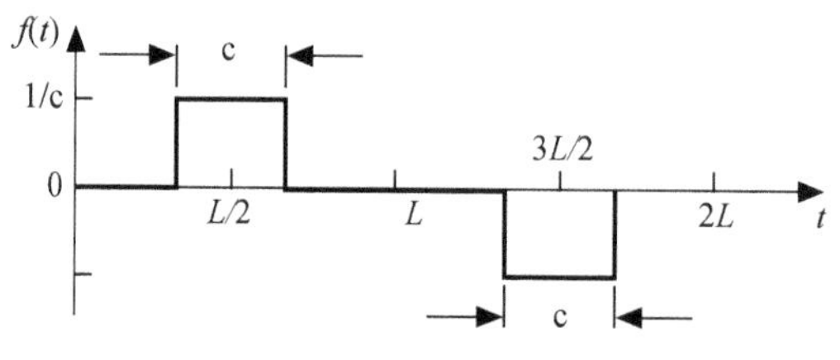

5. $$f(t)=\frac{2}{\pi}\sum_{n=1}^{\infty}\frac{(-1)^{n+1}}{n}\sin\frac{n\pi t}{L}$$

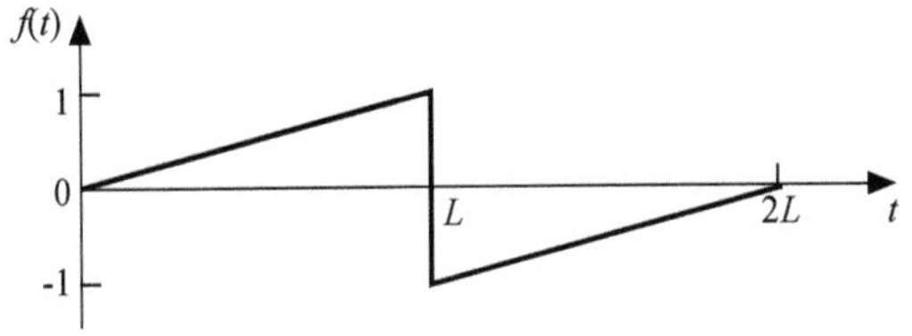

6. $$f(t)=\frac{1}{2}-\frac{4}{\pi^2}\sum_{n=1,3,5\cdots}\frac{1}{n^2}\cos\frac{n\pi t}{L}$$

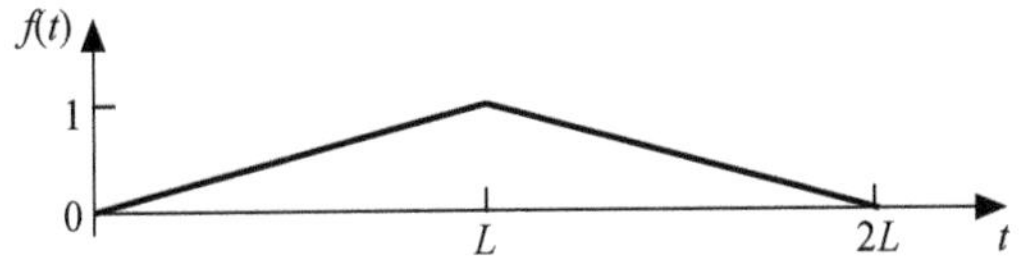

7. $$f(t)=-\frac{2}{\pi}\sum_{n=1}^{\infty}\frac{(-1)^n}{n}\left[1+\frac{1+(-1)^n}{n\pi(1-2a)}\sin n\pi a\right]\sin\frac{n\pi t}{L};\quad a=\frac{c}{2L}$$

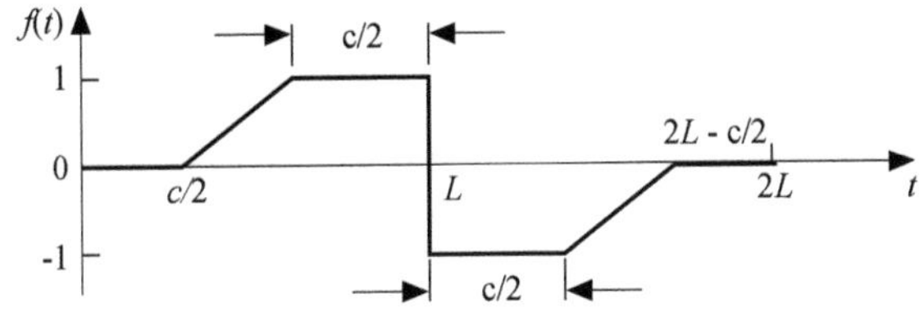

8. $$f(t)=\frac{4}{\pi}\sum_{n=1}^{\infty}\frac{1}{n}\sin\frac{n\pi}{4}\sin n\pi a\sin\frac{n\pi t}{L};\quad a=\frac{c}{2L}$$

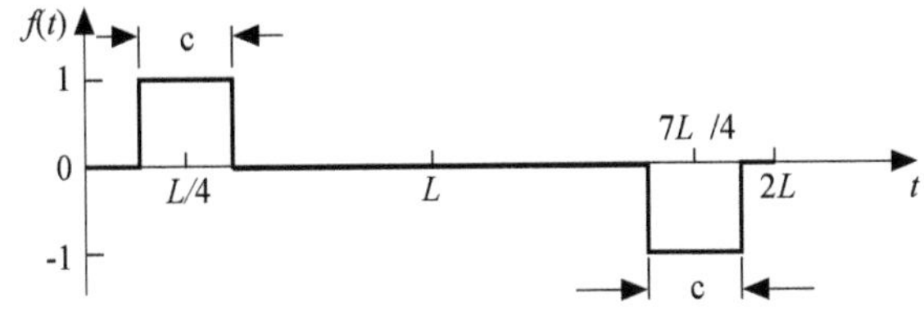

9. $$f(t)=\frac{9}{\pi^2}\sum_{n=1}^{\infty}\frac{1}{n^2}\sin\frac{n\pi}{3}\sin\frac{n\pi t}{L}$$

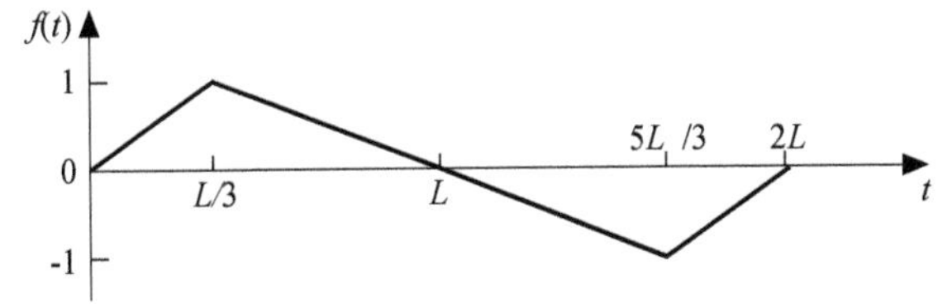

10. $f(t) = \frac{32}{3\pi^2}\sum_{n=1}^{\infty}\frac{1}{n^2}\sin\frac{n\pi}{4}\sin\frac{n\pi t}{L}$

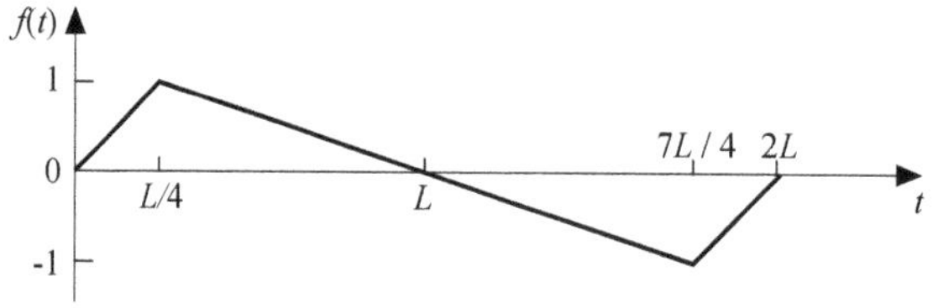

11. $f(t) = \frac{1}{\pi} + \frac{1}{2}\sin\omega t - \frac{2}{\pi}\sum_{n=2,4,6\cdots}\frac{1}{n^2-1}\cos n\omega t$

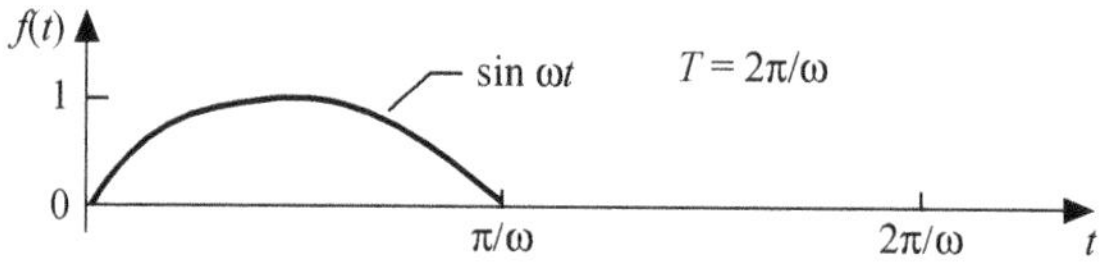

12. $f(t) = \frac{1}{2L} + \frac{1}{L}\sum_{n=1}^{\infty}\cos\frac{n\pi t}{L}$

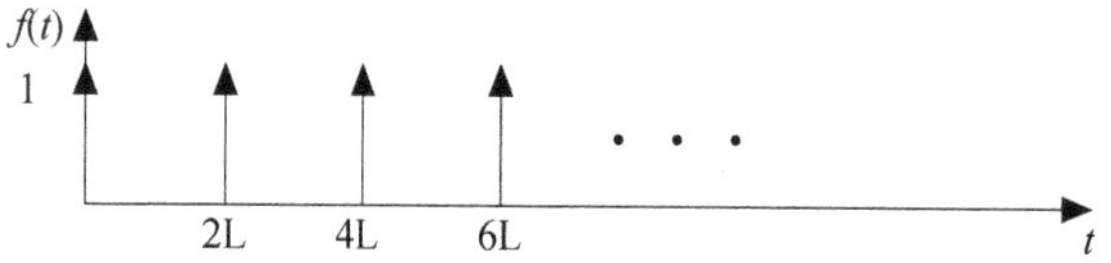

13. $f(t) = \frac{2A}{\pi} - \frac{4A}{\pi}\sum_{n=1}^{\infty}\frac{1}{4n^2-1}\cos 2nt$

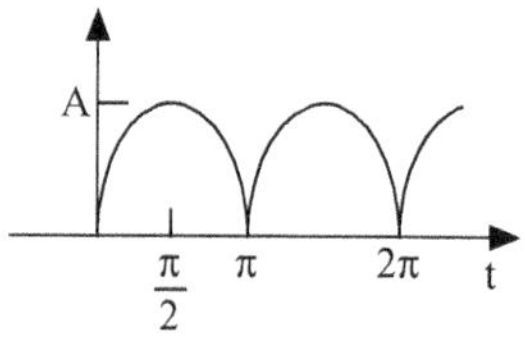

1.2 Orthogonal Systems and Fourier Series

1.2.1 An infinite system of real functions $\varphi_0(t),\varphi_1(t),\varphi_2(t),\ldots,\varphi_n(t),\ldots$ is said to be **orthogonal** on an interval [a,b] if $\int_a^b \varphi_n(t)\varphi_m(t)\,dt = 0$ for $n \neq m$ and $n,m = 0,1,2,\ldots$. It is assumed that

$$\int_a^b \varphi_n^2(t)\,dt \neq 0 \quad \text{for} \quad n = 0,1,2,\cdots.$$

1.2.2 The expansion of a function $f(t)$ in [a,b] is given by

$$f(t) = c_o\varphi_o(t) + c_1\varphi_1(t) + \cdots + c_n\varphi_n(t) + \cdots$$

$$c_n = \frac{\int_a^b f(t)\varphi_n(t)\,dt}{\int_a^b \varphi_n^2(t)\,dt} = \frac{\int_a^b f(t)\varphi_n(t)\,dt}{\|\varphi_n\|^2} \qquad n = 0,1,2,\cdots$$

1.2.3 ***Bessel's*** inequality

$$\int_a^b f^2(t)\,dt \ge \sum_{k=0}^{n} c_k^2 \|\varphi_k\|^2 \qquad n = \text{arbitrary}$$

1.2.4 ***Completeness*** of the system (1.2.1): A necessary and sufficient condition for the system (1.2.1) to be complete is that the Fourier series of any square integrable function $f(t)$ converges to $f(t)$ in the mean.

If the system (1.2.1) is complete, then every square integrable function $f(t)$ is completely determined (except for its values at a finite number of points) by its Fourier series.

1.2.5 The ***limits*** as $n \to \infty$ of the ***trigonometric*** integrals

$$\lim_{n\to\infty} \int_{-T/2}^{T/2} f(t)\cos\frac{2\pi nt}{T}\,dt = \lim_{n\to\infty} \int_{-T/2}^{T/2} f(t)\sin\frac{2\pi nt}{T}\,dt$$

1.2.6 ***Convergence*** in discontinuity: If $f(t)$ is the absolutely integrable function of period T, then at every point of discontinuity where $f(t)$ has a right-hand and left-hand derivative, the Fourier series of $f(t)$ converges to the value $[f(t+0)+f(t-0)]/2$.

1.3 Decreasing Coefficients of Trigonometric Series

1.3.1 ***Abel*** lemma: Let $u_o + u_1 + u_2 + \ldots + u_n + \ldots$ be a numerical series whose partial sums σ_n satisfy the condition $|\sigma_n| \le M$, where M is a constant. Then, if the positive numbers $\alpha_o, \alpha_1, \alpha_2, \ldots, \alpha_n, \ldots$ approach zero monotonically, the series $\alpha_o u_o + \alpha_1 u_1 + \ldots + \alpha_n u_n + \ldots$ converges, and the sum S satisfies the inequality $|S| \le M\alpha_o$.

1.3.2 The ***sum*** of sines

$$\sin t + \sin 2t + \sin 3t + \cdots + \sin nt = \frac{\cos\frac{t}{2} - \cos\left[\left(n+\frac{1}{2}\right)t\right]}{2\cos\frac{t}{2}}$$

$$1 + \frac{\cos t}{p} + \frac{\cos 2t}{p^2} + \cdots + \frac{\cos nt}{p^n} + \cdots = \frac{p(p-\cos t)}{p^2 - 2p\cos t + 1}$$

$$\frac{\sin t}{p} + \frac{\sin 2t}{p^2} + \cdots + \frac{\sin nt}{p^n} + \cdots = \frac{p\sin t}{p^2 - 2p\cos t + 1}$$

1.4 Operations on Fourier Series

1.4.1 ***Integration*** of Fourier series: If the absolutely integrable function $f(t)$ of period T is specified by its Fourier series (1.1.6) then

$$\int_a^b f(t)\,dt$$

can be found by term-by-term integration of the series.

1.4.2 ***Differentiation*** of Fourier series: If $f(t)$ is a continuous function of period T with absolutely integrable derivative, which may not exist at certain points, then the Fourier series of $df(t)/dt$ can be obtained from the Fourier series of $f(t)$ by term-by-term differentiation.

1.5 Two-Dimensional Fourier Series

1.5.1 Complex form

$$f(x,y) = \sum_{m,n=-\infty}^{\infty} c_{mn} e^{j\pi\left(\frac{mx}{l}+\frac{ny}{h}\right)} \qquad R\{-l \le x \le l, \quad -h \le y \le h\}$$

$$c_{mn} = \frac{1}{2l2h}\iint_R f(x,y)\, e^{-j\pi\left(\frac{mx}{l}+\frac{ny}{h}\right)} \qquad m,n = 0, \pm 1, \pm 2, \cdots$$

1.5.2 Trigonometric form

$$f(x,y) = \sum_{m,n=0}^{\infty}\left[A_{mn}\cos\frac{\pi mx}{l}\cos\frac{\pi ny}{h} + B_{mn}\sin\frac{\pi mx}{l}\cos\frac{\pi ny}{h}\right.$$

$$\left.+ C_{mn}\cos\frac{\pi mx}{l}\sin\frac{\pi ny}{h} + D_{mn}\sin\frac{\pi mx}{l}\sin\frac{\pi ny}{h}\right]$$

$$R\{-l \le x \le l, \quad -h \le y \le h\}$$

$$A_{mn} = \frac{1}{lh}\int_{-l}^{l}\int_{-h}^{h} f(x,y)\cos\frac{\pi mx}{l}\cos\frac{\pi ny}{h}\,dxdy$$

$$B_{mn} = \frac{1}{lh}\int_{-l}^{l}\int_{-h}^{h} f(x,y)\sin\frac{\pi mx}{l}\cos\frac{\pi ny}{h}\,dxdy$$

$$C_{mn} = \frac{1}{lh}\int_{-l}^{l}\int_{-h}^{h} f(x,y)\cos\frac{\pi mx}{l}\sin\frac{\pi ny}{h}\,dxdy$$

$$D_{mn} = \frac{1}{lh}\int_{-l}^{l}\int_{-h}^{h} f(x,y)\sin\frac{\pi mx}{l}\sin\frac{\pi ny}{h}\,dxdy$$

1.5.3 Trigonometric form with limits $-\pi \le x \le \pi$, $-\pi \le y \le \pi$

$$f(x,y) = \sum_{m,n=0}^{\infty} \lambda_{mn}[a_{mn} \cos mx \cos ny + b_{mn} \sin mx \cos ny$$
$$+ c_{mn} \cos mx \sin ny + d_{mn} \sin mx \sin ny]$$
$$R\{-\pi \le x \le \pi, \quad -\pi \le y \le \pi\}$$

$$a_{mn} = \frac{1}{\pi^2} \int_{-\pi}^{\pi}\int_{-\pi}^{\pi} f(x,y) \cos mx \cos ny \, dxdy$$

$$b_{mn} = \frac{1}{\pi^2} \int_{-\pi}^{\pi}\int_{-\pi}^{\pi} f(x,y) \sin mx \cos ny \, dxdy$$

$$c_{mn} = \frac{1}{\pi^2} \int_{-\pi}^{\pi}\int_{-\pi}^{\pi} f(x,y) \cos mx \sin ny \, dxdy$$

$$d_{mn} = \frac{1}{\pi^2} \int_{-\pi}^{\pi}\int_{-\pi}^{\pi} f(x,y) \sin mx \sin ny \, dxdy$$

$$\lambda_{mn} = \begin{cases} \frac{1}{4} & m = n = 0 \\ \frac{1}{2} & m > 0,\ n = 0, \text{ or } m = 0,\ n > 0 \\ 1 & m > 0,\ n > 0 \end{cases}$$

$$m,n = 0,1,2,3,4\cdots$$

1.5.4 Parseval's formula

$$\frac{1}{\pi^2} \int_{-\pi}^{\pi}\int_{-\pi}^{\pi} f^2(x,y)\, dxdy = \sum_{m,n=0}^{\infty} \lambda_{mn}(a_{mn}^2 + b_{mn}^2 + c_{mn}^2 + d_{mn}^2)$$

Appendix 1

Examples

Example 1

Expand the function shown in Figure 1.1 in Fourier series and plot the results.

$$\alpha_n = \frac{1}{3.5} \int_{-0.5}^{3} f(t) e^{-jn\omega_o t}\, dt = \frac{1}{3.5}\left[\int_{-0.5}^{1} 1 \cdot e^{-jn\omega_o t} dt + \int_{1}^{3} 0 \cdot e^{-jn\omega_o t}\, dt\right]$$

$$= \frac{1}{3.5(-jn\omega_o)} e^{-jn\omega_o t}\Big|_{-0.5}^{1} = \frac{1}{-j3.5\, n\omega_o}(e^{-jn\omega_o} - e^{j0.5 n\omega_o})$$

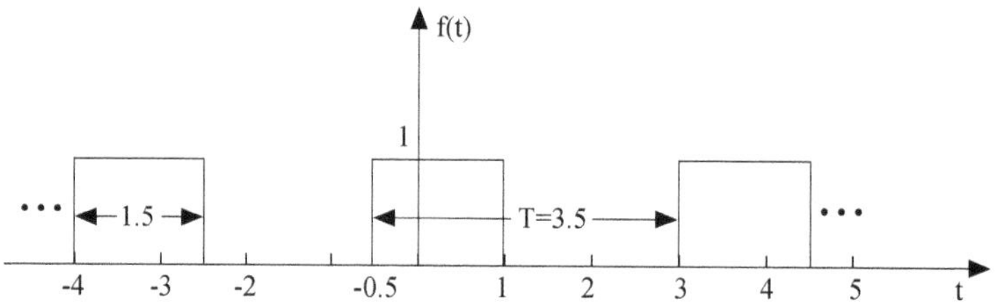

FIGURE 1.1

$$\alpha_o = \frac{1}{3.5}\int_{-0.5}^{1} dt = \frac{3}{7}, \qquad \omega_o = \frac{2\pi}{3.5}$$

$$f(t) = \alpha_o + \sum_{n=-\infty}^{\infty} \alpha_n e^{jn\omega_o t} = \frac{3}{7} + \sum_{n=1}^{\infty} \left\{ \left[\frac{1}{-j3.5n\omega_o}(e^{-jn\omega_o} - e^{j0.5n\omega_o}) + \frac{1}{j3.5n\omega_o}(e^{jn\omega_o} - e^{-j0.5n\omega_o}) \right] \cos n\omega_o t \right.$$

$$\left. + j\left[\frac{1}{-j3.5n\omega_o} \times (e^{-jn\omega_o} - e^{j0.5n\omega_o}) - \frac{1}{j3.5n\omega_o}(e^{jn\omega_o} - e^{-j0.5n\omega_o}) \right] \sin n\omega_o t \right\}$$

$$= \frac{3}{7} + \sum_{n=1}^{\infty} \left[\frac{4}{3.5n\omega_o} [(\sin 0.75n\omega_o \cos 0.25n\omega_o)\cos n\omega_o t + (\sin 0.75n\omega_o \sin 0.25n\omega_o)\sin n\omega_o t] \right]$$

Figure 1.2 shows $f(t)$ for the cases $1 \le n \le 3$ (curve 1) and $1 \le n \le 10$ (curve 2). Figure 1.3 shows $f(t)$ for $10 \le n \le 50$, and Figure 1.4 shows $f(t)$ for $1 \le n \le 60$. Observe the Gibbs phenomenon in Figures 1.2 and 1.4.

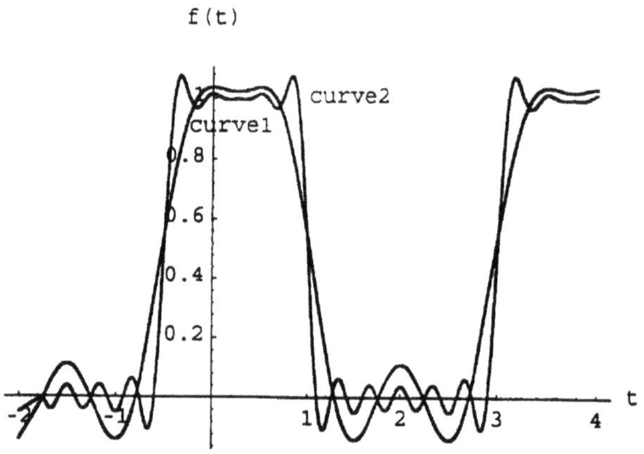

FIGURE 1.2

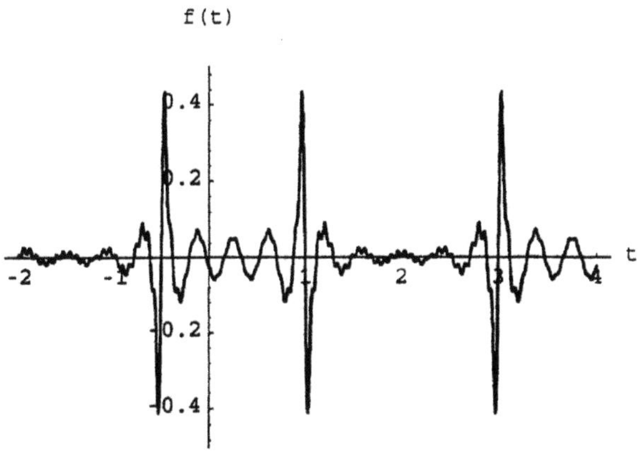

FIGURE 1.3

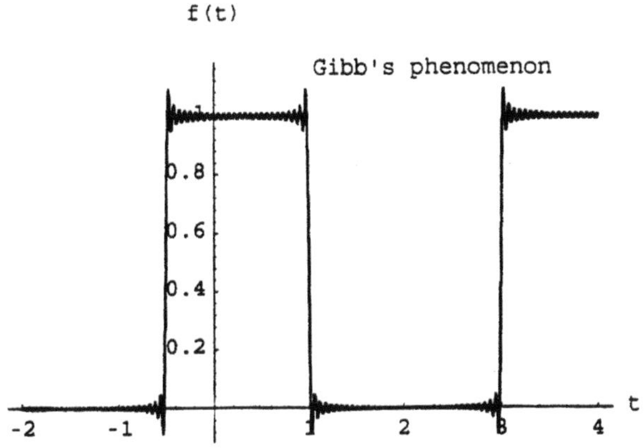

FIGURE 1.4

References

Beyer, W. H., *Standard Mathematical Tables*, 26*th* edition, CRC Press, Boca Raton, Florida, 1981.

Churchill, R. V., *Fourier Series and Boundary Value Problems*, McGraw-Hill Book Co., New York, NY, 1941.

Hsu, H. P., *Outline of Fourier Analysis*, Associate Educational Services, New York, NY, 1967.

Spiegel, M. R., *Fourier Analysis*, McGraw-Hill Book Co., New York, NY, 1974.

Tolstov, G. P., *Fourier Series*, Dover Publications, New York, NY, 1962.

2

Laplace Transforms

2.1 Definitions and Laplace Transform Formulae

2.1.1 One-Sided Laplace Transform

$$F(s) = \int_0^{\infty} f(t) e^{-st}\, dt \qquad s = \sigma + j\omega$$

$f(t)$ = piecewise continuous and of exponential order

2.1.2 One-Sided Inverse Laplace Transform

$$f(t) = \frac{1}{2\pi j} \int_{\sigma - j\infty}^{\sigma + j\infty} F(s) e^{st}\, ds$$

where the integration is within the regions of convergence. The region of convergence is half-plane $\sigma < \mathrm{Re}\{s\}$.

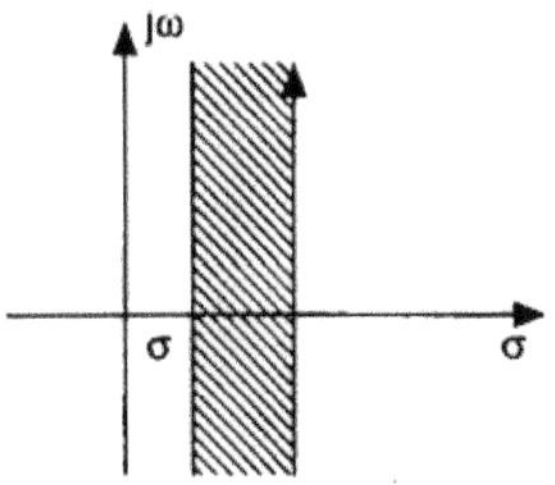

2.1.3 Two-Sided Laplace Transform

$$F(s) = \int_{-\infty}^{\infty} f(t)e^{-st}\,dt \qquad s = \sigma + j\omega$$

$f(t)$ = piecewise continuous and of exponential order

2.1.4 Two-Sided Inverse Laplace Transform

$$f(t) = \frac{1}{2\pi j}\int_{\sigma-j\infty}^{\sigma+j\infty} F(s)e^{st}\,ds$$

where the integration is within the regions of convergence which is a vertical strip $\sigma_1 < \mathrm{Re}\{s\} < \sigma_2$.

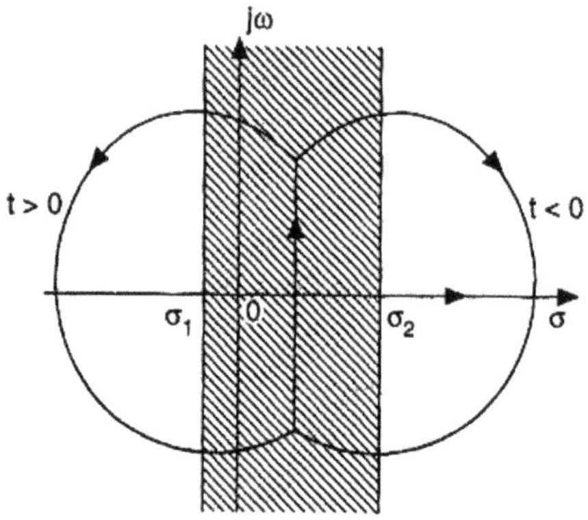

2.2 Properties

2.2.1 Properties of the Laplace Transform (one sided)

TABLE 2.1 Laplace Transform Properties

1. Linearity $L\{K_1 f_1(t) \pm K_2 f_2(t)\} = L\{K_1 f_1(t)\} \pm L\{K_2 f_2(t)\} = K_1 F_1(s) \pm K_2 F_2(s)$
2. Time derivative

$$L\left\{\frac{d}{dt} f(t)\right\} = sF(s) - f(0+)$$

3. Higher time derivative

$$L\left\{\frac{d^n}{dt^n} f(t)\right\} = s^n F(s) - s^{n-1} f(0+) - s^{n-2} f^{(1)}(0+) - \cdots - f^{(n-1)}(0+)$$

where $f^{(i)}(0+)$, $i = 1,2,\ldots,n-1$ is the ith derivative of $f(\cdot)$ at $t = 0+$.

4. Integral with zero initial condition $L\left\{\int_0^t f(\xi)\,d\xi\right\} = \frac{F(s)}{s}$

5. Integral with initial conditions $L\left\{\int_{-\infty}^t f(\xi)\,d\xi\right\} = \frac{F(s)}{s} + \frac{f^{(-1)}(0+)}{s}$ where $f^{(-1)}(0+) = \lim_{t \to 0+} \int_{-\infty}^t f(\xi)\,d\xi$

6. Multiplication by exponential $L\{e^{\pm at} f(t)\} = F(s \mp a)$

7. Multiplication by t $L\{t\,f(t)\} = -\frac{d}{ds} F(s)$; $L\{t^n f(t)\} = (-1)^n \frac{d^n}{ds^s} F(s)$

8. Time shifting $L\{f(t \pm \lambda) u(t \pm \lambda)\} = e^{\pm s\lambda} F(s)$

9. Scaling $L\left\{f\left(\frac{t}{a}\right)\right\} = aF(as)$; $L\{f(ta)\} = \frac{1}{a} F\left(\frac{s}{a}\right)$ $a > 0$

10. Time convolution $L\left\{\int_0^t f_1(t-\tau) f_2(\tau)\,d\tau\right\} \overset{\Delta}{=} L\{f_1(t) * f_2(t)\} = F_1(s) F_2(s)$

11. Frequency convolution

$$L\{f_1(t) f_2(t)\} = \frac{1}{2\pi j} \int_{x-j\infty}^{x+j\infty} F_1(z) F_2(s-z)\,dz = \frac{1}{2\pi j} \{F_1(s) * F_2(s)\}$$

where $z = x + jy$, and where x must be greater than the abscissa of absolute convergence for $f_1(t)$ over the path of integration.

12. Initial value $\lim_{t \to 0+} f(t) = \lim_{s \to \infty} sF(s)$ provided that this limit exists.

13. Final value $\lim_{t \to \infty} f(t) = \lim_{s \to 0+} sF(s)$ provided that sF(s) is analytic on the $j\omega$ axis and in the right half of the s plane

14. Division by t $L\left\{\frac{f(t)}{t}\right\} = \int_s^\infty F(s')\,ds'$

15. $f(t)$ periodic $L\{f(t)\} = \dfrac{\int_0^T e^{-st} f(t)\,dt}{1 - e^{-sT}}$ $f(t) = f(t+T)$

2.2.2 Methods of Finding the Laplace Transform

1. Direct method by solving (2.1.1).
2. Expand $f(t)$ in power series if such an expansion exists.
3. Differentiation with respect to a parameter.
4. Use of tables.

2.3 Inverse Laplace Transforms

2.3.1 Properties

1. Linearity $L^{-1}\{c_1F_1(s) \pm c_2F_2(s)\} = c_1f_1(t) \pm c_2f_2(t)$
2. Shifting $L^{-1}\{F(s-a)\} = e^{at}f(t)$
3. Time shifting $L^{-1}\{e^{-as}F(s)\} = f(t-a) \qquad t > a$
4. Scaling property $L^{-1}\{F(as)\} = \frac{1}{a}f\left(\frac{t}{a}\right) \qquad a > 0$
5. Derivatives $L^{-1}\{F^{(n)}(s)\} = (-1)^n t^n f(t) \qquad F^{(n)}(s) = \frac{d^nF(s)}{ds^n}$
6. Multiplication by s $L^{-1}\{sF(s) - f(0+)\} = L\{sF(s)\} - f(0+)L\{1\} = f^{(1)}(t) + f(0)\delta(t)$
7. Division by s $L^{-1}\left\{\frac{F(s)}{s}\right\} = \int_0^t f(t')\,dt'$
8. Convolution $L^{-1}\{F(s)H(s)\} = \int_0^t F(u)H(t-u)\,du = F(s) * H(s)$

2.3.2 Methods of Finding Inverse Laplace Transforms

1. ***Partial fraction method***: Any rational function $P(s)/Q(s)$ where $P(s)$ and $Q(s)$ are polynomials, with the degree of $P(s)$ less than that of $Q(s)$, can be written as the sum of rational functions, known as partial fractions, having the form $A/(as+b)^r$, $(As+B)/(as^2+bs+c)^r$, $r = 1,2,\ldots$.
2. Expand $F(s)$ in inverse powers of s if such an expansion exists.
3. Differentiation with respect to a parameter.
4. Combination of the above methods.
5. Use of tables.
6. Complex inversion (see Appendix 1).

2.4 Relationship Between Fourier Integrals of Causal Functions and One-Sided Laplace Transforms

2.4.1 $F(\omega)$ from $F(s)$

$$F(\omega) = \int_0^{\infty} e^{-j\omega t} f(t)\,dt \qquad f(t) = \begin{cases} f(t) & t \ge 0 \\ 0 & t < 0 \end{cases}$$

a) The region of convergence of $F(s)$ contains the $j\omega$ axis in its interior, $\sigma < 0$ (see 2.1.2)

$$F(\omega) = F(s)\big|_{s=j\omega}$$

b) If the axis $j\omega$ is outside the region of convergence of $F(s)$, $\sigma > 0$, then $F(\omega)$ does not exist; the function $f(t)$ has no Fourier transform.

c) Let $\sigma = 0$, $F(s)$ is analytic for $s > 0$, and has one singular point on the $j\omega$ axis, hence, $F(s) = \frac{1}{s - j\omega_o}$ or $F(s) = L\{e^{j\omega_o t}u(t)\}$. But $F\{e^{j\omega_o t}u(t)\} = \pi\delta(\omega - \omega_o) + \frac{1}{j\omega - j\omega_o}$ and there we obtain the correspondence

$$F(s) = \frac{1}{s - j\omega_o} \qquad F(\omega) = F(s)\big|_{s=j\omega} = \pi\delta(\omega - \omega_o) + F(s)\big|_{s=j\omega}$$

Also

$$F(s) = \frac{1}{(s - j\omega_o)^n} \qquad F(\omega) = \frac{\pi j^{n-1}}{(n-1)!}\delta^{(n-1)}(\omega - \omega_o) + F(s)\big|_{s=j\omega}$$

$\delta^{(n-1)}(\cdot)$ = the $(n - 1)^{th}$ derivative.

d) $F(s)$ has n simple poles $j\omega_1, j\omega_2, \ldots, j\omega_n$ and no other singularities in the half plane Re $s \ge 0$. $F(s)$ takes the form $F(s) = G(s) + \sum_{n=1}^{n} \frac{a_n}{s - j\omega_n}$ where $G(s)$ is free of singularities for Re $s \ge 0$. The correspondence is

$$F(\omega) = G(s)\big|_{s=j\omega} + \sum_{n=1}^{n} \frac{a_n}{s - j\omega_n}\bigg|_{s=j\omega} + \pi\sum_{n=1}^{n} a_n \delta(\omega - \omega_n)$$

2.5 Table of Laplace Transforms

TABLE 2.2 Table of Laplace Operations

	$F(s)$	$f(t)$
1	$\int_0^\infty e^{-st} f(t)\,dt$	$f(t)$
2	$AF(s) + BG(s)$	$Af(t) + Bg(t)$
3	$sF(s) - f(+0)$	$f'(t)$
4	$s^n F(s) - s^{n-1} f(+0) - s^{n-2} f^{(1)}(+0) - \cdots - f^{(n-1)}(+0)$	$f^{(n)}(t)$
5	$\frac{1}{s} F(s)$	$\int_0^t f(\tau)\,d\tau$
6	$\frac{1}{s^2} F(s)$	$\int_0^t \int_0^\tau f(\lambda)\,d\lambda\,d\tau$
7	$F_1(s)F_2(s)$	$\int_0^t f_1(t-\tau) f_2(\tau)\,d\tau = f_1 * f_2$
8	$-F'(s)$	$tf(t)$
9	$(-1)^n F^{(n)}(s)$	$t^n f(t)$
10	$\int_s^\infty F(x)\,dx$	$\frac{1}{t} f(t)$
11	$F(s-a)$	$e^{at} f(t)$
12	$e^{-bs} F(s)$	$f(t-b)$, where $f(t) = 0;\ t < 0$
13	$F(cs)$	$\frac{1}{c} f\left(\frac{t}{c}\right)$
14	$F(cs - b)$	$\frac{1}{c} e^{(bt)/c} f\left(\frac{t}{c}\right)$
15	$\frac{\int_0^a e^{-st} f(t)\,dt}{1 - e^{-as}}$	$f(t + a) = f(t)$ periodic signal
16	$\frac{\int_0^a e^{-st} f(t)\,dt}{1 + e^{-as}}$	$f(t + a) = -f(t)$
17	$\frac{F(s)}{1 - e^{-as}}$	$f_1(t)$, the half-wave rectification of $f(t)$ in No. 16.
18	$F(s) \coth \frac{as}{2}$	$f_2(t)$, the full-wave rectification of $f(t)$ in No. 16.
19	$\frac{p(s)}{q(s)},\ q(s) = (s-a_1)(s-a_2)\cdots(s-a_m)$	$\sum_1^m \frac{p(a_n)}{q'(a_n)} e^{a_n t}$
20	$\frac{p(s)}{q(s)} = \frac{\phi(s)}{(s-a)^r}$	$e^{at} \sum_{n=1}^r \frac{\phi^{(r-n)}(a)}{(r-n)!} \frac{t^{n-1}}{(n-1)!} + \cdots$

TABLE 2.3 Table of Laplace Transforms

	$F(s)$	$f(t)$
1	s^n	$\delta^{(n)}(t)$ n^{th} derivative of the delta function
2	s	$\frac{d\delta(t)}{dt}$
3	1	$\delta(t)$
4	$\frac{1}{s}$	1
5	$\frac{1}{s^2}$	t
6	$\frac{1}{s^n}\ (n = 1,2,\cdots)$	$\frac{t^{n-1}}{(n-1)!}$
7	$\frac{1}{\sqrt{s}}$	$\frac{1}{\sqrt{\pi t}}$
8	$s^{-3/2}$	$2\sqrt{\frac{t}{\pi}}$
9	$s^{-[n+(1/2)]}\ (n = 1,2,\cdots)$	$\frac{2^n t^{n-(1/2)}}{1\cdot 3\cdot 5\cdots(2n-1)\sqrt{\pi}}$
10	$\frac{\Gamma(k)}{s^k}\ (k \ge 0)$	t^{k-1}
11	$\frac{1}{s-a}$	e^{at}
12	$\frac{1}{(s-a)^2}$	te^{at}
13	$\frac{1}{(s-a)^n}\ (n = 1,2,\cdots)$	$\frac{1}{(n-1)!}t^{n-1}e^{at}$
14	$\frac{\Gamma(k)}{(s-a)^k}\ (k \ge 0)$	$t^{k-1}e^{at}$
15	$\frac{1}{(s-a)(s-b)}$	$\frac{1}{(a-b)}(e^{at} - e^{bt})$
16	$\frac{s}{(s-a)(s-b)}$	$\frac{1}{(a-b)}(ae^{at} - be^{bt})$
17	$\frac{1}{(s-a)(s-b)(s-c)}$	$-\frac{(b-c)e^{at} + (c-a)e^{bt} + (a-b)e^{ct}}{(a-b)(b-c)(c-a)}$
18	$\frac{1}{(s+a)}$	e^{-at} valid for complex a
19	$\frac{1}{s(s+a)}$	$\frac{1}{a}(1-e^{-at})$
20	$\frac{1}{s^2(s+a)}$	$\frac{1}{a^2}(e^{-at} + at - 1)$
21	$\frac{1}{s^3(s+a)}$	$\frac{1}{a^2}\left[\frac{1}{a} - t + \frac{at^2}{2} - \frac{1}{a}e^{-at}\right]$
22	$\frac{1}{(s+a)(s+b)}$	$\frac{1}{(b-a)}(e^{-at} - e^{-bt})$
23	$\frac{1}{s(s+a)(s+b)}$	$\frac{1}{ab}\left[1 + \frac{1}{(a-b)}(be^{-at} - ae^{-bt})\right]$

TABLE 2.3 Table of Laplace Transforms (continued)

	$F(s)$	$f(t)$
24	$\frac{1}{s^2(s+a)(s+b)}$	$\frac{1}{(ab)^2}\left[\frac{1}{(a-b)}(a^2e^{-bt}-b^2e^{-at})+abt-a-b\right]$
25	$\frac{1}{s^3(s+a)(s+b)}$	$\frac{1}{(ab)}\left[\frac{a^3-b^3}{(ab)^2(a-b)}+\frac{1}{2}t^2-\frac{(a+b)}{ab}t+\frac{1}{(a-b)}\left(\frac{b}{a^2}e^{-at}-\frac{a}{b^2}e^{-bt}\right)\right]$
26	$\frac{1}{(s+a)(s+b)(s+c)}$	$\frac{1}{(b-a)(c-a)}e^{-at}+\frac{1}{(a-b)(c-b)}e^{-bt}+\frac{1}{(a-c)(b-c)}e^{-ct}$
27	$\frac{1}{s(s+a)(s+b)(s+c)}$	$\frac{1}{abc}-\frac{1}{a(b-a)(c-a)}e^{-at}-\frac{1}{b(a-b)(c-b)}e^{-bt}-\frac{1}{c(a-c)(b-c)}e^{-ct}$
28	$\frac{1}{s^2(s+a)(s+b)(s+c)}$	$\left\{\frac{ab(ct-1)-ac-bc}{(abc)^2}+\frac{1}{a^2(b-a)(c-a)}e^{-at}+\frac{1}{b^2(a-b)(c-b)}e^{-bt}+\frac{1}{c^2(a-c)(b-c)}e^{-ct}\right.$
29	$\frac{1}{s^3(s+a)(s+b)(s+c)}$	$\left\{\frac{1}{(abc)^3}[(ab+ac+bc)^2-abc(a+b+c)]-\frac{ab+ac+bc}{(abc)^2}t+\frac{1}{2abc}t^2-\frac{1}{a^3(b-a)(c-a)}e^{-at}-\frac{1}{b^3(a-b)(c-b)}e^{-bt}-\frac{1}{c^3(a-c)(b-c)}\varepsilon^{-ct}\right.$
30	$\frac{1}{s^2+a^2}$	$\frac{1}{a}\sin at$
31	$\frac{s}{s^2+a^2}$	$\cos at$
32	$\frac{1}{s^2-a^2}$	$\frac{1}{a}\sinh at$
33	$\frac{s}{s^2-a^2}$	$\cosh at$
34	$\frac{1}{s(s^2+a^2)}$	$\frac{1}{a^2}(1-\cos at)$
35	$\frac{1}{s^2(s^2+a^2)}$	$\frac{1}{a^3}(at-\sin at)$
36	$\frac{1}{(s^2+a^2)^2}$	$\frac{1}{2a^3}(\sin at-at\cos at)$
37	$\frac{s}{(s^2+a^2)^2}$	$\frac{t}{2a}\sin at$
38	$\frac{s^2}{(s^2+a^2)^2}$	$\frac{1}{2a}(\sin at+at\cos at)$
39	$\frac{s^2-a^2}{(s^2+a^2)^2}$	$t\cos at$
40	$\frac{s}{(s^2+a^2)(s^2+b^2)}(a^2\neq b^2)$	$\frac{\cos at-\cos bt}{b^2-a^2}$
41	$\frac{1}{(s-a)^2+b^2}$	$\frac{1}{b}e^{at}\sin bt$
42	$\frac{s-a}{(s-a)^2+b^2}$	$e^{at}\cos bt$
43	$\frac{1}{[(s+a)^2+b^2]^n}$	$\frac{-e^{-at}}{4^{n-1}b^{2n}}\sum_{r=1}^{n}\binom{2n-r-1}{n-1}(-2t)^{r-1}\frac{d^r}{dt^r}[\cos(bt)]$

TABLE 2.3 Table of Laplace Transforms (continued)

	$F(s)$	$f(t)$
44	$\dfrac{s}{[(s+a)^2+b^2]^n}$	$\dfrac{e^{-at}}{4^{n-1}b^{2n}}\left\{\sum_{r=1}^{n}\binom{2n-r-1}{n-1}(-2t)^{r-1}\dfrac{d^r}{dt^r}[a\cos(bt)+b\sin(bt)] - 2b\sum_{r=1}^{n-1} r\binom{2n-r-2}{n-1}(-2t)^{r-1}\dfrac{d^r}{dt^r}[\sin(bt)]\right\}$
45	$\dfrac{3a^2}{s^3+a^3}$	$e^{-at}-e^{(at)/2}\left(\cos\dfrac{at\sqrt{3}}{2}-\sqrt{3}\sin\dfrac{at\sqrt{3}}{2}\right)$
46	$\dfrac{4a^3}{s^4+4a^4}$	$\sin at\cosh at-\cos at\sinh at$
47	$\dfrac{s}{s^4+4a^4}$	$\dfrac{1}{2a^2}(\sin at\sinh at)$
48	$\dfrac{1}{s^4-a^4}$	$\dfrac{1}{2a^3}(\sinh at-\sin at)$
49	$\dfrac{s}{s^4-a^4}$	$\dfrac{1}{2a^2}(\cosh at-\cos at)$
50	$\dfrac{8a^3s^2}{(s^2+a^2)^3}$	$(1+a^2t^2)\sin at-\cos at$
51	$\dfrac{1}{s}\left(\dfrac{s-1}{s}\right)^n$	$L_n(t)=\dfrac{e^t}{n!}\dfrac{d^n}{dt^n}(t^ne^{-t})$ [$L_n(t)$ is the Laguerre polynomial of degree n]
52	$\dfrac{1}{(s+a)^n}$	$\dfrac{t^{(n-1)}e^{-at}}{(n-1)!}$ where n is a positive integer
53	$\dfrac{1}{s(s+a)^2}$	$\dfrac{1}{a^2}[1-e^{-at}-ate^{-at}]$
54	$\dfrac{1}{s^2(s+a)^2}$	$\dfrac{1}{a^3}[at-2+ate^{-at}+2e^{-at}]$
55	$\dfrac{1}{s(s+a)^3}$	$\dfrac{1}{a^3}\left[1-\left(\dfrac{1}{2}a^2t^2+at+1\right)e^{-at}\right]$
56	$\dfrac{1}{(s+a)(s+b)^2}$	$\dfrac{1}{(a-b)^2}\{e^{-at}+[(a-b)t-1]e^{-bt}\}$
57	$\dfrac{1}{s(s+a)(s+b)^2}$	$\dfrac{1}{ab^2}-\dfrac{1}{a(a-b)^2}e^{-at}-\left[\dfrac{1}{b(a-b)}t+\dfrac{a-2b}{b^2(a-b)^2}\right]e^{-bt}$
58	$\dfrac{1}{s^2(s+a)(s+b)^2}$	$\dfrac{1}{a^2(a-b)^2}e^{-at}+\dfrac{1}{ab^2}\left(t-\dfrac{1}{a}-\dfrac{2}{b}\right)+\left[\dfrac{1}{b^2(a-b)}t+\dfrac{2(a-b)-b}{b^3(a-b)^2}\right]e^{-bt}$
59	$\dfrac{1}{(s+a)(s+b)(s+c)^2}$	$\left[\dfrac{1}{(c-b)(c-a)}t+\dfrac{2c-a-b}{(c-a)^2(c-b)^2}\right]e^{-ct}+\dfrac{1}{(b-a)(c-a)^2}e^{-at}+\dfrac{1}{(a-b)(c-b)^2}e^{-bt}$
60	$\dfrac{1}{(s+a)(s^2+\omega^2)}$	$\dfrac{1}{a^2+\omega^2}e^{-at}+\dfrac{1}{\omega\sqrt{a^2+\omega^2}}\sin(\omega t-\phi);\quad \phi=\tan^{-1}\left(\dfrac{\omega}{a}\right)$
61	$\dfrac{1}{s(s+a)(s^2+\omega^2)}$	$\dfrac{1}{a\omega^2}-\dfrac{1}{a^2+\omega^2}\left(\dfrac{1}{\omega}\sin\omega t+\dfrac{a}{\omega^2}\cos\omega t+\dfrac{1}{a}e^{-at}\right)$

TABLE 2.3 Table of Laplace Transforms (continued)

	$F(s)$	$f(t)$
62	$\frac{1}{s^2(s+a)(s^2+\omega^2)}$	$\left\{ \begin{array}{l} \frac{1}{a\omega^2}t - \frac{1}{a^2\omega^2} + \frac{1}{a^2(a^2+\omega^2)}e^{-at} \\ + \frac{1}{\omega^3\sqrt{a^2+\omega^2}}\cos(\omega t+\phi); \quad \phi = \tan^{-1}\left(\frac{a}{\omega}\right) \end{array} \right.$
63	$\frac{1}{[(s+a)^2+\omega^2]^2}$	$\frac{1}{2\omega^3}e^{-at}[\sin\omega t - \omega t\cos\omega t]$
64	$\frac{1}{s^2-a^2}$	$\frac{1}{a}\sinh at$
65	$\frac{1}{s^2(s^2-a^2)}$	$\frac{1}{a^3}\sinh at - \frac{1}{a^2}t$
66	$\frac{1}{s^3(s^2-a^2)}$	$\frac{1}{a^4}(\cosh at - 1) - \frac{1}{2a^2}t^2$
67	$\frac{1}{s^3+a^3}$	$\frac{1}{3a^2}\left[e^{-at} - e^{\frac{a}{2}t}\left(\cos\frac{\sqrt{3}}{2}at - \sqrt{3}\sin\frac{\sqrt{3}}{2}at\right)\right]$
68	$\frac{1}{s^4+4a^4}$	$\frac{1}{4a^3}(\sin at\cosh at - \cos at\sinh at)$
69	$\frac{1}{s^4-a^4}$	$\frac{1}{2a^3}(\sinh at - \sin at)$
70	$\frac{1}{[(s+a)^2-\omega^2]}$	$\frac{1}{\omega}e^{-at}\sinh\omega t$
71	$\frac{s+a}{s[(s+b)^2+\omega^2]}$	$\left\{ \begin{array}{l} \frac{a}{b^2+\omega^2} - \frac{1}{\omega} + \sqrt{\frac{(a-b)^2+\omega^2}{b^2+\omega^2}}e^{-bt}\sin(\omega t+\phi); \\ \phi = \tan^{-1}\left(\frac{\omega}{b}\right) + \tan^{-1}\left(\frac{\omega}{a-b}\right) \end{array} \right.$
72	$\frac{s+a}{s^2[(s+b)^2+\omega^2]}$	$\left\{ \begin{array}{l} \frac{1}{b^2+\omega^2}[1+at] - \frac{2ab}{(b^2+\omega^2)^2} + \frac{\sqrt{(a-b)^2+\omega^2}}{\omega(b^2+\omega^2)}e^{-bt}\sin(\omega t+\phi) \\ \phi = \tan^{-1}\left(\frac{\omega}{a-b}\right) + 2\tan^{-1}\left(\frac{\omega}{b}\right) \end{array} \right.$
73	$\frac{s+a}{(s+c)[(s+b)^2+\omega^2]}$	$\left\{ \begin{array}{l} \frac{a-c}{(c-b)^2+\omega^2}e^{-ct} + \frac{1}{\omega}\sqrt{\frac{(a-b)^2+\omega^2}{(c-b)^2+\omega^2}}e^{-bt}\sin(\omega t+\phi) \\ \phi = \tan^{-1}\left(\frac{\omega}{a-b}\right) - \tan^{-1}\left(\frac{\omega}{c-b}\right) \end{array} \right.$
74	$\frac{s+a}{s(s+c)[(s+b)^2+\omega^2]}$	$\left\{ \begin{array}{l} \frac{a}{c(b^2+\omega^2)} + \frac{(c-a)}{c[(b-c)^2+\omega^2]}e^{-ct} \\ - \frac{1}{\omega\sqrt{b^2+\omega^2}}\sqrt{\frac{(a-b)^2+\omega^2}{(b-c)^2+\omega^2}}e^{-bt}\sin(\omega t+\phi) \\ \phi = \tan^{-1}\left(\frac{\omega}{b}\right) + \tan^{-1}\left(\frac{\omega}{a-b}\right) - \tan^{-1}\left(\frac{\omega}{c-b}\right) \end{array} \right.$
75	$\frac{s+a}{s^2(s+b)^3}$	$\frac{a}{b^3}t + \frac{b-3a}{b^4} + \left[\frac{3a-b}{b^4} + \frac{a-b}{2b^2}t^2 + \frac{2a-b}{b^3}t\right]e^{-bt}$

TABLE 2.3 Table of Laplace Transforms (continued)

	$F(s)$	$f(t)$
76	$\frac{s+a}{(s+c)(s+b)^3}$	$\frac{a-c}{(b-c)^3}e^{-ct}+\left[\frac{a-b}{2(c-b)}t^2+\frac{c-a}{(c-b)^2}t+\frac{a-c}{(c-b)^3}\right]e^{-bt}$
77	$\frac{s^2}{(s+a)(s+b)(s+c)}$	$\frac{a^2}{(b-a)(c-a)}e^{-at}+\frac{b^2}{(a-b)(c-b)}e^{-bt}+\frac{c^2}{(a-c)(b-c)}e^{-ct}$
78	$\frac{s^2}{(s+a)(s+b)^2}$	$\frac{a^2}{(b-a)^2}e^{-at}+\left[\frac{b^2}{(a-b)}t+\frac{b^2-2ab}{(a-b)^2}\right]e^{-bt}$
79	$\frac{s^2}{(s+a)^3}$	$\left[2-2at+\frac{a^2}{2}t^2\right]e^{-at}$
80	$\frac{s^2}{(s+a)(s^2+\omega^2)}$	$\frac{a^2}{(a^2+\omega^2)}e^{-at}-\frac{\omega}{\sqrt{a^2+\omega^2}}\sin(\omega t+\phi);\ \phi=\tan^{-1}\left(\frac{\omega}{a}\right)$
81	$\frac{s^2}{(s+a)^2(s^2+\omega^2)}$	$\left[\frac{a^2}{(a^2+\omega^2)}t-\frac{2a\omega^2}{(a^2+\omega^2)^2}\right]e^{-at}-\frac{\omega}{(a^2+\omega^2)}\sin(\omega t+\phi);$ $\phi=-2\tan^{-1}\left(\frac{\omega}{a}\right)$
82	$\frac{s^2}{(s+a)(s+b)(s^2+\omega^2)}$	$\frac{a^2}{(b-a)(a^2+\omega^2)}e^{-at}+\frac{b^2}{(a-b)(b^2+\omega^2)}e^{-bt}$ $-\frac{\omega}{\sqrt{(a^2+\omega^2)(b^2+\omega^2)}}\sin(\omega t+\phi);\ \phi=-\left[\tan^{-1}\left(\frac{\omega}{a}\right)+\tan^{-1}\left(\frac{\omega}{b}\right)\right]$
83	$\frac{s^2}{(s^2+a^2)(s^2+\omega^2)}$	$-\frac{a}{(\omega^2-a^2)}\sin(at)-\frac{\omega}{(a^2-\omega^2)}\sin(\omega t)$
84	$\frac{s^2}{(s^2+\omega^2)^2}$	$\frac{1}{2\omega}(\sin\omega t+\omega t\cos\omega t)$
85	$\frac{s^2}{(s+a)[(s+b)^2+\omega^2]}$	$\frac{a^2}{(a-b)^2+\omega^2}e^{-at}+\frac{1}{\omega}\sqrt{\frac{(b^2-\omega^2)^2+4b^2\omega^2}{(a-b)^2+\omega^2}}\,e^{-bt}\sin(\omega t+\phi)$ $\phi=\tan^{-1}\left(\frac{-2b\omega}{b^2-\omega^2}\right)-\tan^{-1}\left(\frac{\omega}{a-b}\right)$
86	$\frac{s^2}{(s+a)^2[(s+b)^2+\omega^2]}$	$\frac{a^2}{(a-b)^2+\omega^2}te^{-at}-2\left[\frac{a[(b-a)^2+\omega^2]+a^2(b-a)}{[(b-a)^2+\omega^2]^2}\right]e^{-at}$ $+\frac{\sqrt{(b^2-\omega^2)^2+4b^2\omega^2}}{\omega[(a-b)^2+\omega^2]}e^{-bt}\sin(\omega t+\phi)$ $\phi=\tan^{-1}\left(\frac{-2b\omega}{b^2-\omega^2}\right)-2\tan^{-1}\left(\frac{\omega}{a-b}\right)$
87	$\frac{s^2+a}{s^2(s+b)}$	$\frac{b^2+a}{b^2}e^{-bt}+\frac{a}{b}t-\frac{a}{b^2}$
88	$\frac{s^2+a}{s^3(s+b)}$	$\frac{a}{2b}t^2-\frac{a}{b^2}t+\frac{1}{b^3}[b^2+a-(a+b^2)e^{-bt}]$
89	$\frac{s^2+a}{s(s+b)(s+c)}$	$\frac{a}{bc}+\frac{(b^2+a)}{b(b-c)}e^{-bt}-\frac{(c^2+a)}{c(b-c)}e^{-ct}$
90	$\frac{s^2+a}{s^2(s+b)(s+c)}$	$\frac{b^2+a}{b^2(c-b)}e^{-bt}+\frac{c^2+a}{c^2(b-c)}e^{-ct}+\frac{a}{bc}t-\frac{a(b+c)}{b^2c^2}$

TABLE 2.3 Table of Laplace Transforms (continued)

	$F(s)$	$f(t)$
91	$\frac{s^2+a}{(s+b)(s+c)(s+d)}$	$\frac{b^2+a}{(c-b)(d-b)}e^{-bt}+\frac{c^2+a}{(b-c)(d-c)}e^{-ct}+\frac{d^2+a}{(b-d)(c-d)}e^{-dt}$
92	$\frac{s^2+a}{s(s+b)(s+c)(s+d)}$	$\frac{a}{bcd}+\frac{b^2+a}{b(b-c)(d-b)}e^{-bt}+\frac{c^2+a}{c(b-c)(c-d)}e^{-ct}+\frac{d^2+a}{d(b-d)(d-c)}e^{-dt}$
93	$\frac{s^2+a}{s^2(s+b)(s+c)(s+d)}$	$\left\{\begin{array}{l}\frac{a}{bcd}t-\frac{a}{b^2c^2d^2}(bc+cd+db)+\frac{b^2+a}{b^2(b-c)(b-d)}e^{-bt}\\+\frac{c^2+a}{c^2(c-b)(c-d)}e^{-ct}+\frac{d^2+a}{d^2(d-b)(d-c)}e^{-dt}\end{array}\right.$
94	$\frac{s^2+a}{(s^2+\omega^2)^2}$	$\frac{1}{2\omega^3}(a+\omega^2)\sin\omega t-\frac{1}{2\omega^2}(a-\omega^2)t\cos\omega t$
95	$\frac{s^2-\omega^2}{(s^2+\omega^2)^2}$	$t\cos\omega t$
96	$\frac{s^2+a}{s(s^2+\omega^2)^2}$	$\frac{a}{\omega^4}-\frac{(a-\omega^2)}{2\omega^3}t\sin\omega t-\frac{a}{\omega^4}\cos\omega t$
97	$\frac{s(s+a)}{(s+b)(s+c)^2}$	$\frac{b^2-ab}{(c-b)^2}e^{-bt}+\left[\frac{c^2-ac}{b-c}t+\frac{c^2-2bc+ab}{(b-c)^2}\right]e^{-ct}$
98	$\frac{s(s+a)}{(s+b)(s+c)(s+d)^2}$	$\left\{\begin{array}{l}\frac{b^2-ab}{(c-b)(d-b)^2}e^{-bt}+\frac{c^2-ac}{(b-c)(d-c)^2}e^{-ct}+\frac{d^2-ad}{(b-d)(c-d)}te^{-dt}\\+\frac{a(bc-d^2)+d(db+dc-2bc)}{(b-d)^2(c-d)^2}e^{-dt}\end{array}\right.$
99	$\frac{s^2+a_1s+a_o}{s^2(s+b)}$	$\frac{b^2-a_1b+a_o}{b^2}e^{-bt}+\frac{a_o}{b}t+\frac{a_1b-a_o}{b^2}$
100	$\frac{s^2+a_1s+a_o}{s^3(s+b)}$	$\frac{a_1b-b^2-a_o}{b^3}e^{-bt}+\frac{a_o}{2b}t^2+\frac{a_1b-a_o}{b^2}t+\frac{b^2-a_1b+a_o}{b^3}$
101	$\frac{s^2+a_1s+a_o}{s(s+b)(s+c)}$	$\frac{a_o}{bc}+\frac{b^2-a_1b+a_o}{b(b-c)}e^{-bt}+\frac{c^2-a_1c+a_o}{c(c-b)}e^{-ct}$
102	$\frac{s^2+a_1s+a_o}{s^2(s+b)(s+c)}$	$\frac{a_o}{bc}t+\frac{a_1bc-a_o(b+c)}{b^2c^2}+\frac{b^2-a_1b+a_o}{b^2(c-b)}e^{-bt}+\frac{c^2-a_1c+a_o}{c^2(b-c)}e^{-ct}$
103	$\frac{s^2+a_1s+a_o}{(s+b)(s+c)(s+d)}$	$\frac{b^2-a_1b+a_o}{(c-b)(d-b)}e^{-bt}+\frac{c^2-a_1c+a_o}{(b-c)(d-c)}e^{-ct}+\frac{d^2-a_1d+a_o}{(b-d)(c-d)}e^{-dt}$
104	$\frac{s^2+a_1s+a_o}{s(s+b)(s+c)(s+d)}$	$\frac{a_o}{bcd}-\frac{b^2-a_1b+a_o}{b(c-b)(d-b)}e^{-bt}-\frac{c^2-a_1c+a_o}{c(b-c)(d-c)}e^{-ct}-\frac{d^2-a_1d+a_o}{d(b-d)(c-d)}e^{-dt}$
105	$\frac{s^2+a_1s+a_o}{s(s+b)^2}$	$\frac{a_o}{b^2}-\frac{b^2-a_1b+a_o}{b}te^{-bt}+\frac{b^2-a_o}{b^2}e^{-bt}$
106	$\frac{s^2+a_1s+a_o}{s^2(s+b)^2}$	$\frac{a_o}{b^2}t+\frac{a_1b-2a_o}{b^3}+\frac{b^2-a_1b+a_o}{b^2}t\varepsilon^{-bt}+\frac{2a_o-a_1b}{b^3}e^{-bt}$
107	$\frac{s^2+a_1s+a_o}{(s+b)(s+c)^2}$	$\frac{b^2-a_1b+a_o}{(c-b)^2}e^{-bt}+\frac{c^2-a_1c+a_o}{(b-c)}te^{-ct}+\frac{c^2-2bc+a_1b-a_o}{(b-c)^2}e^{-ct}$
108	$\frac{s^3}{(s+b)(s+c)(s+d)^2}$	$\left\{\begin{array}{l}\frac{b^3}{(b-c)(d-b)^2}e^{-bt}+\frac{c^3}{(c-b)(d-c)^2}e^{-ct}+\frac{d^3}{(d-b)(c-d)}te^{-dt}\\+\frac{d^2[d^2-2d(b+c)+3bc]}{(b-d)^2(c-d)^2}e^{-dt}\end{array}\right.$

TABLE 2.3 Table of Laplace Transforms (continued)

	$F(s)$	$f(t)$
109	$\dfrac{s^3}{(s+b)(s+c)(s+d)(s+f)^2}$	$\begin{cases} \dfrac{b^3}{(b-c)(d-b)(f-b)^2}e^{-bt} + \dfrac{c^3}{(c-b)(d-c)(f-c)^2}e^{-ct} \\ + \dfrac{d^3}{(d-b)(c-d)(f-d)^2}e^{-dt} + \dfrac{f^3}{(f-b)(c-f)(d-f)}te^{-ft} \\ + \left[\dfrac{3f^2}{(b-f)(c-f)(d-f)}\right. \\ \left. + \dfrac{f^3[(b-f)(c-f)+(b-f)(d-f)+(c-f)(d-f)]}{(b-f)^2(c-f)^2(d-f)^2}\right]\varepsilon^{-dt} \end{cases}$
110	$\dfrac{s^3}{(s+b)^2(s+c)^2}$	$-\dfrac{b^3}{(c-b)^2}te^{-bt} + \dfrac{b^2(3c-b)}{(c-b)^3}e^{-bt} - \dfrac{c^3}{(b-c)^2}te^{-ct} + \dfrac{c^2(3b-c)}{(b-c)^3}e^{-ct}$
111	$\dfrac{s^3}{(s+d)(s+b)^2(s+c)^2}$	$\begin{cases} -\dfrac{d^3}{(b-d)^2(c-d)^2}e^{-dt} + \dfrac{b^3}{(c-b)^2(b-d)}te^{-bt} \\ + \left[\dfrac{3b^2}{(c-b)^2(d-b)} + \dfrac{b^3(c+2d-3b)}{(c-b)^3(d-b)^2}\right]e^{-bt} + \dfrac{c^3}{(b-c)^2(c-d)}te^{-ct} \\ + \left[\dfrac{3c^2}{(b-c)^2(d-c)} + \dfrac{c^3(b+2d-3c)}{(b-c)^3(d-c)^2}\right]e^{-ct} \end{cases}$
112	$\dfrac{s^3}{(s+b)(s+c)(s^2+\omega^2)}$	$\begin{cases} \dfrac{b^3}{(b-c)(b^2+\omega^2)}e^{-bt} + \dfrac{c^3}{(c-b)(c^2+\omega^2)}e^{-ct} \\ -\dfrac{\omega^2}{\sqrt{(b^2+\omega^2)(c^2+\omega^2)}}\sin(\omega t+\phi) \\ \phi = \tan^{-1}\left(\dfrac{c}{\omega}\right) - \tan^{-1}\left(\dfrac{\omega}{b}\right) \end{cases}$
113	$\dfrac{s^3}{(s+b)(s+c)(s+d)(s^2+\omega^2)}$	$\begin{cases} \dfrac{b^3}{(b-c)(d-b)(b^2+\omega^2)}e^{-bt} + \dfrac{c^3}{(c-b)(d-c)(c^2+\omega^2)}e^{-ct} \\ + \dfrac{d^3}{(d-b)(c-d)(d^2+\omega^2)}e^{-dt} \\ -\dfrac{\omega^2}{\sqrt{(b^2+\omega^2)(c^2+\omega^2)(d^2+\omega^2)}}\cos(\omega t-\phi) \\ \phi = \tan^{-1}\left(\dfrac{\omega}{b}\right) + \tan^{-1}\left(\dfrac{\omega}{c}\right) + \tan^{-1}\left(\dfrac{\omega}{d}\right) \end{cases}$
114	$\dfrac{s^3}{(s+b)^2(s^2+\omega^2)}$	$\begin{cases} -\dfrac{b^3}{b^2+\omega^2}te^{-bt} + \dfrac{b^2(b^2+3\omega^2)}{(b^2+\omega^2)^2}e^{-bt} - \dfrac{\omega^2}{(b^2+\omega^2)}\sin(\omega t+\phi) \\ \phi = \tan^{-1}\left(\dfrac{b}{\omega}\right) - \tan^{-1}\left(\dfrac{\omega}{b}\right) \end{cases}$
115	$\dfrac{s^3}{s^4+4\omega^4}$	$\cos(\omega t)\cosh(\omega t)$
116	$\dfrac{s^3}{s^4-\omega^4}$	$\tfrac{1}{2}[\cosh(\omega t)+\cos(\omega t)]$

TABLE 2.3 Table of Laplace Transforms (continued)

	$F(s)$	$f(t)$
117	$\frac{s^3+a_2s^2+a_1s+a_o}{s^2(s+b)(s+c)}$	$\frac{a_o}{bc}t-\frac{a_o(b+c)-a_1bc}{b^2c^2}+\frac{-b^3+a_2b^2-a_1b+a_o}{b^2(c-b)}e^{-bt}+\frac{-c^3+a_2c^2-a_1c+a_o}{c^2(b-c)}e^{-ct}$
118	$\frac{s^3+a_2s^2+a_1s+a_o}{s(s+b)(s+c)(s+d)}$	$\frac{a_o}{bcd}-\frac{-b^3+a_2b^2-a_1b+a_o}{b(c-b)(d-b)}e^{-bt}-\frac{-c^3+a_2c^2-a_1c+a_o}{c(b-c)(d-c)}e^{-ct}-\frac{-d^3+a_2d^2-a_1d+a_o}{d(b-d)(c-d)}e^{-dt}$
119	$\frac{s^3+a_2s^2+a_1s+a_o}{s^2(s+b)(s+c)(s+d)}$	$\frac{a_o}{bcd}t+\left[\frac{a_1}{bcd}-\frac{a_o(bc+bd+cd)}{b^2c^2d^2}\right]+\frac{-b^3+a_2b^2-a_1b+a_o}{b^2(c-b)(d-b)}\varepsilon^{-bt}+\frac{-c^3+a_2c^2-a_1c+a_o}{c^2(b-c)(d-c)}e^{-ct}+\frac{-d^3+a_2d^2-a_1d+a_o}{d^2(b-d)(c-d)}e^{-dt}$
120	$\frac{s^3+a_2s^2+a_1s+a_o}{(s+b)(s+c)(s+d)(s+f)}$	$\frac{-b^3+a_2b^2-a_1b+a_o}{(c-b)(d-b)(f-b)}e^{-bt}+\frac{-c^3+a_2c^2-a_1c+a_o}{(b-c)(d-c)(f-c)}e^{-ct}+\frac{-d^3+a_2d^2-a_1d+a_o}{(b-d)(c-d)(f-d)}e^{-dt}+\frac{-f^3+a_2f^2-a_1f+a_o}{(b-f)(c-f)(d-f)}e^{-ft}$
121	$\frac{s^3+a_2s^2+a_1s+a_o}{s(s+b)(s+c)(s+d)(s+f)}$	$\frac{a_o}{bcdf}-\frac{-b^3+a_2b^2-a_1b+a_o}{b(c-b)(d-b)(f-b)}e^{-bt}-\frac{-c^3+a_2c^2-a_1c+a_o}{c(b-c)(d-c)(f-c)}e^{-ct}-\frac{-d^3+a_2d^2-a_1d+a_o}{d(b-d)(c-d)(f-d)}e^{-dt}-\frac{-f^3+a_2f^2-a_1f+a_o}{f(b-f)(c-f)(d-f)}e^{-ft}$
122	$\frac{s^3+a_2s^2+a_1s+a_o}{(s+b)(s+c)(s+d)(s+f)(s+g)}$	$\frac{-b^3+a_2b^2-a_1b+a_o}{(c-b)(d-b)(f-b)(g-b)}e^{-bt}+\frac{-c^3+a_2c^2-a_1c+a_o}{(b-c)(d-c)(f-c)(g-c)}e^{-ct}+\frac{-d^3+a_2d^2-a_1d+a_o}{(b-d)(c-d)(f-d)(g-d)}e^{-dt}+\frac{-f^3+a_2f^2-a_1f+a_o}{(b-f)(c-f)(d-f)(g-f)}e^{-ft}+\frac{-g^3+a_2g^2-a_1g+a_o}{(b-g)(c-g)(d-g)(f-g)}e^{-gt}$
123	$\frac{s^3+a_2s^2+a_1s+a_o}{(s+b)(s+c)(s+d)^2}$	$\frac{-b^3+a_2b^2-a_1b+a_o}{(c-b)(d-b)^2}e^{-bt}+\frac{-c^3+a_2c^2-a_1c+a_o}{(b-c)(d-c)^2}e^{-ct}+\frac{-d^3+a_2d^2-a_1d+a_o}{(b-d)(c-d)}te^{-dt}+\frac{a_o(2d-b-c)+a_1(bc-d^2)+a_2d(db+dc-2bc)+d^2(d^2-2db-2dc+3bc)}{(b-d)^2(c-d)^2}e^{-dt}$
124	$\frac{s^3+a_2s^2+a_1s+a_o}{s(s+b)(s+c)(s+d)^2}$	$\frac{a_o}{bcd^2}-\frac{-b^3+a_2b^2-a_1b+a_o}{b(c-b)(d-b)^2}e^{-bt}-\frac{-c^3+a_2c^2-a_1c+a_o}{c(b-c)(d-c)^2}e^{-ct}-\frac{-d^3+a_2d^2-a_1d+a_o}{d(b-d)(c-d)}te^{-dt}-\frac{3d^2-2a_2d+a_1}{d(b-d)(c-d)}e^{-dt}-\frac{(-d^3+a_2d^2-a_1d+a_o)[(b-d)(c-d)-d(b-d)-d(c-d)]}{d^2(b-d)^2(c-d)^2}e^{-dt}$

TABLE 2.3 Table of Laplace Transforms (continued)

	$F(s)$	$f(t)$
125	$\frac{s^3+a_2s^2+a_1s+a_o}{(s+b)(s+c)(s+d)(s+f)^2}$	$\frac{-b^3+a_2b^2-a_1b+a_o}{(c-b)(d-b)(f-b)^2}e^{-bt}+\frac{-c^3+a_2c^2-a_1c+a_o}{(b-c)(d-c)(f-c)^2}e^{-ct}+\frac{-d^3+a_2d^2-a_1d+a_o}{(b-d)(c-d)(f-d)^2}e^{-dt}+\frac{-f^3+a_2f^2-a_1f+a_o}{(b-f)(c-f)(d-f)}te^{-ft}+\frac{3f^2-2a_2f+a_1}{(b-f)(c-f)(d-f)}e^{-ft}-\frac{(-f^3+a_2f^2-a_1f+a_o)[(b-f)(c-f)+(b-f)(d-f)+(c-f)(d-f)]}{(b-f)^2(c-f)^2(d-f)^2}e^{-ft}$
126	$\frac{s}{(s-a)^{3/2}}$	$\frac{1}{\sqrt{\pi t}}e^{at}(1+2at)$
127	$\sqrt{s-a}-\sqrt{s-b}$	$\frac{1}{2\sqrt{\pi t^3}}(e^{bt}-e^{at})$
128	$\frac{1}{\sqrt{s}+a}$	$\frac{1}{\sqrt{\pi t}}-ae^{a^2t}\operatorname{erfc}(a\sqrt{t})$
129	$\frac{\sqrt{s}}{s-a^2}$	$\frac{1}{\sqrt{\pi t}}+ae^{a^2t}\operatorname{erf}(a\sqrt{t})$
130	$\frac{\sqrt{s}}{s+a^2}$	$\frac{1}{\sqrt{\pi t}}-\frac{2a}{\sqrt{\pi}}e^{-a^2t}\int_0^{a\sqrt{t}}e^{\lambda^2}\,d\lambda$
131	$\frac{1}{\sqrt{s}(s-a^2)}$	$\frac{1}{a}e^{a^2t}\operatorname{erf}(a\sqrt{t})$
132	$\frac{1}{\sqrt{s}(s+a^2)}$	$\frac{2}{a\sqrt{\pi}}e^{-a^2t}\int_0^{a\sqrt{t}}e^{\lambda^2}\,d\pi$
133	$\frac{b^2-a^2}{(s-a^2)(b+\sqrt{s})}$	$e^{a^2t}[b-a\operatorname{erf}(a\sqrt{t})]-be^{b^2t}\operatorname{erfc}(b\sqrt{t})$
134	$\frac{1}{\sqrt{s}(\sqrt{s}+a)}$	$e^{a^2t}\operatorname{erfc}(a\sqrt{t})$
135	$\frac{1}{(s+a)\sqrt{s+b}}$	$\frac{1}{\sqrt{b-a}}e^{-at}\operatorname{erf}(\sqrt{b-a}\sqrt{t})$
136	$\frac{b^2-a^2}{\sqrt{s}(s-a^2)(\sqrt{s}+b)}$	$e^{a^2t}\left[\frac{b}{a}\operatorname{erf}(a\sqrt{t})-1\right]+e^{b^2t}\operatorname{erfc}(b\sqrt{t})$
137	$\frac{(1-s)^n}{s^{n+(1/2)}}$	$\frac{n!}{(2n)!\sqrt{\pi t}}H_{2n}(\sqrt{t})$ $\left[H_n(t)=\text{Hermite polynomial}=e^{x^2}\frac{d^n}{dx^n}(e^{-x^2})\right]$
138	$\frac{(1-s)^n}{s^{n+(3/2)}}$	$-\frac{n!}{\sqrt{\pi}(2n+1)!}H_{2n+1}(\sqrt{t})$
139	$\frac{\sqrt{s+2a}}{\sqrt{s}}-1$	$ae^{-at}[I_1(at)+I_o(at)]$ $[I_n(t)=j^{-n}J_n(jt)$ where J_n is Bessel's function of the first kind$]$
140	$\frac{1}{\sqrt{s+a}\sqrt{s+b}}$	$e^{-(1/2)(a+b)t}I_o\left(\frac{a-b}{2}t\right)$

TABLE 2.3 Table of Laplace Transforms (continued)

	$F(s)$	$f(t)$
141	$\dfrac{\Gamma(k)}{(s+a)^k(s+b)^k}(k \ge 0)$	$\sqrt{\pi}\left(\dfrac{t}{a-b}\right)^{k-(1/2)} e^{-(1/2)(a+b)t} I_{k-(1/2)}\left(\dfrac{a-b}{2}t\right)$
142	$\dfrac{1}{(s+a)^{1/2}(s+b)^{3/2}}$	$te^{-(1/2)(a+b)t}\left[I_o\left(\dfrac{a-b}{2}t\right)+I_1\left(\dfrac{a-b}{2}t\right)\right]$
143	$\dfrac{\sqrt{s+2a}-\sqrt{s}}{\sqrt{s+2a}+\sqrt{s}}$	$\dfrac{1}{t}e^{-at}I_1(at)$
144	$\dfrac{(a-b)^k}{(\sqrt{s+a}+\sqrt{s+b})^{2k}} \quad (k>0)$	$\dfrac{k}{t}e^{-(1/2)(a+b)t}I_k\left(\dfrac{a-b}{2}t\right)$
145	$\dfrac{(\sqrt{s+a}+\sqrt{s})^{-2v}}{\sqrt{s}\sqrt{s+a}}$	$\dfrac{1}{a^v}e^{-(1/2)(at)}I_v\left(\dfrac{1}{2}at\right)$
146	$\dfrac{1}{\sqrt{s^2+a^2}}$	$J_o(at)$
147	$\dfrac{(\sqrt{s^2+a^2}-s)^v}{\sqrt{s^2+a^2}} \quad (v>-1)$	$a^v J_v(at)$
148	$\dfrac{1}{(s^2+a^2)^k} \quad (k>0)$	$\dfrac{\sqrt{\pi}}{\Gamma(k)}\left(\dfrac{t}{2a}\right)^{k-(1/2)} J_{k-(1/2)}(at)$
149	$(\sqrt{s^2+a^2}-s)^k \quad (k>0)$	$\dfrac{ka^k}{t}J_k(at)$
150	$\dfrac{(s-\sqrt{s^2-a^2})^v}{\sqrt{s^2-a^2}} \quad (v>-1)$	$a^v I_v(at)$
151	$\dfrac{1}{(s^2-a^2)^k} \quad (k>0)$	$\dfrac{\sqrt{\pi}}{\Gamma(k)}\left(\dfrac{t}{2a}\right)^{k-(1/2)} I_{k-(1/2)}(at)$
152	$\dfrac{1}{s\sqrt{s+1}}$	$\text{erf}(\sqrt{t})$; $\text{erf}(y) \stackrel{\Delta}{=}$ the error function $= \dfrac{2}{\sqrt{\pi}}\int_o^y e^{-u^2}du$
153	$\dfrac{1}{\sqrt{s^2+a^2}}$	$J_o(at)$; Bessel function of 1st kind, zero order
154	$\dfrac{1}{\sqrt{s^2+a^2}+s}$	$\dfrac{J_1(at)}{at}$; J_1 is the Bessel function of 1st kind, 1st order
155	$\dfrac{1}{[\sqrt{s^2+a^2}+s]^N}$	$\dfrac{N}{a^N}\dfrac{J_N(at)}{t}$; $N=1,2,3,\cdots$, J_N is the Bessel function of 1st kind, Nth order
156	$\dfrac{1}{s[\sqrt{s^2+a^2}+s]^N}$	$\dfrac{N}{a^N}\int_o^t \dfrac{J_N(au)}{u}du$; $N=1,2,3,\cdots$, J_N is the Bessel function of 1st kind, Nth order
157	$\dfrac{1}{\sqrt{s^2+a^2}(\sqrt{s^2+a^2}+s)}$	$\dfrac{1}{a}J_1(at)$; J_1 is the Bessel function of 1st kind, 1st order
158	$\dfrac{1}{\sqrt{s^2+a^2}[\sqrt{s^2+a^2}+s]^N}$	$\dfrac{1}{a^N}J_N(at)$; $N=1,2,3,\cdots$, J_N is the Bessel function of 1st kind, Nth order
159	$\dfrac{1}{\sqrt{s^2-a^2}}$	$I_o(at)$; I_o is the modified Bessel function of 1st kind, zero order
160	$\dfrac{e^{-ks}}{s}$	$S_k(t)=\begin{cases}0 & \text{when } 0<t<k \\ 1 & \text{when } t>k\end{cases}$

TABLE 2.3 Table of Laplace Transforms (continued)

	$F(s)$	$f(t)$
161	$\frac{e^{-ks}}{s^2}$	$\begin{cases} 0 & \text{when } 0<t<k \\ t-k & \text{when } t>k \end{cases}$
162	$\frac{e^{-ks}}{s^\mu}\quad(\mu>0)$	$\begin{cases} 0 & \text{when } 0<t<k \\ \frac{(t-k)^{\mu-1}}{\Gamma(\mu)} & \text{when } t>k \end{cases}$
163	$\frac{1-e^{-ks}}{s}$	$\begin{cases} 1 & \text{when } 0<t<k \\ 0 & \text{when } t>k \end{cases}$
164	$\frac{1}{s(1-e^{-ks})}=\frac{1+\coth\frac{1}{2}ks}{2s}$	$S(k,t)=\begin{cases} n & \text{when} \\ & (n-1)k<t<n\ k(n=1,2,\cdots) \end{cases}$
165	$\frac{1}{s(e^{+ks}-a)}$	$S_k(t)=\begin{cases} 0 & \text{when } 0<t<k \\ 1+a+a^2+\cdots+a^{n-1} \\ \text{when } nk<t<(n+1)k\ (n=1,2,\cdots) \end{cases}$
166	$\frac{1}{s}\tanh ks$	$\begin{cases} M(2k,t)=(-1)^{n-1} \\ \quad\text{when } 2k(n-1)<t<2nk \\ \quad(n=1,2,\cdots) \end{cases}$
167	$\frac{1}{s(1+e^{-ks})}$	$\begin{cases} \frac{1}{2}M(k,t)+\frac{1}{2}=\frac{1-(-1)^n}{2} \\ \quad\text{when } (n-1)k<t<nk \end{cases}$
168	$\frac{1}{s^2}\tanh ks$	$\begin{cases} H(2k,t) & [H(2k,t)=k+(r-k)(-1)^n \text{ where } t=2kn+r; \\ & 0\le r\le 2k;\ n=0,1,2,\cdots] \end{cases}$
169	$\frac{1}{s\sinh ks}$	$\begin{cases} 2S(2k,t+k)-2=2(n-1) \\ \quad\text{when } (2n-3)k<t<(2n-1)k\ (t>0) \end{cases}$
170	$\frac{1}{s\cosh ks}$	$\begin{cases} M(2k,t+3k)+1=1+(-1)^n \\ \quad\text{when } (2n-3)k<t<(2n-1)k\ (t>0) \end{cases}$
171	$\frac{1}{s}\coth ks$	$\begin{cases} 2S(2k,t)-1=2n-1 \\ \quad\text{when } 2k(n-1)<t<2kn \end{cases}$
172	$\frac{k}{s^2+k^2}\coth\frac{\pi s}{2k}$	$\lvert\sin kt\rvert$
173	$\frac{1}{(s^2+1)(1-e^{-\pi s})}$	$\begin{cases} \sin t & \text{when } (2n-2)\pi<t<(2n-1)\pi \\ 0 & \text{when } (2n-1)\pi<t<2n\pi \end{cases}$
174	$\frac{1}{s}e^{-k/s}$	$J_o(2\sqrt{kt})$
175	$\frac{1}{\sqrt{s}}e^{-k/s}$	$\frac{1}{\sqrt{\pi t}}\cos 2\sqrt{kt}$
176	$\frac{1}{\sqrt{s}}e^{k/s}$	$\frac{1}{\sqrt{\pi t}}\cosh 2\sqrt{kt}$
177	$\frac{1}{s^{3/2}}e^{-k/s}$	$\frac{1}{\sqrt{\pi k}}\sin 2\sqrt{kt}$
178	$\frac{1}{s^{3/2}}e^{k/s}$	$\frac{1}{\sqrt{\pi k}}\sinh 2\sqrt{kt}$

TABLE 2.3 Table of Laplace Transforms (continued)

	$F(s)$	$f(t)$
179	$\frac{1}{s^{\mu}} e^{-k/s} \quad (\mu > 0)$	$\left(\frac{t}{k}\right)^{(\mu-1)/2} J_{\mu-1}(2\sqrt{kt})$
180	$\frac{1}{s^{\mu}} e^{k/s} \quad (\mu > 0)$	$\left(\frac{t}{k}\right)^{(\mu-1)/2} I_{\mu-1}(2\sqrt{kt})$
181	$e^{-k\sqrt{s}} \quad (k > 0)$	$\frac{k}{2\sqrt{\pi t^3}} \exp\left(-\frac{k^2}{4t}\right)$
182	$\frac{1}{s} e^{-k\sqrt{s}} \quad (k \ge 0)$	$\operatorname{erfc}\left(\frac{k}{2\sqrt{t}}\right)$
183	$\frac{1}{\sqrt{s}} e^{-k\sqrt{s}} \quad (k \ge 0)$	$\frac{1}{\sqrt{\pi t}} \exp\left(-\frac{k^2}{4t}\right)$
184	$s^{-3/2} e^{-k\sqrt{s}} \quad (k \ge 0)$	$2\sqrt{\frac{t}{\pi}} \exp\left(-\frac{k^2}{4t}\right) - k \operatorname{erfc}\left(\frac{k}{2\sqrt{t}}\right)$
185	$\frac{a e^{-k\sqrt{s}}}{s(a+\sqrt{s})} \quad (k \ge 0)$	$-e^{ak} e^{a^2 t} \operatorname{erfc}\left(a\sqrt{t} + \frac{k}{2\sqrt{t}}\right) + \operatorname{erfc}\left(\frac{k}{2\sqrt{t}}\right)$
186	$\frac{e^{-k\sqrt{s}}}{\sqrt{s}(a+\sqrt{s})} \quad (k \ge 0)$	$e^{ak} e^{a^2 t} \operatorname{erfc}\left(a\sqrt{t} + \frac{k}{2\sqrt{t}}\right)$
187	$\frac{e^{-k\sqrt{s(s+a)}}}{\sqrt{s(s+a)}}$	$\begin{cases} 0 & \text{when } 0 < t < k \\ e^{-(1/2)at} I_o\left(\frac{1}{2} a\sqrt{t^2 - k^2}\right) & \text{when } t > k \end{cases}$
188	$\frac{e^{-k\sqrt{s^2+a^2}}}{\sqrt{(s^2+a^2)}}$	$\begin{cases} 0 & \text{when } 0 < t < k \\ J_o(a\sqrt{t^2 - k^2}) & \text{when } t > k \end{cases}$
189	$\frac{e^{-k\sqrt{s^2-a^2}}}{\sqrt{(s^2-a^2)}}$	$\begin{cases} 0 & \text{when } 0 < t < k \\ I_o(a\sqrt{t^2 - k^2}) & \text{when } t > k \end{cases}$
190	$\frac{e^{-k(\sqrt{s^2+a^2}-s)}}{\sqrt{(s^2+a^2)}} \quad (k \ge 0)$	$J_o(a\sqrt{t^2 + 2kt})$
191	$e^{-ks} - e^{-k\sqrt{s^2+a^2}}$	$\begin{cases} 0 & \text{when } 0 < t < k \\ \frac{ak}{\sqrt{t^2-k^2}} J_1(a\sqrt{t^2 - k^2}) & \text{when } t > k \end{cases}$
192	$e^{-k\sqrt{s^2+a^2}} - e^{-ks}$	$\begin{cases} 0 & \text{when } 0 < t < k \\ \frac{ak}{\sqrt{t^2-k^2}} I_1(a\sqrt{t^2 - k^2}) & \text{when } t > k \end{cases}$
193	$\frac{a^{v} e^{-k\sqrt{s^2-a^2}}}{\sqrt{(s^2+a^2)}\left(\sqrt{s^2+a^2}+s\right)^{v}}$ $(v > -1)$	$\begin{cases} 0 & \text{when } 0 < t < k \\ \left(\frac{t-k}{t+k}\right)^{(1/2)v} J_v(a\sqrt{t^2 - k^2}) & \text{when } t > k \end{cases}$
194	$\frac{1}{s} \log s$	$\Gamma'(1) - \log t \quad [\Gamma'(1) = -0.5772]$
195	$\frac{1}{s^k} \log s \quad (k > 0)$	$t^{k-1}\left\{\frac{\Gamma'(k)}{[\Gamma(k)]^2} \frac{\log t}{\Gamma(k)}\right\}$
196	$\frac{\log s}{s-a} \quad (a > 0)$	$e^{at}[\log a - \operatorname{Ei}(-at)]$

TABLE 2.3 Table of Laplace Transforms (continued)

	$F(s)$	$f(t)$
197	$\frac{\log s}{s^2+1}$	$\cos t\,\mathrm{Si}(t) - \sin t\,\mathrm{Ci}(t)$
198	$\frac{s\log s}{s^2+1}$	$-\sin t\,\mathrm{Si}(t) - \cos t\,\mathrm{Ci}(t)$
199	$\frac{1}{s}\log(1+ks)\ (k>0)$	$-\mathrm{Ei}\left(-\frac{t}{k}\right)$
200	$\log\frac{s-a}{s-b}$	$\frac{1}{t}(e^{bt}-e^{at})$
201	$\frac{1}{s}\log(1+k^2s^2)$	$-2\mathrm{Ci}\left(\frac{t}{k}\right)$
202	$\frac{1}{s}\log(s^2+a^2)\ (a>0)$	$2\log a - 2\mathrm{Ci}(at)$
203	$\frac{1}{s^2}\log(s^2+a^2)\ (a>0)$	$\frac{2}{a}[at\log a + \sin at - at\,\mathrm{Ci}(at)]$
204	$\log\frac{s^2+a^2}{s^2}$	$\frac{2}{t}(1-\cos at)$
205	$\log\frac{s^2-a^2}{s^2}$	$\frac{2}{t}(1-\cosh at)$
206	$\arctan\frac{k}{s}$	$\frac{1}{t}\sin kt$
207	$\frac{1}{s}\arctan\frac{k}{s}$	$\mathrm{Si}(kt)$
208	$e^{k^2s^2}\mathrm{erfc}(ks)\ (k>0)$	$\frac{1}{k\sqrt{\pi}}\exp\left(-\frac{t^2}{4k^2}\right)$
209	$\frac{1}{s}e^{k^2s^2}\mathrm{erfc}(ks)\ (k>0)$	$\mathrm{erf}\left(\frac{t}{2k}\right)$
210	$e^{ks}\mathrm{erfc}(\sqrt{ks})\ (k>0)$	$\frac{\sqrt{k}}{\pi\sqrt{t(t+k)}}$
211	$\frac{1}{\sqrt{s}}\mathrm{erfc}(\sqrt{ks})$	$\begin{cases}0 & \text{when } 0<t<k\\ (\pi t)^{-1/2} & \text{when } t>k\end{cases}$
212	$\frac{1}{\sqrt{s}}e^{ks}\mathrm{erfc}(\sqrt{ks})\ (k>0)$	$\frac{1}{\sqrt{\pi(t+k)}}$
213	$\mathrm{erf}\left(\frac{k}{\sqrt{s}}\right)$	$\frac{1}{\pi t}\sin(2k\sqrt{t})$
214	$\frac{1}{\sqrt{s}}e^{k^2/s}\mathrm{erfc}\left(\frac{k}{\sqrt{s}}\right)$	$\frac{1}{\sqrt{\pi t}}e^{-2k\sqrt{t}}$
215	$-e^{as}\mathrm{Ei}(-as)$	$\frac{1}{t+a};\ (a>0)$
216	$\frac{1}{a}+se^{as}\mathrm{Ei}(-as)$	$\frac{1}{(t+a)^2};\ (a>0)$
217	$\left[\frac{\pi}{2}-\mathrm{Si}(s)\right]\cos s + \mathrm{Ci}(s)\sin s$	$\frac{1}{t^2+1}$

TABLE 2.3 Table of Laplace Transforms (continued)

	$F(s)$	$f(t)$	
218	$K_o(ks)$	$\begin{cases} 0 & \text{when } 0<t<k \\ (t^2-k^2)^{-1/2} & \text{when } t>k \end{cases}$	[$K_n(t)$ is Bessel function of the second kind of imaginary argument]
219	$K_o(k\sqrt{s})$	$\frac{1}{2t}\exp\left(-\frac{k^2}{4t}\right)$	
220	$\frac{1}{s}e^{ks}K_1(ks)$	$\frac{1}{k}\sqrt{t(t+2k)}$	
221	$\frac{1}{\sqrt{s}}K_1(k\sqrt{s})$	$\frac{1}{k}\exp\left(-\frac{k^2}{4t}\right)$	
222	$\frac{1}{\sqrt{s}}e^{k/s}K_o\left(\frac{k}{s}\right)$	$\frac{2}{\sqrt{\pi t}}K_o(2\sqrt{2kt})$	
223	$\pi e^{-ks}I_o(ks)$	$\begin{cases} [t(2k-t)]^{-1/2} & \text{when } 0<t<2k \\ 0 & \text{when } t>2k \end{cases}$	
224	$e^{-ks}I_1(ks)$	$\begin{cases} \frac{k-t}{\pi k\sqrt{t(2k-t)}} & \text{when } 0<t<2k \\ 0 & \text{when } t>2k \end{cases}$	
225	$\frac{1}{s\sinh(as)}$	$2\sum_{k=0}^{\infty} u[t-(2k+1)a]$ 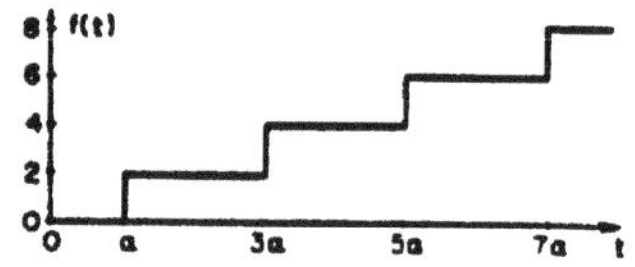	
226	$\frac{1}{s\cosh s}$	$2\sum_{k=0}^{\infty} (-1)^k u(t-2k-1)$ 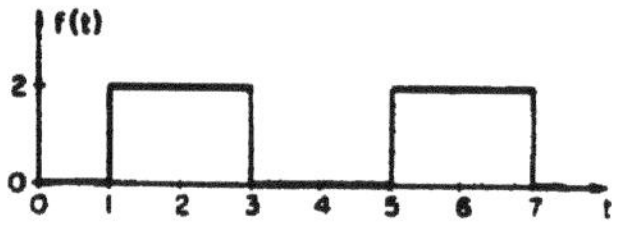	
227	$\frac{1}{s}\tanh\left(\frac{as}{2}\right)$	$u(t)+2\sum_{k=1}^{\infty} (-1)^k u(t-ak)$ square wave 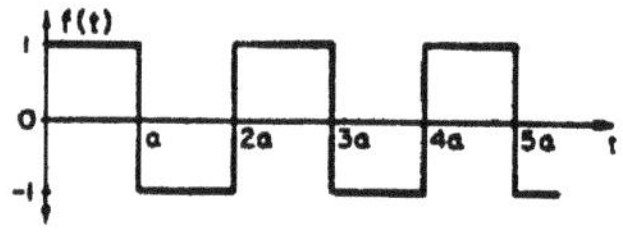	
228	$\frac{1}{2s}\left(1+\coth\frac{as}{2}\right)$	$\sum_{k=0}^{\infty} u(t-ak)$ stepped function 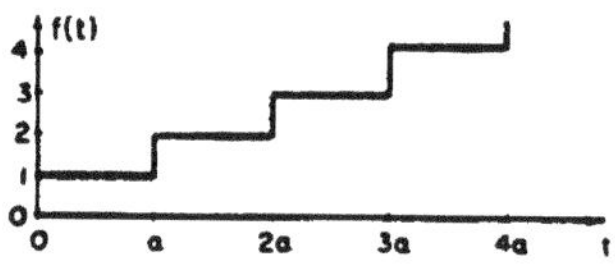	

TABLE 2.3 Table of Laplace Transforms (continued)

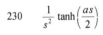

	$F(s)$	$f(t)$
229	$\frac{m}{s^2} - \frac{ma}{2s}\left(\coth\frac{as}{2} - 1\right)$	$mt - ma\sum_{k=1}^{\infty} u(t - ka)$ saw – tooth function 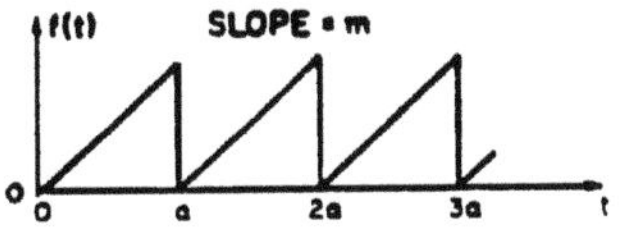
230	$\frac{1}{s^2}\tanh\left(\frac{as}{2}\right)$	$\frac{1}{a}\left[t + 2\sum_{k=1}^{\infty}(-1)^k(t - ka)\cdot u(t - ka)\right]$ triangular wave 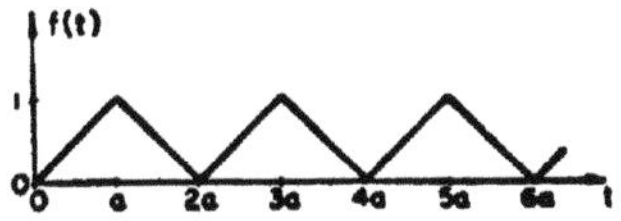
231	$\frac{1}{s(1 + e^{-s})}$	$\sum_{k=0}^{\infty}(-1)^k u(t - k)$ 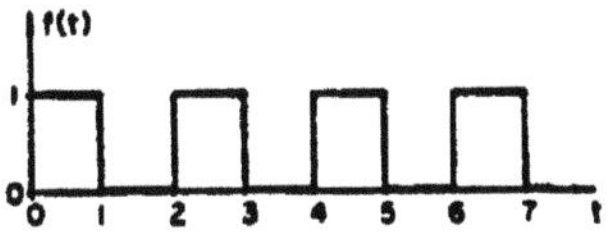
232	$\frac{a}{(s^2 + a^2)(1 - e^{-\frac{\pi}{a}s})}$	$\sum_{k=0}^{\infty}\left[\sin a\left(t - k\frac{\pi}{a}\right)\right]\cdot u\left(t - k\frac{\pi}{a}\right)$ half – wave rectification of sine wave 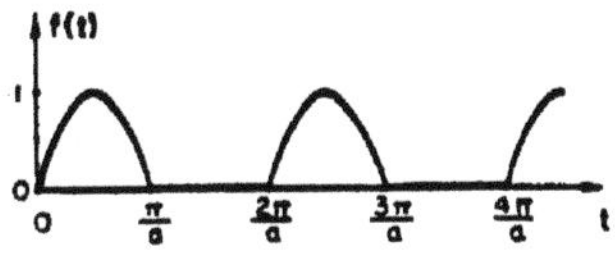
233	$\left[\frac{a}{(s^2 + a^2)}\right]\coth\left(\frac{\pi s}{2a}\right)$	$[\sin(at)]\cdot u(t) + 2\sum_{k=1}^{\infty}\left[\sin a\left(t - k\frac{\pi}{a}\right)\right]\cdot u\left(t - k\frac{\pi}{a}\right)$ full – wave rectification of sine wave 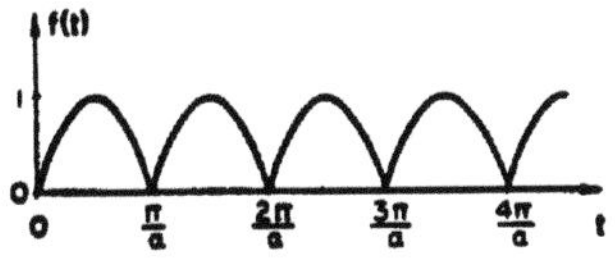
234	$\frac{1}{s}e^{-as}$	$u(t - a)$ 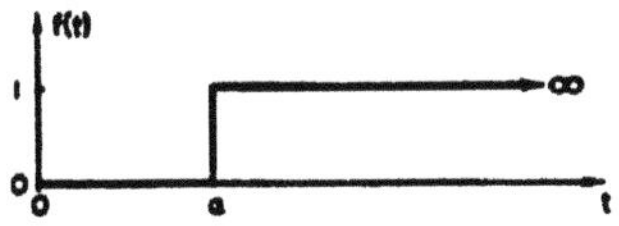

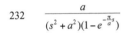

TABLE 2.3 Table of Laplace Transforms (continued)

	$F(s)$	$f(t)$
235	$\frac{1}{s}(e^{-as} - e^{-bs})$	$u(t-a) - u(t-b)$ 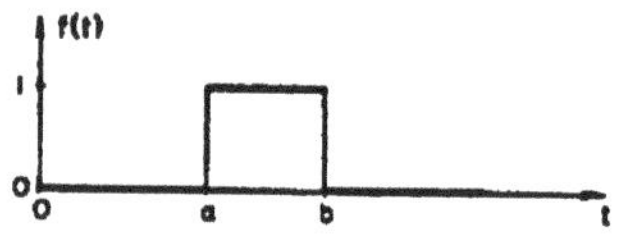
236	$\frac{m}{s^2} e^{-as}$	$m \cdot (t-a) \cdot u(t-a)$ 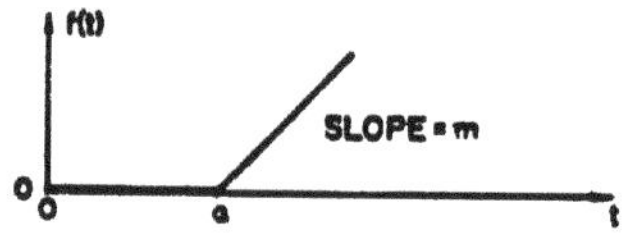
237	$\left[\frac{ma}{s} + \frac{m}{s^2}\right] e^{-as}$	$mt \cdot u(t-a)$ or $\left[ma + m(t-a)\right] \cdot u(t-a)$ 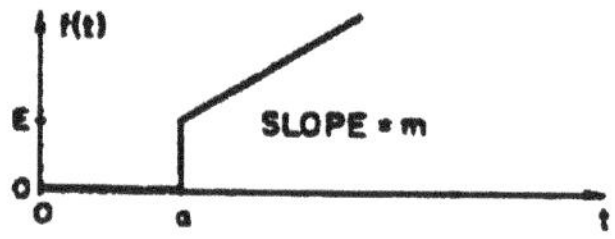
238	$\frac{2}{s^3} e^{-as}$	$(t-a)^2 \cdot u(t-a)$ 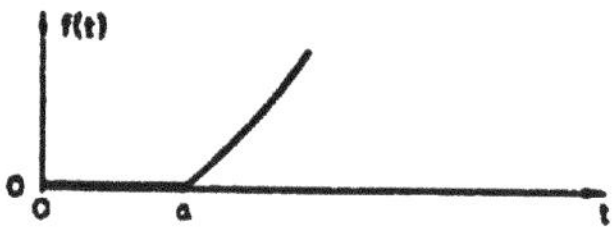
239	$\left[\frac{2}{s^3} + \frac{2a}{s^2} + \frac{a^2}{s}\right] e^{-as}$	$t^2 \cdot u(t-a)$ 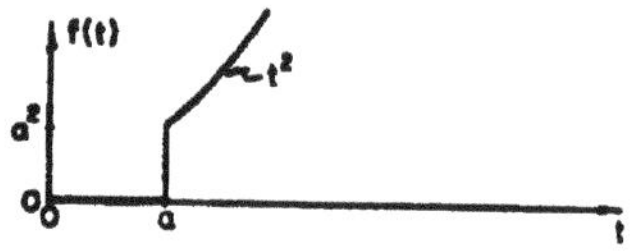
240	$\frac{m}{s^2} - \frac{m}{s^2} e^{-as}$	$mt \cdot u(t) - m(t-a) \cdot u(t-a)$ 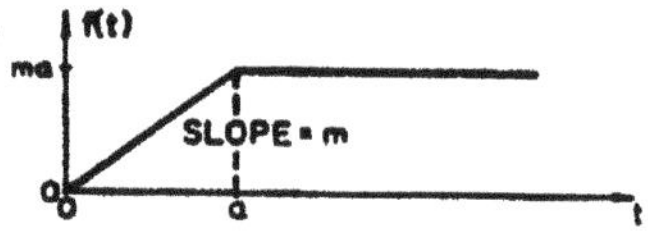
241	$\frac{m}{s^2} - \frac{2m}{s^2} e^{-as} + \frac{m}{s^2} e^{-2as}$	$mt - 2m(t-a) \cdot u(t-a) + m(t-2a) \cdot u(t-2a)$ 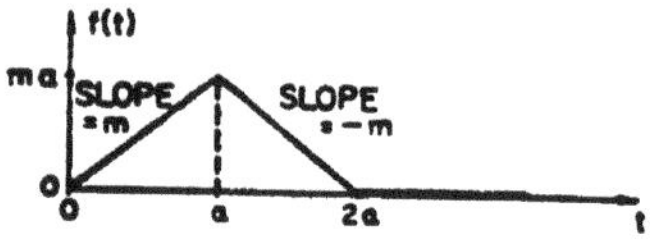
242	$\frac{m}{s^2} - \left(\frac{ma}{s} + \frac{m}{s^2}\right) e^{-as}$	$mt - \left[ma + m(t-a)\right] \cdot u(t-a)$ 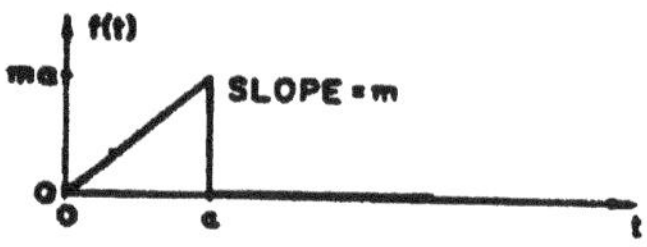

TABLE 2.3 Table of Laplace Transforms (continued)

	$F(s)$	$f(t)$
243	$\frac{(1-e^{-s})^2}{s^3}$	$0.5t^2$ for $0 \le t < 1$ $1-0.5(t-2)^2$ for $0 \le t < 2$ 1 for $2 \le t$ 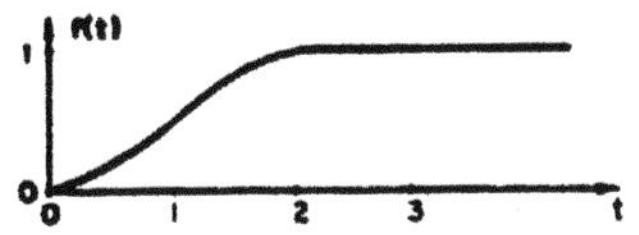
244	$\left[\frac{(1-e^{-s})}{s}\right]^3$	$0.5t^2$ for $0 \le t < 1$ $0.75-(t-1.5)^2$ for $1 \le t < 2$ $0.5(t-3)^2$ for $2 \le t < 3$ 0 for $3 < t$ 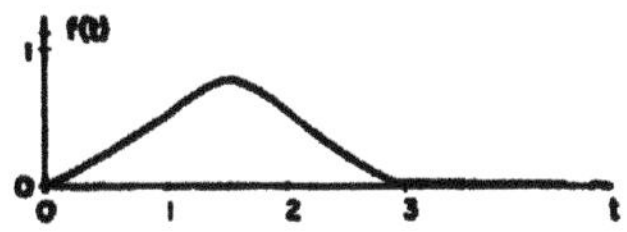
245	$\frac{b}{s(s-b)}+(e^{ba}-1)\left[\frac{1}{s+b}-\frac{s+\frac{b}{e^{ba}-1}}{s(s-b)}\right]e^{-as}$	$(e^{bt}-1)\cdot u(t)-(e^{bt}-1)\cdot u(t-a)+Ke^{-b(t-a)}\cdot u(t-a)$ where $K=(e^{ba}-1)$ 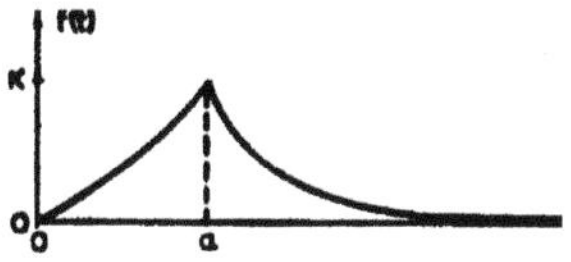

References

W. H. Beyer, *CRC Standard Mathematical Tables*, 2nd Ed., CRC Press, Boca Raton, FL, 1982.

R. V. Churchill, *Modern Operational Mathematics in Engineering*, McGraw-Hill Book Co., New York, NY, 1944.

W. Magnus, F. Oberhettinger, and F. G. Tricom, *Tables of Integral Transforms, Vol. I*, McGraw-Hill Book Co., New York, NY, 1954.

P. A. McCollum and B. F. Brown, *Laplace Transform Tables and Theorems*, Holt Rinehart and Winston, New York, NY, 1965.

Appendix 1

Examples

1.1 Laplace Transformations

Example 2.1 (Inversion)

The inverse of $\dfrac{s^2+a}{s^2(s+b)}$ is found by partial expansion

$$\frac{s^2+a}{s^2(s+b)} = \frac{A}{s} + \frac{B}{s^2} + \frac{c}{s+b}; \qquad B = \left.\frac{s^2+a}{s+b}\right|_{s=0} = \frac{a}{b}, \qquad C = \left.\frac{s^2+a}{s^2}\right|_{s=-b} = \frac{b^2+a}{b^2}.$$

Hence

$$\frac{s^2+a}{s^2(s+b)} = \frac{A}{s} + \frac{a}{b}\frac{1}{s^2} + \frac{b^2+a}{b^2}\frac{1}{s+b}.$$

Set any value of s, e.g., $s = 1$, and solve for $A = -\dfrac{a}{b^2}$.
Hence

$$L^{-1}\left\{\frac{s^2+a}{s^2(s+b)}\right\} = -\frac{a}{b^2}L^{-1}\left\{\frac{1}{s}\right\} + \frac{a}{b}L^{-1}\left\{\frac{1}{s^2}\right\} + \frac{b^2+a}{b^2}L^{-1}\left\{\frac{1}{s+b}\right\} = -\frac{a}{b^2}u(t) + \frac{a}{b}t + \frac{b^2+a}{b^2}e^{-bt}.$$

Example 2.2 (Differential equation)

To solve $y' + by = e^{-t}$ with $y(0) = 1$ we take the Laplace transform of both sides. Hence we obtain $sY(s) - y(0) + bY(s) = \frac{1}{s+1}$ or $Y(s) = \frac{1}{s+b} + \frac{1}{(s+1)(s+b)}$. The inverse transform is $y(t) = e^{-bt} + L^{-1}\{\frac{1}{-1+b}\frac{1}{s+1} + \frac{1}{1-b}\frac{1}{s+b}\} = e^{-bt} + \frac{1}{b-1}e^{-t} + \frac{1}{1-b}e^{-bt} = \frac{2-b}{1-b}e^{-bt} - \frac{1}{1-b}e^{-t}$

1.2 Inversion in the Complex Plane

When the Laplace transform $F(s)$ is known, the function of time can be found by (2.1.2), which is rewritten

$$f(t) = L^{-1}\{F(s)\} = \frac{1}{2\pi j}\int_{\sigma-j\infty}^{\sigma+j\infty} F(s)e^{st}\,ds$$

This equation applies equally well to both the two-sided and the one-sided transforms.

The path of integration is restricted to values of σ for which the direct transform formula converges. In fact, for the two-sided Laplace transform, the region of convergence must be specified in order to determine uniquely the inverse transform. That is, for the two-sided transform, the regions of convergence for functions of time that are zero for $t > 0$, zero for $t < 0$, or in neither category, must be distinguished. For the one-sided transform, the region of convergence is given by σ, where σ is the abscissa of absolute convergence.

The path of integration is usually taken as shown in Figure 2.1 and consists of the straight line ABC displaced to the right of the origin by σ and extending in the limit from $-j\infty$ to $+j\infty$ with connecting semicircles. The evaluation of the integral usually proceeds by using the Cauchy integral theorem (see Chapter 20), which specifies that

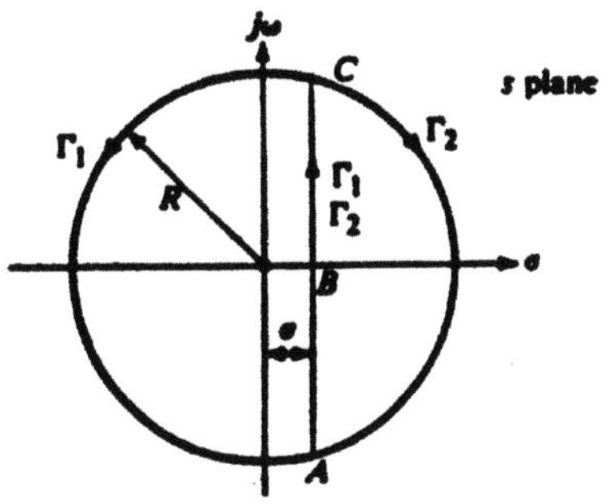

FIGURE 2.1 The Path of Integraton in the s Plane

$$f(t) = \frac{1}{2\pi j} \lim_{R\to\infty} \oint_{\Gamma_1} F(s)e^{st}\, ds$$

$$= \sum [\text{residues of } F(s)e^{st} \text{ at the singularities to the left of } ABC] \text{ for } t > 0$$

As we shall find, the contribution to the integral around the circular path with $R \to \infty$ is zero, leaving the desired integral along path ABC, and

$$f(t) = \frac{1}{2\pi j} \lim_{R\to\infty} \oint_{\Gamma_2} F(s)e^{st}\, ds$$

$$= -\sum [\text{residues of } F(s)e^{st} \text{ at the singularities to the right of } ABC] \text{ for } t < 0$$

Example 2.3

Use the inversion integral to find $f(t)$ for the function

$$F(s) = \frac{1}{s^2 + \omega^2}$$

Note that the inverse of the above formula is $\sin\omega t/\omega$.

Solution

The inversion integral is written in a form that shows the poles of the integrand

$$f(t) = \frac{1}{2\pi j} \oint \frac{e^{st}}{(s + j\omega)(s - j\omega)}\, ds$$

The path chosen is Γ_1 in Figure 2.1. Evaluate the residues

$$\text{Res}\left[(s - j\omega)\frac{e^{st}}{(s^2 + \omega^2)}\right]_{s=j\omega} = \left.\frac{e^{st}}{(s + j\omega)}\right|_{s=j\omega} = \frac{e^{j\omega t}}{2j\omega}$$

$$\text{Res}\left[(s + j\omega)\frac{e^{st}}{(s^2 + \omega^2)}\right]_{s=-j\omega} = \left.\frac{e^{st}}{(s - j\omega)}\right|_{s=-j\omega} = \frac{e^{-j\omega t}}{-2j\omega}$$

Therefore,

$$f(t)=\sum \text{Res}=\frac{e^{j\omega t}-e^{-j\omega t}}{-2j\omega}=\frac{\sin\omega t}{\omega}.$$

Example 2.4
Find $L^{-1}\left\{1/\sqrt{s}\right\}$.

Solution
The function $F(s)=1/\sqrt{s}$ is a double-valued function because of the square root operation. That is, if s is represented in polar form by $re^{j\theta}$, then $re^{j(\theta+2\pi)}$ is a second acceptable representation, and $\sqrt{s}=\sqrt{re^{j(\theta+2\pi)}}=-\sqrt{re^{j\theta}}$, thus showing two different values for $\sqrt{s}$. But a double-valued function is not analytic and requires a special procedure in its solution.

The procedure is to make the function analytic by restricting the angle of s to the range $-\pi<\theta<\pi$ and by excluding the point $s=0$. This is done by constructing a branch cut along the negative real axis, as shown in Figure 2.2. The end of the **branch cut**, which is the origin in this case, is called a **branch point**. Since a branch cut can never be crossed, this essentially ensures that $F(s)$ is single-valued. Now, however, the inversion integral becomes, for $t>0$,

$$f(t)=\lim_{R\to\infty}\frac{1}{2\pi j}\int_{GAB}F(s)e^{st}\,ds=\frac{1}{2\pi j}\int_{\sigma-j\infty}^{\sigma+j\infty}F(s)e^{st}\,ds$$

$$=-\frac{1}{2\pi j}\left[\int_{BC}+\int_{\Gamma_2}+\int_{l-}+\int_{\gamma}+\int_{l+}+\int_{\Gamma_3}+\int_{FG}\right]$$

which does not include any singularity.

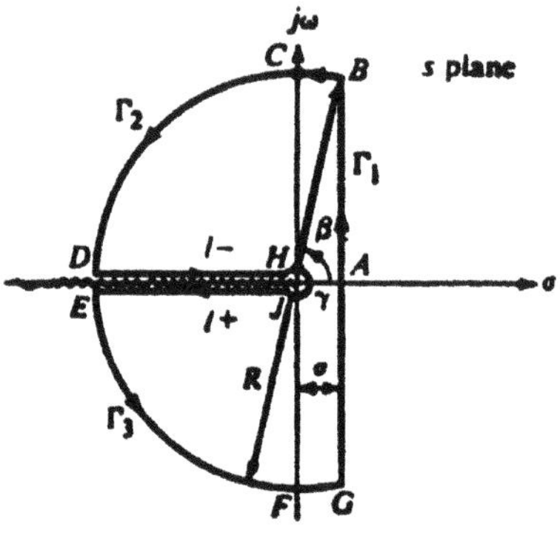

FIGURE 2.2 The Integration Contour for $L^{-1}\{1/\sqrt{s}\}$

First we will show that for $t>0$ the integrals over the contours BC and CD vanish as $R\to\infty$, from which $\int_{\Gamma_2}=\int_{\Gamma_3}=\int_{BC}=\int_{FG}=0$. Note from Figure 2.2 that $\beta=\cos^{-1}(\sigma/R)$ so that the integral over the arc BC is, since $\left|e^{j\theta}\right|=1$,

$$|I|\le\int_{BC}\left|\frac{e^{\sigma t}e^{j\omega t}}{R^{1/2}e^{j\theta/2}}jRe^{j\theta}\,d\theta\right|=e^{\sigma t}R^{1/2}\int_{\beta}^{\pi/2}d\theta=e^{\sigma t}R^{1/2}\left(\frac{\pi}{2}-\cos^{-1}\frac{\sigma}{R}\right)$$

$$=e^{\sigma t}R^{1/2}\sin^{-1}\frac{\sigma}{R}$$

But for small arguments $\sin^{-1}(\sigma/R) = \sigma/R$, and in the limit as $R \to \infty$, $I \to 0$. By a similar approach, we find that the integral over CD is zero. Thus the integrals over the contours Γ_2 and Γ_3 are also zero as $R \to \infty$.

For evaluating the integral over γ, let $s = re^{j\theta} = r(\cos\theta + j\sin\theta)$ and

$$\int_{\gamma} F(s)e^{st}\, ds = \int_{\pi}^{-\pi} \frac{e^{r(\cos\theta + j\sin\theta)}}{\sqrt{r}\, e^{j\theta/2}} jr e^{j\theta}\, d\theta$$

$$= 0 \text{ as } r \to 0$$

The remaining integrals are written

$$f(t) = -\frac{1}{2\pi j}\left[\int_{\ell-} F(s)e^{st}\, ds + \int_{\ell+} F(s)e^{st}\, ds\right]$$

Along path $l-$, let $s = -u$; $\sqrt{s} = j\sqrt{u}$, and $ds = -du$, where u and $\sqrt{u}$ are real positive quantities. Then

$$\int_{l-} F(s)e^{st}\, ds = -\int_{\infty}^{0} \frac{e^{-ut}}{j\sqrt{u}}\, du = \frac{1}{j}\int_{0}^{\infty} \frac{e^{-ut}}{\sqrt{u}}\, du$$

Along path $l+$, $s = -u$; $\sqrt{s} = -j\sqrt{u}$ (not $+j\sqrt{u}$), and $ds = -du$. Then

$$\int_{l+} F(s)e^{st}\, ds = -\int_{0}^{\infty} \frac{e^{-ut}}{-j\sqrt{u}}\, du = \frac{1}{j}\int_{0}^{\infty} \frac{e^{-ut}}{j\sqrt{u}}\, du$$

Combine these results to find

$$f(t) = -\frac{1}{2\pi j}\left[\frac{2}{j}\int_{0}^{\infty} u^{-1/2} e^{-ut}\, du\right] = \frac{1}{\pi}\int_{0}^{\infty} u^{-1/2} e^{-ut}\, du$$

which is a standard form integral listed in most handbooks of mathematical tables, with the result

$$f(t) = \frac{1}{\pi}\sqrt{\frac{\pi}{t}} = \frac{1}{\sqrt{\pi t}} \qquad t > 0\,.$$

Example 2.5

Find the inverse Laplace transform of the given function with an infinite number of poles.

$$F(s) = \frac{1}{s(1 + e^{-s})}$$

Solution

The integrand in the inversion integral $e^{st}/s(1 + e^{-s})$ possesses simple poles at

$$s = 0 \text{ and } s = jn\pi, \quad n = \pm 1, \pm 3, + \cdots \quad \text{(odd values)}$$

These are illustrated in Figure 2.3. This means that the function $e^{st}/s(1 + e^{-s})$ is analytic in the s plane except at the simple poles at $s = 0$ and $s = jn\pi$. Hence, the integral is specified in terms of the residues in the various poles. We thus have:

For $s = 0$

$$\text{Res} = \left\{ \frac{se^{st}}{s(1+e^{-s})} \right\}\Bigg|_{s=0} = \frac{1}{2}$$

For $s = jn\pi$

$$\text{Res} = \left\{ \frac{(s - jn\pi)e^{st}}{s(1+e^{-s})} \right\}\Bigg|_{s=jn\pi} = \frac{0}{0}$$

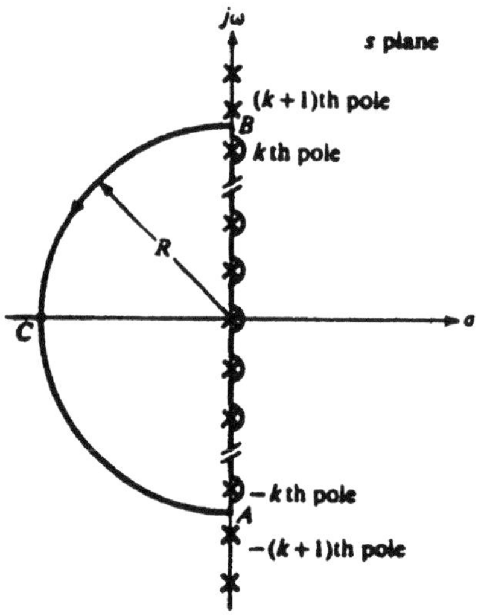

FIGURE 2.3 Illustrating Example 2.5, the Laplace Inversion for the Case of Infinitely Many Poles

The problem we now face in this evaluation is that

$$\text{Res} = \left\{ (s-a)\frac{n(s)}{d(s)} \right\}\Bigg|_{s=a} = \frac{0}{0}$$

where the roots of d(s) are such that $s = a$ cannot be factored. However, we have discussed such a situation in Chapter 20 for complex variables, and we have the following result

$$\frac{d[d(s)]}{ds}\Bigg|_{s=a} = \lim_{s\to a}\frac{d(s)-d(a)}{s-a} = \lim_{s\to a}\frac{d(s)}{s-a} \quad \text{since } d(a) = 0.$$

Combine this expression with the above equation to obtain

$$\text{Res}\left\{ (s-a)\frac{n(s)}{d(s)} \right\}\Bigg|_{s=a} = \frac{n(s)}{\dfrac{d}{ds}[d(s)]}\Bigg|_{s=a}.$$

Therefore, we proceed as follows:

$$\text{Res} = \left\{ \frac{e^{st}}{s\dfrac{a}{ds}(1+e^{-s})} \right\}\Bigg|_{s=jn\pi} = \frac{e^{jn\pi t}}{jn\pi} \qquad (n = \text{odd})$$

We obtain, by adding all of the residues,

$$f(t) = \frac{1}{2} + \sum_{n=-\infty}^{\infty} \frac{e^{jn\pi t}}{jn\pi} \qquad (n = \text{odd})$$

This can be rewritten as follows

$$f(t) = \frac{1}{2} + \left[\cdots + \frac{e^{-j3\pi t}}{-j3\pi} + \frac{e^{-j\pi t}}{-j\pi} + \frac{e^{j\pi t}}{j\pi} + \frac{e^{j3\pi t}}{j3\pi} + \cdots \right]$$

$$= \frac{1}{2} + \sum_{n=1}^{\infty} \frac{2j\sin n\pi t}{jn\pi} \qquad (n = \text{odd})$$

which we write, finally

$$f(t) = \frac{1}{2} + \frac{2}{\pi} \sum_{k=1}^{\infty} \frac{\sin(2k-1)\pi t}{2k-1}$$

As a second approach to a solution to this problem, we will show the details in carrying out the contour integration for this problem. We choose the path shown in Figure 2.3 that includes semicircular hooks around each pole, the vertical connecting line from hook to hook, and the semicircular path as $R \to \infty$. Thus we have

$$f(t) = \frac{1}{2\pi j} \oint \frac{se^{st}\,ds}{s(1+e^{-s})}$$

$$= \frac{1}{2\pi j} \left[\underbrace{\int_{BCA}}_{I_1} + \underbrace{\int_{\text{Vertical connecting lines}}}_{I_2} + \underbrace{\sum \int_{\text{Hooks}}}_{I_3} - \sum \text{Res} \right]$$

We consider the several integrals:

Integral I_1. By setting $s = re^{j\theta}$ and taking into consideration that cos = $-\cos\theta$ for $\theta > \pi/2$, the integral $I_1 \to 0$ as $r \to \infty$.

Integral I_2. Along the *Y*-axis, $s = jy$ and

$$I_2 = j\int_{\substack{-\infty \\ r\to 0}}^{\infty} \frac{e^{jyt}}{jy(1+e^{-jy})}\,dy$$

Note that the integrated is an odd function, whence $I_2 = 0$.

Integral I_3. Consider a typical hook at $s = jn\pi$. Since

$$\lim_{\substack{r\to 0\\ s\to jn\pi}} \left[\frac{(s-jn\pi)e^{st}}{s(1+e^{-s})}\right] = \frac{0}{0},$$

this expression is evaluated and yields $e^{jn\pi t}/jn\pi$. Thus, for all poles,

$$I_3 = \frac{1}{2\pi j}\int_{\substack{-\pi/2\\ r\to 0\\ s\to jn\pi}}^{\pi/2} \frac{e^{st}}{s(1+e^{-s})}\,ds$$

$$= \frac{j\pi}{2\pi j}\left[\sum_{\substack{n=-\infty\\ n\text{ odd}}}^{\infty} \frac{e^{jn\pi t}}{jn\pi} + \frac{1}{2}\right] = \frac{1}{2}\left[\frac{1}{2} + \frac{2}{\pi}\sum_{\substack{n=1\\ n\text{ odd}}}^{\infty} \frac{\sin n\pi t}{n}\right]$$

Finally, the residues enclosed within the contour are

$$\text{Res}\frac{e^{st}}{s(1+e^{-s})} = \frac{1}{2} + \sum_{\substack{n=-\infty\\ n\text{ odd}}}^{\infty} \frac{e^{jn\pi t}}{jn\pi} = \frac{1}{2} + \frac{2}{\pi}\sum_{\substack{n=1\\ n\text{ odd}}}^{\infty} \frac{\sin n\pi t}{n}$$

which is seen to be twice the value around the hooks. Then when all terms are included

$$f(t) = \frac{1}{2} + \frac{2}{\pi}\sum_{\substack{n=1\\ n\text{ odd}}}^{\infty} \frac{\sin n\pi t}{n} = \frac{1}{2} + \frac{2}{\pi}\sum_{k=1}^{\infty} \frac{\sin(2k-1)\pi t}{2k-1}.$$

1.3 Complex Integration and the Bilateral Laplace Transform

We have discussed the fact that the region of absolute convergence of the unilateral Laplace transform is the region to the left of the abscissa of convergence. This is not true for the bilateral Laplace transform: the region of convergence must be specified to invert a function $F(s)$ obtained using the bilateral Laplace transform. This requirement is necessary because different time signals might have the same Laplace transform but different regions of absolute convergence.

To establish the region of convergence, we write the bilateral transform in the form

$$F_2(s) = \int_0^{\infty} e^{-st} f(t)\,dt + \int_{-\infty}^{0} e^{-st} f(t)\,dt$$

If the function $f(t)$ is of exponential order $(e^{\sigma_1 t})$, the region of convergence for $t > 0$ is $\text{Re}\{s\} > \sigma_1$ If the function $f(t)$ for $t < 0$ is of exponential order $\exp(\sigma_2 t)$, then the region of convergence is $\text{Re}\{s\} < \sigma_2$. Hence, the function $F_2(s)$ exists and is analytic in the vertical strip defined by

$$\sigma_1 < \text{Re}\{s\} < \sigma_2$$

Provided, of course, that $\sigma_1 < \sigma_2$. If $\sigma_1 > \sigma_2$, no region of convergence would exist and the inversion process could not be performed. This region of convergence is shown in Figure 2.4.

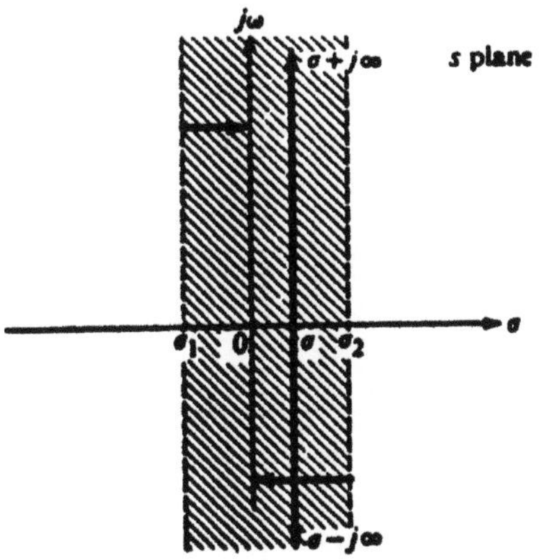

FIGURE 2.4 Region of Convergence for the Bilateral Laplace Transform

Example 2.6

Find the bilateral Laplace transform of the signals $f(t) = e^{-at}u(t)$ and $f(t) = -e^{-at}u(-t)$ and specify their regions of convergence.

Solution

Using the basic definition of the transform, we obtain

a. $$F_2(s) = \int_{-\infty}^{\infty} e^{-at}u(t)e^{-st}\,dt = \int_{0}^{\infty} e^{-(s+a)t}\,dt = \frac{1}{s+a}$$

and its region of convergence is

$$\mathrm{Re}\{s\} > -a$$

For the second signal

b. $$F_2(s) = \int_{-\infty}^{\infty} -e^{-at}u(-t)e^{-st}\,dt = -\int_{-\infty}^{0} e^{-(s+a)t}\,dt = \frac{1}{s+a}$$

and its region of convergence is

$$\mathrm{Re}\{s\} < -a$$

Clearly, the knowledge of the region of convergence is necessary to find the time function unambiguously.

Example 2.7

Find the function, if its Laplace transform is given by

$$F_2(s) = \frac{1}{(s-4)(s+1)(s+2)} \qquad -2 < \mathrm{Re}\{s\} < -1$$

Solution

The region of convergence and the paths of integration are shown in Figure 2.5.

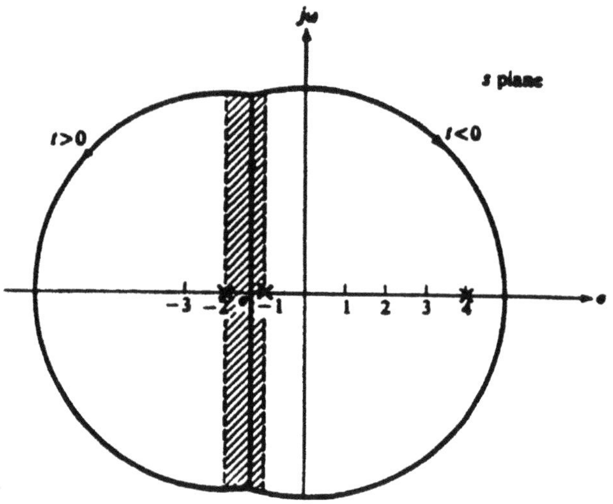

FIGURE 2.5 Illustrating Example 2.7

For $t > 0$, we close the contour to the left, we obtain

$$f(t) = \left.\frac{3e^{st}}{(s-4)(s+1)}\right|_{s=-1} = \frac{1}{2}e^{-2t} \qquad t > 0$$

For $t < 0$, the contour closes to the right, and now

$$f(t) = \left.\frac{3e^{st}}{(s-4)(s+2)}\right|_{s=-1} + \left.\frac{3e^{st}}{(s+1)(s+2)}\right|_{s=4} = -\frac{3}{5}e^{-t} + \frac{e^{4t}}{10} \qquad t < 0$$

These examples confirm that we must know the region of convergence to find the inverse transform.

3

Fourier Transform

3.1 One-Dimensional Fourier Transform

3.1.1 Definitions

3.1.1.1 Fourier Transform

$$F(f) = \int_{-\infty}^{\infty} f(t) e^{-j2\pi ft}\, dt = |F(f)| e^{j\phi(\omega)}$$

$$f(t) = \int_{-\infty}^{\infty} F(f) e^{j2\pi ft}\, df$$

or

$$F(\omega) = \int_{-\infty}^{\infty} f(t) e^{-j\omega t}\, dt = |F(\omega)| e^{j\phi(\omega)} = R(\omega) + jX(\omega)$$

$$f(t) = \frac{1}{2\pi} \int_{-\infty}^{\infty} F(\omega) e^{j\omega t}\, d\omega$$

If $f(t)$ is piecewise continuous and absolutely integrable, then its Fourier transform is a bounded continuous function, bounded by $\int_{-\infty}^{\infty} |F(\omega)|\, d\omega$. If $F(\omega)$ is absolutely integrable $|F(\omega)|$ then its inverse is $f(t)$.

3.1.1.2

$f(t) = f_r(t) + j f_i(t)$ is complex, $f_r(t)$ and $f_i(t)$ are real functions

$$F(\omega) = \int_{-\infty}^{\infty} [f_r(t)\cos\omega t + f_i(t)\sin\omega t]dt - j\int_{-\infty}^{\infty} [f_r(t)\sin\omega t - f_i(t)\cos\omega t]dt$$

$$R(\omega) = \int_{-\infty}^{\infty} [f_r(t)\cos\omega t + f_i(t)\sin\omega t]dt$$

$$X(\omega) = -\int_{-\infty}^{\infty} [f_r(t)\sin\omega t - f_i(t)\cos\omega t]dt$$

$$f_r(t) = \frac{1}{2\pi}\int_{-\infty}^{\infty} [R(\omega)\cos\omega t - X(\omega)\sin\omega t]d\omega$$

$$f_i(t) = \frac{1}{2\pi}\int_{-\infty}^{\infty} [R(\omega)\sin\omega t + X(\omega)\cos\omega t]d\omega$$

$f(t) = \text{real}$ $\quad (f_r(t) = f(t), f_i(t) = 0)$

$$R(\omega) = \int_{-\infty}^{\infty} f(t)\cos\omega t\, dt \qquad X(\omega) = -\int_{-\infty}^{\infty} f(t)\sin\omega t\, dt$$

$$R(\omega) = R(-\omega) \qquad X(-\omega) = -X(\omega) \qquad F(-\omega) = F^*(\omega)$$

$$f(t) = \frac{1}{\pi}\operatorname{Re}\int_{-\infty}^{\infty} F(\omega)e^{j\omega t}\, d\omega$$

$f(t) = j f_i(t)$ is purely imaginary $\quad (f_r(t) = 0)$

$$R(\omega) = \int_{-\infty}^{\infty} f_i(t)\sin\omega t\, dt \qquad X(\omega) = \int_{-\infty}^{\infty} f_i(t)\cos\omega t\, dt$$

$$R(\omega) = -R(-\omega) \qquad X(\omega) = X(-\omega) \qquad F(-\omega) = -F^*(\omega)$$

$f(t)$ is even $[f(t) = f(-t)]$

$$R(\omega) = 2\int_{0}^{\infty} f(t)\cos\omega t\, dt \qquad X(\omega) = 0$$

$$f(t) = \frac{1}{\pi}\int_{0}^{\infty} R(\omega)\cos\omega t\, d\omega$$

$f(t)$ is odd $[f(t) = -f(-t)]$

$$R(\omega) = 0 \qquad X(\omega) = -2\int_{0}^{\infty} f(t)\sin\omega t\, dt$$

$$f(t) = -\frac{1}{\pi}\int_{0}^{\infty} X(\omega)\sin\omega t\, d\omega$$

3.1.2 Properties

3.1.2.1 Properties of Fourier Transform

TABLE 3.1 Properties of Fourier Transform

	Operation	$f(t)$	$F(\omega)$
1.	Transform-direct	$f(t)$	$\int_{-\infty}^{\infty} f(t)e^{-j\omega t}\,dt$
2.	Inverse transform	$\frac{1}{2\pi}\int_{-\infty}^{\infty} F(\omega)e^{j\omega t}\,d\omega$	$F(\omega)$
3.	Linearity	$af_1(t) + bf_2(t)$	$aF_1(\omega) + bF_2(\omega)$
4.	Symmetry	$F(t)$	$2\pi f(-\omega)$
5.	Time shifting	$f(t \pm t_o)$	$e^{\pm j\omega t_o}F(\omega)$
6.	Scaling	$f(at)$	$\frac{1}{\lvert a\rvert}F\left(\frac{\omega}{a}\right)$
7.	Frequency shifting	$e^{\pm j\omega_o t}f(t)$	$F(\omega \mp \omega_o)$
8.	Modulation	$\begin{cases} f(t)\cos\omega_o t \\ f(t)\sin\omega_o t \end{cases}$	$\frac{1}{2}[F(\omega+\omega_o)+F(\omega-\omega_o)]$ $\frac{1}{2j}[F(\omega-\omega_o)-F(\omega+\omega_o)]$
9.	Time differentiation	$\frac{d^n}{dt^n}f(t)$	$(j\omega)^n F(\omega)$
10.	Time convolution	$f(t)*h(t) = \int_{-\infty}^{\infty} f(\tau)h(t-\tau)\,d\tau$	$F(\omega)H(\omega)$
11.	Frequency convolution	$f(t)h(t)$	$\frac{1}{2\pi}F(\omega)*H(\omega) = \frac{1}{2\pi}\int_{-\infty}^{\infty} F(\tau)H(\omega-\tau)\,d\tau$
12.	Autocorrelation	$f(t)\bullet f^*(t) = \int_{-\infty}^{\infty} f(\tau)f^*(\tau - t)\,d\tau$	$F(\omega)F^*(\omega) = \lvert F(\omega)\rvert^2$
13.	Parseval's formula	$E = \int_{-\infty}^{\infty} \lvert f(t)\rvert^2\,dt$	$E = \frac{1}{2\pi}\int_{-\infty}^{\infty} \lvert F(\omega)\rvert^2\,d\omega$
14.	Moments formula	$m_n = \int_{-\infty}^{\infty} t^n f(t)\,dt = \frac{F^{(n)}(0)}{(-j)^n}$ where	$F^{(n)}(0) = \left.\frac{d^n F(\omega)}{d\omega^n}\right\rvert_{\omega=0}, \quad n = 0,1,2\cdots$
15.	Frequency differentiation	$\begin{cases} (-jt)f(t) \\ (-jt)^n f(t) \end{cases}$	$\frac{dF(\omega)}{d\omega}$ $\frac{d^n F(\omega)}{d\omega^n}$
16.	Time reversal	$f(-t)$	$F(-\omega)$
17.	Conjugate function	$f^*(t)$	$F^*(-\omega)$
18.	Integral $(F(0) = 0)$	$\int_{-\infty}^{t} f(t)\,dt$	$\frac{1}{j\omega}F(\omega)$
19.	Integral $(F(0) \neq 0)$	$\int_{-\infty}^{t} f(t)\,dt$	$\frac{1}{j\omega}F(\omega) + \pi F(0)\delta(\omega)$

3.1.3 Tables

3.1.3.1 Graphical Representations of Some Fourier Transforms

TABLE 3.2 Table of Fourier Transforms ($x = t$; $y = w$)

$f(x)$ $\left[f(x) = (1/2\pi)\int_{-\infty}^{+\infty} F(y)e^{+ixy}\,dy\right]$	$F(y)$ $\left[F(y) = \int_{-\infty}^{+\infty} f(x)e^{-ixy}\,dx\right]$	
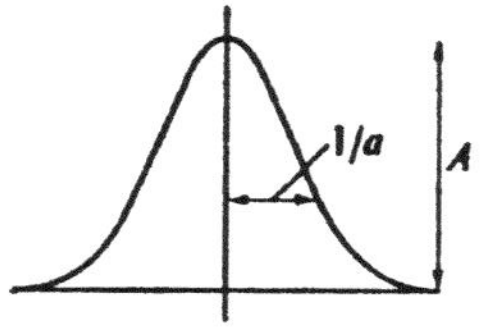 $A\exp(-a^2x^2)$ [Gaussian]	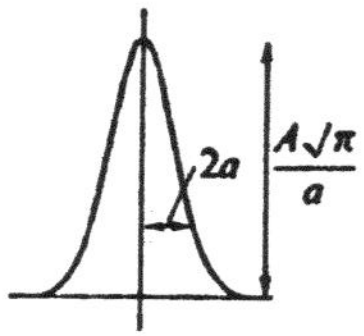 $\frac{A\sqrt{\pi}}{a}\exp(-y^2/4a^2)$ [Gaussian]	(3.1)
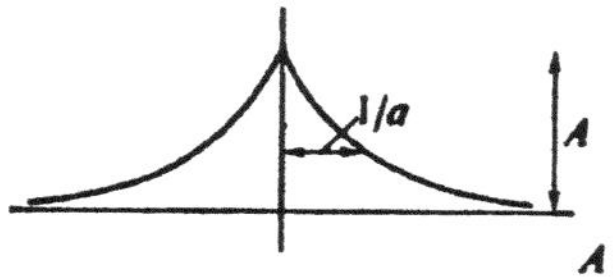 $A\exp(-a\vert x\vert)$	$\frac{2A}{a}\frac{a^2}{a^2+y^2}$ [Lorentzian] 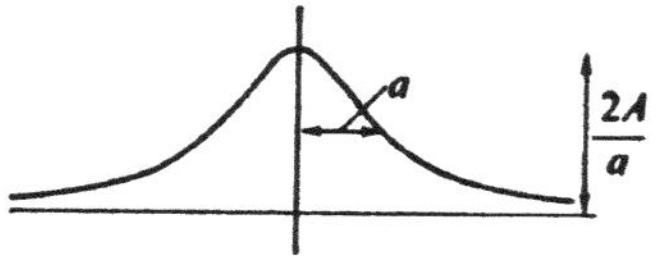	(3.2)
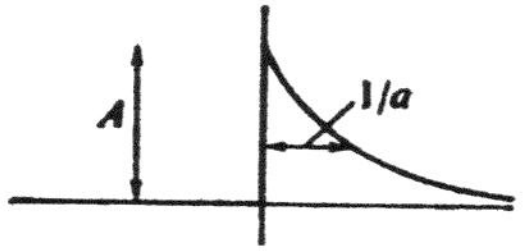 $A\exp(-ax)$ $[x > 0]$; 0 $[x < 0]$	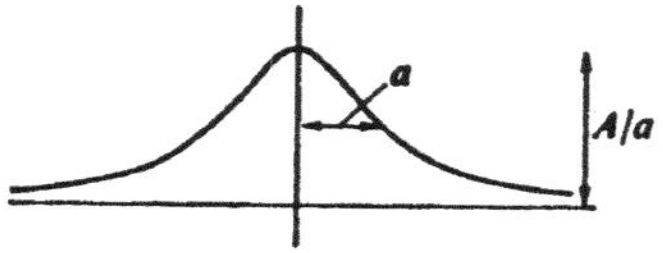 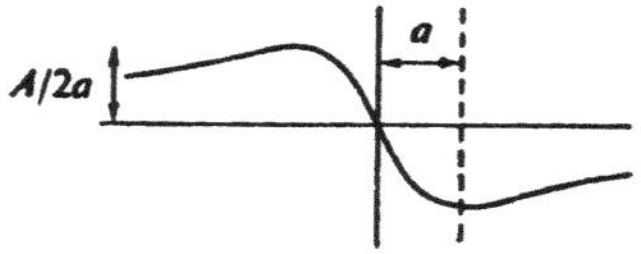 $A\left\{\frac{a-iy}{a^2+y^2}\right\}$	(3.3)

$f(x)$	$F(y)$	
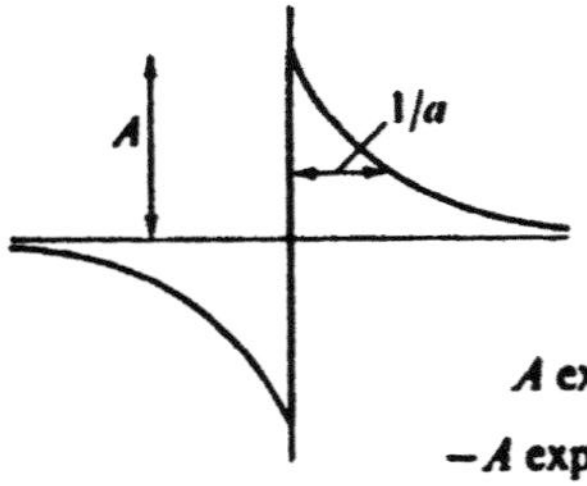 $A \exp(-ax)$ $[x > 0]$ $-A \exp(-a\|x\|)$ $[x < 0]$	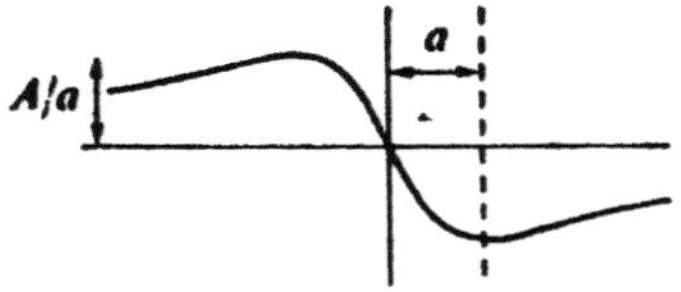 $-2iA \dfrac{y}{a^2 + y^2}$	(3.4)
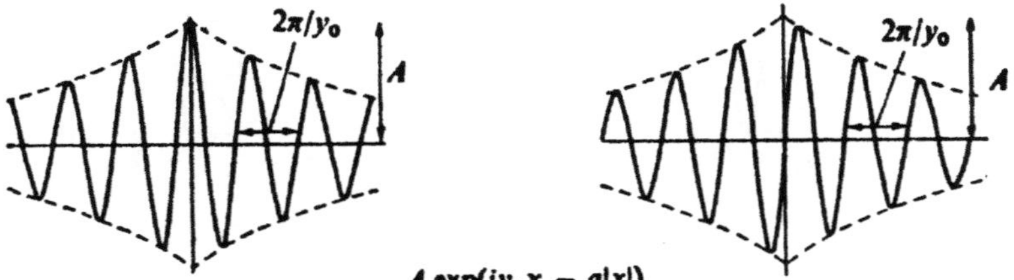 $A \exp(iy_0x - a\|x\|)$	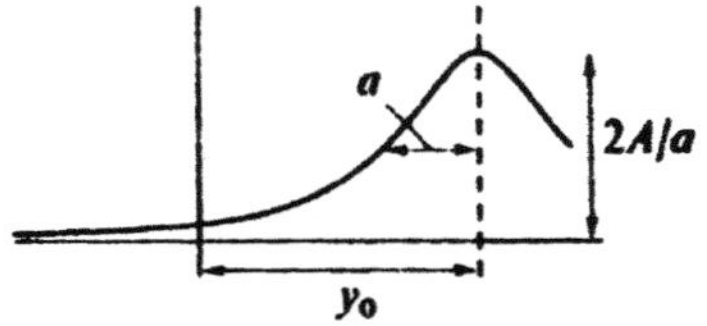 $\dfrac{2A}{a} \dfrac{a^2}{a^2 + (y - y_0)^2}$	(3.5)
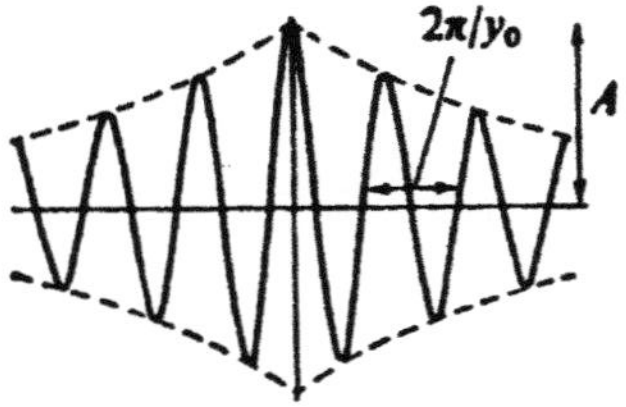 $A \cos y_0 x \exp(-a\|x\|)$	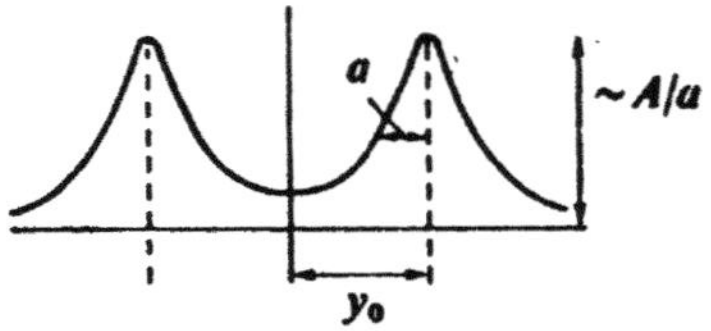 $\dfrac{A}{a}\left\{\dfrac{a^2}{a^2 + (y - y_0)^2} + \dfrac{a^2}{a^2 + (y + y_0)^2}\right\}$ $= \dfrac{A}{a}\left\{\dfrac{2a^2(a^2 + y_0^2 + y^2)}{(a^2 + y_0^2 - y^2)^2 + 4a^2y^2}\right\}$	(3.6)

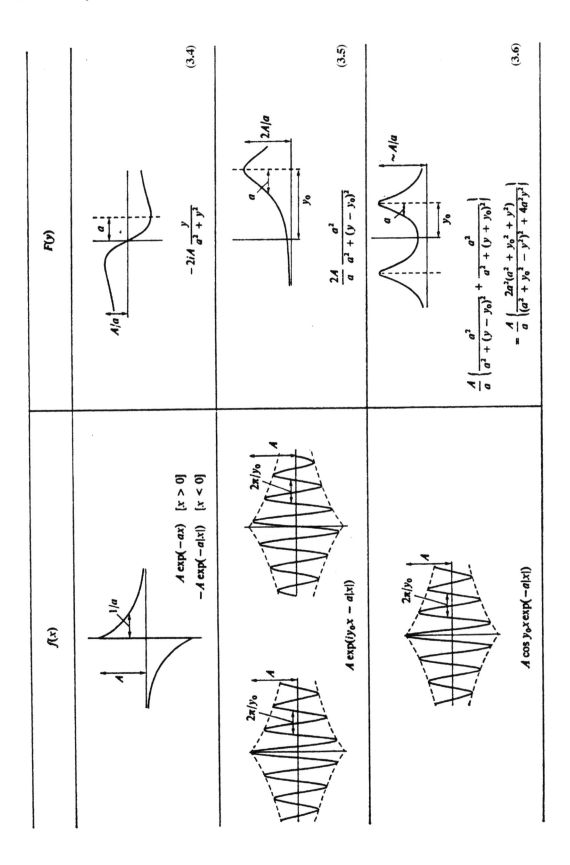

TABLE 3.2 Table of Fourier Transforms ($x = t$; $y = w$) (continued)

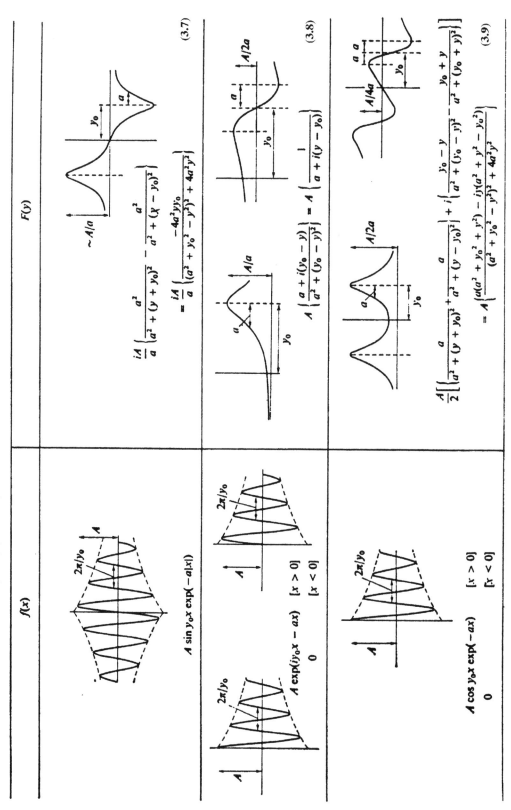

$f(x)$	$F(y)$
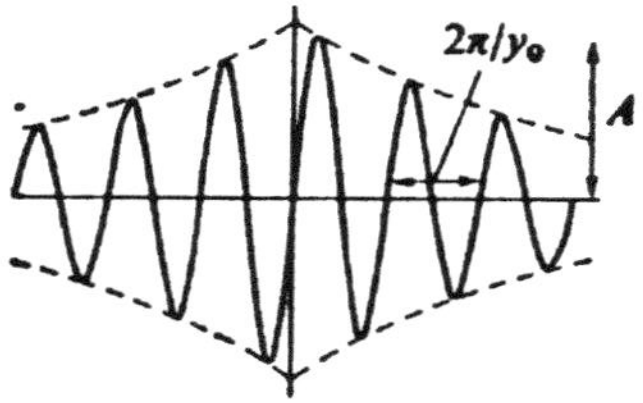 $A \sin y_0x \exp(-a\lvert x\rvert)$	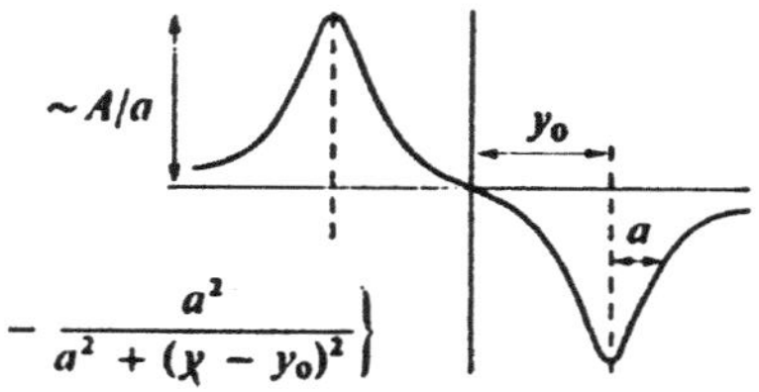 $\frac{iA}{a}\left\{\frac{a^2}{a^2+(y+y_0)^2} - \frac{a^2}{a^2+(y-y_0)^2}\right\}$ $= \frac{iA}{a}\left\{\frac{-4a^2yy_0}{(a^2+y_0^2-y^2)^2+4a^2y^2}\right\}$ (3.7)
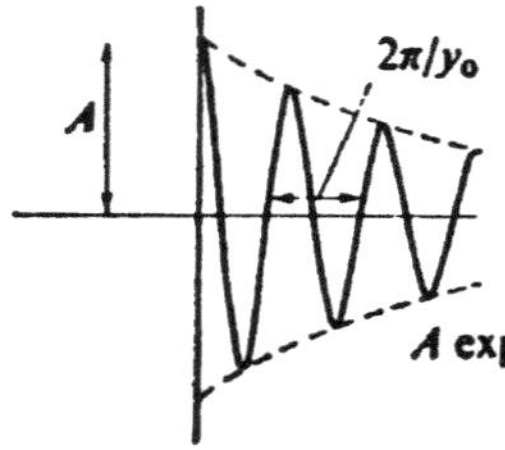 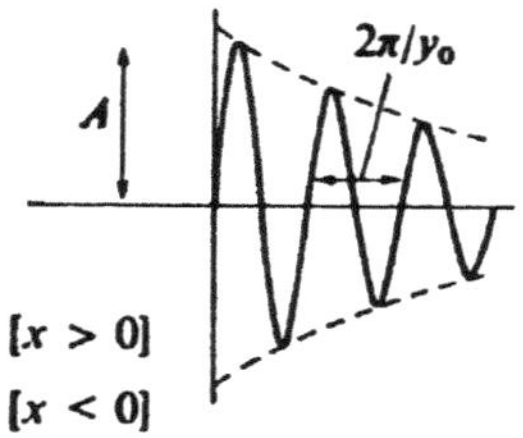 $A \exp(iy_0x - ax)$ $[x > 0]$ 0 $[x < 0]$	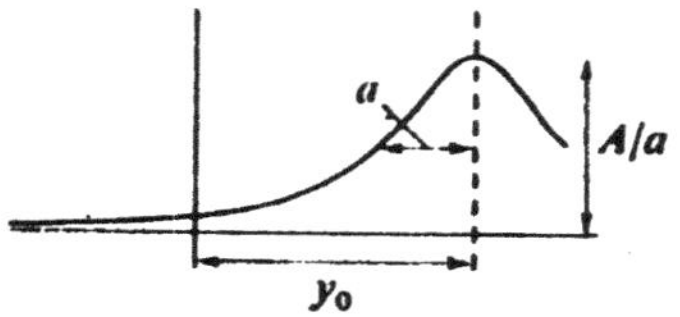 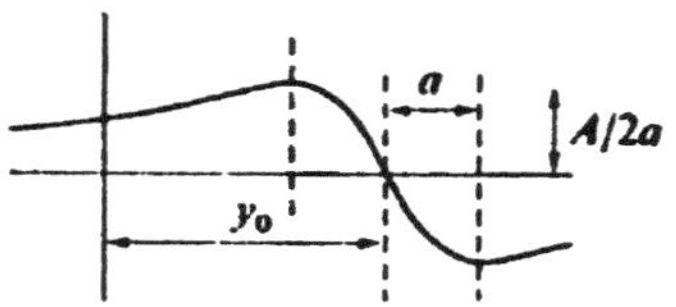 $A\left\{\frac{a+i(y_0-y)}{a^2+(y_0-y)^2}\right\} = A\left\{\frac{1}{a+i(y-y_0)}\right\}$ (3.8)
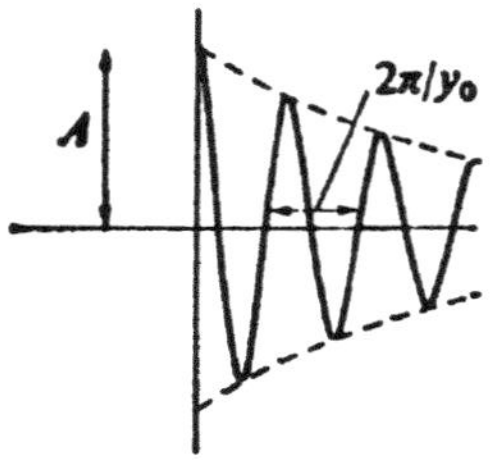 $A \cos y_0x \exp(-ax)$ $[x > 0]$ 0 $[x < 0]$	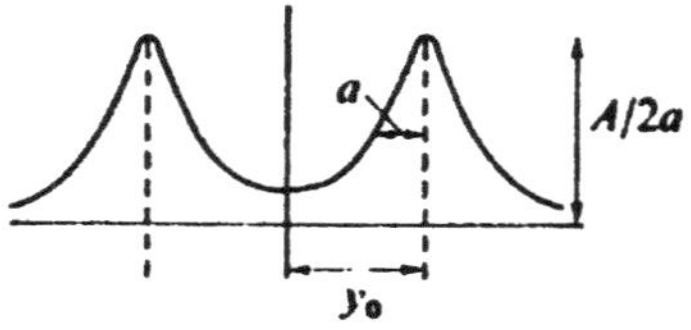 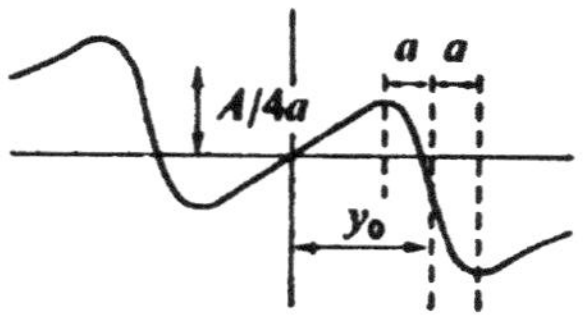 $\frac{A}{2}\left[\left\{\frac{a}{a^2+(y+y_0)^2} + \frac{a}{a^2+(y-y_0)^2}\right\} + i\left\{\frac{y_0-y}{a^2+(y_0-y)^2} - \frac{y_0+y}{a^2+(y_0+y)^2}\right\}\right]$ $= A\left\{\frac{a(a^2+y_0^2+y^2) - iy(a^2+y^2-y_0^2)}{(a^2+y_0^2-y^2)^2+4a^2y^2}\right\}$ (3.9)

$f(x)$	$F(y)$
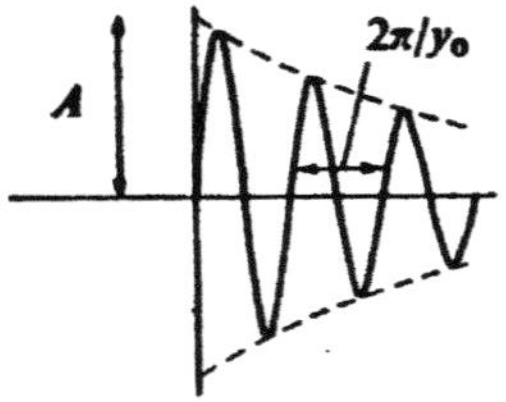 $A \sin y_0 x \exp(-ax)$ $[x > 0]$ 0 $[x < 0]$	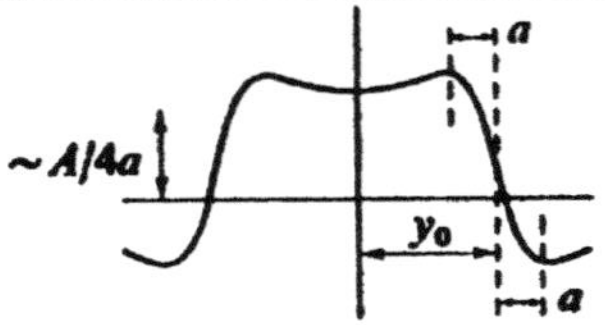 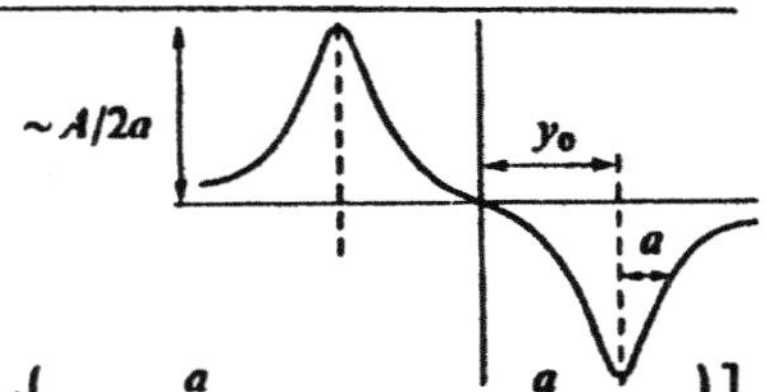 $\frac{A}{2}\left[\left\{\frac{y_0 - y}{a^2 + (y_0 - y)^2} + \frac{y_0 + y}{a^2 + (y_0 + y)^2}\right\} + i\left\{\frac{a}{a^2 + (y_0 + y)^2} - \frac{a}{a^2 + (y_0 - y)^2}\right\}\right]$ $= Ay_0 \left\{\frac{1}{(a^2 + y_0^2 - y^2) + i2ay}\right\}$ (3.10)
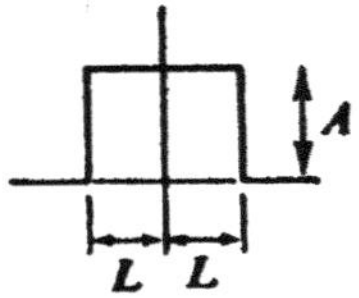 A $[\lvert x\rvert < L]$ 0 $[\lvert x\rvert > L]$	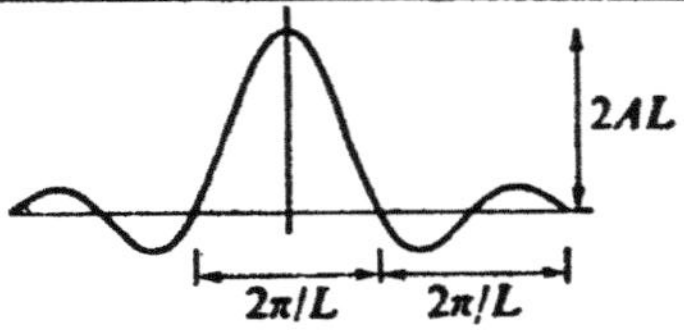 $2A \frac{\sin Ly}{y}$ (3.11)
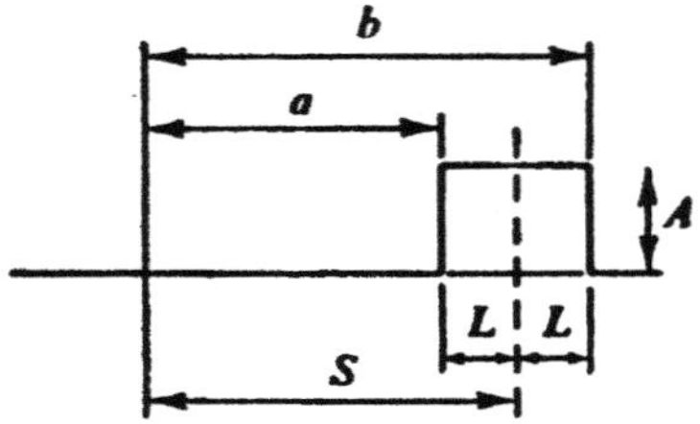 A $[a < x < b]$ 0 $[x < a;\ x > b]$	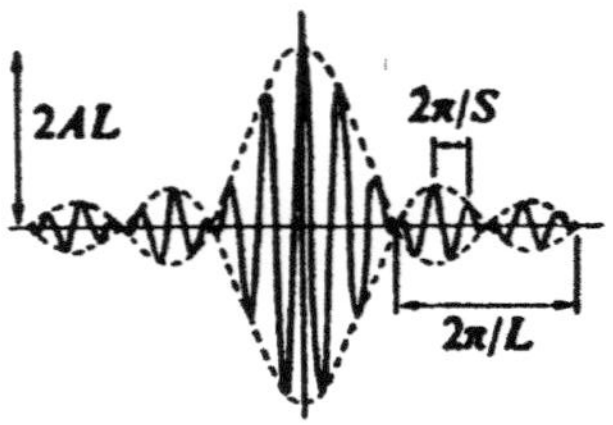 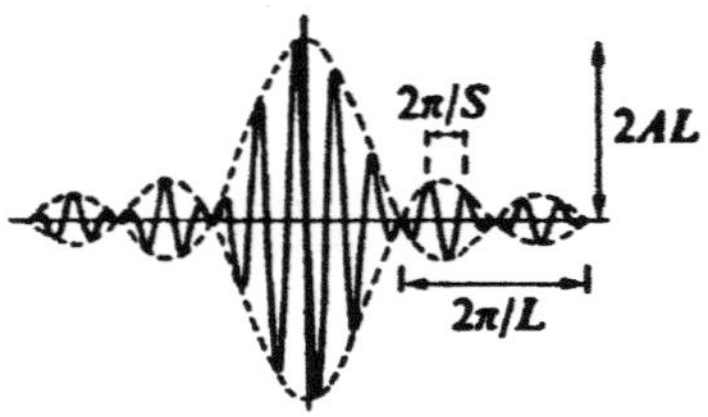 $2A \frac{\sin Ly}{y} \exp(-iSy) = A\left[\frac{(\sin by - \sin ay) - i(\cos ay - \cos by)}{y}\right]$ $= 2A\left[\frac{(\sin Ly \cos Sy) - i(\sin Ly \sin Sy)}{y}\right] = \frac{iA}{y}\left[\exp(-iby) - \exp(-iay)\right]$ (3.12)

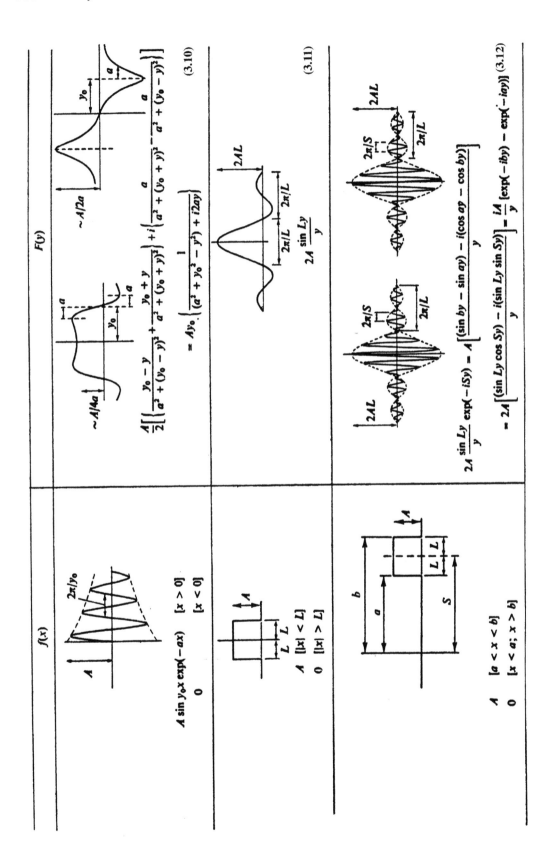

TABLE 3.2 Table of Fourier Transforms ($x = t$; $y = w$) (continued)

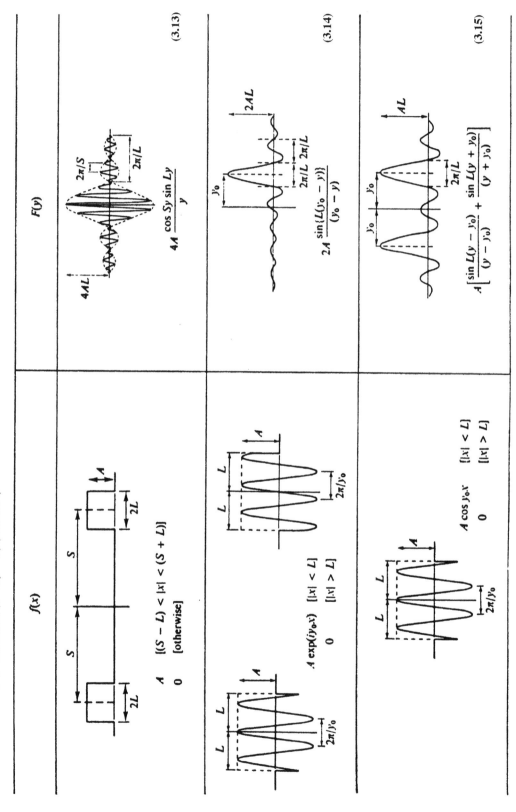

$f(x)$	$F(y)$	
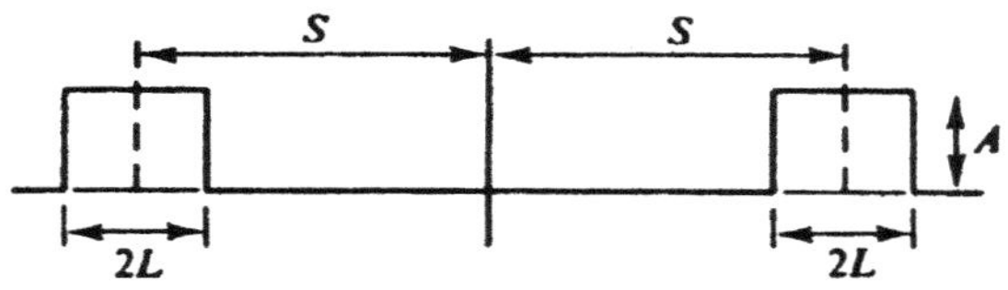 $A \quad [(S - L) < \lvert x\rvert < (S + L)]$ $0 \quad$ [otherwise]	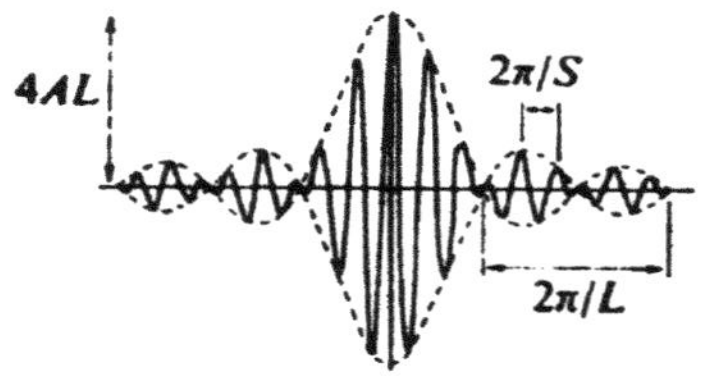 $4A\,\dfrac{\cos Sy \sin Ly}{y}$	(3.13)
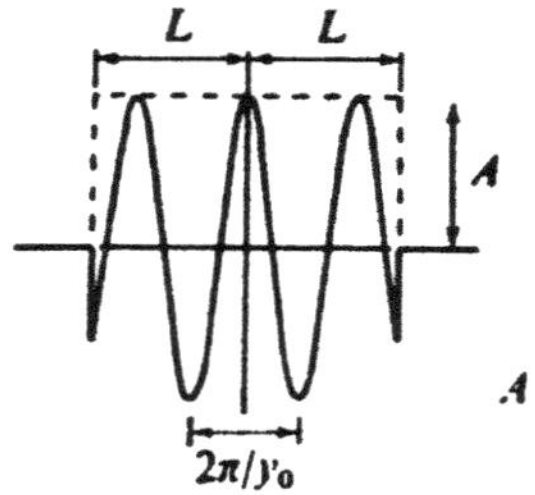 $A\exp(iy_0x) \quad [\lvert x\rvert < L]$ $0 \quad [\lvert x\rvert > L]$ 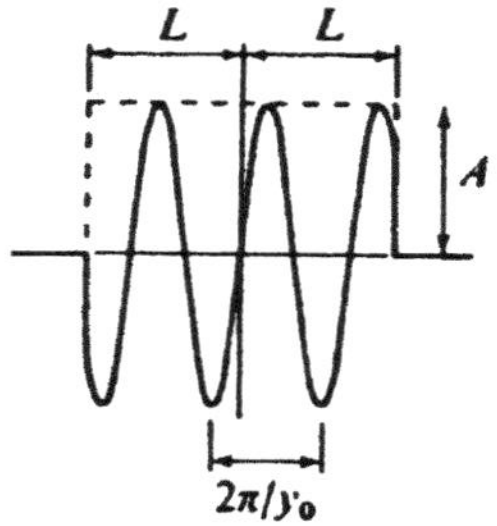	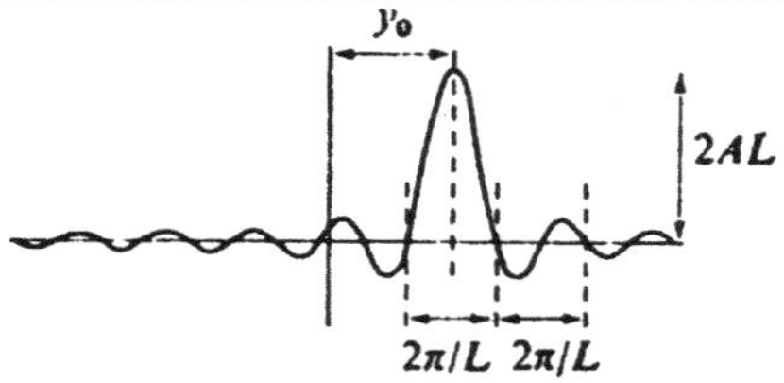 $2A\,\dfrac{\sin\{L(y_0 - y)\}}{(y_0 - y)}$	(3.14)
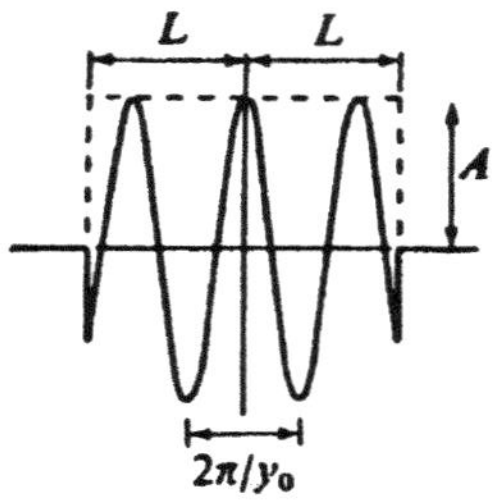 $A\cos y_0x \quad [\lvert x\rvert < L]$ $0 \quad [\lvert x\rvert > L]$	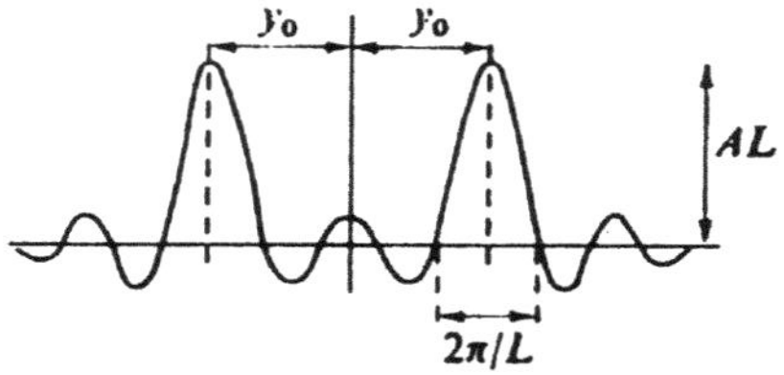 $A\left[\dfrac{\sin L(y - y_0)}{(y - y_0)} + \dfrac{\sin L(y + y_0)}{(y + y_0)}\right]$	(3.15)

$f(x)$	$F(y)$	
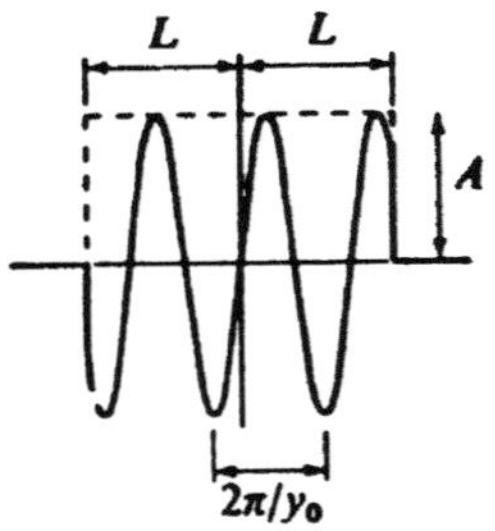 $A \sin y_0 x \quad [\lvert x\rvert < L]$ $0 \quad [\lvert x\rvert > L]$	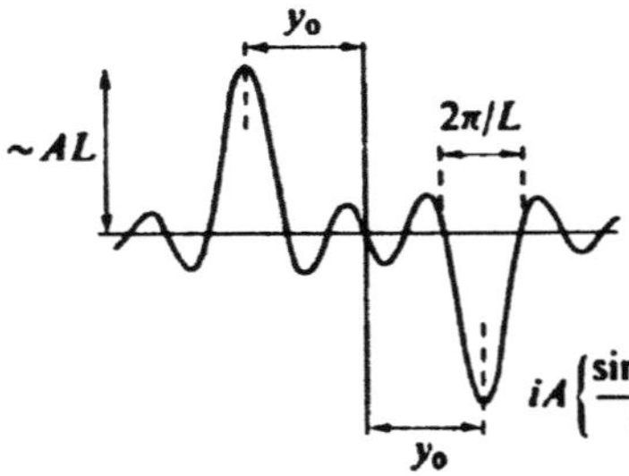 $iA\left\{\frac{\sin L(y+y_0)}{(y+y_0)} - \frac{\sin L(y-y_0)}{(y-y_0)}\right\}$	(3.16)
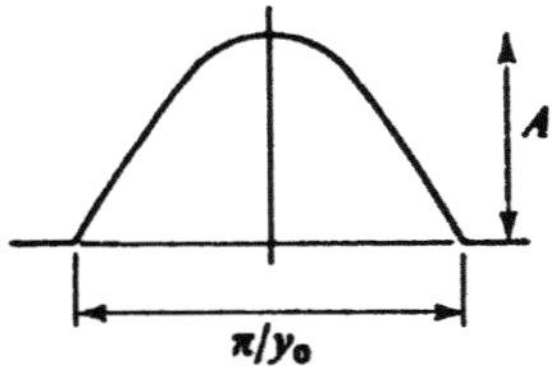 $A \cos y_0 x \quad [\lvert x\rvert < (\pi/2y_0)]$ $0 \quad [\lvert x\rvert > (\pi/2y_0)]$	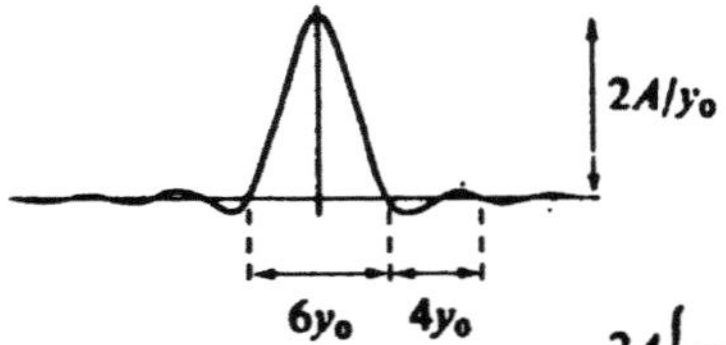 $2A\left\{\frac{y_0}{y_0^2 - y^2}\right\} \cos\left(\frac{\pi y}{2y_0}\right)$ [See (2.52) with $L = \pi/2y_0$].	(3.17)
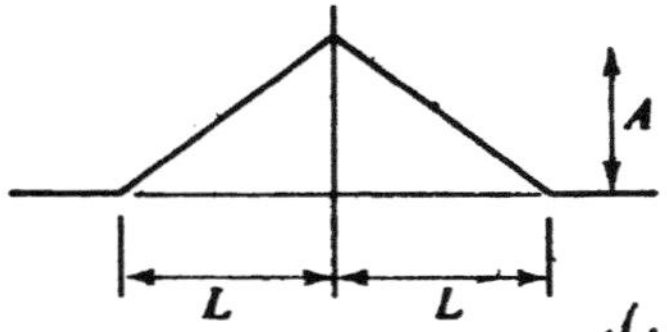 $A\left(1 - \frac{\lvert x\rvert}{L}\right) \quad [\lvert x\rvert < L]$ $0 \quad [\lvert x\rvert > L]$	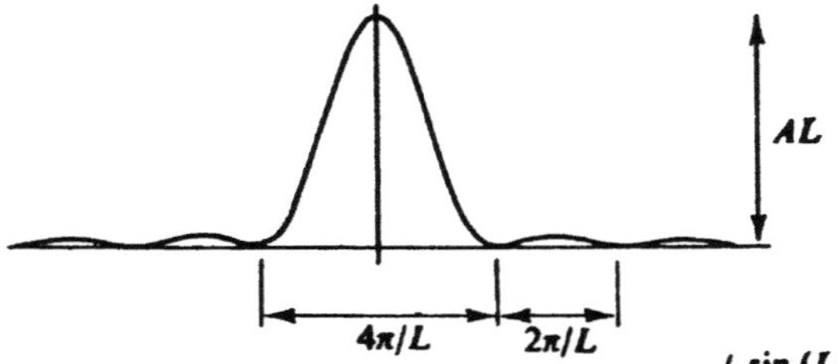 $AL\left\{\frac{\sin (Ly/2)}{(Ly/2)}\right\}^2$	(3.18)

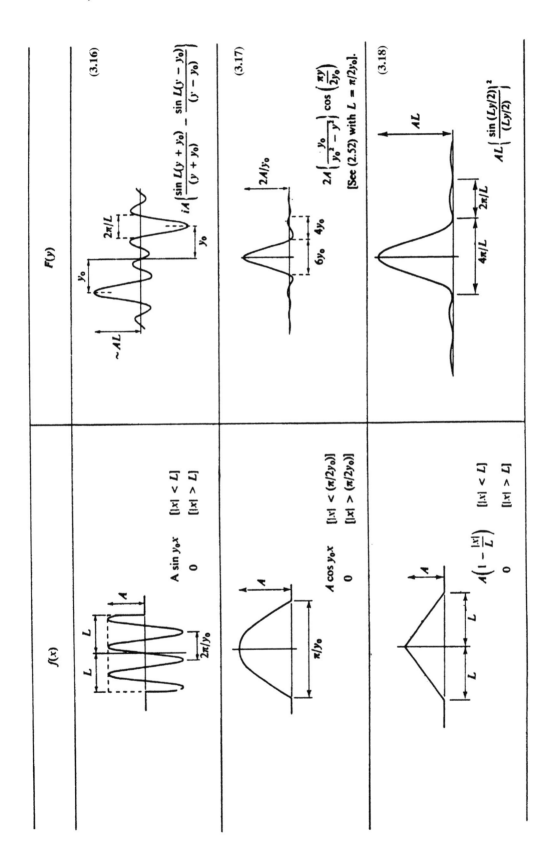

TABLE 3.2 Table of Fourier Transforms ($x = t$; $y = w$) (continued)

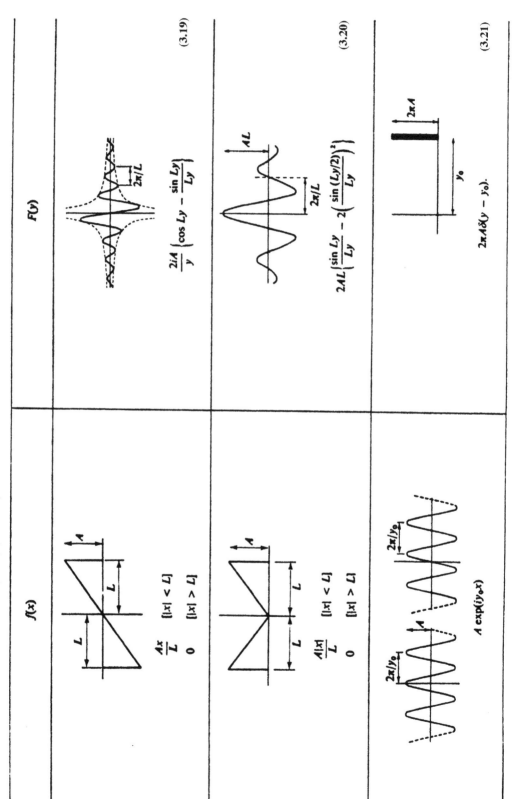

$f(x)$	$F(y)$	
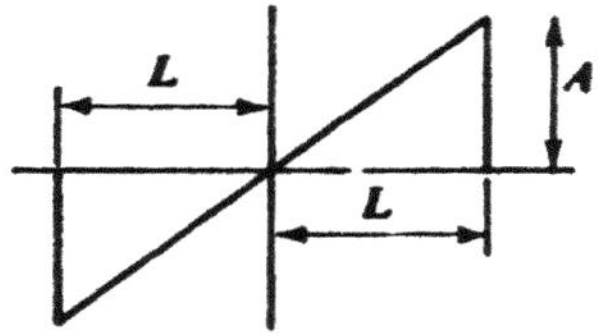 $\frac{Ax}{L}$ $[\lvert x\rvert < L]$ 0 $[\lvert x\rvert > L]$	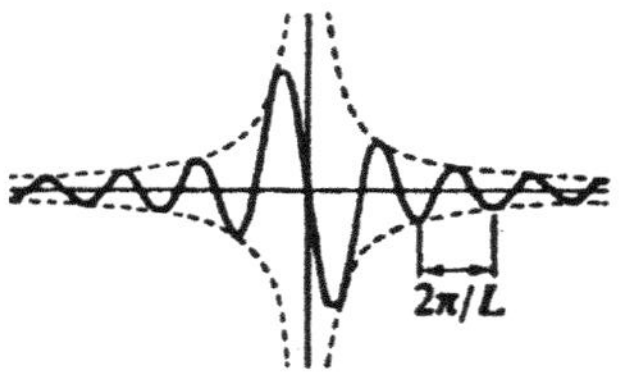 $\frac{2iA}{y}\left\{\cos Ly - \frac{\sin Ly}{Ly}\right\}$	(3.19)
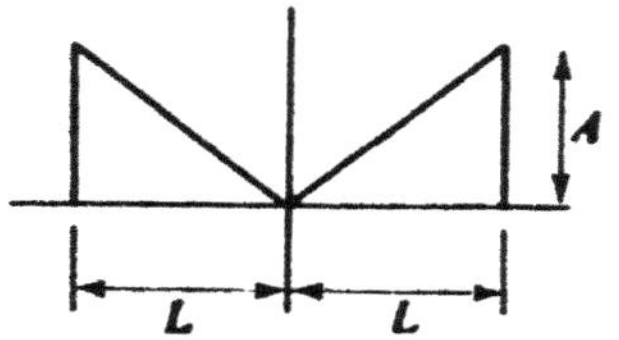 $\frac{A\lvert x\rvert}{L}$ $[\lvert x\rvert < L]$ 0 $[\lvert x\rvert > L]$	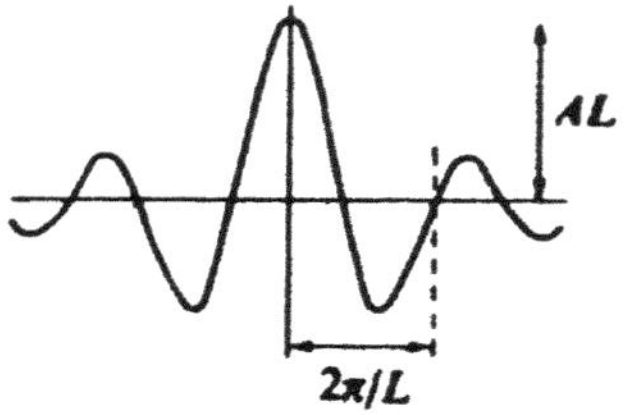 $2AL\left\{\frac{\sin Ly}{Ly} - 2\left(\frac{\sin(Ly/2)}{Ly}\right)^2\right\}$	(3.20)
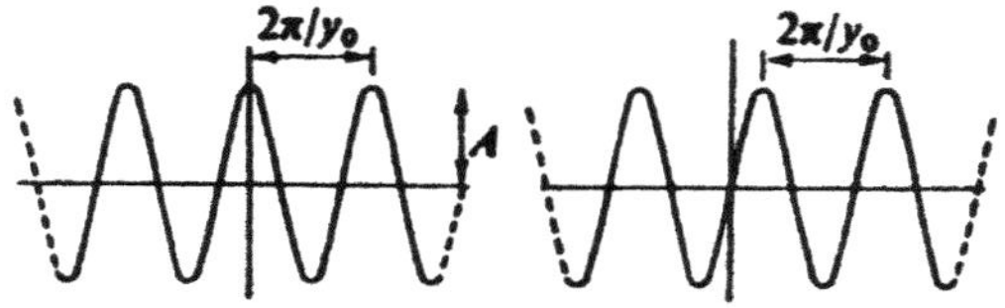 $A\exp(iy_0x)$	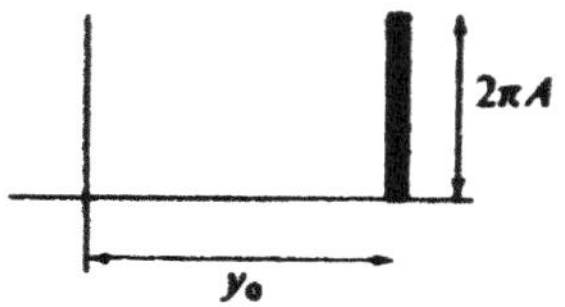 $2\pi A\delta(y - y_0)$.	(3.21)

$f(x)$	$F(y)$	
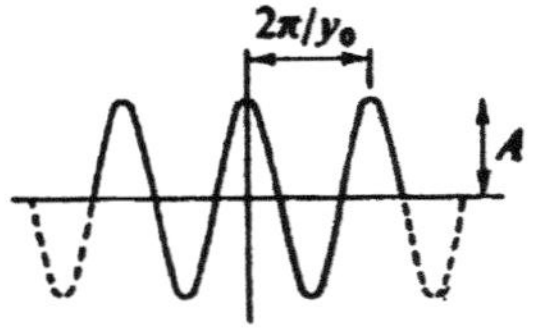 $A \cos y_0 x$	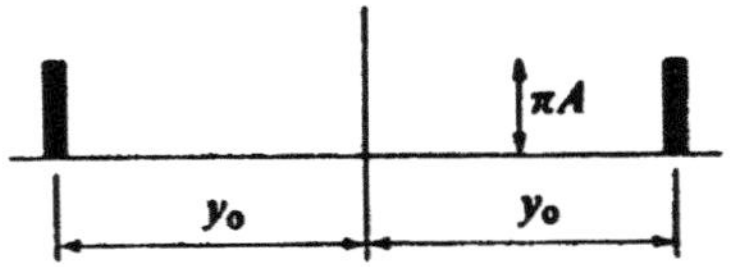 $\pi A\{\delta(y - y_0) + \delta(y + y_0)\}$	(3.22)
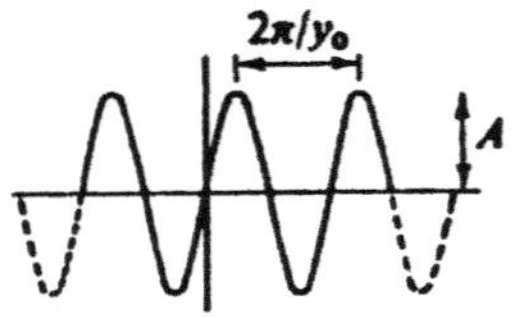 $A \sin y_0 x$	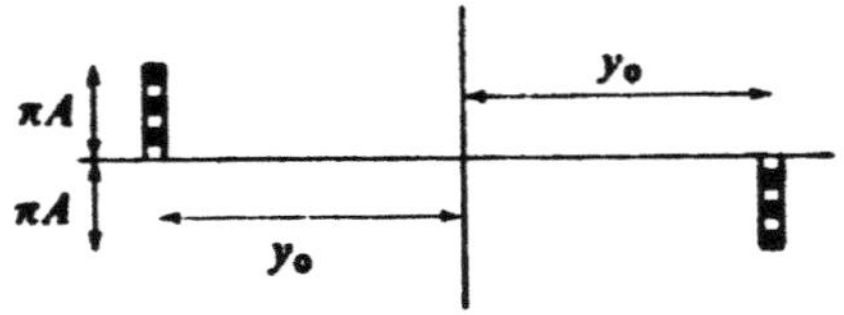 $\pi i A\{\delta(y + y_0) - \delta(y - y_0)\}$	(3.23)
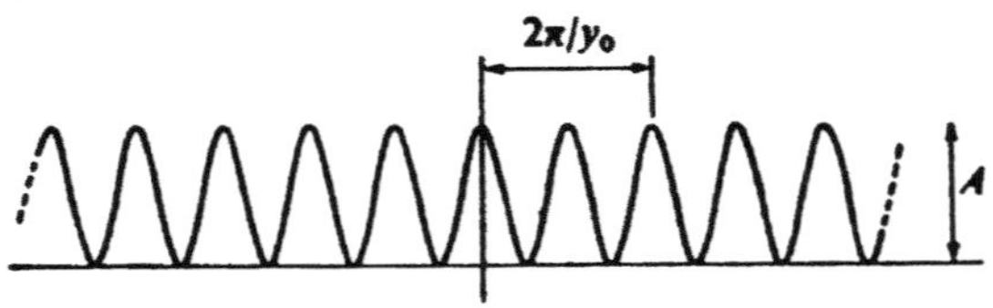 $A \cos^2 y_0 x$	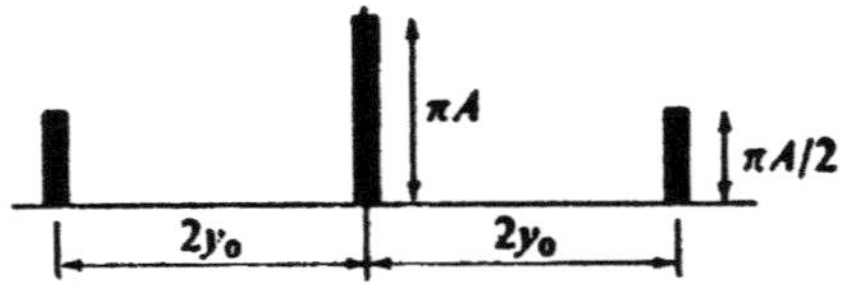 $\pi A\{\frac{1}{2}\delta(y + 2y_0) + \delta(y) + \frac{1}{2}\delta(y - 2y_0)\}$	(3.24)
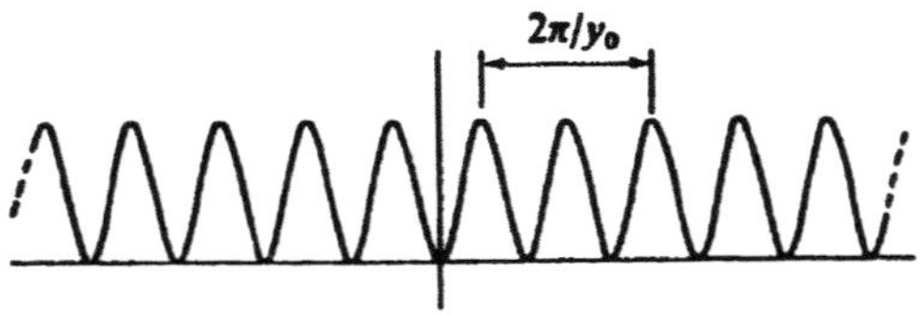 $A \sin^2 y_0 x$	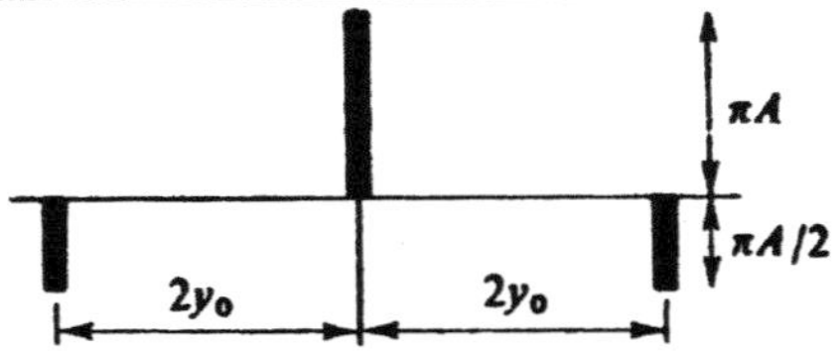 $\pi A\{-\frac{1}{2}\delta(y + 2y_0) + \delta(y) - \frac{1}{2}\delta(y - 2y_0)\}$	(3.25)

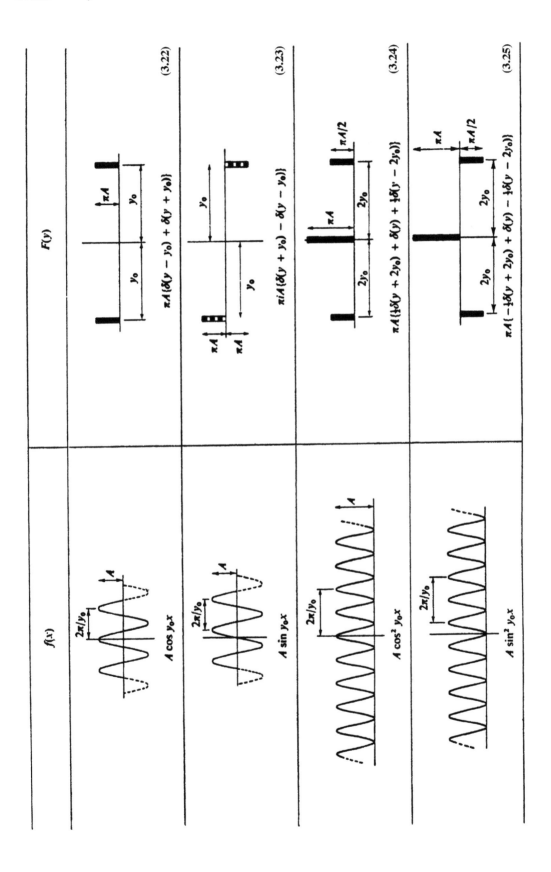

TABLE 3.2 Table of Fourier Transforms ($x = t$; $y = w$) (continued)

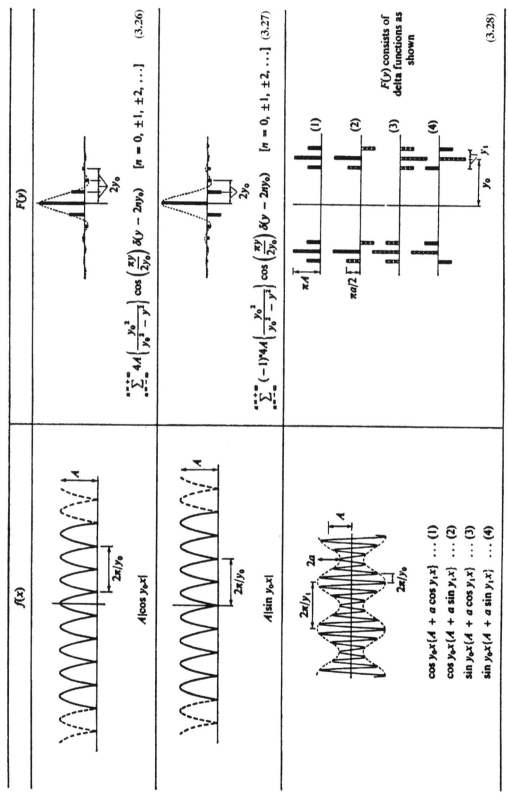

$f(x)$	$F(y)$
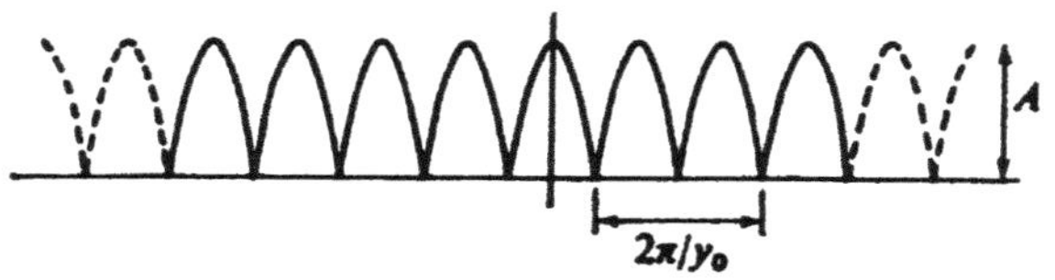 $A\lvert\cos y_0x\rvert$	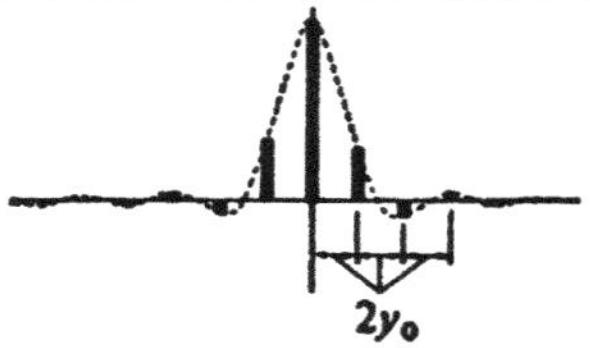 $\sum_{n=-\infty}^{n=+\infty} 4A\left\{\frac{y_0^2}{y_0^2 - y^2}\right\} \cos\left(\frac{\pi y}{2y_0}\right) \delta(y - 2ny_0)$ $[n = 0, \pm 1, \pm 2, \ldots]$ (3.26)
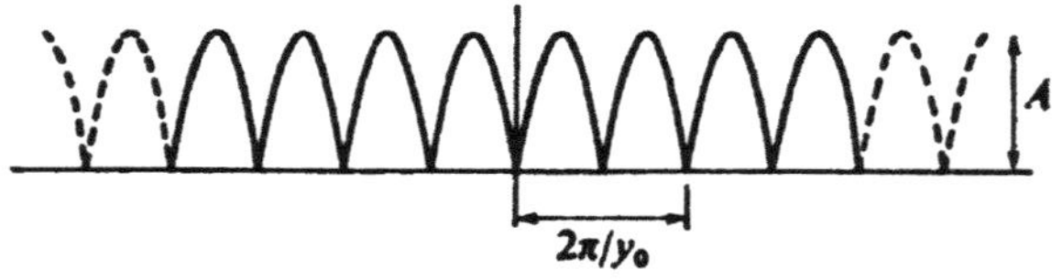 $A\lvert\sin y_0x\rvert$	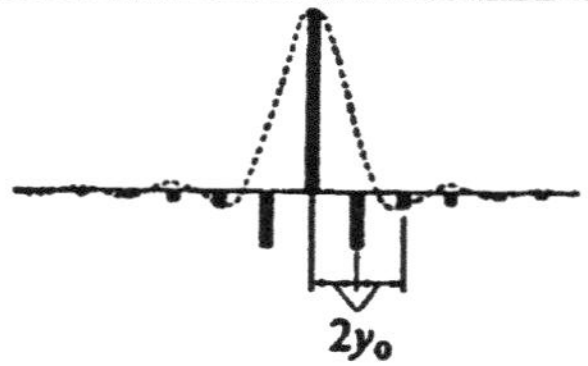 $\sum_{n=-\infty}^{n=+\infty} (-1)^n 4A\left\{\frac{y_0^2}{y_0^2 - y^2}\right\} \cos\left(\frac{\pi y}{2y_0}\right) \delta(y - 2ny_0)$ $[n = 0, \pm 1, \pm 2, \ldots]$ (3.27)
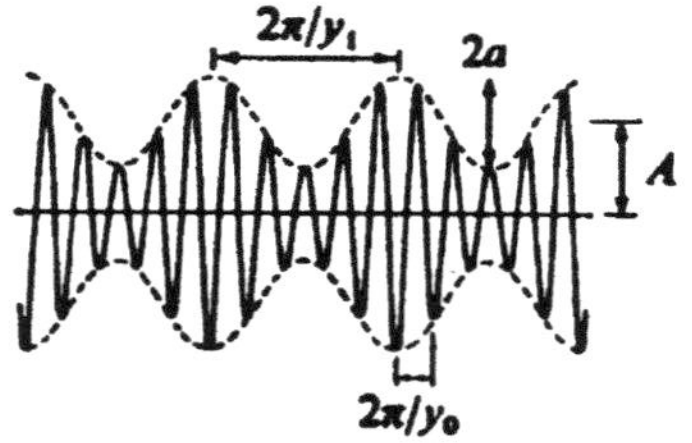 $\cos y_0x\{A + a\cos y_1x\}$ ... (1) $\cos y_0x\{A + a\sin y_1x\}$ ... (2) $\sin y_0x\{A + a\cos y_1x\}$ ... (3) $\sin y_0x\{A + a\sin y_1x\}$ ... (4)	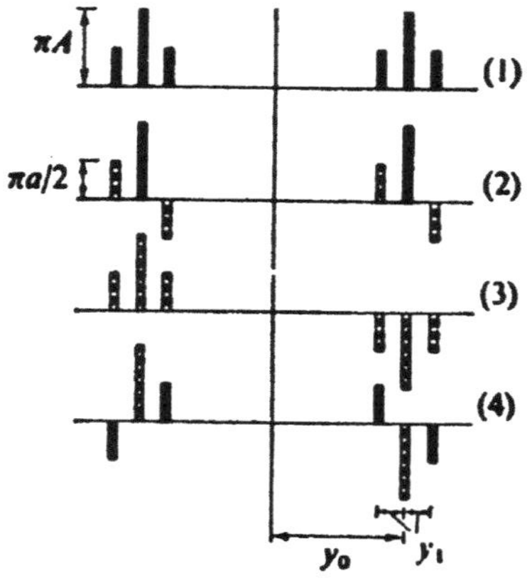 **$F(y)$ consists of delta functions as shown** (3.28)

$f(x)$	$F(y)$	
$\exp(iy_0x)(A + a\cos y_1x)$	$2\pi\left\{A\delta(y - y_0) + \frac{a}{2}\delta(y - y_0 + y_1) + \frac{a}{2}\delta(y - y_0 - y_1)\right\}$ 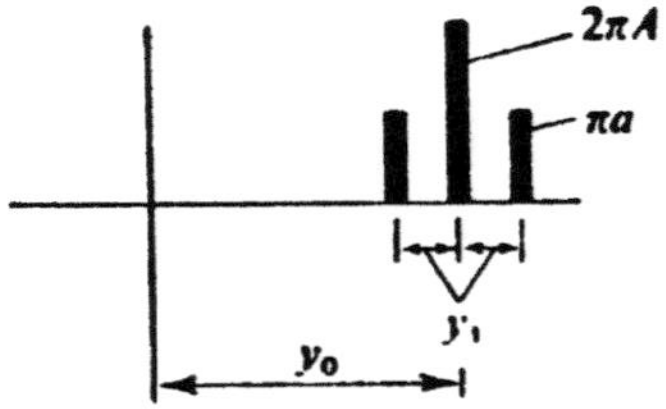	(3.29)
$\exp(iy_0x)(A + a\sin y_1x)$	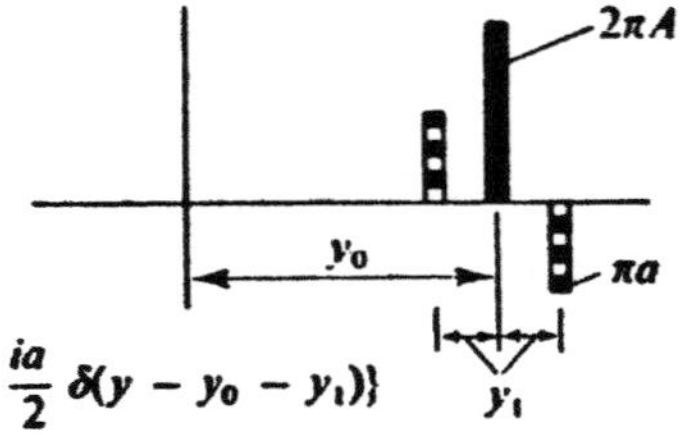 $2\pi\{A\delta(y - y_0) + \frac{ia}{2}\delta(y - y_0 + y_1) - \frac{ia}{2}\delta(y - y_0 - y_1)\}$	(3.30)
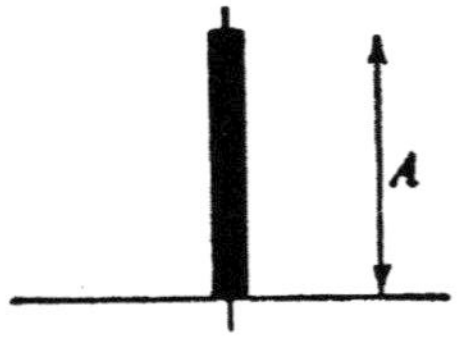 $A\delta(x)$	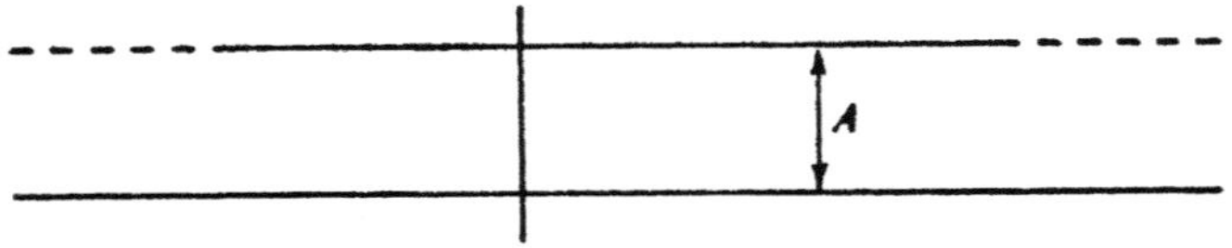 A	(3.31)
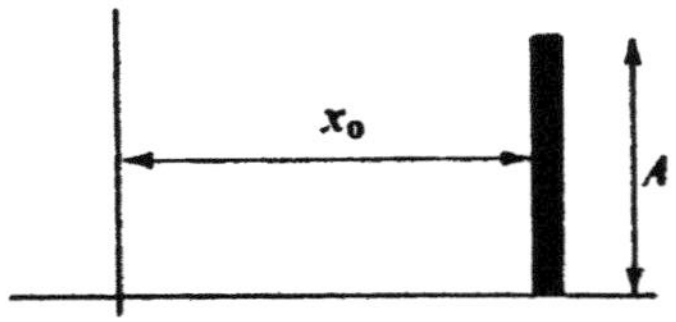 $A\delta(x - x_0)$	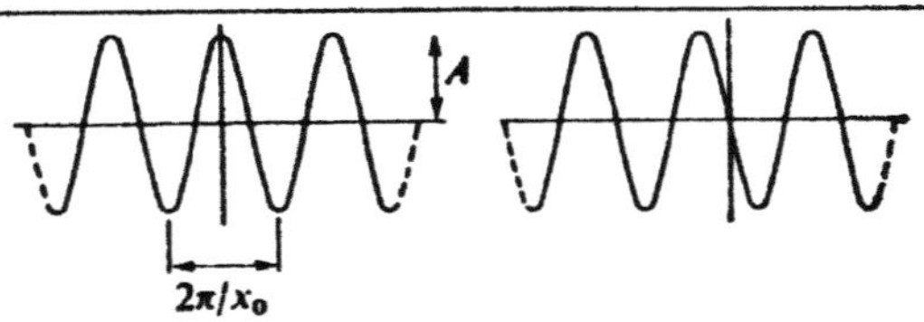 $A\exp(-ix_0y)$	(3.32)

TABLE 3.2 Table of Fourier Transforms ($x = t$; $y = w$) (continued)

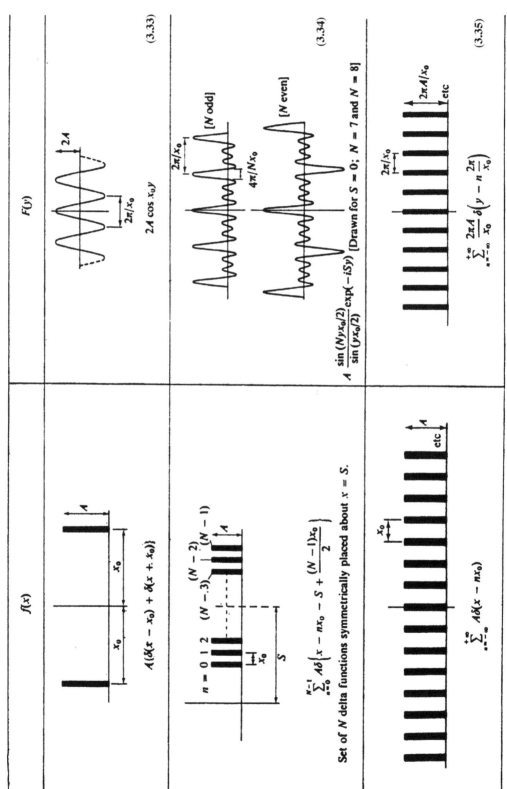

$f(x)$	$F(y)$	
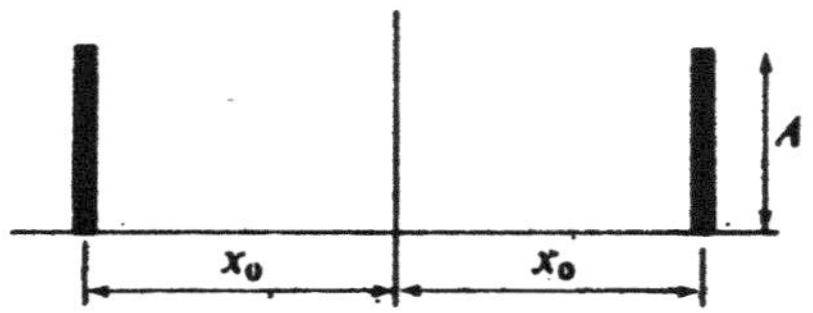 $A\{\delta(x - x_0) + \delta(x + x_0)\}$	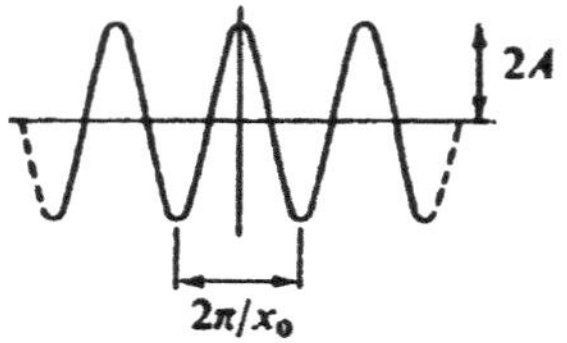 $2A \cos x_0 y$	(3.33)
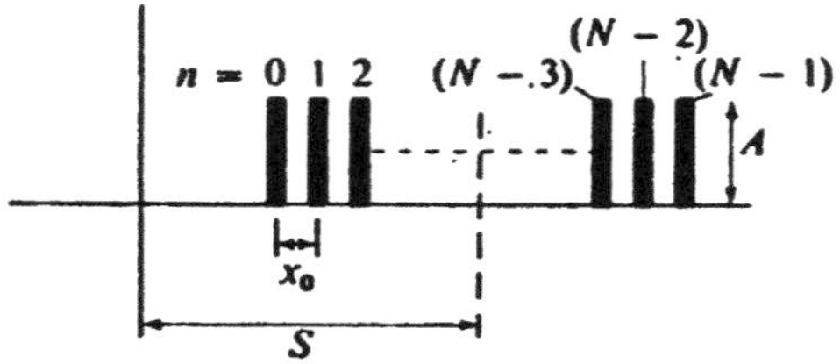 $\sum_{n=0}^{N-1} A\delta\left\{x - nx_0 - S + \frac{(N-1)x_0}{2}\right\}$ Set of N delta functions symmetrically placed about $x = S$.	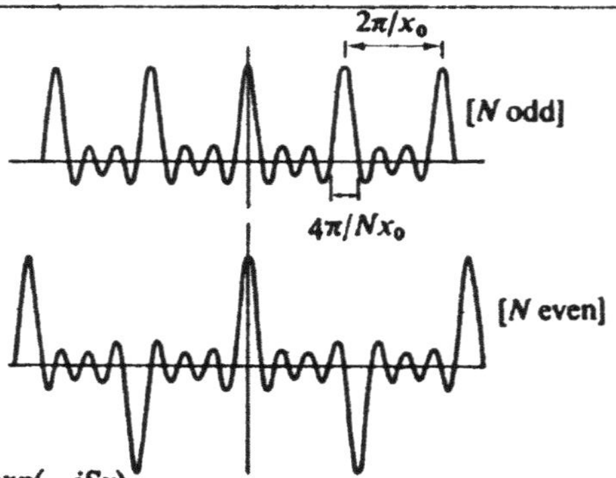 $A \frac{\sin(Nyx_0/2)}{\sin(yx_0/2)} \exp(-iSy)$ [Drawn for $S = 0$; $N = 7$ and $N = 8$]	(3.34)
 $\sum_{n=-\infty}^{+\infty} A\delta(x - nx_0)$	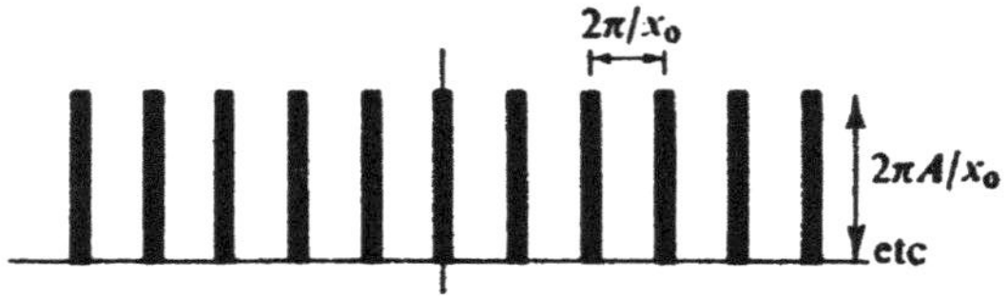 $\sum_{n=-\infty}^{+\infty} \frac{2\pi A}{x_0} \delta\left(y - n\frac{2\pi}{x_0}\right)$	(3.35)

$f(x)$	$F(y)$	
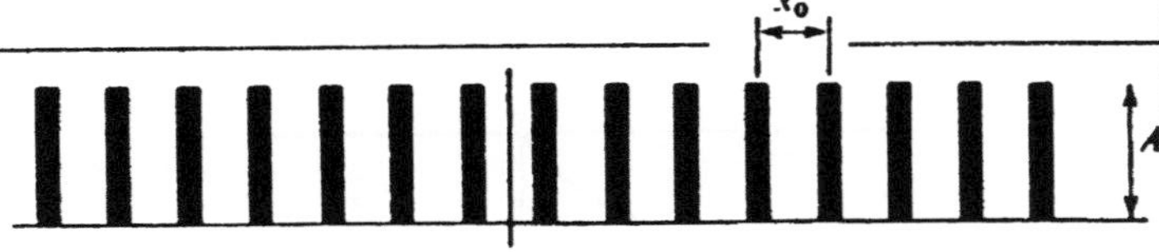 $\sum_{n=-\infty}^{+\infty} A\delta\left(x - \frac{x_0}{2} - nx_0\right)$	 $\sum_{n=-\infty}^{+\infty} (-1)^n \frac{2\pi A}{x_0} \delta\left(y - n\frac{2\pi}{x_0}\right)$	(3.36)
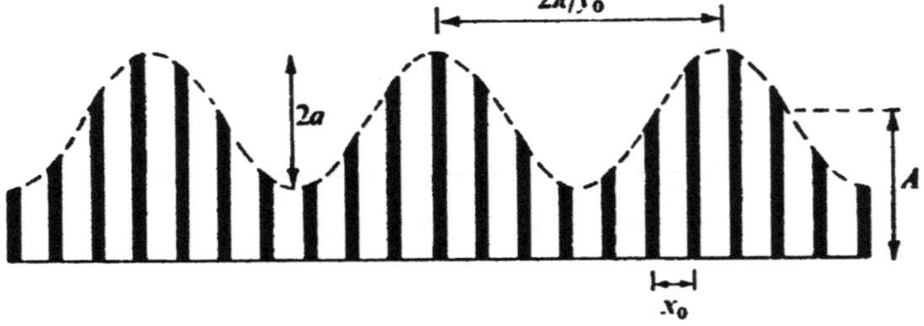 $\sum \delta(x - nx_0)\{A + a\cos y_0 x\}$ (1) $\sum \delta(x - nx_0)\{A + a\sin y_0 x\}$ (2) $[n = 0, \pm 1, \pm 2, \ldots]$	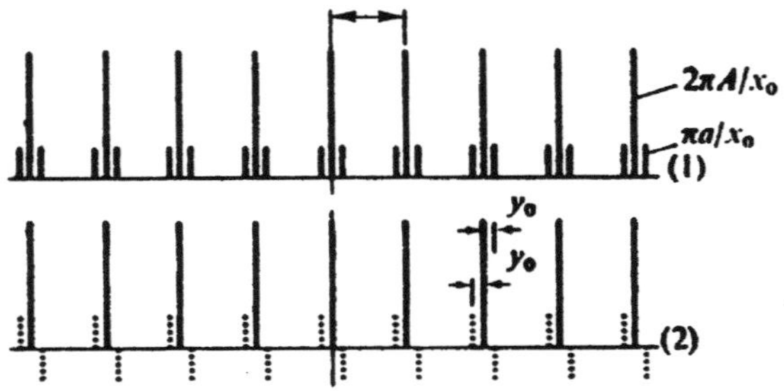 $\sum_n \frac{2\pi}{x_0}\left\{A\delta\left(y - n\frac{2\pi}{x_0}\right) + \frac{a}{2}\delta\left(y - n\frac{2\pi}{x_0} + y_0\right) + \frac{a}{2}\delta\left(y - \frac{2\pi}{x_0} - y_0\right)\right\}$ (1) $\sum_n \frac{2\pi}{x_0}\left\{A\delta\left(y - n\frac{2\pi}{x_0}\right) + \frac{ia}{2}\delta\left(y - n\frac{2\pi}{x_0} + y_0\right) - \frac{ia}{2}\delta\left(y - n\frac{2\pi}{x_0} - y_0\right)\right\}$ (2) $[n = 0, \pm 1, \pm 2, \ldots]$	(3.37)
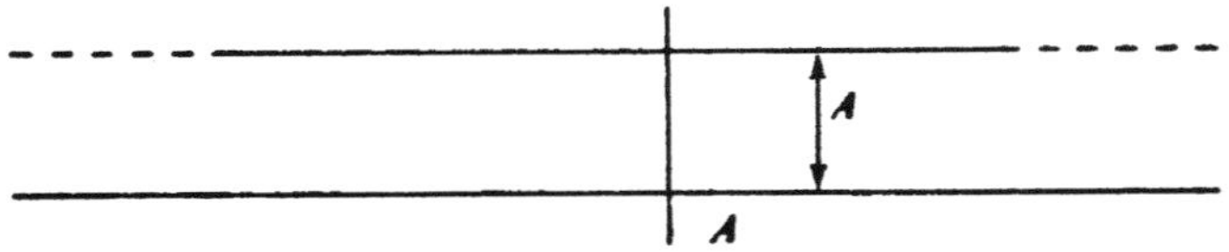 A	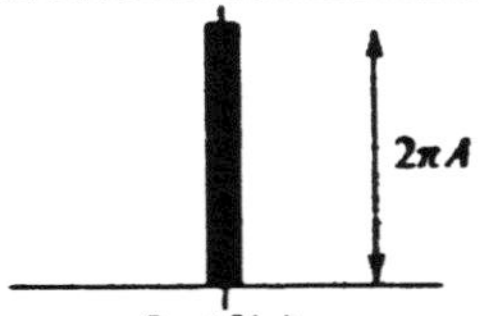 $2\pi A\delta(y)$	(3.38)

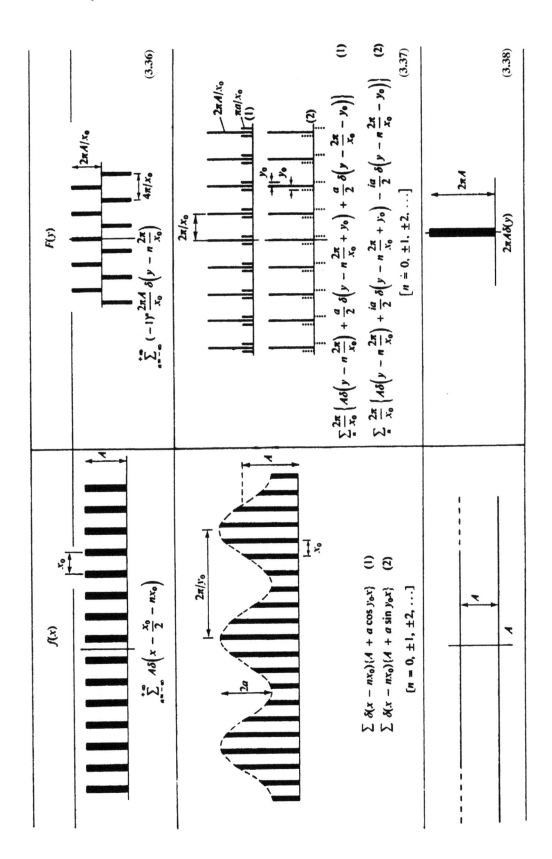

TABLE 3.2 Table of Fourier Transforms ($x = t$; $y = w$) (continued)

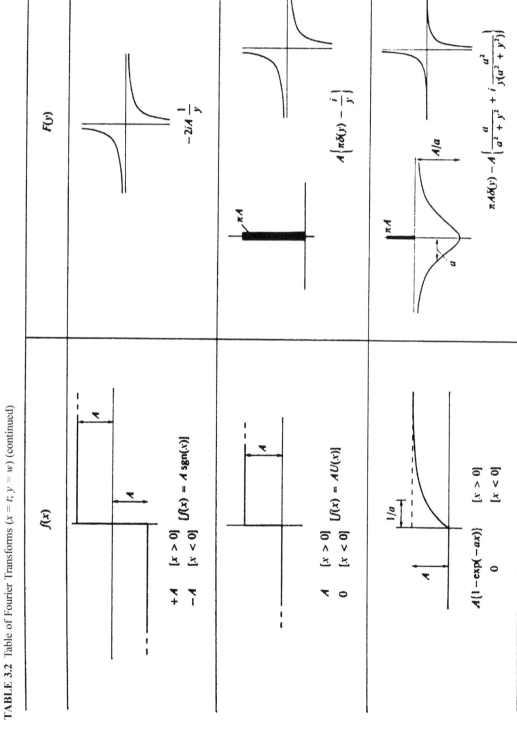

$f(x)$	$F(y)$	
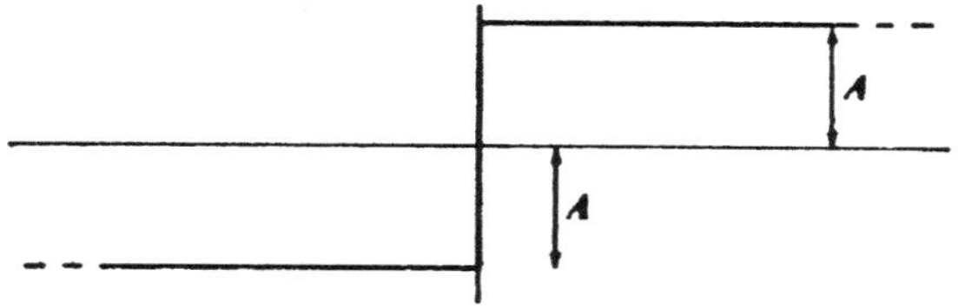 $+A \quad [x > 0]$; $-A \quad [x < 0]$ $[f(x) = A\,\mathrm{sgn}(x)]$	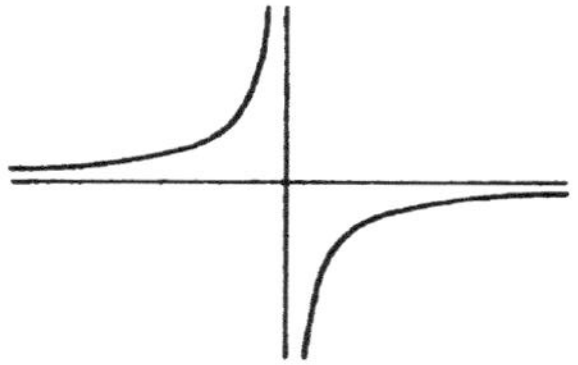 $-2iA\dfrac{1}{y}$	(3.39)
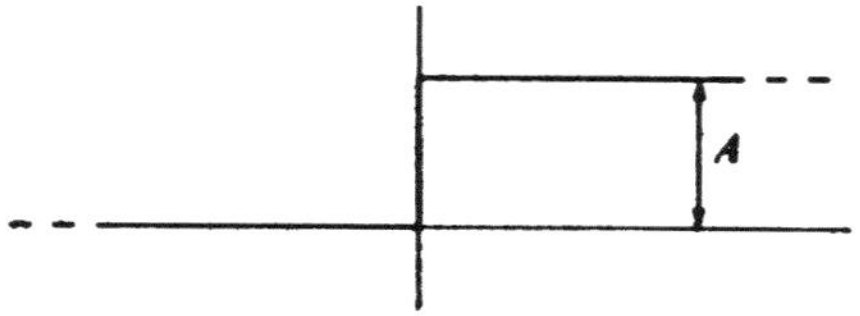 $A \quad [x > 0]$; $0 \quad [x < 0]$ $[f(x) = AU(x)]$	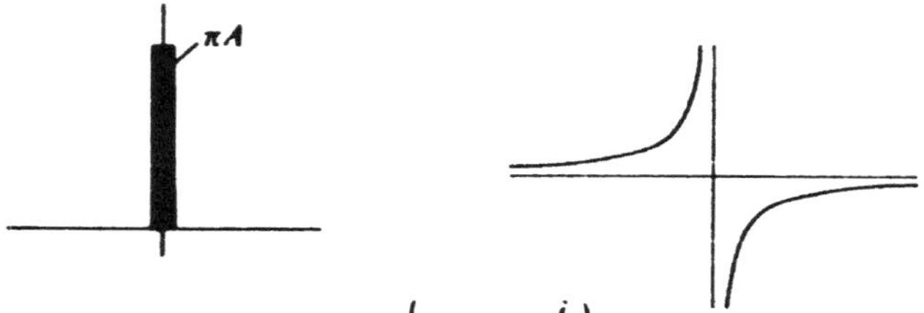 $A\left\{\pi\delta(y) - \dfrac{i}{y}\right\}$	(3.40)
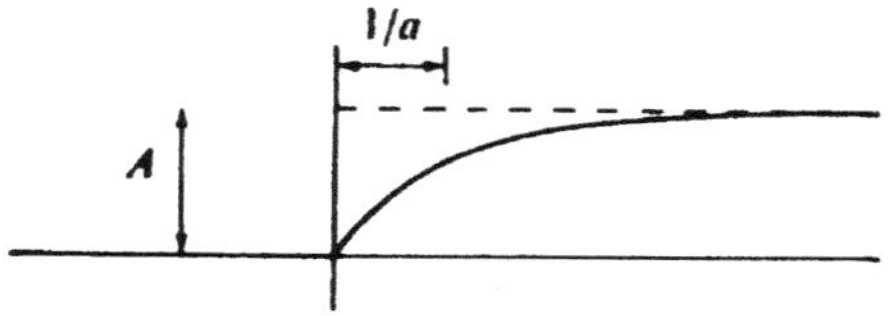 $A\{1 - \exp(-ax)\} \quad [x > 0]$; $0 \quad [x < 0]$	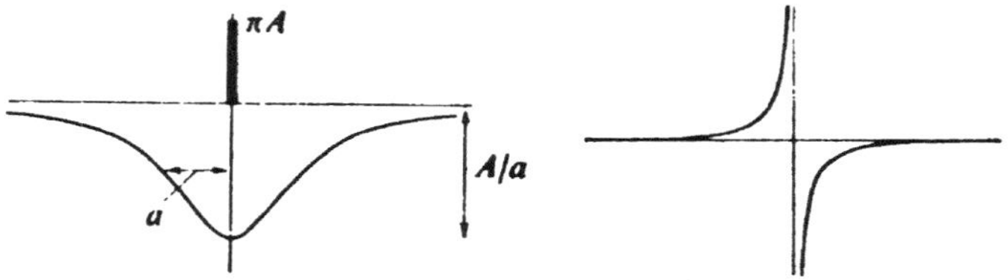 $\pi A\delta(y) - A\left\{\dfrac{a}{a^2 + y^2} + i\dfrac{a^2}{y(a^2 + y^2)}\right\}$	(3.41)

$f(x)$	$F(y)$	
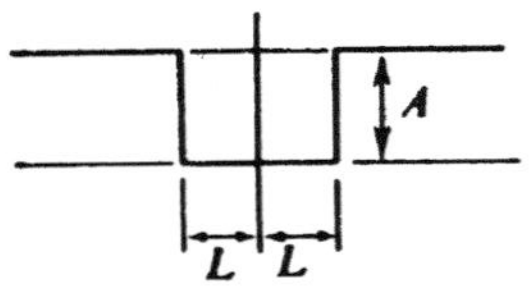 $A \quad [\lvert x\rvert > L]$ $0 \quad [\lvert x\rvert < L]$	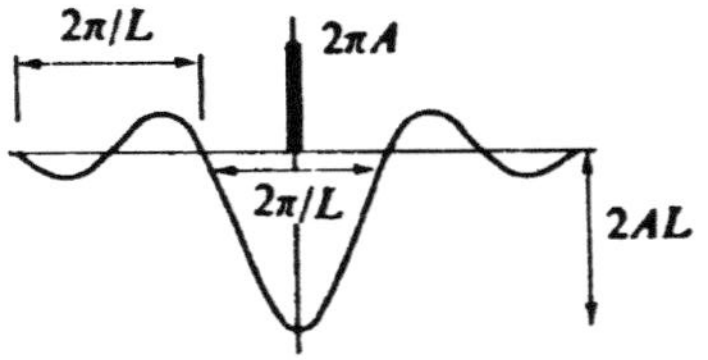 $2\pi A\delta(y) - 2A\dfrac{\sin Ly}{y}$	(3.42)
$A\exp\{i(a\cos y_0 x + bx)\}$	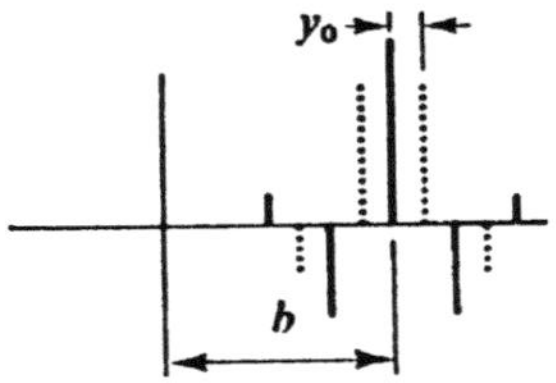 $2\pi A\sum_{n=-\infty}^{+\infty}(i)^n J_n(a)\delta(y - b - ny_0)$	(3.43)
$A\exp\{i(a\sin y_0 x + bx)\}$	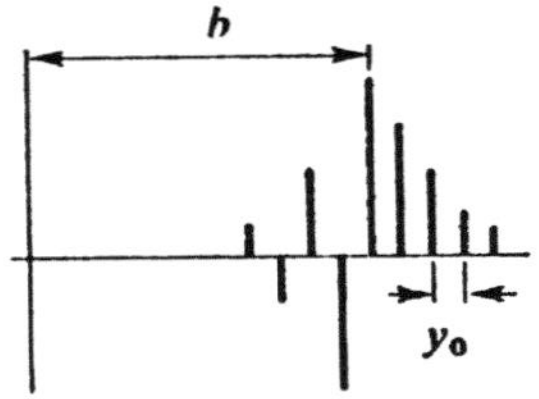 $2\pi A\sum_{n=-\infty}^{+\infty} J_n(a)\delta(y - b - ny_0)\}$	(3.44)

Note: $J_n(-a) = J_{-n}(a) = (-1)^n J_n(a)$. See Appendix H for some properties of Bessel functions.

TABLE 3.2 Table of Fourier Transforms ($x = t$; $y = w$) (continued)

$f(x)$	$F(y)$
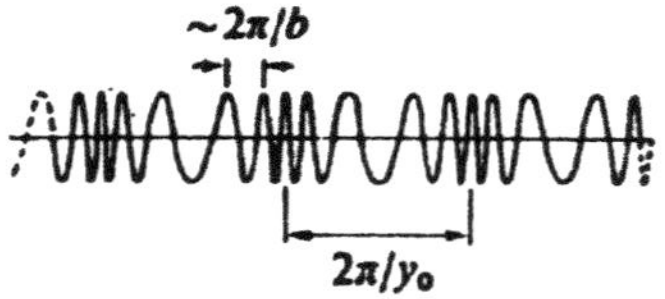 $A\cos(a\sin y_0x + bx)$	 $\pi A\sum_{n=-\infty}^{+\infty}\{J_n(a)\delta(y - b - ny_0) + J_n(a)\delta(y + b + ny_0)\}$ (3.45)
$A\cos(a\cos y_0x + bx)$	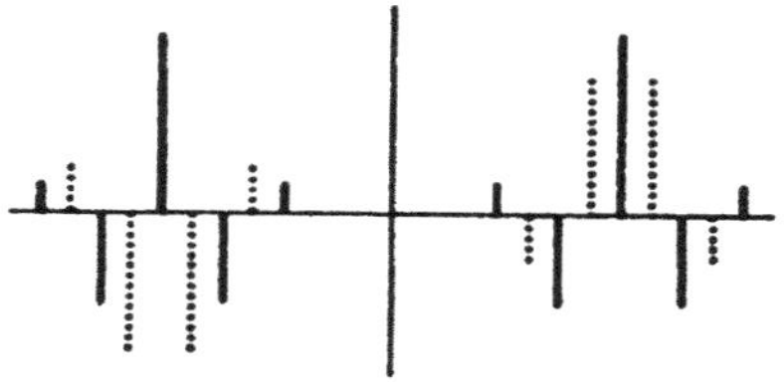 $\pi A\sum_{n=-\infty}^{+\infty}\{(+i)^nJ_n(a)\delta(y - b - ny_0) + (-i)^nJ_n(a)\delta(y + b + ny_0)\}$ (3.46)
$A\sin(a\sin y_0x + bx)$	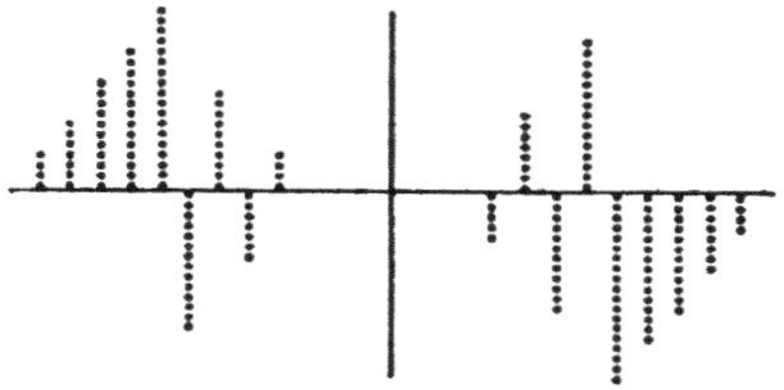 $i\pi A\sum_{n=-\infty}^{+\infty}\{-J_n(a)\delta(y - b - ny_0) + J_n(a)\delta(y + b + ny_0)\}$ (3.47)

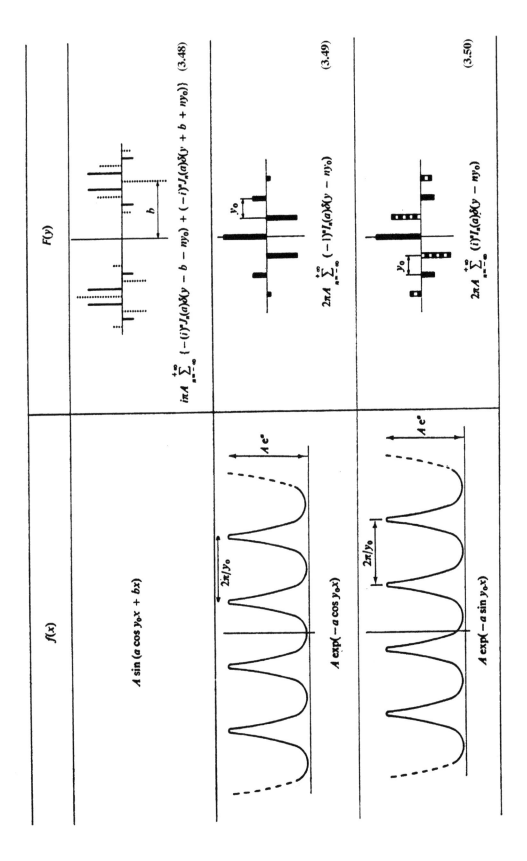

$f(x)$	$F(y)$
$A \sin(a \cos y_0 x + bx)$	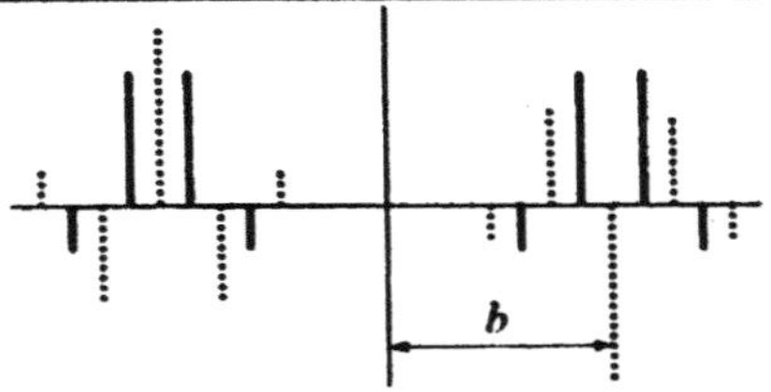 $i\pi A \sum_{n=-\infty}^{+\infty} \{-(i)^n J_n(a)\delta(y - b - ny_0) + (-i)^n J_n(a)\delta(y + b + ny_0)\}$ (3.48)
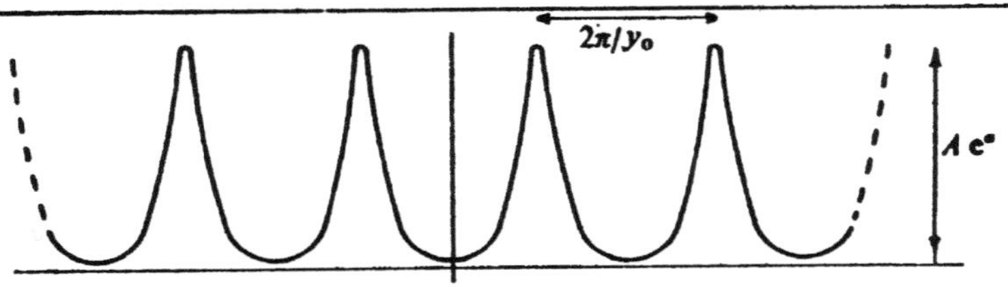 $A \exp(-a \cos y_0 x)$	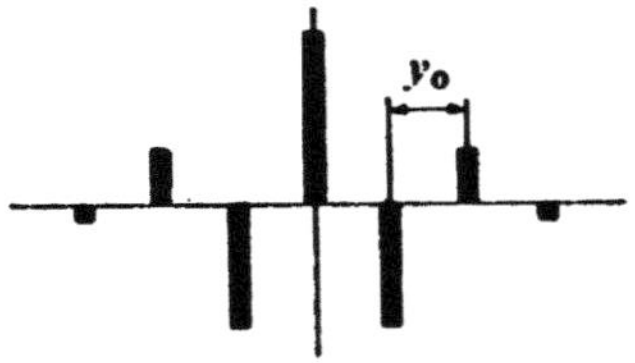 $2\pi A \sum_{n=-\infty}^{+\infty} (-1)^n I_n(a)\delta(y - ny_0)$ (3.49)
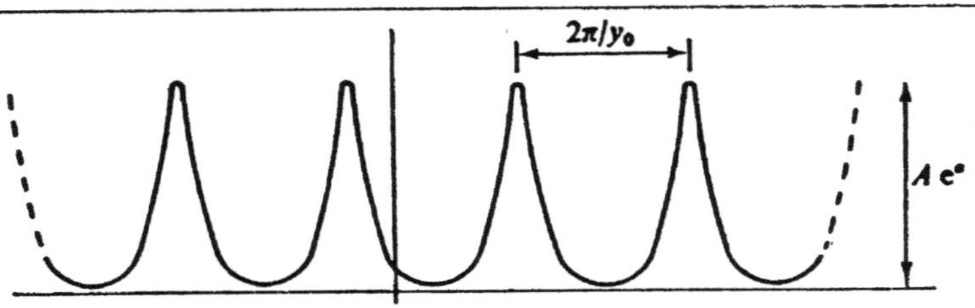 $A \exp(-a \sin y_0 x)$	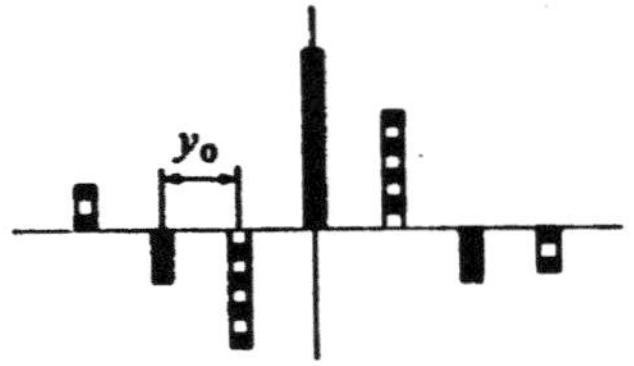 $2\pi A \sum_{n=-\infty}^{+\infty} (i)^n I_n(a)\delta(y - ny_0)$ (3.50)

TABLE 3.2 Table of Fourier Transforms ($x = t$; $y = w$) (continued)

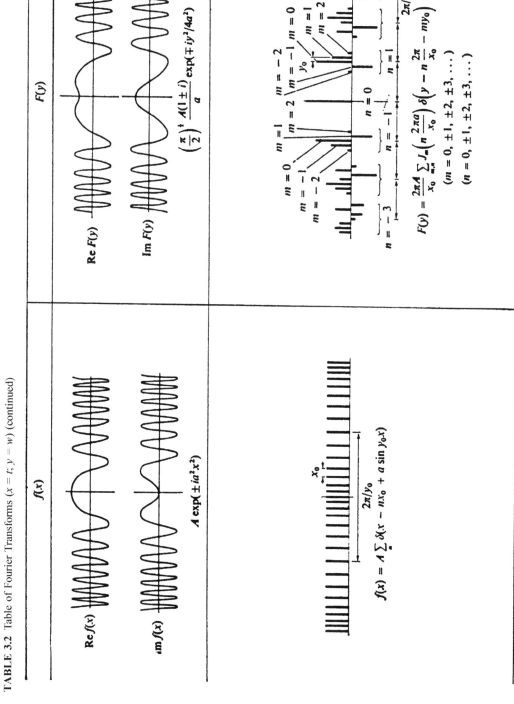

$f(x)$	$F(y)$	
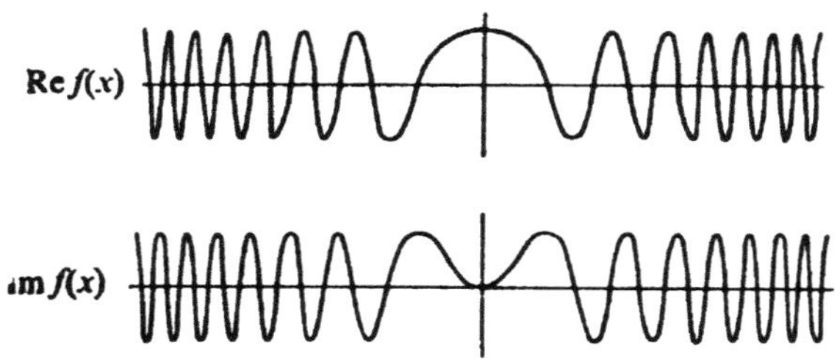 $A\exp(\pm ia^2x^2)$	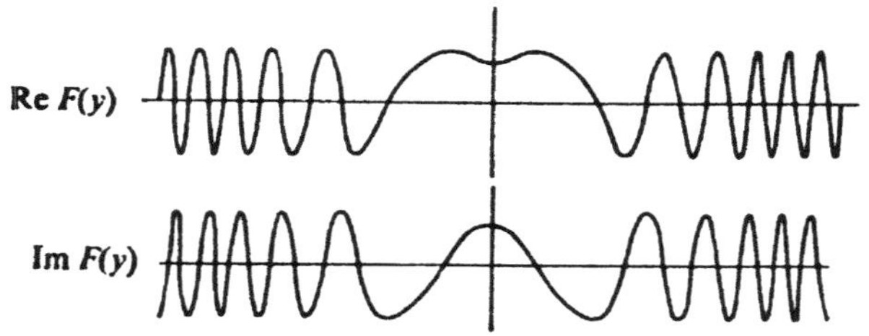 $\left(\frac{\pi}{2}\right)^{1/2}\frac{A(1\pm i)}{a}\exp(\mp iy^2/4a^2)$	(3.51)
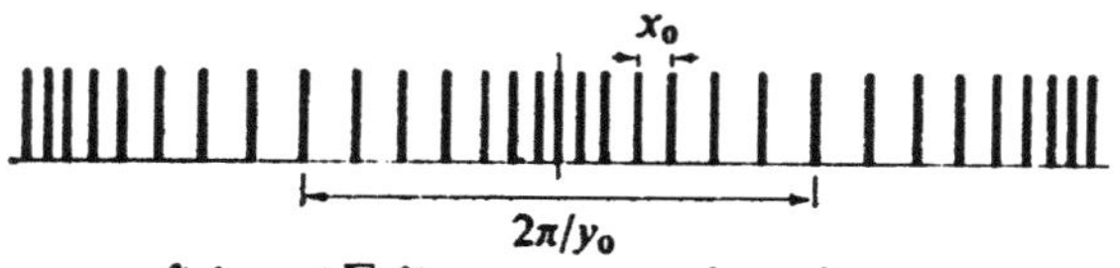 $f(x) = A\sum_{n}\delta(x - nx_0 + a\sin y_0x)$	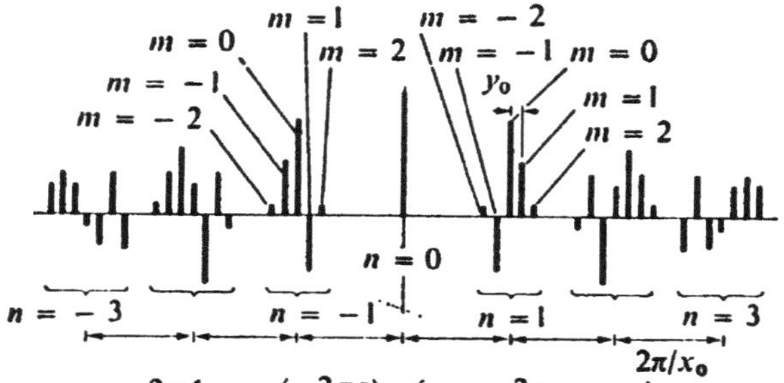 $F(y) = \frac{2\pi A}{x_0}\sum_{m,n}J_m\left(n\frac{2\pi a}{x_0}\right)\delta\left(y - n\frac{2\pi}{x_0} - my_0\right)$ $(m = 0, \pm 1, \pm 2, \pm 3, \ldots)$ $(n = 0, \pm 1, \pm 2, \pm 3, \ldots)$	(3.52)

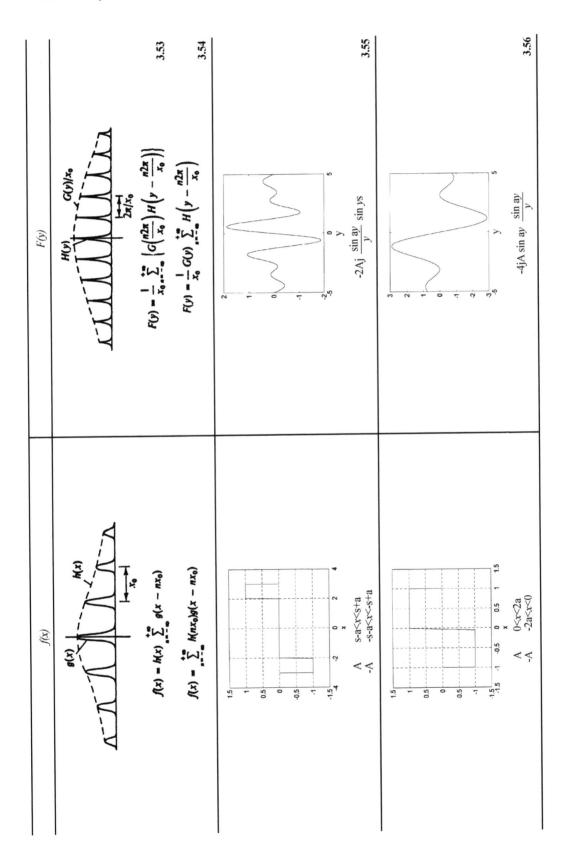

f(x)	*F(y)*	
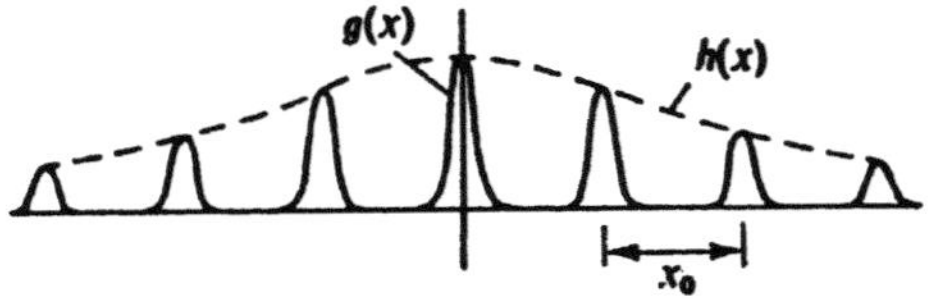 $f(x) = h(x) \sum_{n=-\infty}^{+\infty} g(x - nx_0)$ $f(x) = \sum_{n=-\infty}^{+\infty} h(nx_0) g(x - nx_0)$	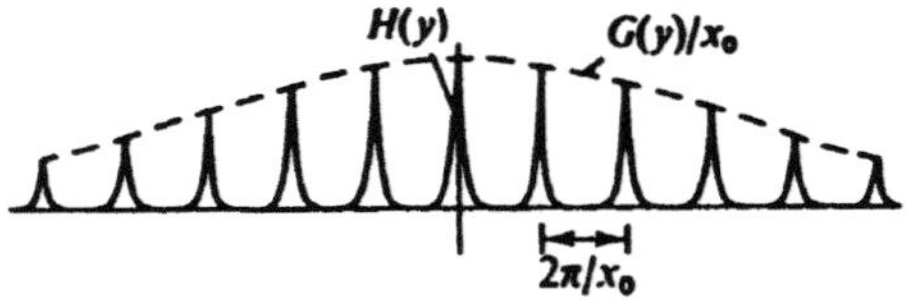 $F(y) = \frac{1}{x_0} \sum_{n=-\infty}^{+\infty} \left\{ G\left(\frac{n2\pi}{x_0}\right) H\left(y - \frac{n2\pi}{x_0}\right) \right\}$ $F(y) = \frac{1}{x_0} G(y) \sum_{n=-\infty}^{+\infty} H\left(y - \frac{n2\pi}{x_0}\right)$	3.53 3.54
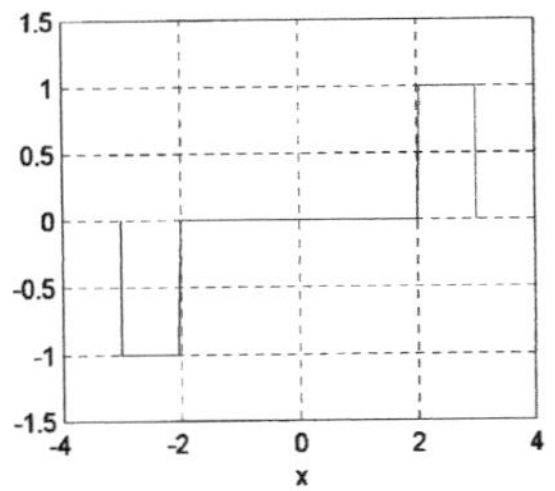 A s-a<*x*<s+a -A -s-a<*x*<-s+a	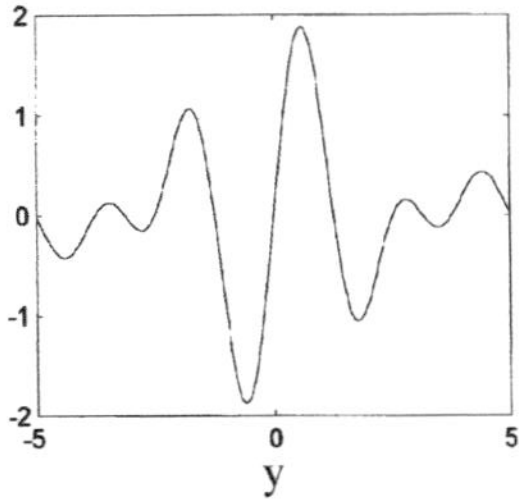 $-2Aj \frac{\sin ay}{y} \sin ys$	3.55
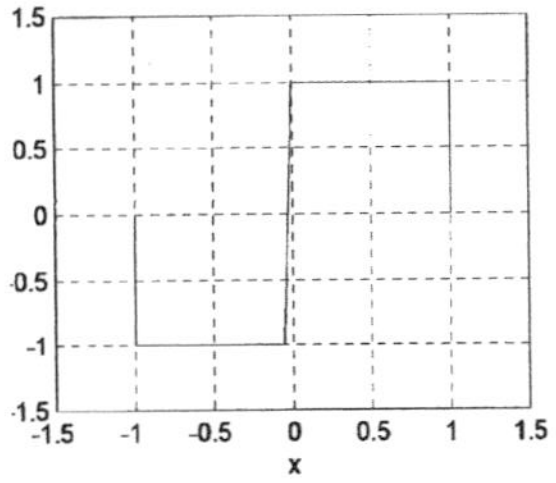 A 0<*x*<2a -A -2a<*x*<0	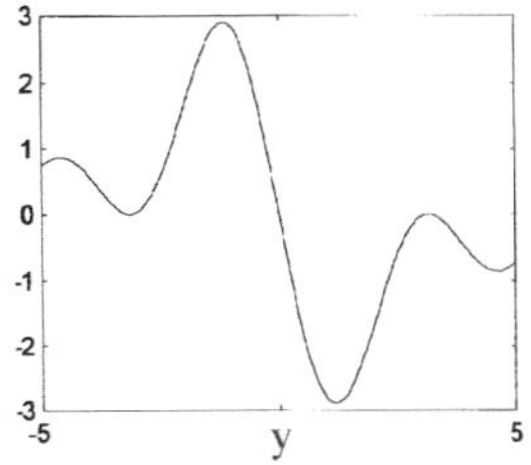 $-4jA \sin ay \frac{\sin ay}{y}$	3.56

TABLE 3.2 Table of Fourier Transforms ($x = t$; $y = w$) (continued)

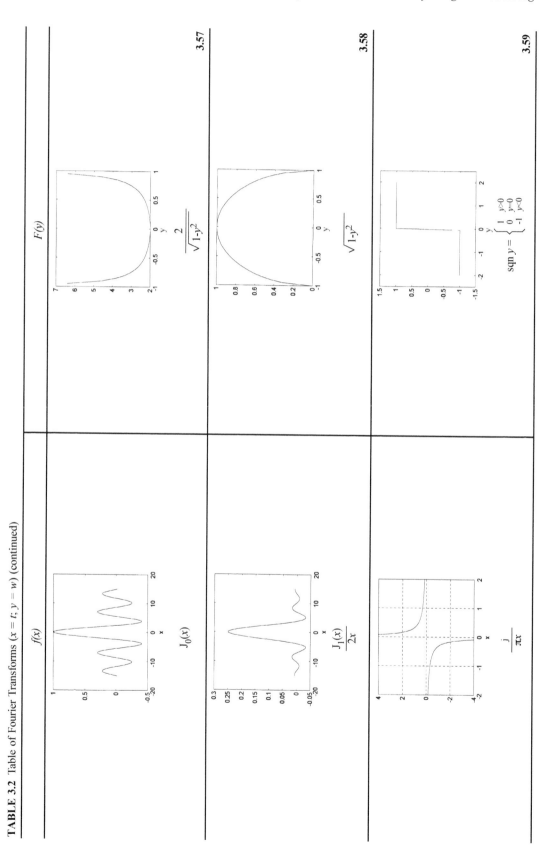

$f(x)$	$F(y)$	
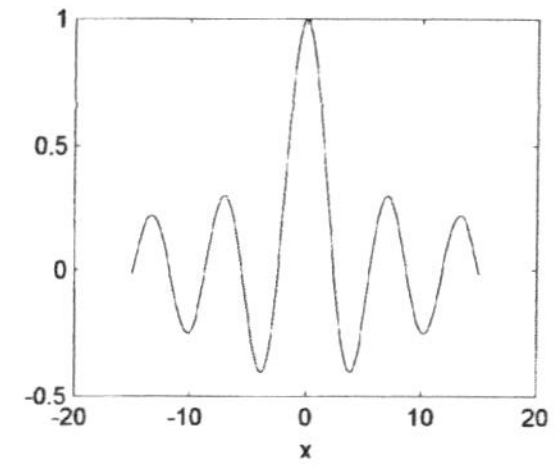 $J_0(x)$	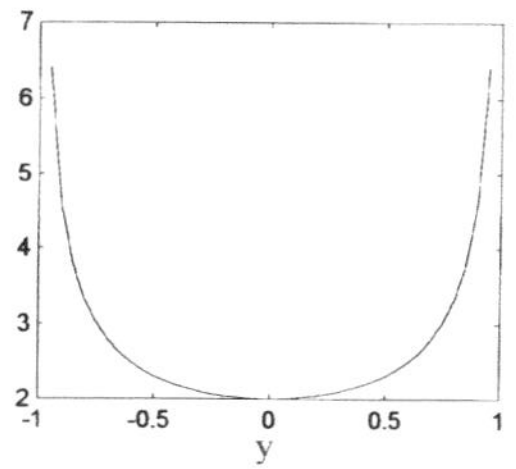 $\frac{2}{\sqrt{1-y^2}}$	**3.57**
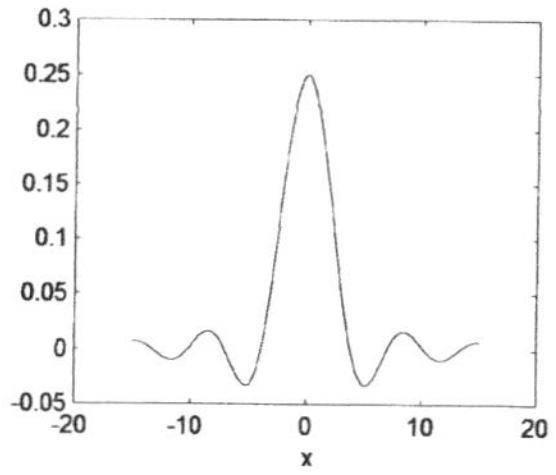 $\frac{J_1(x)}{2x}$	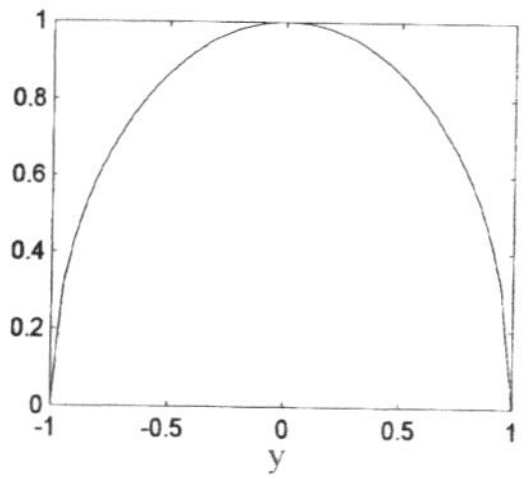 $\sqrt{1-y^2}$	**3.58**
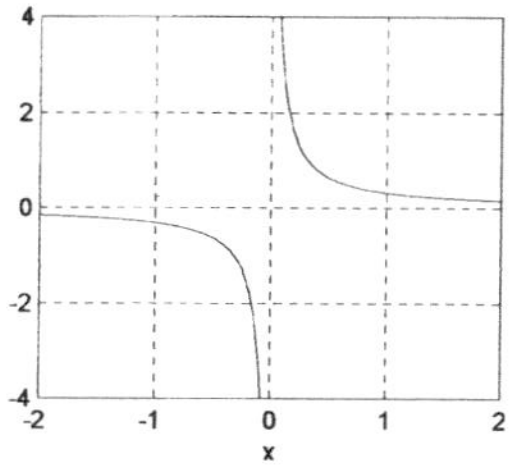 $\frac{j}{\pi x}$	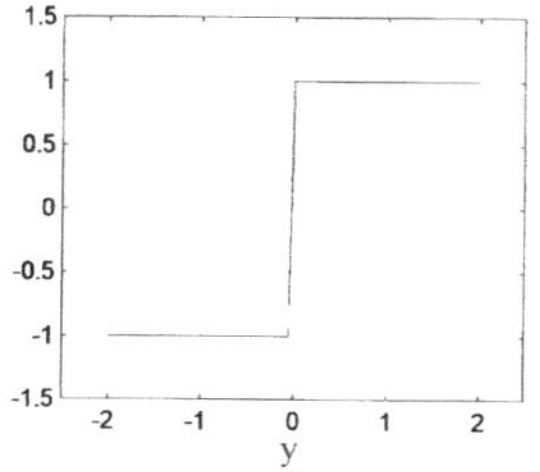 $\text{sqn}\, y = \begin{cases} 1 & y>0 \\ 0 & y=0 \\ -1 & y<0 \end{cases}$	**3.59**

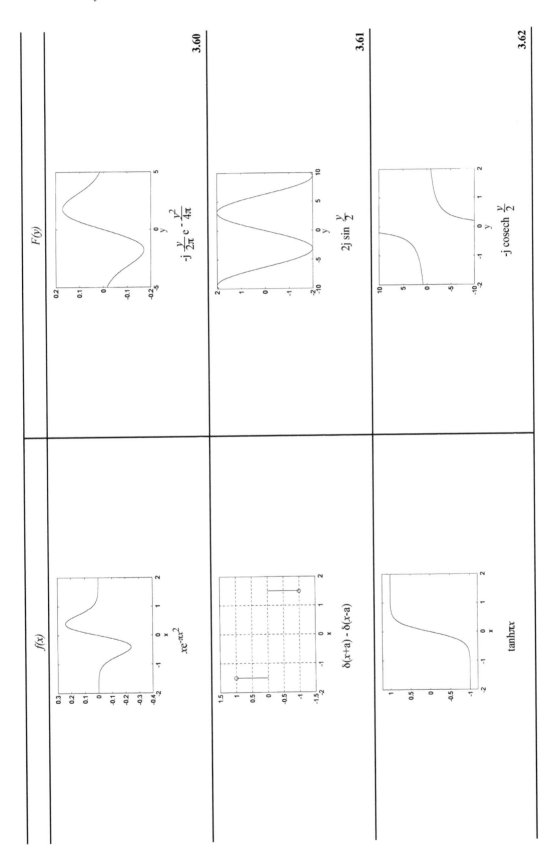

$f(x)$	$F(y)$	
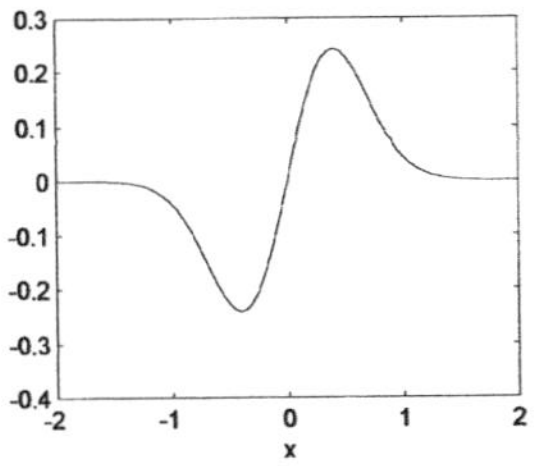 $xe^{-\pi x^2}$	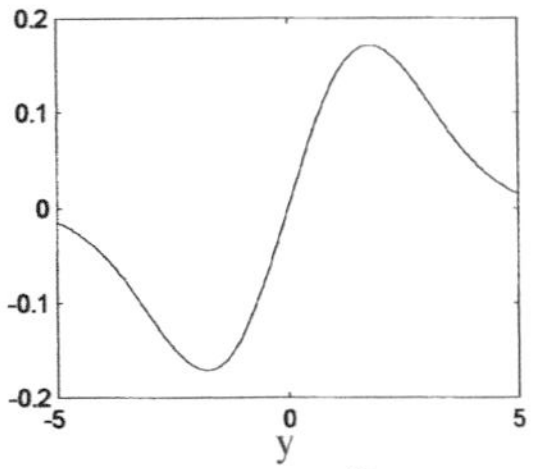 $-j\,\frac{y}{2\pi}\,e^{-\frac{y^2}{4\pi}}$	**3.60**
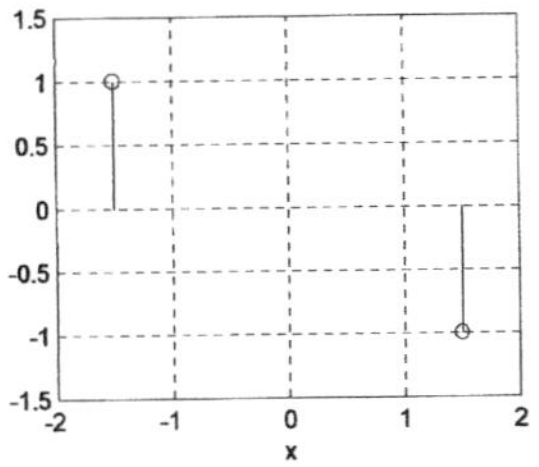 $\delta(x+a) - \delta(x-a)$	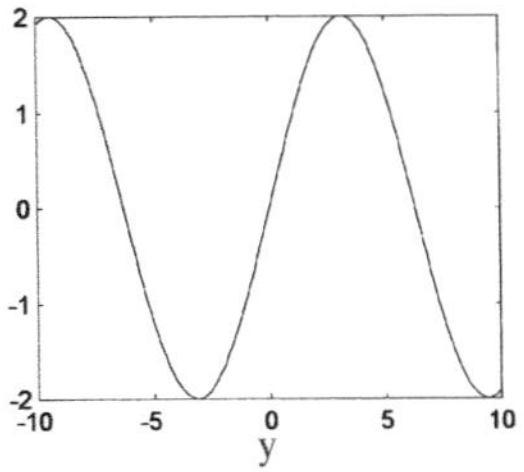 $2j \sin \frac{y}{2}$	**3.61**
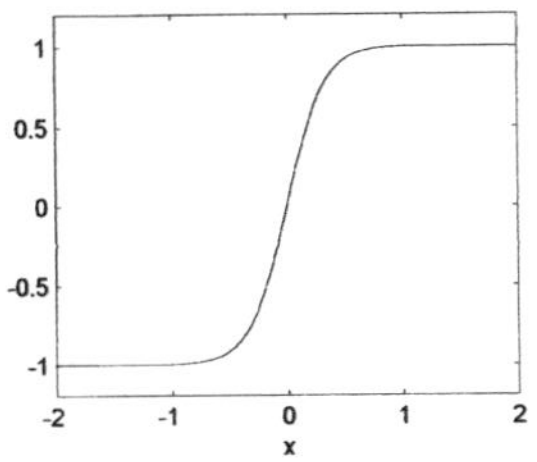 $\tanh \pi x$	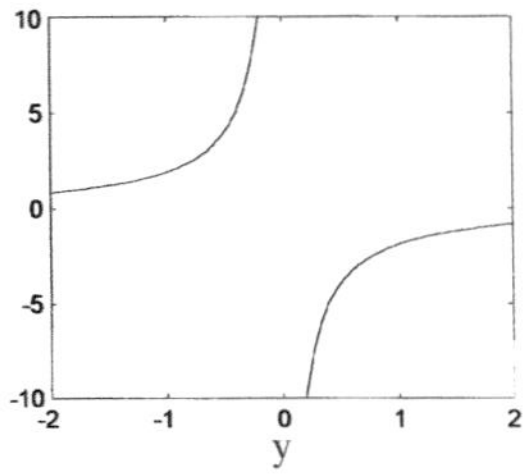 $-j\,\mathrm{cosech}\,\frac{y}{2}$	**3.62**

TABLE 3.2 Table of Fourier Transforms ($x = t$; $y = w$) (continued)

$f(x)$	$F(y)$	
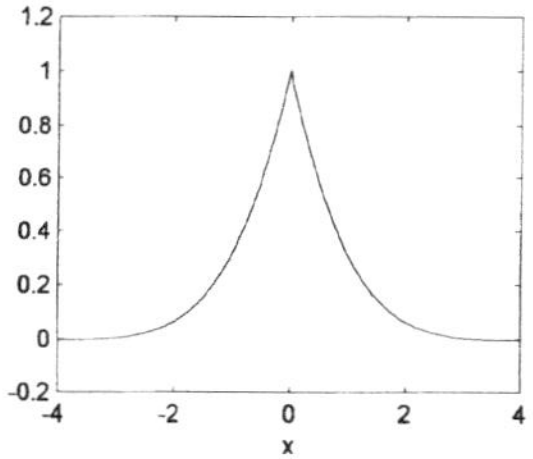 $e^{-\lvert x\rvert}\,\dfrac{\sin x}{x}$	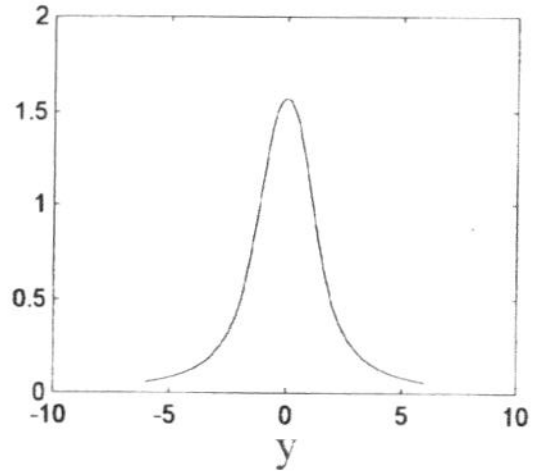 $\tan^{-1}\dfrac{2}{y^2}$	**3.63**
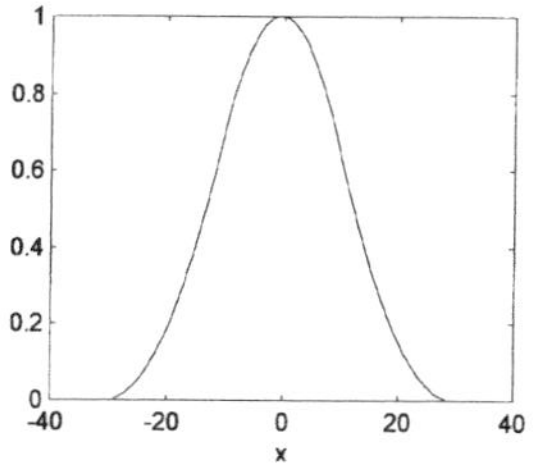 $p(x) * p(x) * p(x)$	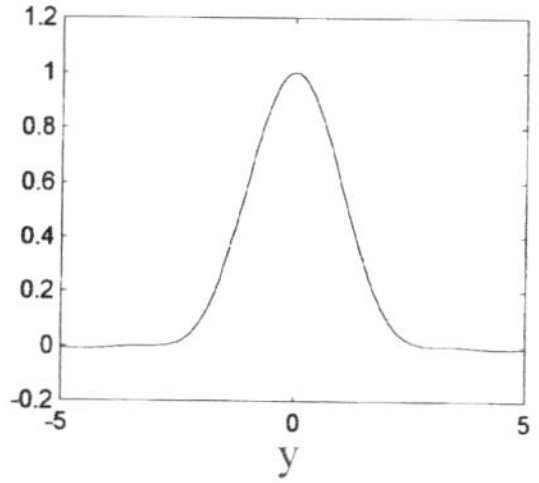 $\left(\dfrac{\sin y}{y}\right)^3$	**3.64**
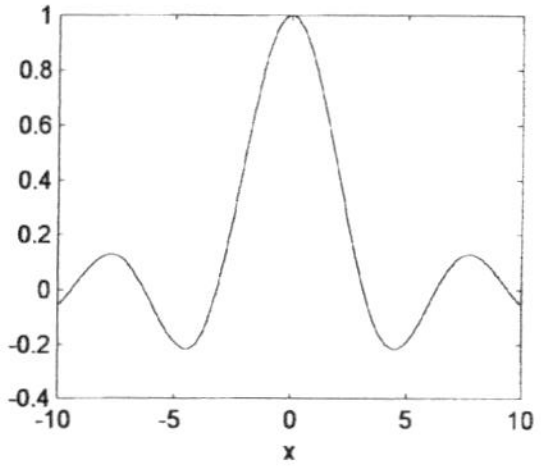 $\dfrac{\sin ax}{\pi x}$	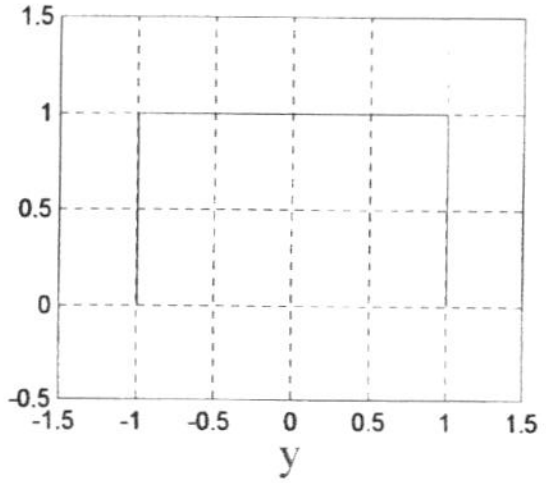 $p_a(x)$	**3.65**

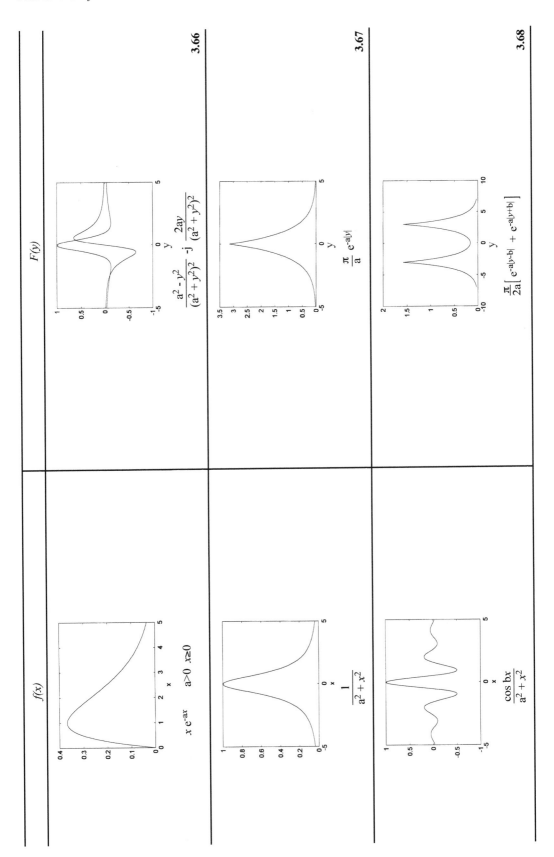

$f(x)$	$F(y)$	
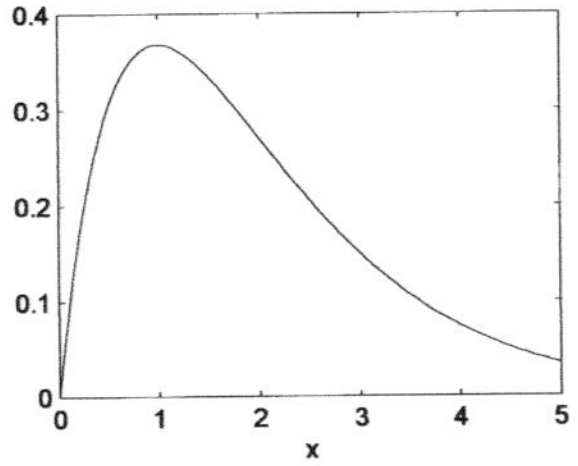 $x\,e^{-ax} \quad a>0 \;\; x\geq 0$	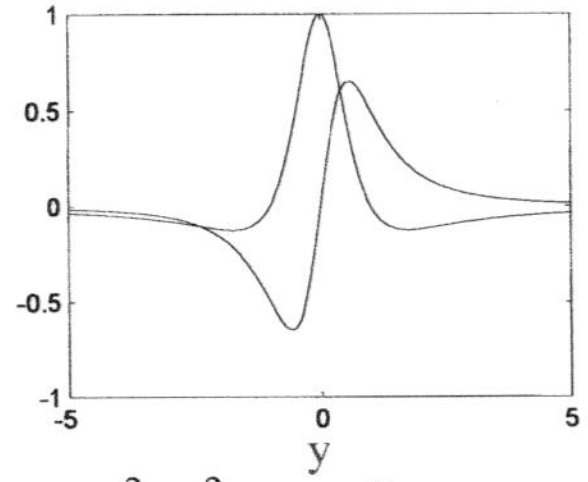 $\frac{a^2 - y^2}{(a^2+y^2)^2} - j\,\frac{2ay}{(a^2+y^2)^2}$	**3.66**
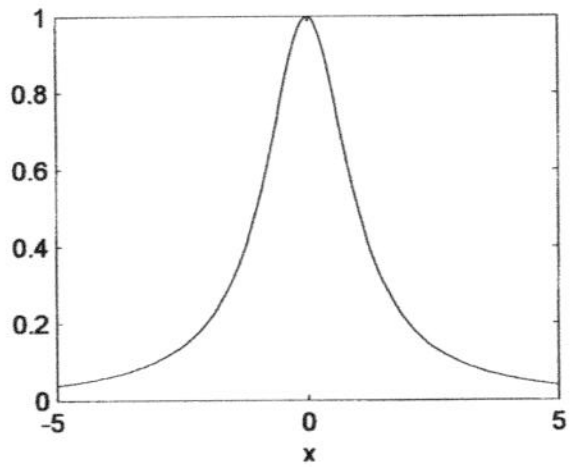 $\frac{1}{a^2 + x^2}$	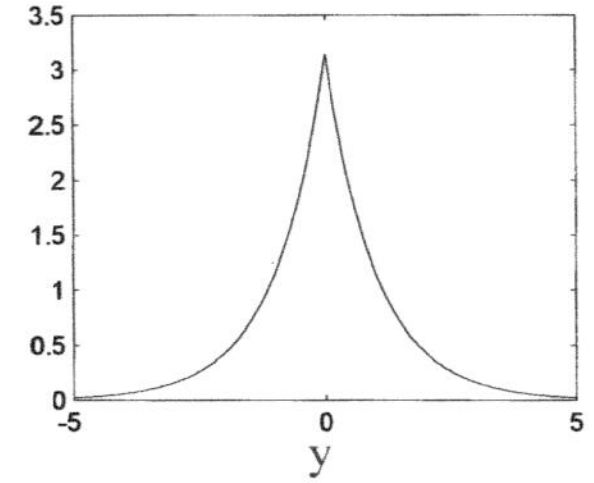 $\frac{\pi}{a}\, e^{-a\lvert y\rvert}$	**3.67**
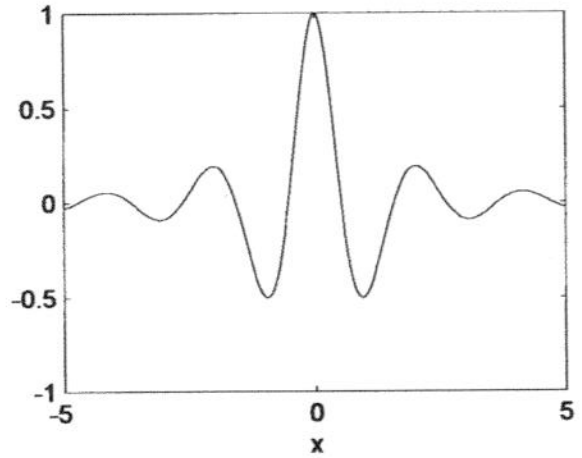 $\frac{\cos bx}{a^2 + x^2}$	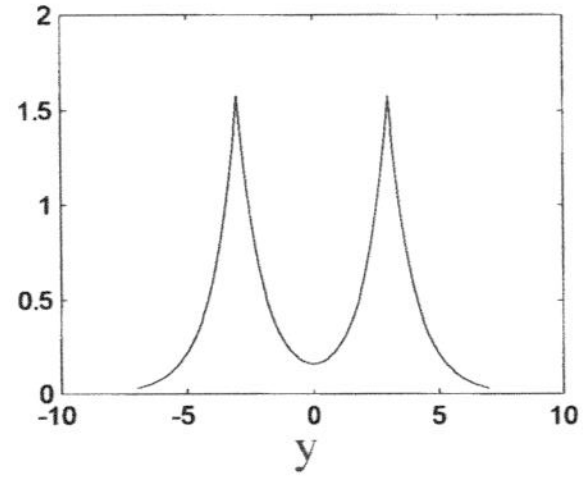 $\frac{\pi}{2a}\left[\, e^{-a\lvert y-b\rvert} + e^{-a\lvert y+b\rvert} \,\right]$	**3.68**

TABLE 3.2 Table of Fourier Transforms ($x = t$; $y = w$) (continued)

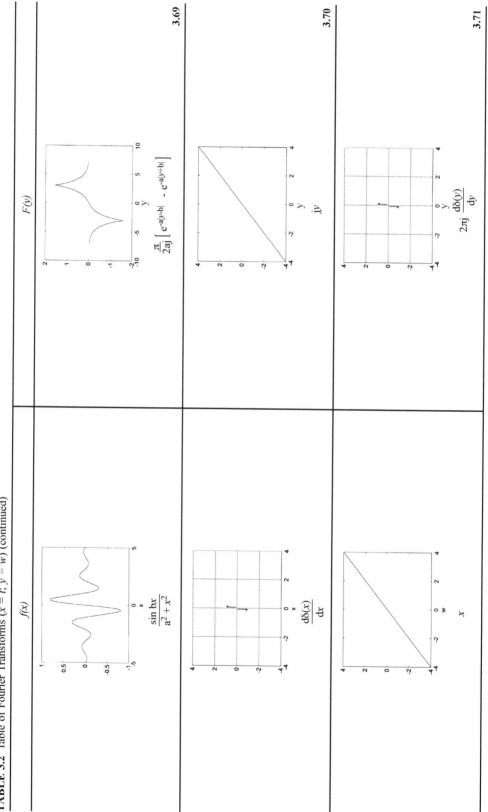

$f(x)$	$F(y)$	
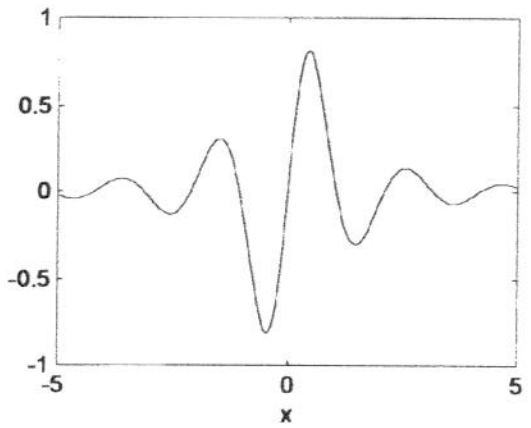 $\frac{\sin bx}{a^2 + x^2}$	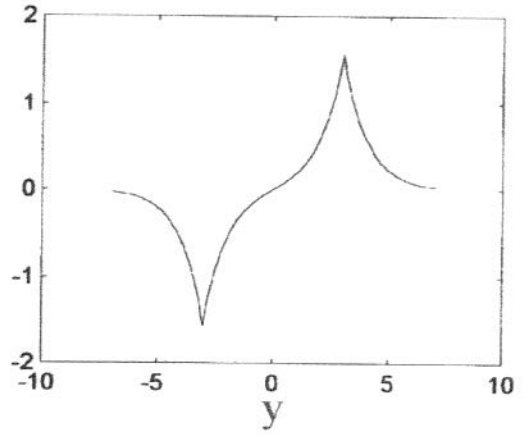 $\frac{\pi}{2aj}\left[e^{-a\lvert y-b\rvert} - e^{-a\lvert y+b\rvert}\right]$	**3.69**
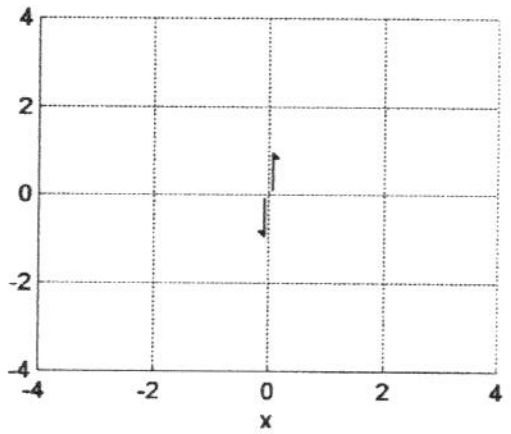 $\frac{d\delta(x)}{dx}$	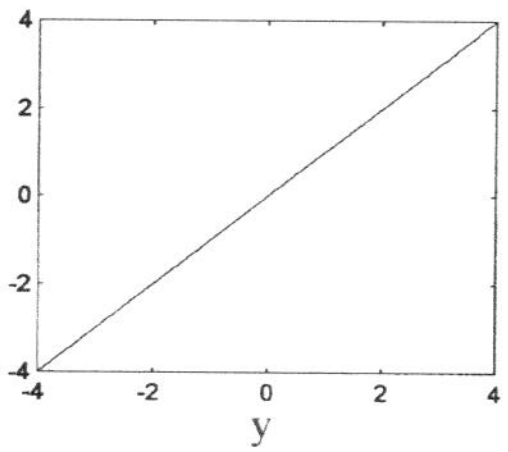 jy	**3.70**
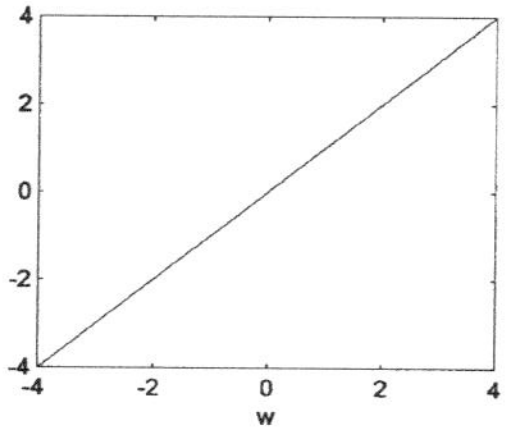 x	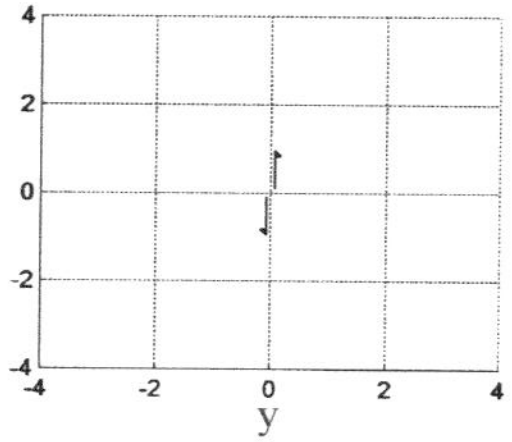 $2\pi j \frac{d\delta(y)}{dy}$	**3.71**

f(x)	F(y)	
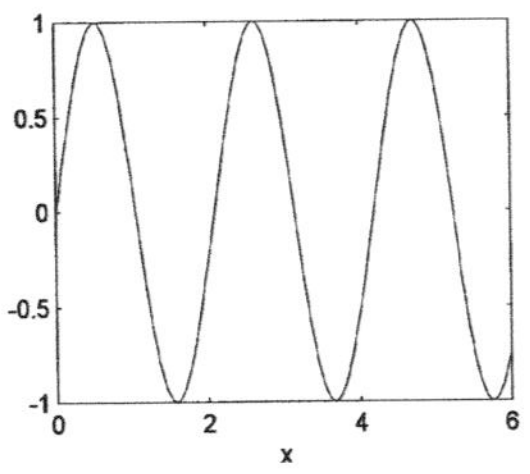 $\begin{matrix} \sin w_0 x & x \ge 0 \\ 0 & x < 0 \end{matrix}$	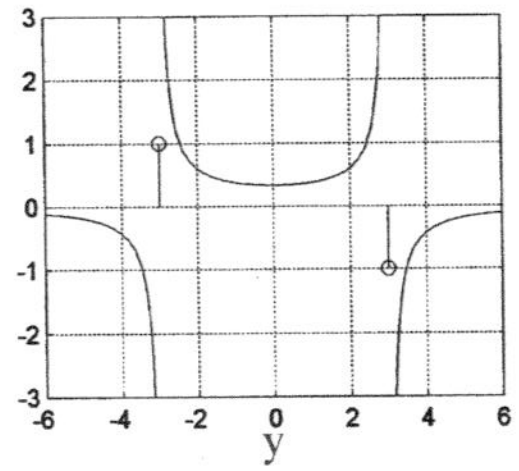 $\frac{w_0}{w_0^2 - y^2} - j\frac{\pi}{2}[\delta(y-w_0) - \delta(y+w_0)]$	**3.72**
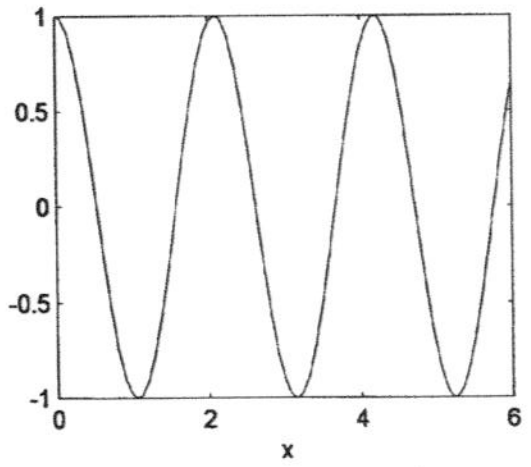 $\begin{matrix} \cos w_0 x & x \ge 0 \\ 0 & x < 0 \end{matrix}$	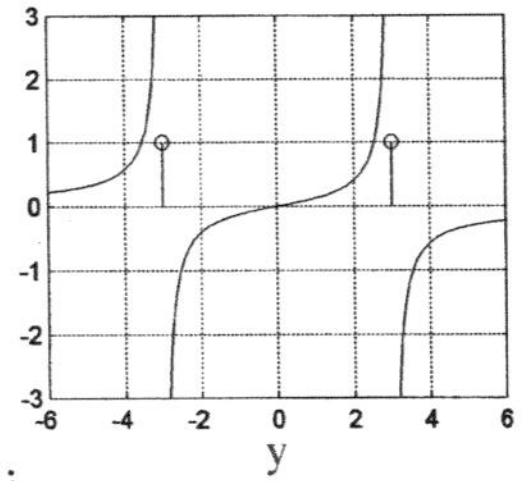 $\frac{jy}{w_0^2 - y^2} + \frac{\pi}{2}[\delta(y-w_0) - \delta(y+w_0)]$	**3.73**
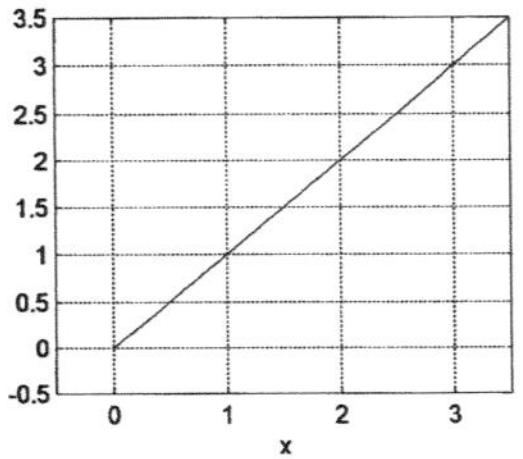 $\begin{matrix} x & x \ge 0 \\ 0 & x < 0 \end{matrix}$	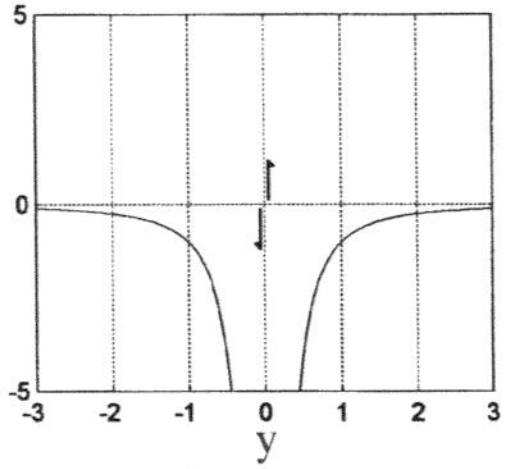 $j\pi\frac{d\delta(y)}{dy} - \frac{1}{y^2}$	**3.74**

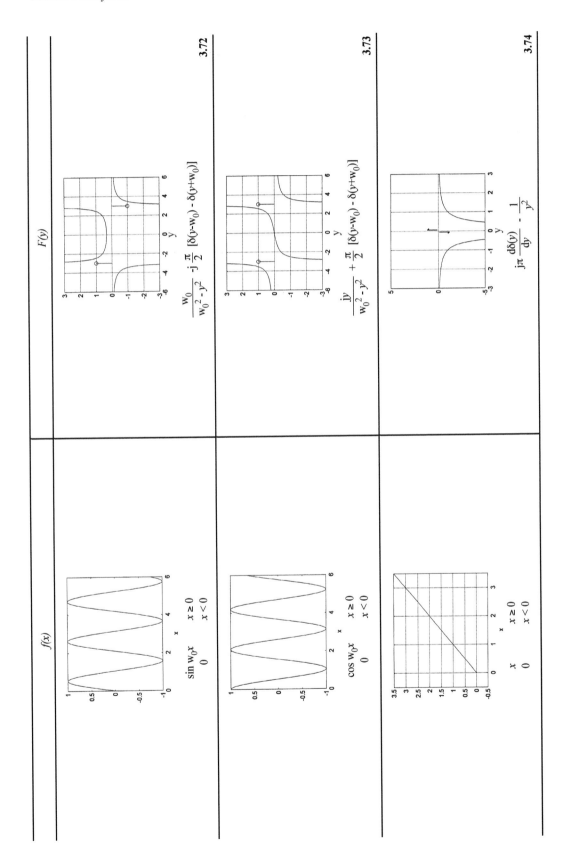

3.2 Two-Dimensional Continuous Fourier Transform

3.2.1 Definitions

3.2.1.1 Two-Dimensional Fourier Transform

$$F\{f(x,y)\} = F(\omega_1,\omega_2) = \iint_{-\infty}^{\infty} f(x,y)e^{-j(x\omega_1+y\omega_2)}\,dx\,dy$$

$$F^{-1}\{F(\omega_1,\omega_2)\} = \frac{1}{(2\pi)^2}\iint_{-\infty}^{\infty} F(\omega_1,\omega_2)e^{j(\omega_1 x+\omega_2 y)}\,d\omega_1\,d\omega_2$$

3.2.1.2 Properties of Two-Dimensional Fourier Transform

TABLE 3.3 Properties of Two-Dimensional Fourier Transform

Rotation	$f(\pm x, \pm y)$	$F(\pm\omega_1, \pm\omega_2)$
Linearity	$a_1 f_1(x,y) + a_2 f_2(x,y)$	$a_1 F_1(\omega_1,\omega_2) + a_2 F_2(\omega_1,\omega_2)$
Conjugation	$f^*(x,y)$	$F^*(-\omega_1,-\omega_2)$
Separability	$f_1(x) f_2(y)$	$F_1(\omega_1) F_2(\omega_2)$
Scaling	$f(ax, by)$	$\frac{1}{\lvert ab\rvert} F\left(\frac{\omega_1}{a}, \frac{\omega_2}{b}\right)$
Shifting	$f(x \pm x_o, y \pm y_o)$	$e^{\pm j(\omega_1 x_o + \omega_2 y_o)} F(\omega_1,\omega_2)$
Modulation	$e^{\pm j(\omega_{c1}x+\omega_{c2}y)} f(x,y)$	$F(\omega_1 \mp \omega_{c1}, \omega_2 \mp \omega_{c2})$
Convolution	$g(x,y) = h(x,y) ** f(x,y)$	$G(\omega_1,\omega_2) = H(\omega_1,\omega_2)F(\omega_1,\omega_2)$
Multiplication	$g(x,y) = h(x,y) f(x,y)$	$G(\omega_1,\omega_2) = \frac{1}{(2\pi)^2} H(\omega_1,\omega_2) ** F(\omega_1,\omega_2)$
Correlation	$c(x,y) = h(x,y) \qquad f(x,y)$	$G(\omega_1,\omega_2) = H(-\omega_1,-\omega_2) F(\omega_1,\omega_2)$
Inner Product	$I = \iint_{-\infty}^{\infty} f(x,y) h^*(x,y)\,dx\,dy$	$I = \frac{1}{(2\pi)^2}\iint_{-\infty}^{\infty} F(\omega_1,\omega_2) H^*(\omega_1,\omega_2)\,d\omega_1\,d\omega_2$
Parseval Formula	$I = \iint_{-\infty}^{\infty} \lvert f(x,y)\rvert^2\,dx\,dy$	$I = \frac{1}{(2\pi)^2}\iint_{-\infty}^{\infty} \lvert F(\omega_1,\omega_2)\rvert^2\,d\omega_1\,d\omega_2$

References

Bracewell, Ron, *The Fourier Transform and Its Applications*, McGraw-Hill Book Company, New York, NY, 1965.

Campbell, G. A., and R. M. Foster, *Fourier Integrals for Practical Applications*, Van Nostrand Company, Princeton, NJ, 1948.

Champency, D. C., *Fourier Transforms and Their Physical Applications*, Academic Press, New York, 1973.

Howell, K., Fourier transform, in *The Transforms and Application Handbook*, Edited by A. D. Poularikas, CRC Press Inc., Boca Raton, FL, 1996.

Papoulis, Athanasios, *The Fourier Integral and Its Applications,* McGraw-Hill Book Company, New York, NY, 1962.

Walker, James S., *Fourier Analysis*, Oxford University Press, New York, NY, 1988.

Appendix 1

Examples

1.1 Gibbs Phenomenon

Example 3.1

Let $U(\omega)$ be the spectrum of unit step function, and let the truncated spectrum $U_{\omega_o}(\omega) = U(\omega)$ for $|\omega| \le \omega_o$ and zero otherwise. We can also write the truncated spectrum as follows: $U_{\omega_o}(\omega) = U(\omega)\, p_{\omega_o}(\omega)$, where $p_{\omega_o}(\omega)$ is a unit pulse of $2\omega_o$ duration and centered at the origin. The approximate step function

$$u_a(t) = F^{-1}\{U(\omega)p_{\omega_o}(\omega)\} = F^{-1}\{U(\omega)\} * F^{-1}\{p_{\omega_o}(\omega)\} = u(t) * \frac{1}{\pi}\frac{\sin\omega_o t}{t} = \frac{1}{\pi}\int_{-\infty}^{t} \frac{\sin\omega_o \tau}{\tau} d\tau$$

and is shown in Figure 3.1.

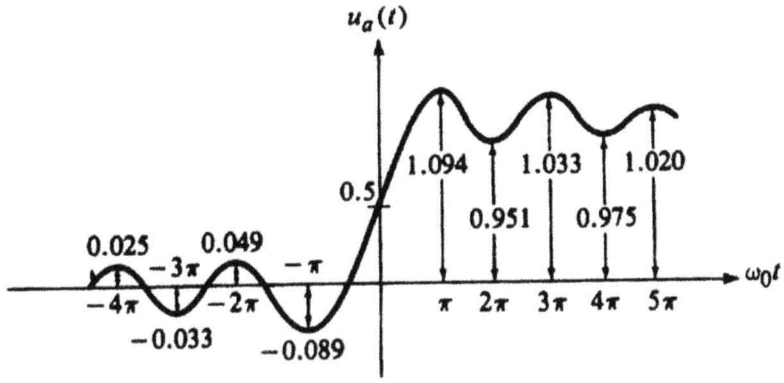

FIGURE 3.1

1.2 Special Functions

Example 3.2

$$F\{\text{sgn}(t)\} = F\{\lim_{\varepsilon\to 0} e^{-\varepsilon|t|}\,\text{sgn}(t)\} = \lim_{\varepsilon\to 0}\int_{-\infty}^{\infty} e^{-\varepsilon|t|}\,\text{sgn}(t) e^{-j\omega t}\, dt$$

$$= \lim_{\varepsilon\to 0}\left[\int_{-\infty}^{o} e^{(\varepsilon-j\omega)t}\, dt + \int_{o}^{\infty} e^{-(\varepsilon+j\omega)t}\, dt\right] = \lim_{\varepsilon\to 0}\left(-\frac{1}{\varepsilon - j\omega} + \frac{1}{\varepsilon + j\omega}\right) = \frac{2}{j\omega}$$

Example 3.3

$$F\{u(t)\} = F\{\tfrac{1}{2} + \tfrac{1}{2}\text{sgn}(t)\} = \tfrac{1}{2}2\pi\delta(\omega) + \tfrac{1}{2}\tfrac{2}{j\omega} = \pi\delta(\omega) + \tfrac{1}{j\omega}$$

Example 3.4

$$F\sum_{n=-\infty}^{\infty}\delta(t - nT)\} \triangleq F\{comb_T(t) = \int_{-\infty}^{\infty}\left[\sum_{n=-\infty}^{\infty}\delta(t - nT)\right]e^{-j\omega t}\, dt$$

But $comb_T(t)$ is periodic with the period $\omega_o = 2\pi/T$, and can be expanded in the complex form of Fourier series $comb_T(t) = comb_T(t) = \frac{1}{T}\sum_{n=-\infty}^{\infty} e^{jn\omega_o t}$. Hence

$$COMB_{\omega_o}(\omega) = \frac{1}{T}\sum_{n=-\infty}^{\infty}\int_{-\infty}^{\infty} e^{-j(\omega - n\omega_o)t}\,dt = \frac{2\pi}{T}\sum_{n=-\infty}^{\infty}\delta(\omega - n\omega_o)$$

4

Discrete-Time Fourier Transform, One- and Two-Dimensional

4.1 One-Dimensional Discrete-Time Fourier Transform

4.1.1 Definitions

$$\mathcal{F}\{x(n)\} \doteq X(\omega) \doteq X(e^{j\omega}) = \sum_{n=-\infty}^{\infty} x(n)e^{-j\omega n} \qquad -\pi \le \omega \le \pi$$

$X(\omega)$ = periodic with period 2π

$$x(n) = \mathcal{F}^{-1}\{X(\omega)\} = \frac{1}{2\pi}\int_{-\pi}^{\pi} X(\omega)e^{-j\omega n}\, d\omega$$

The Fourier transform function appears as a function of $e^{j\omega}$ and, hence, in sequences we will use both representations for convenience.

4.1.2 Properties

TABLE 4.1 Properties of One-Dimensional Discrete-Time Fourier Transform

Properties	Sequence	Transform $[X(\omega) \equiv X(e^{j\omega})]$
Discrete-time Fourier transform	$x(n)$	$X(\omega)$
Linearity	$ax(n) + by(n)$	$aX(\omega) + bY(\omega)$
Time reversal	$x(-n)$	$X(-\omega)$
Complex conjugation	$x^*(n)$	$X^*(-\omega)$
Reversal and complex conjugate	$x^*(-n)$	$X^*(\omega)$

TABLE 4.1 Properties of One-Dimensional Discrete-Time Fourier Transform (continued)

Properties	Sequence	Transform $[X(\omega) \equiv X(e^{j\omega})]$
Time shifting	$x(n \pm m)$	$e^{\pm j\omega m} X(\omega)$
Frequency shift	$e^{\pm j\omega_o n} x(n)$	$X(\omega \mp \omega_o)$
Modulation	$\cos\omega_o n\, x(n)$	$\frac{1}{2}[X(\omega+\omega_o)+X(\omega-\omega_o)]$
	$\sin\omega_o n\, x(n)$	$\frac{1}{2j}[X(\omega+\omega_o)-X(\omega-\omega_o)]$
Convolution	$x(n) * h(n)$	$X(\omega)\,Y(\omega)$
Multiplication	$x(n)\,h(n)$	$\frac{1}{2\pi} X(\omega) * Y(\omega)$
Delta function	$\delta(n-n_o)=\begin{cases}1, & n=n_o\\ 0, & \text{otherwise}\end{cases}$	$e^{-jn_o\omega}$
Frequency domain delta function	$e^{j\omega_o n}$	$\frac{1}{2\pi}\delta(\omega-\omega_o)$
Cosine function	$\cos\omega_o n$	$\pi\delta(\omega+\omega_o)+\pi\delta(\omega-\omega_o)$
Sine function	$\sin\omega_o n$	$\frac{\pi}{j}(\delta\omega+\omega_o)-\frac{1}{2}\delta(\omega-\omega_o)$
N sample step sequence	$u_N(n)=\begin{cases}1, & n=0,1,2,\cdots\\ 0, & \text{otherwise}\end{cases}$	$e^{-j\omega(N-1)/2}\dfrac{\sin(\omega N/2)}{\sin(\omega/2)}$
Symmetric pulse	$p_N(n)=\begin{cases}1, & \lvert n\rvert \le N\\ 0, & \lvert n\rvert \ge N\end{cases}$	$\dfrac{\sin[\omega(N+1/2)]}{\sin(\omega/2)}$
Triangle sequence	$\wedge_N(n)=\begin{cases}N-\lvert n\rvert, & \lvert n\rvert \le N\\ 0, & \lvert n\rvert > N\end{cases}$	$\dfrac{\sin^2(\omega N/2)}{\sin^2(\omega/2)}$
Real sequence	$x(n)$	$X(\omega)=X^*(-\omega)$
Decomposition of real $x(n)$ in even $x_e(n)$ and $x_o(n)$ parts.	$\begin{cases}x(n)=x_e(n)+x_o(n)\\ x_e=\frac{1}{2}[x(n)+x(-n)]\\ x_o=\frac{1}{2}[x(n)-x(-n)]\end{cases}$	$X(\omega)$ $\mathrm{Re}\{X(\omega)\}$ $j\,\mathrm{Im}\{X(\omega)\}$
Decomposition of a complex sequence $x(n)$ into a conjugate symmetric part $x_e(n)$ and conjugate antisymmetric part $x_o(n)$	$\begin{cases}x(n)=x_e(n)+x_o(n)\\ x_e=\frac{1}{2}[x(n)+x^*(-n)]\\ x_o=\frac{1}{2}[x(n)-x^*(-n)]\end{cases}$	$X(\omega)$ $\mathrm{Re}\{X(\omega)\}$ $j\,\mathrm{Im}\{X(\omega)\}$
Decomposition of a complex transform $X(\omega)$	$\begin{cases}x(n)\\ \mathrm{Re}\{x(n)\}\\ j\,\mathrm{Im}\{x(n)\}\end{cases}$	$X(\omega)=X_e(\omega)+X_o(\omega)$ $X_e(\omega)=\frac{1}{2}[X(\omega)+X^*(-\omega)]$ $X_o(\omega)=\frac{1}{2}[X(\omega)-X^*(-\omega)]$
Increasing sampling frequency by m; i.e., transforming a data sequence $x_1(n)$ padded with zeros $x_1(n)$ by a factor of M	$x(n)=\begin{cases}x_1(n), & \text{if } n/M=m\\ 0, & \text{otherwise}\end{cases}$	$X_1(M\omega)$
Reducing sampling frequency by M; i.e., decimating a sequence $x_1(n)$ by a factor of M	$\begin{cases}x(n)=x_1(Mn)\\ n=0,\pm1,\pm2,\cdots\end{cases}$	$\dfrac{1}{M}\sum_{\ell=0}^{M-1} X_1\left(\omega-\dfrac{2\pi\ell}{M}\right)$
Parseval's theorem	$\sum_{n=-\infty}^{\infty} x(n)h^*(n)$	$=\dfrac{1}{2\pi}\int_{-\pi}^{\pi} X(\omega)H^*(\omega)\,d\omega$
	$\sum_{n=-\infty}^{\infty} \lvert x(n)\rvert^2$	$=\dfrac{1}{2\pi}\int_{-\pi}^{\pi} \lvert X(\omega)\rvert^2\,d\omega$
Correlation	$x(n) \star h(n)$	$X(\omega)\,H(\omega)$

4.1.3 Finite Sequence

$$F_N(\omega) = \sum_{n=0}^{N-1} f(n)e^{-j\omega n} = e^{-j\omega(N-1)/2} \frac{\sin\frac{\omega N}{2}}{\sin\frac{\omega}{2}}$$

4.1.4 Approximation to Continuous-Time Fourier Transform

$$F(\omega_c) = \int_{-\infty}^{\infty} f(t)e^{-j\omega_c t}\,dt \qquad \omega_c = \text{frequency for continuous Fourier transform}$$

$$F(\omega_c) \cong \sum_{n=-\infty}^{\infty} T\,f(nt)e^{-j\omega_c nT} \qquad T = \text{sampling time such that } F(\omega_c) \cong 0 \text{ for all } |\omega_c| > \pi/T$$

4.2 Two-Dimensional Discrete-Time Fourier Transform

4.2.1 Definition

$$X(\omega_1,\omega_2) \doteq \sum_{m,n=-\infty}^{\infty}\sum x(m,n)e^{-j(m\omega_1+n\omega_2)} \qquad -\pi \le \omega_1,\omega_2 \le \pi$$

$$x(m,n) = \frac{1}{(2\pi)^2}\iint_{-\pi}^{\pi} X(\omega_1,\omega_2)e^{j(m\omega_1+n\omega_2)}$$

4.2.2 Properties of Two-Dimensional Discrete-Time Fourier Transform

TABLE 4.2 Properties of Two-Dimensional Discrete-Time Fourier Transform

Properties	Sequence	Transform
	$x(m,n), y(m,n), h(m,n), \ldots$	$X(\omega_1,\omega_2), Y(\omega_1,\omega_2), H(\omega_1,\omega_2), \ldots$
Linearity	$a_1x_1(m,n) + a_2x_2(m,n)$	$a_1X_1(\omega_1,\omega_2) + a_2X_2(\omega_1,\omega_2)$
Conjugation	$x^*(m,n)$	$X^*(-\omega_1,-\omega_2)$
Separability	$x_1(m)x_2(m)$	$X_1(\omega_1)X_2(\omega_2)$
Shifting	$x(m \pm m_0, n \pm n_0)$	$\exp[\pm j(m_0\omega_1 + n_0\omega_2)]X(\omega_1,\omega_2)$
Modulation	$\exp[\pm j(\omega_{01}m + \omega_{02}n)]x(m,n)$	$X(\omega_1 \mp \omega_{01}, \omega_2 \mp \omega_{02})$
Convolution	$y(m,n) = h(m,n) ** x(m,n)$	$Y(\omega_1,\omega_2) = H(\omega_1,\omega_2)X(\omega_1,\omega_2)$
Multiplication	$h(m,n)x(m,n)$	$\left(\frac{1}{4\pi^2}\right)H(\omega_1,\omega_2) ** X(\omega_1,\omega_2)$
Spatial correlation	$c(m,n) = h(m,n) .. x(m,n)$	$C(\omega_1,\omega_2) = H(-\omega_1,-\omega_2)X(\omega_1,\omega_2)$
Inner product	$I = \sum_{m,n=-\infty}^{\infty}\sum x(m,n)y^*(m,n)$	$I = \frac{1}{4\pi^2}\int_{-\pi}^{\pi}\int_{-\pi}^{\pi} X(\omega_1,\omega_2)Y^*(\omega_1,\omega_2)\,d\omega_1\,d\omega_2$
Energy conservation	$\mathsf{E} = \sum_{m,n=-\infty}^{\infty}\sum \lvert x(m,n)\rvert^2$	$\mathsf{E} = \frac{1}{4\pi^2}\int_{-\pi}^{\pi}\int_{-\pi}^{\pi} \lvert X(\omega_1,\omega_2)\rvert^2\,d\omega_1\,d\omega_2$
	$\sum_{m,n=-\infty}^{\infty}\sum \exp[j(m\omega_{01} + n\omega_{02})]$	$4\pi^2\delta(\omega_1 - \omega_{01}, \omega_2 - \omega_{02})$

TABLE 4.2 Properties of Two-Dimensional Discrete-Time Fourier Transform (continued)

Properties	Sequence	Transform
	$\delta(m,n)$	$\frac{1}{4\pi^2}\int_{-\pi}^{\pi}\int_{-\pi}^{\pi}\exp[-j(\omega_1 m+\omega_2 n)d\omega_1\, d\omega_2$
Differentiation	$-jmx(m,n)$	$\frac{\partial X(\omega_1,\omega_2)}{\partial\omega_1}$
	$-jnx(m,n)$	$\frac{\partial X(\omega_1,\omega_2)}{\partial\omega_2}$
	$mnx(m,n)$	$\frac{\partial^2 X(\omega_1,\omega_2)}{\partial\omega_1\partial\omega_2}$
Transportation	$x(m,n)$	$X(\omega_2,\omega_1)$
Reflection	$x(-m,n)$	$X(-\omega_1,\omega_2)$
	$x(m,-n)$	$X(\omega_1,-\omega_2)$
	$x(-m,-n)$	$X(-\omega_1,-\omega_2)$
Real and Imaginary	$\mathrm{Re}[x(m,n)]$	$\frac{1}{2}[X(\omega_1,\omega_2)+X^*(-\omega_1,-\omega_2)]$
Parts	$j\mathrm{Im}[x(m,n)]$	$\frac{1}{2}[X(\omega_1,\omega_2)-X^*(-\omega_1,-\omega_2)]$
	$\frac{1}{2}[x(m,n)+x^*(-m,-n)]$	$\mathrm{Re}[X(\omega_1,\omega_2)]$
	$\frac{1}{2}[x(m,n)-x^*(-m,-n)]$	$j\mathrm{Im}[X(\omega_1,\omega_2)]$
Real-valued sequence		$X(\omega_1,\omega_2)=X^*(-\omega_1,-\omega_2)$ $\mathrm{Re}[X(\omega_1,\omega_2)]=\mathrm{Re}[X(-\omega_1,-\omega_2)]$ $\mathrm{Im}[X(\omega_1,\omega_2)]=-\mathrm{Im}[X(-\omega_1,-\omega_2)]$ $\mathrm{Re}[X(\omega_1,\omega_2)],\ \lvert X(\omega_1,\omega_2)\rvert$: even (symmetric with respect to the origin) $\mathrm{Im}[X(\omega_1,\omega_2)]\,;\tan^{-1}\frac{\mathrm{Im}[X(\omega_1,\omega_2)]}{\mathrm{Re}[X(\omega_1,\omega_2)]}$: odd (antisymmetric with respect to the origin)
	$x(m,n)$: real and even	$X(\omega_1,\omega_2)$: real and even
	$x(m,n)$: real and odd	$X(\omega_1,\omega_2)$: pure imaginary and odd

References

Dudgeon, Dan E., and Russell M. Mersereau, *Multidimensional Digital Signal Processing*, Prentice Hall, Englewood Cliffs, NJ, 1984.

Jain, Aril K., *Fundamentals of Digital Image Processing*, Prentice Hall, Englewood Cliffs, NJ, 1989.

Lim, Jae S., *Two-Dimensional Signal and Image Processing*, Prentice Hall, Englewood Cliffs, NJ, 1990.

Appendix 1

1.1 Two-Dimensional Discrete-Time Fourier Transform

Example

The transform of $x(m,n) = \frac{1}{5}\delta(m-1)\delta(n) + \frac{1}{5}\delta(m+1)\delta(n) + \frac{1}{5}\delta(m)\delta(n-1) + \frac{1}{5}\delta(m)\delta(n+1) + \frac{2}{5}\delta(m)\delta(n)$ which is shown in Figure 4.1 is $X(\omega_1,\omega_2) = \sum\sum_{m,n=-\infty}^{\infty} x(m,n)e^{-jm\omega_1 - jn\omega_2} = \frac{1}{5}e^{-j\omega_1} + \frac{1}{5}e^{j\omega_1} + \frac{1}{5}e^{-j\omega_2} + \frac{1}{5}e^{j\omega_2} + \frac{1}{2} = \frac{2}{5}\cos\omega_1 + \frac{2}{5}\cos\omega_2 + \frac{2}{5}$ and is shown in Figure 4.2.

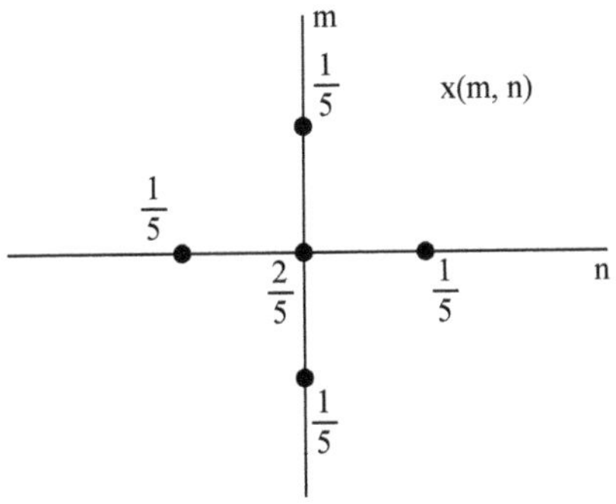

FIGURE 4.1

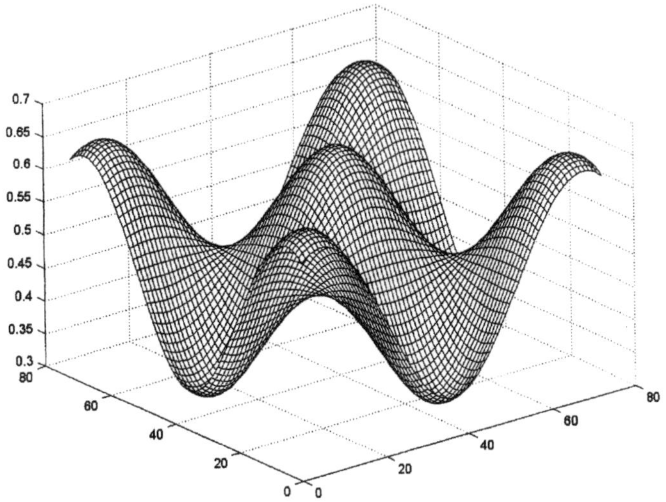

FIGURE 4.2

1.2 Two-Dimensional Discrete-Time Fourier Transform

Example

(Ideal Lowpass Filter). To find the inverse of $H(\omega_1,\omega_2)$ shown in Figure 4.3 we observe that it is equal to the multiplication of two pulse filters whose one direction extends from minus infinity to infinity. Hence (a and b are less than π)

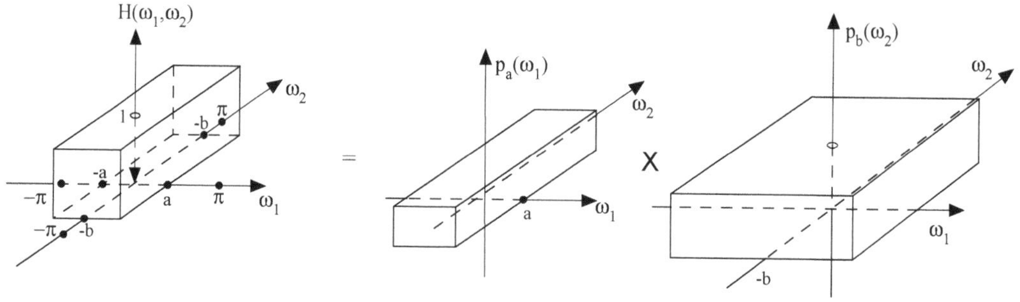

FIGURE 4.3

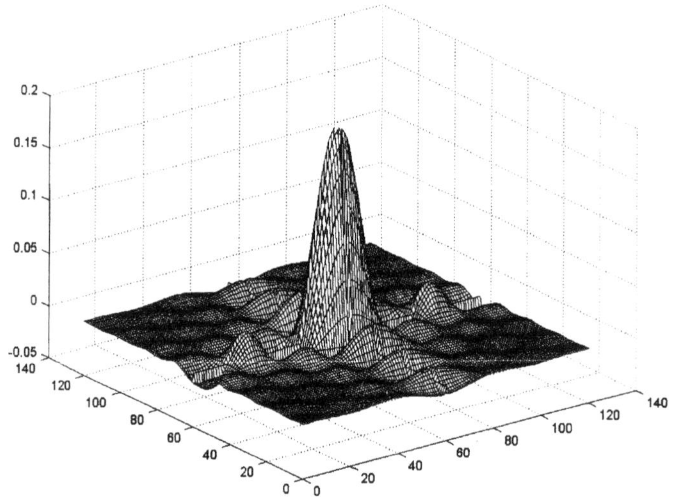

FIGURE 4.4

$$h(m,n)=\frac{1}{(2\pi)^2}\int_{-\pi}^{\pi}\int_{-\pi}^{\pi}H(\omega_1,\omega_2)e^{j\omega_1 m}e^{j\omega_2 n}\,d\omega_1\,d\omega_2=\frac{1}{2\pi}\int_{-\pi}^{\pi}p_a(\omega_1)e^{j\omega_1 m}\,d\omega_1\frac{1}{2\pi}\int_{-\pi}^{\pi}p_b(\omega_2)e^{j\omega_2 n}\,d\omega_2$$

$$=\frac{1}{2\pi}\int_{-a}^{a}e^{j\omega_1 m}\,d\omega_1\frac{1}{2\pi}\int_{-b}^{b}e^{j\omega_2 n}\,d\omega_2=\frac{\sin am}{\pi m}\frac{\sin bn}{\pi n}$$

which is plotted in Figure 4.4.

Example

Let the ideal filter be of the form $H(\omega_1,\omega_2)=1$ for $\omega_1^2+\omega_2^2\le\omega_c^2$ and zero otherwise (see Figure 4.5). Hence

$$h(m,n)=\frac{1}{(2\pi)^2}\iint_{\substack{(\omega_1,\omega_2)\in\\ [\omega_1^2+\omega_2^2\le\omega_c^2]}}1e^{j\omega_1 m}e^{j\omega_2 n}\,d\omega_1\,d\omega_2$$

$$=\frac{1}{(2\pi)^2}\int_{r=0}^{\omega_c}r\,dr\int_{\theta=a}^{a+2\pi}e^{jr(m\cos\theta+n\sin\theta)}\,d\theta\qquad a=\text{real constant}$$

where we set $r\cos\theta=\omega_1$, $r\sin\theta=\omega_2$. Next change to: $m=q\cos\phi$ and $n=q\sin\phi$. Hence

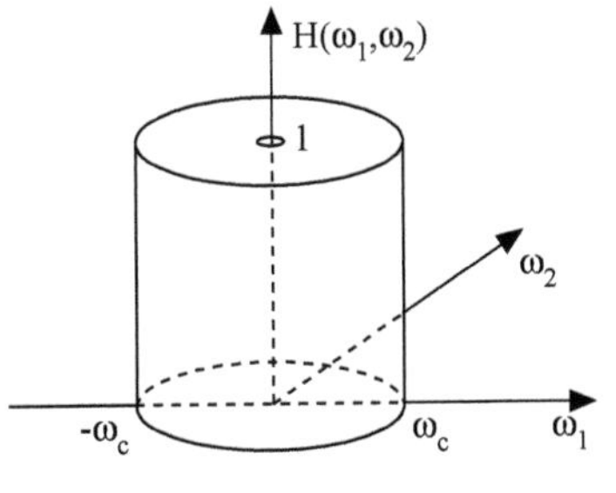

FIGURE 4.5

$$h(m,n)=\frac{1}{(2\pi)^2}\int_{r=0}^{\omega_c} r\int_{\theta=a}^{a+2\pi} e^{jrn\cos(\theta-\phi)}\,d\theta=\frac{1}{(2\pi)^2}\int_{r=0}^{\omega_c} r\,2\pi J_o(r\sqrt{m^2+n^2})\,dr=\frac{\omega_c}{2\pi\sqrt{m^2+n^2}}J_1(\omega_c\sqrt{m^2+n^2})$$

5

Distributions, Delta Function

5.1 Test Function

5.1.1 A Test Function

$\varphi(t)$ is a real-valued function of the real independent variable that can be differentiated an arbitrary number of times, and which is identical to zero outside a finite interval.

Example 5.1

$$\varphi(t,a) = \text{test function} = \begin{cases} \exp\left[-a^2/(a^2-t^2)\right] & |t| < a \\ 0 & |t| \geq a \end{cases}$$

5.1.2 Properties of Test Functions

1. If $f(t)$ can be differentiated arbitrarily often, $\psi(t) = f(t)\varphi(t) =$ test function.
2. If $f(t)$ is zero outside a finite interval, $\psi(t) = \int_{-\infty}^{\infty} f(\tau)\varphi(t-\tau)\,d\tau$, $-\infty < t < \infty$, is a test function.
3. A sequence of test functions, $\{\varphi_n(t)\}$ $1 \leq n < \infty$, converges to zero if all φ_n are identically zero outside some interval independent of n and each φ_n, as well as all of its derivatives, tend uniformly to zero.

Example 5.2

$$\varphi_n(t) = \varphi\left(t + \frac{1}{n}\right) - \varphi(t).$$

4. Test functions belong to a set D, where D is a linear vector space such that if $\varphi_1 \in D$ and $\varphi_2 \in D$, then $\varphi_1 + \varphi_2 \in D$ and $a\varphi_1 \in D$ for any number a.

5.2 Distributions

5.2.1 Definition

A ***distribution*** (or ***generalized*** function) $g(t)$ is a process of assigning our arbitrary test function $\varphi(t)$ a number $N_g[\varphi(t)]$. A distribution is also a ***functional***.

Example 5.3

$$\int_{-\infty}^{\infty} u(\tau)\varphi(t)\,dt = \int_{o}^{\infty} \varphi(t)\,dt = N_f[\varphi(t)]$$

and implies that $u(t)$ is a distribution that assigns a number to each $\varphi(t)$ equal to its area.

5.2.2 Properties

1. Linearity-homogeneity:

$$\int_{-\infty}^{\infty} g(t)[a_1\varphi_1(t)+a_2\varphi_2(t)]dt = a_1\int_{-\infty}^{\infty} g(t)\varphi_1(t)\,dt + a_2\int_{-\infty}^{\infty} g(t)\varphi_2(t)\,dt$$

2. Shifting: $\displaystyle\int_{-\infty}^{\infty} g(t-t_o)\varphi(t)\,dt = \int_{-\infty}^{\infty} g(t)\varphi(t+t_o)\,dt$

3. Scaling: $\displaystyle\int_{-\infty}^{\infty} g(at)\varphi(t)\,dt = \frac{1}{|a|}\int_{-\infty}^{\infty} g(t)\varphi\left(\frac{t}{a}\right)dt$

4. Even distribution: $\displaystyle\int_{-\infty}^{\infty} g(t)\varphi(t)\,dt = 0, \qquad \varphi(t) = \text{odd}$

5. Odd distribution: $\displaystyle\int_{-\infty}^{\infty} g(t)\varphi(t)\,dt = 0, \qquad \varphi(t) = \text{even}$

6. Derivative: $\displaystyle\int_{-\infty}^{\infty} \frac{dg(t)}{dt}\varphi(t)\,dt = -\int_{-\infty}^{\infty} g(t)\frac{d\varphi(t)}{dt}dt$

7. nth derivative: $\displaystyle\int_{-\infty}^{\infty} \frac{d^n g(t)}{dt^n}\varphi(t)\,dt = (-1)^n\int_{-\infty}^{\infty} g(t)\frac{d^n\varphi(t)}{dt^n}dt$

8. Product with ordinary function: $\displaystyle\int_{-\infty}^{\infty} [g(t)f(t)]\varphi(t)\,dt = \int_{-\infty}^{\infty} g(t)[f(t)\varphi(t)]dt$

   provided that $f(t)\varphi(t)$ belongs to the set of test functions.

9. Convolution: $\displaystyle\int_{-\infty}^{\infty}\left[\int_{-\infty}^{\infty} g_1(\tau)g_2(t-\tau)d\tau\right]\varphi(t)\,dt = \int_{-\infty}^{\infty} g_1(\tau)\left[\int_{-\infty}^{\infty} g_2(t-\tau)\varphi(t)\,dt\right]d\tau$

Definition

A sequence of distributions $\{g_n(t)\}_1^\infty$ is said to converge to the distribution $g(t)$ if

$$\lim_{n\to\infty} \int_{-\infty}^{\infty} g_n(t)\varphi(t)\,dt = \int_{-\infty}^{\infty} g(t)\varphi(t)\,dt$$

for all φ belonging to the set of test functions.

10. Every distribution is the limit, in the sense of distributions, of a sequence of infinitely differentiable functions.
11. If $g_n(t) \to g(t)$ and $r_n(t) \to r(t)$ (r being a distribution), and the numbers $a_n \to a$, then $\frac{d}{dt} g_n(t) \to$ $\frac{dg(t)}{dt}$, $g_n(t) + r_n(t) \to g(t) + r(t)$, $a_n g_n(t) \to ag(t)$
12. Any distribution $g(t)$ may be differentiated as many times as desired. The derivative of any distribution always exists, and it is a distribution.

5.3 One-Dimensional Delta Function

5.3.1 Definition

$$\delta(t) = 0 \qquad t \neq 0$$

$$\int_{-\infty}^{\infty} \delta(t)\varphi(t)\,dt = \varphi(0), \qquad \varphi(t) \text{ is a testing function}$$

5.3.2 Properties

TABLE 5.1 Properties of Delta Function

Delta Function Properties
$\delta(at) = \frac{1}{\lvert a\rvert}\delta(t)$
$\delta\left(\frac{t-t_o}{a}\right) = \lvert a\rvert\,\delta(t-t_o)$
$\delta(at-t_o) = \frac{1}{\lvert a\rvert}\delta\left(t \quad \frac{t_o}{a}\right)$
$\delta(-t+t_o) = \delta(t \quad t_o)$
$\delta(-t) = \delta(t)$; $\quad \delta(t) =$ even function
$\int_{-\infty}^{\infty} \delta(t)f(t)\,dt = f(0)$
$\int_{-\infty}^{\infty} \delta(t-t_o)f(t) = f(t_o)$
$f(t)\delta(t) = f(0)\delta(t)$

TABLE 5.1 Properties of Delta Function (continued)

Delta Function Properties

$$f(t)\delta(t-t_o) = f(t_o)\delta(t\quad t_o)$$

$$t\delta(t) = 0$$

$$\int_{-\infty}^{\infty} A\delta(t)\,dt = \int_{-\infty}^{\infty} A\delta(t-t_o)\,dt \quad A$$

$$f(t) * \delta(t) = \text{convolution} = \int_{-\infty}^{\infty} f(t-\tau)\delta(\tau)\,d\tau \quad f(t)$$

$$\delta(t-t_1) * \delta(t-t_2) = \int_{-\infty}^{\infty} \delta(\tau-t_1)\delta(t-\tau-t_2)\,d\tau = \delta[t-(t_1+t_2)]$$

$$\sum_{n=-N}^{N} \delta(t-nT) * \sum_{n\ -N}^{N} \delta(t-nT) = \sum_{n\ 2N}^{2N} (2N+1-|n|)\delta(t-nT)$$

$$\int_{-\infty}^{\infty} \frac{d\delta(t)}{dt} f(t)\,dt = -\frac{df(0)}{dt}$$

$$\int_{-\infty}^{\infty} \frac{d\delta(t-t_o)}{dt} f(t)\,dt = -\frac{df(t_o)}{dt}$$

$$\int_{-\infty}^{\infty} \frac{d^n\delta(t)}{dt^n} f(t)\,dt = (-1)^n \frac{d^n f(0)}{dt^n}$$

$$f(t)\frac{d\delta(t)}{dt} = -\frac{df(0)}{dt}\delta(t) \quad f(0)\frac{d\delta(t)}{dt}$$

$$t\frac{d\delta(t)}{dt} = -\delta(t)$$

$$t^n \frac{d^m\delta(t)}{dt^m} = \begin{cases} (-1)^n\, n!\, \delta(t), & m = n \\ (-1)^n \dfrac{m!}{m-n!} \dfrac{d^{m-n}\delta(t)}{dt^{m-n}}, & m > n \\ 0, & m < n \end{cases}$$

$$\int_{-\infty}^{\infty} \frac{d\delta(t)}{dt}\,dt = 0, \quad \frac{d\delta(t)}{dt} = \text{odd function}$$

$$f(t) * \frac{d\delta(t)}{dt} = \frac{df(t)}{dt}$$

$$f(t)\frac{d^n\delta(t)}{dt^n} = \sum_{k=0}^{n} (-1)^k \frac{n!}{k!(n-k)!} \frac{d^k f(0)}{dt^k} \frac{d^{n-k}\delta(t)}{dt^{n-k}}$$

$$\frac{\partial\delta(yt)}{\partial y} = -\frac{1}{y^2}\delta(t)$$

$$\delta(t) = \frac{du(t)}{dt}$$

$$\frac{d^n\delta(-t)}{dt^n} = (-1)^n \frac{d^n\delta(t)}{dt^n}, \quad \left\{ \frac{d^n\delta(t)}{dt^n} \text{ is even if } n \text{ is even, and odd if } n \text{ is odd.} \right\}$$

$$(\sin at)\frac{d\delta(t)}{dt} = -a\delta(t)$$

TABLE 5.1 Properties of Delta Function (continued)

Delta Function Properties
$\frac{d\delta(t)}{dt} = \frac{d^2u(t)}{dt^2}$
$-\delta(t) = \frac{du(-t)}{dt}$
$\delta(t-t_o) = \frac{du(t-t_o)}{dt}$
$\frac{d\,\mathrm{sgn}(t)}{dt} = 2\delta(t)$
$\delta[r(t)] = \sum_n \frac{\delta(t-t_n)}{\left\lvert \frac{dr(t_n)}{dt} \right\rvert}$ $\quad t_n =$ zeros of $r(t)$, $\frac{dr(t_n)}{dt} \neq 0$
$\frac{d\,\delta[r(t)]}{dt} = \sum_n \frac{\frac{d\delta(t-t_n)}{dt}}{\frac{dr(t)}{dt}\left\lvert \frac{dr(t_n)}{dt} \right\rvert}$ $\quad t_n =$ zeros of $r(t)$, $\frac{dr(t_n)}{dt} \neq 0$, $\frac{dr(t)}{dt} \neq 0$
$\delta(\sin t) = \sum_{n=-\infty}^{\infty} \delta(t - n\pi)$
$\delta(t^2-1) = \frac{1}{2}\,\delta(t-1) + \frac{1}{2}\,\delta(t+1)$
$\delta(t^2-a^2) = \frac{1}{2a}\,[\delta(t+a) + \delta(t-a)]$
$\delta(t) = \lim_{\varepsilon \to o} \frac{e^{-t^2/\varepsilon}}{\sqrt{\varepsilon\pi}}$
$\delta(t) = \lim_{\omega \to \infty} \frac{\sin \omega t}{\pi t}$
$\delta(t) = \lim_{\varepsilon \to o} \frac{1}{\pi} \frac{\varepsilon}{t^2+\varepsilon^2}$
$\frac{d\delta(t)}{dt} = \lim_{\varepsilon \to 0} -\frac{2}{\pi} \frac{\varepsilon t}{(t^2+\varepsilon^2)^2}$
$\delta(t) = \frac{1}{2\pi} \int_{-\infty}^{\infty} \cos \omega t \, d\omega$
$\frac{df(t)}{dt} = \frac{d}{dt}\,[t\,u(t) - (t-1)\,u(t-1) - u(t-1)]$ $= t\,\delta(t) + u(t) - (t-1)\,\delta(t-1) - u(t-1) - \delta(t-1)$
$comb_{T_s}(t) = \sum_{n=-\infty}^{\infty} \delta(t - nT_s)$, $\quad f(t)\,comb_{T_s}(t) = \sum_{n=-\infty}^{\infty} f(nT_s)\,\delta(t\ \ nT_s)$
$COMB_{\omega_s}(\omega) = F\{comb_{T_s}(t)\} = \omega_s \sum_{n=-\infty}^{\infty} \delta(t - n\omega_s)$ $\quad \omega_s = \frac{2\pi}{T_s}$
$\lim_{a \to \infty} \int_{-a}^{a} e^{j\omega t}\, d\omega = 2\pi\delta(t)$

TABLE 5.1 Properties of Delta Function (continued)

Delta Function Properties
$\lim_{a\to\infty}\int_{-a}^{a}(j\omega)e^{j\omega t}\,d\omega = 2\pi\frac{d\delta(t)}{dt}$
$\lim_{a\to\infty}\int_{-a}^{a}(j\omega)^2 e^{j\omega t}\,d\omega = 2\pi\frac{d^2\delta(t)}{dt^2}$

The following examples will elucidate some of the delta properties and the use of the delta function.

5.4 Examples

Example 5.4

Equivalence of expressions involving the delta functions:

a) $(\cos t + \sin t)\,\delta(t) = \delta(t)$ b) $\cos 2t + \sin t\,\delta(t) = \cos 2t$

c) $1 + 2e^{-t}\,\delta(t-1) = 1 + 2e^{-1}\,\delta(t-1)$

Example 5.5

The values of the following integrals are:

$$\int_{-\infty}^{\infty}(t^2+4t+5)\delta(t)\,dt = 0^2+4\cdot 0+5 = 5,$$

$$\int_{-\infty}^{\infty}\frac{(1+\cos t)\delta(t)}{1+2e^t}\,dt = \frac{2}{1+2}$$

$$\int_{-\infty}^{\infty}t^2\sum_{k=1}^{n}\delta(t-k)\,dt = \sum_{k=1}^{n}k^2 = \tfrac{1}{6}\,[n(n+1)(2n+1)]$$

Example 5.6

The first derivative of the functions is:

$$\frac{d}{dt}(2u(t+1)+u(1-t)) = \frac{d}{dt}(2u(t+1)+u[-(t-1)]) = 2\delta(t+1)-\delta(t-1)$$

$$\frac{d}{dt}([2-u(t)]\cos t = \frac{d}{dt}(2\cos t - u(t)\cos t) = 2\sin t - \delta(t)\cos t + u(t)\sin t = (u(t)-2)\sin t - \delta(t)$$

$$\frac{d}{dt}\left(\left[u\left(t-\frac{\pi}{2}\right)-u(t-\pi)\right]\sin t\right) = \left[\delta\left(t-\frac{\pi}{2}\right)-\delta(t-\pi)\right]\sin t + \left[u\left(t-\frac{\pi}{2}\right)-u(t-\pi)\right]\cos t$$

$$= \delta\left(t-\frac{\pi}{2}\right) + [u\left(t-\frac{\pi}{2}\right) - u(t-\pi)]\cos t$$

Example 5.7

The values of the following integrals are:

$$\int_{-\infty}^{\infty} e^{2t}\sin 4t \frac{d^2\delta(t)}{dt^2}\,dt = (-1)^2 \frac{d^2}{dt^2}[e^{2t}\sin 4t]\Big|_{t=o} = 2\times 2\times 4 = 16$$

$$\int_{-\infty}^{\infty} (t^3+2t+3)\left(\frac{d\delta(t-1)}{dt}+2\frac{d^2\delta(t-2)}{dt^2}\right)dt = \int_{-\infty}^{\infty} (t^3+2t+3)\frac{d\delta(t-1)}{dt}\,dt$$

$$+2\int_{-\infty}^{\infty} (t^3+2t+3)\frac{d^2\delta(t-2)}{dt^2}\,dt = (-1)(3t^2+2)\big|_{t=1} + (-1)^2\,2(6t)\big|_{t=2} = -5+24 = 19$$

Example 5.8

The values of the following integrals are:

$$\int_{o}^{4} e^{4t}\delta(2t-3)\,dt = \int_{o}^{4} e^{4t}\delta\left[2\left(t-\frac{3}{2}\right)\right] = \frac{1}{2}\int_{o}^{4} e^{4t}\delta\left(t-\frac{3}{2}\right)dt = \frac{1}{2}e^{4\frac{3}{2}} = \frac{1}{2}e^{6}$$

$$\int_{o}^{4} e^{4t}\delta(3-2t)\,dt = \int_{o}^{4} e^{4t}\delta[-(2t-3)]\,dt = \frac{1}{2}\int_{o}^{4} e^{4t}\delta(2t-3)\,dt = \frac{1}{2}e^{6}$$

5.5 Two-Dimensional Delta Function

5.5.1 Definitions

$$\delta(x,y) = \delta(x)\delta(y)$$

$$\delta(x-x_o, y-y_o) = \delta(x-x_o)\delta(y-y_o)$$

$$\iint_{-\infty}^{\infty} f(\xi,\eta)\delta(x-\xi)\delta(y-\eta)\,d\xi\,d\eta = f(x,y)$$

$$\iint_{A} \delta(x-a)\delta(y-b) = p_A(a,b) \qquad a,b \text{ not on the boundary of } A$$

$$p_A(x,y) = \begin{cases} 1 & x,y\in A \\ 0 & \text{otherwise} \end{cases}$$

5.5.2 Line Masses

The function $\varphi(y)\delta(x-a)$ can be interpreted as a line mass on the line $x = a$ of density $\varphi(y)$.

Example 5.9

$p_a(y)\delta(x)$ is a line mass on the y-axis with density one on the y-axis from $y = -a$ to $y = a$.

Example 5.10

$f(x,y) ** \delta(x) = \iint_{-\infty}^{\infty} f(\xi,\eta)\,\delta(x-\xi)\,d\xi\,d\eta = \int_{-\infty}^{\infty} f(x,\eta)\,d\eta$ which is the x profile of $f(x,y)$.

5.5.3 Line Mass on a Curve

$\delta[\alpha(x,y)]$ is a line mass on the curve $\alpha(x,y) = 0$ with density $\lambda(x,y) = \dfrac{1}{\sqrt{\alpha_x^2+\alpha_y^2}}$ where $\alpha_x = \dfrac{\partial\alpha(x,y)}{\partial x}$, $\alpha_y = \dfrac{\partial\alpha(x,y)}{\partial y}$.

5.5.4 Line Masses Along x- and y-Axes

The line masses have densities along the x- and y-directions given by $\dfrac{dm}{dx} = \dfrac{1}{|\alpha_x|}\left(\alpha_x = \dfrac{\partial\alpha(x,y)}{\partial x}\right)$ and $\dfrac{dm}{dy} = \dfrac{1}{|\alpha_y|}\left(\alpha_y = \dfrac{\partial\alpha(x,y)}{\partial y}\right)$, respectively. $\dfrac{dm}{ds} = \dfrac{dm}{\sqrt{dx^2+dy^2}} = \dfrac{1}{\sqrt{\left(\dfrac{dx}{dm}\right)^2+\left(\dfrac{dy}{dm}\right)^2}} = \dfrac{1}{\sqrt{\alpha_x^2+\alpha_y^2}}$ and hence $\delta[\alpha(x,y)] = \dfrac{1}{\sqrt{\alpha_x^2+\alpha_y^2}}\delta(s)$ and $s = \alpha(x,y) = 0$ is the curve of $\alpha(x,y)$.

5.5.5 Solution of $\alpha(x,y)$

If we solve $\alpha(x,y) = 0$ for x and denote i^{th} root with x_i then we may regard $\delta[\alpha(x,y)]$ as the line mass

$$\delta[\alpha(x,y)] = \sum_i \frac{1}{|\alpha_x|}\delta(x-x_i), \qquad \alpha_x = \frac{\partial\alpha(x,y)}{\partial x}$$

and similarly for the y solution

$$\delta[\alpha(x,y)] = \sum_i \frac{1}{|\alpha_y|}\delta(y-y_i), \qquad \alpha_y = \frac{\partial\alpha(x,y)}{\partial y}$$

Example 5.11

If $\delta[\sqrt{x^2+y^2}-r_o]$ then $\alpha(x,y) = \sqrt{x^2+y^2}-r_o$, $\alpha_x = x/\sqrt{x^2+y^2}$, $\alpha_y = y/\sqrt{x^2+y^2}$, $x_1,x_2 = \pm\sqrt{r_o^2-y^2}$, $y_1,y_2 = \pm\sqrt{r_o^2-x^2}$.

$$\delta[\alpha(x,y)] = \left|\frac{\sqrt{x^2+y^2}}{x^2}\right|\left[\delta(x-\sqrt{r_o^2-y^2})+\delta(x+\sqrt{r_o^2-y^2})\right]$$

$$= \frac{r_o}{\sqrt{r_o^2-y^2}}\left[\delta(x-\sqrt{r_o^2-y^2})+\delta(x+\sqrt{r_o^2-y^2})\right], \quad |y|<r_o. \text{ Also}$$

$$\delta[\alpha(x,y)] = \frac{r_o}{\sqrt{r_o^2-x^2}}\left[\delta(y-\sqrt{r_o^2-x^2})+\delta(y+\sqrt{r_o^2-x^2})\right].$$

Since $\alpha_x^2+\alpha_y^2 = 1$ and $r = \sqrt{x^2+y^2}$ then $\delta[\alpha(x,y)] = \delta(r-r_o)$ is a ring delta function with unit density along $r = r_o$.

Example 5.12

If $\delta(\alpha x + by + c)$, then $\alpha(x,y) = \alpha x + by + c$, $\alpha_x = a$, $\alpha_y = b$, $x = -\frac{b}{a}y - \frac{c}{a}$, $y = -\frac{a}{b}x - \frac{c}{b}$ and hence

$$\delta(\alpha x + by + c) = \frac{1}{|a|}\delta\left(x + \frac{b}{a}y + \frac{c}{a}\right) = \frac{1}{|b|}\delta\left(y + \frac{a}{b}y + \frac{c}{b}\right).$$

5.5.6 Transformation of Coordinates for $\delta(ax + by + c)$ (see Figure 5.1)

$x' = \cos\theta' + y\sin\theta'$, $y' = -x\sin\theta' + y\cos\theta'$, $\theta' = \tan^{-1}\left(\frac{b}{a}\right)$, $k = \sqrt{a^2 + b^2}$, $\cos\theta' = \frac{a}{k}$, $\sin\theta' = \frac{b}{k}$, $x = (ax' - by')/k$, $y = (bx' + ay')/k$.

$$\delta(ax + by + c) = \delta\left(\frac{a^2x' - aby'}{k} + \frac{b^2x' + aby'}{k} + c\right) = \delta(kx' + c) = \frac{1}{k}\delta(x' - x_o) \text{ where } x_o' = -c/k.$$

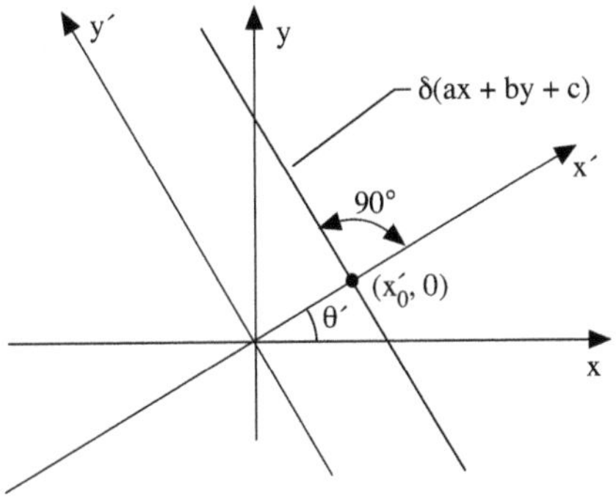

FIGURE 5.1

Example 5.13

$$f(x,y)\delta(ax + by + c) = \frac{1}{k}f\left(\frac{ax_o' - b_1y'}{k}, \frac{bx_o' + ay'}{k}\right)\delta(x' - x_o') \text{ where } k = \sqrt{a^2 + b^2} \text{ and } c/k = -x_o'.$$

The density along this line is $\frac{dm}{dy'} = \left(\frac{1}{k}\right)f\left(\frac{ax_o' - by'}{k}, \frac{bx_o' + ay'}{k}\right).$

5.5.7 The Function $\delta(a_1x + b_1y + c_1, a_2x + b_2y + c_2)$: From (5.5.5)

$$\begin{aligned}\delta(a_1x + b_1y + c_1, a_2x + b_2y + c_2) &= \frac{1}{|a_1|}\delta\left(x + \frac{b_1y}{a_1} + \frac{c_1}{a_1}\right)\delta\left(\frac{-a_2b_1y - a_2c_1}{a_1} + b_2y + c_2\right) \\ &= \frac{1}{|a_1|}\delta\left(x + \frac{b_1}{a_1}y + \frac{c_1}{a_1}\right)\frac{|a_1|}{|a_1b_2 - a_2b_1|}\delta\left(y - \frac{a_2c_1 - a_1c_2}{a_1b_2 - a_2b_1}\right) \\ &= \frac{1}{|a_1b_2 - a_2b_1|}\delta\left(x - \frac{b_1c_2 - b_2c_1}{a_1b_2 - a_2b_1}\right)\delta\left(y - \frac{a_2c_1 - a_1c_2}{a_1b_2 - a_2b_1}\right) \\ &= \frac{1}{|D|}\delta(x - x_o, y - y_o)\end{aligned}$$

5.5.8 The function $f(x,y)\delta(a_1x + b_1y + c_1, a_2x + b_2y + c_2)$

$f(x,y)\delta(a_1x+b_1y+c_1, a_2x+b_2y+c_2) = \frac{1}{|D|}f(x_o,y_o)\delta(x-x_o, y-y_o)$. See (5.5.7) for the values of D, x_o, and y_o.

5.5.9 $comb(a_1x + b_1y + c_1, a_2x + b_2y + c_2)$

$$comb(a_1x+b_1y+c_1, a_2x+b_2y+c_2) = \sum_{n=-\infty}^{\infty}\sum_{m=-\infty}^{\infty}\delta(a_1x+b_2y+c_1-n)\delta(a_2x+b_2y+c_2 - m)$$

$$= \frac{1}{|D|}\sum_{n=-\infty}^{\infty}\sum_{m=-\infty}^{\infty}\delta\left(x-x_o-\frac{b_2n}{D}\quad\frac{b_1m}{D}\right)\delta\left(y+y_o+\frac{a_2n}{D}\quad\frac{a_1m}{D}\right)$$

See (5.5.7) for the values of D, x_o, and y_o.

5.5.10 $comb(a_1x + b_1y, a_2x + b_2y)$

$$comb(a_1x+b_1y, a_2x+b_2y) = \frac{1}{|D|}\sum_{n=-\infty}^{\infty}\sum_{m=-\infty}^{\infty}\delta\left(x-\frac{b_2n}{D}\quad\frac{b_1m}{D}\right)\delta\left(y+\frac{a_2n}{D}-\frac{a_1m}{D}\right)$$

5.5.11 $f(x,y)\ comb(a_1x + b_1y + c_1, a_2x + b_2y + c_2)$

$$f(x,y)comb(a_1x+b_1y+c_1, a_2x+b_2y+c_2) = \frac{1}{|D|}\sum_{n=-\infty}^{\infty}\sum_{m=-\infty}^{\infty}f\left(x_o+\frac{b_2n}{D}-\frac{b_1m}{D}, y_o-\frac{a_2n}{D}+\frac{a_1m}{D}\right)$$

$$\times\delta\left(x-x_o-\frac{b_2n}{D}+\frac{b_1m}{D}\right)\delta\left(y-y_o+\frac{a_2n}{D}\quad\frac{a_1m}{D}\right)$$

5.5.12 $\delta[\alpha_1(x,y)]\ \delta[\alpha_2(x,y)]$

$$\delta[\alpha_1(x,y)]\delta[\alpha_2(x,y)] = \sum_i \frac{\delta(x-x_i)\delta(y-y_i)}{|\alpha_{1x}\alpha_{2y}-\alpha_{1y}\alpha_{2x}|}$$

where x_i, y_i are the coordinates of the intersections of the curves $\alpha_1(x,y)$ and $\alpha_2(x,y)$,

$$\alpha_{1x} = \frac{\partial\alpha_1(x,y)}{\partial x},\ \alpha_{1y} = \frac{\partial\alpha_1(x,y)}{\partial y},\ \alpha_{2x} = \frac{\partial\alpha_2(x,y)}{\partial x},\ \text{and}\ \alpha_{2y} = \frac{\partial\alpha_2(x,y)}{\partial y}.$$

Example 5.14

(See Figure 5.2)

$\delta[\alpha_1(x,y)]\ \delta[\alpha_2(x,y)] = \delta(\sqrt{x^2+y^2}-a)\ \delta(x-x_o)$. Intersect at $(x_o, \sqrt{a^2-x_o^2})$ and at $(x_o, -\sqrt{a^2-x_o^2})$ and $x_o \le a$. $\alpha_1(x,y) = \sqrt{x^2+y^2}-a$, $\alpha_2(x,y) = x - x_o$, $\partial\alpha_1/\partial x = x/\sqrt{x^2+y^2}$, $\partial\alpha_1/\partial y = y/\sqrt{x^2+y^2}$, $\partial\alpha_2/\partial x = 1$, and $\partial\alpha_2/\partial y = 0$.

Hence from (5.5.12)

$$\delta[\alpha_1(x,y)]\delta[\alpha_2(x,y)] = \frac{\sqrt{x^2+y^2}}{y}[\delta(x-x_o)\delta(y+\sqrt{a^2-x_o^2})+\delta(x-x_o)\delta(y-\sqrt{a^2-x_o^2})]$$

and

$$\iint_{-\infty}^{\infty} \delta[\alpha_1(x,y)]\delta[\alpha_2(x,y)]dx\,dy = \iint_{-\infty}^{\infty} \frac{\sqrt{x^2+y^2}}{y}\delta(x-x_o)\delta(y+\sqrt{a^2 \; x_o^2})dx\,dy$$

$$+\iint_{-\infty}^{\infty} \frac{\sqrt{x^2+y^2}}{y}\delta(x-x_o)\delta(y-\sqrt{a^2-x_o^2})dx\,dy$$

$$= \frac{\sqrt{x_o^2+a^2-x_o^2}}{\sqrt{a^2-x_o^2}} + \frac{\sqrt{x_o^2+a^2-x_o^2}}{\sqrt{a^2-x_o^2}} = \frac{2a}{\sqrt{a^2-x_o^2}} \qquad \text{for } x_o < a.$$

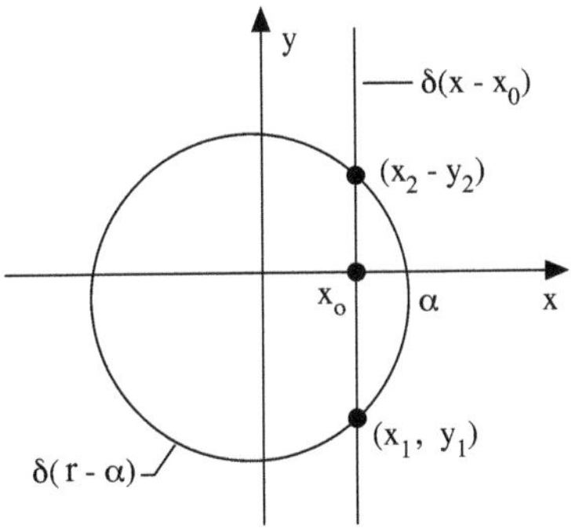

FIGURE 5.2

References

Gelfand, I. M., et al., *Generalized Functions*, Vol. 1-6, Academic Press, New York, NY 1964-69.

Hoskins, R. F., *Generalized Functions*, Chichester, England, 1979.

Lighthill, M. J., *Introduction to Fourier Analysis and Generalized Functions*, Cambridge University Press, New York, NY, 1959.

Poularikas, A. D., Signals and systems, in *The Transforms and Applications Handbook*, Edited by A. D. Poularikas, CRC Press Inc., Boca Raton, Florida, 1996.

6

The Z-Transform

6.1 One-Dimensional Z-Transform (positive-time sequences)

6.1.1 Definitions of One-Sided Z-Transform

$$Z\{f(nT)\} \doteq F(z) = \sum_{n=0}^{\infty} f(nT)z^{-n}$$

$$Z\{f(n)\} \doteq F(z) = \sum_{n=0}^{\infty} f(n)z^{-n}$$

6.1.2 Radius of Convergence for One-Sided Z-Transform

$$|z| > \overline{\lim_{n\to\infty}} \sqrt[n]{|f(nT)|} = R$$

where $\overline{\lim}$ denotes the **greatest** limit points of $\overline{\lim_{n\to\infty}} |f(nT)|^{1/n}$.

6.1.3 Region of Convergence for One-Sided Z-Transform

The series will converge absolutely for all points in the z-plane that lie **outside** the circle of radius R.

For example, the region of convergence of $f(nT) = e^{-aTn}u(nT)$ is: $\left|z^{-n}\,e^{-aTn}\right|^{1/n} < 1$ or $|z| > e^{-aT}$.

6.1.4 Table of One-Sided Z-Transform Properties

TABLE 6.1 Z-Transform Properties for Positive-Time Sequences

1. Linearity

$$Z\{c_i f_i(nT)\} = c_i F_i(z) \qquad |z| > R_i, \quad c_i = \text{constants}$$

$$Z\left\{\sum_{i=0}^{\ell} c_i f_i(nT)\right\} = \sum_{i=0}^{\ell} c_i F_i(z) \qquad |z| > \max R_i$$

2. Shifting Property

$$Z\{f(nT-kT)\} = z^{-k} F(z) \qquad f(-nT) = 0 \qquad n = 1, 2, \cdots$$

$$Z\{f(nT-kT)\} = z^{-k} F(z) + \sum_{n=1}^{k} f(\;nT) z^{-(k-n)}$$

$$Z\{f(nT+kT)\} = z^{k} F(z) \;\; \sum_{n=0}^{k-1} f(nT) z^{k-n}$$

$$Z\{f(nT+T)\} = z[F(z) \;\; f(0)]$$

3. Time Scaling

$$Z\{a^{nT} f(nT)\} = F(a^{-T} z) = \sum_{n=0}^{\infty} f(nT)(a^{-T} z)^{-n} \qquad |z| > a^{T}$$

4. Periodic Sequence

$$Z\{f(nT)\} = \frac{z^N}{z^N - 1} F_{(1)}(z) \qquad |z| > R$$

N = number of time units in a period

R = radius of convergence of $F_{(1)}(z)$

$F_{(1)}(z)$ = Z-transform of the first period

5. Multiplication by n and nT

$$Z\{n f(nT)\} = -z \frac{dF(z)}{dz} \qquad |z| \;\; R$$

$$Z\{nT f(nT)\} = -zT \frac{dF(z)}{dz} \qquad |z| \;\; R$$

R = radius of convergence of $F(z)$

6. Convolution

$$Z\{f(nT)\} = F(z) \qquad |z| > R_1$$

$$Z\{h(nT)\} = H(z) \qquad |z| > R_2$$

$$Z\{f(nT) * h(nT)\} = F(z) H(z) \qquad |z| > \max(R_1, R_2)$$

7. Initial Value

$$f(0T) = \lim_{z\to\infty} F(z) \qquad |z| > R$$

8. Final Value

$$\lim_{n\to\infty} f(nT) = \lim_{z\to 1} (z-1) F(z) \qquad \text{if } f(\infty T) \text{ exists}$$

9. Multiplication by $(nT)^k$

$$Z\{n^k T^k f(nT)\} = -Tz \frac{d}{dz} Z\{(nT)^{k-1} f(nT)\} \qquad k \;\; 0 \text{ and integer}$$

10. Complex Conjugate Signals

$$Z\{f(nT)\} = F(z) \qquad |z| > R$$

$$Z\{f^*(nT)\} = F^*(z^*) \qquad |z| > R$$

TABLE 6.1 Z-Transform Properties for Positive-Time Sequences (continued)

11. Transform of Product

$$Z\{f(nT)\} = F(z) \qquad |z| > R_f$$

$$Z\{h(nT)\} = H(z) \qquad |z| > R_h$$

$$Z\{f(nT)h(nT)\} = \frac{1}{2\pi j}\oint_C F(\tau)H\left(\frac{z}{\tau}\right)\frac{d\tau}{\tau}, \qquad |z| > R_f\, R_h,\; R_f < |\tau| < \frac{|z|}{R_h}$$

counter-clockwise integration

12. Parseval's Theorem

$$Z\{f(nT)\} = F(z) \qquad |z| > R_f$$

$$Z\{h(nT)\} = H(z) \qquad |z| > R_h$$

$$\sum_{n=0}^{\infty} f(nT)h(nT) = \frac{1}{2\pi j}\oint_C F(z)H(z^{-1})\frac{dz}{z} \qquad |z| = 1 > R_f\, R_h$$

counter-clockwise integration

13. Correlation

$$f(nT).\ h(nT) = \sum_{m=0}^{\infty} f(mT)h(mT - nT) \quad \frac{1}{2\pi j}\oint_C F(\tau)H\left(\frac{1}{\tau}\right)\tau^{n-1}\,d\tau, \qquad n \geq 1$$

Both $f(nT)$ and $h(nT)$ must exist for $|z| > 1$. The integration is counter-clockwise.

14. Transforms with Parameters

$$Z\left\{\frac{\partial}{\partial a}f(nT,a)\right\} = \frac{\partial}{\partial a}F(z,a)$$

$$Z\{\lim_{a\to a_o} f(nT,a)\} = \lim_{a\to a_o} F(z,a)$$

$$Z\left\{\int_{a_o}^{a_1} f(nT,a)\,da\right\} = \int_{a_o}^{a_1} F(z,a)\,da \qquad \text{finite interval}$$

6.1.5 Inverse Z-Transform (see Appendix)

1. Use tables.
2. Decompose the expression into simpler partial forms which are included in the tables
3. If the transform is decomposed into a product of partial series, the resulting object function is obtained as the convolution of the partial object function.
4. Use inversion integral.

6.2 One-Dimensional, Two-Sided Z-Transform

6.2.1 Definitions

$$Z_{II}\{f(nT)\} \doteq F(z) = \sum_{n=-\infty}^{\infty} f(nT)z^{-n} \qquad R_+ < |z| < R_-$$

$$Z_{II}\{f(n)\} \doteq F(z) = \sum_{n=-\infty}^{\infty} f(n)z^{-n} \qquad R_+ < |z| < R_-$$

6.2.2 Region of Convergence

Assuming that the algebraic expression for the Z-transform $F(z)$ is a rational function and that $f(nT)$ has finite amplitude, except possibly at infinities, the properties of the region of convergence are:

1. The ROC is a ring or disc in the z-plane and centered at the origin. $0 \le R_+ < |z| < R_- \le \infty$.
2. The Fourier transform converges also absolutely if and only if the ROC of the Z-transform of $f(nT)$ includes the unit circle.
3. No poles exist in the ROC.
4. The ROC of a finite sequence $\{f(nT)\}$ is the entire z-plane except possibly for $z = 0$ or $z = \infty$.
5. If $f(nT)$ is right-handed, $0 \le n < \infty$, the ROC extends outward from the outermost pole of $F(z)$ to infinity.
6. If $f(nT)$ is left-handed, $-\infty \le n < 0$, the ROC extends inward from the innermost pole of $F(z)$ to zero.
7. An infinite-duration two-sided sequence $\{f(nT)\}$ has a ring as its ROC bounded on the interior and exterior by a pole. The ring contains no poles.
8. The ROC must be a connected region.

6.2.3 Properties of Two-Sided Z-Transform

TABLE 6.2 Z-Transform Properties for Positive- and Negative-Time Sequences

1. Linearity

$$Z_{II}\left\{\sum_{i=0}^{\ell} c_i f_i(nT)\right\} = \sum_{i=0}^{\ell} c_i F_i(z) \qquad \max R_{i+} < |z| < \min R_{i-}$$

2. Shifting Property

$$Z_{II}\{f(nT \pm kT)\} = z^{\pm k} F(z) \qquad R_+ < |z| < R_-$$

3. Scaling

$$Z_{II}\{f(nT)\} = F(z) \qquad R_+ < |z| < R_-$$

$$Z_{II}\{a^{nT} f(nT)\} = F(a^{-T} z) \qquad |a^T| R_+ < |z| < |a^T| R_-$$

4. Time Reversal

$$Z_{II}\{f(nT)\} = F(z) \qquad R_+ < |z| < R_-$$

$$Z_{II}\{f(-nT)\} = F(z^{-1}) \qquad \tfrac{1}{R_-} < |z| < \tfrac{1}{R_+}$$

5. Multiplication by nT

$$Z_{II}\{f(nT)\} = F(z) \qquad R_+ < |z| < R_-$$

$$Z_{II}\{nT\, f(nT)\} = -zT\,\frac{dF(z)}{dz} \qquad R_+ < |z| < R_-$$

6. Convolution

$$Z_{II}\{f_1(nT) * f_2(nT)\} = F_1(z) F_2(z)$$

$$ROC\, F_1(z) \cup ROC\, F_2(z), \quad \max(R_{+f_1}, R_{+f_2}) < |z| < \min(R_{-f_1}, R_{-f_2})$$

7. Correlation

$$R_{f_1 f_2}(z) = Z_{II}\{f_1(nT) \cdot f_2(nT)\} = F_1(z) F_2(z^{-1})$$

$$ROC\, F_1(z) \cup ROC\, F_2(z^{-1}), \quad \max(R_{+f_1}, R_{+f_2}) < |z| < \min(R_{-f_1}, R_{-f_2})$$

8. Multiplication by e^{-anT}

$$Z_{II}\{f(nT)\} = F(z) \qquad R_+ < |z| < R_-$$

$$Z_{II}\{e^{-anT} f(nT)\} = F(e^{aT} z) \qquad |e^{-aT}| R_+ < |z| < |e^{-aT}| R_-$$

TABLE 6.2 Z-Transform Properties for Positive- and Negative-Time Sequences (continued)

9. Frequency Translation

$$G(\omega) = Z_{II}\{e^{j\omega_o nT} f(nT)\} = G(z)\Big|_{z=e^{j\omega T}} = F(e^{j(\omega-\omega_o)T}) = F(\omega \quad \omega_o) \qquad -$$

ROC of $F(z)$ must include the unit circle

10. Product

$$Z_{II}\{f(nT)\} = F(z) \qquad R_{+f} < |z| < R_{-f}$$

$$Z_{II}\{h(nT)\} = H(z) \qquad R_{+h} < |z| < R_{-h}$$

$$Z_{II}\{f(nT)h(nT)\} = G(z) = \frac{1}{2\pi j}\oint_C F(\tau)\, H\!\left(\frac{z}{\tau}\right)\frac{d\tau}{\tau}, \qquad R_{+f}\, R_{+h} < |z| < R_{-f}\, R_{-h}$$

$$\max\left(R_{+f}, \frac{|z|}{R_{-h}}\right) < |\tau| < \min\left(R_{-f}, \frac{|z|}{R_{+h}}\right)$$

counter-clockwise integration

11. Parseval's Theorem

$$Z_{II}\{f(nT)\} = F(z) \qquad R_{+f} < |z| < R_{-f}$$

$$Z_{II}\{h(nT)\} = H(z) \qquad R_{+h} < |z| < R_{-h}$$

$$\sum_{n=-\infty}^{\infty} f(nT)h(nT) = \frac{1}{2\pi j}\oint_C F(z)H(z^{-1})\frac{dz}{z} \qquad R_{+f}\, R_{+h} < |z| = 1 < R_{-f}\, R_{-h}$$

$$\max\left(R_{+f}, \frac{1}{R_{-h}}\right) < |z| < \min\left(R_{-f}, \frac{1}{R_{+h}}\right)$$

counter-clockwise integration

12. Complex Conjugate Signals

$$Z_{II}\{f(nT)\} = F(z) \qquad R_{+f} < |z| < R_{-f}$$

$$Z_{II}\{f^*(nT)\} = F^*(z^*) \qquad R_{+f} < |z| < R_{-f}$$

6.3 Inverse Z-Transform

TABLE 6.3 Inverse Transforms of the Partial Fractions of $F(z)$

Partial Fraction Term	Inverse Transform Term If $F(z)$ Converges Absolutely for Some $\|z\| > \|a\|$
$\frac{z}{z-a}$	$a^k, \quad k \geq 0$
$\frac{z^2}{(z-a)^2}$	$(k+1)a^k, \quad k \geq 0$
$\frac{z^3}{(z-a)^3}$	$\frac{1}{2}(k+1)(k+2)a^k \geq \ k \quad 0$
⋮	⋮
$\frac{z^n}{(z-a)^n}$	$\frac{1}{(n-1)!}(k+1)(k+2)\cdots(k+n-1)a^k, \quad k \geq 0$

TABLE 6.3 Inverse Transforms of the Partial Fractions of $F(z)$ (continued)

Partial Fraction Term	Inverse Transform Term If $F(z)$ Converges Absolutely for Some $\|z\| < \|a\|$
$\frac{z}{z-a}$	$-a^k, \quad k \le -1$
$\frac{z^2}{(z-a)^2}$	$-(k+1)a^k, \quad k \le -1$
$\frac{z^3}{(z-a)^3}$	$-\frac{1}{2}(k+1)(k+2)a^k, \quad k \le -1$
$\vdots$	$\vdots$
$\frac{z^n}{(z-a)^n}$	$-\frac{1}{(n-1)!}(k+1)(k+2)\cdots(k+n-1)a^k, \quad k \le -1$

TABLE 6.4 Inverse Transforms of the Partial Fractions of $F_i(z)$[1]

Elementary Transform Term $F_i(z)$	Corresponding Time Sequence: (I) $F_i(z)$ converges for $\|z\| > R_c$	Corresponding Time Sequence: (I) $F_i(z)$ converges for $\|z\| < R_c$
$\frac{1}{z-a}$	$a^{k-1}\big\|_{k\ge 1}$	$-a^{k-1}\big\|_{k\le 0}$
$\frac{z}{(z-a)^2}$	$ka^{k-1}\big\|_{k\ge 1}$	$-ka^{k-1}\big\|_{k\le 0}$
$\frac{z(z+a)}{(z-a)^3}$	$k^2a^{k-1}\big\|_{k\ge 1}$	$-k^2a^{k-1}\big\|_{k\le 0}$
$\frac{z(z^2+4az+a^2)}{(z-a)^4}$	$k^3a^{k-1}\big\|_{k\ge 1}$	$-k^3a^{k-1}\big\|_{k\le 0}$

1. The function must be a proper function.

TABLE 6.5 Total Square Integrals

A solution of the integral

$$I_n = \frac{1}{2\pi j}\oint_{\substack{\text{unit}\\ \text{circle}}} F(z,m)F(z^{-1},m)z^{-1}\,dz$$

is presented in this table.

Let

$$F(z,m) = \frac{B(z)}{A(z)}$$

where

$$A(z) = \sum_{r=0}^{n} a_r z^{n-r}, \qquad a_0 \ne 0$$

$$B(z) = \sum_{r=0}^{n} b_r z^{n-r}, \quad \text{and the } br\text{'s are bounded functions of } m, \ 0 \le m \le 1$$

and the coefficients a_r, $0 < r \le n$, and b_r, $0 \le r \le n$ are not necessarily nonzero.

The integral I_n is equivalent to the ratio of two determinants as follows:

$$I_n = \frac{|\Omega_1|}{a_0 |\Omega|}$$

where Ω is the following matrix:

$$\Omega \triangleq \begin{bmatrix} a_0 & a_1 & a_2 & a_3 & \cdots & a_n \\ a_1 & a_0 + a_2 & a_1 + a_3 & a_2 + a_4 & \cdots & a_{n-1} \\ a_2 & a_3 & a_0 + a_1 & a_1 + a_5 & \cdots & a_{n-2} \\ \vdots & & & & & \\ a_n & 0 & 0 & 0 & \cdots & a_0 \end{bmatrix}$$

and Ω_1 is the maxtrix formed from Ω by replacing the first column by

$$\begin{bmatrix} \sum_{i=0}^{n} b_i^2 \\ 2\sum b_i b_{i+1} \\ 2\sum b_i b_{i+2} \\ \vdots \\ 2\sum b_i b_{i+n-1} \\ 2b_0 b_n \end{bmatrix}$$

Tabulated values of the integral I_n

$$I_n = \frac{1}{2\pi j} \oint_{\substack{\text{unit} \\ \text{circle}}} \frac{B(z)\, B(z^{-1})}{A(z)\, A(z^{-1})} z^{-1}\, dz$$

Counter clockwise integration

1. $F(z) = \dfrac{b_0 z + b_1}{a_0 z + a_1} = \dfrac{B(z)}{A(z)}$

$$I_1 = \frac{(b_0^2 + b_1^2) a_0 - 2 b_0 b_1 a_1}{a_0 (a_0^2 - a_1^2)}$$

2. $F(z) = \dfrac{b_0 z^2 + b_1 z + b_2}{a_0 z^2 + a_1 z + a_2}$

$$I_2 = \frac{B_0 a_0 e_1 - B_1 a_0 a_1 + B_2 (a_1^2 - a_2 e_1)}{a_0 [(a_0^2 - a_2^2) e_1 - (a_0 a_1 - a_1 a_2) a_1]}$$

$$B_0 = b_0^2 + b_1^2 + b_2^2$$

$$B_1 = 2(b_0 b_1 + b_1 b_2)$$

$$B_2 = 2 b_0 b_2$$

$$e_1 = a_0 + a_2$$

3. $F(z) = \dfrac{b_0 z^3 + b_1 z^2 + b_2 z + b_3}{a_0 z^3 + a_1 z^2 + a_2 z + a_3}$

$$I_3 = \frac{a_0 B_0 Q_0 - a_0 B_1 Q_1 + a_0 B_2 Q_2 - B_3 Q_3}{[(a_0^2 - a_3^2)Q_0 - (a_0 a_1 - a_2 a_3)Q_1 + (a_0 a_2 - a_1 a_3)Q_2]a_0}$$

$B_0 = b_0^2 + b_1^2 + b_2^2 + b_3^2$

$B_1 = 2(b_0 b_1 + b_1 b_2 + b_2 b_3)$

$B_2 = 2(b_0 b_2 + b_1 b_3)$

$B_3 = 2 b_0 b_3$

$Q_0 = (a_0 e_1 - a_3 e_2)$

$Q_1 = (a_0 a_1 - a_2 a_3)$

$Q_2 = (a_1 e_2 - a_2 e_1)$

$Q_3 = (a_1 - a_3)(e_2^2 - e_1^2) + a_0(a_0 e_2 - a_3 e_1)$

$e_1 = a_0 + a_2$

$e_2 = a_1 + a_3$

4. $F(z) = \dfrac{b_0 z^4 + b_1 z^3 + b_2 z^2 + b_3 z + b_4}{a_0 z^4 + a_1 z^3 + a_2 z^2 + a_3 z + a_4}$

$$I_4 = \frac{a_0 B_0 Q_0 - a_0 B_1 Q_1 + a_0 B_2 Q_2 - a_0 B_3 Q_3 + B_4 Q_4}{a_0[(a_0^2 - a_4^2)Q_0 - (a_0 a_1 - a_3 a_4)Q_1 + (a_0 a_2 - a_2 a_4)Q_2 - (a_0 a_3 - a_1 a_4)Q_3]}$$

$B_0 = b_0^2 + b_1^2 + b_2^2 + b_3^2 + b_4^2$, $\quad B_3 = 2(b_0 b_3 + b_1 b_4)$

$B_1 = 2(b_0 b_1 + b_1 b_2 + b_2 b_3 + b_3 b_4)$, $\quad B_4 = 2 b_0 b_4$

$B_2 = 2(b_0 b_2 + b_1 b_3 + b_2 b_4)$

$Q_0 = a_0 e_1 e_4 - a_0 a_3 e_2 + a_4(a_1 e_2 - e_3 e_4)$

$Q_1 = a_0 a_1 e_4 - a_0 a_2 a_3 + a_4(a_1 a_2 - a_3 e_4)$

$Q_2 = a_0 a_1 e_2 - a_0 a_2 e_1 + a_4(a_2 e_3 - a_3 e_2)$

$Q_3 = a_1(a_1 e_2 - e_3 e_4) - a_2(a_1 e_1 - a_3 e_3) + a_3(e_1 e_4 - a_3 e_2)$

$Q_4^* = a_0[e_2(a_1 a_4 - a_0 a_3) + e_5(a_0^2 - a_4^2)] + (e_2^2 - e_5^2)[a_1(a_1 - a_3) + (a_0 - a_4)(e_4 - a_2)]$

$e_1 = a_0 + a_2$

$e_2 = a_1 + a_3$

$e_3 = a_2 + a_4$

$e_4 = a_0 + a_4$

$e_5 = a_0 + a_2 + a_4$

For ease of calculations Q_4 and Q_3 of I_3 can be respectively written

$$Q_4 = -a_4 Q_0 + a_3 Q_1 - {+a_2} Q_2 \quad a_1 Q_3$$

$$Q_3 = a_3 Q_0 - a_2 Q_1 + a_1 Q_2$$

TABLE 6.6 Closed Forms of the Function $\sum_{n=0}^{\infty} n^r x^n, \quad x < 1$

r	$\sum_{n=0}^{\infty} n^r x^n$
0	$\frac{1}{1-x}$
1	$\frac{x}{(1-x)^2}$
2	$\frac{x^2+x}{(1-x)^3}$
3	$\frac{x^3+4x^2+x}{(1-x)^4}$
4	$\frac{x^4+11x^3+11x^2+x}{(1-x)^5}$
5	$\frac{x^5+26x^4+66x^3+26x^2+x}{(1-x)^6}$
6	$\frac{x^6+57x^5+302x^3+57x^2+x}{(1-x)^7}$
7	$\frac{x^7+120x^6+1191x^5+2416x^4+1191x^3+120x^2+x}{(1-x)^8}$
8	$\frac{x^8+247x^7+4293x^6+15619x^5+\ldots+x}{(1-x)^9}$
9	$\frac{x^9+502x^8+14608x^7+88234x^6+156190x^5+\ldots+x}{(1-x)^{10}}$
10	$\frac{x^{10}+1013x^9+47840x^8+455192x^7+1310354x^6+\ldots+x}{(1-x)^{11}}$

The missing terms are apparent since the numerator polynomials are symmetric in the coefficients.

6.4 Positive-Time Z-Transform Tables

TABLE 6.7 Positive-Time Z-Transform Tables

Number	Discrete Time-Function $f(n), n \geq 0$	z-Transform $F(z) = Z[f(n)],\ \lvert z\rvert > R$ $= \sum_{n=0}^{\infty} f(n) z^{-n}$
1	$u(n) = \begin{cases} 1, & \text{for } n \geq 0 \\ 0, & \text{otherwise} \end{cases}$	$\frac{z}{z-1}$
2	$e^{-\alpha n}$	$\frac{z}{z-e^{-\alpha}}$
3	n	$\frac{z}{(z-1)^2}$
4	n^2	$\frac{z(z+1)}{(z-1)^3}$

TABLE 6.7 Positive-Time Z-Transform Tables (continued)

Number	Discrete Time-Function $f(n),\ n \geq 0$	z-Transform $F(z) = Z[f(n)],\ \lvert z\rvert > R$ $= \sum_{n=0}^{\infty} f(n) z^{-n}$
5	n^3	$\frac{z(z^2+4z+1)}{(z-1)^4}$
6	n^4	$\frac{z(z^3+11z^2+11z+1)}{(z-1)^5}$
7	n^5	$\frac{z(z^4+26z^3+66z^2+26z+1)}{(z-1)^6}$
8	n^{k*}	$(-1)^k D^k\left(\frac{z}{z-1}\right);\quad D = z\frac{d}{dz}$
9	$u(n-k)$	$\frac{z^{-k+1}}{z-1}$
10	$e^{-\alpha n} f(n)$	$F(e^{\alpha} z)$
11	$n^{(2)} = n(n-1)$	$2\frac{z}{(z-1)^3}$
12	$n^{(3)} = n(n-1)(n-2)$	$3!\frac{z}{(z-1)^4}$
13	$n^{(k)} = n(n-1)(n-2)\ldots(n-k+1)$	$k!\frac{z}{(z-1)^{k+1}}$
14	$n^{[k]} f(n), n^{[k]} = n(n+1)(n+2)\ldots(n+k-1)$	$(-1)^k z^k \frac{d^k}{dz^k}[F(z)]$
15	$(-1)^k n(n-1)(n-2)\ldots(n-k+1) f_{n-k+1}$†	$zF^{(k)}(z),\ F^{(k)}(z) = \frac{d^k}{dz^k}F(z)$
16	$-(n-1) f_{n-1}$	$F^{(1)}(z)$
17	$(-1)^k (n-1)(n-2)\ldots(n-k) f_{n-k}$	$F^{(k)}(z)$
18	$nf(n)$	$-zF^{(1)}(z)$
19	$n^2 f(n)$	$z^2 F^{(2)}(z) + zF^{(1)}(z)$
20	$n^3 f(n)$	$-z^3 F^{(3)}(z) - 3z^2 F^{(2)}(z) - zF^{(1)}(z)$
21	$\frac{c^n}{n!}$	$e^{c/z}$
22	$\frac{(\ln c)^n}{n!}$	$c^{1/z}$
23	$\binom{k}{n} c^n a^{k-n},\quad \binom{k}{n} = \frac{k!}{(k-n)!\,n!},\quad n \leq k$	$\frac{(az+c)^k}{z^k}$
24	$\binom{n+k}{k} c^n$	$\frac{z^{k+1}}{(z-c)^{k+1}}$
25	$\frac{c^n}{n!},\quad (n = 1,3,5,7,\ldots)$	$\sinh\left(\frac{c}{z}\right)$

* Table 6.6 represents entries for k up to 10.

† It may be noted that f_n is the same as $f(n)$.

TABLE 6.7 Positive-Time Z-Transform Tables (continued)

Number	Discrete Time-Function $f(n),\ n \ge 0$	z-Transform $F(z) = Z[f(n)],\ \lvert z\rvert > R$ $= \sum_{n=0}^{\infty} f(n) z^{-n}$
26	$\frac{c^n}{n!},\quad (n = 0, 2, 4, 6, \ldots)$	$\cosh\left(\frac{c}{z}\right)$
27	$\sin(\alpha n)$	$\frac{z \sin\alpha}{z^2 - 2z\cos\alpha + 1}$
28	$\cos(\alpha n)$	$\frac{z(z - \cos\alpha)}{z^2 - 2z\cos\alpha + 1}$
29	$\sin(\alpha n + \psi)$	$\frac{z^2 \sin\psi + z\sin(\alpha - \psi)}{z^2 - 2z\cos\alpha + 1}$
30	$\cosh(\alpha n)$	$\frac{z(z - \cosh\alpha)}{z^2 - 2z\cosh\alpha + 1}$
31	$\sinh(\alpha n)$	$\frac{z\sinh\alpha}{z^2 - 2z\cosh\alpha + 1}$
32	$\frac{1}{n},\quad n > 0$	$\ln\frac{z}{z-1}$
33	$\frac{1 - e^{-\alpha n}}{n}$	$\alpha + \ln\frac{z - e^{-\alpha}}{z - 1},\quad \alpha > 0$
34	$\frac{\sin\alpha n}{n}$	$\alpha + \tan^{-1}\frac{\sin\alpha}{z - \cos\alpha},\quad \alpha > 0$
35	$\frac{\cos\alpha n}{n},\quad n > 0$	$\ln\frac{z}{\sqrt{z^2 - 2z\cos\alpha + 1}}$
36	$\frac{(n+1)(n+2)\ldots(n+k-1)}{(k-1)!}$	$\left(1 - \frac{1}{z}\right)^{-k},\quad k = 2, 3, \ldots$
37	$\sum_{m=1}^{n} \frac{1}{m}$	$\frac{z}{z-1}\ln\frac{z}{z-1}$
38	$\sum_{m=0}^{n-1} \frac{1}{m!}$	$\frac{e^{1/z}}{z-1}$
39	$\begin{cases} \frac{(-1)^{(n-p)/2}}{2^n\left(\frac{n-p}{2}\right)!\left(\frac{n+p}{2}\right)!}, & \text{for } n \ge p \text{ and } n-p \text{ even} = \\ = 0, & \text{for } n < p \text{ or } n-p \text{ odd} \end{cases}$	$J_p(z^{-1})$ =
40	$\begin{cases} \binom{\alpha}{n/k} b^{n/k}, & n = mk,\ (m = 0, 1, 2, \ldots) \\ = 0 & n \ne mk \end{cases}$	$\left(\frac{z^k + b}{z^k}\right)^{\alpha}$
41	$a^n P_n(x) = \frac{a^n}{2^n n!}\left(\frac{d}{dx}\right)^n (x^2 - 1)^n$	$\frac{z}{\sqrt{z^2 - 2xaz + a^2}}$
42	$a^n T_n(x) = a^n \cos(n\cos^{-1} x)$	$\frac{z(z - ax)}{z^2 - 2xaz + a^2}$
43	$\frac{L_n(x)}{n!} = \sum_{r=0}^{\infty}\binom{n}{r}\frac{(-x)^r}{r!}$	$\frac{z}{z-1}e^{-x/(z-1)}$

TABLE 6.7 Positive-Time Z-Transform Tables (continued)

Number	Discrete Time-Function $f(n),\ n \geq 0$	z-Transform $F(z) = Z[f(n)],\ \lvert z \rvert > R$ $= \sum_{n=0}^{\infty} f(n) z^{-n}$
44	$\frac{H_n(x)}{n!} = \sum_{k=0}^{[n/2]} \frac{(-1)^{n-k} x^{n-2k}}{k!(n-2k)!2^k}$	$e^{-x/z-1/2z^2}$
45	$a^n P_n^m(x) = a^n (1-x^2)^{m/2} \left(\frac{d}{dx}\right)^m P_n(x), \quad m = \text{integer}$	$\frac{(2m)!}{2^m m!} \frac{z^{m+1}(1-x^2)^{m/2} a^m}{(z^2 - 2xaz + a^2)^{m+1/2}}$
46	$\frac{L_n^m(x)}{n!} = \left(\frac{d}{dx}\right)^m \frac{L_n(x)}{n!}, \quad m = \text{integer}$	$\frac{(-1)^m z}{(z-1)^{m+1}} e^{-x/(z-1)}$
47	$-\frac{1}{n} Z^{-1}\left[z \frac{F'(z)}{F(z)} - \frac{G'(z)}{G(z)}\right]$, where $F(z)$ and $G(z)$ are rational	$\ln \frac{F(z)}{G(z)}$
48	$\frac{1}{m(m+1)(m+2)\ldots(m+n)}$	$(m-1)!\,z^m \left[e^{1/z} - \sum_{k=0}^{m-1} \frac{1}{k!\,z^k}\right]$
49	$\frac{\sin(\alpha n)}{n!}$	$e^{\cos\alpha/z} \cdot \sin\left(\frac{\sin\alpha}{z}\right)$
50	$\frac{\cos(\alpha n)}{n!}$	$e^{\cos\alpha/z} \cdot \cos\left(\frac{\sin\alpha}{z}\right)$
51	$\sum_{k=0}^{n} f_k g_{n-k}$	$F(z)G(z)$
52	$\sum_{k=0}^{n} k f_k g_{n-k}$	$-F^{(1)}(z)G(z),\ F^{(1)}(z) = \frac{dF(z)}{dz}$
53	$\sum_{k=0}^{n} k^2 f_k g_{n-k}$	$F^{(2)}(z)G(z)$
54	$\frac{\alpha^n + (-\alpha)^n}{2\alpha^2}$	$\frac{1}{\alpha^2} \frac{z^2}{z^2 - \alpha^2}$
55	$\frac{\alpha^n - \beta^n}{\alpha - \beta}$	$\frac{z}{(z-\alpha)(z-\beta)}$
56	$(n+k)^{(k)}$	$k!\,z^k \frac{z}{(z-1)^{k+1}}$
57	$(n-k)^{(k)}$	$k!\,z^{-k} \frac{z}{(z-1)^{k+1}}$
58	$\frac{(n \mp k)^{(m)}}{m!} e^{\alpha(n-k)}$	$\frac{z^{1 \mp k} e^{m\alpha}}{(z - e^{\alpha})^{m+1}}$
59	$\frac{1}{n} \sin \frac{\pi}{2} n$	$\frac{\pi}{2} + \tan^{-1} \frac{1}{z}$
60	$\frac{\cos\alpha(2n-1)}{2n-1}, \quad n > 0$	$\frac{1}{4\sqrt{z}} \ln \frac{z + 2\sqrt{z}\cos\alpha + 1}{z - 2\sqrt{z}\cos\alpha + 1}$
61	$\frac{\gamma^n}{(\gamma-1)^2} + \frac{n}{1-\gamma} - \frac{1}{(1-\gamma)^2}$	$\frac{z}{(z-\gamma)(z-1)^2}$
62	$\frac{\gamma + a_0}{(\gamma-1)^2} \gamma^n + \frac{1+a_0}{1-\gamma} n + \left(\frac{1}{1-\gamma} - \frac{a_0+1}{(1-\gamma)^2}\right)$	$\frac{z(z+a_0)}{(z-\gamma)(z-1)^2}$

TABLE 6.7 Positive-Time Z-Transform Tables (continued)

Number	Discrete Time-Function $f(n),\ n \ge 0$	z-Transform $F(z) = Z[f(n)],\ \lvert z\rvert > R$ $= \sum_{n=0}^{\infty} f(n) z^{-n}$
63	$a^n \cos \pi n$	$\frac{z}{z+a}$
64	$e^{-\alpha n} \cos an$	$\frac{z(z - e^{-\alpha}\cos a)}{z^2 - 2ze^{-\alpha}\cos a + e^{-2\alpha}}$
65	$e^{-\alpha n} \sin(an + \psi)$	$\frac{z^2 \sinh \psi + ze^{-\alpha}\sinh(a-\psi)}{z^2 - 2ze^{-\alpha}\cosh a + e^{-2\alpha}}$
66	$\frac{\gamma^n}{(\gamma-\alpha)^2 + \beta^2} + \frac{(\alpha^2+\beta^2)^{n/2}\sin(n\theta+\psi)}{\beta[(\alpha-\gamma)^2+\beta^2]^{1/2}}$ $\quad \theta = \tan^{-1}\frac{\beta}{\alpha}$, $\psi = \tan^{-1}\frac{\beta}{\alpha-\gamma}$	$\frac{z}{(z-\gamma)[(z-\alpha)^2 + \beta^2]}$
67	$\frac{n\gamma^{n-1}}{(\gamma-1)^3} - \frac{3\gamma^n}{(\gamma-1)^4} + \frac{1}{2}\left[\frac{n(n-1)}{(1-\gamma)^2} - \frac{4n}{(1-\gamma)^3} + \frac{6}{(1-\gamma)^4}\right]$	$\frac{z}{(z-\gamma)^2(z-1)^3}$
68	$\sum_{v=0}^{k} (-1)^v \binom{k}{v} \frac{(n+k-v)^{(k)}}{k!} e^{\alpha(n-v)}$	$\frac{z(z-1)^k}{(z-e^{\alpha})^{k+1}}$
69	$\frac{f(n)}{n}$	$\int_z^{\infty} p^{-1}F(p)\,dp + \lim_{n\to 0}\frac{f(n)}{n}$
70	$\frac{f_{n+2}}{n+1},\quad f_0 = 0,\ f_1 = 0$	$z\int_z^{\infty} F(p)\,dp$
71	$\frac{1+a_0}{(1-\gamma)[(1-\alpha)^2 + \beta^2]} + \frac{(\gamma+a_0)\gamma^n}{(\gamma-1)[(\gamma-\alpha)^2 + \beta^2]}$ $+ \frac{[\alpha^2+\beta^2]^{n/2}[(a_0+\alpha)^2+\beta^2]^{1/2}}{\beta[(\alpha-1)^2+\beta^2]^{1/2}[(\alpha-\gamma)^2+\beta^2]^{1/2}}\sin(n\theta+\psi+\lambda),$ $\psi = \psi_1 + \psi_2,\quad \psi_1 = \tan^{-1}\frac{\beta}{\alpha-1},\quad \theta = \tan^{-1}\frac{\beta}{\alpha}$ $\lambda = \tan^{-1}\frac{\beta}{a_0+\alpha},\quad \psi_2 = -\tan^{-1}\frac{\beta}{\alpha-\gamma}$	$\frac{z(z+a_0)}{(z-1)(z-\gamma)[(z-\alpha)^2+\beta^2]}$
72	$(n+1)e^{\alpha n} - 2ne^{\alpha(n+1)} + e^{\alpha(n-2)}(n \quad 1)$	$\left(\frac{z-1}{z-e^{\alpha}}\right)^2$
73	$(-1)^n \frac{\cos \alpha n}{n},\quad n > 0$	$\ln\frac{z}{\sqrt{z^2 + 2z\cos\alpha + 1}}$
74	$\frac{(n+k)!}{n!} f_{n+k},\quad f_n = 0,$ for $0 \le n \quad k \quad <$	$(-1)^k z^{2k}\frac{d^k}{dz^k}[F(z)]$
75	$\frac{f(n)}{n+h},\quad h > 0$	$z^h \int_z^{\infty} p^{-(1+h)}F(p)\,dp$
76	$-na^n \cos\frac{\pi}{2}n$	$\frac{2a^2z^2}{(z^2+a^2)^2}$
77	$na^n \frac{1+\cos\pi n}{2}$	$\frac{2a^2z^2}{(z^2-a^2)^2}$
78	$a^n \sin\frac{\pi}{4}n \cdot \frac{1+\cos\pi n}{2}$	$\frac{a^2z^2}{z^4+a^4}$

TABLE 6.7 Positive-Time Z-Transform Tables (continued)

Number	Discrete Time-Function $f(n),\ n \ge 0$	z-Transform $F(z) = Z[f(n)],\ \lvert z \rvert > R$ $= \sum_{n=0}^{\infty} f(n) z^{-n}$
79	$a^n \left(\frac{1+\cos \pi n}{2} - \cos \frac{\pi}{2} n \right)$	$\frac{2a^2 z^2}{z^4 - a^4}$
80	$\frac{P_n(x)}{n!}$	$e^{xz^{-1}} J_0(\sqrt{1-x^2}\, z^{-1})$
81	$\frac{P_n^{(m)}(x)}{(n+m)!}, \quad m > 0, \quad P_n^m = 0, \quad \text{for } n < m$	$(-1)^m e^{xz^{-1}} J_m(\sqrt{1-x^2}\, z^{-1})$
82	$\frac{1}{(n+\alpha)^\beta}, \quad \alpha > 0, \quad \mathrm{Re}\beta > 0$	$\Phi(z^{-1}, \alpha, \beta)$, where $\Phi(1, \beta, \alpha) = \zeta(\beta, \alpha)$ = generalized Rieman-Zeta function
83	$a^n \left(\frac{1+\cos \pi n}{2} + \cos \frac{\pi}{2} n \right)$	$\frac{2z^4}{z^4 - a^4}$
84	$\frac{c^n}{n}, \quad (n = 1, 2, 3, 4, \ldots)$	$\ln z - \ln(z - c)$
85	$\frac{c^n}{n}, \quad n = 2, 4, 6, 8, \ldots$	$\ln z - \frac{1}{2}\ln(z^2 - c^2)$
86	$n^2 c^n$	$\frac{cz(z+c)}{(z-c)^3}$
87	$n^3 c^n$	$\frac{cz(z^2 + 4cz + c^2)}{(z-c)^4}$
88	$n^k c^n$	$-\frac{dF(z/c)}{dz}, \quad F(z) = Z[n^{k-1}]$
89	$-\cos \frac{\pi}{2} n \sum_{i=0}^{(n-2)/4} \binom{n/2}{2i+1} a^{n-2-4i} (a^4 - b^4)^i$	$\frac{z^2}{z^4 + 2a^2 z^2 + b^4}$
90	$n^k f(n), \quad k > 0$ and integer	$-z \frac{d}{dz} F_1(z),\ F_1(z) = Z[n^{k-1} f(n)]$
91	$\frac{(n-1)(n-2)(n-3)\ldots(n-k+1)}{(k-1)!} a^{n-k}$	$\frac{1}{(z-a)^k}$
92	$\frac{k(k-1)(k-2)\ldots(k-n+1)}{n!}$	$\left(1 + \frac{1}{z}\right)^k$
93	$na^n \cos bn$	$\frac{[(z/a)^3 + z/a]\cos b - 2(z/a)^2}{[(z/a)^2 - 2(z/a)\cos b + 1]^2}$
94	$na^n \sin bn$	$\frac{(z/a)^3 \sin b - (z/a)\sin b}{[(z/a)^2 - 2(z/a)\cos b + 1]^2}$
95	$\frac{na^n}{(n+1)(n+2)}$	$\frac{z(a-2z)}{a^2} \ln\left(1 - \frac{a}{z}\right) - \frac{2}{a} z$
96	$\frac{(-a)^n}{(n+1)(2n+1)}$	$2\sqrt{z/a} \tan^{-1} \sqrt{a/z} - \frac{z}{a} \ln\left(1 + \frac{a}{z}\right)$

TABLE 6.7 Positive-Time Z-Transform Tables (continued)

Number	Discrete Time-Function $f(n), n \geq 0$	z-Transform $F(z) = Z[f(n)], \lvert z \rvert > R$ $= \sum_{n=0}^{\infty} f(n) z^{-n}$
97	$\frac{a^n \sin \alpha n}{n+1}$	$\frac{z\cos\alpha}{a}\tan^{-1}\frac{a\sin\alpha}{z - a\cos\alpha} + \frac{z\sin\alpha}{2a}\ln\frac{z^2 - 2az\cos\alpha + a^2}{z^2}$
98	$\frac{a^n \cos(\pi/2)n \sin\alpha(n+1)}{n+1}$	$\frac{z}{4a}\ln\frac{z^2 + 2az\sin\alpha + a^2}{z^2 - 2az\sin\alpha + a^2}$
99	$\frac{1}{(2n)!}$	$\cosh(z^{-1/2})$
100	$\binom{-\frac{1}{2}}{n}(-a)^n$	$\sqrt{z/(z-a)}$
101	$\binom{-\frac{1}{2}}{\frac{n}{2}} a^n \cos\frac{\pi}{2}n$	$\frac{z}{\sqrt{z^2 - a^2}}$
102	$\frac{B_n(x)}{n!}$ $B_n(x)$ are Bernoulli polynomials	$\frac{e^{x/z}}{z(e^{1/z} - 1)}$
103	$W_n(x) \triangleq$ Tchebycheff polynomials of the second kind	$\frac{z^2}{z^2 - 2xz + 1}$
104	$\left\lvert \sin\frac{n\pi}{m} \right\rvert, \quad m = 1, 2, \ldots$	$\frac{z\sin\pi/m}{z^2 - 2z\cos\pi/m + 1}\frac{1+z^{-m}}{1-z^{-m}}$
105	$Q_n(x) = \sin(n\cos^{-1} x)$	$\frac{z}{z^2 - 2xz + 1}$

References

Gabel, R. A. and Roberts, R. A., *Signals and Linear Systems*, John Wiley & Sons, Inc., New York, NY, 1980.

Freeman, H., *Discrete-Time Systems*, John Wiley & Sons, Inc., New York, NY, 1965.

Jury, E. I., *Theory and Application of the Z-Transform Method*, Krieger Publishing Co., Melbourne, FL, 1973.

Poularikas, A. D., Ed., *Transforms and Applications Handbook*, CRC Press Inc., Boca Raton, FL, 1996.

Poularikas, A. D. and Seely, S., *Signals and Systems*, Reprinted Second Edition, Krieger Publishing Co., Melbourne, FL, 1994.

Tretter, S. A., *Introduction to Discrete-Time Signal Processing*, John Wiley & Sons, Inc., New York, NY, 1976.

Vich, R., *Z-Transform Theory and Applications*, D. Reidel Publishing Co., Boston, MA, 1987.

Appendix 1

Examples

1.1 One-Sided Z-Transforms

Example 6.1

The radius of convergence of $f(nT) = e^{-anT}u(nT)$, a positive number, is:

$$\left|z^{-1}e^{-aT}\right| < 1 \quad \text{or} \quad |z| > e^{-aT}.$$

The Z-transform of $f(nT) = e^{-anT}u(nT)$ is

$$F(z) = \sum_{n=0}^{\infty} f(nT)z^{-n} = \sum_{n=0}^{\infty} (e^{-aT}z^{-1})^n = \frac{1}{1-e^{-aT}z^{-1}}.$$

If $a = 0$

$$F(z) = \sum_{n=0}^{\infty} u(nT)z^{-n} = \frac{1}{1-z^{-1}} = \frac{z}{z-1}.$$

Example 6.2

The function $f(nT) = a^{nT}\cos nT\omega\, u(nT)$ has the Z-transform

$$F(z) = \sum_{n=0}^{\infty} a^{nT}\frac{e^{jnT\omega}+e^{-jnT\omega}}{2}z^{-n} = \frac{1}{2}\sum_{n=0}^{\infty}(a^T e^{jT\omega}z^{-1})^n + \frac{1}{2}\sum_{n=0}^{\infty}(a^T e^{-jT\omega}z^{-1})^n$$

$$= \frac{1}{2}\frac{1}{1-a^T e^{jT\omega}z^{-1}} + \frac{1}{2}\frac{1}{1-a^T e^{-jT\omega}z^{-1}} = \frac{1-a^T z^{-1}\cos T\omega}{1-2a^T z^{-1}\cos T\omega + a^{2T}z^{-2}}.$$

The ROC is given by the relations

$$\left|a^T e^{jT\omega}z^{-1}\right| < 1 \quad \text{or} \quad |z| > \left|a^T\right|$$

$$\left|a^T e^{-jT\omega}z^{-1}\right| < 1 \quad \text{or} \quad |z| > \left|a^T\right|$$

Therefore the ROC is $|z| > \left|a^T\right|$.

Example 6.3

$$Z\{n\} = -z\frac{d}{dz}\left(\frac{z}{z-1}\right) = \frac{z}{(z-1)^2}, \qquad n \ge 0$$

$$Z\{n^2\} = -z\frac{d}{dz}Z\{n\} \quad z\frac{d}{dz}\frac{z}{(z-1)^2} = \frac{z(z+1)}{(z-1)^3}, \qquad n \ge 0$$

$$Z\{n^3\} = -z\frac{d}{dz}\frac{z(z+1)}{(z-1)^3} = \frac{z(z^2+4z+1)}{(z-1)^4} \qquad n = 0,1,2$$

Example 6.4

See Figure 6.1 for graphical representation of the complex integration ($n \geq 0$).

$$Z\{nT\} \doteq H(z) = \frac{z}{(z-1)^2}T \quad |z| > 1, \quad Z\{e^{-nT}\} \doteq F(z) = \frac{z}{z-e^{-T}} \quad |z| > e^{-T}$$

Hence (counter-clockwise integration)

$$Z\{nTe^{-nT}\} = \frac{1}{2\pi j}\oint_C T\frac{z}{\tau(\tau - e^{-T})\left(\frac{z}{\tau}-1\right)^2}d\tau.$$

The contour must have a radius $|\tau|$ of the value $e^{-T} < |\tau| < |z| = 1$ and we have from complex integration

$$Z\{nTe^{-nT}\} = \underset{\tau=e^{-T}}{Res}\left\{(\tau - e^{-T})T\frac{z\tau}{(\tau-e^{-T})(z-\tau)^2}\right\} = T\frac{ze^{-T}}{(z-e^{-T})^2}$$

But

$$Z\{nTe^{-nT}\} = -Tz\frac{d}{dz}\left(\frac{1}{1-e^{-T}z^{-1}}\right) = T\frac{ze^{-T}}{(z-e^{-T})^2}$$

and verifies the complex integration approach.

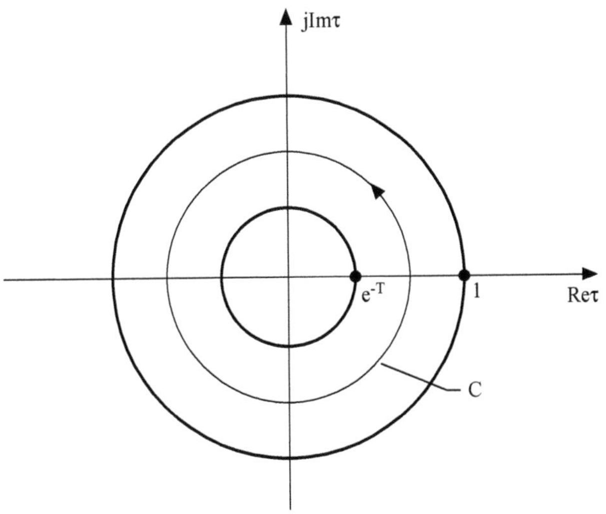

FIGURE 6.1

1.2 One-Sided Inverse Z-Transform

Example 6.5

a) If $F(z) = \dfrac{z^2+1}{(z-1)(z-2)} = A + \dfrac{Bz}{z-1} + \dfrac{Cz}{z-2}$ with $|z| > 2$, then we obtain $A = \dfrac{0+1}{(0-1)(0-2)} = \dfrac{1}{2}$, $B = \left.\dfrac{1}{z}\dfrac{z^2+1}{(z-2)}\right|_{z=1} = -2$ and $C = \left.\dfrac{1}{z}\dfrac{z^2+1}{(z-1)}\right|_{z=2} = \dfrac{5}{2}$. Hence $F(z) = \dfrac{1}{2} - 2\dfrac{z}{z-1} + \dfrac{5}{2}\dfrac{z}{z-2}$ and its inverse is $f(nT) = \dfrac{1}{2}\delta(nT) - 2u(nT) + \dfrac{5}{2}(2)^n u(nT)$.

b) If $F(z) = \dfrac{z+1}{(z-1)(z-2)} = \dfrac{A}{z-1} + \dfrac{B}{z-2}$ then we obtain $A = \left.\dfrac{z+1}{(z-2)}\right|_{z=1} = -2$ and $B = \left.\dfrac{z+1}{(z-1)}\right|_{z=2} = 3$. Hence $F(z) = -2\dfrac{1}{(z-1)} + 3\dfrac{1}{(z-2)}$ and its inverse is $f(nT) = -2u(nT-T) + 3(2)^{n-1}u(nT-T)$ with ROC $|z| > 2$.

Example 6.6

1. By Expansion
   If $F(z)$ has the region of convergence $|z| > 5$, then

$$F(z) = \frac{5z}{(z-5)^2} = \frac{5z}{z^2 - 10z + 25}$$

$$= 5z^{-1} + 50z^{-2} + 375z^{-3} + \cdots = 0\cdot 5^0 z^{-0} + 1\cdot 5z^{-1} + 2\cdot 5^2 z^{-2} + 3\cdot 5^3 z^{-3} + \cdots.$$

   Hence $f(nT) = n5^n$ $n = 0,1,2,\ldots$ which sometimes is difficult to recognize using the expansion method.

2. By Fraction Expansion

$$F(z) = \frac{5z}{(z-5)^2} = \frac{Az}{z-5} + \frac{Bz^2}{(z-5)^2},\quad B = \left.\frac{5}{z}\right|_{z=5} = 1,\quad \frac{5\times 6}{(6-1)^2} = \frac{A\times 6}{6-5} + \frac{6^2}{(6-5)^2}$$ or $A = -1$. Hence

$$F(z) = -\frac{z}{z-5} + \frac{z^2}{(z-5)^2}$$ and $f(nT) = -(5)^n + (n+1)5^n = n5^n$, $n \geq 0$.

3. By Integration

$$\frac{1}{(2-1)!}\frac{d^{2-1}}{dz^{2-1}}\left[(z-5)^2\frac{5z}{(z-5)^2}z^{n-1}\right]\Bigg|_{z=5} = 5nz^{n-1}\Big|_{z=5} = n5^n,\qquad n \geq 0.$$

1.3 Two-Sided Z-Transform

Example 6.7

$$F(z) = Z_{II}\{e^{-|nT|}\} = \sum_{n=-\infty}^{-1} e^{nT}z^{-n} + \sum_{n=0}^{\infty} e^{-nT}z^{-n} = \sum_{n=-\infty}^{0} e^{nT}z^{-n} - 1 \sum_{n=0}^{\infty} e^{-nT}{+}z^{-n}$$

$$= \sum_{n=0}^{\infty} e^{-nT}z^{n} - 1 + \sum_{n=0}^{\infty} e^{-nT}z^{-n} = \frac{1}{1-e^{-nT}z} - 1 + \frac{1}{1-e^{-nT}z^{-1}}$$

The first sum (negative time) converges if $|e^{-T}z| <$ or $|z| < e^{T}$. The second sum (positive time) converges if $|e^{-T}z^{-1}| < 1$ or $e^{-T} < |z|$. Hence the region of convergence is $R_+ = e^{-T} < |z| < R_- = e^{T}$. The two poles of $F(z)$ are $z = e^{T}$ and $z = e^{-T}$.

Example 6.8

The Z-transform of $u(nT)$ is

$$F(z) = \frac{1}{1-z^{-1}} \qquad |z| > 1 = R_{+f},\ R_{-f} = \infty$$

and the Z-transform of $h(nT) = \exp(-|nT|)$ is

$$H(z) = \frac{1-e^{-2T}}{(1-e^{-T}z^{-1})(1-e^{-T}z)} \qquad R_{+h} = e^{-T} < |z| < e^{T} = R_{-h}.$$

But $R_{-f} = \infty$ and hence from the product property $1 \cdot \exp(-T) < |z| < \infty$. The contour must lie in the region $\max(1, |z|e^{-T}) < |\tau| < \min(-\infty, |z|e^{T})$. The pole-zero configuration and the contour are shown in Figure 6.2. If we choose $|z| > e^{T}$ then the contour is that shown in the figure.

Therefore, we obtain

$$Z_{II}\{u(nT)h(nT)\} \doteq G(z) = \frac{1}{2\pi j}\oint_C \frac{1}{1-\tau^{-1}} \frac{1-e^{-2T}}{\left(1-e^{-T}\frac{\tau}{z}\right)\left(1-e^{-T}\frac{z}{\tau}\right)} \frac{d\tau}{\tau}$$

The poles of $H(z/\tau)$ are at $\tau = z\exp(-T)$ and $\tau = z\exp(T)$. Hence the contour encloses the poles $\tau = 1$ and $\tau = z\exp(-T)$. Applying the residue theorem next, we obtain

$$G(z) = \frac{1}{1-e^{-T}z^{-1}} \qquad |z| > e^{-T}$$

which has the inverse function $g(nT) = e^{-nT}\,u(nT)$ as it was expected.

Example 6.9

The Z-transform of $f(nT) = \exp(-nT)u(nT)$ is $F(z) = 1/(1 - e^{-T}z^{-1})$ for $|z| > e^{-T}$. From the Parseval property we obtain (counter-clockwise integration)

$$\sum_{n=-\infty}^{\infty} f^2(nT) = \sum_{n=0}^{\infty} f^2(nT) = \frac{1}{2\pi j}\oint_C \frac{1}{1-e^{-T}z^{-1}} \frac{1}{1-e^{-T}z} \frac{dz}{z}$$

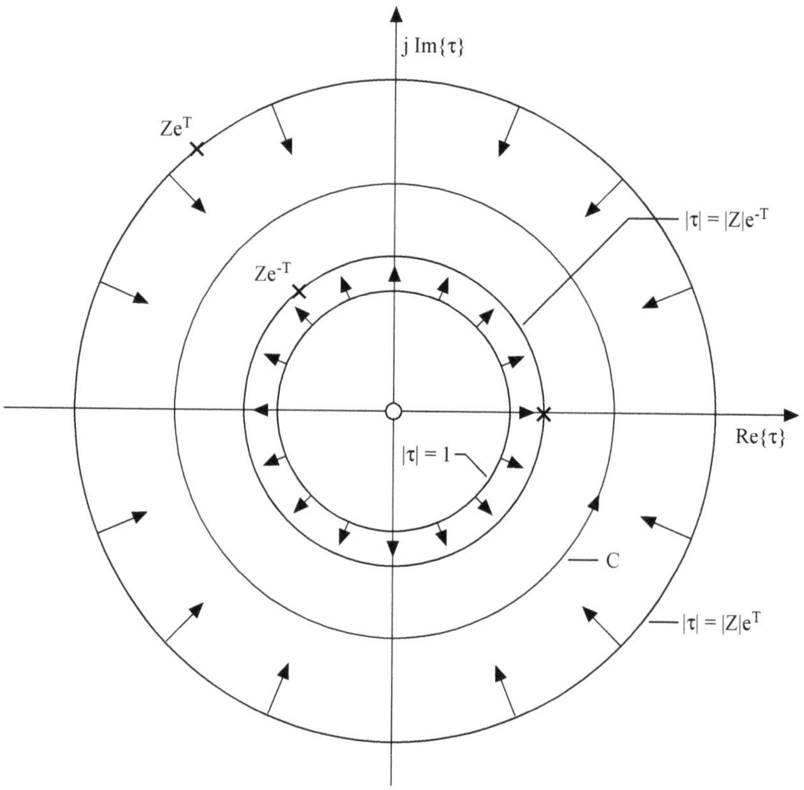

FIGURE 6.2

with $\max(e^{-T},0) < |z| < \min(\ ,e^{T})$. The contour encircles the pole at $z = e^{-T}$ so that

$$\sum_{n=0}^{\infty} f^2(nT) = Res\left[\frac{z - e^{-T}}{(z - e^{-T})(1 - e^{-T}z)}\right]\Bigg|_{z=e^{-T}} = \frac{1}{1 - e^{-2T}}$$

Also we find directly

$$\sum_{n=0}^{\infty} e^{-nT} e^{-nT} = \sum_{n=0}^{\infty} e^{-2nT} = (1 + e^{-2T} + (e^{-2T})^2 + \cdots) = \frac{1}{1 - e^{-2T}}\,.$$

1.4 Two-Sided Inverse Z-Transform

Example 6.10

If $F(z) = [z(z+1)]/(z^2 - 2z + 1) = (1 + z^{-1})/(1 - 2z^{-1} + z^{-2})$ and the ROC is $|z| > 1$, then

$$
\begin{array}{r}
1+3z^{-1}+5z^{-2}+7z^{-3}+\cdots \\
1-2z^{-1}+z^{-2}\ \overline{)\,1+z^{-1}\qquad\qquad\qquad} \\
\underline{1-2z^{-1}+z^{-2}} \\
3z^{-1}-z^{-2} \\
\underline{3z^{-1}-6z^{-2}+3z^{-3}} \\
5z^{-2}-3z^{-3} \\
\cdots
\end{array}
$$

and by continuing the division, we recognize that

$$
f(nT)=\begin{cases} 0 & n<0 \\ (2n+1) & n\geq 0 \end{cases}
$$

If $f(nT)$ is known to be zero for positive n, that the ROC is $|z|<1$, then

$$
\begin{array}{r}
z+3z^{2}+5z^{3}+\cdots \\
z^{-2}-2z^{-1}+1\ \overline{)\,z^{-1}+1\qquad\qquad} \\
\underline{z^{-1}-2+z} \\
3-z \\
\underline{3-6z+3z^{2}} \\
5z-3z^{2} \\
\cdots
\end{array}
$$

This series is recognized as

$$
f(nT)=\begin{cases} -(2n+1) & n<0 \\ 0 & n\geq 0 \end{cases}
$$

Example 6.11

To determine the inverse Z-transform of $F(z) = 1/(1-1.5z^{-1}+0.5z^{-2})$ if a) $ROC: |z|>1$, b) $ROC: |z| < 0.5$, and c) $ROC: 0.5 < |z| < 1$, we proceed as follows:

$$
F(z)=\frac{z^2}{z^2-1.5z+0.5}=\frac{z^2}{(z-1)(z-\frac{1}{2})}=A+\frac{Bz}{z-1}+\frac{Cz}{z-\frac{1}{2}}
$$

or

$$
F(z)=2\frac{z}{z-1}-\frac{z}{z-\frac{1}{2}}
$$

a) $f(nT) = 2(1)^n - (\frac{1}{2})^n$, $n \geq 0$ since both poles are outside the region of convergence $|z| > 1$ (inside the unit circle).
b) $f(nT) = -2(1)^n u(-nT - T) + (\frac{1}{2})^n u(-nT - T)$, $n \leq -1$ since both poles are outside the region of convergence (outside the circle $|z| = 0.5$).
c) Pole at $\frac{1}{2}$ provides the causal part, and the pole at 1 provides the anticausal. Hence

$$f(nT) = -2(1)^n u(-nT - T) - (\tfrac{1}{2})^n u(nT) \qquad -\infty < n < \infty$$

Example 6.12

Let

$$F(z) = \frac{1 - 0.8^2}{(1 - 0.8z)(1 - 0.8z^{-1})} \qquad 0.8 < |z| < 0.8^{-1}$$

For $n \geq 0$ the contour C encloses counter-clockwise only the pole $z = 0.8$ of the function $F(z)z^{n-1}$. Therefore

$$f(nT) = Res\{F(z)z^{n-1}\}\Big|_{z=0.8} = \frac{\left(1 - 0.8^2\right)z^n(z - 0.8)}{(1 - 0.8z)\,(z - 0.8)}\Bigg|_{z=0.8} = 0.8^n \qquad n \geq 0$$

For $n < 0$ only the pole $z = 1/0.8$ is outside C. Hence

$$f(nT) = Res\{F(z)z^{n-1}\}\Big|_{z=1/0.8} = -\frac{\left(1 - 0.8^2\right)0.8^{-1}z^n\left(z - 0.8^{-1}\right)}{-\left(1 - 0.8^{-1}\right)(z - 0.8)}\Bigg|_{z=0.8^{-1}} = 0.8^{-n} \qquad n \leq -1$$

The residue for a multiple pole of order k at z_o is given by

$$Res\{F(z)z^{n-1}\}\Big|_{z=z_o} = \lim_{z \to z_o} \frac{1}{(k-1)!} \frac{d^{k-1}}{dz^{k-1}} \left[(z - z_o)^k F(z)z^{n-1}\right]$$

1.5 Solution of Difference Equation

Example 6.13

The solution of the difference equation $y(n) - ay(n-1) = u(n)$ with initial condition $y(-1) = 2$ and $|a| < 1$ proceeds as follows:

$$Y(z) - ay(-1) - az^{-1}Y(z) = \frac{z}{z-1}$$

$$Y(z) = \frac{2a}{1 - az^{-1}} + \frac{z}{z-1}\frac{1}{1 - az^{-1}} = \frac{2a}{1 - az^{-1}} + \frac{z^2}{(z-1)(z-a)}$$

$$= \frac{2a}{1 - az^{-1}} + \frac{1}{1-a}\frac{1}{1 - z^{-1}} + \frac{a}{a-1}\frac{1}{1 - az^{-1}}$$

Hence, the inverse Z-transform gives

$$y(n) = \underbrace{2a \cdot a^n}_{\text{zero input}} + \underbrace{\frac{1}{1-a}u(n) + \frac{a}{a-1}a^n}_{\text{zero state}} = \underbrace{\frac{1}{1-a}u(n)}_{\text{steady state}} + \underbrace{\frac{2a-1}{a-1}a^{n+1}}_{\text{transient}} \qquad n \geq 0$$

7

Windows

7.1 Introductory Material

7.1.1 Introduction

N = number of samples

T = interval between samples

NT = total time duration of the signal

$\frac{1}{NT}$ = minimum spectral resolution (s^{-1})

DFT = discrete Fourier transform

Leakage = Spectral leakage takes place when the signal has frequencies other than those of the basis set. These other frequencies will exhibit non-zero properties on the entire basis set known as leakage.

7.2 Figures of Merit (see Table 7.1)

7.2.1 Equivalent Noise Bandwidth

$$ENBW = \frac{\sum_n w^2(nT)}{\left|\sum_n w(nT)\right|^2} = \text{equivalent noise bandwidth} = $$

the width of an equivalent ideal rectangular spectral response that will pass the same noise power as the window (filter) under test.

$w(nT)$ = window samples

TABLE 7.1 Figures of Merit for Shaped DFT Filters

Weighting		Highest sidelobe level (dB)	Sidelobe falloff (dB/octave)	Coherent gain	Equivalent noise BW (bins)	3.0-dB BW (bins)	Scallop loss (dB)	Worst-case process loss (dB)	6.0-dB BW (bins)	Overlap correlation (%) 75% OL	Overlap correlation (%) 50% OL
Rectangle		−13	−6	1.00	1.00	0.89	3.92	3.92	1.21	75.0	50.0
Triangle		−27	−12	0.50	1.33	1.28	1.82	3.07	1.78	71.9	25.0
	$\alpha = 1.0$	−23	−12	0.64	1.23	1.20	2.10	3.01	1.65	75.5	31.8
$\cos^{\alpha}(x)$	$\alpha = 2.0$	−32	−18	0.50	1.50	1.44	1.42	3.18	2.00	65.9	16.7
Hann	$\alpha = 3.0$	−39	−24	0.42	1.73	1.66	1.08	3.47	2.32	56.7	8.5
	$\alpha = 4.0$	−47	−30	0.38	1.94	1.86	0.86	3.75	2.59	48.6	4.3
Hamming		−43	−6	0.54	1.36	1.30	1.78	3.10	1.81	70.7	23.5
Parabolic		−21	−12	0.67	1.20	1.16	2.22	3.01	1.59	76.5	34.4
Riemann		−26	−12	0.59	1.30	1.26	1.89	3.03	1.74	73.4	27.4
Cubic		−53	−24	0.38	1.92	1.82	0.90	3.72	2.55	49.3	5.0
Tukey	$\alpha = 0.25$	−14	−18	0.88	1.10	1.01	2.96	3.39	1.38	74.1	44.4
	$\alpha = 0.50$	−15	−18	0.75	1.22	1.15	2.24	3.11	1.57	72.7	36.4
	$\alpha = 0.75$	−19	−18	0.63	1.36	1.31	1.73	3.07	1.80	70.5	25.1
Bohman		−46	−24	0.41	1.79	1.71	1.02	3.54	2.38	54.5	7.4
Poisson	$\alpha = 2.0$	−19	−6	0.44	1.30	1.21	2.09	3.23	1.69	69.9	27.8
	$\alpha = 3.0$	−24	−6	0.32	1.65	1.45	1.46	3.64	2.08	54.8	15.1
	$\alpha = 4.0$	−31	−6	0.25	2.08	1.75	1.03	4.21	2.58	40.4	7.4
Hamming Poisson	$\alpha = 0.5$	−35	−18	0.43	1.61	1.54	1.26	3.33	2.14	61.3	12.6
	$\alpha = 1.0$	−39	−18	0.38	1.73	1.64	1.11	3.50	2.30	56.0	9.2
	$\alpha = 2.0$	none	−18	0.29	2.02	1.87	0.87	3.94	2.65	44.6	4.7
Cauchy	$\alpha = 3.0$	−31	−6	0.42	1.48	1.34	1.71	3.40	1.90	61.6	20.2
	$\alpha = 4.0$	−35	−6	0.33	1.76	1.50	1.36	3.83	2.20	48.8	13.2
	$\alpha = 5.0$	−30	−6	0.28	2.06	1.68	1.13	4.28	2.53	38.3	9.0

Taylor	α = 2.0	–40	–6	0.57	1.30	1.25	1.91	3.06	1.74	75.7	28.3
	α = 2.5	–50	–6	0.51	1.43	1.36	1.60	3.15	1.90	71.3	21.4
	α = 3.0	–60	–6	0.47	1.55	1.47	1.37	3.26	2.06	67.0	16.1
	α = 3.5	–70	–6	0.44	1.66	1.58	1.20	3.40	2.21	62.9	12.1
	α = 4.0	–80	–6	0.41	1.76	1.67	1.06	3.52	2.35	59.1	9.1
Gaussian	α = 2.5	–42	–6	0.51	1.39	1.33	1.69	3.14	1.86	67.7	20.0
	α = 3.0	–55	–6	0.43	1.64	1.55	1.25	3.40	2.18	57.5	10.6
	α = 3.5	–69	–6	0.37	1.90	1.79	0.94	3.73	2.52	47.2	4.9
Dolph-Chebyshev	α = 2.5	–50	0	0.53	1.39	1.33	1.70	3.12	1.85	69.6	22.3
	α = 3.0	–60	0	0.48	1.51	1.44	1.44	3.23	2.01	64.7	16.3
	α = 3.5	–70	0	0.45	1.62	1.55	1.25	3.35	2.17	60.2	11.9
	α = 4.0	–80	0	0.42	1.73	1.65	1.10	3.48	2.31	55.9	8.7
Kaiser-Bessel	α = 2.0	–46	–6	0.49	1.50	1.43	1.46	3.20	1.99	65.7	16.9
	α = 2.5	–57	–6	0.44	1.65	1.57	1.20	3.38	2.20	59.5	11.2
	α = 3.0	–69	–6	0.40	1.80	1.71	1.02	3.56	2.39	53.9	7.4
	α = 3.5	–82	–6	0.37	1.93	1.83	0.89	3.74	2.57	48.8	4.8
Barcilon Temes	α = 3.0	–53	–6	0.47	1.56	1.49	1.34	3.27	2.07	63.0	14.2
	α = 3.5	–58	–6	0.43	1.67	1.59	1.18	3.40	2.23	58.6	10.4
	α = 4.0	–68	–6	0.41	1.77	1.69	1.05	3.52	2.36	54.4	7.6
Exact Blackman		–68	–6	0.46	1.57	1.52	1.33	3.29	2.13	62.7	14.0
Blackman		–58	–18	0.42	1.73	1.68	1.10	3.47	2.35	56.7	9.0
Minimum 3-sample Blackman-Harris		–71	–6	0.42	1.71	1.66	1.13	3.45	1.81	57.2	9.6
Minimum 4-sample Blackman-Harris		–92	–6	0.36	2.00	1.90	0.83	3.85	2.72	46.0	3.8
62-dB 3-sample Blackman-Harris		–62	–6	0.45	1.61	1.56	1.27	3.34	2.19	61.0	12.6
74-dB 4-sample Blackman-Harris		–74	–6	0.40	1.79	1.74	1.03	3.56	2.44	53.9	7.4
4-sample Kaiser–Bessel	α = 3.0	–69	–6	0.40	1.80	1.74	1.02	3.56	2.44	53.9	7.4

7.2.2 Coherent Gain

$$CG = \text{coherent gain} = \frac{1}{N}\sum_{n=0}^{N-1} w(nT) = \text{zero frequency gain (dc gain) of the window}$$

7.2.3 Processing Gain

$$PG = \frac{1}{ENBW} = \frac{\text{output signal-to-noise ratio}}{\text{input signal-to-noise ratio}}$$

7.2.4 Scalloping Loss

$$\text{scalloping loss} = \frac{\left|\sum_n w(nT)\exp\left(-j\frac{\pi}{N}n\right)\right|}{\sum_n w(nT)} =$$

$$= \frac{\text{coherent gain for a tone located half a bin from DFT sample point}}{\text{coherent gain for a tone located at a DFT sample point}}$$

= maximum reduction in PG due to signal frequency

7.2.5 Mainlobe Spectral Response

mainlobe spectral response = spectral interval between the peak gain and the –3.0 dB and –6.0 dB response level

7.2.6 Overlap Correlation

Correlation coefficients represent the degree of correlation of filter output points that are separated by 25% and 50% of the filter length. These terms are useful in quantifying the estimation uncertainty (or variance reduction) related to incoherent averaging of filter (window) data.

7.3 Window (Filter) Descriptions

7.3.1 Introduction

- $T = 1$
- $-\pi \le \omega \le \pi$ or $0 \le \omega \le 2\pi$
- $DFT\ bin = 2\pi/N$
- Windows are even (about the origin) sequences with an odd number of points.
- The right-most point of the window will be discarded.
- N will be taken to be even, and the total points will be *odd*, and hence

$$N = 2 \times (\text{total points}) = \text{even}$$

7.3.2 Rectangle (Dirichlet) Window

$$w(n) = 1.0 \qquad n = -\tfrac{N}{2},\cdots,-1,0,1,\cdots,\tfrac{N}{2}$$

$$W(\omega) = \sum_{n=-N/2}^{N/2} w(n)e^{-jn\omega}$$

To make it realizable shift the sequence by $N/2$ to the right. Hence we obtain

$$w(n) = 1 \qquad n = 0, 1, \cdots, N-1$$

$$W(\omega) = \sum_{n=0}^{N-1} e^{-jn\omega} = e^{-j\frac{N-1}{2}\omega} \frac{\sin\left(\frac{N}{2}\omega\right)}{\sin\frac{\omega}{2}}$$

Figure 7.1 shows the rectangular window and its amplitude spectrum $|W(\omega)|$.

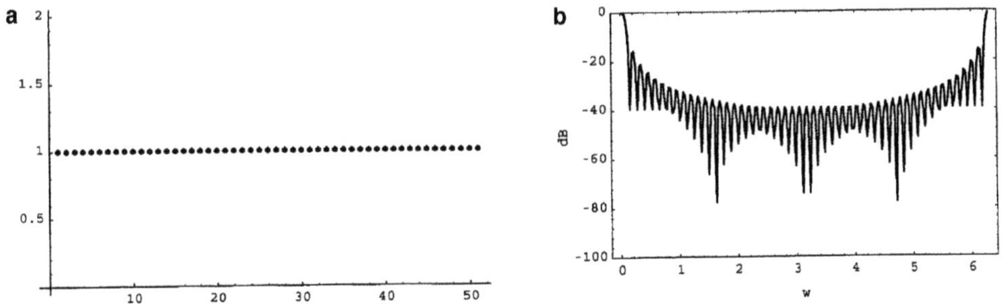

FIGURE 7.1 a) Rectangular window. b) Amplitude spectrum of rectangular window.

7.3.3 Triangle (Fejer, Bartlet) Window

$$w(n) = 1.0 - \frac{|n|}{N/2} \qquad n = -\tfrac{N}{2}, \cdots, -1, 0, 1, \cdots, \tfrac{N}{2}$$

For DFT the window is

$$w(n) = \begin{cases} \dfrac{n}{N/2} & n = 0, 1, \cdots, \tfrac{N}{2} \\ \dfrac{N-n}{N/2} & n = \tfrac{N}{2}+1, \cdots, N-1 \end{cases}$$

and its DFT is

$$W(\omega) = e^{-j(\frac{N}{2}-1)\omega} \left[\frac{\sin\left(\frac{N}{4}\omega\right)}{\sin\frac{\omega}{2}} \right]^2$$

since the symmetric function $w(n)$ is shifted by $\frac{N}{2} - 1$ positions to produce the DFT sequence. Figure 7.2 shows the triangular window and its DFT.

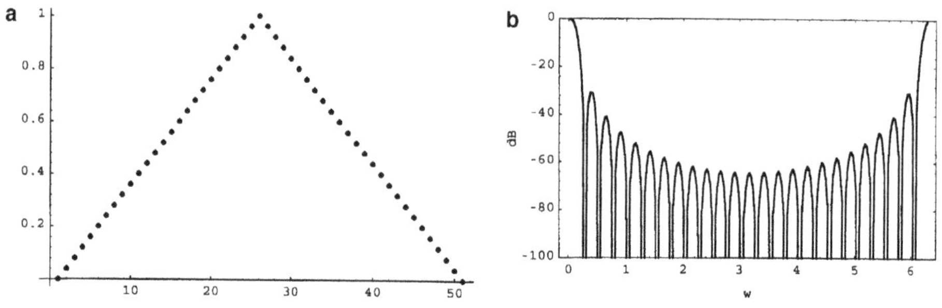

FIGURE 7.2 a) Triangular window. b) Amplitude spectrum of triangular window.

7.3.4 $\cos^{\alpha}$ (t) Windows

$$w(n) = \cos^{\alpha}\left[\left(\frac{n}{N}\right)\pi\right] \qquad n = -\tfrac{N}{2},\cdots,-1,0,1,\cdots,\tfrac{N}{2}$$

$$w(n) = \sin^{\alpha}\left[\left(\frac{n}{N}\right)\pi\right] \qquad n = 0,1,\cdots,N-1$$

Common values of α : $1 \le \alpha \le 4$

Figures 7.3 through 7.5 show the $\cos^2\left(\frac{n\pi}{N}\right)$, $\cos^3\left(\frac{n\pi}{N}\right)$, and $\cos^4\left(\frac{n\pi}{N}\right)$ windows and their Fourier transform.

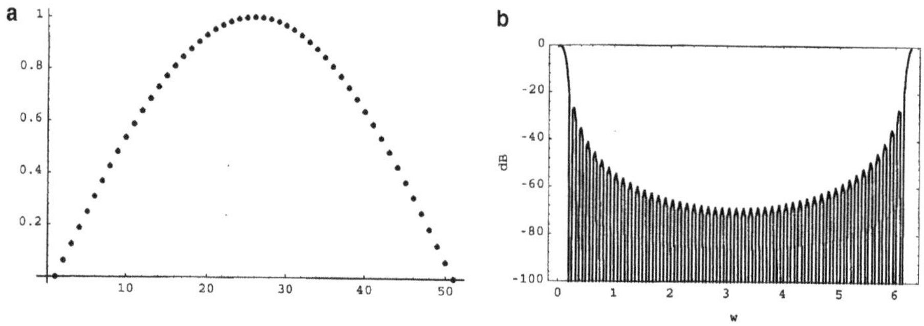

FIGURE 7.3 a) Cosine window with $\alpha = 1$. b) Amplitude spectrum of cosine window with $\alpha = 1$.

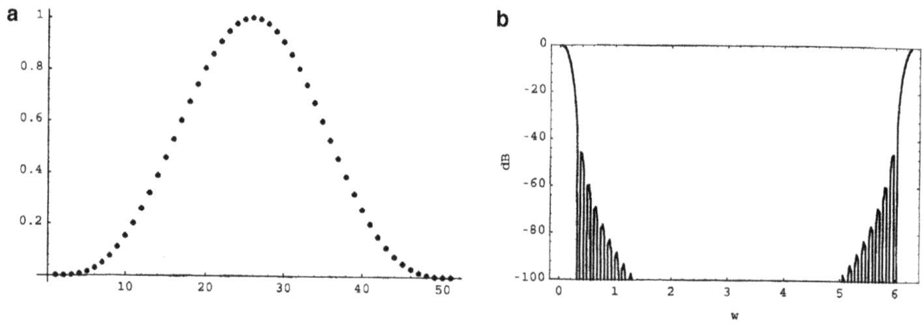

FIGURE 7.4 a) Cosine window with $\alpha = 3$. b) Amplitude spectrum of cosine window with $\alpha = 3$.

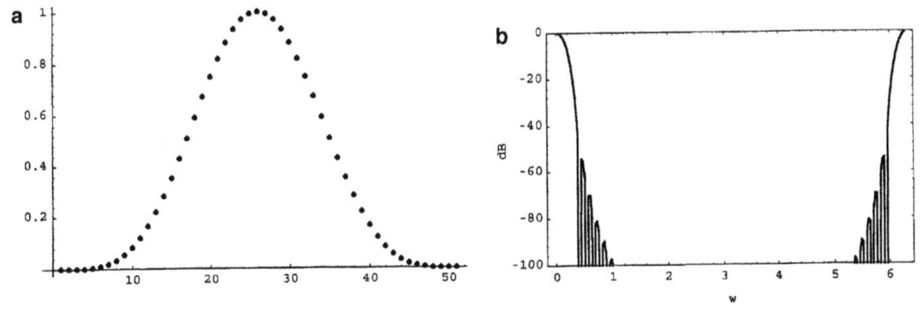

FIGURE 7.5 a) Cosine window with $\alpha = 4$. b) Amplitude spectrum of cosine window with $\alpha = 4$.

7.3.5 Hann Window

$$w(n) = \cos^2\left(\frac{n}{N}\pi\right) = \frac{1}{2}\left[1 + \cos\left(\frac{2n}{N}\pi\right)\right] \qquad n = -\tfrac{N}{2}, \cdots, -1, 0, 1, \cdots, \tfrac{N}{2}$$

$$w(n) = \sin^2\left[\left(\frac{n}{N}\right)\pi\right] = \frac{1}{2}\left[1 - \cos\left[\left(\frac{2n}{N}\right)\pi\right]\right] \qquad n = 0, 1, \cdots, N-1$$

DFT of the window is

$$W(\omega) = \frac{1}{2}D(\omega) + \frac{1}{4}\left[D\left(\omega - \frac{2\pi}{N}\right) + D\left(\omega + \frac{2\pi}{N}\right)\right]$$

$$D(\omega) = e^{j\frac{\omega}{2}} \frac{\sin\left(\frac{N}{2}\omega\right)}{\sin\frac{\omega}{2}} = \text{Dirichlet Kernel} \qquad -\pi \le \omega \le \pi$$

See Figure 7.6 for the Hann window.

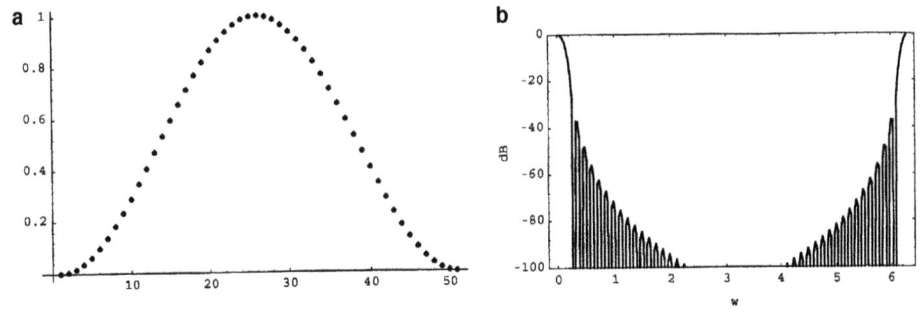

FIGURE 7.6 a) Hann window. b) Amplitude spectrum of Hann window.

7.3.6 Hamming Window

$$w(n) = \alpha + (1 - \alpha)\cos\frac{2\pi}{N}n \qquad n = -\tfrac{N}{2}, \cdots, -1, 0, 1, \cdots, \tfrac{N}{2}$$

$$W(\omega) = \alpha D(\omega) + \frac{1}{2}(1-\alpha)\left[D\left(\omega - \frac{2\pi}{N}\right) + D\left(\omega + \frac{2\pi}{N}\right)\right] \qquad -\pi \le \omega \le \pi$$

D = Dirichlet Kernel (see 7.3.5)

$$w(n) = 0.54 + 0.46\cos\frac{2\pi}{N}n \qquad n = -\tfrac{N}{2},\cdots,-1,0,1,\cdots,\tfrac{N}{2}$$

$$w(n) = 0.54 - 0.46\cos\frac{2\pi}{N}n \qquad n = 0,1,\cdots,N-1$$

Figure 7.7 depicts the Hamming window and its amplitude spectrum.

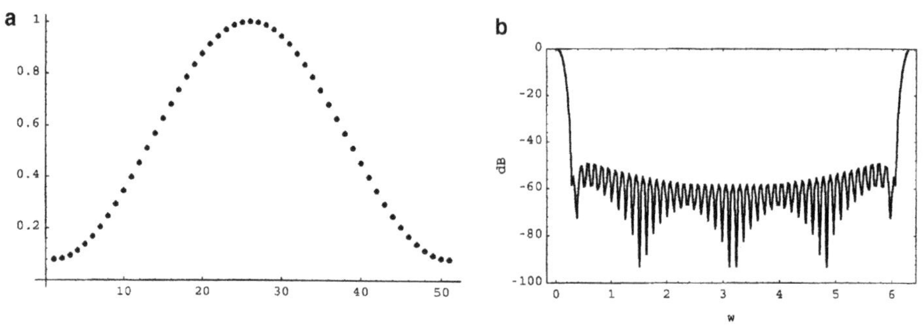

FIGURE 7.7 a) Hamming window. b) Amplitude spectrum of Hamming window.

7.3.7 Short Cosine Series Window

$$w(n) = \sum_{k=0}^{K/2} a(k)\cos\left[\left(\frac{2\pi}{N}\right)kn\right] \qquad n = -\tfrac{N}{2},\cdots,-1,0,1,\cdots,\tfrac{N}{2}$$

$$\sum_{k=0}^{K/2} a(k) = 1 \qquad \text{constraint}$$

$$W(\omega) = a(0)D(\omega) + \sum_{k=0}^{K/2}\frac{a(k)}{2}\left[D\left(\omega - k\frac{2\pi}{N}\right) + D\left(\omega + k\frac{2\pi}{N}\right)\right] \qquad -\pi \le \omega \le \pi$$

7.3.8 Blackman Window

$$a(0) = 0.42659071 \cong 0.42, \quad a(1) = 0.49656062 \cong 0.50, \quad a(2) = 0.07684867 \cong 0.08$$

$$w(n) = 0.42 + 0.5\cos\left(\frac{2\pi}{N}n\right) + 0.08\cos\left(\frac{2\pi}{N}2n\right) \qquad n = -\tfrac{N}{2},\cdots,-1,0,1,\cdots,\tfrac{N}{2}$$

$$w(n) = 0.42 + 0.5\cos\left(\frac{2\pi}{N}(n-25)\right) + 0.08\cos\left(\frac{2\pi}{N}2(n-25)\right) \qquad n = 0,1,\cdots,N-1$$

Figure 7.8 shows the characteristics of the Blackman window.

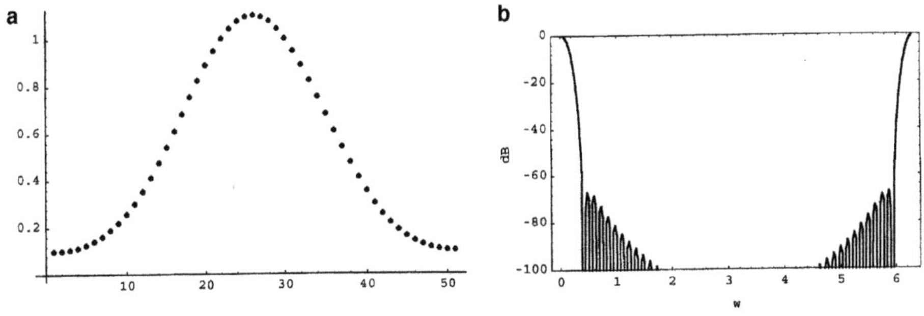

FIGURE 7.8 a) Blackman window. b) Amplitude spectrum of Blackman window.

7.3.9 Harris-Nutall Window

Table 7.2 gives the coefficients for short cosine series windows.

TABLE 7.2 Coefficients of Three- and Four-Term Harris-Nutall Windows

	3-Term (–61 dB)	3-Term (–67 dB)	4-Term (–74 dB)	4-Term (–94 dB)
a(0)	0.44959	0.42323	0.40217	0.35875
a(1)	0.49364	0.49755	0.49703	0.48829
a(2)	0.05677	0.07922	0.09392	0.14128
a(3)	0	0	0.00183	0.01168

Figures 7.9 and 7.10 show the Harris-Nutall window characteristics.

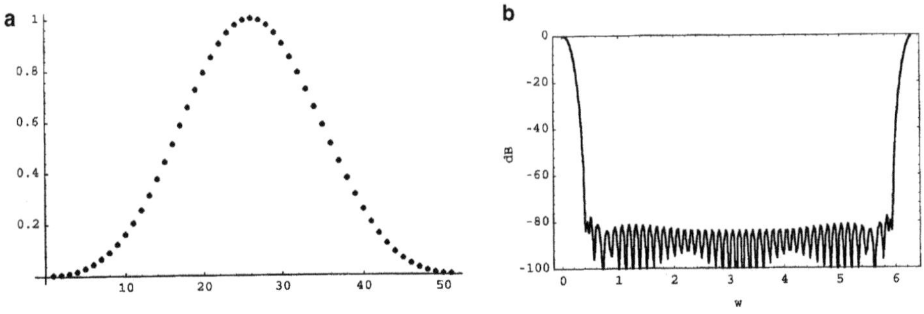

FIGURE 7.9 a) Harris-Nutall window (3-term). b) Amplitude spectrum of Harris-Nutall window (3-term).

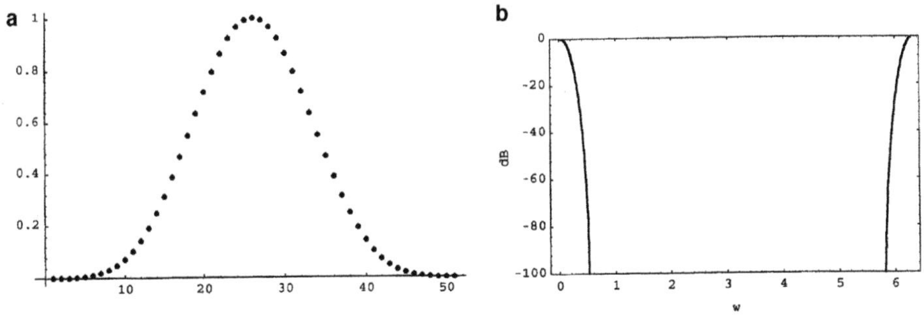

FIGURE 7.10 a) Harris-Nutall window (4-term). b) Amplitude spectrum of Harris-Nutall window (4-term).

7.3.10 Sampled Kaiser-Bessel Window

$$\text{Kaiser-Bessel spectrum window} = W(\omega) = \frac{\sinh\sqrt{\pi^2\alpha^2 - (\omega N/2)^2}}{\sqrt{\pi^2\alpha^2 - (\omega N/2)^2}} \qquad 0 \le \alpha \le 4$$

$$H_1(m) = \frac{\sinh(\pi\sqrt{\alpha^2 - m^2})}{\pi\sqrt{\alpha^2 - m^2}} \qquad \omega = m(2\pi / N)$$

$$c = H_1(0) + 2H_1(1) + 2H_1(2) + [2H_1(3)]$$

$$a(0) = \frac{H_1(0)}{c}, \qquad a(m) = \frac{2H_1(0)}{c}, \qquad m = 1,2,3$$

$$a(0) = 0.40243, \quad a(1) = 0.49804, \quad a(2) = 0.09831, \quad a(3) = 0.00122$$

7.3.11 Parabolic (Riesz, Bochner, Parzen) Window

$$w(n) = 1.0 - \left(\frac{n}{N/2}\right)^2 \qquad 0 \le |n| \le \tfrac{N}{2}$$

$$w(n) = 1.0 - \left(\frac{n - \frac{N}{2}}{N/2}\right)^2 \qquad n = 0,1,2,\cdots,N-1$$

Figure 7.11 shows the parabolic window and its spectrum characteristics.

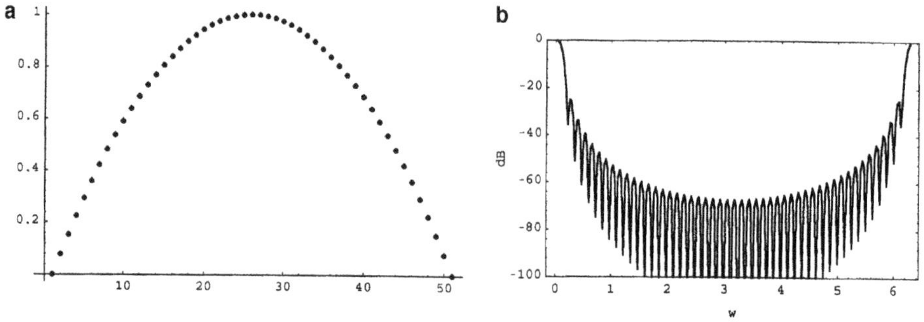

FIGURE 7.11 a) Parabolic window. b) Amplitude spectrum of Parabolic window.

7.3.12 Riemann Window

$$w(n) = \frac{\sin\frac{2\pi n}{N}}{\frac{2\pi n}{N}} \qquad 0 \le |n| \le \tfrac{N}{2}$$

$$w(n) = \frac{\sin\left(\frac{2\pi\left(n-\frac{N}{2}\right)}{N}\right)}{\frac{2\pi\left(n-\frac{N}{2}\right)}{N}} \qquad n = 0,1,2,\cdots,N-1$$

Figure 7.12 shows the window's characteristics.

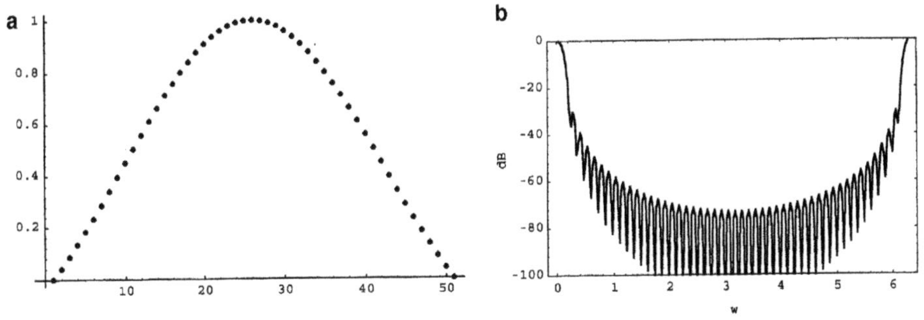

FIGURE 7.12 a) Riemann window. b) Amplitude spectrum of Riemann window.

7.3.13 de la Vallé-Poussin (Jackson, Parzen) Window

$$w(n) = \begin{cases} 1-6\left[\frac{n}{N/2}\right]^2\left[1-\frac{|n|}{N/2}\right] & 0 \le |n| \le \frac{N}{4} \\ 2\left[1-\frac{|n|}{N/2}\right]^3 & \frac{N}{4} \le |n| \le \frac{N}{2} \end{cases}$$

Figure 7.13 shows the window and its frequency response.

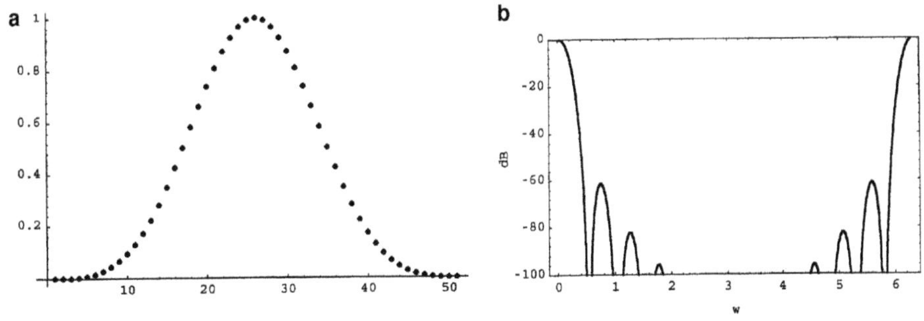

FIGURE 7.13 a) de la Vallé-Poussin window. b) Amplitude spectrum of de la Vallé-Poussin window.

7.3.14 Cosine Taper (Tukey) Window

The Tukey window is equal to one over $(1 - \alpha/2)N$ points, with a cosine taper from one to zero for the remaining points $(\alpha/2)N$.

$$w(n) = \begin{cases} 1 & 0 \le |n| \le \alpha \frac{N}{2} \\ 0.5\left(1 + \cos\left[\pi \dfrac{n - \alpha(N/2)}{(1-\alpha)(N/2)}\right]\right) & \alpha \frac{N}{2} < |n| \le \frac{N}{2} \end{cases}$$

Figures 7.14 and 7.15 show the window and its frequency responses for $\alpha = 8/25$ and $\alpha = 12/25$.

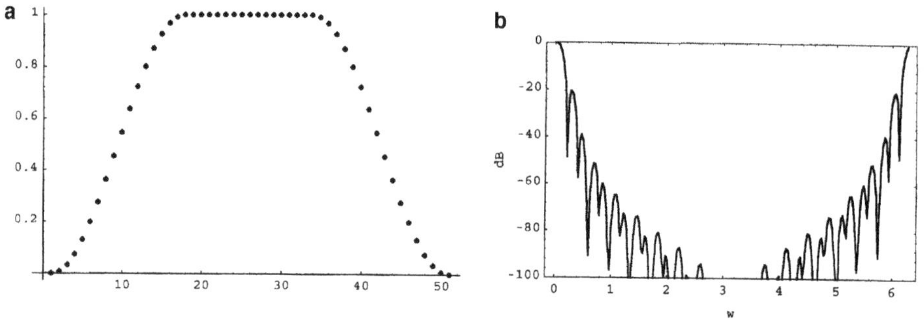

FIGURE 7.14 a) Tukey window with $\alpha = 8/25$. b) Amplitude spectrum of Tukey window with $\alpha = 8/25$.

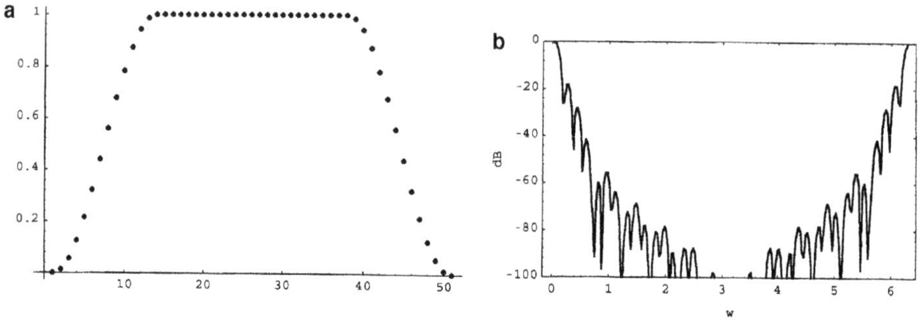

FIGURE 7.15 a) Tukey window with $\alpha = 12/25$. b) Amplitude spectrum of Tukey window with $\alpha = 12/25$.

7.3.15 Bohman Window

$$w(n) = \left(1 - \frac{|n|}{N/2}\right)\cos\left(\pi \frac{|n|}{N/2}\right) + \frac{1}{\pi}\sin\left(\pi \frac{|n|}{N/2}\right), \qquad 0 \le |n| \le \tfrac{N}{2}$$

Figure 7.16 shows the window and its spectrum.

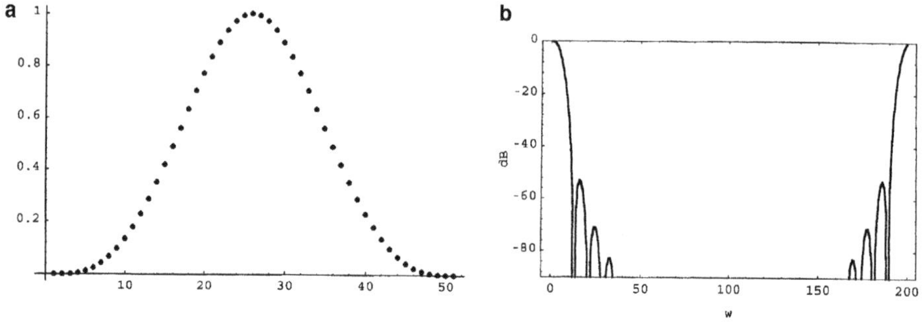

FIGURE 7.16 a) Bohman window. b) Amplitude spectrum of Bohman window.

7.3.16 Poisson Window

$$w(n) = \exp\left(-\alpha \frac{|n|}{N/2}\right), \qquad 0 \le |n| \le \tfrac{N}{2}$$

Figures 7.17 through 7.19 show the window and its spectrum with $\alpha = 1.5$, $\alpha = 3$, and $\alpha = 4$.

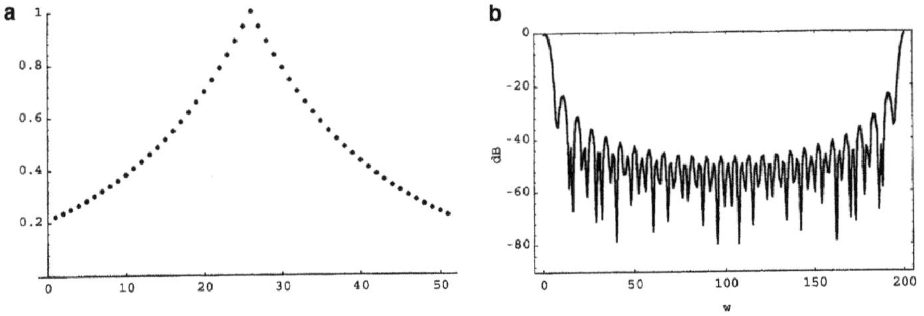

FIGURE 7.17 a) Poisson window with $\alpha = 1.5$. b) Amplitude spectrum of Poisson window with $\alpha = 1.5$.

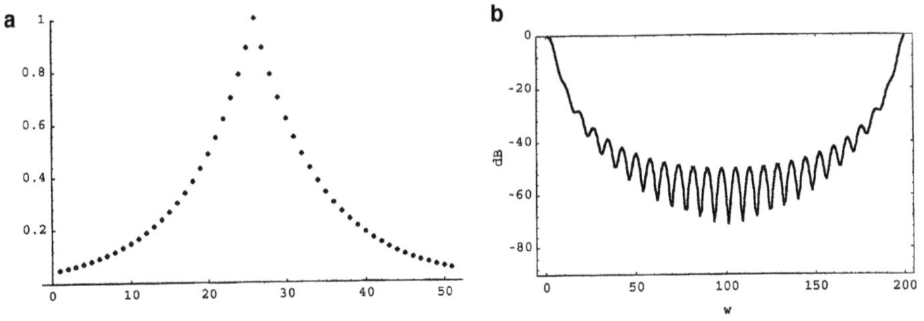

FIGURE 7.18 a) Poisson window with $\alpha = 3.0$. b) Amplitude spectrum of Poisson window with $\alpha = 3.0$.

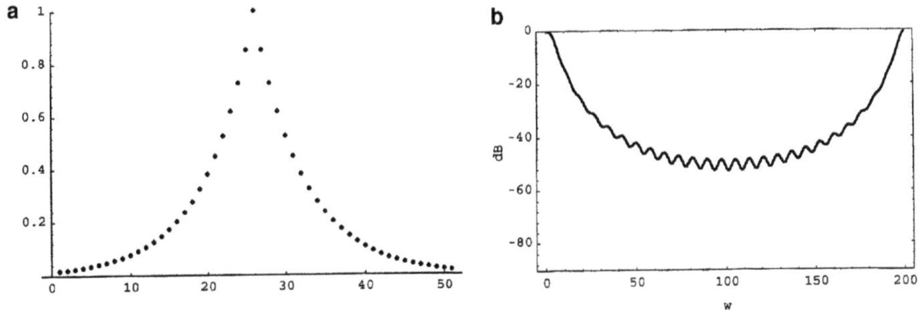

FIGURE 7.19 a) Poisson window with $\alpha = 4.0$. b) Amplitude spectrum of Poisson window with $\alpha = 4.0$.

7.3.17 Hann-Poisson Window

$$w(n) = 0.5\left[1 + \cos\left(\pi \frac{n}{N/2}\right)\right]\exp\left(-\alpha \frac{|n|}{N/2}\right), \qquad 0 \le |n| \le \tfrac{N}{2}$$

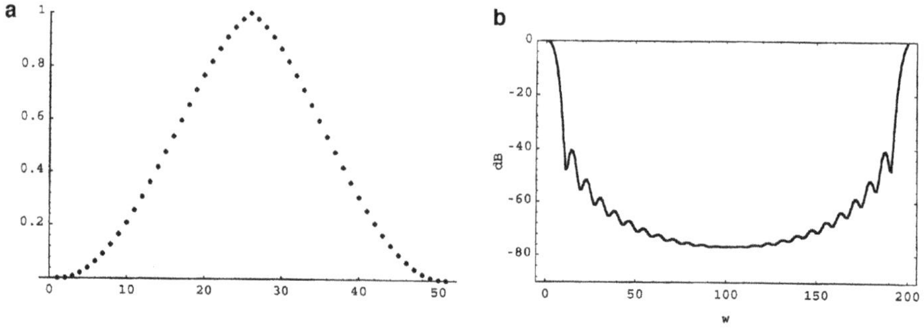

FIGURE 7.20 a) Hann-Poisson window with $\alpha = 0.5$. b) Amplitude spectrum of Hann-Poisson window with $\alpha = 0.5$.

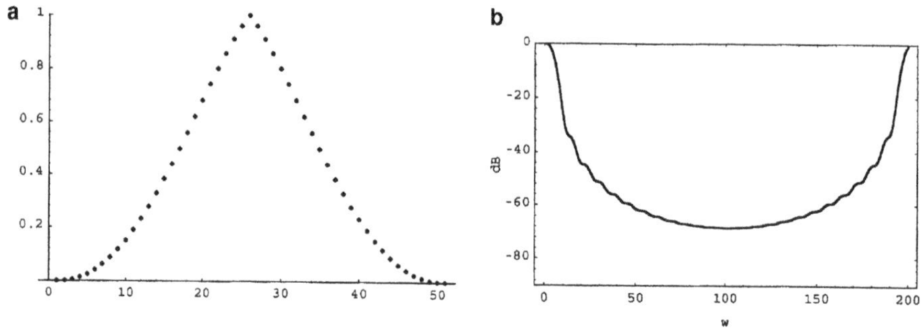

FIGURE 7.21 a) Hann-Poisson window with $\alpha = 1.0$. b) Amplitude spectrum of Hann-Poisson window with $\alpha = 1.0$.

Figures 7.20 and 7.21 show the window and its spectrum with $\alpha = 0.5$ and $\alpha = 1.0$, respectively.

7.3.18 Cauchy (Abel, Poisson) Window

$$w(n) = \frac{1}{1 + \left(\alpha \dfrac{n}{N/2}\right)^2}, \qquad 0 \le |n| \le \tfrac{N}{2}$$

Figures 7.22 through 7.24 show the window and its spectrum with $\alpha = 3.0$, $\alpha = 4.0$, and $\alpha = 6.0$, respectively.

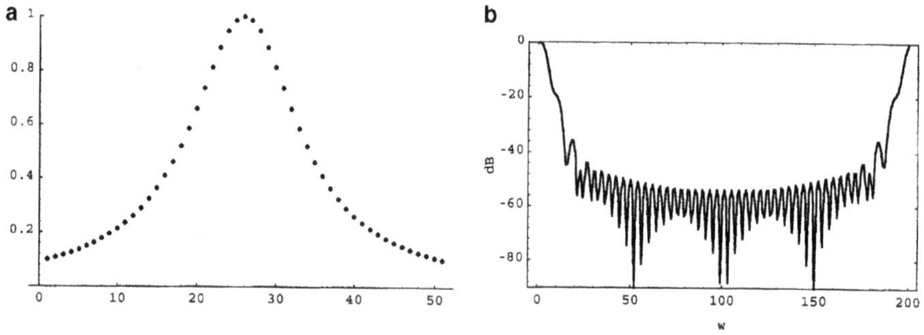

FIGURE 7.22 a) Cauchy window with $\alpha = 3.0$. b) Amplitude spectrum of Cauchy window with $\alpha = 3.0$.

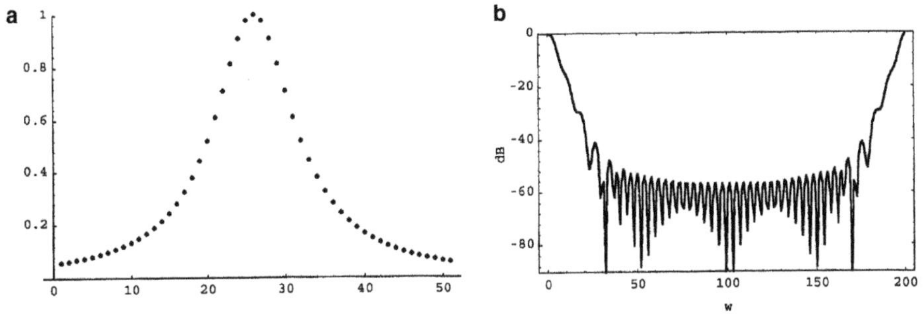

FIGURE 7.23 a) Cauchy window with α = 4.0. b) Amplitude spectrum of Cauchy window with α = 4.0.

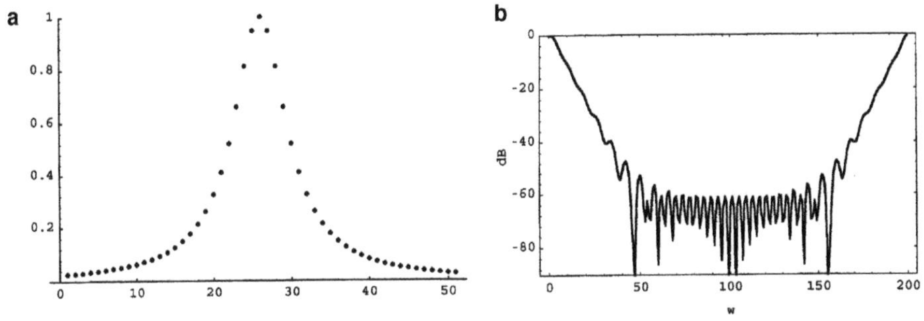

FIGURE 7.24 a) Cauchy window with α = 6.0. b) Amplitude spectrum of Cauchy window with α = 6.0.

7.3.19 Gaussian (Weierstrass) Window

$$w(n) = \exp\left[-\tfrac{1}{2}\left(\alpha \frac{n}{N/2}\right)^2\right], \qquad 0 \le |n| \le \tfrac{N}{2}$$

Figures 7.25 and 7.26 show the window and its spectrum for α = 2.5 and α = 3.5.

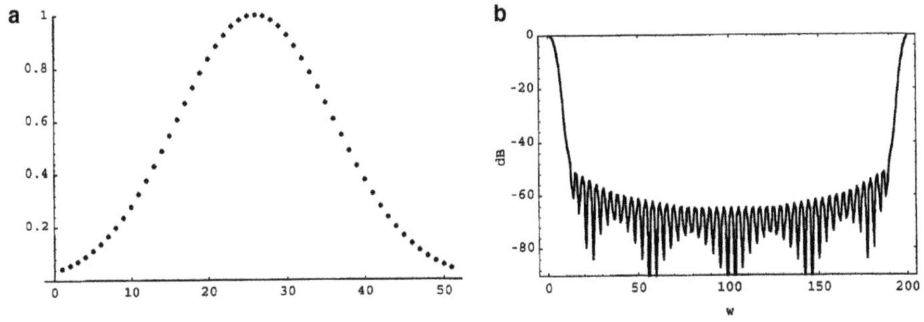

FIGURE 7.25 a) Gaussian window with α = 2.5. b) Amplitude spectrum of Gaussian window with α = 2.5.

7.3.20 Dolph-Chebyshev Window

$$W(k) = \frac{\cosh(N\cosh^{-1}(\beta\cosh(\pi k/N)))}{\cosh(N\cosh^{-1}(\beta))}$$

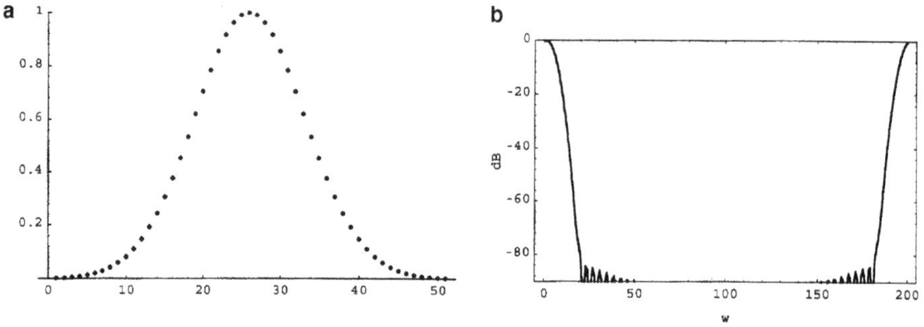

FIGURE 7.26 a) Gaussian window with $\alpha = 3.5$. b) Amplitude spectrum of Gaussian window with $\alpha = 3.5$.

where

$$\cosh^{-1}(X) = \ln\left(X + \sqrt{X^2 - 1.0}\right), \qquad |X| > 1.0$$

$$W(k) = \frac{\cos(N\cos^{-1}(\beta\cos(\pi k / N)))}{\cosh(N\cosh^{-1}(\beta))}$$

where

$$\cos^{-1}(X) = \frac{\pi}{2} - \tan^{-1}\left(\frac{X}{\sqrt{X^2 - 1.0}}\right), \qquad |X| \le 1.0$$

where β satisfies

$$\beta = \cosh\left(\frac{1}{N}\cosh^{-1} 10^{\alpha}\right)$$

and

$$w(n) = \sum_{k=0}^{N-1} W(k)\exp\left(j\frac{2\pi}{N}nk\right)$$

7.3.21 Kaiser-Bessel Window

$$w(n) = \frac{I_o\left[\pi\alpha\sqrt{1.0 - \left(\frac{n}{N/2}\right)^2}\right]}{I_o[\pi\alpha]} \qquad 0 \le |n| \le \tfrac{N}{2}$$

$$I_o(x) = \sum_{k=0}^{\infty}\left[\frac{\left(\frac{x}{2}\right)^k}{k!}\right]^2 = \text{zero-order modified Bessel function}$$

Figures 7.27 and 7.28 show the window and its spectrum for $\alpha = 2$ and $\alpha = 3$.

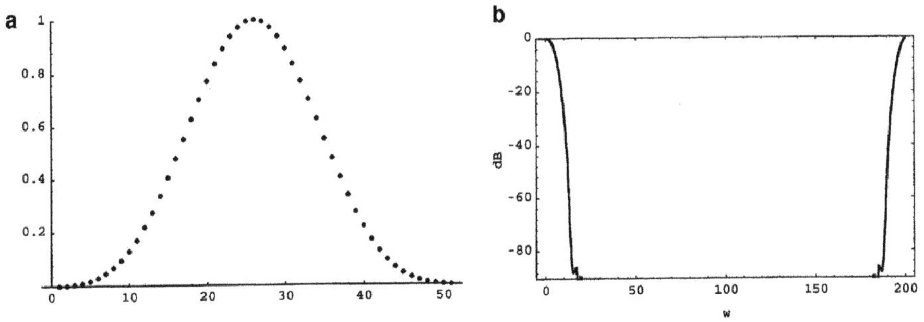

FIGURE 7.27 a) Kaiser-Bessel window with $\alpha = 3.0$. b) Amplitude spectrum of Kaiser-Bessel window with $\alpha = 3.0$.

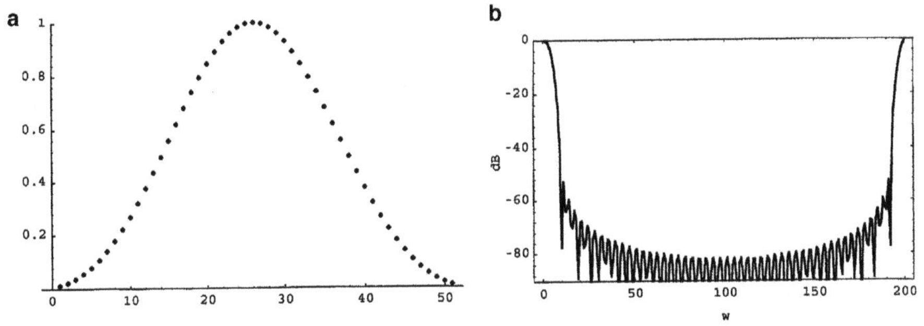

FIGURE 7.28 a) Kaiser-Bessel window with $\alpha = 2.0$. b) Amplitude spectrum of Kaiser-Bessel window with $\alpha = 2.0$.

7.3.22 Barcilon-Themes Window

$$W(k) = \frac{A\cos[y(k)] + B[y(k)/C]\sin[y(k)]}{(C + AB)[(y(k)/C)^2 + 1.0]}$$

where

$$A = \sinh C = \sqrt{10^{2\alpha} - 1.0}$$

$$B = \cosh C = 10^{\alpha}$$

$$C = \cosh^{-1}(10^{\alpha})$$

$$\beta = \cosh(C/N)$$

$$y(k) = N\cos^{-1}\left[\beta\cos\left(\frac{\pi k}{N}\right)\right]$$

7.3.23 Highest Sidelobe Level versus Worst-Case Processing Loss

Figure 7.29 shows the highest sidelobe level versus worst-case processing loss. Shaped DFT filters in the lower left tend to perform well.

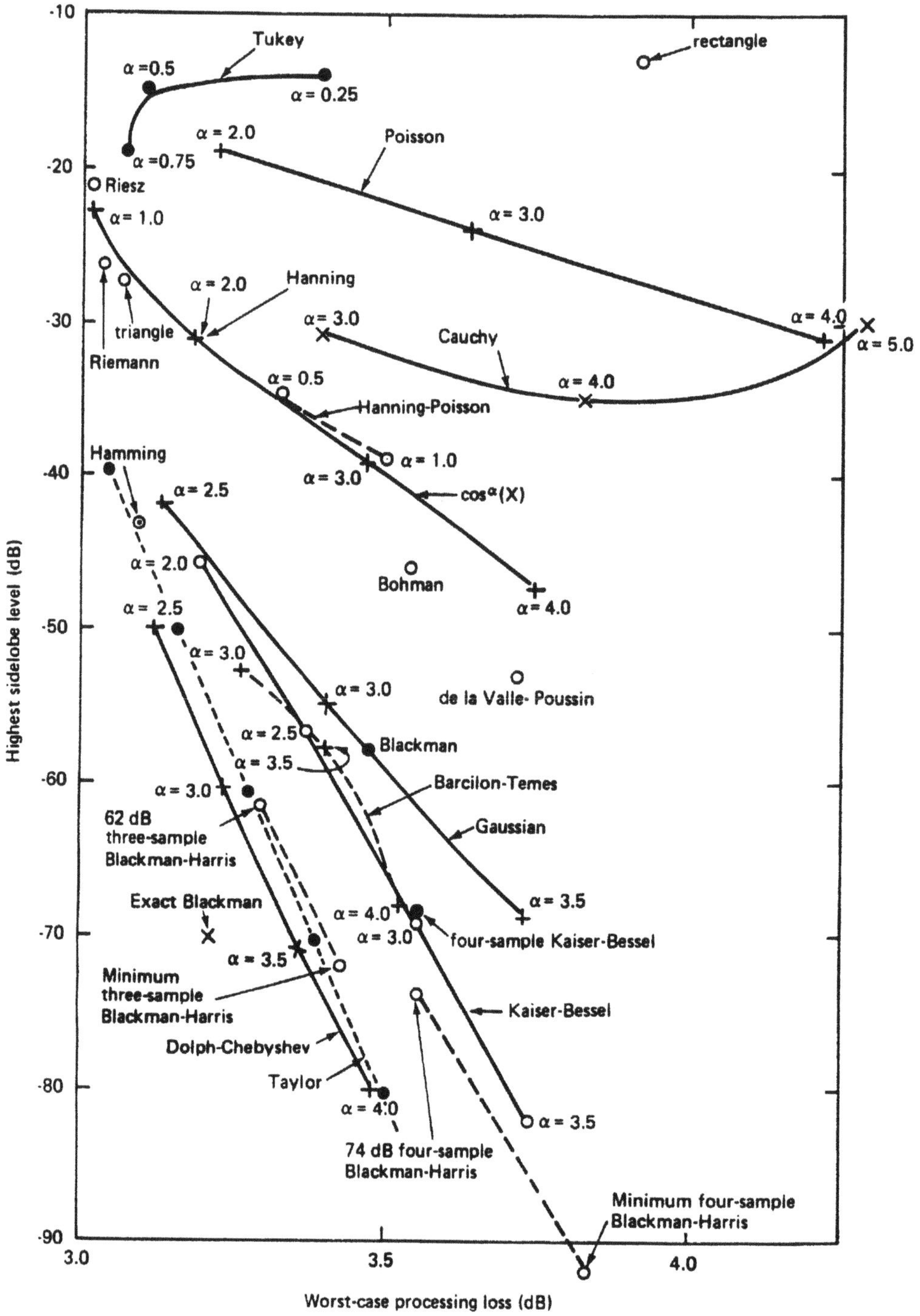

FIGURE 7.29 Highest sidelobe level versus worst-case processing loss. Shaped DFT filters in the lower left tend to perform well.

References

Harris, F. J., On the use of windows for harmonic analysis with the discrete fourier transforms, *Proc. IEEE*, 66, 55-83, January 1978.

8

Two-Dimensional Z-Transform

8.1 The Z-Transform

8.1.1 Definition

$$X(z_1,z_2)=\sum_{n_1=-\infty}^{\infty}\sum_{n_2=-\infty}^{\infty}x(n_1,n_2)z_1^{-n_1}z_2^{-n_2}$$

8.1.2 Relationship to Discrete-Time Fourier Transform

$$X(z_1,z_2)\Big|_{z_1=e^{j\omega_1},z_2=e^{j\omega_2}}=\sum_{n_1=-\infty}^{\infty}\sum_{n_2=-\infty}^{\infty}x(n_1,n_2)e^{-j\omega_1 n_1}e^{-j\omega_2 n_2}=X(\omega_1,\omega_2)$$

evaluated at $z_1=e^{j\omega_1}$ and $z_2=e^{j\omega_2}$.

8.1.3 Region of Convergence (ROC)

Points (z_1,z_2) for which

$$\sum_{n_1}\sum_{n_2}|x(n_1,n_2)||z_1|^{-n_1}|z_2|^{-n_2}<\infty$$

are located in the ROC. This implies that

$$|X(z_1,z_2)|<\infty$$

If (z_{01},z_{02}) point lies in the ROC, then all points (z_1,z_2) that satisfy

$$|z_1| \geq |z_{01}|, \quad |z_2| \geq |z_{02}|$$

also lie in the ROC.

For the first quadrant sequences, the boundary of the ROC must have nonpositive slope.

8.1.4 Sequences with Finite Support

$$X(z_1,z_2) = \sum_{n_1=N_1}^{M_1} \sum_{n_2=N_2}^{M_2} x(n_1,n_2) z_1^{-n_1} z_2^{-n_2}$$

The Z-transform converges for all finite values of z_1 and z_2, except possibly for $z_1 = 0$ and $z_2 = 0$.

8.1.5 Sequences with support on a wedge

If a sequence has support shown in Figure 8.1a, then its ROC is shown in Figure 8.1b (see Dudgeon and Mersereau, 1984). If the point (z_{01},z_{02}) belongs to the ROC, then

$$\ell n|z_1| \geq \ell n|z_{01}| \quad \text{and} \quad \ell n|z_2| \geq L\ell n|z_1| + \{\ell n|z_{02}| - L\ell n|z_{01}|\}$$

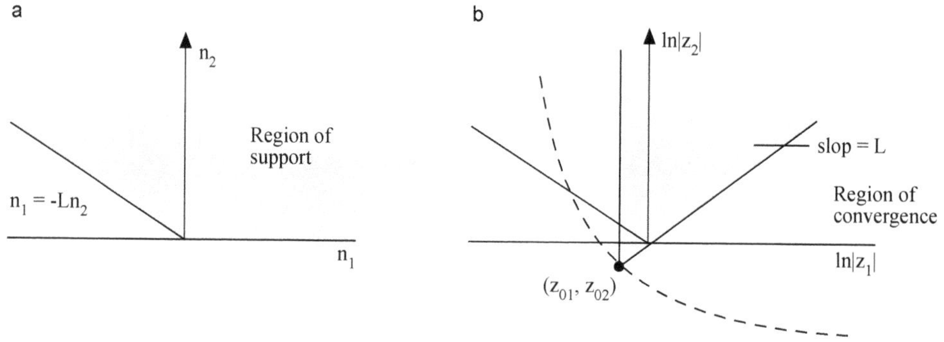

FIGURE 8.1

8.1.6 Sequences with Support on a Half-Plane

The boundary of the region of convergence is constrained to be a single-valued function of $|z_1|$ (or $\ell n|z_1|$) (see Dudgeon and Mersereau, 1984).

8.1.7 Sequences with Support Everywhere

- $x(n_1,n_2) = e^{-n_1^2-n_2^2}$ converges for all values of (z_1,z_2)
- $x(n_1,n_2) = 2^{|n_1|}2^{|n_2|}$ will not converge for any value of (z_1,z_2)
- Often the z-transform of a sequence with support everywhere will converge in a region of finite area.
- A sequence with support everywhere can be split into four quadrant sequences, e. g.,

$$x(n_1,n_2) = x_1(n_1,n_2) + x_2(n_1,n_2) + x_3(n_1,n_2) + x_4(n_1,n_2)$$

where

$$x_1(n_1,n_2)=\begin{cases} x(n_1,n_2) & \text{for } n_1>0,\ n_2>0 \\ \frac{1}{2}x(n_1,n_2) & \text{for } n_1=0,\ n_2>0 \text{ or } n_1>0,\ n_2=0 \\ \frac{1}{4}x(n_1,n_2) & \text{for } n_1=n_2=0 \\ 0 & \text{for } n_1<0 \text{ or } n_2<0 \end{cases}$$

Similarly are defined $x_2(n_1,n_2)$, $x_3(n_1,n_2)$, and $x_4(n_1,n_2)$, . The region of convergence of $x(n_1,n_2)$ is the *intersection* of the region of convergence of the four z-transforms of the four quadrants.

8.1.8 ROCs of Different Supports

Figure 8.2 shows the support of the function and its ROC (Lim, 1990).

8.2 Properties of the Z-Transform

8.2.1 Properties of the Z-Transform

TABLE 8.1 Properties of the 2-D Z-Transform

$$x(n_1,n_2)\longleftrightarrow X(z_1,z_2),\quad ROC: R_x$$

$$y(n_1,n_2)\longleftrightarrow Y(z_1,z_2),\quad ROC: R_y$$

1. Linearity

$$ax(n_1,n_2)+by(n_1,n_2)\longleftrightarrow aX(z_1,z_2)+bY(z_1,z_2),\quad ROC:\ \text{at least } R_x\cap R_y$$

2. Convolution

$$x(n_1,n_2)*y(n_1,n_2)=\sum_{k_1}\sum_{k_2}x(n_1-k_1,n_2-k_2)y(k_1,k_2)\longleftrightarrow X(z_1,z_2)Y(z_1,z_2),\quad ROC:\ R_x\cap R_y$$

3. Separable Signals

$$x(n_1,n_2)=x_1(n_1)x_2(n_2)\longleftrightarrow X(z_1,z_2)=X_1(z_1)X_2(z_2),\quad ROC:\ |z_1|\in ROC\,X_1(z_1) \text{ and } |z_2|\in ROC\,X_2(z_2)$$

4. Shift Property

$$x(n_1\pm m_1,\ n_2\pm m_2)\longleftrightarrow X(z_1,z_2)=z_1^{\pm m_1}z_2^{\pm m_2}X(z_1,z_2),$$

$$ROC: R_x \text{ with possible exceptions } |z_1|=0,\infty \text{ and } |z_2|=0,\infty$$

5. Differentiation Property

$$-n_1x(n_1,n_2)\longleftrightarrow z_1\frac{\partial X(z_1,z_2)}{\partial z_1},\quad ROC: R_x$$

$$-n_2x(n_1,n_2)\longleftrightarrow z_2\frac{\partial X(z_1,z_2)}{\partial z_2},\quad ROC: R_x$$

$$n_1n_2x(n_1,n_2)\longleftrightarrow z_1z_2\frac{\partial^2 X(z_1,z_2)}{\partial z_1\partial z_2},\quad ROC: R_x$$

6. Modulation Property

$$w(n_1,n_2)=a^{n_1}b^{n_2}x(n_1,n_2)\longleftrightarrow X(a^{-1}z_1,b^{-1}z_2),\quad ROC: W(z_1,z_2) \text{ has the same as } X(z_1,z_2),$$

but scaled by $|a|$ in z_1 variable and by $|b|$ in the z_2 variable.

7. Conjugate Properties

$$x(n_1,n_2)=\text{complex function}\longleftrightarrow X(z_1,z_2)$$

TABLE 8.1 Properties of the 2-D Z-Transform (continued)

$$x^*(n_1,n_2) \longleftrightarrow X^*(z_1^*,z_2^*)$$

$$\operatorname{Re}\{x(n_1,n_2)\} \longleftrightarrow \tfrac{1}{2}[X(z_1,z_2)+X^*(z_1^*,z_2^*)]$$

$$\operatorname{Im}\{x(n_1,n_2)\} \longleftrightarrow \tfrac{1}{2j}[X(z_1,z_2)-X^*(z_1^*,z_2^*)]$$

ROC : same as $X(z_1,z_2)$

8. Reflection Properties

$$x(n_1,n_2) \longleftrightarrow X(z_1,z_2)$$

$$x(-n_1,n_2) \longleftrightarrow X(z_1^{-1},z_2)$$

$$x(n_1,-n_2) \longleftrightarrow X(z_1,z_2^{-1})$$

$$x(-n_1,-n_2) \longleftrightarrow X(z_1^{-1},z_2^{-1}) \qquad ROC: |z_1^{-1}|,\ |z_2^{-1}| \text{ in } R_x$$

9. Multiplication Property

$$x(n_1,n_2)y(n_1,n_2) \longleftrightarrow \left(\frac{1}{2\pi j}\right)^2 \oint_{C_2}\oint_{C_1} X\left(\frac{z_i}{\upsilon_1},\frac{z_2}{\upsilon_2}\right) Y(z_1,z_2)\,\frac{d\upsilon_1}{\upsilon_1}\,\frac{d\upsilon_2}{\upsilon_2}$$

10. Parseval's Theorem

$$\sum_{n_1=-\infty}^{\infty}\sum_{n_2=-\infty}^{\infty} x(n_1,n_2)y^*(n_1,n_2) \longleftrightarrow \left(\frac{1}{2\pi j}\right)^2 \oint_{C_2}\oint_{C_1} X(z_1,z_2)Y^*\left(\frac{1}{z_1^*},\frac{1}{z_2^*}\right)\frac{dz_1}{z_1}\,\frac{dz_2}{z_2}$$

Contours must: closed, counter - clockwise, encircle the origin, lie totally within ROC.

11. Initial Value Theorems

$$x(n_1,n_2)=0 \qquad n_1<0,\ n_2<0$$

$$\lim_{z_1\to\infty} X(z_1,z_2) = \sum_{n_2} x(0,n_2)z_2^{-n_2}$$

$$\lim_{z_2\to\infty} X(z_1,z_2) = \sum_{n_1} x(n_1,0)z_1^{-n_1}$$

$$\lim_{\substack{z_1\to\infty\\ z_2\to\infty}} X(z_1,z_2) = x(0,0)$$

12. Linear Mapping of Variables

$$x(n_1,n_2)=y(m_1,m_2) \qquad n_1 = Im_1 + Jm_2\ ,\quad n_2 = Km_1 + Lm_2$$

$I,\ J,\ K,\ L$ are integers

$IL - KJ \neq 0$

$$X(z_1,z_2) = Y(z_1^I z_2^K, z_1^J z_2^L) \qquad ROC: (|z_1^I z_2^K|\ ,\ |z_1^J z_2^L|) \text{ in } R_x$$

8.3. Inverse Z-Transform

8.3.1 Inverse Z-Transform

$$x(n_1,n_2) = \left(\frac{1}{2\pi j}\right)^2 \oint_{C_2}\oint_{C_1} X(z_1,z_2)z_1^{n_1-1}\,z_2^{n_2-1}\,dz_1\,dz_2$$

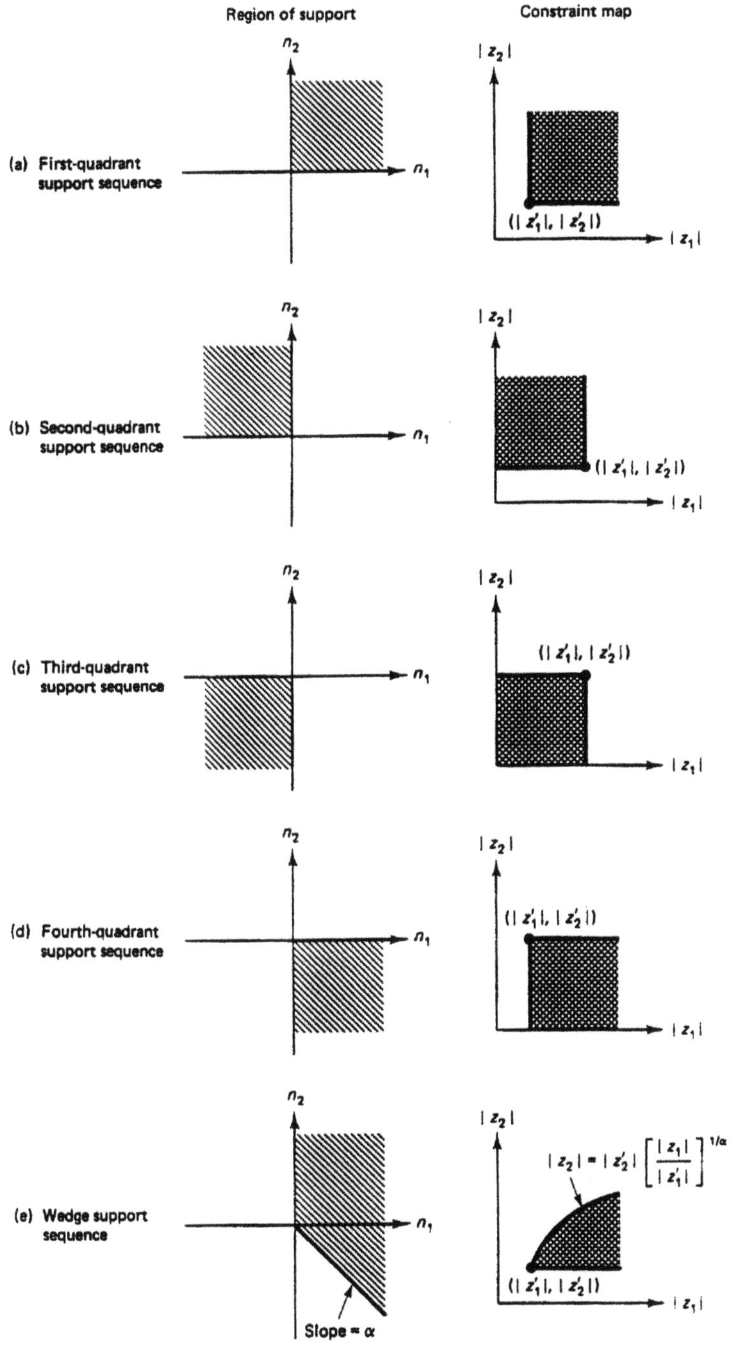

FIGURE 8.2

C_1, C_2 both in the ROC

C_1 counter-clockwise encircling the origin in the z_1 plane with z_2 fixed

C_2 counter-clockwise encircling the origin in the z_2 plane with z_1 fixed

8.4 System Function

8.4.1 System Function

$$\sum_{k_1}\sum_{k_2} a(k_1,k_2)y(n_1-k_1,n_2-k_2)=\sum_{k_1}\sum_{k_2} b(k_1,k_2)x(n_1-k_1,n_2-k_2)$$

$$a(k_1,k_2),\;\; b(k_1,k_2) \equiv \text{finite – extent sequences}$$

$$Y(z_1,z_2)\sum_{k_1}\sum_{k_2} a(k_1,k_2)z_1^{-k_1}\,z_2^{-k_2} = X(z_1,z_2)\sum_{k_1}\sum_{k_2} b(k_1,k_2)z_1^{-k_1}\,z_2^{-k_2}$$

or

$$H(z_1,z_2)=\frac{Y(z_1,z_2)}{X(z_1,z_2)}=\frac{\sum_{k_1}\sum_{k_2} b(k_1,k_2)z_1^{-k_1}\,z_2^{-k_2}}{\sum_{k_1}\sum_{k_2} a(k_1,k_2)z_1^{-k_1}\,z_2^{-k_2}}=\frac{B(z_1,z_2)}{A(z_1,z_2)}$$

8.5 Stability Theorems

8.5.1 Theorem 8.5.1.1 (Shanks, 1972)

Let $H(z_1,z_2) = 1/A(z_1,z_2)$ be a first quadrant recursive filter. This filter is stable if, and only if, $A(z_1,z_2) \neq 0$ for every point (z_1,z_2) such that $|z_1| \geq 1$ or $|z_2| \geq 1$.

8.5.2 Theorem 8.5.1.2 (Shanks, 1972)

Let $H(z_1,z_2) = 1/A(z_1,z_2)$ be a first quadrant recursive filter. Then $H(z_1,z_2)$ is stable if, and only if, the following conditions are true:

a) $A(z_1,z_2) \neq 0,\;\; |z_1| \geq 1,\;\; |z_2| = 1$
b) $A(z_1,z_2) \neq 0,\;\; |z_1| = 1,\;\; |z_2| \geq 1$

8.5.3 Theorem 8.5.1.3 (Huang, 1972)

Let $H(z_1,z_2) = 1/A(z_1,z_2)$ be a first-quadrant recursive filter. The filter is stable if, and only if, $A(z_1,z_2)$ satisfies the following two conditions:

a) $A(z_1,z_2) \neq 0,\;\; |z_1| \geq 1,\;\; |z_2| = 1$
b) $A(a,z_2) \neq 0,\;\; |z_2| \geq 1$ for any a such that $|a| \geq 1$

8.5.4 Theorem 8.5.1.4 (DeCarlo, 1977; Strintzis, 1977)

Let $H(z_1,z_2) = 1/A(z_1,z_2)$ be a first quadrant recursive filter. The filter is stable if, and only if, $A(z_1,z_2)$ satisfies the following three conditions:

a) $A(z_1,z_2) \neq 0,\;\; |z_1| = 1,\;\; |z_2| = 1$
b) $A(a,z_2) \neq 0$, for $|z_2| \geq 1$ for any a such that $|a| = 1$
c) $A(z_1,b) \neq 0$, for $|z_1| \geq 1$ for any b such that $|b| = 1$

References

DeCarlo, R., J. Murray, and R. Saeks, Multivariate Nyquist theory, *Int. J. Control*, 25, 657-75, 1977.

Dudgeon, Dan E. and Russell M. Mersereau, *Multidimensional Digital Signal Processing*, Prentice-Hall, Inc., Englewood Cliffs, NJ, 1984.

Huang, Thomas S., Stability of two-dimensional recursive filters, *IEEE Trans. Audio and Electroacoustics*, AU-20, 158-63, 1972.

Lim, Jae S., *Two-Dimensional Signal and Image Processing,* Prentice-Hall Inc., Englewood Cliffs, NJ, 1990.

Shanks, John L., Sven Treitel, and James H. Justice, Stability and synthesis of two-dimensional recursive filters, *IEEE Trans. Audio and Electroacoustics*, AU-20, 115-28, 1972.

Strintzis, Michael G., Test of stability of multidimensional filters, *IEEE Trans. Circuits and Systems,* CAS-24, 432-37, 1977.

Appendix 1

Examples

1.1 Two-Dimensional Z-transforms

Example 8.1

The Z-transform of $x(n_1,n_2) = a^{n_1} b^{n_2} u(n_1,n_2) = a^{n_1} b^{n_2} u(n_1)u(n_2)$ is

$$X(z_1,z_2) = \sum_{n_1=-\infty}^{\infty} a^{n_1} u(n_1) z_1^{-n_1} \sum_{n_2=-\infty}^{\infty} b^{n_2} u(n_2) z_2^{-n_2} = \sum_{n_1=0}^{\infty} a^{n_1} z_1^{-n_1} \sum_{n_2=0}^{\infty} b^{n_2} z_2^{-n_2}$$

$$= \frac{1}{1-az_1^{-1}} \frac{1}{1-bz_2^{-1}}$$

with region of convergence (ROC) $|az_1^{-1}| < 1$ and $|bz_2^{-1}| < 1$, or $|z_1| > |a|$ and $|z_2| > |b|$ (see Figure 8.3).

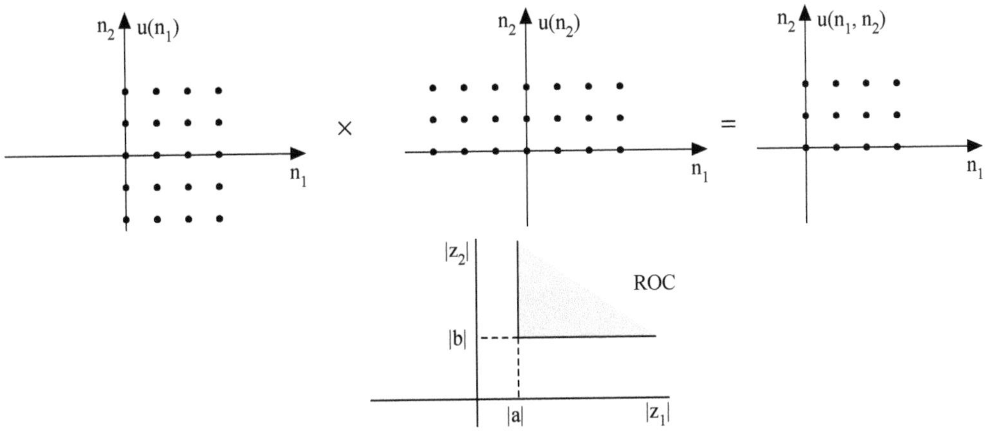

FIGURE 8.3

Example 8.2

The Z-transform of $x(n_1,n_2)=a^{n_1}\delta(n_1-n_2)u(n_1,n_2)$ is

$$X(z_1,z_2)=\sum_{n_1=-\infty}^{\infty}\sum_{n_2=-\infty}^{\infty}a^{n_1}u(n_1)u(n_2)\delta(n_1-n_2)z_1^{-n_1}z_2^{-n_2}=\sum_{n_1=0}^{\infty}\sum_{n_2=0}^{\infty}a^{n_1}\delta(n_1-n_2)z_1^{-n_1}z_2^{-n_2}$$

$$=\sum_{n_2=0}^{\infty}a^{n_2}z_1^{-n_2}z_2^{-n_2}=\frac{1}{1-az_1^{-1}z_2^{-1}}$$

From last summation the ROC is $|az_1^{-1}z_2^{-1}|<1$ or $|a|<|z_1||z_2|$ equivalently

$$\ell n|a|<\ell n|z_1|+\ell n|z_2|$$

The function and its ROC are shown in Figure 8.4.

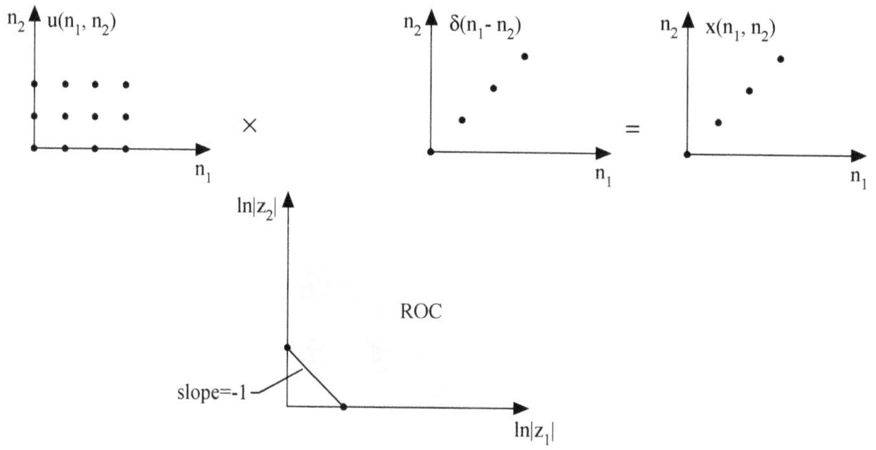

FIGURE 8.4

Example 8.3

The Z-transform of $x(n_1,n_2)=a^{n_1}u(n_1,n_2)u(n_1-n_2)$ is

$$X(z_1,z_2)=\sum_{n_1=-\infty}^{\infty}\sum_{n_2=-\infty}^{\infty}a^{n_1}u(n_1,n_2)u(n_1-n_2)z_1^{-n_1}z_2^{-n_2}$$

$$=\sum_{n_1=0}^{\infty}\sum_{n_2=0}^{\infty}a^{n_1}u(n_1-n_2)z_1^{-n_1}z_2^{-n_2}=\sum_{n_1=0}^{\infty}\sum_{n_2=0}^{\infty}a^{n_1}z_1^{-n_1}z_2^{-n_2}$$

$$=\sum_{n_1=0}^{\infty}a^{n_1}z_1^{-n_1}\frac{1-\left(z_2^{-1}\right)^{n_1+1}}{1-z_2^{-1}}=\frac{1}{\left(1-az_1^{-1}\right)\left(1-az_1^{-1}z_2^{-1}\right)}$$

ROC : $|z_1|>|a|$, $|z_1 z_2|>|a|$

The function and its ROC are shown in Figure 8.5.

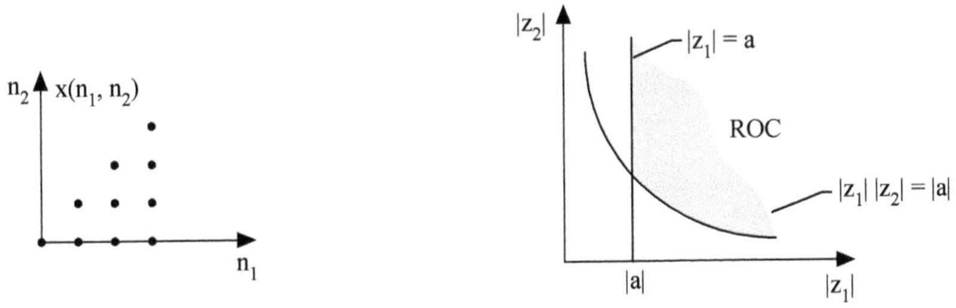

FIGURE 8.5

Example 8.4 (inverse integration)

If $X(z_1,z_2)$ has a region of convergence in the 2-D unit surface,

$$x(n_1,n_2)=\left(\frac{1}{2\pi j}\right)^2 \oint_{C_2}\oint_{C_1} \frac{z_1^{n_1}\, z_2^{n_2}}{1-\frac{1}{2}z_1^{-1}-\frac{1}{4}z_2^{-2}}\, dz_1\, dz_2$$

$$=\left(\frac{1}{2\pi j}\right)^2 \oint_{C_2}\oint_{C_1} \frac{(z_2-\frac{1}{4})^{-1}\, z_1^{n_1}\, z_2^{n_2}}{z_1-[\frac{1}{2}z_2/(z_2-\frac{1}{4})]}\, dz_1\, dz_2$$

$$=\frac{2\pi j}{(2\pi j)^2}\oint_{C_2} (z_2-\tfrac{1}{4})^{-1}\, \frac{(\frac{1}{2})^{n_1}\, z_2^{n_1}}{(z_2-\frac{1}{4})^{n_1}}\, z_2^{n_2}\, dz_2 = (\tfrac{1}{2})^{n_1}\, (\tfrac{1}{4})^{n_2}\, \frac{(n_1+n_2)!}{n_1!\, n_2!}, \quad n_1,n_2 \ge 0$$

9

Analytical Methods

9.1. Binomial Theorem and Binomial Coefficients; Arithmetic and Geometric Progressions; Arithmetic, Geometric, Harmonic, and Generalized Means

Binomial Theorem

9.1.1 $$(a+b)^n = a^n + \binom{n}{1}a^{n-1}b + \binom{n}{2}a^{n-2}b^2 + \binom{n}{3}a^{n-3}b^3 + \cdots + b^n = \sum_{k=0}^{n}\binom{n}{k}a^{n-k}b^k$$

(n is a positive integer)

Binomial Coefficients

9.1.2 $$\binom{n}{k} = \frac{n(n-1)\cdots(n-k+1)}{k!} = \frac{n!}{(n-k)!k!}$$

9.1.3 $$\binom{n}{k} = \binom{n}{n-k} = (-1)^k\binom{k-n-1}{k}$$

9.1.4 $\binom{n+1}{k}=\binom{n}{k}+\binom{n}{k-1}$

9.1.5 $\binom{n}{0}=\binom{n}{n}=1$

9.1.6 $1+\binom{n}{1}+\binom{n}{2}+\cdots+\binom{n}{n}=2^n$

9.1.7 $1-\binom{n}{1}+\binom{n}{2}-\cdots+(-1)^n\binom{n}{n}=0$

9.1.8 **The Multinomial Theorem**

$$(a_1+\cdots+a_m)^n=\sum_{k_1+\cdots+k_m=n}\frac{n!}{k_1!\cdots k_m!}a_1^{k_1}\cdots a_m^{k_m}$$

9.1.9 **Factorials and Binomial Coefficients**

$n!=1\cdot 2\cdot 3\cdots n,\qquad 0!=1$ (factorials)

$(2n-1)!!=1\cdot 3\cdot 5\cdots(2n-1),\qquad (2n)!!=2\cdot 4\cdot 6\cdots 2n$ (semifactorials)

$\binom{n}{k}=\frac{n(n-1)\cdots(n-k+1)}{k!}=\frac{n!}{k!(n-k)!}$ (binomial coefficients)

$n!\sim\sqrt{2\pi}\,n^{n+1/2}e^{-n}$ as $n\to\infty$ (Stirling's formula)

9.1.10 $\binom{n}{n-k}=\binom{n}{k},\quad \binom{n}{k}+\binom{n}{k+1}=\binom{n+1}{k+1}=\binom{n}{k}+\binom{n-1}{k}+\cdots+\binom{k}{k}$

9.1.11 $\binom{n}{0}+\binom{n+1}{1}+\binom{n+2}{2}+\cdots+\binom{n+k}{k}=\binom{n+k+1}{k}$

9.1.12 $\binom{m}{0}\binom{n}{k}+\binom{m}{1}\binom{n}{k-1}+\cdots+\binom{m}{k}\binom{n}{0}=\binom{m+n}{k}$

9.1.13 $\binom{n}{0}+\binom{n}{1}+\cdots+\binom{n}{n}=2^n,\qquad \binom{n}{0}^2+\binom{n}{1}^2+\cdots+\binom{n}{n}^2=\binom{2n}{n}$

9.1.14 $\binom{n}{0}-\binom{n}{1}+\cdots+(-1)^n\binom{n}{n}=0$

9.1.15 $\binom{-r}{k}=(-1)^k\binom{r+k-1}{k}$

9.1.16 $\sum_{k=0}^{n}\binom{n}{k}=2^n$

9.1.17 $\sum_{k=0}^{n}\binom{r+k}{k}=\binom{r+n+1}{n}$

9.1.18 $\sum_{k=0}^{n}\binom{k}{m}=\binom{n+1}{m+1},\qquad m=0,1,2,\cdots$

9.1.19 $\sum_{k=0}^{n} \binom{r}{k}\binom{s}{n-k} = \binom{r+s}{n}, \quad n = 0,1,2,\cdots$

9.1.20 $\sum_{k=0}^{n} \binom{n}{k}\binom{s+k}{m}(-1)^k = (-1)^n \binom{s}{m-n}$

9.1.21 **Cauchy product**: Let $\sum_{n=0}^{\infty} a_n$ and $\sum_{n=0}^{\infty} b_n$ have absolute convergence. Then

$$\sum_{n=0}^{\infty} a_n \sum_{n=0}^{\infty} b_n = \sum_{n=0}^{\infty} c_n = \sum_{n=0}^{\infty}\left[\sum_{k=0}^{n} a_k b_{n-k}\right]$$

9.1.22 $\binom{-1/2}{n} = \frac{(-1)^n (2n)!}{2^{2n}(n!)^2}$

9.1.23 $\binom{-2k-1}{n} = (-1)^n \frac{(n+2k)!}{(2k!)n!}, \quad k = 0,1,2,\cdots$

9.1.24 **Sum of Arithmetic Progression to *n* Terms**

$$a+(a+d)+(a+2d)+\cdots+(a+(n-1)d) = na + \frac{1}{2}n(n-1)d = \frac{n}{2}(a+l),$$

last term in series $= l = a + (n-1)d$

9.1.25 **Sum of Geometric Progression to *n* Terms**

$$s_n = a + ar + ar^2 + \cdots + ar^{n-1} = \frac{a(1-r^n)}{1-r}$$

$$\lim_{n\to\infty} = s_n = a/(1-r), \quad (-1 < r < 1)$$

9.1.26 **Arithmetic Mean of *n* Quantities *A***

$$A = \frac{a_1 + a_2 + \cdots + a_n}{n}$$

9.1.27 **Geometric Mean of *n* Quantities *G***

$$G = (a_1 a_2 \cdots a_n)^{1/n} \qquad (a_k > 0,\ k = 1,2,\cdots n)$$

9.1.28 **Harmonic Mean of *n* Quantities *H***

$$\frac{1}{H} = \frac{1}{n}\left(\frac{1}{a_1} + \frac{1}{a_2} + \cdots + \frac{1}{a_n}\right) \qquad (a_k > 0,\ k = 1,2,\cdots n)$$

Generalized Mean

9.1.29 $M(t) = \left(\frac{1}{n}\sum_{k=1}^{n} a_k^t\right)^{1/t}$

9.1.30 $M(t) = 0 \quad (t < 0,$ some a_k zero)

9.1.31 $\lim_{t\to\infty} M(t) = \max, \qquad (a_1, a_2, \cdots, a_n) = \max a$

9.1.32 $\lim_{t=-\infty} M(t) = \min, \qquad (a_1, a_2, \cdots, a_n) = \min a$

9.1.33 $\lim_{t\to 0} M(t) = G$

9.1.34 $M(1) = A$

9.1.35 $M(-1) = H$

9.2 Inequalities

Relation Between Arithmetic, Geometric, Harmonic, and Generalized Means

9.2.1 $A \ge G \ge H$, equality if and only if $a_1 = a_2 = \dots = a_n$

9.2.2 $\min a < M(t) < \max a$

9.2.3 $\min a < G < \max a$; equality holds if all a_k are equal, or $t < 0$ and an a_k is zero.

9.2.4 $M(t) < M(s)$ if $t < s$ unless all a_k are equal, or $s < 0$ and an a_k is zero.

Triangle Inequalities

9.2.5 $|a_1| - |a_2| \le |a_1 + a_2| \le |a_1| + |a_2|$

9.2.6 $\left|\sum_{k=1}^{n} a_k\right| \le \sum_{k=1}^{n} |a_k|$

9.2.7 **Chebyshev's Inequality**

If $a_1 \ge a_2 \ge a_3 \ge \dots \ge a_n$

$b_1 \ge b_2 \ge b_3 \ge \dots \ge b_n$

$$n\sum_{k=1}^{n} a_k b_k \ge \left(\sum_{k=1}^{n} a_k\right)\left(\sum_{k=1}^{n} b_k\right)$$

9.2.8 **Holder's Inequality for Sums**

If $\frac{1}{p} + \frac{1}{q} = 1, \qquad p > 1, \quad q > 1$

$$\sum_{k=1}^{n} |a_k b_k| \le \left(\sum_{k=1}^{n} |a_k|^p\right)^{1/p} \left(\sum_{k=1}^{n} |b_k|^q\right)^{1/q};$$

equality holds if and only if $|b_k| = c|a_k|^{p-1}$ (c = constant > 0). If $p = q = 2$, we get Cauchy's inequality

9.2.9 **Cauchy's Inequality**

$$\left[\sum_{k=1}^{n} a_k b_k\right]^2 \le \sum_{k=1}^{n} a_k^2 \sum_{k=1}^{n} b_k^2 \quad \text{(equality for } a_k = cb_k,\ c \text{ constant).}$$

9.2.10 **Holder's Inequality for Integrals**

If $\frac{1}{p} + \frac{1}{q} = 1, \qquad p > 1, \quad q > 1$

$$\int_a^b |f(x)g(x)|\,dx \le \left[\int_a^b |f(x)|^p\,dx\right]^{1/p} \left[\int_a^b |g(x)|^q\,dx\right]^{1/q}$$

equality holds if and only if $|g(x)| = c|f(x)|^{p-1}$ (c = constant > 0). If $p = q = 2$, we get Schwarz's inequality.

9.2.11 **Schwarz's Inequality**

$$\left[\int_a^b f(x)g(x)\,dx\right]^2 \le \int_a^b |f(x)|^2\,dx \int_a^b |g(x)|^2\,dx$$

9.2.12 **Minkowski's Inequality for Sums**

If $p > 1$ and $a_k, b_k > 0$ for all k,

$$\left(\sum_{k=1}^{n}(a_k+b_k)^p\right)^{1/p} \le \left(\sum_{k=1}^{n} a_k^p\right)^{1/p} + \left(\sum_{k=1}^{n} b_k^p\right)^{1/p},$$

equality holds if and only if $b_k = ca_k$ (c = constant > 0).

9.2.13 **Minkowski's Inequality for Integrals**

If $p > 1$,

$$\left(\int_a^b |f(x)+g(x)|^p\,dx\right)^{1/p} \le \left(\int_a^b |f(x)|^p\,dx\right)^{1/p} + \left(\int_a^b |g(x)|^p\,dx\right)^{1/p}$$

equality holds if and only if $g(x) = cf(x)$ (c = constant > 0).

9.3 Numbers

9.3.1 Number Systems

N ***natural number***, $N = \{0,1,2,3,\cdots\}$ sometimes 0 is omitted.

Z ***integers***, (Z^+ positive integers), $Z = \{0,\pm1,\pm2,\pm3,\cdots\}$

Q ***rational numbers***, $Q = \{p/q : p,q \in z,\ q \ne 0\}$, Q is ***countable***, i.e., there exists a one-to-one correspondence between Q and N

R ***real numbers***, R = {real numbers} is not countable. Real numbers which are not rational are called ***irrational***. Every irrational number can be represented by an infinite non-periodic decimal expansion. ***Algebraic numbers*** are solutions to an equation of the form$a_n x^n + \ldots + a_o = 0$, $a_k \in z$. ***Transcendental*** are those numbers in **R** which are not algebraic. (Example: 4/7 is rational; $\sqrt{5}$ is algebraic and irrational; e and π are transcendentals.)

C ***Complex numbers***, $C = \{x + jy : x,y \in R\}$ where $j^2 = -1$.

9.3.2 The Supreme Axiom

For any non-empty bounded subset S of R there exist unique numbers $G = \sup S$ and $g = \inf S$ such that:

1. $g \le x \le G$, all $x \in S$
2. For any $\varepsilon > 0$ there exists $x_1 \in S$ and $x_2 \in S$ such that

$$x_1 > G - \varepsilon \text{ and } x_2 < g + \varepsilon$$

9.3.3 Theorems on Prime Numbers

1. For every positive integer n exists a prime factor of $n!+1$ exceeding n.
2. Every prime factor of $p_1p_2\ldots p_n+1$, where $p_1,p_2,\ldots,p_n$ are prime, differs from each of $p_1,p_2,\ldots,p_n$.
3. There are infinitely many primes. (Euclid)
4. For every positive integer $n \ge 2$, there exists a string of n consecutive composite integers.
5. If a and b are relatively prime, then the arithmetic sequence $an+b$, $n = 1,2,\ldots$, contains an infinite number of primes. (Lejeune-Dirichlet)

The following conjectures have not been proved.

6. Every even number ≥ 6 is the sum of two odd primes (the Goldbach conjecture).
7. There exists infinitely many prime twins. Prime twins are pairs like (3,5), (5,7), and (2087,2089).

9.3.4 Unique Factorization Theorem

Every integer > 1 is either a prime or a product of uniquely determined primes.

9.3.5 The Function $\pi(x)$

The function value $\pi(x)$ is the number of primes which are less than or equal to x. Asymptotic behavior:

$$\pi(x) \sim \frac{x}{\ell n x} \quad \text{as } x \to \infty$$

9.3.6 Least Common Multiple (LCM)

Let $[a_1,\ldots,a_n]$ denote the *least common multiple* of the integers $a_1,\cdots,a_n$. One method of finding that number is: Prime number factorize $a_1,\cdots,a_n$. Then form the product of these primes raised to the greatest power in which they appear.

Example: Determine $A = [18,24,30]$. $18 = 2 \cdot 3^2$, $24 = 2^3 \cdot 3$, $30 = 2 \cdot 3 \cdot 5$. Thus, $A = 2^3 \cdot 3^2 \cdot 5 = 360$.

9.3.7 Greatest Common Divisor (GCD)

Let (a,b) denote the *greatest common divisor* of a and b. If $(a,b) = 1$ the numbers are *relatively prime*. One method (Euclid's algorithm) of finding (a,b) is:

Assuming $a > b$ and dividing a by b yields $a = q_1 b_1 + r_1$, $0 \le r_1 < b$. Dividing b by r_1 gives $b = q_2 r_1 + r_2$, $0 \le r_2 < r_1$. Continuing like this, let r_k be the first remainder which equals 0. Then $(a,b) = r_{k-1}$

Example: Determine (112,42). By the above algorithm, $112 = 2 \cdot 42 + 28$, $42 = 1 \cdot 28 + 14$, $28 = 2 \cdot 14 + 0$. Thus, $(112,42) = 14$. ***Note:*** $(a,b) \cdot [a,b] = ab$

9.3.8 Modulo

If m, n, and p are integers, then m and n are *congruent modulo* $p, m = n \bmod(p)$, if $m - n$ is a multiple of p, i.e., m/p and n/p have equal remainders.

$$m_1 = n_1 \mathrm{mod}(p), \quad m_2 = n_2 \mathrm{mod}(p) \Rightarrow$$

$$\text{(i)}\ \ m_1 + m_2 = (n_1 \pm n_2)\mathrm{mod}(p) \qquad \text{(ii)}\ \ m_1 m_2 = (n_1 n_2)\mathrm{mod}(p)$$

9.3.9 Diophantine Equations

A *Diophantine* equation has integer coefficients and integer solutions. As an example, the equation $ax + by = c$, $a,b,c \in \mathbf{Z}$ (a), has integer solutions x and y if and only if (a,b) divides c. In particular, $ax + by = 1$ is solvable $\Leftrightarrow (a,b) = 1$. If x_o, y_o is a particular solution of (a), then the general solution is $x = x_o + nb/(a,b)$, $y = y_o - na/(a,b)$, $n \in \mathbf{Z}$.

9.3.10 Mersenne Numbers

$$M_n = 2^n - 1$$

9.3.11 Mersenne Primes

If $2^p - 1$ is prime, then p is prime.

9.3.12 Fermat Primes

If $2^p + 1$ is prime, then p is a power of 2.

9.3.13 Fibonacci Numbers

$$F_1 = 1 \qquad F_2 = 1 \qquad F_{n+2} = F_n + F_{n+1} \qquad n \geq 1$$

9.3.14 Decimal and Binary Systems

$$x = x_m B^m + x_{m-1} B^{m-1} + \cdots + x_o B^o + x_{-1} B^{-1} + \cdots = (x_m x_{m-1} \cdots x_o . x_{-1}, \cdots)$$

$B > 1$ is base; x_i is one of the numbers $0, 1 \ldots, B-1$.

Example: 36.625_{10}, $x = 3 \cdot 10^1 + 6 \cdot 10^0 + 6 \cdot 10^{-1} + 2 \cdot 10^{-2} + 5 \cdot 10^{-3} = (36.625)_{10}$

$x = 1 \cdot 2^5 + 0 \cdot 2^4 + 0 \cdot 2^3 + 1 \cdot 2^2 + 0 \cdot 2^1 + 0 \cdot 2^0 + 1 \cdot 2^{-1} + 0 \cdot 2^{-2} + 1 \cdot 2^{-3} = (100100.101)_2$

9.3.15 Conversion Bases

1. $B \to 10$: $x_B = (x_m x_{m-1} \cdots x_o . x_{-1}, \cdots)_B$

   $$x_{10} = x_m B^m + x_{m-1} B^{m-1} + \cdots + x_o B^o + x_{-1} B^{-1} + \cdots$$

2. $10 \to B$: (**Example**: $x_{10} = 12545.6789$ to x_8)

$\frac{12545}{8} = 1568 + \frac{1}{8}$	$R_1 = 1$	$0.6789 \times 8 = 5.4312$	$I_1 = 5$
$\frac{1568}{8} = 196 + \frac{0}{8}$	$R_2 = 0$	$0.4312 \times 8 = 3.4496$	$I_2 = 3$
$\frac{196}{8} = 24 + \frac{4}{8}$	$R_3 = 4$	$0.4496 \times 8 = 3.5968$	$I_3 = 3$
$\frac{24}{8} = 3 + \frac{0}{8}$	$R_4 = 0$	$0.5968 \times 8 = 4.7744$	$I_4 = 4$
$\frac{3}{8} = 0 + \frac{3}{8}$	$R_5 = 3$	etc.	

Therefore $x_8 = (30401.5334)_8$

9.3.16 Binary System

Addition: $0 + 0 = 0$, $0 + 1 = 1 + 0 = 1$, $1 + 1 = 10$
Multiplication: $0 \cdot 0 = 0 \cdot 1 = 1 \cdot 0 = 0$, $1 \cdot 1 = 1$

9.3.17 Hexadecimal System

Digits: 0,1,2,3,4,5,6,7,8,9, $A = 10$, $B = 11$, $C = 12$, $D = 13$, $E = 14$, and $F = 15$

Addition Table

	1	2	3	4	5	6	7	8	9	A	B	C	D	E	F	
1	2	3	4	5	6	7	8	9	A	B	C	D	E	F	10	**1**
2	3	4	5	6	7	8	9	A	B	C	D	E	F	10	11	**2**
3	4	5	6	7	8	9	A	B	C	D	E	F	10	11	12	**3**
4	5	6	7	8	9	A	B	C	D	E	F	10	11	12	13	**4**
5	6	7	8	9	A	B	C	D	E	F	10	11	12	13	14	**5**
6	7	8	9	A	B	C	D	E	F	10	11	12	13	14	15	**6**
7	8	9	A	B	C	D	E	F	10	11	12	13	14	15	16	**7**
8	9	A	B	C	D	E	F	10	11	12	13	14	15	16	17	**8**
9	A	B	C	D	E	F	10	11	12	13	14	15	16	17	18	**9**
A	B	C	D	E	F	10	11	12	13	14	15	16	17	18	19	**A**
B	C	D	E	F	10	11	12	13	14	15	16	17	18	19	1A	**B**
C	D	E	F	10	11	12	13	14	15	16	17	18	19	1A	1B	**C**
D	E	F	10	11	12	13	14	15	16	17	18	19	1A	1B	1C	**D**
E	F	10	11	12	13	14	15	16	17	18	19	1A	1B	1C	1D	**E**
F	10	11	12	13	14	15	16	17	18	19	1A	1B	1C	1D	1E	**F**
	1	**2**	**3**	**4**	**5**	**6**	**7**	**8**	**9**	**A**	**B**	**C**	**D**	**E**	**F**	

E.g., B + 6 = 11

Multiplication Table

	1	2	3	4	5	6	7	8	9	A	B	C	D	E	F	
1	1	2	3	4	5	6	7	8	9	A	B	C	D	E	F	**1**
2	2	4	6	8	A	C	E	10	12	14	16	18	1A	1C	1E	**2**
3	3	6	9	C	F	12	15	18	1B	1E	21	24	27	2A	2D	**3**
4	4	8	C	10	14	18	1C	20	24	28	2C	30	34	38	3C	**4**
5	5	A	F	14	19	1E	23	28	2D	32	37	3C	41	46	4B	**5**
6	6	C	12	18	1E	24	2A	30	36	3C	42	48	4E	54	5A	**6**
7	7	E	15	1C	23	2A	31	38	3F	46	4D	54	5B	62	69	**7**
8	8	10	18	20	28	30	38	40	48	50	58	60	68	70	78	**8**
9	9	12	1B	24	2D	36	3F	48	51	5A	63	6C	75	7E	87	**9**
A	A	14	1E	28	32	3C	46	50	5A	64	6E	78	82	8C	96	**A**
B	B	16	21	2C	37	42	4D	58	63	6E	79	84	8F	9A	A5	**B**
C	C	18	24	30	3C	48	54	60	6C	78	84	90	9C	A8	B4	**C**
D	D	1A	27	34	41	4E	5B	68	75	82	8F	9C	A9	B6	C3	**D**
E	E	1C	2A	38	46	54	62	70	7E	8C	9A	A8	B6	C4	D2	**E**
F	F	1E	2D	3C	4B	5A	69	78	87	96	A5	B4	C3	D2	E1	**F**
	1	**2**	**3**	**4**	**5**	**6**	**7**	**8**	**9**	**A**	**B**	**C**	**D**	**E**	**F**	

E.g., B · 6 = 42

9.3.18 Special Numbers in Different Number Bases

B = 2: π = 11.001001 000011 111101 101010 100010 001000 010110 100011···

e = 10.101101 111110 000101 010001 011000 101000 101011 101101···

γ = 0.100100 111100 010001 100111 111000 110111 110110 110110···

$\sqrt{2}$ = 1.011010 100000 100111 100110 011001 111111 001110 111100···

In2 = 0.101100 010111 001000 010111 111101 111101 000111 001111···

B = 3: $\pi = 10.010211\ 012222\cdots$

$e = 2.201101\ 121221\cdots$

$\gamma = 0.120120\ 210100\cdots$

$\sqrt{2} = 1.102011\ 221222\cdots$

$\ln 2 = 0.200201\ 022012\cdots$

B = 8: $\pi = 3.110375\ 524210\ 264302\cdots$

$e = 2.557605\ 213050\ 535512\cdots$

$\gamma = 0.447421\ 477067\ 666061\cdots$

$\sqrt{2} = 1.324047\ 463177\ 167462\cdots$

$\ln 2 = 0.542710\ 277574\ 071736\cdots$

B = 12: $\pi = 3.184809\ 493B91\cdots$

$e = 2.875236\ 069821\cdots$

$\gamma = 0.6B1518\ 8A6760\cdots$

$\sqrt{2} = 1.4B7917\ 0A07B8\cdots$

$\ln 2 = 0.839912\ 483369\cdots$

B = 16: $\pi = 3.243F6A\ 8885A3\cdots$

$e = 2.B7E151\ 628AED\cdots$

$\gamma = 0.93C467\ E37DB0\cdots$

$\sqrt{2} = 1.6A09E6\ 67F3BC\cdots$

$\ln 2 = 0.B17217\ F7D1CF\cdots$

9.4 Complex Numbers ($j^2 = -1$)

9.4.1 Rectangular Form

$z = x + jy,\ z^* = x - jy = \text{conjugate},\ |z| = [zz^*]^{1/2} = [x^2 + y^2]^{1/2} = \text{modulus},$

$|z_1 - z_2|$ = distance between the points z_1 and z_2, $x = \text{Re}\{z\}$, $y = \text{Im}\{z\}$,

$\theta = \tan^{-1}(y/x) + n\pi, \quad (n = 0 \text{ if } x > 0, \quad n = 1 \text{ if } x < 0)$

9.4.2 Polar Form

$z = x + jy = r(\cos\theta + j\sin\theta) = re^{j\theta}$

$x = r\cos\theta \qquad r = \sqrt{x^2 + y^2}$

$y = r\sin\theta \qquad \theta = \tan^{-1}\frac{y}{x} + n\pi \qquad (n = 0 \text{ if } x > 0,\ n = 1 \text{ if } x < 0)$

9.4.3 De Moivre's and Euler Formulas

$$(\cos\theta + j\sin\theta)^n = \cos n\theta + j\sin n\theta, \quad \cos\theta = \frac{e^{j\theta} + e^{-j\theta}}{2}, \quad \sin\theta = \frac{e^{j\theta} - e^{-j\theta}}{2j}$$

9.5 Algebraic Equations

9.5.1 Algebraic Equation

$P(z) = a_n z^n + a_{n-1} z^{n-1} + \cdots + a_1 z + a_o = 0 \quad (a_i = \text{complex numbers})\ n^{th}$ degree equation

9.5.2 Zeros and Roots

If $P(z) = (z - r)^m Q(z)$, r = *zero* of multiplicity m, also a *root* of multiplicity m. If r is a root of multiplicity $m(m \geq 1)$ of Eq. (9.5.1) $P(z) = 0$, then r is a root of multiplicity $m - 1$, of the equation $P'(z) = 0$.

9.5.3 Factor Formula

$P(z)$ contains the factor $z - r \Leftrightarrow P(r) = 0$. $P(z)$ contains the factor $(z - r)^m \Leftrightarrow P'(r) = P(r) = \cdots = P^{m-1}(r) = 0$, then r is a root of multiplicity m–1 of the equation $P'(z) = 0$

9.5.4 Fundamental Theorem of Algebra

Eq. (9.5.1) $P(z) = 0$ of degree n has n roots (including multiplicity). If the roots are $r_1,\cdots,r_n$ then $P(z) = a_n(z - r_1)\cdots(z - r_n)$.

9.5.5 Relationship Between Roots and Coefficients

If $r_1,\cdots,r_n$ are the roots of Eq. (9.5.1) then

$$\begin{cases} r_1 + r_2 + \cdots + r_n = -\dfrac{a_{n-1}}{a_n} \\ r_1r_2 + r_1r_3 + \cdots + r_{n-1}r_n = \sum\limits_{i<j} r_ir_j = \dfrac{a_{n-2}}{a_n} \\ \cdots \\ r_1r_2\cdots r_n = (-1)^n \dfrac{a_o}{a_n} \end{cases}$$

9.5.6 Equations with Real Coefficients

Assume that all a_i of Eq. (9.5.1) are real.

1. If r is a non-real root of Eq. (9.5.1), then so is $\bar{r}$ (conjugate of r), i.e., $P(r) = 0 \Rightarrow P(\bar{r}) = 0$.
2. $P(z)$ can be factorized into real polynomials of degree, at most two.
3. If all a_i are integers and if $r = p/q$ (p and q having no common divisor) is a rational root of Eq. (9.5.1), then p divides a_o and q divides a_n.
4. The number of positive real roots (including multiplicity) of Eq. (9.5.1) either equals the number of sign changes of the sequence $a_0,a_1,\cdots,a_n$ or equals this number minus an even number. If all roots of the equation are real, the first case always applies. (Descartes' rule of signs)

9.5.7 Quadratic Equations

$ax^2 + bx + c = 0$ $\qquad x^2 + px + q = 0$

$$x = \frac{-b \pm \sqrt{b^2 - 4ac}}{2a} \qquad x = -\frac{p}{2} \pm \sqrt{\left(\frac{p}{2}\right)^2 - q}$$

$b^2 - 4ac > 0 \Rightarrow$ two unequal real roots

$b^2 - 4ac < 0 \Rightarrow$ two unequal complex roots $(\pm\sqrt{-d} = \pm j\sqrt{d})$

$b^2 - 4ac = 0 \Rightarrow$ the roots are real and equal

The expression $b^2 - 4ac$ is called the *discriminant*.

Let x_1 and x_2 be roots of the equation $x^2 + px + q = 0$. Then

$$\begin{cases} x_1 + x_2 = -p \\ x_1 x_2 = q \end{cases}$$

9.5.8 Cubic Equations

The equation $az^3 + bz^2 + cz + d = 0$ is by the substitution $z = x - b/3a$ reduced to

$$x^3 + px + q = 0.$$

Set $$D = \left(\frac{p}{3}\right)^3 + \left(\frac{q}{2}\right)^2.$$

Then the cubic equation has (1) one real root if $D > 0$, (2) three real roots of which at least two are equal if $D = 0$, and (3) three distinct real roots if $D < 0$. Put

$$u = \sqrt[3]{-\frac{q}{2} + \sqrt{D}}, \qquad v = \sqrt[3]{-\frac{q}{2} - \sqrt{D}}.$$

The roots of the cubic equation are

$$x_1 = u + v \qquad x_{2,3} = -\frac{u+v}{2} \pm \frac{u-v}{2} j\sqrt{3} \qquad \textit{(Cardano's formula)}$$

If x_1, x_2, x_3 are roots of the equation $x^3 + rx^2 + sx + t = 0$ then

$$\begin{cases} x_1 + x_2 + x_3 = -r \\ x_1 x_2 + x_1 x_3 + x_2 x_3 = s \\ x_1 x_2 x_3 = -t \end{cases}$$

9.5.9 Binomic Equations

A *binomic equation* is of the form

$$z^n = c, \quad c = \text{complex number}$$

1. Special case $n = 2$: $z^2 = a + jb$.
   Roots:

$$z = \pm\sqrt{a + jb} = \begin{cases} \pm\left[\sqrt{\frac{r+a}{2}} + j\sqrt{\frac{r-a}{2}}\right], & b \ge 0 \\ \pm\left[\sqrt{\frac{r+a}{2}} - j\sqrt{\frac{r-a}{2}}\right], & b \le 0 \end{cases}, \qquad r = \sqrt{a^2 + b^2}$$

2. General case: Solution in *polar* form: Set $c = re^{i\theta}$

$$z^n = c = re^{i(\theta + 2k\pi)}$$

   Roots:

$$z = \sqrt[n]{r}\, e^{j(\theta+2k\pi)/n} = \sqrt[n]{r}\left(\cos\frac{\theta + 2k\pi}{n} + j\sin\frac{\theta + 2k\pi}{n}\right),$$

$$k = 0, 1, \cdots, n-1$$

9.6 Differentiation

9.6.1 $\frac{d}{dx}(cu) = c\frac{du}{dx}$, c constant

9.6.2 $\frac{d}{dx}(u+v) = \frac{du}{dx} + \frac{dv}{dx}$

9.6.3 $\frac{d}{dx}(uv) = u\frac{dv}{dx} + v\frac{du}{dx}$

9.6.4 $\frac{d}{dx}(u/v) = \frac{vdu/dx - udv/dx}{v^2}$

9.6.5 $\frac{d}{dx}u(v) = \frac{du}{dv}\frac{dv}{dx}$

9.6.6 $\frac{d^n}{dx^n}(uv) = \sum_{k=0}^{n}\binom{n}{k}u^{(n-k)}v^{(k)}$, where parentheses in exponents mean number of differentiations.

9.6.7 $\frac{d^n}{dx^n}(u^n v^m) = u^{n-1}v^{m-1}\left(nv\frac{du}{dx} + mu\frac{dv}{dx}\right)$

9.6.8 $\frac{d}{dx}\left(\frac{u^n}{v^m}\right) = \frac{u^{n-1}}{v^{m+1}}\left(nv\frac{du}{dx} - mu\frac{dv}{dx}\right)$

9.6.9 $f(x) = u(x)^a\, v(x)^b\, w(x)^c$

$$\frac{\frac{df(x)}{dx}}{f(x)} = a\frac{\frac{du(x)}{dx}}{u(x)} + b\frac{\frac{dv(x)}{dx}}{v(x)} + c\frac{\frac{dw(x)}{dx}}{w(x)}$$

9.6.10 $\frac{d^2}{dx^2}u(v(x)) = \frac{d^2u}{dv^2}\left(\frac{du}{dx}\right)^2 + \frac{du}{dv}\frac{d^2v(x)}{dx^2}$, $\quad \frac{d^2f}{dx^2} = \frac{d^2f}{dy^2}\left(\frac{dy}{dx}\right)^2 + \frac{df}{dy}\cdot\frac{d^2y}{dx^2}$

9.6.11 $\frac{d}{dx}\int_{u(x)}^{v(x)} f(x,t)dt = f(x,v)\frac{dv}{dx} - f(x,u)\frac{du}{dx} + \int_{u(x)}^{v(x)}\frac{\partial}{\partial x}f(x,t)dt$

$$\frac{d}{dx}\int_{a}^{x} f(t)dt = f(x) \qquad \frac{d}{dx}\int_{x}^{a} f(x)dt = -f(x)$$

$$\frac{d}{dx}\int_{u(x)}^{v(x)} f(x)dt = f(v(x))\frac{dv}{dx} - f(u(x))\frac{du}{dx}$$

9.7 Functions

9.7.1 Definitions

$f(x) = f(-x) \equiv$ even; $\quad f(x) = -f(-x) \equiv$ odd; $\quad f(x) = f(x+T) \equiv$ periodic

$x_1 < x_2 \Rightarrow f(x_1) \le f(x_2)$ $\quad [f(x_1) \ge f(x_2)] \equiv$ increasing [decreasing]

Convex (concave): if for any two points the chord lies above (below) the curve.
Inflection point: the point at which the curve changes from convex to concave (or vice versa)
Local maximum (minimum): A function has a local maximum (minimum) at $x = a$ if there is a neighborhood U such that $f(x) \le f(a)$ $[f(x) \ge f(a)]$ for all $x \in U \cap D_f$ (domain of the function).

Strictly increasing: $\frac{df(x)}{dx} > 0$

Increasing: $\frac{df(x)}{dx} \ge 0$; **Constant:** $\frac{df(x)}{dx} = 0$

Decreasing: $\frac{df(x)}{dx} \le 0$; **Strictly Decreasing:** $\frac{df(x)}{dx} < 0$

Stationary (critical) point: $\left.\frac{df(x)}{dx}\right|_{x=x_o} = 0$

Convex: $\frac{d^2 f(x)}{dx^2} \ge 0$; **Concave:** $\frac{d^2 f(x)}{dx^2} \le 0$

Inflection point: $\left.\frac{d^2 f(x)}{dx^2}\right|_{x=x_o} = 0$ and $\frac{d^2 f(x)}{dx^2}$ changes sign at x_o

Jensen's inequality: If $f(x)$ is convex and $a_1 + a_2 + \ldots + a_n = 1$, $a_i > 0$, then $f(a_1x_1 + \ldots + a_nx_n) \le a_1 f(x_1) + \ldots + a_n f(x_n)$.

Continuous at x_o: $\lim_{x \to x_o} f(x) = f(x_o)$

Uniformly continuous in I: If for any $\varepsilon > 0$ there exists a $\delta > 0$ for all $|f(x_1) - f(x_2)| < \varepsilon$ for all $x_1, x_2 \in I$ such that $|x_1 - x_2| < \delta$.

9.8 Limits, Maxima and Minima

9.8.1 Limits

1. $\lim_{x \to a}(f(x) \pm g(x)) = f(a) \pm g(a)$
2. $\lim_{x \to a} \frac{f(x)}{g(x)} = \frac{f(a)}{g(a)}, \quad g(a) \ne 0$
3. $\lim_{x \to a} h(f(x)) = h(f(a)) \quad (h(t) = \text{continuous})$
4. $\lim_{x \to a} f(x)g(x) = f(a)g(a)$
5. $\lim_{x \to a} f(x)^{g(x)} = f(a)^{g(a)} \quad (f(a) > 0)$
6. $f(x) \le g(x) \Rightarrow f(a) \le g(a)$

9.8.2 l'Hospital's Rules

1. $\lim_{x \to a} \frac{f(x)}{g(x)} = \lim_{x \to a} \frac{\frac{df(x)}{dx}}{\frac{dg(x)}{dx}}$, if the latter limit exists.

2. $\lim_{x \to \infty} \frac{f(x)}{g(x)} = \lim_{x \to \infty} \frac{\frac{df(x)}{dx}}{\frac{dg(x)}{dx}}$, if the latter limit exists.

9.8.3 Not Well-Defined Forms

$$\frac{0}{0};\quad \frac{\infty}{\infty};\quad 0\cdot\infty;\quad [0^+]^0;\quad \infty^0;\quad 1^\infty;\quad \infty-\infty$$

9.8.4 Limits

$$\lim_{x\to\pm\infty}\left(1+\frac{1}{x}\right)^x = e;\quad \lim_{x\to\infty} x^{1/x}=1;\quad \lim_{m\to\infty}\frac{a^m}{m!}=0,\quad \lim_{x\to 0}\frac{\sin ax}{x}=a$$

$$\lim_{x\to 0}\frac{\ell n(1+x)}{x}=1;\quad \lim_{x\to\infty}\frac{\ell nx}{\sqrt{x}}=\lim_{x\to\infty}\frac{1/x}{1/2\sqrt{x}}=0$$

9.8.5 Function of Two Variables

The function $f(x,y)$ has a maximum or minimum for those values of (x_o,y_o) for which

$$\frac{\partial f}{\partial x}=0,\qquad \frac{\partial f}{\partial y}=0,$$

and for which $\begin{vmatrix}\partial^2 f/\partial x\partial y & \partial^2 f/\partial x^2\\ \partial^2 f/\partial y^2 & \partial^2 f/\partial x\partial y\end{vmatrix}<0$

(a) $f(x,y)$ has a maximum

$$\text{if } \frac{\partial^2 f}{\partial x^2}<0 \text{ and } \frac{\partial^2 f}{\partial y^2}<0 \text{ at } (x_o,y_o)$$

(b) $f(x,y)$ has a minimum

$$\text{if } \frac{\partial^2 f}{\partial x^2}>0 \text{ and } \frac{\partial^2 f}{\partial y^2}>0 \text{ at } (x_o,y_o)$$

9.9 Integrals

9.9.1 Primitive Function

$F(x)$ if a *primitive* function of $f(x)$ on an interval I if $dF(x)/dx = f(x)$ for all $x \in I$.

$$F(x)=\int f(x)\,dx$$

9.9.2 Integration Properties

Linearity:

$$\int [af(x)+bg(x)]dx = a\int f(x)dx + b\int g(x)dx$$

Integration by Parts:

$$\int f(x)g(x)dx = F(x)g(x) - \int F(x)g'(x)dx$$

Substitution:

$$\int f(g(x))g'(x)dx = \int f(t)dt, \quad [t = g(x)]$$

$$\int f(g(x))g'(x)dx = F(g(x))$$

$$\int f(ax+b)\,dx = \frac{1}{a}F(ax+b)$$

$$\int \frac{f'(x)}{f(x)}dx = \ln|f(x)|$$

$f(x)$ odd $\Rightarrow F(x)$ even

$f(x)$ even $\Rightarrow F(x)$ odd (if $F(0) = 0$)

9.9.3 Useful Integrals

$$\int x^a\,dx = \frac{x^{a+1}}{a+1} \quad (a \neq -1)$$

$$\int \frac{dx}{x} = \ln|x|$$

$$\int e^x\,dx = e^x$$

$$\int \sin x\,dx = -\cos x$$

$$\int \cos x\,dx = -\sin x$$

$$\int \frac{dx}{\sin^2 x} = -\cot x$$

$$\int \frac{dx}{\cos^2 x} = \tan x$$

$$\int \frac{dx}{a^2+x^2} = \frac{1}{a}\arctan\frac{x}{a}$$

$$\int \frac{dx}{\sqrt{a^2-x^2}} = \arcsin\frac{x}{a} \quad (a > 0)$$

$$\int \frac{dx}{\sqrt{a+x^2}} = \ln\left|x + \sqrt{x^2+a}\right|$$

$$\int \sinh x\,dx = \cosh x$$

$$\int \cosh x\,dx = \sinh x$$

9.9.4 Integrals of Rational Algebraic Functions (constants of integration are omitted)

1. $$\int (ax+b)^n\,dx = \frac{(ax+b)^{n+1}}{a(n+1)} \qquad (n \neq -1)$$

2. $$\int \frac{dx}{ax+b} = \frac{1}{a}\ln|ax+b|$$

The following formulas are useful for evaluating

$$\int \frac{P(x)dx}{(ax^2+bx+x)^n}$$

where $P(x)$ is a polynomial and $n > 1$ is an integer.

3. $$\int \frac{dx}{(ax^2+bx+c)} = \frac{2}{(4ac-b^2)^{1/2}} \arctan \frac{2ax+b}{(4ac-b^2)^{1/2}}, \qquad (b^2-4ac<0)$$

4. $$= \frac{1}{(b^2-4ac)^{1/2}} \ln \left| \frac{2ax+b-(b^2-4ac)^{1/2}}{2ax+b+(b^2-4ac)^{1/2}} \right|, \qquad (b^2-4ac>0)$$

5. $$= \frac{-2}{2ax+b}, \qquad (b^2-4ac=0)$$

6. $$\int \frac{x\,dx}{ax^2+bx+c} = \frac{1}{2a} \ln|ax^2+bx+c| - \frac{b}{2a} \int \frac{dx}{ax^2+bx+c}$$

7. $$\int \frac{dx}{(a+bx)(c+dx)} = \frac{1}{ad-bc} \ln \left| \frac{c+dx}{a+bx} \right| \qquad (ad \neq bc)$$

8. $$\int \frac{dx}{a^2+b^2x^2} = \frac{1}{ab} \arctan \frac{bx}{a}$$

9. $$\int \frac{x\,dx}{a^2+b^2x^2} = \frac{1}{2b^2} \ln|a^2+b^2x^2|$$

10. $$\int \frac{dx}{a^2-b^2x^2} = \frac{1}{2ab} \ln \left| \frac{a+bx}{a-bx} \right|$$

11. $$\int \frac{dx}{(x^2+a^2)^2} = \frac{1}{2a^3} \arctan \frac{x}{a} + \frac{x}{2a^2(x^2+a^2)}$$

12. $$\int \frac{dx}{(x^2-a^2)^2} = \frac{-x}{2a^2(x^2-a^2)} + \frac{1}{4a^3} \ln \left| \frac{a+x}{a-x} \right|$$

9.9.5 Integrals of Irrational Algebraic Functions

1. $$\int \frac{dx}{[(a+bx)(c+dx)]^{1/2}} = \frac{2}{(-bd)^{1/2}} \arctan \left[\frac{-d(a+bx)}{b(a+dx)} \right]^{1/2} \qquad (bd<0)$$

2. $$= \frac{-1}{(-bd)^{1/2}} \arcsin \left(\frac{2bdx+ad+bc}{bc-ad} \right) \qquad (b>0,\ d<0)$$

3. $$= \frac{2}{(bd)^{1/2}} \ln \left| [bd(a+bx)]^{1/2} + b(c+dx)^{1/2} \right| \qquad (bd>0)$$

4. $$\int \frac{dx}{(a+bx)^{1/2}(c+dx)} = \frac{2}{[d(bc-ad)]^{1/2}} \arctan \left[\frac{d(a+bx)}{(bc-ad)} \right]^{1/2} \qquad (d(ad-bc)<0)$$

5. $$= \frac{2}{[d(ad-bc)]^{1/2}} \ln \left| \frac{d(a+bx)^{1/2} - [d(ad-bc)]^{1/2}}{d(a+bx)^{1/2} + [d(ad-bc)]^{1/2}} \right| \qquad (d(ad-bc)>0)$$

6. $$\int [(a+bx)(c+dx)]^{1/2}$$

$$= \frac{(ad-bc)+2b(c+dx)}{4bd}[(a+bx)(c+dx)]^{1/2} - \frac{(ad-bc)^2}{8bd}\int \frac{dx}{[(a+bx)(c+dx)]^{1/2}}$$

7. $$\int \left[\frac{c+dx}{a+bx}\right]^{1/2} dx = \frac{1}{b}[(a+bx)(c+dx)]^{1/2} - \frac{(ad-bc)}{2b}\int \frac{dx}{[(a+bx)(c+dx)]^{1/2}}$$

8. $$\int \frac{dx}{(ax^2+bx+c)^{1/2}} = a^{-1/2}\ln\left|2a^{1/2}(ax^2+bx+c)^{1/2}+2ax+b\right| \qquad (a>0)$$

9. $$= a^{-1/2}\operatorname{arcsin h}\frac{(2ax+b)}{(4ac-b^2)^{1/2}} \qquad (a>0, \quad 4ac>b^2)$$

10. $$= a^{-1/2}\ln|2ax+b| \qquad (a>0, \quad b^2=4ac)$$

11. $$= -(-a)^{-1/2}\arcsin\frac{(2ax+b)}{(b^2-4ac)^{1/2}}$$

$$(a<0, \quad b^2>4ac, \quad |2ax+b|<(b^2-4ac)^{1/2})$$

12. $$\int (ax^2+bx+c)^{1/2}dx = \frac{2ax+b}{4a}(ax^2+bx+c)^{1/2} + \frac{4ac-b^2}{8a}\int \frac{dx}{(ax^2+bx+c)^{1/2}}$$

13. $$\int \frac{dx}{x(ax^2+bx+c)^{1/2}} = -\int \frac{dt}{(a+bt+ct^2)^{1/2}} \quad \text{where } t=1/x$$

14. $$\int \frac{x\,dx}{(ax^2+bx+c)^{1/2}} = \frac{1}{a}(ax^2+bx+c)^{1/2} - \frac{b}{2a}\int \frac{dx}{(ax^2+bx+c)^{1/2}}$$

15. $$\int \frac{dx}{(x^2\pm a^2)^{1/2}} = \ln\left|x+(x^2\pm a^2)^{1/2}\right|$$

16. $$\int (x^2\pm a^2)^{1/2}dx = \frac{x}{2}(x^2\pm a^2)^{1/2} \pm \frac{a^2}{2}\ln\left|x+(x^2\pm a^2)^{1/2}\right|$$

17. $$\int \frac{dx}{x(x^2+a^2)^{1/2}} = -\frac{1}{a}\ln\left|\frac{a+(x^2+a^2)^{1/2}}{x}\right|$$

18. $$\int \frac{dx}{x(x^2-a^2)^{1/2}} = \frac{1}{a}\arccos\frac{a}{x}$$

19. $$\int \frac{dx}{(a^2-x^2)^{1/2}} = \arcsin\frac{x}{a}$$

20. $$\int (a^2-x^2)^{1/2}dx = \frac{x}{2}(a^2-x^2)^{1/2} + \frac{a^2}{2}\arcsin\frac{x}{a}$$

21. $$\int \frac{dx}{x(a^2-x^2)^{1/2}} = -\frac{1}{a}\ln\left|\frac{a+(a^2-x^2)^{1/2}}{x}\right|$$

22. $$\int \frac{dx}{(2ax - x^2)^{1/2}} = \arcsin \frac{x-a}{a}$$

23. $$\int (2ax - x^2)^{1/2}\,dx = \frac{x-a}{a}(2ax - x^2)^{1/2} + \frac{a^2}{2}\arcsin\frac{x-a}{a}$$

24. $$\int \frac{dx}{(ax^2+b)(cx^2+d)^{1/2}} = \frac{1}{[b(ad-bc)]^{1/2}}\arctan\frac{x(ad-bc)^{1/2}}{[b(cx^2+d)]^{1/2}} \qquad (ad > bc)$$

25. $$= \frac{1}{2[b(bc-ad)]^{1/2}}\ln\left|\frac{[b(cx^2+d)]^{1/2} + x(bc-ad)^{1/2}}{[b(cx^2+d)]^{1/2} - x(bc-ad)^{1/2}}\right| \qquad (bc > ad)$$

9.9.6 Exponential, Logarithmic, and Trigonometric Functions

1. $\int R(e^{ax})\,dx$ Substitution: $e^{ax} = t,\ x = \frac{1}{a}\ln t,\qquad dx = \frac{dt}{at}$

2. $\int P(x)e^{ax}\,dx = [\text{integration by parts}] = \frac{1}{a}P(x)e^{ax} - \frac{1}{a}\int P'(x)e^{ax}\,dx$, etc.

   ($P(x)$ polynomial

3. $$\int x^a(\ln x)^n\,dx = \begin{cases} [\text{integration by parts}] = \dfrac{x^a+1}{a+1}(\ln x)^n - \\ \dfrac{n}{a+1}\displaystyle\int x^a(\ln x)^{n-1}\,dx, \text{ etc.} (a \neq -1) \\ \dfrac{(\ln x)^{n+1}}{n+1} \quad (a=-1) \end{cases}$$

   or $t = \ln x$ and use #2

4. $\int \frac{1}{x}f(\ln x)\,dx$ Substitution: $\ln x = t,\qquad \frac{dx}{x} = dt$

5. $\int f(\sin x)\cos x\,dx$ Substitution: $\sin x = t,\qquad \cos x\,dx = dt$

6. $\int f(\cos x)\sin x\,dx$ Substitution: $\cos x = t,\qquad -\sin x\,dx = dt$

7. $\int f(\tan x)\,dx$ Substitution: $\tan x = t,\qquad dx = \frac{dt}{1+t^2}$

8. $\int R(\cos x, \sin x)\,dx$ Substitution:

   $$\tan\frac{x}{2} = t,\qquad \sin x = \frac{2t}{1+t^2},\qquad \cos x = \frac{1-t^2}{1+t^2},\qquad dx = \frac{2dt}{1+t^2}$$

9. $\int \sin^n x\,dx \quad (n \geq 1)$ $\qquad$ n odd: Use $\sin^2 x = 1 - \cos^2 x$ and #6

   n even: Use $\sin^2 x = \frac{1}{2}(1 - \cos 2x)$ etc.

10. $\int \cos^n x\,dx \quad (n \geq 1)$ $\qquad$ n odd : Use $\cos^2 x = 1 - \sin^2 x$ and #5

n even : Use $\cos^2 x = \frac{1}{2}(1 + \cos 2x)$ etc.

11. $\int P(x)\begin{Bmatrix}\cos x \\ \sin x\end{Bmatrix} dx$ = Integration by parts, differentiating the polynomial #2.

($P(x)$ polynomial)

12. $\int P(x)e^{ax}\cos bx\,dx = \mathrm{Re}\int P(x)e^{(a+jb)x}\,dx$. Use #2.

13. $\int P(x)e^{ax}\sin bx\,dx = \mathrm{Im}\int P(x)e^{(a+jb)x}\,dx$. Use #2.

14. $\int x^n \arctan x\,dx$ = [integration by parts]

$$= \frac{x^{n+1}}{n+1}\arctan x - \frac{1}{n+1}\int \frac{x^{n+1}}{1+x^2}dx$$

15. $\int x^n \arcsin x\,dx$ = [integration by parts]

$$= \frac{x^{n+1}}{n+1}\arcsin x - \frac{1}{n+1}\int \frac{x^{n+1}}{\sqrt{1-x^2}}dx$$

16. $\int f(\arcsin x)dx$ $\quad$ Substitution: $\arcsin x = t,\ x = \sin t$

17. $\int f(\arctan x)dx$ $\quad$ Substitution: $\arctan x = t,\ x = \tan t$

9.9.7 Definite Integrals

$$I = \lim_{\max|x_i - x_{i-1}| \to 0} \sum_{i=1}^{m} f(\xi_i)(x_i - x_{i-1}) = \int_a^b f(x)dx$$

is the *definite integral* of $f(x)$ over (a,b) in the sense of Riemann integration.

9.9.8 Mean Value Theorem

$$\int_a^b f(x)dx = f(\xi)(b-a), \qquad \int_a^b f(x)g(x)dx = f(\xi)\int_a^b g(x)dx$$

where 1) $f(x)$, $g(x)$ are continuous in $[a,b]$, and 2) $g(x)$ does not change sign.

9.9.9 Improper Integrals

$$\int_a^\infty f(x)dx = \lim_{b\to\infty}\int_a^b f(x)dx$$

9.9.10 Cauchy Principal Value

$$\int_a^b f(x)\,dx = \lim_{\varepsilon \to 0^+}\left(\int_a^{c-\varepsilon} f(x)\,dx + \int_{c+\varepsilon}^b f(x)\,dx\right)$$

9.9.11 Convergence Test

$$\int_a^b g(x) \text{ converges} \Rightarrow \int_a^b f(x)\,dx \text{ converges}; \qquad 0 \le f(x) \le g(x)$$

$$\int_a^b |f(x)| \text{ converges} \Rightarrow \int_a^b f(x)\,dx \text{ converges}$$

9.9.12 Stieltjes Integral

The *Riemann-Stieltjes integral* of $f(x)$ with respect to $g(x)$ over the bounded interval $[a,b]$ is defined as

$$\int_a^b f(x)\,dg(x) = \lim_{\max|x_i - x_{i-1}|} \sum_{i=1}^{m} f(\xi_i)[g(x_i) - g(x_{i-1})]$$

for an arbitrary sequence of partitions

$$a = x_0 < \xi_1 < x_1 < \xi_2 < x_2 < \cdots < \xi_m < x_m = b$$

The limit exists whenever $g(x)$ is of bounded variation and $f(x)$ is continuous on $[a,b]$. (A real function $f(x)$ is of *bounded variation* in (a,b) if and only if there exists a real number M such that $\sum_{i=1}^{m} |f(x_i) - f(x_{i-1})| < M$ for all partitions $a = x_0 < x_1 < \cdots < x_m = b$. If $f(x)$ and $g(x)$ are of bounded variation so is $f(x) + g(x)$ and $f(x)g(x)$.)

9.9.13 Properties of Stieltjes Integrals

1. $\int_a^b f\,dg = -\int_b^a f\,dg \qquad \int_a^b f\,dg = \int_a^c f\,dg + \int_c^b f\,dg$

2. $\int_a^b (f_1 + f_2)\,dg = \int_a^b f_1\,dg + \int_a^b f_2\,dg$ and $\int_a^b f\,d(g_1 + g_2) = \int_a^b f\,dg_1 + \int_a^b f\,dg_2$

3. $\int_a^b (\alpha f)\,dg = \int_b^a f\,d(\alpha g) = \alpha \int_a^b f\,dg$

4. $\int_a^b f\,dg = fg\Big|_a^b - \int_a^b g\,df$

5. $\left|\int_a^b f\,dg\right| \le \int_b^a |f|\,dg, \quad g(x) = \text{nondecreasing}$

6. $\int_a^b f dg \le \int_a^b F dg, \quad g(x) = \text{nondecreasing}, \quad f(x) \le F(x)$

7. $\int_a^b f(x) dg(x) = \int_a^b f(x) \frac{dg(x)}{dx} dx, \quad g(x) = \text{continuous}$

9.10 Sequences and Series

9.10.1 Convergence

A sequence of real or complex numbers $s_0, s_1, s_2, \ldots,$ converges if and only if, for every positive real number ε, there exists a real integer N such that $m > N$, $n > N$ implies $|s_n - s_m| < \varepsilon$.

9.10.2 Test of Convergence

An infinite series, $a_0 + a_1 + \ldots,$ of real positive terms converges if there exists a real number N such $n > N$ implies one or more of the following:

1. $a_n \le M_n$ and/or $\frac{a_{n+1}}{a_n} \le \frac{m_{n+1}}{m_n}$ where $m_0 + m_1 + \ldots$ is a convergence comparison series with real positive terms.
2. At least one of the quantities

$$\frac{a_{n+1}}{a_n}, \quad \sqrt[n]{a_n}, \quad n\left(\frac{a_{n+1}}{a_n} - 1\right) + 2, \quad \left[n\left(\frac{a_{n+1}}{a_n} - 1\right) + 1\right] \ln n + 2$$ has an upper bound $A < 1$.

3. $a_n \le f(n)$, where $f(x)$ is a real positive decreasing function whose (improper) integral $\int_{N+1}^{\infty} f(x) dx$ exists.
4. An infinite series, $a_0 + a_1 + \ldots,$ of real terms converges
   a) If successive terms are alternatively positive and negative (alternating series), decrease in absolute value, and $\lim_{n\to\infty} a_n = 0$.
   b) If the sequence $s_0, s_1, s_2, \ldots,$ of the partial sums is bounded and monotonic.
5. Given a decreasing sequence of real positive numbers $\alpha_0, \alpha_1, \alpha_2, \ldots$ the infinite series $\alpha_0 a_0 + \alpha_1 a_1 + \alpha_2 a_2, \ldots$ converges
   a) If the series $a_0 + a_1 + a_2 + \ldots,$ converges (Abel's test)
   b) If $\lim_{n\to\infty} a_n = 0$ and $\sum_{k=0}^{n} a_k$ is bounded for all n. (Dirichlet's test)
6. If $\{a_n\}$ is decreasing and $a_n \to 0$ as $n \to \infty$, then $\sum_{n=1}^{\infty} (-1)^n a_n$ converges. (Leibniz's test)

Example

1. $\sum_{n=1}^{\infty} \left(\frac{x}{2}\right)^n \begin{cases} \text{conv. for } |x| < 2 \\ \text{div. for } |x| > 2 \end{cases}$ (Root test)

2. $\sum_{n=1}^{\infty} \frac{1}{n^p}$ and $\sum_{n=1}^{\infty} \frac{1}{n(\ln n)^p}$ $\begin{cases} \text{conv. for } p > 1 \\ \text{div. for } p \le 1 \end{cases}$ (integral set)

3. $\sum_{n=1}^{\infty}\left(1-\cos\frac{1}{n}\right)$ conv.

comparison test : $1-\cos\frac{1}{n}=\frac{1}{2n^2}+O\left(\frac{1}{n^4}\right)$ and $\sum_{1}^{\infty}\frac{1}{n^2}$ conv.

4. $\sum_{1}^{\infty}\frac{(-1)^n}{\sqrt{n}}$ conv. (Leibniz' test)

5. $\sum_{k=1}^{\infty}\frac{1}{\sqrt{k}}e^{jkx}$ conv. for $x \neq 2m\pi$. (Dirichlet's test)

9.11 Absolute and Relative Errors

If x_0 is an approximation to the true value of x, then

9.11.1 The *absolute error* of x_0 is $\Delta x = x_0 - x$, $x - x_0$ is the correction to x.

9.11.2 The *relative error* x_0 is $\delta x = \frac{\Delta x}{x} \approx \frac{\Delta x}{x_0}$

9.11.3 The *percentage error* is 100 times the relative error. The absolute error of the sum or difference of several numbers is at most equal to the sum of the absolute errors of the individual numbers.

9.11.4 If $f(x_1,x_2,\ldots,x_n)$ is a function of $x_1,x_2,\ldots,x_n$ and the absolute error in $x_i (i = 1,2,\ldots,n)$ is Δx_i, then the absolute error in f is

$$\Delta f \approx \frac{\partial f}{\partial x_1}\Delta x_1 + \frac{\partial f}{\partial x_2}\Delta x_2 + \cdots + \frac{\partial f}{\partial x_n}\Delta x_n$$

9.11.5 The relative error of the product or quotient of several factors is at most equal to the sum of the relative errors of the individual factors.

9.11.6 If $y = f(x)$, the relative error $\delta y = \frac{\Delta y}{y} \approx \frac{f'(x)}{f(x)}\Delta x$

Approximate Values:

If $|\varepsilon| << 1$, $|\eta| << 1$, $b << a$,

9.11.7 $(a+b)^k \approx a^k + ka^{k-1}b$

9.11.8 $(1+\varepsilon)(1+\eta) \approx 1+\varepsilon+\eta$

9.11.9 $\frac{1+\varepsilon}{1+\eta} \approx 1+\varepsilon-\eta$

9.12 Convergence of Infinite Series

9.12.1 A sequence $s_0(x)$, $s_1(x)$,… of real or complex function *converges uniformly* on a set S of values of x if and only if for every positive real number ε there exists a real number N independent of x such that $m > N$, $n > N$ implies $|s_n(x) - s_m(x)| < \varepsilon$ for x in S. (Cauchy's test)

9.12.2 An infinite series $a_0(x) + a_1(x) + a_2(x) + \ldots$ of real or complex functions converges uniformly and absolutely on every set S of values of x such that $|a_n(x)| \le M_n$ for all n, where $M_0 + M_1 + M_2 + \ldots$ is a convergent comparison series of real positive terms. (Weierstrass' test)

9.12.3 Given a decreasing sequence of real positive terms $\alpha_0, \alpha_1, \alpha_2, \ldots$, the infinite series $\alpha_0 a_0(x) + \alpha_1 a_1(x) + \alpha_2 a_2(x) + \ldots$ converges uniformly on a set S of values x

1. If the infinite series $a_0(x) + a_1(x) + a_2(x) + \ldots$ converges uniformly on S (Abel's test)

2. If $\lim_{n\to\infty} \alpha_n = 0$ and there exists a real number $A \geq \left|\sum_{k=0}^{n} a_k(x)\right|$ for all n and all x in S. (Dirichlet's test)

9.12.4 Assume (i) $f_n(x) \to f(x)$ pointwise on $[a,b]$

(ii) $|f_n(x)| < M$, all n and $x \in [a,b]$

(iii) $f_n(x), f(x)$ integrable

Then

$$\lim_{n\to\infty} \int_a^b f_n(x)dx = \int_a^b f(x)dx \quad \text{(Arzela's theorem)}$$

9.12.5 Assume (i) $\{f_n(x)\}_1^\infty$ increasing, i.e., $f_n(x) \leq f_{n+1}(x)$, all n, x (or decreasing)

(ii) $f_n(x) \to f(x)$ pointwise on $[a,b]$

(iii) $f_n(x), f(x)$ continuous on $[a,b]$

Then the convergence is uniform. (Dini's theorem)

9.13 Series of Functions

9.13.1 Representation of Functions by Infinite Series and Integrals

A function $f(x)$ is often represented by a corresponding infinite series

$$\sum_{k=1}^{\infty} \alpha_k \varphi_k(x)$$

because

1. A sequence of partial sums may yield useful numerical approximations to $f(x)$.
2. It may be possible to describe operations on $f(x)$ in terms of simpler operations on the functions $\varphi_k(x)$ or on the coefficients α_k (transform methods). The functions $\varphi_k(x)$ and the coefficients α_k may have an intuitive (physical) meaning.

9.13.2 Power Series

$$f(x) = \sum_{n=0}^{\infty} a_n x^n = a_0 x + a_1 x + a_2 x^2 + \cdots, \quad a_n = \frac{1}{n!}\frac{d^n f(0)}{dx^n}$$

$$f(x) = \sum_{n=0}^{\infty} a_n (x - x_0)^n, \quad a_n = \frac{1}{n!}\frac{d^n f(x_0)}{dx^n}$$

where x and a can be real or complex.

9.13.3 Taylor's Expansion

Given a real function $f(x)$ such that $\frac{d^n f(x)}{dx^n} \doteq f^{(n)}$ exists $a \le x < b$

$$\left.\begin{aligned} f(x) &= f(a) + f'(a)(x-a) + \frac{1}{2!} f''(a)(x-a)^2 + \cdots \\ &\quad + \frac{1}{(n-1)!} f^{n-1}(a)(x-a)^{n-1} + R_n(x) \\ \text{with } |R_n(x)| &\le \frac{|x-a|^n}{n!} \sup_{a<\xi<x} \left|f^{(n)}(\xi)\right| \end{aligned}\right\} \quad (a \le x < b)$$

9.13.4 Taylor-Series Expansion

Given a function $f(x)$ such that all derivatives $f^{(k)}(x)$ exist and $\lim_{n\to\infty} R_n(x) = 0$ for $a \le x < b$,

$$f(x) = \sum_{k=0}^{\infty} \frac{1}{k!} f^{(k)}(a)(x-a)^k, \quad (a \le x < b)$$

and the series converges uniformly to $f(x)$ for $a \le x < b$. (Taylor-series expansion of $f(x)$ about $x = a$.)

9.13.5 Order Concepts

a) $f(x) = O(x^a)$ as $x \to 0$ means: $f(x) = x^a F(x)$, where $F(x)$ is bounded in a neighborhood of $x = 0$.
b) $f(x) = o(x^a)$ as $x \to 0$ means: $f(x)/x^a \to 0$ as $x \to 0$.

$$O(x^5) \pm O(x^5) = O(x^5); \qquad O(x^3) \pm O(x^4) = O(x^3);$$

$$x^2 O(x^3) = O(x^5) = x^5 O(1); \qquad O(x^2)\,O(x^3) = O(x^5)$$

c) $f(x) - \sum_{k=0}^{n-1} a_k x^{-k} = O(x^{-n})$ as $x \to \infty$

Example $f(x) = e^{-4x^2}$, $a = 0$, $n = 6$

$$f(x) = [t = -4x^2] = e^t = 1 + t + \frac{t^2}{2} + O(t^3) = 1 - 4x^2 + 8x^4 + O(x^6)$$

Example $f(x) = e^x \sin x$, $a = 0$, $n = 5$

$$f(x) = \left(1 + x + \frac{x^2}{2} + \frac{x^3}{6} + O(x^4)\right)\left(x - \frac{x^3}{6} + O(x^5)\right) = x + x^2 + \frac{x^3}{3} + O(x^5)$$

Example $f(x) = e^{(x-1)^2}$, $a = 0$, $n = 3$

$$f(x) = e^{x^2-2x+1} = ee^{x^2-2x} = (t = x^2 - 2x) = ee^t = e\left(1 + t + \frac{t^2}{2} + O(t^3)\right)$$

$$= e\left(1 - 2x + 3x^2\right) + O(x^3)$$

Example $$\frac{1}{(1-x)^2} = \frac{d}{dx}\left(\frac{1}{1-x}\right) = \frac{d}{dx}(1 + x + x^2 + \cdots) = 1 + 2x + 3x^2 + \cdots$$

9.13.6 Multiple Taylor Expansion

$$f(x,y)=f(0,0)+x\frac{\partial f(0,0)}{\partial x}+y\frac{\partial f(0,0)}{\partial y}+\cdots+\frac{1}{n!}\left[x^n\frac{\partial^n f(0,0)}{\partial x^n}+\cdots\right]$$

$$+\frac{1}{(n+1)!}\left[x^{n+1}\frac{\partial^{n+1} f(0,0)}{\partial x^{n+1}}+(n+1)x^n y\frac{\partial^{n+1} f(0,0)}{\partial x^n\partial y}+\cdots\right]$$

$$f(x,y)=f(0,0)+x\frac{\partial f(0,0)}{\partial x}+y\frac{\partial f(0,0)}{\partial y}+\frac{1}{2!}\left[x^2\frac{\partial^2 f(0,0)}{\partial x^2}+2xy\frac{\partial^2 f(0,0)}{\partial x\partial y}+y^2\frac{\partial^2 f(0,0)}{\partial y^2}\right]$$

$$+\frac{1}{3!}\left[x^3\frac{\partial^3 f(0,0)}{\partial x^3}+3x^2y\frac{\partial^3 f(0,0)}{\partial x^2\partial y}+3xy^2\frac{\partial^2 f(0,0)}{\partial x\partial y^2}+y^3\frac{\partial^3 f(0,0)}{\partial y^3}\right]+\cdots$$

9.14 Sums and Series

9.14.1 $$(1+x)^\alpha=\sum_{k=0}^{\infty}\binom{\alpha}{k}x^k \qquad (-1<x<1)$$

9.14.2 $$\sum_{k=1}^{\infty}kx^k=\frac{x}{(1-x)^2} \qquad (-1<x<1)$$

9.14.3 $$\sum_{k=1}^{\infty}\frac{x^k}{k}=-\ln(1-x) \qquad (-1\le x<1)$$

9.14.4 $$\sum_{k=1}^{n}e^{kx}=e^x\cdot\frac{e^{nx}-1}{e^x-1}=\frac{\sinh\frac{nx}{2}}{\sinh\frac{x}{2}}e^{(n+1)x/2} \qquad (x\neq 0)$$

9.14.5 $$\sum_{k=0}^{\infty}e^{-kx}=\frac{1}{1-e^{-x}} \qquad (x>0)$$

9.14.6 $$\sum_{k=1}^{n}e^{jkx}=e^{jx}\cdot\frac{1-e^{jnx}}{1-e^{jx}}=\frac{\sin\frac{nx}{2}}{\sin\frac{x}{2}}\cdot e^{j(n+1)x/2} \qquad (x\neq 2m\pi)$$

9.14.7 $$\sum_{k=1}^{n}\sin kx=\operatorname{Im}\sum_{k=0}^{n}e^{ikx}=\frac{\sin\frac{nx}{2}\sin\frac{(n+1)x}{2}}{\sin\frac{x}{2}} \qquad \text{(cf. 9.14.6)}$$

9.14.8 $$\sum_{k=0}^{n}\cos kx=\operatorname{Re}\sum_{k=0}^{n}e^{jkx}=\frac{\cos\frac{nx}{2}\sin\frac{(n+1)x}{2}}{\sin\frac{x}{2}} \qquad \text{(cf. 9.14.6)}$$

9.14.9 $\sum_{k=1}^{n-1} r^k \sin kx = \operatorname{Im} \sum_{k=0}^{n-1} (re^{jx})^k = \frac{r\sin x(1-r^n \cos nx)-(1-r\cos x)r^n \sin nx}{1-2r\cos x + r^2}$

9.14.10 $\sum_{k=0}^{n-1} r^k \cos kx = \operatorname{Re} \sum_{k=0}^{n-1} (re^{jx})^k = \frac{(1-r\cos x)(1-r^n \cos nx)+r^{n+1}\sin x \sin nx}{1-2r\cos x + r^2}$

9.14.11 $\sum_{k=1}^{n-1} \sin\frac{k\pi}{n} = \cot\frac{\pi}{2n}$

9.14.12 $e = \sum_{k=0}^{\infty} \frac{1}{k!} = 2.7182818284\cdots$ (transcendental)

9.14.13 $\pi = 4\arctan 1 = 4\sum_{k=0}^{\infty} \frac{(-1)^k}{2k+1} = 3.1415926535\cdots$ (transcendental)

9.14.14 $\ln 2 = \sum_{k=1}^{\infty} \frac{(-1)^{k-1}}{k} = \sum_{k=1}^{\infty} \frac{1}{k\cdot 2^k} = 0.69315\cdots$ (transcendental)

9.14.15 $\gamma = \lim_{n\to\infty}\left(\sum_{k=1}^{n} \frac{1}{k} - \ln n\right) = 0.577215665\cdots$ (Euler's constant, irrational)

9.14.16 $\sum_{k=1}^{\infty} kx^{k-1} = \frac{1}{(1-x)^2}$ $(-1 < x < 1)$

9.14.17 $\sum_{k=0}^{\infty} \frac{x^k}{k!} = e^x,\ 0! = 1,\ R_n(x) = \frac{e^{\theta x}}{n!}x^n$ $(0 < \theta < 1)$ $-\infty < x < \infty$

9.14.18 $\sum_{k=0}^{\infty} \frac{(x\ln a)^k}{k!} = a^x$ $-\infty < x < \infty$

9.14.19 $\sum_{k=0}^{\infty} \frac{x^{2k+1}}{(2k+1)!} = \sinh x$ $-\infty < x < \infty$

9.14.20 $\sum_{k=0}^{\infty} \frac{x^{2n}}{(2n)!} = \cosh x$ $-\infty < x < \infty$

9.14.21 $\sum_{k=1}^{\infty} (-1)^{k-1}\frac{x^k}{k} = \ln(1+x)$ $-1 < x < 1$

9.14.22 $\ln a + \sum_{k=1}^{\infty} \frac{(-1)^{k-1}}{k}\left(\frac{x}{a}\right)^k = \ln(a+x)$ $-a < x < a$

9.14.23 $\sum_{k=1}^{\infty} \frac{1}{k}\left(\frac{x}{1+x}\right)^k = \ln(1+x)$

9.14.24 $\sum_{k=0}^{\infty} (-1)^k \frac{x^{2k+1}}{(2k+1)!} = \sin x, \quad R_{2k+1}(x) = (-1)^k \frac{\cos\theta x}{(2n+1)!} x^{2k+1}, \qquad 0<\theta<1, \qquad -\infty < x < \infty$

9.14.25 $\sum_{k=0}^{\infty} (-1)^k \frac{x^{2k}}{(2k)!} = \cos x, \quad R_{2k}(x) = (-1)^k \frac{\cos\theta x}{(2k)!} x^{2k}, \qquad 0<\theta<1, \qquad -\infty < x < \infty$

9.14.26 $\sum_{k=0}^{n} \binom{n}{k} a^{n-k} b^k = (a+b)^n$

9.14.27 $\sum_{k=0}^{n} \binom{n}{k} = 2^n$

9.14.28 $\sum_{k=0}^{n} k\binom{n}{k} = n2^{n-1}$

9.14.29 $\sum_{k=0}^{n} k^2\binom{n}{k} = (n^2+n)2^{n-2}$

9.14.30 $\sum_{k=1}^{n} k = \frac{n(n+1)}{2}$

9.14.31 $\sum_{k=1}^{n} k^2 = \frac{n(n+1)(2n+1)}{6}$

9.14.32 $\sum_{k=1}^{n} k^3 = \frac{n^2(n+1)^2}{4}$

9.15 Lagrange's Expansion

9.15.1 If $y = f(x)$, $y_0 = f(x_0)$, $f'(x_0) \neq 0$, then

$$x = x_0 + \sum_{k=1}^{\infty} \frac{(y-y_0)^k}{k!}\left[\frac{d^{k-1}}{dx^{k-1}}\left\{\frac{x-x_0}{f(x)-y_0}\right\}^k\right]_{x=x_0}$$

9.15.2 $g(x) = g(x_0) + \sum_{k=1}^{\infty} \frac{(y-y_0)^k}{k!}\left[\frac{d^{k-1}}{dx^{k-1}}\left(g'(x)\left\{\frac{x-x_0}{f(x)-y_0}\right\}^k\right)\right]_{x=x_0}$

where $g(x)$ is any function indefinitely differentiable.

9.16 Orthogonal Polynomial

9.16.1 Gram-Schmidt Orthogonalization Process

If the system $\{f_i\}$is linear independent, we can replace by an equivalent orthonormal system ($\|f\|$ = norm = $\langle f,f\rangle^{1/2}$ = inner product)

$$\varphi_1 = \frac{f_1}{\|f_1\|};\ \varphi_2 = \frac{f_2 - \langle f_2,\varphi_1\rangle\varphi_1}{\|f_2 - \langle f_2,\varphi_1\rangle\varphi_1\|};\ \varphi_k = \frac{f_k - \sum_{i=1}^{k-1}\langle f_k,\varphi_i\rangle\varphi_i}{\left\| f_k - \sum_{i=1}^{k-1}\langle f_k,\varphi_i\rangle\varphi_i \right\|}, \qquad k = 1,2,\cdots,n$$

9.16.2 Gram-Schmidt Orthogonalization Matrix Form Process

$$d_k = \begin{vmatrix} \langle f_1,f_1\rangle & \langle f_1,f_2\rangle & \cdots & \langle f_1,f_k\rangle \\ \langle f_2,f_1\rangle & \langle f_2,f_2\rangle & \cdots & \langle f_2,f_k\rangle \\ \vdots & & & \\ \langle f_{k-1},f_1\rangle & \langle f_{k-1},f_2\rangle & \cdots & \langle f_{k-1},f_k\rangle \\ f_1 & f_2 & \cdots & f_k \end{vmatrix}, \qquad \varphi_k = \frac{d_k}{\langle d_k,d_k\rangle^{1/2}}$$

Example If $f_1(x) = 1$ and $f_2(x) = x$ and $f_3(x) = x^2$ in the range $(-1,1)$, we obtain

$$\varphi_0(x) = \frac{1}{\left[\int_{-1}^{1} dx\right]^{1/2}} = \frac{1}{\sqrt{2}}; \qquad d_1(x) = \begin{vmatrix} \langle 1,1\rangle & \langle 1,x\rangle \\ 1 & x \end{vmatrix} = \begin{vmatrix} 2 & 0 \\ 1 & x \end{vmatrix} = 2x$$

$$\varphi_1(x) = \frac{2x}{\|2x\|} = \frac{2x}{\left[\int_{-1}^{1} 4x^2 dx\right]^{1/2}} = \frac{x}{[2/3]^{1/2}} = \sqrt{\frac{3}{2}}\,x;$$

$$d_2(x) = \begin{vmatrix} \langle 1,1\rangle & \langle 1,x\rangle & \langle 1,x^2\rangle \\ \langle x,1\rangle & \langle x,x\rangle & \langle x,x^2\rangle \\ 1 & x & x^2 \end{vmatrix} = \begin{vmatrix} 2 & 0 & 2/3 \\ 0 & 2/3 & 0 \\ 1 & x & x^2 \end{vmatrix} = \frac{4}{3}x^2 - \frac{4}{3}; \qquad \varphi_2(x) = \frac{\frac{4}{3}x^2 - \frac{4}{9}}{\left\|\frac{4}{3}x^2 - \frac{4}{9}\right\|} = \sqrt{\frac{5}{8}}\,(3x^2 - 1)$$

9.16.3 Recurrence Formula

If $\varphi_n(x) = k_n x^n + k_{n-1}x^{n-1} + \ldots$, then $\{\varphi_n\}$ satisfy

$$\varphi_{n+1} - (A_n x + B_n)\varphi_n + C_n\varphi_{n-1} = 0 \qquad n = 0,1,2,\cdots$$

$$A_n = \frac{k_{n+1}}{k_n}, \quad C_n = \frac{A_n}{A_{n-1}}, \quad C_o = 0$$

9.16.4 Christoffel-Darboux Formula

If $K_n(x,y) = \sum_{k=0}^{n} \varphi_k(x)\varphi_k(y)$ then

$$K_n(x,y)=\sum_{k=0}^{n}\varphi_k(x)\varphi_k(y)=\frac{k_n}{k_{n+1}}\left[\frac{\varphi_n(y)\varphi_{n+1}(x)-\varphi_n(x)\varphi_{n+1}(y)}{x-y}\right]$$

In the limit as $y\to x$

$$K_n(x,y)=\sum_{k=0}^{n}\varphi_k^2(x)\geq 0$$

9.16.5 Weierstrass Approximation Theorem

If $f(x)$ is a continuous function defined on the closed and bounded interval $[a,b]$, then it is possible to find a polynomial $p(x)$ such that $|f(x)-p(x)|<\varepsilon$ for all $x\in[a,b]$, $\varepsilon>0$.

9.16.6 Zeros of the Orthogonal Polynomials

The polynomial of $\varphi_n(x)$ has n real, simple zeros, all in the interval (a,b).

9.16.7 Zeros of $\varphi_n(x)$ and $\varphi_{n+1}(x)$

The zeros of $\varphi_n(x)$ and $\varphi_{n+1}(x)$ alternate on the interval (a,b), and $\varphi_n(x)$ and $\varphi_{n+1}(x)$ do not vanish simultaneously.

9.16.8 Minimization of a Function

Among all polynomials of degree n, there is precisely one for which $\|f(x)-p_n(x)\|$ is minimized. It is given by

$$p_n(x)=\sum_{k=0}^{n}\alpha_k\varphi_k(x),\qquad \alpha_k=\langle f,\varphi_k\rangle=\int_a^b w(x)f(x)\varphi_k(x)\,dx$$

$$\left(\int_a^b w(x)f^2(x)\,dx<\infty,\qquad w(x)=\text{weighting function}\right)$$

9.16.9 Zeros of $\varphi_n(x)$ and $\varphi_m(x)$

If $m>n$, then between any two zeros of $\varphi_n(x)$, $\varphi_m(x)$ has to vanish at least once.

9.16.10 Zeros in an Interval

Let (α,β) be a subinterval of the finite and bounded interval (a,b). Then, for sufficiently large n, $\varphi_n(x)$ vanishes at least once in (α,β).

9.17 Completeness of Orthonormal Polynomials

9.17.1 Hilbert Space

The space $L_2(w)$ consisting of real function $f(x)$ for which $\int_a^b w(x)f^2(x)\,dx<\infty$ holds, with the inner product

$$\langle f,g\rangle = \int_a^b w(x)f(x)g(x)dx$$

and the norm $\|f\| = \int_a^b [w(x)f^2(x)dx]^{1/2} = \langle f,f\rangle^{1/2}$ is a Hilbert space.

9.17.2 Completeness

An orthonormal set $\{\varphi_n\}$ is complete in $L_2(w)$ if for every $f(x)$ in $L_2(w)$ we have $\|f\|^2 = \sum_{n=0}^{\infty} \alpha_k^2$, $\alpha_k = \langle f,\varphi_k\rangle$. (*Parseval's* inequality)
The set of orthonormal polynomials $\{\varphi_n\}$ is complete in the space $L_2(w)$.

9.17.3 Closed Space

The set $\{\varphi_n\}$ in a Hilbert space is closed if from $\langle f,\varphi_k\rangle = 0,\ n \ge 0$, we conclude that $f = 0$.

9.17.4 Countably Infinite

An orthonormal set $\{\varphi_n\}$ in $L_2(w)$ is countably infinite.

10

Signals and Their Characterization

10.1 Common One-Dimensional Continuous Signals

10.1.1 Step Function

$$u(t)=\begin{cases}1 & t>0\\ 0 & t<0\end{cases}$$

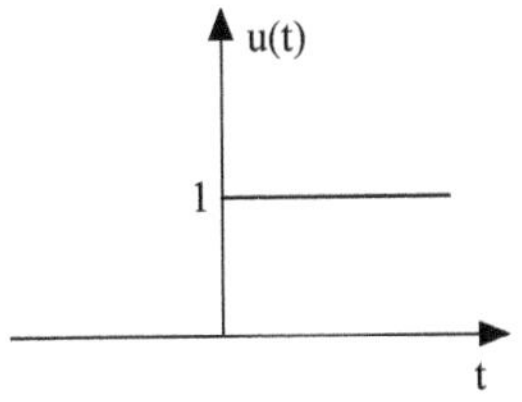

10.1.2 Pulse Function

$$p_a(t)=\begin{cases}1 & -a<t<a\\ 0 & \text{otherwise}\end{cases}$$

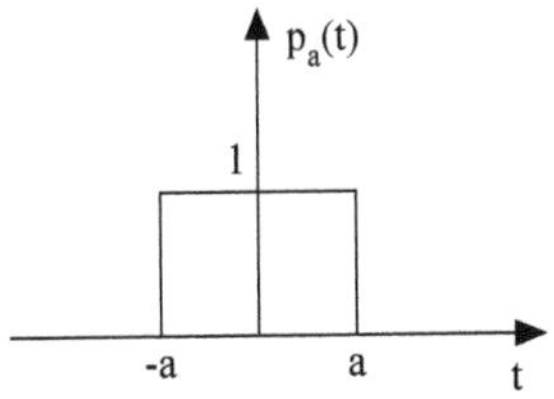

10.1.3 Ramp Function

$$r(at) \doteq ar(t) = \begin{cases} at & t \geq 0 \\ 0 & \text{otherwise} \end{cases}$$

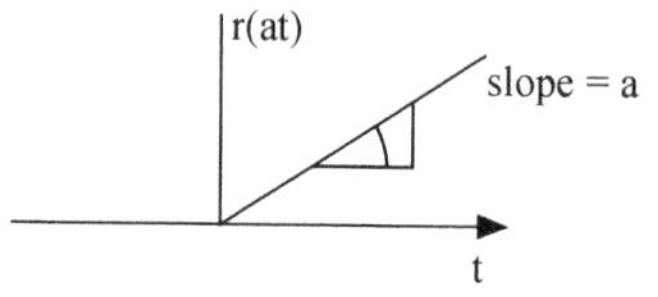

10.1.4 Triangular Function

$$\wedge_a(t) = \begin{cases} 1 - \dfrac{|t|}{a} & |t| \leq a \\ 0 & \text{otherwise} \end{cases}$$

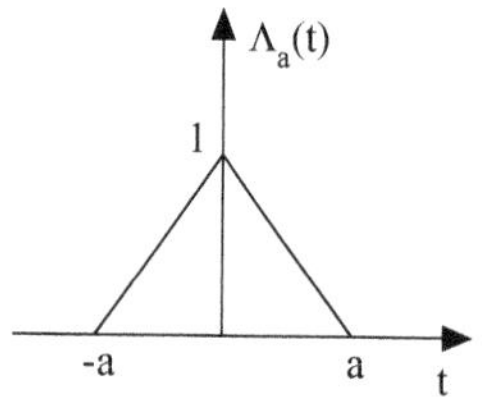

10.1.5 Signum Function

$$\text{sgn}(t) = \begin{cases} 1 & t > 0 \\ 0 & t = 0 \\ -1 & t < 0 \end{cases}$$

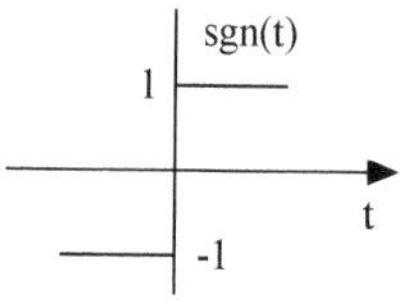

10.1.6 Sinc Function

$$\text{sinc}_a(t) = \frac{\sin at}{t} \qquad -\infty < t < \infty$$

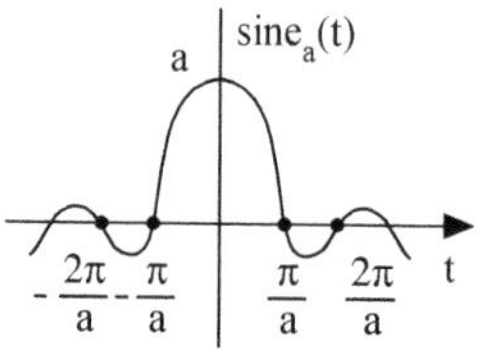

10.1.7 Gaussian Function

$$g_a(t) = e^{-at^2} \qquad -\infty < t < \infty$$

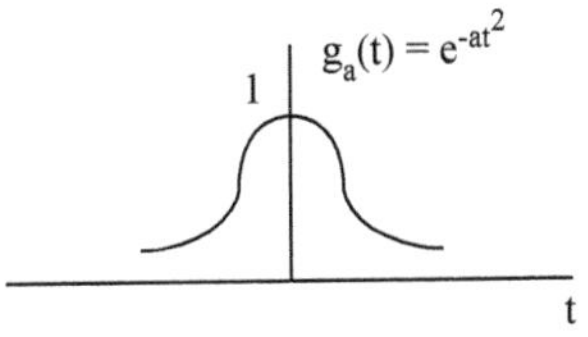

10.1.8 Exponential Function

$$x(t) = e^{-at}u(t) \qquad 0 \le t < \infty$$

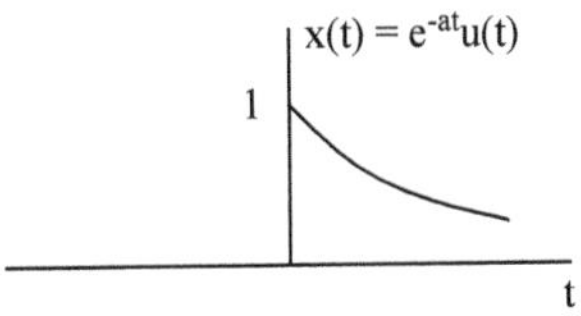

10.1.9 Double Exponential Function

$$x(t) = e^{-a|t|} \qquad -\infty < t < \infty$$

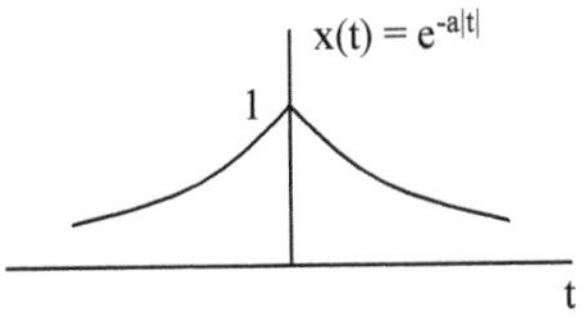

10.1.10 Exponentially Decaying Cosine Function

$$x(t) = \begin{cases} e^{-at}\cos\omega_o t u(t) & t \ge 0 \\ 0 & \text{otherwise} \end{cases}$$

10.1.11 Hyperbolic Cosine Function

$$x(t) = \frac{a}{2}(e^{t/a} + e^{-t/a}) = a\cosh\frac{t}{a} \qquad -\infty < t < \infty$$

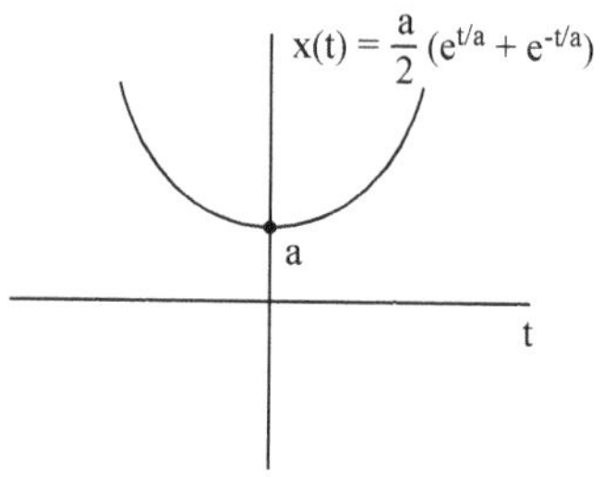

10.1.12 Cosecant Function

$$x(t) = \csc t = \frac{1}{\sin t} \qquad -\infty < t < \infty$$

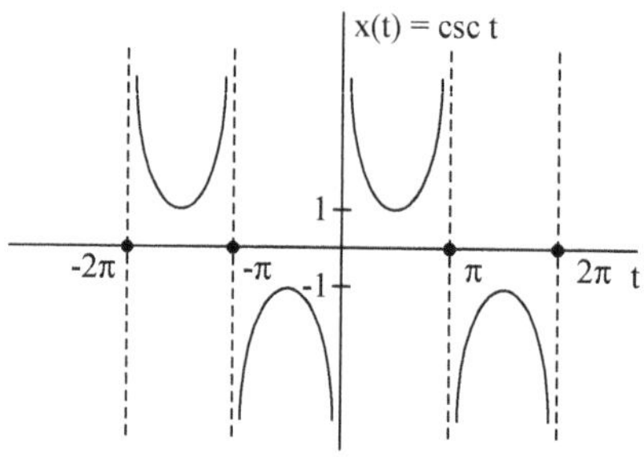

10.1.13 Cotangent Function

$$x(t) = \cot t = \frac{\cos t}{\sin t} \qquad -\infty < t < \infty$$

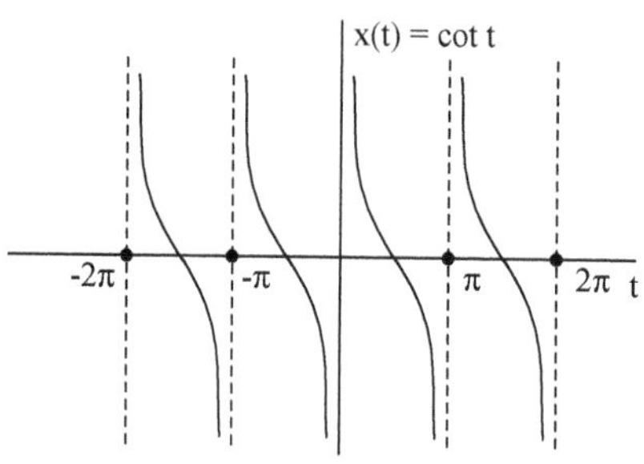

10.1.14 Cubical parabola

$$x(t) = at^3 \qquad a > \qquad -\infty < t < \infty$$

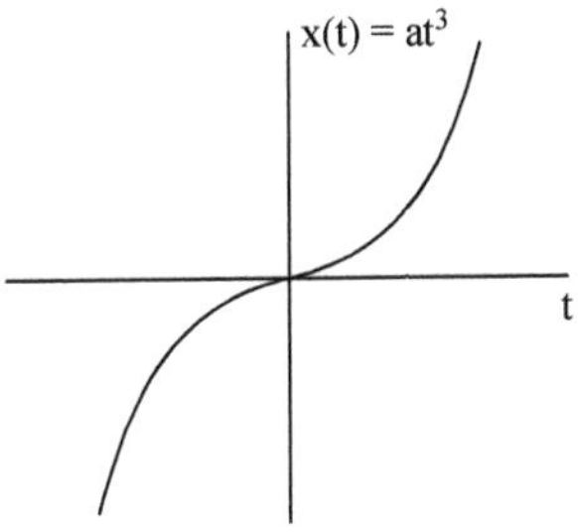

10.1.15 Ellipse

$$\frac{t^2}{a^2} + \frac{y^2}{b^2} = 1 \qquad t = a\cos\phi, \quad y = b\sin\phi$$

$$BF' = BF = a, \qquad PF' + PF = 2a$$

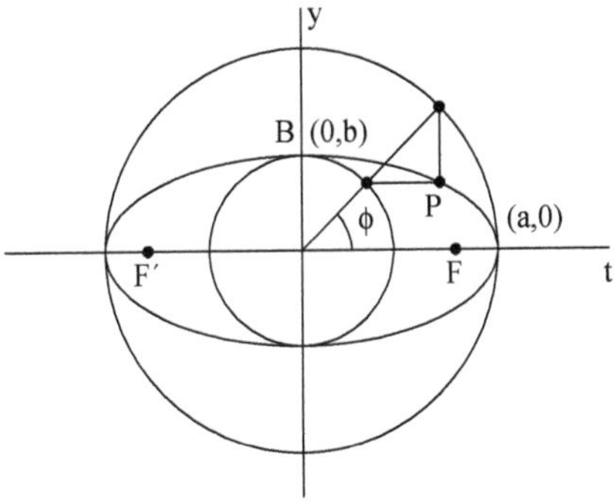

10.1.16 Arcsine Function

$$x = \arcsin t \qquad -1 \le t \le 1$$

$$\sin x = t$$

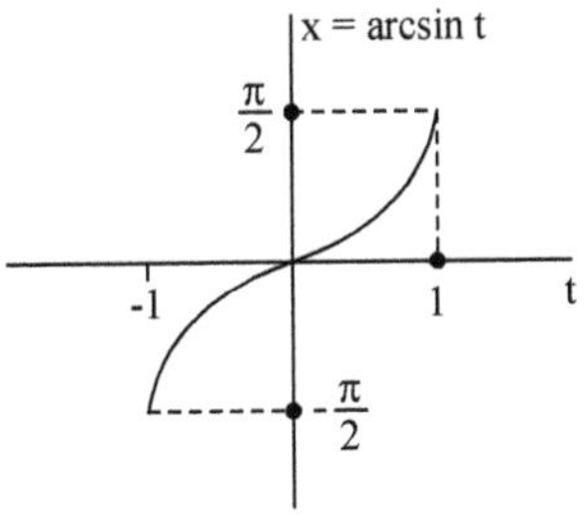

10.1.17 Arctangent Function

$$x = \arctan t \qquad -\infty < t < \infty$$

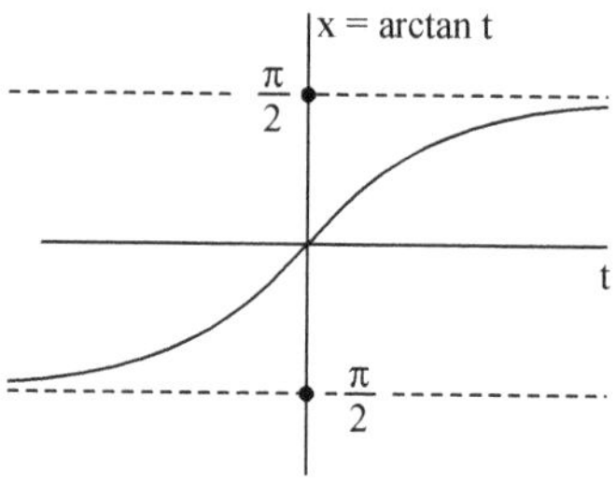

10.1.18 Logarithmic Function

$$x(t) = \log_a t \qquad 0 < t < \infty$$

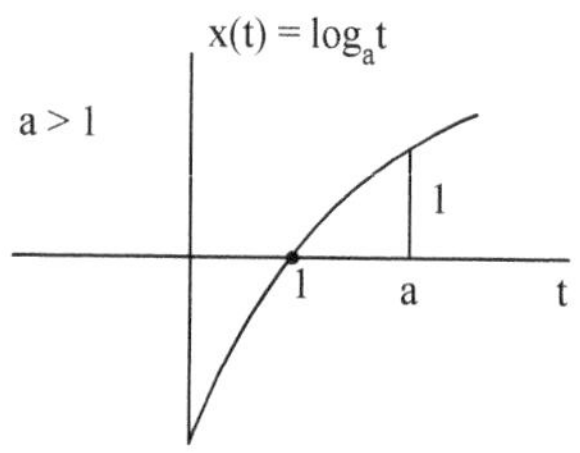

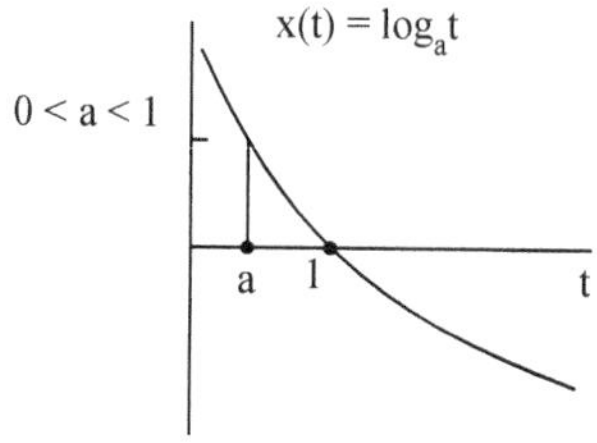

10.1.19 Parabola Function

$$x(t) = at^2 + bt + c \qquad a > 0$$

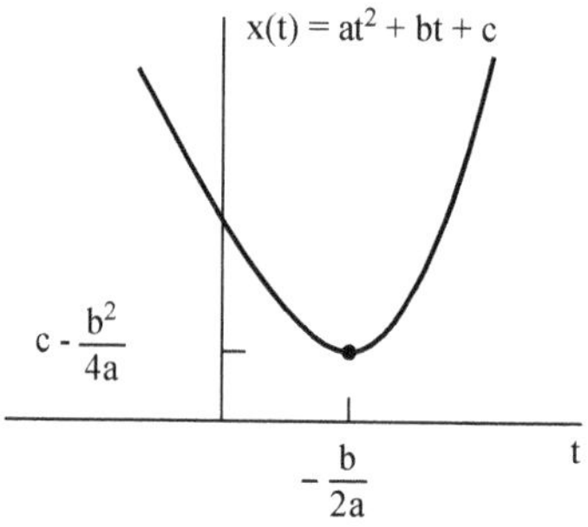

10.1.20 Power Function

$$x(t) = t^{-2} \qquad -\infty < t < \infty$$

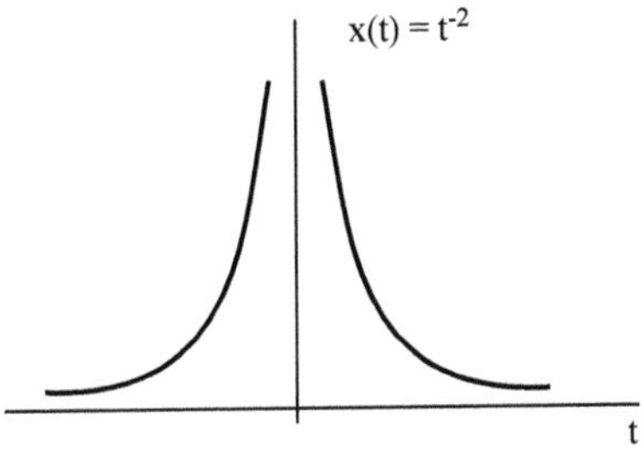

10.1.21 Equilateral Hyperbola

$$x(t) = t^{-1} \qquad -\infty < t < \infty$$

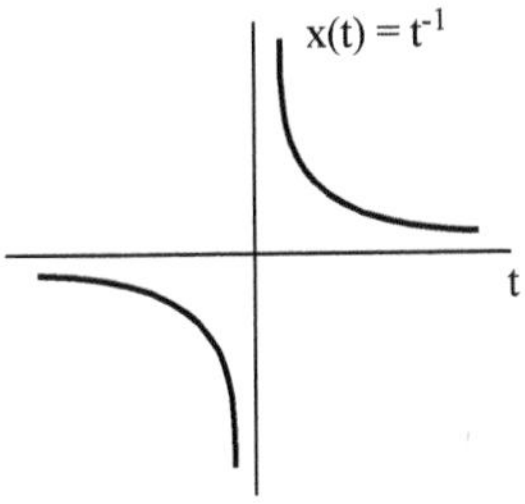

10.1.22 Cubical Parabola

$$x(t) = t^{1/3} \qquad -\infty < t < \infty$$

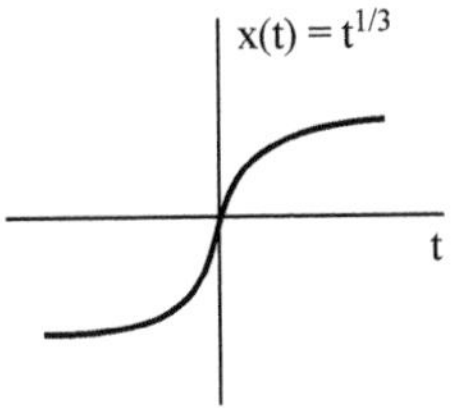

10.1.23 Secant Function

$$x(t) = \sec t = \frac{1}{\cos t} \qquad -\infty < t < \infty$$

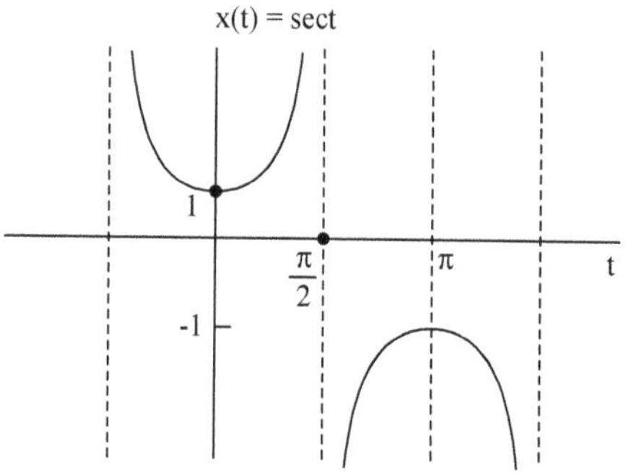

10.2 One-Dimensional Discrete Signal

10.2.1 Unit Sample Sequence

$$\delta(nT_s) = \begin{cases} 0 & n \neq 0 \\ 1 & n = 0 \end{cases}$$

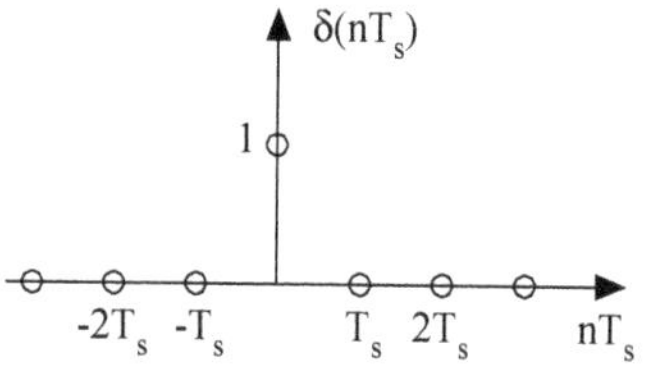

10.2.2 Unit Step Sequence

$$u(nT_s) = \begin{cases} 1 & n \geq 0 \\ 0 & n < 0 \end{cases}$$

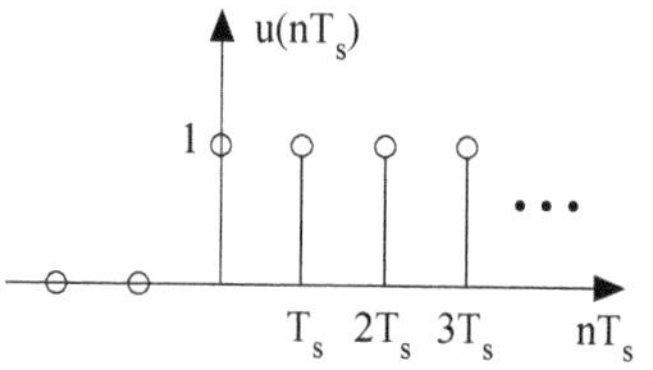

10.2.3 Real Exponential Sequence

$$f(nT_s) = \begin{cases} e^{-anT_s} & n \geq 0 \\ 0 & n < 0 \end{cases} \qquad a = \text{positive constant}$$

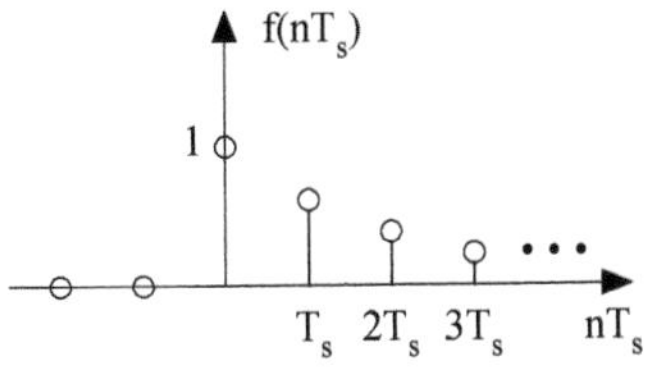

10.2.4 Sinc Function

$$\operatorname{sinc}_a(nT_s) = \frac{\sin anT_s}{nT_s} \qquad -\infty < n < \infty$$

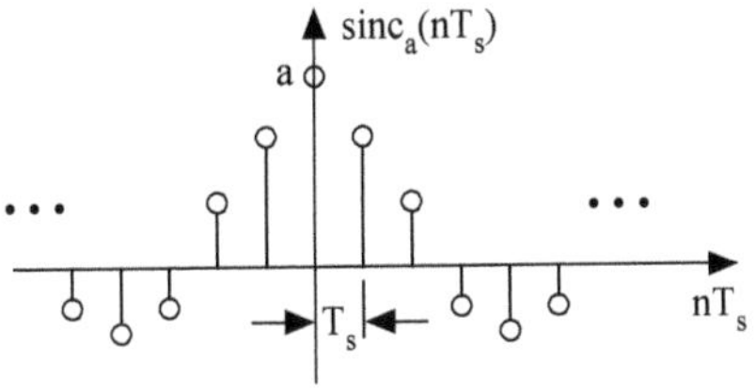

10.2.5 Double Exponential

$$f(nT_s) = e^{-a|nT_s|} \qquad -\infty < n < \infty$$

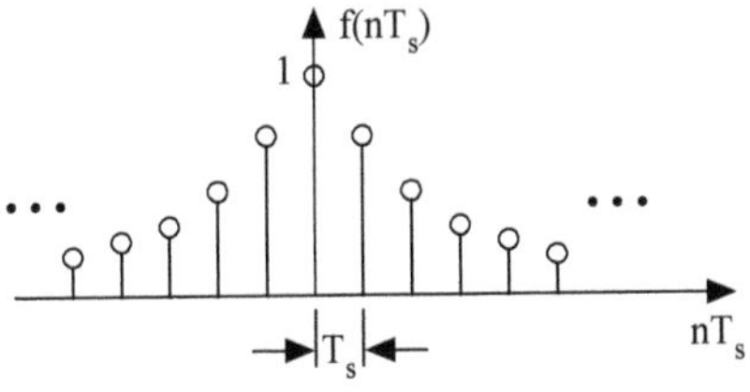

10.2.6 Exponentially Decaying Cosine Function

$$f(nT_s) = e^{-anT_s} \cos \omega_o nT_s \, u(nT_s) \quad \text{for } n \geq 0$$

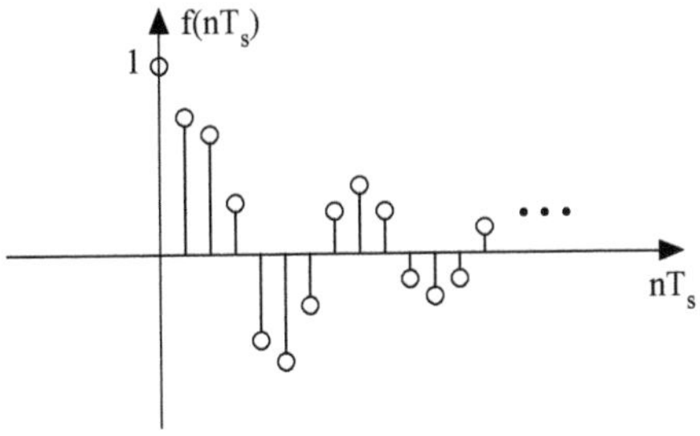

and zero otherwise. a and ω_o are positive constants.

10.2.7 Comb Function

$$comb_{T_s}(t) = \sum_{n=-\infty}^{\infty} \delta(t - nT_s)$$

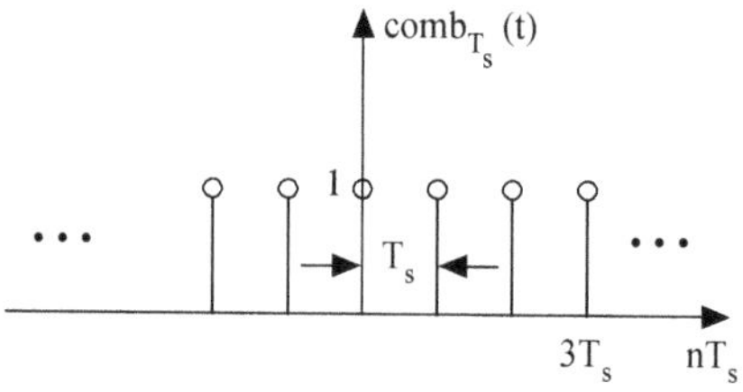

10.3 Two-Dimensional Continuous Signal

10.3.1 The Rectangle Function

$$f(x,y) = p_a(x - x_o) = \begin{cases} 1 & x_o - a \le x \le x_o + a \\ 0 & \text{otherwise} \end{cases}$$

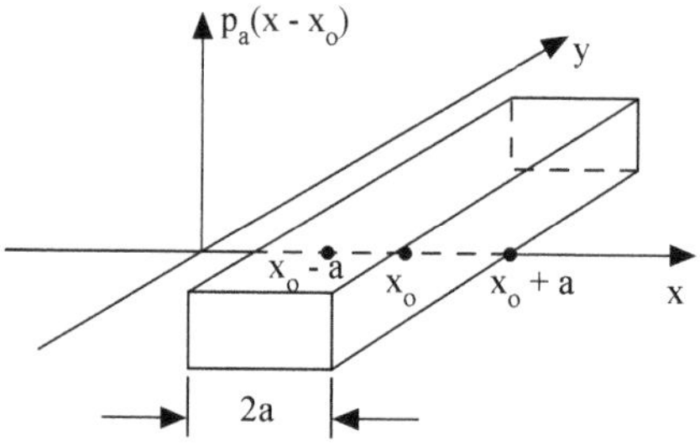

10.3.2 The Pulse Function

$$f(x,y) = p_a(x - x_o)p_b(y - y_o) = \begin{cases} 1 & x_o - a \le x \le x_o + a,\ y_o - b \le y \le y_o + b \\ 0 & \text{otherwise} \end{cases}$$

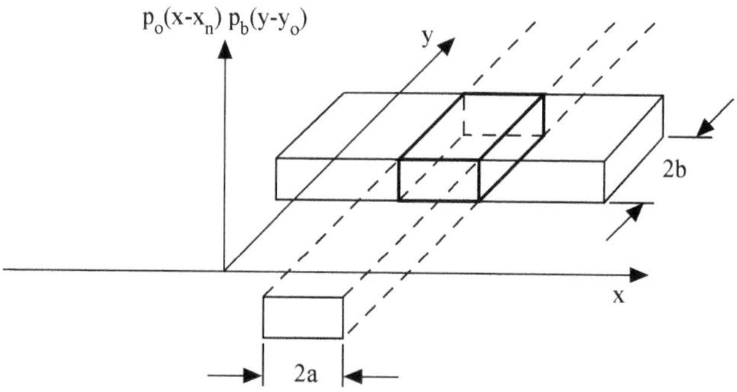

10.3.3 The Cosine Function

$$f(x,y) = \cos\omega_0 x$$

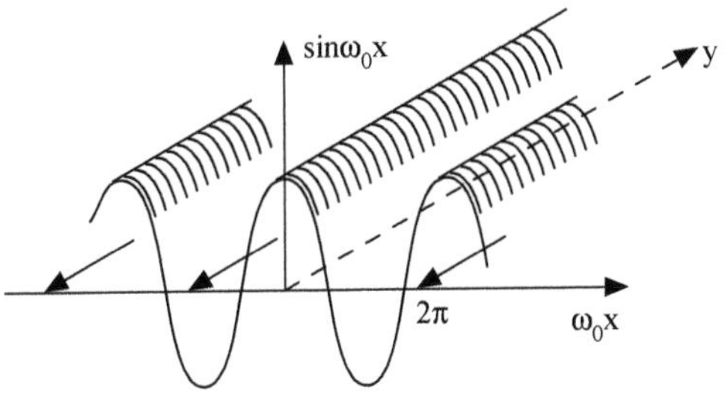

10.3.4 The Cosine-Cosine Function

$$f(x,y) = \cos\omega_1 x \cos\omega_2 x$$

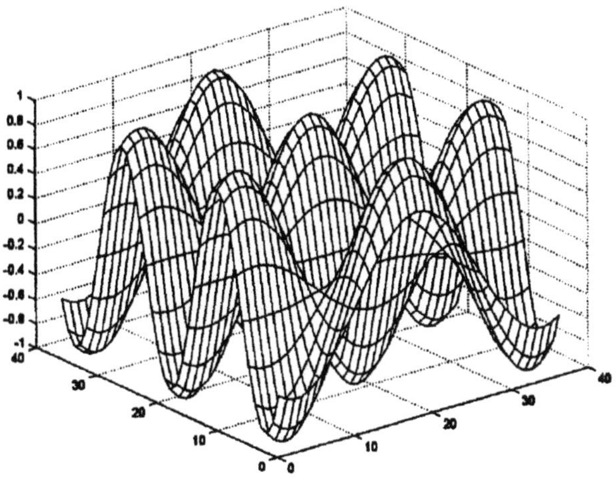

10.3.5 The Gaussian Function

$$f(x,y) = e^{-ax^2-by^2}$$

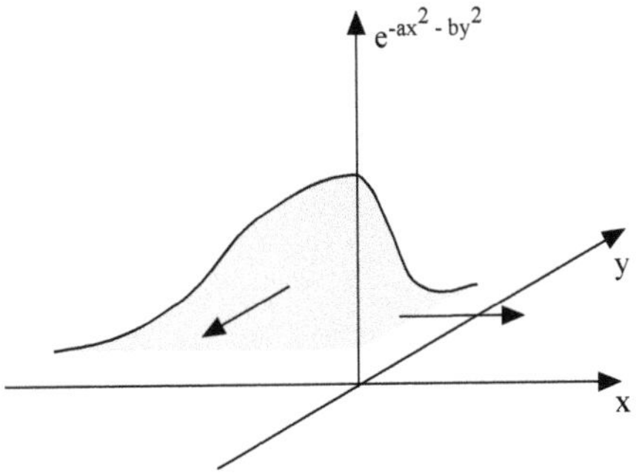

10.3.6 The Sinc Function

$$f(x,y) = \frac{\sin ax}{x}$$

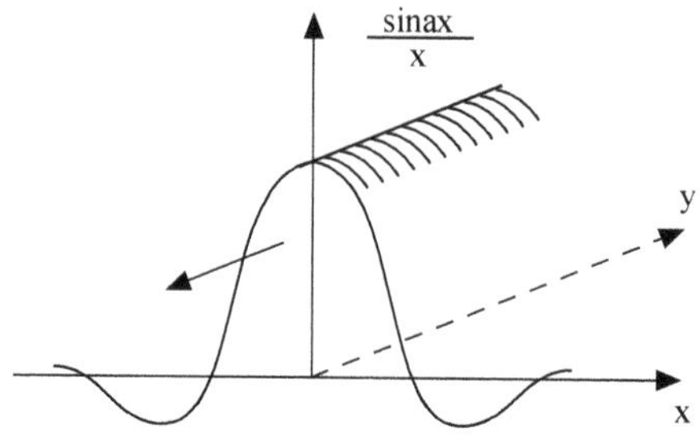

11

Discrete Fourier Transform

11.1 Definitions

11.1.1 Discrete Fourier Transform of Sequences

$$X(k) = DFT\{x(n)\} = \sum_{n=0}^{N-1} x(n) e^{-j\frac{2\pi}{N}kn} = \sum_{n=0}^{N-1} x(n) W^{kn}, \qquad k = 0,1,2,\cdots,N-1$$

$$x(n) = IDFT\{X(k)\} = \tfrac{1}{N}\sum_{k=0}^{N-1} X(k) e^{j\frac{2\pi}{N}nk} = \sum_{k=0}^{N-1} X(k) W^{-kn}, \qquad n = 0,1,2,\cdots,N-1$$

where $W = e^{-j\frac{2\pi}{N}}$. The spectrum has symmetry at N/2 or at frequency $\frac{2\pi}{N} \times \frac{N}{2} = \pi$ radians and is periodic every N or every 2π radians. Both $x(n)$ and $x(k)$ are periodic with period N.

11.1.2 Discrete Fourier Transform of Sampled Functions

$$X(k) = T\sum_{n=0}^{N-1} x(n) e^{-j\frac{2\pi}{N}kn} = T\sum_{n=0}^{N-1} x(n) W^{kn}, \qquad k = 0,1,2,\cdots,N-1$$

$$x(n) = \frac{1}{NT}\sum_{k=0}^{N-1} X(k) e^{j\frac{2\pi}{N}kn} = \frac{1}{NT}\sum_{k=0}^{N-1} X(k) W^{-kn}, \qquad n = 0,1,2,\cdots,N-1$$

where $x(n) \equiv x(nT)$ and $X(k) \equiv X(k2\pi/NT)$. The spectrum has symmetry at N/2 or at $\frac{2\pi}{NT} \times \frac{N}{2} = \pi/T$ radians and is periodic every N or every $2\pi/T$ radians. Both $X(k)$ and $x(n)$ are periodic because of the definition by their transform.

$$X(k) = X(k + Ni) \qquad i = 0, \pm 1, \pm 2, \cdots$$

$$x(n) = x(n + Ni) \qquad i = 0, \pm 1, \pm 2, \cdots$$

11.2 Properties

11.2.1 Table 11.1 Presents the Properties of Discrete Fourier Transform.

TABLE 11.1 Discrete Fourier Transform Properties

Sequence (period N)	DFT (period N)
$x(n),\ x_1(n),\ x_2(n)$	$X(k),\ X_1(k),\ X_2(k)$
$ax_1(n)+\ bx_2(n)$ (linearity)	$aX_1(k)+\ bX_2(k)$
$X(n)$ (symmetry)	$Nx(-k)$
$x(n-m)$ (time shifting)	$W^{km}X(k)$
$W^{-\ell n}x(n)$ (frequency shifting)	$X(k-\ell)$
$\sum_{m=0}^{N-1} x_1(m)x_2(n-m)$ (periodic time convolution)	$X_1(k)X_2(k)$
$x_1(n)x_2(n)$ (periodic frequency convolution)	$\frac{1}{N}\sum_{\ell=0}^{N-1} X_1(\ell)X_2(k-\ell)$
$x^*(n)$ (conjugate)	$X^*(-k)$
$x^*(-n)$ (Hermitian conjugate)	$X^*(k)$
$\mathrm{Re}\{x(n)\}$ (real part)	$X_e(k)=\frac{1}{2}[X(k)+X^*(-k)]$
$j\mathrm{Im}\{x(n)\}$ (imaginary part)	$X_o(k)=\frac{1}{2}[X(k)-X^*(-k)]$
$x_e(k)=\frac{1}{2}[x(n)+x^*(-n)]$ (even)	$\mathrm{Re}\{X(k)\}$
$x_o(n)=\frac{1}{2}[x(n)-x^*(-n)]$ (odd)	$j\mathrm{Im}\{X(k)\}$
$x(n)$ (real)	$X(k)=X^*(-k)$
$x(n)$ (real)	$\mathrm{Re}\{X(k)\}=\mathrm{Re}\{X(-k)\}$
$x(n)$ (real)	$\mathrm{Im}\{X(k)\}=-\mathrm{Im}\{X(-k)\}$
$x(n)$ (real)	$\lvert X(k)\rvert=\lvert X(-k)\rvert$
$x(n)$ (real)	$\arg\{X(k)\}=-\arg\{X(-k)\}$
$x_e(k)=\frac{1}{2}[x(n)+x(-n)]$ (even and real)	$\mathrm{Re}\{X(k)\}$
$x_o(n)=\frac{1}{2}[x(n)-x(-n)]$ (odd and real)	$j\mathrm{Im}\{X(k)\}$

11.3 Convolutions

11.3.1 Cyclic Convolution

$$y(n)=\sum_{m=0}^{N-1} x(m)h([n-m]_N)=x(n)*_N h(n) \qquad n=0,1,2,\cdots,N-1$$

where $y(n)$, $h(n)$ are sequences of length N. The symbol $[n-m]_N$ indicates the residue of $n-m$ evaluated modulo N, and $*_N$ denotes ***cyclic*** convolution of length N.

11.3.2 DFT of Cyclic Convolution (see Section 11.3.1)

$$Y(k)=X(k)H(k) \qquad k=0,1,2,\cdots,N-1$$

$$y(n)=IDFT\{X(k)H(k)\}=\frac{1}{N}\sum_{k=0}^{N-1} X(k)H(k)W^{-kn} \qquad n=0,1,2,\cdots,N-1$$

11.3.3 Polynomial of *X*(*z*) of *x*(*n*)

$$X(z) = \sum_{n=0}^{N-1} x(n) z^n$$

11.3.4 Noncylic Convolution of *x*(*n*) and *h*(*n*)

$$Y(z) = X(z)H(z)$$

The elements of the output sequence $y(n)$ is of $M + N - 1$ length and can be found as the coefficients of the polynomial $Y(z)$. M is the length of $h(n)$, and N is the length of $x(n)$.

11.3.5 Cyclic Convolution

$$Y(z) = H(z)H(z) \bmod(Z^N - 1)$$

The above indicates that the product polynomial is reduced modulo the polynomial $(Z^N - 1)$.

Example 11.1: Cyclic Convolution

$$x(n) = \{1,2,-1\},\ h(n) = \{5,1,4\},\ Y(z) = X(z)H(z) = (1 + 2z - z^2)(5 + z + 4z^2)$$

$$= 5 + 10z - 5z^2 + z + 2z^2 - z^3 + 4z^2 + 8z^3 - 4z^4 \bmod(z^3 - 1) = 5 + 11z + z^2 + 7z^3 - 4z^4$$

$$= 5 + 11z + z^2 + 7z^0 z^3 - 4zz^3 = 12 + 7z + z^2,\ y(n) = \{12,7,1\}.$$

Example 11.2: Noncyclic Convolution

From 3.5.1 $Y(z) = 5 + 11z + z^2 + 7z^3 - 4z^4$, $y(n) = \{5,11,1,7,-4\}$. Length of $y(n)$ is $3 + 3 - 1 = 5$.

11.4 Fast Fourier Transform Programs

Program 11.1: Radix-2 DIF FFT

```
CC ================================================================= CC
CC                                                                   CC
CC     Subroutine CFFT2DF (X,Y,M):                                   CC
CC            A Cooley-Tukey in-place, radix-2 complex FFT program   CC
CC            Decimation-in-frequency, cos/sin in second loop        CC
CC                                                                   CC
CC     Input/output:                                                 CC
CC            X    Array of real part of input/output (length > = N) CC
CC            Y    Array of imaginary part of input/output (length > = N) CC
CC            M    Transform length is N = 2**M                      CC
CC                                                                   CC
CC     Author:                                                       CC
CC            H. V. Sorensen, University of Pennsylvania, Dec. 1984  CC
CC                 Internet address: hvs@ee.upenn.edu                CC
CC                                                                   CC
CC         This program may be used and distributed freely provided  CC
CC         this header is included and left intact.                  CC
CC                                                                   CC
```

```
CC ======================================================================= CC
          SUBROUTINE CFFT2DF (X,Y,M)
          Real X(1), Y(1)
C ------------ Main FFT loops ---------------------------------------------CC
        N = 2**M
        N2 = N
        DO 10 K = 1, M
              N1 = N2
              N2 = N2/2
              E = 6.283185307179586/N1
              A = 0
              DO 20 J = 1, N2
                    C = cos(A)
                    S = sin(A)
                    A = J*E
                    DO 30 I = J, N, N1
                          L = I + N2
                          XT = X(I) – X(L)
                          X(I) = X(I) + X(L)
                          YT = Y(I) – Y(L)
                          Y(I) = Y(I) + Y(L)
                          X(L) = C*XT + S*YT
                          Y(L) = C*YT – S*XT
30                  CONTINUE
20            CONTINUE
10      CONTINUE
C ------------ Digit reverse counter ----------------------------------------CC
100     J = 1
        DO 104 I = 1, N-1
              IF (I . GE . J) GOTO 101
                    XT = X(J)
                    X(J) = X(I)
                    X(I) = XT
                    XT = Y(J)
                    Y(J) = Y(I)
                    Y(I) = XT
101           K = N/2
102           IF (K.GE.J) GOTO 103
                    J = J - K
                    K = K/2
                    GOTO 102
103           J = J + K
104     CONTINUE
        RETURN
        END
```

Program 11.2: Radix-2 DIT FFT

```
CC ======================================================================= CC
CC                                                                         CC
CC      Subroutine CFFT2DF (X,Y,M):                                        CC
CC              A Cooley-Tukey in-place, radix-2 complex FFT program       CC
```

```
CC              Decimation-in-time, cos/sin in second loop                        CC
CC                                                                                CC
CC      Input/output:                                                             CC
CC              X     Array of real part of input/output (length > = N)           CC
CC              Y     Array of imaginary part of input/output (length > = N)      CC
CC              M     Transform length is N = 2**M                                CC
CC                                                                                CC
CC      Author:                                                                   CC
CC              H. V. Sorensen, University of Pennsylvania, Dec. 1984             CC
CC                                 Internet address: hvs@ee.upenn.edu             CC
CC                                                                                CC
CC              This program may be used and distributed freely provided          CC
CC              this header is included and left intact.                          CC
CC                                                                                CC
CC ============================================================================== CC
        SUBROUTINE CFFT2DF (X,Y,M)
        Real X(1), Y(1)
        N = 2**M
C -------------Digit reverse counter -------------------------------------------------CC
100     J = 1
        DO 104 I = 1, N-1
                IF (I . GE . J) GOTO 101
                        XT = X(J)
                        X(J) = X(I)
                        X(I) = XT
                        XT = Y(J)
                        Y(J) = Y(I)
                        Y(I) = XT
101             K = N/2
102             IF (K.GE.J) GOTO 103
                        J = J – K
                        K = K/2
                        GOTO 102
103             J = J + K
104     CONTINUE
C -------------Main FFT loops ---------------------------------------------------------CC
        N1 = 1
        DO 10 K = 1, M
                N2 = N1
                N1 = N2*2
                E = 6.283185307179586/N1
                A = 0
                DO 20 J = 1, N2
                        C = cos (A)
                        S = sin (A)
                        A = J*E
                        DO 30 I = J, N, N1
                                L = I + N2
                                X(T) = C*X(L) + S*Y(L)
                                Y(T) = C*Y(L) – S*X(L)
                                X(L) = X(I) – XT
```

```
                    X(I) = X(I) + XT
                    Y(L) = Y(I) * YT
                    Y(I) = Y(I) + YT
30              CONTINUE
20          CONTINUE
10      CONTINUE
        RETURN
        END
```

Program 11.3: Split-Radix FFT Without Table Look-up

```
CC ================================================================ CC
CC                                                                  CC
CC    Subroutine CFFTSR (X,Y,M):                                    CC
CC        An in-place, split-radix complex FFT program              CC
CC        Decimation-in-frequency, cos/sin in second loop           CC
CC        and is computed recursively                               CC
CC                                                                  CC
CC    Input/output:                                                 CC
CC        X    Array of real part of input/output (length > = N)    CC
CC        Y    Array of imaginary part of input/output (length > = N) CC
CC        M    Transform length is N = 2**M                         CC
CC                                                                  CC
CC    Calls:                                                        CC
CC        CSTAGE, CBITREV                                           CC
CC                                                                  CC
CC    Author:                                                       CC
CC        H. V. Sorensen, University of Pennsylvania, Dec. 1984     CC
CC                         Internet address: hvs@ee.upenn.edu       CC
CC    Modified:                                                     CC
CC        H. V. Sorensen, University of Pennsylvania, July 1987     CC
CC                                                                  CC
CC    Reference:                                                    CC
CC        Sorensen, Heideman, Burrus: "On computing the split-radix CC
CC        FFT," IEEE Trans. ASSP, Vol. ASSP-34, No. 1, pp. 152-156, CC
CC        Feb. 1986                                                 CC
CC                                                                  CC
CC        This program may be used and distributed freely provided  CC
CC        this header is included and left intact.                  CC
CC                                                                  CC
CC ================================================================ CC
        SUBROUTINE CFFTSR (X,Y,M)
        Real X(1), Y(1)
        N = 2**M
C ------------L shaped butterflies--------------------------------CC
        N2 = 2*N
        DO 10 K = 1, M-1
            N2 = N2/2
            N4 = N2/4
            CALL CSTAGE (N, N2, N4, X(1), X(N4+1), X(2*N4+1), X(3*N4+1),
     $                          Y(1), Y(N4+1), Y(2*N4+1), Y(3*N4+1) )
```

```
10        CONTINUE
C -------------Length two butterflies ---------------------------------------------------------------------------CC
          IS = 1
          ID = 4
20              DO 30 I1          = IS, N, ID
                     T1           = X(I1)
                     X(I1)        = T1 + X(I1+1)
                     X(I1+1)      = T1 – X(I1+1)
                     T1           = Y(I1)
                     Y(I1)        = TI + Y(I1+1)
                     Y(I1+1)      = TI – Y(I1+1)
30              CONTINUE
                IS = 2*ID – 1
                ID = 4*ID
            IF (IS.LT.N) GOTO 20
C -------------Digit reverse counter ----------------------------------------------------------------------------CC
          CALL CBITREV (X,Y,M)
          RETURN
          END
CC ============================================================================================= CC
CC                                                                                               CC
CC        Subroutine CSTAGE – the workhorse of the CFFTSR                                        CC
CC                computes one stage of a complex split-radix length N                           CC
CC                transform.                                                                     CC
CC                                                                                               CC
CC        Author:                                                                                CC
CC                H. V. Sorensen, University of Pennsylvania, July 1987                          CC
CC                                                                                               CC
CC                This program may be used and distributed freely provided                       CC
CC                this header is included and left intact.                                       CC
CC                                                                                               CC
CC ============================================================================================= CC
          SUBROUTINE CSTAGE (N, N2, N4, X1, X2, X3, X4, Y1, Y2, Y3, Y4)
          REAL X1(1), X2(1), X3(1), X4(1), Y1(1), Y2(1), Y3(1), Y4(1)
          N8 = N4/2
C -------------zero butterfly-------------------------------------------------------------------------------------CC
          IS = 0
          ID = 2*N2
10              DO 20 I1= IS+1, N, ID
                     T1           = X1(I1) - X3(I1)
                     X1(I1)       = X1(I1) + X3(I1)
                     T2           = Y2(I1) – Y4(I1)
                     Y2(I1)       = Y2(I1) + Y4(I1)
                     X3(I1)       = TI + T2
                     T2           = T1 – T2
                     T1           = X2(I1) – X4(I1)
                     X2 (I1)      = X2(I1) + X4(I1)
                     X4(I1)       = T2
                     T2           = Y1(I1) – Y3(I1)
                     Y1(I1)       = Y1(I1) + Y3(I1)
                     Y3(I1)       = T2 – T1
```

```
                  Y4(I1)  = T2 + T1
30            CONTINUE
              IS = 2*ID – N2
              ID = 4*ID
         IF (IS. LT. N) GOTO 10
C
         IF (N8–1) 100, 100, 60
C -------------N/8 butterfly ---------------------------------------------------------------------------------------C
30       IS = 0
         ID = 2*N2
40            DO 50 I1 = IS + 1 + N8, N, ID
                  T1     = X1(I1) – X3(I1)
                  X1(I1) = X1(I1) + X3(I1)
                  T2     = X2(I1) – X4(I1)
                  X2(I1) = X2(I1) + X4(I1)
                  T3     = Y1(I1) – Y3(I1)
                  Y1(I1) = Y1(I1) + Y3(I1)
                  T4     = Y2(I1) – Y4(I1)
                  Y2(I1) = Y2(I1) + Y4(I1)
                  T5     = (T4 – T1)*0.707106778
                  T1     = (T4 + T1)*0.707106778
                  T4     = (T3 – T2)*0.707106778
                  T2     = (T3 + T2)*0.707106778
                  X3(I1) = T4 + T1
                  Y3(I1) = T4 – T1
                  X4(I1) = T5 + T2
                  Y4(I1) = T5 – T2
50            CONTINUE
              IS = 2*ID – N2
              ID = 4*ID
         IF (IS . LT. N–1) GOTO 40
C
         IF (N8–1) 100, 100, 60
C -------------General butterfly. Two at a time -----------------------------------------------------------------C
60       E = 6.283185307179586/N2
         SS1 = SIN(E)
         SD1 = SS1
         SD3 = 3 . *SD1–4 . *SD1**3
         SS3 = SD3
         CC1 = COS (E)
         CD1 = CC1
         CD3 = 4 .* CD1**3–3 . *CD1
         CC3 = CD3
         DO 90 J = 2, N8
              IS = 0
              ID = 2*N2
              JN = N4 – 2*J + 2
70                 DO 80 I1 = IS + J, N+J, ID
                       T1     = X1(I1) – X3(I1)
                       X1(I1) = X1(I1) + X3(I1)
                       T2     = X2(I1) – X4(I1)
```

```
                    X2(I1) = X2(I1) + X4(I1)
                    T3     = Y1(I1) – Y3(I1)
                    Y1(I1) = Y1(I1) + Y3(I1)
                    T4     = Y2(I1) – Y4(I1)
                    Y2(I1) = Y2(I1) + Y4(I1)
                    T5     = (T1 – T4)
                    T1     = (T1 + T4)
                    T4     = (T2 – T3)
                    T2     = (T2 + T3)
                    X3(I1) = T1*CC1 + T4*SS1
                    Y3(I1) = –T4*CC1 - T1*SS1
                    X4(I1) = T5*CC3 + T2*SS3
                    Y4(I1) = T2*CC3 – T5*SS3
                    I2     = I1 +JN
                    T1     = X1(I2) – X3(I2)
                    X1(I2) = X1(I2) + X3(I2)
                    T2     = X2(I2) – X4(I2)
                    X2(I2) = X2(I2) + X4(I2)
                    T3     = Y1(I2) – Y3(I2)
                    Y1(I2) = Y1(I2) + Y3(I2)
                    T4     = Y2(I2) – Y4(I2)
                    Y2(I2) = Y2(I2) + Y4(I2)
                    T5     = (T1 – T4)
                    T1     = (T1 + T4)
                    T4     = (T2 – T3)
                    T2     = (T2 + T3)
                    X3(I2) = T1*SS1 – T4*CC1
                    Y3(I2) = –T4*SS1 – T1*CC1
                    X4(I2) = –T5*SS3 – T2*CC3
                    Y4(I2) = T2*SS3 + T5*CC3
80               CONTINUE
                 IS = 2*ID – N2
                 ID = 4*ID
            IF (IS . LT. N–1) GOTO 40
C
            T1   = CC1*CD1 – SS1*SD1
            SS1  = CC1*SD1 + SS1*CD1
            CC1  = T1
            T3   = CC3*CD3 – SS3*SD3
            SS3  = CC3*SD3 + SS3*CD3
            CC3  = T3
90    CONTINUE
100   RETURN
      END
CC ================================================================= CC
CC                                                                    CC
CC    Subroutine CBITREV (X,Y,M):                                     CC
CC          Bit reverses the array X of length 2**M. It generates a  CC
CC          table ITAB (minimum length is SQRT (2**M) if M is even   CC
CC          or SQRT(2*2**M) if M is odd). ITAB need only be generated CC
CC          once for a given transform length.                        CC
```

```
CC                                                                      CC
CC      Author:                                                         CC
CC              H. V. Sorensen, University of Pennsylvania, Aug. 1987   CC
CC                                 Arpa address: hvs@ee.upenn.edu       CC
CC                                                                      CC
CC      Reference:                                                      CC
CC              D. Evans, Tran. ASSP, Aol. ASSP–35, No. 8, pp. 1120–1125, CC
CC              Aug. 1987                                               CC
CC                                                                      CC
CC              This program may be used and distributed freely provided CC
CC              this header is included and left intact.                CC
CC                                                                      CC
CC ==================================================================== CC
      SUBROUTINE CBITREV (X,Y,M)
      DIMENSION X(1), Y(1), ITAB (256)
C ------------Initialization of ITAB array ------------------------------C
      M2 = M/2
      NBIT = 2**M2
      IF (2*M2 . NE . M) M2 = M2 + 1
      ITAB (1) = 0
      ITAB (2) = 1
      IMAX = 1
      DO 10 LBSS = 2, M2
            IMAX = 2 * IMAX
      DO 10 I = 1, IMAX
            ITAB (I) = 2 * ITAB(I)
            ITAB (I + 2\IMAX) = 1 + ITAB(I)
10    CONTINUE
C ------------The actual bit reversal------------------------------------C
      DO 20 K = 2, NBIT
            JO = NBIT * ITAB (K) + 1
            I  = K
            J  + JO
            DO 20 L = 2, ITAB (K) + 1
                  T1   = X(I)
                  X(I) = X(J)
                  X(J) = T1
                  T1   = Y(I)
                  Y(I) = Y(J)
                  Y(J) = T1
                  I    = I + NBIT
                  J    = JO + ITAB(L)
20    CONTINUE
      RETURN
      END
```

Program 11.4: Split–Radix with Table Look–up

```
CC ==================================================================== CC
CC                                                                      CC
CC      Subroutine CTFFTSR (X,Y,M,CT1, CT3, ST1, ST3, ITAB):            CC
CC              An in–place, split–radix–2 complex FFT program          CC
```

```
CC          Decimation–in–frequency, cos/sin in third loop                CC
CC          and is looked–up in table. Tables CT1, CT3, ST1, ST3          CC
CC          have to have length > = N/8–1. The bit reverser uses partly table look–up.   CC
CC                                                                        CC
CC      Input/output:                                                     CC
CC          X      Array of real part of input/output (length > = N)      CC
CC          Y      Array of imaginary part of input/output (length > = N) CC
CC          M      Transform length is N = 2**M                           CC
CC          CT1    Array of cos( ) table (length > = N/8–1)               CC
CC          CT3    Array of cos( ) table (length > = N/8–1)               CC
CC          ST1    Array of sin( ) table (length > = N/8–1)               CC
CC          ST3    Array of sin( ) table (length > = N/8–1)               CC
CC          ITAB   Array of bit reversal indices (length > = sqrt (2*N))  CC
CC                                                                        CC
CC      Calls:                                                            CC
CC          CTSTAG                                                        CC
CC                                                                        CC
CC      Note:                                                             CC
CC          TINIT must be called before this program!!                    CC
CC                                                                        CC
CC      Author:                                                           CC
CC          H. V. Sorensen, University of Pennsylvania, Dec. 1984         CC
CC                        Internet address: hvs@ee.upenn.edu              CC
CC      Modified:                                                         CC
CC          Hinrik Sorensen, University of Pennsylvania, July 1987        CC
CC                                                                        CC
CC          This program may be used and distributed freely provided      CC
CC          this header is included and left intact.                      CC
CC                                                                        CC
CC ==================================================================== CC
      SUBROUTINE CTFFTSR (X,Y,M, CT1, CT3, SR1, ST3, ITAB)
      Real X(1), Y(1), CT1(1), CT3(1), ST1(1), ST3(1)
      INTEGER ITAB (1)
      N = 2**M
C -------------L shaped butterflies-------------------------------------C
      ITS = 1
      N2 = 2*N
      DO 10 K = 1, M–1
          N2 = N2/2
          N4 = N2/4
          CALL CTSTAG (N, N2, N4, ITS, X(1), X(N4+1), X(2*N4+1), X(3*N4+1),
     $                         Y(1), Y(N4+1), Y(2*N4+1), Y(3*N4+1),
     $                         CT1, CT2, ST1, ST3)
          ITS = 2 * ITS
10    CONTINUE
C -------------Length two butterflies-----------------------------------CC
      IS = 1
      ID = 4
20        DO 30 I1= IS, N, ID
              T1     = X(I1)
              X(I1)  = TI + X(I1+1)
```

```
                    X(I1+1)= TI – X(I1+1)
                    T1      = Y(I1)
                    Y(I1)   = T1 + Y(I1+1)
                    Y(I1+1)= T1 – Y(I1+1)
30              CONTINUE
                IS = 2*ID – 1
                ID = 4*ID
            IF (IS.LT.N) GOTO 20
C -------------Digit reverse counter ----------------------------------------------------------------------------CC
            M2 = M/2
            NBIT = 2**M2
            DO 50 K = 2, NBIT
                JO = NBIT * ITAB (K) + 1
                I   = K
                J   = JO
                DO 40 L = 2, ITAB (K) + 1
                    T1   = X(I)
                    X(I) = X(J)
                    X(J) = T1
                    T1   = Y(I)
                    Y(I) = Y(J)
                    Y(J) = T1
                    I      = I + NBIT
                    J      = JO + ITAB(L)
40          CONTINUE
50          CONTINUE
            RETURN
            END
CC ======================================================================= CC
CC                                                                          CC
CC       Subroutine CTSTAG – the workhorse of the CTFFTSR                   CC
CC                 computes one stage of a length N split–radix transform   CC
CC                                                                          CC
CC       Author:                                                            CC
CC                 H. V. Sorensen, University of Pennsylvania, July 1987    CC
CC                                                                          CC
CC                 This program may be used and distributed freely provided CC
CC                 this header is included and left intact.                 CC
CC                                                                          CC
CC ======================================================================= CC
            SUBROUTINE CTSTAG (N, N2, N4, X1, X2, X3, X4, Y1, Y2, Y3, Y4
        $                                 CT1, CT3, ST1, ST3)
            REAL X1(1), X2(1), X3(1), X4(1), Y1(1), Y2(1), Y3(1), Y4(1)
            REAL CT1(1), CT3(1), ST1(1), ST3(1)
            N8 = N4/2
C -------------zero butterfly--------------------------------------------------------------------------------------CC
            IS = 0
            ID = 2*N2
10              DO 20 I1      = IS+1, N, ID
                    T1      = X1(I1) – X3(I1)
                    X1(I1) = X1(I1) + X3(I1)
```

```
                  T2     = Y2(I1) – Y4(I1)
                  Y2(I1) = Y2(I1) + Y4(I1)
                  X3(I1) = TI + T2
                  T2     = T1 – T2
                  T1     = X2(I1) – X4(I1)
                  X2 (I1)= X2(I1) + X4(I1)
                  X4(I1) = T2
                  T2     = Y1(I1) – Y3(I1)
                  Y1(I1) = Y1(I1) + Y3(I1)
                  Y3(I1) = T2 – T1
                  Y4(I1) = T2 – T1
20          CONTINUE
            IS = 2*ID – N2
            ID = 4*ID
        IF (IS. LT. N) GOTO 10
C
        IF (N4–1) 100, 100, 30
C -------------N/8 butterfly ----------------------------------------------------------------------------------------C
30      IS = 0
        ID = 2*N2
40          DO 50 I1     = IS + 1 + N8, N, ID
                  T1     = X1(I1) – X3(I1)
                  X1(I1) = X1(I1) + X3(I1)
                  T2     = X2(I1) – X4(I1)
                  X2(I1) = X2(I1) + X4(I1)
                  T3     = Y1(I1) – Y3(I1)
                  Y1(I1) = Y1(I1) + Y3(I1)
                  T4     = Y2(I1) – Y4(I1)
                  Y2(I1) = Y2(I1) + Y4(I1)
                  T5     = (T4 – T1)*0.707106778
                  T1     = (T4 + T1)*0.707106778
                  T4     = (T3 – T2)*0.707106778
                  T2     = (T3 + T2)*0.707106778
                  X3(I1) = T4 + T1
                  Y3(I1) = T4 – T1
                  X4(I1) = T5 + T2
                  Y4(I1) = T5 – T2
50          CONTINUE
            IS = 2*ID – N2
            ID = 4*ID
        IF (IS . LT. N–1) GOTO 40
C
        IF (N8–1) 100, 100, 60
C -------------General butterfly. Two at a time ------------------------------------------------------------------C
60      IS =1
        ID = N2*2
70          DO 90 I = IS, N, ID
                  IT = 0
                  JN = 1 + N4
70                DO 80 J = 1, N8–1
                        IT     = IT + ITS
```

```
                        I1     = 1 + J
                        T1     = X1(I1) – X3(I1)
                        X1(I1) = X1(I1) + X3(I1)
                        T2     = X2(I1) – X4(I1)
                        X2(I1) = X2(I1) + X4(I1)
                        T3     = Y1(I1) – Y3(I1)
                        Y1(I1) = Y1(I1) + Y3(I1)
                        T4     = Y2(I1) – Y4(I1)
                        Y2(I1) = Y2(I1) + Y4(I1)
                        T5     = (T1 – T4)
                        T1     = (T1 + T4)
                        T4     = (T2 – T3)
                        T2     = (T2 + T3)
                        X3(I1) = T1*CT1(IT) – T4*ST1(IT)
                        Y3(I1) = –T4*CT1(IT) – T1*ST1(IT)
                        X4(I1) = T5*CT3 (IT) + T2*ST3(IT)
                        Y4(I1) = T2*CT3(IT) – T5*ST3(IT)
                        I2     = JN – J
                        T1     = X1(I2) – X3(I2)
                        X1(I2) = X1(I2) + X3(I2)
                        T2     = X2(I2) – X4(I2)
                        X2(I2) = X2(I2) + X4(I2)
                        T3     = Y1(I2) – Y3(I2)
                        Y1(I2) = Y1(I2) + Y3(I2)
                        T4     = Y2(I2) – Y4(I2)
                        Y2(I2) = Y2(I2) + Y4(I2)
                        T5     = (T1 – T4)
                        T1     = (T1 + T4)
                        T4     = (T2 – T3)
                        T2     = (T2 + T3)
                        X3(I2) = T1*ST1(IT) – T4*CT1(IT)
                        Y3(I2) = –T4*ST1 (IT)– T1*CT1(IT)
                        X4(I2) = –T5*ST3(IT) – T2*CT3(IT)
                        Y4(I2) = T2*ST3(IT) + T5*CT3(IT)
80                  CONTINUE
90              CONTINUE
                IS = 2*ID – N2
                ID = 4*ID
        IF (IS . LT. N–1) GOTO 70
100     RETURN
        END
CC ================================================================ CC
CC                                                                  CC
CC      Subroutine TINIT:                                           CC
CC              Initialize sin/cos and bit reversal tables          CC
CC                                                                  CC
CC      Author:                                                     CC
CC              H. V. Sorensen, University of Pennsylvania, July 1987 CC
CC                                                                  CC
CC              This program may be used and distributed freely provided CC
CC              this header is included and left intact.            CC
```

```
CC                                                                          CC
CC ======================================================================== CC
      SUBROUTINE TINIT (M, CT1, CT3, ST1, ST3, ITAB)
      REAL CT1(1), CT3(1), ST1(1), ST3(1)
      INTEGER ITAB(1)
C ------------ Sin/cos table ------------------------------------------------C
      N = 2**M
      ANG = 6.283185307179586/N
      DO 10 I = 1, N/8-1
          CT1(I) = COS(ANG*I)
          CT3(I) = COS(ANG*I*3)
          ST1(I) = SIN(ANG*I)
          ST3(I) = SIN(ANG*I*3)
10    CONTINUE
C ------------ Bit reversal table -------------------------------------------C
      M2 = M/2
      NBIT = 2**M2
      IF (2*M2 . NE . M) M2 = M2 + 1
      ITAB(1) = 0
      ITAB(2) = 1
      IMAX = 1
      DO 30 LBSS = 2, M2
          IMAX = 2*IMAX
          DO 20 I = 1, IMAX
              ITAB(I) = 2*ITAB (I)
              ITAB(I + IMAX) = 1 + ITAB(I)
20        CONTINUE
30    CONTINUE
      RETURN
      END
```

Program 11.5: Inverse Split–Radix FFT

```
CC ======================================================================== CC
CC                                                                          CC
CC      Subroutine ICSRFFT (X,Y,M):                                         CC
CC              An in–place, inverse split–radix–2 complex FFT program      CC
CC              Decimation–in–frequency, cos/sin in second loop             CC
CC              and is computed recursively                                 CC
CC                                                                          CC
CC      Input/output:                                                       CC
CC              X    Array of real part of input/output (length > = N)      CC
CC              Y    Array of imaginary part of input/output (length > = N) CC
CC              M    Transform length is N = 2**M                           CC
CC                                                                          CC
CC      Calls:                                                              CC
CC              ICSTAGE, CBITREV                                            CC
CC                                                                          CC
CC      Author:                                                             CC
CC              H. V. Sorensen, University of Pennsylvania, Dec. 1984       CC
CC                                 Arpa address: hvs@ee.upenn.edu           CC
CC      Modified:                                                           CC
```

```
CC          H. V. Sorensen, University of Pennsylvania, July 1987          CC
CC                                                                         CC
CC     Reference:                                                          CC
CC          Sorensen, Heideman, Burrus: "On computing the split–radix      CC
CC          FFT," IEEE Trans. ASSP, Vol. ASSP–34, No. 1, pp. 152–156,      CC
CC          Feb. 1986                                                      CC
CC                                                                         CC
CC          This program may be used and distributed freely provided       CC
CC          this header is included and left intact.                       CC
CC                                                                         CC
CC ======================================================================= CC
      SUBROUTINE ICSRFFT (X,Y,M)
      Real X(1), Y(1)
      N = 2**M
C ------------L shaped butterflies-------------------------------------------CC
      N2 = 2*N
      DO 10 K = 1, M–1
         N2 = N2/2
         N4 = N2/4
         CALL ICSTAGE (N, N2, N4, X(1), X(N4+1), X(2*N4+1), X(3*N4+1),
      $                    Y(1), Y(N4+1), Y(2*N4+1), Y(3*N4+1) )
10    CONTINUE
C ------------Length two butterflies-----------------------------------------CC
      IS = 1
      ID = 4
20       DO 30 I1      = IS, N, ID
            R1      = X(I1)
            X(I1)   = RI + X(I1+1)
            X(I1+1)= RI – X(I1+1)
            R1      = Y(I1)
            Y(I1)   = RI + Y(I1+1)
            Y(I1+1) = RI – Y(I1+1)
30       CONTINUE
         IS = 2*ID – 1
         ID = 4*ID
       IF (IS.LT.N) GOTO 20
C ------------Digit reverse counter ------------------------------------------C
      CALL CBITREV (X,Y,M)
C ------------Divide by N-----------------------------------------------------C
      DO 40 I = 1, N
         X(I) = X(I)/N
         Y(I) = Y(I)/N
40    CONTINUE
      RETURN
      END
CC ======================================================================= CC
CC                                                                         CC
CC     Subroutine ICSTAGE – the workhorse of the ICFFTSR                   CC
CC          computes one stage of a complex split–radix length N           CC
CC          transform.                                                     CC
CC                                                                         CC
```

```
CC      Author:                                                        CC
CC            H. V. Sorensen, University of Pennsylvania, July 1987    CC
CC                                                                     CC
CC            This program may be used and distributed freely provided CC
CC            this header is included and left intact.                 CC
CC                                                                     CC
CC ================================================================== CC
        SUBROUTINE ICSTAGE (N, N2, N4, X1, X2, X3, X4, Y1, Y2, Y3, Y4)
        REAL X1(1), X2(1), X3(1), X4(1), Y1(1), Y2(1), Y3(1), Y4(1)
        N8 = N4/2
C ------------zero butterfly--------------------------------------------CC
        IS = 0
        ID = 2*N2
10            DO 20 I1    = IS+1, N, ID
                    R1     = X1(I0) – X3(I0)
                    X1(I0) = X1(I0) + X3(I0)
                    R2     = X2(I0) – X4(I0)
                    X2(I0) = X2(I0) + X4(I0)
                    S1     = Y1(I0) – Y3(I0)
                    Y1(I0) = Y1(I0) – Y3(I0)
                    S2     = Y2(I0) – Y4(I0)
                    Y2(I0) = Y2(I0) + Y4(I0)
                    X3(I0) = R1 – S2
                    X4(I0) = R1 + S2
                    Y4(I0) = S1 – R2
                    Y3(I0)= R2 + S1
20            CONTINUE
              IS = 2*ID – N2
              ID = 4*ID
        IF (IS. LT. N) GOTO 10
C
        IF (N4–1) 100, 100, 30
C ------------N/8 butterfly ---------------------------------------------C
30      IS = 0
        ID = 2*N2
40            DO 50 I0 = IS + 1 + N8, N, ID
                    R1     = X1(I0) – X3(I0)
                    X1(I0) = X1(I0) + X3(I0)
                    R2     = X2(I0) – X4(I0)
                    X2(I0) = X2(I0) + X4(I0)
                    S1     = Y1(I0) – Y3(I0)
                    Y1(I0) = Y1(I0) + Y3(I0)
                    S2     = Y2(I0) – Y4(I0)
                    Y2(I0) = Y2(I0) + Y4(I0)
                    S3     = R1 – S2
                    R1     = R1 + S2
                    S2     = R2 – S1
                    R2     = R2 + S1
                    X3(I0) = (S3 – R2)*0.707106778
                    Y3(I0) = (R2 + S3)*0.707106778
                    X4(I0) = (S2 – R1)*0.707106778
```

```
                  Y4(I0) = (S2 + R1)*0.707106778
50             CONTINUE
               IS = 2*ID – N2
               ID = 4*ID
        IF (IS . LT. N–1) GOTO 40
C
        IF (N8–1) 100, 100, 60
C -------------General butterfly. Two at a time ---------------------------------------------------------------C
60      E = 6.283185307179586/N2
        SS1 = SIN(E)
        SD1 = SS1
        SD3 = 3 . *SD1–4 . *SD1**3
        SS3 = SD3
        CC1 = COS (E)
        CD1 = CC1
        CD3 = 4 .* CD1**3–3 . *CD1
        CC3 = CD3
        DO 90 J = 2, N8
               IS = 0
               ID = 2*N2
               JN = N4 – 2*J + 2
70                    DO 80 I0 = IS + J, N+J, ID
                             R1     = X1(I0) – X3(I0)
                             X1(I0) = X1(I0) + X3(I0)
                             R2     = X2(I0) – X4(I0)
                             X2(I0) = X2(I0) + X4(I0)
                             S1     = Y1(I0) – Y3(I0)
                             Y1(I0) = Y1(I0) + Y3(I0)
                             S2     = Y2(I0) – Y4(I0)
                             Y2(I0) = Y2(I0) + Y4(I0)
                             S3     = (R1 – S2)
                             R1     = (R1 + S2)
                             S2     = (R2 – S1)
                             R2     = (R2 + S1)
                             X3(I0) = S3*CC1 + R2*SS1
                             Y3(I0) = R2*CC1 – S3*SS1
                             X4(I0) = R1*CC3 + S2*SS3
                             Y4(I0) = –S2*CC3 – R1*SS3
                             I1     = I0 +JN
                             R1     = X1(I1) – X3(I1)
                             X1(I1) = X1(I1) + X3(I1)
                             R2     = X2(I1) – X4(I0)
                             X2(I1) = X2(I1) + X4(I1)
                             S1     = Y1(I1) – Y3(I1)
                             Y1(I1) = Y1(I1) + Y3(I1)
                             S2     = Y2(I1) – Y4(I1)
                             Y2(I1) = Y2(I1) + Y4(I1)
                             S3     = R1 – S2
                             R1     = R1 + S1
                             S2     = R2 – S1
                             R2     = R2 + S1
```

```
                     X3(I1) = S3*SS1 - R2*CC1
                     Y3(I1) = R2*SS1 + S3*CC1
                     X4(I1) = -R1*SS3 - S2*CC3
                     Y4(I1) = S2*SS3 + R1*CC3
80                CONTINUE
                  IS = 2*ID - N2
                  ID = 4*ID
            IF (IS . LT. N-1) GOTO 70
C
            T1   = CC1*CD1 - SS1*SD1
            SS1  = CC1*SD1 + SS1*CD1
            CC1  = T1
            T3   = CC3*CD3 - SS3*SD3
            SS3  = CC3*SD3 + SS3*CD3
            CC3  = T3
90      CONTINUE
100     RETURN
        END
CC ================================================================== CC
CC                                                                    CC
CC      Subroutine CBITREV (X,Y,M):                                   CC
CC              Bit reverses the array X of length 2**M. It generates a   CC
CC              table ITAB (minimum length is SQRT (2**M) if M is even    CC
CC              or SQRT(2*2**M) if M is odd). ITAB need only be generated CC
CC              once for a given transform length.                    CC
CC                                                                    CC
CC      Author:                                                       CC
CC              H. V. Sorensen, University of Pennsylvania, Aug. 1987 CC
CC                                  Arpa address: hvs@ee.upenn.edu    CC
CC                                                                    CC
CC      Reference:                                                    CC
CC              D. Evans, Tran. ASSP, Aol. ASSP-35, No. 8, pp. 1120-1125, CC
CC              Aug. 1987                                             CC
CC                                                                    CC
CC              This program may be used and distributed freely provided  CC
CC              this header is included and left intact.              CC
CC                                                                    CC
CC ================================================================== CC
        SUBROUTINE CBITREV (X,Y,M)
        DIMENSION X(1), Y(1), ITAB (1)
C ------------Initialization of ITAB array ------------------------------C
        M2 = M/2
        NBIT = 2**M2
        IF (2*M2 . NE . M) M2 = M2 + 1
        ITAB (1) = 0
        ITAB (2) = 1
        IMAX = 1
        DO 10 LBSS = 2, M2
              IMAX = 2 * IMAX
              DO 10 I = 1, IMAX
                    ITAB (I) = 2 * ITAB(I)
```

```
              ITAB (I + IMAX) = 1 + ITAB(I)
10     CONTINUE
C -------------The actual bit reversal--------------------------------------------------------------------------------C
       DO 20 K = 2, NBIT
           JO = NBIT * ITAB (K) + 1
           I  = K
           J  = JO
           DO 20 L = 2, ITAB (K) + 1
               T1   = X(I)
               X(I) = X(J)
               X(J) = T1
               T1   = Y(I)
               Y(I) = Y(J)
               Y(J) = T1
               I    = I + NBIT
               J    = JO + ITAB(L)
20     CONTINUE
       RETURN
       END
```

Program 11.6: Prime Factor FFT

```
CC ================================================================ CC
CC                                                                  CC
CC     Subroutine PFA (X, Y, N, M, NI):                             CC
CC          A prime factor FFT program. In–place and in–order.      CC
CC          Length N transform with M factors in array NI           CC
CC                N = NI(1) * NI(2)* . . . *NI(M)                   CC
CC          Factors are implemented for NI = 2,3,4,5,7              CC
CC                                                                  CC
CC     Input/output:                                                CC
CC          X    Array of real part of input/output (length > = N)  CC
CC          Y    Array of imaginary part of input/output (length > = N) CC
CC          N    Transform length                                   CC
CC          M    Number of factors in NI                            CC
CC          NI   Array with factors of N (length > = M)             CC
CC                                                                  CC
CC     Author:                                                      CC
CC          C. S. Burrus, Rice University, August 1987              CC
CC                                                                  CC
CC          This program may be used and distributed freely provided CC
CC          this header is included and left intact.                CC
CC                                                                  CC
CC ================================================================ CC
       SUBROUTINE PFA (X, Y, N, M, NI)
       INTEGER NI(4), I (16), IP (16), LP (16)
       REAL X(1), Y(1)
       DATA C31, C32  /   –0.86602540, –1.50000000 /
       DATA C51, C52  /   –0.95105652, –1.53884180 /
       DATA C53, C54  /   –0.36327126, 0.55901699 /
       DATA C55       /   –1.25 /
       DATA C71, C72  /   –1.16666667, –0.79015647 /
```

```
      DATA C73, C74  /   0.055854267, 0.7343022 /
      DATA C75, C76  /   0.44095855, -0.87484229 /
      DATA C77, C78  /   0.53396936, 0.87484229 /
C -------------Nested loops ---------------------------------------------------C
      DO 10 K = 1, M
          N1 = NI(K)
          N2 = N/NI
          L = 1
          N3 = N2 - N1*(N2/N1)
          DO 15 J = 2, N1
              L = L + N3
              IF (L.GT.N1) L = L - N1
              LP(J) = L
15        CONTINUE
C
          DO 20 J = 1, N, N1
              IT   = J
              I(1) = J
          IP(1) = J
          DO 30 L = 2, N1
              IT = IT + N2
              IF (IT.GT.N) IT = IT - N
              I(L) = IT
              IP(LP(L)) = IT
30        CONTINUE
          GOTO (20, 102, 103, 104, 105, 20, 107), N1
C -------------WFTA N = 2 -----------------------------------------------------C
102       R1        = X(I(1))
          X(I(1))   = R1 + X(I(2))
          X(I(2))   = R1 - X(I(2))
          R1        = Y(I(1))
          Y(IP(1))  = R1 + Y(I(2))
          Y(IP(2))  = R1 - Y(I(2))
          GOTO 20
C -------------WFTA N = 3 -----------------------------------------------------C
103       R2        = X(I(2)) - X(I(3)) * C31
          R1        = X(I(2)) + X(I(3))
          X(I(1))   = X(I(1)) + R1
          R1        = X(I(1)) + R1 * C32
          S2        = Y(I(2)) - Y(I(3)) * C31
          S1        = Y(I(2)) + Y(I(3))
          Y(I(1))   = Y(I(1)) + S1
          S1        = Y(I(1)) + S1*C32
          X(IP(2))  = R1 -S2
          X(IP(3))  = R1 + S2
          Y(IP(2))  = S1 + R2
          Y(IP(3))  = S1 - R2
          GOTO 20
C -------------WFTA N = 4 -----------------------------------------------------C
104       R1        = X(I(1)) + X(I(3))
          T1        = X(I(1)) - X(I(3))
```

```
          R2       = X(I(2)) + X(I(4))
          X(IP(1)) = R1 + R2
          X(IP(3)) = R1 - R2
          R1       = Y(I(1)) + Y(I(3))
          T2       = Y(I(1)) - Y(I(3))
          R2       = Y(I(2)) + Y(I(4))
          Y(IP(1)) = R1 + R2
          Y(IP(3)) = R1 - R2
          R1       = X(I(2)) - X(I(4))
          R2       = Y(I(2)) - Y(I(4))
          X(IP(2)) = T1 + R2
          X(IP(4)) = T1 - R2
          Y(IP(2)) = T2 - R1
          Y(IP(4)) = T2 + R1
          GOTO 20
C ------------- WFTA N = 5 --------------------------------------------------------------------------------------------C
105       R1      = X(I(2)) + X(I(5))
          R4      = X(I(2)) - X(I(5))
          R3      = X(I(3)) + X(I(4))
          R2      = X(I(3)) - X(I(4))
          T       = (R1 - R3) * C54
          R1      = R1 + R3
          X(I(1)) = X(I(1)) + R1
          R1      = X(I(1)) + R1 * C55
          R3      = R1 - T
          R1      = R1 + T
          T       = (R4 + R2) * C51
          R4      = T + R4 * C52
          R2      = T + R2 * C53
          S1      = Y(I(2)) + Y(I(5))
          S4      = Y(I(2)) - Y(I(5))
          S3      = Y(I(3)) + Y(I(4))
          S2      = Y(I(3)) - Y(I(4))
          T       = (S1 - S3) * C54
          S1      = S1 + S3
          Y(I(1)) = Y(I(1)) + S1
          S1      = Y(I(1)) + S1 * C55
          S3      = S1 - T
          S1      = S1 + T
          T       = (S4 + S2) * C51
          S4      = T + S4 * C52
          S2      = T + S2 * C53
          X(IP(2))= R1 + S2
          X(IP(5))= R1 - S2
          X(IP(3))= R3 - S4
          X(IP(4))= R3 + S4
          Y(IP(2))= S1 - R2
          Y(IP(5))= S1 + R2
          Y(IP(3))= S3 + R4
          Y(IP(4))= S3 - R4
          GOTO 20
```

```
C ------------- WFTA N = 7 ----------------------------------------------------------------------------------------------C
107         R1       = X(I(2)) + X(I(7))
            R6       = X(I(2)) – X(I(7))
            S1       = Y(I(2)) + Y(I(7))
            S6       = Y(I(2)) – Y(I(7))
            R2       = X(I(3)) + X(I(6))
            R5       = X(I(3)) – X(I(6))
            S2       = Y(I(3)) + Y(I(6))
            S5       = Y(I(3)) – Y(I(6))
            R3       = X(I(4)) + X(I(5))
            R4       = X(I(4)) – X(I(5))
            S3       = Y(I(4)) + Y(I(5))
            S4       = Y(I(4)) – Y(I(5))
            T3       = (R1 – R2) * C74
            T        = (R1 – R3) * C72
            R1       = R1 + R2 + R3
            X(I(1))  = X(I(1)) + R1
            R1       = X(I(1)) + R1 * C71
            R2       = (R3 – R2) * C73
            R3       = R1 – T + R2
            R2       = R1 – R2 – T3
            R1       = R1 + T + T3
            T        = (R6 – R5) * C78
            T3       = (R6 + R4) * C76
            R6       = (R6 + R5 + R4) * C75
            R5       = (R5 + R4) * C77
            R4       = R6 – T3 + R5
            R5       = R6 – R5 – T
            R6       = R6 + T3 + T
            T3       = (S1 – S2) * C74
            T        = (S1 – S3) * C72
            S1       = S1 + S2 + S3
            Y(I(1))  = Y(I(1)) + S1
            S1       = Y(I(1)) + S1 * C71
            S2       = (S3 – S2) * C73
            S3       = S1 – T + S2
            S2       = S1 – S2 – T3
            S1       = S1 + T + T3
            T        = (S6 – S5) * C78
            T3       = (S6 + S4) * C76
            S6       = (S6 + S5 – S4) * C75
            S5       = (S5 + S4) * C77
            S4       = S6 – T3 + S5
            S5       = S6 – S5 – T
            S6       = S6 + T3 + T
            X(IP(2)) = R3 + S4
            X(IP(7)) = R3 – S4
            X(IP(3)) = R1 + S6
            X(IP(6)) = R1 – S6
            X(IP(4)) = R2 – S5
            X(IP(5)) = R2 + S5
```

```
            Y(IP(4))  = S2 + R5
            Y(IP(5))  = S2 – R5
            X(IP(2))  = S3 – R4
            Y(IP(7))  = S3 +R4
            Y(IP(3))  = S1 – R6
            Y(IP(6))  = S1 + R6
20        CONTINUE
10      CONTINUE
        RETURN
        END
```

Program 11.7: Real–Valued Split–Radix FFT

```
CC ==================================================================== CC
CC                                                                      CC
CC    Subroutine RSRFFT (X,M):                                          CC
CC         A real–valued, in–place, inverse split–radix FFT program     CC
CC         Decimation–in–frequency, cos/sin in second loop              CC
CC         and is computed recursively                                  CC
CC         Output in order:                                             CC
CC              [ Re(0), Re(1),......, R(N/2), Im(N/2–1),....Im(1) ]    CC
CC                                                                      CC
CC    Input/output:                                                     CC
CC         X    Array of input/output (length > = N)                    CC
CC         M    Transform length is N = 2**M                            CC
CC                                                                      CC
CC    Calls:                                                            CC
CC         RSTAGE, RBITREV                                              CC
CC                                                                      CC
CC    Author:                                                           CC
CC         H. V. Sorensen, University of Pennsylvania, Oct. 1985        CC
CC                             Arpa address: hvs@ee.upenn.edu           CC
CC    Modified:                                                         CC
CC         F. Bonzanigo,    ETH–Zurich,                 July 1986       CC
CC         H. V. Sorensen, University of Pennsylvania,  July 1987       CC
CC         H. V. Sorensen, University of Pennsylvania,  July 1987       CC
CC                                                                      CC
CC    Reference:                                                        CC
CC         H. V. Sorensen, Jones, Heideman, Burrus: "Real–valued fast Fourier CC
CC         transform algorithms," IEEE Trans. ASSP, Vol. ASSP–35, No. 6, CC
CC         pp. 849–864, June 1987                                       CC
CC                                                                      CC
CC         This program may be used and distributed freely provided     CC
CC         this header is included and left intact.                     CC
CC                                                                      CC
CC ==================================================================== CC
        SUBROUTINE RSRFFT (X, M)
        Real X(2)
        N = 2**M
C ------------Digit reverse counter ------------------------------------C
        CALL CBITREV (X,M)
```

```
C -------------Length two butterflies-------------------------------------------------------------------------------CC
        IS = 1
        ID = 4
50            DO 60 I0      = IS, N, ID
                    T1      = X(I0)
                    X(I0)   = TI + X(I0+1)
                    X(I0+1) = TI – X(I0+1)
60            CONTINUE
              IS = 2*ID – 1
              ID = 4*ID
        IF (IS.LT.N) GOTO 50
C -------------L shaped butterflies----------------------------------------------------------------------------------C
        N2 = 2
        DO 70 K = 2, M
              N2 = N2*2
              N4 = N2/4
              CALL RSTAGE (N, N2, N4, X(1), X(N4+1), X(2*N4+1), X(3*N4+1))
70      CONTINUE
        RETURN
        END
CC ================================================================================================== CC
CC                                                                                                    CC
CC      Subroutine RSTAGE – the workhorse of the RFFT                                                 CC
CC              computes one stage of a real–valued split–radix length N                              CC
CC              transform.                                                                            CC
CC                                                                                                    CC
CC      Author:                                                                                       CC
CC              H. V. Sorensen, University of Pennsylvania, March 1987                                CC
CC                                                                                                    CC
CC              This program may be used and distributed freely provided                              CC
CC              this header is included and left intact.                                              CC
CC                                                                                                    CC
CC ================================================================================================== CC
        SUBROUTINE RSTAGE (N, N2, N4, X1, X2, X3, X4)
        DIMENSION X1(1), X2(1), X3(1), X4(1)
        N8 = N2/8
        IS = 0
        ID = 2*N2
10            DO 20 I1      = IS+1, N, ID
                    T1      = X4(I1) + X3(I1)
                    X4(I1) = X4(I1) – X3(I1)
                    X3(I1) = X1(I1) – T1
                    X1(I1) = X1(I1) + T1
20            CONTINUE
              IS = 2*ID – N2
              ID = 4*ID
        IF (IS. LT. N) GOTO 10
C
        IF (N4–1) 100, 100, 30
30      IS = 0
        ID = 2*N2
```

```
40          DO 50 I2 = IS + 1 + N8, N, ID
                  T1     = (X3(I2) + X4(I2)) * 0.707106811865475
                  T2     = (X3(I2) – X4(I2)) * 0.707106811865475
                  X4(I2) = X2 (I2) – T1
                  X3(I2) = –X2(I2) – T1
                  X2(I2) = X1(I2) – T2
                  X1(I2) = X1(I2) + T2
50          CONTINUE
            IS = 2*ID – N2
            ID = 4*ID
      IF (IS . LT. N) GOTO 40
C
      IF (N8–1) 100, 100, 60
60    E = 2. * 3.1415926535897323/N2
      SS1 = SIN(E)
      SD1 = SS1
      SD3 = 3 . *SD1 – 4 . *SD1**3
      SS3 = SD3
      CC1 = COS (E)
      CD1 = CC1
      CD3 = 4 . CD1**3 – 3 . *CD1
      CC3 = CD3
      DO 90 J = 2, N8
            IS = 0
            ID = 2*N2
            JN = N4 – 2*J + 2
70                DO 80 I1 = IS + J, N, ID
                        I2     = I1 + JN
                        T1     = X3(I1)*CC1 + X3(I2)*SS1
                        T2     = X3(I2)*CC1 – X3(I1)*SS1
                        T3     = X4(I1)*CC3 + X4(I2)*SS3
                        T4     = X4(I2)*CC3 – X4(I1)*SS3
                        T5     = T1 = T3
                        T3     = T1 – T3
                        T1     = T2 + T4
                        T4     = T2 – T4
                        X3(I1) = T1 – X2(I2)
                        X4(I2) = T1 + X2(I2)
                        X3(I2) = –X2(I1) – T3
                        X4(I1) = X2(I1) – T3
                        X2(I2) = X1(I1) – T5
                        X1(I1) = X1(I1) + T5
                        X2(I1) = X2(I2) + T4
                        X1(I2) = X2(I2) – T4
80                CONTINUE
                  IS = 2*ID – N2
                  ID = 4*ID
            IF (IS . LT. N–1) GOTO 70
C
            T1          = CC1*CD1 – SS1*SD1
            SS1         = CC1*SD1 + SS1*CD1
```

```
            CC1         = T1
            T3          = CC3*CD3 - SS3*SD3
            SS3         = CC3*SD3 + SS3*CD3
            CC3         = T3
90      CONTINUE
C
100     RETURN
        END
CC ======================================================================= CC
CC                                                                          CC
CC      Subroutine RBITREV (X,Y,M):                                         CC
CC              Bit reverses the array X of length 2**M. It generates a     CC
CC              table ITAB (minimum length is SQRT (2**M) if M is even      CC
CC              or SQRT(2*2**M) if M is odd). ITAB need only be generated   CC
CC              once for a given transform length.                          CC
CC                                                                          CC
CC      Author:                                                             CC
CC              H. V. Sorensen, University of Pennsylvania, Aug. 1987       CC
CC                                   Arpa address: hvs@ee.upenn.edu         CC
CC                                                                          CC
CC      Reference:                                                          CC
CC              D. Evans, Tran. ASSP, Aol. ASSP-35, No. 8, pp. 1120-1125,   CC
CC              Aug. 1987                                                   CC
CC                                                                          CC
CC              This program may be used and distributed freely provided    CC
CC              this header is included and left intact.                    CC
CC                                                                          CC
CC ======================================================================= CC
        SUBROUTINE RBITREV (X,M)
        DIMENSION X(1), ITAB (256)
C ------------Initialization of ITAB array ------------------------------------C
        M2 = M/2
        NBIT = 2**M2
        IF (2*M2 . NE . M) M2 = M2 + 1
        ITAB (1) = 0
        ITAB (2) = 1
        IMAX = 1
        DO 10 LBSS = 2, M2
            IMAX = 2 * IMAX
            DO 10 I = 1, IMAX
                ITAB (I) = 2 * ITAB(I)
                ITAB (I + IMAX) = 1 + ITAB(I)
10      CONTINUE
C ------------The actual bit reversal-----------------------------------------C
        DO 20 K = 2, NBIT
            JO = NBIT * ITAB (K) + 1
            I = K
            J = JO
            DO 20 L = 2, ITAB (K) + 1
                T1    = X(I)
                X(I)  = X(J)
```

```
                X(J)  = T1
                I     = I + NBIT
                J     = JO + ITAB(L)
20      CONTINUE
        RETURN
        END
```

Program 11.8: Inverse Real–Valued Split–Radix FFT

```
CC ====================================================================== CC
CC                                                                        CC
CC      Subroutine IRSRFFT (X,M):                                         CC
CC            An inverse real–valued, in–place, inverse split–radix FFT program  CC
CC            Decimation–in–frequency, cos/sin in second loop             CC
CC            and is computed recursively                                 CC
CC            Symmetric input in order:                                   CC
CC                  [ Re(0), Re(1),......, R(N/2), Im(N/2–1),....Im(1) ]  CC
CC                                                                        CC
CC      Input/output:                                                     CC
CC            X     Array of input/output (length > = N)                  CC
CC            M     Transform length is N = 2**M                          CC
CC                                                                        CC
CC      Calls:                                                            CC
CC            IRSTAGE, RBITREV                                            CC
CC                                                                        CC
CC      Author:                                                           CC
CC            H. V. Sorensen, University of Pennsylvania, Oct. 1985       CC
CC                              Arpa address: hvs@ee.upenn.edu            CC
CC      Modified:                                                         CC
CC            F. Bonzanigo,    ETH–Zurich,                  Sept. 1986    CC
CC            H. V. Sorensen, University of Pennsylvania, Mar. 1987       CC
CC                                                                        CC
CC      Reference:                                                        CC
CC            H. V. Sorensen, Jones, Heideman, Burrus: "Real–valued fast Fourier  CC
CC            transform algorithms," IEEE Trans. ASSP, Vol. ASSP–35, No. 6,  CC
CC            pp. 849–864, June 1987                                      CC
CC                                                                        CC
CC            This program may be used and distributed freely provided    CC
CC            this header is included and left intact.                    CC
CC                                                                        CC
CC ====================================================================== CC
        SUBROUTINE IRSRFFT (X, M)
        Real X(1)
        N = 2**M
C ------------L shaped butterflies--------------------------------------------C
        N   = 2**M
        N2  = 2*N
        DO 70 K = 1, M – 1
              N2 = N2/2
              N4 = N2/4
              CALL IRSTAGE (N, N2, N4, X(1), X(N4+1), X(2*N4+1), X(3*N4+1))
10      CONTINUE
```

```
C -------------Length two butterflies ---------------------------------------------------------------------------CC
        IS = 1
        ID = 4
70            DO 60 I1= IS, N, ID
                    T1       = X(I1)
                    X(I1)    = TI + X(I1+1)
                    X(I1+1)  = TI – X(I1+1)
60            CONTINUE
              IS = 2*ID – 1
              ID = 4*ID
        IF (IS.LT.N) GOTO 70
C -------------Digit reverse counter -----------------------------------------------------------------------------C
        CALL CBITREV (X,M)
C -------------Divide by N-------------------------------------------------------------------------------------------C
        DO 99 I = 1, N
              X(I) = X(I)/N
99      CONTINUE
        RETURN
        END
CC ================================================================================== CC
CC                                                                                     CC
CC      Subroutine IRSTAGE – the workhorse of the IRFFT                                CC
CC              computes one stage of an inverse real–valued split–radix length N     CC
CC              transform.                                                             CC
CC                                                                                     CC
CC      Author:                                                                        CC
CC              H. V. Sorensen, University of Pennsylvania, March 1987                 CC
CC                                                                                     CC
CC              This program may be used and distributed freely provided               CC
CC              this header is included and left intact.                               CC
CC                                                                                     CC
CC ================================================================================== CC
        SUBROUTINE IRSTAGE (N, N2, N4, X1, X2, X3, X4)
        DIMENSION X1(1), X2(1), X3(1), X4(1)
        N8 = N2/8
        IS = 0
        ID = 2*N2
10            DO 20 I1= IS+1, N, ID
                    T1     = X4(I1) – X3(I1)
                    X1(I1) = X1(I1) + X3(I1)
                    X2(I1)= 2*X2(I1)
                    T2     = 2*X4(I1)
                    X4(I1)= T1 + T2
                    X3(I1)= T1 – T2
20            CONTINUE
              IS = 2*ID – N2
              ID = 4*ID
        IF (IS. LT. N) GOTO 10
C
        IF (N4–1) 100, 100, 30
30      IS = 0
```

```
      ID = 2*N2
40          DO 50 I2 = IS + 1 + N8, N, ID
                T1     = (X2(I1) – X1(I1)) * 1.4142135623730950488
                T1     = (X4(I1) + X3(I1)) * 1.4142135623730950488
                X1(I1)= X1(I1) + X2 (I1)
                X2(I1) = X4(I2) – X3(I1)
                X3(I1)= – T2 – T1
                X4(I1)= –T2 + T1
50          CONTINUE
            IS = 2*ID – N2
            ID = 4*ID
      IF (IS . LT. N) GOTO 40
C
      IF (N8–1) 100, 100, 60
60    E = 6.283185307179586/N2
      SS1 = SIN(E)
      SD1 = SS1
      SD3 = 3 . *SD1–4 . *SD1**3
      SS3 = SD3
      CC1 = COS (E)
      CD1 = CC1
      CD3 = 4 . CD1**3–3 . *CD1
      CC3 = CD3
      DO 90 J = 2, N8
            IS = 0
            ID = 2*N2
            JN = N4 – 2*J + 2
70                DO 80 I0 = IS + J, N, ID
                      I2     = I1 + JN
                      T1     = X1(I1) – X2(I2)
                      X1(I1) = X1(I1) + X2(I2)
                      T2     = X1(I2) – X2(I1)
                      X1(I2)= X2(I1) + X1(I2)
                      T3     = X4(I2) + X3(I1)
                      X2(I2) = X4(I2) – X3(I1)
                      T4     = X4(I1) + X3(I2)
                      X2(I1) = X4(I1) – X3(I2)
                      T5     = T1 – T4
                      T1     = T1 + T4
                      T4     = T2 – T3
                      T2     = T2 + T3
                      X3(I1) = T5*CC1 + T4*SS1
                      X3(I2) = T4*CC1 + T5*SS1
                      X4(I1) = T1*CC3 – T2*SS3
                      X4(I2) = T2*CC3 + T1*SS3
80                CONTINUE
                  IS = 2*ID – N2
                  ID = 4*ID
            IF (IS . LT. N–1) GOTO 70
C
            T1   = CC1*CD1 – SS1*SD1
```

```
          SS1  = CC1*SD1 + SS1*CD1
          CC1 = T1
          T3   = CC3*CD3 – SS3*SD3
          SS3  = CC3*SD3 + SS3*CD3
          CC3 = T3
90     CONTINUE
C
100    RETURN
       END
CC ================================================================= CC
CC                                                                   CC
CC     Subroutine RBITREV (X, M):                                    CC
CC            Bit reverses the array X of length 2**M. It generates a   CC
CC            table ITAB (minimum length is SQRT (2**M) if M is even    CC
CC            or SQRT(2*2**M) if M is odd). ITAB need only be generated CC
CC            once for a given transform length.                     CC
CC                                                                   CC
CC     Author:                                                       CC
CC            H. V. Sorensen, University of Pennsylvania, Aug. 1987  CC
CC                                Arpa address: hvs@ee.upenn.edu     CC
CC                                                                   CC
CC     Reference:                                                    CC
CC            D. Evans, Tran. ASSP, Aol. ASSP–35, No. 8, pp. 1120–1125, CC
CC            Aug. 1987                                              CC
CC                                                                   CC
CC            This program may be used and distributed freely provided CC
CC            this header is included and left intact.               CC
CC                                                                   CC
CC ================================================================= CC
       SUBROUTINE RBITREV (X, M)
       DIMENSION X(1), ITAB (256)
C ------------Initialization of ITAB array --------------------------------C
       M2 = M/2
       NBIT = 2**M2
       IF (2*M2 . NE . M) M2 = M2 + 1
       ITAB (1) = 0
       ITAB (2) = 1
       IMAX = 1
       DO 10 LBSS = 2, M2
            IMAX = 2 * IMAX
            DO 10 I = 1, IMAX
                 ITAB (I) = 2 * ITAB(I)
                 ITAB (I + IMAX) = 1 + ITAB(I)
10     CONTINUE
C ------------The actual bit reversal--------------------------------------C
       DO 20 K = 2, NBIT
            JO = NBIT * ITAB (K) + 1
            I   = K
            J   = JO
            DO 20 L = 2, ITAB (K) + 1
                 T1   = X(I)
```

```
              X(I) = X(J)
              X(J) = T1
              I    = I + NBIT
              J    = JO + ITAB(L)
20          CONTINUE
        RETURN
        END
```

References

Mitra, S. K., and J. F. Kaiser, Editors, *Handbook for Digital Signal Processing*, John Wiley & Sons, New York, NY, 1993.

Proakis, J. G., and D. G. Manolakis, *Digital Signal Processing*, Prentice-Hall, 2nd Edition, Upper Saddle River, NJ, 1996.

Poularikas, A. D. and S. Seely, *Signals and Systems*, 2nd edition, Krieger Publishing Co., Melbourne, FL, 1994.

12

Analog Filter Approximations

12.1 Filter Definitions

12.1.1 Normalized *Ideal Low-pass Filter* (see Figure 12.1a)

$$H(j\omega) = e^{-j\omega} \qquad 0 \le |\omega| \le 1$$

$$= 0 \qquad |\omega| > 1$$

12.1.2 Filter Transfer Function

$$H(j\omega) = |H(j\omega)| e^{j\theta(\omega)}$$

$$\theta(\omega) = Arg\, H(j\omega)$$

$$\tau(\omega) = -\frac{d\theta(\omega)}{d\omega} = \text{group delay}$$

$$\omega_c = \text{cutoff frequency at which } |H(j\omega_c)|^2 = \frac{1}{2}$$

or

$$20\log|H(j\omega)|_{\omega=\omega_c} = -3\,dB$$

$$A(\omega) = -10\log|H(j\omega)|^2 \ (= \text{attenuation})\ dB$$

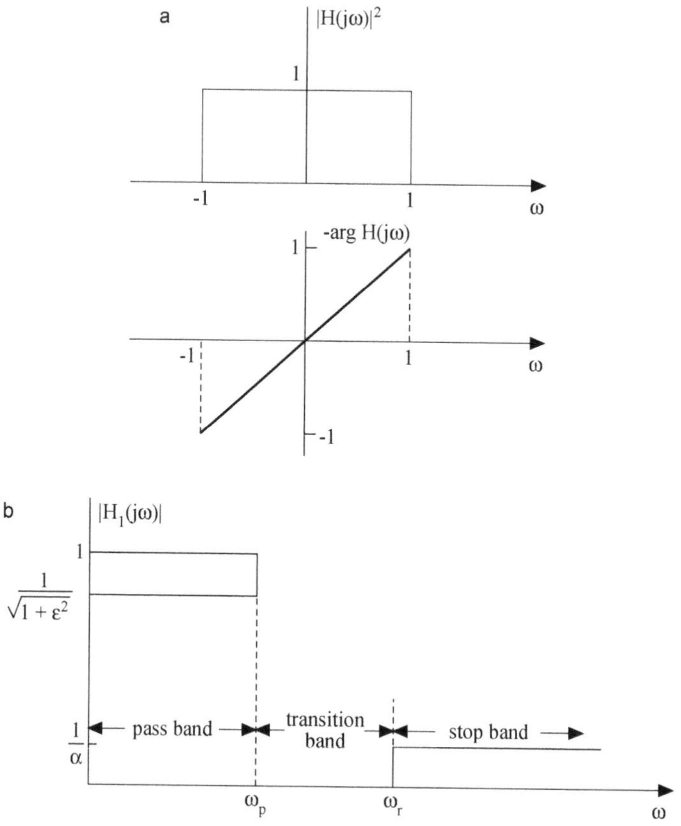

FIGURE 12.1

$$|H(j\omega)| = \left[H(j\omega)H^*(j\omega)\right]^{1/2} = \left[H(j\omega)H(-j\omega)\right]^{1/2} = \left[H(s)H(-s)\Big|_{s=j\omega}\right]^{1/2}$$

$$|H(j\omega)|^2 = \frac{K(\omega^2+z_1^2)(\omega^2+z_2^2)\cdots}{(\omega^2+p_1^2)(\omega^2+p_2^2)\cdots}$$

Complex poles and zeros occur in conjugate pairs. Both the numerator and denominator polynomials of the magnitude squared function of a transfer function are polynomials of ω^2 with real coefficients, and these polynomials are greater than zero for all ω.

12.2 Butterworth Approximation

12.2.1 Definition of Butterworth Low-Pass Filter

$$|H(j\omega)|^2 = \frac{1}{1+\left(\dfrac{\omega}{\omega_c}\right)^{2n}}; \qquad |H(j\omega_c)|^2 = \frac{1}{2}$$

$$10\log|H(j\omega)|^2\Big|_{\omega=\omega_c} = -3.01 \cong -3.0\ dB$$

Normalized

$$|H(j\omega)|^2 = \frac{1}{1+\omega^{2n}}; \qquad |H(j1)|^2 = \frac{1}{2}$$

12.3 Properties of Butterworth Approximation

12.3.1 $|H(j0)|^2 = 1;$ $\quad |H(j1)|^2 = \frac{1}{2};$ $\quad |H(j\infty)|^2 = 0$

$$-10\log|H(j1)|^2 = -10\log 0.5 = 3.01 \cong 3.0\,dB$$

12.3.2 $|H(j\omega)|^2$ monotonically decreasing for $\omega \geq 0$. Its maximum value is at $\omega = 0$.

12.3.3 The first $(2n - 1)$ derivatives of an nth -order low-pass Butterworth filter are zero at $\omega = 0$ *maximally flat* magnitude).

12.3.4 The high-frequency roll-off of an nth -order filter is $20n$ *dB/decade*

$$-10\log|H(j\omega)|^2 = -\log\frac{1}{1+\omega^{2n}} \cong -\log\frac{1}{\omega^{2n}} = 10\log\omega^{2n} = 20n\log\omega\ dB$$

12.4 Transfer Function of Butterworth Approximation

12.4.1 $|H(j\omega)|^2 = H(s)H(-s)\Big|_{s=j\omega} = \dfrac{1}{1+\left(\dfrac{\omega}{\omega_c}\right)^{2n}} = \dfrac{1}{1+(-1)^n\left(\dfrac{s}{\omega_c}\right)^{2n}}$

Poles:

$$1+(-1)^n\left(\frac{s}{\omega_c}\right)^{2n} = 0 \quad \text{or} \quad s_k = \omega_c e^{j\pi(1-n+2K)/2n}, \qquad K = 0,1,\cdots,2n-1$$

12.4.2 Stable Function

Left-half-plane poles are used

$$s_K = \omega_c\left[-\sin\frac{(2K+1)\pi}{2n} + j\cos\frac{(2K+1)\pi}{2n}\right], \qquad K = 0,1,\cdots,n-1$$

12.4.3 Transfer Function

$$H(s) = (-1)^n \prod_{K=0}^{n-1} \frac{s_K}{s - s_K}$$

12.4.4 Butterworth Normalized Low-Pass Filter

Table 12.1 gives the Butterworth polynomials ($\omega_c = 1$) to be used for normalized filters.

TABLE 12.1 Butterworth Normalized and Factored Polynomials

n	Butterworth Polynomials
1	$s + 1$
2	$s^2 + 1.41421s + 1$
3	$(s + 1)(s^2 + s + 1)$
4	$(s^2 + 0.76537s + 1)(s^2 + 1.84776s + 1)$
5	$(s + 1)(s^2 + 0.61803s + 1)(s^2 + 1.61803s + 1)$
6	$(s^2 + 0.51764s + 1)(s^2 + 1.41421s + 1)(s^2 + 1.93185s + 1)$
7	$(s + 1)(s^2 + 0.44504s + 1)(s^2 + 1.24798s + 1)(s^2 + 1.80194s + 1)$
8	$(s^2 + 0.39018s + 1)(s^2 + 1.11114s + 1)(s^2 + 1.66294s + 1)(s^2 + 1.96157s + 1)$
9	$(s + 1)(s^2 + 0.34730s + 1)(s^2 + s + 1)(s^2 + 1.53209s + 1)(s^2 + 1.87939s + 1)$
10	$(s^2 + 0.31287s + 1)(s^2 + 0.90798s + 1)(s^2 + 1.41421s + 1)(s^2 + 1.78201s + 1)(s^2 + 1.97538s + 1)$

12.4.5 Butterworth Filter Specifications (see also Figure 12.1)

A_p = maximum passband attenuation

f_p = passband edge frequency

Maximum allowable attenuation in the stopband

f_r = stopband edge frequency

$$A_p = 10\log\left[1+\left(\frac{\omega_p}{\omega_c}\right)^{2n}\right] \quad \text{(see also 12.3.4)}$$

$$A_r = 10\log\left[1+\left(\frac{\omega_r}{\omega_c}\right)^{2n}\right]$$

$$\omega_p = 2\pi f_p$$

$$\omega_r = 2\pi f_r$$

Solve A_p and A_r to find

$$n = \frac{\left|\log[(10^{0.1A_p} - 1)/(10^{0.1A_r} - 1)]\right|}{\left|\log(\omega_p/\omega_r)\right|}$$

$$k = \text{selectivity parameter} = \frac{\omega_p}{\omega_r} = \frac{f_p}{f_r} < 1$$

$$d = \text{discrimination factor} = \left(\frac{(10^{0.1A_p} - 1)}{(10^{0.1A_r} - 1)}\right)$$

Note: a) larger values of k imply steeper roll off, b) smaller d values imply greater difference between A_p and A_r

$$n \ge \frac{|\log d|}{|\log k|} \quad \text{(accept next higher integer to noninteger } n\text{)}$$

$$\omega_c = \frac{\omega_p}{(10^{0.1A_p} - 1)^{1/2n}}$$

$$\omega_c = \frac{\omega_r}{(10^{0.1A_r} - 1)^{1/2n}} \equiv \text{meets stopband attenuation exactly and exceeds the requirement of passband specification}$$

Figure 12.2 shows magnitude-squared characteristics of the Butterworth low-pass filter.

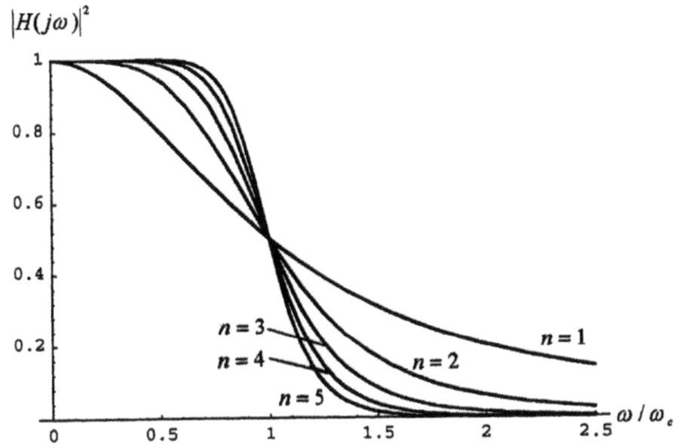

FIGURE 12.2

Example 12.1 Butterworth Filter Design

Filter requirements: a) no more than 1.5 dB deviation from ideal filter at 1300 Hz; b) at least 35 dB for frequencies above 6000 Hz.

Solution:

$$A_p = 1.5\,dB \qquad \omega_p = 2\pi \times 1300 \text{ rad s}^{-1}$$

$$A_r = 35\,dB \qquad \omega_r = 2\pi \times 6000 \text{ rad s}^{-1}$$

$$d = \frac{\sqrt{10^{0.1A_p} - 1}}{\sqrt{10^{0.1A_r} - 1}} = \frac{\sqrt{10^{0.15} - 1}}{\sqrt{10^{3.5} - 1}} = \frac{0.6423}{56.2252} = 1.1424 \times 10^{-2}$$

$$n \geq \frac{|\log d|}{|\log k|} = \frac{1.9422}{0.6576} = 2.953 \qquad \Rightarrow \qquad n = 3$$

$$s_k = -\sin\frac{(2K+1)\pi}{2n} + j\cos\frac{(2K+1)\pi}{2n} \qquad K = 0, 1, \cdots, n-1$$

$$s_o = -\sin\frac{\pi}{6} + j\cos\frac{\pi}{6} = -\frac{1}{2} + j\frac{\sqrt{3}}{2}$$

$$s_1 = -\sin\frac{3\pi}{6} + j\cos\frac{3\pi}{6} = -1$$

$$s_2 = -\sin\frac{5\pi}{6} + j\cos\frac{5\pi}{6} = -\frac{1}{2} - j\frac{\sqrt{3}}{2}$$

$$H(s) = (-1)^n \prod_{K=0}^{n-1} \frac{s_k}{s - s_k} = -\frac{\left(-\frac{1}{2} + j\frac{\sqrt{3}}{2}\right)}{s - \left(-\frac{1}{2} + j\frac{\sqrt{3}}{2}\right)} \frac{-1}{s-(-1)} \frac{-\frac{1}{2} - j\frac{\sqrt{3}}{2}}{s - \left(-\frac{1}{2} - j\frac{\sqrt{3}}{2}\right)}$$

$$= \frac{1}{(s+1)(s^2+s+1)} = \text{normalized}$$

$$\omega_c = \omega_p (10^{0.1A_p} - 1)^{-1/2n} = 2\pi \times 1300 (10^{0.15} - 1)^{-1/6} = 9416 \text{ rad s}^{-1}$$

$$H\left(\frac{s}{\omega_c}\right) = \frac{1}{\left(\frac{s}{9461} + 1\right)\left[\left(\frac{s}{9461}\right)^2 + \frac{s}{9461} + 1\right]}$$

12.5 Chebyshev Filter Approximation

12.5.1 Definition of Chebyshev Filters (equi-ripple passband)

$$C_o(\omega) = 1 \text{ and } C_1(\omega) = \omega. \quad |H(j\omega)|^2 = \frac{1}{1 + \varepsilon C_n^2(\omega)} = \text{normalized}$$

$$C_n(\omega) = \text{Chebyshev polynomials} = \cos(n \cos^{-1} \omega) \qquad 0 \le \omega \le 1$$

$$= \cosh(n \cosh^{-1} \omega) \qquad \omega > 1$$

$\varepsilon = \textit{ripple factor}$

If we set $u = \cos^{-1}\omega$, then $C_n(\omega) = \cos nu$ and thus

$$C_o(\omega) = \cos 0 = 1, C_1(\omega) = \cos u = \cos(\cos^{-1}\omega) = \omega, \quad C_2(\omega) = \cos 2u = 2\cos^2 u - 1 = 2\omega^2 - 1,$$

$$C_3(\omega) = \cos 3u = 4\cos^3 u - 3\cos u = 4\omega^3 - 3\omega, \text{ etc.}$$

12.5.2 Recursive Formula for Chebyshev Polynomials

From $\cos[(n+1)u] = 2\cos nu \cos u - \cos[(n-1)u]$, we get

$$C_{n+1}(\omega) = 2\omega C_n(\omega) - C_{n-1}(\omega) \qquad n = 0,1,2,\cdots$$

with $C_o(\omega) = 1$ and $C_1(\omega) = \omega$. Figure 12.3 shows the first five Chebyshev polynomials.

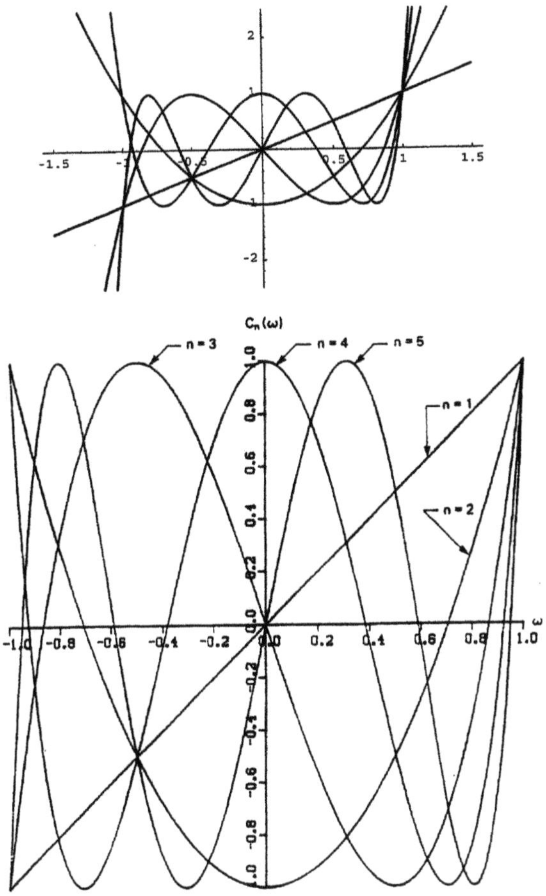

FIGURE 12.3 Chebyshev polynomials.

12.5.3 Table 12.2 gives the first ten Chebyshev polynomials

TABLE 12.2 Chebyshev Polynomials $C_n(\omega)$

n	Chebyshev Polynomials $C_n(\omega)$
0	1
1	ω
2	$2\omega^2 - 1$
3	$4\omega^3 - 3\omega$
4	$8\omega^4 - 8\omega^2 + 1$
5	$16\omega^5 - 20\omega^3 + 5\omega$
6	$32\omega^6 - 48\omega^4 + 18\omega^2 - 1$
7	$64\omega^7 - 112\omega^5 + 56\omega^3 - 7\omega$
8	$128\omega^8 - 256\omega^6 + 160\omega^4 - 32\omega^2 + 1$
9	$256\omega^9 - 576\omega^7 + 432\omega^5 - 120\omega^3 + 9\omega$
10	$512\omega^{10} - 1280\omega^8 + 1120\omega^6 - 400\omega^4 + 50\omega^2 - 1$

12.5.4 Properties of the Chebyshev Polynomials

1. For any n

$$\begin{aligned} 0 \le |C_n(\omega)| \le 1 \quad & \text{for } 0 \le |\omega| \le 1 \\ |C_n(\omega)| > 1 \quad & \text{for } \quad |\omega| > 1 \end{aligned}$$

2. $C_n(1) = 1$ for any n
3. $|C_n(\omega)|$ increases monotonically for $|\omega| > 1$
4. $C_n(\omega)$ is an even (odd) polynomial if n is even (odd)
5. $|C_n(0)| = 0$ for odd n
6. $|C_n(0)| = 1$ for even n

12.5.5 Chebyshev Magnitude Response Properties

1. $|H(j\omega)|_{\omega=0} = 1$ when n is odd

$$= \frac{1}{\sqrt{1+\varepsilon^2}} \quad \text{when } n \text{ is even}$$

2. Since $C_n(1) = 1$ for any n

$$|H(j1)| = \frac{1}{\sqrt{1+\varepsilon^2}} \qquad \text{for any } n$$

3. $|H(j\omega)|$ decreases monotonically

12.5.6 Pole Location of Chebyshev Filters

$$|H(j\omega)|^2 = \frac{1}{1+\varepsilon C_n^2(\omega)} = \left.\frac{1}{1+\varepsilon C_n^2(-js)}\right|_{s=j\omega}$$

$$s = \sigma + j\omega$$

$$\sigma_K = \pm \sin\left[(2K+1)\frac{\pi}{2n}\right] \sinh\left[\frac{1}{n}\sinh^{-1}\frac{1}{\varepsilon}\right]$$

$$\omega_K = \cos\left[(2K+1)\frac{\pi}{2n}\right] \cosh\left[\frac{1}{n}\sinh^{-1}\frac{1}{\varepsilon}\right] \qquad K = 0,1,\cdots,2n-1$$

$$\frac{\sigma_K}{\sinh^2 y} + \frac{\omega_K}{\cosh^2 y} = 1 \qquad \text{an ellipse on the } \sigma-\omega \text{ plane}$$

$$y = \frac{1}{n}\sinh^{-1}\frac{1}{\varepsilon}$$

12.5.7 Design Relations of Chebyshev Filters

$$|H(j\omega)|^2 = \frac{1}{1+\varepsilon C_n^2\left(\dfrac{\omega}{\omega_p}\right)}$$

$$|H(j\omega_p)|^2 = \left.\frac{1}{1+\varepsilon C_n^2\left(\dfrac{\omega}{\omega_p}\right)}\right|_{\omega=\omega_p} = \frac{1}{1+\varepsilon^2}$$

$$A_p = 10\log(1+\varepsilon^2)$$

$$\varepsilon = \sqrt{10^{0.1A_p}-1}$$

$$A_r = 10\log\left[1+\varepsilon^2 C_n^2\left(\frac{\omega_r}{\omega_p}\right)\right]$$

$$= 10\log\left[1+\varepsilon^2\cosh^2\left[n\cosh^{-1}\left(\frac{\omega_r}{\omega_p}\right)\right]\right]$$

$$n \geq \frac{\cosh^{-1}\left(\dfrac{10^{0.1A_r}-1}{\varepsilon^2}\right)^{1/2}}{\cosh^{-1}\left(\dfrac{\omega_r}{\omega_p}\right)}$$

$$k = \frac{\omega_p}{\omega_r} = \frac{f_p}{f_r}, \qquad d = \left(\frac{10^{0.1A_p}-1}{10^{0.1A_r}-1}\right)^{1/2}$$

or

$$n \geq \frac{\cosh^{-1}\left(\dfrac{1}{d}\right)}{\cosh^{-1}\left(\dfrac{1}{k}\right)}$$

Left-Hand Poles for the Transfer Function

$$s_K = \sin\left[(2K+1)\frac{\pi}{2n}\right]\sinh\left[\frac{1}{n}\sinh^{-1}\frac{1}{\varepsilon}\right] + j\cos\left[(2K+1)\frac{\pi}{2n}\right]\cosh\left[\frac{1}{n}\sinh^{-1}\frac{1}{\varepsilon}\right]$$

$$H(s) = -\prod_{K=0}^{n-1}\frac{s_K}{s-s_K}, \quad n \text{ odd}$$

$$H(s) = \frac{1}{\sqrt{1+\varepsilon^2}} \prod_{K=0}^{n-1} \frac{s_K}{s - s_K}, \qquad n \text{ even}$$

For non-normalized transfer function set s/ω_p in place of s

$$|H(j\omega_c)|^2 = \frac{1}{2} = \frac{1}{1+\varepsilon^2 C_n^2(\omega_c)}, \qquad 3-dB \text{ cutoff}$$

$$\omega_c = \cosh\left(\frac{1}{n}\cosh^{-1}\frac{1}{\varepsilon}\right)$$

Example 12.2 (Chebyshev Filter Design):

Filter requirements: a) ripple not to exceed 2 dB up to ω_p; b) 50 dB rejection above 5 ω_p.

Solution

$$A_p \le 2dB \quad \text{at} \quad \omega = \omega_p$$

$$A_r \ge 50\,dB \quad \text{at} \quad \omega = \omega_r = 5\omega_p$$

$$\varepsilon = (10^{0.1A_p} - 1)^{1/2} = (10^{0.2} - 1)^{1/2} = 0.765$$

$$k = \frac{\omega_p}{\omega_r} = \frac{\omega_p}{5\omega_p} = 0.2$$

$$d = \frac{0.765}{(10^{0.1A_r} - 1)^{1/2}} = \frac{0.765}{(10^5 - 1)^{1/2}} = 2.42 \times 10^{-3}$$

$$n \ge \frac{\cosh^{-1}(1/d)}{\cosh^{-1}(1/k)} = \frac{\ln(1/d + \sqrt{1/d^2 - 1})}{\ln(1/k + \sqrt{1/k^2 - 1})} = \frac{2.718}{2.312} = 2.91$$

accept $n = 3$

From 12.5.6

$$y = \frac{1}{n}\sinh^{-1}\frac{1}{\varepsilon} = \frac{1}{n}\ell n\left(\frac{1}{\varepsilon} + \sqrt{\frac{1}{\varepsilon^2} + 1}\right) = 0.361$$

$$\sinh y = \frac{e^y - e^{-y}}{2} = 0.3689 \qquad \cosh y = \frac{e^y + e^{-y}}{2} = 1.0659$$

$$s_0 = \sin\left(\frac{\pi}{6}\right)(0.3689) + j\cos\left(\frac{\pi}{6}\right)(1.0659) = -0.1844 + j0.9231$$

$$s_1 = \sin\left(\frac{\pi}{2}\right)(0.3689) + j\cos\left(\frac{\pi}{2}\right)(1.0659) = -0.3689$$

$$s_2 = -\sin\left(\frac{\pi}{6}\right)(0.3689) + j\cos\left(\frac{5\pi}{6}\right)(1.0659) = -0.1844 - j0.9231 = s_0^*$$

$$H(s) = \frac{0.3289}{(s + 0.3689)(s^2 + 0.3689s + 0.8861)}$$

To denormalize $H(s)$ we set $\omega_p = 2\pi f_p$ we set s/ω_p in place of s. Table 12.3 gives the denominator or the normalized Chebyshev low-pass filters. Figure 12.4 shows the third-order filter with $\omega_p = 2\pi \times 2$.

TABLE 12.3 Factors of the Denominator Polynomials Normalized Chebyshev Low-Pass Filters

n	0.1-dB Ripple (ε = 0.15262)
1	s + 6.55220
2	s² + 2.37236s + 3.31403
3	(s + 0.96941) (s² + 0.96941s + 1.68975)
4	(s² + 0.52831s + 1.33003) (s² + 1.27546s + 0.62292)
5	(s + 0.53891) (s² + 0.33307s + 1.19494) (s² + 0.87198s + 0.63592)
6	(s² + 0.22939s + 1.12939) (s² + 0.62670s + 0.69637) (s² + 0.85608s + 0.26336)
7	(s + 0.37678) (s² + 0.16768s + 1.09245) (s² + 0.46983s + 0.75322) (s² + 0.67893s + 0.33022)
8	(s² + 0.12796s + 1.06949) (s² + 0.36440s + 0.79889) (s² + 0.54536s + 0.41621) (s² + 0.64330s + 0.14561)
9	(s + 0.29046) (s² + 0.10088s + 1.05421) (s² + 0.29046s + 0.83437) · (s² + 0.44501s + 0.49754) (s² + 0.54589s + 0.20134)
10	(s² + 0.08158s + 1.04351) (s² + 0.23675s + 0.86188) (s² + 0.36874s + 0.56799) · (s² + 0.46464s + 0.27409) (s² + 0.51506s + 0.09246)
n	**0.2-dB Ripple (ε = 0.21709)**
1	s + 4.60636
2	s² + 1.92709s + 2.35683
3	(s + 0.81463) (s² + 0.81463s + 1.41363)
4	(s² + 0.44962s + 1.19866) (s² + 1.08548s + 0.49155)
5	(s + 0.46141) (s² + 0.28517s + 1.11741) (s² + 0.74658s + 0.55839)
6	(s² + 0.19705s + 1.07792) (s² + 0.53835s + 0.64491) (s² + 0.73540s + 0.21190)
7	(s + 0.32431) (s² + 0.14433s + 1.05566) (s² + 0.40441s + 0.71644) (s² + 0.58439s + 0.29343)
8	(s² + 0.11028s + 1.04183) (s² + 0.31407s + 0.77124) (s² + 0.47004s + 0.38855) (s² + 0.55445s + 0.11795)
9	(s + 0.25057) (s² + 0.08702s + 1.03263) (s² + 0.25057s + 0.81278) · (s² + 0.38389s + 0.47596) (s² + 0.47092s + 0.17976)
10	(s² + 0.44461s + 0.07513) (s² + 0.40109s + 0.25677) (s² + 0.31830s + 0.55066) · (s² + 0.20436s + 0.84455) (s² + 0.07042s + 1.02619)
n	**0.5-dB Ripple (ε = 0.34931)**
1	s + 2.86278
2	s² + 1.42562s + 1.51620
3	(s + 0.62646) (s² + 0.62646s + 1.14245)
4	(s² + 0.35071s + 1.06352) (s² + 0.84668s + 0.35641)
5	(s + 0.36232) (s² + 0.22393s + 1.03578) (s² + 0.58625s + 0.47677)
6	(s² + 0.15530s + 1.02302) (s² + 0.42429s + 0.59001) (s² + 0.57959s + 0.15610)
7	(s + 0.25617) (s² + 0.11401s + 1.01611) (s² + 0.31944s + 0.67688) (s² + 0.46160s + 0.25388)
8	(s² + 0.08724s + 1.01193) (s² + 0.24844s + 0.74133) (s² + 0.37182s + 0.35865) (s² + 0.43859s + 0.08805)
9	(s + 0.19841) (s² + 0.06891s + 1.00921) (s² + 0.19841s + 0.78937) · (s² + 0.30398s + 0.45254) (s² + 0.37288s + 0.15634)
10	(s² + 0.05580s + 1.00734) (s² + 0.161934s + 0.82570) (s² + 0.25222s + 0.53181) · (s² + 0.31781s + 0.23791) (s² + 0.35230s + 0.05628)
n	**1-dB Ripple (ε = 0.50885)**
1	s + 1.96523
2	s² + 1.09773s + 1.10251
3	(s + 0.49417) (s² + 0.49417s + 0.99421)
4	(s² + 0.27907s + 0.98651) (s² + 0.67374s + 0.27940)
5	(s + 0.28949) (s² + 0.17892s + 0.98832) (s² + 0.46841s + 0.42930)

TABLE 12.3 Factors of the Denominator Polynomials Normalized Chebyshev Low-Pass Filters (continued)

n	1-dB Ripple ($\varepsilon = 0.50885$)
6	$(s^2 + 0.12436s + 0.99073)\ (s^2 + 0.33976s + 0.55772)\ (s^2 + 0.46413s + 0.12471)$
7	$(s + 0.20541)\ (s^2 + 0.09142s + 0.99268)\ (s^2 + 0.25615s + 0.65346)\ (s^2 + 0.37014s + 0.23045)$
8	$(s^2 + 0.07002s + 0.99414)\ (s^2 + 0.19939s + 0.72354)\ (s^2 + 0.29841s + 0.34086)\ (s^2 + 0.35110s + 0.07026)$
9	$(s + 0.15933)\ (s^2 + 0.05533s + 0.99523)\ (s^2 + 0.15933s + 0.77539)$
	$\cdot\ (s^2 + 0.24411s + 0.43856)\ (s^2 + 0.29944s + 0.14236)$
10	$(s^2 + 0.04483s + 0.99606)\ (s^2 + 0.13010s + 0.81442)\ (s^2 + 0.20263s + 0.52053)$
	$\cdot\ (s^2 + 0.25533s + 0.22664)\ (s^2 + 0.28304s + 0.04500)$

n	1.5-dB Ripple ($\varepsilon = 0.64229$)
1	$s + 1.55693$
2	$s^2 + 0.92218s + 0.92521$
3	$(s + 0.42011)\ (s^2 + 0.42011s + 0.92649)$
4	$(s^2 + 0.23826s + 0.95046)\ (s^2 + 0.57521s + 0.24336)$
5	$(s + 0.24765)\ (s^2 + 0.15306s + 0.96584)\ (s^2 + 0.40071s + 0.40682)$
6	$(s^2 + 0.10650s + 0.97534)\ (s^2 + 0.29097s + 0.54233)\ (s^2 + 0.39747s + 0.10932)$
7	$(s + 0.17603)\ (s^2 + 0.07834s + 0.98147)\ (s^2 + 0.21951s + 0.64225)\ (s^2 + 0.31720s + 0.21924)$
8	$(s^2 + 0.06003s + 0.98561)\ (s^2 + 0.17094s + 0.71501)\ (s^2 + 0.25583s + 0.33233)\ (s^2 + 0.30177s + 0.06173)$
9	$(s + 0.13667)\ (s^2 + 0.04745s + 0.98852)\ (s^2 + 0.13664s + 0.76867)$
	$\cdot\ (s^2 + 0.20934s + 0.43185)\ (s^2 + 0.25679s + 0.13565)$
10	$(s^2 + 0.03845s + 0.99063)\ (s^2 + 0.11159s + 0.80900)\ (s^2 + 0.17381s + 0.51510)$
	$\cdot\ (s^2 + 0.21901s + 0.22121)\ (s^2 + 0.24277s + 0.03958)$

n	2-dB Ripple ($\varepsilon = 0.76478$)
1	$s + 1.30756$
2	$s^2 + 0.80382s + 0.82306$
3	$(s + 0.36891)\ (s^2 + 0.36891s + 0.88610)$
4	$(s^2 + 0.20978s + 0.92868)\ (s^2 + 0.50644s + 0.22157)$
5	$(s + 0.21831)\ (s^2 + 0.13492s + 0.95217)\ (s^2 + 0.35323s + 0.39315)$
6	$(s^2 + 0.09395s + 0.96595)\ (s^2 + 0.25667s + 0.53294)\ (s^2 + 0.35061s + 0.09993)$
7	$(s + 0.15533)\ (s^2 + 0.06913s + 0.97462)\ (s^2 + 0.19371s + 0.63539)\ (s^2 + 0.27991s + 0.21239)$
8	$(s^2 + 0.05298s + 0.98038)\ (s^2 + 0.15089s + 0.70978)\ (s^2 + 0.22582s + 0.32710)\ (s^2 + 0.26637s + 0.05650)$
9	$(s + 0.12063)\ (s^2 + 0.04189s + 0.98440)\ (s^2 + 0.12063s + 0.76455)$
	$\cdot\ (s^2 + 0.18482s + 0.42773)\ (s^2 + 0.22671s + 0.13153)$
10	$(s^2 + 0.03395s + 0.98730)\ (s^2 + 0.09853s + 0.80567)\ (s^2 + 0.15347s + 0.51178)$
	$\cdot\ (s^2 + 0.19338s + 0.21788)\ (s^2 + 0.21436s + 0.03625)$

n	2.5-dB Ripple ($\varepsilon = 0.88220$)
1	$(s + 1.13353)$
2	$(s^2 + 0.71525s + 0.75579)$
3	$(s + 0.32995)\ (s^2 + 0.32995s + 0.85887)$
4	$(s^2 + 0.18796s + 0.91386)\ (s^2 + 0.45378s + 0.20676)$
5	$(s + 0.19577)\ (s^2 + 0.12099s + 0.94284)\ (s^2 + 0.31677s + 0.38382)$
6	$(s^2 + 0.08429s + 0.95953)\ (s^2 + 0.23028s + 0.52651)\ (s^2 + 0.31456s + 0.09350)$
7	$(s + 0.13941)\ (s^2\ 0.06204s + 0.96992)\ (s^2 + 0.17384 + 0.63070)\ (s^2 + 0.25120s + 0.20769)$
8	$(s^2 + 0.04756s + 0.97680)\ (s^2 + 0.13054s + 0.70620)\ (s^2 + 0.20269s + 0.32352)\ (s^2 + 0.23909s + 0.05292)$
9	$(s + 0.10829)\ (s^2 + 0.03761s + 0.98157)\ (s^2 + 0.10829s + 0.76173)$
	$\cdot\ (s^2 + 0.16591s + 0.42490)\ (s^2 + 0.20352s + 0.12870)$
10	$(s^2 + 0.19245s + 0.03396)\ (s^2 + 0.17361s + 0.21560)\ (s^2 + 0.13778s + 0.50949)$
	$\cdot\ (s^2 + 0.08846s + 0.80338)\ (s^2 + 0.03048s + 0.98502)$

TABLE 12.3 Factors of the Denominator Polynomials Normalized Chebyshev Low-Pass Filters (continued)

n	3-dB Ripple (ε = 0.99763)
1	$(s + 1.00238)$
2	$(s^2 + 0.64490s + 0.70795)$
3	$(s + 0.29862)\ (s^2 + 0.29862s + 0.83917)$
4	$(s^2 + 0.17034s + 0.90309)\ (s^2 + 0.41124s + 0.19598)$
5	$(s + 0.17753)\ (s^2 + 0.10970s + 0.93603)\ (s^2 + 0.28725s + 0.37701)$
6	$(s^2 + 0.07646s + 0.95483)\ (s^2 + 0.20889s + 0.52182)\ (s^2 + 0.28535s + 0.08880)$
7	$(s + 0.12649)\ (s^2\ 0.05629s + 0.96648)\ (s^2 + 0.15773 + 0.62726)\ (s^2 + 0.22792s + 0.20425)$
8	$(s^2 + 0.04316s + 0.97417)\ (s^2 + 0.12290s + 0.70358)\ (s^2 + 0.18393s + 0.32089)\ (s^2 + 0.21696s + 0.05029)$
9	$(s + 0.09827)\ (s^2 + 0.03413s + 0.97950)\ (s^2 + 0.09827s + 0.75966)$
	$\cdot\ (s^2 + 0.15057s + 0.42283)\ (s^2 + 0.18470s + 0.12664)$
10	$(s^2 + 0.02766s + 0.98335)\ (s^2 + 0.08028s + 0.80171)\ (s^2 + 0.12504s + 0.50782)$
	$\cdot\ (s^2 + 0.15757s + 0.21393)\ (s^2 + 0.17466s + 0.03229)$

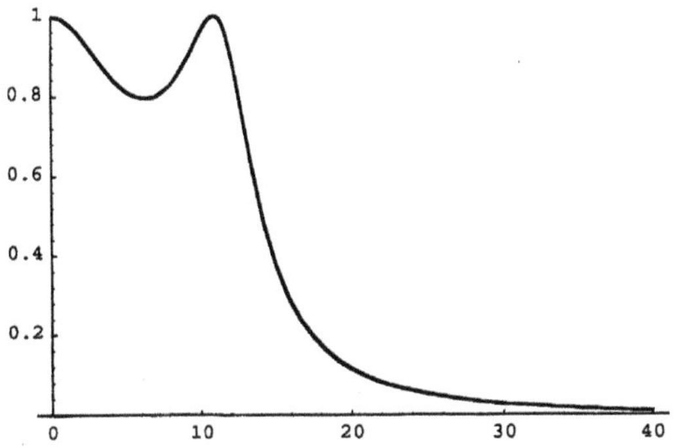

FIGURE 12.4 $f_p = 2H_z$; $\omega_p = 12.5664$

12.6 Inverse-Chebyshev Approximation

12.6.1 Definition

The inverse-Chebyshev filter is flat in the passband and equi-ripple in the stopband.

12.6.2 The Magnitude-Squared Transfer Function

$$|H(j\omega)|^2 = \frac{\varepsilon^2 C_n^2(\omega_r/\omega)}{1 + \varepsilon^2 C_n^2(\omega_r/\omega)}$$

$C_n(\omega)$ = Chebyshev polynomial; ω_r = stopband edge frequency

12.6.3 Attenuation

$$A(\omega) = 10 \log\left(1 + \frac{1}{\varepsilon^2 C_n^2(\omega_r/\omega)}\right) dB$$

ε = ripple factor calculated at $\omega = \omega_r$

$$A_r(\omega) = 10 \log\left(1 + \frac{1}{\varepsilon^2 C_n^2(1)}\right), \quad C_n^2(1) = 1$$

$$\varepsilon = \frac{1}{\sqrt{10^{0.1A_r} - 1}}$$

12.6.4 Filter Order

$$n \geq \frac{\cosh^{-1}(1/d)}{\cosh^{-1}(1/k)}$$

$$k = \frac{f_p}{f_r}, \quad d = \left(\frac{10^{0.1A_p} - 1}{10^{0.1A_r} - 1}\right)^{1/2}$$

12.6.5 Poles and Zeros

$$H(s)H(-s) = \frac{\varepsilon^2 C_n^2(j\omega_r/s)}{1 + \varepsilon^2 C_n^2(j\omega_r/s)}$$

Zeros

$$C_n(j\omega_r/s) = 0 = \cos(n\cos^{-1}(j\omega_r/s))$$

$$\cos^{-1}(j\omega_r/s) = m\pi/2n, \quad m \text{ odd}$$

$$s_m = zeros = j\omega_r \sec(m\pi/2n), \quad m = 1,3,\cdots,2n-1$$

Poles

$$1 + \varepsilon^2 C_n^2(j\omega_r/s) = 0$$

same poles as in 12.5.6 except that $-s$ is replaced by $1/s$.

Denormalization is accomplished with respect o stopband edge frequency ω_r.

12.7 Elliptic Filters

12.7.1 Square Magnitude Response Function for Elliptic Filters

$$|H(j\omega)|^2 = \frac{1}{1 + \varepsilon^2 R_n^2(\omega)}$$

$$R_n(\omega) = \text{rational function}; \quad \varepsilon = \text{ripple factor}$$

12.7.2 Properties of the Rational Function $R_n(\omega)$

1. $R_n(\omega)$ = even for n even. $R_n(\omega)$ = odd for n odd.
2. The zeros of $R_n(\omega)$ are in the range $|\omega| < 1$.
   The poles of $R_n(\omega)$ are in the range $|\omega| > 1$.
3. The function $R_n(\omega)$ oscillates between ± 1 in the passband.
4. $R_n(\omega) = 1$ at $\omega = 1$.
5. $R_n(\omega)$ oscillates between $\pm 1/d$ and infinity in the stopband, where d is given in 12.5.7.

12.7.3 The Rational Normalized Function $R_n(\omega)$ with Respect to Center Frequency $\omega_0 = 1$

$$R_n(\omega) = \omega \prod_{i=1}^{(n-1)/2} \frac{\omega_i^2 - \omega^2}{1 - \omega_i^2 \omega^2} \qquad \text{for } n \text{ odd}$$

$$R_n(\omega) = \prod_{i=1}^{n/2} \frac{\omega_i^2 - \omega^2}{1 - \omega_i^2 \omega^2} \qquad \text{for } n \text{ even}$$

12.7.4 Steps to Calculate the Elliptic Filter

1. Find the selectivity factor k

$$k = \frac{\omega_p}{\omega_r}, \quad \omega_p = \text{passband frequency}, \ \omega_r = \text{stopband frequency}$$

2. Define

$$q_o = \frac{1}{2} \frac{1 - (1 - k^2)^{1/4}}{1 + (1 - k^2)^{1/4}}$$

3. Find the expression

$$q = q_o + 2q_o^5 + 15q_o^9 + 150q_o^{13}$$

4. Find d

$$d = \left(\frac{10^{0.1A_p} - 1}{10^{0.1A_r} - 1} \right)^{1/2}$$

5. Find the filter order n

$$n \geq \frac{\log(16/d^2)}{\log(1/q)}$$

6. Calculate ε

$$\varepsilon = \left(10^{0.1A_p} - 1\right)^{1/2}$$

$$A_p = 10 \log(1 + \varepsilon^2)$$

7. Define

$$\beta = \frac{1}{2n} \ell n \frac{(1 + \varepsilon^2)^{1/2} + 1}{(1 + \varepsilon^2)^{1/2} - 1}$$

8. Calculate

$$a = \frac{2q^{1/4}\sum_{m=0}^{\infty}(-1)^m q^{m(m+1)}\sinh[(2m+1)\beta]}{1+\sum_{m=1}^{\infty}(-1)^m q^{m^2}\cosh(2m\beta)}$$

9. Define

$$U = \left[(1+ka^2)\left(1+\frac{a^2}{k}\right)\right]^{1/2}$$

10. Calculate

$$\omega_i = \frac{2q^{1/4}\sum_{m=0}^{\infty}(-1)^m q^{m(m+1)}\sin[(2m+1)\pi\ell/n]}{1+2\sum_{m=1}^{\infty}(-1)^m q^{m^2}\cos(2m\pi\ell/n)}$$

$\ell = i-\frac{1}{2},\ i=1,2,\cdots,\frac{n}{2},\ n=\text{even},\ \ell=i,\ i=1,2,\cdots,(n-1)/2,\ n=\text{odd}$

11. Define

$$V_i = \left[(1-k\omega_i^2)(1-\tfrac{\omega_i^2}{k})\right]^{1/2}$$

12. Set

$$a_i = \frac{1}{\omega_i^2}$$

13. Set

$$b_i = \frac{2aV_i}{1+a^2\omega_i^2}$$

14. Set

$$c_i = \frac{(aV_i)^2+(\omega_i U)^2}{(1+a^2\omega_i^2)^2}$$

15. Find

$$H_o = a\prod_{i=1}^{(n-1)/2}\frac{c_i}{a_i} \qquad \text{for } n=\text{odd}$$

$$H_o = \frac{1}{\sqrt{1+\varepsilon^2}} \prod_{i=1}^{n/2} \frac{c_i}{a_i} \qquad \text{for } n = \text{even}$$

$$H(s) = H_o \prod_{i=1}^{n/2} \frac{s^2 + a_i}{s^2 + b_i s + c_i} \qquad \text{for } n = \text{even}$$

$$H(s) = \frac{H_o}{s+a} \prod_{i=1}^{(n-1)/2} \frac{s^2 + a_i}{s^2 + b_i s + c_i} \qquad \text{for } n \text{ odd}$$

12.7.5 Unormalized Transfer Function

Replace s in the 15th step above (in Section 12.7.4) with s/ω_0 where $\omega_o = \sqrt{\omega_p \omega_r}$.

Note: Summations in steps 8 and 10 above converge fast, and up to four or five terms of the series will provide good accuracy.

12.8 Elliptic Filters (Second Approach*)

12.8.1 Transfer Function

$$H(s) = \frac{H_o}{D_o(s)} \prod_{i=1}^{r} \frac{s^2 + a_{oi}}{s^2 + b_{1i} s + b_{oi}}$$

$$r = \begin{cases} \dfrac{n-1}{2} & \text{for odd } n \\ \dfrac{n}{2} & \text{for even } n \end{cases}$$

$$D_o(s) = \begin{cases} s + \sigma_o & \text{for odd } n \\ 1 & \text{for even } n \end{cases}$$

12.8.2 Steps of Implementation

Given

ω_p = passband frequency; ω_r = stopband frequency;
A_p = maximum passband loss (dB); A_r = minimum stopband loss (dB)
k = selectivity factor = ω_p/ω_r

Steps

1. $k^1 = \sqrt{1-k^2}$

2. $q_o = \dfrac{1}{2}\left(\dfrac{1-\sqrt{k'}}{1+\sqrt{k'}}\right)$

3. $q = q_o + 2q_o^5 + 15q_o^9 + 150q_o^{13}$

* Antoniou (1993)

4. $D = \dfrac{10^{0.1A_r} - 1}{10^{0.1A_p} - 1}$

5. $n \geq \dfrac{\log 16D}{\log(1/q)}$

6. $\Lambda = \dfrac{1}{2n} \ell n \dfrac{10^{0.05A_p} + 1}{10^{0.05A_p} - 1}$

7. $\sigma_o = \left| \dfrac{2q^{1/4} \sum_{m=0}^{\infty} (-1)^m q^{m(m+1)} \sinh[(2m+1)\Lambda]}{1 + 2\sum_{m=1}^{\infty} (-1)^m q^{m^2} \cosh 2m\Lambda} \right|$

8. $W = \sqrt{(1 + k\sigma_o^2)\left(1 + \dfrac{\sigma_o^2}{k}\right)}$

9. $\Omega_i = \dfrac{2q^{1/4} \sum_{m=0}^{\infty} (-1)^m q^{m(m+1)} \sin\left[\dfrac{(2m+1)\pi\mu}{n}\right]}{1 + 2\sum_{m=1}^{\infty} (-1)^m q^{m^2} \cos\left[\dfrac{2m\pi\mu}{n}\right]}$

$$\mu = \begin{cases} i & \text{for odd } n \\ i - \dfrac{1}{2} & \text{for even } n \end{cases} \qquad i = 1, 2, \cdots, r$$

10. $V_i = \sqrt{(1 - k\Omega_i^2)\left(1 - \dfrac{\Omega_i^2}{k}\right)}$

11. $a_{oi} = \dfrac{1}{\Omega_i^2}$

$$b_{oi} = \frac{(\sigma_o V_i)^2 + (\Omega_i W)^2}{(1 + \sigma_o^2 \Omega_i^2)^2}$$

$$b_{1i} = \frac{2\sigma_o V_i}{1 + \sigma_o^2 \Omega_i^2}$$

12. $H_o = \begin{cases} \sigma_o \prod_{i=1}^{r} \dfrac{b_{oi}}{a_{oi}} & \text{for odd } n \\ 10^{-0.05A_p} \prod_{i=1}^{r} \dfrac{b_{oi}}{a_{oi}} & \text{for even } n \end{cases}$

The series in steps 7 and 9 converge rapidly, and three to four terms are sufficient for most purposes.

Example 12.3 Requirements for an Elliptic Filter:

$$\omega_p = \sqrt{0.9} \text{ rad/s}, \ \ \omega_r = 1/\sqrt{0.9} \text{ rad/s}, \ \ A_p = 0.1 \text{ dB}, \text{ and } \ A_r \geq 50.0 \text{ dB}$$

Results from steps 1 through 5 above:

$$k = 0.9, \ k' = 0.43589, \ q_o = 0.10233, \ q = 0.102352, \ D = 4{,}293{,}090, \ n \geq 7.92 \text{ or } n = 8$$

The coefficients for $H(s)$ are found from the rest of the steps and are:

i	a_{oi}	b_{oi}	b_{1i}
1	1.434825×10	2.914919×10^{-1}	8.711574×10^{-1}
2	2.231643	6.123726×10^{-1}	4.729136×10^{-1}
3	1.320447	8.397386×10^{-1}	1.825141×10^{-1}
4	1.128832	9.264592×10^{-1}	4.471442×10^{-2}

$H_o = 2.876332 \times 10^{-3}$

Hence from 12.8.1

$$H(s) = 2.87633 \times 10^{-3}\left(\frac{s^2 + 14.34825}{s^2 + 0.8711574s + 0.2914919}\right)\left(\frac{s^2 + 2.231643}{s^2 + 0.4729136s + 0.6123726}\right)$$

$$\times\left(\frac{s^2 + 1.320447}{s^2 + 0.1825141s + 0.8397386}\right)\left(\frac{s^2 + 1.128832}{s^2 + 0.04471442s + 0.9264592}\right)$$

and $|H(j\omega)|$ is plotted in Figure 12.5.

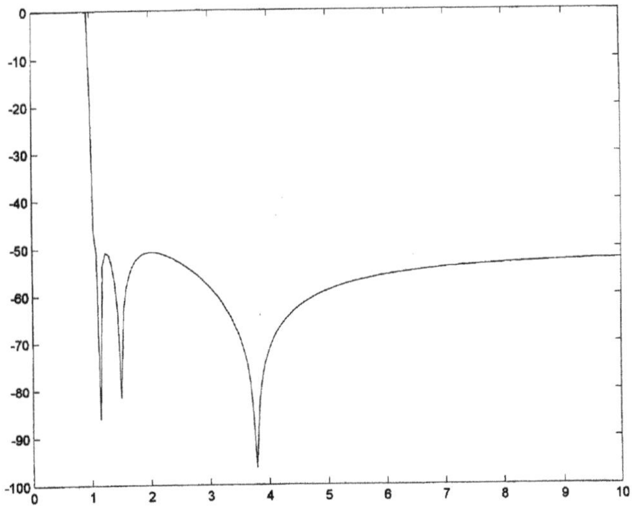

FIGURE 12.5

12.9 Transformations

12.9.1 Lowpass to Lowpass

$$\text{Set } s \to \frac{s}{\omega_p}, \quad \omega_p = \text{new passband frequency}$$

12.9.2 Lowpass to Highpass

$$\text{Set } s \to \frac{\omega_p}{s}, \quad \omega_p = \text{new passband frequency}$$

12.9.3 Lowpass to Bandpass

$$\text{Set } s \to \frac{s^2 + \omega_m^2}{Bs}$$

ω_m = geometric mean of the upper band edge frequency ω_u

and the lower band edge frequency $\omega_\ell = \sqrt{\omega_u \omega_\ell}$

$B = \omega_u - \omega_\ell$ = filter bandwidth

12.9.4 Lowpass to Bandstop

$$s \to \frac{Bs}{s^2 + \omega_m^2}$$

ω_m and B are the same as in 12.9.3

References

Antoniou, A., *Digital Filters*, Second Edition, McGraw-Hill, New York, 1993.

Elliott, D. F., Editor, *Handbook of Digital Signal Processing*, Academic Press, New York, 1987.

Mitra, S. K. and J. F. Kaiser, *Handbook for Digital Signal Processing*, John Wiley & Sons, New York, 1937.

Parks, T. W. and C. S. Burrus, *Digital Filter Design*, John Wiley & Sons, New York, 1987.

Zuerev, A. I., *Handbook of Filter Synthesis*, Wiley, New York, 1967.

13

Sine and Cosine Transforms

13.1 Fourier Cosine Transform (FCT)

13.1.1 Definitions of FCT

$$F_c\{f(t)\} = F_c(\omega) = \int_0^\infty f(t)\cos\omega t\, dt \qquad \omega \quad 0$$

$$F_c^{-1}\{F_c(\omega)\} = f(t) = \frac{2}{\pi}\int_0^\infty F_c(\omega)\cos\omega t\, dt \qquad t \quad 0$$

The sufficient conditions for the inversion formula are that $f(t)$ be absolutely integrable in $[0,\infty)$ and that $f(t)$ be piece-wise continuous in each bounded subinterval of $[0,\infty)$. At the point t_o where $f(t)$ has a jump discontinuity $f(t) = [f(t_o + 0) + f(t_o \; -0)]/2$.

13.1.2 Properties of the FCT

13.1.2.1 Transform of Derivatives

$$F_c\{f''(t)\} = \int_0^\infty f''(t)\cos\omega t\, dt = -\omega^2 F_c(\omega) - f'(0)$$

$f(t)$ and $f'(t)$ vanish as $t \to \infty$ and are continuous in $[0,\infty)$.

If $f(t)$ and $f'(t)$ have a jump discontinuity at t_o of d and d', respectively,

$$F_c\{f''(t)\} = -\omega^2 F_c(\omega) - f'(0) - \omega d \sin \omega t_o - d' \cos \omega t_o$$

$$d = f(t_o + 0) - f(t_o - 0), \qquad d' = f'(t_o + 0) - f'(t_o - 0)$$

13.1.2.2 Scaling

$$F_c\{f(at)\} = \frac{1}{a} F_c\left(\frac{\omega}{a}\right) \qquad a > 0$$

13.1.2.3 Shifting in t-domain

If $f_e(t) = f(|t|)$ and $f(t)$ is piece-wise continuous and absolutely integrable $[0,\infty)$, then

$$F_c\{f_e(t+a) + f_e(t-a)\} = 2F_c(\omega)\cos a\omega \qquad a > 0$$

$$F_c\{f(t+a) + f(|t-a|)\} = 2F_c(\omega)\cos a\omega$$

13.1.2.4 Shifting in the ω-domain

$$F_c(\omega + \beta) = F_c\{f(t)\cos \beta t\} - F_s\{f(t)\sin \beta t\} \qquad \beta \quad 0$$

where $\beta > 0$. F_s means the sine transform.

$$F_c(\omega - \beta) = F_c\{f(t)\cos \beta t\} + F_s\{f(t)\sin \beta t\} \qquad \beta > 0$$

$$F_c\{f(t)\cos \beta t\} = \frac{1}{2}\left[F_c(\omega + \beta) + F_c(\omega \quad \beta)\right]$$

$$F_c\{f(at)\cos \beta t\} = \frac{1}{2a}\left[F_c\left(\frac{\omega + \beta}{a}\right) + F_c\left(\frac{\omega - \beta}{a}\right)\right] \qquad a, \beta > 0$$

13.1.2.4.1 Differentiation in the ω -Domain

$$F_c^{(2n)}(\omega) = F_c\{(\quad 1)^n t^{2n} f(t)\}$$

13.1.2.5 Asymptotic Behavior

$$\lim_{\omega \to \infty} F_c(\omega) = 0$$

13.1.2.6 Integration in the t-Domain

$$F_c\left\{\int_t^{\infty} f(\tau)\, d\tau\right\} = \frac{1}{\omega} F_s(\omega), \qquad F_s(\cdot) \quad \text{sine transform} \;=$$

13.1.2.7 Convolution

For $f_e(t) = f(|t|)$ and $g_e(t) = g(|t|)$ then

$$f_e(t) * g_e(t) = \int_0^{\infty} f(\tau)[g(t + \tau) + g(|t \quad \tau|)]\, d\tau$$

$$F_c\left\{\int_0^{\infty} f(\tau)[g(t + \tau) + g(|t \quad \tau|)]\, d\tau\right\} = 2F_c(\omega) G_c(\omega)$$

13.1.3 Examples of FCT

Pulse Function

$$F_c\{p_a(t-1)\} = \int_0^{\infty} p_a(t-a)\cos\omega t\,dt = \int_0^{2a}\cos\omega t\,dt = \frac{\sin 2a\omega}{\omega}$$

Lambda Function

$$f(t) = \begin{cases} t/a\;, & 0<t<a \\ (2a-t)/a, & a<t<2a \\ 0\;, & t>2a \end{cases}$$

$$F_c\{f(t)\} = \int_0^{a}\frac{t}{a}\cos\omega t\,dt + \int_0^{2a}\frac{2a-t}{a}\cos\omega t\,dt = \frac{1}{a\omega^2}(2\cos a\omega - \cos 2a\omega - 1)$$

Inverse Function

$$f(t) = \frac{1}{t} \qquad\qquad t>a$$

$$F_c\{f(t)\} = \int_a^{\infty}\frac{1}{t}\cos\omega t\,dt = \int_{a\omega}^{\infty}\frac{1}{\tau}\cos\tau\,d\tau = \;Ci(a\omega) \qquad -$$

$$Ci(y) = -\int_y^{\infty}\frac{1}{\tau}\cos\tau\,d\tau \quad \text{cosine integral function}$$

Exponential Function

$$f(t) = e^{-at} \qquad a>0,\; t \geq 0$$

$$F_c\{f(t)\} = \int_0^{\infty} e^{-at}\cos\omega t\,dt = \frac{a}{a^2+\omega^2}$$

Decaying Cosine

$$f(t) = e^{-bt}\cos at, \qquad a,b>0,\; t \geq 0$$

$$F_c\{f(t)\} = \int_0^{\infty} e^{-bt}\cos at\,\cos\omega t\,dt = \frac{b}{2}\left[\frac{1}{b^2+(a-\omega)^2} + \frac{1}{b^2+(a+\omega)^2}\right]$$

by setting $\cos at = \dfrac{e^{jat}+e^{-jat}}{2}$ and $\cos\omega t = \dfrac{e^{j\omega t}+e^{-j\omega t}}{2}$.

13.2 Fourier Sine Transform (FST)

13.2.1 Definition FST

$$F_s(\omega) = F_s\{f(t)\} = \int_0^{\infty} f(t)\sin\omega t\,dt \qquad \omega>0$$

$$f(t) = F_s^{-1}\{F_s(\omega)\} = \frac{2}{\pi}\int_0^{\infty} F_s(\omega)\sin\omega t\,d\omega \qquad t \quad 0$$

13.2.2 Properties of FST

13.2.2.1 Transforms of Derivatives

$$F_s\{f''(t)\} = -\omega^2 F_s(\omega) \quad \omega f(0)$$

$$F_s\{f'(t)\} = -\omega F_c(\omega)$$

13.2.2.2 Scaling

$$F_s\{f(at)\} = \frac{1}{a} F_s(\omega / a) \qquad a > 0$$

13.2.2.3 Shifting in t-Domain

$$f_e(t) = f(|t|) \qquad f_0(t) = \frac{t}{|t|} f(|t|)$$

$$F_s\{f_0(t+a) + f_0(t-a)\}= 2F_s(\omega)\cos a\omega$$

$$F_c\{f_0(t+a) - f_0(t-a)\} = 2F_s(\omega)\sin a\omega \qquad a > 0$$

13.2.2.4 Shifting in ω-Domain

$$F_s(\omega+\beta) = F_s\{f(t)\cos\beta t\} + F_c\{f(t)\sin\beta t\}$$

$$F_s\{f(t)\cos\beta t\} = \frac{1}{2}\left[F_s(\omega+\beta) + F_s(\omega \quad \beta)\right]$$

$$F_s\{f(at)\cos\beta t\} = \frac{1}{2a}\left[F_s\left(\frac{\omega+\beta}{a}\right) + F_s\left(\frac{\omega-\beta}{a}\right)\right]$$

$$F_s\{f(at)\sin\beta t\} = -\frac{1}{2a}\left[F_c\left(\frac{\omega+\beta}{a}\right) - F_c\left(\frac{\omega-\beta}{a}\right)\right]$$

13.2.2.5 Differentiation in the ω-Domain

$$F_s^{(2n)}(\omega) = F_s\{(-1)^n t^{2n} f(t)\}$$

$$F_s^{(2n+1)}(\omega) = F_c\{(-1)^n t^{2n+1} f(t)\}$$

13.2.2.6 Asymptotic Behavior

$$\lim_{\omega\to\infty} F_s(\omega) = 0$$

13.2.2.7 Integration in the t-Domain

$$F_s\left\{\int_o^t f(\tau)\,d\tau\right\} = \frac{1}{\omega} F_c(\omega)$$

13.2.2.8 Integration in the ω-Domain

$$F_c^{\pm 1}\left\{\int_\omega^\infty F_s(\beta)\,d\beta\right\} = \frac{1}{t}f(t)$$

13.2.2.9 The Convolution Property

$$2F_s(\omega)G_c(\omega) = F_s\left\{\int_0^\infty f(\tau)[g(|t-\tau|) - g(t+\tau)]\,d\tau\right\}$$

13.2.3 Examples of FST

Pulse Function

$$F_s\left\{p_{a/2}\left(t-\frac{a}{2}\right)\right\} = \int_0^a \sin\omega t\,dt = \frac{1-\cos\omega a}{\omega} \qquad a>0$$

Lambda Function

$$f(t) = \begin{cases} t/a \quad , & 0<t<a \\ (2a-t)/a, & a<t<2a \\ 0 \quad , & \text{otherwise} \end{cases}$$

$$F_s\{f(t)\} = \int_0^a \frac{t}{a}\sin\omega t\,dt + \int_a^{2a}\frac{2a-t}{a}\sin\omega t\,dt = \frac{1}{a\omega^2}(2\sin a\omega - \sin 2a\omega)$$

Inverse Function

$$f(t) = \frac{1}{t} \qquad t>a$$

$$F_s\{f(t)\} = \int_a^\infty \frac{1}{t}\sin\omega t\,dt = \int_{a\omega}^\infty \frac{1}{\tau}\sin\tau\,d\tau = -si(a\omega)$$

$$si(y) = -\int_y^\infty \frac{\sin x}{x}dx = \int_0^y \frac{\sin x}{x}dx - \int_0^\infty \frac{\sin x}{x}dx = Si(y) - \frac{\pi}{2}$$

Exponential Function

$$f(t) = e^{-at} \qquad a>0,\ t \geq 0$$

$$F_s\{f(t)\} = \int_0^\infty e^{-at}\sin\omega t\,dt = \frac{\omega}{a^2+\omega^2}$$

by setting $\sin\omega t = (e^{j\omega t} - e^{-j\omega t})/2j$

Decaying Cosine

$$f(t) = e^{-bt}\cos at, \qquad a,b>0,\ t \quad 0 \qquad \geq$$

$$F_s\{f(t)\} = \int_0^{\infty} e^{-bt}\cos at \,\sin \omega t\, dt = \frac{1}{2}\left[\frac{\omega - a}{b^2 + (\omega - a)^2} + \frac{\omega + a}{b^2 + (\omega + a)^2}\right]$$

by setting $\cos at = (e^{jat} + e^{-jat})/2$ and $\sin \omega t = (e^{j\omega t} - e^{-j\omega t})/2j$.

13.3 Discrete Cosine Transform (DCT)

13.3.1 Transform Kernel

$$K_c(\omega, t) = \cos \omega t$$

$$K_c(m,n) = K_c(\omega_m, t_n) = \cos(2\pi mn\,\Delta f\,\Delta t)$$

$$\omega_m = 2\pi m\,\Delta f = \text{sampled angular frequency}$$

$$t_n = \text{sampled time}$$

$$\Delta f, \Delta t = \text{sample intervals of frequency and time}$$

$$m,n = \text{positive integers}$$

If we set

$$\Delta f\,\Delta t = 1/2N$$

$$K_c(m,n) = \cos(\pi mn/N)$$

13.3.2 Discrete Cosine Transform (DCT)

If a finite duration signal is divided into N intervals of Δt each, there exists $N+1$ sample points. If these $N+1$ points are represented by vector $\underline{x}$, the DCT $\underline{X}_c$ is

$$\underline{X}_c = [C]\underline{x}, \qquad [C] = (N+1)\times(N \quad 1)\ \text{matrix} \quad +$$

where

$$[C]_{mn} = \text{matrix element of } [C] = \sqrt{2/N}\left[k_m k_n \cos\frac{mn\pi}{N}\right] \qquad m,n = 0,1,\cdots,N$$

$$k_i = 1 \ \text{ for } i \neq 0 \text{ or } N, \qquad k_i = \frac{1}{\sqrt{2}} \ \text{ for } i = 0 \text{ or } N$$

$$X_c(m) = \text{element of } \underline{X}_c = \sqrt{2/N}\sum_{n=0}^{N} k_m k_n \cos\left(\frac{mn\pi}{N}\right)x(n)$$

$$x(n) = \text{element of } \underline{x} = \sqrt{2/N}\sum_{m=0}^{N} k_m k_n \cos\left(\frac{mn\pi}{N}\right)X_c(m)$$

13.4 Discrete Sine Transform (DST)

13.4.1 Discrete Sine Transform (DST)

$$[S]_{mn} = \text{matrix element of } [S] = \sqrt{2/N}\sin\left(\frac{mn\pi}{N}\right) \qquad m,n = 1,2,\cdots,N-1$$

$$\underline{X}_s = [S]\underline{x}, \qquad [S] = (N-1)\times(N-1) \text{ matrix (boundary points are zero)}$$

$\underline{x} = N-1$ data vector, $\underline{X}_s$ N 1 DST vector

$$X_s(m) = \text{element of } \underline{X}_s = \sqrt{2/N}\sum_{n=1}^{N-1}\sin\left(\frac{mn\pi}{N}\right)x(n)$$

$X_s(m), x(n)$ DST pairs

$$x(n) = \text{element of } \underline{x} = \sqrt{2/N}\sum_{m=1}^{N-1}\sin\left(\frac{mn\pi}{N}\right)X_s(m)$$

13.5 Properties of DCT and DST

13.5.1 The Unitary Property

If $\underline{c}_m$ denotes the mth column vector of matrix $[C]$ then

$$\underline{c}_m^T\underline{c}_n = 0 \qquad \text{for } m \neq n$$

$$\underline{c}_m^T\underline{c}_n = 1 \qquad \text{for } m = n \quad 0 \qquad \neq$$

$$\underline{c}_m^T\underline{c}_n = 1 \qquad \text{for } m = n = 0$$

or

$$\underline{c}_m^T\underline{c}_n = \delta_{mn}, \qquad \delta_{mn} = \text{Kronecker delta}$$

The same applies for matrix $[S]$.

13.5.2 Inverse Transformation

$$[C]^{-1} = [C], \quad [S]^{-1} = [S]$$

These are **unitary** symmetric matrices.

13.5.3 Scaling

$$\Delta f\,\Delta t = 1/2N \qquad \text{or} \qquad \Delta f = 1/2N\,\Delta t$$

A change of Δt to $a\Delta t$ changes Δf to $\Delta f/a$, provided N remains the same. If we set $T = N\Delta t$, the time duration of the data, then

$\Delta f = 1/2T$.

13.6 FCT and FST Algorithm Based on FFT

13.6.1 FCT of Real Data Sequence

$\{x(n), n = 0,1,\cdots,N\}$ = sequence with $N+1$ points

$$X_c(m) = \sqrt{2/N}\sum_{n=0}^{N} k_m k_n \cos\left(\frac{mn\pi}{N}\right)x(n)$$

$$k_n = 1 \quad \text{for } n \neq 0 \ or \ N$$

$$k_n = 1/\sqrt{2} \quad \text{for } n = 0 \ or \ N$$

Construct an even or symmetric sequence using the sequence as follows:

$$\begin{aligned} s(n) &= x(n), && 0 < n < N \\ &= 2x(n), && n = 0, N \\ &= x(2N-n), && N < n \leq 2N-1 \end{aligned}$$

DFT of $\{s(n)\}$ is

$$S_F(m) = 2\left[x(0) + (-1)^m x(N) + \sum_{n=1}^{N-1} \cos\left(\frac{mn\pi}{N}\right)x(n)\right]$$

The $(N+1)$ – point DCT of $\{x(n)\}$ is the same as the 2N-point DFT of the sequence $\{s(n)\}$. Hence the DCT of $\{x(n)\}$ can be computed using a 2N-point FFT of $\{s(n)\}$.

13.6.2 FST of Real Data Sequence

Let $\{x(n), n = 1,\cdots,N-1\}$ be an $(N-1)$ – point data sequence. Its DST is (see 13.4.1)

$$X_s(m) = \sqrt{2/N}\sum_{n=1}^{N-1} \sin\left(\frac{mn\pi}{N}\right)x(n)$$

Construct a $(2N-1)$ – point odd or skew-symmetric sequence $\{s(n)\}$ using $\{x(n)\}$.

$$\begin{aligned} s(n) &= x(n), && 0 < n < N \\ &= 0, && n = 0, N \\ &= -x(2N-n), && N < n \leq 2N-1 \end{aligned}$$

The $2N$-point DFT of $\{s(n)\}$ is

$$S_F(m) = -2j\sum_{n=1}^{N-1} \sin\left(\frac{mn\pi}{N}\right)x(n)$$

13.7 Fourier Cosine Transform Pairs

13.7.1 Fourier Cosine Transform Properties

TABLE 13.1 Properties of FCT

	$f(t)$	$F_c(\omega) = \int_o^\infty f(t)\cos\omega t\,dt \quad \omega > 0$
1.	$F_c(t)$	$(\pi/2)f(\omega)$
2.	$f(at) \quad a > 0$	$(1/a)F_c(\omega/a)$
3.	$f(at)\cos bt \quad a,b > 0$	$(1/2a)\left[F_c\left(\frac{\omega+b}{a}\right) + F_c\left(\frac{\omega-b}{a}\right)\right]$
4.	$f(at)\sin bt \quad a,b > 0$	$(1/2a)\left[F_s\left(\frac{\omega+b}{a}\right) - F_s\left(\frac{\omega-b}{a}\right)\right]$
5.	$t^{2n} f(t)$	$(-1)^n \frac{d^{2n}}{d\omega^{2n}} F_c(\omega)$
6.	$t^{2n+1} f(t)$	$(-1)^n \frac{d^{2n+1}}{d\omega^{2n+1}} F_s(\omega)$
7.	$\int_o^\infty f(r)[g(t+r) + g(\lvert t-r \rvert)]\,dr$	$2F_c(\omega)G_c(\omega)$
8.	$\int_t^\infty f(r)\,dr$	$(1/\omega)F_s(\omega)$
9.	$f(t+a) - f_o(t-a)$	$2F_s(\omega)\sin a\omega \quad a > 0$
10.	$\int_o^\infty f(r)[g(t+r) - g_o(t-r)]\,dr$	$2F_s(\omega)G_s(\omega)$

13.7.2 Fourier Cosine Transform Pairs (see Section 13.9 for notation and definitions)

TABLE 13.2 Fourier Cosine Transform Pairs

	$f(t)$	$F_c(\omega) \quad \omega \quad 0 >$
1.	$(1/\sqrt{t})$	$\sqrt{(\pi/2)}\,(1/\omega)^{1/2}$
2.	$(1/\sqrt{t})[1 - U(t-1)]$	$(2\pi/\omega)^{1/2}\, C(\omega)$
3.	$(1/\sqrt{t})U(t-1)$	$(2\pi/\omega)^{1/2}\,[1/2 - C(\omega)]$
4.	$(t+a)^{-1/2} \quad \lvert \arg a \rvert < \pi$	$(\pi/2\omega)^{1/2}\{\cos a\omega[1 - 2C(a\omega)] + \sin a\omega[1 - 2S(a\omega)]\}$
5.	$(t-a)^{-1/2}U(t-a)$	$(\pi/2\omega)^{1/2}[\cos a\omega - \sin a\omega]$
6.	$a(t^2+a^2)^{-1} \quad a > 0$	$(\pi/2)\exp(-a\omega)$
7.	$t(t^2+a^2)^{-1} \quad a > 0$	$-1/2[e^{-a\omega}\overline{\mathrm{Ei}}(a\omega) + e^{a\omega}\mathrm{Ei}(a\omega)]$
8.	$(1-t^2)(1+t^2)^{-2}$	$(\pi/2)\omega\exp(-\omega)$
9.	$-t(t^2-a^2)^{-1} > a \quad 0$	$\cos a\omega\mathrm{Ci}(a\omega) + \sin a\omega\mathrm{Si}(a\omega)$
10.	$1 \quad 0 < t < a$ $0 \quad a < t < \infty$	$\frac{1}{\omega}\sin(a\omega)$

TABLE 13.2 Fourier Cosine Transform Pairs (continued)

	$f(t)$	$F_c(\omega)$ ω 0 $>$
11.	$1 \quad 0<t<a$ $1/t \quad a<t<\infty$	$-\mathrm{Ci}(a\omega)$
12.	$\dfrac{[(a^2+t^2)^{1/2}+a]^{1/2}}{[a^2+t^2]^{1/2}} \quad a>0$	$\left(\dfrac{2\omega}{\pi}\right)^{-1/2} e^{-a\omega}$
13.	$\dfrac{1}{(a+jt)^{\nu}}+\dfrac{1}{(a-jt)^{\nu}},\ a>0,\ \nu>0$	$\pi[\Gamma(\nu)]^{-1}\omega^{\nu-1}e^{-a\omega}$
14.	$e^{-at} \quad \mathrm{Re}\,a>0$	$a(a^2+\omega^2)^{-1}$
15.	$(1+t)e^{-t}$	$2(1+\omega^2)^{-2}$
16.	$\sqrt{t}\,e^{-at} \quad \mathrm{Re}\,a>0$	$\dfrac{\sqrt{\pi}}{2}(a^2+\omega^2)^{-3/4}\cos[3/2\tan^{-1}(\omega/a)]$
17.	$e^{-at}/\sqrt{t} \quad \mathrm{Re}\,a>0$	$\sqrt{(\pi/2)}(a^2+\omega^2)^{-1/2}[(a^2+\omega^2)^{1/2}+a]^{1/2}$
18.	$t^n e^{-at} \quad \mathrm{Re}\,a>0$	$n![a/(a^2+\omega^2)]^{n+1}\sum_{2m=0}^{n+1}(-1)^m\binom{n+1}{2m}\left(\dfrac{\omega}{a}\right)^{2m}$
19.	$\exp(-at^2)/\sqrt{t},\ \mathrm{Re}\,a>0$	$\pi(\omega/8a)^{1/2}\exp(-\omega^2/8a)I_{-1/4}(-\omega^2/8a)$
20.	$t^{2n}\exp(-a^2t^2),\ \lvert\arg a\rvert<\ /4$	$\pi\ (-1)^n\sqrt{\pi}\,2^{-n-1}a^{-2n-1}$ $\cdot\exp[-(\omega/2a)^2]\mathrm{He}_{2n}(2^{-1/2}\omega/a)$
21.	$t^{-3/2}\exp(-a/t),\ \mathrm{Re}\,a>0$	$(\pi/a)^{1/2}\exp[-(2a\omega)^{1/2}]\cos(2a\omega)^{1/2}$
22.	$t^{-1/2}\exp(-a/\sqrt{t}),\ \mathrm{Re}\,a>0$	$(\pi/2\omega)^{1/2}[\cos(2a\sqrt{\omega})-\sin(2a\sqrt{\omega})]$
23.	$e^{-at^2} \quad \mathrm{Re}\,a>0$	$\dfrac{\sqrt{\pi}}{2}\dfrac{1}{\sqrt{a}}e^{-\omega^2/(4a)}$
24.	$t^{\nu-1}e^{-at} \quad \mathrm{Re}\,a>0,\ \mathrm{Re}\,\nu>0$	$\Gamma(\nu)(a^2+\omega^2)^{-\nu/2}\cos[\nu\tan^{-1}(\omega/a)]$
25.	$t^{-1/2}\ln t$	$-(\pi/2\omega)^{1/2}[\ln(4\omega)+C+\pi/2]$
26.	$(t^2-a^2)^{-1}\ln t,\ a>0$	$(\pi/2\omega)\{\sin(a\omega)[\mathrm{ci}(a\omega)-\ln a]$ $-\cos(a\omega)[\mathrm{si}(a\omega)-\pi/2]\}$
27.	$t^{-1}\ln(1+t)$	$(1/2)\{[\mathrm{ci}(\omega)]^2+[\mathrm{si}(\omega)]^2\}$
28.	$\exp(-t/\sqrt{2})\sin(\pi/4+t/\sqrt{2})$	$(1+\omega^4)^{-1}$
29.	$\exp(-t/\sqrt{2})\cos(\pi/4+t/\sqrt{2})$	$\omega^2(1+\omega^4)^{-1}$
30.	$\ln\dfrac{a^2+t^2}{1+t^2},\ a>0$	$(\pi/\omega)[\exp(-\omega)-\exp(-a\omega)]$
31.	$\ln[1+(a/t)^2],\ a>0$	$(\pi/\omega)[1-\exp(-a\omega)]$
32.	$\ln(1+e^{-at}),\ \mathrm{Re}\,a>0$	$\dfrac{1}{2}a\dfrac{1}{\omega^2}-\dfrac{1}{2}\dfrac{\pi}{\omega}\operatorname{csc}h\left(\pi\dfrac{\omega}{a}\right)$
33.	$t^{-1}e^{-t}\sin t$	$(1/2)\tan^{-1}(2\omega^{-2})$
34.	$t^{-2}\sin^2(at)$	$\begin{cases}(\pi/2)(a-\omega/2) & \omega<2a\\ 0 & \omega>2a\end{cases}$
35.	$\left(\dfrac{\sin t}{t}\right)^n \quad n=2,3,\cdots$	$\begin{cases}\dfrac{n\pi}{2^n}\displaystyle\sum_{r>0}^{r<(\omega+n)/2}\dfrac{(-1)^r(\omega+n\ \ 2r)^{n-1}}{r!(n-r)!}, & 0<\omega<n\\ 0 & n\le\omega\end{cases}$

TABLE 13.2 Fourier Cosine Transform Pairs (continued)

	$f(t)$	$F_c(\omega)\quad \omega\quad 0 >$
36.	$\exp(-\beta t^2)\cos at,\ \ \mathrm{Re}\,\beta > 0$	$(1/2)(\pi/\beta)^{1/2}\exp\left(-\frac{a^2+\omega^2}{4\beta}\right)\cosh\left(\frac{a\omega}{2\beta}\right)$
37.	$(a^2+t^2)^{-1}(1-2\beta\cos t + \beta^2)^{-1}$ $\mathrm{Re}\,a > 0,\ \ \lvert\beta\rvert < 1$	$(1/2)(\pi/a)(1-\beta^2)^{-1}(e^a-\beta)^{-1}$ $\cdot(e^{a-a\omega}+\beta e^{a\omega}),\ \ 0 \le \omega\ \ 1\ \ <$
38.	$\sin(at^2),\ \ a > 0$	$(1/4)(2\pi/a)^{1/2}\left[\cos\left(\frac{\omega^2}{4a}\right)-\sin\left(-\frac{\omega^2}{4a}\right)\right]$
39.	$\sin[a(1-t^2)],\ \ a > 0$	$-(1/2)(\pi/a)^{1/2}\cos[a+\pi/4+\omega^2/(4a)]$
40.	$\cos(at^2),\ \ a > 0$	$(1/4)(2\pi/a)^{1/2}\left[\cos\left(\frac{\omega^2}{4a}\right)+\sin\left(\frac{\omega^2}{4a}\right)\right]$
41.	$\cos[a(1-t^2)],\ \ a > 0$	$(1/2)(\pi/a)^{1/2}\sin[a+\pi/4+\omega^2/(4a)]$
42.	$\frac{\sin at}{t},\ \ a > 0$	$\begin{cases}\frac{\pi}{2} & \omega < a\\ \frac{\pi}{4} & \omega = a\\ 0 & \omega > a\end{cases}$
43.	$e^{-\beta t}\sin at,\ \ a > 0,\ \mathrm{Re}\,\beta > 0$	$\frac{\frac{1}{2}(a+\omega)}{\beta^2+(a+\omega)^2}+\frac{\frac{1}{2}(a-\omega)}{\beta^2+(a-\omega)^2}$
44.	$\frac{\sin t}{te^{-t}}$	$\frac{1}{2}\tan^{-1}\left(\frac{2}{\omega^2}\right)$
45.	$\frac{\sin^2(at)}{t^2},\ \ a > 0$	$\begin{cases}\frac{\pi}{2}(a-\frac{1}{2}\omega) & \omega < 2a\\ 0 & 2a < \omega\end{cases}$
46.	$\frac{1-\cos at}{t^2},\ \ a > 0$	$\begin{cases}\frac{\pi}{2}(a-\omega) & \omega < a\\ 0 & a < \omega\end{cases}$
47.	$e^{-\beta t}\cos at,\ \ \mathrm{Re}\,\beta > \lvert\mathrm{Im}\,a\rvert$	$\frac{\beta}{2}\left[\frac{1}{\beta^2+(a-\omega)^2}+\frac{1}{\beta^2+(a+\omega)^2}\right]$
48.	$\begin{cases}\frac{\cos[b(a^2-t^2)^{1/2}]}{(a^2-t^2)^{1/2}} & 0 < t < a\\ 0 & a < t < \ \infty\end{cases}$	$\frac{\pi}{2}J_o[a(b^2+\omega^2)^{1/2}]$
49.	$\frac{\tan^{-1}\left(\frac{t}{a}\right)}{t}$	$-\frac{\pi}{2}Ei(-a\omega)$
50.	$e^{-t^2}He_{2n}(2t)$	$\frac{\sqrt{\pi}}{2}(-1)^n e^{-\omega^2/4}He_{2n}(\omega)$
51.	$e^{-t^2/2}[He_n(t)]^2$	$\sqrt{\frac{\pi}{2}}\ n!\ e^{-\omega^2/2}L_n(\omega^2)$
52.	$J_o(at),\ \ a > 0$	$\begin{cases}(a^2-\omega^2)^{-1/2} & 0 < \omega < a\\ \infty & \omega = a\\ 0 & a < \omega < \ \infty\end{cases}$

13.8 Fourier Sine Transform Pairs

13.8.1 Fourier Sine Transform Properties

TABLE 13.3 Fourier Sine Transform Properties

	$f(t)$	$F_s(\omega) = \int_o^\infty f(t)\sin\omega t\, dt \quad \omega > 0$
1.	$F_s(t)$	$(\pi/2)f(\omega)$
2.	$f(at),\ a>0$	$(1/a)F_s(\omega/a)$
3.	$f(at)\cos bt,\ a,b>0$	$(1/2a)\left[F_s\left(\frac{\omega+b}{a}\right)+F_s\left(\frac{\omega-b}{a}\right)\right]$
4.	$f(at)\sin bt,\ a,b>0$	$-(1/2a)\left[F_c\left(\frac{\omega+b}{a}\right)-F_c\left(\frac{\omega-b}{a}\right)\right]$
5.	$t^{2n}f(t)$	$(-1)^n \frac{d^{2n}}{d\omega^{2n}}F_s(\omega)$
6.	$t^{2n+1}f(t)$	$(-1)^{n+1} \frac{d^{2n+1}}{d\omega^{2n+1}}F_c(\omega)$
7.	$\int_o^\infty f(r)\int_{\lvert t-r\rvert}^{t+r} g(s)\,ds\,dr$	$(2/\omega)F_s(\omega)G_s(\omega)$
8.	$f_o(t+a)+f_o(t-a)$	$2F_s(\omega)\cos a\omega$
9.	$f_e(t-a)-f_e(t+a)$	$2F_c(\omega)\sin a\omega$
10.	$\int_o^\infty f(r)[g(\lvert t-r\rvert)-g(t+r)]\,dr$	$2F_s(\omega)G_c(\omega)$

13.8.2 Fourier Sine Transform Pairs

TABLE 13.4 Fourier Sine Transform Pairs

	$f(t)$	$F_s(\omega)$
1.	$1/t$	$\pi/2$
2.	$1/\sqrt{t}$	$(\pi/2\omega)^{1/2}$
3.	$1/\sqrt{t}[1-U(t-1)]$	$(2\pi/\omega)^{1/2}S(\omega)$
4.	$(1/\sqrt{t})U(t-1)$	$(2\pi/\omega)^{1/2}[1/2-S(\omega)]$
5.	$(t+a)^{-1/2},\ \lvert\arg a\rvert<\pi$	$\begin{cases}(\pi/2\omega)^{1/2}\{\cos a\omega[1-2S(a\omega)] \\ \quad -\sin a\omega[1-2C(a\omega)]\}\end{cases}$
6.	$(t-a)^{-1/2}U(t-a)$	$(\pi/2\omega)^{1/2}(\sin a\omega+\cos a\omega)$
7.	$t(t^2+a^2)^{-1},\quad a>0$	$(\pi/2)\exp(-a\omega)$
8.	$t(a^2-t^2)^{-1},\quad a>0$	$-(\pi/2)\cos a\omega$
9.	$t(t^2+a^2)^{-2},\quad a>0$	$(\pi\omega/4a)\exp(-a\omega)$
10.	$a^2[t(t^2+a^2)]^{-1},\quad a>0$	$(\pi/2)[1-\exp(-a\omega)]$

TABLE 13.4 Fourier Sine Transform Pairs (continued)

	$f(t)$	$F_s(\omega)$
11.	$t(4+t^4)^{-1}$	$(\pi/4)\exp(-\omega)\sin\omega$
12.	$\begin{cases} 1 & 0<t<a \\ 0 & a<t<\ \infty \end{cases}$	$\dfrac{1-\cos a\omega}{\omega}$
13.	$\dfrac{\left[\sqrt{t^2+a^2}\ -a\right]^{1/2}}{(t^2+a^2)^{1/2}}$	$\pi\ \dfrac{e^{-a\omega}}{\sqrt{2\omega}}$
14.	$e^{-at},\ \mathrm{Re}\,a>0$	$\omega(a^2+\omega^2)^{-1}$
15.	$te^{-at},\ \mathrm{Re}\,a>0$	$(2a\omega)(a^2+\omega^2)^{-2}$
16.	$t(1+at)e^{-at},\ \mathrm{Re}\,a>0$	$(8a^3\omega)(a^2+\omega^2)^{-3}$
17.	$e^{-at}\sqrt{t},\ \mathrm{Re}\,a>0$	$\sqrt{(\pi/2)}\,(a^2+\omega^2)^{-1/2}[(a^2+\omega^2)^{1/2}\quad a]^{1/2}$
18.	$t^{-3/2}e^{-at},\ \mathrm{Re}\,a>0$	$(2\pi)^{1/2}[(a^2+\omega^2)^{1/2}-a]^{1/2}$
19.	$\exp(-at^2),\ \mathrm{Re}\,a>0$	$-j(1/2)(\pi/a)^{1/2}\exp(-\omega^2/4a)\,\mathrm{Erf}\left(\dfrac{j\omega}{2\sqrt{a}}\right)$
20.	$t\exp(-t^2/4a),\ \mathrm{Re}\,a>0$	$2a\omega\sqrt{(\pi a)}\exp(-a\omega^2)$
21.	$t^{-3/2}\exp(-a/t),\ \lvert\arg a\rvert<\pi/2$	$(\pi/a)^{1/2}\exp[-(2a\omega)^{1/2}]\sin(2a\omega)^{1/2}$
22.	$t^{-3/4}\exp(-a\sqrt{t}),\ \lvert\arg a\rvert<\pi/2$	$-(\pi/2)(a/\omega)^{1/2}[J_{1/4}(a^2/8\omega)$
		$\cdot\cos(\pi/8+a^2/8\omega)+Y_{1/4}(a^2/8\omega)$
		$\cdot\sin(\pi/8+a^2/8\omega)]$
23.	$e^{-t/2}(1-e^{-t})^{-1}$	$-\tfrac{1}{2}\tanh(\pi\omega)$
24.	$t^{-1}e^{-at^2},\ \lvert\arg a\rvert<\pi/2$	$\tfrac{1}{2}\pi\ \mathrm{Erf}\left(\dfrac{\omega}{2\sqrt{a}}\right)$
25.	$t^{-1}\ln t$	$-(\pi/2)[C+\ln\omega]$
26.	$t(t^2-a^2)^{-1}\ln t,\ a>0$	$-(\pi/2)\{\cos a\omega[\mathrm{Ci}(a\omega)-\ln a]$
		$+\sin a\omega[\mathrm{Si}(a\omega)-\pi/2]\}$
27.	$t^{-1}\ln(1+a^2t^2),\ a>0$	$-\pi\,\mathrm{Ei}(-\omega/a)$
28.	$\ln\dfrac{t+a}{\lvert t-a\rvert},\ a>0$	$(\pi/\omega)\sin a\omega$
29.	$t^{-1}\sin^2(at),\ a>0$	$\begin{cases} \pi/4 & 0<\omega<2a \\ \pi/8 & \omega=2a \\ 0 & \omega>2a \end{cases}$
30.	$t^{-2}\sin^2(at),\ a>0$	$(1/4)(\omega+2a)\ln\lvert\omega+2a\rvert$
		$+(1/4)(\omega-2a)\ln\lvert\omega-2a\rvert-(1/2)\omega\ln\omega$
31.	$t^{-2}[1-\cos at],\ a>0$	$(\omega/2)\ln\lvert(\omega^2-a^2)/\omega^2\rvert$
		$+(a/2)\ln\lvert(\omega+a)/(\omega\quad a)\rvert\ -$
32.	$\sin(at^2),\ a>0$	$(\pi/2a)^{1/2}\{\cos(\omega^2/4a)C[\omega/(2\pi a)^{1/2}]$
		$+\sin(\omega^2/4a)S[\omega/(2\pi a)^{1/2}]\}$
33.	$\cos(at^2),\ a>0$	$(\pi/2a)^{1/2}\{\sin(\omega^2/4a)C[\omega/(2\pi a)^{1/2}]$

TABLE 13.4 Fourier Sine Transform Pairs (continued)

	$f(t)$	$F_s(\omega)$
		$-\cos(\omega^2/4a)S[\omega/(2\pi a)^{1/2}]\}$
34.	$\tan^{-1}(a/t),\ a>0$	$(\pi/2\omega)[1-\exp(-a\omega)]$
35.	$\frac{\sin at}{t},\ a>0$	$\frac{1}{2}\ln\left\lvert\frac{\omega+a}{\omega-a}\right\rvert$
36.	$\frac{\sin \pi t}{1-t^2}$	$\begin{cases}\sin\omega & 0\le\omega\le\pi \\ 0 & \pi\le\omega\end{cases}$
37.	$\sin\left(\frac{a^2}{t}\right),\ a>0$	$\left(\frac{\pi}{2}\right)\frac{a}{\sqrt{\omega}}J_1(2a\sqrt{\omega})$
38.	$\tan^{-1}(t/a),\ a>0$	$\frac{\pi}{2}\frac{e^{-a\omega}}{\omega}$
39.	$\tan^{-1}(2a/t),\ \mathrm{Re}\,a>0$	$\frac{\pi}{\omega}e^{-a\omega}\sinh(a\omega)$
40.	$\mathrm{Erfc}(at),\ a>0$	$(1-e^{-\omega^2/4a^2})/\omega$
41.	$J_o(at),\ a>0$	$\begin{cases}0 & 0<\omega<a \\ 1/\sqrt{\omega^2-a^2} & a<\omega<\infty\end{cases}$
42.	$J_o(at)/t,\ a>0$	$\begin{cases}\sin^{-1}(\omega/a) & 0<\omega<a \\ \pi/2 & a<\omega<\infty\end{cases}$

13.9 Notations and Definitions

1. $f(t)$ Piece-wise smooth and absolutely integrable function on the positive real line.
2. $F_c(\omega)$ The Fourier cosine transform of $f(t)$.
3. $F_s(\omega)$ The Fourier sine transform of $f(t)$.
4. $f_o(t)$ The odd extension of the function f over the entire real line.
5. $f_e(t)$ The even extension of the function f over the entire real line.
6. $C(\omega)$ is defined as the integral:

$$(2\pi)^{-1/2}\int_o^{\omega} t^{-1/2}\cos t\,dt\,.$$

7. $S(\omega)$ is defined as the integral:

$$(2\pi)^{-1/2}\int_o^{\omega} t^{-1/2}\sin t\,dt\,.$$

8. $\mathrm{Ei}(x)$ is the exponential integral function defined as:

$$-\int_{-x}^{\infty} t^{-1}e^{-t}\,dt, \qquad |\arg(x)|<\pi\,.$$

9. $\overline{\mathrm{Ei}}(x)$ is defined as $(1/2)[(\mathrm{Ei}(x+j0)+\mathrm{Ei}(x-j0)]$.
10. $\mathrm{Ci}(x)$ is the cosine integral function defined as:

$$-\int_x^{\infty} t^{-1}\cos t\,dt\,.$$

11. Si(x) is the sine integral function defined as:

$$\int_o^x t^{-1}\sin t\,dt$$

12. $I_v(z)$ is the modified Bessell function of the first kind defined as:

$$\sum_{m=0}^{\infty}\frac{(z/2)^{v+2m}}{m!\Gamma(v+m+1)}, \qquad |z|<\infty,< \qquad |\arg(z)| \quad \pi\,.$$

13. $He_n(x)$ is the Hermite polynomial function defined as

$$(-1)^n\exp(x^2/2)\frac{d^n}{dx^n}[\exp(-x^2/2)]\,.$$

14. C is the Euler constant defined as

$$\lim_{m\to\infty}\left[\sum_{n=1}^{m}(1/n)-\ln m\right]=0.5772156649...$$

15. ci(x) and si(x) are related Ci(x) and Si(x) by the equations:

$$\text{ci}(x)=-\text{Ci}(x), \qquad \text{si}(x)=\text{Si}(x)-\pi/2\,.$$

16. Erf(x) is the error function defined by

$$(2/\sqrt{\pi})\int_o^x \exp(-t^2)\,dt\,.$$

17. $J_v(x)$ and $Y_v(x)$ are the Bessel functions for the first and second kind, respectively,

$$J_v(x)=\sum_{m=0}^{\infty}(-1)^m\frac{(x/2)^{v+2m}}{m!\Gamma(v+m+1)}$$

and

$$Y_v(x)=\text{cosec}\{v\pi[J_v(x)\cos v\pi-J_{-v}(x)]\}\,.$$

18. $U(t)$ is the Heaviside step function defined as

$$U(t)=0 \qquad t<0$$
$$U(t)=1 \qquad t>0$$

19. $\binom{m}{n}$ is the binomial coefficient defined as $\frac{m!}{n!(m-n)!}$.

20. $\Gamma(x)$ is the Gamma function defined as

$$\Gamma(x) = \int_0^\infty e^{-t}\, t^{x-1}\, dt\,.$$

References

Churchill, R. V., 1958. *Operational Mathematics*, 3rd ed. New York: McGraw-Hill.

Erdelyi, A. 1954. *Bateman Manuscript*, Vol. 1. New York: McGraw-Hill.

Rao, K. R. and Yip, P. 1990. *Discrete Cosine Transform: Algorithms, Advantages, Applications*. Boston: Academic Press.

Sneddon, I. N. 1972. *The Uses of Integral Transforms*. New York: McGraw-Hill.

Yip, P., 1996. The discrete sine and cosine transforms, Chapter 3 in *The Transforms and Applications Handbook*, Ed. A. D. Poularikas, Boca Raton, Florida: CRC Press.

14

The Hartley Transform

14.1 Introduction to the Hartley Transform

14.1.1 Definition of the Pair with Symmetry and the Use of ω (units: rad s^{-1})

$$H_x(\omega) = \frac{1}{\sqrt{2\pi}} \int_{-\infty}^{\infty} x(t)\operatorname{cas}(\omega t)\,dt$$

$$x(t) = \frac{1}{\sqrt{2\pi}} \int_{-\infty}^{\infty} H_x(\omega)\operatorname{cas}(\omega t)\,d\omega$$

14.1.2 Definition of the Pair with Use of f (units: s^{-1})

$$H_x(f) = \int_{-\infty}^{\infty} x(t)\operatorname{cas}(2\pi f t)\,dt$$

$$x(t) = \int_{-\infty}^{\infty} H_x(f)\operatorname{cas}(2\pi f t)\,df$$

$$\operatorname{cas}(\omega t) = \cos(\omega t) + \sin(\omega t)$$

$$\operatorname{cas}(\omega t) = \sqrt{2}\sin\left(\omega t + \frac{\pi}{4}\right)$$

$$\operatorname{cas}(\omega t) = \sqrt{2}\cos\left(\omega t - \frac{\pi}{4}\right)$$

$$\omega = 2\pi f$$

14.1.3 Odd and Even Parts of the Hartley Transform

$$H_x^e(f) = \frac{H_x(f) + H_x(-f)}{2} = \int_{-\infty}^{\infty} x(t)\cos(2\pi f t)\,dt$$

$$H_x^o(f) = \frac{H_x(f) - H_x(-f)}{2} = \int_{-\infty}^{\infty} x(t)\sin(2\pi f t)\,dt$$

14.1.4 Properties of the cas Function

TABLE 14.1 Trigonometric Properties of the cas Function

The cas function	$\operatorname{cas}\xi = \cos\xi + \sin\xi$
The cas function	$\operatorname{cas}\xi = \frac{1}{2}[(1+j)\exp(-j\xi) + (1-j)\exp(j\xi)]$
The complementary cas function	$\operatorname{cas}'\xi = \operatorname{cas}(-\xi) = \cos\xi - \sin\xi$
The complementary cas function	$\sqrt{2}\,\sin\left(\xi + \frac{3\pi}{4}\right) = \sqrt{2}\,\cos\left(\xi + \frac{\pi}{4}\right)$
Relation to cos	$\cos\xi = \frac{1}{2}[\operatorname{cas}\xi + \operatorname{cas}(-\xi)]$
Relation to sin	$\sin\xi = \frac{1}{2}[\operatorname{cas}\xi - \operatorname{cas}(-\xi)]$
Reciprocal relation	$\operatorname{cas}\xi = \frac{\csc\xi + \sec\xi}{\sec\xi\csc\xi}$
Product relation	$\operatorname{cas}\xi = \cot\xi\sin\xi + \tan\xi\cos\xi$
Function product relation	$\operatorname{cas}\tau\,\operatorname{cas}\nu = \cos(\tau-\nu) + \sin(\tau+\nu)$
Quotient relation	$\operatorname{cas}\xi = \frac{\cot\xi\sec\xi + \tan\xi\csc\xi}{\csc\xi\sec\xi}$
Double angle relation	$\operatorname{cas}2\xi = \operatorname{cas}^2\xi - \operatorname{cas}^2(-\xi)$
Indefinite integral relation	$\int \operatorname{cas}(\tau)\,d\tau = -\operatorname{cas}(-\tau) = -\operatorname{cas}'\tau$
Derivative relation	$\frac{d}{dt}\operatorname{cas}\tau = \operatorname{cas}(-\tau) = \operatorname{cas}'\tau$
Angle-sum relation	$\operatorname{cas}(\tau+\nu) = \cos\tau\,\operatorname{cas}\nu + \sin\tau\,\operatorname{cas}'\nu$
Angle-difference relation	$\operatorname{cas}(\tau-\nu) = \cos\tau\,\operatorname{cas}'\nu + \sin\tau\,\operatorname{cas}\nu$
Function-sum relation	$\operatorname{cas}\tau + \operatorname{cas}\nu = 2\operatorname{cas}\frac{1}{2}(\tau+\nu)\cos\frac{1}{2}(\tau-\nu)$
Function-difference relation	$\operatorname{cas}\tau - \operatorname{cas}\nu = 2\operatorname{cas}'\frac{1}{2}(\tau+\nu)\sin\frac{1}{2}(\tau-\nu)$

14.1.5 Signs of the cas Function

TABLE 14.2 Signs of the cas Function

Quadrant	cas
1st	+
2nd	+ and –
3rd	–
4th	+ and –

14.1.6 Variations of the cas Function

TABLE 14.3 Variations of the cas Function

Quadrant	cas
1st	$+1 \to +1$ with a maximum at $\pi/4$
2nd	$+1 \to -1$
3rd	$-1 \to -1$ with a maximum at $5\pi/4$
4th	$-1 \to +1$

14.1.7 Values of the cas at Special Angles

TABLE 14.4 Values of the cas at Special Angles

Angle	cas
$0° = 0$	0
$30° = \frac{\pi}{6}$	$\frac{1}{2}(\sqrt{3}+1)$
$45° = \frac{\pi}{4}$	$\sqrt{2}$
$60° = \frac{\pi}{3}$	$\frac{1}{2}(1+\sqrt{3})$
$90° = \frac{\pi}{2}$	1
$120° = \frac{2\pi}{3}$	$\frac{1}{2}(-1+\sqrt{3})$
$150° = \frac{5\pi}{6}$	$\frac{1}{2}(1-\sqrt{3})$
$180° = \pi$	-1
$270° = \frac{3\pi}{2}$	-1

14.2 Relationship with the Fourier Transform

14.2.1 Relationship to Fourier Transform

$$X(f) = \int_{-\infty}^{\infty} x(t)e^{-j2\pi f}\,dt = \int_{-\infty}^{\infty} x(t)\cos(2\pi ft)\,dt \quad j\int_{-\infty}^{\infty} x(t)\sin(2\pi ft)\,dt$$

$$= H_x^e(f) - jH_x^o(f) = \frac{H_x(f)+H_x(-f)}{2} - j\,\frac{H_x(f)-H_x(-f)}{2}$$

$$\mathrm{Re}\{X(f)\} = H_x^e(f), \quad \mathrm{Im}\{X(f)\} = \quad H_x^o(f) \qquad -$$

$$H_x(f) = H_x^e(f) + H_x^o(f) = \mathrm{Re}\{X(f)\} \quad \mathrm{Im}\{X(f)\}$$

***Note*:** The Fourier transform is the even part of the Hartley transform minus j times the odd part. The Hartley transform is the real part of the Fourier transform minus the imaginary part.

Example (Exponential Function):

$$H_x(f) = \int_{-\infty}^{\infty} e^{-at}u(t)\text{cas}(2\pi ft)\,dt = \int_{o}^{\infty} e^{-at}\,\text{cas}(2\pi ft)\,dt = \int_{o}^{\infty} e^{-at}\frac{e^{j2\pi ft}+e^{-j2\pi ft}}{2}dt$$

$$+\int_{o}^{\infty} e^{-at}\frac{e^{j2\pi ft}-e^{-j2\pi ft}}{2j}dt = \frac{1}{2}\frac{1}{a-j2\pi ft}+\frac{1}{2}\frac{1}{a+j2\pi ft}+\frac{1}{2j}\frac{1}{a-j2\pi ft}$$

$$-\frac{1}{2j}\frac{1}{a+j2\pi ft} = \frac{a+2\pi f}{a^2+4\pi^2 f^2} = \frac{a}{a^2+4\pi^2 f^2}+\frac{2\pi f}{a^2+4\pi^2 f^2} = H_x^e(f)+H_x^o(f)$$

$$X(f) = F\{e^{-at}u(t)\} = \frac{a}{a^2+4\pi^2 f^2} - j\frac{2\pi f}{a^2+4\pi^2 f^2}$$

Example (Delta Function):

$$H_x(f) = \int_{-\infty}^{\infty}\delta(t-t_o)\text{cas}(2\pi ft)\,dt \quad \text{cas}2\pi ft_o$$

Example (Pulse Function):

$$H_x(f) = \int_{-\infty}^{\infty} p_a(t)\text{cas}(2\pi ft)\,dt = \int_{-a}^{a}\text{cas}(2\pi ft)\,dt = \int_{-a}^{a}\frac{e^{j2\pi ft}+e^{-j2\pi ft}}{2}dt$$

$$+\int_{-a}^{a}\frac{e^{j2\pi ft}-e^{-j2\pi ft}}{2j}dt = \frac{\sin 2\pi fa}{\pi f}$$

14.3 Power and Phase Spectra

14.3.1 Power Spectrum

If $X(f)$ is the Fourier transform of $x(t)$, then

$$P_x(f) = \text{power spectrum} = X(f)X^*(f) = [\text{Re}\{X(f)\}]^2 + [\text{Im}\{X(f)\}]^2$$

From (14.2.1)

$$P_x(f) = [H_x^e(f)]^2 + [-H_x^o(f)]^2 = \left[\frac{H_x(f)+H_x(-f)}{2}\right]^2 + \left[\frac{H_x(f)-H_x(-f)}{2}\right]^2$$

$$= \frac{[H_x(f)]^2+[H_x(-f)]^2}{2}$$

14.3.2 Phase Spectrum

If $X(f)$ is the Fourier transform of $x(t)$, then (see ([14.2.1])

$$\text{pha}\{X(f)\} = \tan^{-1}\left[\frac{\text{Im}\{X(f)\}}{\text{Re}\{X(f)\}}\right] = \tan^{-1}\left[-\frac{H_x^o(f)}{H_x^e(f)}\right] = \tan^{-1}\left[\frac{H_x(-f)-H_x(f)}{H_x(f)+H_x(-f)}\right]$$

14.4 Properties of the Hartley Transform

14.4.1 Linearity

$$\int_{-\infty}^{\infty} [ax_1(t) + bx_2(t)]\text{cas}(2\pi ft)\,dt = aH_{x_1}(f) + bH_{x_2}(f)$$

where $H_{x_1}(f)$ and $H_{x_2}(f)$ are the Hartley transforms of $x_1(t)$ and $x_2(t)$.

14.4.2 Power and Phase Spectra (see 14.3.1 and 14.3.2)

$$P_x(f) = \frac{[H_x(f)]^2 + [H_x(-f)]^2}{2}, \quad \text{phase}\{X(f)\} = \tan^{-1}\left[\frac{H_x(-f) - H_x(f)}{H_x(f) + H_x(-f)}\right]$$

14.4.3 Scaling

If $H_x(f)$ is the Hartley transform of $x(t)$, then

$$\int_{-\infty}^{\infty} x(at)\text{cas}(2\pi ft)\,dt = \frac{1}{|a|}\int_{-\infty}^{\infty} x(t')\text{cas}\left(\frac{2\pi ft'}{a}\right)dt' = \frac{1}{|a|}H_x\left(\frac{f}{a}\right)$$

Similarly

$$\int_{-\infty}^{\infty} x\left(\frac{t}{a}\right)\text{cas}(2\pi ft)\,dt = |a|\,H_x(af)$$

14.4.4 Function Reversal

If $x(t)$ and $H_x(f)$ are a Hartley transform pair, then

$$\int_{-\infty}^{\infty} x(-t)\text{cas}(2\pi ft)\,dt = H_x(\ f) \quad \text{(see 14.4.3 with a = -1)}$$

14.4.5 Shifting

$$H_x(f) = \int_{-\infty}^{\infty} x(t - t_o)\text{cas}(2\pi ft)\,dt \quad \int_{-\infty}^{\infty} x(t')\text{cas}[2\pi f(t' + t_o)]\,dt'$$

$$= \cos(2\pi ft_o)H_x(f) + \cos(2\pi ft_o)H_x(\ f)$$

where we set $t - t_o = t'$ and expanded $\text{cas}[2\pi f(t' + t_o)]$.

14.4.6 Modulation

The Hartley transform of $x(t)\cos 2\pi f_o t$ is

$$H_x(f) = \int_{-\infty}^{\infty} x(t)\cos(2\pi f_o t)\text{cas}(2\pi ft)\,dt = \int_{-\infty}^{\infty} x(t)\cos(2\pi f_o t)\cos(2\pi ft)\,dt + \int_{-\infty}^{\infty} x(t)\cos(2\pi f_o t)\sin(2\pi ft)\,dt$$

$$= \tfrac{1}{2}H_x(f - f_o) + \tfrac{1}{2}H_x(f + f_o)$$

Example:

$$\int_{-\infty}^{\infty} p_a(t)\cos(2\pi f_o t)\operatorname{cas}(2\pi f t)\,dt = \frac{1}{2}\frac{\sin[2\pi(f-f_o)a]}{\pi(f-f_o)} + \frac{1}{2}\frac{\sin[2\pi(f+f_o)a]}{\pi(f+f_o)}$$

See (14.2.1), the third example.

14.4.7 Convolution

If $g(t) = x_1(t) * x_2(t)$, where $*$ stands for convolution, then [see (14.2.1)]

$$F\{g(t)\} = G(f) = X_1(f)X_2(f) = [H_{x_1}^e(f) - jH_{x_1}^o(f)][H_{x_2}^e(f) - jH_{x_2}^o(f)]$$

After separating and replacing the even and odd Hartley transform with their Hartley transform equivalent, e.g., $H_{x_1}^e(f) = [H_{x_1}(f) + H_{x_1}(\ f)]/2$, we obtain

$$G(f) = \operatorname{Re}\{G(f)\} - j\operatorname{Im}\{G(f)\}$$

$$H_g(f) = \operatorname{Re}\{G(f)\} - \operatorname{Im}\{G(f)\}$$

$$= \tfrac{1}{2}[H_{x_1}(f)H_{x_2}(f) + H_{x_1}(-f)H_{x_2}(f) + H_{x_1}(f)H_{x_2}(-f) - H_{x_1}(-f)H_{x_2}(-f)]$$

Note:

1. If $x_1(t)$ is even and $x_2(t)$ is odd, or $x_1(t)$ is odd and $x_2(t)$ is even, then

$$\int_{-\infty}^{\infty} [x_1(t) * x_2(t)]\operatorname{cas}(2\pi f t)\,dt = H_{x_1}(f)H_{x_2}(f)$$

2. If $x_1(t)$ is odd, then $\int_{-\infty}^{\infty} [x_1(t) * x_2(t)]\operatorname{cas}(2\pi f t)\,dt = H_{x_1}(f)H_{x_2}(\ f)\ \ -$

3. If $x_2(t)$ is odd, then $\int_{-\infty}^{\infty} [x_1(t) * x_2(t)]\operatorname{cas}(2\pi f t)\,dt = H_{x_1}(\ f)H_{x_2}(f)\ \ -$

4. If both functions are odd, then $\int_{-\infty}^{\infty} [x_1(t) * x_2(t)]\operatorname{cas}(2\pi f t)\,dt = \ \ H_{x_1}(f)H_{x_2}(f)\ \ -$

14.4.8 Autocorrelation

$$\int_{-\infty}^{\infty}\left[\int_{-\infty}^{\infty} x(\tau)\star x_2(\tau - t)\,d\tau\right]\operatorname{cas}(2\pi f t)\,dt = \tfrac{1}{2}[H_x^2(f) + H_x^2(-f)] = [H_x^e(f)]^2 + [H_x^o(f)]^2$$

14.4.9 Product

The Hartley transform of the product $x_1(t)x_2(t)$ is given by

$$\tfrac{1}{2}[H_{x_1}(f) * H_{x_2}(f) + H_{x_1}(-f) * H_{x_2}(f) + H_{x_1}(f) * H_{x_2}(-f) - H_{x_1}(-f) * H_{x_2}(-f)]$$

$$= H_{x_1}^e(f) * H_{x_2}^e(f) - H_{x_1}^o(f) * H_{x_2}^o(f) + H_{x_1}^e(f) * H_{x_2}^o(f) + H_{x_1}^o(f) * H_{x_2}^e(f)$$

14.4.10 Derivative of x(t)

$$F\left\{\frac{dx(t)}{dt}\right\} = j2\pi f\, X(f) = j2\pi f\, \mathrm{Re}\{X(f)\} + 2\pi f\, \mathrm{Im}\{X(f)\}$$

Hence

$$H_{x'(t)} = 2\pi f\, \mathrm{Im}\{X(f)\} + 2\pi f\, \mathrm{Re}\{X(f)\} = 2\pi f[-H_x^o(f) \quad H_x^e(f)]$$
$$= -2\pi f\, H_x(-f)$$

By revision we can write

$$\int_{-\infty}^{\infty} \frac{d^n x(t)}{dt^n} \mathrm{cas}\, 2\pi ft\, dt = \left(\mathrm{cas}' \frac{n\pi}{2} \right) (2\pi f)^n H_x[(-1)^n f]$$

14.4.11 Second Moment

$$H(f) = \int_{-\infty}^{\infty} x(t)[\cos 2\pi ft + \sin 2\pi ft]\, dt$$

Differentiating both sides with respect to f and then setting $f = 0$, we easily obtain

$$\left. \frac{d^2 H(f)}{df^2} \right|_{f=0} = -4\pi^2 \int_{-\infty}^{\infty} t^2 x(t)\, \mathrm{cas}\, 2\pi ft\, dt$$

14.4.12 Fourier Transform from Hartley Transform

$$X(f) = \frac{1}{\sqrt{2}} e^{-j\pi/4} H(f) + \frac{1}{\sqrt{2}} e^{j\pi/4} H(\ f)-$$

14.4.13 Hartley Transform from Fourier Transform

$$H(f) = \frac{1}{\sqrt{2}} e^{j\pi/4} X(f) + \frac{1}{\sqrt{2}} e^{-j\pi/4} X^*(\ f)$$

14.4.14 Hartley Transform Properties

TABLE 14.5 Hartley Transform Properties

Property	$x(t)$	$H(f) = \int_{-\infty}^{\infty} x(t)\, \mathrm{cas}\, 2\pi ft\, dt$
Linearity	$ax_1(t) + bx_2(t)$	$aH_{x_1}(f) + bH_{x_2}(f)$
Scaling	$x(at)$	$\frac{1}{\lvert a \rvert} H_x\left(\frac{f}{a}\right)$
	$x(t/a)$	$\lvert a \rvert H_x(af)$
Reversal	$x(-t)$	$H_x(-f)$

TABLE 14.5 Hartley Transform Properties (continued)

Property	$x(t)$	$H(f)=\int_{-\infty}^{\infty} x(t)\,\text{cas}2\pi ft\,dt$
Shift	$x(t-t_o)$	$\cos 2\pi ft_o H_x(f)+\sin 2\pi ft_o H_x(-f)$
Modulation	$\cos 2\pi f_o t\, x(t)$	$\frac{1}{2}H_x(f-f_o)+\frac{1}{2}H_x(f+f_o)$
Convolution	$x_1(t)*x_2(t)$	$\frac{1}{2}[H_{x_1}(f)H_{x_2}(f)+H_{x_1}(-f)H_{x_2}(f)$ $+H_{x_1}(f)H_{x_2}(-f)-H_{x_1}(-f)H_{x_2}(-f)]$
Autocorrelation	$x(t)\bigstar x(t)$	$\frac{1}{2}[H_x^2(f)+H_x^2(-f)]=[H_x^e(f)]^2+[H_x^o(f)]^2$
Product	$x_1(t)x_2(t)$	$\frac{1}{2}[H_{x_1}(f)*H_{x_2}(f)+H_{x_1}(-f)*H_{x_2}(f)$ $+H_{x_1}(f)*H_{x_2}(-f)-H_{x_1}(-f)*H_{x_2}(-f)]$ $=H_{x_1}^e(f)*H_{x_2}^e(f)-H_{x_1}^o(f)*H_{x_2}^o(f)$ $+H_{x_1}^e(f)*H_{x_2}^o(f)+H_{x_1}^o(f)*H_{x_2}^e(f)$
Derivative	$\frac{dx(t)}{dt}$	$-2\pi f\,H_x(-f)$
	$\frac{d^n x(t)}{dt^n}$	$\left(\text{cas}'\frac{n\pi}{2}\right)(2\pi f)^n H_x[(-1)^n f]$
FT from HT	$X(f)=\frac{1}{\sqrt{2}}e^{-j\pi/4}H(f)+\frac{1}{\sqrt{2}}e^{j\pi/4}H(\ f)-$	
HT from FT	$H(f)=\frac{1}{\sqrt{2}}e^{j\pi/4}X(f)+\frac{1}{\sqrt{2}}e^{-j\pi/4}X^*(\ -f)$	
Power spectrum	$P_x(f)=\lvert X(f)\rvert^2=$	$\frac{[H_x(f)]^2+[H_x(-f)]^2}{2}$
Phase spectrum	$\text{Phas}\,X(f)=\tan^{-1}$	$\left[\frac{H_x(-f)-H_x(f)}{H_x(f)+H_x(-f)}\right]$

14.5 Examples of Hartley Transforms

14.5.1 Example (Gaussian Function)

$$F\{e^{-at^2}\}=\sqrt{\frac{\pi}{a}}\,e^{-\frac{\pi^2 f^2}{a}},\qquad a>0.$$

Since HT is the even part of FT implies that

$$H_x(f)=\sqrt{\frac{\pi}{a}}\,e^{-\frac{\pi^2 f^2}{a}}.$$

If $a=\pi$, then $H_x(f)=e^{-\pi^2 f^2}$, which is identical with the function.

14.5.2 Example (Shifted Gaussian)

$$F\{e^{-a(t-t_o)^2}\}=e^{-j\omega t_o}F\{e^{-at^2}\}=\sqrt{\frac{\pi}{a}}\,e^{-j\omega t_o}e^{-\frac{\pi^2 f^2}{a}}=\sqrt{\frac{\pi}{a}}\,e^{-\frac{\pi^2 f^2}{a}}(\cos 2\pi ft_o\quad j\sin 2\pi ft_o).$$

Since the $H_x(f) = \mathrm{Re}\{X(f)\} - \mathrm{Im}\{X(f)\}$ (see 14.2.1), we obtain $H_x(f) = \sqrt{\frac{\pi}{a}}\, e^{-\frac{\pi^2 f^2}{a}} \operatorname{cas} 2\pi f t_o$.

14.5.3 Example (Modulated Gaussian):

$$F\{e^{-\pi t^2}\cos \pi t\} = F\left\{e^{-\pi t^2}\frac{e^{j\pi t}+e^{-j\pi t}}{2}\right\} = \int_{-\infty}^{\infty} e^{-\pi t^2}\frac{e^{-j2\pi(f-\frac{1}{2})t}+e^{-j2\pi(f+\frac{1}{2})t}}{2}dt = \tfrac{1}{2}e^{-\pi(f-\frac{1}{2})^2}+\tfrac{1}{2}e^{-\pi(f+\frac{1}{2})^2}.$$

Since this is an even function implies from (14.2.1) that $H_x(f) = F\{e^{-\pi t^2}\cos \pi t\} = X(f)$.

14.5.4 Example (Truncated Cosine)

If $x(t) = \cos\omega_o t\, p_a(t)$, then

$$F\{x(t)\} = \int_{-a}^{a} e^{-j2\pi f t}\tfrac{1}{2}(e^{j\omega_o t}+e^{-j\omega_o t})dt = \tfrac{1}{2}\int_{-a}^{a} e^{-j(2\pi f-\omega_o)t}dt + \tfrac{1}{2}\int_{-a}^{a} e^{-j(2\pi f+\omega_o)t}dt$$

$$= \frac{\sin[(2\pi f-\omega_o)a]}{2\pi f-\omega_o} + \frac{\sin[(2\pi f+\omega_o)a]}{2\pi f+\omega_o}.$$

Since $X(f)$ is even, then $H_x(f) = X(f)$.

14.5.5 Example (Signum)

If $x(t) = \operatorname{sgn}(t)$, then

$$F\{\operatorname{sgn}(t)\} = \frac{1}{j\pi f} = j\frac{1}{\pi f}.$$

Hence, $H_x(f) = -\mathrm{Im}\{X(f)\}\ \frac{1}{\pi f}$.

14.5.6 Example (Unit Step)

If $x(t) = u(t)$, then

$$F\{u(t)\} = \tfrac{1}{2}\delta(f) - j\frac{1}{2\pi f}..$$

Hence, $H_x(f) = \mathrm{Re}\{X(f)\} - \mathrm{Im}\{X(f)\} = \tfrac{1}{2}\delta(f) + \frac{1}{2\pi f}$.

14.5.7 Example (Cosine)

If $x(t) = \cos\omega_o t$, the $F\{\cos\omega_o t\} = \tfrac{1}{2}[\delta(f-f_o)+\delta(f+f_o)]$. Since the delta functions are even, this implies the $H_x(f) = \tfrac{1}{2}[\delta(f-f_o)+\delta(f+f_o)]$.

14.5.8 Example (Shifted Exponential)

If $x(t) = e^{-a(t-t_o)} u(t-t_o)$, then

$$F\{e^{-a(t-t_o)} u(t-t_o)\} = \int_{-\infty}^{\infty} e^{-a(t-t_o)} u(t-t_o) e^{-j2\pi ft}\, dt \quad \int_{-\infty}^{\infty} e^{-a(\tau)} u(\tau) e^{-j2\pi ft_o} e^{-j2\pi f\tau}\, d\tau$$

$$= e^{-j2\pi ft_o} \int_{o}^{\infty} e^{-(a+j2\pi f)\tau}\, d\tau = e^{-j2\pi ft_o} \frac{1}{a+j2\pi f}.$$

Hence, $H_x(f) = \mathrm{Re}\{X(f)\} - \mathrm{Im}\{X(f)\} = \dfrac{(a+2\pi f)\cos 2\pi ft_o + (a \quad 2\pi f)\sin 2\pi ft_o}{a^2 + (2\pi f)^2}.$

14.5.9 Example

Because the $F\{p_a(t)\} = \dfrac{2\sin \omega a}{\omega}$, the symmetry property of Fourier transform (see 3.1.2) gives $F\left\{\dfrac{2\sin at}{t}\right\} = 2\pi\, p_a(\omega)$ or equivalently $F\left\{\dfrac{\sin at}{\pi t}\right\} = p_a(\omega)$. Hence, on the f axis $F\left\{\dfrac{\sin at}{\pi t}\right\} = p_{a/2\pi}(f)$, and since the pulse is an even function $H_x(f) = p_{a/2\pi}(f)$.

14.5.10 Example (Shifted Pulse)

The FT of $x(t) = p_a(t-a)$ is

$$X(f) = \int_{o}^{2a} e^{-j2\pi ft}\, dt = \frac{(\cos 2\pi fa - j\sin 2\pi fa)\sin 2\pi fa}{\pi f}.$$

Hence, $H_x(f) = \mathrm{Re}\{X(f)\} - \mathrm{Im}\{X(f)\} = \mathrm{cas} 2\pi fa \dfrac{\sin 2\pi fa}{\pi f}.$

14.5.11 Example

The FT of $x(t) = te^{-at} u(t)$, $a > 0$, is

$$X(f) = \int_{o}^{\infty} te^{-at} e^{-j2\pi ft}\, dt = \int_{o}^{\infty} te^{-(a+j2\pi f)t}\, dt = \frac{1}{(a+j2\pi f)(a+j2\pi f)}.$$

Hence, $H_x(f) = \dfrac{a^2 - (2\pi f)^2 + 4a\pi f}{[a^2 + (2\pi f)^2]^2}.$

14.5.12 Example

The FT of $x(t) = \sin \pi t\, p(t)$ is

$$X(f) = \int_{-1}^{1} \sin\pi t\, e^{-j2\pi ft}\, dt = \int_{-1}^{1} \frac{e^{j\pi t} - e^{-j\pi t}}{2j} e^{-j2\pi ft}\, dt = -j\frac{\sin(2\pi f - \pi)}{2\pi f - \pi} + j\frac{\sin(2\pi f + \pi)}{2\pi f + \pi}..$$

Hence, $H_x(f) = -\mathrm{Im}\{X(f)\} \quad \dfrac{\sin(2\pi f - \pi)}{2\pi f - \pi} - \dfrac{\sin(2\pi f + \pi)}{2\pi f + \pi}.$

14.6 Hartley Series

14.6.1 Orthonormal Set of Functions

The set $\{\phi_n(t)\} = \frac{1}{\sqrt{2\pi}} \text{cas}(n2\pi f_o t)$, $n = 0, \pm 1, \pm 2, \cdots$; $f_o = 2\pi / \text{T}$; T = period, is orthonormal in the interval

$$t_o \leq t \leq t_o + \text{T}, \qquad (\phi_i, \phi_k) = \begin{cases} 1 & i = k \\ 0 & i \neq k \end{cases}.$$

14.6.2 Expansion of a Periodic Function

$$x(t + \text{T}) = x(t) \text{ for all } t$$

$$x(t) = \sum_{i=-\infty}^{\infty} \gamma_i \phi_i(t)$$

$$\phi_i(t) = \text{cas}(i2\pi f_o t)$$

$$\gamma_i = \frac{1}{T} \int_{T_o}^{t_o + \text{T}} x(t) \text{cas}(i2\pi f_o t)\, dt$$

14.6.3 Relation to Fourier Series

$$x(t) = \sum_{n=-\infty}^{\infty} \alpha_n e^{jn2\pi f_o t} = \text{complex Fourier series}$$

$$\alpha_n = \frac{1}{T} \int_{T_o}^{t_o + \text{T}} x(t) e^{-jn2\pi f_o t}\, dt$$

$$\alpha^*_{-n} = \alpha_n$$

$$\gamma_k = \begin{cases} \text{Re}\{\alpha_k\} - \text{Im}\{\alpha_k\} & k \neq 0 \\ \alpha_k & k = 0 \end{cases}$$

$$\alpha_k = E\{\gamma_k\} - jO\{\gamma_k\}$$

$$E\{\gamma_k\} = \text{even part} = \frac{\gamma_k + \gamma_{-k}}{2}; \qquad O\{\gamma_k\} = \text{odd part} = \frac{\gamma_k - \gamma_{-k}}{2}$$

14.7 Tables of Fourier and Hartley Transforms

TABLE 14.6 Table of Fourier and Hartley Transform Pairs

$x(t)$	$X(t)$	$H_x(f)$
$u\left(t + \frac{T}{2}\right) - u\left(t - \frac{T}{2}\right)$	$T \frac{\sin \pi T f}{\pi T f} = T \,\text{sinc}\, Tf$	Because $x(t)$ is even, $H_x(f) = X(f)$
$\beta e^{-\alpha t} u(t)$	$\frac{\beta}{j\omega + \alpha}$	$\frac{\beta(\alpha + 2\pi f)}{\alpha^2 + (2\pi f)^2}$

TABLE 14.6 Table of Fourier and Hartley Transform Pairs (continued)

$x(t)$	$X(t)$	$H_x(f)$
$1-2\frac{\lvert t\rvert}{T},\ \lvert t\rvert<\frac{T}{2}$	$\frac{T}{2}\text{sinc}^2\left(\frac{Tf}{2}\right)=\frac{1-\cos\pi f T}{T\pi^2 f^2}$	Because $x(t)$ is even, $H_x(f)=X(f)$
$e^{-\alpha^2 t^2}$	$\frac{\sqrt{\pi}}{\alpha}e^{-(\pi^2 f^2/\alpha^2)}$	Because $x(t)$ is even, $H_x(f)=X(f)$
$e^{-\alpha\lvert t\rvert}$	$\frac{2\alpha}{\alpha^2+4\pi^2 f^2}$	Because $x(t)$ is even, $H_x(f)=X(f)$
$e^{-\alpha t}\sin(\omega_o t)u(t)$	$\frac{\omega_o}{(\alpha+j2\pi f)^2+\omega_o^2}$	$\frac{\omega_o(\alpha^2+\omega_o^2-4\pi^2 f^2 \quad 4\pi f\alpha)}{(\alpha^2+\omega_o^2-4\pi^2 f^2)^2 \quad (4\pi f\alpha)^2}$
$e^{-\alpha t}\cos(\omega_o t)u(t)$	$\frac{\alpha+j2\pi f}{(\alpha+j2\pi f)^2+\omega_o^2}$	$\frac{(\alpha-2\pi f)(\alpha^2+\omega_o^2-4\pi^2 f^2)+(4\pi f\alpha)(\alpha+2\pi f)}{(\alpha^2+\omega_o^2-4\pi^2 f^2)^2 \quad (4\pi f\alpha)^2}$
$\frac{1}{\beta-\alpha}(e^{-\alpha t}-e^{-\beta t})u(t)$	$\frac{1}{(\alpha+j2\pi f)(\beta+j2\pi f)}$	$\frac{\alpha\beta-2\pi f(\alpha+\beta+2\pi f)}{[\alpha\beta-(2\pi f)^2]^2+[2\pi f(\alpha+\beta)]^2}$
$\cos\omega_o t\left[u\left(t+\frac{T}{2}\right)-u\left(t-\frac{T}{2}\right)\right]$	$\frac{T}{2}\left[\frac{\sin\pi T(f-f_o)}{\pi T(f-f_o)}+\frac{\sin\pi T(f+f_o)}{\pi T(f+f_o)}\right]$	Because $x(t)$ is even, $H_x(f)=X(f)$
$K\delta(t)$	K	K
K	$K\delta(f)$	$K\delta(f)$
$u(t)$	$\frac{1}{2}\delta(f)+\frac{1}{j2\pi f}$	$\frac{1}{2}\delta(f)+\frac{1}{2\pi f}$
$\text{sgn}\,t=\frac{t}{\lvert t\rvert}$	$\frac{1}{j\pi f}$	$\frac{1}{\pi f}$
$\cos\omega_o t$	$\frac{1}{2}[\delta(f-f_o)+\delta(f+f_o)]$	Because $x(t)$ is even, $H_x(f)=X(f)$
$\sin\omega_o t$	$\frac{-j}{2}[\delta(f-f_o)-\delta(f+f_o)]$	$\frac{1}{2}[\delta(f-f_o)-\delta(f+f_o)]$
$\sum_{-\infty}^{\infty}\delta(t-nT)$	$\frac{1}{T}\sum_{-\infty}^{\infty}\delta\left(f-\frac{n}{T}\right)$	Because $x(t)$ is even, $H_x(f)=X(f)$
$\sum_{-\infty}^{\infty}\alpha_n e^{jn2\pi f_o t}$	$\sum_{-\infty}^{\infty}\alpha_n\delta\left(f-\frac{n}{T}\right)$	$\sum_{-\infty}^{\infty}\gamma_n\delta\left(f-\frac{n}{T}\right)$
$Ae^{j\omega_o t}$	$A\delta(f-f_o)$	$H_x(f)=X(f)$ ($\delta(f)$ is even)
$t\,u(t)$	$\frac{j}{4\pi}\delta'(f)-\frac{1}{4\pi^2 f^2}$	$\frac{-1}{4\pi}\delta'(f)-\frac{1}{4\pi^2 f^2}$
$e^{-\alpha(t-t_o)^2}$	$\sqrt{\frac{\pi}{a}}e^{-j\omega t_o}e^{-\frac{\pi^2 f^2}{a}}$	$\sqrt{\frac{\pi}{a}}e^{-\frac{\pi^2 f^2}{a}}\text{cas}2\pi f t_o$
$e^{-\pi t^2}\cos\pi t$	$\frac{1}{2}e^{-\pi(f-\frac{1}{2})^2}+\frac{1}{2}e^{-\pi(f+\frac{1}{2})^2}$	$H_x(f)=X(f)$
$\cos\omega_o t\,p_a(t)$	$\left\{\frac{\sin[(2\pi f-\omega_o)a]}{2\pi f-\omega_o}+\frac{\sin(2\pi f+\omega_o)}{2\pi f+\omega_o}\right.$	$H_x(f)=X(f)$

TABLE 14.6 Table of Fourier and Hartley Transform Pairs (continued)

$x(t)$	$X(t)$	$H_x(f)$
$e^{-\alpha(t-t_o)}u(t)$	$e^{-j2\pi f t_o}\dfrac{1}{a+j2\pi f}$	$\dfrac{(\alpha+2\pi f)\cos 2\pi f t_o}{\alpha^2+(2\pi f)^2}+\dfrac{(\alpha-2\pi f)\sin 2\pi f t_o}{\alpha^2+(2\pi f)^2}$
$p_a(t)$	$\dfrac{\sin 2\pi f a}{\pi f}$	$H_x(f)=X(f)$
$\dfrac{\sin at}{\pi t}$	$p_{a/2\pi}(f)$	$H_x(f)=X(f)$
$p_a(t-a)$	$e^{-j2\pi f a}\dfrac{\sin 2\pi f a}{\pi f}$	$\text{cas} 2\pi f a \dfrac{\sin 2\pi f a}{\pi f}$
$te^{-t}u(t)$	$\dfrac{1}{(a+j2\pi f)(a+j2\pi f)}$	$\dfrac{a^2+4a\pi f-(2\pi f)^2}{[a^2+(2\pi f)^2]^2}$
$\sin \pi t p(t)$	$-j\dfrac{\sin(2\pi f-\pi)}{2\pi f-\pi}+j\dfrac{\sin(2\pi f+\pi)}{2\pi f+\pi}$	$\dfrac{\sin(2\pi f-\pi)}{2\pi f-\pi}-\dfrac{\sin(2\pi f+\pi)}{2\pi f+\pi}$

14.8 Two-Dimensional Hartley Transform

14.8.1 Two-Dimensional

$$H_f(u,v)=\iint_{-\infty}^{\infty} f(x,y)\,\text{cas}[2\pi(ux+vy)]\,dx\,dy$$

$$f(x,y)=\iint_{-\infty}^{\infty} H_f(u,v)\,\text{cas}[2\pi(ux+vy)]\,du\,dv$$

14.8.2 Relation to Fourier Transform

$$F(u,v)=\iint_{-\infty}^{\infty} f(x,y)e^{-j2\pi(ux+vy)}\,dx\,dy$$

$$=R_f(u,v)-jI_f(u,v)$$

$$f(x,y)=\iint_{-\infty}^{\infty} F(u,v)e^{j2\pi(ux+vy)}\,du\,dv$$

$$R_f(u,v)=\text{Re}\{F(u,v)\}=\text{even}$$

$$I_f(u,v)=\text{Im}\{F(u,v)\}=\text{odd}$$

$$H_f(u,v)=R_f(u,v)-I_f(u,v)$$

14.9 Discrete Hartley Transform

14.9.1 The Discrete Hartley Transform (DHT)

$$H_h(k\Omega_n) = \frac{1}{\sqrt{N}} \sum_{n=0}^{N-1} h(nT) cas(k\Omega_h nT)$$

$$h(nT) = \frac{1}{\sqrt{N}} \sum_{k=0}^{N-1} H_h(k\Omega_n) cas(k\Omega_n nT)$$

$$\Omega_h = \frac{2\pi}{NT} = \text{frequency resolution (rad s}^{-1}\text{)}$$

Similarly the discrete Fourier transform is given by

$$F(k\Omega_f) = \sum_{n=0}^{N-1} f(nT) e^{-jk\Omega_f nT}$$

$$f(nT) = \frac{1}{N} \sum_{k=0}^{N-1} F(k\Omega_f) e^{jk\Omega_f nT}$$

$$\Omega_f = \frac{2\pi}{NT} = \text{frequency resolution (rad s}^{-1}\text{)}$$

14.9.2 Relation to Discrete Fourier Transform

$$H(k\Omega_h) = \mathrm{Re}\{F(k\Omega_f)\} - \mathrm{Im}\{F(k\Omega_f)\}$$

Note:

- DHT avoids complex arithmetic
- DHT requires half the memory storage
- DHT requires $N \log_2 N$ real operation instead of $N \log_2 N$ complex operations required for DFT
- DHT fewer operation may help in truncation and rounding errors
- DHT is its own inverse

14.9.3 A C Program for Fast Hartley Transforms

TABLE 14.7

```
/*****************************************************************************
/* Program FHT.C ***********************************************************
/*
/* This FHT algorithm utilizes an efficient permutation algorithm
/* developed by David M. W. Evans. Additional /* details may be found
/* in: IEEE Transaction on Acoustics, Speech, and Signal Processing,
/* vol. ASSP-35, n. 8, pp. 1120-1125. August 1987.
/*
```

```
/* This FHT algorithm, authored by Lakshmikantha S. Prabhu, is
/* optimized for the SPARC RISC platform. Additional details may
/* be found in his M.S.E.E. thesis referenced below.
/*
/* L. S. Prabhu, "A Complexity-Based Timing Analysis of Fast
/* Real Transform Algorithms," Master's Thesis, University of
/* Arkansas, Fayetteville, AR, 72701-1201, 1993.
/***********************************************************************/
/* This program assumes a maximum array length of 2^M = N where         */
/* M=9 and N=512.                                                        */
/* See line 52 if the array length is increased.                        */
# include <stdio.h>
# include <math.h>
# define M 3
# define N 8
float * myFht () ;
main ()
{
/* Read the integer values 1, . . . , N into the vector X(N).            */
        int i;
        float X[N];
        for ( i = 0 ; i < N; i++)
           X[i] = i+1 ;
        for ( i = 0; i < N; i++)
           printf ("%f\n", X[i]) ;
           myFht(X,N,M) ;
           printf("\n")
        for ( i = 0 ; i < N; i++)
           printf("%d: %f\n , i,X[i]/N;
/* It is assumed that the user divides by the integer N.                 */
}
float*
myFht (x,n,m)
float* x;
int n,m;
{
int i, j, k, kk, 1, 10, 11, 12, 13, 14, 15, m1, n1, n2, NN, s;
int diff = 0, diff2, gamma, gamma2=2, n2_2, n2_4, n_2, n_4,n_8, n_16;
int itemp, ntemp, phi, theta_by_2;
float ee, temp1, temp2, xtemp1, xtemp2;
double ccl, cc2, ss1, ss2;
double since [257] ;
/***********************************************************************/
/* digit reverse counter.                                                */
/***********************************************************************/
int powers_of_2[16] , seed[256] ;
int firstj, log2_n, log2_seed_size ;
int group_no, nn, offset ;
log2_n = m >> 1 ;
nn = 2 << {log2_n -1) ;
if ( (m % 2) == 1)
```

```
  log2_n = log2_n + 1 ;
  seed[0] = 0 ; seed[1] = 1 ;
  for (log2_seed_size = 2; log2_seed_size < = log2_n; log2_seed_size++)
     {
     for ( i = 0; i < 2 << (log2_seed_size -2); i++)
        {
        seed [i] = 2 * seed [i]
        for (k = 1; k < 2; k++)
           seed [ i + k * (2 << (log2_seed_size - 1) >> 1] = seed [i] ;
        }
     }
  for (offset = 1; offset < nn; offset++)
     {
     firstj - nn * seed [offset];
     i = offset ; j = firstj ;
     xtemp = x[i] ;
     x[i] = x[j] ;
     x[j] = xtemp ;
     for ( group_no = 1; group_no < seed [offset] ; group_no++)
        {
        i = i + nn ; j = firstj + seed [group_no] ;
        xtemp = x[i];
        x[i] = x[j] ;
        x[j] = xtemp ;
        }
     }
j = 0
n1 = n - 1 ;
n_16 = n >> 4 ;
n_8 = n >> 3;
n_4 = n >> 2;
n_2 = n >> 1;
/****************************************************************************/
/* Start the transform computation with 2-point butterflies.               */
/****************************************************************************/
for (i = 0 ; i < n ; i + -2)
   {
   s = 1i + 1
   xtemp = x[i] ;
   x[i] = x[s] ;
   x[s] = xtemp - x[s] ;
   }
/****************************************************************************/
/* Now the 4-point butterflies.                                             */
/****************************************************************************/
for (i = 0 ; i < n ; i + -2)
   {
   xtemp = x[i] ;
   x[i] += x[i+2] ;
   x[i+2] = xtemp - x[i+2]
   xtemp = x[i+1]
```

```
  x[i+1] += x[i+3] ;
  x[i+3] += xtemp - x[i+3] ;
  }
/***************************************************************************/
/* Sine table initialization.                                              */
/***************************************************************************/
NN = n_4;
sine[0] = 0
sine [n_16] = 0.382683432 ;
sine [n_8] = 0.707106781 ;
sine [3*n_16] = 0.923879533 ;
sine -n_4] = 1.000000000 ;
h_sec_b = 0.509795579 ;
diff = n_16 ;
theta_by_2 = n_4 >> 3 ;
j = 0 ;
while (theta_by_2 >= 1)
    {
    for ( i = 0 ; i <= n_4 ; i += diff)
       {
       sine[j + theta_by_2] = h_sec_b * (sine[j] + sine[j + diff]) ;
       j = j + diff
       }
    j = 0
    diff = diff >> 1 ;
    theta_by_2 = theta_by_2 >> 1 ;
    h_sec_b = 1 / sqrt)2 + 1/h_sec_b) ;
    }
/***************************************************************************/
/* Other butterflies.                                                      */
/***************************************************************************/
for (i = 3 ; i < m ; i ++)
   {
   diff = 1 ; gamma = 0 ;
   ntemp = 0 ; phi = 2 << (m-1) >> 1 ;
   ss1 = sine [phi] ;
   ccl = sine [n_4 - phi] ;
   n2 = 2 << (i-1) ;
   n2_2 = n2 >> 1 ;
   n2_4 = n2 >> 2 ;
   gamma2 = n2_4 ;
   diff2 - gamma2 + gamma2 - 1 ;
   itemp = n2_4 ;
   k = 0 ;
/***************************************************************************/
/* Initial section of stages 3,4,... for which sines & cosines are not required.  */
/***************************************************************************/
for (k = 0 ; k < (2 << (m-i) >>1) ; k+ +)
    {
    10 = gamma ;
    11 = 10 + n2_2 ;
```

```
    13 = gamma2 ;
    14 = gamma2 + n2_2 ;
    15 = 11 + itemp ;
    x0 = x[10] ;
    x1 = x[11] ;
    x3 = x[13] ;
    x5 = x[15] ;
    x[10] = x0 + x1 ;
    x[11] = x0 - x1 ;
    x[13] = x3 + x5 ;
      x[14] = x3 - x5 ;
      gamma = gamma + n2 ;
      gamma2 = gamma2 + n2 ;
      }
    gamma = diff ;
    gamma2 = diff2 ;
/*******************************************************************************/
/* Next sections of stages 3,4,...                                              */
/*******************************************************************************/
    for (k = 0 ; k < (2 << (m-i) >>1) ; k+ +)
      {
      for (k = 0 ; k < (2 << (m-i) >>1) ; k+ +)
        {
        10 = gamma ;
        11 = 10 + n2_2 ;
        13 = gamma2 ;
        14 = 13+ n2_2 ;
        x0 = x[10] ;
        x1 = x[11] ;
        x3 = x[13] ;
        x4 = x[14] ;
        x[10] = x0 + x1 * cc1 + x4 * ss1 ;
        x[10] = x0 + x1 * cc1 + x4 * ss1 ;
        x[13] = x3 - x4 * cc1 + x1 * ss1 ;
        x[14] = x3 + x4 * cc1 + x1 * ss1 ;
        gamma = gamma + n2 ;
        gamma2 = gamma2 + n2 ;
        }
      itemp = 0 ;
      phi = phi + (2 << (m-i) >> 1) ;
      ntemp = (phi < n_4) ? 0 : n_4 ;
      ss1 = sine [phi - ntemp] ;
      cc1 = sine [n_4 - phi + ntemp] ;
      diff++ ; diff2- ;
      gamma = diff ;
      gamma2 = diff2 ;
      }
    }
}
```

References

Bracewell, R. N., *The Hartley Transform*, Oxford University Press, New York, 1986.

Olejniczak, K. J., The Hartley Transform, Chapter 4, in *The Transforms and Applications Handbook*, Ed. A. D. Poularikas, CRC Press, Boca Raton, Florida, 1996.

Wang, Z., Harmonic analysis with a real frequency function - I: Aperiodic case, *Appl. Math. and Comput.*, 9, 53-73, 1981.

Wang, Z., Harmonic analysis with a real frequency function - II: Periodic and bounded case, *Appl. Math. and Comput.*, 9, 153-156, 1981.

Wang, Z., Harmonic analysis with a real frequency function - III: Data Sequence, *Appl. Math. and Comput.*, 9, 245-255, 1981.

15

The Hilbert Transform

15.1 The Hilbert Transform

15.1.1 Definition of Hilbert Transform

$$\upsilon(t) = \mathrm{H}\{x(t)\} = \frac{-1}{\pi} P\int_{-\infty}^{\infty} \frac{x(\eta)}{\eta - t} d\eta = \frac{1}{\pi} P\int_{-\infty}^{\infty} \frac{x(\eta)}{t - \eta} d\eta$$

$$x(t) = \mathrm{H}^{-1}\{\upsilon(t)\} = \frac{1}{\pi} P\int_{-\infty}^{\infty} \frac{\upsilon(\eta)}{\eta - t} d\eta = \frac{-1}{\pi} P\int_{-\infty}^{\infty} \frac{\upsilon(\eta)}{t - \eta} d\eta$$

where P stands for the Cauchy principal value of the integral.

Convolution form representation

$$v(t) = x(t) * \frac{1}{\pi t}$$

$$x(t) = -\upsilon(t) * \frac{1}{\pi t}$$

Fourier transform of $v(t)$ and $x(t)$ and $1/\pi t$ (see Table 3.1.3)

$$V(\omega) = X(\omega)[-j\,\mathrm{sgn}(\omega)]$$

$$\mathcal{F}^{-1}\{-j\,\mathrm{sgn}(\omega)X(\omega)\} = \upsilon(t)$$

$$\mathcal{F}\left\{\frac{1}{\pi t}\right\} = -j\,\mathrm{sgn}(\omega)$$

Example

If $x(t) = \cos\omega t$, then

$$\begin{aligned}\mathcal{H}\{\cos\omega t\} &= \upsilon(t)\\ &= \frac{-1}{\pi}P\int_{-\infty}^{\infty}\frac{\cos\omega\eta}{\eta - t}d\eta\\ &= \frac{-1}{\pi}P\int_{-\infty}^{\infty}\frac{\cos[\omega(y+t)]}{y}dy\\ &= \frac{-1}{\pi}\left\{\cos\omega t\,P\int_{-\infty}^{\infty}\frac{\cos\omega y}{y}dy - \sin\omega t\,P\int_{-\infty}^{\infty}\frac{\sin\omega y}{y}dy\right\}\\ &= \sin\omega t.\end{aligned}$$

The result is due to the fact that $\cos\omega y/y$ is an odd function and $P\int_{-\infty}^{\infty}\frac{\sin\omega y}{y}dy = \pi$.

Example

If $x(t) = p_a(t)$ then

$$v(t) = \mathcal{H}\{p_a(t)\} = \frac{-1}{\pi}P\int_{-a}^{t-\varepsilon}\frac{d\eta}{\eta - t} - \frac{1}{\pi t}P\int_{t+\varepsilon}^{a}\frac{d\eta}{\eta - t}$$

$$= \lim_{\varepsilon\to 0}\left[-\frac{1}{\pi}\ln(\eta - t)\Big|_{-a}^{t-\varepsilon} - \frac{1}{\pi}\ln(\eta - t)\Big|_{t+\varepsilon}^{a}\right] = \upsilon(t) = \frac{1}{\pi}\ln\left|\frac{t+a}{t-a}\right|$$

Example

If $x(t) = a$ then

$$a\mathcal{H}\{1\} = a\lim_{a\to\infty}\frac{1}{\pi}\ln\left|\frac{t+a}{t-a}\right| = 0\,.$$

Hence, if $x_o = constant$ is the mean value of a function, then $x(t) = x_o + x_1(t)$. Therefore $\mathcal{H}\{x_o + x_1(t)\} = \mathcal{H}\{x_1(t)\}$. This implies that the Hilbert transform cancels the mean value or the DC term in electrical engineering terminology.

15.1.2 Analytic Signal

A complex signal whose imaginary part is the Hilbert transform of its real part is called the *analytic signal.*

$$\psi(z) = \psi(t,\tau) = x(t,\tau) + j\upsilon(t,\tau), \quad x \text{ and } \upsilon \text{ are real functions}$$

$$z = t + j\tau$$

$$\upsilon(t,\tau) = \mathrm{H}\{x(t,\tau)\}$$

The function $\psi(z) = x(t,\tau) + j\upsilon(t,\tau)$ is analytic if the Cauchy-Riemann conditions

$$\frac{\partial x}{\partial t} = \frac{\partial \upsilon}{\partial \tau} \quad \text{and} \quad \frac{\partial x}{\partial \tau} = -\frac{\partial \upsilon}{\partial t}$$

are satisfied.

Example

The real and imaginary parts of the analytic function

$$\psi(z) = 1/(\alpha - jz) = \frac{\alpha + \tau}{(\alpha + \tau)^2 + t^2} + j\frac{t}{(\alpha + \tau)^2 + t^2}$$

satisfy Cauchy-Riemann conditions and, hence, they are Hilbert transform pairs.

$$x(t) = \frac{\psi(t) + \psi^*(t)}{2} \qquad \upsilon(t) = \frac{\psi(t) - \psi^*(t)}{2j} \qquad (\tau = 0)$$

15.2 Spectra of Hilbert Transformation

15.2.1 One-Sided Spectrum of the Analytic Signal

$$x(t) = x_e(t) + x_o(t) = \frac{x(t) + x(-t)}{2} + \frac{x(t) - x(-t)}{2}$$

$$X(\omega) = X_r(\omega) + jX_i(\omega) = \int_{-\infty}^{\infty} x_e(t)\cos\omega t\, dt + j\left(-\int_{-\infty}^{\infty} x_o(t)\sin\omega t\, dt\right)$$

$V(\omega) = V_r(\omega) + jV_i(\omega) =$ Spectrum of the Hilbert transform

$V_r(\omega) = -j\,\mathrm{sgn}(\omega)[jX_i(\omega)] = \mathrm{sgn}(\omega)X_i(\omega)$ (see also 15.1.1)

$$V_i(\omega) = -\mathrm{sgn}(\omega)X_r(\omega)$$

Example

$\mathrm{H}\{\cos\omega t\} = \sin\omega t,\ \mathrm{H}\{\sin\omega t\} = -\cos\omega t$ and, therefore,

$$\mathrm{H}\{e^{j\omega t}\} = \sin\omega t - j\cos\omega t = -j\,\mathrm{sgn}(\omega)e^{j\omega t} = \mathrm{sgn}(\omega)e^{j(\omega t - \frac{\pi}{2})}$$

Note: The operator $-j\,\mathrm{sgn}(\omega)$ provides a $\pi/2$ phase lag for all positive frequencies and $\pi/2$ lead for all negative frequencies.

15.2.2 Fourier Spectrum of the Analytic Signal

$$\mathrm{H}\{x(t)\} = \upsilon(t); \quad \mathrm{F}\{x(t)\} = X(\omega); \quad \mathrm{F}\{\upsilon(t)\} = -j\,\mathrm{sgn}(\omega)X(\omega)$$

$$\mathsf{F}\{\psi(t)\} = x(t) + j\upsilon(t) = \Psi(\omega) = X(\omega) + jV(\omega) = [1 + \operatorname{sgn}(\omega)]X(\omega)$$

$$1 + \operatorname{sgn}(\omega) = \begin{cases} 2 & \omega > 0 \\ 1 & \omega = 0 \\ 0 & \omega < 0 \end{cases}$$

Note: The spectrum of the analytic signal is twice that of its Fourier transform at the positive frequency range $0 < \omega < \infty$.

Example

If $\psi(t) = \frac{1}{1+t^2} + j\frac{t}{1+t^2}$ then $\mathsf{F}\{\psi(t)\} = [1 + \operatorname{sgn}(\omega)]\pi e^{-|\omega|}$ where

$$\mathsf{H}\{1/(1+t^2)\} = t/(1+t^2) \text{ and } \mathsf{F}\{1/(1+t^2)\} = \pi e^{-|\omega|}.$$

15.3 Hilbert Transform and Delta Function

15.3.1 Complex Delta Function

If we define $2 \cdot 1(f) = 1(f) + \operatorname{sgn}(f)$, then the function (see Fourier transform properties [symmetry] and function, Chapter 3).

$$\psi_\delta(t) = \int_{-\infty}^{\infty} 2 \cdot 1(f) e^{j\omega t}\, df = \int_{-\infty}^{\infty} 1(f) e^{j\omega t}\, df + \int_{-\infty}^{\infty} \operatorname{sgn}(f) e^{j\omega t}\, df$$

$$= \delta(t) + j\frac{1}{\pi t}$$

15.3.2 Hilbert Transform of the Delta Function

From (15.3.1) implies

$$\mathsf{H}\{\delta(t)\} = \frac{1}{\pi t}$$

15.4 Hilbert Transform of Periodic Signals

15.4.1 Hilbert Transform of Period Functions

A periodic function can be written in trigonometric form

$$x_p(t) = C_o + \sum_{n=1}^{\infty} C_n \cos(n\omega_o t + \varphi_n), \quad \omega_o = 2\pi/T, \quad T = \text{period}$$

Therefore we obtain

$$v_p(t) = \mathsf{H}\{x_p(t)\} = \sum_{n=1}^{\infty} C_n \sin(n\omega_o t + \varphi_n)$$

because the Hilbert transform of a constant is zero (see 15.1.1).

A periodic function can also be written in *complex* form

$$x_p(t) = \sum_{n=-\infty}^{\infty} \alpha_n e^{jn\omega_o t}$$

Therefore,

$$v_p(t) = \mathrm{H}\{x_p(t)\} = \sum_{n=-\infty}^{\infty} \alpha_n \mathrm{H}\{e^{jn\omega_o t}\} = \sum_{n=-\infty}^{\infty} -j\,\mathrm{sgn}(n)e^{jn\omega_o t}$$

15.5 Hilbert Transform Properties and Pairs

15.5.1 Hilbert Transform Properties

TABLE 15.1 Properties of the Hilbert transformation

No.	Name	Original or Inverse Hilbert Transform	Hilbert Transform
1	Notations	$x(t)$ or $\mathrm{H}^{-1}[\upsilon]$	$\upsilon(t)$ or $\hat{x}(t)$ or $\mathrm{H}[\upsilon]$
2	Time domain definitions	$x(t) = \frac{1}{\pi}\int_{-\infty}^{\infty}\frac{\upsilon(\eta)}{\eta - t}d\eta$; $x(t) = \frac{-1}{\pi t} * \upsilon(t)$ or	$\upsilon(t) = \frac{-1}{\pi}\int_{-\infty}^{\infty}\frac{x(\eta)}{\eta - t}d\eta$; $\upsilon(t) = \frac{1}{\pi t} * x(t)$
3	Change of symmetry	$x(t) = x_{1e}(t) + x_{2o}(t)^*$;	$\upsilon(t) = \upsilon_{1o}(t) + \upsilon_{2e}(t)$
4	Fourier spectra	$x(t) = \overset{F}{\Longleftrightarrow} X(\omega) = X_e(\omega) + jX_o(\omega)$;	$\upsilon(t) = \overset{F}{\Longleftrightarrow} V(\omega) = V_e(\omega) + jV_o(\omega)$
		$X(\omega) = j\,\mathrm{sgn}(\omega)\,V(\omega)$;	$V(\omega) = -j\,\mathrm{sgn}(\omega)\,X(\omega)$
	For even functions the Hilbert transform is odd:		
		$X_e(\omega) = 2\int_o^{\infty} x_{1e}(t)\cos(\omega t)\,dt$	$\upsilon_o(t) = 2\int_o^{\infty} X_e(\omega)\sin(\omega t)\,df$
	For odd functions the Hilbert transform is even:		
		$X_o(\omega) = -2\int_o^{\infty} x_{2o}(t)\sin(\omega t)\,dt$	$\upsilon_e(t) = 2\int_o^{\infty} X_o(\omega)\cos(\omega t)\,df$
5	Linearity	$ax_1(t) + bx_2(t)$	$a\upsilon_1(t) + b\upsilon_2(t)$
6	Scaling and	$x(at);\ a > 0$	$\upsilon(at)$
	time reversal	$x(-at)$	$-\upsilon(-at)$
7	Time shift	$x(t-a)$	$\upsilon(t-a)$
8	Scaling and time shift	$x(bt-a)$	$\upsilon(bt-a)$
			<u>Fourier image</u>
9	Iteration	$\mathrm{H}[x(t)] = \upsilon(t)$	$-j\,\mathrm{sgn}(\omega)\,X(\omega)$
		$\mathrm{H}[\mathrm{H}[x]] = -x(t)$	$[-j\,\mathrm{sgn}(\omega)]^2 X(\omega)$
		$\mathrm{H}[\mathrm{H}[\mathrm{H}[x]]] = -\upsilon(t)$	$[-j\,\mathrm{sgn}(\omega)]^3 X(\omega)$
		$\mathrm{H}[\mathrm{H}[\mathrm{H}[\mathrm{H}[x]]]] = x(t)$	$[-j\,\mathrm{sgn}(\omega)]^4 X(\omega)$

e = even; o = odd

TABLE 15.1 Properties of the Hilbert transformation (continued)

No.	Name	Original or Inverse Hilbert Transform	Hilbert Transform
			First option
10	Time derivatives	$\dot{x}(t) = \frac{-1}{\pi t} * \dot{\upsilon}(t)$	$\dot{\upsilon}(t) = \frac{1}{\pi t} * \dot{x}(t)$
			Second option
		$\dot{x}(t) = \left[\frac{d}{dt}\frac{-1}{\pi t}\right] * \upsilon(t)$	$\dot{\upsilon}(t) = \left[\frac{d}{dt}\frac{1}{\pi t}\right] * x(t)$
11	Convolution	$\begin{cases} x_1(t) * x_2(t) = \\ -\upsilon_1(t) * \upsilon_2(t) \end{cases}$	$x_1(t) * \upsilon_2(t) = \upsilon_1(t) * x_2(t)$
12	Autoconvolution equality	$\int x(\tau)x(t-\tau)\,d\tau = -\int \upsilon(\tau)\upsilon(t-\tau)\,d\tau$ for $\tau = 0$ energy equality	
13	Multiplication by t	$t x(t)$	$t\upsilon(t) - \int_{-\infty}^{\infty} x(\tau)\,d\tau$
14	Multiplication of signals with non-overlapping spectra	$x_1(t)$ (low-pass signal)	$x_2(t)$ (high-pass signal)
		$x_1(t)x_2(t)$	$x_1(t)\upsilon_2(t)$
15	Analytic signal	$\psi(t) = x(t) + j\mathsf{H}[x(t)]$	$\mathsf{H}[\psi(t)] = -j\,\psi(t)$
16	Product of analytic signals	$\psi(t) = \psi_1(t)\psi_2(t)$	$\mathsf{H}[\psi(t)] = \psi_1(t)\mathsf{H}[\psi_2(t)]$ $= \mathsf{H}[\psi_1(t)]\psi_2(t)$
17	Nonlinear transformations	$x(x)$	$\upsilon(x)$
17a	$y = \frac{c}{bt+a}$	$x_1(t) = x\left[\frac{c}{bt+a}\right]$	$\upsilon_1(t) = \upsilon\left[\frac{c}{bt+a}\right] - \frac{1}{\pi}P\int_{-\infty}^{\infty}\frac{x(t)}{t}dt$
17 b	$y = a + \frac{b}{t}$	$x_1(t) = x\left[a + \frac{b}{t}\right]$	$\upsilon_1(t) = \frac{b}{a}\left\{\upsilon\left[a + \frac{b}{t}\right] - \upsilon(a)\right\}$
	Notice that the nonlinear transformation may change the signal $x(t)$ of finite energy to a signal $x_1(t)$ of infinite energy. P is the Cauchy Principal Value.		
18	Asymptotic value as $t \Rightarrow \infty$ for even functions of finite support:		
	$x_e(t) = x_e(-t)$		$\lim_{t\Rightarrow\infty}\lvert\upsilon_o(t)\rvert = \frac{1}{\pi t}\int_S x_e(t)\,dt$ [a]

[a] S is support of $x_e(t)$

15.5.2 Iteration

- Iteration of the HT two times yields the original signal with reverse sign.
- Iteration of the HT four times restores the original signal
- In Fourier frequency domain, n-time iteration translates the n-time multiplication by $-j\text{sgn}(\omega)$

15.5.3 Parseval's Theorem

$$v(t) = \mathsf{H}\{x(t)\}$$

$$\mathsf{F}\{\upsilon(t)\} = V(\omega) = -j\,\mathrm{sgn}(\omega)X(\omega)$$

$$|V(\omega)|^2 = |-j\,\mathrm{sgn}(\omega)X(\omega)|^2 = |X(\omega)|^2$$

since

$$E_x = \int_{-\infty}^{\infty} x^2(t)\,dt = \int_{-\infty}^{\infty} |X(\omega)|^2\,df = \text{energy of } x(t)$$

$$E_\upsilon = \int_{-\infty}^{\infty} |V(\omega)|^2\,df = \int_{-\infty}^{\infty} |X(\omega)|^2\,df = E_x$$

15.5.4 Orthogonality

$$\int_{-\infty}^{\infty} \upsilon(t)x(t)\,dt = 0$$

15.5.5 Fourier Transform of the Autoconvolution of the Hilbert Pairs

$$\mathsf{F}\{x(t) * x(t)\} = X^2(\omega)$$

$$\mathsf{F}\{\upsilon(t) * \upsilon(t)\} = [-j\,\mathrm{sgn}(\omega)X(\omega)]^2 = -X^2(\omega)$$

$$x(t) * x(t) = \int_{-\infty}^{\infty} x(\tau)x(t-\tau)\,d\tau = -\int_{-\infty}^{\infty} \upsilon(\tau)\upsilon(t-\tau)\,d\tau = -\upsilon(t) * \upsilon(t)$$

$$x_1(t) * x_2(t) = -\upsilon_1(t) * \upsilon_2(t)$$

15.5.6 Hilbert Transform Pairs

TABLE 15.2 Selected Useful Hilbert Pairs

No.	Name	Function	Hilbert Transform
1	sine	$\sin(\omega t)$	$-\cos(\omega t)$
2	cosine	$\cos(\omega t)$	$\sin(\omega t)$
3	Exponential	$e^{j\omega t}$	$-j\,\mathrm{sgn}(\omega)e^{j\omega t}$
4	Square pulse	$\prod_{2a}(t)$	$\frac{1}{\pi}\ln\left\|\frac{t+a}{t-a}\right\|$
5	Bipolar pulse	$\prod_{2a}(t)\mathrm{sgn}(t)$	$-\frac{1}{\pi}\ln\left\|1-(a/t)^2\right\|$
6	Double triangle	$t\prod_{2a}(t)\mathrm{sgn}(t)$	$-\frac{1}{\pi}\ln\left\|1-(a/t)^2\right\|$
7	Triangle, tri(t)	$1-\|t/a\|,\ \|t\| \le a$	$\frac{-1}{\pi}\left\{\ln\left\|\frac{t-a}{t+a}\right\| + \frac{t}{a}\ln\left\|\frac{t^2}{t^2-a^2}\right\|\right\}$
		$0,\ \|t\| > a$	
8	One-sided triangle		$\frac{1}{\pi}\left\{(1-t/a)\ln\left\|\frac{t}{t-a}\right\| + 1\right\}$

TABLE 15.2 Selected Useful Hilbert Pairs (continued)

No.	Name	Function	Hilbert Transform
9	Trapezoid		$\frac{-1}{\pi}\left\{\frac{b}{b-a}\ln\left\|\frac{(a+t)(b-t)}{(a-t)(b+t)}\right\| + \frac{t}{b-a}\ln\left\|\frac{a^2-t^2}{b^2-t^2}\right\| + \ln\left\|\frac{(a-t)}{(a+t)}\right\|\right\}$
10	Cauchy pulse	$\frac{a}{a^2+t^2}$	$\frac{t}{a^2+t^2}$
11	Gaussian pulse	$e^{-\pi t^2}$	$2\int_0^\infty e^{-\pi f^2}\sin(\omega t)\,df;\ \omega = 2\pi f$
12	Parabolic pulse	$1-(t/a)^2,\ \|t\|\le a$	$\frac{-1}{\pi}\left\{\left[1-(t/a)^2\right]\ \ln\left\|\frac{t-a}{t+a}\right\| - \frac{2t}{a}\right\}$
13	Symmetric exponential	$e^{-a\|t\|}$	$2\int_0^\infty \frac{2a}{a^2-\omega^2}\sin(\omega t)\,df$
14	Antisymmetric exponential	$\mathrm{sgn}(t)e^{-a\|t\|}$	$-2\int_0^\infty \frac{2a}{a^2-\omega^2}\cos(\omega t)\,df$
15	One-sided exponential	$1(t)e^{-a\|t\|}$	$2\int_0^\infty \frac{a\sin(\omega t)-\omega\cos(\omega t)}{a^2-\omega^2}\,df$
16	Sinc pulse	$\frac{\sin(at)}{at}$	$\frac{\sin^2(at/2)}{(at/2)} = \frac{1-\cos(at)}{at}$
17	Video test pulse	$\begin{cases}\cos^2(\pi t/2a); & \|t\|\le a\\ 0, & \|t\|>a\end{cases}$	$2\int_0^\infty \frac{2a^2}{4a^2-\omega^2}\frac{\sin[\pi\omega/(2a)]}{\omega}\sin(\omega t)\,df$
18	{Spectra of $a(t)$ and $\cos(\omega_0 t)$ overlapping	$a(t)\cos(\omega_0 t)$	$\left[a(t) * \frac{\sin(\omega_0 t)}{\pi t}\right]\sin(\omega_0 t) + \left[a(t) * \frac{\cos(\omega_0 t)}{\pi t}\right]\cos(\omega_0 t)$
19	Bedrosian's theorem	$a(t)\cos(\omega_0 t)$	$a(t)\sin(\omega_0 t)$
20	A constant	a	zero

Hyperbolic Functions: Approximation by Summation of Cauchy Functions (see Hilbert Pairs No. 10 and 45)

No.	Name	Function	Hilbert Transform
21	Tangent hyp.	$\tanh(t) = 2\sum_{\eta=0}^{\infty}\frac{t}{(\eta+0.5)^2\pi^2+t^2}$	$-2\pi\sum_{\eta=0}^{\infty}\frac{(\eta+0.5)}{(\eta+0.5)^2\pi^2+t^2}$
22	Part of finite energy of tanh	$\mathrm{sgn}(t)-\tanh(t)$	$\pi\delta(t)+2\pi\sum_{\eta=0}^{\infty}\frac{(\eta+0.5)}{(\eta+0.5)^2\pi^2+t^2}$
23	Cotangent hyp.	$\coth(t) = \frac{1}{t}+2\sum_{\eta=1}^{\infty}\frac{t}{(\eta\pi)^2+t^2}$	$-\pi\delta(t)+2\pi\sum_{\eta=1}^{\infty}\frac{\eta}{(\eta\pi)^2+t^2}$
24	Secans hyp.	$\mathrm{sech}(t) = -2\pi\sum_{\eta=0}^{\infty}(-1)^{(\eta-1)}\frac{(\eta+0.5)}{(\eta+0.5)^2\pi^2+t^2}$	$-2\sum_{\eta=0}^{\infty}(-1)^{(\eta-1)}\frac{t}{(\eta+0.5)^2\pi^2+t^2}$
25	Cosecans hyp.	$\mathrm{cosech}(t) = \frac{1}{t}-2\sum_{\eta=1}^{\infty}(-1)^{(\eta-1)}\frac{t}{(\eta\pi)^2+t^2}$	$-\pi\delta(t)+2\pi\sum_{\eta=1}^{\infty}(-1)^{(\eta-1)}\frac{\eta}{(\eta\pi)^2+t^2}$

TABLE 15.2 Selected Useful Hilbert Pairs (continued)

No.	Name	Function	Hilbert Transform
		Hyperbolic Functions by Inverse Fourier Transformation; $\omega = 2\pi f$	
26		$\operatorname{sgn}(t) - \tanh(at/2)$	$2\int_0^{\infty}\left[\frac{2\pi}{a\sinh(\pi\omega/a)} - \frac{2}{\omega}\right]\cos(\omega t)\,df$
		$\operatorname{Re} a > 0$	
27		$\coth(t) - \operatorname{sgn}(t)$	$2\int_0^{\infty}\left[\frac{2\pi}{a}\coth(\pi\omega/a) - \frac{2}{\omega}\right]\cos(\omega t)\,df$
28		$\operatorname{sec h}(at/2)$	$2\int_0^{\infty}\frac{2\pi}{a\cosh(\pi\omega/(2a))}\sin(\omega t)\,df$
29		$\operatorname{cos ech}(at/2)$	$-2\int_0^{\infty}\frac{2\pi}{a}\tanh(\pi\omega/(2a))\cos(\omega t)\,df$
30		$\operatorname{sec h}^2(at/2)$	$2\int\frac{2\pi\omega}{a\sinh(\pi\omega/(2a))}\sin(\omega t)\,df$
		Delta Distribution, $1/(\pi t)$ Distribution and its Derivatives: Derivation Using Successive Iteration and	
		Differentiation	Iteration
		If $x(t) \overset{H}{\Longleftrightarrow} v(t)$ then $\dot{x}(t) \overset{H}{\Longleftrightarrow} \dot{v}(t)$	$\mathrm{H}[v(t)] = \mathrm{HH}[u(t)] = -x(t)$
No.	**Operation**	$x(t)$	$v(t)$
31		$\delta(t)$	$1/(\pi t)$
32	Iteration	$1/(\pi t)$	$-\delta(t)$
33	Differentiation	$\dot{\delta}(t)$	$-1/(\pi t^2)$
34	Iteration	$1/(\pi t^2)$	$\dot{\delta}(t)$
35	Differentiation	$\ddot{\delta}(t)$	$2/(\pi t^3)$
36	Iteration	$1/(\pi t^3)$	$-0.5\ddot{\delta}(t)$
37	Differentiation	$\dddot{\delta}(t)$	$-6/(\pi t^4)$
38	Iteration	$1/(\pi t^4)$	$(1/6)\dddot{\delta}(t)$
39		$x(t)\delta(t)$	$x(0)/(\pi t)$
		The procedure could be continued.	
		Equality of Convolution	
40		$\delta(t) * \delta(t) * \delta(t)$	$\frac{1}{\pi t} * \frac{1}{\pi t} = -\delta(t)$
41		$\dot{\delta}(t) * \delta(t) = \dot{\delta}(t)$	$\frac{1}{\pi t^2} * \frac{1}{\pi t} = \dot{\delta}(t)$
42		$\dot{\delta}(t) * \dot{\delta}(t) = \ddot{\delta}(t)$	$\frac{1}{\pi t^2} * \frac{1}{\pi t^2} = -\ddot{\delta}(t)$
43		$\dddot{\delta}(t) * \delta(t) = \dddot{\delta}(t) = \ddot{\delta}(t) * \dot{\delta}(t)$	$\frac{6}{\pi t^4} * \frac{1}{\pi t} = \dddot{\delta}(t) = \frac{2}{\pi t^3} * \frac{1}{\pi t^2}$
		Approximating Functions of Distributions (see No. 31 to 37 of this table)	
		$x(t)$	$v(t)$
44		$\int \delta(a,t)\,dt = \frac{1}{\pi}\tan^{-1}(t/a)$	$\int \theta(a,t)\,dt = \frac{\ln(a^2 + t^2)}{2\pi}$

TABLE 15.2 Selected Useful Hilbert Pairs (continued)

No.	Name	Function	Hilbert Transform
45		$\delta(a,t)=\frac{1}{\pi}\frac{a}{a^2+t^2}$	$\theta(a,t)=\frac{1}{\pi}\frac{t}{a^2+t^2}$
46		$\dot{\delta}(a,t)=\frac{1}{\pi}\frac{-2at}{(a^2+t^2)^2}$	$\dot{\theta}(a,t)=\frac{1}{\pi}\frac{a^2-t^2}{(a^2+t^2)^2}$
47		$\ddot{\delta}(a,t)=\frac{1}{\pi}\frac{6at^2-2a^2}{(a^2+t^2)^3}$	$\ddot{\theta}(a,t)=\frac{1}{\pi}\frac{2t^2-6at^2}{(a^2+t^2)^3}$
48		$\dddot{\delta}(a,t)=\frac{1}{\pi}\frac{24a^3t-24at^2}{(a^2+t^2)^4}$	$\dddot{\theta}(a,t)=\frac{1}{\pi}\frac{-6t^2+36a^2t^2-6a^4}{(a^2+t^2)^4}$

Derivation Using Successive Iteration and Differentiation (see the information above No. 31)

Trigonometric Expressions

	Operation	$x(t)$	$v(t)$
49		$\frac{\sin(at)}{t}$	$\frac{1-\cos(at)}{t}=\frac{2\sin^2(at/2)}{t}$
50	Iteration	$\frac{\cos(at)}{t}$	$-\pi\delta(t)+\frac{\sin(at)}{t}$
51	Differentiation	$\frac{\sin(at)}{t^2}$	$-a\pi\delta(t)+\frac{1-\cos(at)}{t^2}$
52	Iteration	$\frac{\cos(at)}{t^2}$	$\pi\dot{\delta}(t)-\frac{a}{t}+\frac{\sin(at)}{t^2}$
53	Differentiation	$\frac{\sin(at)}{t^3}$	$\pi a\dot{\delta}(t)-\frac{a^2}{2t}+\frac{1-\cos(at)}{t^3}$
54	Iteration	$\frac{\cos(at)}{t^3}$	$-\frac{\pi}{2}\ddot{\delta}(t)+\frac{\pi a^2}{2}\delta(t)-\frac{a}{t^2}+\frac{\sin(at)}{t^3}$

Selected Useful Hilbert Pairs of Periodic Signals

	Name	$x_p(t)$	$v_p(t)$
55	Sampling sequence	$\sum_{n=-\infty}^{\infty}\delta(t-nT)$	$\frac{1}{T}\sum_{n=-\infty}^{\infty}\cos[(\pi/T)(t-nT)]$
56	Even square wave	$\mathrm{sgn}[\cos(\omega t)],\ \omega=2\pi/T$	$(2/\pi)\ln\lvert\tan(\omega t/2+\pi/4)\rvert$
57	Odd square wave	$\mathrm{sgn}[\sin(\omega t)],\ \omega=2\pi/T$	$(2/\pi)\ln\lvert\tan(\omega t/2)\rvert$
58	Squared cosine	$\cos^2(\omega t)$	$0.5\sin(2\omega t)$
59	Squared sine	$\sin^2(\omega t)$	$-0.5\sin(2\omega t)$
60	Cube cosine	$\cos^3(\omega t)$	$\frac{3}{4}\sin(\omega t)+\frac{1}{4}\sin(3\omega t)$
61	Cube sine	$\sin^3(\omega t)$	$\frac{-3}{4}\cos(\omega t)+\frac{1}{4}\cos(3\omega t)$
62		$\cos^4(\omega t)$	$\frac{1}{2}\sin(2\omega t)+\frac{1}{8}\sin(4\omega t)$
63		$\sin^4(\omega t)$	$-\frac{1}{2}\sin(2\omega t)+\frac{1}{8}\sin(4\omega t)$
64		$\cos^5(\omega t)$	$\frac{5}{8}\sin(2\omega t)+\frac{5}{16}\sin(3\omega t)+\frac{1}{16}\sin(5\omega t)$
65		$\cos^6(\omega t)$	$\frac{15}{32}\sin(2\omega t)+\frac{6}{32}\sin(4\omega t)+\frac{1}{32}\sin(6\omega t)$
66		$\cos(at+\varphi)\cos(bt+\Psi)$ $0<a<b$ $\varphi,\Psi=$ constants	$\cos(at+\varphi)\sin(bt+\Psi)$

TABLE 15.2 Selected Useful Hilbert Pairs (continued)

No.	Name	Function	Hilbert Transform
67	Fourier Series	$X_o + \sum_{n=1}^{\infty} X_n \cos(n\omega t + \varphi_n)$	$\sum_{n=1}^{\infty} X_n \sin(n\omega t + \varphi_n)$
68	Any periodic function	x_T = generating function	$x_T(t) \cdot \frac{1}{T} \sum_{k=-\infty}^{\infty} \cot[(\pi / T)(t - kT)]$
		$x_T(t) * \sum_{k=-\infty}^{\infty} \delta(t - kT)$	

15.6 Differentiation of Hilbert Pairs

15.6.1 Differentiation Pairs

$$\mathsf{H}\{\dot{x}(t)\} = \dot{v}(t)$$

$$\mathsf{H}\left\{\frac{d^n x(t)}{dt^n}\right\} = \frac{d^n v(t)}{dt^n}$$

Example

$$\mathsf{H}\{\delta(t)\} = \frac{1}{\pi t};\quad \mathsf{H}\{\dot{\delta}(t)\} = \frac{d}{dt}\left(\frac{1}{\pi t}\right) = -\frac{1}{\pi t^2}$$

15.6.2 Derivative of Convolution

$$\mathsf{H}\{x(t)\} = \mathsf{H}\left\{\frac{-1}{\pi t} * v(t)\right\} \Rightarrow v(t) = \frac{1}{\pi t} * x(t)$$

$$\mathsf{H}\{\dot{x}(t)\} = \mathsf{H}\left\{-\frac{d}{dt}\left(\frac{1}{\pi t}\right) * v(t)\right\} \Rightarrow \dot{v}(t) = \frac{d}{dt}\left(\frac{1}{\pi t}\right) * x(t) \quad \text{(see 15.6.1 and 15.5.5)}$$

$$= \left(-\frac{-1}{\pi t^2}\right) * x(t) = \frac{1}{\pi t^2} * v(t)$$

$$\mathsf{H}\{\dot{x}(t)\} = \mathsf{H}\left\{-\frac{1}{\pi t} * \dot{v}(t)\right\} \Rightarrow \dot{v}(t) = \frac{1}{\pi t} * \dot{x}(t)$$

15.6.3 Fourier Transform of Hilbert Transform

$$v(t) = \frac{1}{\pi t} * x(t), \qquad \mathsf{F}\{v(t)\} = -j\,\mathrm{sgn}(\omega) X(\omega)$$

$$\mathsf{F}\{\dot{v}(t)\} = j\omega[-j\,\mathrm{sgn}(\omega) X(\omega)] = \omega\,\mathrm{sgn}(\omega) X(\omega)$$

15.7 Hilbert Transform of Hermite Polynomials

15.7.1 Hermite Polynomials and their Hilbert Transform

$$H_n(t) = (-1)^n e^{t^2} \frac{d^n}{dt^n} e^{-t^2} \qquad n = 0,1,2,\cdots, \; -\infty < t < \infty$$

$$H_n(t) = 2t\, H_{n-1}(t) - 2(n-1) H_{n-2}(t) \qquad n = 1,2,\cdots$$

$$\mathsf{F}\{e^{-t^2}\} = \sqrt{\pi}\, e^{-\pi^2 f^2} = \sqrt{\pi}\, e^{-\omega^2/4}$$

$$v(t) = \mathsf{H}\{x(t)\} \doteq \mathsf{H}\{e^{-t^2}\} = \mathsf{F}^{-1}\{V(\omega)\} = \int_{-\infty}^{\infty} -j\,\mathrm{sgn}(\omega)\sqrt{\pi}\, e^{-\pi^2 f^2} e^{j\omega t}\, df$$

$$= 2\sqrt{\pi} \int_{-\infty}^{\infty} e^{-\pi^2 f^2} \sin \omega t\, df$$

$$\mathsf{H}\{2te^{-t^2}\} = -2\sqrt{\pi} \int_{-\infty}^{\infty} \omega e^{-\pi^2 f^2} \cos \omega t\, df$$

15.7.2 Table of Hilbert Transform of Hermite Polynomials

TABLE 15.3 Hilbert Transform of Weighted Hermite Polynomials [Notation: $x = \exp(-t^2)$]

	Hermite Polynomial	Hilbert Transform	Energy
n	$H_n x$	$\mathsf{H}(H_n x)$	E
0	$(1)x$	$2\sqrt{\pi}\int_0^{\infty} \exp(-\pi^2 f^2)\sin(\omega t)\, df$	$\sqrt{\pi/2}$
1	$(2t)x$	$-2\sqrt{\pi}\int_0^{\infty} \omega \exp(-\pi^2 f^2)\cos(\omega t)\, df$	$\sqrt{\pi/2}$
2	$(4t^2-2)x$	$-2\sqrt{\pi}\int_0^{\infty} \omega^2 \exp(-\pi^2 f^2)\sin(\omega t)\, df$	$3\sqrt{\pi/2}$
3	$(8t^3-12t)x$	$2\sqrt{\pi}\int_0^{\infty} \omega^3 \exp(-\pi^2 f^2)\cos(\omega t)\, df$	$15\sqrt{\pi/2}$
4	$(16t^4-48t^2+12)x$	$2\sqrt{\pi}\int_0^{\infty} \omega^4 \exp(-\pi^2 f^2)\sin(\omega t)\, df$	$105\sqrt{\pi/2}$
5	$(32t^5-160t^3+120t)x$	$-2\sqrt{\pi}\int_0^{\infty} \omega^5 \exp(-\pi^2 f^2)\cos(\omega t)\, df$	$945\sqrt{\pi/2}$
n	$H_n x = (-1)^n[2tH_{n-1}(t)$ $-2(n-1)H_{n-2}(t)]$	$(-1)^n 2\sqrt{\pi}\int_0^{\infty} \omega^n \exp(-\pi^2 f^2)\sin\left(\omega t + \frac{n\pi}{2}\right) df$	

Energy $= \int_{-\infty}^{\infty} x^2 H_n^2\, dt = \int_{-\infty}^{\infty} [\mathsf{H}(xH_n)]^2\, dt = 1\times 3\times 5\times\cdots\times(2n-1)\times\sqrt{\pi/2},\; n \geq 1$

15.7.3 Hilbert Transform of Orthonormal Hermite Functions (see Chapter 22)

$$h_n(t) = (2^n n!)^{-1/2}\,\pi^{-1/2}\,e^{-t^2}\,H_n(t) \qquad n = 0,1,2,\cdots$$

$$\mathsf{H}\{h_n(t)\} = \nu_n(t)$$

$$= \left[\frac{2(n-1)!}{n!}\right]^{1/2}\left[t\nu_{n-1}(t) - \frac{1}{\pi}\int_{-\infty}^{\infty} h_{n-1}(\tau)\,d\tau\right] - (n-1)\left[\frac{(n-2)!}{n!}\right]^{1/2}\nu_{n-2}(t)$$

15.7.4 Hilbert Transform of Orthonormal Hermite Functions

TABLE 15.4 Hilbert Transforms of Orthonormal Hermite Functions (Energy = 1).

Notations: $h_o(t), h_1(t), \cdots \Rightarrow h_o, h_1, \cdots;$ $\quad \upsilon_o(t), \cdots \Rightarrow \upsilon_o, \upsilon_1, \cdots$

$$g(t) = \int_0^{\infty} e^{-2\pi^2 f^2}\sin(2\pi f t)\,df; \qquad a = \pi^{-0.25}e^{-t^2/2}; \qquad b = \pi^{0.25}$$

Hermite Functions $h_n(t)$	Hilbert Transforms $\upsilon_n(t)$
	Recurrent Notation
$h_0 = a$	$\upsilon_0 = 2\sqrt{2}\ bg(t)$
$h_1 = \sqrt{2}\ th_0$	$\upsilon_1 = \sqrt{2}\left[t\upsilon_0 \frac{\sqrt{2}\,b}{\pi}\right]$
$h_2 = th_1 - \sqrt{1/2}\ h_0$	$\upsilon_2 = t\upsilon_1 - \sqrt{1/2}\ \upsilon_0$
$h_3 = \sqrt{2/3}\ [th_2 - h_1]$	$\upsilon_3 = \sqrt{2/3}\left[t\upsilon_2 - \frac{b}{\pi} - \upsilon_1\right]$
$h_4 = \sqrt{1/2}\ th_3 - \sqrt{3/4}\ h_2$	$\upsilon_4 = \sqrt{1/2}\ t\upsilon_3 - \sqrt{3/4}\ \upsilon_2$
$h_5 = \sqrt{2/5}\ th_4 - \sqrt{4/5}\ h_3$	$\upsilon_5 = \sqrt{2/5}\left[t\upsilon_4 - \frac{\sqrt{3}\,b}{2\pi}\right] - \sqrt{4/5}\ \upsilon_3$
$h_n = \sqrt{\frac{2(n-1)!}{n!}}\ th_{n-1} +$	$\upsilon_n = \sqrt{\frac{2(n-1)!}{n!}}\ [t\upsilon_{n-1}$
$(n-1)\sqrt{\frac{(n-2)!}{n!}}\ h_{n-2}$	$-\frac{1}{\pi}\int h_{n-1}(\tau)\,d\tau] - (n-1)\sqrt{\frac{(n-2)!}{n!}}\ \upsilon_{n-2}$
	Nonrecurrent Notation
$h_0 = a1$	$2\sqrt{2}\ bg(t)$
$h_1 = \sqrt{2}\ at$	$2b[2tg(t) - \pi^{-1}]$
$h_2 = \frac{a}{\sqrt{8}}(4t^2 - 2)$	$2b[(2t^2 - 1)\,g(t) - t\pi^{-1}]$
$h_3 = \frac{a}{\sqrt{48}}(8t^3 - 12t)$	$\sqrt{8/3b}\left[(2t^3 - 3t)\,g(t) - \frac{t^2}{\pi} + \frac{1}{2\pi}\right]$
$h_4 = \frac{a}{\sqrt{384}}(16t^4 - 48t^2 + 12)$	$\sqrt{4/3b}\left[(2t^4 - 6t^2 + 1.5)\,g(t) - \frac{t^3}{\pi} + \frac{2t}{2\pi}\right]$

TABLE 15.4 Hilbert Transforms of Orthonormal Hermite Functions (Energy = 1). (continued)

Notations: $h_o(t), h_1(t), \cdots \Rightarrow h_o, h_1, \cdots$; $\quad \upsilon_o(t), \cdots \Rightarrow \upsilon_o, \upsilon_1, \cdots$

$$g(t) = \int_0^\infty e^{-2\pi^2 f^2} \sin(2\pi f t)\, df; \qquad a = \pi^{-0.25} e^{-t^2/2}; \qquad b = \pi^{0.25}$$

Hermite Functions $h_n(t)$	Hilbert Transforms $\upsilon_n(t)$
$h_5 = \frac{a}{\sqrt{3840}}(32t^5 - 160t^3 + 120t)$	$\sqrt{8/15}\, b\left[(2t^5 - 10t^3 + 7.5)g(t) - \frac{(t^4 - 4t^2) + 1.75}{\pi}\right]$
$h_n(t) = \frac{a}{\sqrt{2^n n!}} H_n(t),$	$H_n(t) = 2tH_{n-1}(t) - 2(n-1)H_{n-2}(t)$

n	0	1	2	3	4	5	...
$\int_{-\infty}^{\infty} h_n(\tau)\, d\tau$	$\sqrt{2}\, b$	0	b	0	$\sqrt{3/4}\, b$	0	...

15.8 Hilbert Transform of Product of Analytic Signals

15.8.1 Hilbert Transform of Product of Analytic Signals:

From

$$\mathrm{H}\{\psi(t)\} = \mathrm{H}\{x(t) + j\upsilon(t)\} = \mathrm{H}\{x(t) + j\mathrm{H}\{x(t)\}\} = \mathrm{H}\{x(t)\} - jx(t)$$
$$= \upsilon(t) - jx(t) = -j(x(t) + j\upsilon(t)) = -j\psi(t)$$

we obtain $\mathrm{H}\{\psi_1(t)\psi_2(t)\} = -j\psi_1(t)\psi_2(t) = \psi_1(t)\mathrm{H}\{\psi_2(t)\} = \psi_2(t)\mathrm{H}\{\psi_1(t)\}$ since the product can be considered as an analytic function $\psi(t)$.

15.8.2 The nth Product of an Analytic Signal

$$\mathrm{H}\{\psi^2(t)\} = \psi(t)\mathrm{H}\{\psi(t)\} = -j\psi^2(t)$$

$$\mathrm{H}\{\psi^n(t)\} = \psi^{n-1}(t)\mathrm{H}\{\psi(t)\} = -j\psi^n(t)$$

Example

Because $\mathrm{H}\{(1 - jt)^{-1}\} = -j(1 - jt)^{-2}$, we obtain

$$\mathrm{H}\{(1 - jt)^{-2}\} = (1 - jt)^{-1}(-j(1 - jt)^{-1}) = -j(1 - jt)^{-2}$$

15.9 Hilbert Transform of Bessel Functions

15.9.1 Hilbert Transform of Bessel Function:

$$\mathrm{H}\{J_n(t)\} = \hat{J}_n(t) = \frac{1}{\pi}\int_0^\pi \sin(t\sin\varphi - n\varphi)\, d\varphi = \sum_{n=0}^{\infty} \frac{\hat{J}_n^{(n)}(t=0)}{n!} t^n$$

$$\hat{J}_0(t) = \frac{1}{\pi}\int_0^1 \frac{2}{(1-\omega^2)^{1/2}} \sin\omega t \, d\omega$$

$$\psi_0(t) = J_0(t) + j\hat{J}_0(t)$$

$$\hat{J}_0(0) = \frac{1}{\pi}\int_0^1 \frac{2\,d\omega}{(1-\omega^2)^{1/2}} \sin(0) = 0, \qquad \hat{J}_0^{(1)}(t) = \frac{1}{\pi}\int_0^1 \frac{2\omega\,d\omega}{(1-\omega^2)^{1/2}} \cos(0) = \frac{2}{\pi}$$

The parenthesis in the exponent indicates number of differentiations with respect to time.

15.9.2 Hilbert Transform Pairs of Bessel Functions:

TABLE 15.5 Hilbert Transform of Bessel Functions of the First Kind

Bessel Function	Fourier Transform	Hilbert Transform
$J_n(t)$	$C_n(f)$	$\hat{J}_n(t) = \mathrm{H}[J_n(t)]$
$J_0(t)$	$C_0 = \frac{2}{(1-\omega^2)^{0.5}};\ \lvert\omega\rvert < 1$ $= 0;\ \lvert\omega\rvert > 0$	$\frac{1}{\pi}\int_0^1 C_0(f)\sin(\omega t)\,d\omega$
$J_1(t)$	$C_1 = -j\omega C_0$	$-\frac{1}{\pi}\int_0^1 \lvert C_1(f)\rvert\cos(\omega t)\,d\omega$
$J_2(t)$	$C_2 = -(2\omega^2 - 1)C_0$	$-\frac{1}{\pi}\int_0^1 \lvert C_2(f)\rvert\sin(\omega t)\,d\omega$
$J_3(t)$	$C_3 = j(4\omega^3 - 3\omega)C_0$	$\frac{1}{\pi}\int_0^1 \lvert C_3(f)\rvert\cos(\omega t)\,d\omega$
$J_4(t)$	$C_4 = (8\omega^4 - 8\omega^2 + 1)C_0$	$\frac{1}{\pi}\int_0^1 \lvert C_4(f)\rvert\sin(\omega t)\,d\omega$
$J_5(t)$	$C_5 = -j(16\omega^5 - 20\omega^3 + 5\omega)C_0$	$-\frac{1}{\pi}\int_0^1 \lvert C_5(f)\rvert\cos(\omega t)\,d\omega$
$J_6(t)$	$C_6 = -(32\omega^6 - 48\omega^4 + 18\omega^2 - 1)C_0$	$-\frac{1}{\pi}\int_0^1 \lvert C_6(f)\rvert\sin(\omega t)\,d\omega$
$J_n(t)$	$C_n = (-j)^n 2^{n-1} T_n(\omega) C_0$	$\frac{(-1)^{n/2}}{\pi}\int_0^1 \lvert C_n(f)\rvert\sin(\omega t)\,d\omega$ for $n = 0,2,4,\ldots$ $\frac{(-1)^{(n+1)/2}}{\pi}\int_0^1 \lvert C_n(f)\rvert\cos(\omega t)\,d\omega$ for $n = 1,3,5,\ldots$

$T_n(\omega) = \cos[n\cos^{-1}(\omega)]$ is the Chebyshev polynomial

15.10 Instantaneous Amplitude, Phase, and Frequency

15.10.1 Instantaneous Angular Frequency

$$\psi(t) = x(t) + j\upsilon(t) = A(t)e^{j\varphi(t)} = A(t)\cos\varphi(t) + A(t)\sin\varphi(t)$$

$$A(t)=\sqrt{x^2(t)+\upsilon^2(t)}, \quad \varphi(t)=\tan^{-1}\frac{\upsilon(t)}{x(t)}$$

$$\dot{\varphi}(t)=\Omega(t)=2\pi F(t)\equiv \text{instantaneous angular frequency}$$

$$F(t)=\text{instantaneous frequency}=\frac{\Omega(t)}{2\pi}=\frac{\dot{\varphi}(t)}{2\pi}$$

$$\Omega(t)=\frac{d}{dt}\tan^{-1}\frac{\upsilon(t)}{x(t)}=\frac{x(t)\dot{\upsilon}(t)-\upsilon(t)\dot{x}(t)}{x^2(t)+\upsilon^2(t)}$$

15.11 Hilbert Transform and Modulation

15.11.1 Modulated Signal (see 15.10.1)

$$\psi(t)=A_o\,\gamma(t)e^{j\Phi_0}\,e^{j\Omega_0 t}$$

$$\psi_x(t)=x(t)+j\hat{x}(t)$$

$$x(t)=\frac{\psi_x(t)+\psi_x^*(t)}{2}$$

15.11.2 Instantaneous Amplitude and Angular Frequency (see 15.10.1)

$$A(t)=\frac{m}{2}\left|\psi_x(t)\right|=\frac{m}{2}[x^2(t)+\hat{x}^2(t)]^{1/2}$$

$$\omega_x(t)=\pm\frac{d}{dt}\tan^{-1}\left[\frac{\hat{x}(t)}{x(t)}\right]$$

15.11.3 High-Frequency Analytic Signals ($\Phi_0 = 0$)

$$\psi_{upper}(t)=\text{upper sideband}=\psi_x(t)e^{j\Omega_0 t}$$

$$\psi_{lower}(t)=\text{lower sideband}=\psi_x^*(t)e^{j\Omega_0 t}$$

$$x_{SSB}(t)=x(t)\cos\Omega_0 t\mp\hat{x}(t)\sin\Omega_0 t$$

where $x(t)\cos(\Omega_0 t)$ and $\hat{x}(t)\sin\Omega_0 t$ represent double sideband (DSB) compressed carrier AM signals.

15.12 Hilbert Transform and Transfer Functions of Linear Systems

15.12.1 Causal Systems

$$H(s)=A(\alpha,\omega)+jB(\alpha,\omega), \qquad \sigma=\alpha+j\omega$$

$$A(\omega)=-\frac{1}{\pi}P\int_{-\infty}^{\infty}\frac{B(\lambda)}{\lambda-\omega}d\lambda$$

$$B(\omega) = \frac{1}{\pi} P \int_{-\infty}^{\infty} \frac{A(\lambda)}{\lambda - \omega} d\lambda$$

15.12.2 Minimum Phase Transfer Function

$$H(j\omega) = H_{\varphi}(j\omega) H_{ap}(j\omega)$$

$$H_{\varphi}(j\omega) = \text{minimum phase transfer function}$$

$$H_{ap}(j\omega) = \text{all-pass transfer function}$$

$$H_{\varphi}(j\omega) = |H(j\omega)| e^{j\varphi(\omega)} = A_{\varphi}(\omega) + jB_{\varphi}(\omega)$$

$H_{\varphi}(j\omega)$ has all the zeros lying in the left half-plane of the s-plane. The minimum phase transfer function is analytic and its real and imaginary parts for a Hilbert pair

$$\mathsf{H}\{A(\omega)\} = -B_{\varphi}(\omega)$$

15.13 The Discrete Hilbert Filter

15.13.1 Discrete Hilbert Filter

$$H(k) = \begin{cases} -j & k = 1, 2, \cdots, \frac{N}{2} - 1 \\ 0 & k = 0 \text{ and } k = \frac{N}{2} \\ j & k = \frac{N}{2} + 1, \frac{N}{2} + 2, \cdots, N - 1 \end{cases} \qquad (N = \text{even})$$

$$H(k) = -j \operatorname{sgn}\left(\frac{N}{2} - k\right) \operatorname{sgn}(k), \qquad k = 0, 1, \cdots, N-1 \quad (N = \text{even})$$

15.13.2 Impulse Response of the Hilbert Filter

$$h(i) = \frac{1}{N} \sum_{k=0}^{N-1} H(k) e^{jw} = \frac{1}{N} \sum_{k=0}^{N-1} -j \operatorname{sgn}\left(\frac{N}{2} - k\right) \operatorname{sgn}(k) e^{jw}$$

$$= \frac{1}{N} \sum_{k=0}^{N-1} \sin(w) = \frac{2}{N} \sin^2\left(\frac{\pi i}{2}\right) \cot\left(\frac{\pi i}{N}\right), \qquad i = 0, 1, \cdots, N-1, \; w = \frac{2\pi k i}{N} \quad (N \text{ even})$$

15.13.3 DHT of a Sequence $x(i)$ in the Form of Convolution

$$\upsilon(i) = -x(i) \otimes h(i) = -x(i) \otimes \frac{2}{N} \sin^2\left(\frac{\pi i}{2}\right) \cot\left(\frac{\pi i}{N}\right), \qquad i = 0, 1, \cdots, N-1$$

$\otimes$ = circular convolution

$$\upsilon(i) = \sum_{r=0}^{N-1} h(i - r) x(r), \qquad i = 0, 1, \cdots, N-1 \quad (N \text{ even})$$

15.13.4 DHT of a Sequence $x(i)$ via DFT

$$\mathsf{F}_D\{x(i)\} = X(k)$$

$$V(k) = -j\,\mathrm{sgn}\left(\frac{N}{2} - k\right)\mathrm{sgn}(k)X(k)$$

$$\upsilon(i) = \mathsf{F}_D^{-1}\{V(k)\}, \qquad i,k = 0,1,2,\cdots,N-1 \ (N \text{ even})$$

$\mathsf{F}_D \equiv$ discrete Fourier transform, $\mathsf{F}_D^{-1} \equiv$ inverse discrete Fourier transform

15.13.5 Discrete Hilbert Filter when N is odd

$$H(k) = \begin{cases} -j & k = 1,2,\cdots,\frac{N-1}{2} \\ 0 & k = 0 \\ j & k = \frac{N}{2}+1,\frac{N}{2}+2,\cdots,N-1 \end{cases}$$

$$h(i) = \frac{2}{N}\sum_{k=1}^{(N-1)/2} \sin(2\pi ki/N), \qquad i = 0,1,\cdots,N-1$$

Also

$$h(i) = \frac{1}{N}\left[1 - \frac{\cos(\pi i)}{\cos(\pi i/N)}\cot\left(\frac{\pi i}{N}\right)\right]$$

15.14 Properties of Discrete Hilbert Transform

15.14.1 Parseval's Theorem

$$E\{x(i)\} = \sum_{i=0}^{N-1} |x(i)|^2 = \frac{1}{N}\sum_{k=0}^{N-1} |X(k)|^2$$

$$E\{x(i)\} \neq E\{\upsilon(i)\}$$

The reason is that the DC term (average value of $x(i)$) is eliminated in the DHT.

$$x_{DC} = \frac{1}{N}\sum_{i=0}^{N-1} x(i) = X(0)$$

15.14.2 Discrete Hilbert Transform

$$\mathsf{H}_D\{x_{AC}(i)\} = \upsilon(i)$$

$$x_{AC}(i) = x(i) - x_{DC}$$

where $x_{AC}(i)$ is the alternating part of $x(i)$.

15.14.3 Energies (powers) of x_{AC} and $\upsilon(i)$

$$\sum_{i=0}^{N-1} |x_{AC}(i)|^2 = \sum_{i=1}^{N-1} |\upsilon(i)|^2 + \frac{|X(N/2)|^2}{N} \qquad (N \text{ even})$$

where the special term $X\left(\frac{N}{2}\right)$ is zero, the two energies are equal.

Example

If $x(i) = \delta(i)$ and $N = 8$ we obtain (see 15.13.3)

$$\upsilon(i) = -\delta(i) \otimes \tfrac{1}{4}\sin^2(\pi i/2)\cot(\pi i/N)$$

Figure 15.1 shows the desired components and transforms. The $x_{DC} = 1/8 = 0.125$ and the energies are:

$$E\{x(i)\} = 1, \quad E\{x_{AC}(i)\} = 1 - \frac{1^2}{N} = 0.875, \text{ and } E\{\upsilon(i)\} = 1 - \frac{1^2}{N} - \frac{1^2}{N} = 1 - \frac{2}{8} = 0.75.$$

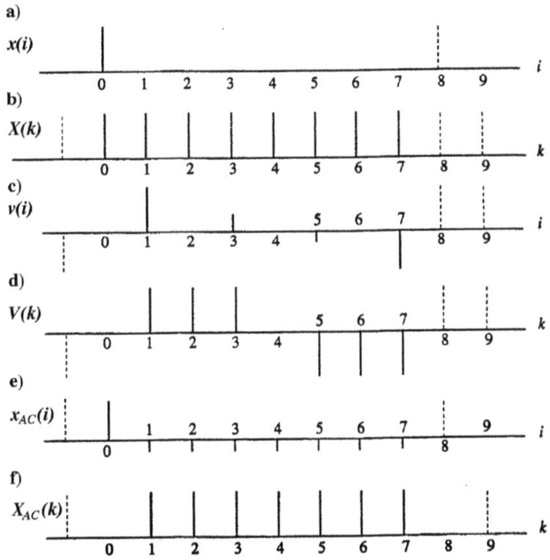

FIGURE 15.1 (a) The sequence $x(i)$ consisting of a single sample $\delta(i)$, (b) its spectrum $X(k)$ given by the DFT, (c) the samples of the discrete Hilbert transform, (d) the corresponding spectrum $V(k)$, (e) the samples of the AC component of $x(i)$, and (f) the corresponding spectrum $X_{AC}(k)$.

15.14.4 Shifting Property:

$$F_D\{x(i \pm m)\} = e^{\pm j2\pi mk/N} X(k)$$

See 15.13.4

$$\upsilon(i) = F_D^{-1}\{-j\,\text{sgn}\left(\frac{N}{2} - k\right)\text{sgn}(k)e^{\pm j2\pi mk/N} X(k)\}$$

15.14.5 Linearity:

$$\mathsf{H}_D\{ax_1(i)+bx_2(i)\}=a\upsilon_1(i)+b\upsilon_2(i)$$

15.14.6 Complex Analytic Discrete Sequence:

$$\psi(i)=x(i)+j\upsilon(i),\qquad \upsilon(i)=\mathsf{H}_D\{x(i)\}$$

$$\mathsf{H}_D\{\psi(i)\}=X(k)+j[-j\,\mathrm{sgn}\left(\frac{N}{2}-k\right)\mathrm{sgn}(k)]X(k),\qquad k=0,1,\cdots,N-1\quad (N\text{ even})$$

15.15 Hilbert Transformers (continuous)

15.15.1 Hilbert Transformer (quadratic filter)

$$H(jf)=\mathsf{F}\left\{\frac{1}{\pi t}\right\}=|H(f)|e^{j\varphi(f)}=-j\,\mathrm{sgn}\,f$$

$$H(jf)=\begin{cases}-j & f>0\\ 0 & f=0\\ j & f<0\end{cases}$$

$$\varphi(f)=\arg H(jf)=-\frac{\pi}{2}\mathrm{sgn}\,f$$

15.15.2 Phase-Splitter Hilbert Transformers

Analog Hilbert transformers are mostly implemented in the form of a phase splitter consisting of two parallel all-pass filters with a common input pot and separated output ports, each having the following transfer function respectively.

$$Y_1(jf)=e^{j\varphi_1(f)},\qquad Y_2(jf)=e^{j\varphi_2(f)}$$

with

$$\delta(f)=\varphi_1(f)-\varphi_2(f)=-\pi/2\quad\text{for all}\quad f>0$$

15.15.3 All-Pass Filters

$$H(j\omega)=\frac{R-jX(\omega)}{R+jX(\omega)}\qquad \omega=2\pi f$$

$$\varphi(\omega)=\arg\{(R-jX(\omega))^2\}=\tan^{-1}\left[\frac{-2RX(\omega)}{R^2-X^2(\omega)}\right]$$

See Figure 15.2a.

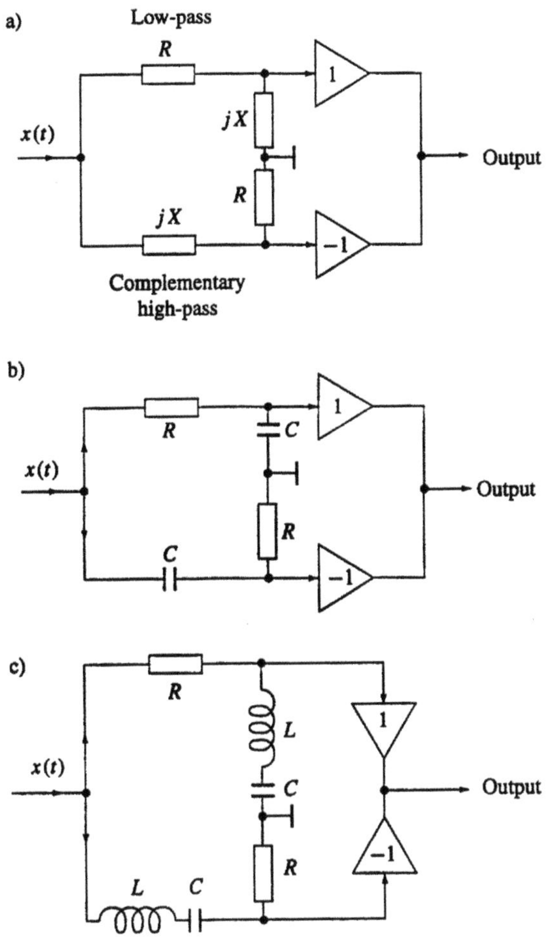

FIGURE 15.2 An all-pass consisting of (a) a low-pass and a complementary high-pass, (b) a first-order RC low-pass and complementary CR high-pass, and (c) a second-order RLC low-pass and complementary RLC high-pass.

If $X(\omega) = \dfrac{1}{\omega C}$, then (see Figure 15.2b)

$$\varphi(y) = \tan^{-1}\left[\frac{-2y}{1-y^2}\right], \qquad y = \omega RC = \omega\tau$$

If $X(\omega) = \omega L - 1/\omega C$, then (see Figure 15.2c)

$$\varphi(y) = \tan^{-1}\left[\frac{2(1-y^2)qy}{(1-y^2)^2 - q^2y^2}\right], \qquad y = \omega/\omega_r, \qquad \omega_r = 1/\sqrt{LC}$$

$$q = \omega_r RC = R\sqrt{C/L}$$

15.15.4 Design Hilbert Phase Splitters

Example

Filter with two first-order all-pass filters in each branch. The phase function for the first branch is (see Figure 15.3)

$$\varphi_1(f) = \tan^{-1}\left[\frac{-2y}{y^2-1}\right] + \tan^{-1}\left[\frac{-2ay}{a^2y^2-1}\right], \qquad y = 2\pi fRC$$

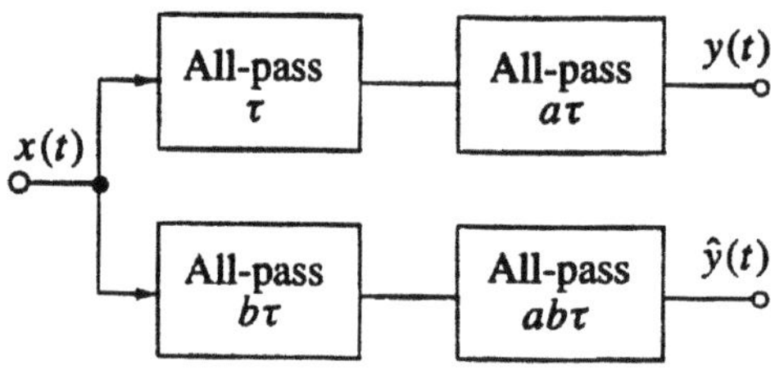

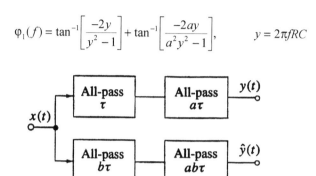

FIGURE 15.3 Phase Hilbert splitter with two all-pass filters.

Find a to get the best linearity of $\varphi_1(f)$ in the logarithmic scale. Small changes of a introduce a trade-off between the RMS phase error and the pass-band of the Hilbert transformer. Find shift parameter b to yield the minimum RMS phase error

$$\varphi_2(f) = \tan^{-1}\left[\frac{2by}{b^2y^2-1}\right] + \tan^{-1}\left[\frac{2aby}{a^2b^2y^2-1}\right]$$

Figure 15.4 shows an example with a = 0.08 and b = 0.24 giving the normalized edge frequencies $y_1 = 1.6$ and $y_2 = 30$ ($f_2/f_1 = 18.75$, or more than 4 octaves) with $\varepsilon_{RMS} = 0.016$.

15.16 Digital Hilbert Transformers

15.16.1 Digital Hilbert Transformers

Ideal discrete-time Hilbert transformer is defined as an all-pass with a pure imaginary transfer function.

$$H(e^{j\psi}) = H_r(\psi) + jH_i(\psi)$$

$$H_r(\psi) = 0 \qquad \text{for all } f$$

$$H(e^{j\psi}) = jH_i(\psi) = \begin{cases} -j & 0 < \psi < \frac{\pi}{2} \\ 0 & \psi = 0, |\psi| = \pi \\ j & -\pi < \psi < 0 \end{cases}$$

Equivalent Notation

$$H(e^{j\psi}) = -j\,\text{sgn}(\sin\psi) = -\text{sgn}(\sin\psi)e^{j\pi/2} = |H(\psi)|e^{j\arg H(\psi)}$$

$$|H(\psi)| = |\text{sgn}(\sin\psi)| = \begin{cases} 1 & 0 < |\psi| < \pi \\ 0 & \psi = 0, |\psi| = \pi \end{cases}$$

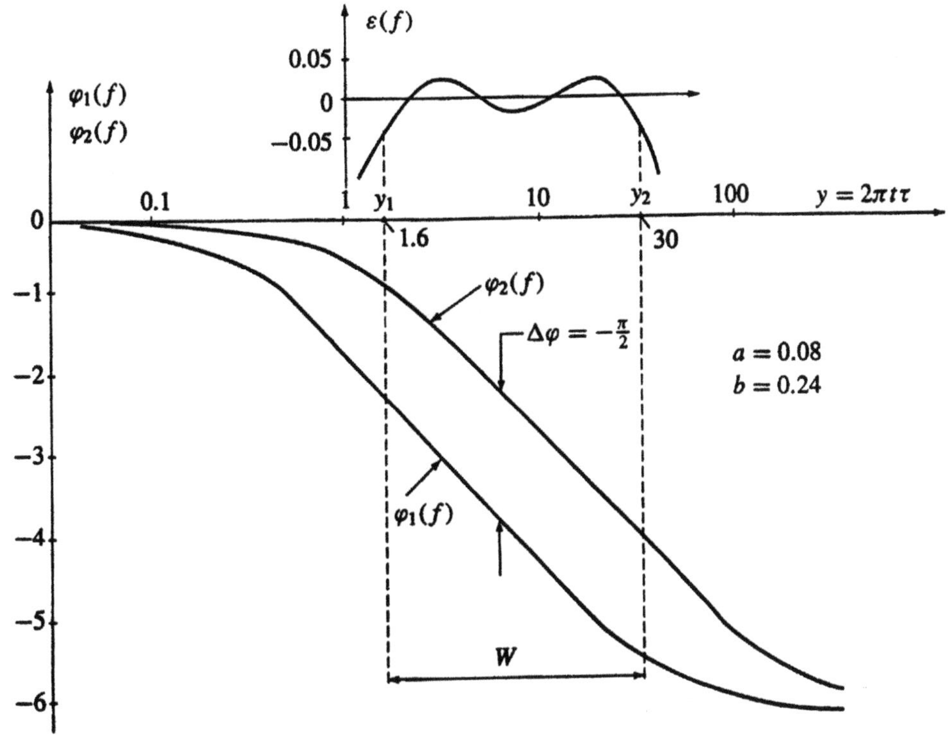

FIGURE 15.4 The phase functions and the phase error of the Hilbert transformer of Figure 15.3.

$$\arg[H(\psi)] = -\frac{\pi}{2}\operatorname{sgn}(\sin\psi)$$

$$\psi = 2\pi f_n, \qquad f_n = f/f_s, \qquad f_s = \text{sampling frequency}$$

Noncausal impulse response of the ideal Hilbert transformer is

$$h(i) = \frac{2}{\pi i}\sin^2\left(\frac{i\pi}{2}\right) \qquad i = 0, \pm 1, \pm 2, \cdots$$

15.16.2 Ideal Hilbert Transformer With Linear Phase Term

$$H(e^{j\psi}) = \begin{cases} -je^{j\psi\tau} & 0 < \psi < \pi \\ 0 & \psi = 0, \ |\psi| = \pi \\ je^{-j(\psi - 2\pi)\tau} & \pi < \psi < 2\pi \end{cases}$$

$$h(i) = \frac{2}{\pi}\frac{\sin^2\left[\frac{\pi}{2}(i-\tau)\right]}{i-\tau} \qquad i = 0, \pm 1, \pm 2, \cdots$$

$$h(i) = -h(-i) \qquad i = 0, 1, 2, \cdots$$

15.16.3 FIR Hilbert Transformers:

Figure 15.5 shows a noncausal impulse response Hilbert transformer and its truncated and shifted version so that a causal one is generated.

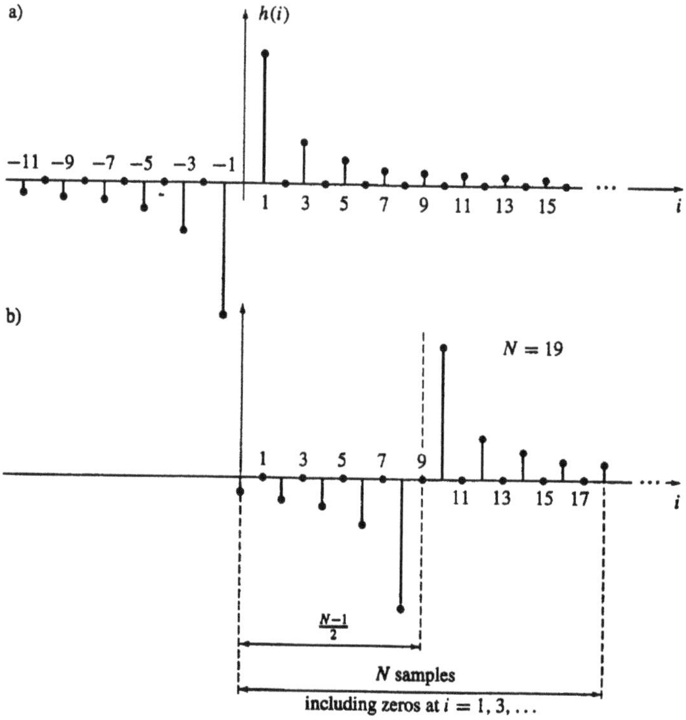

FIGURE 15.5 Impulse responses of (a) the ideal discrete time Hilbert transformer (see 15.16.1) and (b) a FIR Hilbert transformer given by the truncation and shifting of the impulse response shown in (a).

Causal Filter Impulse Response

$$H(i_1) = \sum_{i_1=0}^{N-1} h_1(i_1) z^{-i_1}$$

$$h_1\left(i + \frac{N-1}{2}\right) = h(i) \qquad i_1 = i + \tfrac{N-1}{2},\quad i = -\tfrac{N-1}{2}, \cdots 0, \cdots, \tfrac{N-1}{2}$$

Transfer function =

$$H(e^{j\psi}) = e^{-j\psi\frac{N-1}{2}} \sum_{i=-\frac{N-1}{2}}^{N-1} h(i) e^{-j\psi i} = e^{-j\psi\frac{N-1}{2}} \sum_{i=1}^{\frac{N-1}{2}} -j2h(i)\sin(\psi i), \quad \psi = \frac{2\pi f}{f_s}$$

Amplitude of Hilbert Transformer (see Figure 15.6)

$$G(e^{j\psi}) = -\sum_{i=1}^{\frac{N-1}{2}} 2h(i)\sin(\psi i)$$

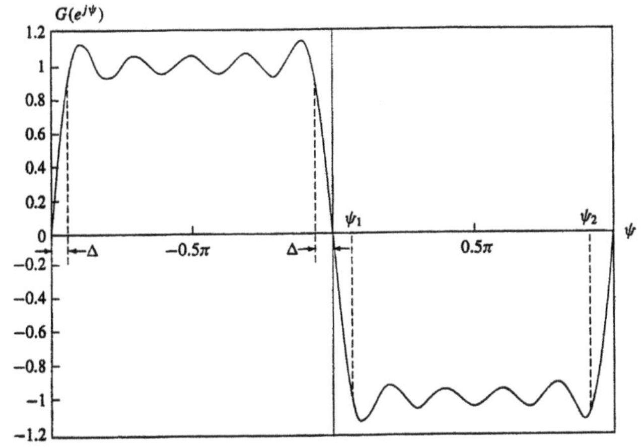

FIGURE 15.6 The $G(e^{j\psi})$ function of an FIR Hilbert transformer (amplitude).

Normalized Dimensionless Pass-band Hilbert Transformer

$$W_\psi = \psi_2 - \psi_1 = \pi - 2\Delta, \quad \psi_1, \psi_2 = \text{edge frequencies}$$

$$W_f[H_z] = \frac{\pi - 2\Delta}{2\pi} f_s$$

15.17 IIR Hilbert Transformers

15.17.1 IIR Ideal Hilbert Transformer (see Figure 15.7)

$$H_{HB}(z) = 1 + z^{-1} G(z^2) \equiv \text{ideal half-band filter} \quad \text{(see Figure 15.7a)}$$

$$G(z^2) = \text{all-pass filter with unit magnitude}$$

$$H_H(z) = z^{-1} G(-z^2) \equiv \text{ideal IIR Hilbert transformer}$$

$$F(z) = z^{-1} G(z^2), \quad z = e^{j\psi} \qquad \text{(see Figure 15.7b)}$$

$$F(e^{j\psi}) = e^{-j\psi} e^{j\Phi_G(\psi)} = e^{j\Phi(\psi)}$$

$$\Phi(\psi) = 0.5\pi[\mathrm{sgn}(\sin(2\psi)) - \mathrm{sgn}\,\psi]$$

$$\Phi_G(\psi) = \Phi(\psi) + \psi \qquad \text{(see Figure 15.7e)}$$

$$H_H(e^{j\psi}) = e^{-j\psi} e^{j\Phi_G(0.5\pi+\psi)}, \quad z^2 = e^{j2\psi}, \; -z^2 = e^{j2(0.5\pi+\psi)}$$

$$\arg\{z^{-1}G(-z^2)\} = -\psi + \Phi_G(0.5\pi + \psi) \qquad \text{(see Figure 15.7g)}$$

IIR Hilbert transformer has an equi-ripple phase function and exact amplitude. A noncausal transfer function may have the form

$$H(z) = z^{-1} \sum_{i=1}^{N} \frac{1 - a_i z^2}{z^2 - a_i}$$

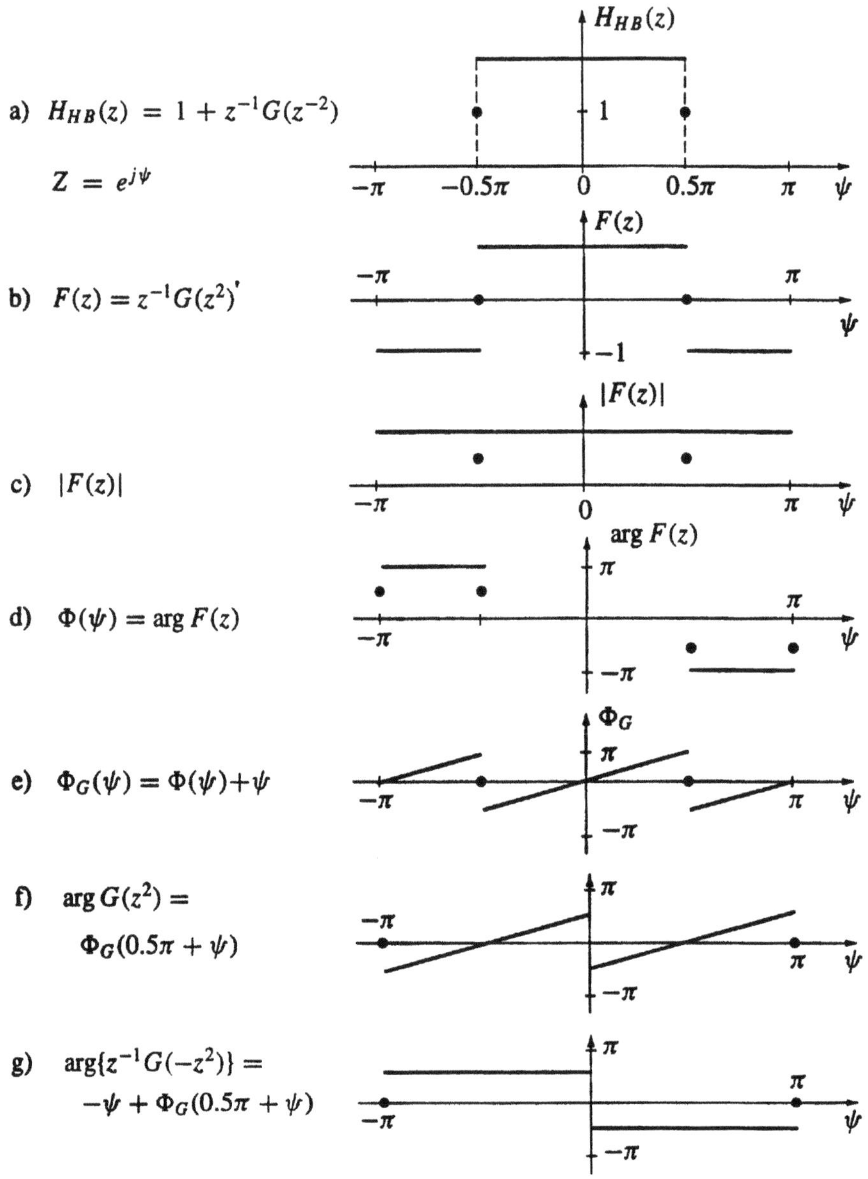

FIGURE 15.7 Step-by-step derivation of the IIR transfer function of a Hilbert transformer $Z^{-1}G(-z^2)$, starting from the transfer function of the ideal half-band filter given by $1 + Z^1G(z^2)$

Example

Let $\psi_1 = 0.02\pi \equiv$ low-frequency edge, $\psi_2 = 0.98\pi$ = high-frequency edge ($\Delta = 0.02\pi$), phase equiripple amplitude $|\Delta\Phi| \le 0.01\pi$. Because $\delta = \sin(0.5\Delta\Phi)$, $\delta = 0.0157$. Using the procedure from Ansari (1985), we find $a(1) = 5.36078$, $a(2) = 1.2655$, $a(3) = 0.94167$, and $a(4) = 0.53239$. Inserting a_i's, in $H(z)$ above, we find the phase function.

References

Ansari, R., IIR discrete-time Hilbert transformers, *IEEE Trans.*, ASSP-33, 1146-1150, 1985.
Erdelyi, A., *Tables of Integral Transform*, McGraw-Hill Book Co. Inc., New York, NY, 1954.
Hahn, Stefan L., Hilbert Transforms, in *Transforms and Applications Handbook*, Ed. Alexander D. Poularikas, CRC Press Inc., Boca Raton, FL, 1996.

16
The Radon and Abel Transform

16.1 The Radon Transform

16.1.1 Definition in two dimensions (see Figure 16.1)

$$\mathrm{R}\{f(\bar{r})\} = \breve{f}(\bar{\xi};p) = \iint_{-\infty}^{\infty} f(\bar{r})\delta(p - \bar{\xi}\cdot\bar{r})\,dx\,dy$$

$$\bar{\xi} = (\xi_1,\xi_2), \quad \bar{r} = (x_1,x_2)$$

$$\bar{\xi}\cdot\bar{r} = \xi_1 x + \xi_2 y = x\cos\phi + y\cos\phi \equiv \text{line in the } (x,y) \text{ plane}$$

Radon space = surface of a half-cylinder (see Figure 16.2)

$$p = \xi_1 x + \xi_2 y$$

$$\bar{\xi} = \text{unit vector} = (\cos\phi, \sin\phi)$$

$$\bar{r} = \text{vector}$$

16.1.2 Other Interpretation

$$M(p,\bar{\xi})\iint_{\bar{\xi}\cdot\bar{r}<p} f(x,y)\,dx\,dy = \iint f(x,y)u(p - \bar{\xi}\cdot\bar{r})\,dx\,dy$$

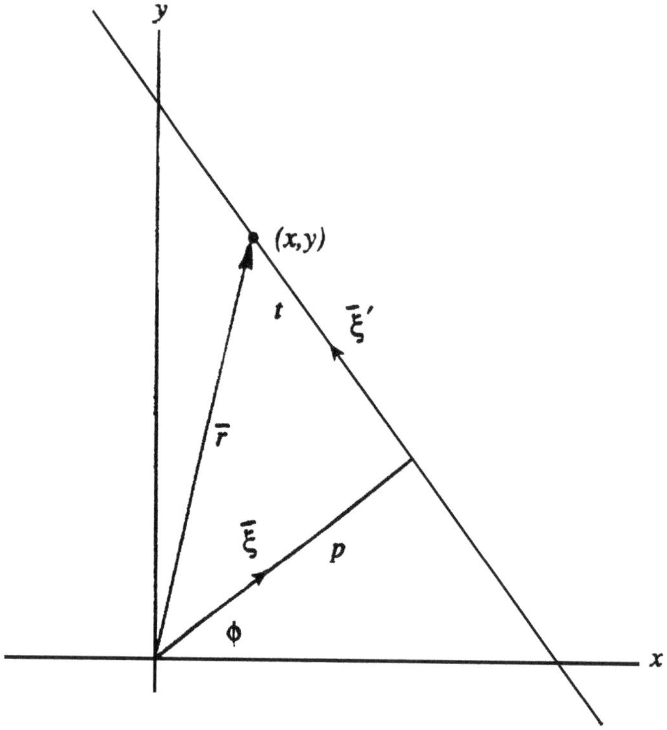

FIGURE 16.1 Coordinates in feature space used to define the Radon transform. The equation of the line is given by $p = x\cos\phi + y\sin\phi$.

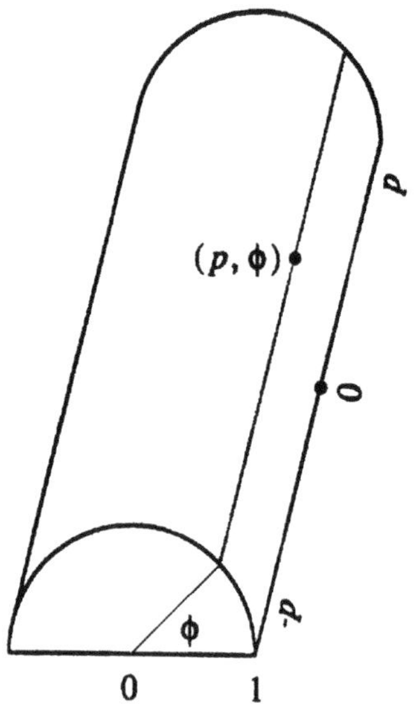

FIGURE 16.2 Coordinates in Radon space on the surface of a cylinder.

$$u(\cdot) = \text{unit step function}$$

$$\frac{\partial u(p)}{\partial p} = \delta(p)$$

$$M(p,\bar{\xi}) = \text{total mass in the region } \bar{\xi}\cdot\bar{r} < p$$

$$\frac{\partial M(p,\bar{\xi})}{\partial p} = \iint_{-\infty}^{\infty} f(\bar{r})\delta(p-\bar{\xi}\cdot\bar{r})\,dx\,dy = \mathcal{R}\{f(x,y)\}$$

16.1.3 Sample of Radon Transform

If the transform is found for only selected values of these variables, we call the result a *sample* of the Radon transform.

Example (see Figure 16.3)

The equation of line is $x = p$ and $\phi = 0$. Hence (see 16.1.2)

$$\check{f}(p) = \frac{\partial M(p)}{\partial p} = \frac{\partial}{\partial p}\left[2\int_0^p \sqrt{1-x^2}\,dx\right] = 2\sqrt{1-p^2}$$

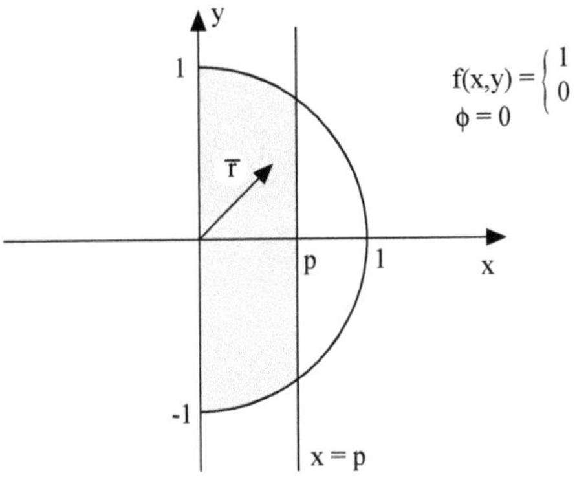

FIGURE 16.3 The sample Radon transform.

16.1.4 Rotated Coordinate System (see Figure 16.4)

$$\begin{bmatrix} x \\ y \end{bmatrix} = \begin{bmatrix} \cos\phi & -\sin\phi \\ \sin\phi & \cos\phi \end{bmatrix}\begin{bmatrix} p \\ t \end{bmatrix} \equiv \text{transformation}$$

$$\mathcal{R}\{f(x,y)\} = \check{f}(p,\phi) = \int_{-\infty}^{\infty} f(p\cos\phi - t\sin\phi,\ p\sin\phi + t\cos\phi)\,dt \quad \text{(integration along } t\text{)}$$

$$\check{f}_\phi(p) = \int_{-\infty}^{\infty} f_\phi(p,t) \equiv \text{integral of } f_\phi(p,t) \text{ with respect to } t \text{ for fixed } \phi;$$

$$f_\phi(p,t) = f(p\cos\phi - t\sin\phi,\ p\sin\phi + t\cos\phi)$$

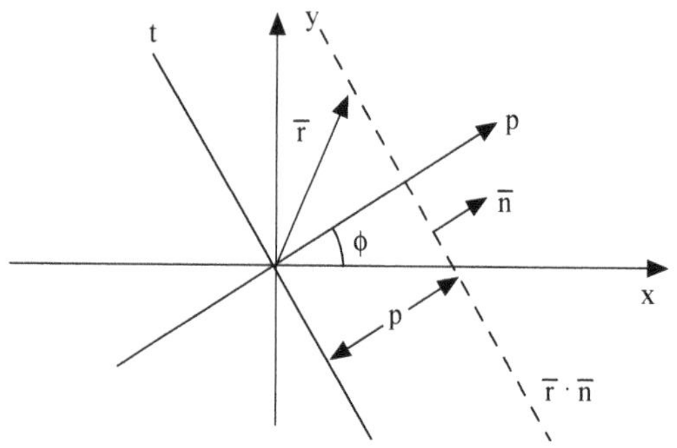

FIGURE 16.4 Rotated coordinate system.

Functions of p for various values of $\phi = \Phi$ are called *projections* of $f(x,y)$ at an angle Φ.

Example (see Figure 16.5)

The function is $f(x,y) = 1$, $0 \le r \le 1$, (or $f(x,y) = cyl_1(r)$) and zero otherwise. When $p = 1$, the length $t = 0$ and when $p = p$, $t = 2\sqrt{1-p^2}$. As ϕ varies, t assumes values in the range $-1 \le p \le p$. Hence

$$\check{f}(p,\phi) = \mathsf{R}\{f(x,y)\} = \int_0^{2\sqrt{1-p^2}} dt = \begin{cases} 2\sqrt{1-p^2} & |p| \le 1 \\ 0 & \text{otherwise} \end{cases}$$

The function $\check{f}(p,\phi)$ is plotted in Figure 16.5.

16.2 Transforms Between Spaces

16.2.1 Central-Slice Theorem

If $f(\bar{r})$ is a function of n variables

$$\mathsf{F}_1\mathsf{R}\{f\} = \mathsf{F}_1\{\check{f}\} = \mathsf{F}_n\{f\} = F(u,\upsilon) = F(q,\phi)$$

Example

$$\mathsf{F}\{f(x,y) = e^{-x^2-y^2}\} = F(u,\upsilon) = \int_0^\infty dr\, r e^{-r^2} \int_0^{2\pi} d\theta\, e^{[-j2\pi qr\cos(\theta-\phi)]}$$

where we let $x = r\cos\theta$, $y = r\sin\theta$, $u = q\cos\phi$, and $\upsilon = q\sin\phi$. The integral over θ is $2\pi J_0(2\pi qr)$. The remaining integral is a Hankel transform of order zero. Hence $F(q,\phi) = 2\pi\int_0^\infty re^{-r^2} J_0(2\pi qr)dr = \pi e^{-\pi^2 q^2}$ and therefore

$$\check{f}(p,\xi) = \pi\int_{-\infty}^{\infty} e^{-\pi^2 q^2} e^{j2\pi qp}\, dq = \sqrt{\pi}\, e^{-p^2}.$$

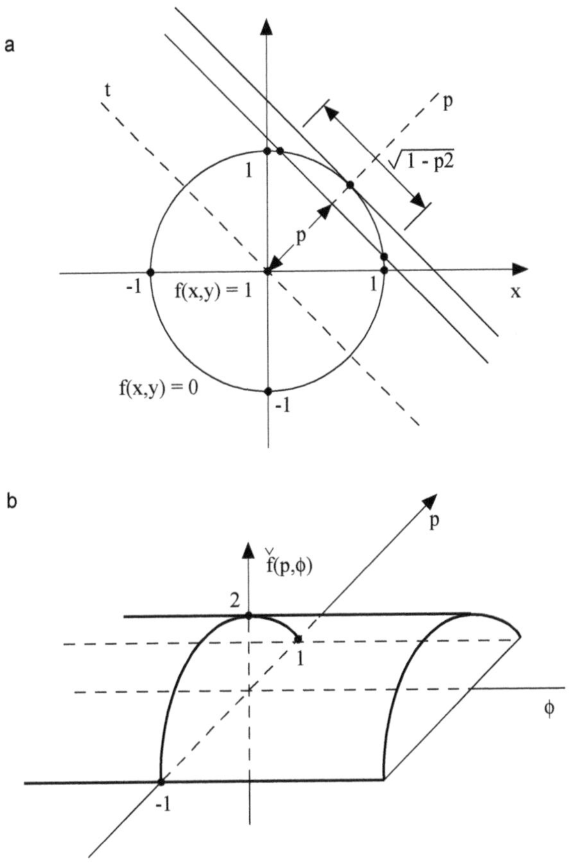

FIGURE 16.5 Radon transform of a cylinder function.

16.3 Basic Properties of the Radon Transform

16.3.1 Notation

$$\xi_1 = \cos\phi,\ \xi_2 = \sin\phi,\ \xi_1^2 + \xi_1^2 = 1,\ \check{f}(p,\xi_1,\xi_2) = \iint_{-\infty}^{\infty} f(x,y)\delta(p - x\xi_1 - y\xi_2)\,dx\,dy$$

$$F(u,\upsilon) = \iint_{-\infty}^{\infty} f(x,y)e^{-j2\pi(ux+\upsilon y)}\,dx\,dy$$

16.3.2 Linearity

$$\mathrm{R}\{af + bg\} = a\check{f} + b\check{g}$$

16.3.3 Similarity

If

$$\mathrm{R}\{f(x,y)\} = \check{f}(p,\xi_1,\xi_2),$$

then

$$\mathsf{R}\{f(ax,by)\} = \frac{1}{|ab|}\check{f}\left(p, \frac{\xi_1}{a}, \frac{\xi_2}{b}\right)$$

$$\mathsf{F}\{f(ax,by)\} = \frac{1}{|ab|}F\left(\frac{u}{a}, \frac{\upsilon}{b}\right)$$

16.3.4 Symmetry

$$\check{f}(ap, a\bar{\xi}) = \frac{1}{|a|}\check{f}\left(p, \bar{\xi}\right)$$

$$\check{f}(-p, -\bar{\xi}) = \check{f}\left(p, \bar{\xi}\right) \equiv \text{even homogeneous function}$$

$$\check{f}(p, s\bar{\xi}) = \frac{1}{|s|}\check{f}\left(\frac{p}{s}, \bar{\xi}\right)$$

16.3.5 Shifting

$$\mathsf{R}\{f(x-a, y-b)\} = \check{f}(p - a\xi_1 - b\xi_2, \bar{\xi})$$

$$\mathsf{F}\{f(x-a, y-b)\} = e^{-j2\pi(au+b\upsilon)}\,F(u,\upsilon)$$

16.3.6 Differentiation

$$\mathsf{R}\left\{\frac{\partial f}{\partial x}\right\} = \xi_1\left\{\frac{\partial \check{f}(p,\bar{\xi})}{\partial p}\right\}, \quad \mathsf{R}\left\{\frac{\partial f}{\partial y}\right\} = \xi_2\left\{\frac{\partial \check{f}(p,\bar{\xi})}{\partial p}\right\}, \quad \mathsf{R}\left\{\frac{\partial^2 f}{\partial x^2}\right\} = \xi_1^2\left\{\frac{\partial^2 \check{f}(p,\bar{\xi})}{\partial p^2}\right\}$$

$$\mathsf{R}\left\{\frac{\partial^2 f}{\partial y^2}\right\} = \xi_2^2\left\{\frac{\partial^2 \check{f}(p,\bar{\xi})}{\partial p^2}\right\}, \quad \mathsf{R}\left\{\frac{\partial^2 f}{\partial x \partial y}\right\} = \xi_1\xi_2\left\{\frac{\partial^2 \check{f}(p,\bar{\xi})}{\partial p^2}\right\}$$

16.3.7 Convolution

$$\check{f}(p,\bar{\xi}) = \mathsf{R}\{g(x,y) ** h(x,y)\} = \check{g} * \check{h} = \int_{-\infty}^{\infty} \check{g}(\tau,\bar{\xi})\check{h}(p-\tau,\bar{\xi})\,d\tau$$

16.3.8 Linear Transformation

$A = n \times n$ nonsingular matrix, $\bar{y} = A\bar{x}$, $\bar{x} = A^{-1}\bar{y} \doteq B\bar{y}$, $A^{-1} = B$,

$$\bar{\xi} \cdot \bar{x} = \xi_1 x_1 + \xi_2 x_2 + \cdots + \xi_n x_n$$

$$\mathcal{R}\{f(A\bar{x})\} = \mathcal{R}\{f(B^{-1}\bar{x})\} = \int f(A\bar{x})\delta(p-\bar{\xi}\cdot\bar{x})d\bar{x} = |\det B|\int f(\bar{y})\delta(p-\bar{\xi}\cdot B\bar{y})d\bar{y}$$

$$= |\det B|\int f(\bar{y})\delta(p-B^T\bar{\xi}\cdot\bar{y})d\bar{y} = |\det B|\check{f}(p,B^T\bar{\xi})$$

If $B^{-1} = B^T = A =$ orthogonal with $|\det B| = 1$, then

$$\mathcal{R}\{f(A\bar{x})\} = \check{f}(p,A\bar{\xi}), \qquad A\bar{\xi} = \text{unit vector}$$

If $A = cI$ with c real, $B = A^{-1} = c^{-1}I$, then

$$\mathcal{R}\{f(c\bar{x})\} = \frac{1}{|c|^n}\check{f}\left(p,\frac{\bar{\xi}}{c}\right) = \frac{1}{|c|^{n-1}}\check{f}(cp,\bar{\xi})$$

If $B^T\bar{\xi}$ is not a unit vector, we define s equal to the magnitude of the vector $B^T\bar{\xi}$, and hence $\bar{\mu} = \frac{B^T\bar{\xi}}{s}$ is a unit vector. Hence we obtain

$$|\det B|\check{f}(p,B^T\bar{\xi}) = |\det B|\check{f}(p,s\bar{\mu}) = \frac{|\det B|}{s}\check{f}\left(\frac{p}{s},\bar{\mu}\right)$$

and

$$\mathcal{R}\{f(B^{-1}\bar{x})\} = \frac{|\det B|}{s}\check{f}\left(\frac{p}{s},\bar{\mu}\right), \quad s = \left|B^T\bar{\xi}\right|$$

16.3.9 Examples:

Example

$$\bar{\xi} = (\cos\phi,\sin\phi), \quad A = \begin{bmatrix} \xi_1 & \xi_2 \\ -\xi_2 & \xi_1 \end{bmatrix}, \quad \begin{bmatrix} u \\ \upsilon \end{bmatrix} = A\begin{bmatrix} x \\ y \end{bmatrix} = \begin{bmatrix} \xi_1 x + \xi_2 y \\ -\xi_2 x + \xi_1 y \end{bmatrix}.$$

A is orthogonal (see 16.3.7) and $u^2 + \upsilon^2 = x^2 + y^2$, $u = \xi_1 x + \xi_2 y$.

$$\mathcal{R}\{f(x,y) = e^{-x^2-y^2}\} = \mathcal{R}\{f(A\bar{x})\} = \mathcal{R}\{f(u,\upsilon)\} = \iint_{-\infty}^{\infty} e^{-u^2-\upsilon^2}\delta(p-u)\,du\,d\upsilon$$

$$= e^{-p^2}\int_{-\infty}^{\infty} e^{-\upsilon^2}d\upsilon = \sqrt{\pi}\,e^{-p^2}$$

(See 16.2.1)

Also, $\mathcal{R}\{e^{-x^2-y^2-z^2}\} = \left(\sqrt{\pi}\right)^{3-1}e^{-p^2} = \pi e^{-p^2}$

Example

Because $\mathcal{R}\{f(\bar{x}) = e^{-x^2-y^2}\} = \check{f}(p,\bar{\xi}) = \sqrt{\pi}\,e^{-p^2}$ we can set

$$B=\begin{bmatrix} a & 0\\ 0 & b\end{bmatrix},\quad B^{-1}=\begin{bmatrix} \frac{1}{a} & 0\\ 0 & \frac{1}{b}\end{bmatrix},\quad |\det B|=|ab|,\quad B^T\bar{\xi}=\begin{bmatrix} a\cos\phi\\ b\sin\phi\end{bmatrix}\ \text{(not a unit vector)},$$

$s=\left|B^T\bar{\xi}\right|=(a^2\cos^2\phi+b^2\sin^2\phi)^{1/2}$ and from 16.3.7 $\mathcal{R}\{e^{-\frac{x^2}{a^2}-\frac{y^2}{b^2}}\}=\frac{|ab|\sqrt{\pi}}{s}e^{-(p/s)^2}$.

Example (see Figure 16.4)

If $(p,t)\to(u,\upsilon)$ from the transformation $[\bar{\xi}=(\cos\phi,\sin\phi)]$

$$\begin{bmatrix} u\\ \upsilon\end{bmatrix}=A\begin{bmatrix} x\\ y\end{bmatrix}=\begin{bmatrix} \xi_1 & \xi_2\\ -\xi_2 & \xi_1\end{bmatrix}\begin{bmatrix} x\\ y\end{bmatrix},$$

we obtain $x=u\cos\phi-\upsilon\sin\phi$, and $y=u\sin\phi+\upsilon\cos\phi$. Therefore the RT of a function in a unit disk is

$$\mathcal{R}\{f(x,y)\}=\check{f}(p,\phi)=\int_{disk} f(x,y)\delta(p-x\cos\phi-y\sin\phi)\,dx\,dy$$

$$=\int_{disk} f(u\cos\phi-\upsilon\sin\phi,\ u\sin\phi+\upsilon\cos\phi)\delta(p-u)\,du\,d\upsilon$$

$$=\int_{-\sqrt{1-p^2}}^{\sqrt{1-p^2}} f(p\cos\phi-\upsilon\sin\phi,\ p\sin\phi+\upsilon\cos\phi)\,d\upsilon$$

Example

Because $\delta(x,y)=\delta(p,t)$, then $\mathcal{R}\{\delta(x,y)=\int_{-\infty}^{\infty}\delta(p,t)\,dt=\delta(p)$ (see Figure 16.6 and Section 16.1.3).

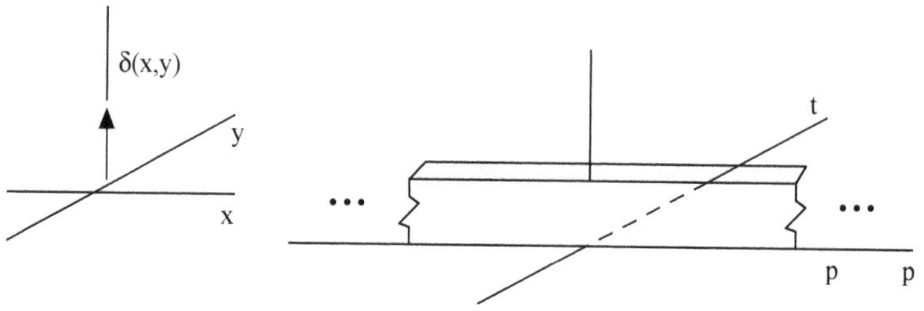

FIGURE 16.6 Delta function.

Example

$$\mathcal{R}\{f(x,y)=\delta(x-a,y-b)=\delta(x-a)\delta(x-b)\}=\iint_{-\infty}^{\infty}\delta(x-a)\delta(x-b)\delta(p-\bar{\xi}\cdot\bar{r})\,dx\,dy$$

$$=\iint_{-\infty}^{\infty}\delta(x-a)\delta(x-b)\delta(p-\xi_1x-\xi_2y)\,dx\,dy=\delta(p-\xi_1a-\xi_2b)=\delta(p-p_0)$$

$p_0=a\cos\phi+b\sin\phi$. Therefore as impulse function at the (x,y) point creates an impulse function along a sinusoidal curve in the Radon space (p,ϕ) which is $p=a\cos\phi+b\sin\phi=\sqrt{a^2+b^2}\cos[\phi-\tan^{-1}(b/a)]$ (see Figure 16.7)

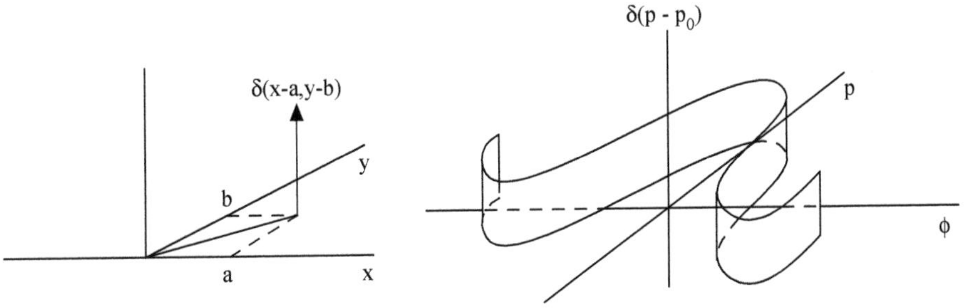

FIGURE 16.7 Displaced delta function.

Example

The Radon transform of a finite-extent delta function

$$f(x,y)=\begin{cases}\delta(x-p_0,0) & |y|<L/2\\ 0 & |y|\geq L/2\end{cases}$$

is given by (see 16.1.3) and Chapter 5 on delta function properties

$$\check{f}(p,\phi)=\int_{-L/2}^{L/2}\delta(p\cos\phi-t\sin\phi-p_0)\,dt=\begin{cases}L\delta(p-p_0) & \phi=2n\pi\\ L\delta(p+p_0) & \phi=(2n+1)\pi\end{cases}$$

If ϕ is different than multiple π (see Chapter 5 on delta function properties)

$$\check{f}(p,\phi)=\int_{p=t}\delta(p\cos\phi-t\sin\phi-p_0)\,dt=\begin{cases}\dfrac{1}{|\sin\phi|} & |p-p_0\cos\phi|\leq|\tfrac{L}{2}\sin\phi|\\ 0 & \text{otherwise}\end{cases}$$

The inequality can be derived from the geometry of Figure 16.8b. The region of support is shown in Figure 16.8c, and the transform is shown in Figure 16.8d.

If $f(x,y)$ is a finite length delta function (see Figure 16.9) such that

$$f(x,y)=\begin{cases}\delta[p-(x\cos\theta_0+y\sin\theta_0)] & x\cos\theta_0+y\sin\theta_0\in(-\tfrac{L}{2},\tfrac{L}{2})\\ 0 & \text{otherwise}\end{cases}$$

then

$$\check{f}(p,\phi)=\begin{cases}L\delta(p-p_0) & \phi=\phi_0+2n\pi\\ L\delta(p+p_0) & \phi=\phi_0+(2n+1)\pi\\ \dfrac{1}{|\sin(\phi-\phi_0)|} & |p-p_0\cos(\phi-\phi_0)|\leq\tfrac{L}{2}|\sin(\phi-\phi_0)|\\ 0 & \text{otherwise}\end{cases}$$

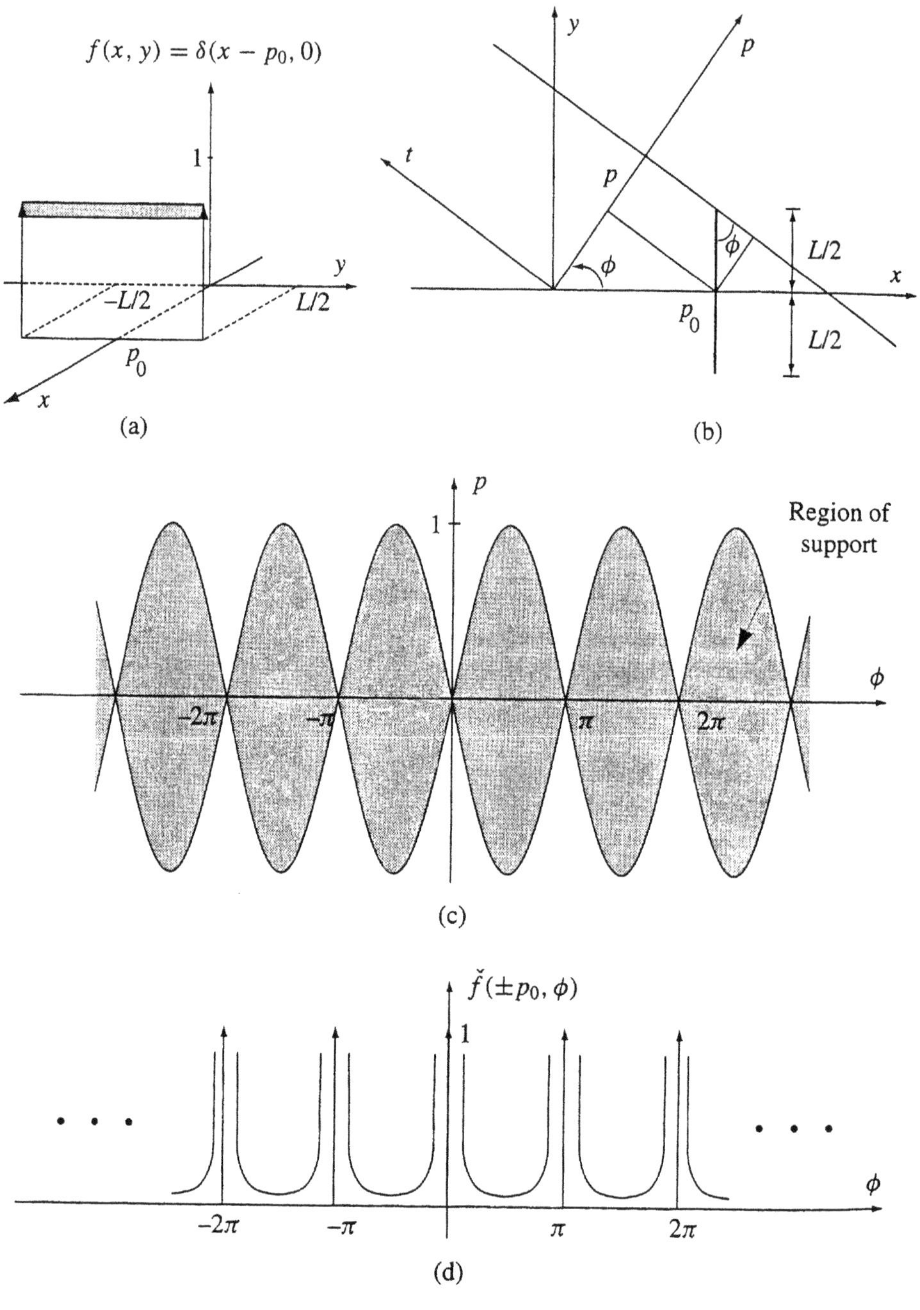

FIGURE 16.8 Radon transform of finite-extended delta function.

Example

Find the Radon transform of the cylinder defined in the Example above displaced at the point (x_o, y_o) as shown in Figure 16.10a. The solution follows immediately from the solution above combined with the shifting property. Also, the solution can be deduced from the geometry in Figure 16.10a. When $d = 1$, the length $t = 0$; also, for p such that the line of integrated passes through the cylinder,

$$t = 2\sqrt{1-[p-r_0\cos(\phi_0-\phi)]^2}\,.$$

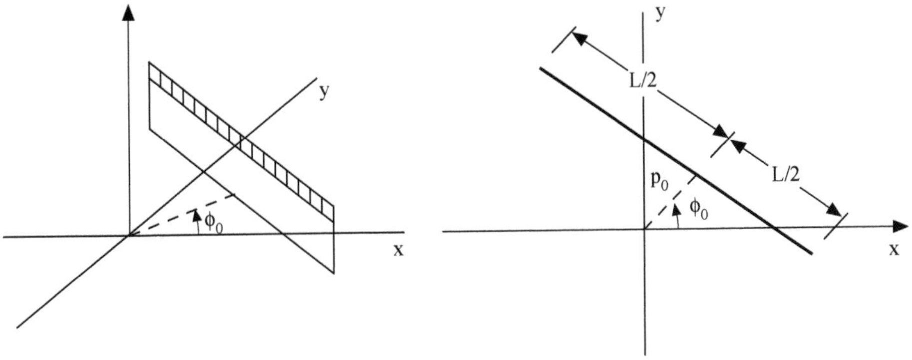

FIGURE 16.9 Finite delta in general direction.

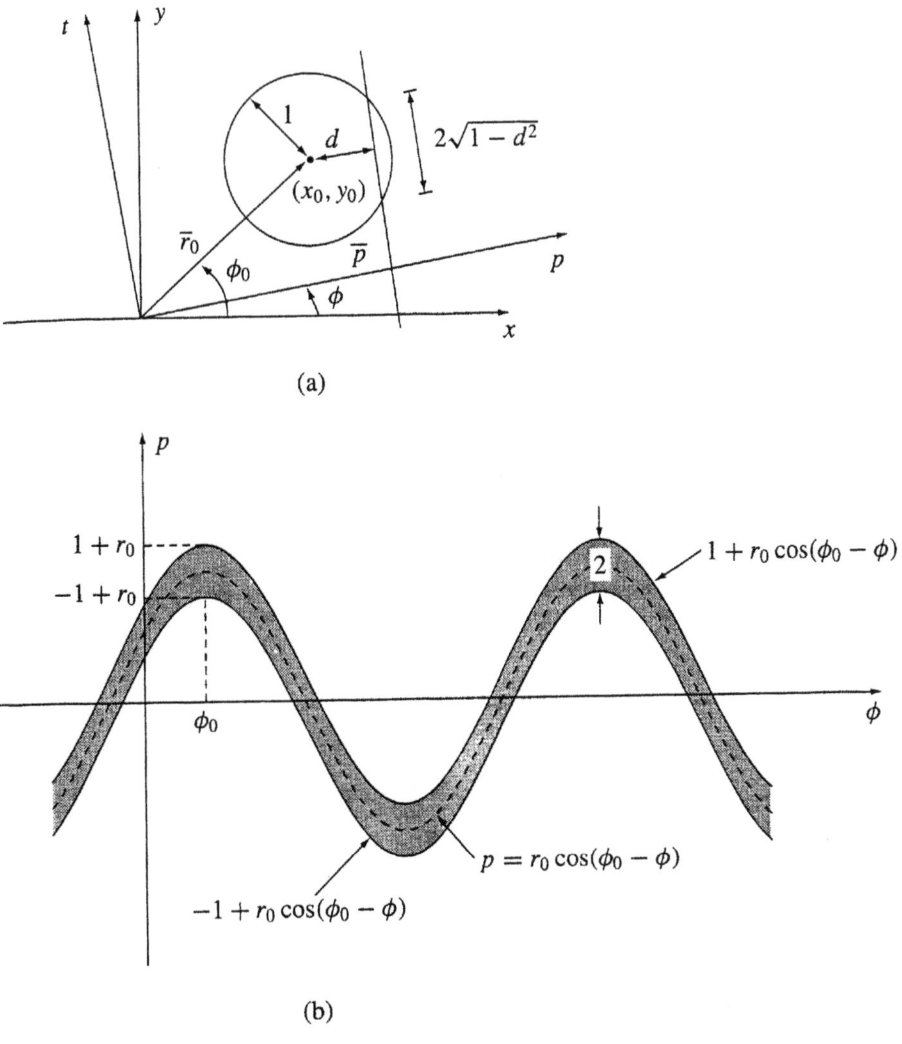

FIGURE 16.10 Displaced cylinder function and region of support of the transform.

Further, when ϕ varies, the values p can assume follow from the geometry. The transform is

$$\check{f}(p,\phi) = \begin{cases} 2\sqrt{1-[p-r_0\cos(\phi_0-\phi)]^2}, & -1+r_0\cos(\phi_0-\phi) \le p \le 1+r_0\cos(\phi_0-\phi) \\ 0, & \text{otherwise} \end{cases}$$

Figure 16.10b shows the (sinusoidal) region of support of the transform.

16.4 Additional Properties of Radon Transforms

16.4.1 Transform of Derivatives

$$\mathsf{R}\left\{\frac{\partial f}{\partial x_k}\right\} = \xi_k \left\{\frac{\partial \check{f}(p,\bar{\xi})}{\partial p}\right\}, \qquad f(x_1,\cdots,x_n)$$

$$\mathsf{R}\left\{\frac{\partial^2 f}{\partial x_k \partial y_k}\right\} = \xi_\ell \xi_k \frac{\partial^2 \check{f}(p,\bar{\xi})}{\partial p^2}$$

Example

$$\mathsf{R}\{\nabla^2 f(\bar{x})\} = |\bar{\xi}| \frac{\partial^2 \check{f}(p,\bar{\xi})}{\partial p^2} = \frac{\partial^2 \check{f}(p,\bar{\xi})}{\partial p^2}$$

Example

$$\mathsf{R}\{\nabla^2 f(\bar{x},t)\} = \mathsf{R}\left\{\frac{\partial^2 \check{f}(\bar{x},t)}{\partial t^2}\right\} \quad \text{or} \quad \frac{\partial^2 \check{f}(p,\bar{\xi})}{\partial p^2} = -\frac{\partial^2 \check{f}(p,\bar{\xi})}{\partial t^2}$$

16.4.2 Derivative of the Radon Transform

$$\frac{\partial}{\partial \eta_j}\delta(p-\bar{\eta}\cdot\bar{x}) = -x_j \frac{\partial}{p}\delta(p-\bar{\eta}\cdot\bar{x})$$

$$\frac{\partial \check{f}}{\partial \xi_k} = \left[\frac{\partial}{\partial \eta_k}\mathsf{R}\{f(\bar{x})\}\right]_{\bar{\eta}=\bar{\xi}} = -\frac{\partial}{\partial p}\mathsf{R}\{x_k f(\bar{x})\}$$

Example

$$\mathsf{R}\{f(x,y) = e^{-x^2-y^2}\} = \check{f}(p,\bar{\xi}) = \sqrt{\pi}\, e^{-p^2}.$$

Set $\bar{\eta} = s\bar{\xi}$ $(s = (\eta_1^2 + \eta_2^2)^{1/2})$ to obtain $\check{f}(p,\bar{\eta}) = \check{f}(p,s\bar{\xi}) = \frac{\sqrt{\pi}}{s} e^{-p^2/s^2}$ (scaling relation). But $\frac{\partial}{\partial \eta_k} = \frac{\partial s}{\partial \eta_k}\frac{\partial}{\partial s}$ $(k = 1,2)$, and thus

$$\frac{\partial \check{f}}{\partial \eta_k} = \sqrt{\pi}\,(\eta_k|s)\frac{\partial}{\partial s}(s^{-1}e^{-p^2/s^2}) = \sqrt{\pi}\,\frac{\eta_k}{s^5}(2p^2 - s^2)e^{-p^2/s^2}$$

To find the derivative $\partial \check{f}/\partial \xi_k$ we calculate the above expression at $\bar{\eta} = \bar{\xi}$ or equivalent setting s = 1. Hence,

$$\frac{\partial \check{f}}{\partial \xi_k} = \sqrt{\pi}\,\xi_k(2p^2 - 1)e^{-p^2}$$

For 2-D function we also have

$$\frac{\partial^{\ell+k}\check{f}(p,\bar{\xi})}{\partial \xi_1^{\ell}\partial \xi_2^{k}} = \left(-\frac{\partial}{\partial p}\right)^{\ell+k} \mathrm{R}\{x^{\ell}y^{k}f(x,y)\}$$

16.5 Hermite Polynomials and Radon Transforms

16.5.1 Hermite Polynomials

From 16.4.1

$$\mathrm{R}\left\{\left(\frac{\partial}{\partial x}\right)^{\ell}\left(\frac{\partial}{\partial y}\right)^{k} f(x,y)\right\} = (\cos\phi)^{\ell}(\sin\phi)^{k}\left(\frac{\partial}{\partial p}\right)^{\ell+k}\check{f}(p,\bar{\xi})$$

and hence

$$\mathrm{R}\left\{H_{\ell}(x)H_k(y)e^{-x^2-y^2} = (-1)^{\ell+k}\left(\frac{\partial}{\partial x}\right)^{\ell}\left(\frac{\partial}{\partial y}\right)^{k} e^{-x^2-y^2}\right\}$$

$$= (-1)^{\ell+k}(\cos\phi)^{\ell}(\sin\phi)^{k}\left(\frac{\partial}{\partial p}\right)^{\ell+k}\sqrt{\pi}\,e^{-p^2} = (-1)^{\ell+k}\sqrt{\pi}\,(\cos\phi)^{\ell}(\sin\phi)^{k}\,e^{-p^2}H_{\ell+k}(p)$$

Example

$$x = \tfrac{1}{2}H_1$$

and hence

$$\mathcal{R}\{xe^{-x^2-y^2}\} = \mathcal{R}\{\tfrac{1}{2}H_1(x)e^{-x^2-y^2}\}$$

$$= (-1)^{1+0}\frac{1}{2}\left((\cos\phi)^1(\sin\phi)^0\frac{\partial}{\partial p}\sqrt{\pi}\,e^{-p^2}\right) = p\sqrt{\pi}\,e^{-p^2}\cos\phi$$

16.6 Inversion of Radon Transforms

16.6.1 2-D Inverse Radon Transform

$$f(x,y) = \frac{-1}{2\pi^2}P\int_0^{\pi}d\phi\int_{-\infty}^{\infty}\frac{\check{f}(p,\bar{\xi})}{p-\bar{\xi}\cdot\bar{x}}dp$$

P stands for principal value of the integral.

16.7 N-Dimensional Radon Transform

16.7.1 N-Dimensional Radon Transform with its Properties

TABLE 16.1 Properties the N-Dimensional Radon Transform

1	Linearity	$\mathcal{R}\{a_1f_1 + a_2f_2\} = a_1\check{f}_1 + a_2\check{f}_2$
2	Homogeneity	$\check{f}(\alpha\bar{\xi},\alpha p) = \lvert\alpha\rvert^{-1}\check{f}(\bar{\xi},p)$
3	Shifting	$\check{f}_{\bar{a}}(\bar{\xi},p) = \check{f}(\bar{\xi},p+\bar{\xi}\cdot\bar{a}),\quad f_{\bar{a}}(\bar{x}) = f(\bar{x}+\bar{a})$
4	Coordinate transformation	$\check{f}_A(\bar{\xi},p) = \lvert\det A\rvert\,\check{f}(A^T\bar{\xi},p)$ $f_A(\bar{x}) = f(A^{-1}\bar{x})$, A is nonsingular matrix
5	Differentiation	$\mathcal{R}\left\{\sum_{k=1}^{n}a_k\frac{\partial f}{\partial x_k}\right\} = \sum_{k=1}^{n}a_k\xi_k\frac{\partial}{\partial p}\check{f}(\bar{\xi},p)$
6	Convolution	$\check{f}(\bar{\xi},p) = \int_{-\infty}^{\infty}\check{f}_1(\bar{\xi},t)\check{f}_2(\bar{\xi},p-t)\,dt$ where $f(\bar{x}) = f_1(\bar{x}) *^n f_2(\bar{x})$
7	Relationship to Fourier transform of $f(\bar{x})$	with $F(\bar{\xi}) = \int f(\bar{x})e^{-j\bar{\xi}\cdot\bar{x}}\,d\bar{x}$ $\check{f}(\bar{\xi},p) = \frac{1}{2\pi}\int_{-\infty}^{\infty}F(\alpha\bar{\xi})e^{j\alpha p}\,d\alpha$ $F(\bar{\xi}) = \int_{-\infty}^{\infty}\check{f}(\bar{\xi},p)e^{-jp}\,dp$
8	Inverse Radon transform	n = odd $\quad f(\bar{x}) = \frac{(-1)^{(n-1)/2}}{2(2\pi)^{n-1}}\int_{\Gamma}\check{f}_p^{(n-1)}(\bar{\xi},\bar{\xi}\cdot\bar{x})\omega(\bar{\xi})\,dp$ n = even $\quad f(\bar{x}) = \frac{(-1)^{n/2}(n-1)!}{(2\pi)^n}$ $\int_{\Gamma}\left[\int_{-\infty}^{\infty}\check{f}(\bar{\xi},p)(p-\bar{\xi}\cdot\bar{x})^{-n}\,dp\right]\omega(\bar{\xi})$

TABLE 16.1 Properties the N-Dimensional Radon Transform (continued)

where $\omega(\bar{\xi}) = \sum_{k=1}^{n} (-1)^{k-1}\xi_k d\xi_1 \cdots d\xi_{k-1} d\xi_{k+1} \cdots d\xi_n$ Γ is any surface enclosing the origin in $\bar{\xi}$ space and $\check{f}_p^{(n-1)}$ is the $(n-1)^{th}$ derivative of $\check{f}(\bar{\xi},p)$ with respect to p.

16.8 Abel Transforms

16.8.1 Definition of Abel Transform

$$g(x) = \int_o^x \frac{f(y)}{(x-y)^\alpha} dy = f(x) * \frac{1}{x^\alpha} \qquad x > 0,\ y < x,\ 0 < \alpha < 1$$

16.8.2 Laplace Transform of Abel Transformation

$$G(s) = F(s)K(s)$$

$$F(s) = \frac{G(s)}{K(s)} = [sG(s)]\left[\frac{1}{sK(s)}\right] = sG(s)H(s)$$

16.8.3 Inverse Solution

$$f(x) = \frac{\sin\alpha\pi}{\pi}\frac{d}{dx}\int_0^x (x-y)^{\alpha-1} g(y)dy$$

$$f(x) = \frac{\sin\alpha\pi}{\pi}\left[\frac{g(0+)}{x^{1-\alpha}} + \int_0^x \frac{g'(y)}{(x-y)^{1-\alpha}} dy\right]$$

16.8.4 Abel Transform Pairs

$$\hat{f}_1(x) = \mathrm{A}_1\{f_1(r);x\} = \int_0^x \frac{f_1(r)dr}{(x^2-r^2)^{1/2}}, \qquad x > 0$$

$$\hat{f}_2(x) = \mathrm{A}_2\{f_2(r);x\} = \int_x^\infty \frac{f_2(r)dr}{(r^2-x^2)^{1/2}}, \qquad x > 0$$

$$\hat{f}_3(x) = \mathrm{A}_3\{f_3(r);x\} = 2\int_x^\infty \frac{rf_3(r)dr}{(r^2-x^2)^{1/2}}, \qquad x > 0$$

$$\hat{f}_4(x) = \mathrm{A}_4\{f_4(r);x\} = 2\int_0^x \frac{rf_4(r)dr}{(x^2-r^2)^{1/2}}, \qquad x > 0$$

16.8.5 Inverse of the Four Types of Abel Transform (see 16.8.3)

$$f_1(r) = \frac{2}{\pi}\frac{d}{dr}\int_0^r \frac{x\hat{f}_1(x)\,dx}{(r^2-x^2)^{1/2}} = \frac{2\hat{f}_1(0)}{\pi} + \frac{2r}{\pi}\int_0^r \frac{\hat{f}_1'(x)\,dx}{(r^2-x^2)^{1/2}}$$

$$f_2(r) = -\frac{2}{\pi}\frac{d}{dr}\int_r^\infty \frac{x\hat{f}_2(x)\,dx}{(x^2-r^2)^{1/2}} = -\frac{2r}{\pi}\int_r^\infty \frac{\hat{f}_2'(x)\,dx}{(x^2-r^2)^{1/2}}$$

$$f_3(r) = -\frac{1}{\pi r}\frac{d}{dr}\int_r^\infty \frac{x\hat{f}_3(x)\,dx}{(x^2-r^2)^{1/2}} = -\frac{1}{\pi}\int_r^\infty \frac{\hat{f}_3'(x)\,dx}{(x^2-r^2)^{1/2}}$$

$$f_4(r) = \frac{1}{\pi r}\frac{d}{dr}\int_0^r \frac{x\hat{f}_4(x)\,dx}{(r^2-x^2)^{1/2}} = \frac{\hat{f}_4(0)}{\pi r} + \frac{1}{\pi}\int_0^r \frac{\hat{f}_4'(x)\,dx}{(r^2-x^2)^{1/2}}$$

16.8.6 Relationships Among the Four Types of Abel Transforms

$$\mathsf{A}_3\{f(r)\} = 2\mathsf{A}_2\{r\,f(r)\}$$

$$\mathsf{A}_4\{f(r)\} = 2\mathsf{A}_1\{r\,f(r)\}$$

$$\mathsf{A}_4\{r^{-1}f_1(r)\} = 2\hat{f}_1(x)$$

$$\mathsf{A}_3\{r^{-1}f_2(r)\} = 2\hat{f}_2(x)$$

$$f_1(r) = \mathsf{A}_1^{-1}\{\hat{f}_1(x)\} = \frac{2}{\pi}\frac{d}{dr}\mathsf{A}_1\{x\,\hat{f}_1(x)\}$$

$$f_2(r) = \mathsf{A}^{-1}\{\hat{f}_2(x)\} = -\frac{2}{\pi}\frac{d}{dr}\mathsf{A}_2\{x\,\hat{f}_2(x)\}$$

Example

$$\hat{f}_1(x) = \mathsf{A}_1\{(a-r)\} = \int_0^x \frac{(a-r)\,dr}{(x^2-r^2)^{1/2}} = a\int_0^x \frac{dr}{(x^2-r^2)^{1/2}} - \int_0^x \frac{r\,dr}{(x^2-r^2)^{1/2}}$$

$$= a\sin^{-1}\frac{r}{x}\Big|_0^x - \frac{1}{2}\int_0^{y^2} \frac{dy}{(x^2-y)^{1/2}} = \frac{\pi a}{2} - \frac{1}{2}\left[\frac{2}{-1}(x^2-y)^{1/2}\Big|_0^{x^2}\right] = \frac{\pi a}{2} - x$$

(see 16.8.4)

$$f_1(r) = \mathsf{A}_1^{-1}\{\hat{f}_1(x)\} = \frac{2}{\pi}\frac{\pi a}{2} + \frac{2r}{\pi}\int_0^r \frac{-dx}{(r^2-x^2)^{1/2}} = a - r$$

16.9 Inverse Radon Transform

16.9.1 Back Projection

$$f(x,y) = 2\mathsf{B}\{\check{f}(t,\phi)\}$$

where $\mathrm{B}\{F(p,\phi)\} = \int_o^\pi F(x\cos\phi + y\sin\phi, \phi)\,d\phi$ is the back projection operation which is defined by replacing p by $x\cos\phi + y\sin\phi$ and integrating over the angle ϕ. The over bar defines the Hilbert transform of the derivative of some function g as follows

$$\bar{g}(t) = -\frac{1}{4\pi}\mathrm{H}\{g_p(p);\ p \to t\} \quad \text{for } n = 2$$

16.9.2 Backprojection of the Filtered Projection

$$f = 2\mathrm{B}\{\mathrm{F}^{-1}\mathrm{F}\bar{\check{f}}\} = \frac{2}{4\pi^2}\mathrm{B}\left\{\mathrm{F}^{-1}\left\{\mathrm{F}\frac{\partial}{\partial p}\left[\frac{1}{p} * \check{f}(p,\phi)\right]\right\}\right\}$$

$$= \frac{1}{2\pi^2}\mathrm{B}\left\{\mathrm{F}^{-1}\left\{(j2\pi k)\left[\mathrm{F}\left\{\frac{1}{p}\right\}\right]\left[\mathrm{F}\{\check{f}(p,\phi)\}\right]\right\}\right\}$$

$$= \frac{1}{2\pi^2}\mathrm{B}\{\mathrm{F}^{-1}\{(j2\pi k)(j\pi gnk)\mathrm{F}\{\check{f}(p,\phi)\}\}\} = \mathrm{B}\{\mathrm{F}^{-1}\{|k|\mathrm{F}\{\check{f}(p,\phi)\}\}\}$$

If we define $F(s,\phi) = \mathrm{F}^{-1}\{|k|\mathrm{F}\{\check{f}(p,\phi)\}\} = \mathrm{F}^{-1}\{|k|\mathrm{F}\{\tilde{\check{f}}(k,\phi)\}$ then the $f(x,y)$ is found by the backprojection of F,

$$f(x,y) = \mathrm{B}\{F(s,\phi)\} = \int_0^\pi F(x\cos\phi + y\sin\phi, \phi)\,d\phi$$

16.9.3 Convolution Method

If $\mathrm{F}\{g\} = |k|w(k)$, $w(k)$ is a window function, then $g(s) = \mathrm{F}^{-1}\{|k|w(k)\}$ and then $F(s,\phi)$ of 16.9.2 becomes

$$F(s,\phi) = \mathrm{F}^{-1}\{\mathrm{F}\{g\}\mathrm{F}\{\check{f}\}\} = \check{f} * g = \int_{-\infty}^{\infty}\check{f}(p,\phi)g(s-p)\,dp$$

16.9.4 Frequency Space Implementation

From 16.9.2 we obtain

$$F(s,\phi) = \mathrm{F}^{-1}\{|k|w(k)\mathrm{F}\{\check{f}(p,\phi)\}\} = \mathrm{F}^{-1}\{|k|w(k)\tilde{\check{f}}(k,\phi)\}$$

Figure 16.11 shows a flow chart of filtered backprojection and convolution.

16.9.5 Filter of Backprojection

$$b(x,y) = \mathrm{B}\{\check{f}(p,\phi)\} = \int_0^\pi \check{f}(x\cos\phi + y\sin\phi, \phi)\,d\phi$$

$$\tilde{g}(u,\upsilon) = |g|w(u,\upsilon), \qquad q = \sqrt{u^2+\upsilon^2}, \qquad w(u,\upsilon) = FT \text{ of window}$$

$$f(x,y) = g(x,y) ** b(x,y)$$

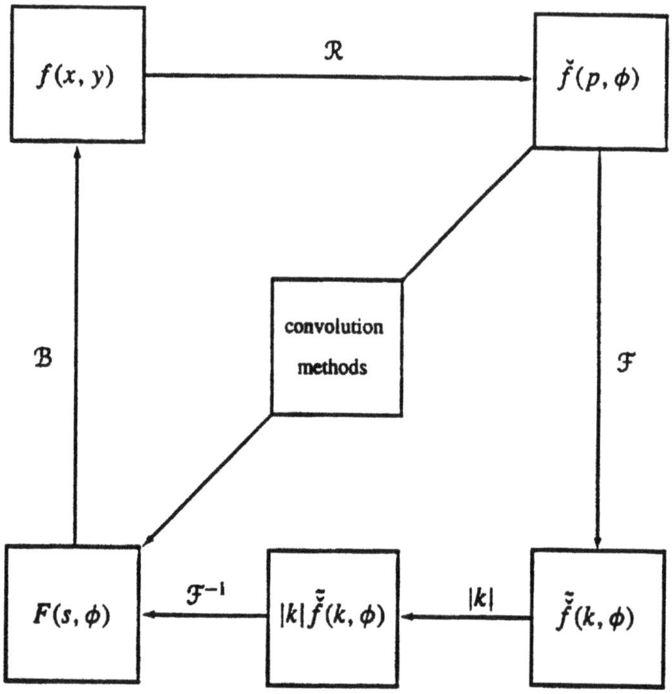

FIGURE 16.11 Filtered backprojection, convolution.

16.10 Tables of Abel and Radon Pairs

16.10.1 Abel and Radon Pairs

Just to remind the user of these tables, the sinc function is defined by

$$\operatorname{sinc} x = \frac{\sin \pi x}{\pi x}$$

and the characteristic function for the unit disk, designated by $\chi(r)$ is defined by

$$\chi(r) = \begin{cases} 1, & \text{for } 0 \le r \le 1 \\ 0, & \text{for } r > 1 \end{cases}$$

The complete elliptic integral of the first kind is designated by $F(\frac{1}{2}\pi, t)$, and the complete elliptic integral of the second kind is designated by $E(\frac{1}{2}\pi, t)$. A good source of these is the tabulation by Gradshteyn et al. The constant $C(n)$ in the table for A_3 is $C(n) = 2\int_0^{\pi/2} \cos^n x\, dx$, with $n \ge 1$; Bessel functions of the first kind J_ν, and second kind N_ν (Neumann functions) conform to the standard definitions in Gradshteyn et al. In these tables, $a > 0$ and $b > 0$.

TABLE 16.2 Abel and Radon Transforms

Abel Transforms A_1

$f(r)$	$A_1\{f(r); x\}$
$\chi(r/a)$	$\sin^{-1}\left(\frac{a}{x}\right),\ x > a$
$\delta(r-a)$	$(x^2-a^2)^{-1/2},\ x > a$
$(a^2-r^2)^{-1/2}$	$a^{-1}F\left(\frac{\pi}{2},\frac{x}{a}\right),\ x < a$
$(a^2-r^2)^{1/2}$	$aE\left(\frac{\pi}{2},\frac{x}{a}\right),\ x < a$
$r^2(a^2-r^2)^{-1/2}$	$a\left[F\left(\frac{\pi}{2},\frac{x}{a}\right)-E\left(\frac{\pi}{2},\frac{x}{a}\right)\right],\ x < a$
$a-r$	$\frac{1}{2}\pi a - x,\ x < a$
$\cos br$	$\frac{1}{2}\pi J_0(bx)$
$r\sin br$	$\frac{1}{2}\pi x J_1(bx)$
$rJ_0(br)$	$b^{-1}\sin bx$
$J_\nu(br)$	$\frac{1}{2}\pi\left[J_{\frac{\nu}{2}}\left(\frac{bx}{2}\right)\right]^2$
$r^{\nu+1}J_\nu(br)$	$\pi^{\frac{1}{2}}(2b)^{-\frac{1}{2}}x^{\nu+\frac{1}{2}}J_{\nu+\frac{1}{2}}(bx)$

Abel Transforms A_2

$f(r)$	$A_2\{f(r); x\}$
$\chi(r/a)$	$\log\left(\frac{a+\sqrt{a^2-x^2}}{x}\right),\ x < a$
$\delta(r-a)$	$(a^2-x^2)^{-1/2},\ x < a$
$(a^2-r^2)^{-1/2}\chi(r/a)$	$a^{-1}F(\frac{1}{2}\pi,t),\ x < a$
$(a^2-r^2)^{1/2}\chi(r/a)$	$a[F(\frac{1}{2}\pi,t)-E(\frac{1}{2}\pi,t)],\ x < a$
$r^2(a^2-r^2)^{-1/2}\chi(r/a)$	$aE(\frac{1}{2}\pi,t),\ x < a$
$(a-r)\chi(r/a)$	$\log\left(\frac{a+\sqrt{a^2-x^2}}{x}\right)-\sqrt{a^2-x^2},\ x < a$
$\sin br$	$\frac{1}{2}\pi J_0(bx)$
$r\cos br$	$-\frac{1}{2}\pi x J_1(bx)$
$rJ_0(br)$	$b^{-1}\cos bx$

Note: $t = a^{-1}\sqrt{a^2-x^2}$.

TABLE 16.2 Abel and Radon Transforms (continued)

Abel Transforms A_3

$f(r)$	$A_3\{f(r); x\}$
$(a^2-r^2)^{-1/2}\chi(r/a)$	$\pi\chi(r/a)$
$\chi(r/a)$	$2(a^2-x^2)^{1/2}\chi(x/a)$
$(a^2-r^2)^{1/2}\chi(r/a)$	$\frac{1}{2}\pi(a^2-x^2)\chi(x/a)$
$(a^2-r^2)\chi(r/a)$	$\frac{4}{3}(a^2-x^2)^{3/2}\chi(x/a)$
$(a^2-r^2)^{\frac{3}{2}}\chi(r/a)$	$\frac{3\pi}{8}(a^2-x^2)^2\chi(x/a)$
$(a^2-r^2)^2\chi(r/a)$	$\frac{15}{16}(a^2-x^2)^{5/2}\chi(x/a)$
$(a^2-r^2)^{\frac{n-1}{2}}\chi(r/a)$	$C(n)(a^2-x^2)^{\frac{n}{2}}\chi(x/a)$
$(a^2+r^2)^{-1}$	$\pi(a^2+x^2)^{-1/2}$
$(a^2+r^2)^{-\frac{3}{2}}$	$2(a^2+x^2)^{-1}$
e^{-r^2}	$\sqrt{\pi}\ e^{-x^2}$
$r^2e^{-r^2}$	$\frac{1}{2}\sqrt{\pi}\ (2x^2+1)e^{-x^2}$
$\operatorname{sinc} 2ar$	$\frac{1}{2a}J_0(2\pi ax)$
$\cos br$	$-\pi x J_1(bx)$
$J_0(br)$	$2b^{-1}\cos bx$
$r^{-1}J_1(br)$	$2(bx)^{-1}\sin bx$
$r^{-1}J_\nu(br)$	$-\pi J_{\frac{\nu}{2}}\left(\frac{bx}{2}\right)N_{\frac{\nu}{2}}\left(\frac{bx}{2}\right)$
$r^{-1}N_\nu(br)$	$\frac{1}{2}\pi\left[J_{\frac{\nu}{2}}\left(\frac{bx}{2}\right)\right]^2-\frac{1}{2}\pi\left[N_{\frac{\nu}{2}}\left(\frac{bx}{2}\right)\right]^2$

Radon Transforms

$f(x,y)$	$\check{f}(p,\phi)$
$e^{-x^2-y^2}$	$\sqrt{\pi}\ e^{-p^2}$
$(x^2+y^2)e^{-x^2-y^2}$	$\frac{1}{2}\sqrt{\pi}\ (2p^2+1)e^{-p^2}$
$xe^{-x^2-y^2}$	$\sqrt{\pi}\ e^{-p^2}\cos\phi$
$ye^{-x^2-y^2}$	$\sqrt{\pi}\ e^{-p^2}\sin\phi$
$x^2e^{-x^2-y^2}$	$\frac{1}{2}\sqrt{\pi}\ (2p^2\cos^2\phi+\sin^2\phi)e^{-p^2}$
$y^2e^{-x^2-y^2}$	$\frac{1}{2}\sqrt{\pi}\ (2p^2\sin^2\phi+\cos^2\phi)e^{-p^2}$
$\exp\left[-\left(\frac{x}{a}\right)^2-\left(\frac{y}{a}\right)^2\right]$	$\frac{\lvert ab\rvert\sqrt{\pi}}{s}\exp\left[-\left(\frac{p}{s}\right)^2\right]$
$\delta(x-a)\delta(y-b)$	$\delta(p-p_0)$
$\chi(r)$	$2(1-p^2)^{\frac{1}{2}}\chi(p)$
$x^2\chi(r)$	$(1-p^2)^{\frac{1}{2}}[2p^2\cos^2\phi+\frac{2}{3}(1-p^2)\sin^2\phi)]$

TABLE 16.2 Abel and Radon Transforms (continued)

Radon Transforms	
$f(x,y)$	$\check{f}(p,\phi)$
$y^2\chi(r)$	$(1-p^2)^{\frac{1}{2}}[2p^2\sin^2\phi+\frac{2}{3}(1-p^2)\cos^2\phi)]$
$(x^2+y^2)\chi(r)$	$\frac{4}{3}(1-p^2)^{\frac{1}{2}}(2p^2+1)$

The following notation issued in the above table,

$$s=(a^2\cos^2\phi+b^2\sin^2\phi)^{\frac{1}{2}},\ r=\sqrt{a^2+y^2},\ p_o=a\cos\phi+b\sin\phi.$$

Formulas for Radon transforms involving Hermite polynomials, Laguerre polynomials, and Zernike polynomials appear in Chapters 22, 23, and 26, respectively.

References

Deans, Stanley R., Radon and Abel Transforms, in *The Transforms and Applications Handbook*, edited by Alexander D. Poularikas, CRC Press, Boca Raton, FL, 1996.

Deans, Stanley R., *The Radon Transform and Some of Its Applications*, John Wiley & Sons, New York, NY, 1983.

Gradshteyn, I. S., I. M. Ryzhik, Yu. V. Geronimus, and M. Yu. Tseytlin, *Tables of Integrals, Series, and Products*, Academic Press, New York, NY, 1965.

17

The Hankel Transform

17.1 The Hankel Transform

17.1.1 Definition of the νth Order Hankel Transform

$$F_\nu(s) \equiv \mathcal{H}a_\nu\{f(r)\} = \int_0^\infty r f(r) J_\nu(sr)\,dr, \qquad r = \sqrt{x^2 + y^2}$$

$$f(r) \equiv \mathcal{H}a_\nu^{-1}\{F_\nu(s)\} = \int_0^\infty s F_\nu(s) J_\nu(sr)\,ds$$

17.1.2 The Zero-Order Hankel Transform

$$F(s) \equiv \mathcal{H}a_0\{f(r)\} = \int_0^\infty r f(r) J_0(sr)\,dr$$

$$f(r) \equiv \mathcal{H}a_0^{-1}\{F(s)\} = \int_0^\infty s F(s) J_0(sr)\,ds$$

17.1.3 Relation to Fourier Transform with Function of Circular Symmetry

$$\mathcal{F}\{f(\sqrt{x^2 + y^2})\} = F(u,v)$$

$$F(u,v) = 2\pi F(s) = 2\pi F(\sqrt{u^2 + v^2})$$

with $\mathcal{F}\{f(x,y)\} = \iint_{-\infty}^{\infty} f(x,y)\exp[-j(xu + yv)]\,dx\,dy$

Example

$\mathcal{F}\{\exp[-a(x^2 + y^2)]\} = \frac{\pi}{a}\exp[-(u^2 + v^2)/4a]$ and, therefore,

$$F(s) = \frac{1}{2\pi} F(u,v) = \frac{1}{2a} \exp[-s^2 / 4a], \quad s^2 = (u^2 + v^2), \quad a > 0$$

17.2 Properties of Hankel Transform

17.2.1 Derivatives

$$F_v(s) = \mathcal{H}a_v\{f(r)\}$$

$$G_v(s) = \mathcal{H}a_v\{f'(r)\} = s\left[\frac{v+1}{2v} F_{v-1}(s) - \frac{v-1}{2v} F_{v+1}(s)\right]$$

$$\mathcal{H}a_v\left\{\frac{d^2 f(r)}{dr^2} + \frac{1}{r}\frac{df(r)}{dr} - \left(\frac{v}{r}\right)^2 f(r)\right\} = -s^2 \mathcal{H}a_v\{f(r)\}$$

Note: $\frac{d}{dr}[rJ_v(sr)] = \frac{sr}{2v}[(v+1)J_{v-1}(sr) - (v-1)J_{v+1}(sr)]$

Example

$$\mathcal{F}\left\{\frac{1}{r}\frac{d}{dr}\left[r\frac{df(r)}{dr}\right]\right\} = -(u^2 \quad v^2)F(u,v).$$

But from 17.1.3

$$\mathcal{F}\left\{\frac{1}{r}\frac{d}{dr}\left[r\frac{df(r)}{dr}\right] = \frac{d^2 f(r)}{dr^2} + \frac{1}{r}\frac{df(r)}{dr}\right\} = -2\pi s^2 F(s)$$

and hence

$$\mathcal{H}a_0\left\{\frac{d^2 f(r)}{dr^2} + \frac{1}{r}\frac{df(r)}{dr}\right\} = -s^2 F(s) \qquad s^2 \mathcal{H}a_0\{f(r)\}$$

17.2.2 Similarity

$$\mathcal{H}a_v\{f(ar)\} = \frac{1}{a^2} F_v\left(\frac{s}{a}\right)$$

Example (see 17.1.3)

$$\mathcal{F}\left\{f(a\sqrt{x^2+y^2})\right\} = \mathcal{F}\left\{f(\sqrt{(ax)^2+(ay)^2})\right\} = \iint f(\sqrt{(ax)^2+(ay)^2})\exp[-jux - jvy]\,dx\,dy$$

$$= \frac{1}{a^2}\iint f(\sqrt{t^2+\tau^2})\exp[-j\tfrac{u}{a}t - j\tfrac{v}{a}\tau]\,dt\,d\tau \quad \frac{1}{a^2} 2\pi F\left(\frac{s}{a}\right).$$

Hence

$$\mathcal{H}a_0\{f(ar)\} = \frac{1}{a^2} F\left(\frac{s}{a}\right)$$

17.2.3 Division by r

1. $\mathcal{H}a_v\{r^{-1}f(r)\} = \dfrac{s}{2v}\left[F_{v-1}(s) + F_{v+1}(s)\right]$

2. $\mathcal{H}a_v\{r^{v-1}\dfrac{d}{dr}[r^{1-v}f(r)]\} = sF_{v-1}(s)$

3. $\mathcal{H}a_v\left\{r^{-v-1}\dfrac{d}{dr}\left[r^{v+1}f(r)\right]\right\} = sF_{v+1}(s)$

17.2.4 Parseval's Theorem

$$F_v(s) = \mathcal{H}a_v\{f(r)\}, \quad G_v(s) = \mathcal{H}a_v\{g(r)\}$$

1. $\int_0^\infty F_v(s)G_v(s)s\,ds = \int_0^\infty rg(r)f(r)\,dr$

2. $\int_0^\infty F_v(s)G_v^*(s)s\,ds = \int_0^\infty rf(r)g^*(r)\,dr$ for complex signals

3. $\int_0^\infty r|f(r)|^2\,dr = \int_0^\infty s|F(s)|^2\,ds$

17.2.5 Convolution Identity

$$\mathcal{F}_{(2)}\left\{\iint_{-\infty}^{\infty} f_1(\sqrt{x_1^2+y_1^2})f_2(\sqrt{(x-x_1)^2+(y-y_1)^2})\,dx_1\,dy_1\right\} = 4\pi^2F_1(s)F_2(s)$$

Hence $\mathcal{H}a_0\{f_1(r)**f_2(r)\} = \dfrac{1}{2\pi}\mathcal{F}_{(2)}\{f_1(r)**f_2(r)\} = 2\pi F_1(s)F_2(s)$

Also $\mathcal{H}a_0\{2\pi f_1(r)f_2(r)\} = F_1(s)**F_2(s)$

17.2.6 Moment

$$m_n = \int_0^\infty r^n f(r)\,dr$$

But

$$J_0(sr) = 1 - \left(\frac{sr}{2}\right)^2 + \frac{1}{(2!)^2}\left(\frac{sr}{2}\right)^4 - \cdots = \sum_{n=0}^{\infty}\frac{(-1)^n}{(n!)^2}\left(\frac{sr}{2}\right)^{2n}$$

hence

$$F(s) = \mathcal{H}a_0\{f(r)\} = \sum_{n=0}^{\infty}\frac{(-1)^n}{(n!)^2}\left(\frac{s}{2}\right)^{2n}\int_0^\infty r^{2n+1}f(r)\,dr = \sum_{n=0}^{\infty}\frac{(-1)^n m_{2n+1}}{(n!)^2 2^{2n}}s^{2n}$$

17.3 Examples of Hankel Transform

17.3.1 Example

From

$$\int_0^a rJ_0(sr)\,dr = \int_0^a \frac{1}{s}\frac{d}{dr}[rJ_1(sr)] = [aJ_1(as)]/s$$

implies that $\mathcal{H}a_0\{p_a(r)\} = \frac{a}{s}J_1(as)$, where $p_a(r) = 1$ for $r < a$ and zero otherwise.

17.3.2 Example

From $\int_0^a J_0(sr)dr = \frac{1}{s}$, $s > 0$ we obtain $\mathcal{H}a_0\{1/r\} = 1/s$.

17.3.3 Example

From $\int_0^a r\delta(r-a)J_0(sr)dr = aJ_0(as)$ we obtain $\mathcal{H}a_0\{\delta(r-a)\} = aJ_0(as)$, $s > 0$ and because of symmetry $\mathcal{H}a_0\{aJ_0(ar)\} = \delta(s-a)$, $>a \;\; 0$

17.3.4 Example

If $f_1(r) = f_2(r) = [J_1(ar)]/r$ then from 17.2.5 $\mathcal{H}a_0\{2\pi J_1^2(ar)/r^2\} = \frac{1}{a^2}p_a(s)**p_a(s)$ where

$$p_a(s)**p_a(s) = \left(2\cos^{-1}\frac{s}{2a} - \frac{s}{a}\sqrt{1-\frac{s^2}{4a^2}}\right)a^2.$$

Hence

$$\mathcal{H}a_0\{2\pi J_1^2(ar)/r^2\} = \left(2\cos^{-1}\frac{s}{2a} - \frac{s}{a}\sqrt{1-\frac{s^2}{4a^2}}\right)p_{2a}(s),$$

where $p_{2a}(s) = 1$ for $|s| \le 2a$ and 0 otherwise.

17.3.5 Example

From the relationship

$$\int_0^a rJ_0(br)J_0(sr)\,dr = a[bJ_1(ab)J_0(as) - sJ_0(ab)J_1(as)]/(b^2 - s^2)$$

we find

$$\mathcal{H}a_0\{J_0(br)p_a(r)\} = [abJ_1(ab)J_0(as) - asJ_0(ab)J_1(as)]/(b^2 - s^2).$$

17.3.6 Example

From $\delta(s-a)**\delta(s-a) = 4a^2/(s\sqrt{4a^2-s^2})$ for $s \quad 2a$ and equals zero for $s > 2a$, 17.2.5 and 17.3.3 we obtain $\mathcal{H}a_0\{J_0(ar)J_0(ar)\} = 2p_{2a}(s)/(\pi s\sqrt{4a^2-s^2})$.

17.3.7 Example

From $p_a(s) ** \delta(s-a) = 2a\cos^{-1}(s/2a)$ for s $2a$ and cycles zero for $s > 2a$, from $\mathcal{H}a_0\{J_0(ar)\} = \delta(s-a)/a$; from $\mathcal{H}a_0\{J_1(ar)/r\} = p_a(s)/a$ and 17.2.5 we obtain $\mathcal{H}a_0\{J_0(ar)J_1(ar)/r\} = p_{2a}(s)\ \cos^{-1}(s/2a)/(a\pi)$.

17.3.8 Example

$\mathcal{H}a_v\{r^v u(a-r)\} = \int_0^a r^{v+1} J_v(sr)\,dr = \frac{1}{s^{v+2}} \int_0^{as} x^{v+1} J_v(x)\,dx,\ a > 0$ where $u(a-r) = 1$ for r a and 0 for $r > a$ is the unit step function. But $\int x^v J_{v-1}(x)\,dx = x^v J_v(x) + C$ (see 25.3.2) and, hence,

$$\mathcal{H}a_v\{r^v h(a-r)\} = \frac{(as)^{v+1}}{s^{v+2}} J_{v+1}(as) = a^{v+1} J_{v+1}(as),\ a > 0,\ v > \ 1/2.$$

17.3.9 Example

$\mathcal{H}a_0\{e^{-ar}\} = L\{rJ_0(sr)\} = -\frac{d}{da}[(s^2+a^2)^{-1/2}] = a/[s^2+a^2]^{3/2},\ \ a > 0$

17.4 Relation to Fourier Transform

17.4.1 Relationship Between Fourier and Hankel Transform

If $F(s)$ is the Hankel transform of $f(r)$, then

1. $2\pi F(\sqrt{u^2+v^2}) = F(u,v) = \mathcal{F}\{f\sqrt{x^2+y^2})$
2. $\Phi(\omega) = \int_{-\infty}^{\infty} e^{-j\omega x} \varphi(x)\,dx$

   $\varphi(x) = \int_{-\infty}^{\infty} f(\sqrt{x^2+y^2})\,dy$

   $\mathcal{F}\{\varphi(x)\} = 2\pi F(s)\big|_{s=\omega}$

17.4.2 Example

If $f(r) = p_a(r)$, then

$$\varphi(x) = \int_{-\sqrt{a^2-x^2}}^{\sqrt{a^2-x^2}} dy = 2\sqrt{a^2\ \ x^2}$$

for $|x| < a$, and $\varphi(x) = 0$ for $|x| > a$. But $\mathcal{H}a_0\{p_a(r)\} = aJ_1(as)/s$ (see 17.3.1) and, hence,

$$\mathcal{F}\{2\sqrt{a^2-x^2}\ p_a(x)\} = 2\pi J_1(a\omega)/\omega\,.$$

17.4.3 Example

If $f(r) = p_a(r)/\sqrt{a^2-x^2}$, then

$$\varphi(x) = \int_{-\sqrt{a^2-x^2}}^{\sqrt{a^2-x^2}} \frac{dy}{\sqrt{a^2-(x^2+y^2)}} = \int_{-\pi/2}^{\pi/2} d\theta = \pi$$

for $|x| < a$ and equals zero for $|x| > a$.
Hence

$$\Phi(\omega) = \int_{-a}^{a} \pi e^{-j\omega x} d = \frac{2\pi \sin a\omega}{\omega} = 2\pi F(s)\Big|_{s=\omega}$$

which implies that $\mathcal{H}a_0\{p_a(r)/\sqrt{a^2-r^2}\} = \sin as / s$.

17.5 Hankel Transforms of Order Zero

Table 17.1 lists the Hankel transforms of some particular functions for the important special case $\nu = 0$. Table 17.2 lists Hankel transforms of general order ν. In these tables, $u(x)$ is the unit step function, I_ν and K_ν are modified Bessel functions, $\mathbf{L}_0$ and $\mathbf{H}_0$ are Struve functions, and Ker and Kei are Kelvin functions as defined in Abramowitz and Stegun.

TABLE 17.1 Hankel Transform of Order Zero

$f(r)$	$F_0(s) = \mathcal{H}a_0\{f(r);s\}$
Algebraic function	
$1/r$	$1/s$
$1/r^{\mu}, \quad \frac{1}{2} < \mu < 2$	$2^{1-\mu}\dfrac{\Gamma\left(1-\frac{\mu}{2}\right)}{\Gamma\left(\frac{\mu}{2}\right)}\dfrac{1}{s^{2-\mu}}$
$\dfrac{1}{(a^2+r^2)^{1/2}}, \quad \mathrm{Re}(a) > 0$	$\dfrac{e^{-as}}{s}$
$\begin{cases} \dfrac{1}{(a^2-r^2)^{1/2}}, & 0 < r < a \\ 0, & a < r < \infty \end{cases}$	$\dfrac{\sin(as)}{s}$
$\dfrac{1}{(r^2+a^2)^{3/2}}, \quad \mathrm{Re}(a) > 0$	$a^{-1}e^{-as}$
$\dfrac{1}{r(r+a)}$	$\dfrac{\pi}{2}[\mathbf{H}_0(as) - Y_0(as)]$
$\dfrac{1}{r^2+a^2}$	$K_0(as)$
$\dfrac{1}{r(r^2+a^2)}$	$\dfrac{\pi}{2a}[I_0(as) - \mathbf{L}_0(as)]$
$\dfrac{1}{1+r^4}$	$-\mathrm{Kei}(s)$
$\dfrac{1}{a^4+r^4} \quad \lvert\arg a\rvert < \pi/4$	$-a^{-2}\mathrm{Kei}(as)$

TABLE 17.1 Hankel Transform of Order Zero (continued)

$f(r)$	$F_0(s) = \mathcal{H}a_0\{f(r);s\}$
$\dfrac{r^2}{1+r^4}$	$\mathrm{Ker}(s)$
$\dfrac{1}{\sqrt{r^4+a^4}}$	$K_0(as/\sqrt{2})J_0(as/\sqrt{2})$
Exponential function	
e^{-ar}, $\mathrm{Re}(a)>0$	$\dfrac{s}{(s^2+a^2)^{3/2}}$
$\dfrac{e^{-ar}}{r}$	$\dfrac{1}{(s^2+a^2)^{1/2}}$
$(1-e^{-ar})r^{-2}$, $\mathrm{Re}(a)>0$	$\sinh^{-1}(a/s)$
$r^{n-\frac{1}{2}}e^{-ar}$, $\mathrm{Re}(a)>0$	$\dfrac{n!}{(a^2+s^2)^{\frac{1}{2}n+\frac{1}{2}}}P_n\left[\dfrac{a}{(a^2+s^2)^{1/2}}\right]$
$e^{-a^2r^2}$	$\dfrac{e^{-s^2/4a^2}}{2a^2}$
Trigonometric function	
$\dfrac{\sin ar}{r}$, $a>0$	$\begin{cases}\dfrac{1}{(a^2-s^2)^{1/2}}, & 0<s<a\\ 0 & a<s<\infty\end{cases}$
$\dfrac{\sin ar}{r^2}$, $a>0$	$\begin{cases}\frac{1}{2}\pi, & 0<s<a\\ \sin^{-1}(a/s), & a<s<\infty\end{cases}$
$\dfrac{\sin ar}{b^2+r^2}$, $a>0$ $\mathrm{Re}(b)>0$	$\dfrac{\pi}{2}e^{-ab}I_0(sb)$, $0<s<a$
$\dfrac{\cos ar}{r}$, $a>0$	$\begin{cases}0, & 0<s<a\\ (s^2-a^2)^{-1/2}, & a<s<\infty\end{cases}$
$\dfrac{1-\cos ar}{r^2}$, $a>0$	$\begin{cases}\cosh^{-1}(a/s), & 0<s<a\\ 0, & a<s<\infty\end{cases}$
$\dfrac{\cos ar}{b^2+r^2}$, $a>0$ $\mathrm{Re}(b)>0$	$\cosh(ab)K_0(bs)$, $a<s<\infty$
$\cos(a^2r^2/2)$, $a>0$	$a^{-2}\sin(a^{-2}s^2/2)$
Other functions	
$\dfrac{1-J_0(ar)}{r^2}$, $a>0$	$\begin{cases}\ln\dfrac{a}{s}, & s<a\\ 0, & s>a\end{cases}$
$\dfrac{J_1(ar)}{r}$, $a>0$	$\begin{cases}a^{-1}, & 0<s<a\\ 0, & a<s<\infty\end{cases}$

TABLE 17.2 Hankel Transform of Order ν^{th}

$f(r)$	$F_\nu(s) = \mathcal{H}a_\nu\{f(r);s\}$
Algebraic functions	
$1/r, \quad \mathrm{Re}(\nu) > -1$	$1/s$
$1/r^{\mu}, \quad \frac{1}{2} < \mu < \nu + 2$	$\dfrac{2^{1-\mu}\Gamma\left(\frac{\nu+2-\mu}{2}\right)}{s^{2-\mu}\Gamma\left(\frac{\nu+\mu}{2}\right)}$
$\begin{cases} r^{\nu}, & 0 < r < 1, \ \mathrm{Re}(\nu) > -1 \\ 0, & 1 < r < \infty \end{cases}$	$\dfrac{J_{\nu+1}(s)}{s}$
$\dfrac{r^{\nu}}{r^2 + a^2}, \quad \mathrm{Re}(a) > 0, \ -1 < \mathrm{Re}(\nu) < \frac{3}{2}$	$a^{\nu} K_{\nu}(as)$
$r^{\nu}(a^2 - r^2)^{\mu}\, u(a-r), \quad \mathrm{Re}(\nu) > 1, \quad -$ $0 < r < a, \quad \mathrm{Re}(\mu) > -1$	$2^{\mu}\Gamma(\mu+1) s^{-\mu-1} a^{\nu+\ +1} J_{\nu+\mu+1}(as)$
$\dfrac{r^{\nu}}{(r^2 + a^2)^{\nu+\frac{1}{2}}}, \quad \mathrm{Re}(a) > 0, \ \mathrm{Re}(\nu) > \frac{1}{2} \quad -$	$\dfrac{\sqrt{\pi}\, s^{\nu-1}}{2^{\nu} e^{as}\, \Gamma(\nu + \frac{1}{2})}$
$\dfrac{r^{\nu}}{(r^2 + a^2)^{\mu+1}}, \quad \mathrm{Re}(a) > 0$ $-1 < \mathrm{Re}(\nu) < 2\,\mathrm{Re}(\mu) + \frac{3}{2}$	$\dfrac{a^{\nu-\mu} s^{\mu} K_{\nu-\mu}(as)}{2^{\mu}\, \Gamma(\mu+1)}$
$\dfrac{r^{\nu}}{(a^2 - r^2)^{\nu+\frac{1}{2}}} u(a-r), \quad 0 < r < a,$ $\lvert\mathrm{Re}(\nu)\rvert < \frac{1}{2}$	$\pi^{-1/2} 2^{-\nu}\, \Gamma(\frac{1}{2} - \nu) s^{\nu-1} \sin(as)$
$\dfrac{r^{\nu}}{(r^4 + 4a^4)^{\nu+\frac{1}{2}}}, \quad \lvert\arg(a)\rvert < \pi/4$ $\lvert\mathrm{Re}(\nu)\rvert > -\frac{1}{2}$	$\dfrac{s^{\nu}\sqrt{\pi}\, J_{\nu}(as) K_{\nu}(as)}{a^{2\nu} 2^{3\nu} e^{as}\, \Gamma(\nu + \frac{1}{2})}$
$\dfrac{r^{\nu+2}}{(r^4 + 4a^4)^{\nu+\frac{1}{2}}}, \quad \lvert\arg(a)\rvert < \pi/4$ $\lvert\mathrm{Re}(\nu)\rvert > \frac{1}{6}$	$\dfrac{\sqrt{\pi}\, s^{\nu} J_{\nu-1}(as) K_{\nu-1}(as)}{2^{3\nu-1} a^{2\nu-1}\, \Gamma(\nu + \frac{1}{2})}$
Exponential function	
$\dfrac{e^{-ar}}{r}, \quad \mathrm{Re}(a) > 0, \ \mathrm{Re}(\nu) > 1 \quad -$	$s^{-\nu}(s^2 + a^2)^{-\frac{1}{2}}[(a^2 + s^2)^{\frac{1}{2}} \quad a]^{\nu} -$
$\dfrac{e^{-ar}}{r^2}, \quad \mathrm{Re}(a) > 0, \ \mathrm{Re}(\nu) > 0$	$\nu^{-1} s^{-\nu}[(a^2 + s^2)^{\frac{1}{2}} - a]^{\nu}$
$r^{\nu} e^{-ar}, \quad \mathrm{Re}(a) > 0, \ \mathrm{Re}(\nu) > 1 \quad -$	$\dfrac{1}{\sqrt{\pi}} 2^{\nu+1}\Gamma(\nu + \frac{3}{2}) a s^{\nu} \dfrac{1}{(a^2 + s^2)^{\nu+\frac{3}{2}}}$
$r^{\nu-1} e^{-ar}, \quad \mathrm{Re}(a) > 0, \ \mathrm{Re}(\nu) > \frac{1}{2} \quad -$	$\dfrac{1}{\sqrt{\pi}} 2^{\nu}\Gamma(\nu + \frac{1}{2}) s^{\nu} \dfrac{1}{(a^2 + s^2)^{\nu+\frac{1}{2}}}$
$r^{\nu} e^{-ar^2}, \quad \mathrm{Re}(a) > 0, \ \mathrm{Re}(\nu) > 1 \quad -$	$\dfrac{s^{\nu}}{(2a)^{\nu+1}} \exp\left(-\dfrac{s^2}{4a}\right)$
Trigonometric functions	
$\dfrac{\sin ar}{r}, \quad a > 0, \ \mathrm{Re}(\nu) > 2 \quad -$	$\cos(\pi\nu/2) s^{\nu}(a^2 - s^2)^{-\frac{1}{2}}[a + (a^2 - s^2)^{\frac{1}{2}}]^{-\nu}, \quad 0 < s < a$ $(s^2 - a^2)^{-\frac{1}{2}} \sin[\nu \sin^{-1}(a/s)], \quad a < s <$

TABLE 17.2 Hankel Transform of Order ν^{th} (continued)

$f(r)$	$F_\nu(s) = \mathcal{H}a_\nu\{f(r);s\}$
$\frac{\sin ar}{r^2}$, $a>0$, $\mathrm{Re}(\nu) > 1$ –	$\nu^{-1}\sin(\nu\pi/2)s^\nu[a+(a^2-s^2)^{\frac{1}{2}}]^{-\nu}$, $0<s<a$ $\nu^{-1}\sin[\nu\sin^{-1}(a/s)]$, $a<s<\infty$
$r^\nu \sin ar$, $a>0$, $-\frac{3}{2}<\mathrm{Re}(\nu)<\frac{1}{2}$ –	$\frac{-2^{\nu+1}\sin\nu\pi\,\Gamma(\nu+\frac{3}{2})as^\nu}{\sqrt{\pi}\,(a^2-s^2)^{-\nu-\frac{3}{2}}}$, $0<s<a$ $\frac{-2^{\nu+1}\Gamma(\nu+\frac{3}{2})as^\nu}{\sqrt{\pi}\,(s^2-a^2)^{-\nu-\frac{3}{2}}}$, $a<s<\infty$
$r^{\nu-1}\sin ar$, $a>0$, $-1<\mathrm{Re}(\nu)<\frac{1}{2}$	$\frac{\sqrt{\pi}\;2^\nu\;s^\nu}{\Gamma(\frac{1}{2}-\nu)\;(a^2-s^2)^{-\nu-\frac{1}{2}}}$, $0<s<a$ 0, $a<s<\infty$
$r^\nu \cos ar$, $a>0$, $-1<\mathrm{Re}(\nu)<\frac{1}{2}$ –	$\frac{2^{1+\nu}\sqrt{\pi}\;a\,s^\nu}{\Gamma(-\frac{1}{2}-\nu)\;(a^2-s^2)^{-\nu-\frac{3}{2}}}$, $0<s<a$ 0, $a<s<\infty$
Other functions	
$\frac{J_{\nu-1}(ar)}{r}$, $a>0$, $\mathrm{Re}(\nu)>1$ –	0, $0<s<a$ $a^{\nu-1}s^{-\nu}$, $a<s<\infty$
$\frac{J_\nu(ar)}{r^2}$, $a>0$, $\mathrm{Re}(\nu)>0$	$\frac{1}{2\nu}\frac{s^\nu}{a^\nu}$, $0<s<a$ $\frac{1}{2\nu}\frac{a^\nu}{s^\nu}$, $a\le s<\infty$
$\frac{J_{\nu+1}(ar)}{r}$, $a>0$, $\mathrm{Re}(\nu)>\frac{3}{2}$ –	$a^{-\nu-1}s^\nu$, $0<s<a$ 0, $a<s<\infty$

References

Bracewell, R. N., *The Fourier Transform and Its Applications*, McGraw-Hill Book Co., New York, 1978.

Davies, B., Integral *Transforms and Their Applications*, 2nd ed., Springer-Verlag, New York, 1984.

Erdelyi, A., W. Magnus, F. Oberhettinger, and F. G. Tricomi, *Tables of Integral Transforms*, McGraw-Hill Book Co., New York, 1954.

Papoulis, A., *Systems and Transforms with Applications in Optics*, McGraw-Hill Book Co., New York, 1968.

Piessens, R., The Hankel Transforms, in *The Transforms and Applications Handbook*, ed. Alexander D. Poularikas, CRC Press, Boca Raton, FL, 1996.

Sneddon, Ian N., *The Use of Integral Transforms*, McGraw-Hill Book Co., New York, 1972.

Wolf, K., B., *Integral Transforms in Science Engineering*, Plenum Press, New York, 1979.

18

The Mellin Transform

18.1 The Mellin Transform

18.1.1 Definition

$$M\{f(t);s\} = F(s) = \int_0^{\infty} f(t)t^{s-1}\,dt$$

Example

$$M\{e^{-at}u(t)\} = \int_0^{\infty} e^{-at}\,t^{s-1}\,dt = a^{-s}\Gamma(s)$$

18.1.2 Relation to Laplace Transform

By letting $t = e^{-x}$, $dt = \; e^{-x}\,dx$,– the transform becomes

$$M\{f(t)\} = F(s) = \int_{-\infty}^{\infty} f(e^{-x})e^{-sx}dx = L\,\{f(e^{-x})\}$$

18.1.3 Relation to Fourier Transform

By setting $s = \sigma + j2\pi\beta$ in 18.1.2 we obtain

$$F(s) = \int_{-\infty}^{\infty} f(e^{-x})e^{-ax}e^{-j2\pi\beta x}\,dx$$

which implies that

$$M\{f(t);s = \sigma + j2\pi\beta\} = F\{f(e^{-x})e^{-ax};\beta\}$$

where

$$\mathcal{F}\{f(x);\beta\} = \int_{-\infty}^{\infty} f(x)e^{-j2\pi\beta x}\,dx = \text{ Fourier Transform}$$

18.1.4 Inversion Formula

$$f(t) = \frac{1}{2\pi j}\int_{c-j\infty}^{c+j\infty} F(s)t^{-s}\,ds$$

where c is within the strip of analyticity $a < \operatorname{Re} s < b$.

18.2 Properties of Mellin Transform

18.2.1 Scaling Property

$$\mathcal{M}\{f(at);s\} = \int_0^{\infty} f(at)t^{s-1}\,dt = a^{-s}\int_0^{\infty} f(x)x^{s-1}\,dx = a^{-s}F(s)$$

18.2.2 Multiplication by t^a

$$\mathcal{M}\{t^a f(t);s\} = \int_0^{\infty} f(t)t^{(s+a)-1}\,dt = F(s+a)$$

18.2.3 Raising the Independent Variable to a Real Power

$$\mathcal{M}\{f(t^a);s\} = \int_0^{\infty} f(t^a)t^{s-1}\,dt = \int_0^{\infty} f(x)x^{\frac{s}{a}-\frac{1}{a}}\left(\frac{1}{a}x^{\frac{1}{a}-1}\,dx\right) = a^{-1}F\left(\frac{s}{a}\right),\quad a>0$$

18.2.4 Inverse of Independent Variable

$$\mathcal{M}\{t^{-1}f(t^{-1});s\} = F(1-s)$$

18.2.5 Multiplication by ln t

$$\mathcal{M}\{\ln t\, f(t);s\} = \frac{d}{ds}F(s)$$

18.2.6 Multiplication by a Power of ln t

$$\mathcal{M}\{(\ln t)^k f(t);s\} = \frac{d^k}{ds^k}F(s)$$

18.2.7 Derivative

$$\mathrm{M}\left\{\frac{d^k}{ds^k} f(t);s\right\} = (-1)^k (s-k)_k F(s-k)$$

$$(s-k)_k \equiv (s-k)(s-k+1)\cdots(s-1) = \frac{(s-1)!}{(s-k-1)!} = \frac{\Gamma(s)}{\Gamma(s-k)}$$

18.2.8 Derivative Multiplied by Independent Variable

$$\mathrm{M}\left\{t^k \frac{d^k}{ds^k} f(t);s\right\} = (-1)^k (s)_k F(s) = (-1)^k \frac{\Gamma(s+k)}{\Gamma(s)} F(s),\ (s)_k = s(s+1)\cdots(s+k-1)$$

Example

$$\mathrm{M}\left\{t^2 \frac{d^2 f(t)}{dt^2} + t\frac{df(t)}{dt};s\right\} = s^2 F(s)$$

18.2.9 Convolution

$$\mathrm{M}\{f(t)g(t);s\} = \frac{1}{2\pi j}\int_{c-j\infty}^{c+j\infty} F(z)G(s-z)\,dz$$

18.2.10 Multiplicative Convolution

$$\mathrm{M}\{f \vee g\} = \mathrm{M}\left\{\int_0^\infty f\left(\frac{t}{u}\right) g(u)\frac{du}{u};s\right\} = F(s)G(s)$$

$$\mathrm{M}^{-1}\{F(s)G(s)\} = \int_0^\infty f\left(\frac{t}{u}\right) g(u)\frac{du}{u}$$

Properties of the Multiplicative Convolution

$$\int_0^\infty f\left(\frac{t}{u}\right) g(u)\frac{du}{u}$$

1. $f \vee g = g \vee f$ commutative
2. $(f \vee g) \vee h = f \vee (g \vee h)$ associative
3. $f \vee \delta(t-1) = f$ unit element
4. $\left(t\frac{d}{dt}\right)^k (f \vee g) = \left[\left(t\frac{d}{dt}\right)^k f\right] \vee g = f \vee \left[\left(t\frac{d}{dt}\right)^k g\right]$
5. $(\ln t)(f \vee g) = [(\ln t) f] \vee g + f \vee [(\ln t) g]$
6. $\delta(t-a) \vee f = a^{-1} f(a^{-1} t)$

$$\delta(t-p) \vee \delta(t-p') = \delta(t-pp'), \quad p, p' > 0$$

$$\frac{d^k \delta(t-1)}{dt^k} \vee f = \left(\frac{d}{ds}\right)^k (t^k f)$$

18.2.11 Parseval's Formulas

$$\int_0^\infty f(t) g(t) = \frac{1}{2\pi j} \int_{c-j\infty}^{c+j\infty} M\{f;s\} M\{g;1-s\}\, ds$$

$$\int_0^\infty f(t) g^*(t) t^{2r+1}\, dt = \int_{-\infty}^{\infty} M\{f\}(\beta) M\{g\}(\beta)\, d\beta$$

where $$M\{f\}(\beta) = \int_0^\infty f(t) t^{2\pi j\beta + r}\, dt$$

18.3 Examples of Mellin Transform

18.3.1 Example

$$M\{t^a u(t-t_0)\} = \int_{t_0}^\infty t^{a+s-1}\, dt = \frac{t_o^{a+s}}{a+s}, \quad \text{Re}\{s\} < -a$$

18.3.2 Example

$$M\left\{\frac{1}{1+t}; s\right\} = \int_0^\infty \frac{1}{1+t} t^{s-1}\, dt.$$

Setting $t+1 = \frac{1}{1-x}$ we obtain: $x = \frac{t}{t+1}$, $dx = \frac{dt}{(1+t)^2}$. Hence ,

$$M\{f;s\} = \int_0^1 (1-x) \frac{x^{s-1}}{(1-x)^{s-1}} \frac{dx}{(1-x)^2} = \int_0^1 x^{s-1}(1-x)^{-s}\, dx = B(s, 1-s) = \Gamma(s)\Gamma(1-s) \quad \frac{\pi}{\sin \pi s}$$

$0 < \text{Re}\{s\} < 1.$

18.3.3 Example

From $\int_o^1 (1-u)^{m-1} u^{s-1}\, du = \frac{\Gamma(m)\Gamma(s)}{\Gamma(m+s)}$, $\text{Re}\{s\} > 0$, $\text{Re}\{m\} > 0$, with the setting $u = x/(1+x)$, we obtain

$$\int_0^\infty \frac{x^{s-1}}{(1+x)^{m+s}} dx = \frac{\Gamma(m)\Gamma(s)}{\Gamma(m+s)}.$$

Hence,

$$M\{(1+t)^{-a}; s\} = \frac{\Gamma(s)\Gamma(a-s)}{\Gamma(a)}, \quad 0 < \text{Re}\{s\} < \text{Re}\{a\}.$$

18.3.4 Example

Using 18.2.3 and 18.3.3 we obtain

$$\mathcal{M}\{(1+t^a)^{-b};s\} = \frac{\Gamma(s/a)\Gamma\left(b-\frac{s}{a}\right)}{a\Gamma(b)}, \quad 0 < \text{Re}\{s\} < \text{Re}\{ab\}.$$

18.3.5 Example

$$\mathcal{M}\{\delta(t-t_o);s\} = \int_0^\infty \delta(t-t_0)t^{s-1}dt = t_0^{s-1} \quad \text{(see 5.3.1)}$$

18.3.6 Example

From 18.3.1 and 18.3.5 $\mathcal{M}\{t^a u(t-t_0)\} = \dfrac{t_0^{a+s}}{a+s}$ and hence,

$$\mathcal{M}\left\{\frac{df}{dt};s\right\} = -(s-1)F(s-1) = -(s-1)\frac{t_0^{a+s-1}}{a+s-1} = -a\frac{t_0^{a+s-1}}{a+s-1} + t_0^{a+s-1}$$

$$= \mathcal{M}\{au(t-t_0)t^{a+s-1};s\} + \mathcal{M}\{t_0^a\,\delta(t-t_0);s\}$$

18.4 Special Functions Frequency Occurring in Mellin Transforms

18.4.1 Definition

The gamma function $\Gamma(s)$ is defined on the complex half-plane $\text{Re}(s) > 0$ by the integral

$$\Gamma(s) = \int_0^\infty e^{-t}t^{s-1}\,dt$$

18.4.2 Analytic Continuation

The analytical continued gamma function is holomorphic in the whole plane except at the points $s = -n, n = 0,1,2,\cdots$, where it has a simple pole.

18.4.3 Residues at the Poles

$$\text{Res}_{s=-n}(\Gamma(s)) = \frac{(-1)^n}{n!}$$

18.4.4 Relation to the Factorial

$$\Gamma(n+1) = n!$$

18.4.5 Functional Relations

$$\Gamma(s+1) = s\Gamma(s)$$

$$\Gamma(s)\Gamma(1-s) = \frac{\pi}{\sin(\pi s)}$$

$$\Gamma\left(\frac{1}{2}\right) = \sqrt{\pi}$$

$$\Gamma(2s) = \pi^{-1/2} 2^{2s-1} \Gamma(s)\Gamma(s+1/2)$$

(Legendre's duplication formula)

$$\Gamma(ms) = m^{ms-1/2} (2\pi)^{(1-m)/2} \prod_{k=0}^{m-1} \Gamma(s+k/m), \quad m = 2,3,\cdots,$$

(Gauss-Legendre's multiplication formula)

$$\Gamma(s) \sim \sqrt{2\pi}\, s^{s-1/2} \exp\left[-s\left(1+\frac{1}{12s}\right) + O(s^{-2})\right], \quad s \to \infty, < |\arg(s)| \quad \pi$$

(Stirling asymptotic formula)

18.4.6 The Beta Function

Definition: $B(x,y) \equiv \int_0^1 t^{x-1}(1-t)^{y-1}\, dt$

Relation to the gamma function: $B(x,y) = \dfrac{\Gamma(x)\Gamma(y)}{\Gamma(x+y)}$

18.4.7 The psi Function (logarithmic derivative of the gamma function)

Definition: $\psi(s) \equiv \dfrac{d}{ds} \ln \Gamma(s)$

$$= -\gamma + \sum_{n=0}^{\infty} \left(\frac{1}{n+1} - \frac{1}{s+n}\right)$$

Euler's constant γ, also called C, is defined by

$$\gamma \equiv -\Gamma'(1)/\Gamma(1)$$

and has value $\gamma \cong 0.577 \ldots.$

18.4.8 Riemann's Zeta Function

$$\zeta(z,q) \equiv \sum_{n=0}^{\infty} \frac{1}{(q+n)^z}, \quad \mathrm{Re}(z) > 1, \; q \neq 0,-1,-2,\cdots$$

$$\zeta(z) \equiv \sum_{n=0}^{\infty} \frac{1}{n^z}, \quad \mathrm{Re}(z) > 1$$

The function $\zeta(z)$ is analytic in the whole complex z-plane except in $z = 1$ where it has a simple pole with residue equal to +1.

18.5 Tables of Mellin Transform

18.5.1 Tables of Mellin Transform

TABLE 18.1 Some Standard Mellin Transform Pairs

Original Function	Mellin Transform	
$f(t),\ t>0$	$M[f;s]\equiv\int_0^\infty f(t)t^{s-1}\,dt$	Strip of holomorphy
Algebraic Functions		
$u(t-a)t^b,\ a>0$	$-\dfrac{a^{b+s}}{b+s}$	$\mathrm{Re}(s)<-\mathrm{Re}(b)$
$(u(t-a)-u(t))t^b$	$-\dfrac{a^{b+s}}{b+s}$	$\mathrm{Re}(s)>-\mathrm{Re}(b)$
$(1+t)^{-1}$	$\dfrac{\pi}{\sin(\pi s)}$	$0<\mathrm{Re}(s)<1$
$(a+t)^{-1},\ \lvert\arg a\rvert<\pi$	$\pi a^{s-1}\csc(\pi s)$	$0<\mathrm{Re}(s)<1$
$(1+t)^{-a}$	$\dfrac{\Gamma(s)\Gamma(a-s)}{\Gamma(a)}$	$0<\mathrm{Re}(s)<\mathrm{Re}(a)$
$(1-t)^{-1}$	$\pi\cot(\pi s)$	$0<\mathrm{Re}(s)<1$
$(a-t)^{-1},\ a>0$	$\pi a^{s-1}\cot(\pi s)$	$0<\mathrm{Re}(s)<1$
$u(1-t)(1-t)^{b-1}$, $\mathrm{Re}(b)\ \ 0$	$\dfrac{\Gamma(s)\Gamma(b)}{\Gamma(s+b)}$	$\mathrm{Re}(s)>0$
$u(t-1)(t-1)^{-b}$	$\dfrac{\Gamma(b-s)\Gamma(1-b)}{\Gamma(1-s)}$	$\mathrm{Re}(s)<\mathrm{Re}(b)<1$
$(t^2+a^2)^{-1},\ \mathrm{Re}(a)>0$	$\frac{1}{2}\pi a^{s-2}\csc\left(\dfrac{\pi s}{2}\right)$	$0<\mathrm{Re}(s)<2$
$(t^n+a),\ \lvert\arg a\rvert<\ \ \pi$	$(\pi/n)\csc(\pi s/n)a^{s/n-1}$	$0<\mathrm{Re}(s)<n$
$\begin{cases} t^\nu & 0<t<1 \\ 0 & t>1 \end{cases}$	$(s+\nu)^{-1}$	$\mathrm{Re}(s)>-\mathrm{Re}(\nu)$
$\begin{cases} (1-t^h)^{\nu-1} & 0<t<1 \\ 0 & t\geq 0 \\ h>0 & \mathrm{Re}(\nu)>0 \end{cases}$	$h^{-1}\dfrac{\Gamma(\nu)\Gamma\left(\dfrac{s}{h}\right)}{\Gamma\left(\nu+\dfrac{s}{h}\right)}$	$\mathrm{Re}(s)>0$
$(1-t^a)(1-t^{na})^{-1}$	$\dfrac{\pi}{na}\sin\left(\dfrac{\pi}{n}\right)\csc\left(\dfrac{\pi s}{na}\right)\csc\left(\dfrac{\pi s+\pi a}{na}\right)$	$0<\mathrm{Re}(s)<(n\ \ 1)a$
Exponential Functions		
$e^{-pt},\ p>0$	$p^{-s}\Gamma(s)$	$\mathrm{Re}(s)>0$
$(e^t-1)^{-1}$	$\Gamma(s)\zeta(s)$ ($\zeta(s)$ = zeta function)	$\mathrm{Re}(s)>1$
$(e^{at}+1)^{-1},\ \mathrm{Re}(a)>0$	$a^{-s}\Gamma(s)(1-2^{1-s})\zeta(s)$	$\mathrm{Re}(s)>0$
$(e^{at}-1)^{-1},\ \mathrm{Re}(a)>0$	$a^{-s}\Gamma(s)\zeta(s)$	$\mathrm{Re}(s)>1$
$(e^{-at})(1-e^{-t})^{-1},\ \mathrm{Re}(a)>0$	$\Gamma(s)\zeta(s,a)$	$\mathrm{Re}(s)>1$
$(e^t-1)^{-2}$	$\Gamma(s)[\zeta(s-1)-\zeta(s)]$	$\mathrm{Re}(s)>2$

TABLE 18.1 Some Standard Mellin Transform Pairs (continued)

Original Function	Mellin Transform	
$f(t),\ t>0$	$M[f;s]\equiv\int_0^\infty f(t)t^{s-1}\,dt$	Strip of holomorphy
$e^{-at^h},\ \mathrm{Re}(a)>0,\ h>0$	$h^{-1}a^{-s/h}\Gamma(s/h)$	$\mathrm{Re}(s)>0$
$t^{-1}e^{-t^{-1}}$	$\Gamma(1-s)$	$-\infty<\mathrm{Re}(s)<1$
e^{-t^2}	$\frac{1}{2}\Gamma(s/2)$	$0<\mathrm{Re}(s)<\quad\infty$
$1-e^{-at^h},\ \mathrm{Re}(a)>0,\ h>0$	$-h^{-1}a^{-s/h}\Gamma(s/h)$	$-h<\mathrm{Re}(s)<0$
e^{jat}	$a^{-s}\Gamma(s)e^{j\pi(s/2)}$	$0<\mathrm{Re}(s)<1$
Logarithmic Functions		
$\ln(1+t)$	$\dfrac{\pi}{s\sin(\pi s)}$	$-1<\mathrm{Re}(s)<0$
$\ln(1+at),\ \lvert\arg a\rvert<\pi$	$\pi s^{-1}a^{-s}\csc(\pi s)$	$-1<\mathrm{Re}(s)<0$
$u(p-1)\ln(p-t)$	$-p^s s^{-1}[\psi(s+1)+p^{-1}\ln\gamma]$	$\mathrm{Re}(s)>0$
$t^{-1}\ln(1+t)$	$\dfrac{\pi}{(1-s)\sin(\pi s)}$	$0<\mathrm{Re}(s)<1$
$\ln\left\lvert\dfrac{1+t}{1-t}\right\rvert$	$(\pi/s)\tan(\pi s/2)$	$-1<\mathrm{Re}(s)<1$
$\begin{cases}\ln t & 0<t<a\\ 0 & t>a\end{cases}$	$s^{-1}a^{-s}(\ln a-s^{-1})$	$\mathrm{Re}(s)>0$
$\begin{cases}t^{\nu}\ln t & 0<t<1\\ 0 & 1<t<\infty\end{cases}$	$-(s+\nu)^{-2}$	$\mathrm{Re}(s)>-\mathrm{Re}(\nu)$
$e^{-t}(\ln t)^n$	$\dfrac{d^n\Gamma(s)}{ds^n}$	$\mathrm{Re}(s)>0$
$u(t-1)\sin(a\ln t)$	$\dfrac{a}{s^2+a^2}$	$\mathrm{Re}(s)<-\lvert\mathrm{Im}(a)\rvert$
$u(1-t)\sin(-a\ln t)$	$\dfrac{a}{s^2+a^2}$	$\mathrm{Re}(s)>\lvert\mathrm{Im}(a)\rvert$
$(u(t)-u(t-p))\ln(p/t),\ p\quad 0$	$\dfrac{p^s}{s^2}$	$\mathrm{Re}(s)>0$
Trigonometric Functions		
$\sin(at),\ a>0$	$a^{-s}\Gamma(s)\sin(\pi s/2)$	$-1<\mathrm{Re}(s)<1$
$e^{-at}\sin(pt),\ \mathrm{Re}(a)>\lvert\mathrm{Im}\,\beta\rvert$	$(a^2+\beta^2)^{-s/2}\Gamma(s)\sin\left(s\tan^{-1}\dfrac{\beta}{a}\right)$	$\mathrm{Re}(s)>-1$
$\sin^2(at),\ a>0$	$-2^{-s-1}a^{-s}\Gamma(s)\cos(\pi s/2)$	$-2<\mathrm{Re}(s)<0$
$\cos(at),\ a>0$	$a^{-s}\Gamma(s)\cos(\pi s/2)$	$0<\mathrm{Re}(s)<1$
$\tan^{-1}(t)$	$\dfrac{-\pi}{2s\cos(\pi s/2)}$	$-1<\mathrm{Re}(s)<0$
$\mathrm{co}\tan^{-1}(t)$	$\dfrac{\pi}{2s\cos(\pi s/2)}$	$0<\mathrm{Re}(s)<1$

TABLE 18.1 Some Standard Mellin Transform Pairs (continued)

Original Function	Mellin Transform	
$f(t),\ t>0$	$M[f;s]\equiv\int_0^\infty f(t)t^{s-1}\,dt$	Strip of holomorphy
Other Functions		
$J_\nu(at),\ a>0$	$\dfrac{2^{s-1}\Gamma\left(\frac{s}{2}+\frac{\nu}{2}\right)}{a^s\Gamma\left(\frac{\nu}{2}-\frac{s}{2}+1\right)}$	$-\mathrm{Re}(\nu)<\mathrm{Re}(s)<3/2$
$\sin at\,J_\nu(at),\ a>0$	$\dfrac{2^{\nu-1}\Gamma\left(\frac{1}{2}-s\right)\Gamma\left(\frac{1}{2}+\frac{\nu}{2}+\frac{s}{2}\right)}{a^s\Gamma(1+\nu-s)\Gamma\left(1-\frac{\nu}{2}-\frac{s}{2}\right)}$	$-1<\mathrm{Re}(\nu)<\mathrm{Re}(s)<1/2$
$\delta(t-p),\ p>0$	p^{s-1}	whole plane
$\sum_{n=1}^{\infty}\delta(t-pn),\ p>0$	$p^{s-1}\zeta(1-s)$	$\mathrm{Re}(s)<0$
$J_\nu(t)$	$\dfrac{2^{s-1}\Gamma(s+\nu)/2}{\Gamma[(1/2)(\nu-s)+1]}$	$-\nu<\mathrm{Re}(s)<3/2$
$\sum_{n=-\infty}^{\infty}p^{-nr}\delta(t-p^n)$,	$\dfrac{1}{\ln p}\sum_{n=-\infty}^{\infty}\delta\left(\beta-\dfrac{n}{\ln p}\right)$,	$s=r+j\beta$
$p>0,\ r=\text{real}$	$\beta=\mathrm{Im}(s)$	
t^b	$\delta(b+s)$	none (analytic functional)

References

Bertrand, Jacqueline, Pierre Bertrand, and Jean-Philippe Ovarlez, The Mellin Transform, in *Transforms and Applications Handbook*, ed. Alexander Poularikas, CRC Press, Boca Raton, Florida, 1996.

Davies, G., *Integral Transforms and Their Applications*, 2nd ed., Springer-Verlag, New York, NY, 1984.

Erdelyi, A., W. Magnus, F. Oberhettinger, and F. G. Tricomi, *Tables of Integral Transfer*, McGraw-Hill Book Co., New York, NY, 1954.

Oberhettinger, F., *Tables of Mellin Transform*, 2nd ed., Springer-Verlag, New York, NY, 1974.

Sneddon, Ian N., *The Use of Integral Transform*, McGraw-Hill Book Co., New York, NY, 1972.

19

Time-Frequency Transformations

19.1 The Wigner Distribution

19.1.1 Definition of WD in Time Domain

$$WD_{x,g}(t;t,f)=\int_{-\infty}^{\infty}e^{-j2\pi f\tau}x\left(t+\frac{\tau}{2}\right)g^{*}\left(t-\frac{\tau}{2}\right)d\tau$$

$$WD_{x}(t;t,f)=WD_{x,x}(t;t,f)=\int_{-\infty}^{\infty}e^{-j2\pi f\tau}x\left(t+\frac{\tau}{2}\right)x^{*}\left(t-\frac{\tau}{2}\right)d\tau$$

19.1.2 Definition of WD in Frequency Domain

$$WD_{X,G}(f;f,t)=\int_{-\infty}^{\infty}e^{j2\pi t\nu}X\left(f+\frac{\nu}{2}\right)G^{*}\left(f-\frac{\nu}{2}\right)d\nu$$

$$WD_{X}(f;f,t)=\int_{-\infty}^{\infty}e^{j2\pi t\nu}X\left(f+\frac{\nu}{2}\right)X^{*}\left(f-\frac{\nu}{2}\right)d\nu$$

$$=\int\int_{-\infty}^{\infty} AF_{x}(t;\tau,\nu)e^{j2\pi(t\nu-f\tau)}\,d\tau\,d\nu \quad \text{[see 19.1.3 for } AF_{x}(t;\tau,\nu)]$$

$WD_{X,G}(f;f,t) = WD_{x,g}(t;t,f)$ which means that the WD of the spectra of two signals can be determined from that of time functions by an interchange of frequency and time variables.

19.1.3 Definition of Ambiguity Function (AF)

$$AF_x(t;\tau,\nu) = \int_{-\infty}^{\infty} x\left(t+\frac{\tau}{2}\right)x^*\left(t-\frac{\tau}{2}\right)e^{-j2\pi\nu t}\,dt$$

$$AF_X(t;\tau,\nu) = \int_{-\infty}^{\infty} X\left(f+\frac{\nu}{2}\right)X^*\left(f-\frac{\nu}{2}\right)e^{j2\pi\tau f}\,df$$

$$= \iint_{-\infty}^{\infty} WD_x(t;f)e^{-j2\pi(\nu t-\tau f)}\,dt\,df$$

19.1.4 Example [WD of the Delta Function $x(t) = \delta(t-t_o)$]

$$WD_x(t;f) = \int_{-\infty}^{\infty} e^{-j2\pi f\tau}\delta\left(t+\frac{\tau}{2}-t_0\right)\delta\left(t-\frac{\tau}{2}-t_0\right)d\tau$$

$$= \int_{-\infty}^{\infty} e^{-j2\pi f\tau}\delta\left(\frac{2t+\tau-2t_0}{2}\right)\delta\left(\frac{2t-\tau-2t_0}{2}\right)d\tau = 4e^{-j2\pi f2(t-t_0)}\delta[4(t-t_0)] = \delta(t-t_0)$$

19.1.5 Example [WD of the Delta Function $X(f) = \delta(f-f_0)$]

$$WD_X(f;t,f) = \int_{-\infty}^{\infty} e^{j2\pi\nu t}\delta\left(f+\frac{\nu}{2}-f_0\right)\delta\left(f-\frac{\nu}{2}-f_0\right)d\nu = \delta(f-f_0)$$

19.1.6 Example [WD of the Linear FM Signal $x(t) = e^{j\pi\alpha t^2}$, α = sweep rate]

$$WD_x(t;t,f) = \int_{-\infty}^{\infty} e^{j\pi\alpha(t+\frac{\tau}{2})^2}e^{-j\pi\alpha(t-\frac{\tau}{2})^2}e^{-j2\pi f\tau}\,d\tau = \int_{-\infty}^{\infty} e^{-j2\pi(f-\alpha t)\tau}\,d\tau = \delta(f-\alpha t)$$

19.1.7 Example [WD of the Gaussian Function $x(t) = \frac{1}{\sqrt{\sigma}}e^{-\pi(t/\sigma)^2}$]

$$WD_x(t;t,f) = \frac{1}{\sigma}\int_{-\infty}^{\infty} e^{-\pi[(t+\frac{\tau}{2})/\sigma]^2}e^{-\pi[(t-\frac{\tau}{2})/\sigma]^2}e^{-j2\pi f\tau}\,d\tau$$

$$= \frac{1}{\sigma}e^{-2\pi(t/\sigma)^2}\int_{-\infty}^{\infty} e^{-\pi\tau^2/(2\sigma^2)}e^{-j2\pi f\tau}\,d\tau = \sqrt{2}\,e^{-2\pi[(t/\sigma)^2+(\sigma f)^2]}$$

19.2 Properties of the Wigner Distribution

19.2.1 Conjugation

$$WD^*_{g,x}(t;t,f) = \int_{-\infty}^{\infty} e^{j2\pi f\tau}g^*(t+\tfrac{\tau}{2})x(t-\tfrac{\tau}{2})\,d\tau \quad \text{set } \tau=-\tau \text{ to find}$$

$$WD_{x,g}(t;t,f) = WD^*_{g,x}(t;t,f)$$

19.2.2 Real-Valued

From 19.2.1 $\quad WD_x(t;t,f) = WD_x^*(t;t,f) \equiv \text{real} \;\; (WD_{x,x} \equiv WD_x)$

19.2.3 Even in Frequency (real function)

$$WD_{x^*}(t;t,f) = \int_{-\infty}^{\infty} e^{-j2\pi f\tau} x^*\left(t+\frac{\tau}{2}\right) x\left(t-\frac{\tau}{2}\right) d\tau$$

$$= -\int_{\infty}^{-\infty} e^{-j2\pi(-f)\tau} x^*\left(t-\frac{\tau}{2}\right) x\left(t+\frac{\tau}{2}\right) d\tau = WD_x(t;t,-f)$$

Also

$$WD_x(t;t,f) = \int_{-\infty}^{\infty} e^{-j2\pi f\tau} x\left(t+\frac{\tau}{2}\right) x^*\left(t-\frac{\tau}{2}\right) d\tau$$

$$= -\int_{\infty}^{-\infty} e^{-j2\pi(-f)\tau} x^*\left(t+\frac{\tau}{2}\right) x\left(t-\frac{\tau}{2}\right) dt = WD_{x^*}(t;t,-f)$$

19.2.4 Time Shift

If $x_s(t) = x(t-t_0)$ and $g_s(t) = g(t-t_0)$, then

$$WD_{x_s,g_s}(t;t,f) = \int_{-\infty}^{\infty} e^{-j2\pi f\tau} x\left(t-t_0+\frac{\tau}{2}\right) g^*\left(t-t_0-\frac{\tau}{2}\right) d(t-t_0) = WD_{x,g}(t-t_0;t-t_0,f)$$

19.2.5 Frequency Shift

If $x_s(t) = x(t)\, e^{j2\pi f_0 t}$ and $g_s(t) = g(t)\, e^{j2\pi f_0 t}$ are modulated, then

$$WD_{x_s,g_s}(t;t,f) = \int_{-\infty}^{\infty} e^{-j2\pi f\tau} x\left(t+\frac{\tau}{2}\right) e^{j2\pi f_0\left(t+\frac{\tau}{2}\right)} g^*\left(t-\frac{\tau}{2}\right) e^{-j2\pi f_0\left(t-\frac{\tau}{2}\right)} d\tau$$

$$= \int_{-\infty}^{\infty} e^{-j2\pi(f-f_0)\tau} x\left(t+\frac{\tau}{2}\right) g^*\left(t-\frac{\tau}{2}\right) d\tau = WD_{x,g}(t;t,f-f_0)$$

19.2.6 Time and Frequency Shifts

If $x_{sm}(t) = x(t-t_0)\, e^{j2\pi f_0(t-t_0)}$ and $g_{sm}(t) = g(t-t_0)\, e^{j2\pi f_0(t-t_0)}$, then

$$WD_{x_{sm}g_{sm}}(t;t,f) = WD_{x,g}(t;t-t_0,f-f_0)$$

19.2.7 Ordinates

$$WD_{x,g}(t;0,0) = \int_{-\infty}^{\infty} x\left(\frac{\tau}{2}\right) g^*\left(-\frac{\tau}{2}\right) d\tau$$

19.2.8 Sum of Functions

$$W_{g+x}(t;t,f)=\int_{-\infty}^{\infty} e^{-j2\pi f\tau}\left[x\left(t+\frac{\tau}{2}\right)g\left(t+\frac{\tau}{2}\right)\right]\left[x^{*}\left(t-\frac{\tau}{2}\right)g^{*}\left(t-\frac{\tau}{2}\right)\right]d\tau$$

$$=WD_{x}(t;t,f)+WD_{x,g}(t;t,f)+WD_{g,x}(t;t,f)+WD_{g}(t;t,f)$$

$$=WD_{x}(t;t,f)+WD_{g}(t;t,f)+2\,\mathrm{Re}\{W_{x,g}(t;t,f)\}\quad \text{since } W^{*}_{x,g}=W_{gx}$$

19.2.9 Multiplication by *t*

$$WD_{tx,g}(t;t,f)+WD_{x,tg}(t;t,f)=\int_{-\infty}^{\infty} e^{-j2\pi f\tau}\left(t+\frac{\tau}{2}\right)x\left(t+\frac{\tau}{2}\right)g^{*}\left(t-\frac{\tau}{2}\right)d\tau$$

$$+\int_{-\infty}^{\infty} e^{-j2\pi f\tau}x\left(t+\frac{\tau}{2}\right)\left(t-\frac{\tau}{2}\right)g^{*}\left(t-\frac{\tau}{2}\right)d\tau$$

$$=2t\int_{-\infty}^{\infty} e^{-j2\pi f\tau}x\left(t+\frac{\tau}{2}\right)g^{*}\left(t-\frac{\tau}{2}\right)d\tau=2tWD_{x,g}(t;t,f)$$

19.2.10 Multiplication by *f*

$$2f\,WD_{x,g}(t;t,f)=WD_{fx,g}(t;t,f)+WD_{x,fg}(t;t,f)$$

19.2.11 Multiplication by *t* and *f*

$$2ft\,WD_{x,g}(t;t,f)=WD_{tx,fg}(t;t,f)+WD_{fx,tg}(t;t,f)$$

19.2.12 Fourier Transform

Since for a specific t the WD is the Fourier transform of $x\left(t+\frac{\tau}{2}\right)g^{*}\left(t-\frac{\tau}{2}\right)$ implies that

$$x\left(t+\frac{\tau}{2}\right)g^{*}\left(t-\frac{\tau}{2}\right)=F^{-1}\{WD_{x,g}(t;t,f)\}=\int_{-\infty}^{\infty} e^{j2\pi f\tau}WD_{x,g}(t;t,f)\,df$$

19.2.13 Time Marginal

In 19.2.12 set $t+\frac{\tau}{2}=t_1$ and $t-\frac{\tau}{2}=t_2$ which implies that $t=\frac{t_1+t_2}{2}$ and $\tau=t_1-t_2$, and hence

$$\int_{-\infty}^{\infty} e^{j2\pi f(t_1-t_2)}WD_{x,g}\left(t;\frac{t_1+t_2}{2},f\right)df=x(t_1)g^{*}(t_2).$$

If we further set $t_1=t_2=t$, we obtain

1. $\int_{-\infty}^{\infty} WD_{x,g}(t;t,f)\,df=x(t)g^{*}(t).$

If x = g, we obtain

2. $\int_{-\infty}^{\infty} WD_x(t;t,f)df = |x(t)|^2$.

19.2.14 Recovery of Time Function

If we set $t_1 = t$ and $t_2 = 0$ the 19.2.13 becomes

$$\int_{-\infty}^{\infty} e^{j2\pi ft} WD_{x,g}\left(t;\frac{t}{2},f\right)df = F^{-1}\left\{WD_{x,g}\left(t;\frac{t}{2},f\right)\right\} = x(t)g^*(0)$$

$$\int_{-\infty}^{\infty} e^{j2\pi ft} WD_x\left(t;\frac{t}{2},f\right)df = F^{-1}\left\{WD_{x,g}\left(t;\frac{t}{2},f\right)\right\} = x(t)x^*(0)$$

which indicate that we can retrieve the function from its WD within a constant.

19.2.15 Frequency Marginal

From 19.1.2

$$\int_{-\infty}^{\infty} WD_{X,G}(f;t,f)dt = \int_{-\infty}^{\infty}\int_{-\infty}^{\infty} e^{j2\pi t\nu} X\left(f+\frac{\nu}{2}\right)G^*\left(f-\frac{\nu}{2}\right)d\nu\, dt$$

$$= \int_{-\infty}^{\infty} X\left(f+\frac{\nu}{2}\right)G^*\left(f-\frac{\nu}{2}\right)d\nu \int_{-\infty}^{\infty} e^{j2\pi t\nu}\, dt$$

$$= \int_{-\infty}^{\infty} \delta(\nu) X\left(f+\frac{\nu}{2}\right)G^*\left(f-\frac{\nu}{2}\right)d\nu = X(f)G^*(f)$$

For $X(f) = G(f)$ implies that $\int_{-\infty}^{\infty} WD_X(f;t,f)dt = |X(f)|^2 = \int_{-\infty}^{\infty} WD_x(t;t,f)dt$.

19.2.16 Total Energy

From 19.2.15

$$\int_{-\infty}^{\infty}\int_{-\infty}^{\infty} WD_{X,G}(f;f,t)dt\,df = \int_{-\infty}^{\infty} X(f)G^*(f)df$$

and for $X(f) = G(f)$ we obtain

$$\int\int_{-\infty}^{\infty} WD_X(f;f,t)dt\,df = \int_{-\infty}^{\infty} |X(f)|^2 df = \|X(f)\| = E_x$$

where $\|\cdot\|$ is known as the norm. We can also write

$$\int\int_{-\infty}^{\infty} WD_x(f;t,f)dt\,df = \int_{-\infty}^{\infty} |X(f)|^2 df.$$

19.2.17 Time Moments

$$\int_{-\infty}^{\infty}\int t^n WD_{x,g}(t;t,f)\,dt\,df = \int\int_{-\infty}^{\infty}\int t^n x\left(t+\frac{\tau}{2}\right)g^*\left(t-\frac{\tau}{2}\right)e^{-j2\pi f\tau}\,d\tau\,dt\,df$$

$$= \int_{-\infty}^{\infty}\int t^n x\left(t+\frac{\tau}{2}\right)g^*\left(t-\frac{\tau}{2}\right)\delta(\tau)\,d\tau\,dt = \int_{-\infty}^{\infty} t^n x(t)g^*(t)\,dt.$$

Similarly we obtain

$$\int_{-\infty}^{\infty}\int t^n WD_x(t;t,f)\,dt\,df = \int_{-\infty}^{\infty} t^n |x(t)|^2\,dt$$

19.2.18 Frequency Moments

$$\int_{-\infty}^{\infty}\int f^n WD_{X,G}(f;f,t)\,dt\,df = \int\int_{-\infty}^{\infty}\int f^n X\left(f+\frac{\nu}{2}\right)G^*\left(f-\frac{\nu}{2}\right)e^{j2\pi t\nu}\,d\nu\,dt\,df$$

$$= \int_{-\infty}^{\infty}\int f^n X\left(f+\frac{\nu}{2}\right)G^*\left(f-\frac{\nu}{2}\right)\delta(\nu)\,d\nu\,df = \int_{-\infty}^{\infty} f^n X(f)G^*(f)\,df.$$

Because of 19.1.2 we also write

$$\int_{-\infty}^{\infty}\int f^n WD_{x,g}(t;t,f)\,dt\,df = \int_{-\infty}^{\infty} f^n X(f)G^*(f)\,df$$

similarly we write

$$\int_{-\infty}^{\infty}\int f^n WD_x(t;t,f)\,dt\,df = \int_{-\infty}^{\infty} f^n |X(f)|^2\,df$$

19.2.19 Scale Covariance

If $y_1(t) = \sqrt{|a|}\,x(at)$ and $y_2(t) = \sqrt{|a|}\,g(at)$, then

$$WD_{y_1,y_2}(t;t,f) = \int_{-\infty}^{\infty} e^{-j2\pi f\tau}\sqrt{|a|}\,x\left(at+\frac{a\tau}{2}\right)\sqrt{|a|}\,g^*\left(at-\frac{a\tau}{2}\right)d\tau$$

$$= \int_{-\infty}^{\infty} e^{-j2\pi\frac{f}{a}r}x\left(at+\frac{r}{2}\right)g^*\left(at-\frac{r}{2}\right)dr = WD_{x,g}(t;at,f/a)$$

19.2.20 Convolution Covariance

If $y(t) = \int_{-\infty}^{\infty} h(t-\tau)x(\tau)\,d\tau$, then

$$WD_y(t;t,f) = \int_{-\infty}^{\infty}\left[\int_{-\infty}^{\infty} h\left(t+\frac{t'}{2}-\alpha\right)x(\alpha)d\alpha\right]\left[\int_{-\infty}^{\infty} h^*\left(t-\frac{t'}{2}-\gamma\right)x^*(\gamma)d\gamma\right]e^{-j2\pi f t'}\,dt'.$$

Set $\alpha = \tau + p/2$, $\gamma = \tau - p/2$ and $t' = q + p$ in the above equation. Hence we obtain

$$WD_y(t;t,f) = \int\int\int_{-\infty}^{\infty} h\left((t-\tau)+\frac{q}{2}\right)h^*\left((t-\tau)-\frac{q}{2}\right)x(\tau+\tfrac{p}{2})x^*\left(\tau-\frac{p}{2}\right)e^{-j2\pi f(q+p)}\,dq\,d\tau\,dp$$

$$= \int_{-\infty}^{\infty} WD_h(t;t-\tau,f)WD_x(t;\tau,f)d\tau$$

The above indicates that convolution of two signals in the time domain produces a WD that is their convolution of their WD's. If

$$x_c(t) = x(t) * h_1(t) = \int_{-\infty}^{\infty} x(\tau)h_1(t-\tau)d\tau$$

and

$$g_c(t) = g(t) * h_2(t) = \int_{-\infty}^{\infty} g(\tau)h_2(t-\tau)d\tau,$$

then

$$WD_{x_c,g_c}(t;t,f) = \int_{-\infty}^{\infty} WD_{x,g}(\tau;\tau,f)WD_{h_1,h_2}(t;t-\tau,f)d\tau$$

19.2.21 Modulation Covariance

If $y(t) = h(t)x(t)$, then

$$\mathrm{F}\{y(t)\} = Y(f) = \mathrm{F}\{h(t)x(t)\} = \int_{-\infty}^{\infty} H(f-f')X(f')df'.$$

Hence from 19.2.20 and 19.1.2 we obtain

$$WD_Y(f;f,t) = WD_y(t;t,f) = \int_{-\infty}^{\infty} WD_h(t;t,f-f')WD_x(t;t,f')df'$$

The above indicates that WD of the product of two functions is equal to the convolution of their WD's in the frequency domain. If $x_m(t) = x(t)m_1(t)$ and $g_m(t) = g(t)m_2(t)$, then

$$WD_{x_m,g_m}(t;t,f)=\int_{-\infty}^{\infty} WD_{x,g}(t;t,\eta)\,WD_{m_1,m_2}(t;t,f-\eta)\,d\eta .$$

19.2.22 Finite Time Support

$WD_{x,g}(t;t,f)=0$ for $t\notin(t_1,t_2)$ if $x(t)$ or $g(t)$ is zero in $t\notin(t_1,t_2)$.

19.2.23 Finite Frequency Support

$WD_{x,g}(t;t,f)=0$ for $f\notin(f_1,f_2)$ if $X(f)$ and $G(f)$ are zero for $f\notin(f_1,f_2)$.

19.2.24 Instantaneous Frequency

$$\int_{-\infty}^{\infty} f\,WD_x(t;t,f)\,df \Big/ \int_{-\infty}^{\infty} WD_x(t;t,f)\,df=\frac{1}{2\pi}\frac{d\arg\{x(t)\}}{dt}$$

Proof: Write $x(t)$ in its polar form $x(t)=A(t)e^{j2\pi\phi(t)}$ where $A(t)>0$. Then

$$\int_{-\infty}^{\infty} f\,WD_x(t;t,f)\,df=\int_{-\infty}^{\infty} f\int_{-\infty}^{\infty} A\left(t+\frac{\tau}{2}\right)e^{j2\pi\phi(t+\frac{\tau}{2})}A\left(t-\frac{\tau}{2}\right)e^{-j2\pi\phi(t-\frac{\tau}{2})}e^{-j2\pi f\tau}\,d\tau\,df$$

$$=\int_{-\infty}^{\infty} A\left(t+\frac{\tau}{2}\right)A\left(t-\frac{\tau}{2}\right)e^{-j2\pi[\phi(t-\frac{\tau}{2})-\phi(t+\frac{\tau}{2})]}\left[\int_{-\infty}^{\infty} f e^{-j2\pi f\tau}\,df\right]d\tau$$

$$=\int_{-\infty}^{\infty} A\left(t+\frac{\tau}{2}\right)A\left(t-\frac{\tau}{2}\right)e^{-j2\pi[\phi(t-\frac{\tau}{2})-\phi(t+\frac{\tau}{2})]}\frac{1}{j2\pi}\frac{\partial}{\partial\tau}\delta(\tau)\,d\tau$$

$$=\frac{1}{j2\pi}\frac{\partial}{\partial\tau}\left[A\left(t+\frac{\tau}{2}\right)A\left(t-\frac{\tau}{2}\right)e^{-j2\pi[\phi(t-\frac{\tau}{2})-\phi(t+\frac{\tau}{2})]}\right]_{\tau=0}$$

$$=\frac{1}{j2\pi}\left[\frac{\dot{A}(t)A(t)}{2}-\frac{A(t)\dot{A}(t)}{2}+A^2(t)\,j2\pi\dot{\phi}(t)\right]=A^2(t)\dot{\phi}(t)$$

From 19.2.13 $\int_{-\infty}^{\infty} WD_x(t;t,f)\,df=|x(t)|^2=A^2(t)$ and the assertion above is shown to be correct.

19.2.25 Group Delay

$$\int_{-\infty}^{\infty} t\,WD_x(t;t,f)\,dt \Big/ \int_{-\infty}^{\infty} WD_x(t;t,f)\,dt=-\frac{1}{2\pi}\frac{d}{df}\arg\{X(f)\}$$

This property is the dual of property 19.2.24. The proof is similar to that of 19.2.24 except that the signal's Fourier transform is expressed in polar form, and the frequency-domain formulation of the WD 19.1.2 is used.

19.2.26 Fourier Transform of X(t)

$y(t) = X(t)$ where $F\{y(t)\} = F\{X(t)\} = x(-f)$ then

$$WD_y(t;t,f) = \int_{-\infty}^{\infty} X\left(t+\frac{\tau}{2}\right)X^*\left(t-\frac{\tau}{2}\right)e^{-j2\pi f\tau}\,d\tau = \int_{-\infty}^{\infty} x\left(t+\frac{v}{2}\right)x^*\left(t-\frac{v}{2}\right)e^{j2\pi(-f)v}\,dv = WD_x(t;-f,t)$$

where from 19.1.2 we see that t substitutes f and f substitutes t.

Example

If $x(t) = p_a(t)$ is a pulse with a 2a width centered at the origin, its WD is:

$$WD_x(t;t,f) = \frac{\sin[4\pi(a-|t|)f]}{\pi f}p_a(t).$$

Therefore if $y(t) = X(t) = \dfrac{\sin 2\pi at}{\pi t}$ implies that

$$WD_y(t;t,f) = WD_x(t;-f,t) = \frac{\sin[4\pi(a-|-f|)t]}{\pi t}p_a(f).$$

19.2.27 Frequency Localization

If $X(f) = \delta(f-f_0)$, then from 19.1.2 we obtain

$$WD_X(f;f,t) = \int_{-\infty} e^{j2\pi vt}\delta\left(f+\frac{v}{2}-f_0\right)\delta\left(f-\frac{v}{2}-f_0\right)dv$$

$$= 2\int_{-\infty} e^{j4\pi tw}\delta(t+w-f_0)\delta(f-w-f_0)dw = e^{j4\pi t(f-f_0)}\delta[2(f-f_0)] = \delta(f-f_0)$$

19.2.28 Time Localization

If $x(t) = \delta(t-t_0)$, then from 19.1.2 we obtain

$$WD_x(t;t,f) = \int_{-\infty} e^{-j2\pi f\tau}\delta\left(t+\frac{\tau}{2}-t_0\right)\delta\left(t-\frac{\tau}{2}-t_0\right)dt = \delta(t-t_0)$$

by following steps similar to those in 19.2.27.

19.2.29 Linear Chirp Localization

If $X(f) = e^{-j\pi cf^2}$ then from 19.1.5 we obtain

$$WD_X(f;f,t) = \int_{-\infty}^{\infty} e^{j2\pi vt}e^{-j\pi c(f+\frac{v}{2})^2}e^{-j\pi c(f-\frac{v}{2})^2}\,dv = \int_{-\infty} e^{j2\pi v(t-cf)}\,dv = \delta(t-cf)$$

19.2.30 Chirp Convolution

If $y(t) = \int_{-\infty}^{\infty} x(t-\tau)\sqrt{|c|}\,e^{+j\pi c\tau^2}\,d\tau$, then (see FT property) $F\{y(t)\} = X(f)\sqrt{j}\,e^{-j\pi f^2/c}$ and from 19.1.5 we obtain

$$WD_Y(f;f,t) = \int_{-\infty}^{\infty} X\left(f+\frac{v}{2}\right)\sqrt{j}\, e^{-j\pi(f+\frac{v}{2})^2/c}\, X^*\left(f-\frac{v}{2}\right)\sqrt{j}\, e^{-j\pi(f-\frac{v}{2})^2/c}\, e^{j2\pi t v}\, dv$$

$$= \int_{-\infty}^{\infty} X\left(f+\frac{v}{2}\right) X^*\left(f-\frac{v}{2}\right) e^{j2\pi v(t-\frac{f}{c})}\, dv = WD_x\left(t;t-\frac{f}{c},f\right)$$

19.2.31 Chirp Multiplication

If $y(t) = x(t)e^{j\pi ct^2}$, then $WD_y(t;t,f) = WD_x(t;t,f-ct)$.

Proof: If $z(t) = e^{j\pi ct^2}$, then

$$WD_z(t;t,f) = \int_{-\infty}^{\infty} e^{+j\pi c(t+\frac{\tau}{2})^2} e^{-j\pi c(t-\frac{\tau}{2})^2} e^{-j2\pi f\tau}\, d\tau = \int_{-\infty}^{\infty} e^{-j2\pi(f-ct)\tau}\, d\tau = \delta(f-ct).$$

Hence, from 19.2.21 we obtain $WD_y(t;t,f) = \int_{-\infty}^{\infty} WD_y(t;t,f')\delta[(f-f')-ct]\,df' = WD_x(t;t,f-ct)$.

Example

If $y(t) = p_a(t)e^{j\pi ct^2}$, then

$$WD_y(t;t,f) = WD_x(t;t,f-ct) = \frac{\sin[4\pi(a-|t|)(f-ct)]}{\pi(f-ct)}\, p_a(t),$$

where $p_a(t)$ is a pulse of 2a width centered at the origin (see 10.1.2). Observe that the WD is a sine function centered along the chirp's linear instantaneous frequency, $f = ct$ in the time-frequency plane.

19.2.32 Moyal's Formula

1. $$\int_{-\infty}^{\infty}\!\!\int WD_x(t;t,f)WD_y^*(t;t,f)\,dt\,df = \int_{-\infty}^{\infty}\!\!\int \left[\int_{-\infty}^{\infty} x\left(t+\frac{u}{2}\right)x^*\left(t-\frac{u}{2}\right)e^{-j2\pi fu}\,du\right]$$

$$\left[\int_{-\infty}^{\infty} y^*\left(t+\frac{t'}{2}\right)y\left(t-\frac{t'}{2}\right)e^{j2\pi ft'}\,dt'\right]dt\,df$$

$$= \int_{-\infty}^{\infty}\!\!\int\!\!\int x\left(t+\frac{u}{2}\right)y^*\left(t+\frac{t'}{2}\right)x^*\left(t-\frac{u}{2}\right)y\left(t-\frac{t'}{2}\right)\delta(t'-u)\,du\,dt\,dt'$$

$$= \int_{-\infty}^{\infty}\!\!\int \left[x\left(t+\frac{u}{2}\right)y^*\left(t+\frac{u}{2}\right)\right]\left[x\left(t-\frac{u}{2}\right)y^*\left(t-\frac{u}{2}\right)\right]^* du\,dt$$

$$= \left|\int_{-\infty}^{\infty} x(t)y^*(t)\,dt\right|^2$$

2. Similarly: $$\int_{-\infty}^{\infty}\!\!\int WD_{x_1,x_2}(t;t,f)WD_{y_1,y_2}^*(t;t,f)\,dt\,df = \int_{-\infty}^{\infty} x_1(t)y_1^*(t)\,dt \int_{-\infty}^{\infty} x_2(t)y_2^*(t)\,dt$$

3. Similarly: $$\int_{-\infty}^{\infty}\!\!\int WD_x(t;t,f)WD_x^*(t;t,f)\,dt\,df = \left|\int_{-\infty}^{\infty} |x(t)|^2\,dt\right|^4$$

19.2.33 Energy of the Time Functions in a Time Range $t_1 < t < t_2$

From 19.2.13 we obtain

$$\int_{-t_1}^{t_2}\left[\int_{-\infty}^{\infty}[WD_x(t;t,f)df\right]dt=\int_{t_1}^{t_2}|x(t)|^2\,dt$$

19.2.34 Energy of the Function Spectrum in the Range $f_1 < f < f_2$

From 19.2.15 we obtain

$$\int_{f_1}^{f_2}\left[\int_{-\infty}^{\infty}[WD_x(t;t,f)dt\right]df=\int_{f_1}^{f_2}|X(f)|^2\,df \quad \text{where } F\{x(f)\}=X(f)$$

19.2.35 Time-Domain Windowing; the Pseudo-Wigner Distribution (PWD)

If $x_t(\tau)=x(\tau)w_x(\tau-t)$ and $g_t(\tau)=g(\tau)w_g(\tau-t)$, where w_x and w_g and are windows, then from 19.2.2

$WD_{x_t,g_t}(\tau;\tau,f) = \int_{-\infty}^{\infty}WD_{x,g}(\tau;\tau,\eta)WD_{w_x,w_g}(\tau;\tau-t,f-\eta)d\eta$ where t appears as a parameter that indicates the position of the window. For $\tau=t$ we obtain

$$WD_{x_t,g_t}(\tau;\tau,f)\big|_{\tau=t}=\int_{-\infty}^{\infty}WD_{x,g}(t;t,\eta)WD_{w_x,w_g}(t;0,f-\eta)d\eta.$$

Hence we can write $\tilde{W}D_{x,g}(t;t,f)=WD_{x_t,g_t}(\tau;\tau,f)\big|_{\tau=t}$ which is a function of t and f that resembles, but in general is <u>not</u> a WD. This distribution is called <u>pseudo-Wigner distribution</u> (PWD).

19.2.36 Analytic Signals

If $x_a(t)=x(t)+j\hat{x}(t)$ is the analytic signal and $\hat{x}(t)$ is the Hilbert transform of $x(t)$, then

$$X_a(f)=F\{x_a(t)\}=\begin{cases}2X(f) & f>0\\ X(0) & f=0\\ 0 & f<0\end{cases}$$

From 19.2.23 $WD_{x_a}(t;t,f)=0$ for $\omega<0$. From 19.1.2 and 19.2.1 and the definition of WD we obtain

$$WD_{X_a}(f;f,t)=\int_{-\infty}^{\infty}X_a\left(f+\frac{v}{2}\right)X_a^*\left(f-\frac{v}{2}\right)e^{j2\pi vt}\,dv$$

$$=4\int_{-\infty}^{\infty}e^{j2\pi vt}X\left(f+\frac{v}{2}\right)X^*\left(f-\frac{v}{2}\right)dv$$

$$=4\int_{-2f}^{2f}X\left(f+\frac{v}{2}\right)X^*\left(f-\frac{v}{2}\right)e^{j2\pi vt}\,dv \text{ for } f>0.$$

From the definition of WD in 19.1.2 we obtain its inverse Fourier transform as follows:

$$\int_{-\infty}^{\infty} e^{-j2\pi vt} WD_X(f;f,t)dt = X\left(f+\frac{v}{2}\right)X^*\left(f-\frac{v}{2}\right)$$

and, hence, the above equation becomes

$$WD_{x_a}(f;f,t) = 4\int_{-\infty}^{\infty} WD_X(f;f,t)\int_{-2f}^{2f} e^{j2\pi v(t-\tau)}\,dv\,d\tau = \frac{4}{\pi}\int_{-\infty}^{\infty} WD_X(f;f,t)\frac{\sin 4\pi f(t-\tau)}{t-\tau}d\tau$$

$$= \frac{4}{\pi}\int_{-\infty}^{\infty} WD_X(f;f,t-\tau)\frac{\sin 4\pi f\tau}{\tau}d\tau = \frac{4}{\pi}\int_{-\infty}^{\infty} WD_x(t;t-\tau,f)\frac{\sin 4\pi f\tau}{\tau}d\tau$$

for $f > 0$ and 0 for $f < 0$. The last equation indicates that the WD of the analytic signal at a fixed frequency value $f > 0$ can be obtained by considering $WD_x(t;t,f)$ for this frequency value as a function of time and passing this time function through an ideal low-pass filter with cut-off frequency 2f. This means that $WD_{x_a}(t;t,f)$ with f fixed is a time function whose highest frequency is at most 2f.

19.3 Instantaneous Frequency and Group Delay

19.3.1 Instantaneous Frequency $f_x(t)$ of $x(t)$

$$f_x(t) = \frac{1}{2\pi}\frac{d}{dt}\arg x(t)$$

19.3.2 Group Delay of Linear Time-Invariant Filters

$$\tau_h(f) = -\frac{1}{2\pi}\frac{d}{dt}\arg H(f)$$

$$H(f) = \mathrm{F}\{h(t)\},\quad h(t) = \text{impulse response of the filter}$$

19.4 Table of WD Properties

19.4.1 WD Properties and Ideal Time-Frequency Representations

TABLE 19.1 Table of WD Properties and Ideal Properties

Property Name	WD Property	Ideal
1. Conjugation	$WD_{x,g}(t;t,f) = WD^*_{g,x}(t;t,f)$	same
2. Real-Valued	$WD_x(t;t,f) = WD^*_x(t;t,f) \equiv \text{real } (WD_{x,x} \equiv WD_x)$	
3. Even in Frequency	$WD_{x^*}(t;t,f) = WD_x(t;t,-f)$	
(real $x(t)$)	$WD_x(t;t,f) = WD_{x^*}(t;t,-f)$	
4. Time Shift	$WD_{x_s,g_s}(t;t,f) = WD_{x,g}(t;t-t_0,f)$	
	$x_s(t) = x(t-t_0),\quad g_s(t) = g(t-t_0)$	

TABLE 19.1 Table of WD Properties and Ideal Properties (continued)

Property Name	WD Property	Ideal
5. Frequency Shift	$WD_{x_s,g_s}(t;t,f) = WD_{x,g}(t;f-f_0)$ $x_s(t)=x(t)e^{j2\pi f_0 t},\quad g_s(t)=g(t)\,e^{j2\pi f_0 t}$	
6. Time and Frequency Shift	$WD_{x_{sm},g_{sm}}(t;t,f) = WD_{x,g}(t;t-t_0,f-f_0)$ $x_{sm}(t)=x(t-t_0)e^{j2\pi f_0(t-t_0)},$ $g_{sm}(t)=g(t-t_0)e^{j2\pi f_0(t-t_0)}$	
7. Ordinates	$WD_{x,g}(t;0,0)=\int_{-\infty}^{\infty} x\left(\frac{\tau}{2}\right)g^*\left(-\frac{\tau}{2}\right)d\tau$	
8. Sum of Functions	$WD_{x+g}(t;t,f)=WD_x(t;t,f)+WD_g(t;t,f)$ $+2\,\mathrm{Re}\{WD_{x,g}(t;t,f)\}$ $WD_y(t;t,f)=\sum_{i=1}^{N} W_{x_i}(t;t,f)$ $+2\sum_{i=1}^{N-1}\sum_{k=i+1}^{N}\mathrm{Re}\{W_{x_i x_k}(t;t,f)\}$ $y(t)=\sum_{i=1}^{N} x_i(t),$ N auto terms $+\frac{N(N-1)}{2}$ cross term $WD_y(t;t,f)=\sum_{i=1}^{N} WD_x(t;t-t_i,f-f_i)$ $+2\sum_{i=1}^{N-1}\sum_{m=i+1}^{N} WD_x\left(t;t-\frac{t_i+t_m}{2},f-\frac{f_i+f_m}{2}\right)$ $\times\cos 2\pi\left[(f_i-f_m)t-(t_i-t_m)f+\frac{f_i+f_m}{2}(t_i-t_m)\right]$	
9. Multiplication by t	$WD_{tx,g}(t;t,f)+WD_{x,tg}(t;t,f)=2t\,WD_{x,g}(t;t,f)$	
10. Multiplication by f	$WD_{fx,g}(t;t,f)+WD_{x,fg}(t;t,f)=2f\,W_{x,g}(t;t,f)$	
11. Multiplication by t and f	$WD_{tx,fg}(t;t,f)+WD_{fx,tg}(t;t,f)=2ft\,WD_{x,g}(t;t,f)$	
12. Fourier Transform	$F^{-1}\{WD_{x,g}(t;t,f)\}=x\left(t+\frac{\tau}{2}\right)g^*\left(t-\frac{\tau}{2}\right)$	
13. Time Marginal	$\int_{-\infty}^{\infty} WD_{x,g}(t;t,f)df=x(t)g^*(t)$ $\int_{-\infty}^{\infty} WD_x(t;t,f)df=\lvert x(t)\rvert^2 \quad (x(t)=g(t))$	
14. Recovery of Time Function	$F^{-1}\left\{WD_{x,g}\left(t;\frac{t}{2},f\right)\right\}=x(t)g^*(0)$ $F^{-1}\left\{WD_x\left(t;\frac{t}{2},f\right)\right\}=x(t)x^*(0)$	
15. Frequency Marginal	$\int_{-\infty}^{\infty} WD_{X,G}(f;t,f)dt=X(f)G^*(f)$ $\int_{-\infty}^{\infty} W_X(f;t,f)dt=\lvert X(f)\rvert^2=\int_{-\infty}^{\infty} WD_X(f;t,f)dt.$	

TABLE 19.1 Table of WD Properties and Ideal Properties (continued)

Property Name	WD Property	Ideal
16. Total Energy	$\iint_{-\infty}^{\infty} WD_{X,G}(f;f,t)\,dt\,df = \int_{-\infty}^{\infty} X(f)G^*(f)\,df$	
	$\iint_{-\infty}^{\infty} WD_X(f;f,t)\,dt\,df = \int_{-\infty}^{\infty} \lvert X(f)\rvert^2\,df$	
	$\iint_{-\infty}^{\infty} WD_x(t;t,f)\,dt\,df = \int_{-\infty}^{\infty} \lvert X(f)\rvert^2\,df$	
17. Time Moments	$\iint_{-\infty}^{\infty} t^n WD_{x,g}(t;t,f)\,dt\,df = \iint_{-\infty}^{\infty} t^n x(t)g^*(t)\,dt$	
	$\iint_{-\infty}^{\infty} t^n WD_x(t;t,f)\,dt\,df = \iint_{-\infty}^{\infty} t^n \lvert x(t)\rvert^2\,dt$	
18. Frequency Moments	$\iint_{-\infty}^{\infty} f^n WD_{X,G}(f;f,t)\,dt\,df = \int_{-\infty}^{\infty} f^n X(f)G^*(f)\,df$	
	$\iint_{-\infty}^{\infty} f^n WD_{x,g}(t;t,f)\,dt\,df = \int_{-\infty}^{\infty} f^n X(f)G^*(f)\,df$	
	$\iint_{-\infty}^{\infty} f^n WD_x(t;t,f)\,dt\,df = \int_{-\infty}^{\infty} f^n \lvert X(f)\rvert^2\,df$	
19. Scale Covariance	$WD_{y_1,y_2}(t;t,f) = WD_{x,g}(t;at,f/a)$	
	$y_1(t) = \sqrt{\lvert a\rvert}\,x(at), \quad y_2(t) = \sqrt{\lvert a\rvert}\,g(at)$	
20. Convolution Covariance	$WD_y(t;t,f) = \int_{-\infty}^{\infty} WD_h(t;t-\tau,f)\,WD_x(t;\tau,f)\,d\tau$	
	$y(t) = \int_{-\infty}^{\infty} h(t-\tau)x(\tau)\,d\tau$	
	$WD_{x_c,g_c}(t;t,f) = \int_{-\infty}^{\infty} WD_{x,g}(\tau;\tau,f)\,WD_{h_1,h_2}(t;t-\tau,f)\,d\tau$	
	$x_c(t) = \int_{-\infty}^{\infty} x(\tau)h_1(t-\tau)\,d\tau,$	
	$g_c(t) = \int_{-\infty}^{\infty} g(\tau)h_2(t-\tau)\,d\tau$	
21. Modulation Covariance	$WD_Y(f;f,t) = WD_y(t;t,f)$	
	$= \int_{-\infty}^{\infty} WD_h(t;t,f-f')\,WD_x(t;t,f')\,df'$	
	$Y(f) = \mathrm{F}\{h(t)x(t)\} = \int_{-\infty}^{\infty} H(f-f')X(f')\,df'$	

TABLE 19.1 Table of WD Properties and Ideal Properties (continued)

Property Name	WD Property	Ideal
	$WD_{x_m,g_m}(t;t,f) =$ $\int_{-\infty}^{\infty} WD_{x,g}(t;t,\eta)\, WD_{m_1,m_2}(t;t,f-\eta)\, d\eta$	
22. Finite Time Support	$WD_{x,g}(t;t,f)=0$ for $t \notin (t_1,t_2)$ if $x(t)$ or $g(t)$ is zero in $t \notin (t_1,t_2)$	
23. Finite Frequency Support	$WD_{x,g}(t;t,f)=0$ for $f \notin (f_1,f_2)$ if $X(f)$ and $G(f)$ are zero for $f \notin (f_1,f_2)$	
24. Instantaneous Frequency	$\dfrac{\int_{-\infty}^{\infty} f\, WD_x(t;t,f)\, df}{\int_{-\infty}^{\infty} WD_x(t;t,f)\, df} = \dfrac{1}{2\pi}\dfrac{d}{dt}\arg\{x(t)\}$	
25. Group Delay	$\dfrac{\int_{-\infty}^{\infty} t\, WD_x(t;t,f)\, dt}{\int_{-\infty}^{\infty} WD_x(t;t,f)\, dt} = -\dfrac{1}{2\pi}\dfrac{d}{df}\arg\{X(f)\}$	
26. Fourier Transform of $X(t)$	$WD_y(t;t,f) = WD_x(t;-f,t)$ $y(t) = X(t),\ F\{y(t)\} = x(-f)$	
27. Frequency Localization	$WD_X(f;f,t) = \delta(f-f_0)$ $X(f) = \delta(f-f_0)$	
28. Time Localization	$WD_x(t;t,f) = \delta(t-t_0)$ $x(t) = \delta(t-t_0)$	
29. Linear Chirp Localization	$WD_X(f;f,t) = \delta(t-cf)$ $X(f) = e^{-j\pi c f^2}$	
30. Chirp Convolution	$WD_Y(f;f,t) = WD_x\left(t;t-\dfrac{f}{c},f\right)$ $y(t) = \int_{-\infty}^{\infty} x(t-\tau)\sqrt{\lvert c\rvert}\, e^{j\pi c\tau^2}\, d\tau$ $F\{y(t)\} = X(f)\sqrt{j}\, e^{-j\pi f^2/c}$	
31. Chirp Multiplication	$WD_y(t;t,f) = WD_x(t;t,f-ct)$ $y(t) = x(t)e^{j\pi c t^2}$	
32. Moyal's Formula	$\iint_{-\infty}^{\infty} WD_x(t;t,f)\, WD_y^*(t;t,f)\, dt\, df = \left\lvert \int_{-\infty}^{\infty} x(t)\, y^*(t)\, dt \right\rvert^2$ $\iint_{-\infty}^{\infty} WD_{x_1,x_2}(t;t,f)\, WD_{y_1,y_2}(t;t,f)\, dt\, df =$ $\int_{-\infty}^{\infty} x_1(t)\, y_1^*(t)\, dt \int_{-\infty}^{\infty} x_2(t)\, y_2^*(t)\, dt$	

TABLE 19.1 Table of WD Properties and Ideal Properties (continued)

Property Name	WD Property	Ideal
	$\iint_{-\infty}^{\infty} WD_x(t;t,f)WD_x^*(t;t,f)\,dt\,df = \left\lvert \int_{-\infty}^{\infty} \lvert x(t)\rvert^2\,dt \right\rvert^4$	
33. Energy of the Time Functions in a Time Range $t_1 < t < t_2$	$\int_{t_1}^{t_2}\left[\int_{-\infty}^{\infty} [WD_x(t;t,f)df\right]dt = \int_{t_1}^{t_2} \lvert x(t)\rvert^2\,dt$	
34. Energy of the Function Spectrum in the Range $f_1 < f < f_2$	$\int_{f_1}^{f_2}\left[\int_{-\infty}^{\infty} [WD_x(t;t,f)dt\right]df = \int_{f_1}^{f_2} \lvert X(f)\rvert^2\,df$ $F\{x(t)\} = X(f)$	
35. Positivity		$W_x(t;t,f) \ge 0$ $-\infty < t < \infty$ $-\infty < f < \infty$
36. Time-Domain Windowing; the Pseudo-Wigner Distribution (PWD)	$\widetilde{WD}_{x,g}(t;t,f) = W_{x_t,g_t}(\tau;\tau,f)\rvert_{\tau=t} = PWD$ $WD_{x_t,g_t}(\tau;\tau,f) =$ $\int_{-\infty} W_{x,g}(\tau;\tau,\eta)\, WD_{w_x,w_g}(\tau;\tau-t,f-\eta)\,d\eta$ $x_t(\tau) = x(\tau)w_x(\tau-t), \quad g_t(\tau) = g(\tau)w_g(\tau-t)$	
37. Analytic Signals	$WD_{x_a}(f;f,t) =$ $\frac{4}{\pi}\int_{-\infty}^{\infty} WD_x(t;t-\tau,f)\frac{\sin 4\pi f\tau}{\tau}d\tau, \quad f>0$ $X_a(f) = F\{x_a(t)\} = \begin{cases} 2X(f) & f>0 \\ X(0) & f=0 \\ 0 & f<0 \end{cases}$	
38. Hyperbolic Shift		$T_y(t;t,f) = T_x\left(t;t-\frac{c}{f},f\right)$ $Y(f) = \exp\left(-j2\pi c \ln\frac{f}{f_r}\right)X(f)$ for $f>0$
39. Hyperbolic Localization		$T_x(t;t,f) = \frac{1}{f}\delta\left(t-\frac{c}{f}\right), \quad f>0$ $X_c(f) = \frac{1}{\sqrt{f}}e^{-j2\pi c \ln\frac{f}{f_r}}, \quad f>0$

19.5 Tables of Wigner Distribution and Ambiguity Function

19.5.1 Table Signals with Closed-Form Wigner Distributions (WD) and Ambiguity Functions (AF) (See Table 19.2)

TABLE 19.2* Signals with Closed-Form Equations for Their Wigner Distribution and Ambiguity Function

Signal, $x(t)$	Fourier Transform, $X(f)$	Wigner Distribution, $WD_x(t;t,f)$	Ambiguity Function $AF_x(t;\tau,\nu)$
$\delta(t-t_i)$	$e^{-j2\pi f t_i}$	$\delta(t-t_i)$	$e^{-j2\pi\nu t_i}\,\delta(\tau)$
$e^{j2\pi f_i t}$	$\delta(f-f_i)$	$\delta(f-f_i)$	$e^{j2\pi f_i \tau}\,\delta(\nu)$
$e^{+j\pi\alpha t^2}$	$\frac{1}{\sqrt{-j\alpha}}e^{-j\pi f^2/\alpha}$	$\delta(f-\alpha t)$	$\delta(\nu-\alpha t)$
$\frac{1}{\sqrt{j\alpha}}e^{j\pi t^2/\alpha}$	$e^{-j\pi\alpha f^2}$	$\delta(t-\alpha f)$	$\delta(\tau-\alpha\nu)$
$e^{j\pi(\alpha t^2+2f_i t+c)}$	$\frac{1}{\sqrt{-j\alpha}}e^{j\pi[c-(f-f_i)^2/\alpha]}$	$\delta(f-f_i-\alpha t)$	$\delta(\nu-\alpha\tau)\,e^{j2\pi f_i\tau}$
$\frac{1}{\sqrt{\sigma}}e^{-\pi(t/\sigma)^2}$	$\sqrt{\sigma}\,e^{-\pi(\sigma f)^2}$	$\sqrt{2}\,e^{-2\pi[(t/\sigma)^2+(\sigma f)^2]}$	$\frac{1}{\sqrt{2}}e^{-(\pi/2)[(\tau/\sigma)^2+(\sigma\nu)^2]}$
$\frac{1}{\sqrt{\sigma}}e^{-\pi(t/\sigma)^2}e^{j\pi\alpha t^2}$	$\frac{1}{\sqrt{\sigma[\sigma^{-2}-j\alpha]}}\exp\left[-\pi f^2\frac{\sigma^{-2}+j\alpha}{\sigma^{-4}+\alpha^2}\right]$	$\sqrt{2}\,e^{-2\pi[(t/\sigma)^2+(\sigma)^2(f-\alpha t)^2]}$	$\frac{1}{\sqrt{2}}e^{-(\pi/2)[(\tau/\sigma)^2+(\sigma)^2(\nu-\alpha\tau)^2]}$
$\frac{1}{\sqrt{\sigma}}e^{-\pi((t-t_i)/\sigma)^2}e^{j2\pi f_i t}$	$\sqrt{\sigma}e^{-\pi\sigma^2(f-f_i)^2}e^{-j2\pi(f-f_i)t_i}$	$\sqrt{2}\,e^{-2\pi[((t-t_i)/\sigma)^2+\sigma^2(f-f_i)^2]}$	$\frac{1}{\sqrt{2}}e^{-(\pi/2)[(\tau/\sigma)^2+(\sigma\nu)^2]}e^{j2\pi(f_i\tau-t_i\nu)}$
$p_a(t)$	$\frac{\sin(2\pi af)}{\pi f}$	$\frac{\sin[4\pi(a-\lvert t\rvert)f]}{\pi f}p_a(t)$	$\frac{\sin[\pi\nu(2a-\lvert\tau\rvert)]}{\pi f}p_{2a}(\tau)$
$\frac{\sin(2\pi at)}{\pi t}$	$p_a(f)$	$\frac{\sin[4\pi(a-\lvert f\rvert)t]}{\pi t}p_a(t)$	$\frac{\sin[\pi\tau(2a-\lvert\nu\rvert)]}{\pi\tau}p_{2a}(\nu)$
$e^{j\pi\alpha t^2}p_a(t)$	$\frac{1}{\sqrt{-j\alpha}}\int e^{-j\frac{\pi}{\alpha}(f-\beta)^2}\frac{\sin(2\pi a\beta)}{\pi\beta}d\beta$	$\frac{\sin[4\pi(a-\lvert t\rvert)(f-\alpha t)]}{\pi(f-\alpha t)}p_a(t)$	$\frac{\sin[\pi(\nu-\alpha\tau)(2a-\lvert\tau\rvert)]}{\pi(\nu-\alpha\tau)}p_{2a}(\tau)$
$u(t)=\begin{cases}1, & t>0\\ 0, & t<0\end{cases}$	$\frac{\delta(f)}{2}-\frac{j}{2\pi f}$	$\frac{\sin(4\pi ft)}{\pi f}u(t)$	$\left[\frac{\delta(\nu)}{2}-\frac{j}{2\pi\nu}\right]e^{-j\pi\nu\lvert\tau\rvert}$
$e^{-\sigma t}u(t)$	$\frac{1}{\sigma+j2\pi f}$	$e^{-2\sigma t}\frac{\sin 4\pi ft}{\pi f}u(t)$	$\frac{e^{-(\sigma+j\pi\nu)\lvert\tau\rvert}-1}{2\sigma+j2\pi\nu}$
$h_n(t),\ n=0,1,\cdots$	$(-j)^n h_n(f)$	$2e^{-2\pi(t^2+f^2)}L_n(4\pi(t^2+f^2))$	$e^{-\pi(\tau^2+\nu^2)/2}L_n(\pi(\tau^2+\nu^2))$

TABLE 19.2* Signals with Closed-Form Equations for Their Wigner Distribution and Ambiguity Function (continued)

Signal, $x(t)$	Fourier Transform, $X(f)$	Wigner Distribution, $WD_x(t;t,f)$	Ambiguity Function $AF_x(t;\tau,\nu)$
$\cos(2\pi f_i t)$	$[\delta(f+f_i)+\delta(f-f_i)]/2$	$[\delta(f+f_i)+\delta(f-f_i)+2\delta(f)\cos(4\pi f_i t)]/4$	$[\delta(\nu+2f_i)+\delta(\nu-2f_i)+2\delta(\nu)\cos(2\pi f_i\tau)]/4$
$\sin(2\pi f_i t)$	$j[\delta(f+f_i)-\delta(f-f_i)]/2$	$[\delta(f+f_i)+\delta(f-f_i)-2\delta(f)\cos(4\pi f_i t)]/4$	$-[\delta(\nu+2f_i)+\delta(\nu-2f_i)-2\delta(\nu)\cos(2\pi f_i\tau)]/4$
$\delta(t-t_i)+\delta(t-t_m)$	$e^{-j2\pi f t_i}+e^{-j2\pi f t_m}$	$\delta(t-t_i)+\delta(t-t_m)$ $+2\delta\left(t-\frac{t_i+t_m}{2}\right)\cos(2\pi(t_i-t_m)f)$	$[e^{-j2\pi\nu t_i}+e^{-j2\pi\nu t_m}]\delta(\tau)$ $+[e^{-j\pi(t_i+t_m)\nu}[\delta(\tau-(t_i-t_m))+\delta(\tau+(t_i-t_m))]]$
$e^{j2\pi f_i t}+e^{j2\pi f_m t}$	$\delta(f-f_i)+\delta(f-f_m)$	$\delta(f-f_i)+\delta(f-f_m)$ $+2\delta\left(f-\frac{f_i+f_m}{2}\right)\cos(2\pi(f_i-f_m)t)$	$[e^{j2\pi f_i\tau}+e^{j2\pi f_m\tau}]\delta(\nu)$ $+[e^{j\pi(f_i+f_m)\tau}[\delta(\nu-(f_i-f_m))+\delta(\nu+(f_i-f_m))]]$
$\sum_k c_k e^{j2\pi k f_0 t}$	$\sum_k c_k\delta(f-kf_0)$	$\sum_k \lvert c_k\rvert^2\delta(f-kf_0)$ $+\sum_k\sum_{m\neq k} c_k c_m^*\delta\left(f-\frac{k+m}{2}f_0\right)e^{j2\pi(k-m)f_0 t}$	$\sum_k \lvert c_k\rvert^2 e^{j2\pi k f_0\tau}\delta(\tau)$ $+\sum_k\sum_{m\neq k} c_k c_m^* e^{j\pi(k+m)f_0\tau}\delta(\nu-(k-m)f_0)$

* $\sigma>;a,\alpha,c\in\Re$, $\operatorname{sgn}(t)=\begin{cases}1 & t>0\\ 0 & t=0\\ -1 & t<0\end{cases}$, $p_a(t)=\begin{cases}1 & -a<\lvert t\rvert<a\\ 0 & \text{otherwise}\end{cases}$, $h_n(t)=\frac{2^{1/4}}{\sqrt{n!}}e^{-\pi t^2}H_n(2\sqrt{\pi}t)$

$H_n(t)=(-1)^n\exp(t^2/2)\frac{d^n}{dt^n}\exp(-t^2/2)=n^{th}$order Hermite Polynomial, $L_n(t)=\frac{1}{n!}e^t\frac{d^n}{dt^n}(t^n e^{-t})=\sum_{k=0}^{n}\frac{n!}{k!(n-k)!}(-t)^k=n^{th}$order Laguerre polynomials

19.6 Effects of WD and AF on Functions

19.6.1 Effects of WD and AF on Functions (See Table 19.3)

19.7 Other Time-Frequency Representation (TFRs)

19.7.1 Cohen's Class

$$C_x(t;t,f;\Psi_c)=\iint_{-\infty}^{\infty}\varphi_c(t-t',\tau)x\left(t'+\frac{\tau}{2}\right)x^*\left(t'-\frac{\tau}{2}\right)e^{-j2\pi f\tau}\,dt'\,d\tau$$

$$=\iint_{-\infty}^{\infty}\Phi_c(f-f',\nu)X\left(f'+\frac{\nu}{2}\right)X^*\left(f'-\frac{\nu}{2}\right)e^{j2\pi t\nu}\,df'\,d\nu$$

$$=\iint_{-\infty}^{\infty}\psi_c(t-t',f-f')WD_x(t;t',f')dt'\,df'$$

$$=\iint_{-\infty}^{\infty}\Psi_c(\tau,\nu)AF_x(t;\tau,\nu)e^{j2\pi(t\nu-f\tau)}\,d\tau\,d\nu$$

$$C_x(t;t,f;\Psi_c)=\iint_{-\infty}^{\infty}\Gamma_c(f-f_1,f-f_2)X(f_1)X^*(f_2)e^{j2\pi t(f-f_2)}\,df_1\,df_2$$

where

$$\varphi_c(t,\tau)=\iint_{-\infty}^{\infty}\Phi_c(f,\nu)e^{j2\pi(f\tau+\nu t)}\,df\,d\nu=\int_{-\infty}^{\infty}\Psi_c(\tau,\nu)e^{j2\pi\nu t}\,d\nu$$

$$\varphi_c(t,\tau)\leftrightarrow\Psi_c(f,\nu)$$

$$\psi_c(t,f)=\iint_{-\infty}^{\infty}\Psi_c(\tau,\nu)e^{j2\pi(\nu t-f\tau)}\,d\tau\,d\nu=\int_{-\infty}^{\infty}\Phi_c(f,\nu)e^{j2\pi\nu t}\,d\nu$$

$$\psi_c(t,f)\leftrightarrow\Psi_c(\tau,\nu)$$

$$\Gamma_c(f_1,f_2)=\Phi_c\left(\frac{f_1+f_2}{2},f_2-f_1\right)$$

TABLE 19.3 Effects of signal operations of the Wigner distribution and ambiguity function. Here $\sigma > 0,\ a,\ \alpha,\ c \in \Re$, and $\mathrm{sgn}(a)$

Signal, $y(t)$	Fourier Transform, $Y(f)$	Wigner Distribution, $WD_y(t;t,f)$	Ambiguity Function $AF_y(t;\tau,\nu)$
$Ax(t)$	$AX(f)$	$\lvert A\rvert^2\, WD_x(t;t,f)$	$\lvert A\rvert^2 AF_x(t;\tau,\nu)$
$x(-t)$	$X(-f)$	$WD_x(t;-t,-f)$	$AF_x(t;-\tau,-\nu)$
$\sqrt{\lvert a\rvert}\, x(at)$	$\frac{1}{\sqrt{\lvert a\rvert}} X(f/a)$	$WD_x(t;at,f/a)$	$AF_x(t;a\tau,\nu/a)$
$\sqrt{\lvert a\rvert}\, X(at)$	$\frac{1}{\sqrt{\lvert a\rvert}} x(-f/a)$	$WD_x(t;-f/a,at)$	$AF_x(t;-\nu/a,a\tau)$
$x(t) = \pm x(\pm t)$	$X(f) = \pm X(\pm f)$	$\pm 2AF_x(t;2t,2f)$	$\pm\frac{1}{2} WD_x(t;\tau/2,\nu/2)$
$x^*(t)$	$X^*(-f)$	$WD_x(t;t,-f)$	$AF_x^*(t;\tau,-\nu)$
$x(t-t_i)e^{j2\pi f_i t}$	$X(f-f_i)e^{-j2\pi(f-f_i)t_i}$	$WD_x(t;t-t_i,f-f_i)$	$AF_x(t;\tau,\nu)e^{j2\pi(f_i\tau-t_i\nu)}$
$x(t)h(t)$	$\int X(f')H(f-f')df'$	$\int WD_x(t;t,f')WD_h(t;t,f-f')df'$	$\int AF_x(t;\tau,\nu')AF_h(t;\tau,\nu-\nu')d\nu'$
$\int x(t')h(t-t')dt'$	$X(f)H(f)$	$\int WD_x(t;t',f)WD_h(t;t-t',f)dt'$	$\int AF_x(t;\tau',\nu)AF_h(t;\tau-\tau',\nu)d\tau'$
$x(t)e^{j\pi\alpha t^2}$	$\frac{1}{\sqrt{-j\alpha}}\int X(f-f')e^{-j\pi f'^2/\alpha}\, df'$	$WD_x(t;t,f-\alpha t)$	$AF_x(t;\tau,\nu-\alpha\tau)$
$\int \sqrt{\lvert\alpha\rvert}\ e^{j\pi\alpha u^2}x(t-u)du$	$\sqrt{j\,\mathrm{sgn}(\alpha)}\ X(f)e^{-j\pi f^2/\alpha}$	$WD_x(t;t-f/\alpha,f)$	$AF_x(t;\tau-\nu/\alpha,\nu)$

$\sum_{i=0}^{N-1} x\left(t-\left(i-\frac{N-1}{2}\right)T_r\right),$ $T_r > 0$	$X(f)\frac{\sin(\pi T_r Nf)}{\sin(\pi T_r f)}$	$\sum_{i=0}^{N-1} WD_x\left(t;t-\left(i-\frac{N-1}{2}\right)T_r,f\right)$ $+2\sum_{i=0}^{N-2}\sum_{m=i+1}^{N-1} WD_x\left(t;t-\left(\frac{(i+m)-(N-1)}{2}\right)T_r,f\right)$ $\times\cos[2\pi T_r(i-m)f]$	$\sum_{i=-N+1}^{N-1} AF_x(t;\tau-nT_r,\nu)\frac{\sin\pi\nu T_r(N-\lvert n\rvert)}{\sin(\pi\nu T_r)}$
$\sum_{i=1}^{N} x(t-t_i)e^{j2\pi f_i t}$	$\sum_{i=1}^{N} X(f-f_i)e^{-j2\pi(f-f_i)t_i}$	$\sum_{i=1}^{N} WD_x(t;t-t_i,f-f_i)$ $+2\sum_{i=1}^{N-1}\sum_{m=i+1}^{N} WD_x\left(t;t-\frac{(t_i+t_m)}{2},f-\frac{(f_i+f_m)}{2}\right)$ $\times\cos 2\pi\left[(f_i-f_m)t-(t_i-t_m)f+\frac{(f_i+f_m)}{2}(t_i-t_m)\right]$	$AF_x(t;\tau,\nu)\sum_{i=1}^{N} e^{j2\pi(f_i\tau-\nu t_i)}$ $+\sum_{i=1}^{N}\sum_{m=1, m\neq i}^{N} AF_x(t;\tau-(t_i-t_m),\nu-(f_i-f_m))$ $\exp\left[j2\pi\left(\frac{f_i+f_m}{2}\right)\tau-\frac{t_i+t_m}{2}\nu+(f_i-f_m)\frac{t_i+t_m}{2}\right]$

19.7.2 Choi-Williams Distribution

The function x is a function of time and $CWD_x(t,f;\sigma)$ means that the distribution is a function of t and f with a parameter σ.

$$CWD_x(t,f;\sigma) = \sqrt{\frac{\sigma}{4\pi}} \iint_{-\infty}^{\infty} \frac{1}{|\tau|} \exp\left(\frac{-\sigma}{4}\left[\frac{t-t'}{\tau}\right]^2\right) x\left(t'+\frac{\tau}{2}\right) x^*\left(t'-\frac{\tau}{2}\right) e^{-j2\pi f\tau}\, dt'\, d\tau$$

$$= \sqrt{\frac{\sigma}{4\pi}} \iint_{-\infty}^{\infty} \frac{1}{|\nu|} \exp\left(\frac{-\sigma}{4}\left[\frac{f-f'}{\nu}\right]^2\right) X\left(f'+\frac{\nu}{2}\right) X^*\left(f'-\frac{\nu}{2}\right) e^{j2\pi t\nu}\, df'\, d\nu$$

$$= \sqrt{\frac{\sigma}{4\pi}} \iiint_{-\infty}^{\infty} \frac{1}{|u|} \exp\left(\frac{-\sigma}{4}\left[\frac{t-t'}{u}\right]^2\right) e^{-j2\pi(f-f')u}\, WD_x(t',f')\, du\, dt'\, df'$$

$$= \iint_{-\infty}^{\infty} e^{-(2\pi\tau\nu)^2/\sigma} AF_x(\tau,\nu) e^{j2\pi(t\nu - f\tau)}\, d\tau\, d\nu$$

$$= \sqrt{\frac{\sigma}{4\pi}} \iint_{-\infty}^{\infty} \frac{1}{|f_1-f_2|} \exp\left(\frac{-\sigma}{4}\left[\frac{f-(f_1+f_2)/2}{f_1-f_2}\right]^2\right) X(f_1) X^*(f_2) e^{j2\pi t(f_1-f_2)}\, df_1\, df_2$$

where $WD_x(t,f) \equiv WD_x(t;t,f)$ and $AF_x(\tau,\nu) = AF_x(t;\tau,\nu)$.

19.7.3 Table of Time-Frequency Representations of Cohen's Class

Cohen's-class TFRs. Here,

$$rect_a(t) = \begin{cases} 1, & |t| < |a| \\ 0, & |t| > |a| \end{cases}, \qquad AF_x(\tau,\nu)$$

is the ambiguity function, and $\tilde{\mu}(\tilde{\tau},\tilde{\nu};\alpha,r,\beta,\gamma) = \tilde{\tau}^2(\tilde{\nu}^2)^\alpha + (\tilde{\tau})^\alpha \tilde{\nu}^2 + 2r(((\tilde{\tau},\tilde{\nu})^\beta)^\gamma)^2$. Functions with lower- and uppercase letters, e.g., $\gamma(t)$ and $\Gamma(f)$, are Fourier transform pairs. The subscript x implies a function of time.

TABLE 19.4

Cohen's-Class Distribution	Formula
Ackroyd	$ACK_x(t,f) = Re\{x^*(t)X(f)e^{j2\pi ft}\}$
Affine-Cohen Subclass	$AC_x(t,f;S_{AC}) = \iint \frac{1}{\lvert\tau\rvert} S_{AC}\left(\frac{t-t'}{\tau}\right) x\left(t'+\frac{\tau}{2}\right) x^*\left(t'-\frac{\tau}{2}\right) e^{-j2\pi f\tau}\, dt'\, d\tau$
Born-Jordan	$BJD_x(t,f) = \iint \frac{\sin(\pi\tau\nu)}{\pi\tau\nu} AF_x(\tau,\nu) e^{j2\pi(t\nu - f\tau)}\, d\tau\, d\nu$
	$= \int \frac{1}{\tau}\left[\int_{t-\lvert\tau\rvert/2}^{t+\lvert\tau\rvert/2} x\left(t'+\frac{\tau}{2}\right) x^*\left(t'-\frac{\tau}{2}\right) dt'\right] e^{-j2\pi f\tau}\, d\tau$

TABLE 19.4 (continued)

Cohen's-Class Distribution	Formula
Butterworth	$BUD_x(t,f;M,N)=\iint\left(1+\left(\frac{\tau}{\tau_0}\right)^{2M}\left(\frac{\nu}{\nu_0}\right)^{2N}\right)^{-1}AF_x(\tau,\nu)e^{j2\pi(t\nu-f\tau)}\,d\tau\,d\nu$
Choi-Williams (Exponential)	$CWD_x(t,f;\sigma)=\iint e^{-(2\pi\tau\nu)^2/\sigma}AF_x(\tau,\nu)e^{j2\pi(t\nu-f\tau)}\,d\tau\,d\nu$ $=\iint\sqrt{\frac{\sigma}{4\pi}}\frac{1}{\lvert\tau\rvert}\exp\left[-\frac{\sigma}{4}\left(\frac{t-t'}{\tau}\right)^2\right]x\left(t'+\frac{\tau}{2}\right)x^*\left(t'-\frac{\tau}{2}\right)e^{-j2\pi f\tau}\,dt'\,d\tau$
Cone Kernel	$CKD_x(t,f)=\iint g(\tau)\lvert\tau\rvert\frac{\sin(\pi\tau\nu)}{\pi\tau\nu}AF_x(\tau,\nu)e^{j2\pi(t\nu-f\tau)}\,d\tau\,d\nu$
Cumulative Attack Spectrum	$CAS_x(t,f)=\left\lvert\int_{-\infty}^{t}x(\tau)e^{-j2\pi f\tau}\,d\tau\right\rvert^2$
Cumulative Decay Spectrum	$CDS_x(t,f)=\left\lvert\int_{t}^{\infty}x(\tau)e^{-j2\pi f\tau}\,d\tau\right\rvert^2$
Generalized Exponential	$GED_x(t,f)=\iint\exp\left[-\left(\frac{\tau}{\tau_0}\right)^{2M}\left(\frac{\nu}{\nu_0}\right)^{2N}\right]AF_x(\tau,\nu)e^{j2\pi(t\nu-f\tau)}\,d\tau\,d\nu$
Generalized Rectangular	$GRD_x(t,f)=\iint rect_1\left(\lvert\tau\rvert^{M/N}\lvert\nu\rvert/\sigma\right)AF_x(\tau,\nu)e^{j2\pi(t\nu-f\tau)}\,d\tau\,d\nu$
Generalized Wigner	$GWD_x(t,f;\tilde{\alpha})=\int x(t+(\tfrac{1}{2}+\tilde{\alpha})\tau)x^*(t-(\tfrac{1}{2}-\tilde{\alpha})\tau)e^{-j2\pi f\tau}\,d\tau$
Levin	$LD_x(t,f)=-\frac{d}{dt}\left\lvert\int_{t}^{\infty}x(\tau)e^{-j2\pi f\tau}\,d\tau\right\rvert^2$
Margineau-Hill	$MH_x(t,f)=Re\{x(t)X^*(f)e^{-j2\pi ft}\}$
Multiform Tiltable Kernel	$MT_x(t,f;S)=\iint S\left(\tilde{\mu}\left(\frac{\tau}{\tau_0},\frac{\nu}{\nu_0};\alpha,r,\beta,\gamma\right)^{2\lambda}\right)AF_x(\tau,\nu)e^{j2\pi(t\nu-f\tau)}\,d\tau\,d\nu$ $S_{MTED}(\beta)=e^{-\pi\beta},\quad S_{MTBUD}(\beta)=[1+\beta]^{-1}$
Nutall	$ND_x(t,f)=\iint\exp\left\{-\pi\left[\left(\frac{\tau}{\tau_0}\right)^2+\left(\frac{\nu}{\nu_0}\right)^2+2r\left(\frac{\tau\nu}{\tau_0\nu_0}\right)\right]\right\}AF_x(\tau,\nu)e^{j2\pi(t\nu-f\tau)}\,d\tau\,d\nu$
Page	$PD_x(t,f)=2Re\left\{x^*(t)e^{j2\pi ft}\int_{-\infty}^{t}x(\tau)e^{-j2\pi f\tau}\,d\tau\right\}$
Pseudo Wigner	$PWD_x(t,f;\Gamma)=\int x\left(t+\frac{\tau}{2}\right)x^*\left(t'-\frac{\tau}{2}\right)\gamma\left(\frac{\tau}{2}\right)\gamma^*\left(-\frac{\tau}{2}\right)e^{-j2\pi f\tau}\,d\tau$ $=\int WD_\gamma(0,f-f')WD_x(t,f')\,df'$

TABLE 19.4 (continued)

Cohen's-Class Distribution	Formula
Reduced Interference	$RID_x(t,f) = \iint \frac{1}{\lvert\tau\rvert} s\left(\frac{t-t'}{\tau}\right) x\left(t'+\frac{\tau}{2}\right) x^*\left(t'-\frac{\tau}{2}\right) e^{-j2\pi f\tau}\, dt'\, d\tau$
	with $S(\beta) \in \Re,\ S(0)=1,\ \frac{d}{d\beta}S(\beta)\big\rvert_{\beta=0} = 0,\ \{S(\alpha)=0 \text{ for } \lvert\alpha\rvert > \frac{1}{2}\}$
Rihaczek	$RD_x(t,f) = x(t)X^*(f)e^{-j2\pi ft}$
Smoothed Pseudo Wigner	$SPWD_x(t,f;\Gamma,s) = \int s(t-t')PWD_x(t',f;\Gamma)\,dt'$
	$= \iint s(t-t')WD_\gamma(0,f-f')WD_x(t',f')\,dt'\,df'$
Spectrogram	$SPEC_x(t,f;\Gamma) = \left\lvert \int x(\tau)\gamma^*(\tau-t)e^{-j2\pi f\tau}\,d\tau \right\rvert^2 = \left\lvert \int X(f')\Gamma^*(f'-f)e^{j2\pi f't}\,df' \right\rvert^2$
Wigner	$WD_x(t,f) = \int x\left(t+\frac{\tau}{2}\right)x^*\left(t-\frac{\tau}{2}\right)e^{-j2\pi f\tau}\,d\tau = \int X\left(f+\frac{\nu}{2}\right)X^*\left(f-\frac{\nu}{2}\right)e^{j2\pi t\nu}\,d\nu$

* From Mixed Time-Frequency Signal Transformations by G. Faye Boudreaux-Bartels in *The Transforms and Applications Handbook*, edited by Alexander D. Poularikas, CRC Press, 1996. Reprinted with permission.

19.8 Kernels of Cohen's Class

19.8.1 Kernels of Cohen's Shift Covariant Class (See Table 19.5)

19.9 Affine and Hyperbolic TFRs

19.9.1 For the affine and hyperbolic classes of TFRs see Mixed Time-Frequency Signal Transformations by G. Faye Boudreaux-Bartels in *The Transforms and Applications Handbook*, edited by Alexander D. Poularikas, CRC Press, 1996.

19.10 Wigner Distribution of Discrete-Time Signals

19.10.1 WD of Discrete-Time Signals x(n) and g(n), ($\omega \in R,\ n \in Z$)

$$WD_{x,g}(n,\omega) \equiv WD_{x,g}(n;n,\omega) = 2\sum_{k=-\infty}^{\infty} x(n+k)g^*(n-k)e^{-j2k\omega} \qquad \lvert\omega\rvert < \tfrac{\pi}{2T}$$

$$WD_{x,g}(n,\omega) = 2T\sum_{k=-\infty}^{\infty} x(n+k)g^*(n-k)e^{-j2\omega kT} \qquad \lvert\omega\rvert < \tfrac{\pi}{2T}$$

$$WD_x(n,\omega) \equiv W_{x,x}(n,\omega) = 2\sum_{k=-\infty}^{\infty} x(n+k)x^*(n-k)e^{-j2k\omega} \qquad \lvert\omega\rvert < \tfrac{\pi}{2T}$$

TABLE 19.5 Kernels of Cohen's Shift Covariant Class of Time-Frequency Representations (TFRs) Defined in Table 19.4. (Here, $\tilde{\mu}(\tilde{\tau},\tilde{\nu};\alpha,r,\beta,\gamma) = \left((\tilde{\tau})^2((\tilde{\nu})^2)^\alpha + ((\tilde{\tau})^2)^\alpha(\tilde{\nu})^2 + 2r((\tilde{\tau}\tilde{\nu})^\beta)^\gamma\right)$. Function with lower case and upper case letters, e.g., $\gamma(t)$ and $\Gamma(f)$, indicate Fourier transform pairs.)

TFR	$\psi_C(t,f)$	$\Psi_C(\tau,\nu)$	$\varphi_C(t,\tau)$	$\Phi_C(f,\nu)$
AC	$\int \frac{1}{\lvert\tau\rvert} s_{AC}\left(\frac{t}{\tau}\right) e^{-j2\pi f\tau}\, d\tau$	$S_{AC}(\tau\nu)$	$\frac{1}{\lvert\tau\rvert} s_{AC}\left(\frac{t}{\tau}\right)$	$\frac{1}{\lvert\nu\rvert} S_{AC}\left(-\frac{f}{\nu}\right)$
ACK	$2\cos(4\pi tf)$	$\cos(\pi\tau\nu)$	$\frac{\delta(t+\tau/2)+\delta(t-\tau/2)}{2}$	$\frac{\delta(f-\nu/2)+\delta(f+\nu/2)}{2}$
BJD		$\frac{\sin(\pi\tau\nu)}{\pi\tau\nu}$	$\begin{cases} \frac{1}{\lvert\tau\rvert}, & \lvert t/\tau\rvert < 1/2 \\ 0, & \lvert t/\tau\rvert > 1/2 \end{cases}$	$\begin{cases} \frac{1}{\lvert\nu\rvert}, & \lvert f/\nu\rvert < 1/2 \\ 0, & \lvert f/\nu\rvert > 1/2 \end{cases}$
BUD		$\left(1+\left(\frac{\tau}{\tau_0}\right)^{2M}\left(\frac{\nu}{\nu_0}\right)^{2N}\right)^{-1}$		
CWD	$\sqrt{\frac{\sigma}{4\pi}} \int \frac{1}{\lvert\beta\rvert} \exp\left[-\frac{\sigma}{4}\left(\frac{t}{\beta}\right)^2\right] e^{-j2\pi f\beta}\, d\beta$	$e^{-(2\pi\tau\nu)^2/\sigma}$	$\sqrt{\frac{\sigma}{4\pi}} \frac{1}{\lvert\tau\rvert} \exp\left[-\frac{\sigma}{4}\left(\frac{t}{\tau}\right)^2\right]$	$\sqrt{\frac{\sigma}{4\pi}} \frac{1}{\lvert\nu\rvert} \exp\left[-\frac{\sigma}{4}\left(\frac{f}{\nu}\right)^2\right]$
CKD		$g(\tau)\lvert\tau\rvert \frac{\sin(\pi\tau\nu)}{\pi\tau\nu}$	$\begin{cases} g(\tau), & \lvert t/\tau\rvert < 1/2 \\ 0, & \lvert t/\tau\rvert > 1/2 \end{cases}$	
CAS		$\left[\frac{1}{2}\delta(\nu) + \frac{1}{j\nu}\right] e^{-j\pi\lvert\tau\rvert\nu}$		
CDS		$\left[\frac{1}{2}\delta(-\nu) - \frac{1}{j\nu}\right] e^{j\pi\lvert\tau\rvert\nu}$		
GED		$\exp\left[-\left(\frac{\tau}{\tau_0}\right)^{2M}\left(\frac{\nu}{\nu_0}\right)^{2N}\right]$	$\frac{\nu_0}{2\sqrt{\pi}} \left\lvert\frac{\tau_0}{\tau}\right\rvert^{M} \exp\left[\frac{-\nu_0^2\tau_0^{2M}t^2}{4\tau^{2M}}\right]$ N = 1 only	$\frac{\tau_0}{2\sqrt{\pi}} \left\lvert\frac{\nu_0}{\nu}\right\rvert^{N} \exp\left[\frac{-\tau_0^2\nu_0^{2N}f^2}{4\nu^{2N}}\right]$ M = 1 only
GRD		$\begin{cases} 1, & \lvert\tau\rvert^{M/N}\lvert\nu\rvert/\sigma < 1 \\ 0, & \lvert\tau\rvert^{M/N}\lvert\nu\rvert/\sigma > 1 \end{cases}$	$\frac{\sin\left(2\pi\sigma t/\lvert\tau\rvert^{M/N}\right)}{\pi t}$	

TABLE 19.5 Kernels of Cohen's Shift Covariant Class of Time-Frequency Representations (TFRs) defined in Table 19.4. (Here, $\tilde{\mu}(\tilde{\tau},\tilde{\nu};\alpha,r,\beta,\gamma) = \left((\tilde{\tau})^2((\tilde{\nu})^2)^{\alpha} + ((\tilde{\tau})^2)^{\alpha}(\tilde{\nu})^2 + 2r((\tilde{\tau}\tilde{\nu})^{\beta})^{\gamma}\right)$. Function with lower case and upper case letters, e.g., $\gamma(t)$ and $\Gamma(f)$, indicate Fourier transform pairs.) (continued)

TFR	$\psi_C(t,f)$	$\Psi_C(\tau,\nu)$	$\varphi_C(t,\tau)$	$\Phi_C(f,\nu)$
GWD	$\frac{1}{\lvert\tilde{\alpha}\rvert}e^{j2\pi tf/\tilde{\alpha}}$	$e^{j2\pi\tilde{\alpha}\tau\nu}$	$\delta(t+\tilde{\alpha}\tau)$	$\delta(f-\tilde{\alpha}\nu)$
LD		$e^{j\pi\lvert\tau\rvert\nu}$	$\delta(t+\lvert\tau\rvert/2)$	
MH	$2\cos(4\pi tf)$	$\cos(\pi\tau\nu)$	$\frac{\delta(t+\tau/2)+\delta(t-\tau/2)}{2}$	$\frac{\delta(f-\nu/2)+\delta(f+\nu/2)}{2}$
MT		$S\left(\tilde{\mu}\left(\frac{\tau}{\tau_0},\frac{\nu}{\nu_0};\alpha,r,\beta,\gamma\right)^{2\lambda}\right)$		
ND		$\exp\left[-\pi\tilde{\mu}\left(\frac{\tau}{\tau_0},\frac{\nu}{\nu_0};0,r,1,1\right)\right]$		
PD		$e^{-j\pi\lvert\tau\rvert\nu}$	$\delta(t-\lvert\tau\rvert/2)$	$\left[\delta\left(f+\frac{\nu}{2}\right)+\delta\left(f-\frac{\nu}{2}\right)+j\frac{\nu}{\pi(f^2-\nu^2/4)}\right]\Big/2$
PWD	$\delta(t)\,WD_\gamma(0,f)$	$\gamma(\tau/2)\gamma^*(-\tau/2)$	$\delta(t)\gamma(\tau/2)\gamma^*(-\tau/2)$	$WD_\gamma(0,f)$
RGWD	$\frac{1}{\lvert\tilde{\alpha}\rvert}\cos(2\pi tf/\tilde{\alpha})$	$\cos(2\pi\tilde{\alpha}\tau\nu)$	$\frac{\delta(t+\tilde{\alpha}\tau)+\delta(t-\tilde{\alpha}\tau)}{2}$	$\frac{\delta(f-\tilde{\alpha}\nu)+\delta(f+\tilde{\alpha}\nu)}{2}$
RID	$\int\frac{1}{\lvert\beta\rvert}s\left(\frac{t}{\beta}\right)e^{-j2\pi f\beta}d\beta$	$S(\tau\nu)$	$\frac{1}{\lvert\tau\rvert}s\left(\frac{t}{\tau}\right)$	$\frac{1}{\lvert\nu\rvert}s\left(-\frac{f}{\nu}\right)$
		$S(\beta)\in\Re,\ S(0)=1,\ \frac{d}{d\beta}S(\beta)\big\vert_{\beta=0}=0$	$s(\alpha)=0,\ \lvert\alpha\rvert>\frac{1}{2}$	$s(\alpha)=0,\ \lvert\alpha\rvert>\frac{1}{2}$
RD	$2e^{-j4\pi tf}$	$e^{-j2\pi\tau\nu}$	$\delta(t-\tau/2)$	$\delta(f+\nu/2)$
SPWD	$s(t)\,WD_\gamma(0,f)$	$S(\nu)\gamma(\tau/2)\gamma^*(-\tau/2)$	$s(t)\gamma(\tau/2)\gamma^*(-\tau/2)$	$S(\nu)\,WD_\gamma(0,f)$
SPEC	$WD_\gamma(-t,-f)$	$AF_\gamma(-\tau,-\nu)$	$\gamma\left(-t-\frac{\tau}{2}\right)\gamma^*\left(-t+\frac{\tau}{2}\right)$	$\Gamma\left(-f-\frac{\nu}{2}\right)\Gamma^*\left(-f+\frac{\nu}{2}\right)$
WD	$\delta(t)\delta(f)$	1	$\delta(t)$	$\delta(f)$

* From Mixed Time-Frequency Signal Transformations by G. Faye Boudreaux-Bartels in *The Transforms and Applications Handbook*, edited by Alexander D. Poularikas, CRC Press, 1996. Reprinted with permission.

19.10.2 WD Using the Signal Spectrums

$$WD_{X,G}(\omega,n) \equiv WD_{X,G}(\omega;\omega,n) = \frac{1}{\pi}\int_{-\pi}^{\pi} X(\omega+\xi)G^*(\omega-\xi)e^{j2n\xi}\,d\xi$$

$$WD_{X,G}(\omega,n) = WD_{x,g}(n,\omega)$$

$$X(\omega) = F_d\{x(n)\} = \sum_{n=-\infty}^{\infty} x(n)e^{-jn\omega} = \sum_{n=-\infty}^{\infty} x(n)e^{-jn2\pi f}$$

$$x(n) \equiv F_d^{-1}\{X(\omega)\} = \frac{1}{2\pi}\int_{-\pi}^{\pi} X(\omega)e^{jn\omega}\,d\omega = \int_{-1/2}^{1/2} X(f)e^{jn2\pi f}\,df$$

19.11 WD Properties Involving Discrete-Time Signals

19.11.1 Periodicity

$WD_{x,g}(n,\omega) = WD_{x,g}(n,\omega+\pi)$ for all (n,ω). Observe that a factor of 2 is added on the exponential $e^{j2k\omega}$ so that the frequency components occur at ω rather than 2ω.

19.11.2 Symmetry

$$WD_{x,g}(n,\omega) = WD^*_{g,x}(n,\omega)$$

$$WD_x(n,\omega) = WD_x^*(n,\omega) = \text{real}$$

$$WD_x(n,\omega) = WD_{x^*}(n,-\omega)$$

19.11.3 Time Shift

$$WD_{x(n-k),g(n-k)}(n,\omega) = WD_{x,g}(n-k,\omega)$$

19.11.4 Modulation by $e^{jn\omega}$

If $y_1(n) = x(n)e^{jn\omega_c}$ and $y_2(n) = g(n)e^{jn\omega_c}$, then

$$WD_{y_1,y_2}(n,\omega) = 2\sum_{k=-\infty}^{\infty} e^{-j2k\omega}\, x(n+k)e^{jn\omega_c}e^{jk\omega_c}\, g^*(n-k)e^{-jn\omega_c}e^{jk\omega_c}$$

$$= 2\sum_{k=-\infty}^{\infty} x(n+k)g^*(n-k)e^{-j2k(\omega-\omega_c)} = WD_{x,g}(n,\omega-\omega_c)$$

19.11.5 Inner Product

$$WD_{x,g}(0,0) = 2\sum_{k=-\infty}^{\infty} x(k)g^*(-k)$$

19.11.6 Sum Formula

$$WD_{x+g}(n,\omega) = 2\sum_{k=-\infty}^{\infty}[x(n+k)+g(n+k)][x^*(n-k)+g^*(n-k)]e^{-j2k\omega}$$

$$= 2\sum_{k=-\infty}^{\infty}[x(n+k)x^*(n-k)+x(n+k)g^*(n-k)+g(n+k)x^*(n-k)$$

$$+g(n+k)g^*(n-k)]e^{-j2k\omega}$$

$$= WD_x(n,\omega)+WD_g(n,\omega)+2Re\{WD_{x,g}(n,\omega)\}$$

19.11.7 Multiplication by n

$$2\sum_{k=-\infty}^{\infty}(n+k)x(n+k)g^*(n-k)e^{-j2k\omega}+2\sum_{k=-\infty}^{\infty}x(n+k)(n-k)g^*(n-k)e^{-j2k\omega}$$

$$= 2n\sum_{k=-\infty}^{\infty}x(n+k)g^*(n-k)e^{-j2k\omega}$$

Hence, $2n\,WD_{x,g}(n,\omega) = WD_{nx,g}(n,\omega)+WD_{x,ng}(n,\omega)$

19.11.8 Multiplication by $e^{j2\omega}$

$$2\sum_{k=-\infty}^{\infty}x(n-1+k)g^*(n+1-k)e^{j2k\omega} = 2\sum_{r=-\infty}^{\infty}x(n+r)g^*(n-r)e^{j2r\omega}e^{j2\omega} \quad \text{where we set } k-1=r.$$

Hence, $WD_{x(n-1),g(n+1)}(n,\omega) = e^{j2\omega}\,WD_{x,g}(n,\omega)$.

19.11.9 Inverse Transform in Time

$$\frac{1}{2\pi}\int_{-\pi}^{\pi}e^{jk\omega}\,WD_{x,g}\left(n,\frac{\omega}{2}\right)d\omega = 2x(n+k)g^*(n-k)$$

where ω was substituted with ω/2. Hence, the WD evaluated at ω/2 can be considered as the Fourier transform of the sequence $2x(n+k)g^*(n-k)$. Set ω/2 = ω in the above equation with $n+k=n_1$ and $n-k=n_2$, we obtain

$$\frac{1}{2\pi}\int_{-\pi/2}^{\pi/2}e^{j(n_1-n_2)\omega}\,WD_{x,g}\left(\frac{n_1+n_2}{2},\omega\right)d\omega = x(n_1)g^*(n_2)$$

Similarly

$$\frac{1}{2\pi}\int_{-\pi/2}^{\pi/2}e^{j(n_1-n_2)\omega}\,WD_x\left(\frac{n_1+n_2}{2},\omega\right)d\omega = x(n_1)x^*(n_2)$$

19.11.10 Inverse Transform of the Product $x(n)g^*(n)$

Setting $n_1 = n_2 = n$ in 19.11.9

$$\frac{1}{2\pi}\int_{-\pi/2}^{\pi/2} WD_{x,g}(n,\omega)\,d\omega = x(n)g^*(n)$$

$$\frac{1}{2\pi}\int_{-\pi/2}^{\pi/2} WD_{x}(n,\omega)\,d\omega = |x(n)|^2$$

The last equation shows that the integral over one period of WD in its frequency variable is equal to the instantaneous signal power.

19.11.11 Recovery

Set $n_1 = 2n$ and $n_2 = 0$ in 19.11.9 we obtain

$$\frac{1}{2\pi}\int_{-\pi/2}^{\pi/2} e^{j2n\omega}\, WD_{x,g}(n,\omega)\,d\omega = x(2n)g^*(0)$$

Also

$$\frac{1}{2\pi}\int_{-\pi/2}^{\pi/2} e^{j2n\omega}\, WD_{x}(n,\omega)\,d\omega = x(2n)x^*(0)$$

Set $n_1 = 2n-1$ and $n_2 = 1$ in 19.11.9 we obtain

$$\frac{1}{2\pi}\int_{-\pi/2}^{\pi/2} e^{j2(n-1)\omega}\, WD_{x,g}(n,\omega)\,d\omega = x(2n-1)g^*(1)$$

19.11.12 Inner Product of Signals

Summing 19.11.10 over n we obtain

$$\frac{1}{2\pi}\sum_{n=-\infty}^{\infty}\int_{-\pi/2}^{\pi/2} WD_{x,g}(n,\omega)\,d\omega = \sum_{n=-\infty}^{\infty} x(n)g^*(n)$$

$$\frac{1}{2\pi}\sum_{n=-\infty}^{\infty}\int_{-\pi/2}^{\pi/2} WD_{x}(n,\omega)\,d\omega = \sum_{n=-\infty}^{\infty} |x(n)|^2$$

19.11.13 Moyal's Formula

$$\frac{1}{2\pi}\int_{-\pi/2}^{\pi/2}\sum_{n=-\infty}^{\infty} WD_{x_1,g_1}(n,\omega)WD^*_{x_2,g_2}(n,\omega)d\omega$$

$$=\left[\sum_{n=-\infty}^{\infty} x_1(n)x_2^*(n)\right]\left[\sum_{n=-\infty}^{\infty} g_1(n)g_2^*(n)\right]^* + \left[\sum_{n=-\infty}^{\infty} x_1(n)x_2^*(n)e^{-jn\pi}\right]\left[\sum_{n=-\infty}^{\infty} g_1(n)g_2^*(n)e^{-jn\pi}\right]^*$$

$$=\left[\frac{1}{2\pi}\int_{-\pi}^{\pi} X_1(\omega)X_2^*(\omega)d\omega\right]\left[\frac{1}{2\pi}\int_{-\pi}^{\pi} G_1(\omega)G_2^*(\omega)d\omega\right]^*$$

$$+\left[\frac{1}{2\pi}\int_{-\pi}^{\pi} X_1(\omega)X_2^*(\omega-\pi)d\omega\right]\left[\frac{1}{2\pi}\int_{-\pi}^{\pi} G_1(\omega)G_2^*(\omega-\pi)d\omega\right]^*$$

19.11.14 Time-Limited Signals

If $x(n)$ and $g(n)$ are time-limited (finite duration signal $x(n)=g(n)=0,\ n_b<n<n_a$, then

$$WD_{x,g}(n,\omega)=0, \qquad n_b<n<n_a$$

19.11.15 Sampled Analog Signals

$$WD_x(n,\omega)=2T\sum_{m=-\infty}^{\infty} x(n+m)x^*(n-m)e^{-j2\omega mT} \qquad |\omega|<\frac{\pi}{2T}$$

$$=2T\sum_{r=-\infty}^{\infty} x(2r)x^*[2(n-r)]e^{-j2\omega(2r-n)T}$$

$$=2T\sum_{r=-\infty}^{\infty} x(2r-1)x^*[2(n-r)+1]e^{-j2\omega(2r-1-n)T} \qquad |\omega|<\frac{\pi}{2T}$$

Example

If $x(n)=1$ for $|n|<N$ and zero otherwise, then

$$WD_x(n,\omega)=\begin{cases}2\dfrac{\sin[2\omega(N-|n|+\frac{1}{2})]}{\sin\omega} & |n|<N\\ 0 \text{ otherwise} & |n|\ge N\end{cases}$$

19.11.16 Multiplication in the Time Domain

If x and g modulate the carriers m_x and m_g, respectively, we obtain: $x_m(n)=x(n)m_x(n)$ and $g_m(n) = g(n)m_g(n)$. The WD is

$$WD_{x_m,g_m}(n,\omega)=\frac{1}{2\pi}\int_{-\pi/2}^{\pi/2} WD_{x,g}(n,\xi)WD_{m_x,m_g}(n,\omega-\xi)d\xi$$

19.11.17 Pseudo-Wigner Distribution (PWD)

$$x_n(\nu) = x(\nu)w_x(\nu - n), \quad g_n(\nu) = g(\nu)w_g(\nu - n)$$

$$PWD_{x_n,g_n}(\nu,\omega) = \frac{1}{2\pi}\int_{-\pi/2}^{\pi/2} WD_{x,g}(\nu,\xi)WD_{w_x,w_g}(\nu - n,\omega - \xi)d\xi$$

The discrete PWD is given by

$$PWD_{x,g}(n,\omega) = PWD_{x_n,g_n}(\nu,\omega)\big|_{\nu=n} = \frac{1}{2\pi}\int_{-\pi/2}^{\pi/2} WD_{x,g}(n,\xi)WD_{w_x,w_g}(0,\omega - \xi)d\xi$$

19.11.18 Analytic Signals

If $x_a(n) = x(n) + j\hat{x}(n)$ where $x(n)$ is real and $\hat{x}(n)$ is the discrete Hilbert transform of $x(n)$, defined by

$$\hat{x}(n) = \mathsf{H}_d\{x(n)\} = \sum_{m\neq n} x(m)\frac{\sin^2 \pi(m-n)/2}{\pi(m-n)/2}$$

$$X_a(\omega) = \begin{cases} 2X(\omega) & 0 < \omega < \pi \\ X(\omega) & \omega = 0 \\ 0 & -\pi < \omega < 0 \end{cases}$$

19.11.19 Central Moments

If $\sum_n k(n) = m_0 > 0$, then the central moments are found by minimization of the expression

$$i(n_0) = \sum_n (n - n_0)^2 k(n)/m_0$$

which gives $n_0 = n_k = \sum_n nk(n)/m_0$ and the minimum is

$$i(n_k) = \sum_n (n - n_k)^2 k(n)/m_0 = \sum_n n^2 k(n)/m_0 - n_k^2$$

19.12 Table of WD of Discrete-Time Function

19.12.1 Table of WD of Discrete-Time Functions

TABLE 19.6 WD Properties of Discrete-Time Signals $WD_{x,g}(n,\omega) = 2\sum_{k=-\infty}^{\infty} x(n+k)g^*(n-k)e^{-j2k\omega}$

	Property Name	Property
1.	Periodicity	$WD_{x,g}(n,\omega) = WD_{x,g}(n,\omega + \pi)$

TABLE 19.6 WD Properties of Discrete-Time Signals $WD_{x,g}(n,\omega)=2\sum_{k=-\infty}^{\infty} x(n+k)g^*(n-k)e^{-j2k\omega}$ (continued)

	Property Name	Property
2.	Symmetry	$WD_{x,g}(n,\omega)=WD^*_{g,x}(n,\omega)$; $WD_x(n,\omega)=WD^*_x(n,\omega)=\text{real}$;
		$WD_x(n,\omega)=WD_{x^*}(n,-\omega)$
3.	Time Shift	$WD_{x(n-k),g(n-k)}(n,\omega)=WD_{x,g}(n-k,\omega)$
4.	Modulation by $e^{jn\omega}$	$WD_{y_1,y_2}(n,\omega)=2\sum_{k=-\infty}^{\infty} x(n+k)g^*(n-k)e^{-j2k(\omega-\omega_c)}=WD_{x,g}(n,\omega-\omega_c)$
		$y_1(n)=x(n)\exp(jn\omega_c)$, $y_2(n)=g(n)\exp(jn\omega_c)$
5.	Inner Product	$WD_{x,g}(0,0)=2\sum_{k=-\infty}^{\infty} x(k)g^*(-k)$; $WD_x(0,0)=2\sum_{k=-\infty}^{\infty} x(k)x^*(-k)$
6.	Sum Formula	$WD_{x+g}(n,\omega)=WD_x(n,\omega)+WD_g(n,\omega)+2Re\{WD_{x,g}(n,\omega)\}$
7.	Multiplication by n	$2n\,WD_{x,g}(n,\omega)=WD_{nx,g}(n,\omega)+WD_{x,ng}(n,\omega)$
8.	Multiplication by $e^{j2\omega}$	$WD_{x(n-1),g(n+1)}(n,\omega)=e^{j2\omega}\,WD_{x,g}(n,\omega)$.
9.	Inverse Transform in Time	$\frac{1}{2\pi}\int_{-\pi}^{\pi} e^{jk\omega}WD_{x,g}\left(n,\frac{\omega}{2}\right)d\omega=2x(n+k)g^*(n-k)$
10.	Inverse Transform of the Product	$\frac{1}{2\pi}\int_{-\pi/2}^{\pi/2} e^{j(n_1-n_2)\omega}\,WD_{x,g}\left(\frac{n_1+n_2}{2},\omega\right)d\omega=x(n_1)g^*(n_2)$
		$\frac{1}{2\pi}\int_{-\pi/2}^{\pi/2} WD_{x,g}(n,\omega)\,d\omega=x(n)g^*(n)$
		$\frac{1}{2\pi}\int_{-\pi/2}^{\pi/2} WD_x(n,\omega)\,d\omega=\lvert x(n)\rvert^2$
11.	Recovery of signals	$\frac{1}{2\pi}\int_{-\pi/2}^{\pi/2} e^{j2n\omega}\,WD_{x,g}(n,\omega)\,d\omega=x(2n)g^*(0)$
		$\frac{1}{2\pi}\int_{-\pi/2}^{\pi/2} e^{j2n\omega}\,WD_x(n,\omega)\,d\omega=x(2n)x^*(0)$
		$\frac{1}{2\pi}\int_{-\pi/2}^{\pi/2} e^{j2(n-1)\omega}\,WD_{x,g}(n,\omega)\,d\omega=x(2n-1)g^*(1)$
12.	Inner Product of Signals	$\frac{1}{2\pi}\sum_{n=-\infty}^{\infty}\int_{-\pi/2}^{\pi/2} WD_{x,g}(n,\omega)\,d\omega=\sum_{n=-\infty}^{\infty} x(n)g^*(n)$
		$\frac{1}{2\pi}\sum_{n=-\infty}^{\infty}\int_{-\pi/2}^{\pi/2} WD_x(n,\omega)\,d\omega=\sum_{n=-\infty}^{\infty}\lvert x(n)\rvert^2$
13.	Moyal's Formula	$\frac{1}{2\pi}\int_{-\pi/2}^{\pi/2}\sum_{n=-\infty}^{\infty} WD_{x_1,g_1}(n,\omega)\,WD^*_{x_2,g_2}(n,\omega)\,d\omega$

TABLE 19.6 WD Properties of Discrete-Time Signals $WD_{x,g}(n,\omega)=2\sum_{k=-\infty}^{\infty} x(n+k)g^*(n-k)e^{-j2k\omega}$ (continued)

Property Name	Property
	$=\left[\sum_{n=-\infty}^{\infty} x_1(n)x_2^*(n)\right]\left[\sum_{n=-\infty}^{\infty} g_1(n)g_2^*(n)\right]^*$
	$+\left[\sum_{n=-\infty}^{\infty} x_1(n)x_2^*(n)e^{-jn\pi}\right]\left[\sum_{n=-\infty}^{\infty} g_1(n)g_2^*(n)e^{-jn\pi}\right]^*$
	$=\left[\frac{1}{2\pi}\int_{-\pi}^{\pi} X_1(\omega)X_2^*(\omega)d\omega\right]\left[\frac{1}{2\pi}\int_{-\pi}^{\pi} G_1(\omega)G_2^*(\omega)d\omega\right]^*$
	$+\left[\frac{1}{2\pi}\int_{-\pi}^{\pi} X_1(\omega)X_2^*(\omega-\pi)d\omega\right]\left[\frac{1}{2\pi}\int_{-\pi}^{\pi} G_1(\omega)G_2^*(\omega-\pi)d\omega\right]^*$
14. Time-Limited Signals	If $x(n)=g(n)=0$, for $n_b<n<n_a$, then $WD_{x,g}(n,\omega)=0, \quad n_b<n<n_a$
15. Sampled Analog Signals	$WD_x(n,\omega)=2T\sum_{m=-\infty}^{\infty} x(n+m)x^*(n-m)e^{-j2\omega mT} \quad \lvert\omega\rvert<\frac{\pi}{2T}$
	$=2T\sum_{r=-\infty}^{\infty} x(2r)x^*[2(n-r)]e^{-j2\omega(2r-n)T}$
	$=2T\sum_{r=-\infty}^{\infty} x(2r-1)x^*[2(n-r)+1]e^{-j2\omega(r-1-n)T} \quad \lvert\omega\rvert<\frac{\pi}{2T}$
16. Multiplication in the Time Domain:	If $x_m(n)=x(n)m_x(n)$ and $g_m(n)=g(n)m_g(n)$, then
	$WD_{x_m,g_m}(n,\omega)=\frac{1}{2\pi}\int_{-\pi/2}^{\pi/2} WD_{x,g}(n,\xi)WD_{m_x,m_g}(n,\omega-\xi)d\xi$
17. Pseudo-Wigner Distribution (PWD):	If $x_n(\nu)=x(\nu)w_x(\nu-n), \quad g_n(\nu)=g(\nu)w_g(\nu-n)$
	$PWD_{x_n,g_n}(\nu,\omega)=\frac{1}{2\pi}\int_{-\pi/2}^{\pi/2} WD_{x,g}(\nu,\xi)WD_{w_x,w_g}(\nu-n,\omega-\xi)d\xi$
	$PWD_{x,g}(n,\omega)\equiv$ discrete $PWD=PWD_{x_n,g_n}(\nu,\omega)\rvert_{\nu=n}$
	$=\frac{1}{2\pi}\int_{-\pi/2}^{\pi/2} WD_{x,g}(n,\xi)WD_{w_x,w_g}(0,\omega-\xi)d\xi$
18. Analytic Signals	If $x_a(n)=x(n)+j\hat{x}(n)$ and $\hat{x}(n)=H_d\{x(n)\}=$ Hilbert transform $=$
	$\sum_{m\neq n} x(m)\frac{\sin^2\pi(m-n)/2}{\pi(m-n)/2}, \quad X_a(\omega)=\begin{cases}2X(\omega) & 0<\omega<\pi\\ X(\omega) & \omega=0\\ 0 & -\pi<\omega<0\end{cases}$
19. Central Moments	If $\sum_n k(n)=m_0>0$, then

TABLE 19.6 WD Properties of Discrete-Time Signals $WD_{x,g}(n,\omega)=2\sum_{k=-\infty}^{\infty} x(n+k)g^*(n-k)e^{-j2k\omega}$ (continued)

Property Name	Property
	$i(n_k)=\text{central moment } =\sum_n (n-n_0)^2\,k(n)/m_0$
	$=\sum_n n^2 k(n)/m_0-n_k^2$ where
	$n_0=n_k=\sum_n nk(n)/m_0$

References

Boudreaux-Bartels, G. Faye, Mixed Time-Frequency Signal Transformations in *The Transforms and Applications Handbook,* edited by Alexander D. Poularikas, CRC Press Inc., Boca Raton, FL, 1996.

Classen, T.A.C.M., and W.F.G. Mecklenbrauker, The Wigner Distribution — A tool for time-frequency signal analysis, Part I: Continuous-time signals, *Phillips Journal of Research,* 35, 217-250, 1980.

Classen, T.A.C.M., and W.F.G. Mecklenbrauker, The Wigner Distribution — A tool for time-frequency signal analysis, Part II: Continuous-time signals, *Phillips Journal of Research,* 35, 276-300, 1980.

Classen, T.A.C.M., and W.F.G. Mecklenbrauker, The Wigner Distribution — A tool for time-frequency signal analysis, Part III: Continuous-time signals, *Phillips Journal of Research,* 35, 372-389, 1980.

Cohen, L., *Time-Frequency Analysis,* Prentice-Hall Inc., Englewood Cliffs, NJ, 1995.

Cohen, L., Time-frequency distribution — A review, *Proceedings IEEE,* 77, 941-981, July 1989.

20

Functions of a Complex Variable

20.1 Basic Concepts

Complex variable: $z = x + jy$

Complex conjugate: $z^* = x - jy,\ (z_1 + z_2)^* = z_1^* + z_2^*,\ (z_1 z_2)^* = z_1^* z_2^*,\ (z_1 - z_2)^* = z_1^* - z_2^*,$

$$(z_1 / z_2)^* = z_1^* / z_2^*,\ \mathrm{Re}\{z\} = \frac{z + z^*}{2},\ \mathrm{Im}\{z\} = \frac{z - z^*}{2j}$$

Polynomial: If z_0 is the root of a polynomial equation $a_n z^n + \cdots + a_1 z + a_0 = 0$ then z_0^* is also a root.

Absolute value: $|z|^2 = x^2 + y^2 = zz^*,\ |z| = \sqrt{x^2 + y^2},\ |z| = |z^*|,\ z^{-1} = z^* / |z|^2,\ z \neq 0,$

$$|z_1 z_2| = |z_1|\,|z_2|,\ |z_1 / z_2| = |z_1| / |z_2|\ \ z_2 \neq 0$$

Triangle inequality: $|z_1 + z_2| \leq |z_1| + |z_2|,$

$$\big||z_1| - |z_2|\big| \leq |z_1 - z_2|$$

Lagrange's identity:

$$\left|\sum_{j=1}^{n} a_j b_j\right|^2 = \sum_{j=1}^{n} |a_j|^2 \sum_{j=1}^{n} |b_j|^2 - \sum_{1\le j<k\le n} |a_j b_k^* - a_k b_j^*|^2$$

Polar form:

$$z = x + jy = r\cos\theta + jr\sin\theta = r(\cos\theta + j\sin\theta) \text{ where } \cos\theta = \frac{x}{|z|},\ \sin\theta = \frac{y}{|z|}$$

$$\theta = \arg\{z\} = \{\theta + 2\pi n;\ n = 0, \pm 1, \pm 2, \cdots\} \equiv \arg\{z\} = \theta \bmod 2\pi$$

$$\arg\{z_1 z_2\} = \arg\{z_1\} + \arg\{z_2\} \bmod 2\pi$$

$$\arg\{z_1 / z_2\} = \arg\{z_1\} - \arg\{z_2\} \bmod 2\pi$$

DeMoivre identity: $(\cos\theta + j\sin\theta)^n = \cos n\theta + j\sin n\theta = e^{jn\theta}$

20.2 Roots of Complex Numbers

20.2.1 Roots of Complex Numbers

If $z^n = a$ then

$$z = a^{1/n} = \sqrt[n]{|a|}\left(\cos\frac{\theta + 2\pi k}{n} + j\sin\frac{\theta + 2\pi k}{n}\right),$$

$$k = 0, 1, 2, \cdots, n-1 \text{ (true also for negative integer)}$$

Example

$$(-1)^{2/3} = [(-1)^2]^{1/3} = 1^{1/3} = \left\{1,\ \cos\frac{2\pi}{3} + j\sin\frac{2\pi}{3},\ \cos\frac{4\pi}{3} + j\sin\frac{4\pi}{3}\right\}.$$

Roots when *n* and *m* are relatively prime

If $z^{n/m} = a$ then,

$$z = (a^m)^{1/n} = a^{m/n} = (a^{1/n})^m = \sqrt[n]{|a|^m}\left(\cos\frac{m\theta + 2\pi k}{n} + j\sin\frac{m\theta + 2\pi k}{n}\right).$$

20.3 Functions, Continuity, and Analyticity

20.3.1 Continuous Function

A function $W(z)$ is *continuous* at a point $z = \lambda$ of R_z if, for each number $\varepsilon > 0$, however small, there exists another number $\delta > 0$ such that whenever

$$|z - \lambda| < \delta \text{ then } |W(z) - W(\lambda)| < \varepsilon$$

20.3.2 Limit of Sum and Difference

If
$$\lim_{z\to z_0} W_1(z) = a \text{ and } \lim_{z\to z_0} W_2(z) = b, \text{ then}$$
$$\lim_{z\to z_0}[W_1(z) \pm W_2(z)] = a + b$$

20.3.3 Limit of Product and Ratio

$$\lim_{z\to z_0} W_1(z)W_2(z) = ab, \quad \lim_{z\to z_0} \frac{W_1(z)}{W_2(z)} = \frac{a}{b}$$

Example

Let D be the punctured disc $0 < |z| < 1$, and let $W(z) = z^{*2}$. Then if z_1 and $z_2 \in D$, we obtain $|W(z_1) - W(z_2)| = |z_1^{*2} - z_2^{*2}| = |z_1^* + z_2^*||z_1^* - z_2^*| \le (|z_1| + |z_2|)\ |z_1 - z_2| \le 2|z_1 - z_2|$. Let $\varepsilon > 0$ be given. Choose $\delta > 0$ such that $|z_1 - z_2| < \delta$, then $|W(z_1) - W(z_2)| < \varepsilon$. Clearly, choosing $\delta = \varepsilon/2$ we obtain $|W(z_1) - W(z_2)| < 2\varepsilon/2 = \varepsilon$ which shows that $f(z)$ is uniformly continuous.

20.3.4 Analytic Function

A function $W(z)$ is *analytic* at a point z if, for each number $\varepsilon > 0$, however small, there exists another number $\delta > 0$ such that whenever

$$|z - \lambda| < \delta \text{ then } \left|\frac{W(z) - W(\lambda)}{z - \lambda} - \frac{dW(\lambda)}{dz}\right| < \varepsilon$$

20.3.5 Cauchy-Riemann Conditions

The function $W(z) = u(x,y) + j\upsilon(x,y)$ is analytic at a point if $\frac{\partial u}{\partial x} = \frac{\partial \upsilon}{\partial y}$ and $\frac{\partial \upsilon}{\partial x} = -\frac{\partial u}{\partial y}$.

20.3.6 Domain of Analyticity

The set of all points where a function is analytic is called the domain of analyticity. A function whose domain of analyticity is the whole complex plane is called entire.

20.3.7 Rules of Differentiation

If F, G, and H are analytic in D, then

$$\frac{d}{dz}(F(z) + G(z)) = \frac{dF(z)}{dz} + \frac{dG(z)}{dz}, \quad \frac{d}{dz}\left[\frac{F(z)}{G(z)}\right] = \frac{F'(z)G(z) - F(z)G'(z)}{G^2(z)}$$

$$\frac{d}{dz}H(z) = \frac{d}{dz}(F(z)G(z)) = F'(z)G(z) + F(z)G'(z).$$

20.4 Power Series

20.4.1 Power Series

$$\sum_{n=0}^{\infty} a_n(z-z_0)^n = a_0 + a_1(z-z_0) + \cdots + a_n(z-z_0)^n + \cdots$$

20.4.2 Convergence of Power Series

If (20.4.1) converges at some point z_1 and diverges at some point z_2, then it converges absolutely for all z such that $|z-z_0| < |z_1-z_0|$, and diverges for all z such that $|z-z_0| > |z_2-z_0|$.

20.4.3 Radius of Convergence

$$R = \frac{1}{\lim_{n\to\infty} \sup_{k\geq n} \sqrt[k]{|a_k|}}$$

20.4.4 Cauchy-Hadamard Rule

If $R = 0$, (20.4.1) converges only for $z = z_0$. If $R = \infty$, (20.4.1) converges absolutely for all z. If $0 < R < \infty$, (20.4.1) converges absolutely if $|z-z_0| < R$ and diverges if $|z-z_0| > R$.

20.4.5 Uniform Convergence

If $0 < r < R$ then (20.4.1) converges uniformly in the set $|z-z_0| \leq r$.

20.4.6 Representation of a Function

When (20.4.1) converges to a complex number $W(z)$ for each point z in a set S, we say that the series represents the function W in S.

20.4.7 Analyticity of Power Series

In the interior of its circle of convergence, the power series $W(z) = \sum_{n=0}^{\infty} a_n(z-z_0)^n$ is an analytic function.

20.4.8 Infinite Differentiable

In the interior of its circle of convergence, a power series is infinitely differentiable.

20.5 Exponential, Trigonometric, and Hyperbolic Functions

20.5.1 Complex Exponential Function

$$e^z = \sum_{n=0}^{\infty} \frac{z^n}{n!} \quad \text{for } z \in C$$

20.5.2 Complex Sine Function

$$\sin z = \sum_{n=0}^{\infty} (-1)^n \frac{z^{2n+1}}{(2n+1)!} \quad \text{for } z \in C \ (R = \infty)$$

20.5.3 Complex Cosine Function

$$\cos z = \sum_{n=0}^{\infty} (-1)^n \frac{z^{2n}}{(2n)!} \quad \text{for } z \in C \ (R = \infty)$$

20.5.4 Euler's Formula

$$e^{jz} = \cos z + j\sin z, \quad \cos z = \frac{e^{jz} + e^{-jz}}{2}, \quad \sin z = \frac{e^{jz} - e^{-jz}}{2j}$$

20.5.5 Periodic

A function is periodic in D if there exists a non-zero constant ω, called period, such that $f(z+\omega) = f(z)$.

20.5.6 Trigonometric Functions

$$\tan z = \frac{\sin z}{\cos z} \quad \text{for } z \in C - \left\{\frac{\pi}{2} + n\pi : n \in Z\right\}; \qquad \cot z = \frac{\cos z}{\sin z} \quad \text{for } z \in C - \{n\pi : n \in Z\}$$

$$\sec z = \frac{1}{\cos z} \quad \text{for } z \in C - \left\{\frac{\pi}{2} + n\pi : n \in Z\right\}; \qquad \csc z = \frac{1}{\sin z} \quad \text{for } z \in C - \{n\pi : n \in Z\}$$

20.5.7 Hyperbolic Functions

$$\cosh z = \frac{e^z + e^{-z}}{2}, \quad \sinh z = \frac{e^z - e^{-z}}{2} \quad \text{for } z \in C$$

$$\tanh z = \frac{\sinh z}{\cosh z} \quad \text{for } z \in C - \{(n+\tfrac{1}{2})\pi j : n \in Z\}; \qquad \coth z = \frac{\cosh z}{\sinh z} \quad \text{for } z \in C - \{n\pi j : n \in Z\}$$

$$\sec hz = \frac{1}{\cosh z} \quad \text{for } z \in C - \{(n+\tfrac{1}{2})\pi j : n \in Z\}; \qquad \csc hz = \frac{1}{\sinh z} \quad \text{for } z \in C - \{n\pi j : n \in Z\}$$

20.5.8 Other Hyperbolic Relations

$\tanh' z = \text{sech}^2 z$; $\coth' z = -\text{csch}^2 z$; $\text{sech}' z = -\text{sech}\, z \tanh z$; $\cosh' z = -\cosh z \coth z$; $\cosh z = \cos jz$; $\sinh z = -j\sin jz$; $\sin(x+jy) = \sin x \cosh y + j\cos x \sinh y$; $\cos(x+jy) = \cos x \cosh y - j\sin x \sinh y$; $|\sin z|^2 = \sin^2 x + \sinh^2 y = -\cos^2 x + \cosh^2 y$; $|\cos z|^2 = \cos^2 x + \sinh^2 y = -\sin^2 x + \cosh^2 y$; $|\sin x| \le |\sin z|$; $|\cos x| \le |\cos z|$; $|\sin z| \le \cosh y$ and $|\sin z| \ge |\sinh y|$; $\cos\left(\frac{\pi}{2} - z\right) = \sin z$; $\cos(\pi - z) = -\cos z$; $\tan(\pi + z) = \tan z$;

$\sin\left(\frac{\pi}{2}-z\right)=\cos z$; $\sin(\pi-z)=\sin z$; $\cot\left(\frac{\pi}{2}-z\right)=\tan z$; $\tan z=(\sin 2x+j\sinh 2y)/(\cos 2x+\cosh 2y)$;

$\cosh^2 z-\sinh^2 z=1$; $\cosh 2z=\cosh^2 z+\sinh^2 z$; $\sinh 2z=2\sinh z\cosh z$; $\sinh\left(\frac{j\pi}{2}-z\right)=j\cosh z$.

20.6 Complex Logarithm

20.6.1 Definitions

Determine all complex numbers q such that $e^q = z$. Hence if $z \in C-\{0\}$, we define $\ell n z=\{q : e^q = z\}$.

20.6.2 If $z=r(\cos\theta+j\sin\theta)$, then $z=|z|\,e^{j\theta}=e^{\ell n|z|+j\arg z}$ $(\arg z=\theta)$. Also

$$\ell n z=\ell n|z|+j\arg z$$

20.6.3 Principal Value

$$Lnz=\ell n|z|+j\mathrm{Arg}z,\quad -\pi<\mathrm{Arg}z\le\pi$$

20.6.4 Additional Properties

$e^{\ln z}=z$; $\ell n e^z=z \bmod 2\pi j$; $\ell n z_1z_2=\ln z_1+\ell n z_2 \bmod 2\pi j$; $\ell n(z_1/z_2)=\ln z_1-\ln z_2 \bmod 2\pi j$; $\ell n z^n = n\ell n z \bmod 2\pi j$ for all $n\in Z$.

20.6.5 Principal Value

The principal value of $z^a = e^{a\mathrm{Log}z}$ (z^a has many distinct elements).

Example

$j^j=e^{j\log j}=\left\{\exp\left[j\left(j\frac{\pi}{2}+2\pi jk\right)\right] : k\in Z\right\}=\{e^{-\frac{\pi}{2}-2\pi k} : k\in Z\}$. Hence the principal value of j^j is $e^{-\pi/2}$.

20.6.6 Other Relationships of Principal Values in General

$$z^{a_1}z^{a_2}=z^{a_1+a_2};\ (z_1z_2)^a\ne z_1^az_2^a;\ (z_1/z_2)^2\ne(z_1^a/z_2^a);\ \mathrm{Log}z^a\ne a\mathrm{Log}z;\ (z^a)^b\ne z^{ab}$$

20.7 Integration

20.7.1 Definition

The sum $\sum_{s=1}^{n} W_s\Delta z_s$ with overall values of s from a to b, and taking the limit $\Delta z_s\to 0$ and $n\to\infty$,

we obtain the integral $I=\int_a^b W(z)dz$ where the path of integration in the z-plane must be specified.

Example

The integration of $W(z)=1/z$ over a circle centered at the origin is given by

$$I=\oint \frac{1}{z}dz=\oint \frac{1}{re^{j\theta}}jre^{j\theta}d\theta=\oint j\,d\theta=\int_1 j\,d\theta+\int_2 j\,d\theta=j\pi+j\pi=2j\pi\,.$$

where $\oint$ integrates counterclockwise, $\int_1$ integrates from 0 to π, and $\int_2$ integrates from $-\pi$ to 0.

Example

The integration of $W(z)=1/z^2,\ W(z)=1/z^3,\cdots,W(z)=1/z^n$ around a contour encircling the origin is equal to zero.

Example

Find the value of the integral $\int_0^{z_0} z\,dz$ from the point (0,0) to (2,j4).

Solution

Because z is an analytic function along any path, then

$$\int_0^{z_0} z\,dz=\left.\frac{z^2}{2}\right|_0^{2+j4}=-6+j8$$

Equivalently, we could write

$$\int_0^{z_0} z\,dz=\int_0^2 x\,dx-\int_0^4 y\,dy+j\int_0^4 x\,dy=\left.\frac{x^2}{2}\right|_0^2-\left.\frac{y^2}{2}\right|_0^4+jxy|_0^4=2-\frac{16}{2}+j2\times 4=-6+j8$$

20.7.2 Properties of Integration

1. $\int_C [kW(z)+\ell G(z)]dz=k\int_C W(z)dz+\ell\int_C G(z)dz$, k and ℓ are complex numbers

2. $\int_C W(z)dz=-\int_{C'} W(z)dz$, C' has opposite orientation to C

3. $\int_C W(z)dz=\int_{C_1} W(z)dz+\int_{C_2} W(z)dz$, $C=C_1+C_2$

4. $\left|\int_C W(z)dz\right|\le ML$ if $|f(t)|\le M$ for $a\le t\le b$ and L is the length of the contour C.

5. $\int_C W(z)dz=\int_a^b W(C(t))C'(t)dt=F(b)-F(a),\quad F'(t)=W(C(t))C'(t)$

20.7.3 Cauchy First Integral Theorem

Given a region of the complex plane within which $W(z)$ is analytic and any closed curve that lies entirely within this region, then

$$\oint_C W(z)dz=0$$

where the contour C is taken counterclockwise.

20.7.4 Corollary 1

If the contour C_2 completely encloses C_1, and if $W(z)$ is analytic in the region between C_1 and C_2, and also on C_1 and C_2, then

$$\oint_{C_1} W(z)\,dz = \oint_{C_2} W(z)\,dz$$

The integration is done in a counterclockwise direction.

20.7.5 Corollary 2

If $W(z)$ has a finite number n of isolated singularities within a region G bounded by a curve C, then

$$\oint_{C} W(z)\,dz = \sum_{s=1}^{N} \oint_{C_s} W(z)\,dz$$

(see Figure 20.1.)

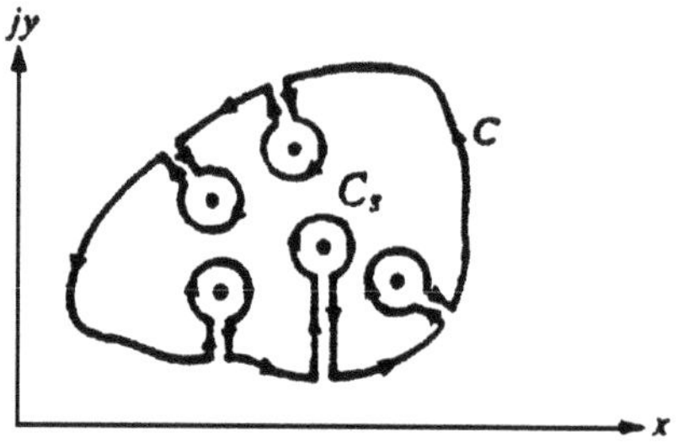

FIGURE 20.1 A countour enclosing n isolated singularities.

20.7.6 Corollary 3

The integral $\int_A^B W(z)\,dz$ depends only upon the end points A and B, and does not depend on the path of integration, provided that this path lies entirely within the region in which $W(z)$ is analytic.

20.7.7 The Cauchy Second Integral Theorem

If $W(z)$ is the function $W(z) = f(z)/(z - z_0)$ and the contour encloses the singularity at z_0, then

$$\oint_C \frac{f(z)}{z - z_0}\,dz = j2\pi f(z_0)$$

or

$$f(z_0) = \frac{1}{2\pi j}\oint_C \frac{f(z)}{z - z_0 d_z}\,dz$$

20.7.8 Derivative of an Analytic Function $W(z)$

The derivative of an analytic function is also analytic, and consequently itself possesses a derivative. Let C be a contour within and upon which $W(z)$ is analytic. Then a is a point inside the contour (the prime indicates first-order derivative)

$$W'(a) = \lim_{|h|\to 0} \frac{W(a+h) - W(a)}{h}$$

and can be shown that

$$W'(a) = \frac{1}{2\pi j} \oint_C \frac{W(z)\,dz}{(z-a)^2}$$

where the contour C is taken in a counterclockwise direction. Proceeding, it can be shown that

$$W^{(n)}(a) = \frac{n!}{2\pi j} \oint_C \frac{W(z)\,dz}{(z-a)^{n+1}}$$

The exponent (n) indicates the nth derivative and the contour is taken counterclockwise.

Example

$$\int_C \frac{\sin z}{(z-\pi)^3} dz = \pi j W''(\pi) = 0 \text{ where } C \text{ was the circle } |z| = 4.$$

20.7.9 Cauchy's Inequality

If $W(z)$ is analytic in the disk $|z-a| < R$ and if $|W(z)| \le M$ in this disk, then

$$\left|W^{(n)}(a)\right| \le \frac{Mn!}{R^n}$$

(see 20.7.8).

20.7.10 Liouville's Theorem

A bounded entire function $W(z)$ is identically constant (a function whose domain of analyticity is the whole complex plane is called <u>entire</u>).

20.7.11 Taylor's Theorem

If $W(z)$ is analytic in the disk $|z - z_0| < R$, then

$$W(z) = \sum_{n=0}^{\infty} \frac{W^{(n)}(z_0)}{n!}(z-z_0)^n \text{ whenever } |z-z_0| < R$$

Example

The Taylor series of ℓnz around $z_0 = 1$ is found by first identifying its derivatives $\ell nz, z^{-1}, -z^{-2}, 2z^{-3}, -3\times 2z^{-4}$, etc. In general $d^r \ell nz / dz^r = (-1)^{r+1}(r-1)! z^{-r}$ $(r = 1,2,\cdots)$.
Evaluating the derivatives at $z = 1$ we obtain,

$$\ell nz = 0 + (z-1) - \frac{(z-1)^2}{2!} + 2!\frac{(z-1)^3}{3!} - 3!\frac{(z-1)^4}{4!} + \cdots = \sum_{r=1}^{\infty} \frac{(-1)^{r+1}(z-1)^r}{r!}$$

which is valid $|z-1| < 1$.

20.7.12 Maclaurin Series

When $z_0 = 0$ in (20.7.11) the series is known as a Maclaurin series.

20.7.13 Cauchy Product

The Cauchy product of two Taylor series

$$\sum_{i=0}^{\infty} a_i(z-z_0)^i \text{ and } \sum_{i=0}^{\infty} b_i(z-z_0)^i$$

is defined to be the series

$$\sum_{i=0}^{\infty} c_i(z-z_0)^i \text{ where } c_i = \sum_{i=0}^{k} a_{k-i} b_i.$$

20.7.14 Product of Taylor Series

Let f and g be analytic functions with Taylor series $f(z) = \sum_{i=0}^{\infty} a_i(z-z_0)^i$ and $g(z) = \sum_{i=0}^{\infty} b_i(z-z_0)^i$ around the point z_0. Then the Taylor series for $f(z)g(z)$ around z_0 is given by the Cauchy product of these two series.

Example

$$\sin z \cos z = \left(z - \frac{z^3}{3!} + \frac{z^5}{5!} - \frac{z^7}{7!} + \cdots\right)\left(1 - \frac{z^2}{2!} + \frac{z^4}{4!} - \frac{z^6}{6!} + \cdots\right) = z - \left(\frac{1}{3!} + \frac{1}{2!}\right)z^3$$

$$+\left(\frac{1}{5!} + \frac{1}{3!}\frac{1}{2!} + \frac{1}{4!}\right)z^5 - \left(\frac{1}{7!} + \frac{1}{5!}\frac{1}{2!} + \frac{1}{3!}\frac{1}{4!} + \frac{1}{6!}\right)z^7 + \cdots = z - \frac{4}{3!}z^3 + \frac{16}{5!}z^5 - \frac{64}{7!}z^7 + \cdots$$

20.7.15 Taylor Expansions

1. $e^z = \sum_{k=0}^{\infty} \frac{z^k}{k!}$ for $z_0 = 0$

2. $\cosh z = \sum_{k=0}^{\infty} \frac{z^{2k}}{(2k)!}$ for $z_0 = 0$

3. $\sinh z = \sum_{k=0}^{\infty} \frac{z^{2k+1}}{(2k+1)!}$ for $z_0 = 0$

4. $\frac{1}{1-z} = \sum_{k=0}^{\infty} \frac{(z-j)^k}{(1-j)^{k+1}}$ for $z_0 = j$

5. $\ln(1-z) = \sum_{k=1}^{\infty} \frac{-z^k}{k!}$ for $z_0 = 0$

20.8 The Laurent Expansion

20.8.1 Laurent Theorem

Let C_1 and C_2 be two concentric circles, as shown in Figure 20.2, with their center at a. The function $f(z)$ is analytic with the ring and $(a+h)$ is any point in it. From the figure and Cauchy's theorem we obtain

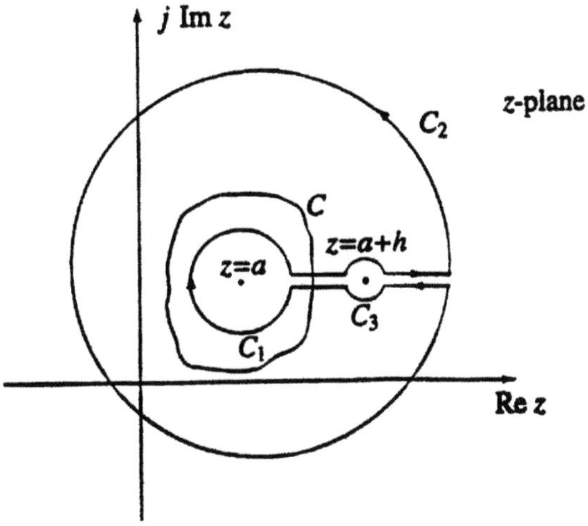

FIGURE 20.2 Explaining Laurent's theorem.

$$\frac{1}{2\pi j}\oint_{C_2}\frac{f(z)dz}{(z-a-h)}+\frac{1}{2\pi j}\oint_{C_1}\frac{f(z)dz}{(z-a-h)}+\frac{1}{2\pi j}\oint_{C_3}\frac{f(z)dz}{(z-a-h)}=0$$

where the first contour is counterclockwise and the last two are clockwise. The above equation becomes

$$f(a+h)=\frac{1}{2\pi j}\oint_{C_2}\frac{f(z)dz}{(z-a-h)}-\frac{1}{2\pi j}\oint_{C_1}\frac{f(z)dz}{(z-a-h)}$$

where both the contours are taken counterclockwise. For the C_2 contour $h<|(z-a)|$ and for the C_1 contour $h>|(z-a)|$. Hence we expand the above integral as follows:

$$f(a+h)=\frac{1}{2\pi j}\oint_{C_2}f(z)\left\{\frac{1}{(z-a)}+\frac{h}{(z-a)^2}+\cdots+\frac{h^n}{(z-a)^{n+1}}+\frac{h^{n+1}}{(z-a)^{n+1}(z-a-h)}\right\}dz$$
$$+\frac{1}{2\pi j}\oint_{C_1}f(z)\left\{\frac{1}{h}+\frac{z-a}{h^2}+\cdots+\frac{(z-a)^{n+1}}{h^{n+1}}+\frac{(z-a)^{n+1}}{h^{n+1}(z-a-h)}\right\}dz$$

From Taylor's theorem it was shown that the integrals of the last term in the two brackets tend to zero as n tends to infinity. Therefore, we have

$$f(a+h)=a_0+a_1h+a_2h^2+\cdots+\frac{b_1}{h}+\frac{b_2}{h^2}+\cdots \tag{20.1}$$

where

$$a_n=\frac{1}{2\pi j}\oint_{C_2}\frac{f(z)dz}{(z-a)^{n+1}}\qquad b_n=\frac{1}{2\pi j}\oint_{C_1}(z-a)^{n-1}f(z)dz$$

The above expansion can be put in more convenient form by substituting $h = z - a$, which gives

$$f(z)=c_0+c_1(z-a)+c_2(z-a)^2+\cdots+\frac{d_1}{(z-a)}+\frac{d_2}{(z-a)^2}+\cdots+\frac{d_n}{(z-a)^n}+\cdots \tag{20.2}$$

Because $z = a + h$, it means that z now is any point within the ring-shaped space between C_1 and C_2 where $f(z)$ is analytic. Equation (20.2) is the Laurent's expansion of $f(z)$ at a point $z + h$ within the ring. The coefficients c_n and d_n are obtained from (20.1) by replacing a_n, b_n, z by c_n, d_n, ζ, respectively. Here ζ is the variable on the contours, and z is inside the ring. When $f(z)$ has a simple pole at $z = a$, there is only one term, namely, $d_1/(z-a)$. If there exists an nth-order term, there are n terms of which the last is $d_n/(z-a)^n$; some of the d_n's may be zero.

If m is the highest index of the inverse power of $f(z)$ in (20.2) it is said that $f(z)$ has a pole of order m at $z = a$. Then

$$f(z) = \sum_{n=0}^{\infty} c_n (z-a)^n + \sum_{n=1}^{m} \frac{d_n}{(z-a)^n}$$

The coefficient d_1 is the *residue* at the pole.

If the series in the *inverse powers of (z – a)* in (20.2) does not terminate, the function $f(z)$ is said to have an *essential singularity* at $z = a$. Thus

$$f(z) = \sum_{n=0}^{\infty} c_n (z-a)^n + \sum_{n=1}^{m} \frac{d_n}{(z-a)^n}$$

The coefficient d_1 is the *residue* of the singularity.

Example

Find the Laurent expansion of $f(z) = 1/[(z-a)(z-b)^n]\,(n \ge 1,\ a \ne b \ne 0)$ near each pole.

Solution

First remove the origin to $z = a$ by the transformation $\zeta = (z-a)$. Hence we obtain

$$f(z) = \frac{1}{\zeta}\frac{1}{(\zeta + c)^n} = \frac{1}{c^n \zeta}\frac{1}{\left(1+\dfrac{\zeta}{c}\right)^n}, \quad c = a - b$$

If $|\zeta/c| < 1$ then we have

$$f(z) = \frac{1}{c^n \zeta}\left[1 - \frac{n\zeta}{c} + \frac{n(n+1)}{2!}\frac{\zeta^2}{c^2} - \cdots\right]$$

$$= \left[-\frac{n}{c^{n+1}} + \frac{n(n+1)\zeta}{2!c^{n+2}} - \cdots\right] + \frac{1}{c^n \zeta}$$

which is the Laurent series expansion near the pole at $z = a$. The residue is $1/c^n = 1/(a-b)^n$.

For the second pole set $\zeta = (z-a)$ and expand as above to find

$$f(z) = -\left(\frac{1}{c^{n+1}} + \frac{\zeta}{z^{n+2}} + \frac{\zeta^2}{c^{n+3}} + \cdots\right) - \left(\frac{1}{c^n \zeta} + \frac{\zeta}{c^{n-1}\zeta^2} + \cdots + \frac{1}{c\zeta^n}\right)$$

The second part of the expansion is the principal expansion near $z = b$, and the residue is $-1/c^n = -1/(a-b)^n$.

Example

Prove that

$$f(z)=\exp\left[\frac{x}{2}\left(z-\frac{1}{2}\right)\right]=J_0(x)+zJ_1(x)+z^2J_2(x)+\cdots+z^nJ_n(x)+\cdots$$

$$-\frac{1}{z}J_1(x)+\frac{1}{z^2}J_2(x)-\cdots+\frac{(-1)^n}{z^n}J_n(x)+\cdots$$

where

$$J_n(x)=\frac{1}{2\pi}\int_0^{2\pi}\cos(n\theta-x\sin\theta)\,d\theta$$

Solution

The function $f(z)$ is analytic except the point $z = a$. Hence by the Laurent's theorem we obtain

$$f(z)=a_0+a_1z+a_2z^2+\cdots+\frac{b_1}{z}+\frac{b_2}{z^2}+\cdots$$

where

$$a_n=\frac{1}{2\pi j}\oint_{C_2}\exp\left[\frac{x}{2}\left(z-\frac{1}{z}\right)\right]\frac{dz}{z^{n+1}},\qquad b_n=\frac{1}{2\pi j}\oint_{C_1}\exp\left[\frac{x}{2}\left(z-\frac{1}{z}\right)\right]z^{n-1}\,dz$$

where the contours are circles with center at the origin and are taken counterclockwise. Set C_2 equal to a circle of unit radius and write $z = \exp(j\theta)$. Then we have

$$a_n=\frac{1}{2\pi j}\int_0^{2\pi}e^{x\sin\theta}e^{-jn\theta}\,jd\theta=\frac{1}{2\pi}\int_0^{2\pi}\cos(n\theta-x\sin\theta)\,d\theta$$

because the last integral vanishes, as can be seen by writing $2\pi-\varphi$ for θ. Thus $a_n=J_n(x)$, and $b_n=(-1)^n a_n$, because the function is unaltered if $-z^{-1}$ is substituted for z, so that $b_n=(-1)^nJ_n(x)$.

20.9 Zeros and Singularities

20.9.1 Zero of Order m

A point z_0 is a *zero* of order m of $f(z)$ if $f(z)$ is analytic at z_0 and $f(z)$ and its $m-1$ derivatives vanish at z_0, but $f^{(m)}(z_0)\neq 0$.

20.9.2 Essential Singularity

A function has an *essential singularity* at $z = z_0$ if its Laurent expansion about the point z_0 contains an infinite number of terms in inverse powers of $(z - z_0)$.

20.9.3 Nonessential Singularity (pole of order m)

A function has a *nonessential singularity* or *pole of order m* if its Laurent expansion can be expressed in the form

$$W(z) = \sum_{n=-m}^{\infty} a_n (z - z_0)^n$$

Note that the summation extends from $-m$ to infinity and not from minus infinity to infinity; that is, the highest inverse power of $(z - z_0)$ is m.

An alternative definition that is equivalent to this but somewhat simpler to apply is the following: If $\lim_{z \to z_0} [(z - z_0)^m W(z)] = c$, a nonzero constant (here m is a positive number), then $W(z)$ is said to possess a pole of order m at z_0. The following examples illustrate these definitions:

Example

1. $\exp(1/z)$ has an essential singularity at the origin.
2. $\cos z/z$ has a pole of order 1 at the origin.
3. Consider the function

$$W(z) = \frac{e^z}{(z-4)^2(z^2+1)}$$

Note that functions of this general type exist frequently in the Laplace inversion integral. Because e^z is regular at all finite points of the z-plane, the singularities of $W(z)$ must occur at the points for which the denominator vanishes; that is, for

$$(z-4)^2(z^2+1) = 0 \quad \text{or} \quad z = 4, +j, -j$$

By the second definition above, it is easily shown that $W(z)$ has a second-order pole at $z = 4$, and first-order poles at the two points $+j$ and $-j$. That is,

$$\lim_{z \to 4} (z-4)^2 \left[\frac{e^z}{(z-4)^2(z^2+1)} \right] = \frac{e^4}{17} \neq 0$$

$$\lim_{z \to j} (z-j) \left[\frac{e^z}{(z-4)^2(z^2+1)} \right] = \frac{e^j}{(j-4)^2 2j} \neq 0$$

20.9.4 Picard's Theorem

A junction with an essential singularity assumes every complex number, with possibly one exception, as a value in any neighborhood of this singularity.

Example

The zeros of $\sin(1 - z^{-1})$ are given by $1 - z^{-1} = n\pi$ or at $z = 1/(1 - n\pi)$ for $n = 0, \pm 1, \pm 2, \cdots$. Furthermore, the zeros are simple because the derivative at these points is

$$\frac{d}{dz} \sin(1 - z^{-1}) \Big|_{z=(1-n\pi)^{-1}} = \frac{1}{z^2} \cos(1 - z^{-1}) \Big|_{z=(1-n\pi)^{-1}} = (1 - n\pi)^2 \cos \pi \neq 0$$

The only singularity of $\sin(1 - z^{-1})$ appears at $z = 0$. Since zero is the limit point of the sequence $(1 - n\pi)^{-1}$, $n = 1, 2, \cdots$, we observe that this function has a zero in every neighborhood of the origin. Hence $z = 0$ is not a pole. This point is not a removable singularity because $\sin(1 - z^{-1})$ does not approach 0 as $z \to 0$ ($\sin(1 - z_p^{-1}) = 1$ for $z_p = \left(1 - 2p\pi - \frac{\pi}{2}\right)^{-1}$, $p = 1,2,\cdots$). Hence by elimination $z = 0$ is an essential singularity.

20.10 Theory of Residues

20.10.1 Residue

$$\frac{1}{2\pi}\oint_C W(z)\,dz = \text{residue of } W(z) \text{ at } z_0 \text{ singularity} \equiv \text{Res}(W)$$

20.10.2 Theorem

If the $\lim_{z\to z_0}[(z - z_0)W(z)]$ is finite, this limit is the residue of $W(z)$ at $z = z_0$. If the limit is not finite, then $W(z)$ has a pole of at least second order at $z = z_0$ (it may possess an essential singularity here). If the limit is zero, then $W(z)$ is regular at $z = z_0$.

Example

Evaluate the following integral

$$\frac{1}{2\pi j}\oint_C \frac{e^{zt}}{(z^2 + \omega^2)}\,dz$$

when the contour C encloses both first-order poles at $z = \pm j\omega$. Note that this is precisely the Laplace inversion integral of the function $1/(z^2 + \omega^2)$.

Solution

This involves finding the following residues

$$\text{Res}\left(\frac{e^{zt}}{(z^2 + \omega^2)}\right)_{z=j\omega} = \frac{e^{j\omega t}}{2j\omega} \qquad \text{Res}\left(\frac{e^{zt}}{(z^2 + \omega^2)}\right)_{z=-j\omega} = \frac{e^{-j\omega t}}{2j\omega}$$

Hence,

$$\frac{1}{2\pi j}\oint_C \frac{e^{zt}}{(z^2 + \omega^2)}\,dz = \sum \text{Res}\left(\frac{e^{j\omega t} - e^{-j\omega t}}{2j\omega}\right) = \frac{\sin \omega t}{\omega}$$

A slight modification of the method for finding residues of simple poles

$$\text{Res}W(z_0) = \lim_{z\to z_0}[(z - z_0)W(z)]$$

makes the process even simpler. This is specified by the following theorem.

20.10.3 Theorem

Suppose that $f(z)$ is analytic at $z = z_0$ and suppose that $g(z)$ is divisible by $z - z_0$ but not by $(z - z_0)^2$. Then

$$\text{Res}\left[\frac{f(z)}{g(z)}\right]_{z=z_0} = \frac{f(z_0)}{g'(z_0)} \quad \text{where } g'(z) = \frac{dg(z)}{dz}$$

Example

If

$$W(z)=\frac{e^z}{(z-4)^2(z^2+1)}$$

then we take

$$f(z)=\frac{e^z}{(z-4)^2},\qquad g(z)=z^2+1$$

thus, $g'(z)=2z$ and the previous result follows immediately with

$$\operatorname{Res}\left[\frac{e^z}{(z-4)^2(z^2+1)}\right]=\frac{e^j}{(j-4)^2 2j}$$

20.10.4 Residue of Pole Order n

If $W(z)=f(z)/(z-z_0)^n$ where $f(z)$ is analytic at $z=z_0$

$$\operatorname{Res}(W(z))\Big|_{z=z_0}=\frac{1}{2\pi j}\oint W(z)dz=\frac{1}{(n-1)!}\frac{d^{n-1}}{dz^{n-1}}[(z-z_0)^n W(z)]\Big|_{z=z_0}$$

20.10.5 Residue with Nonfactorable Denominator

Sometimes the function takes the form

$$W(z)=\frac{f(z)}{zg(z)}$$

where the numerator and denominator are prime to each other, $g(z)$ has not zero at $z=0$ and cannot be factored readily. The residue due to the pole at zero is given by

$$\operatorname{Res} W(z)=\frac{f(z)}{g(z)}\Big|_{z=0}=\frac{f(0)}{g(0)}$$

If $z=a$ is the zero of $g(z)$, then the residue at $z=a$ is given by

$$\operatorname{Res} W(z)=\frac{f(a)}{ag'(a)}$$

If there are N poles of $g(z)$, then the residues at all simple poles of $W(z)$ is given by

$$\Sigma\operatorname{Res}=\frac{f(z)}{g(z)}\Big|_{z=0}+\sum_{m=1}^{N}\left[f(z)/z\frac{dg(z)}{dz}\right]_{z=a_m}$$

If $W(z)$ takes the form $W(z)=f(z)/[h(z)g(z)]$ and the simple poles to the two functions are not common, then the residues at all simple poles are given by

$$\sum \text{Res} = \sum_{m=1}^{N} \frac{f(a_m)}{h(a_m)g'(a_m)} + \sum_{r=1}^{R} \frac{f(b_r)}{h'(b_r)g(b_r)}$$

Example

Find the sum of the residues $e^{2z}/\sin mz$ at the first $N+1$ pole on the negative axis.

Example

The simple poles occur at $z = -n\pi/m,\ n = 0,1,2,\cdots$. Thus

$$\sum \text{Res} = \sum_{n=0}^{N} \left[\frac{e^{2z}}{m\cos mz}\right]_{z=-n\pi/m} = \frac{1}{m}\sum_{n=0}^{N} (-1)^n e^{-2n\pi/m}$$

Example

Find the sum of the residues of $e^{2z}/(z\cosh mz)$ at the origin of the first N pole on each side of it.

Solution

The zeros of $\cosh mz$ are $z = -j(n+1/2)\pi/m$, n integral. Because $\cosh mz$ has no zero at $z = 0$, then we obtain

$$\sum \text{Res} = 1 + \sum_{n=-N}^{N-1} \left[\frac{e^{2z}}{mz\sinh mz}\right]_{z=-(n+\frac{1}{2})\pi j/m}$$

20.11 Aids to Complex Integration

20.11.1 Integration of an Arc (R → ∞) Theorem

If AB is the arc of a circle of radius $|z| = R$ for which $\theta_1 \le \theta \le \theta_2$ and if $\lim_{R\to\infty}(zW(z)) = k$, a constant that may be zero, then

$$\lim_{R\to\infty} \int_{AB} W(z)dz = jk(\theta_2 - \theta_1)$$

20.11.2 Integration of an Arc (r → 0); Theorem

If AB is the arc of a circle of radius $|z - z_0| = r$ for which $\varphi_1 \le \varphi \le \varphi_2$ and if $\lim_{z\to z_0}[(z - z_0)W(z)] = k$, a constant that may be zero, then

$$\lim_{r\to 0} \int_{AB} W(z)dz = jk(\varphi_2 - \varphi_1)$$

where r and φ are introduced polar coordinates, with the point $z = z_0$ as origin.

20.11.3 Maximum Value Over a Path, Theorem

If the maximum value of $W(z)$ along a path C (not necessarily closed) is M, the maximum value of the integral of $W(z)$ along C is Ml, where l is the length of C. When expressed analytically, this specifies that

$$\left|\int_C W(z)dz\right| \le Ml$$

20.11.4 Jordan's Lemma

If $t < 0$ and

$$f(z) \to 0 \quad \text{as } z \to \infty$$

then

$$\int_C e^{tz} f(z)\,dz \to 0 \quad \text{as } r \to \infty$$

where C is the arc shown in Figure 20.3.

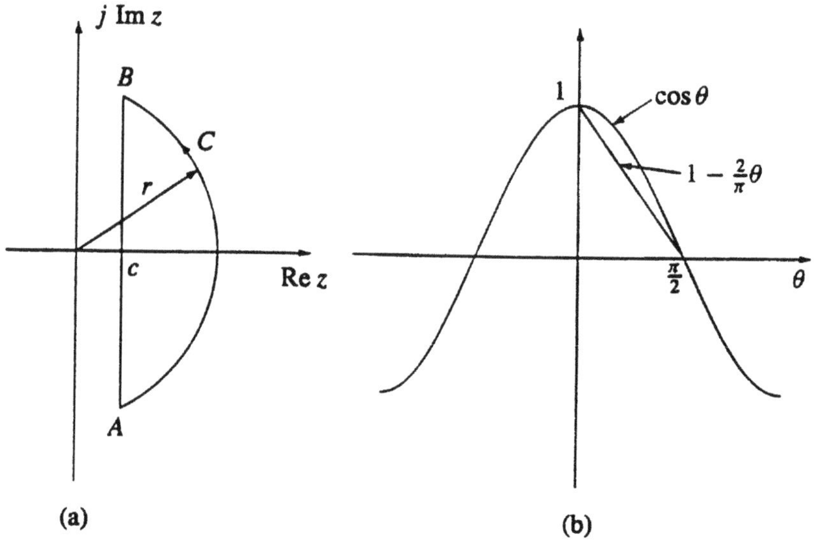

FIGURE 20.3

20.11.5 Theorem (Mellin 1)

Let

a. $\phi(z)$ be analytic in the strip $\alpha < x < \beta$, both alpha and beta being real
b. $\int_{x-j\infty}^{x+j\infty} |\phi(z)|\,dz = \int_{-\infty}^{\infty} |\phi(x+jy)|\,dy$ converges
c. $\phi(z) \to 0$ uniformly as $|y| \to \infty$ in the strip $\alpha < x < \beta$
d. θ = real and positive: If

$$f(\theta) = \frac{1}{2\pi j}\int_{c-j\infty}^{c+j\infty} \theta^{-z}\phi(z)\,dz \tag{20.11.5.1}$$

then

$$\phi(z) = \int_0^{\infty} \theta^{z-1} f(\theta)\,d\theta \tag{20.11.5.2}$$

20.11.6 Theorem (Mellin 2)

For θ real and positive, $\alpha < \text{Re}\, z < \beta$, let $f(\theta)$ be continuous or piecewise continuous, and integral (20.11.5.2) be absolutely convergent. Then (20.11.5.1) follows from (20.11.5.2).

20.11.7 Theorem (Mellin 3)

If in (20.11.5.1) and (20.11.5.2) we write $\theta = e^{-t}$, t being real, and in (20.11.5.2) put p for z and $g(t)$ for $f(e^{-t})$, we get

$$g(t) = \frac{1}{2\pi j}\int_{c-j\infty}^{c+j\infty} e^{zt}\phi(z)\,dz$$

$$\phi(p) = \int_0^{\infty} e^{-pt} g(t)\,dt$$

20.11.8 Transformation of Contour

To evaluate formally the integral

$$I = \int_0^a \cos xt\, dt$$

we set $\upsilon = xt$ that gives $dx = d\upsilon / t$ and, thus,

$$I = \frac{1}{t}\int_0^{at} \cos\upsilon\, d\upsilon = \frac{\sin at}{t}$$

Regarding this as a contour integral along the real axis for $x = 0$ to a, the change to $\upsilon = xt$ does not change the real axis. However, the contour is unaltered except in length.

Let t be real and positive. If we set $z = \zeta t$ or $\zeta = z/t$, the contour in the ζ-plane is identical in type with that in the z-plane. If it were a circle of radius r in the z-plane, the contour in the ζ-plane would be a circle of radius r/t. When t is complex $z = r_1 e^{j\theta_1}$, $z = r_2 e^{j\theta_2}$, so $\zeta = (r_1/r_2)e^{j(\theta_1-\theta_2)}$, r_1 and θ_1 being variables, while r_2 and θ_2 are fixed. If $z = jy = |z|e^{j\theta_1} = |z|e^{j\pi/2}$ and if the phase of t was $\theta_2 = \pi/4$ then the contour in the ζ-plane would be a straight line at 45 degrees with respect to the real axis. In effect, any figure in the z-plane transforms into a similar figure in the ζ-plane, whose orientation and dimensions are governed by the factor $1/t = e^{-j\theta_2}/r_2$.

Example

Make the transformation $z = \zeta t$ to the integral $I = \oint_C e^{z/t}\frac{dz}{z}$, where C is a circle of radius r_0 around the same origin.

Solution

$dz/z = d\zeta/\zeta$ so $I = \oint_{C'} e^{\zeta}\frac{d\zeta}{\zeta}$, where C' is a circle around the origin of radius $r_0/r (r = |t|)$.

Example

Discuss the transformation $z = (\zeta - a)$, a being complex and finite.

Solution

This is equivalent to a shift of the origin to point $z = -a$. Neither the contour nor the position of the singularities is affected in relation to each other, so the transformation can be made without any alteration in technique.

Example

Find the new contour due to transformation $z = \zeta^2$ if the contour was the imaginary axis, $z = jy$.

Solution

Choosing the positive square root we have $\zeta = (jy)^{1/2}$ above and $\zeta = (-jy)^{1/2}$ below the origin. Because

$$\sqrt{j} = (e^{j\pi/2})^{1/2} = e^{j\pi/4} \text{ and } \sqrt{-j} = (e^{-j\pi/2})^{1/2} = e^{-j\pi/4}$$

the imaginary axis of the z-plane transforms to that in Figure 20.4.

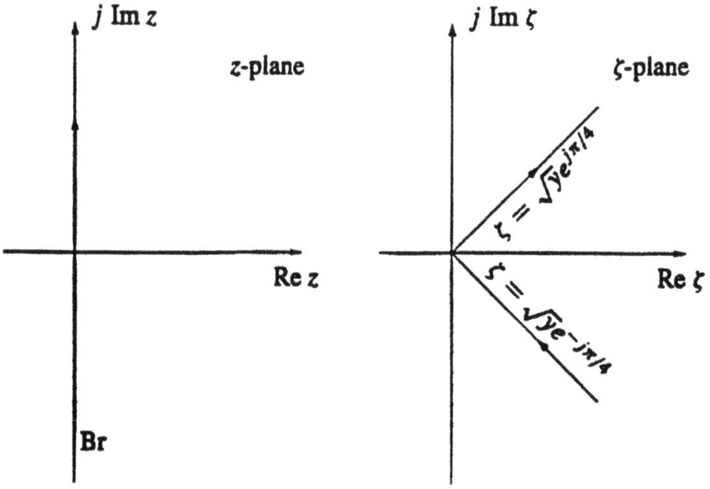

FIGURE 20.4

Example

Evaluate the integral $\int_C \frac{dz}{z}$, where C is a circle of radius 4 units around the origin, under the transformation $z = \zeta^2$.

Solution

The integral has a pole at $z = 0$ and its value is $2\pi j$. If we apply the transformation $z = \zeta^2$ then $dz = 2\,\zeta d\zeta$. Also, $\zeta = \sqrt{z} = \sqrt{r}\,e^{j\theta/2}$ if we choose the positive root. From this relation we observe that as the z traces a circle around the origin, the ζ traces a half-circle from 0 to π. Hence, the integral becomes

$$2\int_C \frac{d\zeta}{\zeta} = 2\int_0^{\pi} \frac{\rho j e^{j\theta}}{\rho e^{j\theta}} d\theta = 2\pi j$$

as we expected.

20.12 Bromwich Contour

20.12.1 Definition of the Bromwich Contour

The Bromwich contour takes the form

$$f(t) = \frac{1}{2\pi j} \int_{c-j\infty}^{c+j\infty} e^{zt} F(z)\, dz$$

where $F(z)$ is a function of z, all of whose singularities lie on the left of the path, and t is the time, which is always real and positive, $t > 0$

20.12.2 Finite Number of Poles

Let us assume that $F(z)$ has n poles at $p_1, p_2, \cdots, p_n$ and no other singularities; this case includes the important case of *rational transforms*. To utilize the Cauchy's integral theorem, we must express $f(t)$ as an integral along a closed contour. Figure 20.5 shows such a situation. We know from Jordan's lemma that if $F(z) \to 0$ as $|z| \to \infty$ on the contour C then for $t > 0$

$$\lim_{R\to\infty} \int_C e^{tz} F(z)dz \to 0, \quad t > 0$$

and because

$$\int_{c-jy}^{c+jy} e^{tz} F(z)dz \to \int_{\mathrm{Br}} e^{tz} F(z)dz, \quad y \to \infty$$

we conclude that $f(t)$ can be written as a limit,

$$f(t) \xrightarrow[R\to\infty]{} \frac{1}{2\pi j} \int_C e^{zt} F(z)dz$$

of an integral along the closed path as shown in Figure 20.5. If we take R large enough to contain all the poles of $F(z)$, then the integral along C is independent of R. Therefore we write

$$f(t) = \frac{1}{2\pi j} \int_C e^{zt} F(z)dz$$

Using Cauchy's theorem it follows that

$$\int_C e^{zt} F(z)dz = \sum_{k=1}^{n} \int_{C_k} e^{zt} F(z)dz$$

where C_k's are the contours around each pole.

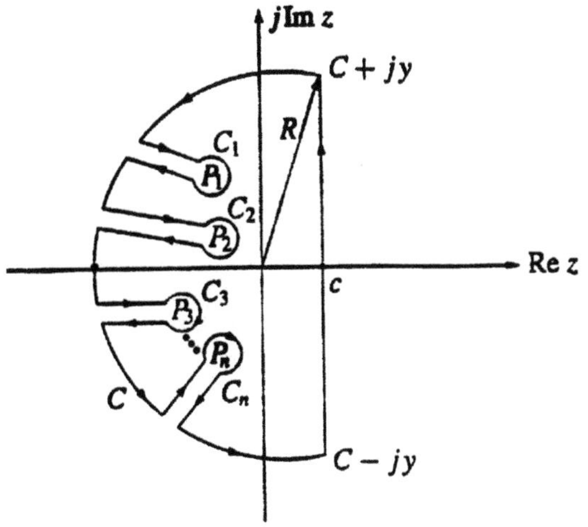

FIGURE 20.5

20.12.3 Simple Poles

$$f(t)=\sum_{k=1}^{n}F_k(z_k)e^{z_kt},\quad t>0$$

$$F_k(z_k)=F(z)(z-z_k)\Big|_{z=z_k}$$

20.12.4 Multiple Pole of m + 1 Multiplicity

$$\int_{C_k}e^{zt}F(z)dz=\int_{C_k}\frac{e^{zt}F_k(z)}{(z-z_k)^{m+1}}dz=\frac{2\pi j}{m!}\frac{d^m}{dz^m}[e^{zt}F_k(z)]\Big|_{z=z_k}$$

20.12.5 Infinitely Many Poles

(See Figure 20.6.) If we can find circular arcs with radii tending to infinity such that

$$F(z)\to 0 \text{ as } z\to\infty \qquad \text{on } C_n$$

Applying Jordan's lemma to the integral along those arcs, we obtain

$$\int_{C_n}e^{zt}F(z)dz\xrightarrow[n\to\infty]{}0,\quad t>0$$

and with C_n' the closed curve, consisting of C_n and the vertical line $\mathrm{Re}\,z=c$, we obtain

$$f(t)=\lim_{n\to\infty}\frac{1}{2\pi j}\int_{C_n'}e^{zt}F(z)dz,\quad t>0$$

Hence, for simple poles $z_1,z_2,\cdots,z_n$ of $F(z)$ we obtain

$$f(t)=\sum_{k=1}^{\infty}F_k(z_k)e^{z_kt}$$

where $F_k(z)=F(z)(z-z_k)$.

Example

Find $f(t)$ from its transformed value $F(z)=1/(z\cosh az),\ a>0$

Solution

The poles of the above function are

$$z_0=0,\quad z_k=\pm j\frac{(2k-1)\pi}{2a},\quad k=1,2,3,\cdots$$

We select the arcs C_n and their radii are $R_n=jn\pi$. It can be shown that $1/\cosh az$ is bounded on C_n and, therefore, $1/(\cosh az)\to 0$ as $z\to\infty$ on C_n. Hence,

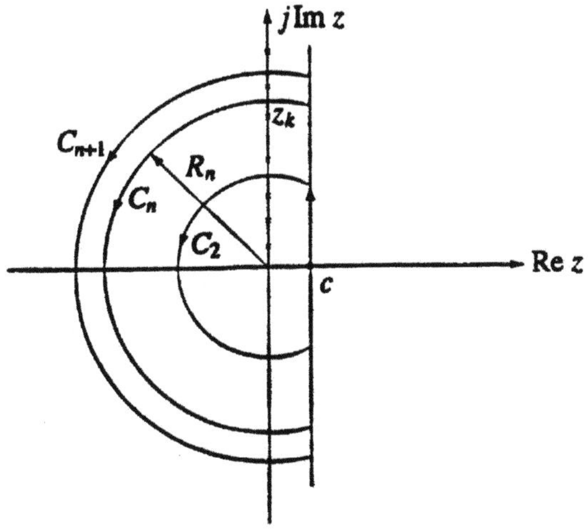

FIGURE 20.6

$$zF(z)\Big|_{z=0}=1, \quad (z-z_k)F(z)\Big|_{z=z_k}=\frac{(-1)^k 2}{(2k-1)\pi}$$

and from (20.12.5) we obtain

$$f(t)=1+\frac{2}{\pi}\sum_{k=1}^{\infty}\frac{(-1)^k}{2k-1}e^{z_k t}+\frac{2}{\pi}\sum_{k=1}^{\infty}\frac{(-1)^k}{2k-1}e^{-z_k t}$$

$$=1+\frac{4}{\pi}\sum_{k=1}^{\infty}\frac{(-1)^k}{2k-1}\cos\frac{(2k-1)\pi t}{2a}$$

20.13 Branch Points and Branch Cuts

20.13.1 Definition of Branch Points and Branch Cuts

The singularities that have been considered are those points at which $|W(z)|$ ceases to be finite. At a branch point the absolute value of $W(z)$ may be finite but $W(z)$ is not single valued, and hence is not regular. One of the simplest functions with these properties is

$$W_1(z)=z^{1/2}=\sqrt{r}\,e^{j\theta/2}$$

which takes on two values for each value of z, one the negative of the other depending on the choice of θ. This follows because we can write an equally valid form for $z^{1/2}$ as

$$W_2(z)=\sqrt{r}\,e^{j(\theta+2\pi)/2}=-\sqrt{r}\,e^{j\theta/2}=-W_1(z)$$

Clearly, $W_1(z)$ is not continuous at points on the positive real axis because

$$\lim_{\theta\to 2\pi}(\sqrt{r}\,e^{j\theta/2})=-\sqrt{r} \quad \text{while} \quad \lim_{\theta\to 0}(\sqrt{r}\,e^{j\theta/2})=\sqrt{r}$$

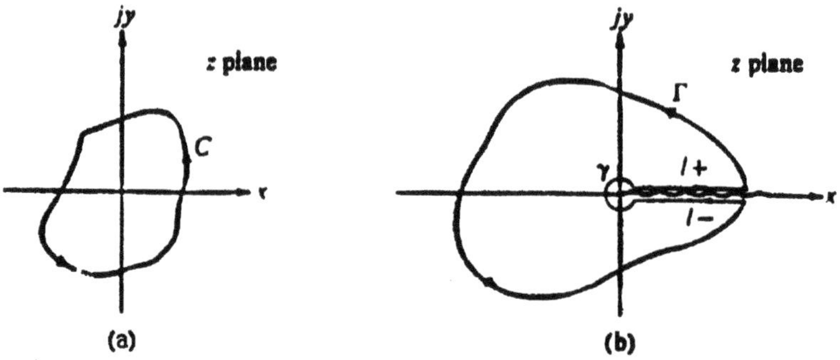

FIGURE 20.7

Hence, $W'(z)$ does not exist when z is real and positive. However, the branch $W_1(z)$ is analytic in the region $0 \leq \theta \leq 2\pi,\ r \to 0$. The part of the real axis where $x \geq 0$ is called a *branch cut* for the branch $W_1(z)$ and the branch is analytic except at points on the cut. Hence, the cut is a boundary introduced so that the corresponding branch is single valued and analytic throughout the open region bounded by the cut.

Suppose that we consider the function $W(z) = z^{1/2}$ and contour C, as shown Figure 20.7a, which encloses the origin. Clearly, after one complete circle in the positive direction enclosing the origin, θ is increased by 2π, given a value of $W(z)$ that changes from $W_1(z)$ to $W_2(z)$; that is, the function has changed from one branch to the second. To avoid this and to make the function analytic, the contour C is replaced by a contour Γ, which consists of a small circle γ surrounding the branch point, a semi-infinite cut connecting the small circle and C, and C itself (as shown in Figure 20.7b). Such a contour, which avoids crossing the branch cut, ensures that $W(z)$ is singled valued. Because $W(z)$ is single valued and excludes the origin, we would write for this composite contour C

$$\int_C W(z)dz = \int_\Gamma + \int_{l-} + \int_\gamma + \int_{l+} = 2\pi j \sum \text{Res}$$

The evaluation of the function along the various segments of C proceeds as before.

Example

If $0 < a < 1$, show that

$$\int_0^\infty \frac{x^{a-1}}{1+x} dx = \frac{\pi}{\sin a\pi}$$

Solution

Consider the integral

$$\oint_C \frac{z^{a-1}}{1+z} dz = \int_\Gamma + \int_{l-} + \int_\gamma + \int_{l+} = I_1 + I_2 + I_3 + I_4 = \sum \text{Res}$$

which we will evaluate using the contour shown in Figure 20.8. Under the conditions

$$\left|\frac{z^a}{1+z}\right| \to 0 \quad \text{as } |z| \to 0 \quad \text{if } a > 0$$

$$\left|\frac{z^a}{z+1}\right| \to 0 \quad \text{as } |z| \to \infty \quad \text{if } a < 0$$

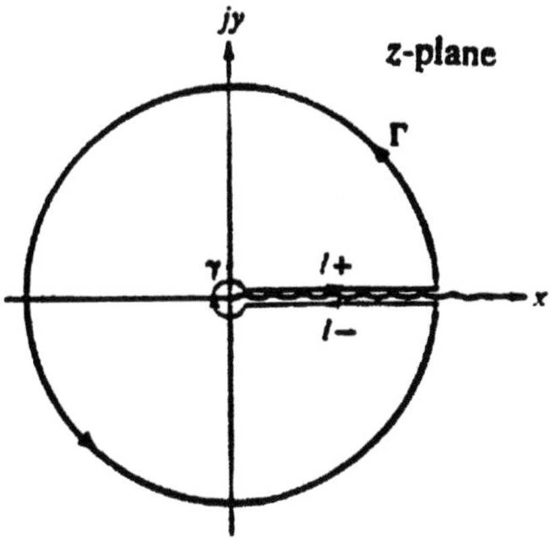

FIGURE 20.8

the integral becomes by (20.11.1)

$$\int_{\Gamma} \to 0, \int_{l-} = -e^{2\pi ja} \int_0^{\infty}$$

by (20.11.2)

$$\int_{\gamma} \to 0, \qquad \int_{l+} = 1\int_0^{\infty}$$

Thus

$$(1 - e^{2\pi ja}) \int_0^{\infty} \frac{x^{a-1}}{1+x} dx = 2\pi j \sum \text{Res}$$

Further, the residue at the pole $z = -1$, which is enclosed, is

$$\lim_{z=e^{j\pi}} (1+z) \frac{z^{a-1}}{1+z} = e^{j\pi(a-1)} = -e^{\pi ja}$$

Therefore,

$$\int_0^{\infty} \frac{x^{a-1}}{1+x} dx = 2\pi j \frac{e^{j\pi a}}{e^{j\pi a} - 1} = \frac{\pi}{\sin \pi a}$$

If, for example, we have the integral $(1/2\pi j)\int_{\text{Br}_1} \frac{e^{zt}\,dz}{z^{\upsilon+1}}$ to evaluate with $\text{Re}\,\upsilon > -1$ and t real and positive, we observe that the integral has a branch point at the origin if υ is a nonintegral constant. Because the integral vanishes along the arcs as $R \to \infty$, the equivalent contour can assume the form depicted in Figure 20.9a and marked Br_2. For the contour made up of Br_1, Br_2, the arc is closed and contains no singularities and, hence, the integral around the contour is zero. Because the arcs do not contribute any value, provided $\text{Re}\,\upsilon > -1$, the integral along Br_1 is equal to that along Br_2, both being described positively. The angle γ between the barrier and the positive real axis may have any value

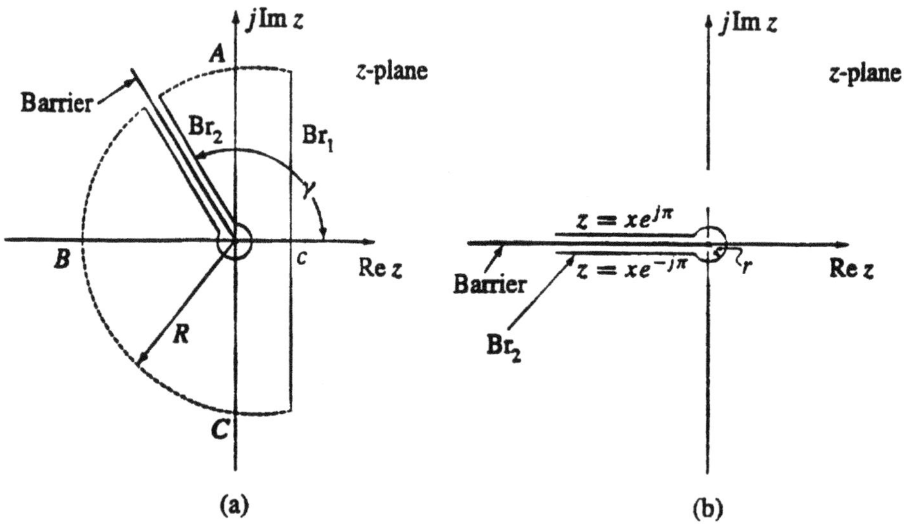

FIGURE 20.9

between $\pi/2$ and $3\pi/2$. When the only singularity is a branch point at the origin, the contour of Figure 20.9b is an approximate one.

Example

Evaluate the integral $I = \frac{1}{2\pi j}\int_{\mathrm{Br}_2} \frac{e^{zt}\,dz}{\sqrt{z}}$, where Br_2 is the contour shown in Figure 20.9b.

Solution

1. Write $z = e^{j\theta}$ on the circle. Hence we get

$$I_1 = \frac{1}{2\pi j}\int_{-\pi}^{\pi} \frac{e^{re^{j\theta}}\,d(re^{j\theta})}{\sqrt{r}\;e^{j\theta/2}} = \frac{\sqrt{r}}{2\pi}\int_{-\pi}^{\pi} e^{r(\cos\theta + j\sin\theta) + j\theta/2}\,d\theta$$

2. On the line below the barrier $z = x\exp(-j\pi)$ where $x = |x|$. Hence the integral becomes

$$I_2 = \frac{1}{2\pi j}\int_{\infty}^{r} \frac{e^{xe^{-j\pi}}\,d(xe^{-j\pi})}{\sqrt{x}\;e^{-j\pi/2}} = \frac{1}{2\pi}\int_{r}^{\infty} e^{-x}x^{-1/2}\,dx$$

3. On the line above the barrier $z = x\exp(j\pi)$ and, hence,

$$I_3 = \frac{1}{2\pi j}\int_{r}^{\infty} \frac{e^{xe^{j\pi}}\,d(xe^{j\pi})}{\sqrt{x}\;e^{j\pi/2}} = \frac{1}{2\pi}\int_{r}^{\infty} e^{-x}x^{-1/2}\,dx$$

Hence we have

$$I_2 + I_3 = \frac{1}{\pi}\int_{r}^{\infty} e^{-x}x^{-1/2}\,dx$$

As $r \to 0$, $I_1 \to 0$ and, hence,

$$I = I_1 + I_2 + I_3 = \frac{1}{\pi}\int_{0}^{\infty} e^{-x}x^{-1/2}\,dx = \frac{\sqrt{\pi}}{\pi} = \frac{1}{\sqrt{\pi}}$$

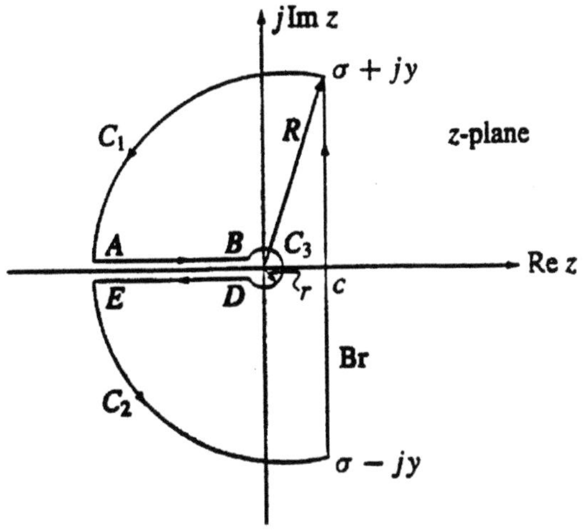

FIGURE 20.10

Example

Evaluate the integral $f(t) = \int_{\text{Br}} \frac{e^{zt} e^{-a\sqrt{z}}}{\sqrt{z}} dz,\ a > 0$ (see Figure 20.10).

Solution

The origin is a point branch and we select the negative axis as the barrier. We select the positive value of $\sqrt{z}$ when z takes positive real values in order that the integral vanishes as z approaches infinity in the region $\text{Re}\, z > \gamma$, where γ indicates the region of convergence, $\gamma \le c$. Hence we obtain

$$z = re^{j\theta} \qquad -\pi < \theta \le \pi \qquad \sqrt{z} = \sqrt{r}\, e^{j\theta/2}$$

The curve $C = \text{Br} + C_1 + C_2 + C_3$ encloses a region with no singularities and, therefore, Cauchy's theorem applies (the integrand is analytic in the region). Hence,

$$\int_C e^{zt} \frac{e^{-a\sqrt{z}}}{\sqrt{z}} dz = 0$$

It is easy to see that the given function converges to zero as R approaches infinity and, therefore, the integration over $C_1 + C_2$ does not contribute any value. For z on the circle we obtain

$$\left| \frac{e^{zt} e^{-a\sqrt{z}}}{\sqrt{z}} \right| \le \frac{e^{rt}}{\sqrt{r}}$$

Therefore, for fixed $t > 0$ we obtain

$$\left| \int_{C_3} e^{zt} \frac{e^{-a\sqrt{z}}}{\sqrt{z}} dz \right| \le 2\pi r \frac{e^{rt}}{\sqrt{r}} = \lim_{r \to 0} 2\pi r \frac{e^{rt}}{\sqrt{r}} = 0$$

because

$$\left| \int_C f(z)\, dz \right| \le ML$$

where L is the length of the contour and $|f(z)| < M$ for z on C.

On AB, $z=-x$, $\sqrt{z}=j\sqrt{x}$, and on DE, $z=-x$, $\sqrt{z}=-j\sqrt{x}$. Therefore, we obtain

$$\int_{AB+DC} e^{zt}\frac{e^{-a\sqrt{z}}}{\sqrt{z}}dz \xrightarrow[R\to\infty]{r\to 0} -\int_{\infty}^{0} e^{-xt}\frac{e^{ja\sqrt{x}}}{j\sqrt{x}}dx - \int_{0}^{\infty} e^{-xt}\frac{e^{-ja\sqrt{x}}}{-j\sqrt{x}}dx$$

But from (20.12.1)

$$\int_{\text{Br}} \frac{e^{zt}e^{-a\sqrt{z}}}{\sqrt{z}}dz = 2\pi jf(t)$$

and, hence,

$$f(t)+\frac{1}{2\pi j}\int_0^{\infty} e^{-xt}\frac{e^{ja\sqrt{x}}+e^{-ja\sqrt{x}}}{j\sqrt{x}}dx = 0$$

If we set $x = y^2$ we have

$$\int_0^{\infty} e^{-xt}\frac{\cos a\sqrt{x}}{\sqrt{x}}dx = 2\int_0^{\infty} e^{-y^2 t}\cos ay\,dy$$

But (see Fourier transform of Gaussian function, see Table 3.2),

$$2\int_0^{\infty} e^{-y^2 t}\cos ay\,dy = \sqrt{\frac{x}{t}}\,e^{-a^2/4t}$$

and hence

$$f(t) = \frac{1}{\sqrt{\pi t}}\,e^{-a^2/4t}$$

Example

Evaluate the integral $f(t)=\dfrac{1}{2\pi j}\displaystyle\int_C \frac{e^{zt}\,dz}{\sqrt{z^2-1}}$ where C is the contour shown in Figure 20.11.

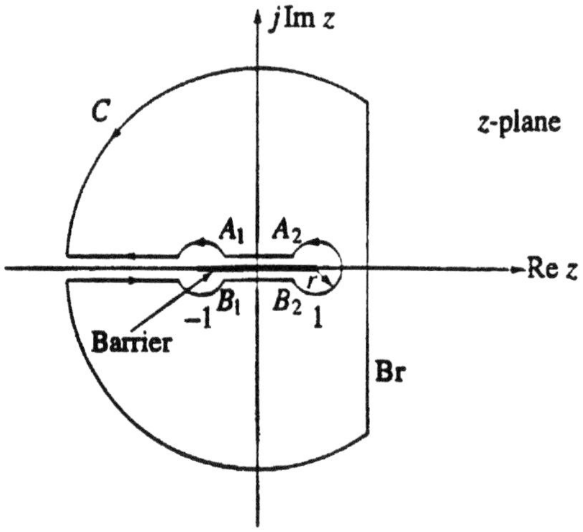

FIGURE 20.11

Solution

The Br contour is equivalent to the dumbbell-type contour shown in Figure 20.11 $B_1B_2A_2A_1B_1$. Set the phase along the line A_2A_1 equal to zero (it can also be set equal to π). Then on A_2A_1 $z = x$ from +1 to −1. Hence we have

$$I_1 = \frac{1}{2\pi j}\int_1^{-1} \frac{e^{xt}\,dx}{\sqrt{x^2-1}} = \frac{1}{2\pi}\int_{-1}^{1} \frac{e^{xt}\,dx}{\sqrt{1-x^2}}, \quad |x| < 1$$

By passing around the $z = -1$ point the phase changes by π and hence on B_1B_2 $z = x\exp(2\pi j)$. The change by 2π is due to the complete transversal of the contour that contains two branch points. Hence we obtain

$$I_2 = -\frac{1}{2\pi j}\int_{-1}^{1} \frac{e^{xt}\,dx}{\sqrt{x^2-1}} = \frac{1}{2\pi}\int_{-1}^{1} \frac{e^{xt}dx}{\sqrt{1-x^2}}, \quad |x| < 1$$

Changing the origin to −1, we set $\zeta = z + 1$ or $z = \zeta - 1$, which gives

$$I_3 = \frac{e^{-t}}{2\pi j}\int \frac{e^{\zeta t}\,d\zeta}{\sqrt{[(\zeta-2)\zeta]}}$$

On the small circle with $z = -1$ as center, $\zeta = r\exp(j\theta)$ and we get

$$I_3 = \frac{e^{-t}}{2\pi}\int_{\pi}^{-\pi} \frac{e^{rt(\cos\theta + j\sin\theta)+(j\theta/2)}\sqrt{r}\,d\theta}{\sqrt{re^{j\theta}-2}}$$

When $\theta = 0$ the integrand has the value $+\sqrt{r}\,e^{rt}/\sqrt{r-2}$, and for $\theta = 2\pi$ the value is $-\sqrt{r}\,e^{rt}/\sqrt{r-2}$. Therefore, the integrand changes sign in rounding the branch point at $z = -1$. Similarly for the branch point at $z = 1$, where the change is from – to +. As $r \to 0$, $I_3 \to 0$, and thus I_3 vanishes. The same is true for the branch point at $z = -1$. Therefore, by setting $x = \cos\theta$ we obtain

$$I = I_1 + I_2 = \frac{1}{\pi}\int_{-1}^{1} \frac{e^{xt}}{\sqrt{1-x^2}}dx = \frac{1}{\pi}\int_0^{\pi} e^{t\cos\theta}d\theta = \frac{1}{\pi}\int_0^{\pi}\sum_{k=0}^{\infty}\frac{(t\cos\theta)^k}{k!}d\theta$$

$$= \frac{1}{\pi}\left[\pi + \pi\frac{1}{2}\frac{t^2}{2!} + \pi\frac{3}{4}\frac{1}{2}\frac{t^4}{4!} + \pi\frac{5}{6}\frac{3}{4}\frac{1}{2}\frac{t^6}{6!} + \cdots\right]$$

$$= 1 + \frac{t^2}{2^2} + \frac{t^4}{2^2 4^2} + \frac{t^6}{2^2 4^2 6^2} + \cdots = \sum_{k=0}^{\infty}\frac{\left(\frac{1}{2}t\right)^{2k}}{(k!)^2} = I_0(t)$$

where $I_0(t)$ is the modified Bessel function of the first kind and zero order.

Example

Evaluate the integral $I = \int_C \frac{e^{zt}}{\sqrt{z^2+1}}dz$ where C is the close contour shown in Figure 20.12.

Solution

The Br contour is equal to the dumbbell-type contour as shown in Figure 20.12, $ABGDA = C_1$. Hence we have

$$f(t) = \frac{1}{2\pi j}\int_{C_1} \frac{e^{zt}}{\sqrt{z^2+1}}dz$$

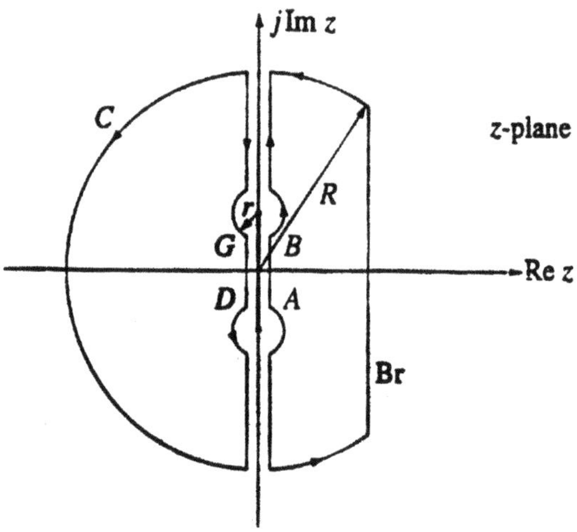

FIGURE 20.12

But

$$\left|\frac{e^{zt}}{\sqrt{z^2+1}}\right| < \frac{e^{rt}}{\sqrt{r}\ \sqrt{2-r}}$$

on the circle on the $+j$ branch point and therefore for $t > 0$

$$\left|\int \frac{e^{zt}}{\sqrt{z^2+1}} dz\right| < \frac{2\pi\sqrt{r}\,e^{rt}}{\sqrt{2-r}} \to 0 \quad \text{as } r \to 0$$

We obtain similar results for the contour around the $-j$ branch point. However,

$$\text{on } AB, z = j\omega,\ \sqrt{1+z^2} = \sqrt{1-\omega^2}\ ;\ \text{on } GD,\ z = j\omega,\ \sqrt{1+z^2} = -\sqrt{1-\omega^2}$$

and, therefore, for $t > 0$ we obtain

$$f(t) = \frac{j}{2\pi j}\int_{-1}^{1} \frac{e^{j\omega t}}{\sqrt{1-\omega^2}} d\omega + \frac{j}{2\pi j}\int_{1}^{-1} \frac{e^{j\omega t}}{-\sqrt{1-\omega^2}} d\omega = \frac{1}{\pi}\int_{-1}^{1} \frac{\cos\omega t}{\sqrt{1-\omega^2}} d\omega$$

If we set $\omega = \sin\theta$

$$f(t) = \frac{1}{\pi}\int_{-\pi/2}^{\pi/2} \cos(t\sin\theta)\,d\theta = J_0(t)$$

where $J_0(t)$ is the Bessel function of the first kind.

20.14 Evaluation of Definite Integrals

20.14.1 Evaluation of the Integrals of Certain Periodic Functions (0 to 2π)

An integral of the form

$$I = \int_0^{\pi/2} F(\cos\theta, \sin\theta)\, d\theta$$

where the integral is a *rational function* of cosθ and sinθ finite on the range of integration, and can be integrated by setting $z = \exp(j\theta)$,

$$\cos\theta = \frac{1}{2}(z + z^{-1}), \qquad \sin\theta = \frac{1}{2j}(z - z^{-1})$$

The above integral takes the form

$$I = \int_C F(z)\, dz$$

where $F(z)$ is a rational function of z finite on C, which is a circle of radius unity with center at the origin.

Example

If $0 < a < 1$, find the value of the integral

$$I = \int_0^{2\pi} \frac{d\theta}{1 - 2a\cos\theta + a^2}$$

Solution

Transforming the above integral, it becomes

$$I = \int_C \frac{dz}{j(1 - az)(z - a)}$$

The only pole inside the unit circle is at a. Therefore, by residue theory we have

$$I = 2\pi j \lim_{z \to a} \frac{z - a}{j(1 - az)(z - a)} = \frac{2\pi}{1 - a^2}$$

20.14.2 Evaluation of Integrals with Limits $-\infty$ and $+\infty$

We can now evaluate the integral $I = \int_{-\infty}^{\infty} F(x)\, dx$ provided that the function $F(z)$ satisfies the following properties:

1. It is analytic when the imaginary part of z is positive or zero (except at a finite number of poles).
2. It has no poles on the real axis.
3. As $|z| \to \infty$, $zF(z) \to 0$ uniformly for all values of arg z such that $0 \le \arg z \le \pi$, provided that when x is real, $xF(x) \to 0$ as $x \to \pm\infty$, in such a way that $\int_0^{\infty} F(x)\, dx$ and $\int_{-\infty}^{0} F(x)\, dx$ both converge.

The integral is given by

$$I = \int_C F(z)\, dz = 2\pi j \sum \text{Res}$$

where the contour is the real axis and a semicircle having its center in the origin and lying above the real axis.

Example

Evaluate the integral $I = \int_{-\infty}^{\infty} \frac{dx}{(x^2+1)^3}$.

Solution

The integral becomes

$$I = \int_C \frac{dz}{(z^2+1)^3} = \int_C \frac{dz}{(z+j)^3(z-j)^3}$$

which has one pole at j of order three. Hence we obtain

$$I = \frac{1}{2!}\frac{d^2}{dz^2}\left[\frac{1}{(z+j)^3}\right]_{z=j} = -j\frac{3}{16}$$

Example

Evaluate the integral $I = \int_0^{\infty} \frac{dx}{x^2+1}$.

Solution

Consider the integral

$$I = \int_C \frac{dz}{z^2+1}$$

where C is the contour of the real axis and the upper semicircle. From $z^2 + 1 = 0$ we obtain $z = \exp(j\pi/2)$ and $z = \exp(-j\pi/2)$. Only the pole $z = \exp(j\pi/2)$ exists inside the contour. Hence we obtain

$$2\pi j \lim_{z \to e^{j\pi/2}} \left(\frac{z - e^{j\pi/2}}{(z - e^{j\pi/2})(z - e^{-j\pi/2})} \right) = \pi$$

Therefore, we obtain

$$\int_{-\infty}^{\infty} \frac{dx}{x^2+1} = 2\int_0^{\infty} \frac{dx}{x^2+1} = \pi \text{ or } I = \frac{\pi}{2}$$

20.14.3 Certain Infinite Integrals Involving Sines and Cosines

If $F(z)$ satisfies conditions (1), (2), and (3), of 20.14.2 and if $m > 0$, then $F(z)e^{jmz}$ also satisfies the same conditions. Hence $\int_0^{\infty} [F(x)e^{jmx} + F(-x)e^{-jmx}]dx$ is equal to $2\pi j \sum \text{Res}$, where $\sum \text{Res}$ means the sum of the residues of $F(z)e^{jmz}$ at its poles in the upper half-plane. Therefore,

1. If $F(x)$ is an even function, that is, $F(x) = -F(-x)$, then

$$\int_0^{\infty} F(x) \cos mx \, dx = \pi j \sum \text{Res}.$$

2. If $F(x)$ is an odd function, that is, $F(x) = -F(-x)$, then

$$\int_o^\infty F(x)\sin mx\,dx = \pi\sum \text{Res}$$

Example

Evaluate the integral $I = \int_0^\infty \frac{\cos x}{x^2 + a^2}dx,\ a > 0.$

Solution

Consider the integral

$$I_1 = \int_C \frac{e^{jz}}{z^2 + a^2}dz$$

where the contour is the real axis and the infinite semicircle on the upper side with respect to the real axis. The contour encircles the pole ja. Hence,

$$I_1 = \int_C \frac{e^{jz}}{z^2 + a^2}dz = 2\pi j\frac{e^{jja}}{2ja} = \frac{\pi}{a}e^{-a}$$

However,

$$I_1 = \int_{-\infty}^\infty \frac{e^{jz}}{z^2 + a^2}dz = \int_{-\infty}^\infty \frac{\cos x}{x^2 + a^2}dx + j\int_{-\infty}^\infty \frac{\sin x}{x^2 + a^2}dx = \int_{-\infty}^\infty \frac{\cos x}{x^2 + a^2}dx$$

because the integrand of the third integral is odd and therefore is equal to zero. From the last two equations we find that

$$I = \int_0^\infty \frac{\cos x}{x^2 + a^2}dx = \frac{\pi}{2a}e^{-a}$$

because the integrand is an even function.

Example

Evaluate the integral $I = \int_0^\infty \frac{x\sin ax}{x^2 + b^2}dx,\ k > 0$ and $a > 0.$

Solution

Consider the integral

$$I_1 = \int_C \frac{ze^{jaz}}{z^2 + b^2}dz$$

where C is the same type of contour as in the previous example. Because there is only one pole at $z = jb$ in the upper half of the z-plane, then

$$I_1 = \int_{-\infty}^\infty \frac{ze^{jaz}}{z^2 + b^2}dz = 2\pi j\frac{jbe^{jajb}}{2jb} = j\pi e^{-ab}$$

Because the integrand $x\sin ax/(x^2 + b^2)$ is an even function, we obtain

$$I_1 = j\int_{-\infty}^\infty \frac{x\sin ax}{x^2 + b^2}dx = j\pi e^{-ab} \text{ or } I = \frac{\pi}{2}e^{-ab}$$

Example

Show that $I_1 = \int_{-\infty}^{\infty} \frac{x \sin \pi x}{x^2 + 2x + 5} dx = -\pi e^{-2\pi}$.

Integrals of the Form $\int_0^{\infty} x^{\alpha-1} f(x)dx,\ 0 < \alpha < 1$:

It can be shown that the above integral has the value

$$I = \int_0^{\infty} x^{\alpha-1} f(x)dx = \frac{2\pi j}{1 - e^{j2\pi\alpha}} \sum_{k=1}^{N} \mathrm{Res}[z^{\alpha-1} f(x)]\Big|_{z=z_k}$$

where $f(z)$ has N singularities and $z^{\alpha-1}f(z)$ has a branch point at the origin.

Example

Evaluate the integral $I = \int_0^{\infty} \frac{x^{-1/2}}{x+1} dx$.

Solution

Because $x^{-1/2} = x^{1/2-1}$ it is implied that $\alpha = 1/2$. From the integrand we observe that the origin is a branch point and the $f(x) = 1/(x+1)$ has a pole at -1. Hence from the previous example we obtain

$$I = \frac{2\pi j}{1 - e^{j2\pi/2}} \mathrm{Res}\left[\frac{z^{-1/2}}{z+1}\right]\Bigg|_{z=-1} = \frac{2\pi j}{j(1 - e^{j\pi})} = \pi$$

We can also proceed by considering the integral $I = \int_C \frac{z^{-1/2}}{z+1} dz$. Because $z = 0$ is a branch point we choose the contour C as shown in Figure 20.13. The integrand has a simple pole at $z = -1$ inside the contour C. Hence the residue at $z = -1 = \exp(j\pi)$ and is

$$\mathrm{Res}\big|_{z=-1} = \lim_{z\to-1} (z+1)\frac{z^{-1/2}}{z+1} = e^{-j\frac{\pi}{2}}$$

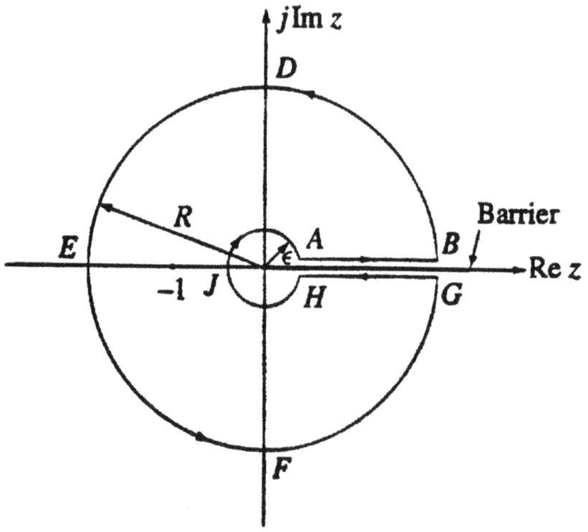

FIGURE 20.13

Therefore we write

$$\oint_C \frac{z^{-1/2}}{z+1}dz = \int_{AB} + \int_{BDEFG} + \int_{GH} + \int_{HJA} = e^{-j\pi/2}$$

The above integrals take the following form:

$$\int_{\varepsilon}^{R} \frac{x^{-1/2}}{x+1}dx + \int_0^{2\pi} \frac{(Re^{j\theta})^{-1/2} jRe^{j\theta}d\theta}{1+Re^{j\theta}} + \int_R^{\varepsilon} \frac{(xe^{j2\pi})^{-1/2}}{1+xe^{j2\pi}}dx + \int_{2\pi}^{0} \frac{(\varepsilon e^{j\theta})^{-1/2} j\varepsilon e^{j\theta}d\theta}{1+\varepsilon e^{j\theta}} = j2\pi e^{-j\pi/2}$$

where we have used $z = x\exp(j2\pi)$ for the integral along GH, because the argument of z is increased by 2π in going around the circle $BDEFG$.

Taking the limit as $\varepsilon \to 0$ and $R \to \infty$ and noting that the second and fourth integrals approach zero, we find

$$\int_0^{\infty} \frac{x^{-1/2}}{x+1}dx + \int_{\infty}^{0} \frac{e^{-j2\pi}x^{-1/2}}{x+1}dx = j2\pi e^{-j\pi/2}$$

or

$$(1-e^{-j\pi})\int_0^{\infty} \frac{x^{-1/2}}{x+1}dx = j2\pi e^{-j\pi/2} \qquad \text{or} \qquad \int_0^{\infty} \frac{x^{-1/2}}{x+1}dx = \frac{j2\pi(-j)}{2} = \pi$$

20.14.4 Miscellaneous Definite Integrals

The following examples will elucidate some of the approaches that have been used to find the values of definite integrals.

Example

Evaluate the integral $I = \int_{-\infty}^{\infty} \frac{1}{x^2+a^2}dx, \; a > 0.$

Solution

We write (see Figure 20.14)

$$\oint_C \frac{dz}{z^2+a^2} = \int_{AB} + \int_{BDA} = 2\pi j \sum \text{Res}$$

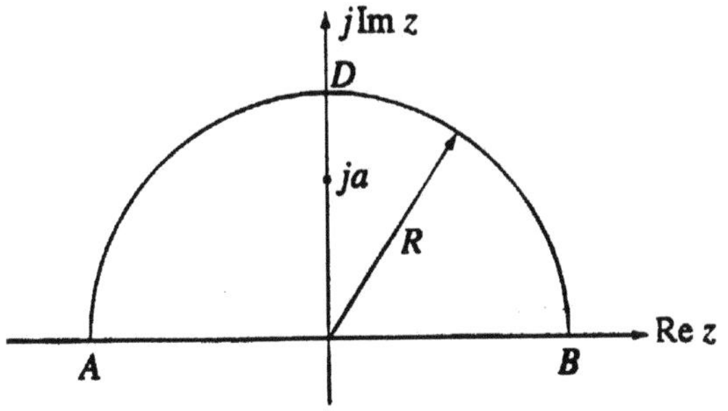

FIGURE 20.14

As $R \to \infty$

$$\int_{BDA} \frac{dz}{z^2 + a^2} = \int_0^{\pi} \frac{Rje^{j\theta} d\theta}{R^2 e^{j2\theta} + a^2} \xrightarrow[R \to \infty]{} 0$$

and, therefore, we have

$$\int_{AB} \frac{dx}{x^2 + a^2} = \int_{-\infty}^{\infty} \frac{dx}{x^2 + a^2} = 2\pi j \frac{z - ja}{z^2 + a^2}\bigg|_{z=ja} = 2\pi j \frac{1}{2ja} = \frac{\pi}{a}$$

Example

Evaluate the integral $I = \int_0^{\infty} \frac{\sin ax}{x} dx.$

Solution

Because $\sin az/z$ is analytic near $z = 0$, we indent the contour around the origin as shown in Figure 20.15. With a positive we write

$$\int_0^{\infty} \frac{\sin ax}{x} dx = \frac{1}{2} \int_{ABCD} \frac{\sin az}{z} dz = \frac{1}{4j} \int_{ABCD} \left[\frac{e^{jaz}}{z} - \frac{e^{-jaz}}{z} \right] dz$$

$$= \frac{1}{4j} \int_{ABCDA} \frac{e^{jaz}}{z} dz - \frac{1}{4j} \int_{ABCFA} \frac{e^{-jaz}}{z} dz = \frac{1}{4j} \left[2\pi j \frac{1}{1} - 0 \right] = \frac{\pi}{2}$$

because the lower contour does not include any singularity. Because $\sin ax$ is an odd function of a and $\sin 0 = 0$, we obtain

$$\int_0^{\infty} \frac{\sin x}{x} dx = \begin{cases} \frac{\pi}{2} & a > 0 \\ 0 & a = 0 \\ -\frac{\pi}{2} & a < 0 \end{cases}$$

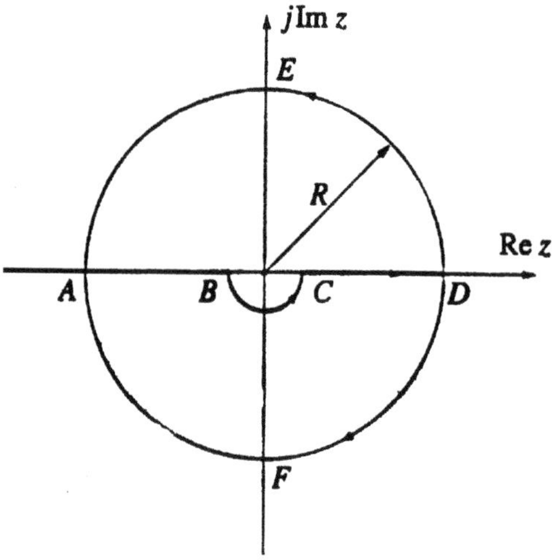

FIGURE 20.15

Example

Evaluate the integral $I = \int_0^{\infty} \frac{dx}{1+x^3}$.

Solution

Because the integrand $f(x)$ is odd, we introduce the ln z. Taking a branch cut along the positive real axis we obtain

$$\ln z = \ln r + j\theta, \qquad 0 \le \theta < 2\pi$$

the discontinuity of ln z across the cut is (see Figure 20.16a)

$$\ln z_1 - \ln z_2 = -2\pi j$$

Therefore, if $f(z)$ is analytic along the real axis and the contribution around an infinitesimal circle at the origin is vanishing, we obtain

$$\int_0^{\infty} f(x)dx = -\frac{1}{2\pi j}\int_{ABC} f(z)\ln(z)\,dz$$

If further $f(z) \to 0$ as $|z| \to \infty$, the contour can be completed with *CDA* (see Figure 20.16b). If $f(z)$ has simple poles of order one at points z_k, with residues Res(f, z_k), we obtain

$$\int_0^{\infty} f(x)dx = -\sum_k \mathrm{Res}(f, z_k)\ln z_k$$

Hence, because $z^3 + 1 = 0$ has poles at $z_1 = e^{j\pi/3}$, $z_2 = e^{j\pi}$, $z_3 = e^{j5\pi/3}$, then the integral is given by

$$I = \int_0^{\infty} \frac{dx}{x^3+1} = -\left[\frac{j\pi/3}{3e^{2j\pi/3}} + \frac{j\pi}{3e^{j2\pi}} + \frac{j5\pi/3}{3e^{j10\pi/3}}\right] = \frac{2\pi\sqrt{3}}{9}$$

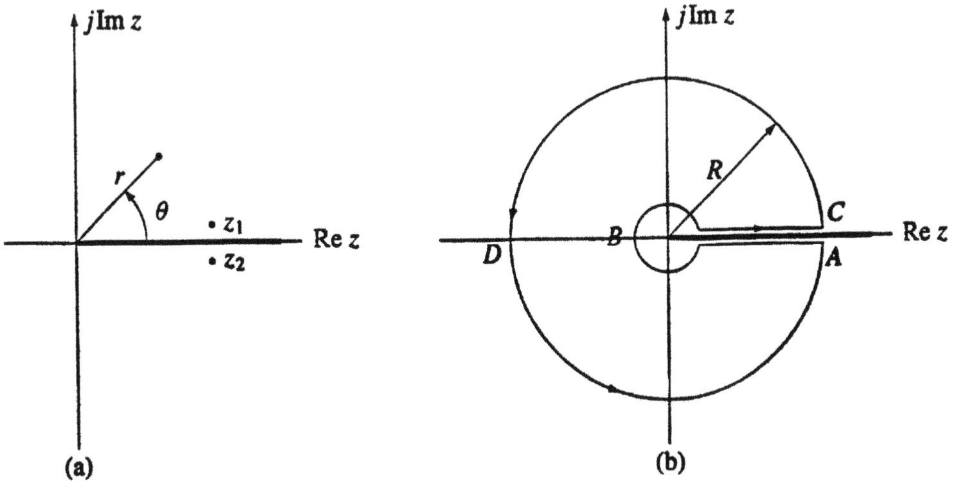

FIGURE 20.16

Example

Show that $\int_0^\infty \cos ax^2\,dx = \int_0^\infty \sin ax^2\,dx = \frac{1}{2}\sqrt{\frac{\pi}{2a}}, \quad a > 0.$

Solution

We first form the integral

$$F = \int_0^\infty \cos ax^2\,dx + j\int_0^\infty \sin ax^2\,dx = \int_0^\infty e^{jax^2}\,dx$$

Because $\exp(jaz^2)$ is analytic in the entire z-plane, we can use Cauchy's theorem and write (see Figure 20.17)

$$F = \int_{AB} a^{jaz^2}\,dz = \int_{AC} e^{jaz^2}\,dz + \int_{CB} e^{jaz^2}\,dz$$

Along the contour CB we obtain

$$\left| -\int_0^{\pi/4} e^{jR^2\cos 2\theta - R^2\sin 2\theta}\, jRe^{j\theta}\,d\theta \right| \le \int_0^{\pi/4} e^{-R^2\sin 2\theta} R\,d\theta = \frac{R}{2}\int_0^{\pi/2} e^{-R^2\sin 2\phi}\,d\phi$$

$$\le \frac{R}{2}\int_0^{\pi/2} e^{-R^2\phi/\pi}\,d\phi = \frac{\pi}{4R}(1 - e^{-R^2})$$

where the transformation $2\theta = \phi$ and the inequality $\sin\phi \ge 2\phi/\pi$ were used $(0 \le \phi \le \pi/2)$. Hence, as R approaches infinity the contribution from CB contour vanishes. Hence,

$$F = \int_{AB} e^{jaz^2}\,dz = e^{j\pi/4}\int_0^\infty a^{-ar^2}\,dr = \frac{1+j}{\sqrt{2}}\frac{1}{2}\sqrt{\frac{\pi}{a}}$$

from which we obtain the desired result.

Example

Evaluate the integral $I = \int_{-1}^{1} \frac{dx}{\sqrt{1-x^2}\,(1+x^2)}$

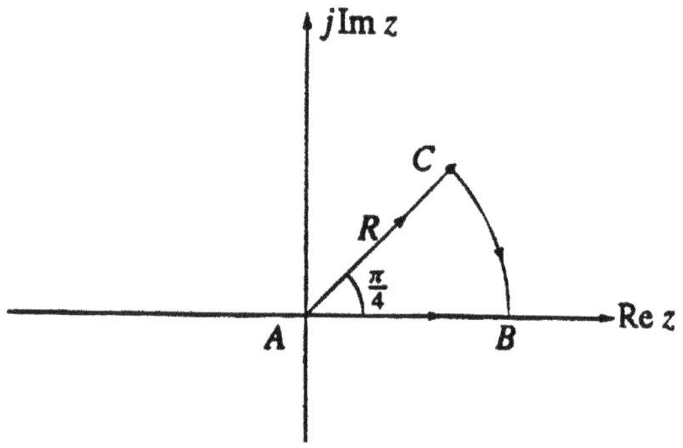

FIGURE 20.17

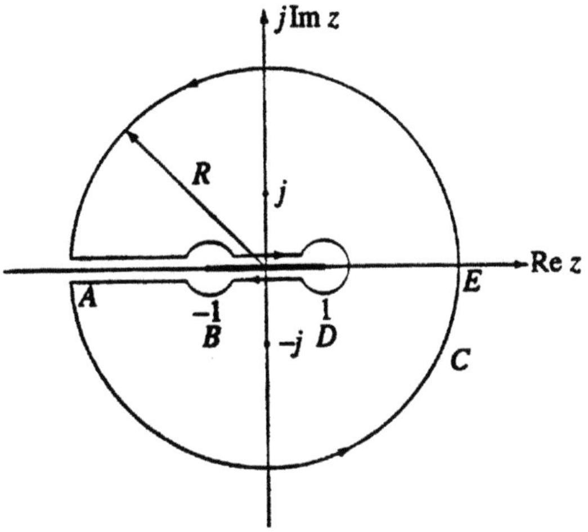

FIGURE 20.18

Solution

Consider the integral

$$I = \oint_C \frac{dx}{\sqrt{1-z^2}\,(1+z^2)}$$

whose contour C is that shown in Figure 20.18 On the top side of the branch cut we obtain *I*, and from the bottom we also get *I*. The contribution of the integral on the outer circle as R approaches infinity vanishes. Hence, due to two poles we obtain

$$2I = 2\pi j\left[\frac{1}{2j\sqrt{2}} + \frac{1}{2j\sqrt{2}}\right] = \pi\sqrt{2} \quad \text{or} \quad I = \frac{\sqrt{2}}{2}\pi$$

Example

Evaluate the integral $I = \int_{-\infty}^{\infty} \frac{e^{ax}}{e^{bx}+1}dx, \quad a,b > 0.$

Solution

From Figure 20.19 we find

$$I = \int_C \frac{e^{az}}{e^{bz}+1}dz = \int_C \frac{e^{az/b}}{e^{z}+1}dz = 2\pi j\sum \text{Res}$$

There is an infinite number of poles at $z = j\pi/b$, residue is $-\exp(j\pi a/b)$; at $z = 3j\pi/b$, residue is $-\exp(3j\pi a/b)$and so on. The sum of residue forms a geometric series and because we assume a small imaginary part of a, $|\exp(j2\pi a/b)| < 1$. Hence, by considering the common factor $\exp(j\pi a/b)$, we obtain

$$I = -\frac{2\pi}{b}j\frac{e^{j\pi a/b}}{1-e^{j2\pi a/b}} = \frac{1}{b}\frac{\pi}{b\sin(\pi a/b)}$$

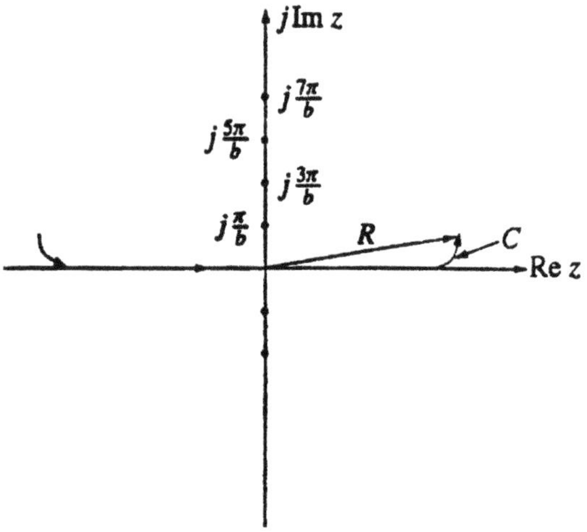

FIGURE 20.19

The integral is of the form $\int e^{j\omega x} f(x)dx$ whose evaluation can be simplified by Jordan's lemma

$$\int_C e^{j\omega x} f(x)dx = 0$$

for the contour semicircle C at infinity for which $\text{Im}(\omega x) > 0$, provided $|f(Re^{j\theta})| < \varepsilon(R) \to 0$ as $R \to \infty$ (note that the bound on $|f(x)|$ must be independent of θ).

Example

A relaxed RL series circuit with an input voltage source $\upsilon(t)$ is described by the equation $Ldi/dt + Ri = \upsilon(t)$. Find the current in the circuit using the inverse Fourier transform when the input voltage is a delta function.

Solution

The Fourier transform of the differential equation with delta input voltage function is

$$Lj\omega I(\omega) + RI(\omega) = 1 \quad \text{or} \quad I(\omega) = \frac{1}{R + j\omega L}$$

and hence

$$i(t) = \frac{1}{2\pi}\int_{-\infty}^{\infty} \frac{e^{j\omega t}}{R + j\omega L} d\omega$$

If $t < 0$ the integral is exponentially small for $\text{Im}\omega \to -\infty$. If we complete the contour by a large semicircle in the lower ω-plane, the integral vanishes by Jordan's lemma. Because the contour does not include any singularities, $i(t) = 0, t < 0$. For $t > 0$, we complete the contour in the upper ω-plane. Similarly no contribution exists from the semicircle. Because there is only one pole at $\omega = jR/L$ inside the contour the value of the integral is

$$i(t) = 2\pi j \frac{1}{2\pi}\frac{1}{jL} e^{j(jR/L)t} = \frac{1}{L} e^{-\frac{R}{L}t}$$

which is known as the *impulse response of the system.*

20.15 Principal Value of an Integral

20.15.1 Cauchy Principal Value

Refer to the limiting process employed in a previous example in section 20.14 for the integral $\int_0^\infty \frac{\sin ax}{x} dx$, which can be written in the form

$$\lim_{R\to\infty} \int_{-R}^{R} \frac{e^{jx}}{x} dx = j\pi$$

The limit is called the *Cauchy principal value* of the integral in the equation

$$\int_{-\infty}^{\infty} \frac{e^{jx}}{x} dx = j\pi$$

In general, if $f(x)$ becomes infinite at a point $x = c$ inside the range of integration, and if

$$\lim_{\varepsilon\to 0} \int_{-R}^{R} f(x)dx = \lim_{\varepsilon\to 0} \left[\int_{-R}^{c-\varepsilon} f(x)dx + \int_{c+\varepsilon}^{R} f(x)dx \right]$$

and if the separate limits on the right also exist, then the integral is convergent and the integral is written as $P\int$ where P indicates the principal value. Whenever each of the integrals

$$\int_{-\infty}^{0} f(x)dx \qquad \int_{0}^{\infty} f(x)dx$$

has a value, here $R \to \infty$, the principal value is the same as the integral. For example, if $f(x) = x$, the principal value of the integral is zero, although the value of the integral itself does not exist.

As another example, consider the integral

$$\int_a^b \frac{dx}{x} = \log \frac{b}{a}$$

If a is negative and b is positive, the integral diverges at $x = 0$. However, we can still define

$$P\int_a^b \frac{dx}{x} = \lim_{\varepsilon\to 0} \left[\int_a^{-\varepsilon} \frac{dx}{x} + \int_{\varepsilon+}^{b} \frac{dx}{x} \right] = \lim_{\varepsilon\to 0} \left(\log \frac{\varepsilon}{-a} + \log \frac{b}{\varepsilon} \right) = \log \frac{b}{|a|}$$

This principal value integral is unambiguous. The condition that the same value of ε must be used in both sides is essential; otherwise, the limit could be almost anything by taking the first integral from a to $-\varepsilon$ and the second from k to b and making these two quantities tend to zero in a suitable ratio.

If the complex variables were used, we could complete the path by a semicircle from $-\varepsilon$ to $+\varepsilon$ about the origin, either above or below the real axis. If the upper semicircle were chosen, there would be a contribution $-j\pi$, whereas if the lower semicircle were chosen, the contribution to the integral would be $+j\pi$. Thus, according to the path permitted in the complex plane we should have

$$\int_a^b \frac{dz}{z} = \log \frac{b}{|a|} \pm j\pi$$

The principal value is the mean of these alternatives.

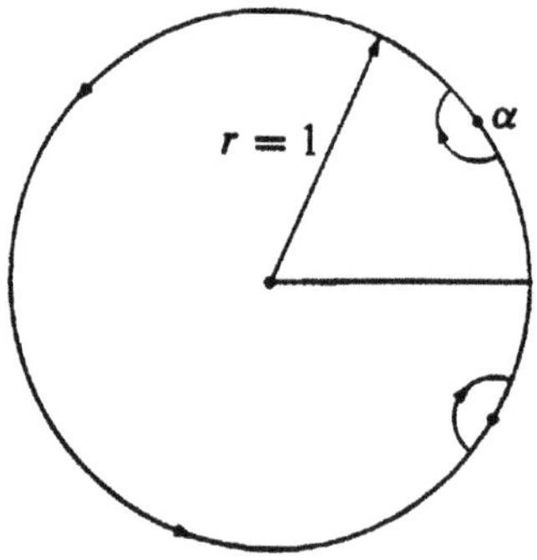

FIGURE 20.20

If a path in the complex plane passes through a simple pole at $z = a$, we can define a principal value of the integral along the path by using a hook of small radius ε about the point a and then making ε tend to zero, as already discussed. If we change the variable z to ζ, and $dz/d\zeta$ is finite and not equal to zero at the pole, this procedure will define an integral in the ζ-plane, but the values of the integrals will be the same. Suppose that the hook in the z-plane cuts the path at $a-\varepsilon$ and $a+\varepsilon'$, where $|\varepsilon| = |\varepsilon'|$, and in the ζ-plane the hook cuts the path at $\alpha - k$ and $\alpha + k'$. Then, if k and k' tend to zero so that $\varepsilon/\varepsilon' \to 1$, k and k' will tend to zero so that $k/k' \to 1$.

To illustrate this discussion, suppose we want to evaluate the integral

$$I = \int_0^{\pi} \frac{d\theta}{a - b\cos\theta}$$

where a and b are real and $a > b > 0$. A change of variable by writing $z = \exp(j\theta)$ transforms the integral to (where a new constant α is introduced)

$$I = \int_0^{\pi} \frac{2e^{j\theta}\, d\theta}{2ae^{j\theta} - b(e^{j2\theta}+1)} = -\frac{1}{j}\int_C \frac{2\,dz}{bz^2 - 2az + b} = -\frac{1}{j}\int_C \frac{2\,dz}{b(z-\alpha)\left(z - \frac{1}{\alpha}\right)}$$

where the path of integration is around the unit circle. Because the contour would pass through the poles, hooks are used to isolate the poles as shown in Figure 20.20. Because no singularities are closed by the path, the integral is zero. The contributions of the hooks are $-j\pi$ times the residue, where the residues are

$$-\frac{1}{j}\frac{\frac{2}{b}}{\alpha - \frac{1}{\alpha}} \qquad -\frac{1}{j}\frac{\frac{2}{b}}{\frac{1}{\alpha} - \alpha}$$

These are equal and opposite and cancel each other. Therefore, the principal value of the integral around the unit circle is zero. This approach for finding principal values succeeds only at simple poles.

20.16 Integral of the Logarithmic Derivative

Of importance in the study of mapping from z-plane to $W(z)$-plane is the integral of the logarithmic derivative. Consider, therefore, the function

$$F(z) = \log W(z)$$

Then

$$\frac{dF(z)}{dz} = \frac{1}{W(z)}\frac{dW(z)}{dz} = \frac{W'(z)}{W(z)}$$

The function to be examined is the following:

$$\int_C \frac{dF(z)}{dz}dz = \int_C \frac{W'(z)}{W(z)}dz$$

The integrand of this expression will be analytic within the contour C except for the points at which $W(z)$ is either zero or infinity.

Suppose that $W(z)$ has a pole of order n at z_0. This means that $W(z)$ can be written

$$W(z) = (z - z_0)^n g(z)$$

with n positive for a zero and n negative for a pole. We differentiate this expression to get

$$W'(z) = n(z - z_0)^{n-1} g(z) + (z - z_0)^n g'(z)$$

and so

$$\frac{W'(z)}{W(z)} = \frac{n}{z - z_0} + \frac{g'(z)}{g(z)}$$

For n positive, $W'(z)/W(z)$ will possess a pole of order one. Similarly, for n negative $W'(z)/W(z)$ will possess a pole of order one, but with a negative sign. Thus, for the case of n positive or negative, the contour integral in the positive sense yields

$$\int_C \frac{W'(z)}{W(z)}dz = \pm\int_C \frac{n}{z - z_0}dz + \int_z \frac{g'(z)}{g(z)}dz$$

But because $g(z)$ is analytic at the point z_0, then $\int_C [g'(z)/g(z)]dz = 0$, and since

$$\int_C \frac{W'(z)}{W(z)}dz = \pm 2\pi jn$$

Thus the existence of a zero of $W(z)$ introduces a contribution $2\pi jn_z$ to the contour integral, where n_z is the multiplicity of the zero of $W(z)$ at z_0. Clearly, if a number of zeros of $W(z)$ exist, the total contribution to the contour integral is $2\pi jN$, where N is the weighted value of the zeros of $W(z)$ (weight 1 to a first-order zero, weight 2 to a second-order zero, and so on).

For the case where n is negative, which specifies that $W(z)$ has a pole of order n at z_0, then in the last equation n is negative and the contribution to the contour integral is now $-2\pi jn_p$ for each pole of $W(z)$; the total contribution is $-2\pi jP$, where P is the weighted number of poles. Clearly, because both zeros and poles of $F(z)$ cause poles of $W'(z)/W(z)$ with opposite signs, then the total value of the integral is

$$\int_C \frac{W'(z)}{W(z)}dz = \pm 2\pi j(N - P)$$

Note further that

$$\int_C W'(z)dz = \int_C \frac{dW(z)}{dz}dz = \int d[\log W(z)] = \int d[\log|W(z)| + j\arg W(z)]$$

$$= \log|W(z)|\Big|_o^{2\pi} + j[\arg W(2\pi) - \arg W(0)]$$

$$= 0 + j[\arg W(2\pi) - \arg W(0)]$$

so that

$$[\arg W(0) - \arg W(2\pi)] = 2\pi(N - P)$$

This relation can be given simple graphical interpretation. Suppose that the function $W(z)$ is represented by its pole and zero configuration on the z-plane. As z traverses the prescribed contour on the z-plane, $W(z)$ will move on the $W(z)$-plane according to its functional dependence on z. But the left-hand side of this equation denotes the total change in the phase angle of $W(z)$ as z traverses around the complete contour. Therefore the number of times that the moving point representing $W(z)$ revolves around the origin in the $W(z)$-plane as z moves around the specified contour is given by $N - P$.

The foregoing is conveniently illustrated graphically. Figure 20.21a shows the prescribed contour in the z-plane, and Figure 20.21b shows a possible form for the variation of $W(z)$. For this particular case, the contour in the x-plane encloses one zero and no poles; hence, $W(z)$ encloses the origin once in the clockwise direction in the $W(z)$-plane.

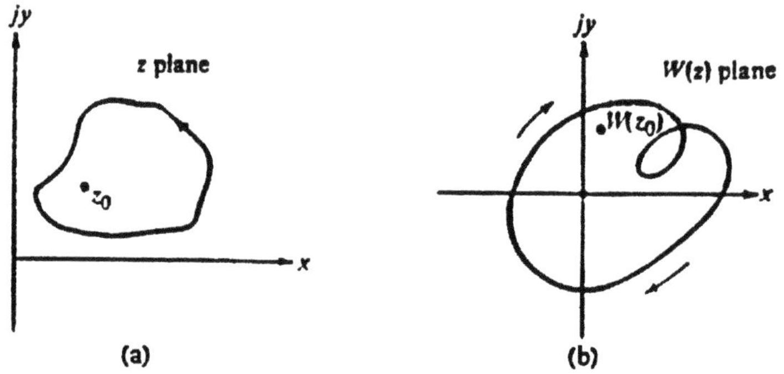

FIGURE 20.21

On the other hand, if the contour includes a pole but no zeros, it can be shown by a similar argument that any point in the interior of the z-contour must correspond to a corresponding point outside the $W(z)$-contour in the $W(z)$-plane. This is manifested by the fact that the $W(z)$-contour is traversed in a counterclockwise direction. With both zeros and poles present, the situation depends on the value of N and P.

Of special interest is the locus of the network function that contains no poles in the right-hand plane or on the $j\omega$-axis. In this case the frequency locus is completely traced as z varies along the ω-axis from $-j\infty$ to $+j\infty$. To show this, because $W(z)$ is analytic along this path, $W(z)$ can be written for the neighborhood of a point z_0 in a Taylor series

$$W(z) = \alpha_o + \alpha_1(z - z_o) + \alpha_2(z - z_o)^2 + \cdots$$

For the neighborhood $z \to \infty$, we examine $W'(z)$, where $z' = 1/z$. Because $W(z)$ does not have a pole at infinity, then $W(z')$ does not have a pole at zero. Therefore, we can expand $W(z')$ in a Maclaurin series.

$$W(z') = \alpha_0 + \alpha_1 z' + \alpha_2 (z')^2 + \cdots$$

which means that

$$W(z') = \alpha_0 + \frac{\alpha_1}{z} + \frac{\alpha_2}{z^2} + \cdots$$

But as z approaches infinity, $W(\infty)$ approaches infinity. In a real network function when z* is written for z, then $W(z^*) = W^*(z)$. This condition requires that $\alpha_0 = a_0 + j0$ be a real number irrespective of how z approaches infinity; that is, as z approaches infinity, $W(z)$ approaches a fixed point in the $W(z)$-plane. This shows that a z varies around the specified contour in the z-plane, W varies from $W(-j\infty)$ to $W(+j\infty)$ as z varies along the imaginary axis. However, $W(-j\infty) = W(+j\infty)$ from the above, which thereby shows that the locus is completely determined. This is illustrated in Figure 20.22.

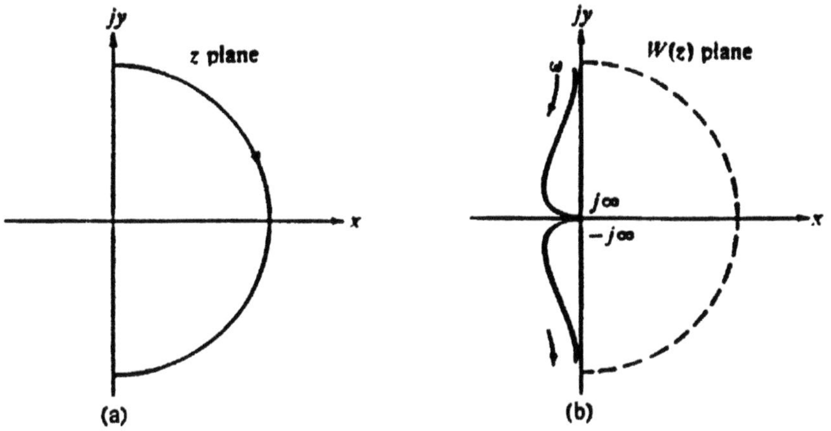

FIGURE 20.22

References

Ahlfors, L. V., *Complex Analysis*, 2nd Edition, McGraw-Hill Book Co., New York, NY, 1966.

Copson, E. T., *An Introduction to the Theory of Functions of a Complex Variable*, Oxford University Press, London, 1935.

Marsden, J. E., *Basic Complex Analysis*, W. H. Freeman and Co., San Francisco, CA, 1973.

21

Legendre Polynomials

21.1 Legendre Polynomials

21.1.1 Definition

$$P_n(t) = \sum_{k=0}^{[n/2]} \frac{(-1)^k (2n-2k)!\, t^{n-2k}}{2^n\, k!(n-k)!(n-2k)!}$$

$$[n/2] = \begin{cases} n/2 & n \text{ even} \\ (n-1)/2 & n \text{ odd} \end{cases}$$

21.1.2 Generating Function

$$w(t,s) = \frac{1}{\sqrt{1-2st+s^2}} = \begin{cases} \sum_{n=0}^{\infty} P_n(t)s^n & |s|<1 \\ \sum_{n=0}^{\infty} P_n(t)s^{-n-1} & |s|>1 \end{cases} \quad \text{generating function}$$

$$w(-t,-s) = w(t,s)$$

21.1.3 Rodrigues Formula

$$P_n(t) = \frac{1}{2^n\, n!}\frac{d^n}{dt^n}(t^2-1)^n \qquad n = 0,1,2\cdots$$

21.1.4 Recursive Formulas

1. $(n+1)P_{n+1}(t) - (2n+1)t\,P_n(t) + n\,P_{n-1}(t) = 0 \qquad n = 1,2,\cdots$

2. $P'_{n+1}(t) - t\,P'_n(t) = (n+1)P_n(t) \quad (P'(t) \doteq \text{derivative of } P(t) \qquad n = 0,1,2,\cdots$

3. $t P_n'(t) - P_{n-1}'(t) = n P_n(t)$ $\quad n = 1,2,\cdots$

4. $P_{n+1}'(t) - P_{n-1}'(t) = (2n+1) P_n(t)$ $\quad n = 1,2,\cdots$

5. $(t^2 - 1) P_n'(t) = nt P_n(t) - n P_{n-1}(t)$

6. $P_0(t) = 1 \qquad P_1(t) = t$

TABLE 21.1 Legendre Polynomials

$P_0 = 1$
$P_1 = t$
$P_2 = \frac{3}{2}t^2 - \frac{1}{2}$
$P_3 = \frac{5}{2}t^3 - \frac{3}{2}t$
$P_4 = \frac{35}{8}t^4 - \frac{30}{8}t^2 + \frac{3}{8}$
$P_5 = \frac{63}{8}t^5 - \frac{70}{8}t^3 + \frac{15}{8}t$
$P_6 = \frac{231}{16}t^6 - \frac{315}{16}t^4 + \frac{105}{16}t^2 - \frac{5}{16}$
$P_7 = \frac{429}{16}t^7 - \frac{693}{16}t^5 + \frac{315}{16}t^3 - \frac{35}{16}t$

Figure 21.1 shows a few Legendre functions.

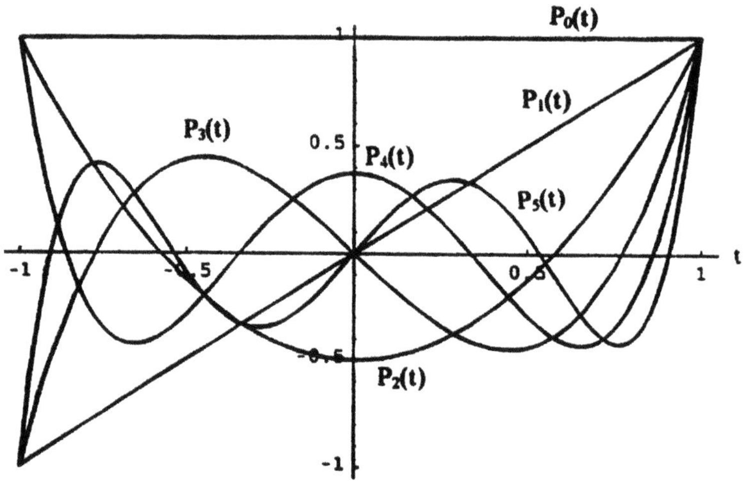

FIGURE 21.1

21.1.5 Legendre Differential Equation

If $y = P_n(x)$ $(n = 0,1,2,\cdots)$ is a solution to the second-order DE

$$(1 - t^2)y'' - 2ty' + n(n+1)y = 0$$

For $t = \cos\varphi$: $\quad \frac{1}{\sin\varphi}\frac{d}{d\varphi}\left(\sin\varphi \frac{dy}{d\varphi}\right) + n(n+1)y = 0$

Example

From (21.1.4.4) and $t = 1$ implies $0 = nP_n(1) - nP_{n-1}(1)$ or $P_n(1) = P_{n-1}(1)$. For $n = 1$, $P_1(1) = P_0(1) = 1$. For $n = 2$, $P_2(1) = P_1(1) = 1$ and so forth. Hence $P_n(1) = 1$.

21.1.6 Integral Representation

1. Laplace integral: $$P_n(t) = \frac{1}{\pi}\int_0^{\pi} [t + \sqrt{t^2 - 1}\cos\varphi]^n \, d\varphi$$

2. Mehler-Dirichlet formula: $$P_n(\cos\theta) = \frac{2}{\pi}\int_0^{\theta} \frac{\cos(n+\frac{1}{2})\psi}{\sqrt{2\cos\psi - \cos\theta}} d\psi \qquad 0 < \theta < \pi, n = 0,1,2,\ldots$$

3. Schläfli integral: $$P_n(t) = \frac{1}{2\pi j}\int_C \frac{(z^2-1)^n}{2^n (z-t)^{n+1}} dz$$

   where C is any regular, simple, closed curve surrounding t.

21.1.7 Complete Orthonormal System

$$\{[\tfrac{1}{2}(2n+1)]^{1/2} P_n(t)\}$$

The Legendre polynomials are orthogonal in [–1,1]

$$\int_{-1}^{1} P_n(t) P_m(t) dt = 0$$

$$\int_{-1}^{1} [P_n(t)]^2 \, dt = \frac{2}{2n+1} \qquad n = 0,1,2\cdots$$

and therefore the set

$$\varphi_n(t) = \sqrt{\frac{2n+1}{2}} P_n(t) \qquad n = 0,1,2\cdots$$

is orthonormal.

21.1.8 Asymptotic Representation:

$$P_n(\cos\theta) \cong \sqrt{\frac{2}{\pi n \sin\theta}} \sin\left[\left(n + \frac{1}{2}\right)\theta + \frac{\pi}{4}\right], \qquad n \to \infty, \quad \delta \le \theta \le \pi - \delta$$

δ = fixed positive number

21.1.9 Series Expansion

If $f(t)$ is integrable in [–1,1] then

$$f(t) = \sum_{n=0}^{\infty} a_n P_n(t) \qquad -1 < t < 1$$

$$a_n = \frac{2n+1}{2}\int_{-1}^{1} f(t)P_n(t)\,dt \qquad n = 0,1,2\cdots$$

For even $f(t)$, the series will contain term $P_n(t)$ of even index; if $f(t)$ is odd, the term of odd index only.

If the real function $f(t)$ is piecewise smooth in $(-1,1)$ and if it is square integrable in $(-1,1)$, then the series converges to $f(t)$ at every continuity point of $f(t)$. If there is a discontinuity at t then the series converges at $[f(t+0)+f(t-0)]/2$.

21.1.10 Change of Range

If a function $f(t)$ is defined in $[a,b]$, it is sometimes necessary in the applications to expand the function in a series in the applications to expand the function in a series of orthogonal polynomials in this interval. Clearly the substitution

$$t = \frac{2}{b-a}\left[x - \frac{b+a}{2}\right], \qquad a<b, \qquad \left[x = \frac{b-a}{2}t + \frac{b+a}{2}\right]$$

transform the interval $[a,b]$ of the x-axis into the interval $[-1,1]$ of the t-axis. It is, therefore, sufficient to consider

$$f\left[\frac{b-a}{2}t + \frac{b+a}{2}\right] = \sum_{n=0}^{\infty} a_n P_n(t)$$

$$a_n = \frac{2n+1}{2}\int_{-1}^{1} f\left[\frac{b-a}{2}t + \frac{b+a}{2}\right]P_n(t)\,dt$$

The above equation can also be accomplished as follows:

$$f(t) = \sum_{n=0}^{\infty} a_n X_n(t)$$

$$X_n(t) = \frac{1}{n!(b-a)^n}\,\frac{d^n(t-a)^n(t-b)^n}{dt^n}$$

$$a_n = \frac{2n+1}{b-a}\int_{b}^{a} f(t)X_n(t)\,dt$$

Example

Suppose $f(t)$ is given by

$$f(t) = \begin{cases} 0 & -1 \le t < a \\ 1 & a < t \le 1 \end{cases}$$

Then

$$a_n = \frac{2n+1}{2}\int_{a}^{1} P_n(t)\,dt$$

Using (21.1.4.4), and noting that $P_n(1) = 1$ we obtain

$$a_n = -\tfrac{1}{2}[P_{n+1}(a) - P_{n-1}(a)], \qquad a_0 = \tfrac{1}{2}(1-a)$$

which leads to the expansion

$$f(t) \cong \frac{1}{2}(1-a) - \frac{1}{2}\sum_{n=1}^{\infty}[P_{n+1}(a) - P_{n-1}(a)]P_n(t), \qquad -1 < t < 1$$

Example

Suppose $f(t)$ is given by

$$f(t) = \begin{cases} -1 & -1 \le t < 0 \\ 1 & 0 < t \le 1 \end{cases}$$

The function is an odd function and, therefore, $f(t)P_n(t)$ is an odd function of $P_n(t)$ with even index. Hence a_n are zero for $n = 0,2,4,\ldots$. For odd index n, the product $f(t)P_n(t)$ is even and hence

$$a_n = \left(n+\frac{1}{2}\right)\int_{-1}^{1} f(t)P_n(t)\,dt = 2\left(n+\frac{1}{2}\right)\int_{0}^{1} P_n(t)\,dt \qquad n = 1,3,5,\cdots$$

Using (21.1.4.4) and setting $n = 2k + 1$, $k = 0,1,2\ldots$ we obtain

$$a_{2k+1} = (4k+3)\int_0^1 P_{2k+1}(t)\,dt = \int_0^1 [P'_{2k+2}(t) - P'_{2k}(t)]\,dt$$

$$= [P_{2k+2}(t) - P_{2k}(t)]\Big|_0^1 = P_{2k}(0) - P_{2k+2}(0)$$

where we have used the property $P_n(1) = 1$ for all n. But

$$P_{2n}(0) = \binom{-\frac{1}{2}}{n} = \frac{(-1)^n (2n)!}{2^{2n}(n!)^2}$$

and, thus, we have

$$a_{2k+1} = \frac{(-1)^k (2k)!}{2^{2k}(k!)^2} - \frac{(-1)^{k+1}(2k+2)}{2^{2k+2}[(k+1)!]^2} = \frac{(-1)^k (2k)!}{2^{2k}(k!)^2}\left[1 + \frac{2k+1}{2k+2}\right]$$

$$= \frac{(-1)^k (2k)!(4k+3)}{2^{2k+1}k!(k+1)!}$$

The expansion is

$$f(t) = \sum_{n=0}^{\infty} \frac{(-1)^k (2k)!(4k+3)}{2^{2k+1}k!(k+1)!}P_{2k+1}(t) \qquad -1 \le t \le 1$$

21.1.11 Expansion of Polynomials

If $q_m(t)=\sum_{k=0}^{m} c_k x^k$ is an arbitrary polynomial, then $q_m(t)=c_0P_0(t)+c_1P_1(t)+\cdots+c_mP_m(t)$ where $c_n = \left(n+\frac{1}{2}\right)\int_{-1}^{1} q_m(t)P_n(t)\,dt = 0,\ n=0,1,2\cdots$. If $q_m(t)$ is a polynomial of degree m and $m < r$, then $\int_{-1}^{1} q_m(t)P_r(t)\,dt = 0,\ m<r.$

Example

To find $P_{2n}(0)$ we use the summation

$$P_{2n}(t)=\frac{(-1)^n}{2^{2n-1}}\sum_{k=0}^{n}\frac{(-1)^k(2n+2k-1)!}{(2k)!(n+k-1)!(n-k)!}t^{2k}$$

with $k = 0$. Hence

$$P_{2n}(0)=\frac{(-1)^n(2n-1)!}{2^{2n-1}(n-1)!n!}=\frac{(-1)^n 2n[(2n-1)!]}{2^{2n}n[(n-1)!]n!}=\frac{(-1)^n(2n)!}{2^{2n}(n!)^2}$$

Example

To evaluate $\int_0^1 P_m(t)\,dt$ for $m \neq 0$ we must consider the two cases: m = odd and m = even.

(a) m = even and $m \neq 0$

$$\int_0^1 P_m(t)\,dt=\frac{1}{2}\int_{-1}^{1}P_m(t)\,dt=\frac{1}{2}\int_{-1}^{1}P_m(t)\cdot 1\,dt=\frac{1}{2}\int_{-1}^{1}P_m(t)P_0(t)\,dt=0$$

The result is due to the orthogonality principle.

(b) m = odd and $m \neq 0$. From the relation (see Table 21.2)

$$\int_{-1}^{1} P_m(t)\,dt=\frac{1}{2m+1}[P_{m-1}(t)-P_{m+1}(t)]$$

with $t = 0$ we obtain

$$\int_0^1 P_m(t)\,dt=\frac{1}{2m+1}[P_{m-1}(0)-P_{m+1}(0)]$$

Using the results of the previous example, we obtain

$$\int_0^1 P_m(t)\,dt=\frac{1}{2m+1}\left[\frac{(-1)^{\frac{m-1}{2}}(m-1)!}{2^{m-1}\left[\left(\frac{m-1}{2}\right)!\right]^2}-\frac{(-1)^{\frac{m+1}{2}}(m+1)!}{2^{m+1}\left[\left(\frac{m+1}{2}\right)!\right]^2}\right]$$

$$=\frac{(-1)^{\frac{m-1}{2}}(m-1)!(2m+1)(m+1)}{(2m+1)2^{m+1}\left(\frac{m+1}{2}\right)!\left(\frac{m+1}{2}\right)\left(\frac{m-1}{2}\right)!}=\frac{(-1)^{\frac{m-1}{2}}(m-1)!}{2^m\left(\frac{m+1}{2}\right)!\left(\frac{m-1}{2}\right)!}\qquad m=\text{odd}$$

21.2 Legendre Functions of the Second Kind (Second Solution)

21.2.1 Second Kind:

1. $Q_0 = \frac{1}{2}\ln\frac{1+t}{1-t}, \quad |t| < 1;$

2. $Q_1(t) = \frac{1}{2}t\ln\frac{1+t}{1-t} - 1, \quad |t| < 1;$

3. $Q_{n+1}(t) = \frac{2n+1}{n+1}tQ_n(t) - \frac{n}{n+1}Q_{n-1}(t), \quad n = 1,2,\cdots$

4. $Q_n(t) = P_n(t)Q_0(t) - \sum_{k=0}^{[\frac{1}{2}(n-1)]}\frac{2n-4k-1}{(2k+1)(n-k)}P_{n-2k-1}(t), \quad |t| < 1, \quad n = 1,2,\cdots$

   for $[\frac{1}{2}(n-1)]$ see 21.1.1.

21.2.2 Recursions

$Q_n(t)$ satisfies all the recurrence relations of $P_n(t)$.

21.2.3 Property

$$\frac{1}{x-t} = \sum_{n=0}^{\infty}(2n+1)P_n(t)Q_n(x)$$

21.2.4 Newman Formula

$$Q_n(t) = \frac{1}{2}\int_{-1}^{1}\frac{P_n(x)}{t-x}dx, \qquad n = 0,1,2\cdots$$

21.3 Associated Legendre Polynomials

21.3.1 Definition

If m is a positive integer and $-1 \le t \le 1$, then

$$P_n^m(t) = (1-t^2)^{m/2}\frac{d^m P_n(t)}{dt^m} \qquad m = 1,2,\cdots,n$$

where $P_n^m(t)$ is known as the *associated Legendre function* or *Ferrers' functions*.

21.3.2 Rodrigues Formula

$$P_n^m(t) = \frac{(1-t^2)^{m/2}}{2^n\,n!}\frac{d^{n+m}}{dt^{n+m}}(t^2-1)^n, \qquad m = 1,2,\cdots,n; \quad n+m \ge 0$$

21.3.3 Properties

1. $P_n^{-m}(t) = (-1)^m\frac{(n-m)!}{(n+m)!}P_n^m(t)$

2. $P_n^0(t) = P_n(t)$

3. $(n-m+1)P_{n+1}^m(t) - (2n+1)t P_n^m(t) + (n+m)P_{n-1}^m(t) = 0$

4. $(1-t^2)^{1/2} P_n^m(t) = \frac{1}{2n+1}[P_{n+1}^{m+1}(t) - P_{n-1}^{m+1}(t)]$

5. $(1-t^2)^{1/2} P_n^m(t) = \frac{1}{2n+1}[(n+m)(n+m-1)P_{n-1}^{m-1}(t) - (n-m+1)(n-m+2)P_{n+1}^{m-1}(t)]$

6. $P_n^m(t) = 2mt(1-t^2)^{-1/2} P_n^m(t) - [n(n+1) - m(m-1)]P_n^{m-1}(t)$

7. $P_n^{m+1}(t) = (t^2-1)^{-1/2}[(n-m)\, t P_n^m(t) - (n+m)P_{n-1}^m(t)]$

8. $P_{n+1}^m(t) = P_{n-1}^m(t) + (2n+1)(t^2-1)^{1/2} P_n^{m-1}(t)$

9. $P_n^m(t) = (t^2-1)^{m/2} \frac{d^m P_n(t)}{dt^m}$

10. $\int_{-1}^{1} P_n^m(t) P_k^m(t)\, dt = 0 \qquad k \neq n$

11. $\int_{-1}^{1} [P_n^m(t)]^2\, dt = \frac{2(n+m)!}{(2n+1)(n-m)!}$

21.3.4 Differential Equation

$$(1-t^2)\frac{d^2 P_n^m(t)}{dt^2} - 2t\frac{dP_n^m(t)}{dt} + \left[n(n+1) - \frac{m^2}{(1-t^2)}\right]P_n^m(t) = 0$$

21.3.5 Schlafli Formula

$$P_n^m(t) = \frac{(n+m)!}{2\pi jn!}(1-t^2)^{m/2}\oint_C \frac{(x^2-1)^n}{2^n(x-t)^{n+m+1}}dx$$

where C is any regular closed curve surrounding the point t and taking it counterclockwise.

21.4 Bounds for Legendre Polynomials

21.4.1 Stieltjes Theorem

$$|P_n(\cos\gamma)| \leq \sqrt{2}\frac{4}{\sqrt{\pi}}\frac{1}{\sqrt{n}\sqrt{\sin\gamma}}, \qquad 0 < \gamma < \pi, \quad n = 1,2,\cdots$$

21.4.2 Second Stieltjes Theorem

$$|P_n(t) - P_{n+2}(t)| < \frac{4}{\sqrt{\pi}\sqrt{n+2}}$$

21.4.3 $\left|\frac{dP_n(t)}{dt}\right| < \frac{2}{\sqrt{\pi}} \frac{\sqrt{n}}{1-t^2}, \qquad |t|<1, \quad n=1,2,\cdots$

21.4.4 $|P_{n+1}(t)+P_n(t)| < \frac{6\sqrt{2}}{\sqrt{\pi}} \frac{1}{\sqrt{n}} \frac{1}{\sqrt{1-t}}, \qquad |t|<1$

21.5 Table of Legendre and Associate Legendre Functions

TABLE 21.2 Properties of Legendre and Associate Legendre Functions [$P_n(t)$ = Legendre Functions, $P_n^m(t)$ = Associate Legendre Functions, $Q_n(t)$ = Legendre Functions of the Second Kind]

1.	$\frac{1}{\sqrt{1-2tx+x^2}} = \sum_{n=0}^{\infty} P_n(t)x^n \quad |t|\le 1 \quad |x|<1$	
2.	$P_n(t)=\sum_{k=0}^{[n/2]} \frac{(-1)^k(2n-2k)!\,t^{n-2k}}{2^n k!(n-k)!(n-2k)!} \quad [n/2]=\frac{n}{2} \quad n=\text{even}; \quad [n/2]=(n-1)/2 \quad n=\text{odd}$	
3.	$P_0(t)=1$	
4.	$P_{2n}(0)=\binom{-\frac{1}{2}}{n}=\frac{(-1)^n(2n)!}{2^{2n}(n!)^2}$	$n=1,2,\cdots$
5.	$P_{2n+1}(0)=0$	$n=0,1,2\cdots$
6.	$P_{2n}(-t)=P_{2n}(t) \qquad P_{2n+1}(-t)=-P_{2n+1}(t)$	$n=0,1,2\cdots$
7.	$P_n(-t)=(-1)^n P_n(t)$	$n=0,1,2\cdots$
8.	$P_n(1)=1 \quad n=0,1,2\cdots; \quad P_n(-1)=(-1)^n$	$n=0,1,2\cdots$
9.	$P_n(t)=\frac{1}{2^n n!}\frac{d^n}{dt^n}(t^2-1)^n$ = Rodrigues formula,	$n=0,1,2\cdots$
10.	$(n+1)P_{n+1}(t)-(2n+1)tP_n(t)+nP_{n-1}(t)=0$	$n=1,2,\cdots$
11.	$P'_{n+1}(t)-2tP'_n(t)+P'_{n-1}(t)-P_n(t)=0$	$n=1,2,\cdots$
12.	$P'_{n-1}(t)=P_n(t)+2tP'_n(t)-P'_{n+1}(t)$	
13.	$P'_{n+1}(t)=P_n(t)+2tP'_n(t)-P'_{n-1}(t)$	
14.	$P'_{n+1}(t)-tP'_n(t)=(n+1)P_n(t)$	
15.	$tP'_n(t)-P'_{n-1}(t)=nP_n(t)$	$n=1,2,\cdots$
16.	$P'_{n+1}(t)-P'_{n-1}(t)=(2n+1)P_n(t)$	$n=1,2,\cdots$
17.	$(1-t^2)P'_n(t)=nP_{n-1}(t)-ntP_n(t)$	
18.	$|P_n(t)|<1, \qquad -1<t<1$	
19.	$P_{2n}(t)=\frac{(-1)^n}{2^{2n-1}}\sum_{k=0}^{n}\frac{(-1)^k(2n+2k-1)!}{(2k)!(n+k-1)!(n-k)!}t^{2k}$	$n=0,1,2\cdots$
20.	$(1-t^2)P'_n(t)=(n+1)[tP_n(t)-P_{n+1}(t)]$	$n=0,1,2\cdots$
21.	$\int_{-1}^{1} P_n(t)dt=0$	$n=1,2,\cdots$
22.	$|P_n(t)|\le 1$	$|t|\le 1$

TABLE 21.2 Properties of Legendre and Associate Legendre Functions [$P_n(t)$ = Legendre Functions, $P_n^m(t)$ = Associate Legendre Functions, $Q_n(t)$ = Legendre Functions of the Second Kind] (continued)

23.	$\int_{-1}^{1} P_n(t)P_m(t)dt = 0$	$n \neq m$
24.	$\int_{-1}^{1} [P_n(t)]^2\, dt = \frac{2}{2n+1}$	$n = 0,1,2\cdots$
25.	$\frac{1}{2}\int_{-1}^{1} t^m\, P_s(t)dt = \frac{m(m-2)\cdots(m-s+2)}{(m+s+1)(m+s-1)\cdots(m+1)}$	$m, s = \text{even}$
26.	$\frac{1}{2}\int_{-1}^{1} t^m\, P_s(t)dt = \frac{(m-1)(m-3)\cdots(m-s+2)}{(m+s+1)(m+s-1)\cdots(m+2)}$	$m, s = \text{odd}$
27.	$\int_{-1}^{1} t\,P_n(t)\,P_{n-1}(t)dt = \frac{2n}{4n^2-1}$	$n = 1,2,\cdots$
28.	$\int_{-1}^{1} P_n(t)P'_{n+1}(t)dt = 2$	$n = 0,1,2\cdots$
29.	$\int_{-1}^{1} t\,P'_n(t)P_n(t)dt = \frac{2n}{2n+1}$	$n = 0,1,2\cdots$
30.	$\int_{-1}^{1} (1-t^2)\,P'_n(t)\,P'_k(t)dt = 0$	$k \neq n$
31.	$\int_{-1}^{1} (1-t)^{-1/2}\, P_n(t)dt = \frac{2\sqrt{2}}{2n+1}$	$n = 0,1,2\cdots$
32.	$\int_{-1}^{1} t^2\, P_{n+1}(t)\,P_{n-1}(t)dt = \frac{2n(n+1)}{(4n^2-1)(2n+3)}$	$n = 1,2,\cdots$
33.	$\int_{-1}^{1} (t^2-1)\,P_{n+1}(t)\,P'_n(t)dt = \frac{2n(n+1)}{(2n+1)(2n+3)}$	$n = 1,2,\cdots$
34.	$\int_{-1}^{1} t^n\, P_n(t)dt = \frac{2^{n+1}(n!)^2}{(2n+1)!}$	$n = 0,1,2\cdots$
35.	$\int_{-1}^{1} t^2\,[P_n(t)]^2\, dt = \frac{2}{(2n+1)^2}\left[\frac{(n+1)^2}{2n+3} + \frac{n^2}{2n-1}\right]$	$n = 0,1,2\cdots$
36.	$P_n^m(t) = (1-t^2)^{m/2}\,\frac{d^m}{dt^m}\,P_n(t)$	$m > 0$
37.	$P_n^m(t) = \frac{1}{2^n\, n!}(1-t^2)^{m/2}\,\frac{d^{n+m}}{dt^{n+m}}[(t^2-1)^n]$	$m+n \geq 0$
38.	$P_n^{-m}(t) = (-1)^m\,\frac{(n-m)!}{(n+m)!}\,P_n^m(t)$	
39.	$P_n^0(t) = P_n(t), \quad P_n^m(t) = 0 \text{ for } m > n$	
40.	$(n-m+1)\,P_{n+1}^m(t) - (2n+1)\,t\,P_n^m(t) + (n+m)\,P_{n-1}^m(t) = 0$	
41.	$(1-t^2)^{1/2}\,P_n^m(t) = \frac{1}{2n+1}[P_{n+1}^{m+1}(t) - P_{n-1}^{m+1}(t)]$	

TABLE 21.2 Properties of Legendre and Associate Legendre Functions [$P_n(t)$ = Legendre Functions, $P_n^m(t)$ = Associate Legendre Functions, $Q_n(t)$ = Legendre Functions of the Second Kind] (continued)

42. $$(1-t^2)^{1/2} P_n^m(t) = \frac{1}{2n+1}[(n+m)(n+m-1)P_{n-1}^{m-1}(t) - (n-m+2)P_{n+1}^{m-1}(t)]$$

43. $$P_n^{m+1}(t) = 2mt(1-t^2)^{-1/2} P_n^m(t) - [n(n+1) - m(m-1)]P_n^{m-1}(t)$$

44. $$\int_{-1}^{1} P_n^m(t) P_k^m(t)\,dt = 0 \qquad k \neq n$$

45. $$\int_{-1}^{1} [P_n^m(t)]^2\,dt = \frac{2}{2n+1}\frac{(n+m)!}{(n-m)!}$$

46. $$P_n^m(-t) = (-1)^{n+m} P_n^m(t)$$

47. $$P_n^m(\pm 1) = 0 \qquad m > 0$$

48. $$P_{2n}^1(0) = 0 \qquad P_{2n+1}^1(0) = \frac{(-1)^n (2n+1)!}{2^{2n}(n!)^2}$$

49. $$P_n^m(0) = 0 \qquad n+m = \text{odd}$$
$$P_n^m(0) = (-1)^{(n-m)/2} \frac{(n+m)!}{2^n[(n-m)/2]![(n+m)/2]!} \qquad n+m = \text{even}$$

50. $$\int_{-1}^{1} P_n^m(t) P_n^k(t)\,(1-t^2)^{-1}\,dt = 0 \qquad k \neq m$$

51. $$\int_{-1}^{1} (1-t^2)^{-1/2} P_{2m}(t)\,dt = \left[\frac{\Gamma(\frac{1}{2}+m)}{m!}\right]^2$$

52. $$\int_{-1}^{1} t(1-t^2)^{-1/2} P_{2m+1}(t)\,dt = \frac{\Gamma(\frac{1}{2}+m)\Gamma(\frac{3}{2}+m)}{m!(m+1)!}$$

53. $$\int_{t}^{1} P_n(t)\,dt = \frac{1}{2n+1}[P_{n-1}(t) - P_{n+1}(t)]$$

54. $$\int_{0}^{1} t^q P_n(t)\,dt = \Gamma(q+1)\sum_{k=0}^{n} \frac{(-1)^k \Gamma(n+k+1)}{2^k k!\Gamma(n-k+1)(q+k+2)} \qquad q > -1$$

55. $$\int_{0}^{1} t^{-1/2} P_n(t)\,dt = \begin{cases} \dfrac{2(-1)^{n/2}}{2n+1} & n = \text{even} \\[2ex] \dfrac{2(-1)^{(n-1)/2}}{2n+1} & n = \text{odd} \end{cases}$$

56. $$\int_{0}^{1} t^{1/2} P_n(t)\,dt = \begin{cases} \dfrac{2(-1)^{(n+2)/2}}{(2n-1)(2n+3)} & n = \text{even} \\[2ex] \dfrac{2(-1)^{(n+3)/2}}{(2n-1)(2n+3)} & n = \text{odd} \end{cases}$$

57. $$Q_0 = \frac{1}{2}\ln\frac{1+t}{1-t}, \qquad |t| < 1$$

TABLE 21.2 Properties of Legendre and Associate Legendre Functions [$P_n(t)$ = Legendre Functions, $P_n^m(t)$ = Associate Legendre Functions, $Q_n(t)$ = Legendre Functions of the Second Kind] (continued)

58. $Q_1(t) = \frac{1}{2} t \ln \frac{1+t}{1-t} - 1 = tQ_0(t) - 1, \quad |t| < 1$

59. $Q_{n+1}(t) = \frac{2n+1}{n+1} tQ_n(t) - \frac{n}{n+1} Q_{n-1}(t), \quad n = 1, 2, \cdots$

60. $Q_n(t) = P_n(t)Q_0(t) - \sum_{k=0}^{[\frac{1}{2}(n-1)]} \frac{2n-4k-1}{(2k+1)(n-k)} P_{n-2k-1}(t), \quad |t| < 1$

61. $Q_n(t) = \frac{1}{2} \int_{-1}^{1} \frac{P_n(x)}{t-x} dx \qquad n = 0, 1, 2 \cdots$

62. $Q'_{n+1}(t) - 2tQ'_n(t) + Q'_{n-1}(t) - Q_n(t) = 0$

63. $Q'_{n+1}(t) - tQ'_n(t) - (n+1)Q_n(t) = 0$

64. $Q'_{n+1}(t) - Q'_{n-1}(t) = (2n+1)Q_n(t)$

65. $Q_0(-t) = -Q_0(t), \quad Q_n(-t) = (-1)^{n+1} Q_n(t), \quad n = 1, 2, \cdots$

References

Abramowitz, M. and I. S. Stegun, *Handbook of Mathematical Functions*, Dover Publications, Inc., New York, NY, 1965.

Andrews, L. C., *Special Functions for Engineers and Applied Mathematicians*, MacMillan Publishing Co. New York, NY. 1985.

Hochstadt, H., *The Functions of Mathematical Physics*, Dover Publications Inc., New York, NY, 1986.

McLachlan, N. W., *Bessel Functions for Engineers,* 2nd Edition, Oxford University Press, London, 1961.

Sansone, G., *Orthogonal Functions*, Interscience Publishers, New York, NY., 1959.

22

Hermite Polynomials

22.1 Hermite Polynomials

22.1.1 Rodrigues Formula

$$H_n(t) = (-1)^n e^{t^2} \frac{d^n e^{-t^2}}{dt^n} \qquad n = 0,1,2,\cdots,\ -\infty < t < \infty$$

The first few Hermite polynomials are:

$$H_0(t) = 1,\ H_1(t) = 2t,\ H_2(t) = 4t^2 - 2,\ H_3(t) = 8t^3 - 12t,\quad H_4(t) = 16t^4 - 48t + 12,$$

$$H_5(t) = 32t^5 - 160t^3 + 120t, \text{ etc.}$$

(See Figure 22.1.)

22.1.2 Expansion Formula

$$H_n(t) = \sum_{k=0}^{[n/2]} \frac{(-1)^k n!}{k!(n-2k)!}(2t)^{n-2k}, \qquad [n/2] = \text{largest integer} \le n/2$$

22.1.3 Generating Function

$$w(t,x) = e^{2tx - x^2} = \sum_{n=0}^{\infty} \frac{H_n(t)}{n!} x^n = \sum_{n=0}^{\infty} \frac{1}{n!}\left[\frac{\partial^n w}{\partial x^n}\right]_{x=0} x^n, \quad |x| < \infty$$

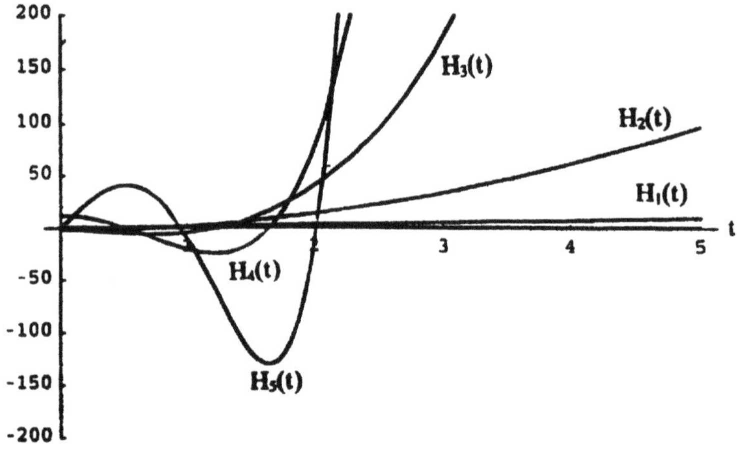

FIGURE 22.1

22.1.4 Even and Odd n

$$H_n(-t) = (-1)^n H_n(t); \; H_{2n}(0) = (-1)^n \frac{(2n)!}{n!}, \quad H_{2n+1}(0) = 0$$

22.2 Recurrence Relation

22.2.1 Recurrence Relation

1. $H_{n+1}(t) - 2t\,H_n(t) + 2n\,H_{n-1}(t)$ $\quad n = 1, 2, \cdots$
2. $H_n'(t) = 2n\,H_{n-1}(t)$ $\quad n = 1, 2, \cdots$

   $H_{n+1}'(t) = 2(n+1)\,H_n(t)$ $\quad n = 1, 2, \cdots$
3. $H_{n+1}(t) - 2t\,H_n(t) + H_n'(t) = 0$ $\quad n = 0, 1, 2, \cdots$

22.2.2 Hermite Differential Equation

$$H_n''(t) - 2t\,H_n'(t) + 2n\,H_n(t) = 0 \qquad n = 0, 1, 2\cdots$$

which implies that the Hermite polynomials are the solution of the second-order ordinary differential equation.

22.3 Integral Representation

$$H_n(t) = \frac{(-j)^n 2^n e^{t^2}}{\sqrt{\pi}} \int_{-\infty}^{\infty} e^{-x^2 + j2tx} x^n \, dx \qquad n = 0, 1, 2\cdots$$

$$e^{-t^2/2} H_n(t) = \frac{1}{j^n \sqrt{2\pi}} \int_{-\infty}^{\infty} e^{jty} e^{-y^2/2} H_n(y)\, dy \qquad n = 0, 1, 2\cdots$$

$$e^{-t^2/2}H_{2m}(t) = (-1)^m \sqrt{\frac{2}{\pi}} \int_0^\infty e^{-y^2/2} H_{2m}(y) \cos ty \, dy$$

$$e^{-t^2/2}H_{2m+1}(t) = (-1)^m \sqrt{\frac{2}{\pi}} \int_0^\infty e^{-y^2/2} H_{2m+1}(y) \sin ty \, dy \qquad m = 0,1,2\cdots$$

22.4 Hermite Series

22.4.1 Orthogonality Property

$$\int_{-\infty}^{\infty} e^{-t^2} H_m(t) H_n(t) dt = 0 \quad \text{if} \quad m \neq n$$

and

$$\int_{-\infty}^{\infty} e^{-t^2} H_n^2(t) dt = 2^n n! \sqrt{\pi} \qquad n = 0,1,2\cdots$$

22.4.2 Orthonormal Hermite Polynomials

$$\varphi_n(t) = (2^n n! \sqrt{\pi})^{-1/2} e^{-t^2/2} H_n(t) \qquad n = 0,1,2,\cdots, \ -\infty < t < \infty$$

22.4.3 Hermite Series

$$f(t) = \sum_{n=0}^{\infty} C_n H_n(t) \qquad -\infty < t < \infty$$

$$C_n = \frac{1}{2^n n! \sqrt{\pi}} \int_{-\infty}^{\infty} e^{-t^2} f(t) H_n(t) dt \qquad n = 0,1,2,\cdots$$

where $f(t)$ is piecewise smooth in every finite interval and $\int_{-\infty}^{\infty} e^{-t^2} f^2(t) dt < \infty$.

Example

$f(t) = t^4 = \sum_{n=0}^{\infty} C_{2n} H_{2n}(t)$, since $f(t)$ is even.

$$C_{2n} = \frac{1}{2^{2n}(2n)!\sqrt{\pi}} \int_{-\infty}^{\infty} e^{-t^2} t^4 H_{2n}(t) dt$$

$$= \frac{1}{2^{2n}(2n)!\sqrt{\pi}} \int_{-\infty}^{\infty} t^4 \frac{d^{2n}}{dt^{2n}}(e^{-t^2}) dt = \frac{1}{2^{2n}(2n)!\sqrt{\pi}} \frac{(4)!}{(4-2n)!} \int_{-\infty}^{\infty} e^{-t^2} t^{4-2n} dt$$

$$= \frac{1}{2^{2n}(2n)!\sqrt{\pi}} \frac{(4)!}{(4-2n)!} \Gamma\left(4 - n + \frac{1}{2}\right)$$

22.5 Properties of the Hermite Polynomials

TABLE 22.1 Properties of the Hermite Polynomial

1. $H_n(t) = (-1)^n e^{t^2} \dfrac{d^n e^{-t^2}}{dt^n}$ $\quad n = 0,1,2\cdots$

2. $H_n(t) = \sum_{k=0}^{\lfloor n/2 \rfloor} \dfrac{(-1)^k n!}{k!(n-2k)!}(2t)^{n-2k}$ $\quad [n/2] =$ largest integer $\le n/2$

3. $e^{2tx-x^2} = \sum_{n=0}^{\infty} H_n(t)\dfrac{x^n}{n!}$

4. $H_{2n}(0) = (-1)^n \dfrac{(2n)!}{n!}$

5. $H_{2n+1}(0) = 0, \quad H'_{2n}(0) = 0, \quad H'_{2n+1}(0) = (-1)^n \dfrac{(2n+2)!}{(n+1)!}$

6. $H_n(-t) = (-1)^n H_n(t)$

7. $H_{2n}(t) =$ even functions, $\quad H_{2n+1}(t) =$ odd functions

8. $H_{n+1}(t) - 2t\,H_n(t) + 2n\,H_{n-1}(t) = 0$ $\quad n = 1,2,\cdots$

9. $H'_n(t) = 2n\,H_{n-1}(t)$ $\quad n = 1,2,\cdots$

10. $H_{n+1}(t) - 2t\,H_n(t) + H'_n(t) = 0$ $\quad n = 0,1,2\cdots$

11. $H''_n(t) - 2t\,H'_n(t) + 2n\,H_n(t) = 0$ $\quad n = 0,1,2\cdots$

12. $H_n(t) = \dfrac{(-j)^n 2^n e^{t^2}}{\sqrt{\pi}} \int_{-\infty}^{\infty} e^{-x^2+j2tx} x^n\,dx$ $\quad n = 0,1,2\cdots$

13. $e^{-t^2/2} H_n(t) = \dfrac{1}{j^n \sqrt{2\pi}} \int_{-\infty}^{\infty} e^{jty} e^{-y^2/2} H_n(y)\,dy =$ integral equation

14. $e^{-t^2/2} H_{2m}(t) = (-1)^m \sqrt{\dfrac{2}{\pi}} \int_0^{\infty} e^{-y^2/2} H_{2m}(y) \cos ty\,dy$

15. $e^{-t^2/2} H_{2m+1}(t) = (-1)^m \sqrt{\dfrac{2}{\pi}} \int_0^{\infty} e^{-y^2/2} H_{2m+1}(y) \sin ty\,dy$

16. $\int_{-\infty}^{\infty} e^{-t^2} H_m(t) H_n(t)\,dt = 0$ if $m \ne n$

17. $\int_{-\infty}^{\infty} e^{-t^2} H_n^2(t)\,dt = 2^n n! \sqrt{\pi}$ $\quad n = 0,1,2\cdots$

18. $f(t) = \sum_{n=0}^{\infty} C_n H_n(t)$ $\quad -\infty < t < \infty$

$$C_n = \frac{1}{2^n n! \sqrt{\pi}} \int_{-\infty}^{\infty} e^{-t^2} f(t) H_n(t)\,dt$$

19. $\int_{-\infty}^{\infty} t^k e^{-t^2} H_n(t)\,dt = 0$ $\quad k = 0,1,\cdots n-1$

TABLE 22.1 Properties of the Hermite Polynomial (continued)

20.	$\int_{-\infty}^{\infty} t^2 e^{-t^2} H_n^2(t)\,dt = \sqrt{\pi}\, 2^n n!\left(n+\frac{1}{2}\right)$	
21.	$\int_{0}^{\infty} x^n e^{-x^2} H_n(tx)\,dx = \frac{\sqrt{\pi}\; n!}{2} P_n(t)$	
22.	$\int_{-\infty}^{\infty} e^{-2t^2} H_n^2(t)\,dt = 2^{n-\frac{1}{2}}\Gamma\left(n+\frac{1}{2}\right)$	
23.	$\frac{d^m H_n(t)}{dt^m} = \frac{2^m n!}{(n-m)!} H_{n-m}(t)$	$m < n$
24.	$\int_{-\infty}^{\infty} e^{-a^2t^2} H_{2n}(t)\,dt = \frac{(2n)!}{n!}\frac{\sqrt{\pi}}{a}\left(\frac{1-a^2}{a^2}\right)^n$	$a > 0$
25.	$\int_{-\infty}^{\infty} e^{-t^2+2bt} H_n(t)\,dt = \sqrt{\pi}\,(2b)^n e^{b^2}$	$n = 0,1,2\cdots$

References

Abramowitz, M. and I. S. Stegun, *Handbook of Mathematical Functions*, Dover Publications, Inc., New York, NY, 1965.

Andrews, L. C., *Special Functions for Engineers and Applied Mathematicians*, MacMillan Publishing Co. New York, NY. 1985.

Hochstadt, H., *The Functions of Mathematical Physics*, Dover Publications Inc., New York, NY, 1986.

McLachlan, N. W., *Bessel Functions for Engineers,* 2nd Edition, Oxford University Press, London 1961.

Sansone, G., *Orthogonal Functions*, Interscience Publishers, New York, NY., 1959.

23

Laguerre Polynomials

23.1 Laguerre Polynomials

23.1.1 Definition

$$L_n(t) = \sum_{k=0}^{n} \frac{(-1)^k n! t^k}{(k!)^2 (n-k)!} \qquad n = 0,1,2,\cdots,\ 0 \le t < \infty$$

Figure 23.1 shows several Laguerre polynomials.

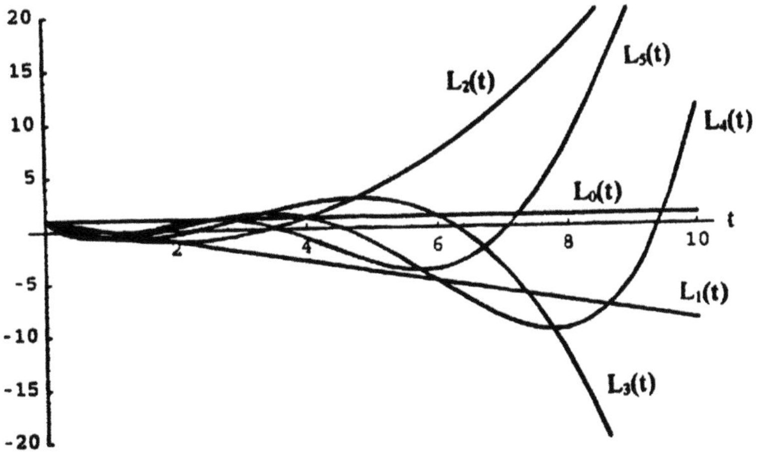

FIGURE 23.1

23.1.2 Rodrigues Formula

$$L_n(t) = \frac{e^t}{n!}\frac{d^n}{dt^n}(t^n e^{-t}) \qquad n = 0,1,2,\cdots$$

23.1.3 Laguerre Polynomials

$$L_0(t) = 1, \quad L_1(t) = -t+1, \quad L_2(t) = \tfrac{1}{2!}(t^2 - 4t + 2),$$

$$L_3(t) = \tfrac{1}{3!}(-t^3 + 9t^2 - 18t + 6), \quad L_4(t) = \tfrac{1}{4!}(t^4 - 16t^3 + 72t^2 - 96t + 24)$$

23.1.4 Generating Function

$$w(t,x) = (1-x)^{-1}\exp\left[-\frac{tx}{1-x}\right] = \sum_{n=0}^{\infty} L_n(t)x^n \qquad |x| < 1, \; 0 \le t < \infty$$

23.2 Recurrence Relations

23.2.1 Recurrence Relations

1. $L_n'(t) - L_{n-1}'(t) + L_{n-1}(t) = 0 \qquad n = 1,2,\cdots$
2. $L_{n+1}'(t) = L_n'(t) - L_n(t) \qquad n = 1,2,\cdots$
3. $L_{n-1}'(t) = L_n'(t) + L_{n-1}(t) \qquad n = 1,2,\cdots$
4. $(n+1)L_{n+1}'(t) + (t-1-2n)L_n'(t) + L_n(t) + nL_{n-1}'(t) = 0 \qquad n = 1,2,\cdots$
5. $tL_n'(t) = nL_n(t) - nL_{n-1}(t) \qquad n = 1,2,\cdots$

23.2.2 Laguerre Equation

$$tL_n''(t) + (1-t)L_n'(t) + nL_n(t) = 0$$

23.3 Laguerre Series

23.3.1 Orthogonality Relation

$$\int_0^\infty e^{-t}L_n(t)L_m(t)\,dt = 0 \qquad n \ne m$$

23.3.2 Orthonormal Functions

$$\varphi_n(t) = e^{-t/2}L_n(t) \qquad n = 0,1,2\cdots$$

23.3.3 Laguerre Series

$$f(t) = \sum_{n=0}^{\infty} C_n L_n(t) \qquad 0 \le t < \infty$$

$$C_n = \int_0^\infty e^{-t} f(t) L_n(t)\,dt \qquad n = 0,1,2\cdots$$

23.4 Associated Laguerre Polynomials (or Generalized)

23.4.1 Definitions

For a real $a > -1$ the general Laguerre polynomials are defined by the formula

$$L_n^a(t) = e^t \frac{t^{-a}}{n!} \frac{d^n}{dt^n}(e^{-t} t^{n+a}) \qquad n = 0,1,2\cdots$$

Using Leibniz's formula

$$L_n^a(t) = \sum_{k=0}^{n} \frac{\Gamma(n+a+1)}{\Gamma(k+a+1} \frac{(-t)^k}{k!(n-k)!} \qquad 0 \le t < \infty$$

For $a = 0$, $L_n^a(t)$ become $L_n(t)$

23.4.2 Polynomials

$$L_0^a(t) = 1,\ L_1^a(t) = 1 + a - t,\ L_2^a(t) = \tfrac{1}{2}[(1+a)(2+a) - 2(2+a)t + t^2],\ldots$$

23.5 Recurrence Relations

23.5.1 Recurence Relations

1. $L_n^{a\prime}(t) - L_{n-1}^{a\prime}(t) + L_{n-1}^a(t) = 0 \qquad n = 1,2,\cdots$
2. $t\,L_n^{a\prime}(t) = n\,L_n^a(t) - (n+a)\,L_{n-1}^a(t) \qquad n = 1,2,\cdots$

23.6 Laguerre Series

23.6.1 Orthogonality Relations

$$\int_0^\infty e^{-t} t^a L_m^a(t) L_n^a(t)\,dt = 0 \qquad n \ne m \quad a > -1$$

$$\int_0^\infty e^{-t} t^a [L_n^a(t)]^2\,dt = \frac{\Gamma(n+a+1)}{n!} \qquad a > -1 \quad n = 0,1,2\cdots$$

23.6.2 Orthonormal Functions

$$\varphi_n^a(t) = \left[\frac{n!}{\Gamma(n+a+1)}\right]^{1/2} e^{-t/2} t^{a/2} L_n^a(t) \qquad n = 0,1,2\cdots 0 \le t < \infty$$

23.6.3 Series

The Laguerre series is given by

$$f(t)=\sum_{n=0}^{\infty} C_n L_n^a(t) \qquad 0\le t<\infty$$

$$C_n=\frac{n!}{\Gamma(n+a+1)}\int_0^\infty e^{-t}t^a f(t)L_n^a(t)\,dt \qquad n=0,1,2\cdots$$

Example

The function t^b can be expanded in series

$$t^b=\sum_{n=0}^{\infty} C_n L_n^a(t) \qquad b>-\tfrac{1}{2}(a+1)$$

$$C_n=\frac{n!}{\Gamma(n+a+1)}\int_0^\infty t^{b+a}e^{-t}L_n^a(t)\,dt=\frac{n!}{\Gamma(n+a+1)}\int_0^\infty e^{-t}t^{b+a}\frac{e^t t^{-a}}{n!}\frac{d^n}{dt^n}(t^{n+a}e^{-t})\,dt$$

$$=\frac{1}{\Gamma(n+a+1)}\int_0^\infty t^b\frac{d^n}{dt^n}(t^{n+a}e^{-t})\,dt=\frac{(-1)^n b(b-1)\cdots(b-n+1)}{\Gamma(n+a+1)}\int_0^\infty e^{-t}t^{b+a}\,dt$$

$$=(-1)^n\frac{\Gamma(b+1)}{\Gamma(n+b+1)\Gamma(b-n+1)}\int_0^\infty e^{-t}t^{(b+a+1)-1}\,dt=(-1)^n\frac{\Gamma(b+1)\Gamma(b+a+1)}{\Gamma(n+b+1)\Gamma(b-n+1)}$$

The steps to find C_n were: a) substitution of (23.4.1), b) integration by parts n times, c) multiplication numerator and denominator by $\Gamma(b-n+1)$. In particular if $b=m=$ *positive integer*

$$t^m=\Gamma(m+a+1)m!\sum_{n=0}^{m}\frac{(-1)^n L_n^a(t)}{\Gamma(n+a+1)(m-n)!} \qquad 0\le t<\infty,\quad a>-1 \quad\text{and}\quad m=0,1,2\cdots$$

If $a=0$ we obtain the expansion

$$t^m=\Gamma(m+1)m!\sum_{n=0}^{m}\frac{(-1)^m L_n(t)}{n!(m-n)!}$$

Example

The function $f(t)=e^{-bt}$, with $b>-\frac{1}{2}$ and $t>0$, is expanded as follows

$$C_n=\frac{n!}{\Gamma(n+a+1)}\int_0^\infty e^{-(b+1)t}t^a L_n^a(t)\,dt=\frac{1}{\Gamma(n+a+1)}\int_0^\infty e^{-bt}\frac{d^n}{dt^n}(e^{-t}t^{n+a})\,dt$$

$$=\frac{b^n}{\Gamma(n+a+1)}\int_0^\infty e^{-(b+1)t}t^{n+a}\,dt=\frac{b^n}{(b+1)^{n+a+1}} \qquad n=0,1,2\cdots$$

and thus

$$e^{-bt} = (b+1)^{-a-1}\sum_{n=0}^{\infty}\left(\frac{b}{b+1}\right)^n L_n^a(t) \qquad 0 \le t < \infty$$

For $a = 0$

$$e^{-bt} = (b+1)^{-1}\sum_{n=0}^{\infty}\left(\frac{b}{b+1}\right)^n L_n(t) \qquad 0 \le t < \infty$$

23.7 Tables of Laguerre Polynomials

TABLE 23.1 Properties of the Laguerre Polynomials

1.	$L_n(t) = \sum_{k=0}^{n} \frac{(-1)^k n! t^k}{(k!)^2 (n-k)!} = \sum_{k=0}^{n} (-1)^k \frac{1}{k!}\binom{n}{k} t^k$	$n = 0,1,2,\cdots,\ 0 \le t < \infty$
2.	$L_n(t) = \frac{e^t}{n!}\frac{d^n}{dt^n}(t^n e^{-t})$	$n = 0,1,2\cdots$
3.	$L_0(t) = 1,\ L_1(t) = -t+1,\ L_2(t) = \frac{1}{2!}(t^2 - 4t + 2),$ $L_3(t) = \frac{1}{3!}(-t^3 + 9t^2 - 18t + 6),\ L_4(t) = \frac{1}{4!}(t^4 - 16t^3 + 72t^2 - 96t + 24)$ $L_n(0) = 1,\ L_n'(0) = -n,\ L_n''(0) = \frac{1}{2}n(n-1)$	
4.	$(n+1)L_{n+1}(t) + (t-1-2n)L_n(t) + nL_{n-1}(t) = 0$	$n = 1,2,3,\cdots$
5.	$L_n'(t) - L_{n-1}'(t) + L_{n-1}(t) = 0$	$n = 1,2,3,\cdots$
6.	$(n+1)L_{n+1}'(t) + (t-1-2n)L_n'(t) + L_n(t) + nL_{n-1}'(t) = 0$	$n = 1,2,3,\cdots$
7.	$L_{n+1}'(t) = L_n'(t) - L_n(t)$	
8.	$tL_n'(t) = nL_n(t) - nL_{n-1}(t)$	$n = 1,2,3,\cdots$
9.	$tL_n''(t) + (1-t)L_n'(t) + nL_n(t) = 0$, Laguerre differential equation	
10.	$w(t,x) = (1-x)^{-1}\exp\left[-\frac{tx}{1-x}\right] = \sum_{n=0}^{\infty} L_n(t)x^n$ generating function	
11.	$\int_0^{\infty} e^{-t} L_n(t) L_k(t)\,dt = 0 \qquad k \ne n$	
12.	$\int_0^{\infty} e^{-t}[L_n(t)]^2\,dt = 1$	
13.	$f(t) = \sum_{n=0}^{\infty} C_n L_n(t)$	$0 \le t < \infty$
	$C_n = \int_0^{\infty} e^{-t} f(t) L_n(t)\,dt$	$n = 0,1,2\cdots$
14.	$L_n^m(t) = (-1)^m \frac{d^m}{dt^m}[L_{n+m}(t)]$	$m = 0,1,2\cdots$
15.	$L_n^m(t) = \sum_{k=0}^{n} \frac{(-1)^k (n+m)! t^k}{(n-k)!(m+k)!k!}$	$m = 0,1,2\cdots$
16.	$(n+1)L_{n+1}^m(t) + (t-1-2n-m)L_n^m(t) + (n+m)L_{n-1}^m(t) = 0$	
17.	$tL_n^{m\prime}(t) - nL_n^m(t) + (n+m)L_{n-1}^m(t) = 0$	

TABLE 23.1 Properties of the Laguerre Polynomials (continued)

18.	$L_n^m(t) = \frac{1}{n!} e^t t^{-m} \frac{d^n}{dt^n}(e^{-t} t^{n+m})$ = Rodrigues formula
19.	$L_{n-1}^m(t) + L_n^{m-1}(t) - L_n^m(t) = 0$
20.	$L_n^{m\prime}(t) = -L_{n-1}^{m+1}(t)$
21.	$L_n^m(0) = \frac{(n+m)!}{n!\,m!}$
22.	$\int_0^\infty e^{-t} t^k L_n(t)\,dt = \begin{cases} 0 & k<n \\ (-1)^n n! & k=n \end{cases}$
23.	$\int_0^t L_k(x) L_n(t-x)\,dx = \int_0^t L_{n+k}(x)\,dx = L_{n+k}(t) - L_{n+k+1}(t)$
24.	$\int_t^\infty e^{-x} L_n^m(x)\,dx = e^{-t}[L_n^m(t) - L_{n-1}^m(t)] \qquad m = 0,1,2.\cdots$
25.	$\int_0^t (t-x)^m L_n(x)\,dx = \frac{m!\,n!}{(m+n+1)!} t^{m+1} L_n^{m+1}(t) \qquad m = 0,1,2.\cdots$
26.	$\int_0^1 x^a (1-x)^{b-1} L_n^a(tx)\,dx = \frac{\Gamma(b)\Gamma(n+a+1)}{\Gamma(n+a+b+1)} L_n^{a+b}(t) \qquad a>-1\ ,\quad b>0$
27.	$\int_0^\infty e^{-t} t^a L_n^a(t) L_k^a(t)\,dt = 0 \qquad k \neq n\ ,\ a>-1$
28.	$\int_0^\infty e^{-t} t^a [L_n^a(t)]^2\,dt = \frac{\Gamma(n+a+1)}{n!} \qquad a>-1$
29.	$\int_0^\infty e^{-t} t^{a+1} [L_n^a(t)]^2\,dt = \frac{\Gamma(n+a+1)}{n!}(2n+a+1) \qquad a>-1$
30.	$L_n^{-1/2}(t) = \frac{(-1)^n}{2^{2n} n!} H_{2n}(\sqrt{t})$
31.	$L_n^{1/2}(t) = \frac{(-1)^n}{2^{2n+1} n!} \frac{H_{2n+1}(\sqrt{t})}{\sqrt{t}}$
32.	$f(t) = \sum_{n=0}^\infty C_n L_n^m(t)$ $C_n = \frac{n!}{\Gamma(n+m+1)} \int_0^\infty e^{-t} t^m f(t) L_n^m(t)\,dt$
33.	$\varphi_n^m = \left[\frac{n!}{\Gamma(n+m+1)}\right]^{1/2} e^{-t/2} t^{m/2} L_n^m(t)$, orthonormal sequence, $n = 0,1,2\cdots$
34.	$t^p = p! \sum_{n=0}^p \binom{p}{n} (-1)^n L_n(t)$
35.	$e^{-at} = (a+1)^{-1} \sum_{n=0}^\infty \left(\frac{a}{a+1}\right)^n L_n(t) \qquad a > -\frac{1}{2}$
36.	$\int_0^\infty \frac{e^{-tx}}{x+1}\,dx = \sum_{n=0}^\infty \frac{L_n(t)}{n+1}$

References

Abramowitz, M. and I. S. Stegun, *Handbook of Mathematical Functions*, Dover Publications, Inc., New York, NY, 1965.

Andrews, L. C., *Special Functions for Engineers and Applied Mathematicians*, MacMillan Publishing Co. New York, NY. 1985.

Hochstadt, H., *The Functions of Mathematical Physics*, Dover Publications Inc., New York, NY, 1986.

McLachlan, N. W., *Bessel Functions for Engineers*, 2nd Edition, Oxford, London, 1961.

Sansone, G., *Orthogonal Functions*, Interscience Publishers, New York, NY., 1959.

24

Chebyshev Polynomials

24.1 Chebyshev Polynomials

24.1.1 Definitions

$$T_0(t) = 1, \quad T_n(t) = \frac{n}{2}\sum_{k=0}^{[n/2]} \frac{(-1)^k (n-k-1)!}{k!(n-2k)!}(2t)^{n-2k} \qquad -1 < t < 1$$

$$U_n(t) = \sum_{k=0}^{[n/2]} \binom{n-k}{k}(-1)^k (2t)^{n-2k}, \quad \text{second kind}$$

$[n/2] = n/2$ for n even and $[n/2] = (n-1)/2$ for n odd.
Figure 24.1 shows several polynomials.

24.2 Recurrence Relations

24.2.1 Recurrence

$$T_{n+1}(t) - 2tT_n(t) + T_{n-1}(t) = 0$$

$$U_{n+1}(t) - 2tU_n(t) + U_{n-1}(t) = 0$$

24.3 Orthogonality Relations

24.3.1 Relations

$$\int_{-1}^{1} (1-t^2)^{-1/2} T_n(t) T_k(t) dt = 0 \qquad k \neq n$$

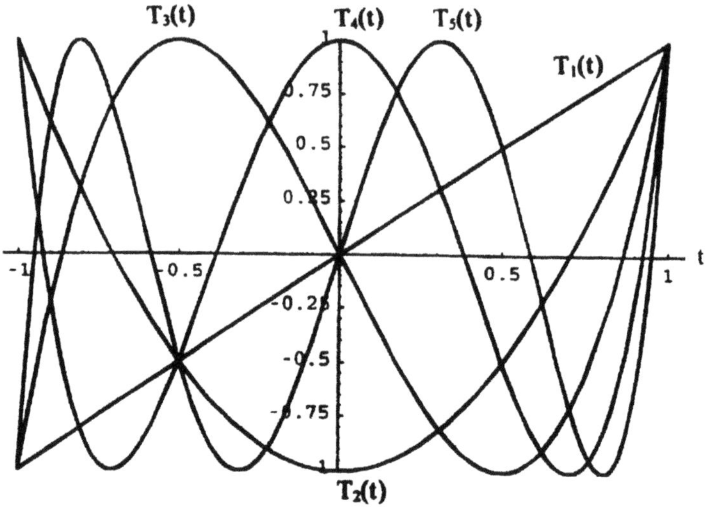

FIGURE 24.1

$$\int_{-1}^{1} (1-t^2)^{-1/2} U_n(t) U_k(t)\, dt = 0 \qquad k \neq n$$

24.4 Differential Equations

24.4.1 For $T_n(t)$: $(1-t^2)y'' - ty' + n^2 y = 0$. For $U_n(t)$: $(1-t^2)y'' - 3ty' + n(n+2)y = 0$.

24.5 Generating Function

$$\frac{1-st}{1-2st+s^2} = \sum_{n=0}^{\infty} T_n(t) s^n$$

24.6 Rodrigues Formula

$$T_n(t) = \frac{(-2)^n n!}{(2n)!} \sqrt{1-t^2} \frac{d^n}{dt^n} (1-t^2)^{n-\frac{1}{2}}$$

24.7 Table of Chebyshev Properties

TABLE 24.1 Properties of the Chebyshev Polynomials

1. $(1-t^2)\dfrac{d^2y}{dt^2} - t\dfrac{dy}{dt} + n^2 y = 0 \quad ; \quad y(t) = T_n(t)$

2. $T_n(t) = \dfrac{n}{2} \sum_{k=0}^{[n/2]} \dfrac{(-1)^k (n-k-1)!}{k!(n-2k)!} (2t)^{n-2k},$ $\quad n = 1, 2\cdots,\ [n/2] = \text{largest integer} \leq n/2$

TABLE 24.1 Properties of the Chebyshev Polynomials (continued)

3. $T_n(t) = \dfrac{(-2)^n n!}{(2n)!}\sqrt{1-t^2}\,\dfrac{d^n}{dt^n}(1-t^2)^{n-\frac{1}{2}}$, Rodrigues formula

4. $T_n(t) = \cos(n\cos^{-1} t)$

5. $\dfrac{1-st}{1-2st+s^2} = \sum_{n=0}^{\infty} T_n(t)s^n$, generating formula

6. $T_{n+1}(t) = 2t\,T_n(t) - T_{n-1}(t)$

7. $\displaystyle\int_{-1}^{1} \frac{T_n(t)T_m(t)}{\sqrt{(1-t^2)}}dt = \begin{cases} 0 & n \neq m \\ \pi/2 & n = m \neq 0 \\ \pi & n = m = 0 \end{cases}$

8. $T_n(1) = 1,\ \ T_n(-1) = (-1)^n,\ \ T_{2n}(0) = (-1)^n,\ \ T_{2n+1}(0) = 0$

References

Abramowitz, M., and I. S. Stegun, *Handbook of Mathematical Functions*, Dover Publications, Inc., New York, NY, 1965.

Andrews, L. C., *Special Functions for Engineers and Applied Mathematicians*, MacMillan Publishing Co. New York, NY. 1985.

Hochstadt, H., *The Functions of Mathematical Physics*, Dover Publications Inc., New York, NY, 1986.

McLachlan, N. W., *Bessel Functions for Engineers*, 2nd Edition, Oxford University Press, London, 1961.

Sansone, G., *Orthogonal Functions*, Interscience Publishers, New York, NY., 1959.

25

Bessel Functions

25.1 Bessel Functions of the First Kind

25.1.1 Definition of Integer Order

$$J_n(t)=\sum_{k=0}^{\infty}\frac{(-1)^k(t/2)^{n+2k}}{k!(n+k)!},\qquad -\infty<t<\infty,\ n=0,1,2,\ldots$$

$$J_{-n}(t)=\sum_{k=0}^{\infty}\frac{(-1)^k(t/2)^{2k-n}}{k!(k-n)!}=\sum_{k=n}^{\infty}\frac{(-1)^k(t/2)^{2k-n}}{k!(k-n)!}$$

$$=\sum_{m=0}^{\infty}\frac{(-1)^{m+n}(t/2)^{2m+n}}{m!(m+n)!}$$

$$J_{-n}(t)=(-1)^n J_n(t)$$

$$J_0(0)=1,\quad J_n(0)=0\qquad n\neq 0$$

Figure 25.1 shows several Bessel functions of the first kind.

25.1.2 Definition of Nonintegral Order

$$J_v(t)=\sum_{k=0}^{\infty}\frac{(-1)^k(t/2)^{2k+v}}{k!\Gamma(k+v+1)}\qquad v\geq 0$$

$$J_{-v}(t)=\sum_{k=0}^{\infty}\frac{(-1)^k(t/2)^{2k-v}}{k!\Gamma(k-v+1)}\qquad v\geq 0$$

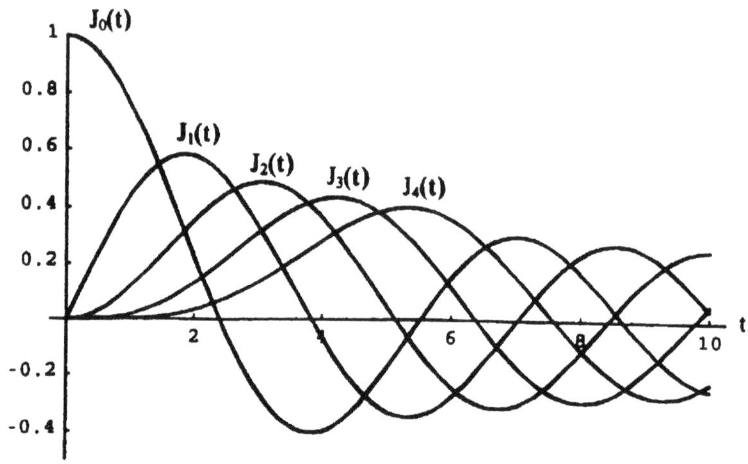

FIGURE 25.1

The two functions $J_{-v}(t)$ and $J_v(t)$ are linear independent for noninteger values of v and they do not satisfy any generating-function relation. The functions $J_{-v}(0) = \infty$ and $J_v(0)$ remain finite. Both these functions share most of the properties of $J_n(t)$ and $J_{-n}(t)$.

25.1.3 Generating Function

$$w(t,x) \doteq e^{\frac{1}{2}t(x-\frac{1}{x})} = \sum_{n=-\infty}^{\infty} J_n(t)x^n \qquad x \neq 0$$

25.1.4 Differential Equation

$$y'' + \frac{1}{t}y' + \left(1 - \frac{n^2}{t^2}\right)y = 0 \qquad n = 0,1,2,\cdots$$

has solution the function $y = J_n(t)$

25.2 Recurrence Relation

25.2.1 Recurrence Relations

1. $$\frac{d}{dt}[t^v J_v(t)] = \frac{d}{dt}\sum_{k=0}^{\infty} \frac{(-1)^k (t)^{2k+2v}}{2^{2k+v} k!\Gamma(k+v+1)} = t^v \sum_{k=0}^{\infty} \frac{(-1)^k (t/2)^{2k+(v-1)}}{k!\Gamma(k+v)}$$
$$= t^v J_{v-1}(t)$$

Similarly,

2. $$\frac{d}{dt}[t^{-v} J_v(t)] = -t^{-v} J_{v+1}(t)$$

Differentiate (1) and (2) and divide by t^v and t^{-v}, respectively, we find

3. $J_\nu'(t) + \frac{\nu}{t} J_\nu(t) = J_{\nu-1}(t)$

4. $J_\nu'(t) - \frac{\nu}{t} J_\nu(t) = -J_{\nu+1}(t)$

Set $\nu = 0$ in (4) to obtain

5. $J_0'(t) = -J_1(t)$

Add and subtract (3) and (4) to find, respectively, the relations

6. $2J_\nu'(t) = J_{\nu-1}(t) - J_{\nu+1}(t)$

7. $\frac{2\nu}{t} J_\nu(t) = J_{\nu-1}(t) + J_{\nu+1}(t)$

The above relation is known as the *three-term recurrence formula*. Repeated operations result in

8. $\left(\frac{d}{t\,dt}\right)^m [t^\nu J_\nu(t)] = t^{\nu-m} J_{\nu-m}(t); \quad \left[\left(\frac{d}{t\,dt}\right)^2 y = \frac{1}{t}\frac{d}{dt}\left(\frac{1}{t}\frac{dy}{dt}\right)\right]$

9. $\left(\frac{d}{t\,dt}\right)^m [t^{-\nu} J_\nu(t)] = (-1)^m t^{-\nu-m} J_{\nu+m}(t) \qquad m = 1,2,\ldots$

Example

We proceed to find the following derivative

$$\frac{d}{dt}[t^\nu J_\nu(at)] = \frac{d}{dt}\left[\left(\frac{u}{a}\right)^\nu J_\nu(u)\right] = \frac{d}{du}\left[\frac{u^\nu}{a^\nu} J_\nu(u)\right]\frac{du}{dt}$$

$$= a^{-\nu}\frac{d}{du}[u^\nu J_\nu(u)]a = a^{1-\nu}[u^\nu J_{\nu-1}(u)]$$

$$= a^{1-\nu}[(at)^\nu J_{\nu-1}(at)] = at^\nu J_{\nu-1}(at)$$

where (1) was used.

25.3 Integral Representation

25.3.1 Integral Representation

Set $x = \exp(-j\varphi)$ in (25.1.3), multiply both sides by $\exp(jn\varphi)$ and integrate the results from 0 to π. Hence

1. $\int_0^\pi e^{j(n\varphi - t\sin\varphi)}\, d\varphi = \sum_{k=-\infty}^{\infty} J_k(t) \int_0^\pi e^{j(n-k)\varphi}\, d\varphi$

Expand on both sides the exponentials in Euler's formula; equate the real and imaginary parts and use the relation

2. $\int_0^\pi \cos(n-k)\varphi\, d\varphi = \begin{cases} 0 & k \neq 0 \\ \pi & k = n \end{cases}$

to find that all terms of the infinite sum vanish except for $k = n$. Hence we obtain

3. $J_n(t) = \frac{1}{\pi}\int_0^{\pi} \cos(n\varphi - t\sin\varphi)\,d\varphi \qquad n = 0,1,2,\ldots$

when $n = 0$, we find

4. $J_0(t) = \frac{1}{\pi}\int_0^{\pi} \cos(t\sin\varphi)\,d\varphi$

For Bessel function with nonintegral order, the Poisson formula is

5. $J_v(t) = \frac{(t/2)^v}{\sqrt{\pi}\,\Gamma(v+\frac{1}{2})}\int_{-1}^{1} (1-x^2)^{v-\frac{1}{2}} e^{jtx}\,dx \qquad v > -\frac{1}{2},\ t > 0$

Set $x = \cos\theta$ to obtain

6. $J_v(t) = \frac{(t/2)^v}{\sqrt{\pi}\,\Gamma(v+\frac{1}{2})}\int_{0}^{\pi} \cos(t\cos\theta)\sin^{2v}\theta\,d\theta \qquad v > -\frac{1}{2},\ t > 0$

25.3.2 Integrals Involving Bessel Functions

Directly integrate (25.2.1.1) and (25.2.1.2) to find

$$\int t^{v} J_{v-1}(t)\,dt = t^{v} J_{v}(t) + C$$

$$\int t^{-v} J_{v+1}(t)\,dt = -t^{-v} J_{v}(t) + C$$

where C is the constant of integration.

Example

We apply the integration procedure to find

$$\int t^2 J_2(t)\,dt = \int t^3[t^{-1}J_2(t)]\,dt = -\int t^3 \frac{d}{dt}[t^{-1}J_1(t)]\,dt$$

$$= -t^2 J_1(t) + 3\int t J_1(t)\,dt = -t^2 J_1(t) - 3\int t[-J_1(t)]\,dt$$

$$= -t^2 J_1(t) - 3\int t\left[\frac{d}{dt}J_0(t)\right]dt = -t^2 J_1(t) - 3tJ_0(t) + 3\int J_0(t)\,dt$$

The last integral has no closed solution.

Example

If $a > 0$ and $b > 0$, then (see 25.3.1.4)

$$\int_0^{\infty} e^{-at} J_0(bt)\,dt = \int_0^{\infty} e^{-at}\,dt \frac{2}{\pi}\int_0^{\pi/2} \cos(bt\sin\varphi)\,d\varphi$$

$$= \frac{2}{\pi}\int_0^{\pi/2} d\varphi \int_0^{\infty} e^{-at}\cos(bt\sin\varphi)\,dt = \frac{2}{\pi}\int_0^{\pi/2} \frac{a\,d\varphi}{a^2 + b^2\sin^2\varphi} = \frac{1}{\sqrt{a^2+b^2}}$$

The usual method to find definite integrals involving Bessel functions is to replace the Bessel function by its series representation.

Example

$$I=\int_0^\infty e^{-at}t^pJ_p(bt)dt \qquad p>-\frac{1}{2},\ a>0,\ b>0$$

$$=\sum_{k=0}^\infty \frac{(-1)^k(b/2)^{2k+p}}{k!\Gamma(k+p+1)}\int_0^\infty e^{-at}t^{2k+2p}\,dt$$

$$=b^p\sum_{k=0}^\infty \frac{(-1)^k\Gamma(2k+2p+1)}{2^{2k+p}k!\Gamma(k+p+1)}(a^2)^{-(p+\frac{1}{2})-k}(b^2)^k$$

where the last integral is in the form of a gamma function. But we know that

$$\binom{-r}{k}=(-1)^k\binom{r+k-1}{k},\quad \binom{n}{k}=\binom{n}{n-k}$$

$$\binom{n+1}{k+1}=\binom{n}{k+1}+\binom{n}{k} \qquad 0\le k\le n-1$$

and thus we obtain

$$\frac{(-1)^k\Gamma(2k+2p+1)}{2^{2k+p}k!\Gamma(k+p+1)}=\frac{(-1)^k2^p\Gamma\left(p+k+\frac{1}{2}\right)}{\sqrt{\pi}k!}$$

$$=\frac{(-1)^k}{\sqrt{\pi}}2^p\Gamma\left(p+\frac{1}{2}\right)\binom{p+k-\frac{1}{2}}{k}=\frac{2^p\Gamma\left(p+\frac{1}{2}\right)}{\sqrt{\pi}}\binom{-\left(p+\frac{1}{2}\right)}{k}$$

Therefore, we obtain

$$I=\int_0^\infty e^{-at}t^pJ_p(bt)dt=\frac{(2b)^p\Gamma\left(p+\frac{1}{2}\right)}{\sqrt{\pi}}\sum_{k=0}^\infty\binom{-\left(p+\frac{1}{2}\right)}{k}(a^2)^{-(p+(1/2))-k}(b^2)^k$$

$$=\frac{(2b)^p\Gamma\left(p+\frac{1}{2}\right)}{\sqrt{\pi}(a^2+b^2)^{p+\frac{1}{2}}} \qquad p>-\frac{1}{2},\ a>0,\ b>0$$

Setting $p=0$ in this equation we find

$$\int_0^\infty e^{-at}J_0(bt)dt=\frac{1}{[a^2+b^2]^{1/2}} \qquad a>0,\quad b>0$$

Set $a = 0+$ in this equation to obtain

$$\int_0^\infty J_0(bt)\,dt = \frac{1}{b} \qquad b > 0$$

By assuming the real part to approach zero and writing a as pure imaginary, the equation before the previous one becomes

$$\int_0^\infty e^{-jat} J_0(bt)\,dt = \begin{cases} \dfrac{1}{(b^2 - a^2)^{1/2}} & b > a \\ \dfrac{-j}{(a^2 - b^2)^{1/2}} & b < a \end{cases}$$

The above integral, by equating real and imaginary parts, becomes

$$\int_0^\infty \cos(at) J_0(bt)\,dt = \frac{1}{(b^2 - a^2)^{1/2}} \qquad b > a$$

$$\int_0^\infty \sin(at) J_0(bt)\,dt = \frac{1}{(a^2 - b^2)^{1/2}} \qquad b < a$$

Example

To evaluate the integral $\int_0^b t J_0(at)\,dt$, we proceed as follows:

$$\int_0^\infty t J_0(at)\,dt = \int_0^\infty \frac{1}{a}\frac{d}{dt}[t J_1(at)]\,dt$$

$$= \frac{1}{a}[t J_1(at)]\Big|_{t=0}^{b} = \frac{b}{a} J_1(ab) \qquad a \neq 0$$

where (25.2.1.1) with $\nu = 1$ was used.

25.4 Fourier-Bessel Series

25.4.1 Fourier-Bessel Series

$f(t) = \sum_{n=1}^{\infty} c_n J_\nu(t_n t) \quad 0 < t < a, \quad \nu > -\frac{1}{2}$, and t_n $(n = 1, 2, \cdots)$ are solutions of $J_\nu(t_n t) = 0 \quad n = 1, 2, \cdots$.

25.4.2 Product Property

$$\int_0^a t J_\nu(\alpha t) J_\nu(\beta t)\,dt = \frac{a\beta J_\nu(\alpha a) J_\nu'(\beta a) - a\alpha J_\nu(\beta a) J_\nu'(a\alpha)}{\alpha^2 - \beta^2}$$

25.4.3 Orthogonality

Setting $\alpha = t_n$, $\beta = t_m$ in (25.4.2), we obtain the orthogonality relation $\int_0^a t J_\nu(t_m t) J_\nu(t_n t)\,dt = 0$ since t_n and t_m are the roots of $J_\nu(t_n a)$ and $J_\nu(t_m a)$.

25.4.4 $t_m = t_n$

$$\int_0^a t[J_\nu(t_n t)]^2\,dt = \frac{a^2}{2}[J_{\nu+1}(t_n a)]^2,$$

which is found from (25.4.2) by limiting process $t_m \to t_n$ (using L' Hopital's rule and treating t_m as the variable).

25.4.5 Fourier Bessel Constants

Multiply (25.4.1) by $t J_\nu(t_m t)$, integrate from 0 to a, and use (25.4.4) to find

$$c_n = \frac{2}{a^2[J_{\nu+1}(t_n a)]^2} \int_0^a t\,f(t) J_\nu(t_n t)\,dt, \quad n = 1,2,\cdots.$$

Note: $f(t)$ must be piecewise continuous in the interval $(0,a)$ and $\int_0^a \sqrt{t}\,|f(t)|\,dt < \infty$.

Example

Find the Fourier-Bessel series for the function

$$f(t) = \begin{cases} t & 0 < t < 1 \\ 0 & 1 < t < 2 \end{cases}$$

corresponding to the set of functions $\{J_1(t_n t)\}$ where t_n satisfies $J_1(2t_n) = 0$ $(n = 1,2,3,\cdots)$.

Solution

We write the solution

$$f(t) = \sum_{n=1}^{\infty} c_n J_1(t_n t) \qquad 0 < t < 2$$

where

$$c_n = \frac{1}{2[J_2(2t_n)]^2} \int_0^2 t f(t) J_1(t_n t)\,dt$$

$$= (\cdot) \int_0^1 t^2 J_1(t_n t)\,dt \qquad (\text{let } r = t_n t)$$

$$= (\cdot) \frac{1}{t_n^3} \int_0^{t_n} r^2 J_1(r)\,dr \qquad (\text{apply (25.2.1.1)})$$

$$= (\cdot) \frac{1}{t_n^3} \int_0^{t_n} \frac{d}{dr}[r^2 J_2(r)]\,dr$$

$$= \frac{1}{2[J_2(2t_n)]^2 t_n^3} t_n^2 J_2(t_n) = \frac{J_2(t_n)}{2t_n[J_2(2t_n)]^2} \qquad n = 1,2,3,\ldots$$

Example

To express the function $f(t) = 1$ on the open interval $0 < t < a$ as an infinite series of Bessel functions of zero order, we proceed as follows (see 25.4.5):

$$c_n = \frac{2}{a^2[J_1(t_n a)]^2}\int_0^a t\cdot 1\cdot J_0(t_n t)\,dt$$

$$= \frac{2}{a^2[J_1(t_n a)]^2}\int_0^a \frac{1}{t_n}\frac{d}{dt}[tJ_1(t_n t)]\,dt \qquad \text{(see (25.2.1.1))}$$

$$= \frac{2}{a^2 t_n[J_1(t_n a)]^2}[tJ_1(t_n t)]\Big|_{t=0}^{a} = \frac{2}{at_n J_1(t_n a)}$$

Hence, the expression is

$$1 = 2\sum_{n=1}^{\infty}\frac{J_n(t_n t)}{t_n J_1(t_n a)} \qquad 0 < t < a$$

25.5 Properties of Bessel Function

TABLE 25.1 Properties of Bessel Functions

1.	$J_n(t) = \sum_{k=0}^{\infty}\frac{(-1)^k(t/2)^{n+2k}}{k!(n+k)!}$	$-\infty < t < \infty, \quad n = 0,1,2,3\cdots$
2.	$J_{-n}(t) = \sum_{m=0}^{\infty}\frac{(-1)^{m+n}(t/2)^{2m+n}}{m!(m+n)!}$	$-\infty < t < \infty, \quad n = 0,1,2,3,\cdots$
3.	$J_{-n}(t) = (-1)^n J_n(t), \quad J_n(-t) = (-1)^n J_n(t) = J_{-n}(t)$	$n = 0,1,2,3,\cdots$
4.	$J_0(0) = 1, \quad J_n(0) = 0$	$n \neq 0$
5.	$J_\nu(t) = \sum_{k=0}^{\infty}\frac{(-1)^k(t/2)^{2k+\nu}}{k!\Gamma(k+\nu+1)}$	$\nu \geq 0, \quad \nu = \text{noninteger}$
6.	$J_{-\nu}(t) = \sum_{k=0}^{\infty}\frac{(-1)^k(t/2)^{2k-\nu}}{k!\Gamma(k-\nu+1)}$	$\nu \geq 0, \quad \nu = \text{noninteger}$
7.	$\frac{d}{dt}[t^\nu J_\nu(t)] = t^\nu J_{\nu-1}(t)$	
8.	$\frac{d}{dt}[t^\nu J_\nu(at)] = at^\nu J_{\nu-1}(at)$	
9.	$\frac{d}{dt}[t^{-\nu} J_\nu(t) = -t^{-\nu} J_{\nu+1}(t)$	
10.	$\frac{d^2 J_\nu(t)}{dt^2} = \frac{1}{2^2}[J_{\nu-2}(t) - 2J_\nu(t) + J_{\nu+2}(t)]$	
11.	$J'_\nu(t) + \frac{\nu}{t}J_\nu(t) = J_{\nu-1}(t)$	

TABLE 25.1 Properties of Bessel Functions (continued)

12.	$J'_\nu(t) - \frac{\nu}{t} J_\nu(t) = -J_{\nu+1}(t), \quad J'_0(t) = -J_1(t)$
13.	$tJ'_\nu(t) = -\nu J_\nu(t) + tJ_{\nu-1}(t)$
14.	$2J'_\nu(t) = J_{\nu-1}(t) - J_{\nu+1}(t)$
15.	$\frac{2\nu}{t} J_\nu(t) = J_{\nu-1}(t) + J_{\nu+1}(t)$
16.	$\left(\frac{d}{t\,dt}\right)^m [t^\nu J_\nu(t)] = t^{\nu-m} J_{\nu-m}(t) \qquad m = 1,2,3,\cdots$
17.	$\left(\frac{d}{tdt}\right)^m [t^{-\nu} J_\nu(t)] = (-1)^m t^{-\nu-m} J_{\nu+m}(t) \qquad m = 1,2,3,\cdots$
18.	$J'_1(0) = \frac{1}{2}, \qquad J'_n(0) = 0 \qquad n > 1$
19.	$J_n(t+r) = \sum_{k=-\infty}^{\infty} J_k(t) J_{n-k}(r)$
20.	$J_0(2t) = [J_0(t)]^2 + 2\sum_{k=1}^{\infty} (-1)^k [J_k(t)]^2$
21.	$\lvert J_0(t)\rvert \le 1, \qquad \lvert J_n(t)\rvert \le \frac{1}{\sqrt{2}} \qquad n = 1,2,3,\cdots$
22.	$e^{jt\sin\theta} = \sum_{n=-\infty}^{\infty} J_n(t) e^{jn\theta}$
23.	$\cos(t\sin\theta) = J_0(t) + 2\sum_{n=1}^{\infty} J_{2n}(t)\cos(2n\theta)$
24.	$\cos(t\cos\theta) = J_0(t) + 2\sum_{n=1}^{\infty} (-1)^n J_{2n}(t)\cos(2n\theta)$
25.	$\sin(t\sin\theta) = 2\sum_{n=1}^{\infty} J_{2n-1}(t)\sin[(2n-1)\theta]$
26.	$\sin(t\cos\theta) = 2\sum_{n=0}^{\infty} (-1)^n J_{2n+1}(t)\cos[(2n+1)\theta]$
27.	$\cos t = J_0(t) + 2\sum_{n=1}^{\infty} (-1)^n J_{2n}(t)$
28.	$\sin t = 2\sum_{n=1}^{\infty} (-1)^n J_{2n-1}(t)$
29.	$J_\nu(t)J_{1-\nu}(t) + J_{-\nu}(t)J_{\nu-1}(t) = \frac{2\sin\nu\pi}{\pi t}$ Lommel's formula
30.	$\frac{d}{dt}[tJ_\nu(t)J_{\nu+1}(t)] = t[[J_\nu(t)]^2 - [J_{\nu+1}(t)]^2]$
31.	$\frac{d}{dt}[t^2 J_{\nu-1}(t)J_{\nu+1}(t)] = 2t^2 J_\nu(t) J'_\nu(t)$
32.	$J_{1/2}(t) = \sqrt{\frac{2}{\pi t}}\sin t, \qquad J_{-1/2}(t) = \sqrt{\frac{2}{\pi t}}\cos t$
33.	$J_{1/2}(t)J_{-1/2}(t) = \frac{\sin 2t}{\pi t}, \qquad [J_{1/2}(t)]^2 + [J_{-1/2}(t)]^2 = \frac{2}{\pi t}$

TABLE 25.1 Properties of Bessel Functions (continued)

34. $[J_0(t)]^2 = \sum_{n=0}^{\infty} \frac{(-1)^n (2n)!}{(n!)^4} \left(\frac{t}{2}\right)^{2n}$

35. $J_n(t) = \frac{1}{\pi} \int_0^{\pi} \cos(n\varphi - t\sin\varphi)\, d\varphi$

36. $J_0(t) = \frac{1}{\pi} \int_0^{\pi} \cos(t\sin\varphi)\, d\varphi$

37. $J_\nu(t) = \frac{(t/2)^\nu}{\sqrt{\pi}\,\Gamma\left(\nu + \frac{1}{2}\right)} \int_{-1}^{1} (1-x^2)^{\nu - \frac{1}{2}} e^{jtx}\, dx, \quad \nu > -\frac{1}{2} \quad t > 0$

38. $J_\nu(t) = \frac{(t/2)^\nu}{\sqrt{\pi}\,\Gamma\left(\nu + \frac{1}{2}\right)} \int_0^{\pi} \cos(t\cos\theta) \sin^{2\nu}\theta\, d\theta \quad \nu > -\frac{1}{2} \quad t > 0$

39. $\int t^\nu J_{\nu-1}(t)\, dt = t^\nu J_\nu(t) + C \quad C = \text{constant}$

40. $\int t^{-\nu} J_{\nu+1}(t)\, dt = -t^{-\nu} J_\nu(t) + C \quad C = \text{constant}$

41. $[1 + (-1)^n] J_n(t) = \frac{2}{\pi} \int_0^{\pi} \cos n\varphi \cos(t\sin\varphi)\, d\varphi \quad n = 0, 1, 2, \cdots$

42. $J_{2k}(t) = \frac{1}{\pi} \int_0^{\pi} \cos 2k\varphi \cos(t\sin\varphi)\, d\varphi \quad k = 0, 1, 2, \cdots$

43. $J_{2k+1}(t) = \frac{1}{\pi} \int_0^{\infty} \sin[(2k+1)\varphi] \sin(t\sin\varphi)\, d\varphi \quad k = 0, 1, 2, \cdots$

44. $\int_0^{\pi} \cos[(2k+1)\varphi] \cos(t\sin\varphi)\, d\varphi = 0 \quad k = 0, 1, 2, \cdots$

45. $\int_0^{\pi} \sin 2k\varphi \sin(t\sin\varphi)\, d\varphi = 0 \quad k = 0, 1, 2, \cdots$

46. $J_0(t) = \frac{2}{\pi} \int_0^{1} \frac{\cos tx}{\sqrt{1-x^2}}\, dx$

47. $\frac{2\sin t}{t} = \sqrt{\frac{2\pi}{t}}\, J_{1/2}(t)$

48. $\int t J_0(t)\, dt = t J_1(t) + C$

49. $\int t^2 J_0(t)\, dt = t^2 J_1(t) + t J_0(t) - \int J_0(t)\, dt + C$

50. $\int t^3 J_0(t)\, dt = (t^3 - 4t) J_1(t) + 2t^2 J_0(t) + C$

51. $\int J_1(t)\, dt = -J_0(t) + C$

52. $\int t J_1(t)\, dt = -t J_0(t) + \int J_0(t)\, dt + C$

TABLE 25.1 Properties of Bessel Functions (continued)

53. $\int t^2 J_1(t)\,dt = 2t J_1(t) - t^2 J_0(t) + C$

54. $\int t^3 J_1(t)\,dt = 3t^2 J_1(t) - (t^3 - 3t) J_0(t) - 3\int J_0(t)\,dt + C$

55. $\int J_3(t)\,dt = -J_2(t) - 2t^{-1} J_1(t) + C$

56. $\int t^{-1} J_1(t)\,dt = -J_1(t) + \int J_0(t)\,dt + C$

57. $\int t^{-2} J_2(t)\,dt = -\frac{2}{3t^2} J_1(t) - \frac{1}{3} J_1(t) + \frac{1}{3t} J_0(t) + \frac{1}{3}\int J_0(t)\,dt + C$

58. $\int J_0(t)\cos t\,dt = t J_0(t)\cos t + t J_1(t)\sin t + C$

59. $\int J_0(t)\sin t\,dt = t J_0(t)\sin t - t J_1(t)\cos t + C$

60. $\int_0^\infty e^{-at} t^\nu J_\nu(bt)\,dt = \frac{(2b)^\nu \Gamma\left(\nu + \frac{1}{2}\right)}{\sqrt{\pi}(a^2+b^2)^{\nu+\frac{1}{2}}}$ $\quad \nu > -\frac{1}{2},\ a>0,\ b>0$

61. $\int_0^\infty e^{-at} J_0(bt)\,dt = \frac{1}{(a^2+b^2)^{1/2}}$ $\quad a>0,\ b>0$

62. $\int_0^\infty J_0(bt)\,dt = \frac{1}{b}$ $\quad b>0$

63. $\int_0^\infty J_{n+1}(t) = \int_0^\infty J_{n-1}(t)\,dt$ $\quad n = 1, 2, \cdots$

64. $\int_0^\infty J_n(at)\,dt = \frac{1}{a}$ $\quad a>0$

65. $\int_0^\infty t^{-1} J_n(t)\,dt = \frac{1}{n}$ $\quad n = 1, 2, \cdots$

66. $\int_0^\infty e^{-at} t^{\nu+1} J_\nu(bt)\,dt = \frac{2^{\nu+1}\Gamma\left(\nu+\frac{3}{2}\right)}{\sqrt{\pi}} \frac{ab^\nu}{(a^2+b^2)^{\nu+\frac{3}{2}}}$ $\quad \nu > -1,\ a>0,\ b>0$

67. $\int_0^\infty t^2 e^{-at} J_0(bt)\,dt = \frac{2a^2 - b^2}{(a^2+b^2)^{5/2}}$ $\quad a>0,\ b>0$

68. $\int_0^\infty e^{-at^2} t^{\nu+1} J_p(bt)\,dt = \frac{b^\nu e^{-b^2/4a}}{(2a)^{\nu+1}}$ $\quad \nu > -1,\ a>0,\ b>0$

69. $\int_0^\infty e^{-at^2} t^{\nu+3} J_\nu(bt)\,dt = \frac{b^\nu}{2^{\nu+1} a^{\nu+2}}\left(\nu + 1 - \frac{b^2}{4a}\right) e^{-b^2/4a}$ $\quad \nu > -1,\ a>0,\ b>0$

70. $\int_0^\infty t^{-1} \sin t\, J_0(bt)\,dt = \arcsin\left(\frac{1}{b}\right)$ $\quad b>1$

TABLE 25.1 Properties of Bessel Functions (continued)

71. $\int_0^{\pi/2} J_0(t\cos\varphi)\cos\varphi\, d\varphi = \frac{\sin t}{t}$

72. $\int_0^{\pi/2} J_1(t\cos\varphi)\, d\varphi = \frac{1-\cos t}{t}$

73. $\int_0^{\infty} e^{-t\cos\varphi} J_0(t\sin\varphi)t^n\, dt = n!\, P_n(\cos\varphi) \quad 0 \le \varphi < \pi$

$P_n(t) = nth$ Legendre polynomial

74. $\int_0^{\infty} t(t^2+a^2)^{-1/2} J_0(bt)\, dt = \frac{e^{-ab}}{b} \quad a \ge 0,\ b > 0$

75. $\int_0^{\infty} \frac{J_\nu(t)}{t^m} dt = \frac{\Gamma((\nu+1-m)/2)}{2^m \Gamma((\nu+1+m)/2)} \quad m > \frac{1}{2},\ \nu - m > -1$

76. $\frac{1}{8}(1-t^2) = \sum_{n=1}^{\infty} \frac{J_0(k_n t)}{k_n^3 J_1(k_n)} \quad 0 \le t \le 1,\ J_0(k_n) = 0 \quad n = 1,2,\cdots$

77. $t^\nu = 2\sum_{n=1}^{\infty} \frac{J_\nu(k_n t)}{k_n J_{\nu+1}(k_n)} \quad 0 < t < 1,\ J_\nu(k_n) = 0,\ n = 1,2,\cdots$

78. $t^{\nu+1} = 2^2(\nu+1)\sum_{n=1}^{\infty} \frac{J_{\nu+1}(k_n t)}{k_n^2 J_{\nu+1}(k_n)} \quad 0 < t < 1,\ \nu > -1/2,\ J_\nu(k_n) = 0,\ n = 1,2,\cdots$

TABLE 25.2

					$J_0(x)$					
x	0	.1	.2	.3	.4	.5	.6	.7	.8	.9
0	1.0000	.9975	.9900	.9776	.9604	.9385	.9120	.8812	.8463	.8075
1	.7652	.7196	.6711	.6201	.5669	.5118	.4554	.3980	.3400	.2818
2	.2239	.1666	.1104	.0555	.0025	−.0484	−.0968	−.1424	−.1850	−.2243
3	−.2601	−.2921	−.3202	−.3443	−.3643	−.3801	−.3918	−.3992	−.4026	−.4018
4	−.3971	−.3887	−.3766	−.3610	−.3423	−.3205	−.2961	−.2693	−.2404	−.2097
5	−.1776	−.1443	−.1103	−.0758	−.0412	−.0068	.0270	.0599	.0917	.1220
6	.1506	.1773	.2017	.2238	.2433	.2601	.2740	.2851	.2931	.2981
7	.3001	.2991	.2951	.2882	.2786	.2663	.2516	.2346	.2154	.1944
8	.1717	.1475	.1222	.0960	.0692	.0419	.0146	−.0125	−.0392	−.0653
9	−.0903	−.1142	−.1376	−.1577	−.1768	−.1939	−.2090	−.2218	−.2323	−.2403
10	−.2459	−.2490	−.2496	−.2477	−.2434	−.2366	−.2276	−.2164	−.2032	−.1881
11	−.1712	−.1528	−.1330	−.1121	−.0902	−.0677	−.0446	−.0213	.0020	.0250
12	.0477	.0697	.0908	.1108	.1296	.1469	.1626	.1766	.1887	.1988
13	.2069	.2129	.2167	.2183	.2177	.2150	.2101	.2032	.1943	.1836
14	.1711	.1570	.1414	.1245	.1065	.0875	.0679	.0476	.0271	.0064
15	−.0142	−.0346	−.0544	−.0736	−.0919	−.1092	−.1253	−.1401	−.1533	−.1650

When $x > 15.9$,

$$J_0(x) \cong \sqrt{\left(\frac{2}{\pi x}\right)}\left\{\sin\left(x+\frac{1}{4}\pi\right)+\frac{1}{8x}\sin\left(x-\frac{1}{4}\pi\right)\right\}$$

$$\cong \frac{.7979}{\sqrt{x}}\left\{\sin(57.296x+45°)+\frac{1}{8x}\sin(57.296x-45°)\right\}$$

					$J_1(x)$					
x	0	.1	.2	.3	.4	.5	.6	.7	.8	.9
0	.0000	.0499	.0995	.1483	.1960	.2423	.2867	.3290	.3688	.4059
1	.4401	.4709	.4983	.5220	.5419	.5579	.5699	.5778	.5815	.5812
2	.5767	.5683	.5560	.5399	.5202	.4971	.4708	.4416	.4097	.3754
3	.3391	.3009	.2613	.2207	.1792	.1374	.0955	.0538	.0128	−.0272
4	−.0660	−.1033	−.1386	−.1719	−.2028	−.2311	−.2566	−.2791	−.2985	−.3147
5	−.3276	−.3371	−.3432	−.3460	−.3453	−.3414	−.3343	−.3241	−.3110	−.2951
6	−.2767	−.2559	−.2329	−.2081	−.1816	−.1538	−.1250	−.0953	−.0652	−.0349
7	−.0047	.0252	.0543	.0826	.1096	.1352	.1592	.1813	.2014	.2192
8	.2346	.2476	.2580	.2657	.2708	.2731	.2728	.2697	.2641	.2559
9	.2453	.2324	.2174	.2004	.1816	.1613	.1395	.1166	.0928	.0684
10	.0435	.0184	−.0066	−.0313	−.0555	−.0789	−.1012	−.1224	−.1422	−.1603
11	−.1768	−.1913	−.2039	−.2143	−.2225	−.2284	−.2320	−.2333	−.2323	−.2290
12	−.2234	−.2157	−.2060	−.1943	−.1807	−.1655	−.1487	−.1307	−.1114	−.0912
13	−.0703	−.0489	−.0271	−.0052	.0166	.0380	.0590	.0791	.0984	.1165
14	.1334	.1488	.1626	.1747	.1850	.1934	.1999	.2043	.2066	.2069
15	.2051	.2013	.1955	.1879	.1784	.1672	.1544	.1402	.1247	.1080

When $x > 15.9$,

$$J_1(x) \cong \sqrt{\left(\frac{2}{\pi x}\right)}\left\{\sin\left(x - \frac{1}{4}\pi\right) + \frac{3}{8x}\sin\left(x + \frac{1}{4}\pi\right)\right\}$$

$$\cong \frac{.7979}{\sqrt{x}}\left\{\sin(57.296x - 45)^\circ + \frac{3}{8x}\sin(57.296x + 45)^\circ\right\}$$

					$J_2(x)$					
x	0	.1	.2	.3	.4	.5	.6	.7	.8	.9
0	.0000	.0012	.0050	.0112	.0197	.0306	.0437	.0588	.0758	.0946
1	.1149	.1366	.1593	.1830	.2074	.2321	.2570	.2817	.3061	.3299
2	.3528	.3746	.3951	.4139	.4310	.4461	.4590	.4696	.4777	.4832
3	.4861	.4862	.4835	.4780	.4697	.4586	.4448	.4283	.4093	.3879
4	.3641	.3383	.3105	.2811	.2501	.2178	.1846	.1506	.1161	.0813

When $0 \le x < 1$,

$$J_2(x) \cong \frac{x^2}{8}\left(1 - \frac{x^2}{12}\right)$$

					$J_3(x)$					
x	0	.1	.2	.3	.4	.5	.6	.7	.8	.9
0	.0000	.0000	.0002	.0006	.0013	.0026	.0044	.0069	.0122	.0144
1	.0196	.0257	.0329	.0411	.0505	.0610	.0725	.0851	.0988	.1134
2	.1289	.1453	.1623	.1800	.1981	.2166	.2353	.2540	.2727	.2911
3	.3091	.3264	.3431	.3588	.3754	.3868	.3988	.4092	.4180	.4250
4	.4302	.4333	.4344	.4333	.4301	.4247	.4171	.4072	.3952	.3811

When $0 \le x < 1$,

$$J_3(x) \cong \frac{x^3}{48}\left(1 - \frac{x^3}{16}\right)$$

					$J_4(x)$					
x	0	.1	.2	.3	.4	.5	.6	.7	.8	.9
0	.0000	..0000	.0000	.0000	.0001	.0002	.0003	.0006	.0010	.0016
1	.0025	.0036	.0050	.0068	.0091	.0118	.0150	.0188	.0232	.0283
2	.0340	.0405	.0476	.0556	.0643	.0738	.0840	.0950	.1067	.1190
3	.1320	.1456	.1597	.1743	.1891	.2044	.2198	.2353	.2507	.2661
4	.2811	.2958	.3100	.3236	.3365	.3484	.3594	.3693	.3780	.3853

When $0 \le x < 1$, $\quad J_4(x) \cong \frac{x^4}{384}\left(1 - \frac{x^2}{20}\right)$

TABLE 25.3 Zeros of $J_o(x)$, $J_1(x)$, $J_2(x)$, $J_3(x)$, $J_4(x)$, $J_5(x)$

m	$j_{0,m}$	$j_{1,m}$	$j_{2,m}$	$j_{3,m}$	$j_{4,m}$	$j_{5,m}$
1	2.4048	3.8317	5.1356	6.3802	7.5883	8.7715
2	5.5201	7.0156	8.4172	9.7610	11.0647	12.3386
3	8.6537	10.1735	11.6198	13.0152	14.3725	15.7002
4	11.7915	13.3237	14.7960	16.2235	17.6160	18.9801
5	14.9309	16.4706	17.9598	19.4094	20.8269	22.2178
6	18.0711	19.6159	21.1170	22.5827	24.0190	25.4303
7	21.2116	22.7601	24.2701	25.7482	27.1991	28.6266
8	24.3525	25.9037	27.4206	28.9084	30.3710	31.8117
9	27.4935	29.0468	30.5692	32.0649	33.5371	34.9888
10	30.6346	32.1897	33.7165	35.2187	36.6990	38.1599

25.6 Bessel Functions of the Second Kind

25.6.1 Definition

$$Y_\nu(t) = \frac{(\cos \nu\pi) J_\nu(t) - J_{-\nu}(t)}{\sin \nu\pi} \qquad \nu \ne n$$

The above function is the second independent solution to the differential equation $t^2 y'' + ty' + (t^2 - \nu^2)y = 0$.

25.6.2 Recurrence Relations

1. $\frac{d}{dt}\left[t^\nu Y_\nu(t)\right] = t^\nu Y_{\nu-1}(t)$

2. $\frac{d}{dt}\left[t^{-\nu} Y_\nu(t)\right] = -t^{-\nu} Y_{\nu+1}(t)$

3. $Y_{\nu-1}(t) + Y_{\nu+1}(t) = \frac{2\nu}{t} Y_\nu(t)$

4. $Y_{\nu-1}(t) - Y_{\nu+1}(t) = 2Y_\nu'(t)$

5. $Y_{-n}(t) = (-1)^n Y_n(t) \qquad n = 0, 1, 2, \cdots, \ \nu \to n$

25.6.3 Approximations

1. $Y_o(t) \cong \frac{2}{\pi} \ln t, \qquad t \to 0+$

2. $Y_\nu(t) \cong -\frac{\Gamma(\nu)}{\pi}\left(\frac{2}{t}\right)^\nu, \qquad \nu > 0,\ t \to 0+$

3. $Y_\nu(t) \cong \sqrt{\frac{2}{\pi t}} \sin\left[t - \frac{(\nu+\frac{1}{2})\pi}{2}\right], \qquad t \to \infty$

25.7 Modified Bessel Function

25.7.1 Definition

$$I_\nu(t) = j^{-\nu} J_\nu(jt) = \sum_{m=0}^{\infty} \frac{(t/2)^{2m+\nu}}{m!\,\Gamma(m+\nu+1)} \qquad \nu \neq \text{integer},\ n = 0,1,2,\cdots$$

$$I_{-n}(t) = I_n(t)$$

$$I_0(0) = 1; \quad I_\nu(0) = 0 \qquad \nu > 0$$

25.7.2 Recurrence Relations

1. $\frac{d}{dt}\left[t^\nu I_\nu(t)\right] = t^\nu I_{\nu-1}(t)$

2. $\frac{d}{dt}\left[t^{-\nu} I_\nu(t)\right] = t^{-\nu} I_{\nu+1}(t)$

3. $I'_\nu(t) + \frac{\nu}{t} I_\nu(t) = I_{\nu-1}(t)$

4. $I'_\nu(t) - \frac{\nu}{t} I_\nu(t) = I_{\nu+1}(t)$

5. $I_{\nu-1}(t) + I_{\nu+1}(t) = 2I'_\nu(t)$

6. $I_{\nu-1}(t) - I_{\nu+1}(t) = \frac{2\nu}{t} I_\nu(t)$

25.7.3 Integral Representation

$$I_\nu(t) = \frac{(t/2)^\nu}{\sqrt{\pi}\,\Gamma(\nu+\frac{1}{2})} \int_{-1}^{1} (1-x^2)^{\nu-\frac{1}{2}} e^{-xt}\, dx \qquad \nu > -\tfrac{1}{2},\ t > 0$$

25.7.4 Expansion Form

$$I_n(t) = \sum_{m=0}^{\infty} \frac{t^m}{m!} J_{n+m}(t), \qquad n = 0,1,2,\cdots$$

25.7.5 Asymptotic Formulas

1. $I_\nu(t) \cong \dfrac{(t/2)^\nu}{\Gamma(\nu+1)}, \qquad t \to 0+, \quad \nu \neq -1,-2,-3,\cdots$

2. $I_\nu(t) \cong \dfrac{e^t}{\sqrt{2\pi t}}, \qquad \nu \geq 0, \quad t \to \infty$

References

Abramowitz, M. and T. A Stegun, editors, *Handbook of Mathematical Functions*, Dover Publications, New York, NY, 1972.

Andrews, L. C., *Special Functions for Engineers and Applied Mathematicians*, Macmillan Publishing Co., New York, NY, 1985.

Hochstadt, H., *The Functions of Mathematical Physics*, Dover Publications, New York, NY, 1971.

McLachlan, N. W., *Bessel Functions for Engineers*, 2nd Edition, Oxford, London, 1961.

Sansone, G., *Orthogonal Functions*, Interscience Publishers Inc., New York, NY, 1959.

26

Zernike Polynomials

26.1 Zernike Polynomials

26.1.1 Definition

$$V_{nl}(x,y) = V_{nl}(r\cos\theta, r\sin\theta) = V_{nl}(r,\theta) = R_{nl}(r)e^{jl\theta}$$

n = nonnegative integer, $n \geq 0$; l = integer subject to constraints: $n - |l|$ = even and $|l| \leq n$; r = length of vector from origin to (x,y) point; θ = angle between r and x axis in counterclockwise direction.

26.1.2 Orthogonality Property

1. $$\iint_{x^2+y^2\leq 1} V_{nl}^*(r,\theta)V_{mk}(r,\theta)r\,dr\,d\theta = \frac{\pi}{n+1}\delta_{mn}\delta_{k\ell}$$

where δ_{ij} is the Kronecker symbol. The real valued radial polynomials satisfy the orthogonality relation

2. $$\int_0^1 R_{nl}(r)R_{ml}(r)r\,dr = \frac{1}{2(n+1)}\delta_{mn}$$

The radial polynomials are given by

3. $$R_{n\pm|l|}(r) = \frac{1}{\left(\frac{n-|l|}{2}\right)!\,r^m}\left[\frac{d}{d(r^2)}\right]^{\frac{n-|l|}{2}}\left[(r^2)^{\frac{n+|l|}{2}}(r^2-1)^{\frac{n-|l|}{2}}\right]$$

$$= \sum_{s=0}^{\frac{n-|l|}{2}}(-1)^s\frac{(n-s)!}{s!\left(\frac{n+|l|}{2}-s\right)!\left(\frac{n-|l|}{2}-s\right)!}r^{n-2s}$$

For all permissible values of n and $|l|$

4. $R_{n\pm|l|}(1) = 1, \qquad R_{n|l|}(r) = R_{n(-|l|)}(r)$

Table 26.1 gives the explicit form of the function $R_{n(|l|)}(r)$.

TABLE 26.1 The Radial Polynomials $R_{n(l)}(r)$ for $|l| \leq 8,\ n \leq 8$

$|l|$ \ n	0	1	2	3	4	5	6	7	8
0	1		$2r^2 - 1$		$6r^4 - 6r^2 + 1$		$20r^6 - 30r^4 + 12r^2 - 1$		$70r^8 - 140r^6 + 90r^4 - 20r^2 + 1$
1		r		$3r^3 - 2r$		$10r^5 - 12r^3 + 3r$		$35r^7 - 60r^5 + 30r^3 - 4r$	
2			r^2		$4r^4 - 3r^2$		$15r^6 - 20r^4 + 6r^2$		$56r^8 - 105r^6 + 60r^4 - 10r^2$
3				r^3		$5r^5 - 4r^3$		$21r^7 - 30r^5 + 10r^3$	
4					r^4		$6r^6 - 5r^4$		$28r^8 - 42r^6 + 15r^4$
5						r^5		$7r^7 - 6r^5$	
6							r^6		$8r^8 - 7r^6$
7								r^7	
8									r^8

26.1.3 Relation to Bessel Function

A relation between radial Zernike polynomials and Bessel functions of the first kind is given by

$$\int_0^1 R_{n|l|}(r) J_n(vr)\, r\, dr = (-1)^{\frac{n-|l|}{2}} \frac{J_{n+1}(v)}{v}$$

26.1.4 Real Zernike Polynomials

$$U_{nl} = \frac{1}{2}[V_{nl} + V_{n(-l)}] = R_{nl}(r)\cos l\theta \qquad l \neq 0$$

$$U_{n(-l)} = \frac{1}{2j}[V_{nl} - V_{n(-l)}] = R_{nl}(r)\sin l\theta \qquad l \neq 0$$

$$V_{n0} = R_{n0}(r)$$

Figure 26.1 shows the function U_{nl} for a few radial modes.

26.2 Expansion in Zernike Polynomials

26.2.1 Zernike Series

If $f(x,y)$ is a piecewise continuous function, we can expand this function in Zernike polynomials in the form

$$f(x,y) = \sum_{n=0}^{\infty} \sum_{l=-\infty}^{\infty} A_{nl} V_{nl}(x,y), \qquad n - |l| = \text{even}, \quad |l| \leq n$$

Multiplying by $V_{nl}^*(x,y)$, integrating over the unit circle and taking into consideration the orthogonality property, we obtain

$$A_{nl} = \frac{n+1}{\pi} \int_0^{2\pi} \int_0^1 V_{nl}^*(r,\theta) f(r\cos\theta, r\sin\theta)\, r\, dr\, d\theta$$

$$= \frac{n+1}{\pi} \iint_{x^2+y^2 \leq 1} V_{nl}^*(x,y) f(x,y)\, dx\, dy = A_{n(-l)}^*$$

with restrictions of the values of n and l as shown above. A_{nl}'s are also known as *Zernike moments.*

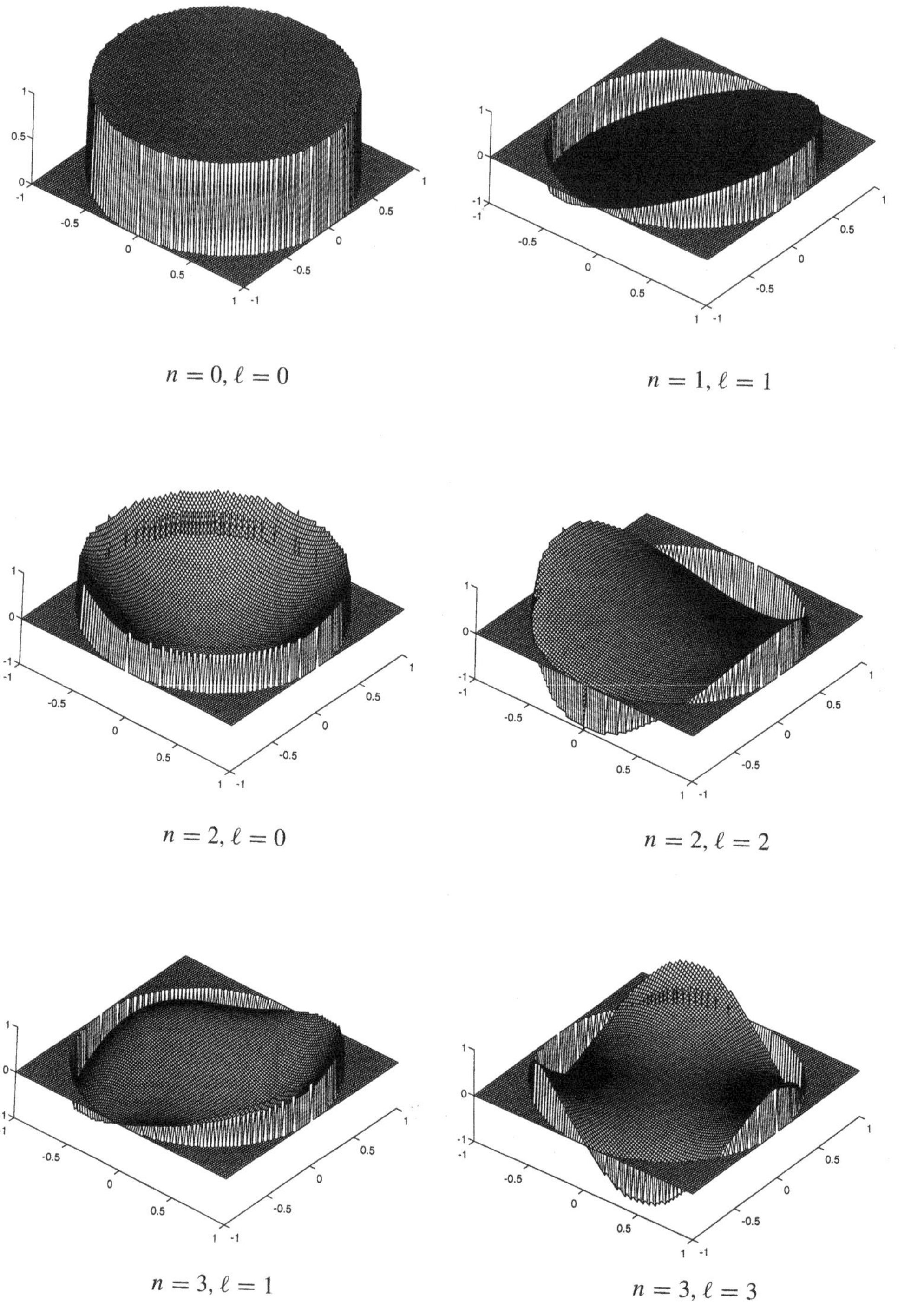

FIGURE 26.1

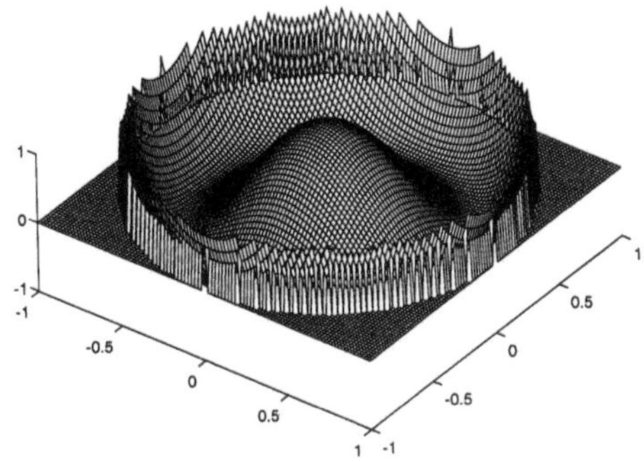

$n = 4, \ell = 0$

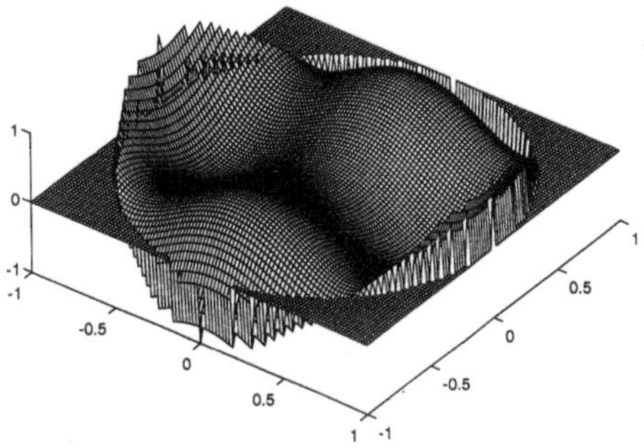

$n = 4, \ell = 2$

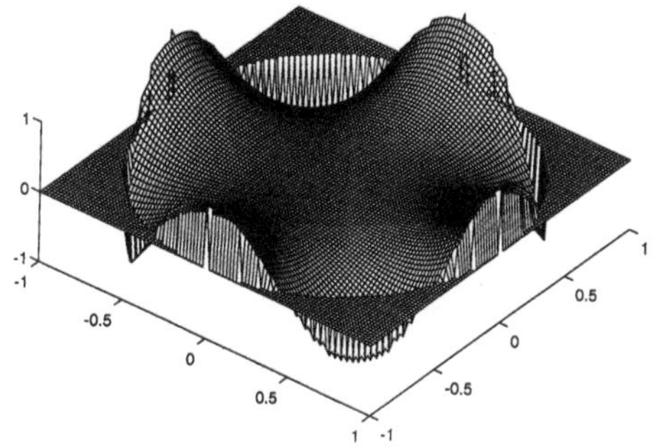

$n = 4, \ell = 4$

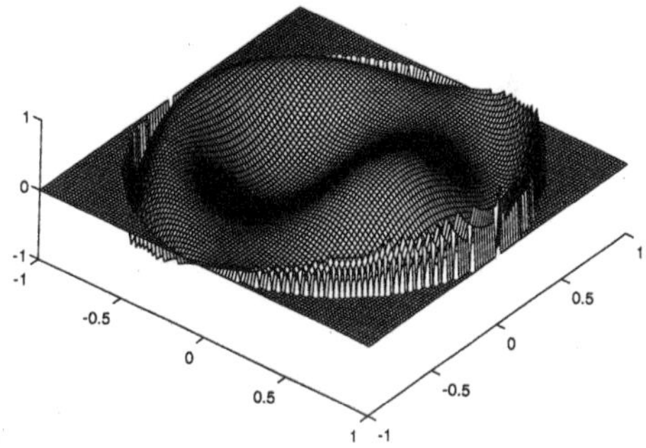

$n = 5, \ell = 1$

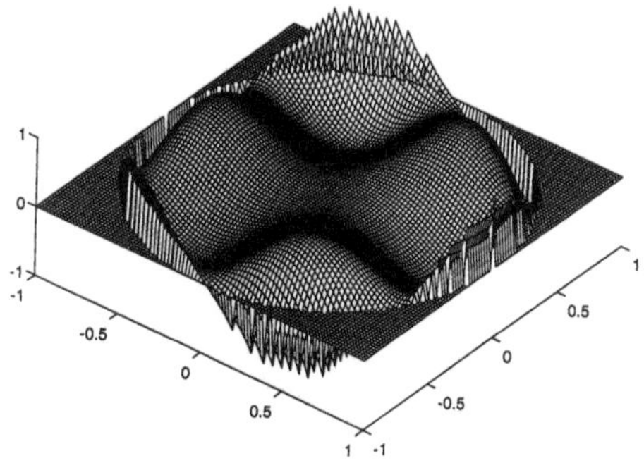

$n = 5, \ell = 3$

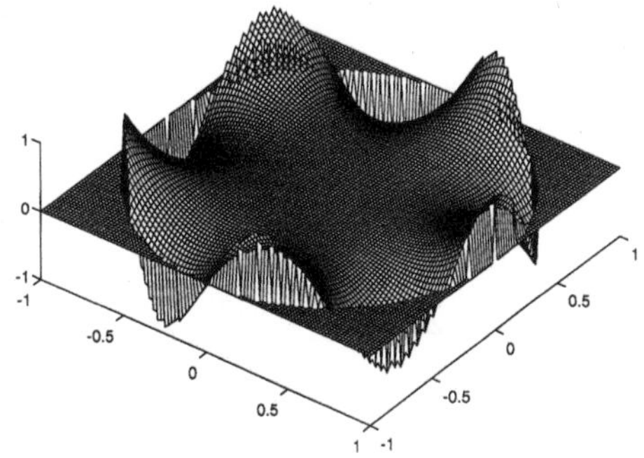

$n = 5, \ell = 5$

FIGURE 26.1(continued)

Example

Expand the function $f(x,y) = x$ in Zernike polynomials.

Solution

We write $f(r\cos\theta, r\sin\theta) = r\cos\theta$ and observe that r has exponent (degree) one. Therefore, the values of n will be 0,1 and since $n - |l|$ must be even, l will take 0,1 and –1 values. We then write

$$f(x,y) = \sum_{n=0}^{\infty}\sum_{l=-\infty}^{\infty} A_{nl}\, R_{nl}(r) e^{jl\theta}$$

$$= \sum_{n=0}^{1} (A_{n(-1)}\, R_{n(-1)}(r) e^{-j\theta} + A_{n0}\, R_{n0}(r) + A_{n1}\, R_{n1}(r) e^{j\theta})$$

$$= A_{00}\, R_{00}(r) + A_{1(-1)}\, R_{1(-1)}(r) e^{-j\theta} + A_{11}\, R_{11}(r) e^{j\theta}$$

where three terms were dropped because they did not obey the relation $n - |l|$ = even. From (26.1.2.4) $R_{1(-1)}(r) = R_{11}(r)$ and hence we obtain

$$A_{00} = \frac{1}{\pi}\int_0^{2\pi}\int_0^{1} R_{00}(r)\, r\cos\theta\, r\, dr\, d\theta = 0$$

$$A_{1(-1)} = \frac{2}{\pi}\int_0^{2\pi}\int_0^{1} R_{11}(r)\, r\cos\theta\, e^{-j\theta}\, r\, dr\, d\theta = \frac{1}{2}$$

$$A_{11} = \frac{2}{\pi}\int_0^{2\pi}\int_0^{1} R_{11}(r)\, r\cos\theta\, e^{j\theta}\, r\, dr\, d\theta = \frac{1}{2}$$

Therefore, the expansion becomes

$$f(x,y) = \frac{1}{2} r e^{j\theta} + \frac{1}{2} r e^{-j\theta} = r\cos\theta = R_{11}(r)\cos\theta = x$$

as was expected.

26.2.2 Expansion of Real Functions

1. $f(x,y) = \sum_{n=0}^{\infty}\sum_{l=0}^{\infty} (C_{nl}\cos l\theta + S_{nl}\sin l\theta) R_{nl}(r)$

where n – l is even and $l <$ n and $f(x,y)$ is real. Observe that l takes only a positive value. The unknown constants are found from

2. $\begin{bmatrix} C_{nl} \\ S_{nl} \end{bmatrix} = \frac{2n+2}{\pi}\int_0^{1}\int_0^{2\pi} r\, dr\, d\theta\, f(r\cos\theta, r\sin\theta) R_{nl}(r) \begin{bmatrix} \cos l\theta \\ \sin l\theta \end{bmatrix} \qquad l \neq 0$

3. $C_{n0} = A_{n0} = \frac{1}{\pi}\int_0^{1}\int_0^{2\pi} r\, dr\, d\theta\, f(r\cos\theta, r\sin\theta) R_{nl}(r) \qquad l = 0$

4. $S_{n0} = 0$

If the function is axially symmetric only the cosine terms are needed. The connection between real and complex Zernike coefficients are:

5. $C_{nl} = 2\,\text{Re}\{A_{nl}\}$

$S_{nl} = -2\,\text{Im}\{A_{nl}\}$

$A_{nl} = (C_{nl} - jS_{nl})/2 = (A_{n(-l)})^*$

References

Abramowitz, M. and I. S. Stegun, *Handbook of Mathematical Functions*, Dover Publications, Inc., New York, NY, 1965.

Andrews, L. C., *Special Functions for Engineers and Applied Mathematicians*, MacMillan Publishing Co., New York, NY, 1985.

Erdelyi, A., Editor, *Tables of Integral Transforms*, McGraw-Hill Book Company, New York, NY, 1954.

Gradshteyn, I. S. and I. M. Ryzhik, *Tables of Integrals, Series and Products*, Academic Press, New York, NY, 1965.

Hochstadt, H., *The Functions of Mathematical Physics*, Dover Publications Inc., New York, NY, 1986.

Lebedev, N. N., *Special Functions and their Applications*, Dover Publications, New York, NY., 1972

Poularikas, A. D., Editor-in-Chief, *The Transform and Application Handbook*, CRC Press Inc., Boca Raton, FL, 1996.

Sansone, G., *Orthogonal Functions*, Interscience Publishers, New York, NY., 1959.

Szegö G., *Orthogonal Polynomials*, 4th Edition, American Mathematical Society, Providence, RI, 1975.

27

Special Functions

27.1 The Gamma and Beta Functions

27.1.1 Gamma Function

$$\Gamma(z) = \int_0^\infty e^{-t} t^{z-1}\, dt \qquad \text{Re}\{z\} > 0$$

$$\Gamma(x) = \int_0^\infty e^{-t} t^{x-1}\, dt \qquad x > 0$$

The gamma function converges for all $x > 0$.

27.1.2 Incomplete Gamma Function

$$\gamma(x,\tau) = \int_0^\tau t^{x-1} e^{-t}\, dt \qquad x > 0, \quad \tau > 0$$

27.1.3 Beta Function

$$B(x,y) = \int_0^1 t^{x-1}(1-t)^{y-1}\, dt \qquad x > 0, \quad y > 0$$

The beta function is related to gamma function as follows:

$$B(x,y) = \frac{\Gamma(x)\Gamma(y)}{\Gamma(x+y)}$$

27.1.4 Properties of $\Gamma(x)$

Setting $x+1$ in place of x we obtain

1. $$\Gamma(x+1)=\int_0^\infty t^{x+1-1}e^{-t}\,dt=\int_0^\infty t^x e^{-t}\,dt$$

$$=-\int_0^\infty t^x d(e^{-t})=-t^x e^{-t}\Big|_0^\infty+\int_0^\infty x t^{x-1}e^{-t}\,dt$$

$$=x\Gamma(x)$$

From the above relation, we also obtain

2. $$\Gamma(x)=\frac{\Gamma(x+1)}{x}$$

3. $$\Gamma(x)=(x-1)\Gamma(x-1)$$

4. $$\Gamma(-x)=\frac{\Gamma(x-1)}{-x}\qquad x\neq 0,1,2,\cdots$$

5. From (27.1.1) with $x=1$, we find that $\Gamma(1)=1$. Using (27.1.4.1) we obtain

$$\Gamma(2)=\Gamma(1+1)=1\Gamma(1)=1\cdot 1=1$$

$$\Gamma(3)=\Gamma(2+1)=2\Gamma(2)=2\cdot 1$$

$$\Gamma(4)=\Gamma(3+1)=3\Gamma(3)=3\cdot 2\cdot 1$$

6. $$\Gamma(n+1)=n\Gamma(n)=n(n-1)!=n!\qquad n=0,1,2,\cdots$$

7. $$\Gamma(n)=(n-1)!\qquad n=0,1,2,\cdots$$

To find $\Gamma(\frac{1}{2})$ we first set $t=u^2$

$$\Gamma\left(\frac{1}{2}\right)=\int_0^\infty t^{-1/2}e^{-t}\,dt=\int_0^\infty 2e^{-u^2}\,du\qquad (t=u^2)$$

Hence, its square value is

$$\Gamma^2\left(\frac{1}{2}\right)=\left[\int_0^\infty 2e^{-x^2}\,dx\right]\left[\int_0^\infty 2e^{-y^2}\,dy\right]$$

$$=4\int_0^\infty\left[\int_0^\infty e^{-y^2}\,dy\right]e^{-x^2}\,dx=4\int_0^{\pi/2}\left[\int_0^\infty e^{-r^2}r\,dr\right]d\theta$$

$$=4\frac{\pi}{2}\cdot\frac{1}{2}=\pi$$

and thus

8. $$\Gamma\left(\frac{1}{2}\right)=\sqrt{\pi}$$

Next, let's find the expression for $\Gamma(n+\frac{1}{2})$ for integer positive value of n. From (27.1.4.3) we obtain

9. $$\Gamma\left(n+\frac{1}{2}\right)=\Gamma\left(\frac{2n+1}{2}\right)=\left(\frac{2n+1}{2}-1\right)\Gamma\left(\frac{2n+1}{2}-1\right)=\frac{2n-1}{2}\Gamma\left(\frac{2n-1}{2}\right)$$
$$=\left(\frac{2n-1}{2}\right)\left(\frac{2n-3}{2}\right)\Gamma\left(\frac{2n-3}{2}\right)$$

If we proceed to apply (27.1.4.3), we finally obtain

10. $$\Gamma\left(n+\frac{1}{2}\right)=\frac{(2n-1)(2n-3)(2n-5)\cdots(3)(1)\sqrt{\pi}}{2^n}$$

Similarly we obtain

11. $$\Gamma\left(n+\frac{3}{2}\right)=\frac{(2n+1)(2n-1)(2n-3)\cdots(3)(1)\sqrt{\pi}}{2^{n+1}}$$

12. $$\Gamma\left(n-\frac{1}{2}\right)=\frac{(2n-3)(2n-5)\cdots(3)(1)\sqrt{\pi}}{2^{n-1}}$$

Example

Applying (27.1.4.3) we find

$$2^n\Gamma(n+1)=2^n n\Gamma(n)=2^n n(n-1)\Gamma(n-1)=\cdots=2^n n(n-1)(n-2)\cdots 2\cdot 1$$
$$=2^n n!=(2\cdot 1)(2\cdot 2)(2\cdot 3)\cdots(2\cdot n)=2\cdot 4\cdot 6\cdots 2n$$

If $n-1$ is substituted in place of n, we obtain

$$2\cdot 4\cdot 6\cdots(2n-2)=2^{n-1}\Gamma(n)$$

27.1.5 Duplication Formula

$$\Gamma(z)\Gamma(z+\tfrac{1}{2})=2^{1-2z}\sqrt{\pi}\,\Gamma(2z)$$

27.1.6 Graph of Gamma Function

Figure 27.1 shows the gamma function.

27.1.7 Definition of Beta Function

$$B(x,y)=\int_0^1 t^{x-1}(1-t)^{y-1}\,dt \qquad x>0,\quad y>0$$

$$B(x,y)=\int_0^{\pi/2} 2\sin^{2x-1}\theta\cos^{y-1}\theta\,d\theta$$

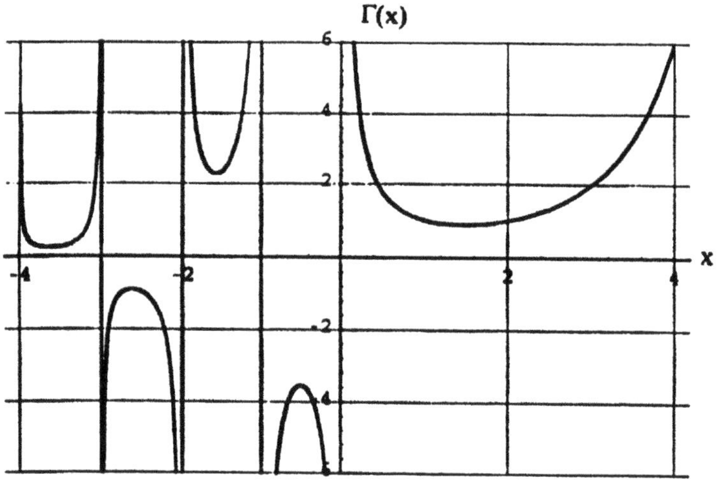

FIGURE 27.1 The gamma function.

$$B(x,y) = \int_0^\infty \frac{u^{x-1}}{(u+1)^{x+y}} du \qquad x > 0, \quad y > 0$$

$$B(x,y) = \frac{\Gamma(x)\Gamma(y)}{\Gamma(x+y)}$$

$$B(p,1-p) = \frac{\pi}{\sin p\pi} \qquad 0 < p < 1$$

Example

$$I = \int_o^\infty x^{-1/2}(1+x)^{-2}\, dx = B\left(\frac{1}{2},\frac{3}{2}\right) = \frac{\Gamma\left(\frac{1}{2}\right)\Gamma\left(\frac{3}{2}\right)}{\Gamma(2)} = \frac{\sqrt{\pi}\,(\sqrt{\pi}/2)}{1} = \frac{\pi}{2}$$

27.1.8 Properties of Beta Function

1. $B(x,y) = B(x+1,y) + B(x,y+1)$
2. $B(x,y) = B(y,x)$

27.1.9 Table of Gamma and Beta Function Relations

TABLE 27.1 Gamma and Beta Function Relations

$\Gamma(x) = \int_0^\infty e^{-t} t^{x-1}\, dt$	$x > 0$
$\Gamma(x) = \int_0^\infty 2u^{2x-1} e^{-u^2}\, du$	$x > 0$

TABLE 27.1 Gamma and Beta Function Relations (continued)

$\Gamma(x)=\int_0^1\left[\log\left(\frac{1}{r}\right)\right]^{x-1}dr$	$x>0$
$\Gamma(x)=\frac{\Gamma(x+1)}{x}$	$x\neq 0,-1,-2,\cdots$
$\Gamma(x)=(x-1)\Gamma(x-1)$	$x\neq 0,-1,-2,\cdots$
$\Gamma(-x)=\frac{\Gamma(1-x)}{-x}$	$x\neq 0,1,2,\cdots$
$\Gamma(n)=(n-1)!$	$n=1,2,3,\cdots\quad,\quad 0!=$
$\Gamma\left(\frac{1}{2}\right)=\sqrt{\pi}$	
$\Gamma\left(n+\frac{1}{2}\right)=\frac{1\cdot 3\cdot 5\cdot\cdots\cdot(2n-1)\sqrt{\pi}}{2^n}$	$n=1,2,\cdots$
$\Gamma\left(n+\frac{3}{2}\right)=\frac{(2n+1)(2n-1)(2n-3)\cdots(3)(1)\sqrt{\pi}}{2^{n+1}}$	$n=1,2,\cdots$
$\Gamma\left(n-\frac{1}{2}\right)=\frac{(2n-3)(2n-5)\cdots(3)(1)\sqrt{\pi}}{2^{n-1}}$	$n=1,2,\cdots$
$\Gamma(n+1)=\frac{2\cdot 4\cdot 6\cdot\cdots\cdot 2n}{2^n}$	$n=1,2,\cdots$
$\Gamma(2n)=1\cdot 3\cdot 5\cdot\cdots\cdot(2n-1)\Gamma(n)2^{1-n}$	$n=1,2,\cdots$
$\frac{\Gamma(2n)}{\Gamma(n)}=\frac{\Gamma\left(n+\frac{1}{2}\right)}{\sqrt{\pi}\;2^{1-2n}}$	$n=1,2,\cdots$
$\Gamma(x)\Gamma(1-x)=\frac{\pi}{\sin x\pi}$	$x\neq 0,\pm 1,\pm 2,\cdots$
$n!=\left(\frac{n}{e}\right)^n\sqrt{2\pi n}+h$	$n=1,2,\cdots\quad,\quad 0<\frac{h}{n!}<\frac{1}{12n}$
$\int_0^\infty t^a e^{-bt^c}\,dt=\frac{\Gamma\left(\frac{a+1}{c}\right)}{c\,b^{(a+1)/c}}$	$a>-1,\quad b>0,\quad c>0$
$B(x,y)=\int_0^1 t^{x-1}(1-t)^{y-1}\,dt$	$x>0,\quad y>0$
$B(x,y)=\int_0^{\pi/2} 2\sin^{2x-1}\theta\cos^{2y-1}\theta\,d\theta$	$x>0,\quad y>0$
$B(x,y)=\int_0^\infty\frac{u^{x-1}}{(u+1)^{x+y}}\,du$	$x>0,\quad y>0$
$B(x,y)=\frac{\Gamma(x)\Gamma(y)}{\Gamma(x+y)}$	
$B(x,y)=B(y,x)$	
$B(x,1-x)=\frac{\pi}{\sin x\pi}$	$0<x<1$
$B(x,y)=B(x+1,y)+B(x,y+1)$	$x>0,\quad y>0$

TABLE 27.1 Gamma and Beta Function Relations (continued)

$B(x,n+1)=\dfrac{1\cdot 2\cdots n}{x(x+1)\cdots(x+n)}$	$x>0$

27.1.10 Table of the Gamma Function

TABLE 27.2 $\Gamma(x)$, $1 \le x \le 1.99$

x	0	1	2	3	4	5	6	7	8	9
1.0	1.0000	.9943	.9888	.9835	.9784	.9735	.9698	.9642	.9597	.9555
.1	.9514	.9474	.9436	.9399	.9364	.9330	.9298	.9267	.9237	.9209
.2	.9182	.9156	.9131	.9108	.9085	.9064	.9044	.9025	.9007	.8990
.3	.8975	.8960	.8946	.8934	.8922	.8912	.8902	.8893	.8885	.8879
.4	.8873	.8868	.8864	.8860	.8858	.8857	8856	.8856	.8857	.8859
.5	.8862	.8866	.8870	.8876	.8882	.8889	.8896	.8905	.8914	.8924
.6	.8935	.8947	.8859	.8972	.8986	.9001	.9017	.9033	.9050	.9068
.7	.9086	.9106	.9126	.9147	.9168	.9191	.9214	.9238	.9262	.9288
.8	.9314	.9341	.9368	.9397	.9426	.9456	.9487	.9518	.9551	.9584
.9	.9618	.9652	.9688	.9724	.9761	.9799	.9837	.9877	.9917	.9958

27.2 Error Function

27.2.1 Error Function

$$erf\,z=\frac{2}{\sqrt{\pi}}\int_0^z e^{-t^2}\,dt$$

27.2.2 Coerror Function

$$erfc\,z=\frac{2}{\sqrt{\pi}}\int_z^\infty e^{-t^2}\,dt=1-erf\,z$$

27.2.3 Series Expansion

$$erf\,z=\frac{2}{\sqrt{\pi}}\sum_{n=0}^{\infty}\frac{(-1)^n z^{2n+1}}{n!(2n+1)} \qquad |z|<\infty$$

27.2.4 Symmetry Relation

$$erf(-z)=-erf\,z,\quad erf\,z^*=(erf\,z)^*$$

27.3 Sine and Cosine Integrals

27.3.1 Sine Integral

$$Si(z)=\int_0^z \frac{\sin t}{t}\,dt$$

27.3.2 Cosine Integral

$$Ci(z) = \gamma + \ln z + \int_0^z \frac{\cos t - 1}{t} dt \qquad |\arg z| < \pi$$

27.3.3

$$si(z) = Si(z) - \frac{\pi}{2}$$

27.3.4 Auxiliary Functions

$$f(z) = Ci(z)\sin z - si(z)\cos z, \qquad g(z) = -Ci(z)\cos z - si(z)\sin z$$

27.3.5 Series Expansion

$$Si(z) = \sum_{n=0}^{\infty} \frac{(-1)^n z^{2n+1}}{(2n+1)(2n+1)!}, \qquad Si(z) = \pi \sum_{n=0}^{\infty} J^2_{n+\frac{1}{2}}\left(\frac{z}{2}\right)$$

$$Ci(z) = \gamma + \ln z + \sum_{n=1}^{\infty} \frac{(-1)^n z^{2n}}{2n(2n)!}$$

27.3.6 Symmetry Relation

$$Si(-z) = -Si(z), \quad Si(z^*) = [Si(z)]^*, \quad Ci(-z) = Ci(z) - j\pi,$$

$$0 < \arg z < \pi, \quad Ci(z^*) = [Ci(z)]^*$$

27.3.7 Value at Infinity

$$\lim_{x\to\infty} Si(x) = \frac{\pi}{2}$$

27.4 Fresnel Integrals

27.4.1 Fresnel Integrals

$$C(z) = \int_0^z \cos\left(\frac{\pi t^2}{2}\right) dt, \quad S(z) = \int_0^z \sin\left(\frac{\pi t^2}{2}\right) dt$$

27.4.2 Extrema

$C(x)$ has extrema at $x = \pm\sqrt{2n+1}$, $S(x)$ has extrema at $x = \pm\sqrt{2n}$ $\qquad (n = 0, 1, 2 \cdots)$

27.4.3 Values at Infinity

$$C(\infty) = S(\infty) = \frac{1}{2}$$

27.4.4 Series Expansion

$$C(z)=\sum_{n=0}^{\infty}\frac{(-1)^n(\pi/2)^{2n}}{(2n)!(4n+1)}z^{4n+1},\quad S(z)=\sum_{n=0}^{\infty}\frac{(-1)^n(\pi/2)^{2n+1}}{(2n+1)!(4n+3)}z^{4n+3}$$

27.4.5 Symmetry Relation

$$C(-z)=-C(z),\quad S(-z)=-S(z),\quad C(jz)=jC(z),\quad S(iz)=-jS(z),$$

$$C(z^*)=[C(z)]^*,\quad S(z^*)=[S(z)]^*$$

27.4.6 Relation to Error Function

$$C(z)+jS(z)=\frac{1+j}{2}erf\left[\frac{\sqrt{\pi}}{2}(1-j)z\right]$$

27.5 Exponential Integrals

27.5.1 $E_1(x)=\int_x^{\infty}\frac{e^{-t}}{t}dt=\int_1^{\infty}\frac{e^{-xt}}{t}dt$

27.5.2 $E_n(x)=\int_1^{\infty}\frac{e^{-xt}}{t^n}dt \qquad (x>0,\ n=0,1,2,\cdots)$

$$E_{n+1}(x)=\frac{1}{n}[e^{-x}-xE_n(x)] \qquad (n=1,2,\cdots)$$

27.5.3 $Ei(x)=\int_{-\infty}^{x}\frac{e^{t}}{t}dt \qquad (x>0,\ \text{Cauchy P.V.})$

27.5.4 $\ell i(x)=\int_0^x\frac{dt}{\ln t}=Ei(\ln x) \qquad (x>1,\ \text{Cauchy P.V.})$

27.5.5 Series Expansions

$$E_1(x)=-\gamma-\ln x-\sum_{n=1}^{\infty}\frac{(-1)^n x^n}{n\,n!} \qquad (x>0)$$

$$Ei(x)=\gamma+\ln x+\sum_{n=1}^{\infty}\frac{x^n}{n\,n!} \qquad (\gamma=\text{Euler's constant})$$

27.5.6 Special Values

$$E_n(0)=\frac{1}{n-1}\ (n>1),\qquad E_0(x)=\frac{e^{-x}}{x}$$

27.5.7 Derivatives

$$\frac{dE_n(x)}{dx} = -E_{n-1}(x) \qquad (n = 1, 2, \cdots)$$

27.6 Elliptic Integrals

27.6.1 Elliptic Integrals of the First Kind

$$F(k,\varphi) = \int_0^{\varphi} \frac{d\theta}{\sqrt{1-k^2\sin^2\theta}} = \int_0^x \frac{dt}{\sqrt{(1-t^2)(1-k^2t^2)}} \qquad (k^2 < 1, \quad x = \sin\varphi)$$

27.6.2 Elliptic Integrals of the Second Kind

$$E(k,\varphi) = \int_0^{\varphi} \sqrt{1-k^2\sin^2\theta}\, d\theta = \int_0^x \sqrt{\left(\frac{1-k^2t^2}{1-t^2}\right)}\, dt \qquad (k^2 < 1, \quad x = \sin\varphi)$$

27.6.3 Elliptic Integrals of the Third Kind

$$\pi(k,n,\varphi) = \int_0^{\varphi} \frac{d\theta}{(1+n\sin^2\theta)\sqrt{1-k^2\sin^2\theta}}$$

$$= \int_0^x \frac{dt}{(1+mt^2)\sqrt{(1-t^2)(1-k^2t^2)}} \qquad (k^2 < 1, \quad x = \sin\varphi)$$

27.6.4 Complete Elliptic Integrals

$$K = K(k) = F\left(k, \frac{\pi}{2}\right) = \int_0^{\pi/2} \frac{d\theta}{\sqrt{1-k^2\sin^2\theta}} \qquad (k^2 < 1)$$

$$E = E(k) = E\left(k, \frac{\pi}{2}\right) = \int_0^{\pi/2} \sqrt{1-k^2\sin^2\theta}\, d\theta \qquad (k^2 < 1)$$

27.6.5 Legendre's Relation

$$E(k)K(k') + E(k')K(k) - K(k)K(k') = \frac{\pi}{2} \qquad (k' = \sqrt{1-k^2})$$

27.6.6 Differential Equations

$$k(1-k^2)\frac{d^2K}{dk^2} + (1-3k^2)\frac{dK}{dk} - kK = 0$$

$$k(1-k^2)\frac{d^2E}{dk^2} + (1-k^2)\frac{dE}{dk} + kE = 0$$

27.6.7 Table of Complete Elliptic Integrals

TABLE 27.3 Numerical tables of complete elliptic integrals $k = \sin a$ (a in degrees)

a	K	E	a	K	E	a	K	E
0°	1.5708	1.5708	50°	1.9356	1.3055	81°.0	3.2553	1.0338
1	1.5709	1.5707	51	1.9539	1.2963	81.2	3.2771	1.0326
2	1.5713	1.5703	52	1.9729	1.2870	81.4	3.2995	1.0314
3	1.5719	1.5697	53	1.9927	1.2776	81.6	3.3223	1.0302
4	1.5727	1.5689	54	2.0133	1.2681	81.8	3.3458	1.0290
5	1.5738	1.5678	55	2.0347	1.2587	82.0	3.3699	1.0278
6	1.5751	1.5665	56	2.0571	1.2492	82.2	3.3946	1.0267
7	1.5767	1.5649	57	2.0804	1.2397	82.4	3.4199	1.0256
8	1.5785	1.5632	58	2.1047	1.2301	82.6	3.4460	1.0245
9	1.5805	1.5611	59	2.1300	1.2206	82.8	3.4728	1.0234
10	1.5828	1.5589	60	2.1565	1.2111	83.0	3.5004	1.0223
11	1.5854	1.5564	61	2.1842	1.2015	83.2	3.5288	1.0213
12	1.5882	1.5537	62	2.2132	1.1920	83.4	3.5581	1.0202
13	1.5913	1.5507	63	2.2435	1.1826	83.6	3.5884	1.0192
14	1.5946	1.5476	64	2.2754	1.1732	83.8	3.6196	1.0182
15	1.5981	1.5442	65	2.3088	1.1638	84.0	3.6519	1.0172
16	1.6020	1.5405	65.5	2.3261	1.1592	84.2	3.6852	1.0163
17	1.6061	1.5367	66.0	2.3439	1.1545	84.4	3.7198	1.0153
18	1.6105	1.5326	66.5	2.3622	1.1499	84.6	3.7557	1.0144
19	1.6151	1.5283	67.0	2.3809	1.1453	84.8	3.7930	1.0135
20	1.6200	1.5238	67.5	2.4001	1.1408	85.0	3.8317	1.0127
21	1.6252	1.5191	68.0	2.4198	1.1362	85.2	3.8721	1.0118
22	1.6307	1.5141	68.5	2.4401	1.1317	85.4	3.9142	1.0110
23	1.6365	1.5090	69.0	2.4610	1.1272	85.6	3.9583	1.0102
24	1.6426	1.5037	69.5	2.4825	1.1228	85.8	4.0044	1.0094
25	1.6490	1.4981	70.0	2.5046	1.1184	86.0	4.0528	1.0086
26	1.6557	1.4924	70.5	2.5273	1.1140	86.2	4.1037	1.0079
27	1.6627	1.4864	71.0	2.5507	1.1096	86.4	4.1574	1.0072
28	1.6701	1.4803	71.5	2.5749	1.1053	86.6	4.2142	1.0065
29	1.6777	1.4740	72.0	2.5998	1.1011	86.8	4.2744	1.0059
30	1.6858	1.4675	72.5	2.6256	1.0968	87.0	4.3387	1.0053
31	1.6941	1.4608	73.0	2.6521	1.0927	87.2	4.4073	1.0047
32	1.7028	1.4539	73.5	2.6796	1.0885	87.4	4.4811	1.0041
33	1.7119	1.4469	74.0	2.7081	1.0844	87.6	4.5609	1.0036
34	1.7214	1.4397	74.5	2.7375	1.0804	87.8	4.6477	1.0031
35	1.7312	1.4323	75.0	2.7681	1.0764	88.0	4.7427	1.0026
36	1.7415	1.4248	75.5	2.7998	1.0725	88.2	4.8478	1.0021
37	1.7522	1.4171	76.0	2.8327	1.0686	88.4	4.9654	1.0017
38	1.7633	1.4092	76.5	2.8669	1.0648	88.6	5.0988	1.0014
39	1.7748	1.4013	77.0	2.9026	1.0611	88.8	5.2527	1.0010
40	1.7868	1.3931	77.5	2.9397	1.0574	89.0	5.4349	1.0008
41	1.7992	1.3849	78.0	2.9786	1.0538	89.1	5.5402	1.0006
42	1.8122	1.3765	78.5	3.0192	1.0502	89.2	5.6579	1.0005
43	1.8256	1.3680	79.0	3.0617	1.0468	89.3	5.7914	1.0004
44	1.8396	1.3594	79.5	3.1064	1.0434	89.4	5.9455	1.0003
45	1.8541	1.3506	80.0	3.1534	1.0401	89.5	6.1278	1.0002

TABLE 27.3 Numerical tables of complete elliptic integrals $k = \sin a$ (a in degrees)

a	K	E	a	K	E	a	K	E
46	1.8691	1.3418	80.2	3.1729	1.0388	89.6	6.3509	1.0001
47	1.8848	1.3329	80.4	3.1928	1.0375	89.7	6.6385	1.0001
48	1.9011	1.3238	80.6	3.2132	1.0363	89.8	7.0440	1.0000
49	1.9180	1.3147	80.8	3.2340	1.0350	89.9	7.7371	1.0000

References

Abramowitz, M. and I. A. Stegun, editors, *Handbook of Mathematical Functions*, Dover Publications, New York, NY, 1972.

Andrews, L. C., *Special Functions for Engineers and Applied Mathematicians*, Macmillan Publishing Co., New York, NY, 1985.

Hochstadt, H., *The Functions of Mathematical Physics*, Dover Publications, New York, NY, 1971.

McLachlan, N. W., *Bessel Functions for Engineers*, 2nd Edition, Oxford, London, 1961.

Samson, G., *Orthogonal Functions*, Interscience Publishers, Inc., New York, NY, 1959.

28

Asymptotic Expansions

28.1 Introduction

28.1.1 Order O ( )

$f(x) = O(g(x))$ as $x \to x_0$ if there exists a constant A such that $|f(x)| \le A|g(x)|$ for all values of x in some neighborhood of x_0.

28.1.2 Order o( )

$f(x) = o(g(x))$ as $x \to x_0$ if $\lim_{x \to x_0} |f(x)/g(x)| = 0$

28.1.3 Order One

$f(x) \sim g(x)$ as $x \to x_0$ if $\lim_{x \to x_0} |f(x)/g(x)| = 1$

28.1.4 Examples

1. $\sin x = O(1),\ x \to \infty,$
2. $(1+x^2)^{-1} = O(1),\ x \to 0,$
3. $(1+x^2)^{-1} = o(x^{-1}),\ x \to \infty,$
4. $(1+x^2)^{-1} = O(x^{-2}),\ x \to \infty,$
5. $(1+x^2)^{-1} \sim x^{-2},\ x \to \infty,$
6. $(1+x^2)^{-1} = x^{-2} + o(x^{-2}),\ x \to \infty,$
7. $(1+x^2)^{-1} = x^{-2} + o(x^{-3}),\ x \to \infty,$
8. $(1+x^2)^{-1} = x^{-2} + O(x^{-4}),\ x \to \infty,$
9. $(1+x^2)^{-1} = x^{-2} - x^{-4} + O(x^{-6}),\ x \to \infty,$
10. $n/(n+1) \sim 1,\ n \to \infty,$
11. $\sin x \sim x,\ x \to 0,$
12. $\cos x = 1 + O(x^2),\ x \to 0,$
13. $\sqrt{n^2+1} \sim n,\ n \to \infty,$
14. $\sqrt{n^2+1} = n + o(1),\ n \to \infty,$
15. $\sqrt{n^2+1} = n + O(n^{-1}),\ n \to \infty,$
16. $(n/e)^n = O(n!),\ n \to \infty,$

17. $\sum_{n=1}^{\infty} x^n = O((1-x)^{-1}),\ x \to 1-,$

18. $\sum_{n=1}^{\infty} n^p x^n = O((1-x)^{-p}),\ x \to 1-,$

19. $\int_2^x \frac{dy}{y} = O(\log x),\ x \to \infty$

28.1.5 Asymptotic Sequence

$\{\varphi_n(x)\}_{n=0}^{\infty}$ is an asymptotic sequence for some fixed point x_0 if for each fixed n we have $\varphi_{n+1}(x) = o(\varphi_n(x))$ as $x \to x_0$.

Example

The sequence $1, x, x^2, \cdots$ is an asymptotic sequence for 0.

28.1.6 Asymptotic Series

An asymptotic series for a given function $f(x)$ at x_0 for each fixed integer n is given by

$$f(x) = a_0\varphi_0(x) + \cdots + a_n\varphi_n(x) + o(\varphi_n(x)) \text{ as } x \to x_0$$

or

$$f(x) \cong \sum_{\nu=0}^{\infty} a_\nu \varphi_\nu(x) \quad \text{as } x \to x_0$$

28.1.7 Example

If $f(x) = \int_0^{\infty} \frac{e^{-t}\,dt}{x+t}$ with x large and positive, then integration by parts yields: $f(x) = \frac{1}{x} - \frac{1}{x^2} + \frac{2!}{x^3} - \cdots + (-1)^n \frac{n!}{x^{n+1}} + (-1)^n (n+1)! \int_0^{\infty} \frac{e^{-t}\,dt}{(x+t)^{n+2}}$. The remainder term is

$$|R_n(x)| = (n+1)! \int_0^{\infty} \frac{e^{-t}\,dt}{(x+t)^{n+2}} = \frac{(n+1)!}{x^{n+1}} \int_0^{\infty} \frac{e^{-xy}}{(1+y)^{n+2}}\,dy \le \frac{(n+1)!}{x^{n+1}} \int_0^{\infty} e^{-xy}\,dy = \frac{(n+1)!}{x^{n+2}}$$

Hence, we terminate the expansion of $f(x)$ after the nth term (ignoring the remainder); the error is of the order of $\varphi_{n+1} = o(\varphi_n(x))$. Hence we write

$$f(x) \cong \sum_{\nu=0}^{\infty} (-1)^{\nu} \frac{\nu!}{x^{n+1}} \quad \text{as } x \to \infty.$$

28.2 Sums

28.2.1 Bernoulli's Numbers B_n (n = 1,2,...):

$$\frac{z}{e^z - 1} = \sum_{n=0}^{\infty} \frac{z^n}{n!} B_n$$

$$B_0 = 1,\ B_1 = -\frac{1}{2},\ B_{2n+1} = 0\ (n = 1,2,3,\cdots),\ B_{2n} = (-1)^{n+1}(2n)!\frac{\zeta(2n)}{(2\pi)^{2n}}2\ (n = 1,2,3,\cdots),$$

where

$$\zeta(2n) = \sum_{n=1}^{\infty} \frac{1}{2^{2n}}\ (n = 1,2,\cdots).$$

Also: $B_2 = \frac{1}{6},\ B_4 = -\frac{1}{30},\ B_6 = \frac{1}{42},\ B_8 = -\frac{1}{30}, B_{10} = \frac{5}{66},\ B_{12} = -\frac{691}{2730},\ B_{14} = \frac{7}{6},\ B_{16} = -\frac{3617}{510},$ $B_{18} = \frac{43867}{798},\ B_{20} = -\frac{174611}{330}$

28.2.2 First-Form of Euler-Maclaurin Sum Formula

If $f(x)$ is continuously differentiable on the interval $[1,n]$ then

$$\sum_{\nu=1}^{\infty} f(\nu) = \int_1^n f(x)dx + \frac{1}{2}[f(1) + f(n)] + \int_1^n \left(x - [x] - \frac{1}{2}\right)f'(x)dx$$

where [x] is the greatest integer contained in x (e.g., $[2] = 2,\ [4.114] = 4,\ [0.315] = 0$).

Example

If $f(x) = \frac{1}{x}$ then

$$\sum_{\nu=1}^{n} \frac{1}{\nu} = \int_1^n \frac{dx}{x} + \frac{1}{2}\left(1 + \frac{1}{n}\right) - \int_1^n \left(x - [x] - \frac{1}{2}\right)\frac{dx}{x^2} = \ln n + \frac{1}{2} + \frac{1}{2n} - \int_1^n \left(x - [x] - \frac{1}{2}\right)\frac{dx}{x^2}.$$

The integral on the right is $o(1)$ as $n \to \infty$ since $(x - [x] - \frac{1}{2})$ is less than $\frac{1}{2}$ in an absolute value and hence $\sum_{\nu=1}^{n} \frac{1}{\nu} = \ln n + O(1)$ as $n \to \infty$.

28.2.3 Euler's Constant

$$\gamma = \frac{1}{2} - \int_1^{\infty} \left(x - [x] - \frac{1}{2}\right)\frac{dx}{x^2} = \lim_{n\to\infty}\left(1 + \frac{1}{2} + \cdots + \frac{1}{n} - \ln n\right) = 0.5772\ldots$$

which comes from the last integral of the last example where the integral is split into two integrals $\int_1^n (\cdot)dx + \int_n^{\infty} (\cdot)dx.$

28.2.4 Second-Form Euler-Maclaurin Sum Formula

If $f(x)$ is $2k+1$ times continuously differentiable in $[1,n]$ then $\sum_{\nu=1}^{\infty} f(\nu) = \int_1^n f(x)dx + \frac{1}{2}(f(1)+f(n))$

$$+\frac{B_2}{2!}(f'(n)-f'(1))+\frac{B_4}{4!}(f'''(n)-f'''(1))+\cdots+\frac{B_{2k}}{(2k)!}(f^{(2k-1)}(n)-f^{(2k-1)}(1))+\int_1^n P_{2k+1}(x)f^{(2k+1)}(x)dx.$$

$P_{2k}(x)=(-1)^{k+1}\sum_{n=1}^{\infty}\frac{2\cos 2n\pi x}{(2n\pi)^{2k}}$ and $P_{2k+1}(x)=(-1)^{k+1}\sum_{n=1}^{\infty}\frac{2\sin 2n\pi x}{(2n\pi)^{2k+1}}$, $k=1,2,\cdots$, and B_i's are the Bernoulli's numbers (see 28.2.1).

28.3 Stirling's Formula

28.3.1 Stirling's Formula

$$\ln(n!)=\left(n+\frac{1}{2}\right)\ln n - n + \ln\sqrt{2\pi}+\frac{B_2}{1\cdot 2}\frac{1}{n}+\frac{B_4}{3\cdot 4}\frac{1}{n^3}+\cdots+\frac{B_{2k}}{2k(2k-1)}\frac{1}{n^{2k-1}}+O(n^{-2k}) \text{ as } n\to\infty.$$

28.3.2 Stirling's Formula

$$\ln(n!)=(n+\tfrac{1}{2})\ln n - n+\ln\sqrt{2\pi}+o(1) \text{ as } n\to\infty$$

$$n!\cong\sqrt{2\pi n}\left(\frac{n}{e}\right)^n \text{ as } n\to\infty$$

28.4 Sums of Powers

28.4.1 Sum of Powers

$$1^p+2^p+\cdots+(n-1)^p=\frac{1}{p+1}\sum_{\nu=0}^{p}\binom{p+1}{\nu}B_\nu n^{p+1-\nu}$$

1. $(p=1)$ $1+2+3+\cdots+n=\frac{n^2}{2}+\frac{n}{2}$
2. $(p=2)$ $1^2+2^2+3^2+\cdots+n^2=\frac{n^3}{3}+\frac{n^2}{2}+\frac{n}{6}$
3. $(p=3)$ $1^3+2^3+3^3+\cdots+n^3=\frac{n^4}{4}+\frac{n^3}{2}+\frac{n^2}{4}$
4. $(p=4)$ $1^4+2^4+3^4+\cdots+n^4=\frac{n^5}{5}+\frac{n^4}{2}+\frac{n^3}{3}-\frac{n}{30}$

28.5 Laplace Method for Integrals

28.5.1 Laplace Theorem

a) If $h(y)$ is real valued and continuous,
b) $h(0) = 0$ and $h(y) < 0$ for $y \neq 0$,
c) there are numbers $\alpha > \beta$ such that $h(y) \leq -\alpha$ when $|y| \geq \beta$,
d) there is a neighborhood of $y = 0$ in which $h(y)$ is twice differentiable and $h''(0) < 0$, then

$$G(x) = \int_{-\infty}^{\infty} e^{xh(y)}\, dy \cong \left[\frac{2\pi}{-xh''(o)}\right]^{1/2} \quad \text{as } x \to \infty.$$

Example

To the Stirling formula $n! = \int_0^{\infty} e^{-t} t^n \, dt$ we replace n with continuous variable x and hence $x! = \Gamma(x + 1) = \int_0^{\infty} e^{-t} t^x \, dt$. Since the integrand has a maximum at $t = x$ we make the substitution $t = x(y + 1)$ to bring it to the above standard form. Hence $x! = x^{x+1} e^{-x} \int_{-1}^{\infty} \exp[x(\log(1+y) - y)]dy$, and thus $h(y) = \log(1+y) - y$ which satisfies all the conditions above. Therefore, $x! = \Gamma(x+1) \cong \sqrt{2\pi x}\left(\frac{x}{e}\right)^x$ as $x \to \infty$, since $h''(0) = -1$.

28.6 The Method of Stationary Phase

28.6.1 Theorem

If the function $r(t)$ is continuous and the derivative of the function $\mu(t)$ vanishes at only a single point $t = t_0$ in the interval $(-\infty, \infty) : \mu'(t_0) = 0,\ \mu''(t_0) \neq 0,$ then for sufficiently large k,

$$\int_{-\infty}^{\infty} r(t) e^{jk\mu(t)} dt \cong e^{jk\mu(t_0)} r(t_0) \sqrt{\frac{2\pi j}{k\mu''(t_0)}}\,.$$

Example

In the relation $J_0(x) = \frac{1}{2\pi}\int_{-\pi}^{\pi} e^{jx\sin t}\, dt$, $r(t) = \frac{1}{2\pi}$, $k = x$ and $\mu(t) = \sin t$. In the interval $(-\pi, \pi)$, $\mu'(t) = \cos t = 0$ for $t = t_1 = \pi/2$ and $t = t_2 = -\pi/2$. Since $\mu(t_1) = 1$, $\mu''(t_1) = -1$, $\mu(t_2) = -1$ and $\mu''(t_2) = 1$, we conclude that

$$J_0(x) \cong \frac{e^{jx}}{2\pi}\sqrt{\frac{2\pi j}{x(-1)}} + \frac{e^{-jx}}{2\pi}\sqrt{\frac{2\pi j}{x}} = \frac{1}{2\pi}\sqrt{\frac{2\pi}{x}}\left(e^{jx - j\frac{\pi}{4}} + e^{-jx + j\frac{\pi}{4}}\right) = \sqrt{\frac{2}{\pi x}}\cos\left(x - \frac{\pi}{4}\right)$$

References

Copson, E., *Asymptotic Expansions*, Cambridge University Press, London, England, 1965.
Erdelyi, A., *Asymptotic Expansions*, Dover Publications, New York, NY, 1956.
Jeffrys, H., *Asymptotic Approximations*, Oxford University Press, London, England, 1962.
Sirovich, L., *Techniques of Asymptotic Analysis*, Springer-Verlag, New York, NY, 1971.

29

Nonrecursive Filters (Finite Impulse Response, Fir)

29.1 Properties of Nonrecursive Filters

29.1.1 Causal Filter

$$H(z) = \sum_{n=0}^{N-1} h(nT_s) z^{-n}$$

$$H(e^{j\omega T_s}) = M(\omega) e^{j\theta(\omega)} = \sum_{n=0}^{N-1} h(nT_s) e^{j\omega n T_s}$$

$$M(\omega) = \left| H(e^{j\omega T_s}) \right|, \; \theta(\omega) = \arg H(e^{j\omega T_s})$$

T_s = sampling time

29.1.2 Phase and Group Delays

$$\tau_p = -\frac{\theta(\omega)}{\omega}, \quad \tau_g = -\frac{d\theta(\omega)}{d\omega}$$

29.1.3 Constant Phase and Group Delays

$$\theta(\omega) = -\tau\omega = \tan^{-1}\frac{-\sum_{n=0}^{N-1} h(nT_s)\sin\omega nT_s}{\sum_{n=0}^{N-1} h(nT_s)\cos\omega nT_s},$$

$$\tau = \frac{(N-1)T_s}{2}, \quad h(nT_s) = h[(N-1-n)T_s] = \text{symmetrical}$$

Impulse response must be symmetrical about the midpoint between samples $(N-2)/2$ and $N/2$ for even N and about samples $(N-1)/2$ for odd N.

29.1.4 Constant Group Delay

$\theta(\omega) = \theta_0 - \tau\omega$; with $\theta_0 = \pm\pi/2$ we must have

$$\tau = \frac{(N-1)T_s}{2} \text{ and } h(nT_s) = -h[(N-1-n)T_s] = \text{antisymmetrical}$$

29.1.5 Frequency Response of Constant-Delay Nonrecursive Filters

$h(nT_s)$	N	$H(e^{j\omega T_s})$
Symmetrical	Odd	$e^{-j\omega(N-1)T_s/2}\sum_{k=0}^{(N-1)/2} a_k\cos\omega kT_s$
	Even	$e^{-j\omega(N-1)T_s/2}\sum_{k=1}^{N/2} b_k\cos\left[\omega\left(k-\frac{1}{2}\right)T_s\right]$
Antisymmetrical	Odd	$e^{-j[\omega(N-1)T_s/2-\frac{\pi}{2}]}\sum_{k=1}^{(N-1)/2} a_k\sin\omega kT_s$
	Even	$e^{-j[\omega(N-1)T_s/2-\frac{\pi}{2}]}\sum_{k=1}^{N-2} b_k\sin\left[\omega\left(k-\frac{1}{2}\right)T_s\right]$

$a_0 = h\left[\frac{(N-1)T_s}{2}\right]$, $a_k = 2h\left[\left(\frac{N-1}{2}-k\right)T_s\right]$, $b_k = 2h\left[\left(\frac{N}{2}-k\right)T_s\right]$

29.2 Fourier Series Design

29.2.1 FIR Filter is periodic function of ω with period $\omega_s = 2\pi/T_s$.

29.2.2 Fourier Series

$$H(e^{j\omega T_s}) = \sum_{n=-\infty}^{\infty} h(nT_s)e^{-j\omega nT_s}$$

$$h(nT_s) = \frac{1}{\omega_s} \int_{-\omega_s/2}^{\omega_s/2} H(e^{j\omega T_s}) e^{j\omega n T_s} \, d\omega$$

29.2.3 Z-Transform Representation

$$H(z) = \sum_{n=-\infty}^{\infty} h(nT_s) z^{-n} \qquad (z = e^{j\omega T_s})$$

29.2.4 Noncausal Finite-Order Filter

$$h(nT_s) = 0 \quad \text{for} \quad |n| > \frac{N-1}{2}$$

$$H(z) = h(0) + \sum_{n=1}^{(N-1)/2} [h(-nT_s) z^n + h(nT_s) z^{-n}] = \text{noncausal}$$

29.2.5 Causal Finite-Order Filter

$$H'(z) = z^{-(N-1)/2} H(z)$$

Example

Design a low-pass filter with a frequency response

$$H(e^{j\omega T_s}) = \begin{cases} 1 & \text{for } |\omega| \le \omega_c \\ 0 & \text{for } \omega_c < |\omega| \le \dfrac{\omega_s}{2}, \end{cases} \qquad \omega_s = \text{sampling frequency}$$

Solution

From (29.2.2)

$$h(nT_s) = \frac{1}{\omega_s} \int_{-\omega_c}^{\omega_c} e^{j\omega n T_s} \, d\omega = \frac{1}{n\pi} \sin \omega_c n T_s$$

From (29.2.4) and (29.2.5)

$$H(z) = z^{-(N-1)/2} \sum_{n=0}^{(N-1)/2} \frac{a_n}{2} (z^n + z^{-n}), \quad a_0 = h(0), \quad a_n = 2h(nT_s)$$

For example, it may be requested that ω_c = cutoff frequency = $10 \, \text{rad s}^{-1}$ and the sampling frequency $\omega_s = 30 \, \text{rad s}^{-1}$. This implies that $T_s = 2\pi/30$, $z = e^{j\omega T_s} = \exp(j\omega 2\pi/30)$ and N is taken to be a relatively small number such as $N = 21, 41, 51$.

29.3 Window Functions in FIR Filters

29.3.1 Window Functions (see also Chapter 7)

$$H(z) = Z\{h(nT_s)\} = \sum_{n=-\infty}^{\infty} h(nT_s)z^{-n}$$

$$W(z) = Z\{w(nT_s)\} = \sum_{n=-\infty}^{\infty} w(nT_s)z^{-n}, \qquad w(nT_s) = \text{window function}$$

$$H_w(z) = Z\{w(nT_s)h(nT_s)\}$$

29.3.2 The Fourier Transform of the Windowed Filter

$$H_w(e^{j\omega T_s}) = \frac{T_s}{2\pi}\int_0^{2\pi/T_s} H(e^{j\xi T_s})W(e^{j(\omega-\xi)T_s})d\xi$$

In the ξ-domain

$$H(e^{j\xi T_s}) = \begin{cases} 1 & \text{for} \quad 0 \le |\xi| \le \omega_c \\ 0 & \text{for} \quad \omega_c < |\xi| \le \dfrac{\omega_s}{2} \end{cases}$$

and let $W(e^{j\xi T_s})$ be real and assume

$$W(e^{j\xi T_s}) = 0 \quad \text{for } \omega_m \le |\xi| \le \frac{\omega_s}{2}$$

29.3.3 Properties of Window Function $w(nT_s)$

1. $w(nT_s)$ for $|n| > \dfrac{N-1}{2}$
2. For odd N, it must be symmetrical about sample $n = 0$
3. Width of main lobe: $k\omega_s / N$, k = constant
4. Sidelobes give Gibbs oscillations in the amplitude response of the filter

29.4 Windows Frequently Used

29.4.1 Rectangular

$$w_R(nT_s) = \begin{cases} 1 & \text{for } |n| \le \dfrac{N-1}{2} \\ 0 & \text{otherwise} \end{cases}$$

$$W(e^{j\omega T_s}) = \sum_{n=-(N-1)/2}^{(N-1)/2} e^{-j\omega n T_s} = \frac{\sin(\omega N T_s/2)}{\sin(\omega T_s/2)}$$

Lobe Widths

$$W(e^{j\omega T_s}) = 0 \text{ at } \omega = m\omega_s / N, \quad m = \pm 1, \pm 2, \cdots$$

$$\text{Main lobe width} = 2\omega_s / N$$

Ripple Ratio

$$r = \frac{100(\text{maximum side-lobe amplitude})}{\text{main-lobe amplitude}}\%$$

29.4.2 Hann and Hamming Windows

$$w_H(nT_s) = \begin{cases} \alpha + (1-\alpha)\cos\dfrac{2\pi n}{N-1} & \text{for } |n| \le \dfrac{N-1}{2} \\ 0 & \text{otherwise} \end{cases}$$

$\alpha = 0.5$ Hann window, $\alpha = 0.54$ Hamming window

29.4.3 Blackman Window

$$w_B(nT_s) = \begin{cases} 0.42 + 0.5\cos\dfrac{2\pi n}{N-1} + 0.08\cos\dfrac{4\pi n}{N-1} & \text{for } |n| \le \dfrac{N-1}{2} \\ 0 & \text{otherwise} \end{cases}$$

Example

Design a low-pass filter with a frequency response

$$H(e^{j\omega T_s}) = \begin{cases} 1 \text{ for } |\omega| \le \omega_c \\ 0 \text{ for } \omega_c < |\omega| \le \dfrac{\omega_s}{2} \end{cases}$$

where ω_s is the sampling frequency using the window approach.

Solution

$$h(nT_s) = \frac{1}{\omega_s}\int_{-\omega_s/2}^{\omega_s/2} e^{j\omega n T_s}\, d\omega = \frac{1}{n\pi}\sin\omega_c nT_s$$

$$H'_w(z) = z^{-(N-1)/2}\sum_{n=0}^{(N-1)/2} \frac{a'_n}{2}(z^n + z^{-n}), \quad a'_0 = w(0)h(0), \; a'_n = 2w(nT_s)h(nT_s)$$

$$|M(\omega)| = \left|\sum_{n=0}^{(N-1)/2} a'_n \cos\omega nT_s\right|.$$

Any of the above windows can be used.

29.4.4 Dolph-Chebyshev Window

$$w_{DC}(nT_s) = \frac{1}{N}\left[\frac{1}{r} + 2\sum_{i=1}^{(N-1)/2} T_{N-1}\left(x_0 \cos\frac{i\pi}{N}\right)\cos\frac{2n\pi i}{N}\right], \qquad n = 0,1,2,\cdots,(N-1)/2$$

r = required ripple ratio (see 29.4.1) and $x_0 = \cosh\left(\frac{1}{N-1}\cosh^{-1}\frac{1}{r}\right)$

$T_{N-1}(x)$ is the $(N-1)^{th}$-order Chebyshev polynomial and is given by

$$T_{N-1}(x) = \begin{cases} \cos((N-1)\cos^{-1}x) & \text{for } |x| \le 1 \\ \cosh(\cosh^{-1}x) & \text{for } |x| > 1 \end{cases}$$

Properties

a) An arbitrary ripple ratio can be achieved,
b) The main-lobe width is controlled by choosing N,
c) With N fixed, the main-lobe width is the smallest that can be achieved for a given ripple ratio, and
d) All the side lobes have the same amplitude.

29.4.5 Kaiser Window

$$w_K(nT) = \begin{cases} \dfrac{I_o(\beta)}{I_o(\alpha)} & \text{for } |n| \le \dfrac{N-1}{2} \\ 0 & \text{otherwise} \end{cases}$$

α = independent parameter, $\qquad \beta = \alpha\sqrt{1-\left(\frac{2n}{N-1}\right)^2}$

$I_0(x) =$ zero order modified Bessel function of the first kind (see Chapter 25)

$$I_0(x) = 1 + \sum_{k=1}^{\infty}\left[\frac{1}{k!}\left(\frac{x}{2}\right)^k\right]^2$$

$$W_K(e^{j\omega T_s}) = w_k(0) + 2\sum_{n=1}^{(N-1)/2} w_K(nT_s)\cos\omega nT_s$$

29.4.6 Window Parameters

		Ripple ratio in %		
Window	Main-lobe width	N = 11	N = 21	N = 101
Rectangular	$2\omega_s/N$	22.34	21.89	21.70
Hann	$4\omega_s/N$	2.62	2.67	2.67
Hamming	$4\omega_s/N$	1.47	0.93	0.74
Blackman	$6\omega_s/N$	0.08	0.12	0.12

29.4.7 Filter Specifications (see Figure 29.1)

A_p = passband ripple $= 20\log\frac{1+\delta}{1-\delta}$ (in dB)

A_a = stopband attenuation $= -20\log\delta$ (in dB)

B_t = transition width (rad s^{-1})

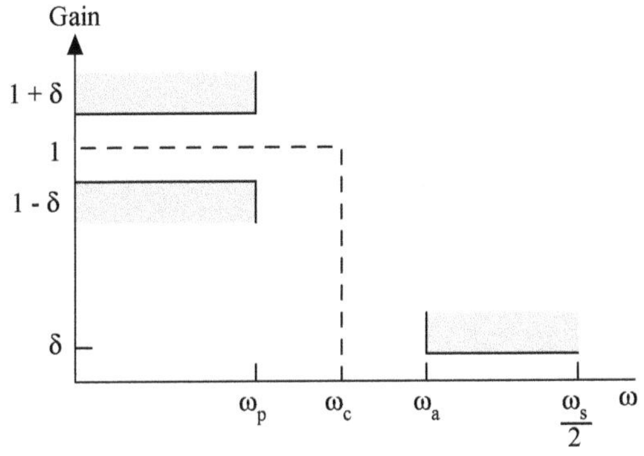

FIGURE 29.1

Steps for Design

1. Determine $h(nT_s)$ using Fourier-series (see 29.2.2), $\omega_c = \frac{1}{2}(\omega_p + \omega_a)$ (see Figure 29.1)
2. Choose δ such that $A_p \le A'_p$ and $A_a \ge A'_a$ where A'_p and A'_p are the desired passband ripple and stopband attenuation, respectively. Choose

$$\delta = \min(\delta_1, \delta_2) \text{ where } \delta_1 = 10^{-0.05A'_a},\ \delta_2 = \frac{10^{0.05A'_p} - 1}{10^{0.05A'_p} + 1}$$

3. Find $A_a = -20\log[\min(\delta_1, \delta_2)]$
4. Choose parameter α as follows:

$$\alpha = \begin{cases} 0 & \text{for } A_a \le 21 \\ 0.5842(A_a - 21)^{0.4} + 0.07886(A_a - 21) & \text{for } 21 < A_a \le 50 \\ 0.1102(A_a - 8.7) & \text{for } A_a > 50 \end{cases}$$

5. Choose D as follows:

$$D = \begin{cases} 0.9222 & \text{for } A_a \le 21 \\ \dfrac{A_a - 7.95}{14.36} & \text{for } A_a > 21 \end{cases}$$

Then select the lowest *odd* value of N satisfying the inequality

$$N \ge \frac{\omega_s D}{B_t} + 1$$

6. Use Kaiser window (see 29.4.5)
7. Form $H'_w(z) = z^{-(N-1)/2} H_w(z), \quad H_w(z) = Z\{w_K(nT_s)h(nT_s)\}$

Example

Design a lowpass filter satisfying the following specifications:

a) Maximum passband ripple to be 0.1 dB in the range 0 to 2 rad s^{-1}
b) Minimum stopband attenuation to be 35 dB in the range from 3 to 4.5 rad s^{-1}
c) Sampling frequency $\omega_s = 10$ rad s^{-1}

Solution

1. $h(nT_s) = \dfrac{\sin \omega_c nT_s}{n\pi}, \quad \omega_c = \tfrac{1}{2}(2+3) = 2.5 \text{ rad s}^{-1}$

2. $\delta_1 = 10^{-0.05\times 35} = 0.0178, \quad \delta_2 = \dfrac{10^{0.05\times 0.1} - 1}{10^{0.05\times 0.1} + 1} = 5.7564 \times 10^{-3},$

   $\min(\delta_1, \delta_2) = 5.7564 \times 10^{-3}$

3. $A_a = -20 \log(5.7564 \times 10^{-3}) = 44.797 \text{ dB}$

4. $\alpha = 0.5842(44.797 - 21)^{0.4} + 0.07886(44.797 - 21) = 3.9524$

   $D = (44.797 - 7.95)/14.36 = 2.5660$

5. $N \geq \dfrac{10(2.566)}{1} + 1 = 26.66 \text{ or } N = 27$

6. $H'_w(z) = z^{-(N-1)/2} \sum_{n=0}^{(N-1)/2} \dfrac{a'_n}{2}(z^n + z^{-n}), \; a'_0 = w_K(0)h(0), \; a'_n = 2w_K(nT_s)h(nT_s)$

29.5 Highpass FIR Filter

29.5.1 Transition Width (see Figure 29.2)

$$B_t = \omega_p - \omega_a$$

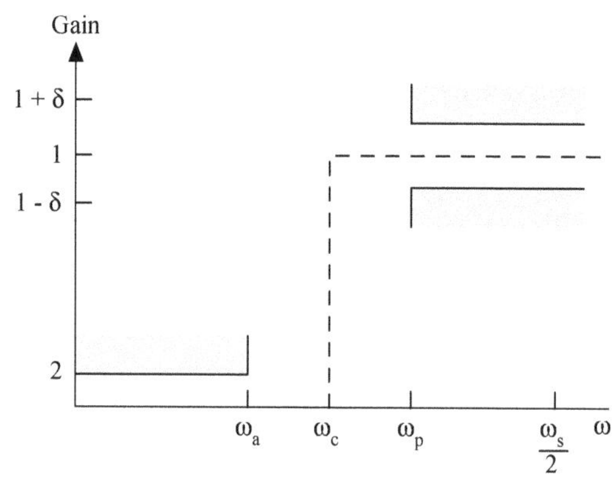

FIGURE 29.2

29.5.2 Ideal Frequency Response

$$H(e^{j\omega T_s}) = \begin{cases} 0 & \text{for } |\omega| \le \omega_c \\ 1 & \text{for } \omega_c < |\omega| \le \dfrac{\omega_s}{2} \end{cases}$$

$$\omega_c = (\omega_a + \omega_p)/2$$

29.6 Bandpass FIR Filter

29.6.1 Transition Width

$$B_t = \min\{(\omega_{p1} - \omega_{a1}),\ (\omega_{a2} - \omega_{p2})\}$$

29.6.2 Ideal Frequency Response (see Figure 29.3)

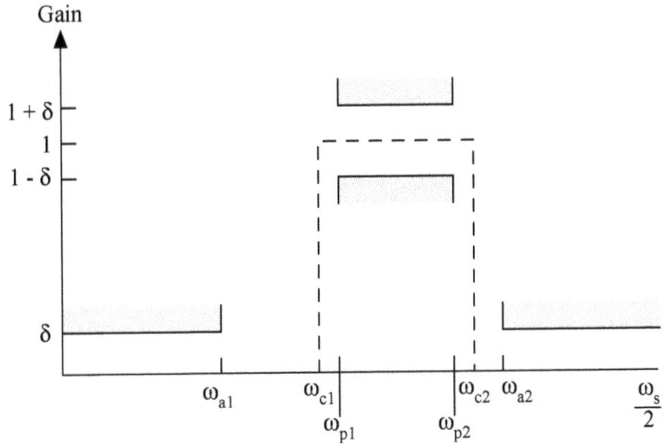

FIGURE 29.3

$$H(e^{j\omega T_s}) = \begin{cases} 0 & \text{for } 0 \le |\omega| \le \omega_{c1} \\ 1 & \text{for } \omega_{c1} \le |\omega| \le \omega_{c2} \\ 0 & \text{for } \omega_{c2} < |\omega| \le \dfrac{\omega_s}{2} \end{cases} \qquad \begin{aligned} \omega_{c1} &= \omega_{p1} - \frac{B_t}{2} \\ \omega_{c2} &= \omega_{p2} + \frac{B_t}{2} \end{aligned}$$

29.7 Bandstop FIR Filter

29.7.1 Transition Width

$$B_t = \min\{(\omega_{a1} - \omega_{p1}),\ (\omega_{p2} - \omega_{a2})\}$$

29.7.2 Ideal Frequency Response (see Figure 29.4)

$$H(e^{j\omega T_s}) = \begin{cases} 1 & \text{for } 0 \le |\omega| \le \omega_{c1} \\ 0 & \text{for } \omega_{c1} \le |\omega| \le \omega_{c2} \\ 1 & \text{for } \omega_{c2} < |\omega| \le \dfrac{\omega_s}{2} \end{cases}$$

$$\omega_{c1} = \omega_{p1} + \frac{B_t}{2}, \quad \omega_{c2} = \omega_{p2} - \frac{B_t}{2}$$

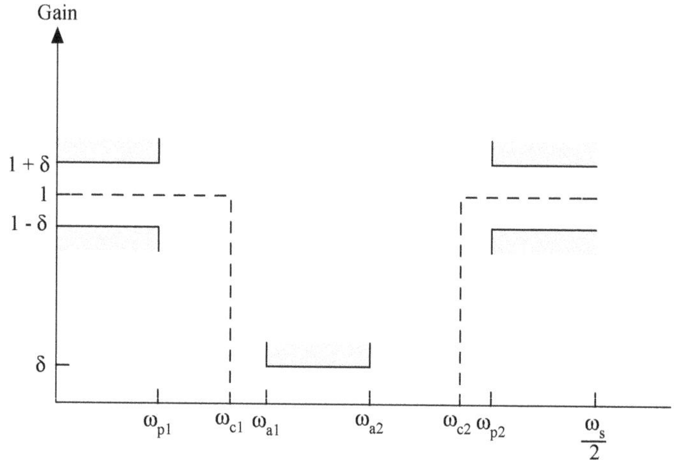

FIGURE 29.4

Example

Design a lowpass filter satisfying the following specifications:

a) Minimum stopband attenuation for $0 \le \omega \le 200$ to be 45 dB
b) Maximum passband ripple for $400 < \omega < 600$ to be 0.2 dB
c) Minimum stopband attenuation for $700 \le \omega \le 1000$ to be 45 dB
d) Sampling frequency $\omega_s = 2000$ rad s^{-1}

Solution

$B_t = \min\{(400-200), (700-600)\} = 100,$, $\omega_{c1} = 400 - 50 = 350,$ $\omega_{c2} = 600 + 50 = 650,$

$$h(nT_s) = \frac{1}{\omega_s}\int_{-\omega_s/2}^{\omega_s/2} H(e^{j\omega T_s})e^{j\omega n T_s}\,d\omega = \frac{1}{\omega_s}\int_o^{\omega_s/2} [H(e^{j\omega T_s})e^{j\omega n T_s}$$

$$+ H(e^{-j\omega T_s})e^{-j\omega n T_s}]d\omega = \frac{1}{\omega_s}\int_{\omega_{c1}}^{\omega_{c2}} 2\cos\omega n T_s\,d\omega = \frac{1}{\pi n}(\sin\omega_{c2} n T_s - \sin\omega_{c1} n T_s)$$

$$T_s = 2\pi/2000, \quad \delta_1 = 10^{-0.05(45)} = 5.6234\times10^{-3}, \quad \delta_2 = \frac{10^{0.05(0.2)}-1}{10^{0.05(0.2)}+1} = 1.1512\times10^{-2},$$

$\min(\delta_1,\delta_2) = 5.6234\times10^{-3}$, $A_a = 45\,dB$, $\alpha = 3.9754$, $D = 2.580$, $N = 53$ and continue as Example 4.2.1.

30

Recursive Filters (Infinite Impulse Response, IIR)

30.1 Introduction

30.1.1 Realizable Filter

The transfer function must a) be a rational function of z with real coefficients, b) have poles that lie within the unit circle of the z plane, and c) have the degree of the numerator equal to or less than that of the denominator polynomial.

30.2 Invariant-Impulse-Response Method

30.2.1 Steps to be taken

1. Deduce $h_A(t) =$ impulse response of the analog filter $= L^{-1}\{H_A(s)\}$, $h(0+) = \lim_{s\to\infty} s\{H_A(s)\}$
2. Replace t by nT_s in $h_A(t)$
3. Form the Z-transform of $h_A(nT_s)$

30.2.2 Conditions

If $H_A(\omega) \cong 0$ for $|\omega| \geq \omega_s/2$ and $h(0+) = 0$, then

$$T_s H_D(e^{j\omega T_s}) = H_A(\omega) \qquad \text{for } |\omega| \leq \frac{\omega_s}{2}$$

Simple poles

$$H_A(s)=\sum_{i=1}^{N}\frac{A_i}{s-p_i},\qquad h_A(t)=\mathcal{L}^{-1}\{H_A(s)\}=\sum_{i=1}^{N}A_i e^{p_i t}$$

$$A_i=[(s-p_i)H_A(s)]\big|_{s=p_i}$$

$$h_A(nT_s)=\sum_{i=1}^{N}A_i e^{p_i nT_s}$$

$$H_D(z)=\mathcal{Z}\{h_A(nT_s)\}=\sum_{i=1}^{N}A_i\sum_{n=0}^{\infty}(e^{p_iT_s}z^{-1})^n=\sum_{i=1}^{N}\frac{A_i}{1-e^{p_iT_s}z^{-1}}=\sum_{i=1}^{N}A_i\frac{z}{z-e^{p_iT_s}}$$

30.2.3 Procedure of Impulse-Invariant of IIR Filters

1. Obtain the transfer function $H_A(s)$ for the desired analog prototype filter (see Chapter 12)
2. For $i=1,2,\cdots,N$ determine the poles of p_i and $H_A(s)$ and compute the coefficients

$$A_i=[(s-p_i)H_A(s)]\big|_{s=p_i}$$

3. Use A_i's from step 2 to generate the digital filter system function

$$H(z)=\sum_{i=1}^{N}\frac{A_i}{1-\exp(p_iT_s)z^{-1}}$$

Example

The normalized transfer function of a second-order Butterworth analog filter with a 3-dB cutoff frequency at 3000 Hz. The sampling frequency is $f_s = 30{,}000\ \text{s}^{-1}$.

Solution

$$H_A(s)=\frac{1}{s^2+\sqrt{2}\,s+1}=\text{normalized analog filter}=\frac{1}{(s-p_1)(s-p_2)}$$

$$\left[p_1=-\frac{\sqrt{2}}{2}+j\frac{\sqrt{2}}{2},\ p_2=-\frac{\sqrt{2}}{2}-j\frac{\sqrt{2}}{2}\right],\qquad \omega_c=2\pi 3000=6000\,\pi,$$

$$H_A\left(\frac{s}{\omega_c}\right)=\text{un-normalized filter}=\frac{1}{\left(\frac{s}{\omega_c}-p_1\right)\left(\frac{s}{\omega_c}-p_2\right)}=\frac{\omega_c^2}{(s-\omega_c p_1)(s-\omega_c p_2)}$$

$$=\frac{A_1}{s-\omega_c p_1}+\frac{A_2}{s-\omega_c p_2},\qquad A_1=\frac{\omega_c^2}{\omega_c p_1-\omega_c p_2}=\frac{\omega_c}{p_1-p_2},\qquad A_2=\frac{\omega_c^2}{\omega_c p_2-\omega_c p_1}=\frac{\omega_c}{p_2-p_1}.$$

But $T_s = 1/f_s = 1/30,000$, $\omega_c T_s = \pi/5$, and hence

$$H_D(z) = \frac{\omega_c}{p_1 - p_2}\frac{1}{1-\exp[\omega_c T_s p_1]z^{-1}} + \frac{\omega_c}{p_2 - p_1}\frac{1}{1-\exp[\omega_c T_s p_2]z^{-1}}$$

$$= \frac{(-j\sqrt{2}\,\omega_c/2)}{1-\exp[\omega_c T_s p_1]z^{-1}} + \frac{(j\sqrt{2}\,\omega_c/2)}{1-\exp[\omega_c T_s p_2]z^{-1}}.$$

Hence $\left|H_D(e^{j\omega T_s})\right| = \left|\frac{(-j\sqrt{2}\,\omega_c/2)}{1-\exp[\omega_c T_s p_1]e^{-j\omega T_s}} + \frac{(j\sqrt{2}\,\omega_c/2)}{1-\exp[\omega_c T_s p_2]e^{-j\omega T_s}}\right|$

30.3 Modified-Invariant-Impulse-Response

30.3.1 Analog Transfer Function

$$H_A(s) = H_0\frac{N(s)}{D(s)} = H_0\frac{\prod_{i=1}^{M}(s-s_i)}{\prod_{i=1}^{N}(s-p_i)}, \quad M \le N;$$

$$H_A(s) = H_0\frac{H_{A1}(s)}{H_{A2}(s)}, \quad H_{A1}(s) = \frac{1}{D(s)}, \quad H_{A2}(s) = \frac{1}{N(s)}$$

Conditions: $h_{A1}(0+) = 0$, $h_{A2}(0+) = 0$, $M, N \ge 2$, $H_{A1}(\omega) = H_{A2}(\omega) \cong 0$ for $|\omega| \ge \frac{\omega_s}{2}$

30.3.2 Digital Filter

$$H_D(z) = H_0\frac{H_{D1}(z)}{H_{D2}(z)}, \quad H_D(e^{j\omega T_s}) = H_0\frac{H_{D1}(e^{j\omega T_s})}{H_{D2}(e^{j\omega T_s})} \cong H_A(\omega) \text{ for } |\omega| \le \frac{\omega_s}{2}$$

30.3.3 Zeroes and poles of $H_A(s)$

$$H_{D1}(z) = \sum_{i=1}^{N}\frac{A_i z}{z - e^{p_i T_s}} = \frac{N_1(z)}{D_1(z)}, \quad H_{D2}(z) = \sum_{i=1}^{M}\frac{B_i z}{z - e^{T_s s_i}} = \frac{N_2(z)}{D_2(z)}, H_D(z) = H_0\frac{N_1(z)D_2(z)}{N_2(z)D_1(z)},$$

30.3.4 Stability

If any pole of $H_D(z)$ is located outside the unit circle it can be replaced by their reciprocals without changing the shape of the loss characteristics, although a constant vertical shift will be introduced.

If any pole is on the unit circle, its magnitude can be reduced slightly.

30.4 Matched-Z-Transform Method

30.4.1 Matched-Z-Transform Method

$$H_D(z) = (z+1)^L H_0 \frac{\prod_{i=1}^{M}(z - e^{s_i T_s})}{\prod_{i=1}^{N}(z - e^{p_i T_s})}, \quad z = e^{j\omega T_s}$$

Values of L

Filter Type	Lowpass	Highpass	Bandpass	Bandstop
All pole	N	0	N/2	0
Elliptic				
N odd	1	0		
N even	0	0	0 for N/2 even	0

30.5 Bilinear-Transformation Method

30.5.1 Analog Integrator

$$H_{AI}(s) = \frac{1}{s}, \quad h_{AI}(t) = \begin{cases} 1 & t \geq 0+ \\ 0 & t \leq 0- \end{cases}$$

30.5.2 Digital Integrator

$$y(nT_s) - y(nT_s - T_s) = \frac{T_s}{2}[x(nT_s - T_s) + x(nT_s)]$$

$$Y(z) - z^{-1}Y(z) = \frac{T_s}{2}[z^{-1}X(z) + X(z)]$$

30.5.3 Transfer Function

$$H_{DI}(z) = \frac{Y(z)}{X(z)} = \frac{T_s}{2}\left(\frac{z+1}{z-1}\right)$$

30.5.4 Bilinear Transformation

$$H_{DI}(z) = H_{AI}(s)\Big|_{s=\frac{2}{T_s}\left(\frac{z-1}{z+1}\right)}$$

30.5.5 Analog Filter Transfer Function

$$H_A(s)=\frac{\sum_{i=0}^{N} a_i s^{N-i}}{s^N+\sum_{i=0}^{N} b_i z^{N-i}}=\frac{\sum_{i=0}^{N} a_i\left[\frac{1}{H_{AI}(s)}\right]^{N-i}}{\left[\frac{1}{H_{AI}(s)}\right]^{N}+\sum_{i=0}^{N} b_i\left[\frac{1}{H_{AI}(s)}\right]^{N-i}} \qquad \text{(see 30.5.1)}$$

30.5.6 Discrete-Time Transfer Function

$$H_D(z)=\frac{\sum_{i=0}^{N} a_i\left[\frac{1}{H_{DI}(s)}\right]^{N-i}}{\left[\frac{1}{H_{DI}(z)}\right]^{N}+\sum_{i=0}^{N} b_i\left[\frac{1}{H_{DI}(z)}\right]^{N-i}}=H_A(s)\Big|_{s=\frac{2}{T_s}\left(\frac{z-1}{z+1}\right)}$$

by replacing $H_{AI}(s)$ by $H_{DI}(z)$.

30.5.7 Mapping Properties of Bilinear Transformation

$$z=\frac{\frac{2}{T_s}+s}{\frac{2}{T_s}-s}$$

a) The open right-half s-plane is mapped onto the region exterior to the unit circle $|z|=1$ of the z-plane;
b) The j axis of the s-plane is mapped onto the unit circle $|z|=1$ of the z-plane;
c) The open left-half s-plane is mapped onto the interior of the unit circle $|z|=1$;
d) The origin of the s-plane maps onto point (1,0) of the z-plane;
e) The positive and negative j axes of the s-plane map onto the upper and lower semicircles of $|z|=1$, respectively;
f) The maxima and minima of $|H_A(\omega)|$ are preserved in $|H_D(e^{j\Omega T_s})|$;
g) If $m_1 \le |H_A(\omega)| \le m_2$ in $\omega_1 \le \omega \le \omega_2$, then $m_1 \le |H_D(e^{j\Omega T_s})| \le m_2$ for a corresponding frequency $\Omega_1 \le \Omega \le \Omega_2$;
h) Passbands or stopbands in the analog filter translate into passbands or stopbands in the digital filter;
i) A stable analog filter will yield a stable digital filter.

30.5.8 The Warping Effect

From (30.5.6) $H_D(e^{j\Omega T_s})=H_A(s)$ provided that $\omega=\frac{2}{T_s}\tan\frac{\Omega T_s}{2}$. For $\Omega<0.3/T_s$, $\omega\cong\Omega$ and hence both filters have the same frequency response. Figure 30.1 shows the warping effect.

Note: If prescribed passband and stopband edges $\tilde{\Omega}_1,\tilde{\Omega}_2,\cdots,\tilde{\Omega}_i$ are to be achieved in the digital filter, the analog filter must be prewarped before application to ensure that

$$\omega_i=\frac{2}{T_s}\tan\frac{\tilde{\Omega}_i T_s}{2}$$

and hence $\Omega_i=\tilde{\Omega}_i$.

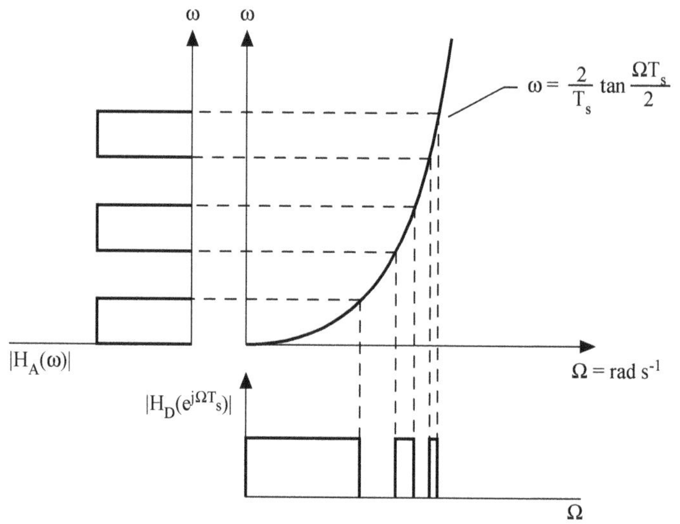

FIGURE 30.1

Note: The phase response of the derived digital filter is nonlinear, although the analog filter has linear phase.

Example

The second-order Butterworth analog filter

$$H_A(s) = \frac{\omega_c^2}{s^2 + \sqrt{2}\,\omega_c s + \omega_c^2}$$

with 3-dB cutoff frequency of 3 kHz and sampling rate of 30,000 samples per second is transformed to an IIR filter using bilinear transformation as follows:

$$\omega_c = 2\pi 3000 = 6000\pi,\ \ s = 2(1 - z^{-1})/[T_s(1 + z^{-1})],\text{ and } T_s = 1/30{,}000$$

$$H_D(z) = \frac{(6000\pi)^2}{(60{,}000)^2\left[\dfrac{1 - z^{-1}}{1 + z^{-1}}\right]^2 + 12000\pi \times 60{,}000\left(\dfrac{1 - z^{-1}}{1 + z^{-1}}\right) + (6000\pi)^2}$$

$$= \frac{0.063964 + 0.127929z^{-1} + 0.063964z^{-2}}{1 - 1.168261z^{-1} + 0.424118z^{-2}}$$

30.6 Digital-Filter Transformations

30.6.1 Constantimides Transformations are given in Table 30.1.

TABLE 30.1 Table of IIR Digital-Filter Transformations

Type	Transformation	α, k
LP to LP	$z = \dfrac{\bar{z} - \alpha}{1 - \alpha\bar{z}}$	$\alpha = \dfrac{\sin[(\Omega_p - \omega_p)T_s/2]}{\sin[(\Omega_p + \omega_p)T_s/2]}$
LP to HP	$z = -\dfrac{\bar{z} - \alpha}{1 - \alpha\bar{z}}$	$\alpha = \dfrac{\cos[(\Omega_p - \omega_p)T_s/2]}{\cos[(\Omega_p + \omega_p)T_s/2]}$
LP to BP	$z = -\dfrac{\bar{z}^2 - \dfrac{2\alpha k}{k+1}\bar{z} + \dfrac{k-1}{k+1}}{1 - \dfrac{2\alpha k}{k+1}\bar{z} + \dfrac{k-1}{k+1}\bar{z}^2}$	$\alpha = \dfrac{\cos[(\omega_{p2} + \omega_{p1})T_s/2]}{\cos[(\omega_{p2} - \omega_{p1})T_s/2]}$
		$k = \tan\dfrac{\Omega_p T_s}{2}\cot\dfrac{(\omega_{p2} - \omega_{p1})T_s}{2}$
LP to BS	$z = \dfrac{\bar{z}^2 - \dfrac{2\alpha}{1+k}\bar{z} + \dfrac{1-k}{1+k}}{1 - \dfrac{2\alpha}{1+k}\bar{z} + \dfrac{1-k}{1+k}\bar{z}^2}$	$\alpha = \dfrac{\cos[(\omega_{p2} + \omega_{p1})T_s/2]}{\cos[(\omega_{p2} - \omega_{p1})T_s/2]}$
		$k = \tan\dfrac{\Omega_p T_s}{2}\tan\dfrac{(\omega_{p2} - \omega_{p1})T_s}{2}$

$z = re^{j\omega T_s}$, $\bar{z} = Re^{j\Omega T_s}$, Ω = digital frequency, ω = analog frequency.

30.6.2 Transformation Applications

1. Obtain a lowpass normalized transfer function $H_N(z)$ using any approximation method.
2. Determine the passband edge Ω_p in $H_N(z)$.
3. Form $H(\bar{z})$ using $H(\bar{z}) = H_N(z)\big|_{z=f(\bar{z})}$.

References

Antoniou, A., *Digital Filters: Analysis Design and Applications*, 2nd Edition, McGraw-Hill Inc., New York, NY, 1993.

Bose, N. K., *Digital Filters: Theory and Applications*, Elsevier Publishers, New York, NY, 1985.

Parks, T. W. and C. S. Burrus, *Digital Filter Design*, John Wiley & Sons Inc., New York, NY, 1987.

31

Recursive Filters Satisfying Prescribed Specifications

31.1 Design Procedure

31.1.1 Design Is Accomplished in Two Steps

1. A normalized lowpass filter transfer function is transformed into a denormalized filter (lowpass, highpass, etc. transfer function) employing the standard analog-filter transformation.
2. A bilinear transformation is applied.

31.1.2 Passband Stopband Edges

If ω_i are analog passband and stopband edges of an analog filter, the corresponding passband and stopband edges in the derived digital filter are given by

$$\Omega_i = \frac{2}{T_s}\tan^{-1}\frac{\omega_i T_s}{2}, \quad i = 1,2,\cdots$$

31.1.3 Prewarping

If desired passband and stopband edges Ω_{di} are to be achieved, the analog filter must be prewarped before the application of the bilinear transformation to ensure $\omega_i = \frac{2}{T_s}\tan\frac{\Omega_{di}T_s}{2}$, so that $\Omega_i = \Omega_{di}$.

31.1.4 Loss Amplitude

$A_N(\omega) = 20\log\frac{1}{|H_N(\omega)|}$ = loss amplitude of an analog normalized lowpass filter $H_N(\omega)$

$$0 \le A_N(\omega) \le A_p \quad \text{for} \quad 0 \le |\omega| \le \omega_p$$

$$A_N(\omega) \ge A_a \quad \text{for} \quad \omega_a \le |\omega| < \infty \qquad \text{(see Figure 31.1)}$$

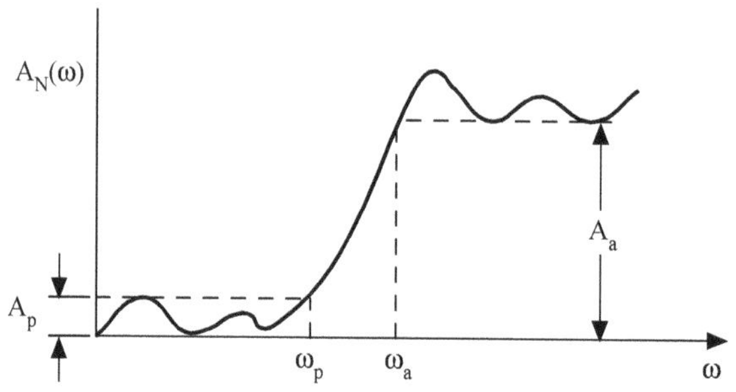

FIGURE 31.1

31.1.5 Transformation

Step 1. Form $H_t(\bar{s}) = H_N(s)\big|_{s=f(\bar{s})}$

Step 2. Form $H_D(s) = H_t(\bar{s})\big|_{\bar{s}=\frac{2}{T_s}\left(\frac{z-1}{z+1}\right)}$

TABLE 31.1 Transformation of Analog Filters

Type	Transformation, $s = f(\bar{s})$
LP to LP	$s = \lambda\bar{s}$
LP to HP	$\lambda/\bar{s}$
LP to BP	$s = \dfrac{1}{B}\left(\bar{s} + \dfrac{\omega_0^2}{\bar{s}}\right)$
LP to BS	$s = \dfrac{B\bar{s}}{\bar{s}^2 + \omega_0^2}$

The parameters λ, ω_0 and B and order $H_N(s)$ must be chosen appropriately (see Section 31.2).

31.2 Analog Filters

31.2.1 Butterworth Filters

$A_N(\omega) = \text{normalized loss} = 10\log(1+\omega^{2n})$

$$n \ge \frac{\log D}{2\log(1/K)},\quad \omega_p = (10^{0.1A_p}-1)^{1/2n},\quad \omega_a = (10^{0.1A_a}-1)^{1/2n},\quad D = \frac{10^{0.1A_a}-1}{10^{0.1A_p}-1}$$

LP	$K = K_o$
HP	$K = 1/K_o$
BP	$K = \begin{cases} K_1 & \text{if } K_c \ge K_B \\ K_2 & \text{if } K_c < K_B \end{cases}$
BS	$K = \begin{cases} \dfrac{1}{K_2} & \text{if } K_c \ge K_B \\ \dfrac{1}{K_1} & \text{if } K_c < K_B \end{cases}$

$$K_A = \tan\frac{\Omega_{dp2}T_s}{2} - \tan\frac{\Omega_{dp1}T_s}{2}, \qquad K_B = \tan\frac{\Omega_{dp1}T_s}{2}\tan\frac{\Omega_{dp2}T_s}{2}$$

$$K_C = \tan\frac{\Omega_{da1}T_s}{2}\tan\frac{\Omega_{da2}T_s}{2}, \qquad K_1 = \frac{K_A\tan(\Omega_{da1}T_s/2)}{K_B - \tan^2(\Omega_{da1}T_s/2)}$$

$$K_2 = \frac{K_A\tan(\Omega_{da2}T_s/2)}{\tan^2(\Omega_{da2}T_s/2) - K_B}, \qquad K_0 = \frac{\tan(\Omega_{dp}T_s/2)}{\tan(\Omega_{da}T_s/2)}$$

$$A_N(\omega_p) = A_p = 10\log(1+\omega_p^{2n}), \quad A_N(\omega_a) = A_a = 10\log(1+\omega_a^{2n})$$

31.2.2 Chebyshev Filter

$A_N(\omega) = \text{normalized loss} = 10\log[1+\varepsilon^2 T_n^2(\omega)]$
$T_n(\omega) = \cosh(n\cosh^{-1}\omega)$ for $\omega_p \le \omega < \infty$, $\varepsilon^2 = 10^{0.1A_p} - 1$, $\omega_p = 1$

$$D = \frac{10^{0.1A_a}-1}{10^{0.1A_p}-1}, \quad n \ge \frac{\cosh^{-1}\sqrt{D}}{\cosh^{-1}(1/K)}, \quad \omega_p = 1, \quad K_0 = \frac{\tan(\Omega_{dp}T_s/2)}{\tan(\Omega_{da}T_s/2)}$$

LP	$K = K_0$
HP	$K = 1/K_0$
BP	$K = \begin{cases} K_1 & \text{if } K_c \ge K_B \\ K_2 & \text{if } K_c < K_B \end{cases}$
BS	$K = \begin{cases} \dfrac{1}{K_2} & \text{if } K_c \ge K_B \\ \dfrac{1}{K_1} & \text{if } K_c < K_B \end{cases}$

31.2.3 Elliptic Filters

$$n \ge \frac{\log 16D}{\log(1/q)}, \quad k' = \sqrt{1-k^2}, \quad q_0 = \frac{1}{2}\left(\frac{1-\sqrt{k'}}{1+\sqrt{k'}}\right)$$

$$q = q_0 + 5q_0^5 - 5q^9 + 10q^{13}, \quad \omega_p = \sqrt{k},, \quad K_0 = \frac{\tan(\Omega_{dp}T_s/2)}{\tan(\Omega_{da}T_s/2)}$$

	k	ω_p
LP	K_0	$\sqrt{K_0}$
HP	$1/K_0$	$1/\sqrt{K_0}$
BP	K_1 if $K_c \ge K_B$	$\sqrt{K_1}$
	K_2 if $K_c < K_B$	$\sqrt{K_2}$

BS	$\frac{1}{K_2}$ if $K_c \geq K_B$	$1/\sqrt{K_2}$
	$\frac{1}{K_1}$ if $K_c < K_B$	$1/\sqrt{K_1}$

31.3 Design Formulas

31.3.1 Lowpass and Highpass Filters (see Figure 31.2)

LP	$\omega_a \leq \frac{\omega_p}{K_0}$	$K_0 = \frac{\tan(\Omega_{dp}T_s/2)}{\tan(\Omega_{da}T_s/2)}$
	$\lambda = \frac{\omega_p T_s}{2\tan(\Omega_{dp}T_s/2)}$	
HP	$\omega_a \leq \omega_p K_0$	
	$\lambda = \frac{2\omega_p \tan(\Omega_{dp}T_s/2)}{T_s}$	

$$A_D(\Omega) = 20\log\frac{1}{\left|H_D(e^{j\Omega T_s})\right|} = \text{loss characteristics}$$

$$0 \leq A_D \leq A_p \quad \text{for} \quad 0 \leq |\Omega| \leq \Omega_p$$

$$A_D \geq A_a \quad \text{for} \quad \Omega_a \leq |\Omega| \leq \frac{\omega_s}{2}$$

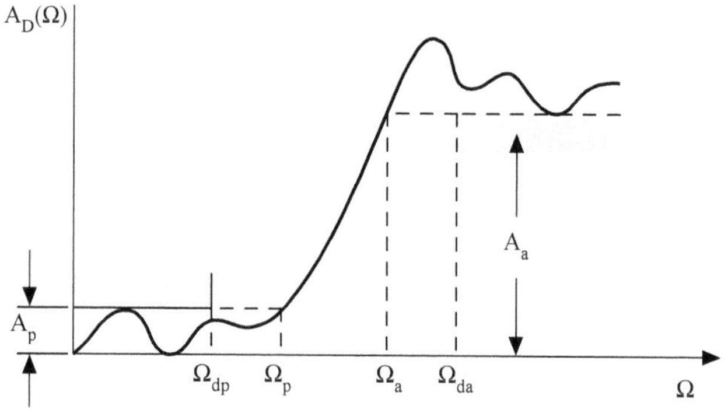

FIGURE 31.2

31.3.2 Bandpass and Bandstop Filters (see Figure 31.3)

BP	$\omega_0 = 2\sqrt{K_B}/T_s$
	$\omega_a \le \begin{cases} \omega_p/K_1 & \text{if } K_c \ge K_B \\ \omega_p/K_2 & \text{if } K_c < K_B \end{cases}$
	$B = 2K_A/(T_s\omega_p)$
BS	$\omega_0 = 2\sqrt{K_B}/T_s$
	$\omega_a \le \begin{cases} \omega_p K_2 & \text{if } K_c \ge K_B \\ \omega_p K_1 & \text{if } K_c < K_B \end{cases}$
	$B = \dfrac{2K_A\omega_p}{T_s}$

$$K_A = \tan\frac{\Omega_{dp2}T_s}{2} - \tan\frac{\Omega_{dp1}T_s}{2}, \qquad K_B = \tan\frac{\Omega_{dp1}T_s}{2}\tan\frac{\Omega_{dp2}T_s}{2}$$

$$K_C = \tan\frac{\Omega_{da1}T_s}{2}\tan\frac{\Omega_{da2}T_s}{2}, \qquad K_1 = \frac{K_A\tan(\Omega_{da1}T_s/2)}{K_B - \tan^2(\Omega_{da1}T_s/2)}$$

$$K_2 = \frac{K_A\tan(\Omega_{da2}T_s/2)}{\tan^2(\Omega_{da2}T_s/2) - K_B}$$

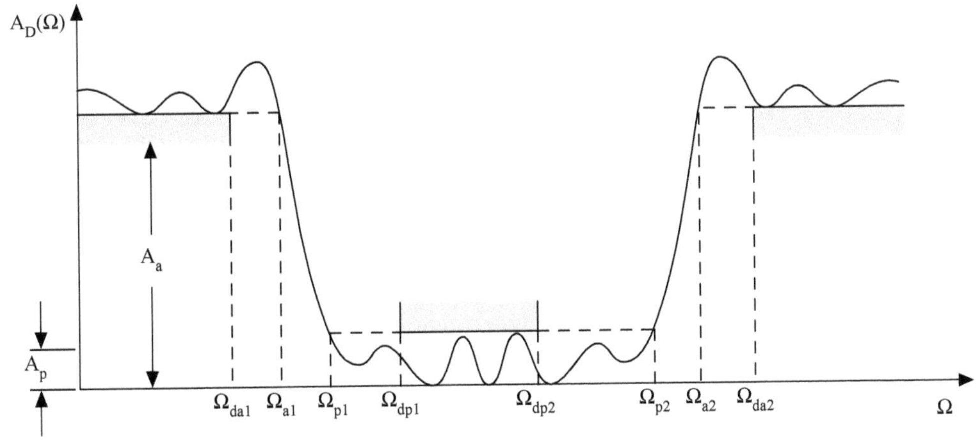

FIGURE 31.3

31.4 Examples

31.4.1 Example

Design a highpass filter using a Butterworth approximation satisfying the following specifications: A_p = 1 dB, $A_a = 15\,\text{dB}$, $\Omega_{dp} = 3.5$ rad s^{-1}, $\Omega_{da} = 1.5$ rad s^{-1} and $\omega_s = 10$ rad s^{-1}.

Solution

$\omega_s = 2\pi / T_s$ or $T_s = 2\pi/10 = 0.2\pi$. From (31.2.1)

$$K_0 = \frac{\tan(3.52\pi/10\times 2)}{\tan(1.52\pi/10\times 2)} = 3.85184; \qquad D = (10^{0.1\times 15}-1)/(10^{0.1\times 1}-1) = 118.268718;$$

$$n \ge \frac{\log D}{2\times \log K_0} = 1.7697 \text{ which implies } n = 2; \qquad \omega_p = (10^{0.1\times 1}-1)^{1/4} = 0.713335.$$

From 31.3.1 $\lambda = \dfrac{2\omega_p \tan(\Omega_{dp}T_s/2)}{T_s} = 4.456334$. From (31.1.5) and Chapter 12 (12.4.4),

$$H_t(\bar{s}) = H_N(s)\Big|_{s=\lambda/\bar{s}} = \frac{1}{s^2+1.41421s+1}\Bigg|_{s=\lambda/\bar{s}} = \frac{\bar{s}^2}{\bar{s}^2+1.41421\lambda\bar{s}+\lambda^2}.$$

From (31.1.5) step #2

$$H_D(z) = H_t(\bar{s})\Big|_{\bar{s}=\frac{2}{T_s}\left(\frac{z-1}{z+1}\right)} = \frac{z^2-2z+1}{4.939888z^2+1.919992z+0.980104}.$$

Figure 31.4 shows the $|H_D(e^{j\omega T_s})|$ versus ω.

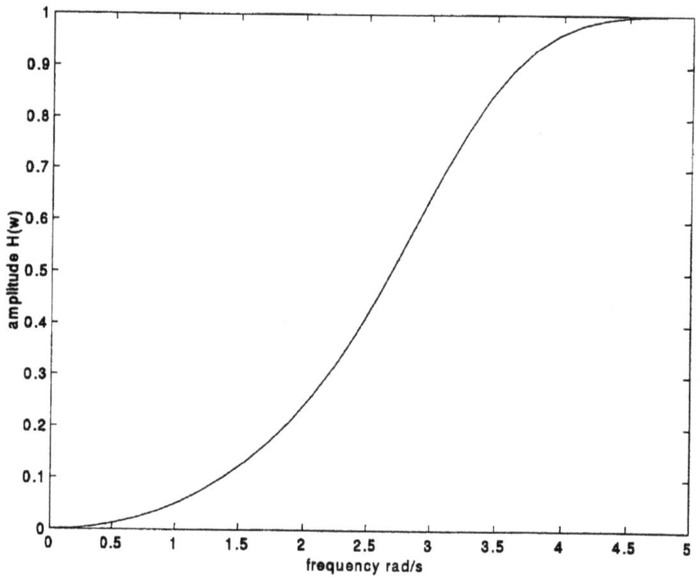

FIGURE 31.4

31.4.2 Example

Design a highpass filter using a Chebyshev approximation satisfying the following specifications: A_p = 1 dB, $A_a = 15\,\text{dB}$, $\Omega_{dp} = 3.5$ rad s^{-1}, $\Omega_{da} = 1.5$ rad s^{-1} and $\omega_s = 10$ rad s^{-1}.

Solution

$\omega_s = 2\pi / T_s$ or $T_s = 2\pi/10 = 0.2\pi$. From (31.2.2) $K_0 = 3.85184$ (see Example 31.4.1), D = 118.268718 (Example 31.4.1),

$$n \geq \frac{\cosh^{-1}\sqrt{D}}{\cosh^{-1} K_0} = \frac{\ln(\sqrt{D}+\sqrt{D-1})}{\ln(K_0+\sqrt{K_0^2-1})} = \frac{3.07963}{2.02439} = 1.521$$

or $n = 2$, $\omega_p = 1$. From (31.3.1) $\lambda = 2\omega_p \tan(\Omega_{dp}T_s/2)/T_s = 6.247183$. From (31.1.5) $s = \lambda/\bar{s}$ and Chapter 12, Table 12.3 of 0.1 dB ripple,

$$H_t(\bar{s}) = H_N(s)\Big|_{s=\lambda/\bar{s}} = \frac{1}{s^2+2.37236s+3.31403}\Bigg|_{s=\lambda/\bar{s}} = \frac{\bar{s}^2}{3.31403\bar{s}^2+2.37236\lambda\bar{s}+\lambda^2}.$$

From (31.1.5) step #2

$$H_D(z) = H_t(\bar{s})\Bigg|_{\bar{s}=\frac{2}{T_s}\left(\frac{z-1}{z+1}\right)} = \frac{z^2-2z+1}{7.911168z^2+1.075619z+1.400869}.$$

31.4.3 Example

Design a Chebyshev bandstop filter satisfying the following specifications: $A_p = 0.5$ dB, $A_a = 15$ dB, $\Omega_{dp1} = 350$ rad s^{-1}, $\Omega_{dp2} = 700$ rad s^{-1}, $\Omega_{da1} = 430$ rad s^{-1}, $\Omega_{da2} = 600$ rad s^{-1}, $\omega_s = 3000$ rad s^{-1}, $T_s = 2\pi/3000$

Solution

From (31.3.2) and (31.2.2)

$$K_B = \tan\frac{\Omega_{dp1}T_s}{2}\tan\frac{\Omega_{dp2}T_s}{2} = 0.34563,$$

$$K_C = \tan\frac{\Omega_{da1}T_s}{2}\tan\frac{\Omega_{da2}T_s}{2} = 0.35122, \qquad K_C > K_B,\ \omega_0 = 2\sqrt{K_B}/T_s = 561.40606,$$

$$K_A = \tan\frac{\Omega_{dp2}T_s}{2} - \tan\frac{\Omega_{dp1}T_s}{2} = 0.51654,$$

$$B = \frac{2K_A\omega_p}{T_s} = 2\times 0.51654\times 1/T_s = 493.2594$$

$$D = (10^{0.1A_a}-1)/(10^{0.1A_p}-1) = 250.968,$$

$$K_2 = \frac{K_A \tan(\Omega_{da2} T_s / 2)}{\tan^2(\Omega_{da2} T_s / 2) - K_B} = 5.28362$$

$$n \geq \frac{\cosh^{-1}\sqrt{D}}{\cosh^{-1} K_2} = \frac{\ln(\sqrt{D} + \sqrt{D-1})}{\ln(K_2 + \sqrt{K_2^2 - 1})} = 1.47$$

or $n = 2$. From Table 12.3 of Chapter 12 and $A_p = 0.5$ ripple,

$$H(s) = \frac{1}{s^2 + 1.42562s + 1.51620}.$$

From (31.1.5) Table 31.1,

$$H_t(\bar{s}) = H(s)\big|_{s=B\bar{s}/(\bar{s}^2+\omega_0^2)} = \frac{\bar{s}^4 + 2\omega_0^2\bar{s}^2 + \omega_0^4}{1.5\bar{s}^4 + 1.42562B\bar{s}^3 + (B^2 + 1.5162 \times 2\omega_0^2)\bar{s}^2 + 1.42562B\omega_0^2\bar{s} + 1.5162\omega_0^4}$$

From (31.1.5) step #2

$$H_D(z) = H_t(\bar{s})\big|_{\bar{s}=\frac{2}{T_s}\left(\frac{z-1}{z+1}\right)} = \frac{\text{Num.}}{\text{Den.}}$$

where

$$\text{Num:}\quad \left(1 + \frac{b_1 T_s^2}{4}\right)z^4 - 4z^3 + \left(6 - 2\frac{b_1 T_s^2}{4} + \frac{b_1 T_s^2}{16}\right)z^2 + \left(\frac{b_2 T_s^2 2}{16} - 4\right)z + \left(1 + \frac{b_1 T_s^2}{4} + \frac{b_2 T_s^2}{16}\right)$$

$$\text{Den:}\quad a_1 z^4 - 4a_1 z^3 + 6a_1 z^2 - 4a_1 z + a_1 + \frac{a_2 T_s}{2} z^4 - 2\frac{a_2 T_s}{2} z^3 + 2\frac{a_2 T_s}{2} z - \frac{a_2 T_s}{2} + \frac{a_3 T_s}{2} z^4$$

$$- 2\frac{a_3 T_s}{4} z^2 + \frac{a_3 T_s}{2}$$

$b_1 = 2\omega_o^2, \quad b_2 = \omega_o^4$

$a_1 = 1.51620, \quad a_2 = 1.42562B, \quad a_3 = B^2 + 1.51620 \times 2\omega_o^2$

$a_4 = 1.42562\omega_o^2, \quad a_5 = 1.51620\omega_o^4$

References

Antonious, A., *Digital Filters*, McGraw-Hill Inc., New York, NY, 1993.

Jackson, L. B., *Digital Filters and Signal Processing*, 3rd Edition, Kluwer Academic Publishers, Boston, MA, 1996.

Mitra, S. K., and J. F. Kaiser, *Handbook for Digital Signal Processing*, John Wiley & Sons, New York, NY, 1993.

32

Statistics

32.1 Estimation

32.1.1 Definitions

32.1.1.1 Sample Data

The experimental values $x(0), x(1), \cdots, x(N-1)$ of the r.v. $X(0), X(1), \cdots, X(N-1)$.

32.1.1.2 Statistic

A function of the observations alone, e.g., $t(x(0), x(1), \cdots, x(N-1))$

32.1.1.3 Estimator-Estimate

The rule or method of estimation is called an estimator, and the value to which it gives rise in a particular is called the estimate.

32.1.1.4 Consistency

Any statistic that converges stochastically to a parameter θ is called a consistent estimator of that parameter θ. $P\{|t_n - \theta| < \varepsilon\} > 1 - N$ for $n > N$ and any positive ε and η, $t \equiv$ statistic (see 32.1.1.2).

32.1.1.5 Biased and Unbiased Estimator

Any statistic whose mathematical expectation is equal to a parameter θ is called the unbiased estimator of the parameter θ. Otherwise, the statistic is said to be biased. For example, $E\{t\} = \theta$, t is an unbiased estimator of θ.

Example 1

$$E\left\{\frac{1}{N}\sum_{n=0}^{N-1} x(n)\right\} = \frac{1}{N}\sum_{n=0}^{N-1} E\{x(n)\} = \frac{N\mu}{N} = \mu,$$

which shows that the sample mean

$$\bar{x} = \left[\sum_{n=0}^{N-1} x(n)\right]\Big/ N$$

is an unbiased estimator of the population mean if it exists.

Example 2

$$E\left\{\sum_{n=0}^{N-1}(x(n)-\bar{x})^2\right\}=E\left\{\sum_{n=0}^{N-1}\left[x(n)-\sum_{n=0}^{N-1}\frac{x(n)}{N}\right]^2\right\}=E\left\{\frac{N-1}{N}\sum_{n=1}^{N-1}x^2(n)-\frac{1}{N}\sum_{n\neq k}\sum x(n)x(k)\right\}$$

$$=(N-1)\mu_2'-(N-1)\mu_2'=(N-1)\mu_2$$

and hence

$$(1/N)\sum_{n=0}^{N-1}(x(n)-\bar{x})^2$$

has the mean value $[(N-1)/N]\mu_2$ which is biased (depends on N). $s^2 \equiv$ unbiased sample variance $=[1/(N-1)]\sum_{n=0}^{N-1}(x(n)-\bar{x})^2$. For μ_i' and μ_i see Chapter 34 Table 34.1. $\bar{x}=$ sample mean.

32.1.1.6 Minimum Variance Unbiased (MVU) Estimator

If the estimator exists whose variance is equal to Cramer-Rao lower bound (CRLB) for each value of the parameter θ then it is an MVU estimator.

32.1.1.7 Vector Parameter

$$\underline{\theta}=[\theta_1,\theta_2,\cdots,\theta_p]^T$$

32.1.1.8 Vector Unbiased Estimator

$$E\{\hat{\theta}_i\}=\theta_i,\quad E\{\hat{\underline{\theta}}\}=[E\{\hat{\theta}_1\},E\{\hat{\theta}_2\},\cdots,E\{\hat{\theta}_p\}]^T=\underline{\theta}$$

32.1.1.9 Likelihood Function

$$L(x(0),x(1),\cdots,x(N-1);\theta)\equiv L(\underline{x};\theta)=p(x(0);\theta),p(x(1);\theta)\cdots p(x(N-1);\theta)\equiv p(\underline{x};\theta)$$

$$\int\cdots\int p(\underline{x};\theta)d\underline{x}=1,\quad \int\cdots\int\frac{\partial p(\underline{x};\theta)}{\partial\theta}d\underline{x}=0,\quad E\left\{\frac{\partial\ln p(\underline{x};\theta)}{\partial\theta}\right\}=\int\cdots\int\frac{1}{p(\underline{x};\theta)}\frac{\partial p(\underline{x};\theta)}{\partial\theta}p(\underline{x};\theta)d\underline{x}=0$$

$$E\left\{\left(\frac{\partial\ln p(\underline{x};\theta)}{\partial\theta}\right)^2\right\}=-E\left\{\left(\frac{\partial^2\ln p(\underline{x};\theta)}{\partial\theta^2}\right)\right\},\quad d\underline{x}=dx(0)dx(1)\cdots dx(N-1)$$

32.1.1.10 Efficient Estimator

If an estimator is unbiased (see 32.1.1.5) and attains the (CRLB), it is said to be efficient.

32.1.2 Cramer-Rao Lower Bound

32.1.2.1 Cramer-Rao Lower Bound (scalar parameter) (CRLB)

$\mathrm{var}(\hat{\theta}) \equiv E\{[\hat{\theta}-\tau(\theta)]^2\} \geq [\tau'(\theta)]^2 \Big/ E\left\{\left(\dfrac{\partial \ln p(\underline{x};\theta)}{\partial\theta}\right)^2\right\} = [\tau'(\theta)]^2 \Big/ \left[-E\left\{\left(\dfrac{\partial^2 \ln p(\underline{x};\theta)}{\partial\theta^2}\right)\right\}\right]$, $\hat{\theta} =$ unbiased estimator, $\tau(\theta) = \int\cdots\int \theta p(\underline{x};\theta)\,d\underline{x}$, $\tau'(\theta) = d\tau(\theta)/d\theta$, $L =$ (see 32.1.1.7) $=$ $p(x(0);\theta)$ $p(x(1);\theta)$ $p(x(N-1);\theta) \equiv p(\underline{x};\theta)$, $d\underline{x} = dx(0)\,dx(1)\cdots dx(N-1)$

Note: $\tau(\theta)=\theta$, $\mathrm{var}(\hat{\theta}) \geq 1 \Big/ \left[-E\left\{\dfrac{\partial^2 \ln p(\underline{x};\theta)}{\partial\theta^2}\right\}\right]$

32.1.2.2 Attainment of CRLB

Attainment is possible if and only if $\dfrac{\partial \ln p(\underline{x};\theta)}{\partial\theta} = I(\theta)\,(t(\underline{x})-\theta)$ and $\hat{\theta} = t(\underline{x})$ is an MVU estimator. *Minimum variance*: $1/I(\theta)$, $t(\underline{x}) = t(x(0),x(1),\cdots,x(N-1))$, $I(\theta) =$ a function of parameter θ. *General formula*: $\dfrac{\partial \ln p(\underline{x};\theta)}{\partial\theta} = I(\theta)\,[t-\tau(\theta)]$, $\mathrm{var}(t) = \tau'(\theta)/I(\theta)$, $I(\theta) =$ independent of observations.

Example 1

$p(\underline{x};\theta) = (1/\sqrt{2\pi})^N \exp\left(-\sum_{i=0}^{N-1}(x(i)-\theta)^2 \Big/ 2\right)$, where $X(0), X(1),\cdots,X(N-1)$ is a random sample from a normal distribution with mean θ and variance 1, $N(\theta,1)$. Since the maximum of $p(\underline{x};\theta)$ and $\ln p(\underline{x};\theta)$ are the same, $d\ln p(\underline{x};\theta)/d\theta = 0 = \sum_{i=0}^{N-1}(x(i)-\theta)$ or $\theta = t(x(0),x(1),\cdots,x(N-1)) = \sum_{i=0}^{N-1} x(i)/N$ maximizes $p(\underline{x};\theta)$.

$\hat{\theta} = t(x(0),\cdots,x(N-1)) = (1/N)\sum_{i=0}^{N-1} x(i) = \bar{x}$ is the maximum likelihood estimator of the mean θ.

Example 2

$x(n) = s(n;\theta) + v(n)$, $n = 0,1,\cdots,N-1$, $s(n;\theta) =$ deterministic signals with unknown parameter θ, $v(n) =$ white Gaussian with variance σ^2, $x(n) =$ observation samples.

$$p(\underline{x};\theta) = [1/(2\pi\sigma^2)^{N/2}]\exp\left[-\frac{1}{2\sigma^2}\sum_{n=0}^{N-1}[x(n)-s(n;\theta)]^2\right],$$

$$\partial \ln p(\underline{x};\theta)/\partial\theta = (1/\sigma^2)\sum_{n=0}^{N-1}(x(n)-s(n;\theta))\frac{\partial s(n;\theta)}{\partial\theta},$$

$$\frac{\partial^2 \ln p(\underline{x};\theta)}{\partial\theta^2} = (1/\sigma^2)\sum_{n=0}^{N-1}\left[[x(n)-s(n;\theta)]\frac{\partial^2 s(n;\theta)}{\partial\theta^2} - \left(\frac{\partial s(n;\theta)}{\partial\theta}\right)^2\right],$$

$$E\left\{\frac{\partial^2 \ln p(\underline{x};\theta)}{\partial\theta^2}\right\} = -(1/\sigma^2)\sum_{n=0}^{N-1}\left(\frac{\partial s(n,\theta)}{\partial\theta}\right)^2, \qquad \operatorname{var}\hat{\theta} \ge -1/E\left\{\frac{\partial^2 \ln p(\underline{x};\theta)}{\partial\theta^2}\right\} = \sigma^2\Big/\sum_{n=0}^{N-1}\left(\frac{\partial s(n;\theta)}{\partial\theta}\right)^2$$

Example 3

$x(n) = A + v(n)$, $v(n) =$ white Gaussian with variance σ^2,

$$p(\underline{x};A) = \prod_{n=0}^{N-1}\frac{1}{\sqrt{2\pi\sigma^2}}\exp\left[-\frac{1}{2\sigma^2}(x(n)-A)^2\right] = \frac{1}{(2\pi\sigma^2)^{N/2}}\exp\left[-\frac{1}{2\sigma^2}\sum_{n=0}^{N-1}(x(n)-A)^2\right],$$

$$\frac{\partial \ln p(\underline{x};A)}{\partial A} = \frac{\partial}{\partial A}\left[-\ln\left[(2\pi\sigma^2)^{N/2}\right] - \frac{1}{2\sigma^2}\sum_{n=0}^{N-1}(x(n)-A)\right] = \frac{1}{\sigma^2}\sum_{n=0}^{N-1}(x(n)-A) = \frac{N}{\sigma^2}(\bar{x}-A),$$

$\bar{x} =$ sample mean, $\dfrac{\partial^2 \ln p(\underline{x};A)}{\partial A^2} = -\dfrac{N}{\sigma^2}$, $\operatorname{var}(\hat{A}) \ge \sigma^2/N$. Since $\dfrac{N}{\sigma^2}(\bar{x}-A)$ is similar to (32.1.2.2) is minimum variance unbiased estimator.

32.1.2.3 CRLB Value

$$\operatorname{var}(\hat{\theta}) = 1/I(\theta)$$

32.1.2.4 Fisher Information

$$I(\theta) = -E\left\{\frac{\partial^2 \ln p(\underline{x};\theta)}{\partial\theta^2}\right\}$$

32.1.2.5 Transformation of Parameter

If $\alpha = g(\theta)$, then the CRLB is

$$\operatorname{var}(\hat{\alpha}) \ge \frac{\left(\dfrac{\partial g(\theta)}{\partial\theta}\right)^2}{-E\left\{\dfrac{\partial^2 \ln p(\underline{x};\theta)}{\partial\theta^2}\right\}}$$

Example 1

From 32.1.2.2 Example 3, $\alpha = g(A) = A^2$ then $\operatorname{var}(\hat{A}^2) \ge \dfrac{(2A)^2}{N/\sigma^2}$

32.1.2.6 CRLB-Vector Parameter

$\underline{\theta} = [\theta_1\ \theta_2 \cdots \theta_p]^T$, $\operatorname{var}(\hat{\theta}_i) \ge [I^{-1}(\underline{\theta})]_{ii}$, $I(\underline{\theta}) =$ Fisher information matrix, $[\ ]_{ii} =$ the ii element of the matrix,

$$[I(\underline{\theta})]_{ij} \equiv I(\underline{\theta})_{ij} = -E\left[\frac{\partial^2 \ln p(\underline{x};\underline{\theta})}{\partial\theta_i\,\partial\theta_j}\right] = E\left\{\frac{\partial \ln p(\underline{x};\underline{\theta})}{\partial\theta_i}\frac{\partial \ln p(\underline{x};\underline{\theta})}{\partial\theta_j}\right\},$$

$i = 1,\cdots,p;\ j = 1,\cdots,p$. $C_{\hat{\theta}} - I^{-1}(\underline{\theta}) \ge \underline{0}$, $C_{\hat{\theta}} =$ covariance matrix of any unbiased estimator $\hat{\underline{\theta}}$.

Example 1

$x(n) = A + Bn + v(n),\ n = 0,1,\cdots,N-1,\ v(n) =$ *WGN*, parameter unknown A and B.
Hence $\underline{\theta} = [A\ B]^T$,

$$I(\underline{\theta})_{11} = -E\left\{\frac{\partial^2 \ln p(\underline{x};\underline{\theta})}{\partial A^2}\right\} = -N/\sigma^2,\quad I(\underline{\theta})_{22} = -E\left\{\frac{\partial^2 \ln p(\underline{x};\underline{\theta})}{\partial B^2}\right\} = -\frac{1}{\sigma^2}\sum_{n=0}^{N-1} n^2$$

$$= -\frac{1}{\sigma^2}\frac{N(N-1)(2N-1)}{6},\quad I(\underline{\theta})_{12} = I(\underline{\theta})_{21} = -E\left\{\frac{\partial^2 \ln p(\underline{x};\underline{\theta})}{\partial A\,\partial B}\right\} = -E\left\{\frac{\partial^2 \ln p(\underline{x};\underline{\theta})}{\partial B\,\partial A}\right\} = -\frac{1}{\sigma^2}\sum_{n=0}^{N-1} n$$

$$= -\frac{1}{\sigma^2}\frac{N(N-1)}{2},\quad p(\underline{x};\underline{\theta}) = \frac{1}{(2\pi\sigma^2)^{N/2}}\exp\left[-\frac{1}{2\sigma^2}\sum_{n=0}^{N-1}(x(n) - A - Bn)^2\right],$$

$$I^{-1}(\underline{\theta}) = \sigma^2\begin{bmatrix} \dfrac{2(2N-1)}{N(N+1)} & -\dfrac{6}{N(N+1)} \\ -\dfrac{6}{N(N+1)} & \dfrac{12}{N(N^2-1)} \end{bmatrix},\quad \operatorname{var}(\hat{A}) \geq \frac{2(2N-1)\sigma^2}{N(N+1)},\quad \operatorname{var}(\hat{B}) \geq \frac{12\sigma^2}{N(N^2-1)}$$

32.1.2.7 CRLB Value-Vector Parameter

$C_{\hat{\underline{\theta}}} = I^{-1}(\underline{\theta})$ if and only if $\dfrac{\partial \ln p(\underline{x};\underline{\theta})}{\partial \underline{\theta}} = I(\underline{\theta})(\underline{t}(\underline{x}) - \underline{\theta}),\ \underline{t} \equiv$ p-dimensional function; $I = p \times p$ matrix,

$\hat{\underline{\theta}} = \underline{t}(\underline{x}) \equiv$ MVU estimator with covariance matrix $I^{-1}(\underline{\theta})$.

Example 1

From Example 32.1.2.6.1,

$$\frac{\partial \ln p(\underline{x};\underline{\theta})}{\partial \underline{\theta}} = \begin{bmatrix} \dfrac{\partial \ln p(\underline{x};\underline{\theta})}{\partial A} \\ \dfrac{\partial \ln p(\underline{x};\underline{\theta})}{\partial B} \end{bmatrix} = \begin{bmatrix} \dfrac{1}{\sigma^2}\sum\limits_{n=0}^{N-1}(x(n) - A - Bn) \\ \dfrac{1}{\sigma^2}\sum\limits_{n=0}^{N-1}(x(n) - A - Bn)n \end{bmatrix} = \begin{bmatrix} \dfrac{N}{\sigma^2} & \dfrac{N(N-1)}{\sigma^2} \\ \dfrac{N(N-1)}{2\sigma^2} & \dfrac{N(N-1)(2N-1)}{6\sigma^2} \end{bmatrix}\begin{bmatrix} \hat{A} - A \\ \hat{B} - B \end{bmatrix},$$

$$\hat{A} = \frac{2(2N-1)}{N(N+1)}\sum_{n=0}^{N-1} x(n) - \frac{6}{N(N+1)}\sum_{n=0}^{N-1} nx(n),\qquad \hat{B} = -\frac{6}{N(N+1)}\sum_{n=0}^{N-1} x(n) + \frac{12}{N(N^2-1)}\sum_{n=0}^{N-1} nx(n).$$

Hence, the conditions for equality are satisfied and $[\hat{A}\ \hat{B}]^T$ is an efficient and therefore MVU estimator.

32.1.2.8 Vector Transformations CRLB

If $\underline{\alpha} = \underline{g}(\underline{\theta}),\ \underline{g} \equiv r-$ dimensional,

$$C_{\hat{\underline{\alpha}}} - \frac{\partial \underline{g}(\underline{\theta})}{\partial \underline{\theta}} I^{-1}(\underline{\theta}) \frac{\partial \underline{g}(\underline{\theta})^T}{\partial \underline{\theta}} \geq \underline{0};\ \partial \underline{g}(\underline{\theta}) / \partial \underline{\theta} \equiv r \times p \text{ Jacobian matrix} = \begin{bmatrix} \dfrac{\partial g_1(\underline{\theta})}{\partial \theta_1} & \cdots & \dfrac{\partial g_1(\underline{\theta})}{\partial \theta_p} \\ \vdots & & \vdots \\ \dfrac{\partial g_r(\underline{\theta})}{\partial \theta_1} & \cdots & \dfrac{\partial g_r(\underline{\theta})}{\partial \theta_p} \end{bmatrix}$$

32.1.2.9 General Gaussian CRLB Case

$x = N(\underline{\mu}(\underline{\theta}), C(\underline{\theta}))$,

$$I(\underline{\theta})_{ij} = \left[\frac{\partial \underline{\mu}(\underline{\theta})}{\partial \theta_i}\right]^T C^{-1}(\underline{\theta})\left[\frac{\partial \underline{\mu}(\underline{\theta})}{\partial \theta_j}\right] + \frac{1}{2} tr\left[C^{-1}(\underline{\theta})\frac{\partial C(\underline{\theta})}{\partial \theta_i} C^{-1}(\underline{\theta})\frac{\partial C(\underline{\theta})}{\partial \theta_j}\right],$$

$$\frac{\partial \underline{\mu}(\underline{\theta})}{\partial \theta_i} = \left[\frac{\partial[\underline{\mu}(\underline{\theta})]_1}{\partial \theta_i}\ \frac{\partial[\underline{\mu}(\underline{\theta})]_2}{\partial \theta_i} \cdots \frac{\partial[\underline{\mu}(\underline{\theta})]_N}{\partial \theta_i}\right]^T, \quad \frac{\partial C(\underline{\theta})}{\partial \theta_i} = \begin{bmatrix} \frac{\partial[C(\underline{\theta})]_{11}}{\partial \theta_i} & \cdots & \frac{\partial[C(\underline{\theta})]_{1N}}{\partial \theta_i} \\ \vdots & & \vdots \\ \frac{\partial[C(\underline{\theta})]_{N1}}{\partial \theta_i} & \cdots & \frac{\partial[C(\underline{\theta})]_{NN}}{\partial \theta_i} \end{bmatrix}$$

If the parameter is scalar θ, $\underline{x} = N(\underline{\mu}(\theta), C(\theta))$,

$$I(\theta) = \left[\frac{\partial \underline{\mu}(\theta)}{\partial \theta}\right]^T C^{-1}(\theta)\left[\frac{\partial \underline{\mu}(\theta)}{\partial \theta}\right] + \frac{1}{2} tr\left[\left(C^{-1}(\theta)\frac{\partial C(\theta)}{\partial \theta}\right)^2\right]$$

Example 1

$x(n) = A + v(n)$, $n = 0,1,\cdots,N-1$, $v(n) =$ WGN, A = Gaussian r.v. with $\mu = 0$ and $\text{var}(A) = \sigma_A^2$, A is independent of $v(n)$, σ_A^2 = unknown.

$[C(\sigma_A^2)]_{ij} = E\{x(i-1)x(j-1)\} = E\{(A + v(i-1))(A + v(j-1))\} = \sigma_A^2 + \sigma^2\delta_{ij}$, $C(\sigma_A^2) = \sigma_A^2 \underline{1}\underline{1}^T + \sigma^2 I$, $\underline{1} = [1\ 1 \cdots 1]^T$,

$$C^{-1}(\sigma_A^2) = \frac{1}{\sigma^2}\left(I - \frac{\sigma_A^2}{\sigma^2 + N\sigma_A^2}\underline{1}\underline{1}^T\right), \quad \frac{\partial C(\sigma_A^2)}{\partial \sigma_A^2} = \underline{1}\underline{1}^T, \quad C^{-1}(\sigma_A^2)\frac{\partial C(\sigma_A^2)}{\partial \sigma_A^2} = \frac{1}{\sigma^2 + N\sigma_A^2}\underline{1}\underline{1}^T,$$

$$I(\theta) = \frac{1}{2} tr\left[\left(\frac{1}{\sigma^2 + N\sigma_A^2}\right)^2 \underline{1}\underline{1}^T\underline{1}\underline{1}^T\right] = \frac{1}{2}\left(\frac{1}{\sigma^2 + N\sigma_A^2}\right)^2, \quad \text{var}(\sigma_A^2) \ge 2\left(\sigma_A^2 + \frac{\sigma^2}{N}\right)^2$$

See (32.1.2.6) and (32.1.2.3).

32.1.3 Linear Models in Estimation

32.1.3.1 MVU Estimation with Gaussian Noise

$\underline{x} = H\underline{\theta} + \underline{v}$, $\underline{x} = N \times 1$ observation vector, $H = N \times p$ known observation matrix and rank p, $\underline{v} = N \times 1$ noise vector with *p.d.f.* $N(\underline{0}, \sigma^2 I)$, $\underline{\theta} = p \times 1$ vector parameter (to be estimated), $\hat{\underline{\theta}} = (H^T H)^{-1} H^T \underline{x} \equiv$ MVU estimator, $C_{\hat{\underline{\theta}}} = \sigma^2 (H^T H)^{-1} \equiv$ covariance matrix of $\hat{\theta}$, for the linear model the MVU estimator is efficient (attains the CRLB).

Example 1

$x(t_n) = \theta_1 + \theta_2 t_n + \theta_3 t_n^2 + v(t_n)$, $n = 0,1,\cdots,N-1$, (to fit a second-order curve to data $x(t_n)$), $\underline{x} = [x(t_0) \cdots x(t_{N-1})]^T$, $\underline{\theta} = [\theta_1\ \theta_2\ \theta_3]^T$,

$$H = \begin{bmatrix} 1 & t_0 & t_0^2 \\ \vdots & \vdots & \vdots \\ 1 & t_{N-1} & t_{N-1}^2 \end{bmatrix},$$

$\underline{\hat{\theta}} = (H^T H)^{-1} H^T \underline{x}$, hence the estimated curve is $\hat{s}(t) = \sum_{i=1}^{3} \hat{\theta}_i t^{i-1}$.

32.1.3.2 MVU Estimator of General Model

$\underline{x} = H\underline{\theta} + \underline{s} + \underline{v}$, $\underline{x} = N \times 1$ observation vector, $H = N \times p$ known observation matrix, $\underline{s} = N \times 1$ known signal, $\underline{v} = N \times 1$ noise with p.d.f. $N(\underline{0}, C)$, $C =$ covariance matrix, $\underline{\hat{\theta}} =$ MVU estimator $= (H^T C^{-1} H)^{-1} H^T C^{-1}(\underline{x} - \underline{s})$, $C_{\underline{\hat{\theta}}} = (H^T C^{-1} H)^{-1}$.

Example 1

$x(n) = A + v(n)$, $n = 0, 1, \cdots, N-1$, $v(n) =$ colored Gaussian noise with $N \times N$ covariance matrix C. $H = \underline{1} = [1\,1 \cdots 1]^T$, then $\hat{A} = (H^T C^{-1} H)^{-1} C^{-1} \underline{x} = \dfrac{\underline{1}^T C^{-1} \underline{x}}{\underline{1}^T C^{-1} \underline{1}}$, $\operatorname{var}(\hat{A}) = (H^T C^{-1} H)^{-1} = 1/[\underline{1}^T C^{-1} \underline{1}]$. If we set $C^{-1} = D^T D$, $D = N \times N$ invertible matrix, then

$$\hat{A} = \frac{\underline{1}^T D^T D \underline{x}}{\underline{1}^T D^T D \underline{1}} = \frac{(D\underline{1})^T \underline{x}'}{\underline{1}^T D^T D \underline{1}} = \sum_{n=0}^{N-1} d_n x'(n),$$

$\underline{x}' = D\underline{x} \equiv$ prewhitened data, $d_n = [D\underline{1}]_n / \underline{1}^T D^T D \underline{1}$.

32.1.4 General MVU Estimation

32.1.4.1 Sufficient Statistic t(x)

If $p(\underline{x};\theta) = g(t(\underline{x}),\theta) h(\underline{x})$, $g(\cdot)$ is a function of $\underline{x}$ through $t(\underline{x})$ only, $h(\underline{x})$ is a function only of $\underline{x}$, then $t(\underline{x})$ is a sufficient statistic.

32.1.4.2 Complete Statistic

A statistic is complete if there is only one function of the statistic that is unbiased.

32.1.4.3 Unbiased Estimator

$\hat{\theta} = E\{\check{\theta} | t(\underline{x})\}$, $\check{\theta} =$ unbiased estimator of θ, $t(\underline{x}) =$ sufficient statistic (see 32.1.4.1) for θ, $\hat{\theta} =$ unbiased and has lesser or equal variance than that of $\check{\theta}$ for all θ. $\hat{\theta} =$ MVU estimator if the sufficient statistic is complete (see 32.1.4.2).

32.1.4.4 Unbiased Estimator (vector parameter)

If $p(\underline{x};\underline{\theta}) = g(\underline{t}(\underline{x}),\underline{\theta}) h(\underline{x})$, then $\underline{t}(\underline{x})$, an $r \times 1$ statistic, is a sufficient statistic for $\underline{\theta}$. $g(\cdot)$ depends only on $\underline{x}$ through $t(\underline{x})$ and on $\underline{\theta}$ and $h(\cdot)$ depends only on $\underline{x}$.

Example

$x(n) = A\cos 2\pi f_0 n + v(n)$, $n = 0, 1, \cdots, N-1$, $\underline{\theta} = [A \;\; \sigma^2]^T$,

$$p(\underline{x};\theta) = [1/(2\pi\sigma^2)^{N/2}]\exp\left[-\frac{1}{2\sigma^2}\sum_{n=0}^{N-1}(x(n) - A\cos 2\pi f_0 n)^2\right]$$

$$= [1/(2\pi\sigma^2)^{N/2}]\exp\left[-\frac{1}{2\sigma^2}\left(\sum_{n=0}^{N-1}x^2(n) - 2A\sum_{n=0}^{N-1}x(n)\cos 2\pi f_0 n + A^2\sum_{n=0}^{N-1}\cos^2 2\pi f_0 n\right)\right]\cdot 1$$

$$= g(\underline{t}(\underline{x}),\underline{\theta})\cdot h(\underline{x}),\quad h(\underline{x}) = 1,\quad \underline{t}(\underline{x}) = \left[\sum_{n=0}^{N-1}x(n)\cos 2\pi f_o n \quad \sum_{n=0}^{N-1}x^2(n)\right]^T$$

32.1.5 Maximum Likelihood Estimation (MLE)

32.1.5.1 Maximum Likelihood Estimation (MLE)

The MLE for a scalar parameter is defined to be the value of θ that maximizes $p(\underline{x};\theta)$ for fixed $\underline{x}$. $\frac{\partial \ln p(\underline{x};\theta)}{\partial\theta} = 0.$

Example 1

$x(n) = A + v(n)$, $n = 0,1,\cdots,N-1$, $v(n) =$ white Gaussian noise (WGN), A = unknown parameter, $p(\underline{x};A)$

$$= [1/(2\pi\sigma^2)^{N/2}]\exp\left[-\frac{1}{2\sigma^2}\sum_{n=0}^{N-1}(x(n) - A)^2\right],\quad \frac{\partial \ln p(\underline{x};\theta)}{\partial A} = \frac{1}{\sigma^2}\sum_{n=0}^{N-1}(x(n) - A) = 0,\quad \hat{A} = \frac{1}{N}\sum_{n=0}^{N-1}x(n)$$

32.1.5.2 MLE-vector Parameters

$\frac{\partial \ln p(\underline{x};\underline{\theta})}{\partial\underline{\theta}} = \underline{0}$ (see Chapters 33, 33.34.5)

Example 1

$x(n) = A + v(n)$, $n = 0,1,\cdots,N-1$, $v(n) = WGN$, $\underline{\theta} = [A\ \sigma^2]^T$, $\frac{\partial \ln p(\underline{x};\theta)}{\partial A} = \left(\frac{1}{\sigma^2}\right)\sum_{n=0}^{N-1}(x(n) - A) = 0$, or $\hat{A} = \bar{x} =$ sample mean,

$$\frac{\partial \ln p(\underline{x};\underline{\theta})}{\partial\sigma^2} = -(N/2\sigma^2) + (1/\sigma^4)\sum_{n=0}^{N-1}(x(n) - A)^2 = 0 \text{ or } \sigma^2 = (1/N)\sum_{n=0}^{N-1}(x(n) - \bar{x})^2$$

since $\hat{A} = \bar{x}$, $\hat{\underline{\theta}} = [\bar{x}\ \ (1/N)\sum_{n=0}^{N-1}(x(n) - \bar{x})^2]^T$.

32.1.5.3 MLE-Linear Model

$\underline{x} = H\underline{\theta} + \underline{v}$, $H = N \times p$ matrix with $N > p$ and rank p, $\underline{\theta} = p\times 1$ parameter vector, $\underline{v}$ = noise vector with p.d.f. $N(\underline{0},C)$, $\hat{\underline{\theta}} = (H^TC^{-1}H)^{-1}H^TC^{-1}\underline{x}$, $\hat{\underline{\theta}} \sim N(\underline{\theta},(H^TC^{-1}H)^{-1})$.

32.1.6 Least Squares (LS)

32.1.6.1 Definition

The Least squares approach attempts to minimize the squared difference between the given data $x(n)$ and the assumed signal (noiseless data). $s(n)$ = deterministic.

32.1.6.2 Error Criterion

$$J(\theta)=\sum_{n=0}^{N-1}[x(n)-s(n)]^2,\ J(\cdot)\ \text{depends on }\ \theta\ \text{via}\ s(n).$$

Example 1

$$s(n)=A,\ J(A)=\sum_{n=0}^{N-1}(x(n)-A)^2,\ \frac{\partial J(A)}{\partial A}=0\ \text{ we obtain }\ \hat{A}=\frac{1}{N}\sum_{n=0}^{N-1}x(n)=\bar{x}$$

32.1.6.3 Linear LS

$$s(n)=\theta h(n),\ h(n)=\text{known sequence},\ J(\theta)=\sum_{n=0}^{N-1}[x(n)-\theta h(n)]^2=\text{ error criterion,}$$

$$\hat{\theta}=\sum_{n=0}^{N-1}x(n)h(n)\Big/\sum_{n=0}^{N-1}h^2(n),\ J_{\min}=J(\hat{\theta})=\sum_{n=0}^{N-1}x^2(\theta)-\hat{\theta}\sum_{n=0}^{N-1}x(n)h(n)=\text{ minimum LS error}$$

32.1.6.4 Linear LS-Vector Parameters

$\underline{s}=H\underline{\theta}$, $\underline{\theta}=p\times 1$ vector parameter, $H=$ observation matrix $=N\times p$ known matrix with rank p, $\underline{s}$ = $[s(0)\ s(1)\cdots s(N-1)]^T$ = signal linear in the unknown parameter,

$$J(\underline{\theta})=\sum_{n=0}^{N-1}(x(n)-s(n))^2=(\underline{x}-H\underline{\theta})^T(\underline{x}-H\underline{\theta})=\underline{x}^T\underline{x}-2\underline{x}^TH\underline{\theta}+\underline{\theta}^TH^TH\underline{\theta},$$

$\dfrac{J(\underline{\theta})}{\partial\underline{\theta}}=-2H^T\underline{x}+2H^TH\underline{\theta}=0$ then $\hat{\underline{\theta}}=(H^TH)^{-1}H^T\underline{x}$, $H^TH\underline{\theta}=H^T\underline{x}\equiv$ normal equations, $J_{\min}$ = $J(\hat{\underline{\theta}})$ = $(\underline{x}-H(H^TH)^{-1}H^T\underline{x})^T(\underline{x}-H(H^TH)^{-1}H^T\underline{x})=\underline{x}^T(I-H(H^TH)^{-1}H^T)\underline{x}$

32.1.6.5 Linear LS Weighted-Vector Parameter

$J(\underline{\theta})=(\underline{x}-H\underline{\theta})^TW(\underline{x}-H\underline{\theta})$, $W=N\times N$ positive definite weighting matrix, $\hat{\theta}=(H^TWH)^{-1}H^TW\underline{x}$, $J_{\min}$ = minimum LS error $=\underline{x}^T(W-WH(H^TWH)^{-1}H^TW)\underline{x}$

32.1.6.6 Order-Recursive LS

$$J_{\min,k+1}=J_{\min,k}-\frac{(\underline{h}_{k+1}^TP_k^{\perp}\underline{x})^2}{\underline{h}_{k+1}^TP_k^{\perp}\underline{h}_{k+1}},$$

$\hat{\underline{\theta}}=(H_k^TH_k)^{-1}H_k^T\underline{x}$, $H_k=N\times k$ observation matrix, $J_{\min,k}$ = minimum LS error based on $H_k=(\underline{x}-H_k\hat{\underline{\theta}}_k)^T(\underline{x}-H_k\hat{\underline{\theta}}_k)$, $H_{k+1}=[H_k\ \ \underline{h}_{k+1}]\equiv[N\times k\ \ N\times 1]$ (add a column),

$$\hat{\underline{\theta}}_{k+1}=\begin{bmatrix}\hat{\underline{\theta}}_k-\dfrac{(H_k^TH_k)^{-1}H_k^T\underline{h}_{k+1}\underline{h}_{k+1}^TP_k^{\perp}\underline{x}}{\underline{h}_{k+1}^TP_k^{\perp}\underline{h}_{k+1}}\\ \dfrac{\underline{h}_{k+1}^TP_k^{\perp}\underline{x}}{\underline{h}_{k+1}^TP_k^{\perp}\underline{h}_{k+1}},\end{bmatrix}\equiv\begin{bmatrix}k\times 1\\ 1\times 1\end{bmatrix}$$

where $P_k^{\perp}=I-H_k(H_k^TH_k)^{-1}H_k^T$ = projection matrix onto the subspace orthogonal to that spanned by the columns of H_k.

Example

(Line fitting) Since $s_1(n) = A_1$ and $s_2(n) = A_2 + B_2 n$ for $n = 0,1,\cdots,N-1$ we have $H_1 = [1\ 1 \cdots 1]^T = \underline{1}^T$, $H_2 = [H_1\ \underline{h}_2]$, $\underline{h}_2 = [0\ 1 \cdots N-1]^T$, $\hat{A}_1 = \hat{\theta}_1 = (H_1^T H_1)^{-1} H_1^T \underline{x} = \bar{x}$,

$$J_{\min 1} = (\underline{x} - H_1\hat{\theta}_1)^T(\underline{x} - H_1\hat{\theta}_1) = \sum_{n=0}^{N-1} (x(n) - \bar{x})^2,\ \bar{x} = \text{sample mean},$$

$$\hat{\underline{\theta}}_2 = [\hat{A}_2\ \ \hat{B}_2]^T = \left[\theta_1 - \frac{(H_1^T H_1)^{-1} H_1^T \underline{h}_2 \underline{h}_2^T P_1^{\perp} \underline{x}}{\underline{h}_2^T P_1^{\perp} \underline{h}_2}\ \ \frac{\underline{h}_2^T P_1^{\perp} \underline{x}}{\underline{h}_2^T P_1^{\perp} \underline{h}_2}\right]^T,$$

$$(H_1^T H_1)^{-1} = 1/N,\ \ P_1^{\perp} = I - H_1(H_1^T H_1)^{-1} - H_1^T = I - \frac{1}{N}\underline{1}\underline{1}^T,$$

$$P_1^{\perp}\underline{x} = \underline{x} - \frac{1}{N}\underline{1}\underline{1}^T\underline{x} = \underline{x} - \bar{x}\underline{1},\ \ \underline{h}_2^T P_1^{\perp}\underline{x} = \underline{h}_2^T\underline{x} - \bar{x}\,\underline{h}_2^T\underline{1} = \sum_{n=0}^{N-1} n x(n) - \bar{x}\sum_{n=0}^{N-1} n,$$

$$\underline{h}_2^T P_1^{\perp}\underline{h}_2 = \underline{h}_2^T\underline{h}_2 - \tfrac{1}{N}(\underline{h}_2^T\underline{1})^2 = \sum_{n=0}^{N-1} n^2 - \frac{1}{N}\left(\sum_{n=0}^{N-1} n\right)^2,$$

$$\hat{\underline{\theta}}_2 = \left[\bar{x} - \frac{\frac{1}{N}\sum_{n=0}^{N-1} n\left[\sum_{n=0}^{N-1} n x(n) - \bar{x}\sum_{n=0}^{N-1} n\right]}{\sum_{n=0}^{N-1} n^2 - \frac{1}{N}\left(\sum_{n=0}^{N-1} n\right)^2}\ \ \frac{\sum_{n=0}^{N-1} n x(n) - \bar{x}\sum_{n=0}^{N-1} n}{\sum_{n=0}^{N-1} n^2 - \frac{1}{N}\left(\sum_{n=0}^{N-1} n\right)^2}\right]^T = \left[\bar{x} - \frac{1}{N}\sum_{n=0}^{N-1} n\hat{B}_2\right],$$

$$\hat{B}_2 = \frac{\sum_{n=0}^{N-1} n x(n) - \frac{N(N-1)}{2}\bar{x}}{N(N^2-1)/12} = -\frac{6}{N(N+1)}\sum_{n=0}^{N-1} x(n) + \frac{12}{N(N^2-1)}\sum_{n=0}^{N-1} n x(n),$$

$$\hat{A} = \bar{x} - \frac{1}{N}\sum_{n=0}^{N-1} n\hat{B}_2 = \bar{x} - \frac{N-1}{2}\hat{B}_2,\ \ J_{\min 2} = J_{\min 1} - \frac{(\underline{h}_2^T P_1^{\perp}\underline{x})^2}{\underline{h}_2^T P_1^{\perp}\underline{h}_2}$$

32.1.6.7 Sequential Least Squares

$\hat{\theta}(N) = \hat{\theta}(N-1) + \frac{1}{N+1}[x(N) - \hat{\theta}(N-1)]$, $\hat{\theta}(N-1) \equiv$ LSE based on $\{x(0), x(1), \cdots, x(N-1)\}$, the argument of $\hat{\theta}$ denotes the index of the most recent data point observed, $[1/(N+1)][x(N) - \hat{\theta}(N-1)] \equiv$ correction term.

32.1.6.8 Sequential Least Squares Error

$$J_{\min}(N) = J_{\min}(N-1) + \frac{N}{N+1}(x(N) - \hat{\theta}(N-1))^2$$

32.1.6.9 Nonlinear LS by Transformation of Parameters

$\underline{\alpha} = g(\underline{\theta})$, $g = p-$ dimensional function of $\underline{\theta}$ whose inverse exists. $\underline{s}(\underline{\theta}(\underline{\alpha})) = \underline{s}(g^{-1}(\underline{\alpha})) = H\underline{\alpha} \equiv$ linear in $\underline{\alpha}$, $\hat{\underline{\theta}} = g^{-1}(\hat{\underline{\alpha}})$ where $\hat{\underline{\alpha}} = (H^T H)^{-1} H^T \underline{x}$, $\hat{\underline{\theta}} = g^{-1}(\hat{\underline{\alpha}})$, J = error criterion $= (\underline{x} - \underline{s}(\underline{\theta}))^T (\underline{x} - \underline{s}(\underline{\theta}))$.

Example 1

$s(n) = A\cos(2\pi f_0 n + \varphi)$, $n = 0,1,\cdots,N-1$, A and φ to be estimated, $f_0 =$ known, $J = \sum_{n=0}^{N-1} [x(n) - A\cos(2\pi f_0 n + \varphi)]^2$, $s(n) = \alpha_1 \cos 2\pi f_0 n + \alpha_2 \sin 2\pi f_0 n$ where $\alpha_1 = A\cos\varphi$ and $\alpha_2 = -A\sin\varphi$, $\underline{s} = H\underline{\alpha}$ where $\underline{\alpha}[\alpha_1 \;\; \alpha_2]^T$ and

$$H = \begin{bmatrix} 1 & 0 \\ \cos 2\pi f_0 & \sin 2\pi f_0 \\ \vdots & \vdots \\ \cos 2\pi f_0 (N-1) & \sin 2\pi f_0 (N-1) \end{bmatrix},$$

$\hat{\alpha} = (H^T H)^{-1} H^T \underline{x}$, $\hat{\theta}$ is found from the inverse transformation $g^{-1}(\underline{\alpha})$. Hence, $A = [\alpha_1^2 + \alpha_2^2]^{1/2}$, $\varphi = \tan^{-1}(-\alpha_2 / \alpha_1)$, and $\hat{\underline{\theta}} = [\hat{A} \;\; \hat{\varphi}]^T = [\sqrt{\hat{\alpha}_1^2 + \hat{\alpha}_2^2} \;\; \tan^{-1}(-\hat{\alpha}_1^2 / \hat{\alpha}_2^2)]^T$.

32.1.6.10 Nonlinear LS by Separation

$\underline{s} = H(\underline{\alpha})\underline{\beta} =$ separable, $\underline{\theta} = [\underline{\alpha} \;\; \underline{\beta}]^T = [(p-q)\times 1 \;\; q\times 1]^T$, $H(\underline{\alpha}) = N\times q$ dependent on $\underline{\alpha}$. Model linear in $\underline{\beta}$ but nonlinear in $\underline{\alpha}$ which implies minimization with respect to $\underline{\beta}$ which results in a function of $\underline{\alpha}$. $J(\underline{\alpha},\underline{\beta}) = (\underline{x} - H(\underline{\alpha})\underline{\beta})^T (\underline{x} - H(\underline{\alpha})\underline{\beta})$, $\hat{\underline{\beta}} = (H^T(\underline{\alpha})H(\underline{\alpha}))^{-1} H^T(\underline{\alpha})\underline{x} \equiv$ minimizes J, $J(\underline{\alpha},\hat{\underline{\beta}})$ = LS error = $\underline{x}^T[I - H(\underline{\alpha})(H^T(\underline{\alpha})H(\underline{\alpha}))^{-1} H^T(\underline{\alpha})]\underline{x}$ which reduces to a minimization of $\underline{x}^T H(\underline{\alpha})(H^T(\underline{\alpha})H(\underline{\alpha}))^{-1} H^T(\underline{\alpha})\underline{x}$.

32.1.7 Method of Moments

32.1.7.1 Scalar Parameter

$$\mu_k' = E\{x^k(n)\} = h(\theta),\; \theta = h^{-1}(\mu_k'),\; \hat{\mu}_k' = \frac{1}{N}\sum_{n=0}^{N-1} x^k(n),\; \hat{\theta} = h^{-1}\left(\frac{1}{N}\sum_{n=0}^{N-1} x^k(n)\right),$$

Example

$x(n) = A + v(n)$, $n = 0,1,\cdots,N-1$, $v(n) =$ WGN with variance σ^2, $A \equiv$ to be estimated. We know

$$\mu_k' = E\{x(n)\} = A \equiv h(\theta) \text{ and } \theta = h^{-1}(\mu_1') = \mu_1' \text{ and hence } \hat{\theta} \equiv \hat{A} = \frac{1}{N}\sum_{n=0}^{N-1} x(n),$$

32.1.7.2 Vector Parameter

$\underline{\mu}' = \underline{h}(\underline{\theta})$ or $[\mu_1' \, \mu_2' \cdots \mu_p']^T = [h_1(\theta_1,\cdots,\theta_p) \;\; h_2(\theta_1,\cdots,\theta_p) \; \cdots \; h_p(\theta_1,\cdots,\theta_p)]^T$

$$\underline{\theta} = \underline{h}^{-1}(\underline{\mu}'),\ \hat{\underline{\theta}} = \underline{h}^{-1}(\hat{\underline{\mu}}')\ \text{where}\ ,\hat{\underline{\mu}}' = \left[\frac{1}{N}\sum_{n=0}^{N-1} x(n) \quad \frac{1}{N}\sum_{n=0}^{N-1} x^2(n)\cdots\frac{1}{N}\sum_{n=0}^{N-1} x^p(n)\right]^T$$

Example

Let

$$p(x(n);\varepsilon) = \frac{1-\varepsilon}{\sqrt{2\pi\sigma_1^2}}\exp\left(-\frac{1}{2}\frac{x^2(n)}{\sigma_1^2}\right) + \frac{\varepsilon}{\sqrt{2\pi\sigma_2^2}}\exp\left(-\frac{1}{2}\frac{x^2(n)}{\sigma_2^2}\right),$$

ε = mixture parameter, $0 < \varepsilon < 1$, σ_1 and σ_2 are unknown variances of the individual Gaussian p.d.f.'s $p(x(n);\varepsilon)$ is thought of as the p.d.f. of r.v. obtained from $N(0,\sigma_1^2)$ with probability $1-\varepsilon$, and from a $N(0,\sigma_2^2)$ p.d.f. with probability ε.

$\mu_2' = E\{x^2(n)\} = (1-\varepsilon)\sigma_1^2 + \varepsilon\sigma_2^2$, $\mu_4' = E\{x^4(n)\} = 3(1-\varepsilon)\sigma_1^4 + 3\varepsilon\sigma_2^4$, $\mu_6' = E\{x^6(n)\} = 15(1-\varepsilon)\sigma_1^6 + 15\varepsilon\sigma_2^6$. Setting $u = \sigma_1^2 + \sigma_2^2$ and $\upsilon = \sigma_1^2\sigma_2^2$ in the above equation we obtain: $u = (\mu_6' - 5\mu_4'\mu_2')/(5\mu_4' - 15\mu_2'^2)$, $\upsilon = \mu_2' u - \frac{\mu_4'}{3}$. We first find u and then υ and, hence, σ_1 and σ_2, which are: $\sigma_1^2 = (u + \sqrt{u^2 - 4\upsilon})/2$, $\sigma_2^2 = \upsilon/\sigma_1^2$. But $E\{x^2(n)\} = \int x^2(n)p(x(n);\varepsilon)dx(n) = (1-\varepsilon)\sigma_1^2 + \varepsilon\sigma_1^2 = \frac{1}{N}\sum_{n=0}^{N-1} x^2(n)$ or $\varepsilon = (\mu_2 - \sigma_1^2)/(\sigma_2^2 - \sigma_1^2)$.

32.1.8 Bayesian MSE

32.1.8.1 Definition:

$B_{mse}(\hat{\theta}) = E\{(\theta-\hat{\theta})^2\} = \iint (\theta-\hat{\theta})^2 p(\underline{x},\theta)d\underline{x}\,d\theta$, the operator E with respect to the joint p.d.f. $p(\underline{x},\theta)$.

32.1.8.2 Prior Knowledge ($\theta \equiv$ random)

$p(\theta)$ = assigned prior $p.d.f.$ of θ, $p(\theta|\underline{x})$ = posterior p.d.f. after data were observed, $\hat{\theta} = E\{\theta|\underline{x}\} = \int \theta p(\theta|\underline{x})d\theta$, $p(\theta|\underline{x}) = p(\underline{x}|\theta)p(\theta)/\int p(\underline{x}|\theta)p(\theta)d\theta$

Example 1

p.d.f. of θ is $1/2A_0$ for $-A_0 \le \theta \le A_0$ and zero everywhere else, $x(n) = \theta + v(n)$ $n = 0,1,\cdots,N-1$, $v(n)$ = GWN with variance σ^2 and independent of θ. Hence $B_{mse}(\hat{\theta}) = \iint (\theta-\hat{\theta})^2 p(\underline{x},\theta)d\underline{x}\,d\theta = \int\left[\int (\theta-\hat{\theta})^2 p(\theta|\underline{x})d\theta\right]p(\underline{x})d\underline{x}$ since $p(\underline{x},\theta) = p(A|\underline{x})p(\underline{x}) \equiv$ Bayes relationship. Since $p(\underline{x}) \ge 0$ implies that the internal integral can be minimized for each $\underline{x}$ and hence $B_{mse}(\hat{\theta})$ will be minimized. $\frac{\partial}{\partial\theta}\int (\theta-\hat{\theta})^2 p(\theta|\underline{x})d\theta = -2\int \theta p(\theta|\underline{x})d\theta + 2\hat{\theta}\int p(\theta|\underline{x})d\theta = 0$ and hence $\hat{\theta} = \int \theta p(\theta|\underline{x})d\theta = E\{\theta|\underline{x}\}$ since the conditional p.d.f. must integrate to one.

$$p_x(x(n)|\theta) = p_v(x(n)-\theta|\theta) = p_v(x(n)-\theta) = \frac{1}{\sqrt{2\pi\sigma^2}}\exp\left[-\frac{1}{2\sigma^2}(x(n)-\theta)^2\right]$$

and hence

$$p_x(\underline{x}|\theta) = \frac{1}{(2\pi\sigma^2)^{N/2}} \exp\left[-\frac{1}{2\sigma^2}\sum_{n=0}^{N-1}(x(n)-\theta)^2\right],$$

$$p(\theta|\underline{x}) = p(\underline{x}|\theta)p(\theta)\Big/\int p(\underline{x}|\theta)p(\theta)d\theta = \frac{\frac{1}{2A_0}p(\underline{x}|\theta)}{\int_{-A_0}^{A_0}(\underline{x}|\theta)p(\theta)d\theta} \quad |\theta| \le A_0$$

and zero for $|\theta| > A_0$. Next we must find c such that

$$p(\theta|\underline{x}) = \frac{1}{c\sqrt{\frac{2\pi\sigma^2}{N}}}\exp\left[-\frac{1}{2\sigma^2/N}(\theta-\bar{x})^2\right] \quad |\theta| \le A_0$$

must integrate to one. Hence,

$$c = \int_{-A_0}^{A}\left[1\Big/\sqrt{\frac{2\pi\sigma^2}{N}}\right]\exp\left[-\frac{1}{2\sigma^2/N}(\theta-\bar{x})^2\right]d\theta$$

and thus MMSE estimator, which is the mean of $p(\theta|\underline{x})$ is

$$\hat{\theta} = E\{\theta|\underline{x}\} = \int_{-\infty}^{\infty}\theta p(\theta|\underline{x})d\theta$$

$$= \left[\int_{-A_0}^{A_0} A\left(1\Big/\left(\sqrt{2\pi\frac{\sigma^2}{N}}\right)\right)\exp\left[-\frac{1}{2\sigma^2/N}(\theta-\bar{x})^2\right]d\theta\right]\Big/\left[\int_{-A_0}^{A_0}\left[1\Big/\left(\sqrt{2\pi\frac{\sigma^2}{N}}\right)\right]\exp[-\frac{1}{2\frac{\sigma^2}{N}}(\theta-\bar{x})^2]d\theta\right]$$

32.1.8.3 Vector Form

$$\underline{\hat{\theta}} = E\{\underline{\theta}|\underline{x}\} = \int \underline{\theta}p(\underline{\theta}|\underline{x})d\underline{\theta},\quad p(\underline{\theta}|\underline{x}) = p(\underline{x}|\underline{\theta})p(\underline{\theta})\Big/\int p(\underline{x}|\underline{\theta})p(\underline{\theta})d\underline{\theta}$$

32.1.8.4 Linear Model (posterior p.d.f. for the general linear model)

$\underline{x} = H\underline{\theta} + \underline{v}$, $\underline{x} = N \times 1$ data vector, $H =$ known $N \times p$ matrix, $\underline{\theta} = p \times 1$ random vector with prior p.d.f. $N(\underline{\mu}_\theta, C_\theta)$ and $\underline{v} = N \times 1$ noise vector with p.d.f. $N(\underline{0}, C_v)$ and independent of $\underline{\theta}$, then the posterior p.d.f. $p(\underline{\theta}|\underline{x})$ is Gaussian with mean ?

$$E\{\underline{\theta}|\underline{x}\} = \underline{\mu}_\theta + C_\theta H^T(H\,C_\theta H^T + C_v)^{-1}(\underline{x} - H\underline{\mu}_\theta)$$

$$C_{\theta|x} = \text{cov}\,ariance = C_\theta - C_\theta H^T(H\,C_\theta H^T + C_v)^{-1}HC_\theta$$

32.2 Statistical Hypotheses

32.2.1 Definitions

32.2.1.1 **Statistical Hypothesis** is a conjecture that a parameter, e.g., θ = mean of a Gaussian process, is larger than a specific value $(\theta > 75)$.

32.2.1.2 **Alternative Hypothesis** is the value of the parameter in 32.2.1.1 is set less than the specific value of 32.2.1.1 $(\theta < 75)$.

32.2.1.3 **Test** is a rule we devise that will tell us what decision to make once the experimental values have been determined. Such a rule is called a test of the statistical hypothesis H_0: $\theta < 75$ against the alternative hypothesis H_1: $\theta > 75$. A test leads to a decision to accept or reject the hypothesis under consideration.

32.2.1.4 **Critical Region** Let C be that subset of the sample space which, in accordance with a prescribed test, leads to the rejection of the hypothesis under consideration. Then C is called the critical region.

32.2.1.5 **Power Function** The power function of a test that yields the probability that the sample point falls in the critical region C of the test; a function that yields the probability of rejecting the hypothesis under consideration.

32.2.1.6 **Power** The value of the power function at a parameter point is called the power of the test at that point.

32.2.1.7 **Significance Level** The significance level of the test (or the *size* of the critical region C) is the maximum value (supremum) of the power function of the test when H_0 is true (H_0 is a hypothesis to be tested against an alternative hypothesis H_1 in accordance with a prescribed test).

Example 1

Let X have p.d.f. $f(x;\theta) = \frac{1}{\theta} e^{-x/\theta}$, $0 < x < \infty$ and $f(x;\theta) = 0$ otherwise. Test $H_0{:}\theta = 2$ (simple hypothesis) against the alternative hypothesis $H_1{:}\theta = 4$. $\{\theta;\theta = 2,4\}$ Random sample X_1,X_2 of size n = 2. C = critical region = $\{(x_1,x_2); 9.5 \le x_1 + x_2 < \infty\}$ will determine the power function and the significance level of the test. $f(x;2)$ specified by H_0 and $f(x;4)$ specified by H_1. The power function is defined at two points $\theta = 2$ and $\theta = 4$. Power function of the test is given by $P\{(X_1,X_2) \in C\}$. If H_0 is true, $\theta = 2$, and the joint p.d.f. of X_1 and X_2 is $f(x_1;2)f(x_2;2) = \frac{1}{4} e^{-(x_1+x_2)/2}$ for $0 < x_1 < \infty$, $0 < x_2 < \infty$, and zero otherwise.
$P\{(X_1,X_2) \in C\} = 1 - P\{(X_1,X_2) \in C^*\} = 1 - \int_0^{9.5}\int_0^{9.5-x_2} \frac{1}{4} e^{-(x_1+x_2)/2} dx_1\, dx_2 \cong 0.05$ (C^* is the complement of C). If H_1 is true, $\theta = 4$, then $f(x_1;4)f(x_2;4) = \frac{1}{16} e^{-(x_1+x_2)/4}$, $0 < x_1 < \infty$, $0 < x_2 < \infty$ and zero otherwise.
$P\{(X_1,X_2) \in C\} = 1 - \int_0^{9.5}\int_0^{9.5-x_2} \frac{1}{16} e^{-(x_1+x_2)/4} dx_1\, dx_2 \cong 0.31$. The power test is 0.05 for $\theta = 2$ and 0.31 for $\theta = 4$. Hence, the probability of rejecting H_0 when H_0 is true is 0.05, and the probability of rejecting H_0

when H_0 is false is 0.31. Since the significance level of this test (size of C) is the power of the test when H_0 is true, the significance level of this test is 0.05.

32.2.2 Neyman-Pearson Theorem

32.2.2.1 Neyman-Pearson Theorem

$X_1,\cdots,X_n$ = random sample from a distribution with p.d.f. $f(x;\theta)$. $L(\theta;x_1,x_2,\cdots x_n)=$ joint *p.d.f.* = $f(x_1;\theta)f(x_2;\theta)\cdots f(x_n;\theta)$. Let θ' and θ'' be distinct fixed vales and k = positive numbers. C = subset of the sample space such that

a) $[L(\theta';x_1,\cdots x_n)/L(\theta'';x_1,\cdots x_n)]\le k$ for each point $(x_1,\cdots,x_n)\in C$,
b) $[L(\theta';x_1,\cdots x_n)/L(\theta'';x_1,\cdots x_n)]\ge k$ for each point $(x_1,\cdots,x_n)\in C^*$ (complement of C)
c) $\alpha = P\{(X_1,\cdots,X_n)\in C; H_0\}$.

Then C is a best critical region of size α for testing the simple hypothesis $H_0:\theta=\theta'$ against the alternative simple hypothesis $H_1:\theta=\theta''$.

Example 1

$X_1,\cdots,X_n$ = random sample from a distribution with p.d.f. $f(x;\theta)=\frac{1}{\sqrt{2\pi}}\exp(-\frac{(x-\theta)^2}{2})$, $-\infty<x<\infty$.

Test $H_0:\theta=\theta'=0$ against the alternative hypothesis $H_1:\theta=\theta''=1$. Now $L(\theta';x_1,\cdots x_n)/L(\theta'';x_1,\cdots x_n)$

$$=\left[\left(\frac{1}{\sqrt{2\pi}}\right)^n \exp\left[-\left(\sum_1^n x_i^2\right)\Big/2\right]\right]\Big/\left[(1/\sqrt{2\pi})^n\exp\left[-\left(\sum_1^n (x_i-1)^2\right)\Big/2\right]\right]=\exp\left[-\sum_1^n x_i+\frac{n}{2}\right].$$ If $k>0$

the set of all points $(x_1,x_2,\cdots,x_n)$ such that $\exp\left[-\sum_1^n x_i+\frac{n}{2}\right]\le k$ is a best critical region. The inequality holds if and only if $-\sum_1^n x_i+\frac{n}{2}\le \ln k$ or $\sum_1^n x_i\ge\frac{n}{2}-\ln k=c$. Hence, the set $C=\{(x_1,x_2,\cdots,x_n);\ \sum_1^n x_i\ge c\}$ where c can be determined so that the size of the critical region is a desired number α. The event $\sum_1^n x_i\ge c$ is equivalent to the event $\overline{X}=\Sigma x_i|n\ge c/n=c_1$, say, so the test may be based upon the statistic $\overline{X}$. If H_0 is true $(\theta=\theta'=0)$, then $\overline{X}$ has a distribution that is $N(0,1/n)$. If $(x_1,x_2,\cdots,x_n)$ are the experimental values, then $\bar{x}=\sum_{i=1}^n x_i|n$. If $\bar{x}\ge c_1$, the simple hypothesis $H_0:\theta=\theta'=0$ would be rejected at the significant level α; if $\bar{x}<c_1$, H_0 would be accepted. The probability of rejecting H_0, when H_0 is true, is α; the probability of rejecting H_0, when H_0 is false, is the value of the power of the test at $\theta=\theta''=1$. That is

$$P\{\overline{X}\ge c_1;H_1\}=\int_{c_1}^{\infty}\frac{1}{\sqrt{2\pi}\sqrt{1/n}}\exp\left(-\frac{(\bar{x}-1)^2}{2(1/n)}\right)d\bar{x}.$$

For example, if n = 25 and if we select $\alpha = 0.05$, then from Table 34.2 in Chapter 34 (interpolate) we find $c_1 = 1.645/\sqrt{25} = 0.329$. Hence the power of this best test of H_0 against H_1 is 0.05, when H_0 is true, and is

$$\int_{0.329}^{\infty} \frac{1}{\sqrt{2\pi}\sqrt{1/25}} \exp\left(-\frac{(\bar{x}-1)^2}{2(1/25)}\right) d\bar{x} = 0.999$$

when H_1 is true.

32.2.2.2 Likelihood Ratio

$\lambda(x_1, x_2, \cdots, x_n) = \lambda = \dfrac{L(\hat{\omega})}{L(\hat{\Omega})}$, $L(\hat{\omega})$ = maximum of $L(\omega)$, $L(\hat{\Omega})$ = maximum of $L(\Omega)$, Ω = the set of all parameter points $(\theta_1, \theta_2, \cdots, \theta_m)$, ω subset of Ω, $(X_1, X_2, \cdots, X_n) = n$ mutually stochastically independent r.v.'s having, respectively, the p.d.f. $f_i(x_i; \theta_1, \theta_2, \cdots, \theta_m)$ $i = 1, 2, \cdots, n$; $L(\omega) = \prod_{i=1}^{n} f_i(x_i; \theta_1, \cdots, \theta_m)$, $(\theta_1, \cdots, \theta_m) \in \omega$, $L(\Omega) = \prod_{i=1}^{n} f_i(x_i; \theta_1, \cdots, \theta_m)$, $(\theta_1, \cdots, \theta_m) \in \Omega$.

32.2.2.3 Likelihood Ration Test Principle

$H_0: (\theta_1, \cdots, \theta_m) \in \omega$, is rejected if and only if $\lambda(x_1, x_2, \cdots, x_n) = \lambda \le \lambda_o$ = positive proper function. The function λ defines an r.v. $\lambda(X_1, \cdots, X_n)$ and the significance level of the test is given by $\alpha = P\{\lambda(X_1, \cdots, X_n) \le \lambda_o; H_o\}$

32.2.3 Hypothesis Testing for the Mean of a Normal Distribution: The Two-Tailed t-test

32.2.3.1 $x_1, x_2, \cdots, x_n = n$ normally distributed observations. We wish to test whether or not the sample supports the hypothesis that $\mu_x = \bar{x}$ is μ_0 = same test value for μ_x which we have specified according to some objectives of our investigation.

Steps

1. Hypothesis specifications: Set μ_0 and α = the size (probability) of the Type I error to be tolerated.

   $H_0: \mu_x = \mu_0$, $H_1: \mu_x \ne \mu_0$; α
2. Test statistic:

   $$t_0 = \frac{\bar{x} - \mu_0}{s_{\bar{x}}} = \text{test statistic},\ \bar{x} = \Sigma x_i / n,\ s_{\bar{x}} = \sqrt{s_x^2 / n},\ s_x^2 = \sum_{1}^{n} (x_i - \bar{x})(x_i - \bar{x}) / (n-1)$$
3. Assumption: $x_i \equiv N(\mu_x, \sigma^2)$. If H_0 is true $t_0 = (\bar{x} - \mu_0)/s_{\bar{x}} = (\bar{x} - \mu_x)/s_{\bar{x}}$ is a member of t-distribution with (n – 1) degrees of freedom.
4. Critical region: Reject H_0 and accept H_1 if $|t_0| = |(\bar{x} - \mu_0)/\sqrt{s_x^2/n}| > t(n-1; \alpha/2)$.

   Accept H_0 if $|t_0| < t(n-1; \alpha/2)$.
5. P-value: P-value = $2P\{t(n-1) > |t_0|\}$

6. Confidence interval: $\bar{x}\left(1 \mp \frac{t_{\alpha/2}}{t_0}\right)$

Example 1

If $x_1 = 17,\ x_2 = 16,\ x_3 = 18,\ x_4 = 21;\ n = 4$. Test whether or not $\mu = 10$ and take $\alpha = 0.05$.

Steps

1. $H_0 : \mu = 10,\ \ H_1 : \mu \neq 10;\ \ \alpha = 0.05$
2. $\bar{x} = 74/4 = 18,\ \sum_1^4 (x_i - \bar{x})(x_i - \bar{x}) = 14,\ s_x^2 = 14/3,\ s_{\bar{x}} = \sqrt{14/12},\ \ t_0 = \dfrac{18-10}{\sqrt{14/12}} = 7.4$
3. Assume $x_i = N(\mu = 10; \sigma^2),\ i = 1, \cdots, 4,$ the test statistic is a member of the t-distribution with $n - 1 = 3$ degrees of freedom.
4. In the 3-degree of freedom row (see Table 34.3, Chapter 34) and column $1 - (0.05/2) = 0.975$ we find $t(3; 0.05/2) = 3.182$. The test statistic value $7.4 > 3.182$ has fallen in the critical region. The test hypothesis is false and prefers the alternative hypothesis.
5. The test statistic value 7.4 is even greater than 5.841 found in the column $1 - 0.005 = 0.995$ in the 3-degree of freedom row. Hence $P < 2(0.005) = 0.01$ which implies that the error is very small.
6. 95% confidence interval for μ_x is $18 \mp (3.182)(\sqrt{14/12}) = 14.56,\ 21.44 \cong 14.5,\ 21.5$.

32.2.4 Hypothesis Testing for the Variance of a Normal Distribution: The Two-Tailed χ^2-Test

32.2.4.1 Steps

1. Hypothesis specification: $H_0 : \sigma_x^2 = \sigma_0^2,\ H_1 : \sigma_x^2 \neq \sigma_0^2;\ \alpha$. Investigator specifies the numerical values of σ_0^2 and α.
2. Test statistic: Given $x_1, x_2, \cdots, x_n$ calculate s_x^2 (see 32.2.3.1) and the test statistic

$$\chi_0^2 = \frac{(n-1)s_x^2}{\sigma_0^2} = \frac{\sum_1^n (x_i - \bar{x})(x_i - \bar{x})}{\sigma_0^2}$$

3. Distribution: If $x_i \equiv N(\mu, \sigma_x^2),\ i = 1, \cdots, n,$ then $\sum_{i=1}^n (x_i - \bar{x})(x_i - \bar{x})/\sigma_x^2 \equiv \chi^2(n-1)$. Hence if H_0 is true and $\sigma_x^2 = \sigma_0^2$, the test statistic is also a $\chi^2(n-1)$-variate; that is $\sum_{i=1}^n (x_i - \bar{x})(x_i - \bar{x})/\sigma_0^2 \equiv \chi^2(n-1)$ so that, when H_0 is true tabulated $\chi^2(n-1)$ probabilities can be used as probabilities for $\sum_{i=1}^n (x_i - \bar{x})(x_i - \bar{x})/\sigma_0^2$

4. Critical region: $n-1 \equiv$ degrees of freedom, $\chi^2(n-1;1-\alpha/2) =$ lower ritical value, $\chi^2(n-1;\alpha/2)$ = upper critical value, $P\{\chi^2 < \chi^2(n-1;1-\alpha/2)\} = \frac{\alpha}{2} = P\{\chi^2 > \chi^2(n-1;\alpha/2)\}$, the two-tail region from 0 to $\chi^2_{1-\alpha/2}$ and from $\chi^2_{\alpha/2}$ to ∞ comprise the critical region. We reject H_0 for H_1 if the test statistic $\chi_0^2 = \sum_{i=1}^{n}(x_i - \bar{x})^2/\sigma_0^2$ fall in the region; we accept H_0 if $\chi^2_{1-\alpha/2} < \chi_0^2 < \chi^2_{\alpha/2}$
5. P-value: $2P\{\chi^2(n-1) > \chi_0^2\}$ or $2P\{\chi^2(n-1) < \chi_0^2\}$, $\chi_0^2 =$ right or left tail of the $\chi^2(n-1)$-distribution.
6. Confidence intervals: A $100(1-\alpha)\%$ confidence interval for the population variance is from $\sum_{i=1}^{n}(x_i - \bar{x})^2/\chi^2_{\alpha/2}$ to $\sum_{i=1}^{n}(x_i - \bar{x})^2/\chi^2_{1-\alpha/2}$

Example 1

If $n = 16$ and $\sum_{i=1}^{n}(x_i - \bar{x}) = 135$, then $s_x^2 = 135/15 = 19$ with 15 degrees of freedom. At $\alpha = 0.05$, is this result statistically compatible with the hypothesis that $\sigma_x^2 = 20$?

1. $H_0{:}\sigma_x^2 = 20,\ H_1{:}\sigma_x^2 \neq 20,\ \alpha = 0.05$
2. $\chi_0^2 = \sum_{i=1}^{n}(x_i - \bar{x})^2/\sigma_0^2 = 135/20 = 6.75 \equiv$ test statistic
3. If $x_i = N(\mu, \sigma_x^2)$, $i = 1,\cdots,16$, the test statistic is a χ_0^2 (15)-variate
4. From Table 34.4 in Chapter 34 and 15 degrees of freedom we obtain $\chi^2(15;1-0.05/2)$ = $\chi^2(15;0.975) = 6.26$ and $\chi^2(15;0.025) = 27.5$ (you should read on $F = 1-0.975$ and F = 1 – 0.025). The test statistic value, $\chi_0^2 = 6.75$, lies between the two critical values and we accept H_0 to the unknown with possible small probability of having made a type II error.

32.2.5 One-Sided Alternative Hypothesis for Means

32.2.5.1 One-tailed t-test for the population mean of a normal distribution.

Steps

1. Hypothesis specification: $H_0{:}\mu_x \geq \mu_0$, $H_1{:}\mu_x < \mu_0$, α
2. Test statistic: n observations, $\bar{x} = \sum_i x_i/n, s_{\bar{x}} = \sqrt{s_x^2/n}$, $t_0 =$ test statistic $= (\bar{x} - \mu_0)/s_{\bar{x}}$ = $(\bar{x} - \mu_0)/\sqrt{s_x^2/n}$
3. Assumptions $x_i \equiv N(\mu_x, \sigma_x^2)$
4. Critical region: $-\infty$ to $-t_\alpha = -t(n-1;\alpha)$. If $t_0 < -t_\alpha$ we reject H_0 in favor of H_1. If $\bar{x} > \mu_0$, H_0 is accepted.

5. P-value: $P = P\{t(n-1) < t_0\} = P\{t(n-1) > -t_0\}$

6. Confidence interval: The $100(1-\alpha)\%$ confidence interval for μ_x is $\bar{x} \mp t_{\alpha/2}\sqrt{s_x^2/n}$

Example 1

Given $\bar{x} = 12.5$, $s_x^2 = 11/3$, $n-1 =$ degrees of freedom = 3, $x_i \equiv$ normally distributed.

Steps

1. $H_0: \mu_x \le 10$, $H_1: \mu_x > 10$, $\alpha = 0.05$
2. Test statistic: $\dfrac{\bar{x}-\mu_0}{s_{\bar{x}}} = \dfrac{12.5-10}{\sqrt{(11/3)/4}} = 2.61$
3. Assuming H_0 is true and $x_i \equiv N(10, \sigma^2)$, the test statistic is a random $t(3)$-variate
4. Since H_1 is right-sided, the critical region comprises those $t(3)$-values exceeding $t_{0.05} = t(3;0.05)$ = 2.353 (see Table 34.3, Chapter 34). Since $t_0 = 2.611 > 2.353$ we accept H_1
5. $P\{t(3) > 2.611\} = 0.05 - \dfrac{2.611-2.353}{3.182-2.353}(0.05-0.025) \cong 0.04$ (by interpolation of the same table)
6. The 95% confidence interval $12.5 \mp 3.182\sqrt{11/12} = 9.4, 15.5$

32.2.6 One-Sided Tests for the Population Variance of a Normal Distribution

32.2.6.1 Steps

1. Hypothesis specifications: $H_0: \sigma_x^2 = \sigma_0^2$, $H_1: \sigma_x^2 < \sigma_0^2$, α

   $H_0: \sigma_x^2 = \sigma_0^2$, $H_1: \sigma_x^2 > \sigma_0^2$, α
2. Test statistic (both cases): $\chi_0^2 = \dfrac{(n-1)s^2}{\sigma_0^2} = \dfrac{\sum_i (x_i - \bar{x})^2}{\sigma_0^2}$
3. If $x_i \equiv N(\mu_x, \sigma_x^2 = \sigma_0^2)$, the test statistic is a $\chi^2(n-1)$-variate
4. Critical region: (a) For the first set of 1) the critical region is from 0 to $\chi_{1-\alpha}^2 = \chi^2(n-1;1-\alpha)$

   (b) For the second set of 1) the critical region is from $\chi_\alpha^2 = \chi^2(n-1;\alpha)$ to ∞.
5. (a) For the first set of 1) P-value is $P\{\chi^2(n-1) < \chi_0^2\}$ see 2, (b) for the second set of 1) P-value is $P\{\chi^2(n-1) > \chi_0^2\}$

Example 1

Given n = 16, x_i normally distributed, $s_x^2 = 9$. Was $\sigma_x^2 > 5$?

1. Specifications: $H_0: \sigma_x^2 \le 5$, $H_1: \sigma_x^2 > 5$, $\alpha = 0.05$
2. Test statistic: $\chi_0^2 = (n-1)s_x^2/\sigma_0^2 = (15)(9)/5 = 27$

3,4. Given normality, since $\chi_0^2 = 27 > 24.50 = \chi^2(15;\, 0.05)$

(see Table 34.4, Chapter 34; note that the table gives 1-tail values), the test statistic lies in the critical region and, hence, we accept H_1.

References

Cox, Phillip, C., *A Handbook of Introductory Statistical Methods*, John Wiley & Sons, New York, NY, 1987.

Hogg, Robert V. and A. T. Craig, *Introduction to Mathematical Statistics*, 4th Edition, Macmillan Publishing Co., New York, NY, 1978.

Kay, Steven M., *Statistical Signal Processing*, Prentice Hall, Upper Saddle River, NJ, 1993.

33

Matrices

33.1 Notation and Some General Properties

33.1.1 Notation

Capital letters will denote matrices, e.g., A, B, N; the lower case of the first few letters of the alphabet wil' denote constants, i.e., a, b, c, d; and the lower case of letters in the last part of the alphabet will denote vectors, e.g., x, y, z. Lower case Greek letters will denote vectors, except for λ which will denote eigenvalues.

33.1.2 Notation

$A = [a_{ij}] =$ matrix where i refers to i^th^ row and j refers to j^th^ column. $A_{n\times m}$ = matrix with n rows and m columns. $a_{ij} \equiv a(i,j)$ the element on the ith row and j^th^ column.

33.1.3 Identity I

Matrix with ones along the diagonal from left to right and zeros for all the other elements.

Example

$$I = \begin{bmatrix} 1 & 0 \\ 0 & 1 \end{bmatrix} = \text{a } 2\times 2 \text{ identity matrix}$$

33.1.4 Diagonal D

Matrix with elements along the diagonal and zeros for all the other elements.

Example

$$D = \begin{bmatrix} a_{11} & 0 \\ 0 & a_{22} \end{bmatrix} = \text{a } 2\times 2 \text{ diagonal matrix}$$

33.1.5 Transpose

$$A^T = [a_{ji}]$$

Example

If

$$A_{3\times 2} = \begin{bmatrix} a_{11} & a_{12} \\ a_{21} & a_{22} \\ a_{31} & a_{32} \end{bmatrix},$$

then

$$A^T = \begin{bmatrix} a_{11} & a_{21} & a_{31} \\ a_{12} & a_{22} & a_{32} \end{bmatrix} \text{ a } 2\times 3 \text{ matrix}$$

33.1.6 Properties of Transposition

1. $(aA)^T = (Aa)^T = A^T a = aA^T$
2. $(aA + bB)^T = aA^T + bB^T$
3. $(A^T)^T = A$
4. $A^T = B^T$ if $A = B$
5. $(AB)^T = B^T A^T$
6. $D = D^T$ if D is diagonal
7. If $A = A^T$, A is *symmetric*
8. If $A = -A^T$, A is *skew-symmetric*
9. $A^T A$ and AA^T are symmetric
10. If A is nonsingular (its determinant is not zero), then A^T and A^{-1} are nonsingular and $(A^T)^{-1}$ $= (A^{-1})^T$. A^{-1} means the inverse of matrix A.

33.1.7 Vector-Vector Multiplication

$z = x^T y$.

Example

$$a = [1 \;\; 2]\begin{bmatrix} 3 \\ 4 \end{bmatrix} = 3\times 1 + 2\times 4 = 11$$

33.1.8 Matrix-Vector Multiplication

$$z = 'Ax\,;\; z_i = \sum_{j=1}^{n} a(i,j)x(j).$$

Example

$$\begin{bmatrix} 1 & 2 \\ 3 & 4 \end{bmatrix}\begin{bmatrix} 5 \\ 6 \end{bmatrix} = \begin{bmatrix} 1\times 5+2\times 6 \\ 3\times 5+4\times 6 \end{bmatrix} = 5\begin{bmatrix} 1 \\ 3 \end{bmatrix} + 6\begin{bmatrix} 2 \\ 4 \end{bmatrix} = \begin{bmatrix} 17 \\ 39 \end{bmatrix}$$

Note: A is 2 × 2, x is a 2 × 1 and the result is 2 × 1. In general, if A is $m \times n$ the vector must be $n \times 1$ and the result will be $m \times 1$ vector.

33.1.9 Matrix-Matrix Multiplication

$$C = AB \text{ or } C_{m\times q} = A_{m\times n}B_{n\times q}.\; c(i,j) = \sum_{k=1}^{n} a(i,k)b(k,j).$$

Example

$$\begin{bmatrix} 1 & 4 & 1 \\ 0 & 2 & 1 \end{bmatrix}\begin{bmatrix} 1 & 2 & 3 \\ 4 & 0 & 0 \\ 1 & 2 & 1 \end{bmatrix} = \begin{bmatrix} 1\times 1+4\times 4+1\times 1 & 1\times 2+4\times 0+1\times 2 & 1\times 3+4\times 0+1\times 1 \\ 0\times 1+2\times 4+1\times 1 & 0\times 2+2\times 0+1\times 2 & 0\times 3+2\times 0+1\times 1 \end{bmatrix}$$

$$= \begin{bmatrix} 18 & 4 & 4 \\ 9 & 2 & 1 \end{bmatrix}$$

$C_{2\times 3} = A_{2\times 3}B_{3\times 3}$

33.1.10 Hermitian

$A^H = (A^T)^* =$ complex conjugate of the transpose of A.

Example

If

$$A = \begin{bmatrix} 1+j2 & 0 \\ 3-j & j4 \end{bmatrix},$$

then

$$A^H = \begin{bmatrix} 1-j2 & 3+j \\ 0 & -j4 \end{bmatrix}$$

Properties

1. $(A+B)^H = A^H + B^H$
2. $(AB)^H = B^H A^H$

33.1.10.1 Inverse of a Hermitian

$(A^H)^{-1} = (A^{-1})^H$

33.1.11 Block Matrix

$$\mathbf{A} = [A(ij)] = \begin{bmatrix} A(1,1) & \cdots & A(1,q) \\ \vdots & \vdots & \vdots \\ A(p,1) & \cdots & A(p,q) \end{bmatrix}.$$

Each matrix $A(i,j)$ has the same dimension $m \times n$. $\mathbf{A}$ has dimensions $p \times q$. The *scalar dimension* of $\mathbf{A}$ $pm \times qn$.

33.1.12 Reflection (or exchange) Matrix J

$$J = \begin{bmatrix} 0 & \cdots & 0 & 1 \\ 0 & \cdots & 1 & 0 \\ 1 & \cdots & 0 & 0 \end{bmatrix}$$

reverses the rows or columns of a matrix.

Example

$$J_{2x2}A_{2x3} = \begin{bmatrix} 0 & 1 \\ 1 & 0 \end{bmatrix} \begin{bmatrix} a(1,1) & a(1,2) & a(1,3) \\ a(2,1) & a(2,2) & a(2,3) \end{bmatrix} = \begin{bmatrix} a(2,1) & a(2,2) & a(2,3) \\ a(1,1) & a(1,2) & a(2,3) \end{bmatrix} = \text{reversed the rows}$$

$$J_{3x3}A_{2x3} = \begin{bmatrix} a(1,1) & a(1,2) & a(1,3) \\ a(2,1) & a(2,2) & a(2,3) \end{bmatrix} \begin{bmatrix} 0 & 0 & 1 \\ 0 & 1 & 0 \\ 1 & 0 & 0 \end{bmatrix} = \begin{bmatrix} a(3,1) & a(1,2) & a(1,1) \\ a(2,3) & a(2,2) & a(2,1) \end{bmatrix} = \text{reversed the columns}$$

33.1.13 Persymmetric P

An $n \times n$ square matrix that is symmetric about its cross diagonal (from right to left). $a(i,j) = a(n-j+1, n-i+1)$

Example

$$P = \begin{bmatrix} a_{11} & a_{12} & a_{13} & a_{14} \\ a_{21} & a_{22} & a_{23} & a_{13} \\ a_{31} & a_{32} & a_{22} & a_{12} \\ a_{41} & a_{31} & a_{21} & a_{11} \end{bmatrix}$$

33.1.13.1 Inverse of Persymmetric

If P is persymmetric, P^{-1} is also persymmetric.

33.1.14 Centrosymmetric

An $n \times n$ square matrix with the property $r(i,j) = r^*(n-i+1, n-j+1)$.

33.1.15 Doubly Symmetric

A square matrix that is Hermitian about the principal diagonal and persymmetric about the cross diagonal:

$$r(i,j) = r^*(j,i) = r(n-j+1, n-i+1) = r^*(n-i+1, n-j+1)$$

Example

$$R = \begin{bmatrix} r(1,1) & r^*(2,1) & r^*(3,1) & r^*(4,1) \\ r(2,1) & r(2,2) & r^*(3,2) & r^*(3,1) \\ r(3,1) & r(3,2) & r(2,2) & r^*(2,1) \\ r(4,1) & r(3,1) & r(2,1) & r(1,1) \end{bmatrix}$$

33.1.15.1 Symmetric

$A = A^T$; $A^{-1} = (A^{-1})^T$ (inverse is also symmetric).

33.1.16 Toeplitz

A matrix A with equal elements along any diagonal, $a(i,j) = a(i-j)$.

33.1.17 Square Toeplitz

A square Toeplitz is Hermitian and a special case of a persymmetric matrix: $a^*(k) = a(-k)$ and $A = JA^*J$. A Hermitian Toeplitz matrix is *centrosymmetric*.

33.1.18 Hankel

A matrix A with equal elements along any cross diagonal, $a(i,j) = a(i+j-n-1)$. A square Hankel matrix A has the property: $A^H = A$.

33.1.19 Right-Circulant

The relationship of its elements of an $n \times n$ right-circulant is matrix given by

$$a(i,j) = \begin{cases} a(j-1) & \text{for } j-i \geq 0 \\ a(n-j+1) & \text{for } j-i < 0 \end{cases}$$

Example

$$A = \begin{bmatrix} a(0) & a(1) & a(2) & a(3) \\ a(3) & a(0) & a(1) & a(2) \\ a(2) & a(3) & a(0) & a(1) \\ a(1) & a(2) & a(3) & a(0) \end{bmatrix}$$

33.1.20 Left-Circulant (n × n matrix)

The relationship of its elements is given by

$$a(i,j) = \begin{cases} a(n+1-i-j) & \text{for } j+i \leq n+1 \\ a(2n+1-i-j) & \text{for } j+i > n+1 \end{cases}$$

Example

$$A = \begin{bmatrix} a(3) & a(2) & a(1) & a(0) \\ a(2) & a(1) & a(0) & a(3) \\ a(1) & a(0) & a(3) & a(2) \\ a(0) & a(3) & a(2) & a(1) \end{bmatrix}$$

33.1.21 Vandermonde (m × n)

$a(i,j) = x_j^{i-1}$ for $1 \leq i < m,\ 1 \leq j \leq n$

Example

$$\begin{bmatrix} 1 & 1 & \cdots & 1 \\ x_1 & x_2 & \cdots & x_n \\ x_1^2 & x_2^2 & \cdots & x_n^2 \\ \vdots & & \vdots & \\ x_1^{m-1} & x_2^{m-1} & \cdots & x_n^{m-1} \end{bmatrix}$$

33.1.22 Upper Triangular (n × m)

$a(i,j) = 0$ for $j < i$

Example

$$\begin{bmatrix} a(1,1) & a(1,2) \\ 0 & a(2,2) \end{bmatrix}$$

33.1.23 Lower Triangular

$a(i,j)=0$ for $j>i$

Example

$$\begin{bmatrix} a(1,1) & 0 \\ a(2,1) & a(2,2) \end{bmatrix}$$

33.1.24 Circulant Permutation (n × n)

$$B = \begin{bmatrix} 0 & 1 & 0 & \cdots & 0 \\ 0 & 0 & 1 & & 0 \\ \vdots & & & & \vdots \\ 0 & 0 & \cdots & & 1 \\ 1 & 0 & \cdots & & 0 \end{bmatrix}$$

If A is circulant then $A=\sum_{k=0}^{n-1} a_{k+1}C^k$ ($C^0=I=C^n$ and a's are the entries of the first row of A).

33.1.25 Upper Hessenberg

$a(i,j)=0$ for $i>j+1$

33.1.26 Lower Hessenberg:

A is lower Hessenberg if A^T is upper Hessenberg.

33.2 Determinants (of square matrices)

33.2.1 Definition

$\det(A)\equiv|A|=\sum_{j=1}^{n} a_{ij}|A_{ij}|$ where A_{ij} is the cofactor of A for any i.

33.2.2 Cofactor

$A_{ij}=(-1)^{i+j}m_{ij}$ where m_{ij} is called the minor of a_{ij}.

Example

$$\det(A)=\det\begin{bmatrix} a_{11} & a_{12} & a_{13} \\ a_{21} & a_{22} & a_{23} \\ a_{31} & a_{32} & a_{33} \end{bmatrix}=a_{11}|A_{11}|+a_{12}|A_{12}|+a_{13}|A_{13}|$$

$$=a_{11}(-1)^{1+1}\begin{vmatrix} a_{22} & a_{23} \\ a_{32} & a_{33} \end{vmatrix}+a_{12}(-1)^{1+2}\begin{vmatrix} a_{21} & a_{23} \\ a_{31} & a_{33} \end{vmatrix}+a_{13}(-1)^{1+3}\begin{vmatrix} a_{21} & a_{22} \\ a_{31} & a_{32} \end{vmatrix}$$

$$=a_{11}(a_{22}a_{23}-a_{23}a_{32})-a_{12}(a_{21}a_{33}-a_{23}a_{31})+a_{13}(a_{21}a_{32}-a_{22}a_{31})$$

33.2.3 Product of Matrices

$\det(AB) = \det(A)\det(B)$

33.2.4 Transpose

$\det(A) = \det(A^T)$

33.2.5 Change of Sign

If two rows (columns) of a matrix are interchanged, the determinant of the matrix changes sign.

33.2.6 If two rows of a matrix A are identical det(A) = 0.

33.2.7 The determinant is not changed if the elements of the ith row are multiplied by a scalar k and the results are added to the corresponding elements of the hth row, $h \neq i$.

33.2.8 If det(A) = 0, A is a singular matrix.

33.3 Rank of Matrices

33.3.1 Elementary Matrix

E_{rs} is an elementary matrix produced by exchanging two rows (or two columns) of an identity matrix I (ones along the diagonal and zeros everywhere else).

33.3.2 Row Exchange of A

$E_{rs}'A$ exchanges two rows of A if E_{rs} has been constructed from an identity matrix by exchanging rows r and s.

33.3.3 Column Exchange of A

A E_{rs} exchanges two columns of A if E_{rs} has been constructed from an identity matrix by exchanging columns r and s.

33.3.4 Rank

rank $r \equiv r_A \equiv r(A)$ is the number of linearly independent rows and columns in the matrix.

1. r is a positive integer
2. r is equal to or less than the smaller of its number of rows or columns
3. When multiplying by elementary matrices, the rank does not change.
4. When $r \neq 0$ of A, there exists at least one nonsingular square submatrix of A.

33.3.5 Calculating Rank

Elementary row operation until the matrix is upper triangular.

Example

$$A = \begin{bmatrix} 1 & 2 & 4 & 3 \\ 3 & -1 & 2 & -2 \\ 5 & -4 & 0 & -7 \end{bmatrix}, \text{ row } 2 - 3\times \text{ (row 1) and row } 3 - 5\times \text{ (row 1)}$$

$$A = \begin{bmatrix} 1 & 2 & 4 & 3 \\ 0 & -7 & -10 & -11 \\ 0 & -14 & -20 & -22 \end{bmatrix}, \text{ row } 3 - 2\times(\text{row } 2),\ A = \begin{bmatrix} 1 & 2 & 4 & 3 \\ 0 & -7 & -10 & -11 \\ 0 & 0 & 0 & 0 \end{bmatrix} \text{ hence the rank r} = 2.$$

33.4 Additional Rank Properties

33.4.1 If $r(A_{nxn}) = n,$ A is nonsingular and full rank.

33.4.2 If $r(A_{nxn}) < n,$ then A is singular.

33.4.3 If $r(A_{pxq}) = p < q,$ A has full row rank.

33.4.4 If $r(A_{pxq}) = q < p,$ A has full column rank.

33.4.5 AC, CB, and ACB have the same rank if A and B are nonsingular and C is any matrix.

33.4.6 $r(AB)$ cannot exceed the rank of either A or B.

33.4.7 $r(A+B) \le r(A) + R(B)$.

33.4.8 If C_{mxn} has rank r then CC^T and C^TC have rank r.

33.5 Vectors

33.5.1 Vector Space

Let V_n be a set of n-component vectors such that for every two vectors in V_n, the sum of the two vectors is also in V_n, and for each vector in V_n and each scalar, the product is in V_n. This set V_n is called a vector space.

33.5.2 Subspace

Let S_n be a subset of vectors in V_n. If S_n is a vector space, then it is called a (vector) subspace.

Example

The set of vectors $x^T = [a_1\ a_2\ 0]$ for all real numbers a_1 and a_2 is a subspace of the three-dimensional space R_3.

33.5.3 Linear Dependent and Independent

The n-dimensional of m vectors $\{x_1, x_2, \cdots, x_m\}$ are linear dependent if we find even one c_i different from zero such that $\sum_{i=1}^{m} c_i x_i = 0.$ If this relation is satisfied only when all c_i's are zeros, the vectors are independent.

33.5.4 If the rank r of the matrix $V = [v_1, v_2, \cdots, v_m]$, where v_i's are $n \times 1$ vectors, is less than m, then the vectors are dependent.

33.5.5 Basis

If $\{v_1, v_2, \cdots, v_m\}$ is a basis of a space, then any vector v has a unique linear combination of the given basis:

$$v = \sum_{i=1}^{m} c_i v_i$$

33.5.6 Orthogonal

If $x^T y = 0,$ then the vectors x and y are orthogonal, $x^T y = \sum_{i=1}^{n} x_i y_i$

33.5.7 Orthogonal Basis

If $\{v_1, v_2, \cdots, v_m\}$ is a basis and $v_i^T v_j = 0$ for all $i \neq j = 1, 2, \cdots, m,$ the basis is orthogonal.

33.5.8 Orthonormalization

If $\{v_1, v_2, \cdots, v_m\}$ is a basis, then $\{z_1, z_2, \cdots, z_m\}$ is an orthonormal basis

$$y_1 = v_1 \qquad z_1 = \frac{y_1}{\sqrt{y_1^T y_1}}$$

$$y_2 = v_2 - \frac{y_1^T v_2}{y_1^T y_1} y_1 \qquad z_2 = \frac{y_2}{\sqrt{y_2^T y_2}}$$

....

$$y_m = v_m - \frac{y_1^T v_m}{y_1^T y_1} y_1 - \frac{y_2^T v_m}{y_2^T y_2} y_2 - \cdots - \frac{y_{m-1}^T v_m}{y_{m-1}^T y_{m-1}} y_{m-1} \qquad z_m = \frac{y_m}{\sqrt{y_m^T y_m}}$$

33.6 Quadratic Form

33.6.1 Definition

$$f(x) = \sum_{j=1}^{n} \sum_{i=1}^{n} a_{ij} x_i x_j = x^T A x \quad (A = n \times n \text{ matrix})$$

33.6.2 Congruent Matrices

$$x^T A x = y^T C^T A C y = y^T B y \quad (B = C^T A C)$$

A and B are congruent. C = $n \times n$ nonsingular matrix and A and B have the same rank.

33.6.3 Positive Definite

When A_{nxn} in 33.4.1 has r = n

33.6.4 Positive Semidefinite

When A_{nxn} in 33.4.1 has r < n

33.7 Orthogonal Matrices

33.7.1 Definition

P_{nxn} matrix is orthogonal if and only if $P^{-1} = P^T$

33.7.2 Partitioned

$P = [p_1, p_2, \cdots, p_n]$, where p_i's are $n \times 1$ matrices (vectors) of the column of P, then $p_i^T p_i = 1$ for $i = 1, 2, \cdots, n,$ and $p_i^T p_j = 0$ for $i \neq j$.

33.7.3 $P^T P = I$ if P is orthogonal.

33.7.4 det (P) is +1 or −1

33.7.5 If A is any $n \times n$ matrix and P an $n \times n$ orthogonal, then det (A) = det $(P^T AP)$.

33.7.6 $P^T AP = D$ where A is an $n \times n$ matrix, P is an $n \times n$ orthogonal matrix, and D is a diagonal matrix.

33.8 Inverse Matrices

33.8.1 Uniqueness

If A^{-1} (inverse of A) is such that $A^{-1}A = AA^{-1} = I$ then A^{-1} is unique for a given A.

33.8.2 Adjoint (or adjugate)

If the elements of a matrix are replaced by their cofactors and then transposed, it is called adjoint.

33.8.3 Inverse ($n \times n$)

$$A^{-1} = \frac{1}{|A|} adj\, A, \quad |A| \neq 0$$

Example

$$A = \begin{bmatrix} 1 & 3 \\ 4 & 6 \end{bmatrix}, \quad adj\, A = \begin{bmatrix} 6 & -4 \\ -3 & 1 \end{bmatrix}^T = \begin{bmatrix} 6 & -3 \\ -4 & 1 \end{bmatrix}, \quad |A| = (6-12) = -6,$$

$$A^{-1} = \frac{1}{-6}\begin{bmatrix} 6 & -3 \\ -4 & 1 \end{bmatrix} = \begin{bmatrix} -1 & +\frac{1}{2} \\ +\frac{2}{3} & -\frac{1}{6} \end{bmatrix}, A^{-1}A = \begin{bmatrix} -1 & \frac{1}{2} \\ \frac{2}{3} & -\frac{1}{6} \end{bmatrix}\begin{bmatrix} 1 & 3 \\ 4 & 6 \end{bmatrix} = \begin{bmatrix} 1 & 0 \\ 0 & 1 \end{bmatrix}$$

33.8.4 Properties

1. $A^{-1}A = AA^{-1} = I$
2. A is unique
3. $|A^{-1}| = 1/|A|$
4. $(A^{-1})^{-1} = A$
5. $(A^T)^{-1} = (A^{-1})^T$
6. $(A^{-1})^T = A^{-T}$ if $A^T = A$
7. $(AB)^{-1} = B^{-1}A^{-1}$
8. $(D[x_i])^{-1} = D[1/x_i]$ for $x_i \neq 0$
9. $(I + AB)^{-1} = I - A(I + BA)^{-1}B$
10. $(A + BCD)^{-1} = A^{-1} - A^{-1}B(DA^{-1}B + C^{-1})^{-1}DA^{-1}$,
    $A = n \times n,\ B = n \times m,\ C = m \times m,\ D = m \times n$
11. $(A + xy^H)^{-1} = A^{-1} - \dfrac{(A^{-1}x)(y^H A^{-1})}{1 + y^H A^{-1}x}$, x and y vectors.
12. a) If $Y = \begin{bmatrix} A & D \\ C & B \end{bmatrix}$ then

$$Y^{-1} = \begin{bmatrix} A^{-1} + A^{-1}D\Delta^{-1}CA^{-1} & -A^{-1}D\Delta^{-1} \\ -\Delta^{-1}CA^{-1} & \Delta^{-1} \end{bmatrix}, \quad \Delta = B - CA^{-1}D$$

$$Y^{-1} = \begin{bmatrix} \Lambda^{-1} & -\Lambda^{-1}DB^{-1} \\ -B^{-1}C\Lambda^{-1} & B^{-1} + B^{-1}C\Lambda^{-1}DB^{-1} \end{bmatrix}, \quad \Lambda = A - DB^{-1}C$$

b) If $D = x$, $C = y^H$, and $B = a$ then

$$Y^{-1} = \begin{bmatrix} A^{-1} + bA^{-1}xy^H A^{-1} & -\beta A^{-1}x \\ -\beta y^H A^{-1} & \beta \end{bmatrix}, \quad \beta = (a - y^H A^{-1}x)^{-1}$$

13. $B = aA + b(I - A)$, $B^{-1} = \frac{1}{a}A + \frac{1}{b}(I - A)$, $A_{k\times k}$ = independent, a , b = constants

14. $A = B^{-1} + CD^{-1}C^H$, $A^{-1} = B - BC(D + C^H BC)^{-1}C^H B$, H stands for Hermitian (conjugate transpose)

33.9 Linear Transformations, Characteristic Roots (eigenvalues)

33.9.1 Definitions

$y = Ax$

If $y_1 = Ax_1$ and $y_2 = Ax_2$ then $y = y_1 + y_2 = A(x_1 + x_2)$
If $y = Ax$ and $z = By$, then $z = BAx$

33.9.2 Characteristic Roots (eigenvalues)

$Ax = \lambda x$ then there exists n complex roots (or real), given by $|A - \lambda I| = 0$ where $|\cdot|$ indicates determinant of the matrix.

33.9.3 If $Ax = \lambda x$, x is the *eigenvector* (characteristic vector).

33.9.4 Properties

1. If A is singular, at least one root is zero.
2. If $C_{n\times n}$ is nonsingular, then $A_{n\times n}$, $C^{-1}AC$, CAC^{-1} have the same eigenvalues.
3. $A_{n\times n}$ has the same eigenvalues of A^T but not necessarily the same eigenvectors.
4. If λ_i is the eigenvalue of $A_{n\times n}$ and x_i its eigenvector, then λ_i^k is the eigenvalue of A^k and x_i is the eigenvector of A^k: $A^k x_i = \lambda_i^k x_i$
5. If λ_i is the eigenvalue of $A_{n\times n}$ then $1/\lambda_i$ is an eigenvalue of A^{-1} (if it exists).
6. If λ_i is the eigenvalue of $A_{n\times n}$ then x_i is real.

Example

$$A = \begin{bmatrix} 1 & 4 \\ 9 & 1 \end{bmatrix}, \ |A - \lambda I| = \begin{vmatrix} 1-\lambda & 4 \\ 9 & 1-\lambda \end{vmatrix} = (1-\lambda)^2 - 36 = 0, \ \lambda_1 = -5, \ \lambda_2 = 7, \ Ax_1 = \lambda_1 x_1,$$

$$A\begin{bmatrix} 2 \\ -3 \end{bmatrix} = -5\begin{bmatrix} 2 \\ -3 \end{bmatrix}, \ Ax_2 = \lambda_2 x_2, \ A\begin{bmatrix} 2 \\ 3 \end{bmatrix} = 7\begin{bmatrix} 2 \\ 3 \end{bmatrix}, \ x_1 = \begin{bmatrix} 2 \\ -3 \end{bmatrix}, \ x_2 = \begin{bmatrix} 2 \\ 3 \end{bmatrix}$$

7. If A has eigenvalue λ then $f(A)$ has eigenvalue $f(\lambda)$.

Example

$f(A) = A^3 + 7A^2 + A + 5I$, then $(A^3 + 7A^2 + A + 5I)x = A^3x + 7A^2x + Ax + 5Ix = \lambda^3x + 7\lambda^2x + \lambda x + 5x$ $= (\lambda^3 + 7\lambda^2 + \lambda + 5)x$. Hence, $f(\lambda) = \lambda^3 + 7\lambda^2 + \lambda + 5$ is an eigenvalue.

Example

$$f(A) = e^A = \sum_{n=0}^{\infty} \frac{A^n}{n!} \text{ has eigenvalue } e^{\lambda}.$$

8. $\sum_{i=1}^{n} \lambda_i = \text{tr}(A)$. (tr stands for trace.)
9. $\prod_{i=1}^{n} \lambda_i = |A|$.
10. The nonzero eigenvectors $x_1, x_2, \cdots, x_n$ corresponding to distinct eigenvalues $\lambda_1, \lambda_2, \cdots, \lambda_n$ are linearly independent.
11. The eigenvalues of a Hermitian matrix are real.
12. A Hermitian matrix is positive definite if and only if the eigenvalues of A are positive $\lambda_k > 0$.
13. The eigenvectors of a Hermitian matrix corresponding to distinct eigenvalues are orthogonal, i.e., $\lambda_i \neq \lambda_j$, then $v_i^T v_j = 0$.
14. Eigenvalue decomposition: $A = V\Lambda V^{-1}$, $A = n \times n$ has n linear independent vectors, V = contain the eigenvectors of A, Λ = diagonal matrix containing the eigenvalues.
15. Any Hermitian matrix A may be decomposed as
    $A = V\Lambda V^H = \lambda_1 v_1 v_1^H + \lambda_2 v_2 v_2^H + \cdots + \lambda_n v_n v_n^H$ where λ_i are the eigenvalues of A and v_i are a set of orthogonal eigenvectors, Λ = diagonal matrix containing the eigenvalues.
16. $B = n \times n$ matrix with eigenvalues λ_i, $A = B + aI$, A and B have the same eigenvectors, and the eigenvalues of A are $\lambda_i + a$.
17. $A = B + aI$, $B = n \times n$ with rank one (u_1 eigenvector of B and λ its eigenvalue),

$$A^{-1} = \frac{1}{a+\lambda} u_1 u_1^H + \frac{1}{a} I - \frac{1}{a} u_1 u_1^H = \frac{1}{a} I - \frac{\lambda}{a(a+\lambda)} u_1 u_1^H$$

18. A = symmetric positive definite matrix, $x^T A x = 1$ differs an ellipse in n dimensions whose axes are in the direction of the eigenvectors v_i of A with the half-length of these axes equal to $1/\sqrt{\lambda_i}$
19. The largest eigenvalue of an $n \times n$ matrix $A = \{a_{ij}\}$ is bounded by $\lambda_{max} \leq \max_i \sum_{j=1}^{n} a_{ij}$ (bounded by the maximum row sum of the matrix).
20. $A = n \times n$ Hermitian with $\lambda_1 \leq \lambda_2 \leq \cdots \leq \lambda_n$, $y =$ arbitrary complex vector, a = arbitrary complex number, if $\tilde{A} = (n+1) \times (n+1)$ Hermitian $= \begin{bmatrix} A & y \\ y^H & a \end{bmatrix}$ then $\tilde{\lambda}_1 \leq \tilde{\lambda}_2 \leq \cdots \leq \tilde{\lambda}_{n+1}$ are interlaced with those of A: $\tilde{\lambda}_1 \leq \lambda_1 \leq \tilde{\lambda}_2 \leq \lambda_2 \leq \cdots \leq \lambda_n \leq \tilde{\lambda}_{n+1}$
21. Eigenvalues are bounded by the maximum and minimum values of the associated power spectral density of the data: $S_{min} \leq \lambda_i \leq S_{max}$, $i = 1, \cdots, m$.
22. Eigenvalue spread: $\chi(R) = \frac{\lambda_{max}}{\lambda_{min}}$, $R \equiv$ correlation matrix of a discrete-time stochastic process.

23. Karhunen-Loeve expansion: $x(n)=\sum_{i=1}^{M} c_i(n)v_i$, $x(n)=[x(n)\ x(n-1)\cdots x(n-M)]^T$, $v_i=$ eigenvector of the matrix R corresponding to eigenvalue λ_i, R = correlation matrix of the wide-sense stationary process $x(n)$, $c_i(n)=$ constants = coefficients of expansion, $c_i(n)=v_i^H x(n)$, $i=1,2,\cdots,M$, $E\{c_i(n)\}=0$, $E\{c_i(n)c_j^*(n)\}=\lambda_i$ if $i=j$ and zero if $i\neq j$, $E\{|c_i(n)|^2\}=\lambda_i$, $\sum_{i=1}^{M}|c_i(n)|^2=\|x(n)\|^2$ where $\|\cdot\|$ is the norm ($\|x(n)\|^2=x^H x$) and H stands for Hermitian (conjugate transpose).

33.9.5 Finding Eigenvectors

$A-\lambda_k I=\begin{bmatrix} R_k & C_k \\ D & E \end{bmatrix}$, $r(R_k)=r(A-\lambda_k I)$, $x_k=\begin{bmatrix} -R_k C_k y \\ y \end{bmatrix}$, y arbitrary of order $n-r(A-\lambda_k I)$, R_k same rank as $A-\lambda_k I$.

Example

$$A=\begin{bmatrix} 2 & 2 & 0 \\ 2 & 1 & 1 \\ -7 & 2 & -3 \end{bmatrix},\ r(A-\lambda_1 I)=2,\ \lambda_1=1,\ \lambda_2=3,\ \lambda_3=-4,\ n-r(A-\lambda_1 I)=3-2=1,$$

$$y=a,\ x_1=\begin{bmatrix} \frac{1}{4}\begin{pmatrix} 0 & -2 \\ -2 & 1 \end{pmatrix}\begin{pmatrix} 0 \\ 1 \end{pmatrix} a \\ a \end{bmatrix}=\begin{bmatrix} -\frac{1}{2}a \\ \frac{1}{4}a \\ a \end{bmatrix} \text{ and for } a=4,\ x_1=\begin{bmatrix} -2 \\ 1 \\ 4 \end{bmatrix}.$$

In similar procedures, we find

$x_2=[-a\ \ -\frac{1}{2}a\ \ a]^T$ and for $a=2$ $x_2=[-2\ \ -1\ \ 2]^T$. $x_3=[a/13\ \ -3a/13\ \ a]^T$ and for $a=13$ $x_3=[-1\ \ -3\ \ 13]^T$.

Example

$$A=\begin{bmatrix} -1 & -2 & -2 \\ 1 & 2 & 1 \\ -1 & -1 & 0 \end{bmatrix},\ \lambda_1=1 \text{ with multiplicity } 2(m_1=2),\ \lambda_2=-1 \text{ with } m_2=1.$$

$$\text{For } \lambda_1=1,\ (A-\lambda_1 I)=\begin{bmatrix} -2 & -2 & -2 \\ 1 & 1 & 1 \\ -1 & -1 & -1 \end{bmatrix},\ r(A-\lambda_1 I)=1,\ n-r(A-\lambda_1 I)=3-1=2,$$

$$x_1=\begin{bmatrix} -(-\frac{1}{2})(-2-2)y \\ y \end{bmatrix}=\begin{bmatrix} -a-b \\ a \\ b \end{bmatrix}.$$

Similarly for $\lambda_2 = -1$, $x_2 = [2a \; -a \; a]^T$. For $a = -b$, $x_1 = [0 \; 1 \; -1]^T$ and for $a = b = 1$ $x_1^* = [-2 \; 1 \; 1]^T$. For $a = 1$, $x_2 = [2 \; -1 \; 1]^T$ and all three eigenvectors are independent.

33.9.6 Similar Matrices

A and B are similar if there exists Q such that $B = Q^{-1}AQ$.

33.9.7 Properties of Similar Matrices

1. $|B| = |A|$
2. If the eigenvalues of λ_i of A are distinct, then there exists Q such that $Q^{-1}AQ = D$.
3. If X is a matrix of eigenvectors x_i corresponding to eigenvalues λ_i then $X^{-1}AX$ = diag $\{\lambda_1, \lambda_2, \cdots \lambda_n\} \equiv D$.
4. If $Q^{-1}AQ = T_u$ (T_u = upper triangular) the diagonal elements of T_u are the eigenvalues of A.
5. Similar matrices have the same set of eigenvalues.
6. Power: $A = XDX^{-1}$ (D = diagonal) implies $A^2 = XDX^{-1}XDX^{-1} = XD^2X^{-1}$ and hence $A^k = XD^kX$.
7. Inverse: $A^{-1} = XD^{-1}X^{-1}$
8. Difference Equations: If A is diagonizable then $x(n) = Ax(n-1) + y$ or

$$x(n) = A(Ax(n-2) + y = A^2x(n-2) + (A + I)y = \cdots$$

$$= A^n x(0) + (A^{n-1} + A^{n-2} + \cdots A + I)y$$

If $A^k \to 0$ as $k \to \infty$ and $(I - A)^{-1}$ exists then

$$x(n) = A^n x(0) + (I - A^n)(I - A)^{-1} y$$

33.10 Symmetric Matrices

33.10.1 Eigenvalues (characteristic roots) of a real symmetric matrix are real.

33.10.2 Real symmetric matrices are diagonizable with eigenvalues along the diagonal $P^T AP = D$, P = orthogonal matrix.

33.10.3 Eigenvectors are orthogonal.

33.10.4 If λ_k is an eigenvalue of $A_{n\times n}$ of multiplicity m_k, then $A - \lambda_k I$ has rank $n - m_k$ and is singular.

33.10.5 If there are m zero eigenvalues of an $n \times n$ real symmetric matrix, then its rank is $n - m$.

Example

$$A = \begin{bmatrix} 1 & 2 & 2 \\ 2 & 1 & 2 \\ 2 & 2 & 1 \end{bmatrix}, \; (\lambda + 1)^2(\lambda - 5) = 0, \; \lambda_1 = 5 \text{ with } m_1 = 1, \; \lambda_2 = -1 \text{ with } m_2 = 2.$$

From (33.9.5)

$$A - \lambda_1 I = \begin{bmatrix} -4 & 2 & 2 \\ 2 & -4 & 2 \\ 2 & 2 & -4 \end{bmatrix}, \; x_1 = \begin{bmatrix} \frac{1}{12}\begin{pmatrix} -4 & -2 \\ -2 & -4 \end{pmatrix}\begin{pmatrix} 2 \\ 2 \end{pmatrix} a \\ a \end{bmatrix} = \begin{bmatrix} a \\ a \\ a \end{bmatrix};$$

for λ_2

$$A-\lambda_2 I=\begin{bmatrix}2&2&2\\2&2&2\\2&2&2\end{bmatrix},\ x_2=\begin{bmatrix}-\frac{1}{2}[2\ 2]y\\ y=\begin{bmatrix}a_1\\a_2\end{bmatrix}\end{bmatrix}=\begin{bmatrix}-(a_1+a_2)\\a_1\\a_2\end{bmatrix},$$

if we set $a_1=a_2=1$, then $x_2^{(1)}=[-2\ 1\ 1]^T$. But $x_2^T x_2^{(1)}=0$ or $2(a_1+a_2)+a_1+a_2=0$ or $a_1=-1$, $a_2=1$ which gives the second vector $x_2^{(2)}=[0\ -1\ 1]$, with $a=1$, $x_1=[1\ 1\ 1]^T$ and the normalized eigenvector matrix is

$$Q=\frac{1}{6}\begin{bmatrix}\sqrt{2}&-2&0\\\sqrt{2}&1&-\sqrt{3}\\\sqrt{2}&1&-\sqrt{3}\end{bmatrix}$$

and gives $Q^TAQ=D=diag\{-5,-1,-1\}$ and $QQ^T=I$.

33.11 Geometric Interpretation

33.11.1 $x=[a_1\ a_2\cdots a_n]^T=\sum_{i=1}^{n}a_i\varepsilon_i$ where $\varepsilon_i=[0\ 0\cdots 1\cdots 0]^T$ (one at the ith place).

33.11.2 Distance in Euclidean E_n space

$$d=\left[\sum_{i=1}^{n}(a_i-b_i)^2\right]^{1/2}=[(x-y)^T(x-y)]^{1/2}\text{ where } x=[a_1\ a_2\cdots a_n]^T \text{ and } y=[b_1\ b_2\cdots b_n]$$

33.11.3 Projection

$P_k^n\equiv$ plane through the origin with $B=[y_1\ y_2\cdots y_k]$ on $n\times k$ matrix of rank k. Then the projection of x on P_k^n is $z=B(B^TB)^{-1}B^Tx$.

Example

$$B=\begin{bmatrix}1&1\\1&0\\0&1\end{bmatrix}\text{ of } y_1=[1\,1\,0]^T$$

and $y_2=[1\,0\,1]$ in P_2^3 plane through the origin. Hence,

$$z=\begin{bmatrix}1&1\\1&0\\0&1\end{bmatrix}\begin{bmatrix}2&1\\1&2\end{bmatrix}^{-1}\begin{bmatrix}1&1&0\\1&0&1\end{bmatrix}\begin{bmatrix}1\\1\\1\end{bmatrix}=\begin{bmatrix}4/3\\2/3\\2/3\end{bmatrix}$$

33.11.4 Distance from Projection

$$d=[(x-B(B^TB)^{-1}B^Tx)^T(x-B(B^TB)^{-1}B^Tx)]^{1/2}$$

33.11.5 Quadratic Form Matrix

$G = I - B(B^T B)^{-1} B$ has the properties $G = G^T$ and $G = G^2$ (G is a symmetric idepotent matrix).

33.12 Relationships of Vector Spaces

33.12.1 Sum of Vector Subspaces

$S = S_1 \oplus S_2 = \{y = x_1 + x_2; x_1 \in S_1 \text{ and } x_2 \in S_2\}$

33.12.2 Sum and Intersection

If S_1 and S_2 belong to subspace V_n of E_n then $S_1 \otimes S_2$ and $S_1 \cap S_2$ are subspaces of V_n

33.12.3 Orthogonal Subspaces

$S_1 \perp S_2$ if for each x_i vector in S_1 and for each vector y_i in S_2 $x_i^T y = 0$.

33.12.4 Orthogonal Complement

Let $S_1 \in E_n$ then $S_1^\perp \in E_n$ is the orthogonal complement of S_1 if $S_1 \perp S_1^T$ and $S_1 \otimes S_2 = E_n$. For each subspace S_1 in E_n the orthogonal complement $S_1^\perp$ always exists and it is unique.

Example

In E_4 let the subspace S be spanned by the linearly independent vectors $x_1 = [1\ 0\ -1\ 1]^T$ and $x_2 = [1\ 1\ 0\ 1]^T$. From (33.5.8) we obtain the orthogonal basis for $S: y_1 = x_1 = [1\ 0\ -1\ 1]^T$ and $y_2 = x_1 - \frac{x_1^T x_2}{x_1^T x_1} y_1 = \frac{1}{3}[1\ 3\ 2\ 1]^T$ with $y_1^T y_2 = 0$ (orthogonal). If we find z_1 and z_2 such that $\{y_1, y_2, z_1, z_2\}$ is an orthogonal basis for E_4 then $\{z_1, z_2\}$ is the basis for $S^\perp$. By inspection $z_1 = [1\ 0\ 0\ -1]^T$ and $z_2 = [1\ -2\ 2\ 1]^T$.

33.12.5 Column Space

If we write $A_{n \times m}$ matrix in the form $A = [x_1 x_2 \dots x_m]$ and consider the columns x_i's as vectors in E_n then the space spanned by the x_i's is called the column space of A. ***Note:*** If A is $n \times n$ and nonsingular, the column space of A is E_n.

33.13 Functions of Matrices

33.13.1 Polynomial of Matrices

$p_q(A) = a_q A^q + \cdots + a_1 A + a_0 I$ $(A = n \times n$ matrix).

33.13.2 Characteristic Polynomial

If $p_n(x) = \det(A - \lambda I) = \sum_{i=0}^{n} c_i \lambda^i$ then $p_n(A) = 0$.

33.13.3 Norm

$\|A\| \equiv$ norm is a real-valued (non-negative) function of A (if the elements a_{ij} of A) that satisfies the following:

1. $\|A\| \geq 0$ and $\|A\| = 0$ if and only if A = 0;
2. $\|cA\| = |c|\,\|A\|$, $c =$ any scalar;
3. $\|A + B\| \leq \|A\| + \|B\|$;
4. $\|AB\| \leq \|A\|\,\|B\|$

Example

Let $f(A) = \sum\sum|a_{ij}|$.

1. $f(A) \geq 0$ and $f(A) = 0$ if $A = 0$
2. $f(cA) = \sum\sum|ca_{ij}| = |c|\sum\sum a_{ij} = |c|f(A)$
3. $f(A+B) = \sum\sum(|a_{ij} + b_{ij}|) \leq \sum\sum(|a_{ij}| + |b_{ij}|) = f(A) + f(B)$
4. $f(AB) = f(C) = \sum_i\sum_j|c_{ij}| = \sum_i\sum_j\left|\sum_t a_{it}b_{tj}\right| \leq \sum_i\sum_j\sum_t|a_{it}||b_{tj}| \leq \sum_i\sum_j\left(\sum_t|a_{it}|\sum_s|b_{sj}|\right) = f(A)f(B)$

Hence, $f(A) = \sum\sum|a_{ij}|$ is a norm.

33.13.4 Special Norms

1. $\|A\|_1 = \max_j\left(\sum_{i=1}^{n}|a_{ij}|\right) =$ maximum of sums of absolute values of column elements.
2. $\|A\|_2 = [\text{maximum eigenvalue root of } A^TA]^{1/2}$ called the *spectrum* norm.
3. $\|A\|_\infty = \max_i\left(\sum_{j=1}^{n}|a_{ij}|\right) =$ maximum of sums of absolute values of row elements.
4. $\|A\|_E = [\sum_i\sum_j(|a_{ij}|)^2]^{1/2} =$ Euclidean norm.

33.13.5 Properties of Norms (all matrices are n × n; refer also to 33.13.4)

1. $\|A^q\| \leq (\|A\|)^q$ for norms in (33.13.4)
2. $0 \leq |(\|A\| - \|B\|)| \leq \|A - B\|$ for norms in (33.13.4)
3. $\|A\|_1 = \|A^T\|_\infty$ for norms in (33.13.4)
4. $\|AB\|_E \leq \|A\|_E\|B\|_2$
5. $\|A\|_E^2 = \text{trace}\,(A^TA) = \|A^T\|_E^2$
6. $\|A\|_E = \|A^T\|_E$
7. $\|A\|_2 \leq \|A\|_E \leq \sqrt{n}\,\|A\|_2$

8. $\|D\|_m = \max|d_{ii}|$ for $m = 1,2,\infty$ and $D = [d_{ij}] =$ diagonal

9. $\|PA\|_m = \|AP\|_m = \|P^TAP\|_m$, $m = E$ or 2 and P is orthogonal

10. $\|A\|_2^2 = \max_{x \in S}[x^TA^TAx / x^Tx]$, $S \equiv$ set of all $n \times 1$ real vectors except the 0 vector.

11. $\|A\|_2 =$ max root of A, A is non-negative

33.14 Generalized Inverse

33.14.1 Generalized Inverse (g-inverse) of $A_{m \times n}$

A^- is generalized inverse of A if:

1. AA^- is symmetric,
2. A^-A is symmetric,
3. $AA^-A = A$,
4. $A^-AA^- = A^-$

33.14.2 Generalized Inverse

If A = 0, then we may write $A_{m \times n} = B_{m \times r}C_{r \times n}$ (A has rank > 0, B and C have rank r), and

$$A^- = C^T(CC^T)^{-1}(B^TB)^{-1}B^T$$

33.14.3 Properties of g-inverse

1. A^- is unique
2. $(A^T)^- = (A^-)^T$
3. $(A^-)^- = A$
4. $r(A) = r(A^-)$
5. If rank A is r, then $A^-, AA^-, A^-A, AA^-A, A^-AA^-$ have rank r
6. $(A^TA)^- = A^-A^{T-}$
7. $(AA^-)^- = AA^-$, $(A^-A)^- = A^-A$
8. $(PAQ)^- = Q^TA^-P^T$, $P_{m \times m} =$ orthogonal, $Q_{n \times n} =$ orthogonal, $A_{m \times n}$
9. If $A = A^T \equiv$ symmetric then $A^- = (A^-)^T =$ symmetric
10. If $A = A^T$ then $AA^- = A^-A$
11. If $A =$ nonsingular then $A^- = A^{-1}$
12. If $A = A^T$ and $A = A^2$ then $A^- = A$
13. If $D_{n \times n}[d_{ii}]$ is diagonal, then $D^- = [d_{ii}^{-1}]$ if $d_{ii} \neq 0$ and zero if $d_{ii} = 0$
14. If $A_{m \times n}$ has rank m, then $A^- = A^T(AA^T)^{-1}$ and $AA^- = I$
15. If $A_{m \times n}$ has rank n, then $A^- = (A^TA)^{-1}A^T$ and $A^-A = I$

16. If $A^- = A^T(AA^T)^- = (A^TA)^-A^T$ whatever the rank of A

17. AA^-, A^-A, $I - AA^-$ and $I - A^-A$ are all symmetric idempotent

18. $(BC)^- = C^-B^-$ if B_{mxr} has rank > 0 and C_{rxm} has rank r

19. $(cA)^- = \frac{1}{c}A^-$, $c =$ nonzero scalar

20. If $A = A_1 + A_2 + \cdots + A_n$ and $A_iA_j^T = 0$ and $A_i^TA_j = 0$ for all $i, j = 1, \cdots, n$

$i \neq j$ then $A^- = A_1^- + \cdots + A_n^-$

21. If A_{nxm} is any matrix, K_{mxm} is nonsingular matrix and $B = AK$, then $BB^- = AA^-$

22. If $A^TA = AA^T$ then $A^-A = AA^-$ and $(A^n)^- = (A^-)^n$

Note: in general $(A^n)^- \neq (A^-)^n$ for $m \times m$ matrix.

23. If $A = \begin{bmatrix} B \\ C \end{bmatrix}$ and $BC^T = 0$ then $A^- = [B^-, C^-]$, $A^-A = B^-B + C^-C$,

$$AA^- = \begin{bmatrix} BB^- & 0 \\ 0 & CC^- \end{bmatrix}$$

24. If $A = \begin{bmatrix} B & 0 \\ 0 & C \end{bmatrix}$, then $A^- = \begin{bmatrix} B^- & 0 \\ 0 & C^- \end{bmatrix}$, $A^-A = \begin{bmatrix} B^-B & 0 \\ 0 & C^-C \end{bmatrix}$,

$$AA^- = \begin{bmatrix} BB^- & 0 \\ 0 & CC^- \end{bmatrix}$$

25. $x^- = (x^Tx)^{-1}x^T$, $x =$ vector

26. $A^- = A^T$ if and only if A^TA is independent

27. The column spaces of A and AA^- are the same.
The column spaces of A^- and A^-A are the same.
The column space of $I - AA^-$ is the orthogonal complement of the column space of A
The column space of $I - A^-A$ is the orthogonal complement of the column space of A^T

33.15 Generalized Matrices Computation

33.15.1 Computation (theorem)

Let X be an m × t matrix and x_t be the tth column vector of X. Then we write $X = [X_{t-1}, x_t]$, and the g-inverse of X is equal to Y where

$$Y = \begin{bmatrix} X_{t-1}^- - X_{t-1}^- x_t y_t^- \\ y_t^- \end{bmatrix}, \quad y_t^- \equiv \text{g-inverse of } y_t \text{ (see 33.14.3.25)}$$

$$y_t = \begin{cases} (I - X_{t-1}X_{t-1}^-)x_t & \text{if } x_t \neq X_{t-1}X_{t-1}^- x_t \text{ (case I)} \\ \dfrac{[1 + x_t^T(X_{t-1}X_{t-1}^T)^- x_t](X_{t-1}X_{t-1}^T)^- x_t}{x_t^T(X_{t-1}X_{t-1}^T)^-(X_{t-1}X_{t-1}^T)^- x_t}, & \text{if } x_t = X_{t-1}X_{t-1}^- x_t \text{ (case II)} \end{cases}$$

33.15.2 Algorithm for 33.15.1

1. Set $Z_2 = [X_1\ x_2] \equiv [x_1\ x_2]$,
2. Compute $X_1^- \equiv x_1^-$ (see Section 33.14.3.25)
3. Compute $X_1^- x_2$
4. Compute $X_1 X_1^- x_2$ (if $x_2 \neq X_1 X_1^- x_2$ use case I, otherwise use case II)
5. Compute y_2^-
6. Compute $Z_2^- = X_2^- = \begin{bmatrix} X_1^- - X_1^- x_2 y_2^- \\ y_2^- \end{bmatrix}$
7. $Z_3^- = X_3^- = \begin{bmatrix} Z_2^- - Z_2^- x_3 y_3^- \\ y_3^- \end{bmatrix}$, $Z_3 = [Z_2\ x_3]$ etc.

Example

$$X = \begin{bmatrix} 1 & 0 & -2 \\ 0 & 1 & -1 \\ -1 & 1 & 1 \\ 2 & -1 & 2 \end{bmatrix},$$

1. $Z_2 = [X_1, x_2] = \begin{bmatrix} 1 & 0 \\ 0 & 1 \\ -1 & 1 \\ 2 & -1 \end{bmatrix}$,
2. $X_1^- \equiv x_1^- = \frac{1}{6}[1 \quad 0 \quad -1 \quad 2]$
3. $X_1^- x_2 = -\frac{3}{6}$
4. $X_1 X_1^- x_2 = \begin{bmatrix} -\frac{3}{6} & 0 & \frac{3}{6} & -\frac{6}{6} \end{bmatrix}^T$
5. since $x_2 \neq X_1 X_1^- x_2$ case I applies and $y_2 = x_2 - X_1 X_1^- x_2 = \begin{bmatrix} \frac{1}{2} & 1 & \frac{1}{2} & 0 \end{bmatrix}^T$ and from (14.3.25) $y_2^- = \begin{bmatrix} \frac{1}{3} & \frac{2}{3} & \frac{1}{3} & 0 \end{bmatrix}$
6. $Z_2^- = X_2^- = \begin{bmatrix} X_1^- - X_1^- x_2 y_2^- \\ y_2^- \end{bmatrix} = \begin{bmatrix} \frac{2}{6} & \frac{2}{6} & 0 & \frac{2}{6} \\ \frac{2}{6} & \frac{4}{6} & \frac{2}{6} & 0 \end{bmatrix}$
7. $Z_3^- = X_3^- = \begin{bmatrix} Z_2^- - Z_2^- x_3 y_3^- \\ y_3^- \end{bmatrix}$, $Z_3 = [Z_2\ x_3] = \begin{bmatrix} 1 & 0 & -2 \\ 0 & 1 & -1 \\ -1 & 1 & 1 \\ 2 & -1 & 2 \end{bmatrix}$

8. $Z_2^- x_3 = X_2^- x_3 = \begin{bmatrix} -\frac{2}{6} & -\frac{6}{6} \end{bmatrix}^T$

9. $X_2 X_2^- x_3 = \frac{1}{6}[-2 \quad -6 \quad -4 \quad 2]^T$

10. $x_3 \neq X_2 X_2^- x_3$ and implies case I, $y_3^- = \begin{bmatrix} -\frac{1}{5} & 0 & \frac{1}{5} & \frac{1}{5} \end{bmatrix}$

$$Z_3^- = X_3^- \equiv X^- = \frac{1}{15}\begin{bmatrix} 4 & 5 & 1 & 6 \\ 2 & 10 & 8 & 3 \\ -3 & 0 & 3 & 3 \end{bmatrix}$$

33.15.3 Block Form

If $A = \begin{bmatrix} A_{11} & A_{12} \\ A_{21} & A_{22} \end{bmatrix}$, where A_{11} is an $r \times r$ matrix of rank r, then

$$A^- = \begin{bmatrix} A_{11}^T B A_{11}^T & A_{11}^T B A_{21}^T \\ A_{12}^T B A_{11}^T & A_{12}^T B A_{21}^T \end{bmatrix}$$

where $B = (A_{11}A_{11}^T + A_{12}A_{12}^T)^{-1} A_{11} (A_{11}^T A_{11} + A_{21}^T A_{21})^{-1}$

33.15.4 Second Form

Steps to find A^- of an $A_{m \times n}$ matrix of rank r

1. Compute $B = A^T A$
2. Let $C_1 = I$
3. Compute $C_{i+1} = I(1/i)\text{tr}(C_i B) - C_i B$ for $i = 1, 2, \cdots r-1$.
4. Compute $rC_r A^T / tr(C_r B) = A^-$ $(C_{r+1}B = 0$ and $\text{tr}(C_r B) \neq 0)$

Example

$A \equiv X$ of example in (33.15.2).

1. $B = AA^T = \begin{bmatrix} 6 & -3 & 1 \\ -3 & 3 & -2 \\ 1 & -2 & 10 \end{bmatrix}$

2. $C_1 = \begin{bmatrix} 1 & 0 & 0 \\ 0 & 1 & 0 \\ 0 & 0 & 1 \end{bmatrix}$

3. $C_2 = Itr(C_1 B) - C_1 B = \begin{bmatrix} 13 & 3 & -1 \\ 3 & 16 & 2 \\ -1 & 2 & 9 \end{bmatrix}$,

$$C_3 = I\frac{1}{2}\text{tr}(C_2B) - C_2B = \begin{bmatrix} 26 & 28 & 3 \\ 28 & 59 & 9 \\ 3 & 9 & 9 \end{bmatrix}, \; r \le 3 \text{ and } C_3B \ne 0 \text{ which implies } r = 3.$$

4. $$A^- = \frac{3C_3A^T}{\text{tr}(C_3B)} = \frac{3}{225}\begin{bmatrix} 20 & 25 & 5 & 30 \\ 10 & 50 & 40 & 15 \\ -15 & 0 & 15 & 15 \end{bmatrix} = \frac{1}{15}\begin{bmatrix} 4 & 5 & 1 & 6 \\ 2 & 10 & 8 & 3 \\ -3 & 0 & 3 & 3 \end{bmatrix} \quad \text{(see 33.15.2)}$$

33.16 Conditional Inverse (or c-inverse)

33.16.1 $A_{m\times n}$ is conditional inverse A^c if $AA^cA = A$. *Note:* A g-inverse is conditional, but the reverse is not always true.

33.16.2 Properties

1. Conditional inverse always exists but may not be unique.
2. If A is $m \times n$ then A^c is $n \times m$.

33.16.3 Hermite Form H

An n × n matrix H is of Hermite form if: 1) H is upper triangular, 2) only zeros and ones are on the diagonal; 3) if a row has a zero on the diagonal, the whole row is zeros; 4) if a row has a one on the diagonal, then every off-diagonal element is zero in the column in which the one appears.

33.16.4 Properties of Hermite Form

1. $H = H^2$.
2. There exists nonsingular B such that BA=H where A is any n × n matrix.
3. B of (2) is conditional inverse of A.
4. If c-inverse of $A_{m\times n}$ is A^c then A^cA and AA^c are idempotent.
5. rank(A) = rank(A^cA) = rank(AA^c) ≤ rank(A^c)
6. rank(A) = rank(H).
7. $A_{n\times n}$ = nonsingular, its Hermite form is $I_{n\times n}$.
8. A^TA, A have the same Hermite form.
9. A^-A and A have the same Hermite form.
10. Rank of A is equal to the number of diagonal elements of H that are equal to one.
11. If $ABA = kA$ then $(1/k)B$ is c-inverse of A.

33.16.5 Steps to find $A^c = B$

$A_{m\times n}$ and $m > n$, $A_0 = [\text{A } 0]$, $0 \equiv m \times m - n$ zero matrix, B_0 = nonsingular such that $B_0A_0 = H$, partition $B_0 = \begin{bmatrix} B \\ B_1 \end{bmatrix}$ with $B = n \times m$ matrix, then $A^c = B$.

Example

$$A=\begin{bmatrix}1 & -1\\ 2 & -1\\ 0 & 1\end{bmatrix},\ A_0=\begin{bmatrix}1 & -1 & 0\\ 2 & -1 & 0\\ 0 & 0 & 0\end{bmatrix},\ B_0=\begin{bmatrix}-1 & 1 & 0\\ -2 & 1 & 0\\ 2 & -1 & 1\end{bmatrix}=\begin{bmatrix}B\\ B_1\end{bmatrix},\ B=\begin{bmatrix}-1 & 1 & 0\\ -2 & 1 & 0\end{bmatrix}=A^c$$

33.17 System of Linear Equation

33.17.1 Definition

$Ax = y$

33.17.2 Existence of Solution of 33.17.1

1) A is an $n \times n$ matrix and nonsingular; 2) solution exists if and only if $\text{rank}(A) = \text{rank}([A\ \ y])$ ($[A\ \ y]$ $\equiv$ augmented matrix); 3) if y is in the column space of A; 4) if there is A^c (c-inverse) of A such that $AA^c y = y$; 5) if there is A^- such that $AA^- y = y$; and 6) the $m \times n$ matrix A has rank m.

33.17.3 Solution of Homogeneous Equation (Ax = 0)

If rank (A) < n, there is a solution different than $x = 0$.

33.17.4 Solutions of Ax = y

1. If a solution exists, then $x_0 = A^c y + (I - A^c A)h$ is a solution where h is any $n \times 1$ vector. A is an $m \times n$ matrix.
2. As in (1) $x_0 = A^- y + (I - A^- A)h$.
3. If $Ax = y$ is consistent (a solution exists) then $x_0 = A^- y$ and $x_1 = A^c y$ are solutions. A is an $m \times n$ matrix.
4. $x_0 = A^- y$ in (3) is unique if and only if $A^- A = I$.
5. The system in (3) has a unique solution if rank (A) = n.
6. If a unique solution exists, then it is $A^- y$ and $A^- y = A^c y$ for any A^c.
7. If the system in (3) has rank $r > 0$ and $y \neq 0$, then there are exactly $n - r + 1$ linearly independent vectors that satisfy the system.
8. The set of solutions of $Ax = 0$ is the orthogonal complement of the column space of A^T.

33.18 Approximate Solutions of Linear Equations

33.18.1 Definition of Approximate System

$Ax - y = \varepsilon(x)$, $\varepsilon(x) \equiv$ error depending on vector x.

33.18.2 Best Approximate Solution

$x_0 = A^- y$ (always exists and is unique).

33.18.3 Minimization of Sum of Squares of Deviations

$\sum \varepsilon_i^2 = \varepsilon^T(x)\varepsilon(x) = (Ax - y)^T(Ax - y)$.

33.18.4 Minimum of $\varepsilon^T(x)\varepsilon(x)$

For $A_{m\times n}$ any $y_{n\times 1}$ the minimum of $\varepsilon^T(x)\varepsilon(x)$ as x varies over E_n is $y^T(I - AA^-)y$.

33.18.5 Linear Model (in statistics)

$z = X\beta + \varepsilon$, $z_{n\times 1}$ = observed quantities, $X_{n\times p}$ = known (correlation matrix), $\beta_{p\times 1}$ = unknown vector, ε = error vector (deviation of observations from the expected value).

33.18.6 Normal Equation

$X^T X\hat{\beta} = X^T z$ or $Ax = y$ ($A_{p\times p} = X^T X, \hat{\beta} = x$ and $X^T z = y$)

33.18.7 Properties of Normal Equation

1. The system in (33.18.6) is consistent.
2. $\hat{\beta} = (X^T X)^- X^T z + [I - (X^T X)^-(X^T X)]h = X^- z + [I - X^- X]h$ = general solution, h is any p × 1 vector.
3. Any vector $\hat{\beta}$ that satisfies (33.18.6) leaves $\hat{\beta}^T X^T z$ invariant.

33.19 Least Squares Solution of Linear Systems

33.19.1 Definition

If x_0 is the least square solution (LSS) to (33.18.1) ($Ax - y = \varepsilon(x)$), then for any other vector x in E_n the following relation holds: $(Ax - y)^T(Ax - y) \ge (Ax_0 - y)^T(Ax_0 - y)$. *Note:* Best approximate solution (BAS) contains LSS, but the reverse is not always true.

33.19.2 Solution

If $x_0 = By \equiv$ least square solution to $Ax - y = \varepsilon(x)$ then a) $ABA = A$, and b) $AB \equiv$ symmetric.

33.19.3 Properties of Least Square Solution

1. If $A_{m\times n}$ and B is such that ABA = A and AB is symmetric then $AB = AA^-$
2. $(Ax_0 - y)^T(Ax_0 - y) = y^T(I - AA^-)y$
3. If x_0 satisfies $Ax = AA^- y$ then x_0 is LSS
4. If x_0 satisfies $A^T Ax = A^T y$ (normal equation) then x_0 is LSS
5. If $A_{m\times n}$ has rank n then the LSS unique

33.19.4 Least Square Inverse

A^L is least square inverse if and only if a) $AA^L A = A$, and b) $AA^L = (AA^L)^T$. *Note:* $A^L \equiv$ *c*-inverse of A, $A^- \equiv$ *c*-inverse and *L*-inverse of A.

33.19.5 Solution with L-Inverse

$x_0 = A^L y + (I - A^L A)h$, h is any $n \times 1$ vector.

33.19.6 Properties of L-Inverse

1. $(A^T A)^c \equiv$ c-inverse of $A^T A$, $A_{m\times n}$ then $B = (A^T A)^c A^T$ is L-inverse of A
2. $AA^L = AA^- \equiv$ symmetric idempotent
3. If $AA^L y = y$ the system is consistent

33.19.7 Solution with Constraint

a) If the system $Ax - \alpha = \varepsilon(x)$, $A_{m\times p}$ of rank p (see 33.18.1) is subject to condition $Bx = \beta$ which is a consistent set of equations, its solution is:

$$x_0 = A^-\alpha - (A^T A)^{-1} B^T [B(A^T A)^{-1} B^T]^{-1} (BA^-\alpha - \beta)$$

b) If $Ax = \alpha$ is not consistent then

$x_0 = B^-\beta + [A(I - B^-B)]^-(\alpha - (AB^-\beta) + (I - B^-B),\ (I - [A(I - B^-B)]^-[A(I - B^-B)]h)$ h = any $p\times 1$ vector, $A \equiv n\times p$ matrix and $B \equiv q\times p$ matrix

33.19.8 Solution with Weighted Least Squares

For the system $y = X\beta + \varepsilon$, if the covariance matrix of ε is not $\sigma^2 I$ but a positive definite matrix V, then the vector β which minimizes $\varepsilon^T V \varepsilon$ is $\hat{\beta} = (X^T V^{-1} X)^c X^T V^{-1} y$.

33.20 Partitioned Matrices

33.20.1 The Inverse of a Partitioned Matrix

$$B_{n\times n} = \begin{bmatrix} B_{11} & B_{12} \\ B_{21} & B_{22} \end{bmatrix},\quad B^{-1} = A = \begin{bmatrix} A_{11} & A_{12} \\ A_{21} & A_{22} \end{bmatrix},\quad |B| \neq 0,\ |B_{11}| \neq 0,\ |B_{22}| \neq 0.$$

1. A_{11}^{-1}, A_{22}^{-1} exist
2. $[B_{11} - B_{12}B_{22}^{-1}B_{21}]^{-1}$ and $[B_{22} - B_{21}B_{11}^{-1}B_{12}]^{-1}$ exist
3. $B^{-1} = \begin{bmatrix} [B_{11} - B_{12}B_{22}^{-1}B_{21}]^{-1} & -B_{11}^{-1}B_{12}[B_{22} - B_{21}B_{11}^{-1}B_{12}]^{-1} \\ -B_{22}^{-1}B_{21}[B_{11} - B_{12}B_{22}^{-1}B_{21}]^{-1} & [B_{22} - B_{21}B_{11}^{-1}B_{12}]^{-1} \end{bmatrix}$
4. $A_{11} = [B_{11} - B_{12}B_{22}^{-1}B_{21}]^{-1} = B_{11}^{-1} + B_{11}^{-1}B_{12}A_{22}B_{21}B_{11}^{-1}$
5. $A_{12} = -B_{11}^{-1}B_{12}[B_{22} - B_{21}B_{11}^{-1}B_{12}]^{-1} = -B_{11}^{-1}B_{12}A_{22}$
6. $A_{22} = [B_{22} - B_{21}B_{11}^{-1}B_{12}]^{-1} = B_{22}^{-1}B_{22}^{-1}B_{21}A_{11}B_{12}B_{22}^{-1}$
7. $A_{21} = -B_{22}^{-1}B_{21}[B_{11} - B_{12}B_{22}^{-1}B_{21}]^{-1} = -B_{22}^{-1}B_{21}A_{11}$
8. $|B| = \dfrac{|B_{11}|}{|A_{22}|} = \dfrac{|B_{22}|}{|A_{11}|}$, $|B_{11}A_{11}| = |B_{22}A_{22}|$
9. $|B| = |B_{22}|\,|B_{11} - B_{12}B_{22}^{-1}B_{21}|$, B_{22} = nonsingular
10. $|B| = |B_{11}|\,|B_{22} - B_{21}B_{11}^{-1}B_{12}|$

33.21 Inverse Patterned Matrices

33.21.1 k × k Lower Triangular

$$C = \begin{bmatrix} a_1b_1 & 0 & \cdots & 0 \\ a_2b_1 & a_2b_2 & \cdots & 0 \\ \vdots \\ a_kb_1 & a_kb_2 & \cdots & a_kb_k \end{bmatrix}$$

$$C^{-1} = \begin{bmatrix} (a_1b_1)^{-1} & 0 & \cdots & \cdots & 0 & 0 \\ -(b_2a_1)^{-1} & (a_2b_2)^{-1} & 0 & \cdots & 0 & 0 \\ 0 & -(b_3a_2)^{-1} & (a_3b_3)^{-1} & \cdots & 0 & 0 \\ \vdots & & & & \vdots & \vdots \\ 0 & 0 & & & (b_{k-1}a_{k-1})^{-1} & \\ 0 & 0 & \cdots & \cdots & -(b_ka_{k-1})^{-1} & (a_kb_k)^{-1} \end{bmatrix}$$

33.21.2 Diagonal Plus Matrix

$C = D + qab^T$, $C^{-1} = D^{-1} + pa^*b^{*T}$, $D =$ nonsingular, $q \neq -[\Sigma a_ib_i / d_{ii}]^{-1}$, $p = -q\left(1 + q\sum_{i=1}^{k} a_ib_id_{ii}^{-1}\right)$

$a_i^* = a_i / d_{ii}$, $b_i^* = b_i / d_{ii}$, $d_{ii} \equiv i^{\text{th}}$ diagonal element of D.

Example

V ≡ variance-covariance of k-dimensional random variable $\left(p_i > 0,\ \sum_{i=1}^{k} p_i < 1\right)$

$$V = \begin{bmatrix} p_1(1-p_1) & -p_1p_2 & -p_1p_3 & \cdots & -p_1p_k \\ -p_1p_2 & p_2(1-p_2) & -p_2p_3 & & -p_2p_k \\ \vdots \\ -p_1p_k & -p_2p_k & -p_3p_k & \cdots & p_k(1-p_k) \end{bmatrix} = D + qab^T$$

$$= \begin{bmatrix} p_1 & 0 & 0 & 0 & \cdots & 0 \\ 0 & p_2 & 0 & 0 & \cdots & 0 \\ 0 & 0 & p_3 & 0 & \cdots & 0 \\ \vdots \\ 0 & 0 & 0 & 0 & \cdots & p_k \end{bmatrix} - \begin{bmatrix} p_1 \\ p_2 \\ \vdots \\ p_k \end{bmatrix} [p_1\ p_2 \cdots p_k],$$

$$V^{-1} = D^{-1} + ba^*b^{*T},\quad a_i^* = b_i^* = 1,\quad a^*b^{*T} = J,\quad q = \left[1 - \sum_{i=1}^{k} p_i\right]^{-1}$$

33.22.3 Equal Diagonal and All Other Elements

$$C_{3x3} = \begin{bmatrix} a & b & b \\ b & a & b \\ b & b & a \end{bmatrix}, \; C_{kxk} = (a-b)I + bJ, \; C^{-1} = \frac{1}{a-b}[I - \frac{b}{a+(k-1)b}J],$$

$a \neq b, \; a \neq -(k-1)b, \; J = 11^T =$ matrix of ones $(1 = [1 \;\; 1 \ldots 1]^T)$

33.22 Determinants of Patterned Matrices

33.22.1 Lower Triangular

This determinant of 33.21.1 is $|C| = a_1 a_2 \cdots a_k b_1 b_2 \cdots b_k$

33.22.2 Diagonal Plus Matrix

The determinant of 33.21.2 is $\det(C) = \left[1 + q\sum_j \frac{a_j b_j}{d_{jj}}\right] \prod_i d_{ii}$

33.22.3 $C_{kxk} = (a-b)I + bJ$ ($J \equiv$ matrix of ones): $\det(C) = (a-b)^{k-1}[a + (k-1)/b]$

33.23 Characteristic Equations of Patterned Matrices

33.23.1 Triangular ($k \times k$): $\prod_{i=1}^{k} (a_{ii} - \lambda) = 0$, eigenvalues: $a_{11}, a_{22}, \cdots, a_{kk}$

33.23.2 $C = D + qab^T$ (see 21.2): $\left(1 + q\sum_{i=1}^{k} \frac{a_i b_i}{d_{ii} - \lambda}\right) \prod_{i=1}^{k} (d_{ii} - \lambda) = 0$

33.23.3 $C = dI + qab^T$: $\left(d + q\sum_{i=1}^{k} a_i b_i - \lambda\right)(d - \lambda)^{k-1} = 0$, $k-1$ eigenvalues are equal to d and one is equal to $d + q\sum_{i=1}^{k} a_i b_i$

33.24 Triangular Matrices

33.24.1 Decomposition to Triangular Matrices

A = RT, R = lower real triangular, T = upper real triangular, $A_{kxk} \equiv$ real with every leading principal minor is nonzero.

33.24.2 Upper Triangular

$A = T^T T$, $T =$ upper triangular, $A_{kxk} =$ positive definite.

Example

$$A=\begin{bmatrix}1 & 2 & 0\\ 2 & 5 & 1\\ 0 & 1 & 17\end{bmatrix}, \quad T=\begin{bmatrix}1 & 2 & 0\\ 0 & 1 & 1\\ 0 & 0 & 4\end{bmatrix}$$

33.24.3 Upper and Diagonal

$A = T^T DT$, $T =$ upper triangular (real and unique) with $t_{ii} = 1$, $D =$ real diagonal, $A_{n\times n} =$ positive definite.

33.24.4 Product of Lower Triangular

$R = R_1 R_2 \cdots R_n$, $R_i =$ lower triangular, R = lower triangular.

33.24.5 Product of Upper Triangular

$T = T_1 T_2 \cdots T_n$, $T_i =$ upper triangular, T = upper triangular.

33.24.6 Inverse

$R^{-1} =$ lower, $T^{-1} =$ upper, $R =$ lower, $T =$ upper

33.24.7 Determinant

$$\det(R \text{ or } T) = \left(\prod_{i=1}^{k} R_{ii} \text{ or } \prod_{i=1}^{k} T_{ii}\right), \quad R_{k\times k} \text{ and } T_{k\times k}$$

33.24.8 Eigenvalues

Eigenvalues $(R \text{ or } T) = (r_{11}, r_{22}, \cdots r_{kk} \text{ or } t_{11}, t_{22}, \cdots t_{kk})$.

33.24.9 Orthogonal Decomposition

$PA = T$, $P =$ real orthogonal matrix, $A_{k\times k} =$ real, T = upper real triangular, $t_{ii} > 0$ for $i = 1,2,\cdots k$.

33.24.10 Eigenvalues

$T = P^{-1}AP$, A = real and its roots are real, $P =$ nonsingular, T = upper triangular, eigenvalues = diagonal elements of T.

33.24.11 Product

i[th] diagonal element of T^n is t_{ii}^n, $T_{k\times k} =$ upper (lower) triangular.

33.24.12 Inverse

If we set $T^{-1} = B$ then $t_{ii} b_{ii} = 1$ for $i = 1,2,\cdots k$, $T_{k\times k} =$ nonsingular triangular.

33.24.13 Orthogonal Decomposition

$P^T AP = T$, $A_{k\times k} =$ real eigenvalues, $P =$ orthogonal, T = upper triangular, eigenvalues of $A =$ diagonal of T.

33.25 Correlation Matrix

33.25.1 Correlation Matrix

$R = [\rho_{ij}],\ \rho_{ij} = \dfrac{\upsilon_{ij}}{\sqrt{\upsilon_{ii}\upsilon_{ij}}}$, V = positive definite covariance matrix $= [\upsilon_{ij}]$, υ_{ij} = covariance between y_i and y_j components of the random vectors y, υ_{ii} = variance between y_i and y_i.

33.25.2 Correlation Matrix

$R = D_\upsilon^{-1/2} V D_\upsilon^{-1/2}$, R = correlation matrix of random vector y, V = positive definite covariance matrix, D_υ = diagonal matrix with ith diagonal element υ_{ii}.

Note: Correlation matrix is positive definite.

33.25.3 Properties of Correlation Matrix

a) $\rho_{ii} = 1,\ i = 1,2,\cdots n$; ; b) $-1 < \rho_{ij} < 1$ all $i \neq j$; c) largest eigenvalue is less than n; d) $0 < |R| \leq 1$; e) positive definite.

33.26 Direct Product and Sum of Matrices

33.26.1 Direct Product

$$C_{m_1 m_2 \times n_1 n_2} = A_{m_2 \times n_2} \otimes B_{m_1 \times n_1} = \begin{bmatrix} Ab_{11} & Ab_{12} & \cdots & Ab_{1n_1} \\ Ab_{21} & Ab_{22} & \cdots & Ab_{2n_1} \\ \vdots & \vdots & & \vdots \\ Ab_{m_1 1} & Ab_{m_1 2} & & Ab_{m_1 n_1} \end{bmatrix}$$

$$= [C_{ij}] = [Ab_{ij}] \quad i = 1,2,\cdots m_1,\ \ j = 1,2,\cdots n_1$$

Example

$$A \otimes B = \begin{bmatrix} 2 & 1 \\ 0 & -1 \end{bmatrix} \otimes [1\ \ 4] = [Ab_{11}\ \ Ab_{12}] = \begin{bmatrix} 2 & 1 & 8 & 4 \\ 0 & -1 & 0 & -4 \end{bmatrix}$$

$$B \otimes A = [1\ \ 4] \begin{bmatrix} 2 & 1 \\ 0 & -1 \end{bmatrix} = \begin{bmatrix} Ba_{11} & Ba_{12} \\ Ba_{21} & Ba_{22} \end{bmatrix} = \begin{bmatrix} 2 & 8 & 1 & 4 \\ 0 & 0 & -1 & -4 \end{bmatrix}$$

33.26.2 Properties of Direct Products

1. $A \otimes B \neq B \otimes A$ in general
2. $A_{m_2 \times n_2} \otimes I_{m_1 \times m_1} = \begin{bmatrix} A & & 0 \\ & \ddots & \\ 0 & & A \end{bmatrix}_{m_1 m_2 \times m_1 n_2} = diag(A)$
3. $(aA) \otimes B = A \otimes (aB) = a(A \otimes B)$, a = any scalar
4. $(A \otimes B) \otimes C = A \otimes (B \otimes C)$
5. $(A \otimes B)^T = A^T \otimes B^T$

Example

$$A^T \otimes B^T = \begin{bmatrix} 3 & -2 \\ 2 & 0 \end{bmatrix} \otimes \begin{bmatrix} 1 \\ 4 \end{bmatrix} = \begin{bmatrix} 3 & -2 \\ 2 & 0 \\ 12 & -8 \\ 8 & 0 \end{bmatrix} = (A \otimes B)^T = \begin{bmatrix} 3 & 2 & 12 & 8 \\ -2 & 0 & -8 & 0 \end{bmatrix}^T$$

6. $tr(A \otimes B) = tr(A)tr(B)$, A and B square matrices
7. $(A \otimes B)(F \otimes G) = (AF) \otimes (BG)$, $A_{m_1 \times n_1}$, $B_{m_2 \times n_2}$, $F_{n_1 \times k_1}$, $G_{n_2 \times k_2}$
8. $(A \otimes B)^{-1} = A^{-1} \otimes B^{-1}$, A and B nonsingular
9. $P \otimes Q =$ orthogonal if P and Q are orthogonal
10. $A \otimes I = diag(A)$
11. $\det(A \otimes B) = \det(B \otimes A) = |A|^n |B|^m$, $A_{m \times m}$, $B_{n \times n}$
12. $\det(A \otimes A^{-1}) = \det(A^{-1} \otimes A) = 1$, A = nonsingular
13. $(I \otimes P)(I \otimes A) = I \otimes T = C$, $A_{m \times m}$, $P_{m \times m}$, $PA = T$, $T =$ upper (lower), C = upper (lower)
14. $A_{m \times m} \equiv \{\lambda_1, \lambda_2, \cdots \lambda_m\} =$ characteristic roots (eigenvalues), the eigenvalues of $I \otimes A$ and $A \otimes I$ are equal and are of multiplicity of n of the eigenvalues of A, $I_{n \times n}$.
15. $(A + B) \otimes C = A \otimes C + B \otimes C$, $A_{m \times m}$, $B_{m \times m}$, $C_{n \times n}$
16. $D_1 \otimes D_2 \equiv$ diagonal, $D_1 =$ diagonal, $D_2 =$ diagonal

    $T_1 \otimes T_2 \equiv$ upper (lower) triangular, $T_1 =$ upper (lower) triangular,

    $T_2 =$ upper (lower) triangular
17. $A \otimes B =$ positive (semi) definite, A and B = positive (semi) definite
18. $rank(A \otimes B) = rank(A)\, rank(B)$

33.27 Direct Sum of Matrices

33.27.1 Definition

$$A = A_1 \oplus A_2 = \begin{bmatrix} A_1 & 0 \\ 0 & A_2 \end{bmatrix}, \; A_{1 n_1 \times n_1}, \; A_{2 n_2 \times n_2}, \; A_{(n_1+n_2)\times(n_1+n_2)}$$

33.27.2 Properties

1. $A_1 \oplus A_2 \neq A_2 \oplus A_1$ in general
2. $(A_1 \oplus A_2) \oplus A_3 = A_1 \oplus (A_2 \oplus A_3)$
3. $A_1 \oplus \cdots \oplus A_k = A_1^T \oplus \cdots \oplus A_K^T = A^T = A$ if and only if A_i is symmetric
4. $\det(A_1 \oplus \cdots \oplus A_K) = \prod_{i=1}^{K} \det(A_i)$

5. $tr(A_1 \oplus \cdots \oplus A_K) = \sum_{i=1}^{K} tr(A_i)$
6. $A_1 \oplus \cdots \oplus A_K$ is upper (lower) triangular if each A_i is upper (lower) triangular
7. $A_1 \oplus \cdots \oplus A_K = A =$ orthogonal, $A^T A = AA^T = I$, if each A_i is orthogonal
8. If $rank(A_i) = r_i$ then $rank(A_1 \oplus \cdots \oplus A_K) = \sum_{i=1}^{K} r_i$
9. $(A_1 \oplus A_2)(A_3 \oplus A_4) = (A_1 A_3) \oplus (A_2 A_4)$ assuming the operations are defined
10. $A \otimes I = A \oplus A \oplus \cdots \oplus A$, A_{nxn}, I_{mxm}

33.28 Circulant Matrices

33.28.1 Definitions

$(j-i)/k \equiv (j-i)$ modulo k, if remainder is positive it is kept, but if it is negative we must add k (Example: 16/5 = 1, –16/5 = –1 hence –1 + 5 = 4).

$$(j-i)/k = k+j-i \quad \text{when} \quad i>j \quad 1 \le i \text{ and } j \le k$$

$$(j-i)/k = j-i \quad \text{when} \quad i \le j \quad 1 \le i \text{ and } j \le k$$

33.28.2 Regular Circulant

$(j-i)/k = (q-p)/k$ which implies that $a_{ij} = a_{pq}, A_{kxk}$

$$A = \begin{bmatrix} a_0 & a_1 & a_2 \\ a_2 & a_0 & a_1 \\ a_1 & a_2 & a_0 \end{bmatrix}, \quad a_{ij} = a_{j-i} \text{ for } j \ge i \text{ (elements above diagonal)}$$

$$a_{ij} = a_{k+j-i} \text{ for } j < i \text{ (elements below the diagonal)}$$

33.28.3 Regular Circulants

1. The zero matrix 0,
2. The identity one I,
3. Matrix with all elements equal to a constant,
4. $aA + bB$, A_{kxk}, B_{kxk} regular circulants

33.28.4 Properties

1. $A^T \equiv$ regular circulant if A is also
2. Diagonal elements of A are equal
3. The diagonals parallel to the main one have their elements equal
4. $P^T AP = A$, $A_{kxk} =$ regular singular, $P = [x_1 x_2 \cdots x_k x_1]$ where $x_i = [0\ 0 \cdots 1 \cdots 0]$ (i^{th} element 1)
5. $AB \equiv$ regular circulant if A_{kxk} and B_{kxk} are regular circulants
6. $C^{-1} \equiv$ regular circulant if C is a nonsingular regular circulant
7. AB = BA if both are $k \times k$ regular circulants
8. $C_{kxk} =$ regular circulant, $[c_0\ c_1 \cdots c_{k-1}] \equiv$ first row,

$\lambda_i =$ eigenvalues of $C = c_0\omega_i^0 + c_1\omega_i + c_2\omega_i^2 + \cdots + c_{k-1}\omega_i^{k-1}$, $\omega_k \equiv k^{th}$ roots of unity,

$x_i = [\omega_i^0\ \omega_i\ \omega_i^2 \cdots \omega_i^{k-1}]^T$ are the eigenvalues

Example

$$C = \begin{bmatrix} 3 & -1 & 2 & 0 \\ 0 & 3 & -1 & 2 \\ 2 & 0 & 3 & -1 \\ -1 & 2 & 0 & 3 \end{bmatrix},\quad e^{j\frac{2\pi k}{4}} \Rightarrow 1, -1, j, -j;$$

$$\lambda_1 = 3\cdot 1 - 1\cdot 1 + 2\cdot 1^2 + 0\cdot 1^3 = 4 \qquad x_1 = [1\ 1\ 1\ 1]^T$$

$$\lambda_2 = 3\cdot 1 - 1(-1) + 2\cdot(-1)^2 + 0(-1)^3 = 6 \qquad x_2 = [1\ -1\ 1\ -1]^T \text{ etc.}$$

9. $C_{k\times k}$ with first row $[c_0\ c_1 \cdots c_{k-1}]$ its

$$\det(C) = \prod_{i=1}^{k} \lambda_i = \prod_{i=1}^{k} (c_0 + c_1\omega_i + c_2\omega_i^2 + \cdots + c_{k-1}\omega_i^{k-1})$$

10. Symmetric regular: $A = \begin{bmatrix} a_0 & a_1 & a_2 & a_1 \\ a_1 & a_0 & a_1 & a_2 \\ a_2 & a_1 & a_0 & a_1 \\ a_1 & a_2 & a_1 & a_0 \end{bmatrix} \equiv$ any regular circulant that is also symmetric

11. Properties of symmetric regular: A and B are symmetric regular
    1. $aA + bB$ ≡ symmetric regular, a and b any real number
    2. A^T, B^T ≡ symmetric regular
    3. AB = BA and AB is symmetric regular circulant
    4. A^{-1} ≡ symmetric regular

12. Symmetric circulant: $A = \begin{bmatrix} a_0 & a_1 & a_2 \\ a_1 & a_2 & a_0 \\ a_2 & a_0 & a_1 \end{bmatrix}$ $(a_{ij} = a_{(i+j-2)/k})$

13. Properties of symmetric circulant: $A_{k\times k}$ and $B_{k\times k}$ are symmetric circulants
    1. aA and bB ≡ symmetric circulants, a and b are any real numbers
    2. I ≡ not symmetric circulant
    3. A^{-1} ≡ symmetric circulant
    4. AB ≡ regular circulant (AB ≠ BA in general)
    5. If $a_0, a_1, \cdots, a_{k-1}$ are the first row elements of A then $s = a_0 + a_1 + \cdots + a_{k-1}$ is an eigenvalue of A
    6. A^- is the g-inverse of B and is symmetric circulant

33.29 Vandermonde and Fourier Matrices

33.29.1 Definition of Vandermonde Matrix (see 33.1.12)

33.29.2 Properties of Vandermonde

1. $\det(A) = \prod_{t=2}^{k} \prod_{i=1}^{t-1} (a_t - a_i)$
2. $\text{rank}(A) = r$, r = number of distinct a_i values
3. A^T = also a Vandermonde matrix if A is one

Example

$y_i = \beta_0 + \beta_1 x_i + \beta_2 x_i^2 + \cdots + \beta_{k-1} x_i^{p-1} + \varepsilon_i,\ \ i = 1,2,\cdots n,\ n \geq p$ Hence, $y = X\beta + \varepsilon$ or

$$\begin{bmatrix} y_1 \\ y_2 \\ \cdots \\ y_n \end{bmatrix} = \begin{bmatrix} 1 & x_1 & x_1^2 \cdots & x_1^{p-1} \\ 1 & x_2 & x_2^2 \cdots & x_2^{p-1} \\ \vdots & & & \\ 1 & x_n & x_n^2 \cdots & x_n^{p-1} \end{bmatrix} \begin{bmatrix} \beta_0 \\ \beta_1 \\ \vdots \\ \beta_{p-1} \end{bmatrix} + \begin{bmatrix} \varepsilon_1 \\ \varepsilon_2 \\ \vdots \\ \varepsilon_n \end{bmatrix}$$

33.29.3 Fourier Matrix

$$F = \frac{1}{\sqrt{k}} \begin{bmatrix} 1 & 1 & 1 & \cdots & 1 \\ 1 & \omega & \omega^2 & \cdots & \omega^{k-1} \\ 1 & \omega^2 & \omega^4 & \cdots & \omega^{2k-2} \\ \vdots & & & & \\ 1 & \omega^{k-1} & \omega^{2k-2} & \cdots & \omega^{(k-1)(k-1)} \end{bmatrix},\ F_{k\times k},\ \omega = e^{j\frac{2\pi}{k}}$$

33.29.4 Properties of the Fourier Matrix

1. $F = F^T,\ F_{k\times k}$
2. $F^{-1} = \bar{F},\ \bar{F} =$ conjugate of F
3. $F^2 = P,\ P =$ permutation matrix $= [\varepsilon_1\ \varepsilon_k\ \varepsilon_{k-1} \cdots \varepsilon_2],\ \varepsilon_j$ j^{th} column of the $k \times k$ identity matrix
4. $F^4 = I$
5. $\sqrt{k}\ F = C + jS$, C and S are real matrices,

$$c_{pq} = \cos\frac{2\pi}{k}(p-1)(q-1),\ \ S_{pq} = \sin\frac{2\pi}{k}(p-1)(q-1)$$

6. $CS = SC,\ C^2 + S^2 = I$

33.30 Permutation Matrices

33.30.1 Definition

a) Column permutation = $n \times n$ matrix resulting from permuting columns of $n \times n$ identity matrix.
b) Row permutation = $n \times n$ matrix resulting from permuting the rows of $n \times n$ identity matrix.

Example

$I = [\varepsilon_1\ \varepsilon_2\ \varepsilon_3\ \varepsilon_4]$, then if $P = [\varepsilon_2\ \varepsilon_4\ \varepsilon_3\ \varepsilon_1]$ we obtain AP = B that moves column 1 of A to column 4 of B, column 2 of A to column 1 of B, column 3 of A to column 3 of B, and column 4 of A to column 2 of B. If we look at P, we observe that row-wise from I row 1 moved to row 2, row 4 moved to row 1, row 3 remains the same, and row 2 moved to row 4. Hence, PA = C moves row 1 of A to row 2 of C, row 2 of A to row 4 of C, row 3 of A to row 3 of C, and row 4 of A to row 1 of C.

33.30.2 Properties

1. $P^T P = PP^T = I,\ P^T = P^{-1}$ which implies P is orthogonal
2. $P^T AP$ has the same diagonal elements (rearranged) as the diagonal elements of A

33.31 Hadamard Matrices

33.31.1 Definitions

a) H consists of +1 and −1 elements, H_{nxn}
b) $H^T H = nI$

Example

$$H_{2x2} = \begin{bmatrix} 1 & 1 \\ -1 & 1 \end{bmatrix},\quad H^T H = 2I$$

33.31.2 Properties

1. $n^{-1/2} H$ is orthogonal, H_{nxn}
2. $HH^T = nI$
3. $H^{-1} = n^{-1} H^T$
4. H^T and nH^{-1} are Hadamard matrices
5. $H_1 \otimes H_2$ = Hadamard matrix, H_1 and H_2 are Hadamard matrices

Example

$$H \otimes H = \begin{bmatrix} 1 & 1 \\ -1 & 1 \end{bmatrix} \otimes \begin{bmatrix} 1 & 1 \\ -1 & 1 \end{bmatrix} = \begin{bmatrix} 1 & 1 & 1 & 1 \\ -1 & 1 & -1 & 1 \\ -1 & -1 & 1 & 1 \\ 1 & -1 & -1 & 1 \end{bmatrix}$$

6. H_{nxn}, $n = 1$ or 2 or multiple of 4
7. $\det(H) = n^{n/2}$

33.32 Toeplitz Matrices (T-matrix)

33.32.1 Definition

All elements on each superdiagonal are equal, and all elements on each subdiagonal are equal.

Example

$$A_1 = \begin{bmatrix} 6 & 2 & 0 \\ 5 & 6 & 2 \\ 0 & 5 & 6 \end{bmatrix},\quad A_2 = \begin{bmatrix} 3 & 1 & 2 \\ 4 & 3 & 1 \\ 2 & 4 & 3 \end{bmatrix}$$

33.32.2 Properties

1. A regular circulant is a T-matrix but a T-matrix is not necessarily a circulant.
2. A linear combination of T-matrices is a T-matrix.
3. If a_{ij} is defined by $a_{ij} = a_{|i-j|}$ then A is a symmetric T-matrix.
4. A^T is a T-matrix.
5. A is symmetric about its secondary diagonals.
6. $A_{n\times n} = \begin{bmatrix} a_0 & a_1 & 0\cdots & 0 & 0 \\ a_2 & a_0 & a_1\cdots & 0 & 0 \\ \vdots & & & & \\ 0 & 0 & 0\cdots & a_2 & a_0 \end{bmatrix}$, eigenvalues: $\lambda_m = a_0 + 2\sqrt{a_1 a_2}\cos\left(\frac{m\pi}{n+1}\right)$

   for $m = 1,2,\cdots,n$
7. If $a_{ij} = a_{|i-j|}$ for $|i-j| \le 1$ and $a_{|i-j|} = 0$ for $|i-j| > 1$ (symmetric matrix) the eigenvalues are:

   $$\lambda_m = a_0 + 2a_1\cos\frac{m\pi}{n+1} \quad \text{for } m = 1,2,\cdots,n.$$
8. A of (7) is positive definite if and only if $a_0 + 2a_1\cos\frac{m\pi}{n+1} > 0$ for $m = 1,2,\cdots,n$.
9. A of (7) is positive for all n if $a_0 > 0$ and $|a_1 / a_0| \le 1/2$
10. The determinant of $A_{n\times n}$ of (6) is $\det(A) = \prod_{m=1}^{n}\left[a_0 + 2\sqrt{a_1 a_2}\cos\left(\frac{m\pi}{n+1}\right)\right]$

33.32.3 Centrosymmetric matrix (or cross-symmetric): $a_{ij} = a_{n+1-i,\,n+1-j}$ for all i and j

Example

$$A = \begin{bmatrix} 3 & 0 & -1 \\ 1 & 2 & 1 \\ -1 & 0 & 3 \end{bmatrix},\quad B = \begin{bmatrix} 1 & 2 & 4 & 0 \\ 2 & 1 & 3 & 6 \\ 6 & 3 & 1 & 2 \\ 0 & 4 & 2 & 1 \end{bmatrix}$$

33.32.4 Properties of Centrosymmetric Matrix (c-matrix)

1. 0 ; I ; J (all ones); aI + bJ are c-matrices.
2. A^T is a c-matrix if A is.
3. $A = \sum_{i=1}^{K} a_i A_i$ is a c-matrix if A_i's are c-matrices.
4. AB is a c-matrix if $A_{n\times n}$ and $B_{n\times n}$ are c-matrices.
5. A^{-1} is c-matrix if A is a c-matrix.

33.33 Trace

33.33.1 Definitions

$$tr(A) = \sum_{i=1}^{n} a_{ii},\quad A_{n\times n}$$

33.33.2 Properties

1. $tr(AB) = tr(BA),\ A_{n\times n},\ B_{n\times n}$

2. $tr(A) = \sum_{i=1}^{n} \lambda_i,\ \lambda_i = \text{eigenvalues}$

3. $tr(A) = tr(P^{-1}AP),\ A_{n\times n},\ P_{n\times n}$

4. $tr(A) = tr(P^T AP),\ P$ is orthogonal

5. $tr(aA + bB) = a\, tr(A) + b\, tr(B)$

6. $tr(A) = m\, rank(A),\ A^2 = mA$

7. $tr(A^T) = tr(A)$

8. $tr[I - A(A^T A)^- A^T] = m - r,\ A_{m\times n},\ rank(A) = r$

9. $tr(AA^T) = tr(A^T A) = \sum_{j=1}^{m} \sum_{i=1}^{n} a_{ij}^2,\ A_{n\times m}$

10. $tr(A^k) = \sum_{i=1}^{n} \lambda_i^k,\ A_{n\times n},\ \lambda_i = \text{eigenvalues}$

11. $tr(A \otimes B) = tr(A)tr(B)$

12. $tr(A^-) = \sum_{i=1}^{r} \lambda_i^{-1},\ A_{n\times n}\ \lambda_1, \lambda_2 \cdots \lambda_r$

13. $tr(I + S) = n,\ tr(S) = 0,\ S_{n\times n} =$ skew symmetric.

14. $tr(A) = 0$ if $A^k = 0$ for some positive k.

15. $x^T Ax = tr(Axx^T)$

16. $tr(A) > 0$ implies A is positive definite.

17. $tr(A) \geq 0$ implies A is positive semidefinite, or non-negative.

18. $tr(AB) \geq 0,\ A_{n\times n}$ and $B_{n\times n}$ are non-negative.

19. $tr(A^T B) = \sum_{i=1}^{n} \sum_{j=1}^{n} A_{ij} B_{ij}$

20. $y^T x = tr(xy^T)$

33.34 Derivatives

33.34.1 Derivative of a Function with Respect to a Vector

$$\frac{\partial f}{\partial x} = \begin{bmatrix} \frac{\partial f}{\partial x_1} \\ \vdots \\ \frac{\partial f}{\partial x_k} \end{bmatrix},\quad x = [x_1, x_2, \cdots x_k]^T,\ x_i = \text{real variables},\ f(x_1, x_2, \cdots x_k)$$

33.34.2 $f(X)=\alpha^T X\beta$, $\frac{\partial f}{\partial X}=\alpha\beta^T$, $X_{m\times n}$, $\alpha_{m\times 1}$, $\beta_{n\times 1}$, α and β are vectors of constants

33.34.3 $f(X)=\alpha^T X\alpha$, $\frac{\partial f}{\partial X}=2\alpha\alpha^T - D_{\alpha\alpha^T}$, $\alpha_{k\times 1}$ = vectors of constants, $X_{k\times k}$ = symmetric matrix of independent real variables (except that $x_{ij}=x_{ji}$), $D_{\alpha\alpha^T}=k\times k$ diagonal matrix (i^th^ diagonal = i^th^ diagonal of $\alpha\alpha^T$)

33.34.4 Derivative of a Matrix with Respect to a Scalar

$\frac{\partial Y}{\partial x_{ij}}=\left[\frac{\partial f_{pq}(x_{11},\cdots,x_{mn})}{\partial x_{ij}}\right]$, $Y_{k\times k}$, y_{pq} = elements of $Y=f_{pq}(x_{11},x_{12}\cdots,x_{mn})$, each y_{pq} is a function of an $m\times n$ matrix of independent real variables x_{ij}.

33.34.5 $$\frac{\partial f(x;\alpha)}{\partial\alpha}=\begin{bmatrix}\frac{\partial f(x;\alpha)}{\partial\alpha_1}\\ \vdots\\ \frac{\partial f(x;\alpha)}{\partial\alpha_n}\end{bmatrix},\quad x=[x_1,x_2,\cdots,x_r]^T,\quad \alpha=[\alpha_1,\alpha_2,\cdots\alpha_n]^T$$

33.34.6 $$\frac{\partial x(\alpha)}{\partial\alpha}=\begin{bmatrix}\frac{\partial x_1(\alpha)}{\partial\alpha_1} & \cdots & \frac{\partial x_1(\alpha)}{\partial\alpha_n}\\ \vdots & & \\ \frac{\partial x_r(\alpha)}{\partial\alpha_1} & \cdots & \frac{\partial x_r(\alpha)}{\partial\alpha_n}\end{bmatrix},\quad x=[x_1(\alpha),x_2(\alpha),\cdots x_r(\alpha)]^T,\quad \alpha=[\alpha_1,\alpha_2,\cdots\alpha_n]^T$$

33.34.7 $$\frac{\partial x(\alpha)}{\partial\alpha_i}=\begin{bmatrix}\frac{\partial x_1(\alpha)}{\partial\alpha_i}\\ \vdots\\ \frac{\partial x_r(\alpha)}{\partial\alpha_i}\end{bmatrix}$$

33.34.8 $\nabla_z(\alpha^H z)=\alpha^*$, $\nabla_z(z^H\alpha)=0$, $\nabla_z(z^H Rz)=(Rz)^*$

33.34.9 $\nabla_{z^*}(\alpha^H z)=0$, $\nabla_{z^*}(z^H\alpha)=\alpha$, $\nabla_{z^*}(z^H Rz)=Rz$

33.34.10 $w_k=x_k+jy_k$, $\frac{\partial}{\partial w_k}=\frac{1}{2}\left(\frac{\partial}{\partial x_k}-j\frac{\partial}{\partial y_k}\right)$, $\frac{\partial}{\partial w_{k^*}}=\frac{1}{2}\left(\frac{\partial}{\partial x_k}+j\frac{\partial}{\partial y_k}\right)$,

$$\frac{\partial w_k}{\partial w_k}=1,\quad \frac{\partial w_k}{\partial w_k^*}=\frac{\partial w_k^*}{\partial w_k}=0,\quad \frac{\partial}{\partial w}=\frac{1}{2}\left[\frac{\partial}{\partial x_0}-j\frac{\partial}{\partial y_0}\ \frac{\partial}{\partial x_1}-j\frac{\partial}{\partial y_1}\cdots\frac{\partial}{\partial x_{m-1}}-j\frac{\partial}{\partial y_{m-1}}\right]^T,$$

$$\frac{\partial}{\partial w^*}=\frac{1}{2}\left[\frac{\partial}{\partial x_0}+j\frac{\partial}{\partial y_0}\ \frac{\partial}{\partial x_1}+j\frac{\partial}{\partial y_1}\cdots\frac{\partial}{\partial x_{m-1}}+j\frac{\partial}{\partial y_{m-1}}\right]^T$$

Example

$$\frac{\partial x^H w}{\partial w^*}=0,\quad \frac{\partial w^H x}{\partial w^*}=x,\quad \frac{\partial w^H R w}{\partial w^*}=Rw,\ \text{if } J=\sigma_d^2-w^H x-x^H w+w^H R w$$

then $\dfrac{\partial J}{\partial w^*}=-x+Rw$

33.35 Positive Definite Matrices

33.35.1 Positive Definite

$A=CC^T$, $C=n\times n$ full rank or the principal minors are all positive. (The ith principal minor is the determinant if the submatrix formed by deleting all rows and columns with an index greater than I.)

$$A^{-1}=(C^{-1})^T(C^{-1})$$

33.35.2 A = positive definite, $B=m\times n$ full rank with $m\le n$, then BAB^T = positive definite

33.35.3 A = positive definite, a) the diagonal elements are positive (non-negative), and b) det A = positive (non-negative)

References

Golub, G. H. and C. F. Van Loan, *Matric Computations*, The Johns Hopkins University Press, Baltimore, Maryland, 1989.

Graybill, F. A., *Matrices with Applications in Statistics*, Wadsworth International Group, Belmont, California, 1983.

Horn, R. A. and C. A. Johnson, *Matrix Analysis*, Cambridge University Press, New York, NY, 1985.

Jennings, A. and J. J. McKeown, *Matrix Computation*, John Wiley & Sons, New York, NY, 1992.

34

Probability and Stochastic Processes

34.1 Axioms of Probability

34.1.1 Axioms of Probability

I. $P(A) \geq 0$, II. $P(S) = 1$, III. If $AB = 0$ then $P(A+B) = P(A) + P(B)$. S = a set of elements of outcomes $\{\zeta_i\}$ of an experiment (certain event), 0 = empty set (impossible event). $\{\zeta_i\}$ = elementary event if $\{\zeta_i\}$ consists of a single element. A + B = *union* of events, AB = *intersection* of events, event = a subset of S, P(A) = probability of event A.

34.1.2 Corollaries of Probability

$$P(0)=0,\ \ P(A)=1-P(\bar{A})\le 1,\ \ (\bar{A}=\text{complement set of } A)$$

$$P(A+B)\ne P(A)+P(B),\ \ P(A+B)=P(A)+P(B)-P(AB)\le P(A)+P(B)$$

Example

$S=\{hh,ht,th,tt\}$ (tossing a coin twice), A = {heads at first tossing} $=\{hh,ht\}$, $B=$ {only one head came up} $=\{ht,th\}$, $G=$ {heads came up at least once} $=\{hh,ht,th\}$, $D=$ {tails at second tossing} $=\{ht,tt\}$

34.2 Conditional Probabilities — Independent Events

34.2.1 Conditional Probabilities

$$P(A|M)=\frac{P(AM)}{P(M)}=\frac{\text{probability of event } AM}{\text{probability of event } M}=\text{conditional probability of } A \text{ given } M.$$

1. $P(A|M)=0$ if $AM=0$
2. $P(A|M)=\dfrac{P(A)}{P(M)}\ge P(A)$ if $AM=A\ (A\subset M)$
3. $P(A|M)=\dfrac{P(M)}{P(M)}=1$ if $M\subset A$
4. $P(A+B|M)=P(A|M)+P(B|M)$ if $AB=0$

Example

$P(f_i)=1/6,\ i=1,\cdots 6.\ M=\{\text{odd}\}=\{f_1,f_3,f_5\},\ A=\{f_1\},\ AM=\{f_1\},\ P(M)=3/6,\ P(AM)=1/6$, then $P(f_1|\text{even})=P(AM)/P(M)=1/3$

34.2.2 Total Probability

$P(B)=P(B|A_1)P(A_1)+\cdots+P(B|A_n)P(A_n)$, B = arbitrary event, $A_iA_j=0\ \ i\ne j=1,2,\cdots n,\ A_1+\cdots+A_n$ $=S=$ certain event

34.2.3 Baye's Theorem

$$P(A_i|B)=\frac{P(B|A_i)P(A_i)}{P(B|A_1)P(A_1)+\cdots+P(B|A_n)P(A_n)}$$

$$A_iA_j=0,\ i\ne j=1,2,\cdots,n,\ \ A_1+A_2+\cdots+A_n=S=\text{certain event},\ \ B=\text{ arbitrary}$$

34.2.4 Independent Events

$P(AB)=P(A)P(B)$ implies A and B are independent events.

34.2.5 Properties

1. $P(A|B)=P(A)$

2. $P(B|A) = P(B)$
3. $P(A_1 A_2 \cdots A_n) = P(A_1)\cdots P(A_n)$, A_i = independent events
4. $P(A+B) = P(A) + P(B) - P(A)P(B)$
5. $\overline{AB} = (\overline{A+B})$, $P(\overline{A+B}) = 1 - P(A+B)$, $P = (\overline{A}\overline{B}) = P(\overline{A})P(\overline{B})$
   If A and B are independent. Overbar means complement set.
6. If A_i, A_2, A_3 are independent and A_1 is independent of $A_2 A_3$ then
   $P(A_1 A_2 A_3) = P(A_1)P(A_2)P(A_3) = P(A_1)P(A_2 A_3)$. Also
   $P[A_1(A_2 + A_3)] = P(A_1 A_2) + P(A_1 A_3) - P(A_1 A_2 A_3) = P(A_1)$
   $[P(A_2) + P(A_3) - P(A_2)P(A_3)] = P(A_1)P(A_2 + A_3)$

34.2.6 $P(A+B+C) = P(A) + P(B) + P(C) - P(AB) - P(AC) - P(BC) + P(ABC)$

34.3 Compound (Combined) Experiments

34.3.1 $S = S_1 \times S_2$ = *Cartesian product*

Example

$S_1 = \{1,2,3\}$, $S_2 = \{\text{heads,tails}\}$, $S = S_1 \times S_2 = \{(1\text{ heads}),(1\text{ tails}),(2\text{ heads}),(2\text{ tails}),(3\text{ heads}),(3\text{ tails})\}$

34.3.2 If $A_1 \subset S_1$, $A_2 \subset S_2$ then $A_1 \times A_2 = (A_1 \times S_2)(A_2 \times S_1)$ (see Figure 34.1)

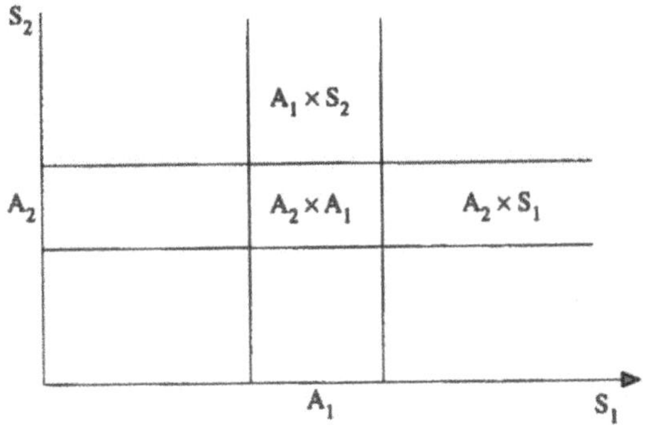

FIGURE 34.1

34.3.3 Probability in Compound Experiments

$$P(A_1) = P(A_1 \times S_2) \text{ where } \zeta_1 \in A_1 \text{ and } \zeta_2 \in A_2$$

34.3.4 Independent Compound Experiments

$P(A_1 \times A_2) = P_1(A_1)P_2(A_2)$

Example

$P(\text{heads}) = p$, $P(\text{tails}) = q$, $p + q = 1$, E = experiment tossing the coin twice = $E_1 \times E_2$ (E_1 = experiment of first tossing), E_2 = experiment of second tossing), $S_1 = \{h,t\}$ $P_1\{h\} = p$ $P_1\{t\} = q$, $E_2 = E_1$ = experiment of the second tossing, $S = S_1 \times S_2 = \{hh, ht, th, tt\}$, $P\{hh\} = P_1\{h\}P_2\{h\} = p^2$ = assume independence, $P\{ht\} = pq$, $P\{th\} = qp$, $P\{t,t\} = q^2$. For heads at the first tossing, $H_1 = \{hh, ht\}$ or $P(H_1)$ = $P\{hh\} + P\{ht\} = p^2 + pq = p$

34.3.5 Sum of More Spaces

$S = S_1 + S_2$, S_1 = outcomes of experiment E_1 and S_2 = outcomes of experiment E_2. S = space of the experiment $E = E_1 + E_2$; $A = A_1 + A_2$ where A_1 and A_2 are events of E_1 and E_2: $A_1 \subset S_1$, $A_2 \subset S_2$; $P(A) = P(A_1) + P(A_2)$.

34.3.6 Bernoulli Trials

$P(A)$ = probability of event A, ExExEx...xE = perform experiment n times = combined experiment.

$p_n(k) = \binom{n}{k} p^k q^{n-k}$ = probability that event A occurs k times in any order $P(A) = p$, $P(\bar{A}) = q$, $p + q = 1$

Example

A fair die was rolled 5 times. $p_5(2) = \frac{5!}{(5-2)!2!}\left(\frac{1}{6}\right)^2\left(\frac{5}{6}\right)^{5-2}$ = probability that "four" will come up twice.

Example

Two fair dice are tossed 10 times. What is the probability that the dice total seven points exactly four times?

Solution

Event $B = \{(1,6),(2,5),(3,4),(4,3),(5,2),(6,1)\}$, $P(B) = 6 \cdot \left(\frac{1}{6}\right)^2 = \frac{1}{6} = p$, $P(\bar{B}) = 1 - p = \frac{5}{6}$. The probability of B occurring four times and $\bar{B}$ six times is $\binom{10}{4}\left(\frac{1}{6}\right)^4\left(\frac{5}{6}\right)^6 = 0.0543$.

34.3.7

$P\{k_1 \le k \le k_2\}$ = probability of success of A (event) will lie between k_1 and k_2

$$P\{k_1 \le k \le k_2\} = \sum_{k=k_1}^{k_2} p_n(k) = \sum_{k=k_1}^{k} \binom{n}{k} p^k q^{n-k}$$

1. Approximate value: $\sum_{k=k_1}^{k_2} \binom{n}{k} p^k q^{n-k} \cong \frac{1}{\sqrt{2\pi npq}} \sum_{k=k_1}^{k_2} e^{-(k-np)^2/2npq}$, $npq >> 1$

34.3.8 DeMoivre-Laplace Theorem

$$p_n(k) = \binom{n}{k} p^k q^{n-k} \cong \frac{1}{\sqrt{2\pi npq}} e^{-(k-np)^2/2npq}, \quad npq >> 1$$

34.3.9 Poisson Theorem

$$\frac{n!}{k!(n-k)!} p^k q^{n-k} \cong e^{-np} \frac{(np)^k}{k!} = e^{-a} \frac{a^k}{k!}, \quad n \to \infty,\ p \to 0,\ np \to a$$

34.3.10 Random Points in Time

$P\{k \text{ in } t_a\} \cong e^{-nt_a/T}\dfrac{(nt_a/T)^k}{k!} = e^{-\lambda t_a}\dfrac{(\lambda t_a)^k}{k!}$, $t_2 - t_1 = t_a << T$, n random points in (0,T), $\lambda = n/T$. If $n \to \infty$, $T \to \infty$, $n/T \to \lambda$ the approximation becomes equality.

1. $P\{\text{one in } t_a\} \cong e^{-\lambda t_a}\lambda t_a \cong \lambda t_a$
2. $\lim\limits_{t_a \to 0} \dfrac{P\{\text{one in } t_a\}}{t_a} = \lambda$
3. $P\{k \text{ in } (t_1,t_2)\} = e^{-\int_{t_1}^{t_2}\lambda(t)dt}\dfrac{\left[\int_{t_1}^{t_2}\lambda(t)dt\right]^k}{k!}$, $p = \int_{t_1}^{t_2}\alpha(t)dt$, $n\alpha(t) = \lambda(t)$

34.4 Random Variables

34.4.1 Random Variable

To every outcome ζ of any experiment we assign a number $X(\zeta) = x$. The function X, whose domain in the space S of all outcomes and its range is a set of numbers, is called a random variable (r.v.).

34.4.2 Distribution Function

$F_X(x) = P\{X \le x\}$ defined on any number $-\infty < x < \infty$. $\{X \le x\}$ is an event for any real number x.

34.4.3 Properties of Distribution Function

1. $F(-\infty) = 0$, $F(+\infty) = 1$
2. $F(x_1) \le F(x_2)$ for $x_1 < x_2$
3. $F(x+) = F(x)$ continuous from the right

34.4.4 Density Function (or Frequency Function)

$f(x) = \dfrac{dF(x)}{dx}$; $f(x) = \lim\limits_{\Delta x \to 0}\dfrac{P\{x \le X \le x + \Delta x\}}{\Delta x}$; $P\{X = x\} = 0$ for continuous distribution function;

$f(x) = \sum_i p_i\delta(x - x_i) =$ density of discrete type, $p_i = F(x_i) - F(x_i -)$.

Example

Poisson distribution: $P\{X = k\} = e^{-\lambda}\dfrac{\lambda^k}{k!}$, $k = 0,1,\cdots$, $\lambda > 0$. Then $f(x) = e^{-\lambda}\sum_{k=0}^{\infty}\dfrac{\lambda^k}{k!}\delta(x-k)$.

Example

If X is normally distributed $\left(f(x) = \dfrac{1}{\sigma\sqrt{2\pi}}e^{-(x-m_x)^2/2\sigma^2}\right)$ with $m_x = 1000$ and $\sigma = 50$, then the probability that X is between 900 and 1,050 is $P\{x_1 \le X \le x_2\} = \int_{-\infty}^{x_2} f(y)d(y) - \int_{-\infty}^{x_1} f(y)d(y) = \dfrac{1}{2} + \operatorname{erf}\dfrac{x_2 - m_x}{\sigma}$ $- \dfrac{1}{2} - \operatorname{erf}\dfrac{x_1 - m_x}{\sigma} = \operatorname{erf}1 + \operatorname{erf}2 = 0.819$ where error function of $x = \operatorname{erf} x = \dfrac{1}{\sqrt{2\pi}}\int_0^x \exp(-y^2/2)dy$

34.4.5 Tables of Distribution Functions (see Table 34.1)

TABLE 34.1 Distribution and Related Quantities

Definitions

1. Distribution function (or cumulative distribution function [c.d.f.]): $F(x)$ = probability that the variate takes values less than or equal to $x = P\{X \le x\} = \int_{-\infty}^{x} f(u)du$
2. Probability density function (p.d.f.):

$$f(x);\; P\{x_\ell < X \le x_u\} = \int_{x_\ell}^{x_u} f(x)dx;\; f(x) = \frac{dF(x)}{dx}$$

3. Probability function (*discrete* variates) $f(x)$ = probability that the variate takes the value $x = P\{X = x\}$
4. Probability generating function (*discrete* variates):

$$P(t) = \sum_{x=0}^{\infty} t^x f(x),\quad f(x) = (1/x!)\left(\frac{\partial^x P(t)}{\partial t^x}\right)_{t=0},\quad x = 0,1,2,\cdots, X > 0$$

5. Moment generating function (m.g.f.):

$$M(t) = \int_{-\infty}^{\infty} t^{tx} f(x)dx,\quad M(t) = 1 + \mu_1' t + \mu_2' \frac{t^2}{2!} + \cdots + \frac{\mu_r' t^r}{r!} + \cdots,$$

$$\mu_r' = r^{th} \text{ moment about the origin} = \int_{-\infty}^{\infty} x^r f(x)dx = \left.\frac{\partial^r M(t)}{\partial t^r}\right|_{t=0}$$

$$M_{X+Y}(t) = M_X(t)M_Y(t)$$

6. Laplace transform of p.d.f.: $f^L(s) = \int_0^{\infty} e^{-sx} f(x)dx,\; X \ge 0$
7. Characteristic function: $\Phi(t) = \int_{-\infty}^{\infty} e^{jtx} f(x)dx,\; \Phi_{X+Y}(t) = \Phi_X(t)\Phi_Y(t)$
8. Cumulant function: $K(t) = \log\Phi(t),\; K_{X+Y}(t) = K_X(t) + K_Y(t)$
9. r^{th} cumulant: the coefficient of $(jt)^r / r!$ in the expansion of K(t)
10. r^{th} moment about the origin:

$$\mu_r' = \int_{-\infty}^{\infty} x^r f(x)dx = \left(\frac{\partial^r M(t)}{\partial t^r}\right)_{t=0} = (-j)^r \left(\frac{\partial^r \Phi(t)}{\partial t^r}\right)_{t=0}$$

11. Mean: μ = first moment about the origin = $\int_{-\infty}^{\infty} xf(x)dx = \mu_1'$
12. r^{th} moment about the mean: $\mu_r = \int_{-\infty}^{\infty} (x-\mu)^r f(x)dx$
13. Variance: σ^2 second moment about the mean $= \int_{-\infty}^{\infty} (x-\mu)^2 f(x)dx = \mu_2$
14. Standard deviation: $\sigma = \sqrt{\sigma^2}$
15. Mean derivation: $\int_{-\infty}^{\infty} |x-\mu| f(x)dx$
16. Mode: A fractile (value of r.v.) for which the p.d.f. is a local maximum
17. Median: m = the fractile which is exceeded with probability 1/2.

18. Standardized r^{th} moment about the mean: $\eta_r = \int_{-\infty}^{\infty}\left(\frac{x-\mu}{\sigma}\right)^r f(x)dx = \frac{\mu_r}{\sigma^r}$

19. Coefficient of skewness: $\eta_3 = \mu_3/\sigma^3$

20. Coefficient of kurtosis: $\eta_4 = \mu_4/\sigma^4$

21. Coefficient of variation: (standard deviation) / mean = σ/μ

22. Information content: $I = -\int_{-\infty}^{\infty} f(x)\log_2(f(x))dx$

23. r^{th} factorial moment about the origin (*discrete* case):

$$\mu'_{(r)} = \sum_{x=0}^{\infty} f(x)x(x-1)\cdots(x-r+1),\ X \ge 0,\ \mu'_{(r)} = \left(\frac{\partial^r P(t)}{\partial t^r}\right)_{t=1}$$

24. r^{th} factorial moment about the mean (discrete case):

$$\mu_{(r)} = \sum_{x=0}^{\infty} f(x-\mu)(x-\mu)(x-\mu-1)\cdots(x-\mu-r+1),\ X \ge 0$$

25. Relationships between moments:

$$\mu'_r = \sum_{i=0}^{r}\binom{r}{i}\mu_{r-i}(\mu'_1)^i;\ \mu_r = \sum_{i=0}^{r}\binom{r}{i}\mu'_{r-i}(-\mu'_1)^i,\ \mu_0 = \mu'_0 = 1,\ \mu_1 = 0$$

26. log is the natural logarithm

Distributions

1. Beta: p.d.f. = $f(x) = x^{v-1}(1-x)^{w-1}/B(v,w)$ $0 \le x \le 1$, $B(v,w) =$ beta function = $\int_0^1 u^{v-1}(1-u)^{w-1}du$; r^{th} moment about the origin: $\prod_{i=0}^{r-1}(v+i)(v+w+i)$; mean = $v/(v+w)$; variance = $vw/(v+w)^2(v+w+1)$; mode $= (v-1)/(v+w+2)$, $v > 1$, $w > 1$; coefficient of skewness: $[2(w-v)(v+w+1)^{1/2}]/[(v+w+2)(vw)^{1/2}]$; coefficient of kurtosis: $([3(v+w)(v+w+1)(v+1)(2w-v)]/[vw(v+w+2)(v+w+3)]) + [v(v-w)]/(v+w)$; coefficient of variation: $[w/[v(v+w+1)]]^{1/2}$; p.d.f. = $f(x) = [(v+w-1)!x^{v-1}(1-x)^{w-1}]/[(v-1)!(w-1)!]$, v and w integers; $B(v,w) = \Gamma(v)\Gamma(w)/\Gamma(v+w) = B(w,v)$, $\Gamma(c) = (c-1)\Gamma(c-1)$

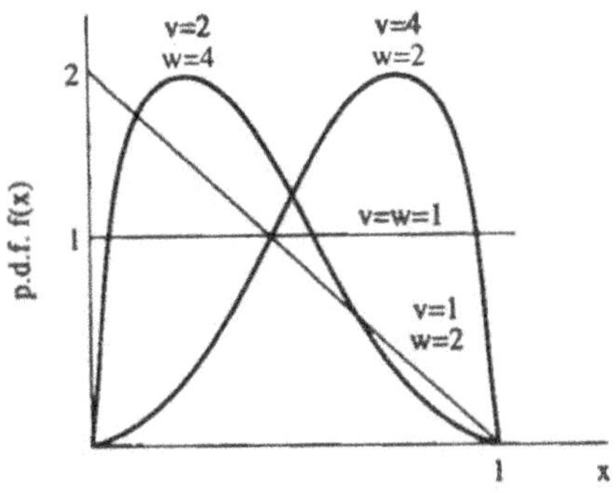

2. Binomial: n , p is the number of successes in n independent Bernoulli trials where the probability of success at each trial is p and the probability of failure is $q = 1-p$, $n =$ positive integer, $0 < p < 1$. c.d.f. = $F(x) = \sum_{i=0}^{x}\binom{n}{i}p^i q^{n-i}$, $x =$

integer; p.d.f. $= f(x) = \binom{n}{x} p^x q^{n-x}$, x = integer; moment generating function: $[p\exp(t)+q]^n$; probability generating function: $(pt+q)^n$; characteristic function: $\Phi(t) = [p\exp(jt)+q]^n$, moments about the origin: mean = np, second = $np(np + q)$, third = $np[(n-1)(n-2)p^2 + 3p(n-1)+1]$; moment about the mean: variance = npq, third = npq(q − p), fourth = $npq[1+3pq(n-2)]$; standard deviation: $(npq)^{1/2}$; mode: $p(n+1)-1 \le x \le p(n+1)$; coefficient of skewness: $(q-p)/(npq)^{1/2}$; coefficient of kurtosis: $3-(6/n)+(1/npq)$; factorial moments about the mean: second = npq, third = $-2npq(1+p)$; coefficient of variation = $(q/np)^{1/2}$

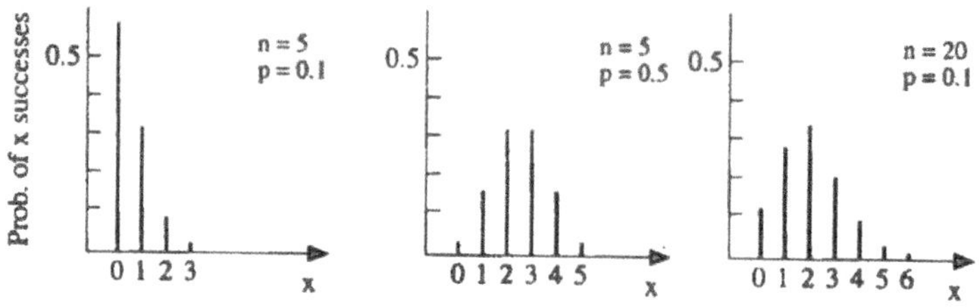

3. Cauchy: p.d.f. $= f(x) = 1/[\pi b[(x-a)/b]^2+1]]$, α = shift parameter, b = scale parameter, $-\infty < x < \infty$; mode = a; median = a.

4. Chi-Squared: p.d.f. $f(x) = [x^{(v-2)/2}\exp(-x/2)]/[2^{v/2}\Gamma(v/2)]$, v (shape parameter) = degrees of freedom, $0 \le x < \infty$; moment generating function: $(1-2t)^{-v/2}$, $t > 1/2$; characteristic function; $\Phi(t) = (1-2jt)^{-v/2}$; cumulant function: $(-v/2)\log(1-2jt)$; rth cumulant: $2^{r-1}v[(r-1)!]$; rth moment about the origin: $2^r \prod_{i=0}^{r-1} [i+(v/2)]$; mean = v; variance: $2v$; standard deviation $(2v)^{1/2}$; mode: $v-2$, $v \ge 2$; coefficient of skewness: $2^{3/2}v^{-1/2}$; coefficient of kurtosis: 3 + 12/v; coefficient of variation: $(2/v)^{1/2}$; Laplace transform of the p.d.f: $(1+2s)^{-v/2}$

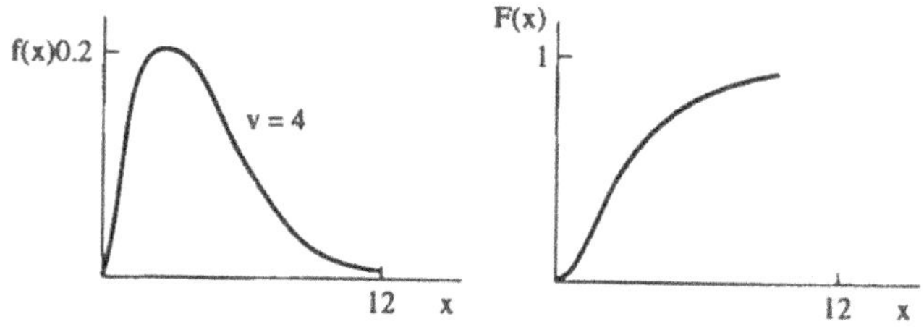

5. Discrete uniform: $a \le x \le a+b-1$, x = integer, a = lower limit of the range, b = scale parameter; c.d.f. = $F(x) = (x-a+1)/b$; p.d.f. $= f(x) = 1/b$; probability generating function: $(t^a - t^{a+b})/(1-t)$; characteristic function: $\exp[j(a-1)t]\sinh(jtb/2)\sinh(jt/2)/b$; mean: $a+(b-1)/2$; variance: $(b^2-1)/12$; coefficient of skewness 0; information content: $\log_2 b$.

6. Exponential: $0 \le x < \infty$, b = scale parameter = mean, $\lambda = 1/b$ = alternative parameter; c.d.f = $F(x) = 1-\exp(-x/b)$; p.d.f. = $f(x) = (1/b)\exp(-x/b)$; moment generating function: $1/(1-bt)$, $t > (1/b)$; Laplace transform of the p.d.f.: $1/(1+bs)$; characteristic function: $1/(1-jbt)$; cumulant function: $-\log(1-jbt)$; rth cumulant: $(r-1)!b^r$; rth moment about the origin: $r!b^r$; mean: b ; variance: b^2; standard deviation: b; mean deviation: 2 b/e (e base of natural log); mode: 0; median: b log 2; coefficient of skewness: 2; coefficient of kurtosis 9; coefficient of variation: 1; information content: $\log_2(eb)$.

7. F-distribution: $0 \le x < \infty$, v and w ≡ positive integers ≡ degrees of freedom; p.d.f = $f(x) = [\Gamma[\frac{1}{2}(v+w)](v/w)^{v/2} x^{(v-2)/2}]/[\Gamma(\frac{1}{2}v)\Gamma(\frac{1}{2}w)(1+xv/w)^{(v+w)/2}]$; rth moment about the origin: $[(w/v)^r\Gamma(\frac{1}{2}v+r)\Gamma(\frac{1}{2}w-r)]/[\Gamma(\frac{1}{2}v)\Gamma(\frac{1}{2}w)]$, $w > 2r$; mean: w/(w − 2), w > 2; variance: $[2w^2(v+w-2)]/[v(w-2)^2(w-4)]$, $w > 4$; mode: $[w(v-2)]/[v(w+2)]$, $v > 1$; coefficient of skewness: $[(2v+w-2)[8(w-4)]^{1/2}]/[(w-6)(v+w-2)^{1/2}]$, $w > 6$; coefficient of variation: $[[2(v+w-2)]/[v(w-4)]]^{1/2}$, $w > 4$.

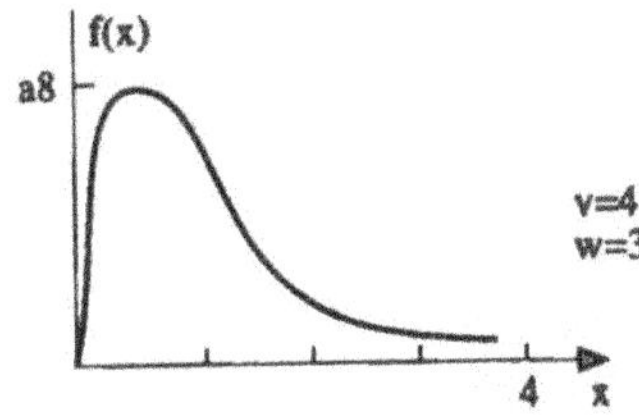

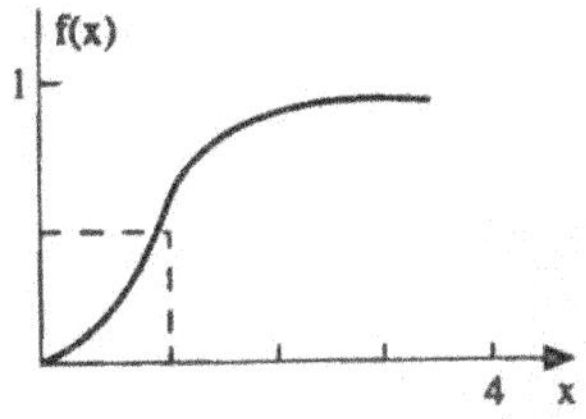

8. Gamma: $0 \le x < \infty$, $b =$ scale parameter > 0 (or $\lambda = 1/b$), $c > 0$ shaper parameter; p.d.f. $= f(x) = (x/b)^{c-1} [\exp(-x/b)]/[b\Gamma(c)], \Gamma(c) = \int_0^{\infty} \exp(-u)u^{c-1}du$; moment generating function: $(1-bt)^{-c}$, $t > 1/b$; Laplace transform of the p.d.f.: $(1+bs)^{-c}$; characteristic function: $(1-jbt)^{-c}$; cumulant function: $-c\log(1-jbt)$; r^{th} cumulant: $(r-1)!cb^r$; r^{th} moment about the origin: $b^r \prod_{i=0}^{r-1}(c+i)$; mean: bc; variance: b^2c; standard deviation: $b\sqrt{c}$; mode: $b(c-1)$, $c \ge 1$; coefficient of skewness: $2c^{-1/2}$; coefficient of kurtosis: 3 + 6/c; coefficient of variation: $c^{-1/2}$

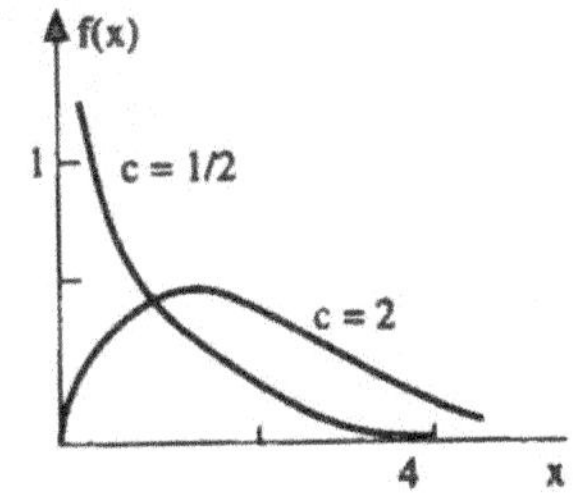

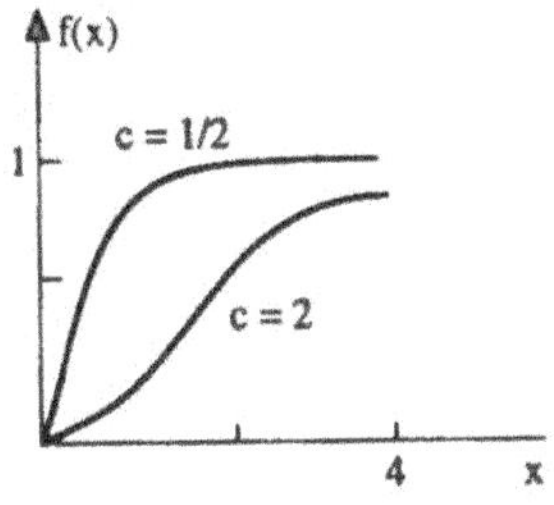

9. Lognormal: $0 \le x < \infty$, $m =$ scale parameter = median > 0, $\mu =$ mean of $\log X > 0, m = \exp\mu$, $\mu = \log m$, $\sigma =$ shape parameter = standard deviation of logX, $w = \exp(\sigma^2)$; p.d.f $= f(x) = [1/x\sigma(2\pi)^{1/2}]\exp[-\{\log(x/m)\}^2/2\sigma^2]$; r^{th} moment about the origin: $m^r \exp(r^2\sigma^2/2)$; mean: $m\exp(\sigma^2/2)$; variance: $m^2w(w-1)$; standard deviation: $m(w^2-w)^{1/2}$; mode: m/w; median; m; coefficient of skewness: $(w+2)(w-1)^{1/2}$; coefficient of kurtosis: $w^4+2w^3+3w^2-3$; coefficient of variation: $(w-1)^{1/2}$.

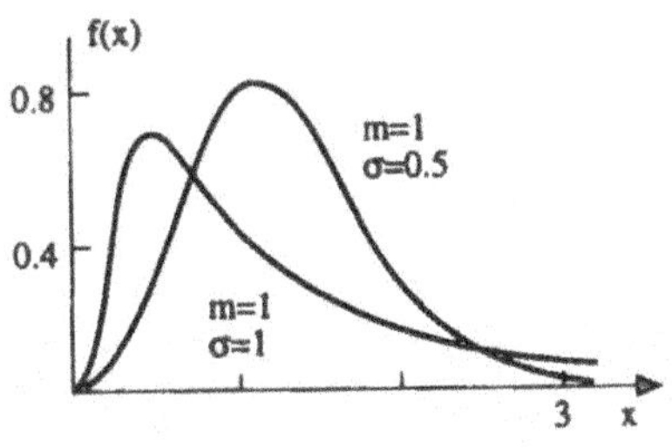

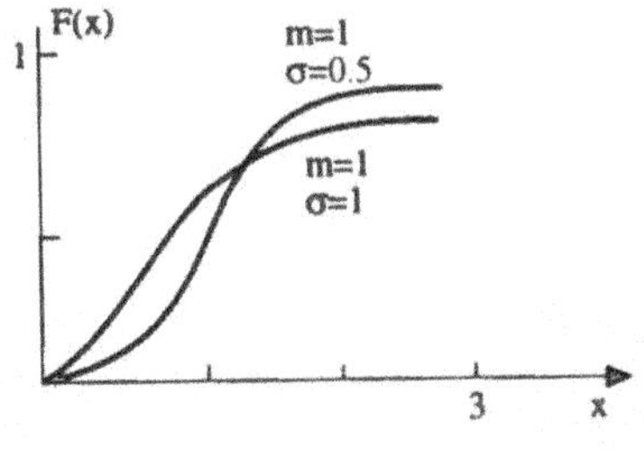

10. Negative bionomial: y = number of failures (integer), x = number of failures before x^{th} success in a sequence of Bernoulli trials; p = probability of success at each trial, $q = 1-p$, $0 \le y < \infty$, $0 < p < 1$; c.d.f. $= F(y) = \sum_{i=1}^{y} \binom{x+i-1}{i} p^x q^i$;

p.d.f. $= f(y) = \binom{x+y-1}{y} p^x q^y$; moment generating function: $p^x(1-q\exp t)^{-x}$; probability generating function: $p^x(1-qt)^{-x}$; characteristic function: $p^x[1-q\exp(jt)]^{-x}$; cumulant function: $x\log(p) - x\log(1-q\exp t)$; Cumulants: first $= xq/p$, second $= xq/p^2$, third $= xq(1+q)/p^3$, fourth $= xq(6q+p^2)/p^4$; mean: xq/p; Moments about the mean: variance $= xq/p^2$, third $= xq(1+q)/p^3$, fourth $= (xq/p^4)(3xq+6q+p^2)$; standard deviation: $(xq)^{1/2}/p$; coefficient of skewness:

$(1+q)(xq)^{-1/2}$; coefficient of kurtosis: $3+\frac{6}{x}+\frac{p^2}{xq}$; factorial moment generating function: $(1-q^t/p)^{-x}$; r^{th} factorial moment about the origin: $(q/p)^r(x+r-1)^r$; coefficient of variation: $(xq)^{-1/2}$.

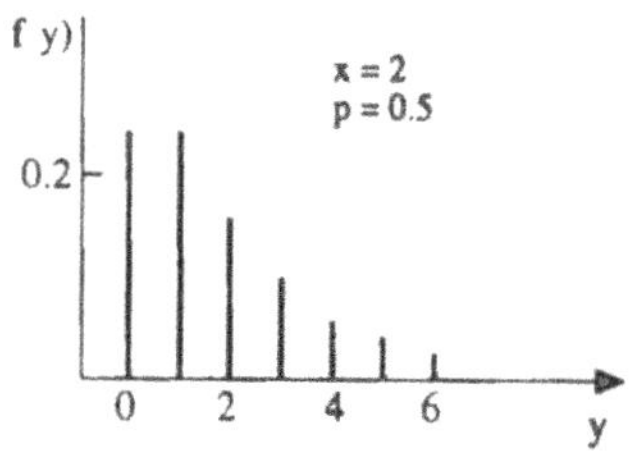

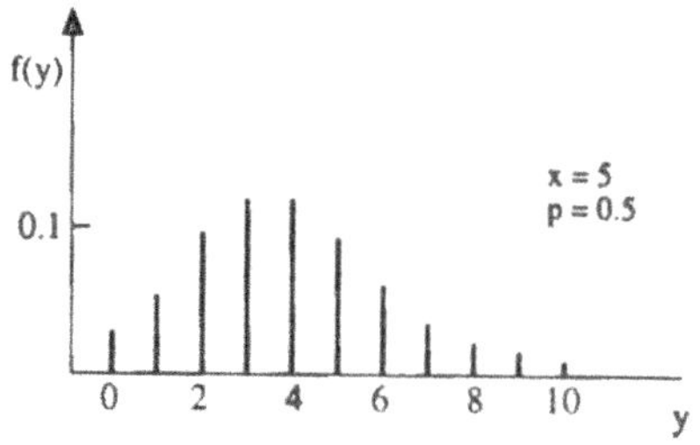

11. Normal: $-8<x<\infty$, $\mu=$ mean = location parameter, σ = standard deviation = scale parameter, $\sigma>0$; p.d.f. = $f(x)=[1/\sigma(2\pi)^{1/2}]\exp[-(x-\mu)^2/2\sigma^2]$; moment generating function: $\exp(\mu t+\frac{1}{2}\sigma^2t^2)$; characteristic function: $\exp(j\mu t-\frac{1}{2}\sigma^2t^2)$; cumulant function: $j\mu t-\frac{1}{2}\sigma^2t^2$; r^{th} cumulant: $K_2=\sigma^2$, $K_r=0$, $r>2$; mean: μ; r^{th} moment about the mean: $\mu_r=0$ for r odd, $\mu_r=(\sigma^r r!)/[2^{r/2}[(r/2)!]]$ for r even; variance: σ^2; standard deviation: σ; mean deviation: $\sigma(2/\pi)^{1/2}$; mode: μ; median: μ; coefficient of skewness: 0; coefficient of kurtosis: 3; information content: $\log_2[\sigma(2\pi e)^{1/2}]$

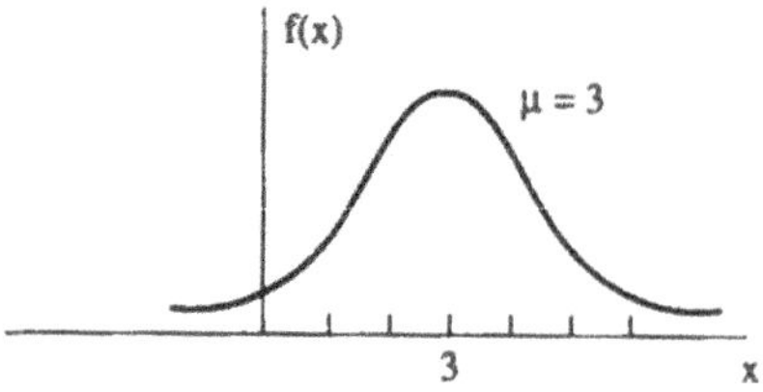

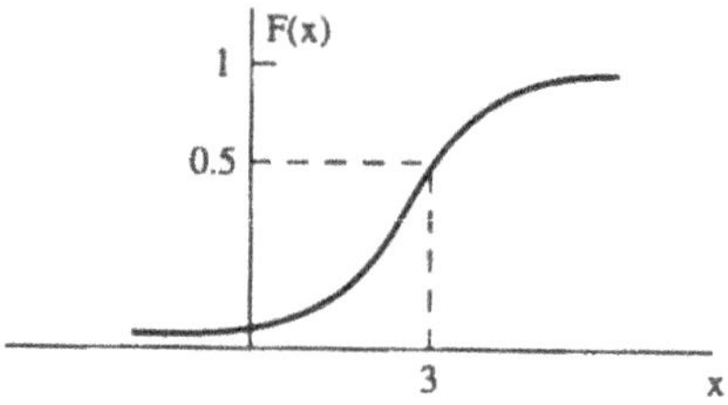

12. Pareto: $1\le x<\infty$, c = shape parameter; c.d.f. = $F(x)=1-x^{-c}$; p.d.f. = $f(x)=cx^{-c-1}$; r^{th} moment about the origin: $c/(c-r)$, $c>r$; mean : $c/(c-1), c>1$; variance: $[c/(c-2)]-[c/(c-1)]^2$, $c>2$; coefficient of variation: $(c-1)/[c(c-1)]^{1/2}$, $c>2$.

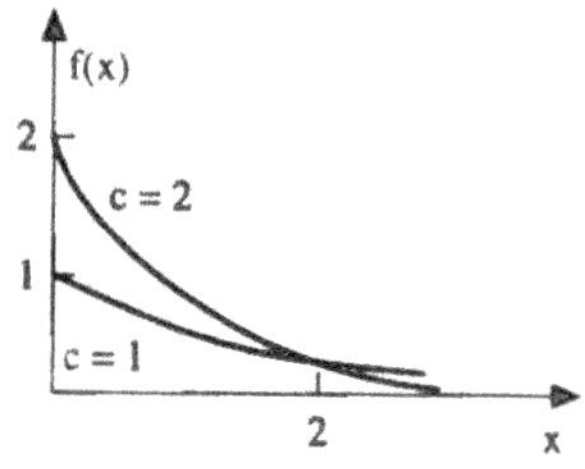

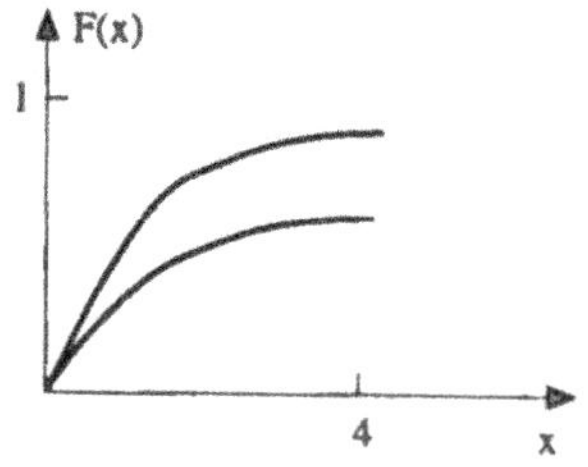

13. Pascal: n = number of trials, n ≥ 1, x = the Bernoulli success parameter = the number of trials up to and including the xth success, p = probability of success at each trial, $0<p<1$, $q=1-p$; p.d.f. = $f(n)=\binom{n-1}{n-x}p^xq^{n-x}$; moment generating function: $p^x\exp(tx)/(1-q\exp t)^x$; probability generating function: $(pt)^x/(1-qt)^x$; characteristic function: $p^x\exp(jtx)/(1-q\exp(jt))^x$; mean: x/p; variance: xq/p^2; standard deviation: $(xq)^{1/2}/p$; coefficient of variation: $(q/x)^{1/2}$.

14. Poisson: $0\le x<\infty$, λ = mean (a parameter); c.d.f. = $F(x)=\sum_{i=0}^{x}\lambda^i\exp(-\lambda)/i!$; p.d.f. = $f(x)=\lambda^x\exp(-\lambda)/x!$; moment generating function: $\exp[\lambda[\exp(t)-1]]$; probability generating function: $\exp[-\lambda(1-t)]$; characteristic function:

$\exp[\lambda[\exp(jt)-1]]$; cumulant function: $\lambda[\exp(t)-1]=\sum_{i=0}^{\infty} t^i/i!$; r^{th} cumulant: λ; moment about the origin: mean $=\lambda$, second $=\lambda+\lambda^2$, third $=\lambda[(\lambda+1)^2+\lambda]$, fourth $=\lambda(\lambda^3+6\lambda^2+7\lambda+1)$; r^{th} moment about the mean, μ_r: $\lambda\sum_{i=0}^{r-2}\binom{r-1}{i}\mu_i$, $r>1$, $\mu_0=1$.

Moments about the mean: variance $=\lambda$, third $=\lambda$, fourth $=\lambda(1+3\lambda)$, fifth $=\lambda(1+10\lambda)$, sixth $=\lambda(1+25\lambda+15\lambda^2)$; standard deviation $=\lambda^{1/2}$; coefficient of skewness: $\lambda^{-1/2}$; coefficient of kurtosis: $3+1/\lambda$; factorial moments about the mean: second $=\lambda$, third $=-2\lambda$, fourth $=3\lambda(\lambda+2)$; coefficient of variation: $\lambda^{-1/2}$

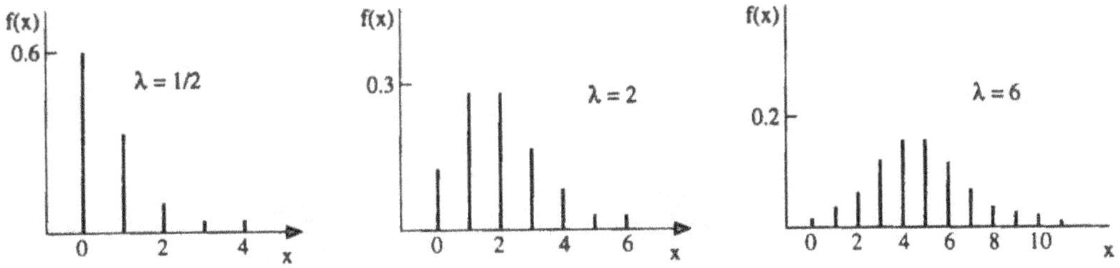

15. Rectangular: $a\le x\le a+b$, $x=$ range, a = lower limit, b = scale parameter; c.d.f. $=F(x)=(x-a)/b$; p.d.f. $=f(x)=1/b$; moment generating function: $\exp(at)[\exp(bt)-1]/bt$; Laplace transform of the p.d.f.: $\exp(-as)[1-\exp(-bs)]/bs$; characteristic function: $\exp(jat)[\exp(jbt)-1]/jbt$; mean: $a+b/2$; r^{th} moment about the mean: $\mu_r=0$ for r odd, $\mu_r=(b/2)^r/(r+1)$ for r even; variance: $b^2/12$; standard deviation: $b/\sqrt{12}$; mean deviation $b/4$; median $a+b/2$; standardized r^{th} moment about the mean: $\mu_r=0$ for r odd, $\mu_r=3^{r/2}/(r+1)$ for r even; coefficient of skewness: 0; coefficient of kurtosis: 9/5; coefficient of variation: $b/[3^{1/2}(2a+b)]$; information content: $\log_2 b$.

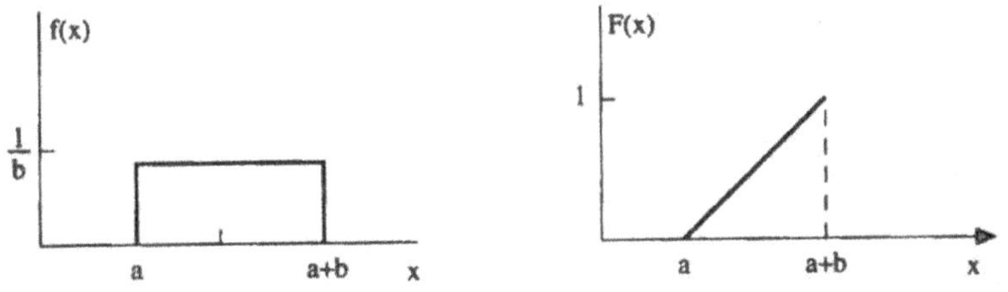

16. Student's t: $-\infty<x<\infty$, $v=$ shape parameter (degrees of freedom), $v\equiv$ positive integer; p.d.f. $=f(x)=[\Gamma[(v+1)/2][1+(x^2/v)]^{-(v+1)/2}]/[(\pi v)^{1/2}\Gamma(v/2)]$; mean: 0; r^{th} moment about the mean: $\mu_r=0$ for r odd, $\mu_r=[1\cdot3\cdot5\cdots(r-1)v^{r/2}]/[(v-2)(v-4)\cdots(v-r)]$ for r even, $r<v$; variance: $v/(v-2)$, $v>2$; mean deviation: $v^{1/2}\Gamma[\frac{1}{2}(v-1)]/\pi^{1/2}\Gamma(\frac{1}{2}v)$; mode: 0; coefficient of skewness and kurtosis: 0

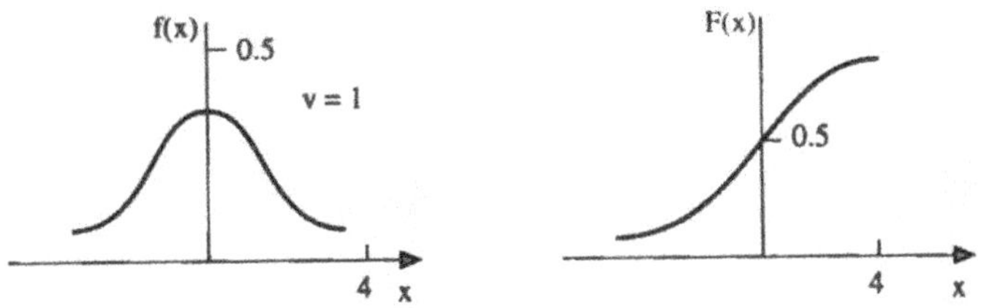

17. Weibull: $0\le x<\infty$, $b>0$ scale parameter, c = shape parameter c > 0; c.d.f. $=F(x)=1-\exp[-(x/b)^c]$; p.d.f. $=f(x)=(cx^{c-1}/b^c)\exp[-(x/b)^c]$; r^{th} moment abut the origin: $b^r\Gamma[(c+r)/c]$; mean: $b\Gamma[(c+1)/c]$.

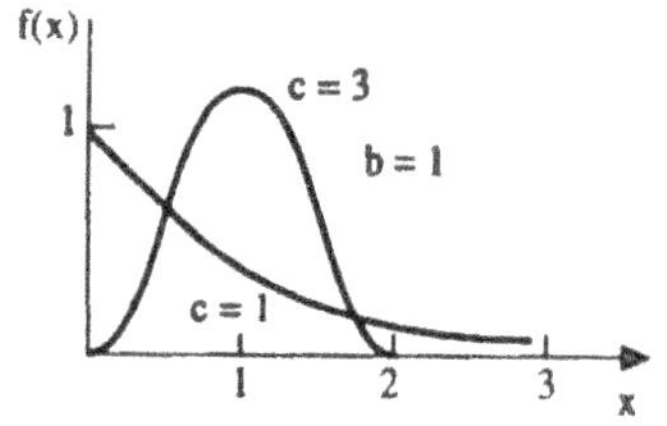

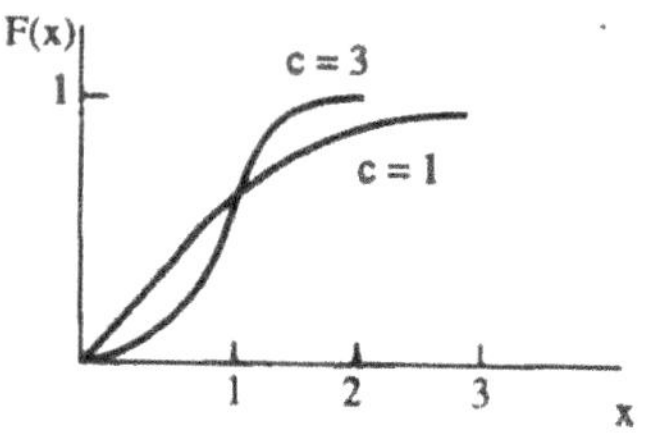

TABLE 34.2 Normal Distribution Tables.

$f(x)$ = distribution density $= (1/\sqrt{2\pi})e^{-x^2/2}$, $F(x)$ = cumulative distribution function $= \int_{-\infty}^{x} (1/\sqrt{2\pi})e^{-\tau^2/2}\,d\tau$,

$f'(x) = -xf(x)$, $f''(x) = (x^2-1)f(x)$, $F(-x) = 1 - F(x)$, $P\{-x < X < x\} = F(x) - F(-x) = 2F(x) - 1$

x	$F(x)$	$f(x)$	$f'(x)$	$f''(x)$	x	$F(x)$	$f(x)$	$f'(x)$	$f''(x)$
.00	.5000	.3989	−.0000	−.3989	.50	.6915	.3521	−.1760	−.2641
.01	.5040	.3989	−.0040	−.3989	.51	.6950	.3503	−.1787	−.2592
.02.	.5080	.3989	−.0080	−.3987	.52	.6985	.3485	−.1812	−.2543
.03	.5120	.3988	−.0120	−.3984	.53	.7019	.3467	−.1837	−.2493
.04	.5160	.3986	−.0159	−.3980	.54	.7054	.3448	−.1862	−.2443
.05	.5199	.3984	−.0199	−.3975	.55	.7088	.3429	−.1886	−.2392
.06	.5239	.3982	−.0239	−.3968	.56	.7123	.3410	−.1920	−.2341
.07	.5279	.3980	−.0279	−.3960	.57	.7157	.3391	−.1933	−.2289
.08	.5319	.3977	−.0318	−.3951	.58	.7190	.3372	−.1956	−.2238
.09	.5359	.3973	−.0358	−.3941	.59	.7224	.3352	−.1978	−.2185
.10	.5398	.3970	−.0397	−.3930	.60	.7257	.3332	−.1999	−.2133
.11	.5438	.3965	−.0436	−.3917	.61	.7291	.3312	−.2020	−.2080
.12	.5478	.3961	−.0475	−.3904	.62	.7324	.3292	−.2041	−.2027
.13	.5517	.3956	−.0514	−.3889	.63	.7357	.3271	−.2061	−.1973
.14	.5557	.3951	−.0553	−.3873	.64	.7389	.3251	−.2080	−.1919
.15	.5596	.3945	−.0592	−.3856	.65	.7422	.3230	−.2099	−.1865
.16	.5636	.3939	−.0630	−.3838	.66	.7454	.3209	−.2118	−.1811
.17	.5675	.3932	−.0668	−.3819	.67	.7486	.3187	−.2136	−.1757
.18	.5714	.3925	−.0707	−.3798	.68	.7517	.3166	−.2153	−.1702
.19	.5753	.3918	−.0744	−.3777	.69	.7549	.3144	−.2170	−.1647
.20	.5793	.3910	−.0782	−.3754	.70	.7580	.3132	−.2186	−.1593
.21	.5832	.3902	−.0820	−.3730	.71	.7611	.3101	−.2201	−.1538
.22	.5871	.3894	−.0857	−.3706	.72	.7642	.3079	−.2217	−.1483
.23	.5910	.3885	−.0894	−.3680	.73	.7673	.3056	−.2231	−.1428
.24	.5948	.3876	−.0930	−.3653	.74	.7704	.3034	−.2245	−.1373
.25	.5987	.3867	−.0967	−.3625	.75	.7734	.3011	−.2259	−.1318
.26	.6026	.3857	−.1003	−.3596	.76	.7764	.2989	−.2271	−.1262
.27	.6064	.3847	−.1039	−.3566	.77	.7794	.2966	−.2284	−.1207
.28	.6103	.3836	−.1074	−.3535	.78	.7823	.2943	−.2296	−.1153
.29	.6141	.3825	−.1109	−.3504	.79	.7852	.2920	−.2307	−.1098
.30	.6179	.3814	−.1144	−.3471	.80	.7881	.2897	−.2318	−.1043
.31	.6217	.3802	−.1179	−.3437	.81	.7910	.2874	−.2328	−.0988
.32	.6255	.3790	−.1213	−.3402	.82	.7939	.2850	−.2337	−.0934
.33	.6293	.3778	−.1247	−.3367	.83	.7967	.2827	−.2346	−.0880
.34	.6331	.3765	−.1280	−.3330	.84	.7995	.2803	−.2355	−.0825

TABLE 34.2 Normal Distribution Tables. (continued)

$f(x) = \text{distribution density} = (1/\sqrt{2\pi})e^{-x^2/2}$, $F(x) = \text{cumulative distribution function} = \int_{-\infty}^{x} (1/\sqrt{2\pi})e^{-t^2/2}\,dt$,

$f'(x) = -xf(x)$, $f''(x) = (x^2-1)f(x)$, $F(-x) = 1 - F(x)$, $P\{-x < X < x\} = F(x) - F(-x) = 2F(x) - 1$

x	$F(x)$	$f(x)$	$f'(x)$	$f''(x)$	x	$F(x)$	$f(x)$	$f'(x)$	$f''(x)$
.35	.6368	.3752	−.1313	−.3293	.85	.8023	.2780	−.2363	−.0771
.36	.6406	.3739	−.1346	−.3255	.86	.8051	.2756	−.2370	−.0718
.37	.6443	.3725	−.1378	−.3216	.87	.8078	.2732	−.2377	−.0664
.38	.6480	.3712	−.1410	−.3176	.88	.8106	.2709	−.2384	−.0611
.39	.6517	.3697	−.1442	−.3135	.89	.8133	.2685	−.2389	−.0558
.40	.6554	.3683	−.1473	−.3094	.90	.8159	.2661	−.2395	−.0506
.41	.6591	.3668	−.1504	−.3051	.91	.8186	.2637	−.2400	−.0453
.42	.6628	.3653	−.1534	−.3008	.92	.8212	.2613	−.2404	−.0401
.43	.6664	.3637	−.1564	−.2965	.93	.8238	.2589	−.2408	−.0350
.44	.6700	.3621	−.1593	−.2920	.94	.8264	.2565	−.2411	−.0299
.45	.6736	.3605	−.1622	−.2875	.95	.8289	.2541	−.2414	−.0248
.46	.6772	.3589	−.1651	−.2830	.96	.8315	.2516	−.2416	−.0197
.47	.6808	.3572	−.1679	−.2783	.97	.8340	.2492	−.2417	−.0147
.48	.6844	.3555	−.1707	−.2736	.98	.8365	.2468	−.2419	−.0098
.49	.6879	.3538	−.1734	−.2689	.99	.8389	.2444	−.2420	−.0049
.50	.6915	.3521	−.1760	−.2641	1.00	.8413	.2420	−.2420	−.0000
1.00	.8413	.2420	−.2420	.0000	1.50	.9332	.1295	−.1943	.1619
1.01	.8438	.2396	−.2420	.0048	1.51	.9345	.1276	−.1927	.1633
1.02	.8461	.2371	−.2419	.0096	1.52	.9357	.1257	−.1910	.1647
1.03	.8485	.2347	−.2418	.0143	1.53	.9370	.1238	−.1894	.1660
1.04	.8508	.2323	−.2416	.0190	1.54	.9382	.1219	−.1877	.1672
1.05	.8531	.2299	−.2414	.0236	1.55	.9394	.1200	−.1860	.1683
1.06	.8554	.2275	−.2411	.0281	1.56	.9406	.1182	−.1843	.1694
1.07	.8577	.2251	−.2408	.0326	1.57	.9418	.1163	−.1826	.1704
1.08	.8599	.2227	−.2405	.0371	1.58	.9429	.1145	−.1809	.1714
1.09	.8621	.2203	−.2401	.0414	1.59	.9441	.1127	−.1792	.1722
1.10	.8643	.2179	−.2396	.0458	1.60	.9452	.1109	−.1775	.1730
1.11	.8665	.2155	−.2392	.0500	1.61	.9463	.1092	−.1757	.1738
1.12	.8686	.2131	−.2386	.0542	1.62	.9474	.1074	−.1740	.1745
1.13	.8708	.2107	−.2381	.0583	1.63	.9484	.1057	−.1723	.1751
1.14	.8729	.1083	−.2375	.0624	1.64	.9495	.1040	−.1705	.1757
1.15	.8749	.2059	−.2368	.0664	1.65	.9505	.1023	−.1687	.1762
1.16	.8770	.2036	−.2361	.0704	1.66	.9515	.1006	−.1670	.1766
1.17	.8790	.2012	−.1354	.0742	1.67	.9525	.0989	−.1652	.1770
1.18	.8810	.1989	−.2347	.0780	1.68	.9535	.0973	−.1634	.1773
1.19	.8830	.1965	−.2339	.0818	1.69	.9545	.0957	−.1617	.1776
1.20	.8849	.1942	−.2330	.0854	1.70	.9554	.0940	−.1599	.1778
1.21	.8869	.1919	−.2322	.0890	1.71	.9564	.0925	−.1581	.1779
1.22	.8888	.1895	−.2312	.0926	1.72	.9573	.0909	−.1563	.1780
1.23	.8907	.1872	−.2303	.0960	1.73	.9582	.0893	−.1546	.1780
1.24	.8925	.1849	−.2293	.0994	1.74	.9591	.0878	−.1528	.1780
1.25	.8944	.1826	−.2283	.1027	1.75	.9599	.0863	−.1510	.1780
1.26	.8962	.1804	−.2273	.1060	1.76	.9608	.0848	−.1492	.1778
1.27	.8980	.1781	−.2262	.1092	1.77	.9616	.0833	−.1474	.1777

TABLE 34.2 Normal Distribution Tables. (continued)

$f(x)$ = distribution density $= (1/\sqrt{2\pi})e^{-x^2/2}$, $F(x)$ = cumulative distribution function $= \int_{-\infty}^{x}(1/\sqrt{2\pi})e^{-t^2/2}\,dt$,

$f'(x) = -xf(x)$, $f''(x) = (x^2-1)f(x)$, $F(-x) = 1 - F(x)$, $P\{-x < X < x\} = F(x) - F(-x) = 2F(x) - 1$

x	$F(x)$	$f(x)$	$f'(x)$	$f''(x)$	x	$F(x)$	$f(x)$	$f'(x)$	$f''(x)$
1.28	.8997	.1758	−.2251	.1123	1.78	.9625	.0818	−.1457	.1774
1.29	.9015	.1736	−.2240	.1153	1.79	.9633	.0804	−.1439	.1772
1.30	.9032	.1714	−.2228	.1182	1.80	.9641	.0790	−.1421	.1769
1.31	.9049	.1691	−.2216	.1211	1.81	.9649	.0775	−.1403	.1765
1.32	.9066	.1669	−.2204	.1239	1.82	.9556	.0761	−.1386	.1761
1.33	.9082	.1647	−.2191	.1267	1.83	.9664	.0748	−.1368	.1756
1.34	.9099	.1626	−.2178	.1293	1.84	.9671	.0734	−.1351	.1751
1.35	.9115	.1604	−.2165	.1319	1.85	.9678	.0721	−.1333	.1746
1.36	.9131	.1582	−.2152	.1344	1.86	.9686	.0707	−.1316	.1740
1.37	.9147	.1561	−.2138	.1369	1.87	.9693	.0694	−.1298	.1734
1.38	.9162	.1539	−.2125	.1392	1.88	.9699	.0681	−.1281	.1727
1.39	.9177	.1518	−.2110	.1415	1.89	.9706	.0689	−.1264	.1720
1.40	.9192	.1479	−.2096	.1437	1.90	.9713	.0656	−.1247	.1713
1.41	.9207	.1476	−.2082	.1459	1.91	.9719	.0644	−.1230	.1705
1.42	.9222	.1456	−.2067	.1480	1.92	.9726	.0632	−.1213	.1697
1.43	.9236	.1435	−.2052	.1500	1.93	.9732	.0620	−.1196	.1688
1.44	.9251	.1415	−.2037	.1519	1.94	.9738	.0608	−.1179	.1679
1.45	.9265	.1394	−.2022	.1537	1.95	.9744	.0596	−.1162	.1670
1.46	.9279	.1374	−.2006	.1555	1.96	.9750	.0584	−.1145	.1661
1.47	.9306	.1354	−.1991	.1572	1.97	.9756	.0573	−.1129	.1651
1.48	.9306	.1334	−.1975	.1588	1.98	.9761	0.562	−.1112	.1641
1.49	.9319	.1315	−.1959	.1604	1.99	.9767	.0551	−.1096	.1630
1.50	.9332	.1295	−.1943	.1619	2.00	.9772	.0540	−.1080	.1620
2.00	.9773	.0540	−.1080	.1620	2.50	.9938	.0175	−.0438	.0920
2.01	.9778	.0529	−.1064	.1609	2.51	.9940	.0171	−.0429	.0906
2.02	.9783	.0519	−.1048	.1598	2.52	.9941	.0167	−.0420	.0892
2.03	.9788	.0508	−.1032	.1586	2.53	.9943	.0163	−.0411	.0868
2.04	.9793	.0498	−.1016	.1575	2.54	.9945	.0158	−.0403	.0878
2.05	.9798	.0488	−.1000	.1563	2.55	.9946	.0155	−.0394	.0850
2.06	.9803	.0478	−.0985	.1550	2.56	.9948	.0151	−.0386	.0836
2.07	.9809	.0468	−.0969	.1538	2.57	.9949	.0147	−.0377	.0823
2.08	.9812	.0459	−.0954	.1526	2.58	.9951	.0143	−.0369	.0809
2.09	.9817	.0449	−.0939	.1513	2.59	.9952	.0139	−.0361	.0796
2.10	.9821	.0440	−.0924	.1500	2.60	.9953	.0136	−.0353	.0782
2.11	.9826	.0431	−.0909	.1487	2.61	.9955	.0132	.0345	.0769
2.12	.9830	.0422	−.0894	.1474	2.62	.9956	.0129	−.0338	.0756
2.13	.9834	.0413	−.0879	.1460	2.63	.9957	.0126	−.0330	.0743
2.14	.9838	.0404	−.0865	.1446	2.64	.9959	.0122	−.0323	.0730
2.15	.9842	.0396	−.0850	.1433	2.65	.9960	.0119	−.0316	.0717
2.16	.9846	.0387	−.0836	.1419	2.66	.9961	.0116	−.0309	.0705
2.17	.9850	.0379	−.0822	.1405	2.67	.9962	.0113	−.0302	.0692
2.18	.9854	.0371	−.0808	.1391	2.68	.9963	.0110	−.0295	.0680
2.19	.9857	.0363	−.0794	.1377	2.69	.9964	.0107	−.0288	.0668

TABLE 34.2 Normal Distribution Tables. (continued)

$f(x)$ = distribution density $= (1/\sqrt{2\pi})e^{-x^2/2}$, $F(x)$ = cumulative distribution function $= \int_{-\infty}^{x} (1/\sqrt{2\pi})e^{-\tau^2/2}\,d\tau$,

$f'(x) = -xf(x)$, $f''(x) = (x^2-1)f(x)$, $F(-x) = 1 - F(x)$, $P\{-x < X < x\} = F(x) - F(-x) = 2F(x) - 1$

x	$F(x)$	$f(x)$	$f'(x)$	$f''(x)$	x	$F(x)$	$f(x)$	$f'(x)$	$f''(x)$
2.20	.9861	.0355	−.0780	.1362	2.70	.9965	.0104	−.0281	.0656
2.21	.9864	.0347	−.0767	.1348	2.71	.9966	.0101	−.0275	.0644
2.22	.9868	.0339	−.0754	.1333	2.72	.9967	.0099	−.0269	.0632
2.23	.9871	.0332	−.0740	.1319	2.73	.9968	.0096	−.0262	.0620
2.24	.9875	.0325	−.0727	.1304	2.74	.9969	.0093	−.0256	.0608
2.25	.9868	.0317	−.0714	.1289	2.75	.9970	.0091	−.0250	.0597
2.26	.9881	.0310	−.0701	.1275	2.76	.9971	.0088	−.0244	.0585
2.27	.9884	.0303	−.0689	.1260	2.77	.9972	.0086	−.0238	.0574
2.28	.9887	.0297	−.0676	.1245	2.78	.9973	.0084	−.0233	.0563
2.29	.9890	.0290	−.0664	.1230	2.79	.9974	.0081	−.0227	.0562
2.30	.9893	.0283	−.0652	.1215	2.80	.9974	.0079	−.0222	.0541
2.31	.9896	.0277	−.0639	.1200	2.81	.9975	.0077	−.0216	.0531
2.32	.9898	.0270	−.0628	.1185	2.82	.9976	.0075	−.0211	.0520
2.33	.9901	.0264	−.0616	.1170	2.83	.9977	.0073	−.0206	.0510
2.34	.9904	.2058	−.0604	.1155	2.84	.9977	.0071	−.0201	.0500
2.35	.9906	.0252	−.0593	.1141	2.85	.9978	.0069	−.0196	.0490
2.36	.9909	.0246	−.0581	.1126	2.86	.9979	.0067	−.0191	.0480
2.37	.9911	.0241	−.0570	.1111	2.87	.9979	.0065	−.0186	.0470
2.38	.9913	.0235	−.0559	.1096	2.88	.9980	.0063	−.0182	.0460
2.39	.9916	.0229	−.0548	.1081	2.89	.9981	.0061	−.0177	.0451
2.40	.9918	.0224	−.0538	.1066	2.90	.9981	.0060	−.0173	.0441
2.41	.9920	.0219	−.0527	.1051	2.91	.9982	.0058	−.0168	.0432
2.42	.9922	.0213	−.0516	.1036	2.92	.9982	.0056	−.0164	.0423
2.43	.9925	.0208	−.0506	.1022	2.93	.9983	.0055	−.0160	.0414
2.44	.9927	.0203	−.0496	.1007	2.94	.9984	.0053	−.0156	.0405
2.45	.9929	.0198	−.0486	.0992	2.95	.9984	.0051	−.0152	.0396
2.46	.9931	.0194	−.0476	.0978	2.96	.9985	.0050	−.0148	.0388
2.47	.9932	.0189	−.0467	.0963	2.97	.9985	.0048	−.0144	.0379
2.48	.9934	.0184	−.0457	.0949	2.98	.9986	.0047	−.0140	.0371
2.49	.9936	.0180	−.0448	.0935	2.99	.9986	.0046	−.0137	.0363
2.50	.9938	.0175	−.0438	.0920	3.00	.9987	.0044	−.0133	.0355
3.00	.9987	.0044	−.0133	.0355	3.50	.9998	.0009	−.0031	.0098
3.05	.9989	.0038	−.0116	.0316	3.55	.9998	.0007	−.0026	.0085
3.10	.9990	.0033	−.0101	.0281	3.60	.9998	.0006	−.0022	.0073
3.15	.9992	.0028	−.0088	.0249	3.65	.9999	.0005	−.0019	.0063
3.20	.9993	.0024	−.0076	.0220	3.70	.9999	.0004	−.0016	.0054
3.25	.9994	.0020	−.0066	.0194	3.75	.9999	.0004	−.0013	.0046
3.30	.9995	.0017	−.0057	.0170	3.80	.9999	.0003	−.0011	.0039
3.35	.9996	.0015	−.0049	.0149	3.85	.9999	.0002	−.0009	.0033
3.40	.9997	.0012	−.0042	.0130	3.90	1.0000	.0002	−.0008	.0028
3.45	.9997	.0010	−.0036	.0113	3.95	1.0000	.0002	−.0006	.0024
3.50	.9998	.0009	−.0031	.0098	4.00	1.0000	.0001	−.0005	.0020

TABLE 34.3 Student t-Distribution Table

$$F(x) = \int_{-\infty}^{x} \frac{\Gamma\left(\frac{n+1}{2}\right)}{\sqrt{n\pi}\,\Gamma(n/2)}\left(1+\frac{y^2}{n}\right)^{-(n+1)/2} dy$$

n = number of degrees of freedom, numbers give x of distribution, e.g., for n = 6 and $F = 0.975,\ x = 2.447,\ F(-x) = 1 - F(x)$

n\F	.60	.75	.90	.95	.975	.99	.995	.9995
1	.325	1.000	3.078	6.314	12.706	31.821	63.657	636.619
2	.289	.816	1.886	2.920	4.303	6.965	9.925	31.598
3	.277	.765	1.638	2.353	3.182	4.541	5.841	12.924
4	.271	.741	1.533	2.132	2.776	3.747	4.604	8.610
5	.267	.727	1.476	2.015	2.571	3.365	4.032	6.869
6	.265	.718	1.440	1.943	2.447	3.143	3.707	5.959
7	.263	.711	1.415	1.895	2.365	2.998	3.499	5.408
8	.262	.706	1.397	1.860	2.306	2.896	3.555	5.041
9	.261	.703	1.383	1.833	2.262	2.821	3.250	4.781
10	.260	.700	1.372	1.812	2.228	2.764	3.169	4.587
11	.260	.697	1.363	1.796	2.201	2.718	3.106	4.437
12	.259	.695	1.356	1.782	2.179	2.681	3.055	4.318
13	.259	.694	1.350	1.771	2.160	2.650	3.012	4.221
14	.258	.692	1.345	1.761	2.145	2.624	2.977	4.140
15	.258	.691	1.341	1.753	2.131	2.602	2.947	4.073
16	.258	.690	1.337	1.746	2.120	2.583	2.921	4.015
17	.257	.689	1.333	1.740	2.110	2.567	2.898	3.965
18	..257	.688	1.330	1.734	2.101	2.552	2.878	3.922
19	.257	.688	1.328	1.729	2.093	2.539	2.861	3.883
20	.257	.687	1.325	1.725	2.086	2.528	2.845	3.850
21	.257	.686	1.323	1.721	2.080	2.518	2.831	3.819
22	.256	.686	1.321	1.717	2.074	2.508	2.819	3.792
23	.256	.685	1.319	1.714	2.069	2.500	2.807	3.767
24	.256	.685	1.318	1.711	2.064	2.492	2.797	3.745
25	.256	.684	1.316	1.708	2.060	2.485	2.787	3.725
26	.256	.684	1.315	1.706	2.056	2.479	2.779	3.707
27	.256	.684	1.314	1.703	2.052	2.473	2.771	3.690
28	.256	.683	1.313	1.701	2.048	2.467	2.763	3.674
29	.256	.683	1.311	1.699	2.045	2.462	2.756	3.659
30	.256	.683	1.310	1.697	2.042	2.457	2.750	3.646
40	.255	.681	1.303	1.684	2.201	2.423	2.704	3.551
60	.254	.679	1.296	1.671	2.000	2.390	2.660	3.460
120	.254	.677	1.289	1.658	1.980	2.358	2.617	3.373
∞	.253	.674	1.282	1.645	1.960	2.326	2.576	3.291

34.4.6 Conditional Distribution

$$F_X(x|M) = P\{X \le x|M\} = \frac{P\{X \le x, M\}}{P\{M\}},$$

$\{X \le x, M\}$ = event of all outcomes ζ such that $X(\zeta) \le x$ and $\zeta \in M$

1. $F(\infty|M) = 1,\ F(-\infty|M) = 0$

2. $F(x_2|M) - F(x_1|M) = P\{x_1 < X \le x_2|M\} = \dfrac{P\{x_1 < X \le x_2, M\}}{P\{M\}}$

TABLE 34.4 The Chi-Squared Distribution

$$F(x) = \int_0^x \frac{1}{2^{n/2} F(n/2)} y^{(n-2)/2} e^{-y/2}\, dy$$

n = number of degrees of freedom

n \ F	.005	.010	.025	.050	.100	.250	.500	.750	.900	.950	.975	.990	.995
1	.0000393	.000157	.000982	.00393	.0158	.102	.455	1.32	2.71	3.84	5.02	6.63	7.88
2	.0100	.0201	.0506	.103	.211	.575	1.39	2.77	4.61	5.99	7.38	9.21	10.6
3	.0717	.115	.216	.352	.584	1.21	2.37	4.11	6.25	7.81	9.35	11.3	12.8
4	.207	.297	.484	.711	1.06	1.92	3.36	5.39	7.78	9.49	11.1	13.3	14.9
5	.412	.554	.831	1.15	1.61	2.67	4.35	6.63	9.24	11.1	12.8	15.1	16.7
6	.676	.872	1.24	2.20	2.20	3.45	5.35	7.84	10.6	12.6	14.4	16.8	18.5
7	.989	1.24	1.69	2.17	2.83	4.25	6.35	9.04	12.0	14.1	16.0	18.5	20.3
8	1.34	1.65	2.18	2.73	3.49	5.07	7.34	10.2	13.4	15.5	17.5	20.1	22.0
9	1.73	2.09	2.70	3.33	4.17	5.90	8.34	11.4	14.7	16.9	19.0	21.7	23.6
10	2.16	2.56	3.25	3.94	4.87	6.74	9.34	12.5	16.0	18.3	20.5	23.2	25.2
11	2.60	3.05	3.82	4.57	5.58	7.58	10.3	13.7	17.3	19.7	21.9	24.7	26.8
12	3.07	3.57	4.40	5.23	6.30	8.44	11.3	14,8	18.5	21.0	23.3	26.2	28.3
13	3.57	4.11	5.01	5.89	7.04	9.30	12.3	16.0	19.8	22.4	24.7	27.7	29.8
14	4.07	4.66	5.63	6.57	7.79	10.2	13.3	17.1	21.1	23.7	26.1	29.1	31.3
15	4.60	5.23	6.26	7.26	8.55	11.0	14.3	18.2	22.3	25.0	27.5	30.6	32.8
16	5.14	5.81	6.91	7.96	9.31	11.9	15.3	19.4	23.5	26.3	28.8	32.0	34.3
17	5.70	6.41	7.56	8.67	10.1	12.8	16.3	20.5	24.8	27.6	30.2	33.4	35.7
18	6.26	7.01	8.23	9.39	10.9	13.7	17.3	21.6	26.0	28.9	31.5	34.8	37.2
19	6.84	7.63	8.91	10.1	11.7	14.6	18.3	22.7	27.2	30.1	32.9	36.2	38.6
20	7.43	8.26	9.59	10.9	12.4	15.5	19.3	23.8	28.4	31.4	34.2	37.6	40.0
21	8.03	8.90	10.3	11.6	13.2	16.3	20.3	24.9	29.6	32.7	35.5	38.9	41.4
22	8.64	9.94	11.0	12.3	14.0	17.2	21.3	26.0	30.8	33.9	36.8	40.3	42.8
23	9.26	10.2	11.7	13.1	14.8	18.1	22.3	27.1	32.0	35.2	38.1	41.6	44.2
24	9.89	10.9	12.4	13.8	15.7	19.0	23.3	28.2	33.2	36.4	39.4	43.0	45.6
25	10.5	11.5	13.1	14.6	16.5	19.9	24.3	29.3	34.4	37.7	40.6	44.3	46.9
26	11.2	12.2	13.8	15.4	17.3	20.8	25.3	30.4	35.6	38.9	41.9	45.6	48.3
27	11.8	12.9	14.6	16.2	18.1	21.7	26.3	31.5	36.7	40.1	43.2	47.0	49.6
28	12.5	13.6	15.3	16.9	18.9	22.7	27.3	32.6	37.9	41.3	44.5	48.3	51.0
29	13.1	14.3	16.0	17.7	19.8	23.6	28.3	33.7	39.1	42.6	45.7	49.6	52.3
30	13.8	15.0	16.8	18.5	20.6	24.5	29.3	34.8	40.3	43.8	47.0	50.9	53.7

TABLE 34.5 The F Distribution

$$F(f)=p\{F\le f\}=\int_0^f \frac{\Gamma[(r_1+r_2)/2](r_1/r_2)^{r_1/2}x^{(r_1/2)-1}}{\Gamma(r_1/2)\Gamma(r_2/2)[1+(r_1x/r_2)]^{(r_1+r_2)/2}}dx$$

$P\{F\le f\}=0.95$

$r_2 \backslash r_1$	1	2	3	4	5	6	7	8	9	10	12	15	20	24	30	40	60
1	161.4	199.5	215.7	224.6	230.2	234.0	236.8	238.9	240.5	241.9	243.9	245.9	248.0	249.1	250.1	251.1	262.2
2	18.51	19.00	19.16	19.25	19.30	19.33	19.35	19.37	19.38	19.40	19.41	19.43	19.45	19.45	19.46	19.47	19.48
3	10.13	9.55	9.28	9.12	9.01	8.94	8.89	8.85	8.81	8.79	8.74	8.70	8.66	8.64	8.62	8.59	8.57
4	7.71	6.94	6.59	6.39	6.26	6.16	6.09	6.04	6.00	5.96	5.91	5.86	5.80	5.77	5.74	5.72	5.69
5	6.61	5.79	5.41	5.19	5.05	4.95	4.88	4.82	4.71	4.74	4.68	4.62	4.56	4.53	4.50	4.46	4.43
6	5.99	5.14	4.76	4.53	4.39	4.28	4.21	4.15	4.10	4.06	4.00	3.94	3.87	3.84	3.81	3.77	3.74
7	5.59	4.74	4.35	4.12	3.97	3.87	3.79	3.73	3.68	3.64	3.57	3.51	3.44	3.41	3.38	3.34	3.30
8	5.32	4.46	4.07	3.84	3.69	3.58	3.50	3.44	3.39	3.35	3.28	3.22	3.15	3.12	3.08	3.04	3.01
9	5.12	4.26	3.86	3.63	3.48	3.37	3.29	3.23	3.18	3.14	3.07	3.01	2.94	2.90	2.86	2.83	2.79
10	4.96	4.10	3.71	3.48	3.33	3.22	3.14	3.07	3.02	2.98	2.91	2.85	2.77	2.74	2.70	2.66	2.62
11	4.84	3.98	3.59	3.36	3.20	3.09	3.01	2.95	2.90	2.85	2.79	2.72	2.65	2.61	2.57	2.53	2.49
12	4.75	3.89	3.49	3.26	3.11	3.00	2.91	2.85	2.80	2.75	2.69	2.62	2.54	2.51	2.47	2.43	2.38
13	4.67	3.81	3.41	3.18	3.03	2.92	1.83	2.77	2.71	2.67	2.60	2.53	2.46	2,41	2.38	2.34	2.30
14	4.60	3.74	3.34	3.11	2.96	2.85	2.76	2.70	2.65	2.60	2.53	2.46	2.39	2.35	2.31	2.27	2.22
15	4.54	3.68	3.29	3.06	2.90	2.79	2.71	2.64	2.59	2.54	2.48	2.40	2.33	2.29	2.25	2.20	2.16
16	4.49	3.63	3.24	3.01	2.85	2.74	2.66	2.59	2.37	2.49	2.42	2.35	2.28	2.24	2.19	2.15	2.11
17	4.45	3.59	3.20	2.96	2.81	2.70	2.61	2.55	2.34	2.45	2.38	2.31	2.23	2.19	2.15	2.10	2.06
18	4.41	3.55	3.16	2.93	2.77	2.66	2.58	2.51	2.32	2.41	2.34	2.27	2.19	2.15	2.11	2.06	2.02
19	4.38	3.52	3.13	2.90	2.74	2.63	2.54	2.48	2.30	2.38	2.31	2.23	2.16	2.11	2.07	2.03	1.98
20	4.35	3.49	3.10	2.87	2.71	2.60	2.51	2.45	2.39	2.35	2.28	2.20	2.12	2.08	2.04	1.99	1.95
21	4.32	3.47	3.07	2.84	2.68	2.57	2.49	2.42	2.37	2.32	2.25	2.18	2.10	2.05	2.01	1.96	1.92
22	4.30	3.44	3.05	2.82	2.66	2.55	2.46	2.40	2.34	2.30	2.23	2.15	2.07	2.03	1.98	1.94	1.89
23	4.28	3.42	3.03	2.80	2.64	2.53	2.44	2.37	2.32	2.27	2.20	2.13	2.05	2.01	1.96	1.91	1.86
24	4.26	3.40	3.01	2.78	2.62	2.51	2.42	2.36	2.30	2.25	2.18	2.11	2.03	1.98	1.94	1.89	1.84

25	4.24	3.39	2.99	2.76	2.60	2.49	2.40	2.34	2.28	2.24	2.16	2.09	2.01	1.96	1.92	1.87	1.82
26	4.23	3.37	2.98	2.74	2.59	2.47	2.39	2.32	2.27	2.22	2.15	2.07	1.99	1.95	1.90	1.85	1.80
27	4.21	3.35	2.96	2.73	2.57	2.46	2.37	2.31	2.25	2.20	2.13	2.06	1.97	1.93	1.88	1.84	1.79
28	1.20	3.34	2.95	2.71	2.56	2.45	2.36	2.29	2.24	2.19	2.12	2.04	1.96	1.91	1.87	1.82	1.77
29	4.18	3.33	2.93	2.70	2.55	2.43	2.35	2.28	2.22	2.18	2.10	2.03	1.94	1.90	1.85	1.81	1.75
30	4.17	3.32	2.92	2.69	2.53	2.42	2.33	2.27	2.21	2.16	2.09	2.01	1.93	1.89	1.84	1.79	1.74
40	4.08	3.23	2.84	2.61	2.45	2.34	2.25	2.18	2.12	2.08	2.00	1.92	1.84	1.79	1.74	1.69	1.64
60	4.00	3.15	2.76	2.53	2.37	2.25	2.17	2.10	2.04	1.99	1.92	1.84	1.75	1.70	1.65	1.59	1.53

$$F(f) = p\{F \le f\} = \int_0^f \frac{\Gamma[(r_1+r_2)/2](r_1/r_2)^{r_1/2} x^{(r_1/2)-1}}{\Gamma(r_1/2)\Gamma(r_2/2)[1+(r_1x/r_2)]^{(r_1+r_2)/2}}\,dx \equiv \text{F Distribution}$$

$P\{F \le f\} = 0.975$

$r_2 \backslash r_1$	1	2	3	4	5	6	7	8	9	10	12	15	20	24	30	40	60
1	647.8	799.5	864.2	899.6	921.8	937.1	948.2	956.7	963.6	968.6	976.7	984.9	993.1	997.2	1001	1006	1010
2	38.51	39.00	39.17	39.25	39.30	39.33	39.36	39.37	39.39	39.40	39.41	39.43	39.45	39.46	39.46	39.47	39.48
3	17.44	16.04	15.44	15.10	14.88	14.73	14.62	14.54	14.47	14.42	14.34	14.25	14.17	14.12	14.08	14.04	13.99
4	12.22	10.65	9.98	9.60	9.36	9.20	9.07	8.98	8.90	8.84	8.85	8.66	8.56	8.51	8.46	8.41	8.36
5	10.01	8.43	7.76	7.39	7.15	6.98	6.85	6.76	6.68	6.62	6.52	6.43	6.33	6.28	6.23	6.18	6.12
6	8.81	7.26	6.60	6.23	5.99	5.82	5.70	5.60	5.52	5.46	5.37	5.27	5.17	5.12	5.07	5.01	4.96
7	8.07	6.54	5.89	5.52	5.29	5.12	4.99	4.90	4.82	4.76	4.67	4.57	4.47	4.42	4.36	4.31	4.25
8	7.57	6.06	5.42	5.05	4.82	4.65	4.53	4.43	4.36	4.30	4.20	4.10	4.00	3.95	3.89	3.84	3.78
9	5.21	5.71	5.08	4.72	4.48	4.32	4.20	4.10	4.03	3.96	3.87	3.77	3.67	3.61	3.56	3.51	3.45
10	6.94	5.46	4.83	4.47	4.24	4.07	3.95	3.85	3.78	3.72	3.62	3.52	3.42	3.37	3.31	3.26	3.20
11	6.72	5.26	4.63	4.28	4.04	3.88	3.76	3.66	3.59	3.53	3.43	3.33	3.23	3.17	3.12	3.06	3.00
12	6.55	5.10	4.47	4.12	3.89	3.73	3.61	3.51	3.44	3.37	3.28	3.18	3.07	3.02	2.96	2.91	2.85
13	6.41	4.97	4.35	4.00	3.77	3.60	3.48	3.39	3.31	3.25	3.15	3.05	2.95	2.89	2.84	2.78	2.72
14	6.30	4.86	4.24	3.89	3.66	3.50	3.38	3.29	3.21	3.15	3.05	2.95	2.84	2.79	2.73	2.67	2.61
15	6.20	4.77	4.15	3.80	3.58	3.41	3.29	3.20	3.12	3.06	2.96	2.86	2.76	2.70	2.64	2.59	2.52
16	6.12	4.69	4.08	3.73	3.50	3.34	3.22	3.12	3.05	2.99	2.89	2.79	2.68	2.63	2.57	2.51	2.45
17	6.04	4.62	4.01	3.66	3.44	3.28	3.26	3.06	2.98	2.92	2.82	2.72	2.62	2.56	2.50	2.44	2.38
18	5.98	4.56	3.95	3.61	3.38	3.22	3.10	3.01	2.93	2.87	2.77	2.67	2.56	2.50	2.44	2.38	2.32
19	5.92	4.51	3.90	3.56	3.33	3.17	3.05	2.96	2.88	2.82	2.72	2.62	2.51	2.45	2.39	2.33	2.27

TABLE 34.5 The F Distribution (continued)

$$F(f) = p\{F \le f\} = \int_0^f \frac{\Gamma[(r_1+r_2)/2](r_1/r_2)^{r_1/2} x^{(r_1/2)-1}}{\Gamma(r_1/2)\Gamma(r_2/2)[1+(r_1 x/r_2)]^{(r_1+r_2)/2}} dx$$

$P\{F \le f\} = 0.975$

$r_2 \backslash r_1$	1	2	3	4	5	6	7	8	9	10	12	15	20	24	30	40	60
20	5.87	4.46	3.86	3.51	3.29	3.13	3.01	2.91	2.84	2.77	2.68	2.57	2.46	2.41	2.35	2.29	2.22
21	5.83	4.42	3.82	3.48	3.25	3.09	2.97	2.87	2.80	2.73	2.64	2.53	2.42	2.37	2.31	2.25	2.18
22	5.79	4.38	3.78	3.44	3.22	3.05	2.93	2.84	2.76	2.70	2.60	2.50	2.39	2.33	2.27	2.21	2.14
23	5.75	4.35	3.75	3.41	3.28	3.02	2.90	2.81	2.73	2.67	2.57	2.47	2.36	2.30	2.24	2.18	2.11
24	5.72	4.32	3.72	3.38	3.15	2.99	2.87	2.78	2.70	2.64	2.54	2.44	2.33	2.27	2.21	2.15	2.08
25	5.69	4.29	3.69	3.35	3.13	2.97	2.85	2.75	2.68	2.61	2.51	2.41	2.30	2.24	2.18	2.12	2.05
26	5.66	4.27	3.67	3.33	3.10	2.94	2.82	2.73	2.65	2.59	2.49	2.39	2.28	2.22	2.16	2.09	2.03
27	5.63	4.24	3.65	3.31	3.08	2.92	2.80	2.71	2.63	2.57	2.47	2.36	2.25	2.19	2.13	2.07	2.00
28	5.61	4.22	3.63	3.29	3.06	2.90	2.78	2.69	2.61	2.55	2.45	2.34	2.23	2.17	2.11	2.05	1.98
29	5.59	4.20	3.61	3.27	3.04	2.88	2.76	2.67	2.59	2.53	2.43	2.32	2.21	2.15	2.09	2.03	1.96
30	5.57	4.18	3.59	3.25	3.03	2.87	2.75	2.65	2.57	2.51	2.41	2.31	2.20	2.14	2.07	2.01	1.94
40	5.42	4.05	3.46	3.13	2.90	2.74	2.62	2.53	2.45	2.39	2.29	2.18	2.07	2.01	1.94	1.88	1.80
60	5.29	3.93	3.34	3.01	2.79	2.63	2.51	2.41	2.33	2.27	2.17	2.06	1.94	1.88	1.82	1.74	1.67

$$F(f) = p\{F \le f\} = \int_0^f \frac{\Gamma[(r_1+r_2)/2](r_1/r_2)^{r_1/2} x^{(r_1/2)-1}}{\Gamma(r_1/2)\Gamma(r_2/2)[1+(r_1 x/r_2)]^{(r_1+r_2)/2}} dx$$

$P\{F \le f\} = 0.99$

$r_2 \backslash r_1$	1	2	3	4	5	6	7	8	9	10	12	15	20	24	30	40	60
1	4052	4999.5	5403	5625	5764	5859	5928	5982	6022	6056	6106	6157	6209	6235	6261	6287	6313
2	98.50	99.00	99.17	99.25	99.30	99.33	99.36	99.37	99.39	99.40	99.42	99.43	99.45	99.46	99.47	99.47	99.48
3	34.12	30.82	29.46	28.71	28.24	27.91	27.67	27.49	27.35	27.23	27.05	26.87	26.69	26.60	26.50	26.41	26.32
4	21.20	18.00	16.69	15.98	15.52	15.21	14.98	14.80	14.66	14.55	14.37	14.20	14.02	13.93	13.84	13.75	13.65
5	16.26	13.27	12.06	11.39	10.97	10.67	10.46	10.29	10.16	10.05	9.89	9.72	9.55	9.47	9.38	9.29	9.20
6	13.75	10.92	9.78	9.15	8.75	8.47	8.26	8.10	7.98	7.87	7.72	7.56	7.40	7.31	7.23	7.14	7.06
7	12.25	9.55	8.45	7.85	7.46	7.19	6.99	6.84	6.72	6.62	6.47	6.31	6.16	6.07	5.99	5.91	5.82

8	11.26	8.65	7.59	7.01	6.63	6.37	6.18	6.03	5.91	5.81	5.67	5.52	5.36	5.28	5.20	5.12	5.03
9	10.56	8.02	6.99	6.42	6.06	5.80	5.61	5.47	5.35	5.26	5.11	4.96	4.81	4.73	4.65	4.57	4.48
10	10.04	7.56	6.55	5.99	5.64	5.39	5.20	5.06	4.94	4.85	4.71	4.56	4.41	4.33	4.25	4.17	4.08
11	9.65	7.21	6.22	5.67	5.32	5.07	4.89	4.74	4.63	4.54	4.40	4.25	4.10	4.02	3.94	3.86	3.78
12	9.33	6.93	5.95	5.41	5.06	4.82	4.64	4.50	4.39	4.30	4.16	4.01	3.86	3.78	3.70	3.62	3.54
13	9.07	6.70	5.74	5.21	4.86	4.62	4.44	4.30	4.19	4.10	3.96	3.82	3.66	3.59	3.51	3.43	3.34
14	8.86	6.51	5.56	5.04	4.69	4.46	4.28	4.14	4.03	3.94	3.80	3.66	3.51	3.43	3.35	3.27	3.18
15	8.68	6.36	5.42	4.89	4.56	4.32	4.14	4.00	3.89	3.80	3.67	3.52	3.37	3.29	3.21	3.13	3.05
16	8.53	6.23	5.29	4.77	4.44	4.20	4.03	3.89	3.78	3.69	3.55	3.41	3.26	3.18	3.10	3.02	2.93
17	8.40	6.11	5.18	4.67	4.34	4.10	3.93	3.79	3.68	3.59	3.46	3.31	3.16	3.08	3.00	2.92	2.83
18	8.29	6.01	5.09	4.58	4.25	4.01	3.84	3.71	3.60	3.51	3.96	3.23	3.08	3.00	2.92	2.84	2.75
19	8.18	5.93	5.01	4.50	4.17	3.94	3.77	3.63	3.32	3.43	3.80	3.15	3.00	2.92	2.84	2.76	2.67
20	8.10	5.85	4.94	4.43	4.10	3.87	3.70	3.56	3.46	3.37	3.23	3.09	2.94	2.86	2.78	2.69	2.61
21	8.02	5.78	4.87	4.37	4.04	3.81	3.64	3.51	3.40	3.31	3.17	3.03	2.88	2.80	2.72	2.64	2.55
22	7.95	5.72	4.82	4.31	3.99	3.76	3.59	3.45	3.35	3.26	3.12	2.98	2.83	2.75	2.67	2.58	2.83
23	7.88	5.66	4.76	4.26	3.94	3.71	3.54	3.41	3.30	3.21	3.07	2.93	2.78	2.70	2.62	2.54	2.75
24	7.82	5.61	4.72	4.22	3.90	3.67	3.77	3.36	3.26	3.17	3.03	2.89	2.74	2.66	2.58	2.49	2.67
25	7.77	5.57	4.68	4.28	3.85	3.63	3.46	3.32	3.22	3.13	2.99	2.85	2.70	2.62	2.54	2.45	2.36
26	7.72	5.53	4.64	4.14	3.82	3.59	3.42	3.29	3.18	3.09	2.96	2.81	2.66	2.58	2.50	2.42	2.33
27	7.68	5.49	4.60	4.11	3.78	3.56	3.39	3.26	3.15	3.06	2.93	2.78	2.63	2.55	2.47	2.38	2.29
28	7.64	5.45	4.57	4.07	3.75	3.53	3.36	3.23	3.12	3.03	2.90	2.75	2.60	1.52	2.44	2.35	2.26
29	7.60	5.42	4.54	4.04	3.73	3.50	3.33	3.20	3.09	3.00	2.87	2.73	2.57	2.49	2.41	2.33	2.23
30	7.56	5.39	4.51	4.02	3.70	3.47	3.30	3.17	3.07	2.98	2.84	2.70	2.55	2.47	2.39	2.30	2.21
40	7.31	5.18	4.31	3.83	3.51	3.29	3.12	2.99	2.89	2.80	2.66	2.52	2.37	2.29	2.20	2.11	2.02
60	7.08	4.98	4.13	3.65	3.34	3.12	2.95	2.82	2.72	2.63	2.50	2.35	2.20	2.12	2.03	1.94	1.84

$$F(f) = p\{F \le f\} = \int_0^f \frac{\Gamma[(r_1+r_2)/2](r_1/r_2)^{r_1/2} x^{(r_1/2)-1}}{\Gamma(r_1/2)\Gamma(r_2/2)[1+(r_1 x/r_2)]^{(r_1+r_2)/2}}\,dx$$

$P\{F \le f\} = 0.995$

$r_2 \backslash r_1$	1	2	3	4	5	6	7	8	9	10	12	15	20	24	30	40	60
1	16211	20000	21615	22500	23056	23437	23715	23925	24091	24224	24426	24630	24836	24920	25044	25148	25253

TABLE 34.5 The F Distribution (continued)

$$F(f) = p\{F \le f\} = \int_0^f \frac{\Gamma[(r_1+r_2)/2](r_1/r_2)^{r_1/2} x^{(r_1/2)-1}}{\Gamma(r_1/2)\Gamma(r_2/2)[1+(r_1x/r_2)]^{(r_1+r_2)/2}} dx$$

$P\{F \le f\} = 0.995$

$r_2 \backslash r_1$	1	2	3	4	5	6	7	8	9	10	12	15	20	24	30	40	60
2	198.5	199.0	199.2	199.2	199.3	199.3	199.4	199.4	199.4	199.4	199.4	199.4	199.4	199.5	199.5	199.5	199.5
3	55.55	49.80	47.47	46.19	45.39	44.84	44.43	44.13	43.88	43.69	43.39	43.08	42.78	42.62	42.47	42.31	42.15
4	31.33	26.28	24.26	23.15	22.46	21.87	21.62	21.35	21.14	20.97	20.70	20.44	20.17	20.03	19.89	19.75	19.61
5	22.78	18.31	16.53	15.56	14.94	14.51	14.20	13.96	13.77	13.62	13.38	13.15	12.90	12.78	12.66	12.53	12.40
6	18.63	14.54	12.92	12.03	11.46	11.07	10.79	10.57	10.39	10.25	10.03	9.81	9.59	9.47	9.36	9.24	9.12
7	16.24	12.40	10.88	10.05	9.52	9.16	8.89	8.68	8.51	8.38	8.18	7.97	7.75	7.65	7.53	7.42	7.31
8	14.69	11.04	9.60	8.81	8.30	7.95	7.69	7.50	7.34	7.21	7.01	6.81	6.61	6.50	6.40	6.29	6.18
9	13.61	10.11	8.72	7.96	7.47	7.13	6.88	6.69	6.54	6.42	6.23	6.03	5.83	5.73	5.62	5.52	5.41
10	12.83	9.43	8.08	7.34	6.87	6.54	6.30	6.12	5.97	5.85	5.66	5.47	5.27	5.17	5.07	4.97	4.86
11	12.23	8.91	7.60	6.88	6.42	6.10	5.86	5.68	5.54	5.42	5.24	5.05	4.86	4.76	4.65	4.55	4.44
12	11.75	8.51	7.23	6.52	6.07	5.76	5.52	5.35	5.20	5.09	4.91	4.72	4.53	4.43	4.33	4.23	4.12
13	11.37	8.19	6.93	6.23	5.79	5.48	5.25	5.08	4.94	4.82	4.64	4.46	4.27	4.17	4.07	3.97	3.87
14	11.06	7.92	6.68	6.00	5.56	5.26	5.03	4.86	4.72	4.60	4.43	4.25	4.06	3.96	3.86	3.76	3.66
15	10.80	7.70	6.48	5.80	5.37	5.07	4.85	4.67	4.54	4.42	4.25	4.07	3.88	3.79	3.69	3.58	3.48
16	10.58	7.51	6.30	5.64	5.21	4.91	4.69	4.52	4.38	4.27	4.10	3.92	3.73	3.64	3.54	3.44	3.33
17	10.38	7.35	6.16	5.50	5.07	4.78	4.56	4.39	4.25	4.14	3.97	3.79	3,61	3.51	3.41	3.31	3.21
18	10.22	7.21	6.03	5.37	4.96	4.66	4.44	4.28	4.14	4.03	3.86	3.68	3.50	3.40	3.30	3.20	3.10
19	10.07	7.09	5.92	5.27	4.85	4.56	4.34	4.18	4.04	3.93	3.76	3.59	3.40	3.31	3.21	3.11	3.00
20	9.94	6.99	5.82	5.17	4.76	4.47	4.26	4.09	3.96	3.85	3.68	3.50	3.32	3.22	3.12	3.02	2.92
21	9.83	6.89	5.73	5.09	4.68	4.39	4.18	4.01	3.88	3.77	3.60	3.43	3.24	3.15	3.05	2.95	2.84
22	9.73	6.81	5.65	5.02	4.61	4.32	4.11	3.94	3.81	3.70	3.54	3.36	3.18	3.08	2.98	2.88	2.77
23	9.63	6.73	5.58	4.95	4.54	4.26	4.05	3.88	3.75	3.64	3.47	3.30	3.12	3.02	2.92	2.82	2.72
24	9.55	6.66	5.52	4.89	4.49	4.20	3.99	3.83	3.69	3.59	3.42	3.25	3.06	2.97	2.87	2.77	2.66

25	9.48	6.60	5.46	4.84	4.43	4.15	3.94	3.78	3.64	3.54	3.37	3.20	3.01	2.92	2.82	2.72	2.61
26	9.41	6.54	5.41	4.79	4.38	4.10	3.89	3.73	3.60	3.49	3.33	3.15	2.97	2.87	2.77	2.67	2.84
27	9.34	6.49	5.36	4.74	4.34	4.06	3.85	3.69	3.56	3.45	3.28	3.11	2.93	2.83	2.73	2.63	2.77
28	9.28	6.44	5.32	4.70	4.30	4.02	3.81	3.65	3.52	3.41	3.25	3.07	2.89	2.79	2.69	2.59	2.71
29	9.23	6.40	5.28	4.66	4.26	3.98	3.77	3.61	3.48	3.48	3.21	3.04	2.86	2.76	2.66	2.56	2.66
30	9.18	6.35	5.24	4.62	4.23	3.95	3.74	3.58	3.45	3.34	3.18	3.01	2.82	2.73	2.63	2.52	2.42
40	8.83	6.07	4.98	4.37	3.99	3.71	3.51	3.35	3.22	3.12	2.95	2.78	2.60	2.50	2.40	2.30	2.18
60	8.49	5.79	4.73	4.14	3.76	3.49	3.29	3.13	3.01	2.90	2.74	2.57	2.39	2.29	2.19	2.08	1.96

TABLE 34.6 The Poisson Density Function

$$f(x) = \frac{e^{-\lambda}\lambda^x}{x!}$$

x \ λ	0.2	0.4	0.6	0.8	1.0	1.5	2.0	2.5	3.0	3.5	4	5	6	7	8	9	10
0	0.9048	0.6703	0.5488	0.4493	0.3679	0.2231	0.1353	0.0821	0.0498	0.0302	0.0183	0.0067	0.0025	0.0009	0.0003	0.0001	0.0000
1	0.0905	0.2681	0.3293	0.3595	0.3679	0.3347	0.2707	0.2052	0.1494	0.1057	0.0733	0.0337	0.0149	0.0064	0.0027	0.0011	0.0005
2	0.0045	0.0536	0.0988	0.1438	0.1839	0.2510	0.2707	0.2565	0.2240	0.1850	0.1465	0.0842	0.0446	0.0223	0.0107	0.0050	0.0023
3	0.0002	0.0072	0.0198	0.0383	0.0613	0.1255	0.1804	0.2138	0.2240	0.2158	0.1954	0.1404	0.0892	0.0521	0.0286	0.0150	0.0076
4		0.0007	0.0030	0.0077	0.0153	0.0471	0.0902	0.1336	0.1680	0.1888	0.1954	0.1755	0.1339	0.0912	0.0573	0.0337	0.0189
5		0.0001	0.0004	0.0012	0.0031	0.0141	0.0361	0.0668	0.1008	0.1322	0.1563	0.1755	0.1606	0.1277	0.0916	0.0607	0.0378
6				0.0002	0.0005	0.0035	0.0120	0.0278	0.0504	0.0771	0.1042	0.1462	0.1606	0.1490	0.1221	0.0911	0.0631
7					0.0001	0.0008	0.0034	0.0099	0.0216	0.0385	0.0595	0.1044	0.0137	0.1490	0.1396	0.1171	0.0901
8						0.0001	0.0009	0.0031	0.0081	0.0169	0.0298	0.0653	0.1033	0.1304	0.1396	0.1318	0.1126
9							0.0002	0.0009	0.0027	0.0066	0.0132	0.0363	0.0688	0.1014	0.1241	0.1318	0.1251
10								0.0002	0.0008	0.0023	0.0053	0.0181	0.0413	0.0710	0.0993	0.1186	0.1251
11									0.0002	0.0007	0.0019	0.0082	0.0225	0.0452	0.0722	0.0970	0.1137
12									0.0001	0.0002	0.0006	0.0034	0.0113	0.0264	0.0481	0.0728	0.0948
13										0.0001	0.0002	0.0013	0.0052	0.0142	0.0296	0.0504	0.0729
14											0.0001	0.0005	0.0022	0.0071	0.0169	0.0324	0.0521
15												0.0002	0.0009	0.0033	0.0090	0.0194	0.0347
16													0.0003	0.0014	0.0045	0.0109	0.0217
17													0.0001	0.0006	0.0021	0.0058	0.0128
18														0.0002	0.0009	0.0029	0.0071
19														0.0001	0.0004	0.0014	0.0037
20															0.0002	0.0006	0.0019
21															0.0001	0.0003	0.0009
22																0.0001	0.0004
23																	0.0002
24																	0.0001

TABLE 34.7 The Poisson Distribution

$$F(x) = \sum_{k=0}^{x} \frac{e^{-\lambda}\lambda^k}{k!}$$

x \ λ	0.2	0.4	0.6	0.8	1.0	1.2	1.4	1.6	1.8	2.0	2.5	3.0	3.5	4.0
0	0.8187	0.6703	0.5488	0.4493	0.3679	0.3012	0.2466	0.2019	0.1653	0.1353	0.0821	0.0489	0.0302	0.0183
1	0.9825	0.9384	0.8781	0.8088	0.7358	0.6626	0.5918	0.5249	0.4268	0.4060	0.2873	0.1991	0.1359	0.0916
2	0.9989	0.9921	0.9769	0.9526	0.9197	0.8795	0.8335	0.7834	0.7306	0.6767	0.5438	0.4232	0.3208	0.2381
3	0.9999	0.9992	0.9966	0.9909	0.9810	0.9662	0.9463	0.9212	0.8913	0.8571	0.7576	0.6472	0.5366	0.4335
4	1.0000	0.9999	0.9996	0.9986	0.9963	0.9923	0.9857	0.9763	0.9636	0.9473	0.8912	0.8153	0.7254	0.6288
5		1.0000	1.0000	0.9998	0.9994	0.9985	0.9968	0.9940	0.9896	0.9834	0.9580	0.9161	0.8576	0.7851
6				1.0000	0.9999	0.9997	0.9994	0.9987	0.9974	0.9955	0.9858	0.9665	0.9347	0.8893
7					1.0000	1.0000	0.9999	0.9997	0.9994	0.9989	0.9958	0.9881	0.9733	0.9489
8							1.0000	1.0000	0.9999	0.9998	0.9989	0.9962	0.9901	0.9786
9									1.0000	1.0000	0.9997	0.9989	0.9967	0.9919
10											0.9999	0.9997	0.9990	0.9972
11											1.0000	0.9999	0.9997	0.9991
12												1.0000	0.9999	0.9997
13													1.0000	0.9999
14														1.0000

34.4.7 Conditional Density

$$f(x|M) = \frac{dF(x|M)}{dx} = \lim_{\Delta x \to 0} \frac{P\{x \le X \le x + \Delta x|M\}}{\Delta x}$$

$$\int_{-\infty}^{\infty} f(x|M)dx = F(\infty|M) - F(-\infty|M) = 1$$

Example

$X(f_i) = 10i$, $i = 1,\cdots 6$ where f_i = face of a die. $M = \{f_2, f_4, f_6\}$ = even event. For $x \ge 60$, $\{X \le x, M\}$ $= \{f_2, f_4, f_6\}$, $F(x|M) = \dfrac{P\{f_2, f_4, f_6\}}{P\{M\}} = 1$; for $40 \le x < 60$, $\{X \le x, M\} = \{f_2, f_4\}$, $F(x|M) = P\{f_2, f_4\}/P\{M\} = (2/6)/(3/6) = 2/3$; for $20 \le x < 40$, $\{X \le x, M\} = \{f_2\}$, $F(x|M) = P\{f_2\}/P\{M\} = (1/6)/(3/6) = 1/3$; for $x < 20$, $\{X \le x, M\} = 0$ and $F(x|M) = 0$.

34.4.8 Total Probability

$F(x) = F(x|A_1)P(A_1) + F(x|A_2)P(A_2) + \cdots + F(x|A_n)P(A_n)$, A_i's are mutually exclusive and their sum is equal to the certain event S.

34.5 Functions of One Random Variable (r.v.)

34.5.1 Random Variable (Definition)

To every experimental outcome ζ we assign a number $X(\zeta)$. The domain of X is the space S, and its range is the set I_X of the real numbers $X(\zeta)$.

34.5.2 Function of r.v.

$Y = g(X) = g[X(\zeta)]$

34.5.3 Distribution Function of Y (see 34.5.2)

$F_Y(y) = P\{Y \le y\} = P\{g(X) \le y\} = P\{X \in I_y\}$

Note: To find $F_Y(y)$ for a given y we must find that set I_y and the probability that X is in I_y. Refer to Figure 34.2: If $y \ge k$ then $g(x) < y$ for any x. Hence $\{Y \le y\}$ = certain event and $F_Y(y) = P\{Y \le y\} = 1$. If $y = y_1$, then $g(x) \le y_1$ for $x \le x_1$ and, hence, $F_Y(y_1) = P\{Y \le y_1\} = P\{X \le x_1\} = F_X(x_1)$ (x_1 depends on y_1). If $y = y_2$, then $g(x) = y_2$ has three solutions x_2', x_2'', x_2''': $g(x_2') = g(x_2'') = g(x_2''') = y_2$ and from Figure 34.2 $g(x) \le y_2$ if $x \le x_2'$ or $x_2'' \le x \le x_2'''$ and hence, $F_Y(y_2) = P\{X \le x_2'\} + P\{x_2'' \le x \le x_2'''\} = F_X(x_2') + F_X(x_2''') - F_X(x_2'')$. If $y < \ell$ no value of x produces $g(x) \le y$ and the event $\{Y \le y\}$ has zero probability: $F_Y(y) = 0$

Example

$Y = 1/X^2$. If $y > 0$, there are two solutions: $x_1 = -\sqrt{y}$, $x_2 = 1/\sqrt{y}$. $g(x) \le y$ if $x \le x_1$ or $x \ge x_2$ and thus

$$F_Y(y) = P\{Y \le y\} = P\{X \le -1/\sqrt{y}\} + P\{X \ge 1/\sqrt{y}\} = F_X(-1/\sqrt{y}) + 1 - F_X(1/\sqrt{y}).$$

If y < 0, no x will produce $g(x) \le y$ and, hence, $F_Y(y) = 0$.

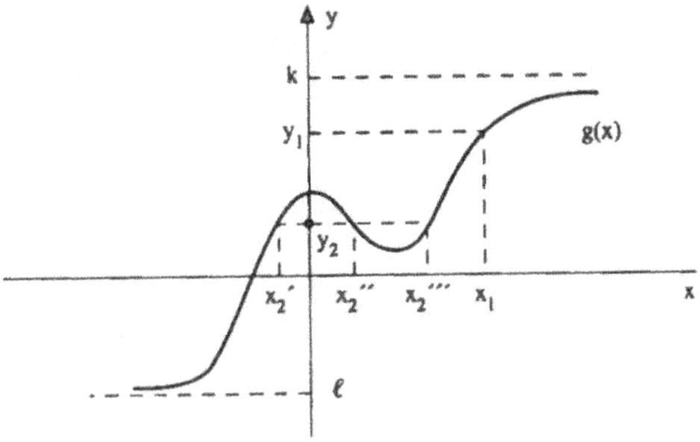

FIGURE 34.2

34.5.4 Density Function of Y = g(X) in Terms of $f_X(x)$ of X

1) Solve $y = g(x)$ for x in terms of y. If $x_1, x_2, \cdots, x_n$ are all its real roots, then $y = g(x_1) = \cdots = g(x_n) = \cdots$, then $f_Y(y) = \frac{f_X(x_1)}{|g'(x_1)|} + \cdots + \frac{f_X(x_n)}{|g'(x_n)|} + \cdots$, $g'(x) = dg(x)/dx$. If $y = g(x)$ has no real roots, then $f_Y(y) = 0$.

Example 1

$g(X) = aX + b$ and $x = (y-b)/a$ for every y. $g'(x) = a$ and hence $f_Y(y) = \frac{1}{|a|} f_X\left(\frac{y-b}{a}\right)$

Example 2

$g(X) = aX^2$ with the r.v. $y = ax^2$, $a > 0$. If $y < 0$ roots are imaginary and $f_Y(y) = 0$. If $y > 0$ then $x_1 = \sqrt{y/a}$ and $x_2 = -\sqrt{y/a}$. Since $g'(x_1) = 2ax_1 = 2\sqrt{ay}$ and $g'(x_2) = 2ax_2 = -2\sqrt{ay}$, then $f_Y(y) = \frac{1}{2\sqrt{ay}}\left[f_X\left(\sqrt{\frac{y}{a}}\right) + f_X\left(-\sqrt{\frac{y}{a}}\right)\right]u(y)$, $u(y) =$ unit step function.

Example 3

$Y = a\sin(X + \theta)$, $a > 0$. If $|y| < a$ then $y = a\sin(x + \theta)$ has infinitely many solutions $x_n = \sin^{-1}\frac{y}{a} - \theta$, $n = \cdots, -1, 0, 1, \cdots$. $dg(x_n)/dx = a\cos(x_n + \theta) = \sqrt{a^2 - y^2}$ and from 34.5.4 $f_Y(y) = \left(1/\sqrt{a^2 - y^2}\right)\sum_{n=-\infty}^{\infty} f_X(x_n)$, $|y| < a$. For $|y| > 0$ there exist no solutions, and $f_Y(y) = 0$.

Example 4

$Y = be^{-aX}u(X)$, $a > 0$, $b > 0$. If $y < 0$ or $y > b$ then the equation $y = b\exp(-ax)u(x)$ has no solution, and hence $f_Y(y) = 0$. If $0 < y < b$, then $x = -(1/a)\ln(y/b)$. $g'(x) = -abe^{-ax} = -ay$ and $f_Y(y) = f_X(-(1/a)\ln(y/b))/ay$, $0 < y < b$.

34.5.5 Conditional Density of Y = g(x)

$$f_Y(y|M) = \frac{f_X(x_1|M)}{|g'(x_1)|} + \cdots + \frac{f_X(x_n|M)}{|g'(x_n)|} + \cdots$$

Example

$Y = aX^2$, $a > 0$, $X \geq 0$, $f_X(x|X \geq 0) = \dfrac{f_X(x)}{1 - F_X(0)} u(x)$ (see 34.5.4 Example 2), and hence $f_Y(y|X \geq 0) =$ $[1/(2\sqrt{ay})] \dfrac{f_X(\sqrt{y/a})}{1 - F_X(0)} u(x)$.

$$\left[f(x|X \geq t) = f(x)/[1 - F(t)] = f(x) \Big/ \int_t^{\infty} f(x)dx, \quad x \geq t \right]$$

34.5.6 Expected Value

$$E\{X\} = \int_{-\infty}^{\infty} xf(x)dx \quad \text{continuous r.v.}$$

$$E\{X\} = \sum_n x_n P\{X = x_n\} = \sum_n x_n p_n \quad \text{discrete r.v.}$$

34.5.7 Expected Value of a Function g(X)

$$E\{Y = g(X)\} = \int_{-\infty}^{\infty} yf_Y(y)dy = \int_{-\infty}^{\infty} g(x)f_X(x)dx \quad \text{continuous r.v.}$$

$$E\{g(X)\} = \sum_k g(x_k)P\{X = x_k\} \quad \text{discrete type of r.v.}$$

34.5.8 Conditional Expected Value

$$E\{X|M\} = \int_{-\infty}^{\infty} xf(x|M)dx \quad \text{continuous r.v.}$$

$$E\{X|M\} = \sum_n x_n P\{X = x_n|M\} \quad \text{discrete r.v.}$$

34.5.9 Variance

$$\sigma^2 = E\{(X - \mu)^2\} = \int_{-\infty}^{\infty} (x - \mu)^2 f(x)dx \quad \text{continuous r.v.}$$

$$\sigma^2 = \sum_n (x_n - \mu)^2 P\{X = x_n\} \quad \text{discrete r.v.}$$

$$\sigma^2 = E\{X^2\} - E^2\{X\}$$

Example

$P\{X = k\} = e^{-\lambda}\frac{\lambda^k}{k!}$, $k = 0,1,\cdots =$ Poisson distribution.

$$E\{X\} = \sum_{k=0}^{\infty} ke^{-\lambda}\frac{\lambda^k}{k!} = e^{-\lambda}\sum_{k=0}^{\infty} k\frac{\lambda^k}{k!} = e^{-\lambda}\sum_{k=1}^{\infty} k\frac{\lambda^k}{k!}.$$

But

$$\frac{d}{d\lambda}e^{\lambda} = \frac{d}{d\lambda}\sum_{k=0}^{\infty}\frac{\lambda^k}{k!} = \sum_{k=1}^{\infty} k\frac{\lambda^{k-1}}{k!} = \frac{1}{\lambda}\sum_{k=1}^{\infty} k\frac{\lambda^k}{k!} = e^{\lambda}$$

or

$$\lambda = e^{-\lambda}\sum_{k=1}^{\infty} k\frac{\lambda^k}{k!}$$

and, hence, $E\{X\} = \lambda$.

34.5.10 Moments About the Origin

$\mu'_k = E\{X^k\} = \int_{-\infty}^{\infty} x^k f(x)dx = \sum_{r=0}^{k}\binom{k}{r}\mu^r\mu_{k-r}$, $\mu'_1 = \mu = E\{X\}$, $\mu'_0 = 1$

34.5.11 Central Moments

$$\mu_k = E\{(X-\mu)^k\} = \int_{-\infty}^{\infty}(x-\mu)^k f(x)dx = E\left\{\sum_{r=0}^{k}\binom{k}{r}(-1)^r\mu^r X^{k-r}\right\} = \sum_{r=0}^{k}\binom{k}{r}(-1)^r\mu^r\mu'_{k-r}$$

$\mu_0 = \mu'_0 = 1$, $\mu_1 = \mu'_1 - \mu = 0$, $\mu_2 = \mu'_2 - 2\mu\mu'_1 + \mu^2 = \mu'_2 - \mu^2$, $\mu_3 = \mu'_3 - 3\mu\mu'_2 + 3\mu^2\mu'_1 - \mu^3 = \mu'_3 - 3\mu\mu'_2 + 2\mu^3$

34.5.12 Absolute Moments

$M_k = E\{|X|^k\} = \int_{-\infty}^{\infty}|x|^k f(x)dx$

34.5.13 Generalized Moments

${}_a\mu'_k = E\{(X-a)^k\}$, ${}_aM'_k = E\{|X-a|^k\}$

Example 1

$$E\{X^{2n}\} = \frac{1}{2a}\int_{-a}^{a} x^{2n}\,dx = \frac{a^{2n}}{2n+1}, \quad \sigma^2 = E\{x^2\} = \frac{a^2}{3}$$

for X uniformly distributed in (–a,a).

Example 2

$$E\{X^n\} = \frac{a^{b+1}}{\Gamma(b+1)}\int_0^{\infty} x^n x^b e^{-ax}\,dx = \frac{a^{b+1}\Gamma(b+n+1)}{a^{b+n+1}\Gamma(b+1)}$$

for a gamma density $f(x) = [a^{b+1}/\Gamma(b+1)]x^b e^{-ax}u(x)$, $u(x) =$ unit step function.

34.5.14 Tchebycheff Inequality

$P\{|X-\mu| \geq k\sigma\} \leq \frac{1}{k^2}$, $\mu = E\{X\}$. Regardless of the shape of $f(x)$, $P\{\mu-\varepsilon < X < \mu+\varepsilon\} \geq 1-\frac{\sigma^2}{\varepsilon^2}$

Generalizations:

1. If $f_Y(y) = 0$ for $y < 0$ then $P\{Y \geq \alpha\} \leq \frac{E\{Y\}}{\alpha}$, $\alpha > 0$

2. $P\{|X-\alpha|^n \geq \varepsilon^n\} \leq \frac{E\{|X-\alpha|^n\}}{\varepsilon^n}$

34.5.15 Characteristic Function

$$\Phi(\omega) = E\{e^{j\omega X}\} = \int_{-\infty}^{\infty} e^{j\omega x} f(x)dx \text{ for continuous r.v.}$$

$$\Phi(\omega) = \sum_k e^{j\omega x_k} P\{X = x_k\} \text{ for discrete type r.v.}$$

$$\Phi(0) = 1, \quad |\Phi(\omega)| \leq 1$$

Example 1

$$\Phi(\omega) = E\{e^{j\omega Y}\} = E\{e^{j\omega(aX+b)}\} = e^{j\omega b}E\{e^{j\omega aX}\}, \text{ if } Y = aX+b$$

Example 2

$$P\{X = k\} = e^{-\lambda}\frac{\lambda^k}{k!}, \; k = 0,1,\cdots \text{ Poisson distribution} \quad \Phi(\omega) = e^{-\lambda}\sum_{k=0}^{\infty} e^{j\omega k}\frac{\lambda^k}{k!} = e^{\lambda(e^{j\omega}-1)}$$

34.5.16 Second Characteristic Function

$\psi(\omega) = \ln\Phi(\omega)$

34.5.17 Inverse of the Characteristic Function

$$f(x)=\frac{1}{2\pi}\int_{-\infty}^{\infty}\Phi(\omega)e^{-j\omega x}\,d\omega$$

34.5.18 Moment Theorem and Characteristic Function

$$\frac{d^n\Phi(0)}{d\omega^n}=j^n\mu'_n,\quad E\{X^n\}=\mu'_n$$

34.5.19 Convolution and Characteristic Function

$\Phi(\omega)=\Phi_1(\omega)\Phi_2(\omega)$, where $\Phi_1(\omega)$ and $\Phi_2(\omega)$ are the characteristic functions of the density functions $f_1(x)$ and $f_2(x)$. $\Phi(\omega)=E\{e^{j\omega X}\}=E\{e^{j\omega(X_1+X_2)}\}$ and $f(x)=f_1(x)*f_2(x)$ where * indicates convolution.

34.5.20 Characteristic Function of Normal r.v.

$$\Phi(\omega)=\exp(j\mu\omega-\tfrac{1}{2}\sigma^2\omega^2)$$

34.6 Two Random Variables

34.6.1 Joint Distribution Function

$$F_{xy}(xy)=P\{X\le x,\,Y\le y\},\; F_{xy}(x,\infty)=F_x(x),\; F_{xy}(\infty,y)=F_y(y),$$

$$F_{xy}(\infty,\infty)=1,\; F_{xy}(-\infty,y)=0,\; F_{xy}(x,-\infty)=0$$

34.6.2 Joint Density Function

$$f(x,y)=\frac{\partial^2F(x,y)}{\partial x\partial y},\; f_x(x)=\int_{-\infty}^{\infty}f(x,y)dy,\; f_y(y)=\int_{-\infty}^{\infty}f(x,y)dx$$

34.6.3 Conditional Distribution Function

$$F_y(y|M)=P\{Y\le y|M\}=\frac{P\{Y\le y,M\}}{P\{M\}},\; F_y\{y|X\le x\}=\frac{P\{X\le x,\,Y\le y\}}{P\{X\le x\}}=\frac{F_{xy}(x,y)}{F_x(x)}$$

$$F_y(y|X\le a,Y\le b)=\frac{P\{X\le a,Y\le b,Y\le y\}}{P\{X\le a,Y\le b\}}=\begin{cases}1 & y\ge b\\ F_{xy}(a,y)/F_{xy}(a,b) & y<b\end{cases}$$

34.6.4 Conditional Density Function

$$f_y(y|X \le x) = \frac{\partial F_{xy}(x,y)/\partial y}{F_x(x)} = \frac{\int_{-\infty}^{x} f_{xy}(\xi,y)d\xi}{\int_{-\infty}^{\infty}\int_{-\infty}^{x} f_{xy}(\xi,y)d\xi dy}, \quad f_y(y|x_1 < X \le x_2) = \frac{\int_{x_1}^{x_2} f_{xy}(x,y)dx}{\Gamma_x(x_2) - F_x(x_1)},$$

$$f_y(y|X = x) = \frac{f_{xy}(x,y)}{f_x(x)}$$

34.6.5 Baye's Theorem

$$f_y(y|X = x) = \frac{f_x(x|Y = y)f_y(y)}{f_x(x)}$$

34.6.6 Joint Conditional Distribution

$$F_{xy}(x,y|a < X \le b) = \frac{P\{X \le x, Y \le y, a < Y \le b\}}{P\{a < X \le b\}} = \begin{cases} \dfrac{F_{xy}(b,y) - F_{xy}(a,y)}{F_x(b) - F_x(a)} & x > b \\ \dfrac{F_{xy}(x,y) - F_{xy}(a,y)}{F_x(b) - F_x(a)} & a < x \le b \\ 0 & x \le a \end{cases}$$

34.6.7 Conditional Expected Value

$$E\{g(Y)|X = x\} = \int_{-\infty}^{\infty} g(y)f_y(y|X = x)dy = \frac{\int_{-\infty}^{\infty} g(y)f_{xy}(x,y)dy}{\int_{-\infty}^{\infty} f_{xy}(x,y)dy}, \quad E\{E\{Y|X\}\} = E\{Y\}$$

34.6.8 Independent r.v.

$$F_{xy}(x,y) = F_x(x)F_y(y); \; f_{xy}(x,y) = f(x)f(y); \; f_y(y|x) = f_y(y); \; f_x(x|y) = f_x(x)$$

34.6.9 Jointly Normal r.v.

$$f(x,y) = \frac{1}{2\pi\sigma_1\sigma_2\sqrt{1-r^2}}\exp\left[-\frac{1}{2(1-r^2)}\left[\frac{(x-\mu_1)^2}{\sigma_1^2} - \frac{2r(x-\mu_1)(y-\mu_2)}{\sigma_1\sigma_2} + \frac{(y-\mu_2)^2}{\sigma_2^2}\right]\right]$$

$E\{X\} = \mu_1$, $E\{Y\} = \mu_2$, $\sigma_x = \sigma_1$, $\sigma_y = \sigma_2$. If $r = 0$, $f(x,y) = f_x(x)f_y(y) \equiv$ independent. $|r| < 1$, $r =$ correlation coefficient.

Conditional Densities

$$f_y(y|X=x)=\frac{1}{\sigma_2\sqrt{2\pi(1-r^2)}}\exp\left[-\frac{1}{2\sigma^2(1-r^2)}\left[y-\mu_2-\frac{r\sigma_2}{\sigma_1}(x-\mu_1)\right]^2\right]$$

$$E\{Y|X=x\}=\mu_2+\frac{r\sigma_2}{\sigma_1}(x-\mu_1),\ \sigma_{y|x=x}=\sigma_2\sqrt{1-r^2}$$

If $\mu_1=\mu_2=0$ then $E\{Y^2|X=x\}=\sigma_2^2(1-r^2)+\frac{r^2\sigma_2^2}{\sigma_1^2}x^2$

34.7 Functions of Two Random Variables

34.7.1 Definitions

$Z=g(X,Y)=g[X(\zeta),Y(\zeta)]$, $F_z(z)=P\{Z\le z\}$, $D_z=$ region of xy-plane such that $g(x,y)\le z$, $\{Z\le z\}$ $=\{(X,Y)\in D_z\}$

34.7.2 Distribution Function

$$F_z(z)=P\{Z\le z\}=P\{(X,Y)\in D_z\}=\iint_{D_z} f_{xy}(x,y)\,dx\,dy$$

34.7.3 Density Function

$$f_z(z)dz=P\{z<Z\le z+dz\}=\iint_{\Delta D_z} f_{xy}(x,y)\,dx\,dy$$

Example 1

$Z=X+Y$, $x+y\le z$, $F_z(z)=\int_{-\infty}^{\infty}\int_{-\infty}^{z-y} f_{xy}(x,y)\,dx\,dy$, $\frac{dF_z(z)}{dz}=f_z(z)=\int_{-\infty}^{\infty} f_{xy}(z-y,y)\,dy$. If the r. v. are independent then $f_{xy}(x,y)=f_x(x)f_y(y)$ and hence $f_z(z)=\int_{-\infty}^{\infty} f_x(z-y)f_y(y)\,dy=\int_{-\infty}^{\infty} f_x(x)f_y(z-x)\,dx=$ $f_x(z)*f_y(z)=$ convolution of densities.

Example 2

$Z=X^2+Y^2$, if $z>0$ so then $x^2+y^2\le z=$ circle with radius $\sqrt{z}, F_z(z)=\iint_{x^2+y^2\le z} f(x,y)\,dx\,dy$, if $z<0$, $F_z(z)=0$. $f_{xy}(x,y)=(1/2\pi\sigma^2)\exp[-(x^2+y^2)/2\sigma^2]$ then $F_z(z)=\frac{1}{2\pi\sigma^2}\int_0^{z} 2\pi r e^{-r^2/2\sigma^2}\,dr=1-e^{-z/2\sigma^2}$, $z>0$, and $f_z(z)=\frac{1}{2\sigma^2}e^{-z/2\sigma^2}$, $z\ge 0$

Example 3

$f_{xy}(x,y) = (1/2\pi\sigma^2)\exp[-(x^2+y^2)/2\sigma^2]$, $Z = +\sqrt{X^2+Y^2}$, $F_z(z) = \frac{1}{2\pi\sigma^2}\int_0^z 2\pi r e^{-r^2/2\sigma^2}\,dr = 1 - e^{-z^2/2\sigma^2}$, $z > 0$, $f_z(z) = (z/\sigma^2)\exp(-z^2/2\sigma^2)$, $z > 0 \equiv$ Rayleigh distributed, $E\{Z\} = \sigma\sqrt{\pi/2}$, $E\{Z^2\} = 2\sigma^2$, $\sigma_z^2 = (2-(\pi/2))\sigma^2$

Example 4

If $f_{xy}(x,y) = f_{xy}(-x,-y)$ then $F_z(z) = 2\int_0^\infty \int_{-\infty}^{yz} f_{xy}(x,y)\,dx\,dy$, $f_z(z) = 2\int_0^\infty y f_{xy}(zy,y)\,dy$. Then for $f_{xy}(x,y) = (1/[2\pi\sigma_1\sigma_2\sqrt{1-r^2}])\exp\left[-\frac{1}{2(1-r^2)}\left(\frac{x^2}{\sigma_1^2} - \frac{2rxy}{\sigma_1\sigma_2} + \frac{y^2}{\sigma_1^2}\right)\right]$ then $f_z(z)$ of $Z = X/Y$ is

$$f_z(z) = [2/(2\pi\sigma_1\sigma_2\sqrt{1-r^2})]\int_0^\infty y\exp\left[-\frac{y^2}{2(1-r^2)}\left[\frac{z^2}{\sigma_1^2} - \frac{2rz}{\sigma_1\sigma_2} + \frac{1}{\sigma_2^2}\right]\right]dy.$$

But $\int_0^\infty y\exp[-y^2/2a^2]\,dy = a^2\int_0^\infty e^{-w}\,dw = a^2$ and hence

$$f_z(z) = [(\sqrt{1-r^2}\,\sigma_1\sigma_2/\pi]/[\sigma_2^2(z - r\sigma_1/\sigma_2)^2 + \sigma_1^2(1-r^2)].$$

If $\mu_2 = \mu_2 = 0$ then $f_z(z)$ is Cauchy density.

34.8 Two Functions of Two Random Variables

34.8.1 Definitions

$Z = g(X,Y)$, $W = h(X,Y)$, D_{zw} region of the xy plane such that $g(x,y) \le z$ and $h(x,y) \le w$, $\{Z \le z, W \le w\} = \{(X,Y) \in D_{xy}\}$, $F_{zw}(z,w) = \iint_{D_{zw}} f_{xy}(x,y)\,dx\,dy$

34.8.2 Density Function $f_{zw}(z,w)$

$f_{zw}(z,w) = \frac{f_{xy}(x_1,y_1)}{|J(x_1,y_1)|} + \cdots + \frac{f_{xy}(x_n,y_{n1})}{|J(x_n,y_n)|} + \cdots$, $z = g(x_i,y_i)$, $w = h(x_i,y_i)$ where (x_i,y_i) are solutions. If there are no real solutions for certain values of (z,w), then $f_{zw}(z,w) = 0$.

Jacobian of transformation

$$J(x,y) = \begin{vmatrix} \frac{\partial g(x,y)}{\partial x} & \frac{\partial g(x,y)}{\partial y} \\ \frac{\partial h(x,y)}{\partial x} & \frac{\partial h(x,y)}{\partial y} \end{vmatrix}.$$

Example 1

If $z = ax + by$, $w = cx + dy$ then $x = a_1 z + b_1 y$, $y = c_1 z + d_1 w$, where a_1, b_1, c_1 and d_1 are functions of a,b,c, and d.

$$J(x,y) = \begin{vmatrix} a & b \\ c & d \end{vmatrix} = ad - bc,\ f_{zw} = (z,w) = 1/[|ad - bc|] f_{xy}(a_1 z + b_1 w, c_1 z + d_1 w)$$

Example 2

$z = +\sqrt{x^2 + y^2}$, $w = x/y$. If $z > 0$ then the system has two solutions: $x_1 = zw/\sqrt{1+w^2}$, $y_1 = z/\sqrt{1+w^2}$ and $x_2 = -x_1$, $y_2 = -y_1$ for any w.

$$J(x,y) = \begin{vmatrix} x/\sqrt{x^2+y^2} & y/\sqrt{x^2+y^2} \\ 1/y & -x/y^2 \end{vmatrix} = (1+w^2)/(-z)$$

and from 34.8.2

$$f_{zw}(z,w) = [z/(1+w^2)]\left[f_{xy}\left(\frac{zw}{\sqrt{1+w^2}}, \frac{z}{\sqrt{1+w^2}} \right) + f_{xy}\left(\frac{-zw}{\sqrt{1+w^2}}, \frac{-z}{\sqrt{1+w^2}} \right) \right].$$

If $z < 0$, $f_{zw}(z,w) = 0$.

34.8.3 Auxiliary Variable

If $z = g(x,y)$ we can introduce an auxiliary function w = x or w = y. $f_z(z) = \int_{-\infty}^{\infty} f_{zw}(z,w)dw$.

Example

If $z = xy$ set auxiliary function w = x. The system has solutions

$$x = w,\ y = z/w.\quad J(x,y) = \begin{vmatrix} y & x \\ 1 & 0 \end{vmatrix} = -x = -w$$

and, hence,

$$f_{zw}(z,w) = (1/|w|) f_{xy}(w, z/w) \text{ and } f_z(z) = \int_{-\infty}^{\infty} (1/|w|) f_{xy}(w, z/w)dw.$$

34.8.4 Functions of Independent r.v.'s

If X and Y are independent then $Z = g(X)$ and $W = h(Y)$ are independent and

$$f_{zw}(z,w) = \frac{f_x(x_1)}{|g'(x_1)|} \frac{f_y(y_1)}{|h'(y_1)|}$$

since

$$J(x,y)=\begin{vmatrix} g'(x) & 0 \\ 0 & h'(x) \end{vmatrix}=g'(x)h'(x)$$

34.9 Expected Value, Moments, and Characteristic Function of Two Random Variables

34.9.1 Expected Value

$$E\{g(X,Y)\}=\iint_{-\infty}^{\infty} g(x,y)f(x,y)\,dx\,dy;$$

$$E\{z\}=\int_{-\infty}^{\infty} zf_z(z)\,dz \text{ if } z=g(x,y);\quad E\{g(X,Y)\}=\sum_{k,n} g(x_k,y_n)p_{kn},$$

$P\{X=x_k, Y=y_n\}=p_{kn}$ discrete case r.v.

34.9.2 Conditional Expected Values

$$E\{g(X,Y|M)\}=\iint_{-\infty}^{\infty} g(x,y)f(x,y|M)\,dx\,dy;$$

$$E\{g(X,Y)|X=x\}=\int_{-\infty}^{\infty} g(x,y)f(x,y)\,dy/f_x(x)=\int_{-\infty}^{\infty} g(x,y)f(y|X=x)\,dy$$

34.9.3 Moments

$$\mu'_{kr}=E\{X^kY^r\}=\iint_{-\infty}^{\infty} x^ky^rf(x,y)\,dx\,dy,\quad \mu'_{11}=R_{xy}=E\{XY\}$$

$$\mu_{kr}=E\{(X-\mu_x)^k(Y-\mu_y)^r\}=\iint_{-\infty}^{\infty}(x-\mu_x)^k(y-\mu_y)^rf(x,y)\,dx\,dy$$

$$\mu_{20}=\sigma_x^2,\ \mu_{02}=\sigma_y^2,\ \mu=\mu'_1$$

34.9.4 Covariance

$$\mu_{11}=E\{(X-\mu_x)(Y-\mu_y)\}=E\{XY\}-\mu_xE\{Y\}-\mu_yE\{X\}+\mu_x\mu_y$$

34.9.5 Correlation Coefficient

$$r=E\{(X-\mu_x)(Y-\mu_y)\}/\sqrt{E\{(X-\mu_x)^2\}E\{(Y-\mu_y)^2\}}=\mu'_{11}/\sigma_x\sigma_y$$

$$\mu_{11}'^2 \le \mu_{20}'\mu_{02}',\ \mu_{11}'^2 \le \mu_{20}\mu_{02},\ |r| = |\mu_{11}'^2| / \sqrt{\mu_{20}\mu_{02}} \le 1$$

34.9.6 Uncorrelated r.v.'s

$E\{XY\} = E\{X\}E\{Y\}$

34.9.7 Orthogonal r.v.'s

$E\{XY\} = 0$

34.9.8 Independent r.v.'s

$f(x,y) = f_x(x)f_y(y)$

Note:

1. If X and Y are independent, $g(X)$ and $h(Y)$ are independent or

$$E\{g(X)h(Y)\} = E\{g(X)\}E\{h(Y)\}$$

2. If X and Y are uncorrelated, then
   a. $E\{(X-\mu_x)(Y-\mu_y)\} = 0,\ r = 0$
   b. $\sigma_{x+y}^2 = \sigma_x^2 + \sigma_y^2$
   c. $E\{(X+Y)^2\} = E\{X^2\} + E\{Y^2\}$
   d. $E\{g(X)h(Y)\} \ne E\{g(X)\}E\{h(Y)\}$ in general

34.9.9 Joint Characteristic Function

$$\Phi_{xy}(\omega_1,\omega_2) = E\{e^{j(\omega_1 X + \omega_2 Y)}\} = \int_{-\infty}^{\infty}\int f_{xy}(x,y)e^{j(\omega_1 x + \omega_2 y)}\,dx\,dy,\ \Psi_{xy}(\omega_1,\omega_2) = \ln\Phi_{xy}(\omega_1,\omega_2)$$

$$f_{xy}(x,y) = \frac{1}{(2\pi)^2}\int_{-\infty}^{\infty}\int e^{-j(\omega_1 x + \omega_2 y)}\,\Psi_{xy}(\omega_1,\omega_2)\,d\omega_1\,d\omega_2$$

$$\Phi_x(\omega) = E\{e^{j\omega X}\} = \Phi_{xy}(\omega,0),\quad \Phi_y(\omega) = \Phi_{xy}(0,\omega)$$

Example

$$\Phi_z(\omega) = E\{e^{j\omega z}\} = E\{e^{j(a\omega X + b\omega Y)}\} = \Phi_{xy}(a\omega, b\omega) \text{ if } Z = aX + bY.$$

$$\Phi_{xy}(\omega_1,\omega_2) = \Phi_x(\omega_1)\Phi_y(\omega_2)$$

if X and Y are independent.

34.9.10 Moment Theorem

$$\frac{\partial^k \partial^r \Phi(0,0)}{\partial\omega_1^k \partial\omega_2^r} = j^{(k+r)} \mu'_{kr}$$

34.9.11 Series Expansion of $\Phi(\omega_1,\omega_2)$

$$\Phi(\omega_1,\omega_2) = 1 + jE\{X\}\omega_1 + jE\{Y\}\omega_2 - \tfrac{1}{2}\{X^2\}\omega_1^2$$

$$-\tfrac{1}{2}E\{Y^2\}\omega_2^2 - E\{XY\}\omega_1\omega_2 + \cdots + \frac{1}{4!}\binom{4}{2}E\{X^2Y^2\}\omega_1^2\omega_2^2 + \cdots,$$

$$\Psi(\omega_1,\omega_2) = \ln\Phi(\omega_1,\omega_2) = j\mu_x\omega_1 + j\mu_y\omega_2 - \tfrac{1}{2}\sigma_x^2\omega_1^2 - r\sigma_x\sigma_y\omega_1\omega_2 - \tfrac{1}{2}\sigma_y^2\omega_2^2 + \cdots$$

34.10 Mean Square Estimation of R.V.'s

34.10.1 Mean Square Estimation of r.v.'s

a. a minimizes $E\{X-a)^2\}$ if $a = E\{X\} = \mu_x$

b. The function $g(X) = E\{Y|X\} =$ regression curve minimizes

$$E\{[Y - g(X)]^2\} = \iint_{-\infty}^{\infty} [y - g(x)]^2 f(x,y)\,dx\,dy$$

c. $a = \dfrac{r\sigma_y}{\sigma_x}$ and $b = E\{Y\} - aE\{X\}$ minimize the m.s. error

$$e = E\{[Y - (aX+b)^2]\} = \iint_{-\infty}^{\infty} [y - (ax-b)^2]\, f(x,y)\,dx\,dy$$

$e_m =$ minimum error $= \sigma_y^2(1-r^2)$, $r =$ correlation coefficient of X and Y.

d) If $E\{X\} = E\{Y\} = 0$ the constant a that minimizes the m.s. error $e = E\{(y-ax)^2\}$ is such that $E\{(Y-aX)X\} = 0$ (orthogonality principle) and the minimum m.s. error is: $e_m = E\{(Y-aX)Y\}$

$a =$ $E\{XY\}/E\{X^2\}$ and hence $e_m = E\{Y^2\} - \dfrac{E^2\{XY\}}{E\{X^2\}}$, also $e_m = E\{Y^2\} - E\{(aX)^2\}$,

$e_m \geq E\{[Y - E\{Y|X\}]^2\}$

34.11 Normal Random Variables

34.11.1 Jointly Normal

If $E\{X\} = E\{Y\} = 0$ the normal joint density is:

$$f(x,y)=\frac{1}{2\pi\sigma_1\sigma_2\sqrt{1-r^2}}\exp\left[-\frac{1}{2(1-r^2)}\left(\frac{x^2}{\sigma_1^2}-\frac{2rxy}{\sigma_1\sigma_2}+\frac{y^2}{\sigma_2^2}\right)\right],\quad E\{X^2\}=\sigma_1^2,\quad E\{Y^2\}=\sigma_2^2$$

34.11.2 Conditional Density

$$f(y|x)=\frac{1}{\sigma_2\sqrt{2\pi(1-r^2)}}\exp\left[-\frac{1}{2\sigma_2^2(1-r^2)}\left(y-\frac{r\sigma_2}{\sigma_1}x\right)^2\right],$$

$$E\{Y|X\}=\frac{r\sigma_2}{\sigma_1}X,\quad E\{Y^2|X\}=\sigma_2^2(1-r^2)+\frac{r^2\sigma_2^2}{\sigma_1^2}x^2$$

$$E\{XY\}=r\sigma_1\sigma_2,\quad E\{X^2Y^2\}=\sigma_1^2\sigma_2^2+2r^2\sigma_1^2\sigma_2^2$$

34.11.3 Mean Value

$$E\{(X-\mu_x)(Y-\mu_y)\}=r\sigma_1\sigma_2$$

34.11.4 Linear Transformations

$$Z=aX+bY,\quad W=cX+dY.$$

If X and Y are jointly normal with zero mean then

$$\sigma_z^2=E\{Z^2\}=E\{(aX+bY)^2\}=a^2\sigma_x^2+b^2\sigma_y^2+2ab\,r_{xy}\sigma_x\sigma_y$$

$$\sigma_w^2=E\{W^2\}=c^2\sigma_x^2+d^2\sigma_y^2+2cd\,r_{xy}\sigma_x\sigma_y,$$

$$r_{zw}\sigma_z\sigma_w=E\{ZW\}=ac\sigma_x^2+bd\sigma_y^2+(ad+bc)r_{xy}\sigma_x\sigma_y$$

34.12 Characteristic Functions of Two Normal Random Variables

34.12.1 Characteristic Function

$\Phi(\omega_1,\omega_2)=E\{\exp[j(\omega_1X+\omega_2Y)]\}=\exp[-\frac{1}{2}(\sigma_1^2\omega_1^2+2r\sigma_1\sigma_2\omega_1\omega_2+\sigma_2^2\omega_2^2)]$ for $E\{X\}=E\{Y\}=0$, and X and Y jointly normal.

34.12.2 Characteristic Function with Means

$\Phi(\omega_1,\omega_2)=\exp[j(\omega_1\mu_x+\omega_2\mu_y)]\exp[-\frac{1}{2}\mu_{20}\omega_1^2+2\mu_{11}\omega_1\omega_2+\mu_{02}\omega_2^2)]$, $\mu_{ij}=$ joint moments about the means.

34.13 Price Theorem for Two R.V.'s

34.13.1 Price Theorem

If X and Y are jointly normal with

$$\mu_{11} = E\{(X-\mu_x)(Y-\mu_y)\} = E\{XY\} - E\{X\}E\{Y\},$$

then,

$$E\{g(X,Y)\} = \iint_{-\infty}^{\infty} g(x,y)f(x,y)\,dx\,dy$$

a. If $\mu_{11} = 0$ (r.v.'s independent) $E\{X^k Y^r\} = E\{X^k\}E\{Y^r\}$

b. $E\{X^k Y^r\} = kr\int_0^{\mu_{11}} E\{X^{k-1}Y^{r-1}\}d\mu_{11} + E\{X^k\}E\{Y^r\}$

c. $E\{X^2Y^2\} = 4\int_0^{\mu_{11}} E\{XY\}d\mu_{11} + E\{X^2\}E\{Y^2\} = 4\int_0^{\mu_{11}} (\mu_{11} + E\{X\}E\{Y\})d\mu_{11}$
$+ E\{X^2\}E\{Y^2\} = 2\mu_{11}^2 + 4\mu_{11}E\{X\}E\{Y\} + E\{X^2\}E\{Y^2\}$

34.14 Sequences of Random Variables

34.14.1 Definitions

34.14.1.1 Definitions

n real r.v. $X_1, X_2, \cdots, X_n$; $F(x_1,x_2,\cdots,x_n) = P\{X_1 \le x_1,\cdots,X_n \le x_n\}$ = distribution function; $f(x_1,\cdots,x_n) = \partial^n F/\partial x_1,\cdots,\partial x_n$ = density function.

34.14.1.2 Marginal Densities

$F(x_1,x_3) = F(x_1,\infty,x_3,\infty) =$ marginal distribution for a sequence of four r.v. ; $f(x_1,x_3) = \iint_{-\infty}^{\infty} f(x_1,x_2,x_3,x_4)dx_2\,dx_4$ marginal density.

34.14.1.3 Functions of r.v.'s

$Y_1 = g_1(X_1,\cdots,X_n),\cdots,\quad Y_n = g_n(X_1,\cdots X_n),$

$$f_{y_1\cdots y_n}(y_1,\cdots,y_n) = f(x_1,x_2,\cdots,x_n)/|J(x_1,\cdots,x_n)|,\quad J(x_1,\cdots,x_n) = \begin{vmatrix} \dfrac{\partial g_1}{\partial x_1} & \cdots & \dfrac{\partial g_1}{\partial x_n} \\ \vdots & & \\ \dfrac{\partial g_n}{\partial x_1} & \cdots & \dfrac{\partial g_n}{\partial x_n} \end{vmatrix}$$

34.14.1.4 Conditional Densities

$f(x_1,\cdots,x_k|\,x_{k+1},\cdots,x_n) = f(x_1,\cdots,x_k,\cdots,x_n)/f(x_{k+1},\cdots,x_n).$

Example

$$f(x_1|\,x_2,x_3) = f(x_1,x_2,x_3)/f(x_2,x_3),\quad F(x_1|\,x_2,x_3) = \int_{-\infty}^{x_1} f(\xi_1,x_2,x_3)d\xi_1/f(x_2,x_3)$$

34.14.1.5 Chain Rule

$f(x_1,\cdots,x_n) = f(x_n | x_{n-1},\cdots,x_1)\cdots f(x_2|x_1)f(x_1)$

34.14.1.6 Removal Rule

$$f(x_1|x_3) = \int_{-\infty}^{\infty} f(x_1,x_2|x_3)dx_2,$$

$$f(x_1|x_4) = \iint_{-\infty}^{\infty} f(x_1|x_2,x_3,x_4)f(x_2,x_3|x_4)dx_2\,dx_3, \quad f(x_1|x_3) = \int_{-\infty}^{\infty} f(x_1|x_2,x_3)f(x_2|x_3)dx_2$$

34.14.1.7 Independent r.v.

$$F(x_1,\cdots,x_n) = F(x_1)\cdots F(x_n);\ f(x_1,\cdots,x_n) = f(x_1)\cdots f(x_n);$$

$$f(x_1,\cdots,x_k,x_{k+1},\cdots,x_n) = f(x_1,\cdots,x_k)f(x_{k+1},\cdots,x_n)$$

if $X_1,\cdots,X_k$ are independent of $X_{k+1},\cdots,X_n$

34.14.2 Mean, Moments, Characteristic Function

34.14.2.1 Expected Value

$$E\{g(X_1,\cdots,X_n)\} = \int_{-\infty}^{\infty}\cdots\int_{-\infty}^{\infty} g(x_1,\cdots,x_n)f(x_1,\cdots,x_n)dx_1\cdots dx_n$$

34.14.2.2 Conditional Expected Values

$$E\{X_1|x_2,\cdots,x_n\} = \int_{-\infty}^{\infty} x_1 f(x_1|x_2,\cdots,x_n)dx_1 = \int_{-\infty}^{\infty} x_1 f(x_1,\cdots,x_n)dx_1 / f(x_2,\cdots,x_n)$$

a. $E\{E\{X_1|X_2,\cdots,X_n\}\} = E\{X_1\}$

b. $E\{X_1X_2|X_3\} = E\{E\{X_1X_2|X_2,X_3\}|X_3\} = E\{X_2E\{X_1|X_2,X_3\}|X_3\}$

c. $E\{X_1|x_2,\cdots,x_n\} = E\{X_1\}$ if X_1 is independent from the remaining r.v.'s

34.14.2.3 Uncorrelated r.v.'s

$X_1,\cdots,X_n$ are uncorrelated if the covariance of any two of them is zero, $E\{X_iX_j\} = E\{X_i\}E\{X_j\}$ for $i \neq j$

34.14.2.4 Orthogonal r.v.'s

$E\{X_iX_j\} = 0$ for any $i \neq j$

34.14.2.5 Variance of Uncorrelated r.v.'s

$\sigma^2_{x_1+\cdots+x_n} = \sigma^2_{x_1} + \cdots + \sigma^2_{x_n}$, $\sigma^2_z = E\{|Z - E\{Z\}|^2\}$ if $Z = X + jY \equiv$ complex r.v., $E\{Z_iZ_j^*\} = E\{Z_i\}E\{Z_j^*\}$ $\equiv$ uncorrelated r.v.'s $i \neq j$, $E\{Z_iZ_j^*\} = 0 \equiv$ orthogonal,

$$f(x_1,y_1,x_2,y_2) = f(x_1,y_1)f(x_2,y_2) \text{ if } Z_1 = X_1 + jY_1 \text{ and } Z_2 = X_2 + jY_2$$

are independent,

$E\{Z_1Z_2^*\} = E\{Z_1\}E\{Z_2^*\} = \iint_{-\infty}^{\infty} z_1 f(x_1,y_1)dx_1\,dy_1 \iint_{-\infty}^{\infty} z_2^* f(x_2,y_2)dx_2\,dy_2$ = uncorrelated if they are independent.

34.14.2.6 Characteristic Functions

$\Phi(\omega_1,\cdots,\omega_n) = E\{e^{j(\omega_1 X + \cdots + \omega_n X_n)}\}$,

$$\frac{\partial^r \Phi(0,0)}{\Phi\omega_1^{k_1},\cdots,\partial\omega_n^{k_n}} = j^r \mu'_{k_1\cdots k_n},\quad \mu'_{k_1\cdots k_n} = E\{X_1^{k_1}\cdots X_n^{k_n}\},\quad r = k_1 + k_2 + \cdots + k_n$$

for dependent variables

34.14.2.7 Characteristic Functions for Independent Variables

$E\{e^{j\omega_1 X_1}\cdots e^{j\omega_n X_n}\} = E\{e^{j\omega_1 X_1}\}\cdots E\{e^{j\omega_n X_n}\}$

Example

$Z = X_1 + X_2 + X_3,\ \Phi_z(\omega) = E\{e^{j\omega(X_1+X_2+X_3)}\} = \Phi_1(\omega)\,\Phi_2(\omega)\,\Phi_3(\omega)$. Hence,

$$f_z(z) = \text{density function} = f_1(z) * f_2(z) * f_3(z)$$

34.14.2.8 Sample Mean

$\bar{X} = (X_1 + \cdots + X_n)/n,\ \bar{X} =$ random variable

34.14.2.9 Sample Variance

$$\bar{S} = [(X_1 - \bar{X})^2 + \cdots + (X_n - \bar{X})^2]/n = \sum_{i=1}^{n} \frac{X_i^2}{n} - \bar{X}^2,\quad \bar{S} = \text{random variable}$$

34.14.2.10 Statistic

A function of one or more variables that does not depend upon any unknown parameter.

Example

$Y = \sum_{i=1}^{n} X_i \equiv$ is a statistic; $Y = (X_1 - \mu)/\sigma \equiv$ is not a statistic unless μ and σ are known; the sample mean is a statistic.

34.14.2.11 Random Sums

$Y = \sum_{k=1}^{n} X_k$ = random sum; $E\{Y\} = \mu E\{n\}$ if n is an r.v. of discrete type and X_i's are independent of n and $E\{X_k\} = \mu$; $E\{Y^2\} = \mu^2 E\{n^2\} + \sigma^2 E(n)$ if the r.v. X_k are uncorrelated with the same variance σ^2.

34.14.3 Normal Random Variables

34.14.3.1 Density Function

$f(x_1,\cdots,x_n) = (1/[(2\pi)^{n/2}\sigma^n])\exp[-(x_1^2 + \cdots + x_n^2)/2\sigma^2]$ where the r.v. X_i are normal, independent with same variance σ^2

34.14.3.2 Density Function of the Sample Mean

$f_{\bar{x}}(x) = (1/\sqrt{2\pi\sigma^2/n})\exp(-nx^2/2\sigma^2)$ (see also 34.14.3.1)

34.14.3.3 Density Function of $\chi = [X_1 + \ldots + X_n]^{1/2}$ (see also 34.14.3.1)

$$f_\chi(\chi) = \frac{2}{2^{n/2}\sigma^n\Gamma(n/2)}\chi^{n-1}\exp(-\chi^2/2\sigma^2)u(\chi),$$ $n = n$ degrees of freedom, $u(\chi)$ = unit step function.

34.14.3.4 Density Fuction of $Y = \chi^2 = X_1^2 + \cdots + X_n^2$ (see also 34.14.3.1)

$$f_y(y) = \frac{1}{2^{n/2}\sigma^n\Gamma(n/2)}y^{(n-2)/2}\exp(-y/2\sigma^2)u(y),$$ $u(y) =$ unit step function

34.14.3.5 Density Function of the Variance $\bar{S}$ (see also 34.14.2.9 and 34.14.3.1)

$$f_{\bar{S}}(s) = \frac{1}{2^{(n-2)/2}(\sigma/\sqrt{n})^{n-1}\Gamma[(n-1)/2]}s^{(n-3)/2}\exp(-ns/2\sigma^2)u(s)$$

34.14.3.6 Characteristic Function

$$\Phi(\omega_1,\omega_2) = E\{e^{j(\omega_1 X_1+\omega_2 X_2)}\} = \exp[-\tfrac{1}{2}(\sigma_1^2\omega_1^2 + r\sigma_1\sigma_2\omega_1\omega_2 + r\sigma_1\sigma_2\omega_2\omega_1 + \sigma_2^2\omega_2^2)]$$

$\sigma_1^2 = \mu'_{11}$, $\sigma_2^2 = \mu'_{22}$, $r\sigma_1\sigma_2 = \mu'_{12}$

34.14.3.7 Matrix Form of Density Function

$$f(x_1,\cdots,x_n) = \exp(-\tfrac{1}{2}x^T\mu'^{-1}x)/\sqrt{(2\pi)^n|\mu'|}$$

$$\mu' = \begin{bmatrix} \mu'_{11} & \mu'_{12} & \cdots & \mu'_{1n} \\ \vdots & & & \vdots \\ \mu'_{n1} & \mu'_{n2} & \cdots & \mu'_{nn} \end{bmatrix},$$ μ'^{-1} = inverse of μ', $x = [x_1,\cdots,x_n]^T$, $|\mu'| =$ determinant

34.14.3.8 Characteristic Function in Matrix Form

$\Phi(\omega_1,\cdots,\omega_n) = \exp(-\tfrac{1}{2}\omega^T\mu'\omega)$, $\omega = [\omega_1,\cdots,\omega_n]^T$, $\mu' \equiv$ see 34.14.3.7

34.14.4 Convergence Concepts, Central Limit Theorem

34.14.4.1 Chebyshev Inequality (see 34.5.14)

$P\{\mu - \varepsilon < X < \mu + \varepsilon\} \geq 1 - \frac{\sigma^2}{\varepsilon^2} \cong 1$ for $\sigma << \varepsilon$. Hence the probability of the event $\{|X - \mu| < \varepsilon\}$ is close to 1. If the observed value of X is $X(\zeta)$ of an experiment, then $\mu - \varepsilon < X(\zeta) < \mu + \varepsilon$ or $X(\zeta) - \varepsilon < \mu < X(\zeta) + \varepsilon$ which estimates the mean.

34.14.4.2 Limiting Distribution

If $\lim_{n\to\infty} F_n(y) = F(y)$ for every point y at which $F(y)$ is continuous, then the r.v. Y_n is said to have a limiting distribution $F(y)$.

34.14.4.3 Stochastic Convergence

The r.v. Y_n converges stochastically to the constant c if and only if, for every $\varepsilon > 0$, $\lim_{n\to\infty} P\{|Y_n - c| < \varepsilon\} = 1$.

Example

$$P\{|\bar{X}_n - \mu| \ge \varepsilon\} = P\left\{|\bar{X}_n - \mu| \ge \frac{k\sigma}{\sqrt{n}}\right\} \le \frac{\sigma^2}{n\varepsilon^2} \quad (k = \varepsilon\sqrt{n}/\sigma)$$

the last inequality is due to Chebyshev inequality, $\bar{X}_n$ = mean of random sample of size n from a distribution with mean μ and variance σ^2, the mean and variance of $\bar{X}_n$ are μ and σ^2/n. Hence, $\lim_{n\to\infty} P\{|\bar{X}_n - \mu| \ge \varepsilon\} \le \lim_{n\to\infty} \frac{\sigma^2}{n\varepsilon^2} = 0$ which implies that $\bar{X}_n$ converges stochastically to μ if σ^2 is finite.

34.14.4.4 Convergence in Probability

$\lim_{n\to\infty} P\{|Y_n - c| < \varepsilon\} = 1$ implies Y_n converges to c in probability (same as stochastic convergence).

34.14.4.5 Convergence with Probability One

$P\{\lim_{n\to\infty} Y_n = c\} = 1$ implies Y_n converges to c with probability one.

34.14.4.6 Limiting Distribution and Moment-Generating Function

If $\lim_{n\to\infty} M(t;n) = M(t)$ then Y_n has a limiting distribution with distribution function $F(Y)$. $M(t;n) =$ moment generating function of Y_n in $-h < t < h$ for all n; $M(t) =$ m.g.f. of Y with d.f. $F(Y)$ in $|t| \le h_1 < h$ (see Table 34.1).

Example

Y_n's have binomial distribution $f(n,p)$. Let $\mu = np$ be the same for every n, that is, $p = \mu/n$. $M(t;n) = E\{e^{tY_n}\} = [(1-p) + pe^t]^n = \left[1 + \frac{\mu(e^t - 1)}{n}\right]^n$ for all real values of t. Hence, $\lim_{n\to\infty} M(t;n) = \exp[\mu(e^t - 1)]$ for all t.

Note: From advanced calculus

$$\lim_{n\to\infty}\left[1 + \frac{b}{n} + \frac{\psi(n)}{n}\right]^{cn} = \lim_{n\to\infty}\left(1 + \frac{b}{n}\right)^{cn} = e^{bc}, \quad \lim_{n\to\infty} \psi(n) = 0,$$

c and b do not depend upon n. Hence, Y_n has a limiting Poisson distribution with mean μ since the moment generating function of Poisson distribution is $\exp[\mu(e^t - 1)]$.

34.14.4.7 Central Limit Theorem

$Y_n = \left(\sum_{n=1}^{n} X_i - n\mu\right)\Big/\sqrt{n}\,\sigma$ has a limiting distribution that is normal with mean zero and variance 1. $X_1, X_2, \cdots, X_n$ is a random sample with mean μ and variance σ^2.

Example

Let $X_1, X_2, \cdots, X_n$ are r.v. from binomial distribution with $\mu = p$ and $\sigma^2 = p(1-p)$. To find the $P\{Y = 48, 49, 50, 51, 52\}$ is equivalent to finding the $P\{47.5 < Y < 52.5\}$. But $(Y_n - np)/\sqrt{np(1-p)}$ where $Y_n = \bar{X}_n = X_1 + X_2 + \cdots + X_n$ has a limiting distribution that is normal with mean zero and variance

1. If $n = 100$ and $p = 1/2$ then $np = 50$ and $\sqrt{np(1-p)} = 5$ and hence $P_r\{47.5 < Y < 52.5\} = P\left\{\frac{47.5-50}{5} < \frac{Y-50}{5} < \frac{52.5-50}{5}\right\} = P\left\{-0.5 < \frac{Y-50}{5} < 0.5\right\}$. But $(Y-50)/5$ has a limiting distribution that is normal with mean zero and variance one. From tables (see 34.2), we find that the probability $(Y-50)/5$ between -0.5 and 0.5 is 0.383.

34.14.4.8 Limiting Distribution of $W_n = U_n/V_n$

If $\lim_{n\to\infty} F_n(u) = F(u)$ and if V_n converges stochastically (see 34.14.4.3) to 1, then the limiting distribution of the r.v. $W_n = U_n/V_n$ is the same as that of U_n (the distribution $F(w)$).

Example

Let the distribution be $N(\mu, \sigma^2)$ (normal with mean μ and variance σ^2) and a sample from it with mean $\bar{X}_n$ and variance S_n^2. $\bar{X}_n$ converges stochastically to μ and S_n^2 converges stochastically to σ^2. Since S_n/σ converges stochastically to 1 (see 34.14.4.9), then $W_n = \bar{X}_n/(S_n/\sigma)$ has the same distribution as the limiting distribution of $\bar{X}_n$.

34.14.4.9 Limiting Distribution of U_n/c

If $\lim_{n\to\infty} F_n(u) = F(u)$ and U_n and U_n converges stochastically (see 34.14.4.3) to $c \neq 0$ then U_n/c converges stochastically to 1.

34.15 General Concepts of Stochastic Processes

34.15.1 Introduction

34.15.1.1 The X(t) Real Function (see Figure 34.3)

$X(t)$ $(X(t,\zeta))$ represents four different things:

1. A family of time functions (t variable, ζ variable)
2. A single time function (t variable, ζ fixed)
3. An r.v. (t fixed, ζ variable)
4. A single number (t , ζ fixed).

($\zeta \equiv$ outcomes of an experiment forming the space S, certain subsets of S are events, and probably probability of these events.)

34.15.1.2 Distribution Function (first order) F(x;t)

$F(x;t) = P\{X(t) \le x\}$. The event $\{X(t) \le x\}$ consists of all outcomes ζ such that at specified t, the functions $X(t)$ of the process do not exceed the given number x.

34.15.1.3 Density Function

$$f(x,t) = \frac{\partial F(x,t)}{\partial x}$$

34.15.1.4 Distribution Function (second order)

$$F(x_1, x_2; t_1, t_2) = P\{X(t_1) \le x_1, X(t_2) \le x_2\}$$

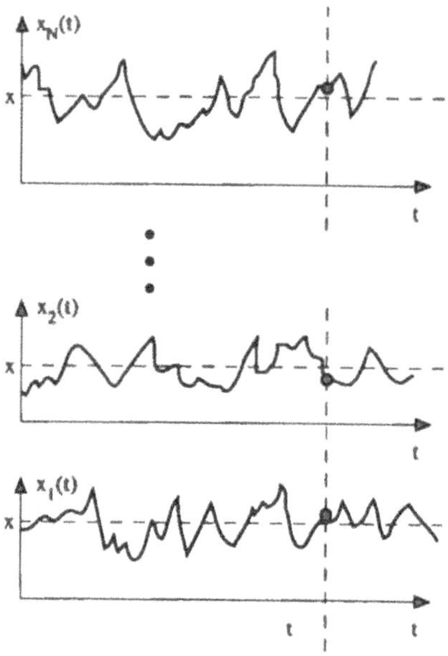

FIGURE 34.3

34.15.1.5 Density Function (second order)

$$f(x_1,x_2;t_1,t_2)=\frac{\partial F(x_1,x_2;t_1,t_2)}{\partial x_1 \partial x_2}$$

Note: $F(x_1,\infty;t_1,t_2)=F(x_1;t_1),\ f(x_1;t_1)=\int_{-\infty}^{\infty} f(x_1,x_2;t_1,t_2)dx_2$

34.15.1.6 Conditional Density

$$f(x_1,t_1|X_2(t_2)=x_2)=f(x_1,x_2;t_1,t_2)/f(x_2;t_2)$$

34.15.1.7 Mean

$$\mu(t)=\mu'(t)=E\{X(t)\}=\int_{-\infty}^{\infty} xf(x;t)dx$$

34.15.1.8 Autocorrelation

$$R(t_1,t_2)=E\{X(t_1)X(t_2)\}=\int_{-\infty}^{\infty} x_1x_2f(x_1,x_2;t_1,t_2)dx_1\,dx_2$$

34.15.1.9 Autocovariance

$$C(t_1,t_2)=E\{[X(t_1)-\mu(t_1)][X(t_2)-\mu(t_2)]\}=R(t_1,t_2)-\mu(t_1)\mu(t_2)$$

34.15.1.10 Variance

$$\sigma^2_{x(t)}=C(t,t)=R(t,t)-\mu^2(t)$$

34.15.1.11 Distribution Function (nth order)

$$F(x_1,\cdots,x_n;t_1,\cdots,t_n)=P\{X(t_1)\le x_1,\cdots,X(t_n)\le x_n\}$$

34.15.1.12 Density Function (n[th] order)

$$f(x_1,\cdots,x_n;t_1,\cdots,t_n)=\frac{\partial F(x_1,\cdots,x_n;t_1,\cdots,t_n)}{\partial x_1,\partial x_2,\cdots,\partial x_n}$$

34.15.1.13 Complex Process

$Z(t)=X(t)+jY(t)$ is a family of complex functions and is statistically determined in terms of the two-dimensional processes $X(t)$ and $Y(t)$.

34.15.1.14 Mean of Complex Process

$\mu_x(t)=E\{X(t)\}$, $X(t)\equiv$ real or complex

34.15.1.15 Autocorrelation of Complex Process

$R(t_1,t_2)=E\{X(t_1)X^*(t_2)\}$, * indicates conjugation.

34.15.1.16 Autocovariance of Complex Process

$$C(t_1,t_2)=E\{[X(t_1)-\mu(t_1)][X^*(t_2)-\mu^*(t_2)]\}=R(t_1,t_2)-\mu(t_1)\mu^*(t_2)$$

34.15.1.17 Crosscorrelation of Complex Processes

$$R_{xy}(t_1,t_2)=E\{X(t_1)Y^*(t_2)\}$$

34.15.1.18 Crosscovariance of Complex Processes

$$C_{xy}(t_1,t_2)=R_{xy}(t_1,t_2)-\mu_x(t_1)\mu_y^*(t_2)$$

34.15.1.19 Uncorrelated Complex Processes

$$R_{xy}(t_1,t_2)=\mu_x(t_1)\mu_y^*(t_2),\quad C_{xy}(t_1,t_2)=0$$

34.15.1.20 Orthogonal

$$R_{xy}(t_1,t_2)=0$$

34.15.1.21 Independent Processes X(t) and Y(t)

The processes are independent if the group $X(t_1),\cdots,X(t_n)$ is independent of the group $Y(t_1'),\cdots,Y(t_n')$ for any $t_1,\cdots,t_y,t_1',\cdots,t_n'$

34.15.1.22 Normal Processes

A process $X(t)$ is normal if r.v. $X(t_1),\cdots,X(t_n)$ are jointly normal for any n.

Example

$X(t)=A\cos at+B\sin at$, A and $B=$ independent normal r.v. with $E\{A\}=E\{B\}=0$ and $E\{A^2\}=E\{B^2\}=\sigma^2$ and $a=$ constant. Hence $E\{X(t)\}=E\{A\}\cos at+E\{B\}\sin at=0$, $R(t_1,t_2)=E\{(A\cos at_1+B\sin at_1)(A\cos at_2+B\sin at_2)\}=E\{A^2\cos at_1\cos at_2+E\{B^2\}\sin at_1\sin at_2=\sigma^2\cos\omega(t_2-t_1)$ which implies that $X(t)$ has mean value zero and variance σ^2 and, hence, $f(x,t)=(1/\sigma\sqrt{2\pi})\exp(-x^2/2\sigma^2)$. But $r=R(t_1,t_2)/\sqrt{R(t_1,t_1)R(t_2,t_2)}=\cos a(t_1-t_2)$ and from (34.11.1) $f(x_1,x_2;t_1,t_2)=[1/(2\pi\sigma^2\sqrt{1-\cos^2 a\tau})]\exp[-(x_1^2-2x_1x_2\cos a\tau+x_2^2)/(2\sigma^2(1-\cos^2 a\tau))]$ where $\tau=t_1-t_2$.

Note: $f(x;t)$ is independent of t and $f(x_1,x_2;t_1,t_2)$ depends only on $t_1-t_2=$ time difference.

34.16 Stationary Processes

34.16.1 Strict and Wide Sense Stationary

34.16.1.1 Strict Sense

$X(t)$ and $X(t+\varepsilon)$ have the same statistics for all shifts ε.

34.16.1.2 Jointly Stationary

$X(t), Y(t)$ the same joint statistics as $X(t+\varepsilon), Y(t+\varepsilon)$ for all shifts ε.

34.16.1.3 Complex Process

$Z(t) = X(t) + jY(t)$ is stationary if $X(t)$ and $Y(t)$ are jointly stationary.

34.1.6.1.4 Density Function of the nth Order

$f(x_1,\cdots,x_n;t_1,\cdots,t_n) = f(x_1,\cdots,x_n;t_1+\varepsilon,\cdots,t_n+\varepsilon)$ for all shifts ε.

1. $f(x,t) = f(x) =$ independent of time,
2. $E\{X(t)\} = \mu =$ constant
3. $f(x_1,x_2;t_1,t_2) = f(x_1,x_2;\tau)$, $\tau = t_1 - t_2 \equiv$ joint density of $X(t+\tau)$ and $X(t)$
4. $R(\tau) = E\{X(t+\tau)X(t)\} = R(-\tau)$,
5. $R_{xy}(\tau) = E\{X(t+\tau)Y(t)\}$

34.16.1.5 Wide Sense Stationary

$X(t)$ is wide sense stationary if $E\{X(t)\} = \mu =$ constant and $E\{X(t+\tau)X(t)\} = R(\tau)$.

Note: A wide sense process may not be a strict one. The converse is always true. A strict sense normal process is also a wide one.

Example

$X(t) = \cos(at+\varphi)$, φ is an r.v. and a is a constant. Then $E\{X(t)\} = E\{\cos(at+\varphi)\} = E\{\cos at \cos\varphi)\} - E\{\sin at \sin\varphi)\} = \cos at\ E\{\cos\varphi\} - \sin at\ E\{\sin\varphi\}$ and to have a stationary process $X(t)$ we must have the mean independent of t or equivalently

$$E\{\cos\varphi\} = E\{\sin\varphi\} = 0.\quad R(t+\tau,t) = E\{\cos(a(t+\tau)+\varphi)\cos(at+\varphi)\}$$

$$= \tfrac{1}{2}\cos a\tau + \tfrac{1}{2}E\{\cos(2at + a\tau + 2\varphi)\}.$$

For the autocorrelation to be independent of t, we must have $E\{\cos 2\varphi\} = E\{\sin 2\varphi\} = 0$. Hence $R(\tau) = \tfrac{1}{2}\cos a\tau$ and the process is wide stationary.

34.17 Stochastic Processes and Linear Deterministic Systems

34.17.1 Memoryless and Time-Invariant System

34.17.1.1 Output System

$Y(t) = g[X(t)]$

34.17.1.2 Mean of Output

$$E\{Y(t)\} = \int_{-\infty}^{\infty} g(x) f_x(x;t)\,dx$$

34.17.1.3 Autocorrelation of Output

$E\{Y(t_1)Y(t_2)\} = \iint_{-\infty}^{\infty} g(x_1)g(x_2)f_x(x_1x_2;t_1,t_2)dx_1\,dx_2$, x_i is the r.v. of the process $X(t)$ at $t = t_i$.

34.17.1.4 Output Density Function

$f_y(y_1,\cdots,y_k;t_1,\cdots,t_k) = f_x(x_1,\cdots,x_k;t_1,\cdots,t_k)/[|g'(x_1)|\cdots|g'(x_k)|]$
$y_1 = g(x_1),\cdots,y_k = g(y_k)$, prime indicates derivative.

Example

If

$$Y(t) = \begin{cases} 1 & X(t) \le x \\ 0 & X(t) > x \end{cases}$$

then $P\{Y(t) = 1\} = P\{X(t) \le x\}$, $P\{Y(t) = 0\} = P\{X(t) \ge x\}$ and hence, $E\{Y(t)\} = 1 \cdot P\{X(t) \le x\} + 0 \cdot P\{X(t) \ge x\} = F_x(x)$

34.17.2 Linear System

34.17.2.1 Linear Operator

$Y(t) = L[X(t)]$, $L[a_1X_1(t) + a_2X_2(t)] = a_1L[X_1(t)] + a_2L[X_2(t)]$, $L[AX(t)] = AL[X(t)]$, $A \equiv$ random variable, $Y(t+\tau) = L[X(t+\tau)]$ implies L (system) is time invariant, L = transformation (system).

34.17.2.2 Linear Transformation

$E\{Y(t)\} = L[E\{X(t)\}]$

34.17.2.3 Autocorrelation

$R_{xy}(t_1,t_2) = L_{t_2}[R_{xx}(t_1,t_2)] = L_{t_2}[E\{X(t_1)X(t_2)\}]$; $R_{yy}(t_1,t_2) = L_{t_1}[R_{xy}(t_1,t_2)] = L_{t_1}[E\{X(t_1)Y(t_2)\}]$;
$E\{Y(t_1)Y(t_2)Y(t_3)\} = L_{t_1}L_{t_2}L_{t_3}[E\{X(t_1)X(t_2)X(t_3)\}]$ which means to evaluate the left-hand side we must know the mean of the right-hand side for every t_1, t_2, t_3.

34.17.3 Stochastic Continuity and Differentiation

34.17.3.1 Continuity in the m.s. Sense

If $E\{[X(t+\tau) - X(t)]^2\} \to 0$ as $\tau \to 0$ the $X(t)$ is continuous in the mean square (m.s.) sense.

34.17.3.2 Continuity with Probability 1

$\lim X(t+\varepsilon) = X(t)$, $\varepsilon \to 0$ for all outcomes

34.17.3.3 Continuity in the m.s.

$E\{[X(t+\tau) - X(t)]^2\} \to 0$ with $\tau \to 0$

34.17.3.4 Continuity in the Mean

$E\{X(t+\tau)\} \to E\{X(t)\}$, $\tau \to 0$

34.17.3.5 Continuity of Stationary Process

$E\{[X(t+\tau) - X(t)]^2\} = 2[R(0) - R(\tau)]$ The process is continuous if its autocorrelation function is continuous at $\tau = 0$.

34.17.3.6 Differentiable Stationary Process

A stationary process $X(t)$ is differentiable in the m.s. sense if its autocorrelation $R(\tau)$ has derivatives of order up to two.

34.17.3.7 Autocorrelation of Derivatives

$$R_{xx'}(\tau) = -\frac{dR_{xx}(\tau)}{d\tau},\quad R_{x'x'}(\tau) = \frac{dR_{xx'}(\tau)}{d\tau},$$

$$R_{x'x'}(\tau) = -\frac{d^2R_{xx}(\tau)}{d\tau^2},\quad E\{[X'(t)]^2\} = R_{x'x'}(0) = -\frac{d^2R_{xx}(0)}{d\tau^2}.$$

Primes indicate differentiation.

34.17.3.8 Taylor Series

If

$$R(\tau) = \sum_{n=0}^{\infty} R^{(n)}(0)\frac{\tau^n}{n!}$$

then $X^{(n)}(t)$ exists in the m.s. sense and

$$X(t+\tau) = \sum_{n=0}^{\infty} X^{(n)}(t)\frac{\tau^n}{n!}$$

where (n) in exponent means n^{th} derivative and $R(\tau)$ must have derivatives of any order.

34.17.4 Stochastic Integrals, Averages, and Ergoticity

34.17.4.1 Integral of a Process

$S = \int_a^b X(t)dt \equiv$ r.v. and defines a number $S(\zeta)$ for each outcome.

34.17.4.2 M.S. Limit of a Sum

$$\lim E\left\{\left[S - \sum_{i=1}^{n} X(t_i)\Delta t_i\right]^2\right\} = 0 \text{ for } \Delta t_i \to 0$$

34.17.4.3 Mean Value of Integrals

$E\{S\} = \int_a^b E\{X(t)\}dt = \int_a^b \mu(t)dt$ (see 34.17.4.1) since the integral can be equated to a sum.

34.17.4.4 The Mean of S^2

$$E\{S^2\} = \int_a^b\int_a^b E\{X(t_1)X(t_2)\}dt_1\,dt_2 = \int_a^b\int_a^b R(t_1,t_2)dt_1\,dt_2$$

34.17.4.5 Variance of S

$$\sigma_s^2 = \int_a^b\int_a^b [R(t_1,t_2) - \mu(t_1)\mu(t_2)]dt_1\,dt_2 = \int_a^b\int_a^b C(t_1,t_2)dt_1\,dt_2;$$

$$\sigma_s^2 = \frac{1}{2T}\int_{-2T}^{2T}\left(1-\frac{|\tau|}{2T}\right)C(\tau)d\tau$$

for $a=-T$ and $b=T$; if $X(t)$ is real $C(\tau)$ is even and

$$\sigma_s^2 = \frac{1}{T}\int_0^{2T}(1-\frac{\tau}{2T})C(\tau)d\tau = \frac{1}{T}\int_0^{2T}\left(1-\frac{\tau}{2T}\right)[R(\tau)-\mu^2]d\tau$$

34.17.4.6 Ergotic Process

A process $X(t)$ is ergotic if all its statistics can be determined from a single function (realization) $X(t,\zeta)$ of the process.

34.17.4.7 Ergoticity of the Mean

$$\lim_{T\to\infty}\overline{X}_T(t) = \lim_{T\to\infty}\frac{1}{2T}\int_{-T}^{T}X(t)dt = E\{X(t)\} = \mu = \text{constant (stationary process) iff}$$

$$\lim_{T\to\infty}\frac{1}{T}\int_0^{2T}\left(1-\frac{\tau}{2T}\right)[R(\tau)-\mu^2]d\tau = 0$$

Example

$$E\{X(t)\},\ R(\tau)=\exp(-\alpha|\tau|),\ E\{S\} = \frac{1}{2T}\int_{-T}^{T}E\{X(t)\}dt = 0,$$

$$\sigma_s^2 = \frac{1}{T}\int_0^{2T}\left(1-\frac{\tau}{2T}\right)\exp(-\alpha\tau)d\tau = \frac{1}{\alpha T}-\frac{1-e^{-2\alpha T}}{2\alpha^2T^2}$$

which approaches zero as $T\to\infty$ and hence $X(t)$ is ergotic in the mean.

34.17.4.8 Ergoticity of the Autocorrelation

$$\lim_{T\to\infty}R_T(\lambda) = \lim_{T\to\infty}\frac{1}{2T}\int_{-T}^{T}X(t+\lambda)X(t)dt = E\{X(t+\lambda)X(t)\} = R(\lambda) \text{ iff}$$

$$\lim_{T\to\infty}\frac{1}{T}\int_0^{2T}\left(1-\frac{\tau}{2T}\right)[R_{yy}(\tau)-R^2(\lambda)]d\tau = 0$$

where $Y(t) = X(t+\lambda)X(t)$, $E\{Y(t)\} = E\{X(t+\lambda)X(t)\} = R(\lambda)$

34.18 Correlation and Power Spectrum of Stationary Processes

34.18.1 Correlation and Covariance

34.18.1.1 Correlation

$R(\tau) = E\{X(t+\tau)X^*(t)\} = R_x(\tau) = R_{xx}(\tau)$, * indicates conjugation; $R(-\tau) = R^*(\tau)$

34.18.1.2 Crosscorrelation

$R_{xy}(\tau) = E\{X(t+\tau)Y^*(t)\} = R^*_{yx}(-\tau)$

34.18.1.3 Auto and Crosscovariance

$C(\tau) = E\{[X(t+\tau)-\mu][X^*(t)-\mu^*]\} = R(\tau) - |\mu|^2 \equiv$ covariance;

$C_{xy}(\tau) = E\{[X(t+\tau)-\mu_x][Y^*(t)-\mu^*_y]\} = R_{xy}(\tau) - \mu_x\mu^*_y \equiv$ crosscovariance

34.18.1.4 Properties of Correlation

1. $R_{zz}(\tau) = R_{xx}(\tau) + R_{yy}(\tau) + R_{xy}(\tau) + R_{yx}(\tau)$ if $Z(t) = X(t) + Y(t)$
2. $R_{ww}(\tau) = R_{xx}(\tau)R_{yy}(\tau)$ if $W(t) = X(t)Y(t)$ and $X(t)$ is independent of $Y(t)$
3. $R(0) = E\{|X(t)|^2\} \geq 0$
4. $E\{[X(t+\tau) \pm X(t)]^2\} = 2[R(0) \pm R(\tau)]$
5. $|R(\tau)| \leq R(0)$, $R(\tau)$ is maximum at the origin,
6. $E\{[X(t+\tau) + aY^*(t)]^2\} = R_{xx}(0) + 2aR_{xy}(\tau) + a^2R_{yy}(0)$, $X(t)$ and $Y(t)$ are real processes
7. $2|R_{xy}(\tau)| \leq R_{xx}(0) + R_{yy}(0)$

34.18.2 Power Spectrum of Stationary Processes

34.18.2.1 Power Spectrum (spectral density; see Table 34.8.)

$$S(\omega) = \int_{-\infty}^{\infty} R(\tau)e^{-j\omega\tau}\,d\tau \equiv \text{real, since } R(-\tau) = R^*(\tau);$$

$$R(\tau) = \frac{1}{2\pi}\int_{-\infty}^{\infty} S(\omega)e^{j\omega\tau}\,d\omega = \int_{-\infty}^{\infty} S(f)e^{j2\pi f\tau}\,df\,; R(0) = \frac{1}{2\pi}\int_{-\infty}^{\infty} S(\omega)\,d\omega = E\{|X(t)|^2\};$$

if $X(t)$ is real implies $R(\tau)$ is real and even and, hence, $S(-\omega) = S(\omega) \equiv$ even, $R(\tau)$ correlation

TABLE 34.8

$R(\tau)=\frac{1}{2\pi}\int_{-\infty}^{\infty} S(\omega)e^{j\omega\tau}\,d\omega$	$S(\omega)=\int_{-\infty}^{\infty} R(\tau)e^{-j\omega\tau}\,d\tau$
$e^{-a\lvert\tau\rvert}$	$\frac{2a}{a^2+\omega^2}$
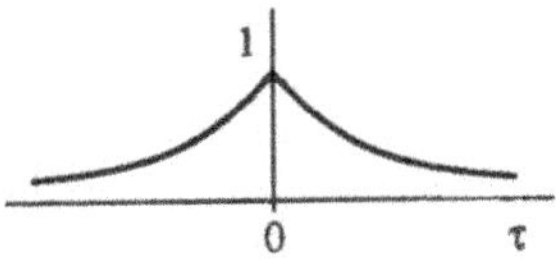	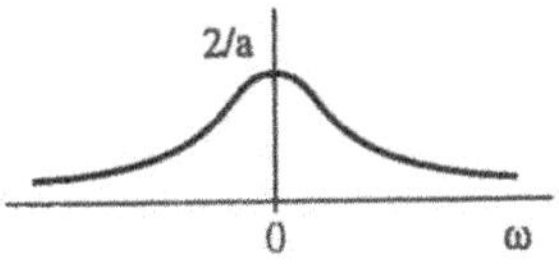
$1-\frac{\lvert\tau\rvert}{T},\ 0\le\tau\le T$ $0\ ,\ \lvert\tau\rvert>0$	$\frac{4\sin^2(\omega T/2)}{T\omega^2}$
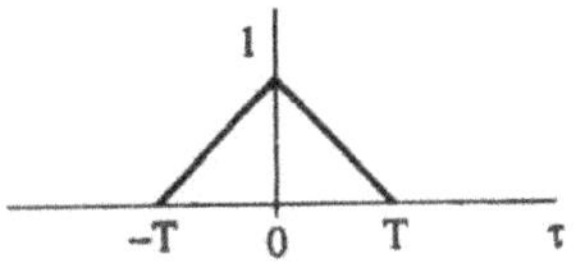	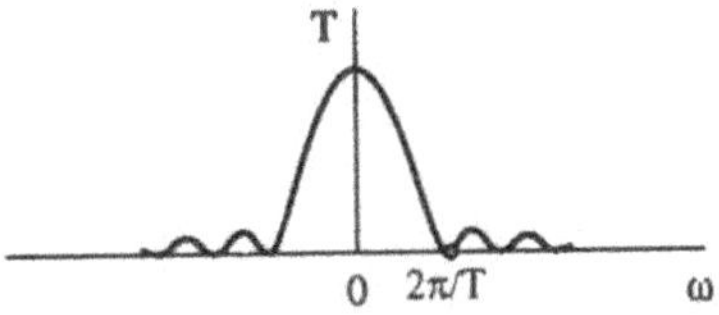
$\sigma^2 e^{-a\lvert\tau\rvert}\cos\omega_0\tau$	$\sigma^2\left[\frac{a}{(\omega-\omega_0)^2+a^2}+\frac{a}{(\omega+\omega_0)^2+a^2}\right]$
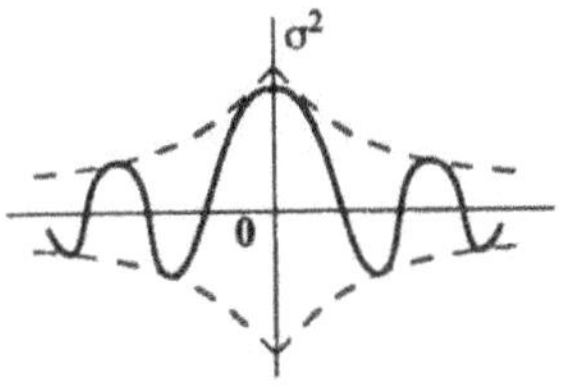	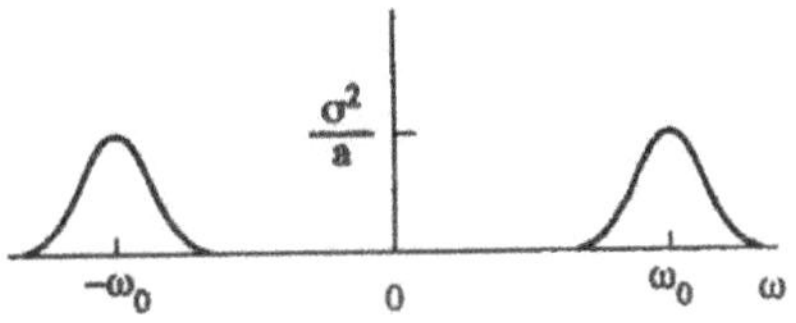
	$2\pi\delta(\omega)$
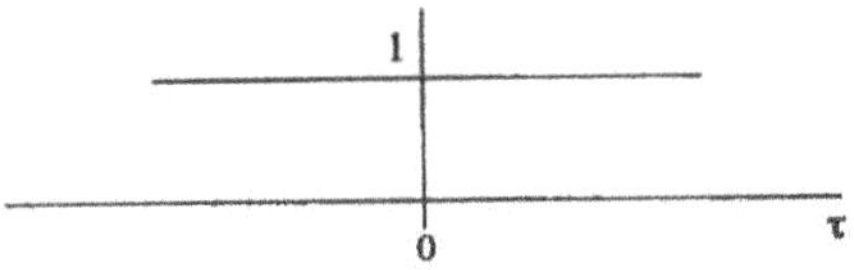	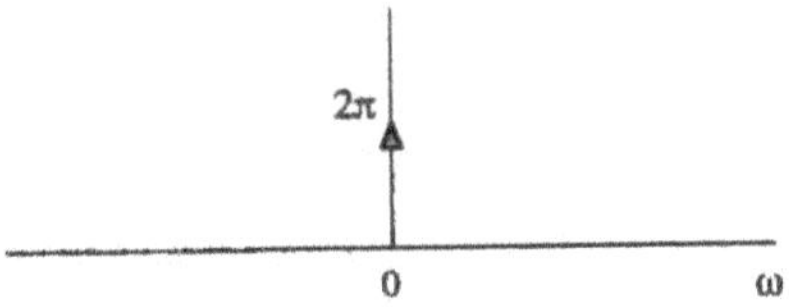
$a\delta(\tau)$	
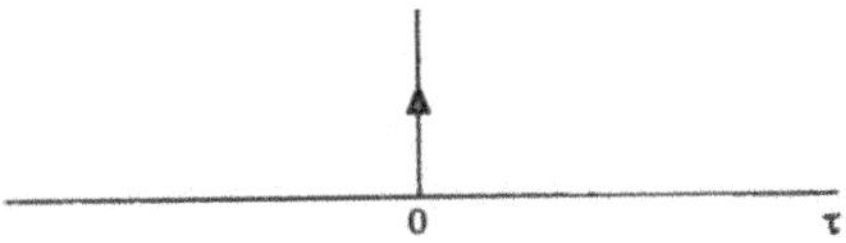	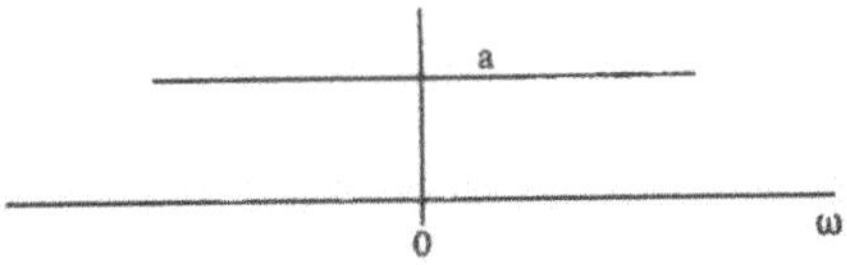
$a\cos\omega_0\tau$	$a\pi[\delta(\omega+\omega_0)+\delta(\omega-\omega_0)]$
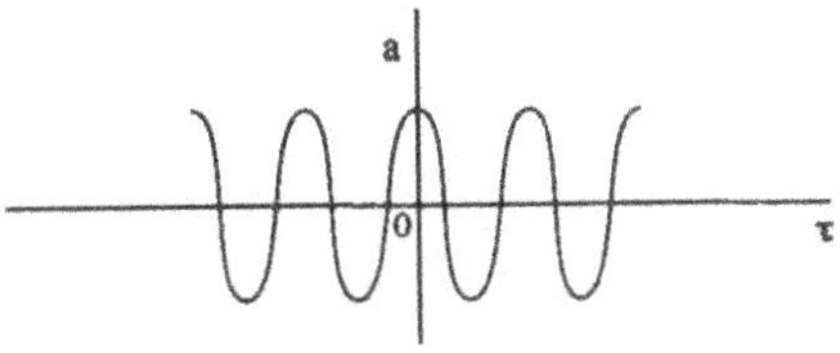	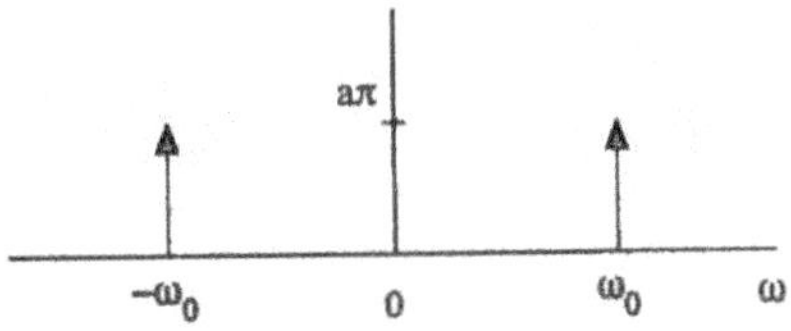

TABLE 34.8 (continued)

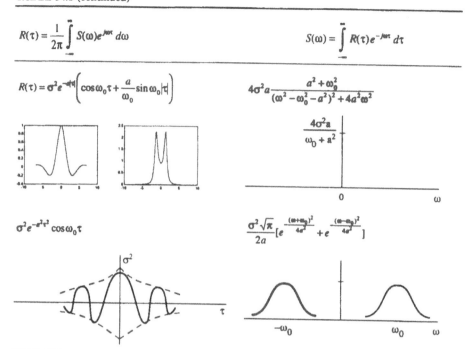

$R(\tau)=\frac{1}{2\pi}\int_{-\infty}^{\infty} S(\omega)e^{j\omega\tau}\,d\omega$	$S(\omega)=\int_{-\infty}^{\infty} R(\tau)e^{-j\omega\tau}\,d\tau$
$R(\tau)=\sigma^2 e^{-a\lvert\tau\rvert}\left(\cos\omega_0\tau+\frac{a}{\omega_0}\sin\omega_0\lvert\tau\rvert\right)$	$4\sigma^2 a\frac{a^2+\omega_0^2}{(\omega^2-\omega_0^2-a^2)^2+4a^2\omega^2}$
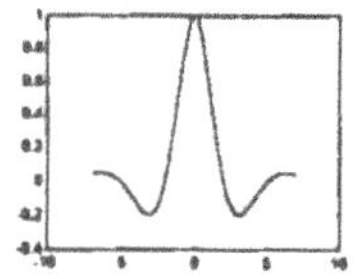 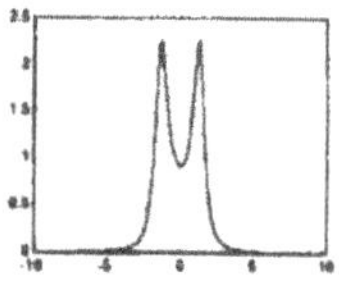	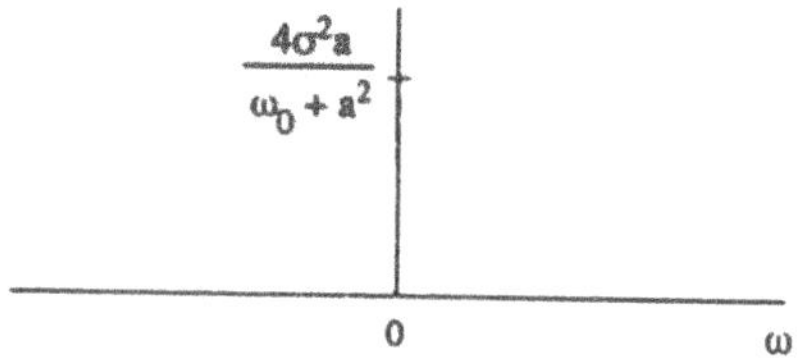
$\sigma^2 e^{-a^2\tau^2}\cos\omega_0\tau$	$\frac{\sigma^2\sqrt{\pi}}{2a}\left[e^{-\frac{(\omega+\omega_0)^2}{4a^2}}+e^{-\frac{(\omega-\omega_0)^2}{4a^2}}\right]$
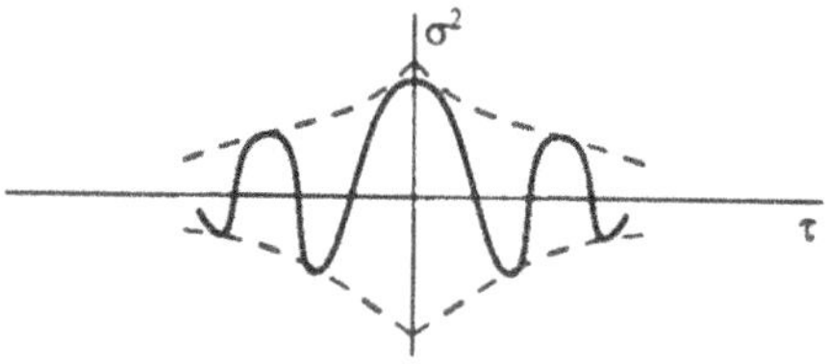	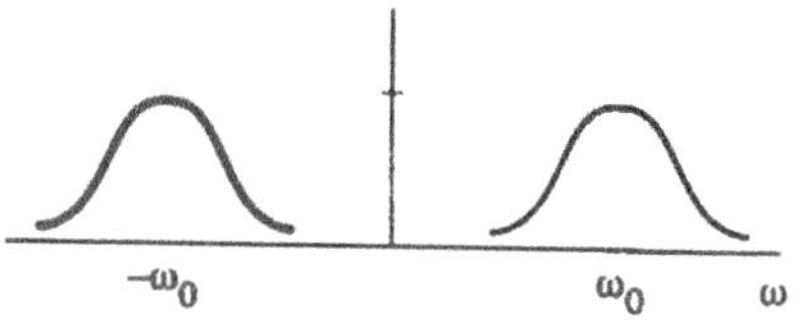

34.18.2.2 Cross-power Spectrum

$$S_{xy}(\omega)=\int_{-\infty}^{\infty} R_{xy}(\tau)e^{-j\omega\tau}\,d\tau=S_{yx}^{*}(\omega),\quad R_{xy}(\tau)=\frac{1}{2\pi}\int_{-\infty}^{\infty} S_{xy}(\omega)e^{j\omega\tau}\,d\omega;$$

$$R_{xy}(0)=\frac{1}{2\pi}\int_{-\infty}^{\infty} S_{xy}(\omega)\,d\omega=E\{X(t)Y^{*}(t)\}$$

34.18.2.3 Orthogonal Processes

$R_{xy}(\tau)=0,\ S_{xy}(\omega)=0,\ R_{x+y}(\tau)=R_x(\tau)+R_y(\tau),\ S_{x+y}(\omega)=S_x(\omega)+S_y(\omega)$ (see 34.14.2.4)

34.18.2.4 Relationships Between Processes (see Table 34.9)

TABLE 34.9

$X(t)$	$R(\tau)=E\{X(t+\tau)X(t)\}$	$S(\omega)=\int_{-\infty}^{\infty} R(\tau)e^{-j\omega\tau}\,d\tau$
$aX(t)$	$\lvert a\rvert^2 R(\tau)$	$\lvert a\rvert^2 S(\omega)$
$\frac{dX(t)}{dt}$	$-\frac{d^2R(\tau)}{d\tau^2}$	$\omega^2 S(\omega)$
$\frac{d^n X(t)}{dt^n}$	$(-1)^n\frac{d^{2n}R(\tau)}{d\tau^{2n}}$	$\omega^{2n}S(\omega)$

TABLE 34.9 (continued)

$X(t)$	$R(\tau)=E\{X(t+\tau)X(t)\}$	$S(\omega)=\int_{-\infty}^{\infty} R(\tau)e^{-j\omega\tau}\,d\tau$
$X(t)e^{\pm j\omega_0 t}$	$e^{\pm j\omega_0 t}R(\tau)$	$S(\omega\mp\omega_0)$
$b_0\frac{d^m X(t)}{dt^m}+b_1\frac{d^{m-1}X(t)}{dt^{m-1}}+\cdots+b_m X(t)$	$\frac{1}{2\pi}\int_{-\infty}^{\infty} e^{-j\omega\tau}\left\|b_0(j\omega)^m+b_1(j\omega)^{m-1}+\cdots+b_m\right\|^2 S(\omega)\,d\omega$	$\left\|b_0(j\omega)^m+b_1(j\omega)^{m-1}+\cdots+b_m\right\|^2 S(\omega)$
$b_0\frac{d^2X(t)}{dt^2}+b_1\frac{dX(t)}{dt}+b_2X(t)$	$\frac{1}{2\pi}\int_{-\infty}^{\infty} e^{-j\omega\tau}\left\|b_0(j\omega)^2+b_1(j\omega)+b_2\right\|S(\omega)\,d\omega$	$\left\|b_0(j\omega)^2+b_1(j\omega)+b_2\right\|S(\omega)$

34.18.2.5 Power Spectrum as Time Average (ergoticity)

$$\lim_{T\to\infty} E\{S_T(\omega)\}=S(\omega)=\int_{-\infty}^{\infty} R(\tau)e^{-j\omega\tau}\,d\tau \text{ where } S_T(\omega)=\frac{1}{2T}\left|\int_{-T}^{T} X(t)e^{-j\omega t}\,dt\right|^2 \text{ and } \int_{-\infty}^{\infty}|\tau R(\tau)|d\tau<\infty.$$

Example

$R(\tau)=e^{-2\lambda|\tau|}\equiv$ autocorrelation of random telegraph signal,

$$S(\omega)=\int_{-\infty}^{\infty} e^{-2\lambda|\tau|}e^{-j\omega\tau}\,d\tau=\int_{-\infty}^{0} e^{2\lambda\tau}e^{-j\omega\tau}\,d\tau+\int_{0}^{\infty} e^{-2\lambda\tau}e^{-j\omega\tau}\,d\tau=\frac{4\lambda}{4\lambda^2+\omega^2}$$

Example

If $X(t)=\sum_{i=1}^{n} A_i e^{j\omega_i t}$ then $R(\tau)=\sum_{i=1}^{n}\sigma_i^2 e^{j\omega_i\tau}$ where $E\{A_i^2\}=\sigma_i^2$ and, hence,

$$S(\omega)=\sum_{i=1}^{n}\sigma_i^2\int_{-\infty}^{\infty} e^{-j(\omega-\omega_i)\tau}\,d\tau=2\pi\sum_{i=1}^{n}\sigma_i^2\,\delta(\omega-\omega_i)$$

34.18.3 Linear Systems and Stationary Processes

34.18.3.1 Transfer Function

$H(\omega)=\int_{-\infty}^{\infty} h(t)e^{-j\omega t}\,dt$; $h(t)$ = impulse response of a linear and time invariant system. $h(t)$ is produced if delta function is the input to the system.

34.18.3.2 Mean

$E\{Y(t)\}=\mu_y=\int_{-\infty}^{\infty} E\{X(t-\alpha)\}h(\alpha)d\alpha=\mu_x\int_{-\infty}^{\infty} h(\alpha)d\alpha=\mu_x H(0)$, $Y(t)\equiv$ output of the system, $X(t)\equiv$ input of the system, $h(t)$ = impulse response of the system

34.18.3.3 Cross-Correlation (outputs — inputs)

$$E\{Y(t)X^*(t-\tau)\} = R_{yx}(\tau) = \int_{-\infty}^{\infty} E\{X(t-\alpha)X^*(t-\tau)\}h(\alpha)d\alpha$$

$$= \int_{-\infty}^{\infty} R_{xx}(\tau-\alpha)h(\alpha)d\alpha = R_{xx}(\tau) * h(\tau)$$

34.18.3.4 Autocorrelation of Input

$E\{X(t-\alpha)X^*(t-\tau)\} = R_{xx}[t-\alpha-(t-\tau)] = R_{xx}(\tau-\alpha)$

34.18.3.5 Autocorrelation of Output

$$R_{yy}(\tau) = E\{Y(t+\tau)Y^*(t)\} = \int_{-\infty}^{\infty} E\{Y(t+\tau)X^*(t-\alpha)\}h^*(\alpha)d\alpha = R_{yx}(\tau) * h^*(-\tau);$$

$$R_{xy}(\tau) = R_{xx}(\tau) * h^*(-\tau),\ \ R_{yy}(\tau) = R_{xy}(\tau) * h(\tau),\ \ R_{yy}(\tau) = R_{xx}(\tau) * h^*(-\tau) * h(\tau)$$

(see 34.18.3.2 and 34.18.3.3)

34.18.3.6 White Noise Input

$X(t)$ = white noise, $R_{xx}(\tau) = \delta(\tau)$, $h(t) = 0$ for $t < 0$ (causal system), then $R_{xy}(\tau) = h(-\tau) = 0$ for $\tau > 0$ which implies $Y(t)$ and $X(t)$ are orthogonal.

34.18.3.7 Power Spectrum

$S_{xy}(\omega) = S_{xx}(\omega)H^*(\omega)$ and $S_{yy}(\omega) = S_{xy}(\omega)H(\omega)$ and hence, $S_{yy}(\omega) = S_{xx}(\omega)|H(\omega)|^2$ ($\mathrm{F}\{h^*(-t)\}$ = $H^*(\omega)$, F stands for Fourier transform)

34.18.3.8 Positiveness of Power Spectrum

$S(\omega) \geq 0$

Example

$Y(t) = dX(t)/dt$, $H(\omega) = j\omega \equiv$ transfer function of a differentiation, and hence, 34.18.3.7 gives $S_{xx'}(\omega) = S_{xx}(\omega)(-j\omega)$ and $S_{x'x'}(\omega) = \omega^2 S_{xx}(\omega)$. Thus $R_{xx'}(\tau) = \mathrm{F}^{-1}\{S_{xx}(\omega)(-j\omega)\} = -\dfrac{dR_{xx}(\tau)}{d\tau}$ and $R_{x'x'}(\tau) = \mathrm{F}^{-1}\{\omega^2 S_{xx}(\omega)\} = -\dfrac{d^2R_{xx}(\tau)}{d\tau^2}$ where F^{-1} stands for inverse Fourier transform.

34.18.3.9 Multiple Terminals Spectra (see Figures 34.4 and 34.5)

$$R_{y_1y_2}(\tau) = \int_{-\infty}^{\infty} R_{x_1y_2}(\tau-\alpha)h_1(\alpha)d\alpha = R_{x_1y_2}(\tau) * h_1(\tau);\ \ R_{x_1y_2}(\tau) = \int_{-\infty}^{\infty} R_{x_1x_2}(\tau+\beta)h_2^*(\beta)d\beta$$

$$= R_{x_1x_2}(\tau) * h_2^*(-\tau);\ \ S_{y_1y_2}(\omega) = S_{x_1y_2}(\omega)H_1(\omega);\ \ S_{x_1y_2}(\omega) = S_{x_1x_2}(\omega)H_2^*(\omega);$$

$$S_{y_1y_2}(\omega) = S_{x_1x_2}(\omega)H_1(\omega)H_2^*(\omega)$$

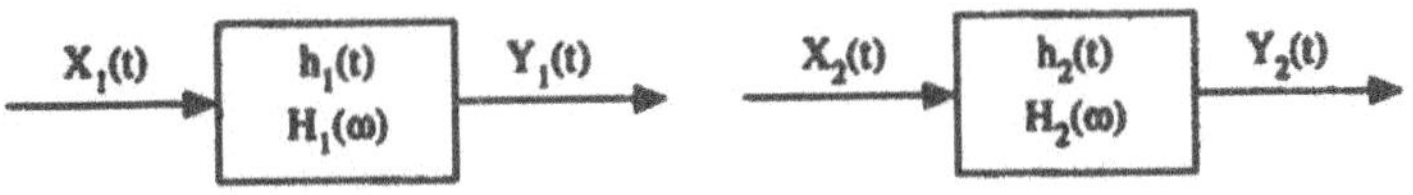

FIGURE 34.4

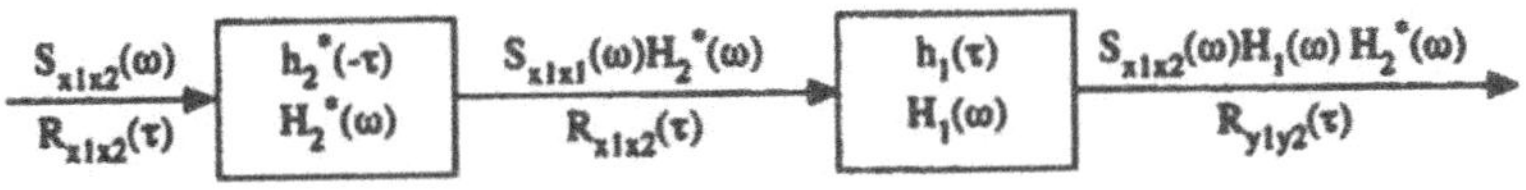

FIGURE 34.5

Note: If $H_1(\omega)H_2(\omega) = 0 \equiv$ disjoint systems, $S_{y_1y_2}(\omega) = 0$; if also $E\{X(t)\} = 0$, $Y_1(t)$ and $Y_2(t)$ are uncorrelated.

Example (see Figure 34.6)

$H_1(\omega) = \dfrac{a + j\omega}{2a + j\omega}$, $H_2(\omega) = \dfrac{a}{2a + j\omega}$, $R_x(\tau) = e^{-2\lambda|\tau|} \equiv$ random telegraph signal, $S_x(\omega) = \dfrac{4\lambda}{4\lambda^2 + \omega^2}$,

$S_{v_1}(\omega) = \dfrac{4\lambda}{4\lambda^2 + \omega^2}\dfrac{a^2 + \omega^2}{4a^2 + \omega^2}$, $S_{v_2}(\omega) = \dfrac{4\lambda}{4\lambda^2 + \omega^2}\dfrac{a^2}{4a^2 + \omega^2}$, $S_{v_1v_2}(\omega) = \dfrac{4\lambda}{4\lambda^2 + \omega^2}\dfrac{a + j\omega}{2a + j\omega}\dfrac{a}{2a - j\omega}$

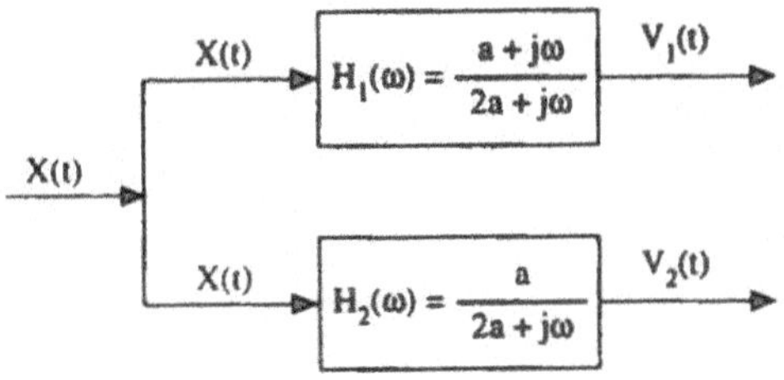

FIGURE 34.6

34.18.3.10 Stochastic Differential Equations

1. $a_nY^{(n)}(t) + \cdots + a_0Y(t) = X(t)$, $E\{Y(t)\} = \frac{1}{a}E\{X(t)\}$ (see (34.18.3.2))

$$H(\omega) = \frac{1}{a_n(j\omega)^n + \cdots + a_0}, \quad S_y(\omega) = S_x(\omega)|H(\omega)|^2$$

2. $a_nY^{(n)}(t) + \cdots + a_0Y(t) = b_mX^{(m)} + \cdots + b_0X(t)$

$$S_y(\omega) = \frac{|b_m(j\omega)^m + \cdots + b_0|^2}{|a_n(j\omega)^n + \cdots + a_0|^2}S_x(\omega)$$

Exponent in parenthesis indicates derivatives.

Example

$Y^{(2)}(t) + 2hY^{(1)}(t) + k^2Y(t) = X(t) \equiv$ white noise process, then $R_x(\tau) = \sigma^2\delta(\tau)$ and $S_x(\omega) = \sigma^2$,

$$S_y(\omega) = \frac{1}{|(j\omega)^2 + 2hj\omega + k^2|^2} S_x(\omega) = \frac{\sigma^2}{(\omega^2 - k^2)^2 + 4h^2\omega^2} = \frac{4\gamma^2\alpha(\alpha^2 + \beta^2)}{[\omega^2 - (\beta^2 + \alpha^2)]^2 + 4\alpha^2\omega^2}$$

where $\alpha = h$, $\beta = \sqrt{k^2 - h^2}$ and $\gamma^2 = \sigma^2/(2hk^2)$. From Table 34.8 $R_y(\tau) = \gamma^2 e^{-\alpha|\tau|}\left(\cos\beta\tau + \frac{\alpha}{\beta}\sin\beta|\tau|\right)$

34.18.3.11 Hilbert Transform

$$F\left\{\frac{1}{\pi t}\right\} = -j\,\mathrm{sgn}(\omega),\ \hat{X}(t) = \frac{1}{\pi}\int_{-\infty}^{\infty} \frac{X(\tau)}{t-\tau}d\tau = X(t) * \frac{1}{\pi t} \equiv \text{Hilbert transform},\ F\{\hat{X}(t)\} = -j\,\mathrm{sgn}(\omega)F\{X(t)\}$$

$$= H(\omega)F\{X(\omega)\},\ S_{\hat{x}}(\omega) = S_x(\omega)|H(\omega)|^2 = S_x(\omega),\quad S_{\hat{x}x}(\omega) = S_x(\omega)H(\omega) = \begin{cases} -jS_x(\omega) & \omega > 0 \\ jS_x(\omega) & \omega < 0 \end{cases}$$

$$\text{(see 34.18.3.7)},\quad R_{\hat{x}}(\tau) = R_x(\tau),\quad R_{\hat{x}x}(\tau) = R_{xx}(\tau) * h(\tau) = \frac{1}{\pi}\int_{-\infty}^{\infty} \frac{R_x(\xi)}{\tau - \xi}d\xi = \hat{R}_x(\tau),\ R_{\hat{x}x}(-\tau) = -R_{\hat{x}x}(\tau),$$

$$R_{\hat{x}x}(0) = 0,\ R_{x\hat{x}}(\tau) = R_{\hat{x}x}(-\tau) = -R_{\hat{x}x}(\tau)$$

34.18.3.12 Analytic Signal

$$Z(t) = X(t) + j\hat{X}(t),\ R_z(\tau) = 2[R_x(\tau) + jR_{\hat{x}x}(\tau)] = 2[R_x(\tau) + j\hat{R}_x(\tau)],$$

$$S_z(\omega) = 2[S_x(\omega) + jS_{\hat{x}x}(\omega)] = \begin{cases} 4S_x(\omega) & \omega > 0 \\ 0 & \omega < 0 \end{cases},\quad X(t) = \mathrm{Re}\{Z(t)\},\ R_x(\tau) = \tfrac{1}{2}\mathrm{Re}\{R_z(\tau)\}$$

34.18.3.13 Periodic Stochastic Process

$R(\tau + T) = R(\tau)$ then $X(t)$ is periodic in the m.s. sense.

$$R(\tau) = \sum_{n=-\infty}^{\infty} \alpha_n e^{jn\omega_0\tau},\ \omega_0 = 2\pi/T,\ \alpha_n = \frac{1}{T}\int_0^T R(\tau)e^{-jn\omega_0\tau}d\tau,\ S(\omega) = F\{R(\tau)\}$$

$$= 2\pi\sum_{n=-\infty}^{\infty} \alpha_n\delta(\omega - n\omega_0)$$

34.18.3.14 Periodic Processes in Linear System (see 34.18.3.13)

$$S_y(\omega) = 2\pi|H(\omega)|^2\sum_{n=-\infty}^{\infty} \alpha_n\delta(\omega - n\omega_0) = 2\pi\sum_{n=-\infty}^{\infty} \alpha_n|H(n\omega_0)|^2\delta(\omega - n\omega_0) \equiv \text{output spectra density.}$$

$$R_{yy}(\tau) = F^{-1}\{S_y(\omega)\} = \sum_{n=-\infty}^{\infty} \alpha_n|H(n\omega_0)|^2 e^{jn\omega_0\tau} \equiv \text{periodic}$$

34.18.3.15 Fourier Series

$$X(t)=\sum_{n=-\infty}^{\infty} A_n e^{jn\omega_0 t},\quad A_n=\frac{1}{T}\int_0^T X(u)e^{-jn\omega_0 u}\,du \equiv \text{uncorrelated (and orthogonal) r.v.,}$$

$$E\{An\}=\begin{cases} E\{X(t)\} & n=0 \\ 0 & n\neq 0 \end{cases}\quad E\{A_k A_n^*\}=\begin{cases} \alpha_n & k=n \\ 0 & k\neq n \end{cases},\quad \alpha_n \equiv \text{coefficients of } R(\tau)$$

(see 34.18.3.13)

34.18.4 Band-Limited Processes

34.18.4.1 Lowpass Process

$$S_x(\omega)=0,\ |\omega|>\omega_N,\quad R_x^{(n)}(\tau)=\frac{1}{2\pi}\int_{-\omega_N}^{\omega_N} (j\omega)^n S_x(\omega)e^{j\omega\tau}\,d\omega<\infty,$$

$$X(t+\tau)=\sum_{n=0}^{\infty} X^{(n)}(t)\frac{\tau^n}{n!}\quad \text{(see 17.3.8).}$$

34.18.4.2 Sampling Theorem

$$R_x(\tau)=\sum_{n=-\infty}^{\infty} R(nT)\frac{\sin(\omega_N\tau-n\pi)}{\omega_N\tau-n\pi},\quad \omega_N=\frac{\pi}{T},\ \text{if } S_x(\omega)$$

is bandlimited (see 34.18.4.1);

$$R_x(\tau-a)=\sum_{n=-\infty}^{\infty} R_x(nT-a)\frac{\sin(\omega_N\tau-n\pi)}{\omega_N\tau-n\pi}$$

or

$$R_x(\tau)=\sum_{n=-\infty}^{\infty} R_x(nT-a)\frac{\sin(\omega_N(\tau+a)-n\pi)}{\omega_N(\tau+a)-n\pi};\quad X(t)=\sum_{n=-\infty}^{\infty} X(nT)\frac{\sin(\omega_N t-n\pi)}{\omega_N t-n\pi},\quad \omega_N=\pi/T$$

if $X(t)$ is bandlimited.

34.18.4.3 Bandpass Process

The narrow-band (quasi-monochromatic $\omega_N \ll \omega_0$) process $W(t)=X(t)\cos\omega_0 t+Y(t)\sin\omega_0 t$ is wide-sense stationary iff $E\{X(t)\}=E\{Y(t)\}=0$ and $R_x(\tau)=R_y(\tau)$ and $R_{xy}(\tau)=-R_{yx}(\tau)$ then $R_w(\tau)=R_x(\tau)\cos\omega_0\tau+R_{yx}(\tau)\sin\omega_0\tau$.

34.19 Linear Mean-Square Estimation

34.19.1 Definitions

34.19.1.1 Estimate

$\hat{G}(t) \equiv$ estimate of $G(t)$

34.19.1.2 Mean Square (m.s.) Error

$e = E\{|G(t) - \hat{G}(t)|^2\}$

34.19.1.3 Minimum m.s. Error, Linear Operation on Data

$\hat{G}(t) = L\{X(t)\}$, L = linear operation

Example

Find a linear operator L such that $e = E\{[G(t) - L\{X(\xi)\}]^2\}$, $\xi \in I$ is minimum. It turns out that $G(t) - L\{X(\xi)\}$ is orthogonal to $X(\xi)$ for every ξ in I and if there is an optimum solution, then L is this optimum. Hence, if $E\{[G(t) - L\{X(\xi)\}]X(\xi_i)\} = 0$, $\xi_i \in I$, then the m.s. is minimum and is given by $e_m = E\{[G(t) - L\{X(\xi)\}]G(t)\}$.

34.19.2 Orthogonality in Linear m.s. Estimation

34.19.2.1 Orthogonality

$E\{X_i X_j^*\} = 0$ implies X_i and X_j are orthogonal r.v. If $E\{X_i Z^*\} = 0$ for $i = 1,2,\cdots,n$ then $E\{a_1 X_1 + \cdots + a_n X_n\}Z^*\} = 0$, a_i's are constants.

34.19.2.2 m.s. Error Estimation

The m.s. error $e = E\{|S_0 - (a_1 S_1 + \cdots + a_n S_n)|^2\}$ is minimum if we find the constants a_i such that the error $S_0 - (a_1 S_1 + \cdots + a_n S_n)$ is orthogonal to the data $S_1,\cdots,S_n$ or $E\{[S_0 - (a_1 S_1 + \cdots + a_n S_n)]S_i^*\} = 0$, $i = 1,2,\cdots,n$ or equivalently $E\{[S_0^* - (a_1^* S_1^* + \cdots + a_n^* S_n^*)]S_i\} = 0$

Example

To estimate $S(t+\tau)$ given $S(t)$ we use 34.19.2.2. $E\{[S(t+\tau) - aS(t)]S(t)\} = 0$ or $R(\tau) - aR(0) = 0$ or $a = R(\tau)/R(0)$ where $S_0 \equiv S(t+\tau)$, $S_1 = S(t)$ and the error is $S(t+\tau) - aS(t)$.

Mean square error (34.19.2.2)
To estimate

$$e = E\{[S(t+\tau) - aS(t)]^2\} = E\{[S(t+\tau) - aS(t)][S(t+\tau) - aS(t)]\}$$

$$= E\{[S(t+\tau) - aS(t)]S(t+\tau)\} = R(0) - aR(\tau) = R(0) - R^2(\tau)/R(0)$$

since $S(t) \equiv$ data is orthogonal to the error $S(t+\tau) - aS(t)$. (This is a prediction problem.)

Example

To estimate $S(t)$ in terms of $X(t)$ we obtain (34.19.2.2) $E\{[S(t) - aX(t)]X(t)\} = 0$ or $R_{sx}(0) - aR_x(0) = 0$ or $a = R_{sx}(0)/R_x(0)$. Mean square error: $e = E\{[S(t) - aX(t)]S(t)\} = R_s(0) - aR_{sx}(0) = R_s(0) - R_{sx}^2(0)/R_x(0)$. If $X(t) = S(t) + n(t)$ and $n(t) \equiv$ noise is orthogonal $S(t)$, then $R_{sx}(\tau) = R_s(\tau)$, $R_x(\tau) = R_s(\tau) + R_n(\tau)$ and $a = R_s(0)/[R_s(0) + R_n(0)]$ and $e = R_s(0)R_n(0)/[R_s(0) + R_n(0)]$. (This is a filtering problem.)

Examples

To estimate $S(t)$ in terms of $S(0)$ and $S(t_0)$ we set the error to be orthogonal to the data $S(0)$ and $S(t_0)$. Hence, $E\{[S(t) - a_1 S(0) - a_2 S(t_0)]S(0)\} = 0$ and $E\{[S(t) - a_1 S(0) - a_2 S(t_0)]S(t_0)\} = 0$. Therefore, we obtain the systems $R(t)=a_1R(0)+a_2R(t_0)$ and $R(t_0-t)=a_1R(t_0)+a_2R(0)$ and solve it for the unknown a_1 and a_2. The m.s. error is $e=E\{[S(t)-a_1S(0)-a_2S(t_0)]S(t)\}=R(0)-[a_1R(t)-a_2R(t_0-t)]$. (This is an interpolation problem.)

34.20 The Filtering Problem for Stationary Processes

34.20.1 Wiener Theory

34.20.1.1 Wiener Integral Equation

$$R_{gx}(t-\xi)=\int_a^b R_x(\alpha-\xi)h(t,\alpha)d\alpha,\quad a\le\xi\le b.$$

If $h(t,\xi)$ is found to satisfy the integral equation for the process $G(t)$ and $X(t)$, then the m.s. error is given by (orthogonality principle):

$$e=E\left\{\left[G(t)-\int_a^b X(\alpha)h(t,\alpha)d\alpha\right]G(t)\right\}=R_g(0)-\int_a^b R_{gx}(t-\alpha)h(t,\alpha)d\alpha$$

34.20.1.2 Wiener Integral Equation for Stationary Processes

$$R_{gx}(\tau)=\int_a^b R_x(\tau-\alpha)h(\alpha)d\alpha$$

34.20.1.3 Solution of 34.20.1.2

$S_{gx}(\omega)=S_x(\omega)H(\omega)$ or $H(\omega)=S_{gx}(\omega)/S_x(\omega)$

Example

$X(t)=S(t)+N(t)$, $X(t)\equiv$ given in $-\infty<t<\infty$, $N(t)\equiv$ noise, $G(t)=S(t)$ and

$$R_{sx}(\tau)=\int_{-\infty}^{\infty} R_x(\tau-\alpha)h(\alpha)d\alpha,\quad S_{sx}(\omega)=S_x(\omega)H(\omega),\quad e=\text{m.s. error}$$

$$=E\{[S(t)-\int_{-\infty}^{\infty} X(t-\alpha)h(\alpha)d\alpha]S(t)\}=R_s(0)-\int_{-\infty}^{\infty} R_{sx}(\alpha)h(\alpha)d\alpha.$$

If $f(\tau)=R_s(\tau)-R_{sx}(-\tau)*h(\tau)$ then $e=f(0)$. But

$$F(\omega)=S_s(\omega)-S_{sx}(-\omega)H(\omega)=S_s(\omega)-\frac{S_{sx}(-\omega)S_{sx}(\omega)}{S_x(\omega)}$$

and from inverse formula with $\omega=0$

$$e = f(0) = \frac{1}{2\pi}\int_{-\infty}^{\infty} [S_s(\omega) - \frac{S_{sx}(-\omega)S_{sx}(\omega)}{S_x(\omega)}]d\omega.$$

34.20.1.4 Signal and Noise Uncorrelated

$S_{sn}(\omega) = 0,\ S_x(\omega) = S_s(\omega) + S_n(\omega),\ S_{sx}(\omega) = S_s(\omega),\ H(\omega) = S_s(\omega)/[S_s(\omega) + S_n(\omega)],$

$$e = \frac{1}{2\pi}\int_{-\infty}^{\infty} \frac{S_s(\omega)S_n(\omega)}{S_s(\omega) + S_n(\omega)} d\omega$$

Example

If $R_s(\tau) = Ae^{-|\tau|}\cos\omega_0\tau,\ \omega_0 >> 1$ and $S_n(\omega) = N,\ S_s(\omega) = A/[1 + (\omega - \omega_0)^2]$ for $\omega > 0,\ H(\omega) = A/[A + N + N(\omega - \omega_0)^2],\ \omega > 0,$

$$e = \frac{1}{\pi}\int_{-\infty}^{\infty} AN\,d\omega/[A + N + N(\omega - \omega_0)^2] = A/[1 + A/N]^{1/2},\ E\{S^2(t)\} = R_s(0) = A.$$

If $N >> A$ the filtering does not improve the estimation of $S(t)$, $e = A$. If $A = 3N$ then $e = R_s(0)/2$.

34.20.1.5 Wiener-Hopf Equation

$$R(\tau + \lambda) = \int_0^{\infty} R(\tau - \alpha)h(\alpha)d\alpha,\ \tau \geq 0$$

$$e = \text{m.s. error} = E\{[S(\tau + \lambda) - \int_0^{\infty} S(\tau - \alpha)h(\alpha)d\alpha]S(\tau + \lambda)\} = R(0) - \int_0^{\infty} R(-\lambda - \alpha)h(\alpha)d\alpha$$

Solution to Wiener-Hopf Equation

$S(\omega)$ = is a rational function of ω.

Step 1

$$S(\omega)\Big|_{\omega = p/j} = \frac{A(p^2)}{B(p^2)} = F(p)F(-p),\ F(p) = \frac{C(p)}{D(p)},\ F(-p) = \frac{C(-p)}{D(-p)},$$

C and D contain all the roots of $A(p^2)$ and $B(p^2)$.

Step 2

$$D(p_i) = 0,\ \text{Re}\,p_i \leq 0,\ F(p) = \frac{C(p)}{D(p)} = \frac{a_1}{p - p_1} + \cdots + \frac{a_n}{p - p_n},\ a_i = \frac{C(p_i)}{D'(p_i)}.$$

Step 3

$$F_1(p) = \frac{a_1 e^{p_1\lambda}}{p - p_1} + \cdots + \frac{a_n e^{p_n\lambda}}{p - p_n} = \frac{C_1(p)}{D(p)},$$

$\lambda =$ the constant in Wiener-Hopf equation $R(\tau+\lambda)$, $C_1(p_i) = C(p_i)e^{p_i\lambda}$.

Step 4

$$H(p) = \frac{F_1(p)}{F(p)} = \frac{C_1(p)}{C(p)}$$

Example

$R(\tau) = Ae^{-|\tau|}$, $S(\omega) = 2A/(1+\omega^2)$.

Step 1

$$S(p/j) = \frac{2A}{1-p^2} = \frac{\sqrt{2A}}{1+p}\frac{\sqrt{2A}}{1-p}$$

Step 2

$$F(p) = \frac{C(p)}{D(p)} = \frac{\sqrt{2A}}{1+p},\ p_1 = -1.$$

Step 3

$$F_1(p) = \frac{\sqrt{2A}}{1+p}e^{-\lambda}$$

Step 4

$$H(p) = \frac{F_1(p)}{F(p)} = e^{-\lambda}\ \ h(t) = e^{-\lambda}\delta(t).$$

Estimate $S(t+\lambda) \sim e^{-\lambda}\ S(t) \equiv$ prediction.

$$e = E\{[S(t+\lambda) - e^{-\lambda}S(t)]S(t+\lambda)\} = R(0) - e^{-\lambda}R(\lambda) = A(1-e^{-2\lambda}).$$

34.21 Harmonic Analysis

34.21.1 Series Expansion

34.21.1.1 $R(\tau) =$ periodic ($R(\tau+T) = R(\tau)$):

$$X(t) = \sum_{n=-\infty}^{\infty} A_n e^{jn\omega_0 t},\ \omega_0 = 2\pi/T,$$

$$A_n = \frac{1}{T}\int_{-T/2}^{T/2} X(t)e^{-jn\omega_0 t}dt,\ E\{A_n A_n^*\} = 0,\ n \neq 0 \text{ orthogonal},$$

$$R(\tau)=\sum_{n=-\infty}^{\infty}\alpha_n e^{jn\omega_0\tau},\quad S(\omega)=2\pi\sum_{n=-\infty}^{\infty}\alpha_n\delta(\omega-n\omega_0),\quad E\{|A_n|^2\}=\alpha_n$$

34.21.1.2 Karhunen–Loève Expansion

If the expansion

$$X(t)\cong\sum_{n=1}^{\infty}B_n\varphi_n(t),\quad |t|<\tfrac{T}{2},$$

has orthogonal coefficients $E\{B_nB_m^*\}=0,\ n\neq m,$ then φ_n must satisfy the integral equation

$$\int_{-T/2}^{T/2}R(t_1,t_2)\varphi(t_2)dt_2=\lambda\varphi(t_1),\quad |t_1|<\tfrac{T}{2}$$

for term $\lambda=\lambda_n$ of λ and the variance of B_n must be equal to $\lambda_n: E\{|B_n|^2\}=\lambda_n$

34.21.1.3 If 34.21.1.2 is satisfied $E\{|X(t)-\hat{X}(t)|^2\}=R(t,t)-\sum_{n=1}^{\infty}\lambda_n|\varphi_n(t)|^2,\ |t|<\tfrac{T}{2}$

34.21.2 Fourier Transforms

34.21.2.1 Fourier Transforms

$$X(\omega)=\int_{-\infty}^{\infty}X(t)e^{-j\omega t}dt,\quad \int_{-\infty}^{\infty}|X(t)|^2dt<\infty,$$

$$X(t)=\frac{1}{2\pi}\int_{-\infty}^{\infty}X(\omega)e^{j\omega t}d\omega,\quad E\{X(\omega)\}=\int_{-\infty}^{\infty}E\{X(t)\}e^{-j\omega t}dt$$

34.21.2.2 Autocorrelation of X(ω)

$$R(t_1,t_2)=E\{X(t_1)X^*(t_2)\}$$

$$\Gamma(\omega_1,\omega_2)=\iint_{-\infty}^{\infty}R(t_1,t_2)e^{-j(\omega_1t_1-\omega_2t_2)}dt_1dt_2=E\{X(\omega_1)X^*(\omega_2)\};$$

$$R(t_1,t_2)\overset{F_{t_1}}{\longleftrightarrow}E\{X(\omega_1)X^*(t_2)\}\overset{F_{t_2}}{\longleftrightarrow}E\{X(\omega_1)X^*(\omega_2)\}$$

34.21.2.3 Linear Systems

$Y(\omega)=\text{output}=X(\omega)H(\omega);\ E\{Y_1(\omega)Y_2^*(\omega)\}=E\{X_1(\omega)H_1(\omega)X_2(\omega)H_2(\omega)\}=$
$E\{X_1(\omega_1)X_2^*(\omega_2)\}H_1(\omega_1)H_2^*(\omega_2);\ \Gamma_{y_1y_2}(\omega_1,\omega_2)=\Gamma_{x_1x_2}(\omega_1,\omega_2)H_1(\omega_1)H_2^*(\omega_2)$

34.22 Markoff Sequences and Processes

34.22.1 Definitions

34.22.1.1 Definition

$F(x_n|x_{n-1},\cdots,x_1)=F(x_n|x_{n-1})$. For continuous r.v. type $f(x_n|x_{n-1},\cdots,x_1)=f(x_n|x_{n-1})$. Also

$$f(x_1,x_2,\cdots,x_n)=f(x_n|x_{n-1})f(x_{n-1}|x_{n-2})\cdots f(x_2|x_1)f(x_1)$$

$$f(x_n|x_{n-1},\cdots,x_1)=f(x_1,x_2,\cdots,x_n)/f(x_1,x_2,\cdots,x_{n-1})=f(x_n|x_{n-1})$$

34.22.1.2 Homogeneous

$f(x_n|x_{n-1})\equiv$ independent of n

34.22.1.3 Stationary

$f_{x_n}(x)\equiv$ the sequence is homogeneous and the r.v. X_n have the same density.

34.22.1.4 Chapman-Kalmogoroff Equation

$$f(x_n|x_s)=\int_{-\infty}^{\infty} f(x_n|x_r)f(x_r|x_s)dx_r \quad \text{where } n>r>s \text{ are any integers.}$$

34.22.2 Markoff Chains

34.22.2.1 Markoff Chains

$P\{X_n=a_{i_n}|X_{n-1}=a_{i_{n-1}},\cdots,X_1=a_{i_1}\}=P\{X_n=a_{i_n}|X_{n-1}=a_{i_{n-1}}\}$ $X_n\equiv$ Markoff chain

34.22.2.2 Conditional Densities

$p_i(n)=P\{X_n=a_i\}$; $P_{ij}(n,s)=P\{X_n=a_i|X_s=a_j\}$, $n>s$; $\quad p_i(n)=\sum_j P_{ij}(n,s)p_j(s)$; $\sum_i p_i(n)=1$,

$\sum_i p_{ij}(n,s)=1$; $P_{ij}(n,s)=\sum_k P_{ik}(n,r)P_{kj}(r,s)$, n > r > s discrete case Chapman-Kolmogoroff equation.

34.22.2.3 Matrices

$P(n,s)=$ square matrix with $P_{ij}(n,s)$ elements; $p(n)$ column matrix with elements $p_i(n)$; $p(n)=P(n,s)p(s)$; $P(n,s)=P(n,r)P(r,s)$

34.22.2.4 Homogeneous Chains

The conditional probabilities $P_{ij}(n,s)$ depend only on the difference n – s , $P_{ij}(n-s)$. Matrix $P(n-s)$ with $P(1)=\Pi$. With $r=s+1$, $n=s+2$, $P(2)=P(1)P(1)=\Pi^2$ and $P(n)=\Pi^n$. Also, $p(n+k)=\Pi^n p(k)$.

34.22.3 Markoff Processes

34.22.3.1 Markoff Process

$P\{X(t_n)\le x_n|X(t_{n-1}),\cdots,X(t_1)\}=P\{X(t_n)\le x_n|X(t_{n-1})\}$ for every n and $t_1<t_2\cdots<t_n$.

Note:

1. For $t_1<t_2$, $P\{X(t_1)\le x_1|X(t)$ for all $t\ge t_2\}=P\{X(t_1)\le x_1|X(t_2)\}$
2. If $X(t_2)-X(t_1)$ is independent of $X(t)$ for every $t\le t_1$ $(t_1<t_2)$ the process $X(t)$ is Markoff.

3. $E\{X(t_n)|X(t_{n-1}),\cdots,X(t_1)\} = E\{X(t_n)|X(t_{n-1})\}$
4. $R(t_3,t_2)R(t_2,t_1) = R(t_3,t_1)R(t_2,t_2)$ for every $t_3 > t_2 > t_1$ and if $X(t)$ is normal Markoff with zero mean.

34.22.3.2 Continuous Process (Chapman-Kolmogoroff equation)

$p(x,t;x_0,t_0) = \int_{-\infty}^{\infty} p(x,t;x_1,t_1)p(x_1,t_1;x_0,t_0)dx_1$ if $t > t_1 > t_0$. $p(x,t;x_0,t_0) = f_{x(t)}(x|X(t_0) = x_0)$ for $t \ge t_0 \equiv$ conditional density of $X(t)$

34.22.3.3 Conditional Mean and Variance

$E\{X(t)|X(t_0) = x_0\} = a(x_0,t,t_0)$; $E\{[X(t) - a(x_0,t,t_0)]^2|X(t_0) = x_0\} = b(x_0,t,t_0)$; $a(x_0,t,t_0) = \int_{-\infty}^{\infty} xp(x,t;x_0,t_0)dx$, $b(x_0,t,t_0) = \int_{-\infty}^{\infty} (x-a)^2 p(x,t;x_0,t_0)dx$; $a(x_0,t_0,t_0) = x_0$, $b(x_0,t_0,t_0) = 0$

References

Larson, H. J. and B. O. Shubert, *Probabilistic Models in Engineering Sciences*, Vol. I and II, John Wiley & Sons, New York, NY, 1979.

Papoulis, A., Probability, *Random Variables, and Stochastic Processes*, McGraw-Hill Book Co., New York, NY, 1965.

Stark, H. and J. W. Woods, *Probability, Random Processes, and Estimation Theory for Engineering*, Prentice Hall, Englewood Cliffs, New Jersey, 1994.

35
Random Digital Signal Processing

35.1 Discrete-Random Processes

35.1.1 Definitions

35.1.1.1 Discrete Stochastic Process

$\{x(1), x(2), \cdots, x(n)\} \equiv$ one realization

35.1.1.2 Cumulative Density Function (c.d.f)

$F_x(x(n)) = P\{X(n) \le x(n)\}$, $X(n) =$ continuous r.v. on time n (over the ensemble), $x(n) =$ the value of $X(n)$ at n

35.1.1.3 Probability Density Function (p.d.f.)

$$f_x(x(n)) = \frac{\partial F_x(x(n))}{\partial x(n)}$$

35.1.1.4 Bivariate c.d.f.

$F_x(x(n_1), x(n_2)) = P\{X(n_1) \le x(n_1), X(n_2) \le x(n_2)\}$

35.1.1.5 Joint p.d.f.

$$f_x(x(n_1), x(n_2)) = \frac{\partial F_x^2(x(n_1), x(n_2))}{\partial x(n_1)\, \partial x(n_2)}.$$

Note: For multivariate forms extend (35.1.1.4) and (35.1.1.5).

35.1.2 Averages (expectation)

35.1.2.1 Average (mean value)

$$\mu(n) = E\{x(n)\} = \int_{-\infty}^{\infty} x(n) f_x(x(n))\, dx(n),\ \mu(n) = \lim_{N\to\infty}\left[\frac{1}{N}\sum_{i=1}^{N} x_i(n)\right] = \text{relative frequency interpretation}$$

35.1.2.2 Correlation

$\mathbf{R}(m,n) = E\{x(m)x(n)\} =$ real r.v., $\mathbf{R}(m,n) = E\{x(m)x^*(n)\} =$ complex r.v.

$$\mathbf{R}(m,n) = \iint_{-\infty}^{\infty} x(m)x^*(n)f(x(n),x(m))dx(m)dx^*(n)$$

$$\mathbf{R}(m,n) = \lim_{N\to\infty}\left[\frac{1}{N}\sum_{i=1}^{N} x_i(m)x_i^*(n)\right] = \text{relative frequency interpretation}$$

$\mathbf{R}(n,n) = E\{x(n)x^*(n)\} =$ autocorrelation

35.1.2.3 Covariance-Variance

$$\mathbf{C}(m,n) = E\{[x(m)-\mu(m)][x(n)-\mu(n)]^*\} = E\{x(m)x^*(n)\} - \mu(m)\mu^*(n)$$

$$\mathbf{C}(m,n) = \iint_{-\infty}^{\infty} [x(m)-\mu(m)][x(n)-\mu(n)]^*\, dx(m)dx^*(n)$$

$$\mathbf{C}(n,n) = \sigma^2(n) = E\{|x(n)-\mu(n)|^2\} = \text{variance}$$

$$\mathbf{R}(m,n) = \mathbf{C}(m,n) \ \text{ if } \ \mu(m) = \mu(n) = 0$$

Example 1

Given $x(n) = Ae^{j(n\omega_0+\phi)}$ with ϕ r.v. uniformly distributed between $-\pi$ and π. Then

$$E\{x(n)\} = E\{A\exp(j(n\omega_0+\phi))\} = 0,\ \ R_x(k,\ell) = E\{x(k),x^*(\ell)\} = E\{Ae^{j(k\omega_0+\phi)}A^*e^{-j(\ell\omega_0+\phi)}\}$$
$$= |A|^2 E\{e^{j(k-\ell)\omega_0}\} = |A|^2 e^{j(k-\ell)\omega_0} = R_x(k-\ell)$$

which implies that the process is at least a wide-sense stationary process.

35.1.2.4 Independent r.v.

$$f(x(m),x(n)) = \text{p.d.f.} = f(x(m))\,f(x(n)),\ E\{f(x(m),x(n))\} = E\{f(x(m))\}\,E\{f(x(n))\},$$

$\mathbf{C}(m,n) = 0 =$ uncorrelated. Independent implies uncorrelated, but the reverse is not always true.

35.1.2.5 Orthogonal r.v.

$E\{x(m)x^*(n)\} = 0$

35.1.3 Stationary Processes

35.1.3.1 Strict Stationary

$F(x(n_1),x(n_2),\cdots,x(n_k)) = F(x(n_1+n_0),x(n_2+n_0),\cdots,x(n_k+n_0))$ for any set $\{n_1,n_2,\cdots,n_k\}$ for any k and any n_0.

35.1.3.2 Wide-sense Stationary (or weak)

$\mu_x(n) = \mu_x =$ constant for any n, $\mathbf{R}_x(k,\ell) = \mathbf{R}_x(k-\ell)$, $C_x(0) =$ variance $< \infty$

35.1.3.3 Correlation Properties

1. Symmetry: $r_x(k) = r_x^*(-k)$ $(r_x(k) = r(-k)$ for real process)
2. Mean Square Value: $r_x(0) = E\{|x(n)|^2\} \geq 0$
3. Maximum Value: $r_x(0) \geq |r_x(k)|$,
4. Periodicity: If $r_x(k_0) = r_x(0)$ for some k_0 then $r_x(k)$ is periodic

35.1.3.4 Autocorrelation Matrix

$$\mathbf{R}_x = E\{\mathbf{x}\mathbf{x}^H\} = \begin{bmatrix} r_x(0) & r_x(-1) & \cdots & r_x(-p) \\ r_x(1) & r_x(0) & \cdots & r_x(p-1) \\ \vdots & & & \\ r_x(p) & r_x(p-1) & \cdots & r_x(0) \end{bmatrix}, \quad \mathbf{x} = [x(0), x(1), \cdots, x(p)]^T$$

The H superscript indicates conjugate transpose quantity (Hermitian)

Properties:

1. $\mathbf{R}^H = \mathbf{R}$ or $r_x(-k) = r_x^*(k)$
2. $\mathbf{R} =$ Toeplitz matrix
3. $\mathbf{x}^H\mathbf{R}\mathbf{x} \geq 0$, $\mathbf{R}$ is non-negative definite
4. $E\{\mathbf{x}^B\mathbf{x}^{BH}\} = \mathbf{R}^T$, $\mathbf{x}^B = [x(p), x(p-1), \cdots, x(0)]^T$

35.1.3.5 Autocovariance Matrix

$\mathbf{C}_x = E\{(\mathbf{x} - \boldsymbol{\mu}_x)(\mathbf{x} - \boldsymbol{\mu}_x)^H] = \mathbf{R}_x - \boldsymbol{\mu}_x\boldsymbol{\mu}_x$, $\boldsymbol{\mu}_x = [\mu_x, \mu_x, \cdots, \mu_x]^T$

35.1.3.6 Ergotic in the Mean

a. $\lim_{N\to\infty} \hat{\mu}_x(N) = \mu_x$, $\hat{\mu}_x(n) = \frac{1}{N}\sum_{n=1}^{N} x(n)$, $\mu_x(n) = E\{x(n)\}$

b. $\lim_{N\to\infty} \frac{1}{N}\sum_{k=1}^{N} c_x(k) = 0$,

c. $\lim_{k\to\infty} c_x(k) = 0$

35.1.3.7 Autocorrelation Ergotic

$$\lim_{N\to\infty} \frac{1}{N}\sum_{k=1}^{N} c_x^2(k) = 0$$

35.1.4 Special Random Signals

35.1.4.1 Independent Identically Distributed

If $f(x(0), x(1), \cdots) = f(x(0)), f(x(1)), \cdots$ for a zero-mean, stationary random signal, it is said that the elements $x(n)$ are independent identically distributed (iid).

35.1.4.2 White Noise (sequence)

If $r(n-m) = E\{v(n)\, v^*(m)\} = \sigma_v^2\, \delta(n-m)$, $\delta(n-m) =$ delta function, $\sigma_v^2 =$ variance of white noise, the sequence is white noise.

35.1.4.3 Correlation Matrix for White Noise

$\mathbf{R} = \sigma_v^2 \mathbf{I}$

35.1.4.4 First-Order Marrkov Signal

$f(x(n)|x(n-1), x(n-2), \cdots, x(0)) = f(x(n)|x(n-1))$

35.1.4.5 Gaussian

1. $f(x(i)) = \dfrac{1}{\sqrt{2\pi}\,\sigma_i} \exp\left[-\dfrac{1}{2\sigma_i^2}(x(i) - \mu(i))^2\right]$

2. $f(x(n_0), x(n_1), \cdots, x(n_{L-1})) = f(\mathbf{x}) = \dfrac{1}{(2\pi)^{L/2}|\mathbf{C}|^{1/2}} \exp[-\tfrac{1}{2}(\mathbf{x} - \boldsymbol{\mu})^T \mathbf{C}^{-1}(\mathbf{x} - \boldsymbol{\mu})]$

   $\mathbf{x} = [x(n_0), x(n_1), \cdots, x(n_{L-1})]^T$, $\boldsymbol{\mu} = [\mu(n_0), \mu(n_1), \cdots, \mu(n_{L-1})]$, $\mathbf{C}$ = covariance matrix for elements of $\mathbf{x}$

3. $f(\mathbf{x}) = \dfrac{1}{(2\pi)^{L/2}|\mathbf{R}|^{1/2}} \exp[-\tfrac{1}{2}\mathbf{x}^T \mathbf{R}^{-1}\mathbf{x}]$, $\mathbf{R} =$ correlation matrix, $E\{\mathbf{x}\} = \mathbf{0}$

4. If $x(n)$ is zero-mean Gaussian iid (Gaussian white noise), then

   $$\mathbf{R}^{-1} = \frac{1}{\sigma_v^2}\mathbf{I},\quad |\mathbf{R}| = \sigma_v^{2L},\quad f(\mathbf{x}) = \frac{1}{(2\pi)^{L/2}\sigma_v^L} \exp\left[-\frac{1}{2\sigma_v^2}\sum_{i=n_0}^{n_{L-1}} x^2(i)\right]$$

5. Linear operation on a Gaussian signal produces a Gaussian signal.

35.1.5 Complex Random Signals

35.1.5.1 Complex Random Signal

$\tilde{x} = u + jv$, u and v real r.v., $\tilde{x}$ = complex r.v

35.1.5.2 Expectation

$E\{\tilde{x}\} = E\{u\} + j\,E\{v\}$

35.1.5.3 Second Moment

$E\{|\tilde{x}|^2\} = E\{\tilde{x}\tilde{x}^*\} = E\{u^2\} + E\{v^2\}$

35.1.5.4 Variance

$\text{var}(\tilde{x}) = E\{|\tilde{x} - E\{\tilde{x}\}|^2\} = E\{|\tilde{x}|^2\} - |E\{\tilde{x}\}|^2$

35.1.5.5 Expectation of Two r.v.

$E\{\tilde{x}_1^*\tilde{x}_2\} = E\{u_1u_2\} + E\{v_1v_2\} + j(E\{u_1v_2\} - E\{u_2v_1\})$, the real vector $[u_1\ u_2\ \upsilon_1\ \upsilon_2]^T$ has a joint p.d.f.

35.1.5.6 Covariance of Two r.v.

$\text{cov}(\tilde{x}_1, \tilde{x}_2) = E\{\tilde{x}_1^*\tilde{x}_2\} - E^*\{\tilde{x}_1\}E\{\tilde{x}_2\}$.

Note: If $\tilde{x}_1$ is independent of $\tilde{x}_2$ (equivalently $[u_1\ \upsilon_1]^T$ is independent of $[u_2\ \upsilon_2]^T$ then $\text{cov}(\tilde{x}_1, \tilde{x}_2) = 0$

35.1.5.7 Expectation of Vectors

If $\tilde{\mathbf{x}} = [\tilde{x}_1\tilde{x}_2\cdots\tilde{x}_n]^T$, then $E\{\tilde{\mathbf{x}}\} = [E\{\tilde{x}_1\}E\{\tilde{x}_2\}\cdots E\{\tilde{x}_n\}]^T$

35.1.5.8 Covariance of $\tilde{\mathbf{x}}$

$$\mathbf{C}_{\tilde{x}} = E\{(\tilde{\mathbf{x}} - E\{\tilde{\mathbf{x}}\})(\tilde{\mathbf{x}} - E\{\tilde{\mathbf{x}}\})^H\} = \begin{bmatrix} \mathrm{var}(\tilde{x}_1) & \mathrm{cov}(\tilde{x}_1,\tilde{x}_2) & \cdots & \mathrm{cov}(\tilde{x}_1,\tilde{x}_n) \\ \mathrm{cov}(\tilde{x}_2,\tilde{x}_1) & \mathrm{var}(\tilde{x}_2) & \cdots & \mathrm{cov}(\tilde{x}_2,\tilde{x}_n) \\ \vdots & & & \\ \mathrm{cov}(\tilde{x}_n,\tilde{x}_1) & \mathrm{cov}(\tilde{x}_n,\tilde{x}_2) & \cdots & \mathrm{var}(\tilde{x}_n) \end{bmatrix}^*$$

H = means complex conjugate of a matrix.

Note: $\mathbf{C}_{\tilde{x}}^H = \mathbf{C}_{\tilde{x}}$ and $\mathbf{C}_{\tilde{x}}$ is positive semi-definite.

Example

$$\tilde{\mathbf{y}} = \mathbf{A}\tilde{\mathbf{x}} + \mathbf{b}, \; E\{\tilde{\mathbf{y}}\} = \mathbf{A}E\{\tilde{\mathbf{x}}\} + \mathbf{b}, \; \tilde{y}_i = \sum_{j=1}^{n} \mathbf{A}_{ij}\tilde{x}_j + b_i, \; E\{\tilde{y}_i\} = \sum_{j=1}^{n} \mathbf{A}_{ij}E\{\tilde{x}_j\} + b_i,$$

$$\mathbf{C}_{\tilde{y}} = E\{(\tilde{\mathbf{y}} - E\{\tilde{\mathbf{y}}\})(\tilde{\mathbf{y}} - E\{\tilde{\mathbf{y}}\})^H\} = E\{\mathbf{A}(\tilde{\mathbf{x}} - E\{\tilde{\mathbf{x}}\})(\tilde{\mathbf{x}} - E\{\tilde{\mathbf{x}}\})^H A^H\},$$

$$\left[\mathbf{C}_{\tilde{y}}\right]_{ij} = E\left\{\sum_{k=1}^{n}\sum_{\ell=1}^{n} \mathbf{A}_{ij}[(\tilde{\mathbf{x}} - E\{\tilde{\mathbf{x}}\})(\tilde{\mathbf{x}} - E\{\tilde{\mathbf{x}}\})^H]_{k\ell}[\mathbf{A}^H]_{\ell j}\right\}$$

35.1.5.9 Complex Gaussian

$\tilde{x} = u + jv$, u and v independent, $u = N(\mu_u, \sigma^2/2)$, $v = N(\mu_v, \sigma^2/2)$,

$$f(u,v) = \frac{1}{\pi\sigma^2}\exp[-\frac{1}{\sigma^2}((u-\mu_u)^2 + (v-\mu_v)^2)].$$

If we write $\tilde{\mu} = E\{\tilde{x}\} = \mu_u + j\mu_v$, then $f(\tilde{x}) = (1/\pi\sigma^2)\exp[-|\tilde{x} - \tilde{\mu}|^2/\sigma^2]$, $\tilde{x} \equiv CN(\tilde{\mu}, \sigma^2) =$ complex Gaussian

35.1.5.10 Complex Gaussian Vector

$\tilde{\mathbf{x}} = [\tilde{x}_1\, \tilde{x}_2 \cdots \tilde{x}_n]^T$, each component of $\tilde{\mathbf{x}}$ is $CN(\tilde{\mu}_i, \sigma_i^2)$, the real random vectors $[u_1\, v_1]^T, [u_2\, v_2]^T, \cdots, [u_n\, v_n]^T$ are independent,

$$f(\tilde{\mathbf{x}}) = \prod_{i=1}^{n} f(\tilde{x}_i) = \left[1 \Big/ \left[\pi^n \prod_{i=1}^{n} \sigma_i^2\right]\right] \exp\left[-\sum_{i=1}^{n} \frac{1}{\sigma_i^2}|\tilde{x}_i - \tilde{\mu}_i|^2\right]$$

$[1/[\pi^n \det(\mathbf{C}_{\tilde{x}})]]\exp[-(\tilde{\mathbf{x}} - \tilde{\boldsymbol{\mu}})^H \mathbf{C}_{\tilde{x}}^{-1}(\tilde{\mathbf{x}} - \tilde{\boldsymbol{\mu}})]$, $\mathbf{C}_{\tilde{x}} = diag(\sigma_1^2, \sigma_2^2, \cdots, \sigma_n^2) =$ multivariate complex Gaussian p.d.f. $CN(\tilde{\boldsymbol{\mu}}, \mathbf{C}_{\tilde{x}})$

Properties

1. Any subvector of $\tilde{\mathbf{x}}$ is complex Gaussian
2. If $\tilde{x}_i$'s of $\tilde{\mathbf{x}}$ are uncorrelated, they are also independent and vice versa
3. Linear transformations again produce complex Gaussian
4. If $\tilde{\mathbf{x}} = [\tilde{x}_1\, \tilde{x}_2\, \tilde{x}_3\, \tilde{x}_4]^T \equiv CN(\mathbf{0}, \mathbf{C}_{\tilde{x}})$, then

   $$E\{\tilde{x}_1^*\, \tilde{x}_2\, \tilde{x}_3^*\, \tilde{x}_4\} = E\{\tilde{x}_1^*\, \tilde{x}_2\}E\{\tilde{x}_3^*\, \tilde{x}_4\} + E\{\tilde{x}_1^*\, \tilde{x}_4\}E\{\tilde{x}_2\, \tilde{x}_3^*\}$$

35.1.6 Complex Wide Sense Stationary Random Processes

35.1.6.1 Mean

$E\{\tilde{x}(n)\} = E\{u(n) + jv(n)\} = E\{u(n)\} + jE\{v(n)\}$

35.1.6.2 Autocorrelation

$$r_{\tilde{x}\tilde{x}}(k) = E\{\tilde{x}^*(n)\tilde{x}(n+k)\} = E\{(u(n) - jv(n))\,(u(n+k) + jv(n+k))\}$$

$$= r_u(k) + jr_{uv}(k) - jr_{vu}(k) + r_v(k) = 2r_u(k) + 2jr_{uv}(k)$$

35.1.7 Derivatives, Gradients, and Optimization

35.1.7.1 Complex Derivative

$\frac{\partial J}{\partial \theta} = \frac{1}{2}\left(\frac{\partial J}{\partial \alpha} - j\frac{\partial J}{\partial \beta}\right)$, J = scalar function with respect to a complex parameter $\theta = \alpha + j\beta$.

Note: $\frac{\partial J}{\partial \theta} = 0$ if and only if $\frac{\partial J}{\partial \alpha} = \frac{\partial J}{\partial \beta} = 0$

Example

$J = |\theta|^2 = \alpha^2 + \beta^2, \quad \frac{\partial J}{\partial \theta} = \frac{1}{2}\left(\frac{\partial J}{\partial \alpha} - j\frac{\partial J}{\partial \beta}\right) = \frac{1}{2}(2\alpha - j2\beta) = \theta^*$

35.1.7.2 Chain Rule

$$\frac{\partial J(\theta,\theta^*)}{\partial \theta} = \frac{\partial \theta}{\partial \theta}\theta^* + \theta\frac{\partial \theta^*}{\partial \theta}, \quad \frac{\partial \theta}{\partial \theta} = \frac{1}{2}\left(\frac{\partial}{\partial \alpha} - j\frac{\partial}{\partial \beta}\right)(\alpha + j\beta)$$

$$= \frac{1}{2}(1 + j0 - j0 + 1) = 1, \quad \frac{\partial \theta^*}{\partial \theta} = 0,$$

and hence

$$\frac{\partial J(\theta,\theta^*)}{\partial \theta} = 1\cdot\theta^* + \theta\cdot 0 = \theta^*$$

35.1.7.3 Complex Conjugate Derivative

$$\frac{\partial J}{\partial \theta^*} = \frac{\partial J}{\partial(\alpha + j(-\beta))} = \frac{1}{2}\left(\frac{\partial J}{\partial \alpha} - j\frac{\partial J}{\partial(-\beta)}\right) = \left(\frac{\partial J}{\partial \theta}\right)^*$$

Note: Setting $\partial J/\partial \theta^* = 0$ will produce the same solutions as $\partial J/\partial \theta = 0$

35.1.7.4 Complex Vector Parameter $\boldsymbol{\theta}$

$$\frac{\partial J}{\partial \boldsymbol{\theta}} = \left[\frac{\partial J}{\partial \theta_1}\ \frac{\partial J}{\partial \theta_2} \cdots \frac{\partial J}{\partial \theta_p}\right]^T,$$

each element is given by (35.1.7.1).

Note: $\partial J/\partial \boldsymbol{\theta} = 0$ if each element is zero and hence if and only if $\partial J/\partial \alpha_i = \partial J/\partial \beta_i = 0$ for all $i = 1,2,\cdots,p$.

35.1.7.5 Hermitian Forms

1. $\ell(\boldsymbol{\theta}) = \mathbf{b}^H\boldsymbol{\theta} = \sum_{i=1}^{p} b_i^*\theta_i$, $\frac{\partial \ell}{\partial \theta_k} = b_k^* \frac{\partial \boldsymbol{\theta}_k}{\partial \boldsymbol{\theta}_k} = b_k^*$ (see 35.1.7.2) and hence $\frac{\partial \mathbf{b}^H\boldsymbol{\theta}}{\partial \boldsymbol{\theta}} = \mathbf{b}^*$

2. Also, $\frac{\partial \boldsymbol{\theta}^H\mathbf{b}}{\partial \boldsymbol{\theta}} = \mathbf{0}$

3. $J = \boldsymbol{\theta}^H \mathbf{A}\, \boldsymbol{\theta} = \text{real}\ (\mathbf{A}^H = \mathbf{A})$,

$$J = \sum_{i=1}^{p}\sum_{j=1}^{p} \theta_i^* \mathbf{A}_{ij}\theta_j,\quad \frac{\partial J}{\partial \theta_k} = \sum_{i=1}^{p}\sum_{j=1}^{p} \theta_i^* \mathbf{A}_{ij}\delta_{ik} = \sum_{i=1}^{p} \theta_i^* \mathbf{A}_{ik} = \sum_{i=1}^{p} \mathbf{A}_{ki}^T\theta_i^*,\quad \frac{\partial J}{\partial \boldsymbol{\theta}} = \mathbf{A}^T\boldsymbol{\theta}^* = (\mathbf{A}\boldsymbol{\theta})^*$$

35.1.8 Power Spectrum Wide-Sense Stationary Processes (WSS)

35.1.8.1 Power Spectrum (power spectral density)

$$S_x(e^{j\omega}) = \sum_{k=-\infty}^{\infty} r_x(k)e^{-jk\omega},\quad r_x(k) = E\{x(n)x^*(n-k)\},\quad x(n) \equiv \text{WSS},\quad r(k) = \frac{1}{2\pi}\int_{-\pi}^{\pi} S_x(e^{j\omega})e^{jk\omega}\,d\omega$$

Properties

1. $S_x(e^{j\omega}) \equiv$ real-valued function
2. $S_x(e^{j\omega}) \geq 0 \equiv$ non-negative,
3. $r(0) = E\{|x(n)|^2\} = \frac{1}{2\pi}\int_{-\pi}^{\pi} S_x(e^{j\omega})\,d\omega$

Example

$$r_x(k) = a^{|k|},\ |a| < 1,\ S_x(e^{j\omega}) = \sum_{k=-\infty}^{\infty} r_x(k)e^{-jk\omega} = \sum_{k=0}^{\infty} a^k e^{-jk\omega} + \sum_{k=0}^{\infty} a^k e^{jk\omega} - 1$$

$$= \frac{1}{1-ae^{-j\omega}} + \frac{1}{1-ae^{j\omega}} - 1 = \frac{1-a^2}{1-2a\cos\omega + a^2}$$

35.1.9 Filtering Wide-Sense Stationary (WSS) Processes and Spectral Factorization

35.1.9.1 Output Autocorrelation

$r_y(k) = r_x(k) * h(k) * h^*(-k)$, $r_y(k) =$ autocorrelation of a linear time invariant system, $h(k)$ = impulse response of at the system, $r_x(k) =$ autocorrelation of the WSS input process $x(n)$.

35.1.9.2 Output Power

$$S_y(e^{j\omega}) = S_x(e^{j\omega})|H(e^{j\omega})|^2,\ S_y(z) = S_x(z)H(z)H^*(1/z^*)\ \text{Z-transforms,}$$

if $h(n)$ is real then $S_y(z) = S_x(z)H(z)H(1/z)$.

Note: If $H(z)$ has a zero at $z = z_0$ then $S_y(z)$ will have a zero at $z = z_0$ and another at $z = 1/z_0^*$.

35.1.9.3 Spectral Factorization

$S_x(z) = \sigma_v^2 Q(z) Q^*(1/z^*)$, $S_x(z) =$ Z-transform of a WSS process $x(n)$, $\sigma_v^2 = \exp\left[\frac{1}{2\pi}\int_{-\pi}^{\pi} \ln S_x(e^{j\omega})\,d\omega\right]$,

$\sigma_v^2 =$ variance of whiter noise.

35.1.9.4 Rational Function

$$S_x(z) = \sigma_v^2 \left[\frac{B(z)}{A(z)}\right]\left[\frac{B^*(1/z^*)}{A^*(1/z^*)}\right],$$

$B(z) = 1 + b(1)z^{-1} + \cdots + b(q)z^{-q}$, $A(z) = 1 + a(1)z^{-1} + \cdots + a(p)z^{-p}$.

Note: $S_x(z) = S_x^*(1/z^*)$ since $S_x(e^{j\omega})$ is real. This implies that for each pole (or zero) in $S_x(z)$ there will be a matching pole (or zero) at the conjugate reciprocal location.

35.1.9.5 Wold Decomposition

Any random process can be written in the form. $x(n) = x_p(n) + x_r(n)$, $x_r(n) =$ regular random process, $x_p(n) =$ predictable process, $x_p(n)$ and $x_r(n)$, are *orthogonal* $E\{x_r(m)x_p^*(n)\} = 0$.

Note: $S_x(e^{j\omega}) = S_{x_r}(e^{j\omega}) + \sum_{k=1}^{N} a_k \delta(\omega - \omega_k) \equiv$ continuous spectrum + line spectrum

35.1.10 Special Types of Random Processes (x(n) WSS process)

35.1.10.1 Autoregressive Moving Average (ARMA)

$$S_x(z) = \sigma_v^2 \frac{B_q(z)B_q^*(1/z^*)}{A_p(z)A_p^*(1/z^*)},$$

$\sigma_v^2 =$ variance of input white noise, $H(z) =$ causal linear time-invariant filter $= \dfrac{B_q(z)}{A_p(z)}$, $S_x(z) =$ output powers spectrum $(z = e^{j\omega})$

35.1.10.2 Power Spectrum of ARMA Process

$$S_x(e^{j\omega}) = \sigma_v^2 \frac{B_q(z)B_q^*(1/z^*)}{A_p(z)A_p^*(1/z^*)} = \sigma_v^2 \frac{\left|B_q(e^{j\omega})\right|^2}{\left|A_p(e^{j\omega})\right|^2}$$

35.1.10.3 Yule-Walker Equations for ARMA Process

$$c_q(k) = \sum_{\ell=k}^{q} b_q(\ell)h^*(\ell - k) = \sum_{\ell=0}^{q-k} b_q(\ell + k)h^*(\ell),$$

$$r_x(k) + \sum_{\ell=1}^{p} a_p(\ell) r_x(k - \ell) = \begin{cases} \sigma_v^2 c_q(k) & 0 \le k \le q \\ 0 & k > q \end{cases},$$ matrix form:

$$\begin{bmatrix} r_x(0) & r_x(-1) & \cdots & r_x(-p) \\ r_x(1) & r_x(0) & \cdots & r_x(-p+1) \\ \vdots & \vdots & & \vdots \\ r_x(q) & r_x(q-1) & \cdots & r_x(q-p) \\ ---- & ---- & - & ---- \\ r_x(q+1) & r_x(q) & \cdots & r_x(q-p+1) \\ \vdots & & & \vdots \\ r_x(q+p) & r_x(q+p-1) & \cdots & r_x(q) \end{bmatrix} \begin{bmatrix} 1 \\ a_p(1) \\ a_p(2) \\ \vdots \\ a_p(p) \end{bmatrix} = \sigma_v^2 \begin{bmatrix} c_q(0) \\ c_q(1) \\ \vdots \\ c_q(q) \\ \hline 0 \\ 0 \\ \vdots \\ 0 \end{bmatrix}$$

35.1.10.4 Extrapolation of Correlation

$$r_x(k) = -\sum_{\ell=1}^{p} a_p(\ell) r_x(k-\ell) \quad \text{for } k \ge p; \text{ given } r_x(0), \cdots, r_x(p-1)$$

35.1.10.5 Autoregressive Process (AR)

$$S_x(z) = \sigma_v^2 \frac{|b(0)|^2}{A_p(z)A_p^*(1/z^*)} = \text{output of the filter}$$

$$H(z) = \frac{b(0)}{1 + \sum_{k=1}^{p} a_p(k) z^{-k}} = \frac{b(0)}{A(z)}$$

35.1.10.6 Power Spectrum of AR Process

$$S_x(e^{j\omega}) = \sigma_v^2 \frac{|b(0)|^2}{|A_p(e^{j\omega})|^2}$$

35.1.10.7 Yule-Walker Equation for AR Process

Set $q = 0$ in (35.1.10.3) and with $c_o(0) = b(0)h^*(0) = |b(0)|^2$ the equations are:

$$r_x(k) + \sum_{\ell=1}^{p} a_p(\ell) r_x(k-\ell) = \sigma_v^2 |b(0)|^2 \delta(k), \quad k \ge 0.$$

In matrix form:

$$\begin{bmatrix} r_x(0) & r_x(-1) & \cdots & r_x(-p) \\ r_x(1) & r_x(0) & \cdots & r_x(-p+1) \\ \vdots & & & \\ r_x(p) & r_x(p-1) & \cdots & r_x(0) \end{bmatrix} \begin{bmatrix} 1 \\ a_p(1) \\ \vdots \\ a_p(p) \end{bmatrix} = \sigma_v^2 |b(0)|^2 \begin{bmatrix} 1 \\ 0 \\ \vdots \\ 0 \end{bmatrix} \equiv \text{linear in the coefficient } a_p(k)$$

35.1.10.8 Moving Average Process (MA)

$$S_x(z) = \sigma_v^2 B_q(z) B_q^*(1/z^*) = \text{output of the filter } H(z) = \sum_{k=0}^{q} b_q(k) z^{-k}$$

35.1.10.9 Power Spectrum of MA Process

$$S_x(e^{j\omega}) = \sigma_v^2 \left|B_q(e^{j\omega})\right|^2$$

35.1.10.10 Yule-Walker Equation of MA Process

From (35.1.10.3) with $a_p(k) = 0$ and noting that $h(n) = b_q(n)$, then

$$c_q(k) = \sum_{\ell=0}^{q-k} b_q(\ell + k) b_q^*(\ell), \quad r_x(k) = \sigma_v^2 b_q(k) * b_q^*(-k) = \sigma_v^2 \sum_{\ell=0}^{q-|k|} b_q(\ell + |k|) b_q^*(\ell),$$

MA(q) process is zero for k outside $[-q, q]$.

35.2 Signal Modeling

35.2.1 The Pade Approximation

35.2.1.1 Pade Approximation

$$x(n) + \sum_{k=1}^{p} a_p(k) x(n-k) = \begin{cases} b_q(n) & n = 0, 1, \cdots, q \\ 0 & n = q+1, \cdots, q+p \end{cases}$$

$h(n) = x(n)$ for $n = 0, 1, \cdots, p+q$, $h(n) = 0$ for $n < 0$ and $b_q(n) = 0$ for $n < 0$ and $n > q$. This means that the data fit exactly to the model over the range $[0, p+q]$. See Figure (35.1) for the model. The system function is $H(z) = B_q(z)/A_p(z)$. (see Section 35.1.10)

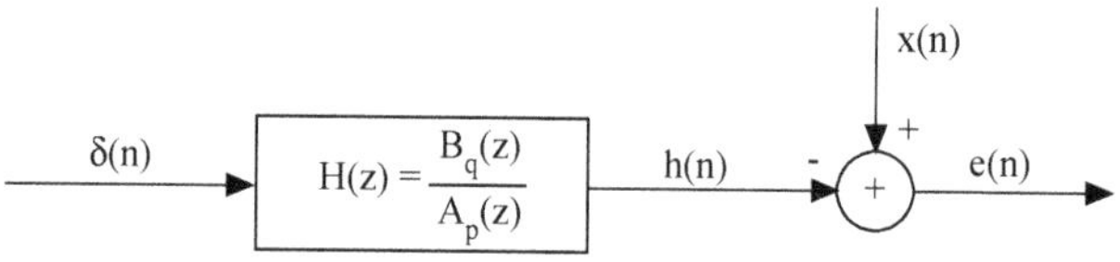

FIGURE 35.1 $x(n)$ modeled as unit sample response of linear shift-invariant system with p poles and q zeros.

35.2.1.2 Matrix Form of Pade Approximation

$$\begin{bmatrix} x(0) & 0 & \cdots & 0 \\ x(1) & x(0) & \cdots & 0 \\ x(2) & x(1) & \cdots & 0 \\ \vdots & & & \\ x(q) & x(q-1) & \cdots & x(q-p) \\ -- & ---- & - & ------ \\ x(q+1) & x(q) & \cdots & x(q-p+1) \\ \vdots & & & \\ x(q+p) & x(q+p-1) & \cdots & x(q) \end{bmatrix} \begin{bmatrix} 1 \\ a_p(1) \\ a_p(2) \\ \vdots \\ a_p(p) \end{bmatrix} = \begin{bmatrix} b_q(0) \\ b_q(1) \\ \vdots \\ b_q(q) \\ --- \\ 0 \\ \vdots \\ 0 \end{bmatrix}$$

Note: Solve for $a_p(p)$'s first (the lower part of the matrix) and then solve for $b_q(q)$'s.

35.2.1.3 Denominator Coefficients $(a_p(p))$

From the lower part of 35.2.1.2 we find

$$\begin{bmatrix} x(q) & x(q-1) & \cdots & x(q-p+1) \\ x(q+1) & x(q) & \cdots & x(q-p+2) \\ \vdots & & & \\ x(q+p-1) & x(q+p-2) & \cdots & x(q) \end{bmatrix} \begin{bmatrix} a_p(1) \\ a_p(2) \\ \vdots \\ a_p(p) \end{bmatrix} = - \begin{bmatrix} x(q+1) \\ x(q+2) \\ \vdots \\ x(q+p) \end{bmatrix}$$

or equivalently $\mathbf{X}_q \bar{\mathbf{a}}_p = -\mathbf{x}_{q+1}$, $\mathbf{X}_q$ = nonsymmetric Toeplitz matrix with $x(q)$ element in the upper left corner, $\mathbf{x}_{q+1}$ = vector with its first element being $x(q + 1)$.

Note:

1. If $\mathbf{X}_q$ is nonsingular, then $\mathbf{X}_q^{-1}$ exists and there is unique solution for the $a_p(p)$'s: $\bar{\mathbf{a}}_p = -\mathbf{X}_q^{-1}\mathbf{x}_{q+1}$.
2. If $\mathbf{X}_q$ is singular and (35.2.1.3) has a solution $\bar{\mathbf{a}}_p$ then $\mathbf{X}_q\mathbf{z} = 0$ has a solution and, hence, there is a solution of the form $\tilde{\mathbf{a}}_p = \bar{\mathbf{a}}_p + \mathbf{z}$.
3. If $\mathbf{X}_q$ is nonsingular and no solution exists, we must set $a_p(0) = 0$ and solve the equation $\mathbf{X}_q\mathbf{a}_p = 0$.

35.2.1.4 All Pole Model

$H(z) = b(0)/\left[1 + \sum_{k=1}^{p} a_p(k)z^{-k}\right]$, with $q = 0$ (35.2.1.3) becomes

$$\begin{bmatrix} x(0) & 0 & \cdots & 0 \\ x(1) & x(0) & \cdots & 0 \\ \vdots & & & \\ x(p-1) & x(p-2) & \cdots & x(0) \end{bmatrix} \begin{bmatrix} a_p(1) \\ a_p(2) \\ \vdots \\ a_p(p) \end{bmatrix} = - \begin{bmatrix} x(1) \\ x(2) \\ \vdots \\ x(p) \end{bmatrix}$$

or $\mathbf{X}_0 \bar{\mathbf{a}}_p = -\mathbf{x}_1$ or

$$a_p(k) = -\frac{1}{x(0)}\left[x(k) + \sum_{\ell=1}^{k-1} a_p(\ell)x(k-\ell)\right]$$

Example 1

$\mathbf{x} = [1\ 1.5\ 0.75\ 0.21\ 0.18\ 0.05]^T$ find the Pade approximation of all-pole and second-order model (p = 2, q = 0). From (35.2.1.4) $a(1) + 0\,a(2) = -1.5$ and $1.5a(1) + a(2) = -0.75$ or $a(1) = -1.5$ and $a(2) =$ 1.5. Also $b(0) = x(0) = 1$ and hence, $H(z) = 1/[1 - 1.5z^{-1} + 1.5z^{-2}]$ and since $h(n) = x(n)$ we obtain $\hat{\mathbf{x}} =$ approximate $= [1, 1.5\ 0.75 - 1.125 - 2.8125 - 2.5312]^T$.

Note: Matches to the second order only.

35.2.2 Prony's Method

35.2.2.1 Prony's Signal Modeling

Figure 35.2 shows the system representation.

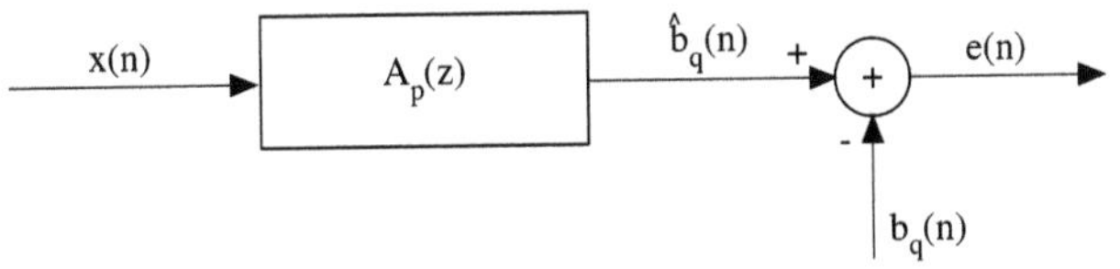

FIGURE 35.2 System interpretation of Prony's; method for signal modeling.

35.2.2.2 Normal Equations

$$\sum_{\ell=1}^{p} a_p(\ell) r_x(k,\ell) = -r_x(k,0) \quad k = 1,2,\cdots,p; \quad r_x(k,\ell) = \sum_{n=q+1}^{\infty} x(n-\ell)x^*(n-k) \quad k,\ell \ge 0 \,.$$

Note: The infinity indicates a nonfinite data sequence.

35.2.2.3 Numerator

$$b_q(n) = x(n) + \sum_{k=1}^{p} a_p(k)x(n-k) \quad n = 0,1,\cdots,q$$

35.2.2.4 Minimum Square Error

$$\varepsilon_{p,q} = r_x(0,0) + \sum_{k=1}^{p} a_p(k) r_x(0,k)$$

Example 1

Let $x(n) = 1$ for $n = 0,1,\cdots,N-1$ and $x(n) = 0$ otherwise. Find Prony's method to model $x(n)$ as the unit sample response of linear time-important filter with one pole and one zero,

$$H(z) = (b(0) + b(1)z^{-1}) / (1 + a(1)z^{-1})$$

Solutions

From (35.2.2.2) and $p = 1$, $a(1)r_x(1,1) = -r_x(1,0)$, $r_x(1,1) = \sum_{n=2}^{\infty} x^2(n-1) = N-1$, $r_x(1,0) = \sum_{n=2}^{\infty} x(n)$ $x(n-1) = N-2$. Hence $a(1) = -r_x(1,0)/r_x(1,1) = -(N-2)/(N-1)$ and the denominator becomes $A(z)$ $= 1 - \frac{N-2}{N-1} z^{-1}$.

Numerator Coefficients

$b(0) = x(0) = 1$, $b(1) = x(1) + a(1)x(0) = 1 - \frac{N-2}{N-1} = \frac{1}{N-1}$. The minimum square error is $\varepsilon_{1,1} = r_x(0,0) +$ $a(1)r_x(0,1)$. But $r_x(0,0) = \sum_{n=2}^{\infty} x(n)^2 = N-2$ and $\varepsilon_{1,1} = (N-2)/(N-1)$. For $N = 21$, $H(z) =$ $(1 + 0.05z^{-1})/(1 - 0.95z^{-1})$ and $h(n) = \delta(n) + (0.95)^{n-1}u(n-1)$, $\varepsilon_{1,1} = 0.95$, $e' =$ comparison error $=$ $x(n) - h(n)$ which gives $\sum_{n=0}^{\infty} [e'(n)]^2 \cong 4.595$.

35.2.3 Shank's Method

35.2.3.1 Shank's Signal Modeling

Figure 35.3 shows the system representation.

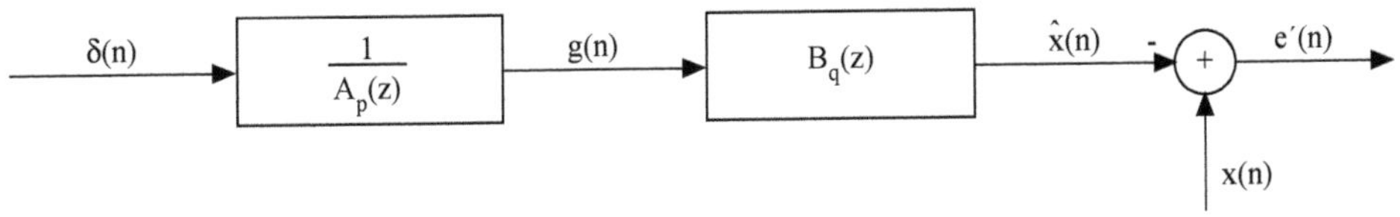

FIGURE 35.3 Shanks method. The denominator $A_p(z)$ is found using Prony's method, $B_q(z)$ is found by minimizing the sum of the squares of the error $e'(n)$.

35.2.3.2 Numerator of H(z)

$$\sum_{\ell=0}^{q} b_q(\ell) r_g(k-\ell) = r_{xg}(k) \quad k = 0,1,\cdots,q; \quad r_g(k-\ell) = \sum_{n=0}^{\infty} g(n-\ell) g^*(n-k),$$

$$r_{xg}(k) = \sum_{n=0}^{\infty} x(n) g^*(n-k), \quad g(n) = \delta(n) - \sum_{k=1}^{p} a_p(k) g(n-k)$$

35.2.4 All-Pole Model-Prony's Method

35.2.4.1 Transfer Function

$$H(z) = b(0) \Big/ \left[1 + \sum_{k=1}^{p} a_p(k) z^{-k}\right]$$

35.2.4.2 Normal Equations

$$\sum_{\ell=1}^{p} a_p(\ell) r_x(k-\ell) = -r_x(k) \quad k = 1,\cdots,p, \ x(n) = 0 \text{ for } n < 0,$$

$$r_x(k) = \sum_{n=0}^{\infty} x(n) x^*(n-k)$$

35.2.4.3 Numerator

$b(0) = \sqrt{\varepsilon_p}$, ε_p = minimum error (see 35.2.4.4)

35.2.4.4 Minimum Error

$$\varepsilon_p = r_x(0) + \sum_{k=1}^{p} a_p(k) r_x^*(k)$$

35.2.5 Finite Data Record Prony All-Pole Model

35.2.5.1 Normal Equations

$$\sum_{\ell=1}^{p} a_p(\ell) r_x(k-\ell) = -r_x(k) \quad k = 1,2,\cdots,p,$$

$$r_x(k) = \sum_{n=k}^{N} x(n) x^*(n-k) \quad k \geq 0, \ x(n) = 0 \text{ for } n < 0 \text{ and } n > N.$$

35.2.5.2 Minimum Error

$$\varepsilon_p = r_x(0) + \sum_{k=1}^{p} a_p(k) r_x^*(k)$$

35.2.6 Finite Data Record-Covariance Method of All-Pole Model

35.2.6.1 Normal Equation

$$\sum_{\ell=1}^{p} a_p(\ell) r_x(k,\ell) = -r_x(k,0) \quad k = 1,2,\cdots,p,$$

$$r_x(k,\ell) = \sum_{n=p}^{N} x(n-\ell)x^*(n-k), \quad k,\ell \ge 0, \quad x(n) = 0 \text{ for } n < 0 \text{ and } n > N.$$

35.2.6.2 Minimum Error

$$[\varepsilon_p]_{\min} = r_x(0) + \sum_{k=1}^{p} a_p(k) r_x(0,k)$$

Example

$\mathbf{x} = [1\ \alpha\ \alpha^2 \cdots \alpha^N]^T$, use first-order model: $H(z) = b(0)/(1 + a(1)z^{-1})$, then $p = 1$ and 35.2.6.1 becomes $\sum_{\ell=1}^{1} a(\ell) r_x(1,\ell) = -r_x(k,0)$. For k = 1, 35.2.6.1 becomes $a(1)r_x(1,1) = -r_x(1,0)$ and must find $r_x(1,1)$ and $r_x(1,0)$. From 35.2.6.1 $r_x(1,1) = \sum_{n=1}^{N} x(n-1)x^*(n-1) = \sum_{n=0}^{N-1} |x(n)|^2 = [1 - |\alpha|^{2N}]/[1 - |\alpha|^2]$, also $r_x(1,0) = \sum_{n=1}^{N} x(n)x^*(n-1) = \alpha[1 - |\alpha|^{2N}]/[1 - |\alpha|^2]$. But $\alpha(1) = -r_x(1,0)/r_x(1,1) = -\alpha$ and with b(0) = 1 so that $x(0) = \hat{x}(0)$ the model is $H(z) = 1/[1 - \alpha z^{-1}]$.

35.3 The Levinson Recursion

35.3.1 The Levinson–Durbin Recursion

35.3.1.1 All-Pole Modeling

$$r_x(k) + \sum_{\ell=1}^{p} a_p(\ell) r_x(k-\ell) = 0 \quad k = 1,2,\cdots,p, \quad \varepsilon_p = r_x(0) + \sum_{\ell=1}^{p} a_p(\ell) r_x^*(\ell),$$

$$r_x(k) = \sum_{n=0}^{N} x(n)x^*(n-k) \quad k \ge 0, \quad x(n) = 0 \text{ for } n < 0 \text{ and } n > N.$$

35.3.1.2 All-Pole Matrix Format

$$\begin{bmatrix} r_x(0) & r_x^*(1) & r_x^*(2) & \cdots & r_x^*(p) \\ r_x(1) & r_x(0) & r_x^*(1) & \cdots & r_x^*(p-1) \\ r_x(2) & r_x(1) & r_x(0) & \cdots & r_x^*(p-2) \\ \vdots & & & & \\ r_x(p) & r_x(p-1) & r_x(p-2) & \cdots & r_x(0) \end{bmatrix} \begin{bmatrix} 1 \\ a_p(1) \\ a_p(2) \\ \vdots \\ a_p(p) \end{bmatrix} = \varepsilon_p \begin{bmatrix} 1 \\ 0 \\ 0 \\ \vdots \\ 0 \end{bmatrix}$$

or $\mathbf{R}_p \mathbf{a}_p = \varepsilon_p \mathbf{u}_1$, $\mathbf{u}_1 = [1,0,\cdots,0]^T$, $\mathbf{R}_p =$ symmetric Toeplitz.

35.3.1.3 Solution of (35.3.1.2)

$a_0(0) = 1$, $\varepsilon_0 = r_x(0)$

1. Initialization of the recursion
   a) $a_0(0) = 1$,
   b) $\varepsilon_0 = r_x(0)$
2. For $j = 0,1,\cdots,p-1$
   a) $\gamma_j = r_x(j+1) + \sum_{i=1}^{j} a_j(i) r_x(j-i+1)$;
   b) $\Gamma_{j+1} = -\gamma_j / \varepsilon_j =$ reflection of coefficient;
   c) For $i = 1,2,\cdots,j$ $\quad a_{j+1}(i) = a_j(i) + \Gamma_{j+1} a_j^*(j-i+1)$;
   d) $a_{j+1}(j+1) = \Gamma_{j+1}$;
   e) $\varepsilon_{j+1} = \varepsilon_j [1 - |\Gamma_{j+1}|^2]$
3. $b(0) = \sqrt{\varepsilon_p}$ (see 35.3.1.1)

35.3.1.4 Properties

1. Γ_j's produced by solving the autocorrelation normal equations (see Section 35.2.4) obey the relation $|\Gamma_j| \le 1$
2. If $|\Gamma_j| < 1$ for all j, then $A_p(z) = 1 + \sum_{k=1}^{p} a_p(k) z^{-k} \equiv$ minimum phase polynomial (all roots lie inside the unit circle)
3. If $\mathbf{a}_p$ is the solution to the Toeplitz normal equation $\mathbf{R}_p \mathbf{a}_p = \varepsilon_p \mathbf{u}_1$ (see 35.3.1.2) and $\mathbf{R}_p > 0$ (positive definite) then $A_p(z) =$ minimum phase
4. If we choose $b(0) = \sqrt{\varepsilon_p}$ (energy matching constraint), then the auto-correlation sequences of $x(n)$ and $h(n)$ are equal for $|k| \le p$.

35.3.2 Step-Up and Step-Down Recursions

35.3.2.1 Step-Up Recursion

The recursion finds $a_p(i)$'s from Γ_j's.

Steps

1. Initialize the recursion: $a_0(0) = 1$
2. For $j = 0, 1, \cdots, p-1$
   a) For $i = 1, 2, \cdots, j$ $a_{j+1}(i) = a_j(i) + \Gamma_{j+1} a_j^*(j-i+1)$
   b) $a_{j+1}(j+1) = \Gamma_{j+1}$
3. $b(0) = \sqrt{\varepsilon_p}$

35.3.2.2 Step-Down Recursion

The recursion finds Γ_j's from $a_p(i)$'s.

Steps

1. Set $\Gamma_p = a_p(p)$
2. For $j = p-1, p-2, \cdots, 1$
   a) For $i = 1, 2, \cdots, j$ $a_j(i) = \frac{1}{1 - |\Gamma_{j+1}|^2}[a_{j+1}(i) - \Gamma_{j+1} a_{j+1}^*(j-i+1)]$
   b) Set $\Gamma_j = a_j(j)$
   c) If $|\Gamma_j| = 1$ quit
3. $\varepsilon_p = b^2(0)$

Example

To implement the third-order filter $H(z) = 1 + 0.5z^{-1} - 0.1z^{-2} - 0.5z^{-3}$ in the form of a lattice structure we proceed as follows. $\mathbf{a}_3 = [1, 0.5, -0.1, -0.5]^T$, $\Gamma_3 = a_3(3) = -0.5$. From step 2) we obtain the second-order polynomial $\mathbf{a}_2 = [1\ a_2(1)\ a_2(2)]^T$ or

$$\begin{bmatrix} a_2(1) \\ a_2(2) \end{bmatrix} = \frac{1}{1-\Gamma_3^2}\left[\begin{bmatrix} a_3(1) \\ a_3(2) \end{bmatrix} - \Gamma_3 \begin{bmatrix} a_3(2) \\ a_3(1) \end{bmatrix}\right] = \frac{1}{1-0.25}\left[\begin{bmatrix} 0.5 \\ -0.1 \end{bmatrix} + 0.5\begin{bmatrix} -0.1 \\ 0.5 \end{bmatrix}\right] = \begin{bmatrix} 0.6 \\ 0.2 \end{bmatrix},$$

$\Gamma_2 = a_2(2) = 0.2$. Next we find $a_1(1) = \frac{1}{1-\Gamma_2^2}[a_2(1) - \Gamma_2 a_2(1)] = 0.5$ and hence $\Gamma_1 = a_1(1) = 0.5$ and $\Gamma = [0.5\ 0.2\ -0.5]^T$. The lattice filter implementation is shown in Figure 35.4.

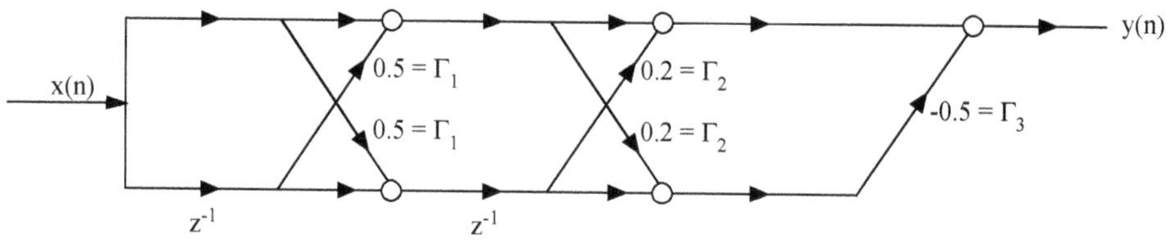

FIGURE 35.4

35.3.3 Cholesky Decomposition

35.3.3.1 Cholesky Decomposition

$\mathbf{R}_p = \mathbf{L}_p \mathbf{D}_p \mathbf{L}_p^H =$ Hermitian Toeplitz autocorrelation matrix,

$$\mathbf{A}_p = \begin{bmatrix} 1 & a_1^*(1) & a_2^*(2) & \cdots & a_p^*(p) \\ 0 & 1 & a_2^*(1) & \cdots & a_p^*(p-1) \\ 0 & 0 & 1 & \cdots & a_p^*(p-2) \\ \vdots & \vdots & \vdots & & \vdots \\ 0 & 0 & 0 & \cdots & 1 \end{bmatrix}, \quad \mathbf{L}_p = \mathbf{A}_p^H = \text{lower triangular},$$

$\mathbf{D}_p = diag\{\varepsilon_0\ \varepsilon_1 \cdots \varepsilon_p\}$, $\det \mathbf{D}_p = \det \mathbf{R}_p$, $\mathbf{R}_p =$ see 35.3.1.2, $\mathbf{A}_p^H \mathbf{R}_p \mathbf{A}_p = \mathbf{D}_p$

35.3.4 Inversion of Toeplitz Matrix

35.3.4.1 Inversion of Toeplitz Matrix

$\mathbf{R}_p \mathbf{x} = \mathbf{b}$, $\mathbf{R}_p^{-1} = \mathbf{A}_p \mathbf{D}_p^{-1} \mathbf{A}_p^H =$ nonsingular Hermitian matrix (see 35.3.1.2), $\mathbf{A}_p =$ see 35.3.3.1, $\mathbf{D}_p =$ see 35.3.3.1, $\mathbf{b} =$ arbitrary vector.

35.3.5 Levinson Recursion for Inverting Toeplitz Matrix

35.3.5.1 Levinson Recursion

$\mathbf{R}_p \mathbf{x} = \mathbf{b}$, $\mathbf{R}_p =$ see 35.3.1.2 (known), $\mathbf{b} =$ arbitrary vector, $\mathbf{x} =$ unknown vector

Recursion:

1. Initialize the recursion
   a) $a_0(0) = 1$
   b) $x_0(0) = b(0)/r_x(0)$
   c) $\varepsilon_0 = r_x(0)$
2. For $j = 0, 1, \cdots, p-1$
   a) $\gamma_j = r_x(j+1) + \sum_{i=1}^{j} a_j(i) r_x(j-i+1)$
   b) $\Gamma_{j+1} = -\gamma_j / \varepsilon_j$
   c) For $i = 1, 2, \cdots, j \quad a_{j+1}(i) = a_j(i) + \Gamma_{j+1} a_j^*(j-i+1)$;
   d) $a_{j+1}(j+1) = \Gamma_{j+1}$
   e) $\varepsilon_{j+1} = \varepsilon_j [1 - |\Gamma_{j+1}|^2]$
   f) $\delta_j = \sum_{i=1}^{j} x_j(i) r_x(j-i+1)$
   g) $q_{j+1} = [b(j+1) - \delta_j] / \varepsilon_{j+1}$
   h) For $i = 0, 1, \cdots, j \quad x_{j+1}(i) = x_j(i) + q_{j+1} a_{j+1}^*(j-i+1)$
   i) $x_{j+1}(j+1) = q_{j+1}$

Example

$$\begin{bmatrix} 8 & 4 & 2 \\ 4 & 8 & 4 \\ 2 & 4 & 8 \end{bmatrix} \begin{bmatrix} x(0) \\ x(1) \\ x(2) \end{bmatrix} = \begin{bmatrix} 18 \\ 12 \\ 24 \end{bmatrix}$$

1. Initialization: $\varepsilon_0 = r(0) = 8,\ x_0(0) = b(0)/r(0) = 18/8 = 9/4$

2. For $j = 0,\ \gamma_0 = r(1) = 4,\ \Gamma_1 = -\gamma_0/\varepsilon_0 = -4/8 = -1/2,\ \mathbf{a}_1 = \begin{bmatrix} 1 \\ \Gamma_1 \end{bmatrix} = \begin{bmatrix} 1 \\ -1/2 \end{bmatrix},\ \varepsilon_1 = \varepsilon_0\left[1 - |\Gamma_1|^2\right] =$ $8[1-1/4] = 6,\ \delta_0 = x_0(0)r(1) = (9/4)4 = 9,\ q_1 = [b(1) - \delta_0]/\varepsilon_1 = (12-9)/6 = 1/2,$ and hence $\mathbf{x}_1 = \begin{bmatrix} x_0(0) \\ 0 \end{bmatrix} + q_1 \begin{bmatrix} a_1(1) \\ 1 \end{bmatrix} = \begin{bmatrix} 9/4 \\ 0 \end{bmatrix} + (1/2)\begin{bmatrix} -1/2 \\ 0 \end{bmatrix} = \begin{bmatrix} 2 \\ 1/2 \end{bmatrix},$

3. For j =1 $\gamma_1 = r(2) + a_1(1)r(1) = 2 + (-1/2)4 = 0,\ \Gamma_2 = 0,\ \mathbf{a}_2 = \begin{bmatrix} \mathbf{a}_1 \\ 0 \end{bmatrix},\ \varepsilon_2 = \varepsilon_1 = 6,\ \delta_1 = [x_1(0)$ $x_1(1)] \begin{bmatrix} r(2) \\ r(1) \end{bmatrix} = [2\ \ 1/2][2\ \ 4]^T = 4 + 2 = 6,\ q_2 = [b(2) - \delta_1]/\varepsilon_2 = [12-6]/6 = 1,\ \mathbf{x}_2 = \begin{bmatrix} x_1(0) \\ x_1(1) \\ 0 \end{bmatrix} +$ $q_2 \begin{bmatrix} a_2(2) \\ a_2(1) \\ 1 \end{bmatrix} = \begin{bmatrix} 2 \\ 1/2 \\ 0 \end{bmatrix} + 6 \begin{bmatrix} 0 \\ -1/2 \\ 1 \end{bmatrix} = \begin{bmatrix} 2 \\ -5/2 \\ 6 \end{bmatrix}$

35.4 Lattice Filters

35.4.1 The FIR Lattice Filter

35.4.1.1 Forward Prediction Error

$e_p^f = x(n) + \sum_{k=1}^{p} a_p(k)x(n-k) =$ forward prediction error, $x(n) =$ data, $a_p(k) =$ all-pole filter coefficients.

35.4.1.2 Square of Error

$$\varepsilon_p^+ = \sum_{n=0}^{\infty} \left|e_p^f(n)\right|^2$$

35.4.1.3 Z-Transform of the pth-Order Error

$E_p^f(z) = A_p(z)X(z),\ A_p(z) = 1 + \sum_{k=1}^{p} a_p(k)z^{-k} =$ forward prediction error filter (all pole), see (35.4.1.1).

Note: The output of the forward prediction filter is e_p^f when the input is $x(n)$.

35.4.1.4 (j+1) Order Coefficient

$a_{j+1}(i) = a_j(i) + \Gamma_{j+1}a_j^*(j-i+1).\ \Gamma_{j+1} \equiv$ see Section 35.3.1

35.4.1.5 (j+1) Order Coefficient in the Z-domain

$$A_{j+1}(z) = A_j(z) + \Gamma_{j+1}[z^{-(j+1)} A_j^*(1/z^*)]$$

35.4.1.6 (j+1) Order Error in the Z-domain

$$E_{j+1}^f(z) = E_j^f(z) + z^{-1}\Gamma_{J+1} E_j^b(z),\ \ E_j^b(z) = z^{-1}X(z)A_j^*(1/z^*),\ \ E_{j+1}^f(z) = A_{j+1}(z)X(z),\ \ E_j^f(z) = A_j(z)X(z)$$

35.4.1.7 (j+1) Order of Error (see 35.4.1.6)

$e_{j+1}^f(n) = e_j^f(n) + \Gamma_{j+1} e_j^b(n-1)$ (inverse Z-transform of (35.4.1.6), e_j^b = j^{th} order of backward prediction error

35.4.1.8 Backward Prediction Error

$$e_j^b(n) = x(n-j) + \sum_{k=1}^{j} a_j^*(k)x(n-j+k)$$

35.4.1.9 (j+1) Backward Prediction Error

$$e_{j+1}^b(n) = e_j^b(n-1) + \Gamma_{j+1}^* e_j^f(n)$$

35.4.1.10 Single Stage of FIR Lattice Filter

See Figure 35.5

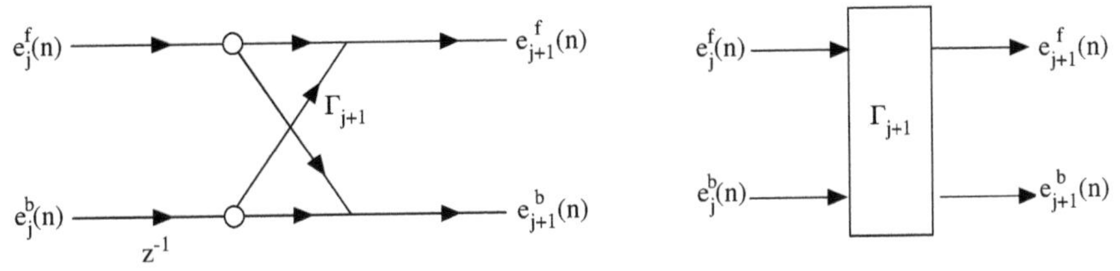

FIGURE 35.5 One-stage FIR lattice filter.

35.4.1.11 p^{th}-Order FIR Lattice Filter

See Figure 35.6

Note: $e_0^f(n) = e_0^b(n) = x(n)$

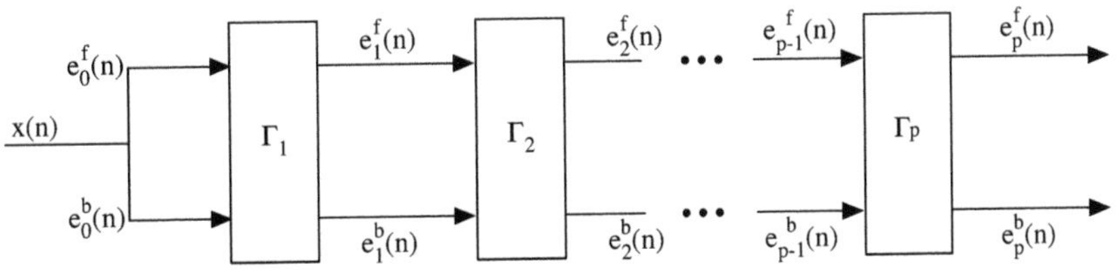

FIGURE 35.6 p^{th}-order FIR lattice filter.

35.4.1.12 All-Pass Filter

$H_{ap}(z) = A_p^R(z)/A_p(z) = [z^{-p}A_p^*(1/z^*)]/\left[1+\sum_{k=1}^{p} a_p(k)z^{-k}\right]$, $E_p^b(z) = H_{ap}(z)E_p^f(z)$ which indicates that $e_p^b(n)$ is the output of an all-pass filter with input $e_p^f(n)$.

35.4.2 IIR Lattice Filters

35.4.2.1 All-pole Filter

$\frac{1}{A_p(z)} = \frac{E_0^f(z)}{E_p^f(z)} = \frac{1}{1+\sum_{k=1}^{p} a_p(k)z^{-k}} \equiv$ produces a response $e_0^f(n)$ to the input $e_p^+(n)$ (see Section 35.4.1 for definition).

Forward and Backward Errors

$e_j^f(n) = e_{j+1}^f(n) - \Gamma_{j+1}\, e_j^b(n-1)$ (see 35.4.1.7), $e_{j+1}^b(n) = e_j^b(n-1) + \Gamma_{j+1}^*\, e_j^f(n)$ (see 35.4.1.9). See Figure 35.7 for a pictorial representation of the single stage of an all-pole lattice filter. Cascading p in such a section we obtain the pth-order all-pole lattice filter.

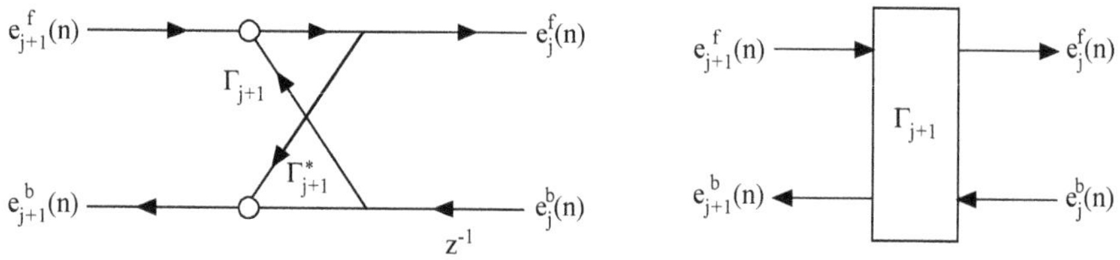

FIGURE 35.7 Single stage of an all-pole lattice filter.

35.4.2.2 All-Pass Filter

$$H_{ap}(z) = z^{-p}\frac{A_p^*(1/z^*)}{A_p(z)}$$

35.4.3 Lattice All-Pole Modeling of Signals

35.4.3.1 Forward Covariance Method

35.4.3.1.1 Reflection Coefficients

$$\Gamma_j^f = -\frac{\sum_{n=j}^{N} e_{j-1}^f(n)[e_{j-1}^b(n-1)]^*}{\sum_{n=j}^{N} \left|e_{j-1}^b(n-1)\right|^2} = -\frac{\langle \mathbf{e}_{j-1}^f, \mathbf{e}_{j-1}^b\rangle}{\left\|\mathbf{e}_{j-1}^b\right\|^2},$$

$<>$ = dot product, $\mathbf{e}_j^f = [e_j^f(j)\; e_j^f(j+1) \cdots e_j^f(N)]^T$,

$$\mathbf{e}_{j-1}^b = [e_{j-1}^b(j-1)\; e_{j-1}^b(j) \cdots e_{j-1}^b(N-1)]^T$$

35.4.3.1.2 Forward Covariance Algorithm

1. Given j – 1 reflection coefficients $\Gamma_{j-1}^{f} = [\Gamma_1^f \ \Gamma_2^f \cdots \Gamma_{j-1}^f]^T$
2. Given the forward and backward prediction errors $e_{j-1}^f(n), e_{j-1}^b(n)$
3. j[th] reflection coefficient is found from 35.4.3.1.1
4. Using lattice filter the (j – 1)[st] order forward and backward prediction errors are updated to form $e_j^f(n)$ and $e_j^b(n)$
5. Repeat the process

Example

Given $x(n) = \alpha^n,\ 0 \le n \le N,\ 0 < \alpha < 1.$

Initialization: $e_0^f(n) = e_0^b(n) = x(n) = \alpha^n,\quad n = 0,1,\cdots,N.$ Next, evaluation of the norm

$$e_0^f(n-1):\ \left\|\mathbf{e}_0^b\right\|^2 = \sum_{n=1}^{N} [e_0^b(n-1)]^2 = \sum_{n=0}^{N-1} \alpha^{2n} = \frac{1-\alpha^{2N}}{1-\alpha^2}\,.$$

Inner product between

$$e_0^f(n) \text{ and } e_0^b(n-1){:}\langle \mathbf{e}_0^f, \mathbf{e}_0^b\rangle = \sum_{n=1}^{N} e_0^f(n) e_0^b(n-1) = \alpha \sum_{n=0}^{N-1} \alpha^{2n} = \alpha \frac{1-\alpha^{2N}}{1-\alpha^2}\,.$$

From 35.4.3.1.1

$$\Gamma_1^f = -\frac{\langle \mathbf{e}_0^f, \mathbf{e}_0^b\rangle}{\left\|\mathbf{e}_0^b\right\|^2} = -\alpha\,.$$

Updating forward prediction error (set $j = j-1$ in 4.1.7): $e_j^f(n) = e_{j-1}^f(n) + \Gamma_j^f e_{j-1}^b(n-1),$ hence, $e_1^f(n) = \alpha^n u(n) - \alpha(\alpha)^{n-1} u(n-1) = \delta(n),\ u(n) =$ unit step function. First-order modeling error (35.4.1.2): $\varepsilon_1^f = \sum_{n=1}^{N} [e_1^f(n)]^2 = 0$ since $\delta(1)\cdots\delta(N)$ are zero. First-order backward prediction error (set j = j – 1 in 4.1.9): $e_j^b(n) = e_{j-1}^b(n-1) + \Gamma_j^* e_{j-1}^f(n)$ or $e_1^b(n) = e_0^b(n-1) + \Gamma_1^* e_0^f(n) = \alpha^{n-1} u(n-1) - \alpha^{n+1} u(n).$ Second reflection coefficient:

$$\Gamma_2^f = -\frac{\langle \mathbf{e}_1^f, \mathbf{e}_1^b\rangle}{\left\|\mathbf{e}_1^b\right\|^2} = 0$$

since $e_1^f(n) = 0$ for $n > 0.$ Similar steps $e_2^f(n) = e_1^b(n) = \delta(n)$ and $\Gamma_3^f = 0.$

Continuing $\Gamma_j^f = 0$ for all $j > 1$ and hence $\boldsymbol{\Gamma}^f = [-\alpha\ 0\ 0 \cdots]^T.$ For finding a_i's see Section 35.4.1.

35.4.3.2 The Backward Covariance Method

35.4.3.2.1 Reflection Coefficients

$$\Gamma_j^b = -\frac{\sum_{n=j}^{N} e_{j-1}^b(n)[e_{j-1}^b(n-1)]^*}{\sum_{n=j}^{N} \left|e_{j-1}^f(n)\right|^2} = -\frac{\langle \mathbf{e}_{j-1}^f, \mathbf{e}_{j-1}^b \rangle}{\left\| \mathbf{e}_{j-1}^f \right\|^2}$$

The steps are similar to those in section 35.4.3.1: Given the first j – 1 reflection coefficients $\mathbf{\Gamma}^b = [\Gamma_1^b\ \Gamma_2^b \cdots \Gamma_{j-1}^b]^T$, and given the forward and backward prediction errors $e_{j-1}^f(n)$ and $e_{j-1}^b(n)$ the jth reflection coefficient is computed. Next, using the lattice filter, the (j – 1)[st]-order forward and backward errors are updated to form the jth-order errors, and the process is repeated.

35.4.3.3 Burg's Method

35.4.3.3.1 Reflection Coefficients

$$\Gamma_j^B = -\frac{2\sum_{n=j}^{N} e_{j-1}^f(n)[e_{j-1}^b(n-1)]^*}{\sum_{n=j}^{N} [\left|e_{j-1}^f(n)\right|^2 + \left|e_{j-1}^b(n-1)\right|^2]} = -\frac{2\langle \mathbf{e}_{j-1}^f, \mathbf{e}_{j-1}^b \rangle}{\left\| \mathbf{e}_{j-1}^f \right\|^2 + \left\| \mathbf{e}_{j-1}^f \right\|^2},$$

< , > = dot product, e's are the forward and backward prediction errors.

35.4.3.3.2 Burg Error

$$\varepsilon_j^B = [\varepsilon_{j-1}^B - \left|e_{j-1}^f(j-1)\right|^2 - \left|e_{j-1}^f(N)\right|^2][1 - \left|\Gamma_j^B\right|^2],\ \ \varepsilon_0^B = \sum_{n=0}^{N} [\left|e_0^f(n)\right|^2 + \left|e_0^b(n)\right|^2] = 2\sum_{n=0}^{N} \left|x(n)\right|^2.$$

The Burg's method has the same steps for computing the necessary unknown as in Section 35.4.3.1 and 35.4.3.2.

35.4.3.3.3 Burg's Algorithm

1. Initialize the recursion:

   a) $e_0^f(n) = e_0^b(n) = x(n)$

   b) $D_1 = 2\sum_{n=1}^{N} [\left|x(n)\right|^2 - \left|x(n-1)\right|^2]$

2. For j = 1 to p:

   a) $\Gamma_j^B = -\frac{2}{D_j}\sum_{n=j}^{N} e_{j-1}^f(n)[e_{j-1}^b(n-1)]^*$

   For n = j to N:

   b) $e_j^f(n) = e_{j-1}^b(n) + \Gamma_j^B e_{j-1}^b(n-1),\ \ e_j^b(n) = e_{j-1}^b(n-1) + (\Gamma_j^B)^* e_{j-1}^f(n)$

   c) $D_{j+1} = D_j(1 - \left|\Gamma_j^B\right|^2) - \left|e_j^f(j)\right|^2 - \left|e_j^b(N)\right|^2,$

   d) $\varepsilon_j^B = D_j[1 - \left|\Gamma_j^B\right|^2]$

Example 1

Given $x(n) = \alpha^n u(n),\ n = 0,1,\cdots N,\ u(n) =$ unit step function. From Example 1 of (35.4.3.1.2)

$$\langle \mathbf{e}_0^f, \mathbf{e}_0^b \rangle = \alpha \frac{1-\alpha^{2N}}{1-\alpha^2} \text{ and } \left\| \mathbf{e}_0^b \right\|^2 = \frac{1-\alpha^{2N}}{1-\alpha^2}.$$

Similarly, from Section 35.4.3.2

$$\left\| \mathbf{e}_0^f \right\|^2 = \alpha^2 \frac{1-\alpha^{2N}}{1-\alpha^2}.$$

Therefore,

$$\Gamma_1^B = -2 \frac{\langle \mathbf{e}_0^f, \mathbf{e}_0^b \rangle}{\left\| \mathbf{e}_0^f \right\|^2 + \left\| \mathbf{e}_0^b \right\|^2} = -\frac{2\alpha}{1+\alpha^2}.$$

Update the errors:

$$e_1^f(n) = e_0^f(n) + \Gamma_1^B e_0^b(n-1) = \alpha^n u(n) + \Gamma_1^B \alpha^{n-1} u(n-1),$$

$$e_1^b(n) = e_0^b(n-1) + \Gamma_1^B e_0^f(n) = \alpha^{n-1} u(n-1) + \Gamma_1^B \alpha^n u(n),$$

$$\left\| \mathbf{e}_1^f \right\|^2 = \sum_{n=2}^{N} [e_1^f(n)]^2 = \alpha^4 (1-\alpha^2) \frac{1-\alpha^{2(N-1)}}{(1+\alpha^2)^2},$$

$$\left\| \mathbf{e}_1^b \right\|^2 = \sum_{n=2}^{N} [e_1^b(n-1)]^2 = (1-\alpha^2) \frac{1-\alpha^{2(N-1)}}{(1+\alpha^2)^2},$$

$$\langle \mathbf{e}_1^f, \mathbf{e}_1^b \rangle = \sum_{n=2}^{N} e_1^f(n) e_1^f(n-1) = -\alpha^2 (1-\alpha^2) \frac{1-\alpha^{2(N-1)}}{(1+\alpha^2)^2}.$$

Hence,

$$\Gamma_2^B = -2 \frac{\langle \mathbf{e}_1^f, \mathbf{e}_1^b \rangle}{\left\| \mathbf{e}_1^f \right\|^2 + \left\| \mathbf{e}_1^b \right\|^2} = \frac{2\alpha^2}{1+\alpha^4}.$$

Zero-order error:

$$\varepsilon_0^B = 2 \sum_{n=0}^{N} x^2(n) = 2 \frac{1-\alpha^{2(N+1)}}{1-\alpha^2} \cong \frac{2}{1-\alpha^2},$$

first-order error

$$\varepsilon_1^B = [\varepsilon_0^B - [e_0^f(0)]^2 - [e_0^b(N)]^2][1 - \left|\Gamma_1^B\right|^2] = [\varepsilon_0^B - 1 - \alpha^{2N}][1 - \left|\Gamma_1^B\right|^2] \cong (1-\alpha^2)/(1+\alpha^2).$$

35.4.3.4 Modified Covariance Method

35.4.3.4.1 Normal Equation

$$\sum_{k=1}^{p}[r_x(\ell,k)+r_x(p-k,p-\ell)]a_p(k)=-[r_x(\ell,0)+r_x(p,p-\ell)],$$

$$\ell=1,\cdots,p,\ r_x(\ell,k)=\sum_{n=p}^{N}x(n-k)x^*(n-\ell)],$$

$$\varepsilon_p^M = \text{modified covariance error } = r_x(0,0)+r_x(p,p)+\sum_{k=1}^{p}a_p(k)[r_x(0,k)+r_x(p,p-k)]$$

Example 1

Given data $x(n)=\alpha^n u(n),\ n=0,\cdots N$. For second-order filter p = 2,

$$r_x(k,\ell)=\sum_{n=2}^{N}\alpha^{n-k}\alpha^{n-\ell}=\alpha^{-k-\ell}\sum_{n=2}^{N}\alpha^{2n}=c\alpha^{4-k-\ell},$$

(a) $c=[1-\alpha^{2(N-1)}]/[1-\alpha^2]$, $\begin{bmatrix} r_x(1,1)+r_x(1,1) & r_x(1,2)+r_x(0,1) \\ r_x(2,1)+r_x(1,0) & r_x(2,2)+r_x(0,0)\end{bmatrix}\begin{bmatrix} a(1) \\ a(2)\end{bmatrix}=-\begin{bmatrix} r_x(1,0)+r_x(2,1) \\ r_x(2,0)+r_x(2,0)\end{bmatrix}$, inserting values of $r_x(k,\ell)$ from (a) and solving for $a(1)$ and $a(2)$ we obtain $a(1)=-(1+\alpha^2)/\alpha$ and $a(2)=1$. Hence, the all-pole model has $A(z)=1-[(1+\alpha^2)/\alpha]z^{-1}+z^{-2}$.

35.4.4 Stochastic Modeling

35.4.4.1 Forward Reflection Coefficients

$$\Gamma_j^f=-\frac{E\{e_{j-1}^f(n)[e_{j-1}^b(n-1)]^*\}}{E\{|e_{j-1}^b(n-1)|^2\}},$$

E stands for expectation.

35.4.4.2 Backward Reflection Coefficients

$$\Gamma_j^b=-\frac{E\{e_{j-1}^f(n)[e_{j-1}^b(n-1)]^*\}}{E\{|e_{j-1}^f(n)|^2\}}$$

35.4.4.3 Burg Reflection Coefficient

$$\Gamma_j^B=-2\frac{E\{e_{j-1}^f(n)[e_{j-1}^b(n-1)]^*\}}{E\{|e_{j-1}^f(n)|^2\}+E\{|e_{j-1}^b(n-1)|^2\}}$$

References

Hayes, M. H., *Statistical Digital Signal Processing and Modeling*, John Wiley & Sons Inc., New York, NY, 1996.

Kay, S., *Modern Spectrum Estimation: Theory and Applications*, Prentice-Hall, Englewood Cliffs, NJ, 1988.

Marple, S. L., *Digital Spectral Analysis with Applications*, Prentice-Hall, Englewood Cliffs, NJ, 1987.

36

Spectrum Estimation of Random Discrete Signals

36.1 Nonparametric Method

36.1.1 Definitions

36.1.1.1 Power Spectrum (unlimited data)

$$S_x(e^{j\omega}) = \sum_{k=-\infty}^{\infty} r_x(k)e^{-jk\omega}, \quad r_x(k) = \lim_{N\to\infty}\frac{1}{2N+1}\sum_{n=-N}^{N} x(n+k)x^*(n)$$

36.1.1.2 Power Spectrum (limited data):

$$\hat{S}_x(e^{j\omega}) = \sum_{k=-N+1}^{N-1} \hat{r}_x(k)e^{-jk\omega}, \quad \hat{r}_x(k) = \frac{1}{N}\sum_{n=0}^{N-1} x(n+k)x^*(k), \quad \hat{r}_x(k) = \frac{1}{N}\sum_{n=0}^{N-1-k} x(n+k)x^*(n)$$

$k = 0,1,\cdots N-1$ ensures that $x(n)$ which fall outside the interval $[0, N-1]$ are excluded, $\hat{r}_x(-k) = \hat{r}_x^*(k) \equiv$ conjugate symmetry, $\hat{r}_x(k) = 0$ for $|k| \geq N$.

36.1.1.3 Power Spectrum Using the Data

$$\hat{S}_x(e^{j\omega}) = \frac{1}{N}X_N(e^{j\omega})X_N^*(e^{j\omega}) = \frac{1}{N}\left|X_N(e^{j\omega})\right|^2,$$

$$X_N(e^{j\omega}) = \sum_{n=-\infty}^{\infty} x_N(n)e^{-jn\omega} = \sum_{n=0}^{N-1} x_N(n)e^{-jn\omega}$$

Figure 36.1 shows the spectrum of many realizations of two sinusoids with noise.

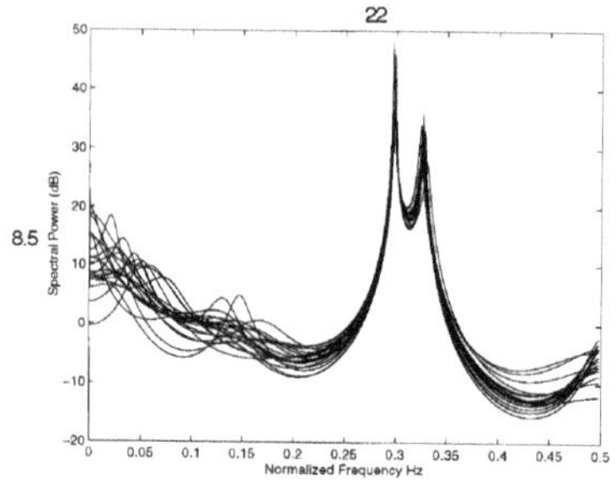

FIGURE 36.1

36.1.1.4 Properties of the Periodogram

$$\hat{S}_x(e^{j\omega}) = \frac{1}{N}\left|\sum_{n=0}^{N-1} x(n)e^{-jn\omega}\right|^2$$

1. Bias: $E\{\hat{S}_{per}(e^{j\omega})\} = \frac{1}{2\pi} S_x(e^{j\omega}) * W_B(e^{j\omega})$, $W_B(e^{j\omega}) =$ spectrum of lag window $= \sum_{k=-N}^{N} \frac{N-|k|}{N}$ $e^{-jk\omega} = \frac{1}{N}\left[\frac{\sin(N\omega/2)}{\sin(\omega/2)}\right]^2$, $*$ = convolution. If $\lim_{N\to\infty} E\{\hat{S}_{per}(e^{j\omega})\} = S_x(e^{j\omega})$, the periodogram is asymptotically unbiased.
2. Variance: $\text{var}\{\hat{S}_{per}(e^{j\omega})\} \cong S_x^2(e^{j\omega})$
3. Resolution: $\Delta\omega = 0.89\frac{2\pi}{N}$

36.1.2 Modified Periodogram

36.1.2.1 Modified Periodogram

$$\hat{S}_m(e^{j\omega}) = \frac{1}{NW}\left|\sum_{n=-\infty}^{\infty} x(n)w(n)e^{-jn\omega}\right|^2,$$

$$w(n) = \text{window (see Chapter 7)}, \quad W = \frac{1}{N}\sum_{n=0}^{N-1} |w(n)|^2 = \frac{1}{2\pi N}\int_{-\pi}^{\pi} |W(e^{-j\omega})|^2\, d\omega$$

36.1.2.2 Properties

$$\text{var}\left(\hat{S}_m(e^{j\omega})\right) \cong S_x^2(e^{j\omega}),\quad \text{Res}\left(\hat{S}_m(e^{j\omega})\right) = (\Delta\omega)_{3dB} = \text{resolution (window dependent)},$$

$$\text{bias } E\left(\hat{S}_m(e^{j\omega})\right) \cong [1/2\pi NW] S_x(e^{j\omega}) * |W(e^{j\omega})|^2$$

36.1.3 Periodogram Averaging (Bartlett's Method)

36.1.3.1 Spectrum

$$\hat{S}_B(e^{j\omega}) = \frac{1}{N}\sum_{i=0}^{K-1}\left|\sum_{n=0}^{L-1} x(n+iL)e^{-jn\omega}\right|^2, \quad n = 0,1,\cdots,L-1,\ i = 0,1,\cdots,K-1,$$

N = length of one realization of data =KL, K = non-overlapping sequences of length L.

36.1.3.2 Properties

1. Bias = $E\{\hat{S}_B(e^{j\omega})\} = \frac{1}{2\pi}S_x(e^{j\omega}) * W_B(e^{j\omega})$,

   $W_B(e^{j\omega}) = FT\{w_B(n)\} = \frac{1}{N}\left[\frac{\sin(N\omega/2)}{\sin(\omega/2)}\right]^2$, $w_B(k) = \frac{N-|k|}{N}$ for $|k| \le N$ and zero for $|k| \ge N$,

   FT stands for Fourier transform.

2. Resolution $\equiv \Delta\omega = \text{Res}\left(\hat{S}_B(e^{j\omega})\right) = 0.89\frac{2\pi}{L} = 0.89K\frac{2\pi}{N}$;

3. Variance: $\text{var}\left(\hat{S}_B(e^{j\omega})\right) \cong \frac{1}{K}S_x^2(e^{j\omega})$

36.1.4 Modified Averaging Periodogram (Welch's Method)

36.1.4.1 Spectrum

$$\hat{S}_w(e^{j\omega}) = \frac{1}{KLW}\sum_{i=0}^{K-1}\left|\sum_{n=0}^{L-1} w(n)x(n+iD)e^{-jn\omega}\right|^2, \quad x_i(n) = x(n+iD),\ n = 0,1,\cdots,L-1,$$

D = offset points for each sequence with length L, $N = L + D(K-1)$, $W = \frac{1}{L}\sum_{n=0}^{L-1}|w(n)|^2$

Example

No overlap $D = L$ and hence $K = N/L$. 50% overlap $D = L/2$ and $K = 2(N/L) - 1$.

36.1.5 Periodogram Smoothing (Blackman-Tukey Method)

36.1.5.1 Spectrum

$$\hat{S}_{BT}(e^{j\omega}) = \sum_{k=-M}^{M} \hat{r}_x(k)w(k)e^{-jk\omega}, \quad w(k) = \text{window},\ 1 << M << N$$

36.1.5.2 Properties

1. Bias: $E\{\hat{S}_{BT}(e^{j\omega})\} \cong \frac{1}{2\pi}S_x(e^{j\omega}) * W(e^{j\omega})$
2. Resolution: window dependent
3. Variance: $\text{var}\left(\hat{S}_{BT}(e^{j\omega})\right) \cong S_x^2(e^{j\omega})\frac{1}{N}\sum_{k=-M}^{M} w^2(k)$

36.1.6 Minimum Variance Spectrum Estimation

36.1.6.1 Minimum Variance Spectrum Estimation

$\hat{S}_{mv}(e^{j\omega}) = \dfrac{p+1}{\underline{e}^H \underline{R}_x^{-1} \underline{e}}$, $\underline{R}_x =$ autocorrelation matrix of WSS process, $\underline{e} = [1, e^{j\omega}, \cdots, e^{jp\omega}]^T$, filter order = P ≤ N, N = number of data

36.1.7 Maximum Entropy Method

36.1.7.1 Property

This method extrapolates $r(k)$ for $|k| \geq N$

36.1.7.2 Spectrum

$$\begin{bmatrix} r_x(0) & r_x^*(1) & \cdots & r_x^*(p) \\ r_x(1) & r_x(0) & \cdots & r_x^*(p-1) \\ \vdots & \vdots & & \\ r_x(p) & r_x(p-1) & \cdots & r_x(0) \end{bmatrix} \begin{bmatrix} 1 \\ a_p(1) \\ \vdots \\ a_p(p) \end{bmatrix} = \varepsilon_p \begin{bmatrix} 1 \\ 0 \\ \vdots \\ 0 \end{bmatrix},$$

$$b(0) = \varepsilon_p = r_x(0) + \sum_{k=1}^{p} a_p(k) r_x^*(k), \quad S_{mem}(e^{j\omega}) = \frac{\varepsilon_p}{\left|\underline{e}^H \underline{a}_p\right|^2},$$

$\underline{a}_p = [1, a_p(1), \cdots, a_p(p)]^T$, $\underline{e} = [1, e^{j\omega}, \cdots, e^{jp\omega}]^T$.

Note: The a's are found from the p lower rows of the matrix equation by excluding the first row.

36.2 Parametric Methods

36.2.1 Introduction

36.2.1.1 *Note:*

Parametric methods are accurate when the model used is appropriate for the processes.

36.2.1.2 Common Models Used

Auto-regressive (AR), moving average (MA), and auto-regressive moving average (ARMA)

36.2.1.3 Steps

a) Select the model
b) Estimate model parameters from the given data
c) Estimate the power spectrum by incorporating the estimated parameters into the model.

36.2.2 Autoregressive Spectrum (AR) Estimation

36.2.2.1 Power Spectrum

$$S_{AR}(e^{j\omega}) = \frac{|b(0)|^2}{\left|1 + \sum_{k=1}^{p} a_p(k)e^{-jk\omega}\right|^2},$$

a's and b's are estimated from the data $x(n)$.

36.2.2.2 Methods to Find Parameters

1. Autocorrelation Method:

$$\begin{bmatrix} r_x(0) & r_x^*(1) & r_x^*(2)\cdots & r_x^*(p) \\ r_x(1) & r_x(0) & r_x^*(1)\cdots & r_x^*(p-1) \\ \vdots & \vdots & & \\ r_x(p) & r_x(p-1) & r_x(p-2)\cdots & r_x(0) \end{bmatrix} \begin{bmatrix} 1 \\ a_p(1) \\ \vdots \\ a_p(p) \end{bmatrix} = \varepsilon_p \begin{bmatrix} 1 \\ 0 \\ \vdots \\ 0 \end{bmatrix}$$

$$r_x(k) = \frac{1}{N}\sum_{n=0}^{N-1-k} x(n+k)x^*(n), \;\; k = 0,1,\cdots,p, \;\; x(n) = \text{data}, \;\; |b(0)|^2 = \varepsilon_p = r_x(0) + \sum_{k=1}^{p} a_p(k)r_x^*(k).$$

Note: The autocorrelation method effectively applies a rectangular window to the data is not usually used. In addition, spectral line splitting may be observed in cases where p is taken to be large.

2. Covariance Method:

$$\begin{bmatrix} r_x(1,1) & r_x(2,1) & \cdots & r_x(p,1) \\ r_x(1,2) & r_x(2,2) & \cdots & r_x(p,2) \\ \vdots & \vdots & & \\ r_x(1,p) & r_x(2,p) & \cdots & r_x(p,p) \end{bmatrix} \begin{bmatrix} a_p(1) \\ a_p(2) \\ \vdots \\ a_p(p) \end{bmatrix} = - \begin{bmatrix} r_x(0,1) \\ r_x(0,2) \\ \vdots \\ r_x(0,p) \end{bmatrix}$$

$r_x(k,\ell) = \sum_{n=p}^{N-1} x(n-\ell)x^*(n-k), \;\; k = 0,1,\cdots,p, \;\; \ell = 0,1,\cdots,p,$ (see also Section 35.4.3 of Chapter 35).

3. Modified Covariance Method: Same as part 2 above with the difference in the autocorrelation function: $r_x(k,\ell) = \sum_{n=p}^{N-1} [x(n-\ell)x^*(n-k) + x(n-p+\ell)x^*(n-p+k)]$ (see also Section 35.4.3 of Chapter 35).
4. Burg Algorithm: see Section 35.4.3 of Chapter 35.

36.2.3 Moving Average (MA)

36.2.3.1 Power Spectrum

$$S_{MA}(e^{j\omega}) = \sum_{k=-q}^{q} r_x(k)e^{-jk\omega}$$

$$r_x(k) = \text{estimate of autocorrelation function} = \sum_{\ell=0}^{q-k} b_q(\ell+k)b_q^*(\ell)$$

$$k = 0,1,\ldots,q, \quad r(-k) = r_x^*(k), \quad r_x(k) = 0 \quad \text{for} \quad |k| > q$$

36.2.3.2 Power Spectrum

$$S_{MA}(e^{j\omega}) = \left| \sum_{k=0}^{q} b_q(k)e^{-jk\omega} \right|^2,$$

$b_q\,(k)$ are estimated from $x(n)$ (see Chapter 35, Section 35.1.10).

36.2.4 Autoregressive Moving Average (ARMA)

36.2.4.1 Power Spectrum

$$S_{ARMA}(e^{j\omega}) = \frac{\left| \sum_{k=0}^{q} b_q(n)e^{-jk\omega} \right|^2}{\left| 1 + \sum_{k=1}^{p} a_p(k)e^{-jk\omega} \right|^2},$$

see Section 35.1.10 of Chapter 35 for evaluating a's and b's.

References

Hays, H. M., *Statistical Digital Signal Processing and Modeling*, John Wiley & Sons Inc., New York, NY, 1996.

Kay, S., *Modern Spectrum Estimation: Theory and Applications*, Prentice-Hall, Englewood Cliffs, NJ, 1988.

Marple, S. L., Jr., *Digital Spectral Analysis with Applications*, Prentice-Hall, Englewood Cliffs, NJ, 1987.

37 Adaptive Filters

37.1 Wiener Filters

37.1.1 Definitions

37.1.1.1 Error

$e(n) = d(n) - \hat{d}(n)$, $d(n) =$ desired signal, $\hat{d}(n) =$ estimated signal

37.1.1.2 Minimum Mean-Square Error

$J_{\min} = E\{|e(n)|^2\}$, $E =$ expectation

37.1.1.3 Filtering

Estimate $d(n)$ from the data $x(n) = d(n) + v(n)$, $v(n) =$ white noise and wide-sense stationary random process, $d(n) =$ wide-sense stationary random process.

37.1.1.4 Smoothing

Same as 37.1.1.3 but with the exception that the filter is non-causal.

37.1.1.5 Prediction

$d(n) = x(n+1)$ and $W(z) =$ Wiener filter is a causal filter.

37.1.1.6 Deconvolution

When $x(n) = d(n) * g(n) + v(n)$ with $g(n)$ being the unit sample response of a linear shift-invariant filter, the Wiener filter becomes a deconvolution filter.

37.1.2 FIR Wiener Filter

37.1.2.1 Estimate

Estimate $\hat{d}(n) = \sum_{\ell=0}^{p-1} w(\ell)x(n-\ell)$, $x(n) = \text{data} = d(n) + v(n)$, $d(n) =$ desired signal, $v(n) =$ noise, $W(z)$

$= \sum_{n=0}^{p-1} w(n)z^{-n} =$ Wiener filter with impulse response $w(n)$

37.1.2.2 Mean Square Error

$J = E\{|e(n)|^2\} = E\{|d(n) - \hat{d}(n)|^2\}$

37.1.2.3 Orthogonality Principle (projection theorem)

$E\{e(n)x^*(n-k)\} = 0, \quad k = 0,1,\cdots p-1$

37.1.2.4 Wiener–Hopf Equations

$\sum_{\ell=0}^{p-1} w(\ell) r_x(k-\ell) = r_{dx}(k), \quad k = 0,1,\cdots,p-1, \; r_x(k-\ell) = E\{x(n-\ell)x^*(n-k)\}, \; r_{dx} = E\{d(n)x^*(n-k)\}$

37.1.2.5 Wiener–Hopf Equation (matrix form)

$$\begin{bmatrix} r_x(0) & r_x^*(1) & \cdots & r_x^*(p-1) \\ r_x(1) & r_x(0) & \cdots & r_x^*(p-2) \\ \vdots & \vdots & & \\ r_x(p-1) & r_x(p-2) & \cdots & r_x(0) \end{bmatrix} \begin{bmatrix} w(0) \\ w(1) \\ \vdots \\ w(p-1) \end{bmatrix} = \begin{bmatrix} r_{dx}(0) \\ r_{dx}(1) \\ \vdots \\ r_{dx}(p-1) \end{bmatrix} \quad \text{or} \quad \underline{R}_x \underline{w} = \underline{r}_{dx}$$

37.1.2.6 Minimum Mean Square Error

$J_{\min} = r_d(0) - \sum_{\ell=0}^{p-1} w(\ell) r_{dx}^*(\ell) = r_d(0) - \underline{r}_{dx}^H \underline{w} = r_d(0) - \underline{r}_{dx}^H \underline{R}_x^{-1} \underline{r}_{dx}$, H stands for Hermitian (transpose conjugate)

37.1.2.7 Filtering

37.1.2.7.1 Wiener–Hopf Equation

$(\underline{R}_d + \underline{R}_v)\underline{w} = \underline{r}_d$, $r_x(k) = E\{x(n+k)x^*(n)\} = E\{[d(n+k) + v(n+k)][d^*(n) + v^*(n)]\} = r_d(k) + r_v(k)$, $\underline{R}_x = \underline{R}_d + \underline{R}_v$, $r_{dx}(k) = E\{d(n)x^*(n-k)\} = E\{d(n)d^*(n-k)\} + E\{d(n)v^*(n-k)\} = r_d(k)$, $\underline{R}_d =$ auto-correlation matrix for $d(n)$, $\underline{R}_v =$ auto-correlation matrix for $v(n)$, $\underline{r}_{dx} = \underline{r}_d = [r_d(0)\cdots r_d(p-1)]^T$, noise $v(n)$ and $d(n)$ are uncorrelated, $E\{v(n)\} = 0$

Example 1

$d(n)$ is a first-order AR process with $r_d(k) = \alpha^{|k|}$, $0 < \alpha < 1$. Let the data be $x(n) = d(n) + v(n)$ with $d(n)$ and noise $v(n)$ uncorrelated. $v(n)$ is white noise with zero mean and σ_v^2 variance.

Solution

The Wiener–Hopf equation is:

$$\begin{bmatrix} r_x(0) & r_x(1) \\ r_x(1) & r_x(0) \end{bmatrix} \begin{bmatrix} w(0) \\ w(1) \end{bmatrix} = \begin{bmatrix} r_{dx}(0) \\ r_{dx}(1) \end{bmatrix}.$$

But $r_{dx}(k) = r_d(k) = \alpha^{|k|}$ and $r_x(k) = r_d(k) + r_v(k) = \alpha^{|k|} + \sigma_v^2 \delta(k)$. Hence the W-H equations become

$$\begin{bmatrix} 1+\sigma_v^2 & \alpha \\ \alpha & 1+\sigma_v^2 \end{bmatrix}\begin{bmatrix} w(0) \\ w(1) \end{bmatrix} = \begin{bmatrix} 1 \\ \alpha \end{bmatrix},$$

and solving for $w(0)$, $w(1)$ we obtain

$$w(0) = \frac{1+\sigma_v^2-\alpha^2}{(1+\sigma_v^2)^2-\alpha^2}, \quad w(1) = \frac{\alpha\sigma_v^2}{(1+\sigma_v^2)^2-\alpha^2}.$$

Hence the filter is $H(z) = w(0) + w(1)z^{-1}$.

37.1.2.8 Linear Prediction

37.1.2.8.1 One Step Linear Prediction

$\hat{x}(n+1) = \sum_{k=0}^{p-1} w(k)x(n-k)$, $w(k) =$ coefficients of the Wiener filter predictor, $r_{dx}(k) = r_x(k+1)$, $d(n) = x(n+1)$, Wiener–Hopf equations:

$$\begin{bmatrix} r_x(0) & r_x^*(1) & \cdots & r_x^*(p-1) \\ r_x(1) & r_x(0) & \cdots & r_x^*(p-2) \\ \vdots & \vdots & & \\ r_x(p-1) & r_x(p-2) & \cdots & r_x(0) \end{bmatrix}\begin{bmatrix} w(0) \\ w(1) \\ \vdots \\ w(p-1) \end{bmatrix} = \begin{bmatrix} r_x(1) \\ r_x(2) \\ \vdots \\ r_x(p) \end{bmatrix}$$

mean-square error: $J_{\min} = r_x(0) - \sum_{k=0}^{p-1} w(k)r_x^*(k+1)$

37.1.2.8.2 q-step Linear Prediction

$\hat{x}(n+q) = \sum_{k=0}^{p-1} w(k)x(n-k)$, $q =$ positive integer, $w(k) =$ filter coefficients, $d(n) = x(n+q)$, $r_{dx}(k) = E\{x(n+q)x^*(n-k)\} = r_x(q+k)$, Wiener–Hopf equations:

$$\begin{bmatrix} r_x(0) & r_x^*(1) & \cdots & r_x^*(p-1) \\ r_x(1) & r_x(0) & \cdots & r_x^*(p-2) \\ \vdots & \vdots & . & \\ r_x(p-1) & r_x(p-2) & \cdots & r_x(0) \end{bmatrix}\begin{bmatrix} w(0) \\ w(1) \\ \vdots \\ w(p-1) \end{bmatrix} = \begin{bmatrix} r_x(q) \\ r_x(q+1) \\ \vdots \\ r_x(q+p-1) \end{bmatrix},$$

mean square error $J_{\min} = r_x(0) - \sum_{k=0}^{p-1} w(k)r_x^*(q+k)$

37.1.2.8.3 Prediction with Noise

$\underline{R}_y \underline{w} = \underline{r}_{dy}$, $y(n) = x(n) + v(n)$, $x(n)$ and $v(n)$ uncorrelated,

$$r_y(k) = E\{y(n)y^*(n-k)\} = r_x(k) + r_v(k),\ r_{dy}(k) = E\{d(n)y^*(n-k)\}$$

$$= E\{x(n+1)y^*(n-k)\} = r_x(k+1)$$

37.1.2.8.4 Noise Cancellation (see Figure 37.1)

$d(n)$ is uncorrelated with $v_1(n)$ and $v_2(n)$, $v_1(n)$ and $v_2(n)$ are correlated, Wiener–Hopf equation (see 37.1.2.5):

$$\underline{R}_{v_2} \underline{w} = \underline{r}_{v_1v_2},\ r_{v_1v_2}(k) = E\{v_1(n)v_2^*(n-k)\} = E\{[x(n) - d(n)]v_2^*(n-k)\} = r_{xv_2}(k),$$

$v_1(n) =$ desired signal, $v_2(n) =$ Wiener filter input, Wiener–Hopf equation becomes $\underline{R}_{v_2} \underline{w} = \underline{r}_{xv_2}$.

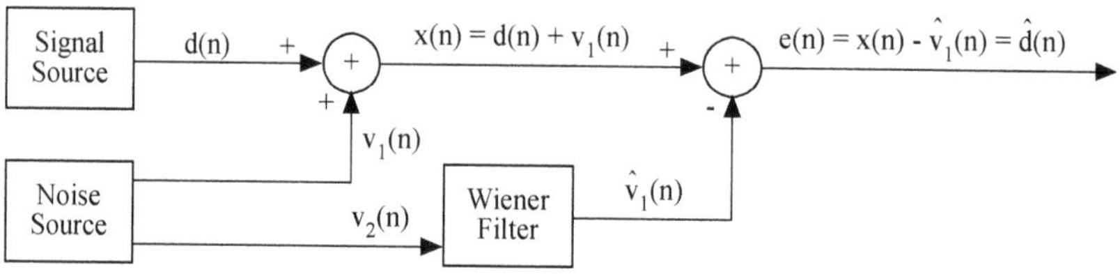

FIGURE 37.1

37.1.3 The IIR Wiener Filter

37.1.3.1 Noncausal IIR Wiener Filter

37.1.3.1.1 Noncausal IIR Wiener Filter

$H(z) = \sum_{n=-\infty}^{\infty} h(n)z^{-n}$, $e(n) =$ error $= d(n) - \hat{d}(n) = d(n) - \sum_{\ell=-\infty}^{\infty} h(\ell)x(n-\ell)$, $x(n) =$ data, Wiener–Hopf equations:

$$\sum_{\ell=-\infty}^{\infty} h(\ell)r_x(k-\ell) = r_{dx}(k),\ -\infty < k < \infty \text{ or } h(k) * r_x(k) = r_{dx}(k),\ r_x(k) = E\{x(n)x^*(n-k)\},$$

$r_{dx}(k) = E\{d(n)x^*(n-k)\}$, Wiener–Hopf equation in frequency and Z-transform form:

$H(e^{j\omega}) = S_{dx}(e^{j\omega})/S_x(e^{j\omega})$ or $H(z) = S_{dx}(z)/S_x(z)$, $J_{\min} =$ minimum mean-square error $= r_d(0) - \sum_{\ell=-\infty}^{\infty} h(\ell)r_{dx}^*(\ell) = \frac{1}{2\pi}\int_{-\pi}^{\pi} [S_d(e^{j\omega}) - H(e^{j\omega})S_{dx}^*(e^{j\omega})]d\omega$, orthogonality principle $E\{e(n)x^*(n-k)\} = 0,\ -\infty<k<\infty$.

Example (smoothing)

$x(n) = d(n) + v(n)$, $d(n) =$ desired signal and $v(n) =$ noise are uncorrelated, $r_x(k) = r_d(k) + r_v(k)$, $S_x(e^{j\omega}) = S_d(e^{j\omega}) + S_v(e^{j\omega})$, $r_{dx}(k) = E\{d(n)x^*(n-k)\} = E\{d(n)d^*(n-k)\} + E\{d(n)r^*(n-k)\} = r_d(k)$, $S_{dx}(e^{j\omega}) = S_d(e^{j\omega})$, IIR Wiener filter: $H(e^{j\omega}) = S_d(e^{j\omega})/[S_d(e^{j\omega}) + S_v(e^{j\omega})]$,

$$J_{\min} = \frac{1}{2\pi}\int_{-\pi}^{\pi}[S_d(e^{j\omega}) - H(e^{j\omega})S_{dx}^*(e^{j\omega})]d\omega = \frac{1}{2\pi}\int_{-\pi}^{\pi}S_d(e^{j\omega})[1 - H(e^{j\omega})]d\omega =$$

$$= \frac{1}{2\pi}\int_{-\pi}^{\pi}S_d(e^{j\omega})\left[1 - \frac{S_d(e^{j\omega})}{S_d(e^{j\omega}) + S_v(e^{j\omega})}\right]d\omega = \frac{1}{2\pi}\int_{-\pi}^{\pi}S_d(e^{j\omega})\frac{S_v(e^{j\omega})}{S_d(e^{j\omega}) + S_v(e^{j\omega})}d\omega$$

$$= \frac{1}{2\pi}\int_{-\pi}^{\pi}S_v(e^{j\omega})H(e^{j\omega})d\omega$$

37.1.3.2 Causal IIR Wiener Filter

37.1.3.2.1 Wiener–Hopf Equation

$\sum_{\ell=0}^{\infty} h(\ell)r_x(k-\ell) = r_{dx}(k),\ 0 \le k < \infty,\ x(n) = \text{data},\ d(n) = \text{desired signal.}$

37.1.3.2.2 Wiener Filter with White Noise Input

$g(n) = r_{d\varepsilon}(n)u(n)$, $u(n) =$ unit step function, $r_\varepsilon(k) = \delta(k)$, $\varepsilon(n) =$ white noise, $g(n) =$ Wiener filter coefficients,

$$r_{d\varepsilon}(k) = E\{d(n)\varepsilon^*(n-k)\} = E\left\{d(n)\left[\sum_{\ell=-\infty}^{\infty} f(\ell)x(n-k-\ell)\right]^*\right\} = \sum_{\ell=-\infty}^{\infty} f^*(\ell)r_{dx}(k+\ell),$$

$f(n) =$ whitening filter coefficients

37.1.3.2.3 Cross-Power Spectral Density

$$S_{d\varepsilon}(z) = S_{dx}(z)F^*(1/z^*) = \frac{P_{dx}(z)}{\sigma_0 Q^*(1/z^*)},$$

$F(z) = 1/(\sigma_0 Q(z)) =$ whitening filter, $x(n) =$ has rational power spectrum with no zeros or poles on the unit circle, $S_x(z) = \sigma_0^2 Q(z)Q^*(1/z^*)$, $S_\varepsilon(e^{j\omega}) = 1$

37.1.3.2.4 Causal Wiener Filter

$G(z) = \frac{1}{\sigma_0}\left[\frac{S_{dx}(z)}{Q^*(1/z^*)}\right]_+ =$ estimates $d(n)$ from $\varepsilon(n)$, "+" = positive-time only.

37.1.3.2.5 Final Causal Wiener Filter

$$H(z) = F(z)G(z) = \frac{1}{\sigma_0^2 Q(z)}\left[\frac{S_{dx}(z)}{Q^*(1/z^*)}\right]_+$$

37.1.3.2.6 Real Process x(n)

$h(n)$ is real and $H(z) = \frac{1}{\sigma_0^2 Q(z)}\left[\frac{S_{dx}(z)}{Q(z^{-1})}\right]_+$

37.1.3.2.7 Mean-Square Error

$$J_{\min} = r_d(0) - \sum_{\ell=0}^{\infty} h(\ell) r_{dx}^*(\ell) = \frac{1}{2\pi}\int_{-\pi}^{\pi} [S_d(e^{j\omega}) - H(e^{j\omega}) S_{dx}^*(e^{j\omega})] d\omega$$

$$= \frac{1}{2\pi}\oint_C [S_d(z) - H(z) S_{dx}^*(1/z^*)] z^{-1}\, dz$$

Example (filtering)

Given $x(n) = d(n) + v(n)(1)$, $r_d(k) = (0.8)^{|k|}$, $v(n) =$ zero mean white noise with variance one, $d(n)$ = desired signal = $0.8\,d(n-1) + w(n)(2)$, $w(n) =$ white noise with variance $\sigma_w^2 = 0.36$, $d(n)$ and $w(n)$ uncorrelated, $d(n)$ and $v(n)$ uncorrelated, $x(n)$ = real data.

Solution

From (2) $D(z) = 1/(1-08z^{-1})$ and hence the output of the filter with $w(n)$ input is

$$S_d(z) = 0.36\,D(z)D(1/z) = 0.36/[(1-0.8z^{-1})(1-0.8z)], \quad \text{from (1)}$$

$$S_x(z) = S_d(z) + S_v(z) = S_d(z) + 1,$$

$$S_x(z) = 1 + S_d(z) = 1.6[(1-0.5z^{-1})(1-0.5z)]/[(1-0.8z^{-1})(1-0.8z)],$$

but

$$S_x(z) = \sigma_0^2 Q(z)Q(1/z) \quad \text{with} \quad \sigma_0^2 = 1.6$$

and

$$Q(z) = [(1-0.5z^{-1})(1-0.8z^{-1})], \quad \left[\frac{S_{dx}(z)}{Q(z^{-1})}\right]_+ = \left[\frac{S_d(z)}{Q(z^{-1})}\right]_+ = \frac{0.6}{1-0.8z^{-1}}$$

since the term $0.3/(z^{-1}-0.5)$ belongs to the negative time. Hence

$$H(z) = \frac{1}{1.6}\frac{1-0.8z^{-1}}{1-0.5z^{-1}}\frac{0.6}{1-0.8z^{-1}} = \frac{0.375}{1-0.5z^{-1}},$$

$$\hat{D}(z) = \text{estimated output} = H(z)X(z) = \frac{0.375}{1-0.5z^{-1}} X(z)$$

or

$$\hat{D}(z)(1-0.5z^{-1}) = 0.375\,X(z)$$

or

$$\hat{d}(n) = 0.5\,d(n-1) + 0.375\,x(n)$$

37.1.3.2.8 Causal IIR Linear Prediction (one step)

$$\hat{x}(n+1)=\sum_{k=0}^{\infty} h(k)x(n-k) \text{ given } x(k) \text{ for all } k\le n,\ d(n)=x(n+1),\ r_{dx}(k)=r_x(k+1),\ S_{dx}(z)=zS_d(z),$$

$$H(z) = \text{causal Wiener filter} = \frac{1}{\sigma_0^2 Q(Z)}\left[\frac{zS_x(z)}{Q^*(1/z^*)}\right]_+, \text{ since } S_x(z)=\sigma_0^2 Q(z)Q^*(1/z^*) \text{ then}$$

$$H(z)=\frac{1}{Q(Z)}[zQ(z)]_+ = z\left[1-\frac{1}{Q(Z)}\right],$$

$$J_{\min}=\frac{1}{2\pi j}\oint_C S_x(z)[1-z^{-1}H(z)]z^{-1}\,dz=\frac{1}{2\pi j}\oint_C \sigma_0^2 Q^*(1/z^*)z^{-1}\,dz=\sigma_0^2 q(0),\ S_x(z)=Z\{r_x(k)\}$$

Example

Given $x(n)=0.9x(n-1)-0.2x(n-2)+w(n)$, $x(n)=$ real-valued AR(2) process, $w(n)=$ zero mean white noise unit variance process.

Solution

$S_x(z)=\dfrac{1}{A(z)A(z^{-1})}$, $A(z)=1-0.9z^{-1}+0.2z^{-2}$, $H(z)=z[1-A(z)]=0.9-0.2z^{-1}$, $\hat{x}(n+1)=0.9x(n)-$ $0.2x(n-1)$

37.2 Discrete Kalman Filter

37.2.1 Definitions

37.2.1.1 Measured Signal

$y(n)=x(n)+v(n)$, $x(n)=$ desired signal, $v(n)=$ noise

37.2.1.2 AR(p) Process

$$x(n)=\sum_{k=1}^{p} a(k)x(n-k)+w(n),\ w(n)=\text{noise uncorrelated with } v(n) \text{ (see 37.2.1.1)}$$

37.2.1.3 Matrix Form of Stationary AR(p) Process

(37.2.1.1) and (37.2.1.2) can be written in the form

$$\underline{x}(n)=\begin{bmatrix} a(1) & a(2)\cdots & a(p-1) & a(p)\\ 1 & 0\cdots & 0 & 0\\ 0 & \vdots & & \\ \vdots & & & \\ 0 & 0\cdots & 1 & 0\end{bmatrix}\underline{x}(n-1)+\begin{bmatrix}1\\0\\0\\\vdots\\0\end{bmatrix}w(n)=\underline{A}\,\underline{x}(n-1)+\underline{w}(n)$$

$y(n)=[1\ 0\ \cdots\ 0]\underline{x}(n)+v(n)=\underline{c}^T\underline{x}(n)+v(n)$

$\underline{x}(n)=[x(n)\ \ x(n-1)\ \cdots\ x(n-p+1)]^T$, $\underline{c}=[1\ \ 0\ \cdots 0]^T$

$\underline{A}=p\times p$ state transition matrix

$\underline{w}(n)=[w(n)\ \ 0\ \cdots\ 0]^T=$ vector noise process

$\underline{c}(n) =$ unit vector of length p

$v(n) =$ measurement noise

37.2.1.4 Nonstationary AR(p) Process

$\underline{x}(n) = \underline{A}(n-1)\underline{x}(n-1) + \underline{w}(n)$, $\underline{A}(n-1) = p \times p$ time varying transition matrix, $\underline{w}(n) =$ vector zero-mean white noise process, $E\{\underline{w}(n)\underline{w}^H(k)\} = \underline{Q}_w(n)$ for $k = n$ and $\underline{0}$ for $k \neq n$, $\underline{y}(n) = \underline{C}(n)\underline{x}(n) + \underline{v}(n)$, $\underline{y}(n) =$ vector of length q, $\underline{C}(n) = q \times p$ time varying matrix, $\underline{v}(n) =$ vector of zero mean white noise process, $\underline{v}(n)$ and $\underline{w}(n)$ statistically independent, $E\{\underline{v}(n)\underline{v}^H(k)\} = \underline{Q}_v(n)$ for $k = n$ and $\underline{0}$ for $k \neq n$

37.2.1.5 Optimum Estimate of State Vector x(n)

$\hat{\underline{x}}(n) = \underline{A}\hat{\underline{x}}(n-1) + \underline{K}[\underline{y}(n) - \underline{C}^T(n)\underline{A}\hat{\underline{x}}(n-1)]$, $\underline{K}(n) =$ Kalman gain vector

37.2.1.6 Optimum Estimate of x (n) (time varying case)

$\hat{\underline{x}}(n) = \underline{A}(n-1)\hat{\underline{x}}(n-1) + \underline{K}(n)[\underline{y}(n) - \underline{C}(n)\underline{A}(n-1)\hat{\underline{x}}(n-1)]$, $\underline{K}(n) =$ Kalman gain matrix, assumed $\underline{A}(n)$, $\underline{C}(n)$, $\underline{Q}_w(n)$ and $\underline{Q}_v(n)$ are known, $\underline{K}(n)$ must be found so that the mean-square error is minimized.

37.2.2 Discrete Kalman Filter

37.2.2.1 Discrete Kalman Filter Equations

Stable Equation	$\underline{x}(n) = \underline{A}(n-1)\underline{x}(n-1) + \underline{w}(n)$
Observation Equation	$\underline{y}(n) = \underline{C}(n)\underline{x}(n) + \underline{v}(n)$
Initialization	$\hat{x}(0\|0) = E\{\underline{x}(0)\}$, $P(0\|0) = E\{\underline{x}(0)\underline{x}^H(0)\}$
Computation	For $n = 1,2,\cdots$ compute
	$\hat{\underline{x}}(n\|n-1) = \underline{A}(n-1)\hat{\underline{x}}(n-1\|n-1)$
	$\underline{P}(n\|n-1) = \underline{A}(n-1)\underline{P}(n-1\|n-1)\underline{A}^H(n-1) + \underline{Q}_w(n)$
	$\underline{K}(n) = \underline{P}(n\|n-1)\underline{C}^H(n)[\underline{C}(n)\underline{P}(n\|n-1)\underline{C}^H(n) + \underline{Q}_v(n)]^{-1}$
	$\hat{\underline{x}}(n\|n) = \hat{\underline{x}}(n\|n-1) + \underline{K}(n)[\underline{y}(n) - \underline{C}(n)\hat{\underline{x}}(n\|n-1)]$
	$\underline{P}(n\|n) = [\underline{I} - \underline{K}(n)\underline{C}(n)]\underline{P}(n\|n-1)$

$\underline{P}(n|n) =$ error covariance matrix $= E\{\underline{e}(n|n)\underline{e}^H(n|n)\}$, $\underline{e}(n|n) = \underline{x}(n) - \hat{\underline{x}}(n|n)$, $\underline{P}(n|n-1) = E\{\underline{e}(n|n-1)$ $\underline{e}^H(n|n-1)\}$, $J(n) = tr\{\underline{P}(n|n)\} =$ mean-square error.

Example

Find unknown constant x given $y(n)$ corrupted by $v(n) =$ zero mean white noise with variance σ_v^2. State equation: $x(n) = x(n-1)$ since x is constant. $y(n) = x(n) + v(n) =$ measurement equation, $\underline{A}(n) = 1$, $\underline{C}(n) = 1$, $\underline{Q}_w(n) = 0$, $\underline{Q}_v(n) = \sigma_v^2$, $P(n|n) = E\{e^2(n|n)\}$ = scalar since $x(n)$ is scalar $= E\{[\underline{x}(n) - \hat{x}(n|n)]^2\}$, $P(n|n-1) = 1 \cdot P(n|n-1) \cdot 1 + 0 = P(n-1|n-1)$, $K(n) = P(n|n-1) \cdot 1 \cdot [1 \cdot P(n|n-1) \cdot 1 + \sigma_v^2]^{-1}$,

$$P(n) = [1 - K(n)]P(n|n-1) = [1 - \frac{P(n|n-1)}{P(n|n-1) + \sigma_v^2}]P(n|n-1) = \frac{P(n|n-1)\sigma_v^2}{P(n|n-1) + \sigma_v^2},$$

$$P(1) = \frac{P(0)\sigma_v^2}{P(0)+\sigma_v^2},\; P(2) = \frac{P(1)\sigma_v^2}{P(1)+\sigma_v^2} = \frac{P(0)\sigma_v^2}{2P(0)+\sigma_v^2},\; P(3) = \frac{P(0)\sigma_v^2}{3P(0)+\sigma_v^2} \text{ etc., } P(n) = \frac{P(0)\sigma_v^2}{nP(0)+\sigma_v^2},$$

$$K(n) = P(n|n-1)[P(n|n-1)+\sigma_v^2]^{-1} =, P(0)/[nP(0)+\sigma_v^2],\; \hat{x}(n) = \hat{x}(n-1) + \frac{P(0)}{nP(0)+\sigma_v^2}[y(n)-\hat{x}(n-1)]$$

Example (estimate AR(1) process)

$x(n) = 0.8x(n-1)+w(n)$, $w(n) =$ white noise with variance $\sigma_w^2 = 0.36$, $y(n) = x(n)+v(n)$, $v(n) =$ white noise uncorrelated with $w(n)$ and variance one, $\underline{A}(n) = 0.8$, $\underline{C}(n) = 1$, $\hat{x}(n) = 0.8\hat{x}(n-1)+K(n)$ $[y(n)-0.8\hat{x}(n-1)]$, $P(n|n-1) = (0.8)^2 P(n-1|n-1)+0.36$, $K(n) = P(n|n-1)[P(n|n-1)+1]^{-1}$, $P(n|n) = [1-K(n)]P(n|n-1)$, $\hat{x}(0) = E\{x(0)\} = 0$, $P(0|0) = E\{|x(0)|^2\} = 1$, for $n = 1$ $P(n|n-1) = 1.00$, $K(n) = 0.5$, $P(n|n) = 0.5$; for n = 2 $P(n|n-1) = 0.68$, $K(n) = 0.40$, $P(n|n) = 0.40$; for $n = 3$ $P(n|n-1) = 0.619$, $K(n) = 0.38$, $P(n|n) = 0.38$, for n = 6 $P(n|n-1) = 0.60$, $K(n) = 0.3751$, $P(n|n) = 0.3751,\cdots$; for n = ∞ $P(n|n-1) = 0.6$, $K(n) = 0.375$, $P(n|n) = 0.375$, $\hat{d}(n) = 0.8\hat{d}(n-1)+0.375[x(n)-0.8\hat{d}(n-1)]$, $J(n) = 0.375$

37.3 Adaptive Filtering

37.3.1 Introduction

37.3.1.1 Filter

$\hat{d}(n) = \sum_{k=0}^{p} w_k(n)x(n-k)$, $w_k(n) =$ k^{th} filter coefficient at time n, $\hat{d}(n) =$ estimate of desired signal, $x(n)$ = data.

37.3.1.2 Vector Form of Filter

$\hat{d}(n) = \underline{w}^T(n)\underline{x}(n)$, $\underline{w}(n) = [w_0(n)\; w_1(n)\cdots w_p(n)]^T$, $\underline{x}(n) = [x(n)\; x(n-1)\cdots x(n-p)]^T$

37.3.1.3 Correction Formula

$\underline{w}(n-1) = \underline{w}(n)+\Delta\underline{w}(n)$, $\Delta\underline{w}(n) =$ correction to $\underline{w}(n)$ at time n to form the set $\underline{w}(n+1)$ at time $n+1$.

37.3.1.4 Autocorrelation

$\hat{r}_x(k) = \frac{1}{N}\sum_{n=0}^{N-1} x(n)x^*(n-k)$, $x(n) =$ data (finite).

37.3.1.5 Cross-Correlation

$\hat{r}_{dx}(k) = \frac{1}{N}\sum_{n=0}^{N-1} d(n)x^*(n-k)$, $d(n) =$ desired signal.

37.3.1.6 Wiener–Hopf Equations

$\underline{R}_x\underline{w} = \underline{r}_{dx}$, $\underline{R}_x =$ autocorrelation matrix.

37.3.1.7 Data

$x(n) = d(n)+v(n)$, $d(n) =$ desired unknown signal, $v(n) =$ noise.

37.3.1.8 Error

$e(n) = d(n) - \hat{d}(n).$

37.3.2 FIR Adaptive Filters

37.3.2.1 Filter

$$\hat{d}(n) = \sum_{k=0}^{p} w_k(n)x(n-k) = \underline{w}^T(n)\underline{x}(n)$$

37.3.2.2 Mean Square Error

$J(n) = E\{|e(n)|^2\},\ e(n) = d(n) - \hat{d}(n) = d(n) - \underline{w}^T(n)\underline{x}(n)$

37.3.2.3 Filter Coefficients

$\underline{R}_x(n)\underline{w}(n) = \underline{r}_{dx}(n),\ \underline{r}_{dx}(n) = [E\{d(n)x^*(n)\}\ \ E\{d(n)x^*(n-1)\}\cdots E\{d(n)x^*(n-p)\}]^T,$

$$\underline{R}_x = \begin{bmatrix} E\{x(n)x^*(n)\} & E\{x(n-1)x^*(n)\} & \cdots & E\{x(n-p)x^*(n)\} \\ E\{x(n)x^*(n-1)\} & E\{x(n-1)x^*(n-1)\} & \cdots & E\{x(n-p)x^*(n-1)\} \\ \vdots & & & \\ E\{x(n)x^*(n-p)\} & E\{x(n-1)x^*(n-p)\} & \cdots & E\{x(n-p)x^*(n-p)\} \end{bmatrix} \equiv (p+1)\times(p+1)$$

Hermitian matrix of autocorrelation, $\underline{r}_{dx}(n) = [E\{d(n)x^*(n)\}\ \ E\{d(n)x^*(n-1)\}\cdots E\{d(n)x^*(n-p)\}]^T$ = cross-correlation between $d(n)$ and $x(n)$.

Note: For jointly wide sense stationary process $\underline{R}_x(n)\underline{w}(n) = \underline{r}_{dx}(n)$ reduces to the Wiener–Hopf equation and the solution $\underline{w}(n)$ becomes independent of time.

37.3.2.4 Steepest Descent Adaptive Filter

37.3.2.4.1 Update Equation

$$\underline{w}(n+1) = \underline{w}(n) - \mu\nabla J(n),\quad \nabla J(n) = \left[\frac{\partial J(n)}{\partial w_1^*(n)}\ \frac{\partial J(n)}{\partial w_2^*(n)} \cdots \frac{\partial J(n)}{\partial w_p^*(n)}\right]^T,$$

$J(n) = E\{|e(n)|^2\},\ \nabla J(n) = \nabla E\{e(n)e^*(n)\} = E\{e(n)\nabla e^*(n)\} = E\{e(n)(-\underline{x}^*(n))\} = -E\{e(n)\underline{x}^*(n)\},\ x(n) =$ data, $d(n) =$ desired signal

37.3.2.4.2 Steepest Descent Algorithm
$\underline{w}(n+1) = \underline{w}(n) + \mu\, E\{e(n)\underline{x}^*(n)\} = \underline{w}(n) + \mu(\underline{r}_{dx} - \underline{R}_x\underline{w}(n)),$ for μ see 37.3.2.1.4

37.3.2.4.3 Steps for Steepest Descent Algorithm

1. Initialize with an initial estimate $\underline{w}(0)$
2. Evaluate the gradient $J(n)$ at the current estimate $\underline{w}(n)$
3. Update the estimate at time n by adding a correction $-\mu\nabla J(n)$, $\underline{w}(n+1) = \underline{w}(n) - \mu\nabla J(n)$
4. Go back to (2) and repeat. For values of μ see below.

37.3.2.4.4 Properties

If $d(n)$ and $x(n)$ are wide sense stationary, at the limit $n \rightarrow \infty$, $\underline{w}(n) = \underline{R}_x^{-1}\underline{r}_{dx}$ = Wiener–Hopf solution and the step size is $0 < \mu < \frac{2}{\lambda_{\max}}$ where $\lambda_{\max}$ is the maximum eigenvalue of the autocorrelation matrix $\underline{R}_x$

37.3.2.4.5 Weight Error Vector

$\underline{c}(n) = \underline{w}(n) - \underline{w}$, shifts the $\underline{w}(w_0\ w_1 \cdots w_p)$ coordinates to point $\underline{w}$

37.3.2.4.6 Steepest Method with Center at w

$\underline{c}(n+1) = (\underline{I} - \mu\underline{R}_x)\underline{c}(n))$, see Figure 37.2 which illustrates the relationships between $\underline{w}$, $\underline{c}$ and $\underline{u}$.

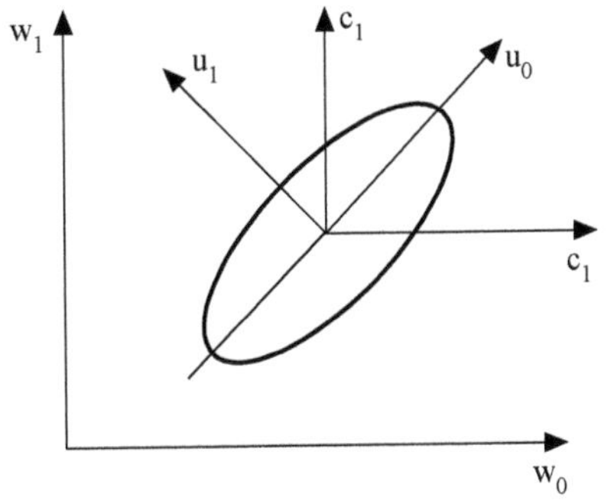

FIGURE 37.2

37.3.2.4.7 Rotated Coordinate System

$\underline{u}(n+1) = (\underline{I} - \mu\underline{\Lambda})\underline{u}(n)$, $\underline{\Lambda} =$ diagonal matrix containing the eigenvalues of $\underline{R}_x$, $\underline{Q} =$ matrix with the columns the eigenvectors $\underline{q}_k$ of $\underline{R}_x$, $\underline{Q}\underline{Q}^H = \underline{I}$ which implies that $\underline{Q}$ is unitary.

37.3.2.4.8 Solution to 3.2.1.7

$$\underline{u}(n) = (\underline{I} - \mu\underline{\Lambda})^n\,\underline{u}(0), \quad k^{\text{th}} \text{ component: } u_k(n) = (1 - \mu\lambda_k)^n u_k(n)$$

37.3.2.4.9 Filter Modes

$$\underline{w}(n) = \underline{w}(n) + \sum_{k=0}^{p} (1 - \mu\lambda_k)^n\, u_k(0)\underline{q}_k$$

37.3.2.4.10 Time Constant for the Mode

$(1-\mu\lambda_k)^{\tau_k} = 1/e$ or $\tau_k = -1/\ln(1-\mu\lambda_k)$, for $\mu\lambda_k \ll 1$, $\tau_k \cong 1/\mu\lambda_k$, $\tau = \max\{\tau_k\} \cong 1/\mu\lambda_{\min}$, $\tau \cong \frac{1}{2\alpha}$

$\frac{\lambda_{\max}}{\lambda_{\min}} = \frac{1}{2\alpha}\chi$ where we set $\mu = \alpha 2/\lambda_{\max}$ and $0 < \alpha < 1$ (see 37.3.2.1.4), χ = condition number

37.3.2.4.11 Minimum Error

$J_{\min} = r_d(0) - \underline{r}_{dx}^H\,\underline{w}$, H stands for Hermitian (transpose complex)

37.3.2.4.12 Error

$J(n) = J_{\min} + \underline{c}^H(n)\underline{R}_x\,\underline{c}(n)$, $J(n) = J_{\min} + \underline{u}^H(n)\underline{\Lambda}_x\,\underline{u}(n)$

37.3.2.4.13 Learning Curve

A plot of $J(n)$ versus n is known as the learning curve (shows how rapidly it learns the Wiener–Hopf solution).

37.3.2.2 The LMS Algorithm

37.3.2.2.1 LMS Algorithm

$$\underline{w}(n+1)=\underline{w}(n)+\mu e(n)\underline{x}^*(n),\ y(n)=\underline{w}^T(n)\underline{x}(n),\ \ e(n)=d(n)-y(n)$$

37.3.2.2.2 LMS Algorithm for p[th] order FIR Adaptive Filter

Parameters: p = filter order

μ = step size (see 37.3.2.2.3 and 37.3.2.2.4)

Initialization: $\underline{w}(0)=\underline{0}$

Computation: For $n = 0,1,2,...$

a) $y(n)=\underline{w}^T(n)\underline{x}(n)$

b) $e(n)=d(n)-y(n)$

c) $\underline{w}(n+1)=\underline{w}(n)+\mu e(n)\underline{x}^*(n)$

37.3.2.2.3 Property

For jointly wide-sense stationary processes, the LMS algorithm converges in the mean if $0<\mu<2/\lambda_{max}$ and the independence assumption is satisfied ($\underline{x}(n)$ and $\underline{w}(n)$) are statistically independent).

37.3.2.2.4 Conservative Bound

$$0<\mu<2/[(p+1)E\{|x(n)|^2\}],\ \ E\{|x(n)|^2\}\cong\hat{E}\{|x(n)|^2\}=\frac{1}{N}\sum_{k=0}^{N-1}|x(n-k)|^2$$

Example (adaptive linear prediction):

$x(n)=\ 1.27x(n-1)-0.80x(n-2)+v(n)$, $v(n)=$ unit variance white noise, $x(n)$ = data of second-order autoregressive process, $\hat{x}(n)=$ optimal causal linear predictor $=1.27x(n-1)-0.80x(n-2)$. Assume the adaptive predictor $\hat{x}(n)=w_1(n)x(n-1)+w_2(n)x(n-2)$. From 37.3.2.2.1 the kth coefficient of the filter is: $w_k(n+1)=w_k(n)+\mu e(n)x^*(n-k)$. For small values of μ, $w_1(n)$ and $w_2(n)$ will converge to 1.27 and –0.8. The error becomes $e(n)=v(n)$ and the minimum mean square error is $J_{min}=\sigma_v^2=1$. Figure 37.3 shows typical converging curves.

37.3.2.2.5 Steady-State Error Value

$$J(\infty)=J_{min}+J_{ex}(\infty)=J_{min}\frac{1}{1-\mu\sum_{k=0}^{p}\frac{\lambda_k}{2-\mu\lambda_k}},\ 0<\mu<\frac{2}{\lambda_{max}},\ \mu\sum_{k=0}^{p}\frac{\lambda_k}{2-\mu\lambda_k}<1,$$

$J_{ex}(\infty)=$ excess mean square error (see 37.3.2.2.6).

37.3.2.2.6 Excess Mean-Square Error $J_{ex}(n)$

$$J_{ex}(\infty)=\mu J_{min}\left[\sum_{k=0}^{p}\lambda_k\Big/(2-\mu\lambda_k)\right]\Bigg/\left[1-\mu\sum_{k=0}^{p}\lambda_k\Big/(2-\mu\lambda_k)\right]$$

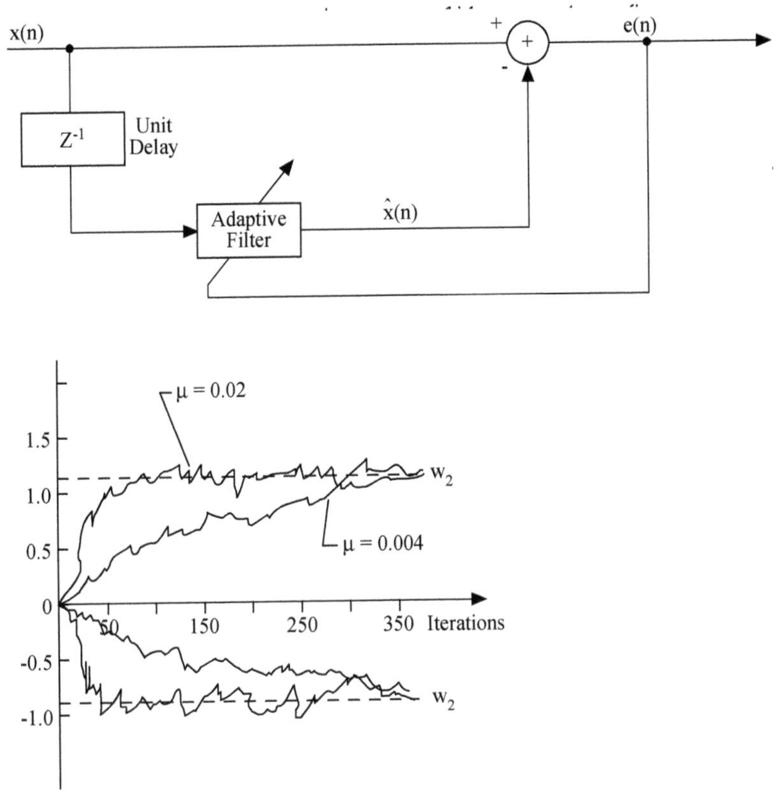

FIGURE 37.3

37.3.2.2.7 Case $\mu << 2/\lambda_{max}$

$$\mu < 2/tr(\underline{R}_x),\ tr = \text{trace of matrix},\ J(\infty) \cong J_{\min}/[1 - \tfrac{1}{2}\mu\, tr(\underline{R}_x)],$$

$$J_{ex}(\infty) \cong \mu J_{\min}[\tfrac{1}{2} tr(\underline{R}_x)]/[1 - \tfrac{1}{2}\mu\, tr(\underline{R}_x)] \cong \tfrac{1}{2}\mu J_{\min}\, tr(\underline{R}_x)$$

37.3.2.2.8 Normalized Mean-Square Error (misadjustment M)

$M = J_{ex}(\infty)/J_{\min}$

37.3.2.2.9 Normalized LMS Algorithm

37.3.2.2.9.1 Normalized LMS Algorithm:

$$\underline{w}(n+1) = \underline{w}(n) + \beta \frac{\underline{x}^*(n)}{\|\underline{x}(n)\|^2} e(n),\ 0 < \beta < 2,\ \|\underline{x}(n)\|^2 = \underline{x}^H(n)\underline{x}(n),\ \mu = \text{step size} = \beta/\|\underline{x}(n)\|^2$$

37.3.2.2.9.2 Modified Normalized LMS:

$$\underline{w}(n+1) = \underline{w}(n) + \beta \frac{\underline{x}^*(n)}{\varepsilon + \|\underline{x}(n)\|^2} e(n),\ \varepsilon = \text{small positive number}$$

37.3.2.2.10 Leaky LMS Algorithm

$\underline{w}(n+1) = (1 - \mu\gamma)\underline{w}(n) + \mu e(n)\underline{x}^*(n),\ 0 < \gamma << 1,\ \gamma =$ leaky coefficient, $0 < \mu < (\gamma + \lambda_{\max})$.

37.3.2.2.11 Block Updating Algorithm

Block LMS Algorithm (filter coefficients are adapted only once for each block of L samples). $\underline{w}[(k+1)L] = \underline{w}(kL) + \mu \frac{1}{L}\sum_{\ell=0}^{L-1} e(kL+\ell)\underline{x}^*(kL+\ell)$, $y(kL+\ell) =$ output of kth block $= \underline{w}^T(kL)\underline{x}(kL+\ell)$, $\ell = 0,1,\cdots,L-1$, $e(kL+\ell) = d(kL+\ell) - \underline{w}^T(kL)\underline{x}(kL+\ell)$, $\ell = 0,1,\cdots,L-1$, k^th^ block updating.

Note: In this approach the filter coefficients are held constant over each block of L samples, and the filter output $y(n)$ and the error $e(n)$ for each value of n within the block are calculated using the filter coefficients for the block. Then, at the end of each block, the coefficients are updated using an average of the L gradient estimates over the block.

37.3.2.2.12 Sign LMS Algorithm

37.3.2.2.12.1 Sign-Error Algorithm: $\underline{w}(n+1) = \underline{w}(n) + \mu\,\mathrm{sgn}[e(n)]\underline{x}(n)$; $\mathrm{sgn}[e(n)] = 1$ for $e(n) > 0$, zero for $e(n) = 0$ and -1 for $e(n) < 0$, $\mathrm{sgn}(\cdot) =$ signum function

37.3.2.2.12.2 Sign-Data Algorithm: $\underline{w}(n+1) = \underline{w}(n) + \mu\, e(n)\mathrm{sgn}[\underline{x}(n)]$, $\mathrm{sgn}(\cdot) =$ signum function

37.3.2.2.12.3 Sign-Data Normalized Algorithm: $w_k(n+1) = w_k(n) + \frac{\mu}{|x(n-k)|} e(n)x(n-k)$

37.3.2.2.12.4 Sign-Data Algorithm: $\underline{w}(n+1) = \underline{w}(n) + \mu\,\mathrm{sgn}[e(n)]\mathrm{sgn}[\underline{x}(n)]$

37.3.2.2.12.5 Sign-Sign Algorithm with Leakage Term: $\underline{w}(n+1) = (1-\mu\gamma)\underline{w}(n) + \mu\,\mathrm{sgn}[e(n)]\mathrm{sgn}[\underline{x}(n)]$

37.3.2.2.12.6 Variable Step-Size Algorithm: $w_k(n+1) = w_k(n) + \mu_k\, e(n)x(n-k)$, $\mu_{\min} < \mu_k(n) < \mu_{\max}$, $\mu_k(n) \equiv$ adjusted independently for each coefficient, when $e(n)x(n-k)$ changes sign it frequently implies that $w_k(n)$ is close to the optimum and $\mu_k(n)$ must be decreased, when $e(n)x(n-k)$ has the same sign for several successive updates it implies that $w_k(n)$ is far from optimum and $\mu_k(n)$ must be increased.

37.3.3 Adaptive Recursive Filters

37.3.3.1 IIR Filter

$y(n) = \sum_{k=1}^{p} a_k(n)y(n-k) + \sum_{k=1}^{q} b_k(n)x(n-k)$, $a_k(n)$ and $b_k(n)$ are coefficients of the adaptive filter at time n, (p,q) filter order.

37.3.3.2 IIR LMS Algorithm

Parameters: p,q = filter order

μ = step size

Initialization: $\underline{a}(0) = \underline{b}(0) = \underline{0}$, $\underline{a}(n) = [a_1(n)\, a_2(n) \cdots a_p(n)]^T$,

$\underline{b}(n) = [b_0(n) b_1(n) \cdots b_q(n)]^T$,

Computation: For n = 0,1,... compute

a) $y(n) = \underline{a}^T(n)\underline{y}(n-1) + \underline{b}^T(n)\underline{x}(n)$

b) $e(n) = d(n) - y(n)$

c) For $k = 1,2,\cdots p$ $\psi_k^a(n) = y^*(n-k) + \sum_{\ell=1}^{p} a_\ell^*(n)\psi_k^a(n-\ell)$

$a_k(n+1) = a_k(n) + \mu\, e(n)\psi_k^a(n)$

d) For $k = 0,1,\cdots,q$

$$\psi_k^b(n) = x^*(n-k) + \sum_{\ell=1}^{p} a_\ell^*(n)\psi_k^b(n-\ell)$$

$$b_k(n+1) = b_k(n) + \mu e(n)\psi_k^b(n)$$

37.3.4 Recursive Least Square (FIR Filter)

37.3.4.1 Lease Square Error

$$\varepsilon(n) = \sum_{i=1}^{n} \lambda^{n-i}|e(i)|^2, \quad 0<\lambda<1$$

37.3.4.2 Data Vector

$\underline{x}(i) = [x(i)\; x(i-1)\cdots x(i-p)]^T$

37.3.4.3 Error

$e(i) = d(i) - y(i) = d(i) - \underline{w}^T(n)\underline{x}(i)$, $y(i) =$ filtered output at time i using the latest set of filter coefficients $w_k(n)$, $\underline{w}(n)$ are assumed constant over the entire observation interval $[0,n]$.

37.3.4.4 Normal Equation (deterministic)

$$\underline{R}_x(n)\,\underline{w}(n) = \underline{r}_{dx}(n),\; \underline{R}_x(n) = \sum_{i=0}^{n} \lambda^{n-i}\underline{x}^*(i)\underline{x}^T(i),\; \underline{x}(i) = [x(i)\cdots x(i-p)]^T,\; \underline{r}_{dx}(n) = \sum_{i=0}^{n}\lambda^{n-i}d(i)\underline{x}^*(i)$$

37.3.4.5 RLS Algorithm

Parameters: p = filter order

λ = exponential weighting factor, $0<\lambda<1$

δ = initializing constant, small positive number, i.e., 0.001, initializes $\underline{P}(0)$

Initialization: $\underline{w}(0) = \underline{0}$

$\underline{R}_x^{-1}(0) \doteq \underline{P}(0) = \delta^{-1}\underline{I}$, $\doteq$ equal by definition

Computation: For $n = 1,2,...$ compute

$$\underline{z}(n) = \underline{P}(n-1)\underline{x}^*(n)$$

$$\underline{g}(n) = [1/[\lambda + \underline{x}^T(n)\underline{z}(n)]]\,\underline{z}(n)$$

$$\alpha(n) = d(n) - \underline{w}^T(n-1)\underline{x}(n)$$

$$\underline{w}(n) = \underline{w}(n-1) + \alpha(n)\underline{g}(n)$$

$$\underline{P}(n) = (1/\lambda)[\underline{P}(n-1) - \underline{g}(n)\underline{z}^H(n)]$$

37.3.5 Frequency Domain Implementation

37.3.5.1 Input Signals

$x_i(n) = x(iL+n) =$ data into blocks of length L, $d_i(n) = d(iL+n) =$ desired signal segmented, $n = 0,1,\cdots,L-1$, $i = 0,1,\cdots$

37.3.5.2 Transformed Input Signals

$X_i(k) = DFT\{x_i(n)\} = \sum_{n=0}^{L-1} x_i(n)e^{-j\frac{2\pi nk}{L}}$, $k = 0,1,\cdots,L-1$; $D_i(k) = DFT\{d_i(n)\} = \sum_{n=0}^{L-1} d_i(n)e^{-j\frac{2\pi nk}{L}}$, $k = 0, 1,\cdots,L-1$.

Note: The filter is updated once per data block.

37.3.5.3 Output

$Y_i(k) = W_i(k)X_i(k)$, $W_i(k) =$ filter coefficients in frequency domain.

37.3.5.4 Error

$E_i(k) = D_i(k) - Y_i(k)$, $k = 0,1,\cdots,L-1$

37.3.5.5 Filter Coefficients

$$W_{i+1}(k) = W_i(k) + \mu E_i(k)X_i^*(k), \quad k = 0,1,\cdots L-1$$

Matrix form: $\underline{W}_{i+1} = \underline{W}_i + \mu \underline{X}_i^* \underline{E}_i$, $\underline{W}_i = [W_i(0)\ W_i(1)\cdots W_i(L-1)]^T$

$\underline{E}_i = [E_i(0)\ E_i(1)\cdots E_i(L-1)]^T$, $\underline{x}_i = diag\{X_i(0)\ X_i(1)\cdots X_i(L-1)\}$

$0 < \mu < \dfrac{2}{E\{|X_i(k)|^2\}}$, for given data we substitute the estimate of the ensemble $E\{|X_i(k)|^2\}$.

37.3.6 Basic Adaptive Filter Configurations

37.3.6.1 Filter Configuration

See Figure 37.4 through 37.11.

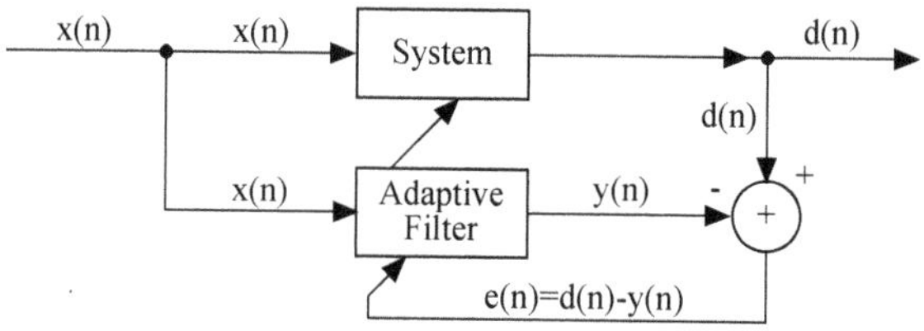

FIGURE 37.4 Identification.

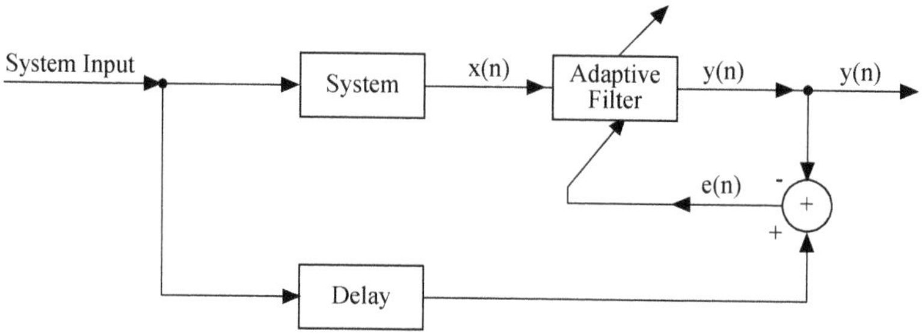

FIGURE 37.5 Inverse modeling.

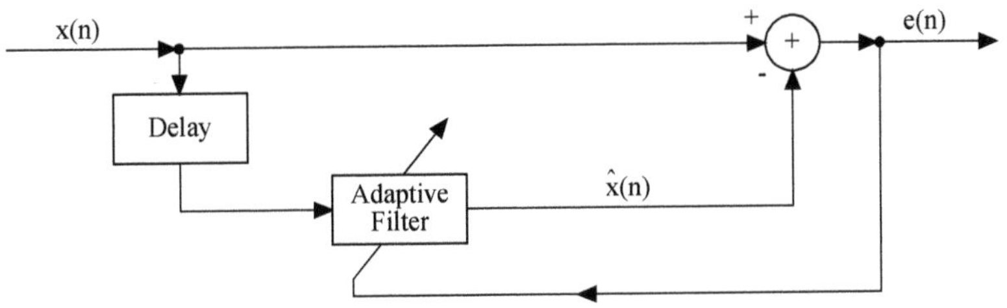

FIGURE 37.6 Linear prediction.

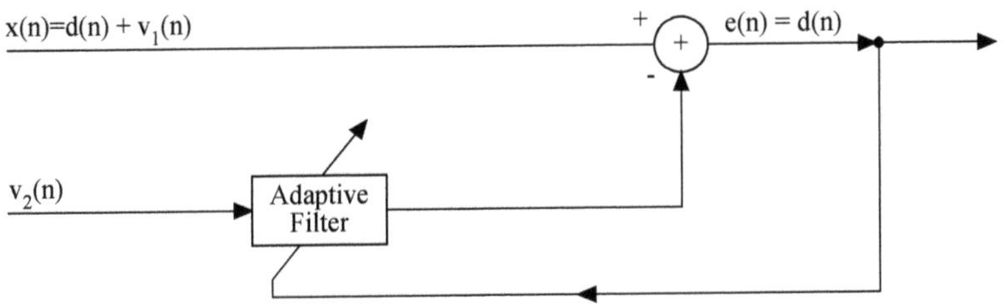

FIGURE 37.7 Noise cancellation with a reference signal.

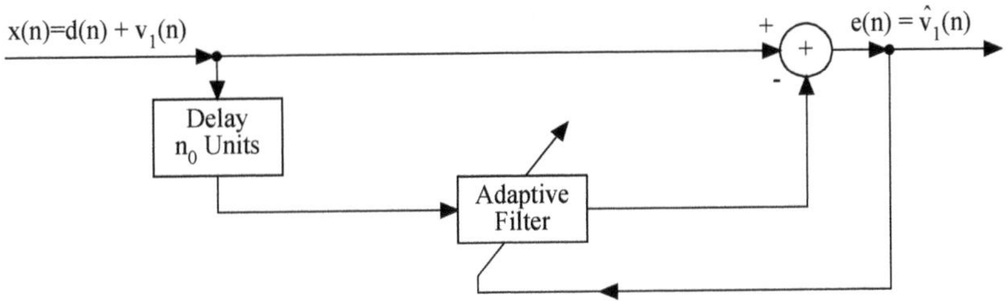

FIGURE 37.8 Noise cancellation with a reference signal.

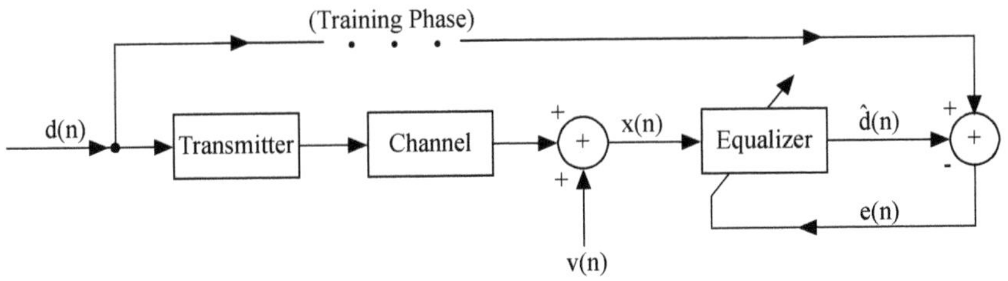

FIGURE 37.9 Channel equilization using training sequence.

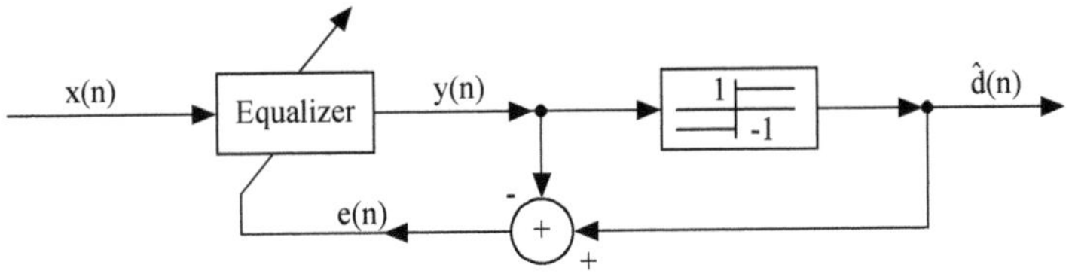

FIGURE 37.10 Equilization in decision-directed mode.

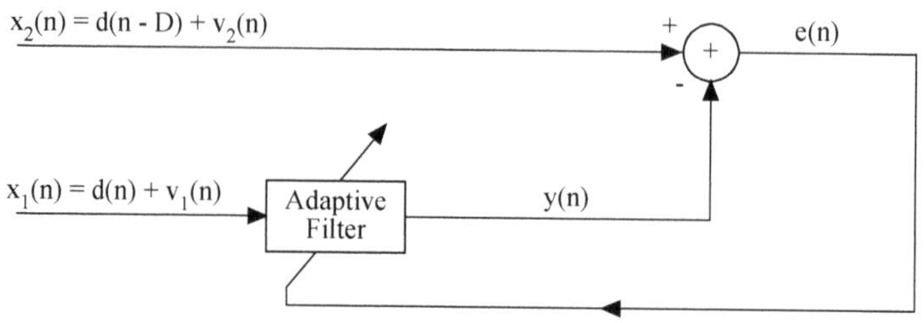

FIGURE 37.11 Dime-delay estimator. Filter cancels the delay between $x_1(n)$ and $x_2(n)$. The peak in w (filter coefficients) gives the delay D which is taken as multiple of the sample interval.

References

Alexander, S. T., *Adaptive Signal Processing: Theory and Applications*, Springer-Verlag, New York, 1986.

Bellanger, M. G., *Adaptive Digital Filters and Signal Analysis*, Marcel Dekker Inc., New York, NY, 1987.

Haykin, S., *Adaptive Filter Theory*, Prentice-Hall, Englewood Cliffs, NJ, 1986.

38

Bandlimited Functions-Sampling

38.1 Bandlimited Functions

38.1.1 Definition (Bandlimited)

$F(\omega) = 0$ for $|\omega| > \sigma \equiv \sigma - BL$ (sigma bandlimited), $E(\equiv$ energy$) < \infty$

38.1.2 Definition (Time-limited)

$f(t) = 0$ for $|t| > \tau$, $E\,(\text{energy}) < \infty \equiv \tau - TL\,(\tau \text{ time-limited})$.

38.1.3 Energy

$$E = \int_{-\infty}^{\infty} |f(t)|^2\, dt = \frac{1}{2\pi} \int_{-\infty}^{\infty} |F(\omega)|^2\, d\omega$$

38.1.4 Properties of Bandlimited Functions

1. $\int_{-\infty}^{\infty} f(t) e^{-j\omega t}\, dt = F(\omega) = T p_\sigma(\omega) \sum_{n=-\infty}^{\infty} f(nT) e^{-jnT\omega}$, T = sampling time, $p_\sigma(\omega) =$ pulse of width of 2σ and centered at $\omega = 0$

2. $\int_{-\infty}^{\infty} |f(t)|^2\, dt = \frac{1}{2\pi} \int_{-\sigma}^{\sigma} |F(\omega)|^2\, d\omega = T \sum_{n=-\infty}^{\infty} |f(nT)|^2$

3. $f(t) = \frac{1}{2\pi} \int_{-\sigma}^{\sigma} F(\omega) e^{j\omega t}\, d\omega$

4. $\int_{-\sigma}^{\sigma} |F(\omega)|\, d\omega < \infty$ ($F(\omega)$ is absolutely integrable)

5. $|f(t)| \le \sqrt{\frac{\sigma E}{\pi}}\, e^{\sigma |t|}$, $E =$ energy (see 38.1.3), $f(t) \equiv \sigma - BL$ (sigma bandlimited)

6. $f'(t) = \frac{1}{2\pi} \int_{-\sigma}^{\sigma} j\omega F(\omega) e^{j\omega t}\, d\omega$, a bandlimited function is analytic in the entire t plane

7. A bandlimited function has finite energy, is analytic in the entire t plane, and is of exponential type $\left(|f(t)| < Ae^{\sigma|t|},\ A \text{ and } \sigma \text{ are constants}\right)$

8. Periodic BL: $y(t) = \sum_{n=-M}^{M} a_n e^{-jn\omega_0 t}$, $E =$ finite in $(0, T)$, $T =$ period,
   $a_n =$ Fourier series coefficients

9. A function $f(t)$ cannot be bandlimited and time-limited

38.1.5 Properties of Time-Limited Functions

1. Derivatives:

$$f^{(k)}(-\tau) = \lim f^{(k)}(-\tau + \varepsilon),\ \ \varepsilon \to 0;\ \ f^{(k)}(\tau) = \lim f^{(k)}(\tau - \varepsilon),\ \ \varepsilon \to 0,\ \varepsilon > 0;$$

$$f^{(k)}(\cdot) \equiv k^{\text{th}} \text{ derivative}$$

2. If $f(t)$ is TL and has bounded derivatives of order up to n for every $|t| < \tau$, then

$$F(\omega) = \frac{1}{2\pi} \int_{-\tau}^{\tau} f(t) e^{-j\omega t}\, dt = \frac{1}{j\omega} [f(-\tau) e^{j\tau\omega} - f(\tau) e^{-j\tau\omega}] + \cdots$$

$$+ \frac{1}{(j\omega)^n} [f^{(n-1)}(-\tau) e^{j\tau\omega} - f^{(n-1)}(\tau) e^{-j\tau\omega}] + O(1/\omega^{n+1})$$

3. If $f(t)$ is TL and has bounded derivatives of order up to n for every $|t| < \tau$ and

$$f^{(k)}(-\tau) = 0,\ f^{(k)}(\tau) = 0,\ k = 0, 1, \cdots, n-1, \text{ then } F(\omega) = O(1/\omega^{n+1}),\ \omega \to \infty.$$

38.2 Sampling and Interpolation

38.2.1 Interpolation Function

$f_k(t) =$ sampled function with $k(t) = \sum_{n=-\infty}^{\infty} Tf(nT) k(t - nT)$, $Tk(0) = 1$, $k(nT) = 0$ for $n \neq 0$ (see Figure 38.1), $T =$ sampling time

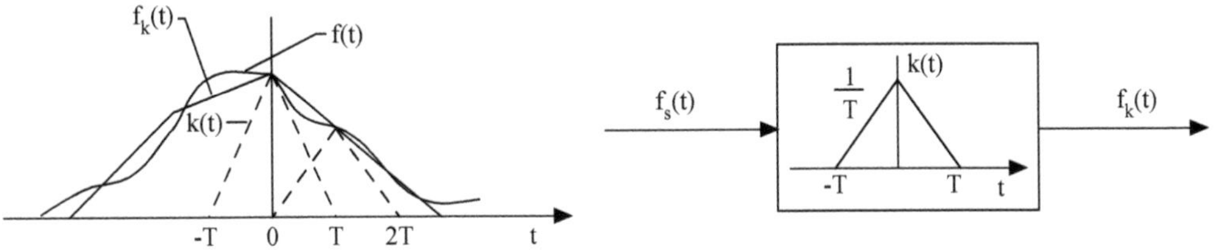

FIGURE 38.1

38.2.2 Sampled Function

$f_s(t) = \sum_{n=-\infty}^{\infty} Tf(nT)\delta(t-nT) = f(t)Tcomb_T(t) =$ sampled function with delta functions, $comb_T(t) = \sum_{n=-\infty}^{\infty} \delta(t-nT)$

38.2.3 Convolution Property

$f_k(t) = f_s(t) * k(t)$

38.2.4 Spectrum of Sampled Function

$$F_s(\omega) = \sum_{n=-\infty}^{\infty} Tf(nT)e^{-jnT\omega} = \sum_{n=-\infty}^{\infty} F(\omega + 2n\sigma),\ \sigma = \frac{\pi}{T};$$

$F_k(\omega) = K(\omega)F_s(\omega) = K(\omega)\sum_{n=-\infty}^{\infty} F(\omega+2n\sigma) = K(\omega)[COMB_{2\sigma}(\omega) * F(\omega)]$ (see 38.2.3 and FT properties), (see Figure 38.2)

38.2.5 Sampling Theorem

$f(t) = \sum_{n=-\infty}^{\infty} Tf(nT)\left[\sin\left[\frac{\omega_s}{2}(t-nT)\right]\Big/[\pi(t-nT)\right]$ for $\omega_s = \frac{2\pi}{T} > 2\sigma,$

$$f(t) = \sigma - BL \text{ function} = \sum_{n=-\infty}^{\infty} f(nT)\frac{\sin\sigma(t-nT)}{\sigma(t-nT)},\quad \sigma = \frac{\pi}{T} \text{ and } \omega_s = 2\sigma,\ F(\omega) = 0 \text{ for } |\omega| > \sigma$$

38.2.6 Sampling Frequency (Nyquist Rate)

$\frac{1}{T} = \frac{\sigma}{\pi}$

38.2.7 Truncation Error

$$f_N(t) = \sum_{n=-N}^{N} f(nT)\frac{\sin\sigma(t-nT)}{\sigma(t-nT)},\quad e_N = \int_{-\infty}^{\infty} |f(t) - f_N(t)|^2\, dt = T\sum_{|n|>N} |f(nT)|^2,$$

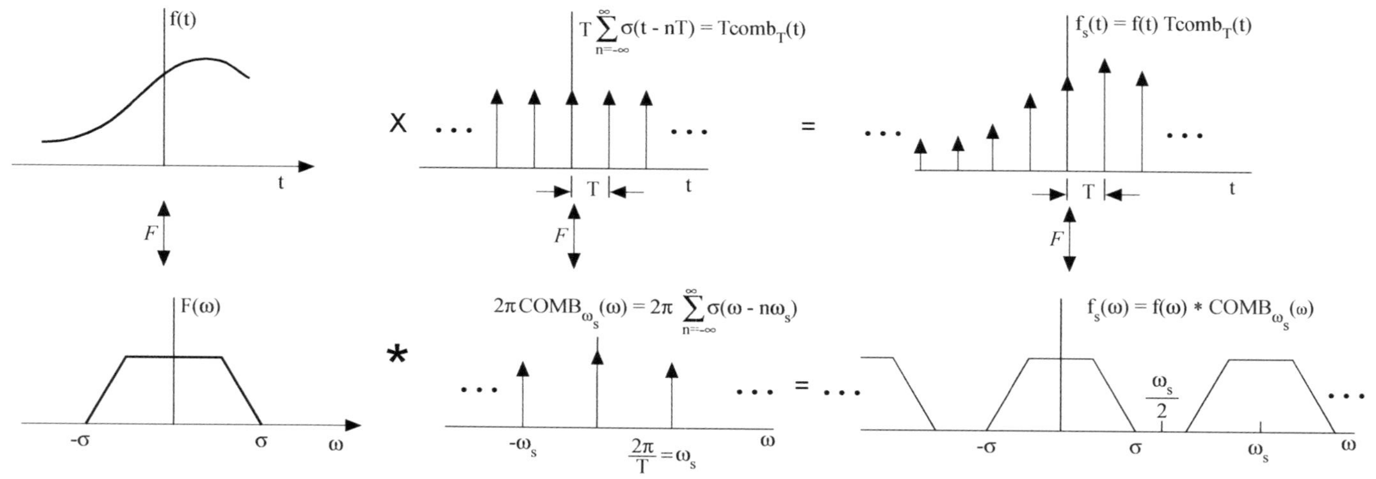

f(t)
t
F
F(ω)
-σ
σ
ω
X
T ∑ σ(t - nT) = Tcomb_T(t)
n=-∞
T
2π COMB_ωs(ω) = 2π ∑ σ(ω - nω_s)
-ω_s
2π/T = ω_s
*
=
f_s(t) = f(t) Tcomb_T(t)
f_s(ω) = f(ω) * COMB_ωs(ω)
ω_s/2
ω_s

FIGURE 38.2

$$|f(t)-f_N(t)| \le \sqrt{\frac{\sigma e_N}{\pi}} \text{ for every } t, \quad f(t) \equiv \sigma - BL$$

38.2.8 Bernstein Inequality

$|f'(t)| \le \sigma M, \; f(t) \equiv \sigma - BL$ and $|f(t)| \le M$

38.2.9 Bandlimited Interpolation

$f(t) \equiv$ arbitrary function, $f_i(t) = \sum_{n=-\infty}^{\infty} f(nT)\frac{\sin\sigma(t-nT)}{\sigma(t-nT)}$ (see also 38.2.1 with $k(t) = \sin\sigma t/\pi t$),

$F_i(\omega) = F_s(\omega)p_\sigma(\omega)$, $p_\sigma(\omega) =$ centered pulse with total width 2σ, $f_i(t) = BL$ obtained by passing $f_s(t)$ (see 38.2.2) through an ideal lowpass filter, $f_i(nT) = f(nT)$, $|f(t)-f_i(t)| \le \frac{1}{\pi}\int_{|\omega|>\sigma} |F(\omega)|\,d\omega$

38.2.10 Bandlimited Mean-Square Approximation

$$|f(t)-f_\sigma(t)| \le \frac{1}{2\pi}\int_{|\omega|>\sigma} |F(\omega)|\,d\omega,$$

$f(t)$ = arbitrary function, $f_\sigma(t) = \frac{1}{2\pi}\int_{-\sigma}^{\sigma} F(\omega)e^{j\omega t}\,d\omega$, $F_\sigma(\omega) = \mathcal{F}\{f_\sigma(t)\} = F(\omega)p_\sigma(\omega)$, $p_\sigma(\omega) =$ centered pulse with width 2σ.

38.3 Sampling with Pulses

38.3.1 Train of Rectangular Pulses

$$F_s(\omega) = \mathcal{F}\{f(t)f_p(t)\} = \sum_{n=-\infty}^{\infty} \frac{\sin\left(\frac{n\omega_s\tau}{2}\right)}{\left(\frac{n\omega_s\tau}{2}\right)} F(\omega - n\omega_s),$$

$f_p(t) = p_{\tau/2}(t) * \sum_{n=-\infty}^{\infty} \delta(n-nT)$, $p_{\tau/2}(t) =$ centered pulse with width τ, $\omega_s = 2\pi/T$, $\mathcal{F}\{f(t)\} = F(\omega)$

38.3.2 Train of Pulses with Flat Tops

$$F_s(\omega) = \frac{2}{T}\frac{\sin\left(\frac{\omega\tau}{2}\right)}{\omega}\sum_{n=-\infty}^{\infty} F(\omega - n\omega_s), \; f_s(t) = [f(t)comb_T(t)] * p_{\tau/2}(t), \; comb_T(t) = \sum_{n=-\infty}^{\infty} \delta(n-nT)$$

38.4 Frequency Sampling

38.4.1 Frequency Sampling

If $f(t)$ = time limited = 0 for $|t| > T_N$,

$$F(\omega) = \sum_{n=-\infty}^{\infty} F\left(n\frac{\pi}{T_N}\right)\frac{\sin(\omega T_N - n\pi)}{\omega T_N - n\pi}$$

38.5 n-Variables Sampling

$$f(t_1,t_2,\cdots,t_n) = \sum_{m_1=-\infty}^{\infty}\cdots\sum_{m_n=-\infty}^{\infty} f\left(\frac{\pi m_1}{\omega_1},\cdots,\frac{\pi m_n}{\omega_n}\right)\times\frac{\sin(\omega_1 t_1 - m_1\pi)}{\omega_1 t_1 - m_1\pi}\cdots\frac{\sin(\omega_n t_n - m_n\pi)}{\omega_n t_n - m_n\pi},$$

$f(t_1,\cdots,t_n) =$ n – variables function, $g(y_1,y_2,\cdots,y_n) = 0$ for $|y_k| > |\omega_k|$, $k = 1,2,\cdots,n$, $g(\cdot) = \mathcal{F}_n\{f(\cdot)\}$

38.6 Sampling and Derivatives

$\mathcal{F}\{f(t)\} = F(\omega) = 0$ for $|\omega| > 2\pi f_N$,

$$f(t) = \sum_{k=-\infty}^{\infty}\left[\xi(kh) + (t-kh)\xi^{(1)}(kh) + \cdots + \frac{(t-kh)^R}{R!}\xi^{(R)}(kh)\right]\left[\frac{\sin\frac{\pi}{h}(t-kh)}{\frac{\pi}{h}(t-kh)}\right]^{R+1},$$

R = highest derivative order, $h = (R+1)/(2f_N)$, $\xi^{(R)}(kh) = R^{\text{th}}$ derivative of the function $\xi(\cdot)$,

$$\xi^{(j)}(kh) = \sum_{i=0}^{j}\binom{j}{i}\left(\frac{\pi}{h}\right)^{j-1}\Gamma_{R+1}^{(j-1)} f^{(i)}(kh),\quad \Gamma_a^{(\beta)} = \frac{d^\beta}{dt^\beta}[(t/\sin t)^\alpha]\Big|_{t=0},\quad \Gamma_\alpha^{(0)} = 1,$$

$$\Gamma_\alpha^{(2)} = \frac{\alpha}{3},\quad \Gamma_\alpha^{(4)} = \frac{\alpha(5\alpha+2)}{15},\quad \Gamma_\alpha^{(6)} = \frac{\alpha(35\alpha^2+42\alpha+16)}{63},\quad \Gamma_\alpha^{(\beta)} = 0 \text{ for odd } \beta.$$

38.7 Papoulis Generalization

38.7.1 One System

$$f(t) = \sigma - BL\text{ signal} = \sum_{n=-\infty}^{\infty} g(nT)y(t-nT),\quad y(t) = \frac{1}{2\sigma}\int_{-\sigma}^{\sigma}\frac{e^{j\omega t}}{H(\omega)}d\omega,\quad g(t) = \frac{1}{2\pi}\int_{-\sigma}^{\sigma}F(\omega)H(\omega)e^{j\omega t}\,d\omega,$$

$1/T = \sigma/\pi$ = Nyquist rate

Example

$g(t) = \alpha \int_{-\sigma}^{\sigma} f(t-\tau)e^{-\alpha\tau}\,d\tau,\ H(\omega) = \mathcal{F}\{\alpha e^{-\alpha t}\} = \dfrac{\alpha}{\alpha + j\omega},\ y(t) = \dfrac{1}{2\sigma}\int_{-\sigma}^{\sigma}\left(1+\dfrac{j\omega}{\alpha}\right)e^{j\omega t}\,d\omega = (\sigma t\cos\sigma t +$ $(\alpha t - 1)\sin\sigma t)/\sigma\alpha t^2$,

38.7.2 Multiple System

$f(t) = \sum_{n=-\infty}^{\infty}[g_1(nT)y_1(t-nT)+\cdots+g_m(nT)y_m(t-nT)]$, where $y_k(t) = \dfrac{1}{C}\int_{-\sigma}^{-\sigma+c} Y_k(\omega,t)e^{j\omega t}\,d\omega$ for $k = 1,2,\cdots,m$; $g_k(t) = \dfrac{1}{2\pi}\int_{-\sigma}^{\sigma} F(\omega)H_k(\omega)e^{j\omega t}\,d\omega$, $1/T = \sigma/m\pi$, $c = 2\sigma/m = 2\pi/T$, $Y_i(\omega,t)$'s are determined from the system.

$$H_1(\omega)Y_1(\omega,t)+\cdots+H_m(\omega)Y_m(\omega,t) = 1$$

$$H_1(\omega+c)Y_1(\omega,t)+\cdots+H_m(\omega+c)Y_m(\omega,t) = e^{jct}$$

$$\cdots$$

$$H_1(\omega+(m-1)c)Y_1(\omega,t)+\cdots+H_m(\omega+(m-1)c)Y_m(\omega,t) = e^{j(m-1)ct},\quad t = \text{arbitrary}$$

ω is in the interval $(-\sigma, -\sigma+c)$

Properties

1. $Y_k(\omega,t) \equiv$ periodic in t with period $T = 2\pi/c$
2. $Y_k(\omega,t)e^{j\omega t} = \sum_{n=-\infty}^{\infty} y_k(t-nT)e^{jnT\omega}$ for every ω in $(-\sigma,-\sigma+c)$ where $Y_k(\omega,t)$ is defined and for every t.
3. For every ω in $(-\sigma, \sigma)$,

$$e^{j\omega t} = H_1(\omega)\sum_{n=-\infty}^{\infty} y_1(t-nT)e^{jnT\omega} + \cdots + H_m(\omega)\sum_{n=-\infty}^{\infty} y_m(t-nT)e^{jnT\omega}$$

38.8 Bounds and Extreme Values of BL Function

38.8.1 Bounds of Output Function

$$|g(t)| \le \left[\frac{E}{2\pi}\int_{-\sigma}^{\sigma} |H(\omega)|^2\,d\omega\right]^{1/2},$$

equality holds for $t = t_0$ only if $F(\omega) = kH^*(\omega)e^{-j\omega t_0}$ for $|\omega| < \sigma$ and k is a constant determined from energy requirements,

$$g(t) = \frac{1}{2\pi}\int_{-\sigma}^{\sigma} F(\omega)H(\omega)e^{j\omega t}\,d\omega,\quad f(t) = \sigma - BL \text{ function},\quad E = \text{energy} = \frac{1}{2\pi}\int_{-\sigma}^{\sigma}|F(\omega)|^2\,d\omega$$

Properties

1. If $H(\omega)=1$ $(g(t)=f(t))$, $|f(t)|\le\sqrt{\frac{E\sigma}{\pi}}$

2 If $H(\omega)=j\omega$ $(g(t)=f'(t))$, $|f'(t)|\le\sigma\sqrt{\frac{E\sigma}{3\pi}}$

38.8.2 Maximum Energy Concentration

$\alpha_y=\frac{1}{E_y}\int_{-\tau}^{\tau}|y(t)|^2\,dt\equiv$ maximum if $y(t)$ is equal to the $y_{eig}(t)=$ eigen function of the integral equation

$$\int_{-\tau}^{\tau} y(x)\frac{\sin\sigma(t-x)}{\pi(t-x)}dx=\lambda y(t) \qquad (1)$$

corresponding to the maximum eigenvalue $\lambda_{\max}$, $y(t)=\sigma-BL$, $E_y=$ energy of $y(t)$, solutions of (1) are known as *prolate spheroidal wave functions.*

References

Jerri, A. J., The Shannon Sampling Theorem — its various extensions and applications: a tutorial review, *Proc. IEEE*, 65, 1565-1596.

Linden, D. A., A discussion of sampling theorems, *Proc. IRE*, 47, 1219-1226, 1959.

Papoulis, A., *Signal Analysis*, McGraw Hill Inc., New York, NY, 1977.

39

High-Order Statistical Analysis

39.1 Moments and Cumulants

39.1.1 Definitions

39.1.1.1 Moments

$$\text{Mom}[x_1^{k_1},x_2^{k_2},\cdots,x_n^{k_n}] \triangleq E\{x_1^{k_1}\,x_2^{k_2}\cdots x_n^{k_n}\}$$

$$=(-j)^r \left.\frac{\partial^r \Phi(\omega_1,\omega_2,\cdots,\omega_n)}{\partial\omega_1^{k_1}\,\partial\omega_2^{k_2}\cdots\partial\omega_n^{k_n}}\right|_{\omega_1=\omega_2=\cdots=\omega_n=0},\quad \{x_1,x_2,\cdots x_n\}\equiv n \text{ random variables},$$

r = order = $k_1+k_2+\cdots+k_n$, $\Phi(\omega_1,\omega_2,\cdots,\omega_n) \triangleq E\{\exp[j(\omega_1x_1+\omega_2x_2+\cdots+\omega_nx_n)]\}$ = characteristic function, $E\{\}$ = expectation operation.

Example

$\text{Mom}[x_1,x_2]=E\{x_1\,x_2\}$, $\text{Mom}[x_1,x_1]=\text{Mom}[x_1^2]=E\{x_1^2\}$

39.1.1.2 Logarithm of $\Phi(\omega_1,\omega_2,\cdots,\omega_n):\tilde{\Psi}(\omega_1,\omega_2,\cdots,\omega_n) \triangleq \ln[\Phi(\omega_1,\omega_2,\cdots,\omega_n)]$

39.1.1.3 Cumulates (semi-invariants)

$$\text{Cum}[x_1^{k_1},x_2^{k_2},\cdots,x_n^{k_n}] \triangleq (-j)^r \left.\frac{\partial^r \tilde{\Psi}(\omega_1,\omega_2,\cdots,\omega_n)}{\partial\omega_1^{k_1}\partial\omega_2^{k_2}\cdots\partial\omega_n^{k_n}}\right|_{\omega_1=\omega_2=\cdots=\omega_n=0},\quad r = \text{order} = k_1^1+k_2^2+\cdots+k_n^n$$

39.1.1.4 Relations of Cumulants and Moments

$$m_1 = \text{Mom}[x_1] = E\{x_1\} = \int_{-\infty}^{\infty} x_1 f(x_1)\,dx_1,\ m_2 = \text{Mom}[x_1,x_1] = E\{x_1^2\} = \int_{-\infty}^{\infty} x_1^2 f(x_1)\,dx_1, \cdots,$$

$$m_n = \text{Mom}[x_1,x_1,\cdots,x_1] = E\{x_1^n\} = \int_{-\infty}^{\infty} x_1^n f(x_1)\,dx_1,\ f(\cdot) = \text{probability density function (pdf)},$$

$$c_1 = \text{Cum}[x_1] = m_1,\ c_2 = \text{Cum}[x_1,x_1] = m_2 - m_1^2,\ c_3 = \text{Cum}[x_1,x_1,x_1] = m_3 - 3m_2m_1 + 2m_1^3,$$

$$c_4 = \text{Cum}[x_1,x_1,x_1,x_1] = m_4 - 4m_3m_1 - 3m_2^2 + 12m_2m_1^2 - 6m_1^4$$

39.1.2 Relationship of Moments Cumulants

39.1.2.1 General Relationship

$$\text{Cum}[x_1,x_2,\cdots,x_n] = \sum (-1)^{p-1}(p-1)!\,E\left\{\prod_{i\in s_1} x_i\right\} E\left\{\prod_{i\in s_2} x_i\right\} \cdots E\left\{\prod_{i\in s_p} x_i\right\},\quad r = n = \text{order},$$

summation extends over all partitions $(s_1,s_2,\cdots,s_p)$, $p = 1,2,\cdots,n$, of the set of integers $(1,2,\cdots,n)$

Example

$$\text{Cum}[x_1,x_2,x_3] = E\{x_1x_2x_3\} - E\{x_1\}E\{x_2x_3\} - E\{x_2\}E\{x_1x_3\} - E\{x_3\}E\{x_1x_2\} + 2E\{x_1\}E\{x_2\}E\{x_3\}$$

because the integers $\{1,2,3\}$ can be partitioned into:

for $p=1$ $s_1 = \{1,2,3\}$, for $p=2$ $s_1 = \{1\}$ $s_2 = \{2,3\}$, and $s_1 = \{2\}$ $s_2 = \{1,3\}$ and

$s_1\{3\}$ $s_2 = \{1,2\}$, for $p=3$ $s_1\{1\}$ $s_2 = \{2\}$ $s_3 = \{3\}$

39.1.2.2 Partitions of Set {1,2,3,4}

p	Partitions	s_1	s_2	s_3	s_4
1		1,2,3,4	0	0	0
2		1,2	3,4	0	0
		1,3	2,4	0	0
		1,4	2,3	0	0
		1	2,3,4	0	0
		2	1,3,4	0	0
		3	1,2,4	0	0
		4	1,2,3	0	0
3		1,2	3	4	0
		1,3	2	4	0
		1,4	2	3	0
		2,4	1	3	0
		3,4	1	2	0
		2,3	1	4	0
4		1	2	3	4

Example

$$\begin{aligned}\text{Cum}[x_1,x_2,x_3,x_4] &= E\{x_1x_2x_3x_4\} - E\{x_1x_2\}E\{x_3x_4\} - E\{x_1x_3\}E\{x_2x_4\} - E\{x_1x_4\}E\{x_2x_3\} \\ &- E\{x_1\}E\{x_2x_3x_4\} - E\{x_2\}E\{x_1x_3x_4\} - E\{x_3\}E\{x_1x_2x_4\} - E\{x_4\}E\{x_1x_2x_3\} + 2E\{x_1x_2\}E\{x_3\} \\ &E\{x_4\} + 2E\{x_1x_3\}E\{x_2\}E\{x_4\} + 2E\{x_1x_4\}E\{x_2\}E\{x_3\} + 2E\{x_2x_4\}E\{x_1\}E\{x_3\} + 2E\{x_3x_4\} \\ &E\{x_1\}E\{x_2\} + 2E\{x_2x_3\}E\{x_1\}E\{x_4\} - 6E\{x_1\}E\{x_2\}E\{x_3\}E\{x_4\}\end{aligned}$$

39.1.3 Properties of Moments and Cumulants

39.1.3.1 Properties

1. $\text{Mom}[a_1x_1,a_2x_2,\cdots,a_nx_n] = a_1\cdots a_n\ \text{Mom}[x_1,\cdots,x_n]$ and

   $\text{Cum}[a_1x_1,a_2x_2,\cdots,a_nx_n] = a_1\cdots a_n\ \text{Cum}[x_1,\cdots,x_n]$

   where $(a_1,a_2,\cdots,a_n)$ are constants. This follows directly from (39.1.4) and (39.1.2.1).

2. Moments and cumulants are symmetric functions in their arguments , e.g.,

   $\text{Mom}[x_1,x_2,x_3] = \text{Mom}[x_2,x_1,x_3] = \text{Mom}[x_3,x_2,x_1]$, and so on.

3. If the random variables $\{x_1,x_2,\cdots,x_n\}$ can be divided into any two or more groups which are statistically independent, their nth-order cumulant is identical to zero; i.e., $\text{Cum}[x_1,x_2,\cdots,x_n]$ = 0, whereas, in general, $\text{Mom}[x_1,x_2,\cdots,x_n] \neq 0$. For example, if the two independent groups are $\{x_1,x_2,\cdots,x_\lambda\}$ and $\{x_{\lambda+1},\cdots,x_n\}$, then their joint characteristic function is $\Phi(\omega_1,\omega_2,\cdots,\omega_n)$ = $\Phi_1(\omega_1,\cdots,\omega_\lambda)\cdot\Phi_2(\omega_{\lambda+1},\cdots,\omega_n)$. On the other hand, their joint second characteristic function is $\tilde{\Psi}(\omega_1,\omega_2,\cdots,\omega_n) = \tilde{\Psi}_1(\omega_1,\cdots,\omega_\lambda) + \tilde{\Psi}_2(\omega_{\lambda+1},\cdots,\omega_n)$.

4. If the sets of random variables $\{x_1,x_2,\cdots,x_n\}$ and $\{y_1,y_2,\cdots,y_n\}$ are independent, then

   $$\text{Cum}[x_1+y_1,x_2+y_2,\cdots,x_n+y_n] = \text{Cum}[x_1,\cdots,x_n\} + \text{Cum}[y_1,\cdots,y_n]$$

   whereas in general

   $$\begin{aligned}\text{Mom}[x_1+y_1,\cdots,x_n+y_n] &= E\{(x_1+y_1)(x_2+y_2)\cdots(x_n+y_n)\} \\ &\neq \text{Mom}[x_1,\cdots,x_n] + \text{Mom}[y_1,\cdots,y_n].\end{aligned}$$

   However, for the random variables $\{y_1,x_1,x_2,\cdots,x_n\}$ we have that

   $$\text{Cum}[x_1+y_1,x_2,\cdots,x_n] = \text{Cum}[x_1,x_2,\cdots,x_n] + \text{Cum}[y_1,x_2,\cdots,x_n]$$

   and

   $$\text{Mom}[x_1+y_1,x_2,\cdots,x_n] = \text{Mom}[x_1,x_2,\cdots,x_n] + \text{Mom}[y_1,x_2,\cdots,x_n].$$

5. If the set of random variables $\{x_1,\cdots,x_n\}$ is jointly Gaussian, then all the information about their distribution is contained in the moments of order $n \leq 2$. Therefore, all moments of order greater than two $n > 2$ have no new information to provide. This leads to the fact that all joint cumulants of order $n > 2$ are identical to zero for Gaussian random vectors. Hence, the cumulants of order

greater than two, in some sense, measure the non-Gaussian nature (or non-normality) of a time series.

Example

$z_i = y_i + x_i$ $i = 1,2,3$, joint p.d.f. of $\{y_1, y_2, y_3\}$ is non-Gaussian, $\{x_1, x_2, x_3\}$ is jointly Gaussian and independent from $\{y_1, y_2, y_3\}$, $E\{y_i\} \neq 0$, $E\{x_i\} \neq 0$ for $i = 1,2,3$, from properties (4) and (5) of cumulants $\text{Cum}[z_1, z_2, z_3] = \text{Cum}[y_1, y_2, y_3]$ since $\text{Cum}[x_1, x_2, x_3] = 0$

39.1.4 Moments and Cumulants of Stationary Processes

39.1.4.1 Moments

$\text{Mom}[X(k), X(k+\tau_1), \cdots, X(k+\tau_{n-1})] \triangleq m_n^x(\tau_1, \tau_2, \cdots, \tau_{n-1}) = E\{X(k)X(k+\tau_1)\cdots X(k+\tau_{n-1})\}$, depend only on time differences $\tau_1, \tau_2, \cdots, \tau_{n-1}$, $\tau_i = 0, \pm 1, \pm 2, \cdots =$ integers for all i, $\{X(k)\} =$ real stationary random process, $k = 0, \pm 1, \pm 2, \cdots$.

39.1.4.2 Cumulants

$$c_n^x(\tau_1, \tau_2, \cdots, \tau_{n-1}) \triangleq \text{Cum}[X(k), X(k+\tau_1), \cdots, X(k+\tau_{n-1})]$$

39.1.4.3 Relationships of Moments and Cumulants

1st order: $c_1^x = m_1^x = E\{X(k)\} =$ mean value, 2nd order: $c_2^x(\tau_1) = m_2^x(\tau_1) - (m_1^x)^2 =$ covariance sequence $= m_2^x(-\tau_1) - (m_1^x)^2 = c_2^x(-\tau_1)$, $m_2^x(\tau_1) =$ autocorrelation sequence, 3rd order: $c_3^x(\tau_1, \tau_2) = m_3^x(\tau_1, \tau_2) - m_1^x[m_2^x(\tau_1) + m_2^x(\tau_2) + m_2^x(\tau_2 - \tau_1)] + 2(m_1^x)^3$, $m_3^x(\tau_1, \tau_2) =$ third-order moment sequence, 4th order: $c_4^x(\tau_1, \tau_2, \tau_3) = m_4^x(\tau_1, \tau_2, \tau_3) - m_2^x(\tau_1)m_2^x(\tau_3 - \tau_2) - m_2^x(\tau_2)m_2^x(\tau_3 - \tau_1) - m_2^x(\tau_3)m_2^x(\tau_2 - \tau_1) - m_1^x[m_3^x(\tau_2 - \tau_1, \tau_3 - \tau_1) + m_3^x(\tau_2, \tau_3) + m_3^x(\tau_2, \tau_4) + m_3^x(\tau_1, \tau_2)] + (m_1^x)^2[m_2^x(\tau_1) + m_2^x(\tau_2) + m_2^x(\tau_3) + m_2^x(\tau_3 - \tau_1) + m_2^x(\tau_3 - \tau_2) + m_2^x(\tau_2 - \tau_1)] - 6(m_1^x)^4$

39.1.4.4 Variance, Skewness, and Kurtosis ($\tau_1 = \tau_2 = \tau_3 = 0$ and $m_1^x = 0$)

$\gamma_2^x = E\{X^2(k)\} = c_2^x(0) \equiv$ variance, $\gamma_3^x = E\{X^3(k)\} = c_3^x(0,0) =$ skewness,

$\gamma_4^x = E\{X^4(k)\} - 3[\gamma_2^x]^2 = c_4^x(0,0,0) \equiv$ kurtosis, $\gamma_4^x / [\gamma_2^x]^2 =$ normalized kurtosis

39.1.4.5 Time-Reversible Process

$c_n^x(\tau_1, \tau_2, \cdots, \tau_{n-1}) = c_n^x(-\tau_1, -\tau_2, \cdots, -\tau_{n-1})$ for all integer values of $\tau_1, \tau_2, \cdots, \tau_{n-1}$ if the probability structure of $\{X(-k)\}$ is the same as that of $\{X(k)\}$

39.1.4.6 Cross-Cumulants: 2nd order

$c_{xy}(\tau_1) = \text{Cum}[X(k), Y(k+\tau_1)] =$ cross-covariance if the processes have zero mean, then $c_{xy}(\tau_1) = E\{X(k)\,Y(k+\tau_1)\}$, $\tau_1 =$ integer value; 3rd order: $c_{xyz}(\tau_1, \tau_2) = \text{Cum}[X(k), Y(k+\tau_1), Z(k+\tau_2)] = E\{(X(k) - m_x)(Y(k+\tau_1) - m_y)(Z(k+\tau_2) - m_z)\}$, $m_x = E\{X(k)\}$, $m_y = E\{Y(k)\}$, $m_z = E\{Z(k)\}$, for zero-mean process $c_{xyz}(\tau_1, \tau_2) = E\{X(k)Y(k+\tau_1)Z(k+\tau_2)\}$, $c_{xyy}(\tau_1, \tau_2) = \text{Cum}[X(k), Y(k+\tau_1), Y(k+\tau_2)]$, $c_{xyx}(\tau_1, \tau_2) = \text{Cum}[X(k), Y(k+\tau_1), X(k+\tau_2)]$; n^{th} *order*: $c_{x_1 x_2 \cdots x_n}(\tau_1, \tau_2, \cdots, \tau_{n-1}) \triangleq \text{Cum}[X_1(k), X_2(k+\tau_1), \cdots, X_n(k+\tau_{n-1})]$

Example

$Z(k) = X(k)\cos(\omega_c k) + Y(k)\sin\omega_c k$, $X(k)$ and $Y(k)$ are independent stationary processes with $E\{X(k)\}$ $E\{Y(k)\} = 0$, $m_2^x(\tau) = E\{X(k)X(k+\tau)\} = m_2^y$, $m_3^x(\tau_1, \tau_2) = E\{X(k)X(k+\tau_1)X(k+\tau_2)\} = m_3^y(\tau_1, \tau_2)$, $\text{Mom}[Z(k), Z(k+\tau)] =$ second moment $= E\{(X(k)\cos(\omega_c k) + Y(k)\sin\omega_c k)(X(k+\tau)\cos[\omega_c(k+\tau)] + Y(k+\tau)\sin[\omega_c(k+\tau)]\} = m_2^x(\tau)\cos\omega_c \tau = m_2^z(\tau) \equiv$ independent of k, and this implies that $\{Z(k)\}$ is a wide-sense stationary process; $\text{Mom}[Z(k), Z(k+\tau_1), Z(k+\tau_2)] = m_3^x(\tau_1, \tau_2)[\cos\omega_c k \cos[\omega_c(k+\tau_1)]\cos[\omega_c(k+\tau_2)] + \sin\omega_c k \sin[\omega_c(k+\tau_1)]\sin[\omega_c(k+\tau_2)] =$ dependent as k, and this implies a non-stationary process in its third-order statistics.

39.1.4.7 Erpodicity and Moments

$$E\{X(k)X(k+\tau_1)\cdots X(k+\tau_{n-1})\} = \langle X(k)\cdots X(k+\tau_{n-1})\rangle = \lim_{M\to\infty}\frac{1}{2M+1}\sum_{k=-M}^{M} X(k)X(k+\tau_1)\cdots X(k+\tau_{n-1}),$$

$\langle\ \rangle \equiv$ time average operator.

Estimate

$$\langle X(k)\cdots X(k+\tau_{n-1})\rangle_M = \frac{1}{2M+1}\sum_{k=-M}^{M} X(k)\cdots X(k+\tau_{n-1})$$

39.2 Cumulant Spectra

39.2.1 Definitions and Properties

39.2.1.1 Cumulant Spectra

$$C_n^x(\omega_1,\omega_2,\cdots,\omega_{n-1}) = \sum_{\tau_1=-\infty}^{\infty}\cdots\sum_{\tau_{n-1}=-\infty}^{\infty} c_n^x(\tau_1,\tau_2,\cdots,\tau_{n-1})\exp[-j(\omega_1\tau_1+\omega_2\tau_2+\cdots+\omega_{n-1}\tau_{n-1})],$$

$|\omega_i| \le \pi$ for $i = 1,2,\cdots,n-1$ and $|\omega_1+\omega_2+\cdots+\omega_{n-1}| \le \pi$, provided that

$$\sum_{\tau_1=-\infty}^{\infty}\cdots\sum_{\tau_{n-1}=-\infty}^{\infty} |c_n^x(\tau_1,\cdots,\tau_{n-1})| < \infty$$

39.2.1.2 Power Spectrum $n = 2$

$C_2^x(\omega) = \sum_{\tau=-\infty}^{\infty} c_2^x(\tau)e^{-j\omega\tau}$ $|\omega| < \pi$, $c_2^x(\tau) =$ covariance of $\{X(k)\}$ (see 39.1.4.3). If the process has zero mean, then the expression is identical with the Wiener-Khintchine identity. From (39.1.4.3) $c_2^x(\tau) = c_2^x(-\tau)$ and hence, $C_2^x(\omega) = C_2^x(-\omega)$ and $C_2^x(\omega) \ge 0$ (real and non-negative).

39.2.1.3 Bispectrum $n = 3$

$$C_3^x(\omega_1,\omega_2) = \sum_{\tau_1=-\infty}^{\infty}\sum_{\tau_2=-\infty}^{\infty} c_3^x(\tau_1,\tau_2)\exp[-j(\omega_1\tau_1+\omega_2\tau_2)]$$

$|\omega_1| \le \pi$, $|\omega_2| \le \pi$, $|\omega_1+\omega_2| \le \pi$, $c_3^x(\tau_1,\tau_2) =$ third-order cumulant (see 39.1.4.3)

Symmetry conditions

$$\begin{aligned} c_3^x(\tau_1,\tau_2) &= c_3^x(\tau_2,\tau_1) = c_3^x(-\tau_2,\tau_1-\tau_2) \\ &= c_3^x(\tau_2-\tau_1,-\tau_1) = c_3^x(\tau_1-\tau_2,-\tau_2) \\ &= c_3^x(-\tau_1,\tau_2-\tau_1) \quad \text{(see Figure 39.1)} \end{aligned}$$

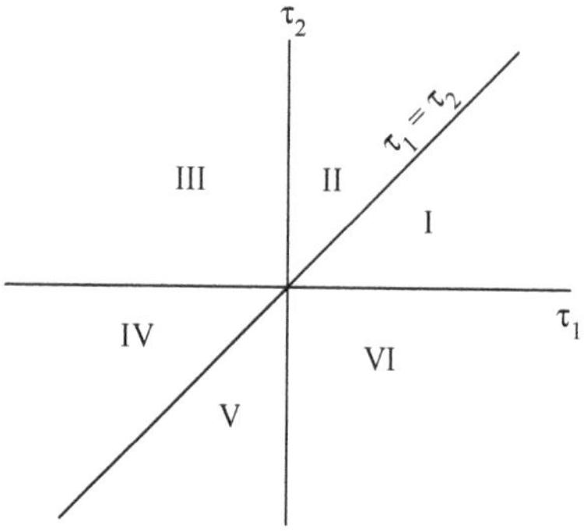

FIGURE 39.1 Symmetry regions of the third-order moments.

39.2.1.4 Symmetry Properties of Bispectrum $n = 3$

$$
\begin{aligned}
C_3^x(\omega_1,\omega_2) &= C_3^x(\omega_2,\omega_1) \\
&= C_3^{x*}(-\omega_2,-\omega_1) = C_3^x(-\omega_1-\omega_2,\omega_2) \\
&= C_3^x(\omega_1,-\omega_1-\omega_2) = C_3^x(-\omega_1-\omega_2,\omega_1) = C_3^x(\omega_2,-\omega_1-\omega_2)
\end{aligned}
$$

Note: Knowledge of the bispectrum in the triangle region $\omega_2 \geq 0,\ \omega_1 \geq \omega_2,\ \omega_1+\omega_2 \leq \pi$ is adequate for complete description (see Figure 39.2.1.4.1)

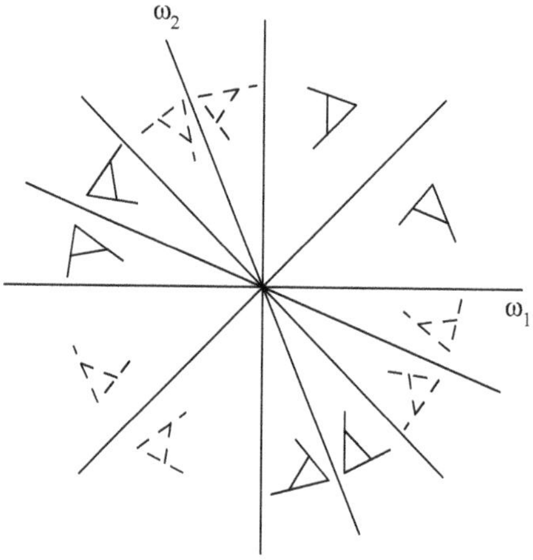

FIGURE 39.2 Symmetry regions of the bispectrum.

39.2.1.5 Triaspectrum $n = 4$

$$
C_4^x(\omega_1,\omega_2,\omega_3) = \sum_{\tau_1=-\infty}^{\infty}\sum_{\tau_2=-\infty}^{\infty}\sum_{\tau_3=-\infty}^{\infty} c_4^x(\tau_1,\tau_2,\tau_3)\exp[-j(\omega_1\tau_1+\omega_2\tau_2+\omega_3\tau_3)]
$$

$|\omega_1| \leq \pi,\ |\omega_2| \leq \pi,\ |\omega_3| \leq \pi,\ |\omega_1+\omega_2+\omega_3| \leq \pi,$

39.2.1.6 Symmetries of Triaspectrum

$C_4^x(\omega_1,\omega_2,\omega_3) = C_4^x(\omega_2,\omega_1,\omega_3) = C_4^x(\omega_3,\omega_2,\omega_1) = C_4^x(\omega_1,\omega_3,\omega_2) = C_4^x(\omega_2,\omega_3,\omega_1) = C_4^x(\omega_3,\omega_1,\omega_2) =$ etc.

39.2.1.7 Inverse of Polyspectrum

$$c_n^x(\tau_1,\tau_2,\cdots,\tau_{n-1}) = \frac{1}{(2\pi)^{n-1}} \int_{-\pi}^{\pi}\cdots\int_{-\pi}^{\pi} C_n^x(\omega_1,\omega_2,\cdots,\omega_{n-1})\exp[j(\omega_1\tau_1 + \cdots + \omega_{n-1}\tau_{n-1})]d\omega_1\cdots d\omega_{n-1}$$

39.2.1.8 Variance

$$c_2^x(0) = \gamma_2^x = \frac{1}{2\pi}\int_{-\pi}^{\pi} C_2^x(\omega)d\omega = \text{variance}, \; n = 2, \; \tau_1 = 0 \; \text{(see 39.2.1.7)}$$

39.2.1.9 Skewness

$$c_3^x(0,0) = \gamma_3^x = \frac{1}{(2\pi)^2}\iint_{-\pi}^{\pi} C_3^x(\omega_1,\omega_2)d\omega_1\, d\omega_2$$

39.2.1.10 Kurtosis

$$c_4^x(0,0,0) = \gamma_4^x = \frac{1}{(2\pi)^2}\iiint_{-\pi}^{\pi} C_4^x(\omega_1,\omega_2,\omega_3)d\omega_1\, d\omega_2\, d\omega_3$$

39.2.1.11 n^{th} Order Coherency Function

$P_n^x(\omega_1,\cdots,\omega_{n-1}) \triangleq C_n^x(\omega_1,\cdots,\omega_{n-1})/[C_2^x(\omega_1)C_2^x(\omega_2)\cdots C_2^x(\omega_{n-1})C_2^x(\omega_1+\omega_2+\cdots+\omega_{n-1})]^{1/2}$, for n = 3 coherency function is also called bicoherency (normalized bispectrum).

39.2.1.12 Cross-Cumulant Spectra

$$C_{x_1,x_2,\cdots,x_{n-1}}(\omega_1,\omega_2,\cdots,\omega_{n-1}) \triangleq \sum_{\tau_1=-\infty}^{\infty}\cdots\sum_{\tau_{n-1}=-\infty}^{\infty} c_{x_1,\cdots,x_{n-1}}(\tau_1,\tau_2,\cdots,\tau_{n-1})\exp[-j(\omega_1\tau_1+\omega_2\tau_2+\cdots+\omega_{n-1}\tau_{n-1})],$$

$\tau_i =$ integers

39.2.1.13 Cross-Bispectrum (see 39.1.4.6)

$$C_{xyy}(\omega_1,\omega_2) = \sum_{\tau_1=-\infty}^{\infty}\sum_{\tau_2=-\infty}^{\infty} c_{xyy}(\tau_1,\tau_2)\exp[-j(\omega_1\tau_1+\omega_2\tau_2)]$$

39.2.1.14 Cross-Spectrum

$$C_{xy}(\omega) = \sum_{\tau=-\infty}^{\infty} c_{xy}(\tau)\exp(-j\omega\tau)$$

39.2.1.15 Linear Phase Shifts

If $Y(k) = X(k-D)$, then $\text{Cum}[X(k),X(k+\tau_1),\cdots,X(k+\tau_{n-1})] = \text{Cum}[Y(k),Y(k+\tau_1),\cdots,Y(k+\tau_{n-1})] = c_n^x(\tau_1,\tau_2,\cdots,\tau_{n-1})$, $X(k) =$ zero mean stationary process

Note: $C_{xyx\cdots x}(\tau_1,\tau_2,\cdots,\tau_{n-1}) = \text{Cum}[X(k),Y(k+\tau_1)\cdots X(k+\tau_{n-1}] = c_n^x(\tau_1,\tau_2-D,\tau_3,\cdots,\tau_{n-1})$ and hence $C_{xyx\cdots x}(\omega_1,\cdots,\omega_{n-1}) = C_n^x(\omega_1,\omega_2,\cdots,\omega_{n-1})\exp(j\omega_2 D)$

39.2.1.16 Complex Regression Coefficients

$R(\omega) \triangleq C_{xy}(\omega)/C_{xx}(\omega)$, $R_1(\omega_1,\omega_2) \triangleq C_{xxy}(\omega_1,\omega_2)/[C_{xy}(\omega_1)C_{xy}(\omega_2)]$,

$R_2(\omega_1,\omega_2) \triangleq C_{yxx}(\omega_1,\omega_2)/C_{xxx}(\omega_1,\omega_2)$

39.2.1.17 Complex Processes (may be defined as)

${}_1c_3(\tau_1,\tau_2) \triangleq \text{Cum}[X(k),X(k+\tau_1),X(k+\tau_2)]$, ${}_2c_3(\tau_1,\tau_2) \triangleq \text{Cum}[X(k),X^*(k+\tau_1),X(k+\tau_2)]$

${}_3c_3(\tau_1,\tau_2) \triangleq \text{Cum}[X(k),X^*(k+\tau_1),X^*(k+\tau_2)]$

39.3 Non-Gaussian White Noise Process

39.3.1 Cumulants

$c_n^w(\tau_1,\tau_2,\cdots,\tau_{n-1}) = \text{Cum}[W(k),W(k+\tau_1),\cdots,W(k+\tau_{n-1})] = \gamma_n^w\delta(\tau_1,\tau_2,\cdots,\tau_{n-1})$, γ_n^w = constant, $W(k)$ = non-Gaussian with zero mean, $\delta(\tau_1,\tau_2,\cdots,\tau_{n-1}) = (n-1)$ — dimensional Kronecker delta function.

39.3.2 Cumulant Spectrum

$C_n^w(\omega_1,\cdots,\omega_{n-1}) = \gamma_n^w$

Example

Let $X(k) = W(k) - W(k-1)$, $\{W(k)\}$ = independent and identically distributed with $E\{W(k)\} = 0$, $E\{W^2(k)\} = 1$ and $E\{W^3(k)\} = 1$. Hence $c_2^x(\tau) = m_2^x(\tau) = E\{X(k)X(k+\tau)\} = 2\delta(\tau) - \delta(\tau-1) - \delta(\tau+1)$, $c_3^x(\tau_1,\tau_2) = m_3^x(\tau_1,\tau_2) = E\{X(k)X(k+\tau_1)X(k+\tau_2)\} = -\delta(\tau_1-1,\tau_2) + \delta(\tau_1-1,\tau_2-1) - \delta(\tau_1,\tau_2-1) +$ $\delta(\tau_1+1,\tau_2) + \delta(\tau_1+1,\tau_2+1) - \delta(\tau_1,\tau_2+1)$, $C_2^x(\omega) = \sum_{\tau=-1}^{1} c_2^x(\tau)\exp(-j\omega) = 2 - 2\cos\omega$, $C_3^x(\omega_1,\omega_2) =$ $-\exp(-j\omega_1) + \exp[-j(\omega_1+\omega_2)] - \exp(-j\omega_2) + \exp(j\omega_1) - \exp[j(\omega_1+\omega_2)] + \exp(j\omega_2) = 2j[\sin\omega_1 +$ $\sin\omega_2 - \sin(\omega_1+\omega_2)]$

39.4 Cumulant Spectra and Linear Systems

39.4.1 Cumulant Spectra of LTI (Linear-Time Invariant) Systems

39.4.1.1 Impulse Response of LTI System

$$h(k) \triangleq \frac{1}{2\pi}\int_{-\pi}^{\pi} H(\omega)\exp(j\omega k)\,d\omega$$

$H(\omega)$ = frequency response function

39.4.1.2 Output of LTI System

$Y(k) = \sum_{i=-\infty}^{\infty} h(k-i)X(i)$, $h(k)$ = impulse response of the system, $\{X(k)\}$ = zero-mean non-Gaussian process with all its moments finite.

39.4.1.3 Cumulant Spectra of the Output

$$C_n^y(\omega_1,\omega_2,\cdots,\omega_{n-1}) = H(\omega_1)H(\omega_2)\cdots H(\omega_{n-1})\cdot H^*(\omega_1+\cdots+\omega_{n-1})C_n^x(\omega_1,\cdots,\omega_{n-1})$$

39.4.1.4 Cumulant Spectra of the Output

$$\left|C_n^y(\omega_1,\omega_2,\cdots,\omega_{n-1})\right| = \left|H(\omega_1)\right|\cdots\left|H(\omega_{n-1})\right|\cdot\left|H^*(\omega_1+\cdots+\omega_{n-1})\right|,\ \Psi_n^y(\omega_1,\cdots,\omega_{n-1}) = \text{phase of}$$

$$C_n^y(\cdot) = \phi_h(\omega_1)+\phi_h(\omega_2)+\cdots+\phi_h(\omega_{n-1})-\phi_h(\omega_1+\omega_2+\cdots+\omega_{n-1})+\Psi_n^x(\omega_1,\omega_2,\cdots,\omega_{n-1}),$$

$$H(\omega) = \left|H(\omega)\right|\exp[j\phi_h(\omega)],\ C_n^x(\cdot) = \left|C_n^x(\cdot)\right|\exp[j\Psi_n^x(\omega_1,\cdots,\omega_{n-1}]$$

39.4.2 LTI Systems Driven by White Noise

39.4.2.1 Cumulant Spectra

$$C_n^y(\omega_1,\cdots,\omega_{n-1}) = \gamma_n^x H(\omega_1)H(\omega_2)\cdots H(\omega_{n-1})\cdot H^*(\omega_1+\omega_2\cdots+\omega_{n-1})$$

$\gamma_n^x = C_n^x(\omega_1,\cdots,\omega_{n-1})$ (see 39.3.2 and 39.4.1.3)

39.4.2.2 Power Spectrum n = 2

$C_2^y(\omega) = \gamma_2^x\left|H(\omega_1)\right|^2$, $\gamma_2^x =$ variance of input noise

39.4.2.3 Bispectrum n = 3

$\left|C_3^y(\omega_1,\omega_2)\right| = \left|\gamma_3^x\right|\left|H(\omega_1)\right|\left|H(\omega_2)\right|\left|H(\omega_1+\omega_2)\right|$ and $\Psi_3^y(\omega_1,\omega_2) = \phi_h(\omega_1)+\phi_h(\omega_2)-\phi_h(\omega_1+\omega_2)$, $\gamma_3^x =$ skewness of the input (see 39.4.1.4)

39.4.2.4 Cumulant Spectra Interrelationship

$$C_n^y(\omega_1,\cdots,\omega_{n-2},0) = C_{n-1}^y(\omega_1,\cdots,\omega_{n-2})H(0)\frac{\gamma_n^y}{\gamma_{n-1}^y}$$

Example

$C_3^y(\omega,0) = C_2^y(\omega)H(0)\gamma_3^y/\gamma_2^y$

Example

$h(k) = \delta(k)-a\delta(k-1)$, $H(\omega) = 1-ae^{-j\omega}$

$$Y(k) = \sum_{i=0}^{1} h(i)X(k-i) = X(k)-aX(k-1),\ C_2^y(\omega) = \text{output power spectrum} = \gamma_2^x\left|H(\omega)\right|^2$$

$$= \gamma_2^x(1-ae^{-j\omega})(1-ae^{j\omega}),\ C_3^y(\omega_1,\omega_2) = \text{bispectrum} = \gamma_3^x(1-ae^{-j\omega_1})(1-ae^{-j\omega_2})(1-ae^{j(\omega_1+\omega_2)}),$$

39.4.3 Minimum and Non-minimum Phase LTI Systems

39.4.3.1 Minimum Phase MA System

$Y_1(k) = W(k)-(a+b)W(k-1)+abW(k-2)$ (second order), $H_1(z) = (1-az^{-1})(1-bz^{-1})$, $0<a<1$, $0<b<1$. Both zeros are inside the unit circle, $W(k) =$ non-Gaussian white noise.

39.4.3.2 Maximum Phase MA System

$Y_2(k) = W(k)-(a+b)W(k+1)+abW(k+2)$, $H_2(z) = (1-az)(1-bz)$, $0<a<1$, $0<b<1$. Both zeros lie outside the unit circle, $W(k)$ (see 39.4.3.1).

39.4.3.3 Non-Minimum or Mixed Phase MA System

$Y_3(k) = -aW(k+1)+(1+ab)W(k)-bW(k-1)$, $H_3(z) = (1-az)(1-bz^{-1})$, $0<a<1$, $0<b<1$, $W(k)$, see (39.4.3.1).

39.4.3.4 ARMA (auto-regressive moving average) System

$$\sum_{i=0}^{p} a_i Y(k-i) = \sum_{i=0}^{q} b_\ell W(k-\ell),\ a_0 = 1,\ H(z) = B(z)/A(z),\ A(z) = \sum_{i=0}^{p} a_i z^{-i},$$

$$B(z) = \sum_{\ell=0}^{q} b_\ell z^{-\ell},\ W(k)\ \text{(see 39.4.3.1).}$$

39.4.3.5 Cumulant Spectrum of ARMA System

See 39.4.2.1 with $H(z)\big|_{z=\exp(j\omega)}$

Note: If $W(k)$ is Gaussian, the real roots z_r can be replaced by their inverse $1/z_r$ and the conjugate roots z_0 can be substituted with their conjugate inverse $(1/z_0^*)$ without changing the power spectrum of the output $\{Y(k)\}$.

39.5 Estimation of Higher-Order Spectra

39.5.1 Estimation of Higher-Order Statistics

39.5.1.1 Higher-Order Statistics Estimates

Steps:

1. Segment the data into K records of M samples each; i.e., $N = K \cdot M$. However, if the data samples correspond to a deterministic energy signal, then the data segmentation is inappropriate. Also, if the process is deterministic periodic, then M should be equal to the period of the signal or multiple integers of the period.
2. Subtract the average value of each record. (This is optional for deterministic signals.)
3. Assuming that $\{x^{(i)}(k),\ k = 0,1,\cdots,M-1\}$ is the data segment $(i = 1,2,\cdots,K)$, natural estimates of higher-order moments are given by

$$m_n^{(i)}(\tau_1,\cdots,\tau_{n-1}) = \frac{1}{M}\sum_{k=s_1}^{s_2} x^{(i)}(k)x^{(i)}(k+\tau_1)\cdots x^{(i)}(k+\tau_{n-1})$$

where $n = 2,3,\cdots,i = 1,2,\cdots,K,\ \tau_k = 0,\pm1,\pm2,\cdots,\ s_1 = \max(0,-\tau_1,\cdots,-\tau_{n-1}),\ s_2 = \min(M-1, M-1-\tau_1,\cdots,M-1-\tau_{n-1}),\ |\tau_k| \le L_n$. Note that L_n determines the region of support of the estimated n^{th}-order moment function.

4. The average over all segments is given by

$$\hat{m}_n^x(\tau_1,\cdots,\tau_{n-1}) = \frac{1}{K}\sum_{i=1}^{K} m_n^{(i)}(\tau_1,\cdots,\tau_{n-1}),\ n = 2,3,\cdots,\ |\tau_k| \le L_n.$$

the $\hat{m}_n^x(\tau_1,\cdots,\tau_{n-1})$ is considered the usual estimate for $m_n^x(\tau_1,\cdots,\tau_{n-1})$. If the signal is deterministic, then $\hat{m}_n^x(\tau_1,\cdots,\tau_{n-1}) = m_n^{(1)}(\tau_1,\cdots,\tau_{n-1})$; i.e., $K = 1$ and thus $N = M$.

5. For stochastic signals, generate the cumulants $\hat{c}_n^x(\tau_1,\cdots,\tau_{n-1})$ using the above equation. If the average value of each record is subtracted (i.e., so each can be considered a zero-mean signal), it follows (see Section 39.1.4) that

$$\hat{c}_2^x(\tau_1) = \hat{m}_2^x(\tau_1)$$

$$\hat{c}_3^x(\tau_1,\tau_2) = \hat{m}_3^x(\tau_1,\tau_2)$$

$$\hat{c}_4^x(\tau_1,\tau_2,\tau_3) = \hat{m}_4^x(\tau_1,\tau_2,\tau_3) - \hat{m}_2^x(\tau_1)\hat{m}_2^x(\tau_3-\tau_2) - \hat{m}_2^x(\tau_2)\cdot\hat{m}_2^x(\tau_3-\tau_1) - \hat{m}_2^x(\tau_3)\cdot\hat{m}_2^x(\tau_2-\tau_1),$$

where $|\tau_k| \le L_n$, $k = 1,2,3$. Data: $\{X(1),\cdots,X(N)\}$

39.5.2 Higher-Order Spectra Estimates

39.5.2.1 Moment Spectra Estimates

$$\hat{M}_n^x(\omega_1,\cdots,\omega_{n-1}) = \frac{1}{K}\sum_{\tau_1=-L_n}^{L_n}\cdots\sum_{\tau_{n-1}=-L_n}^{L_n}\hat{m}_n^x(\tau_1,\cdots,\tau_{n-1})w(\tau_1\Delta_n,\cdots,\tau_{n-1}\Delta_n)\exp[j(\omega_1\tau_1+\cdots+\omega_{n-1}\tau_{n-1})]$$

39.5.2.2 Cumulate Spectra Estimates

$$\hat{C}_n^x(\omega_1,\cdots,\omega_{n-1}) = \sum_{\tau_1=-L_n}^{L_n}\cdots\sum_{\tau_{n-1}=-L_n}^{L_n}\hat{c}_n^x(\tau_1,\cdots,\tau_{n-1})w(\tau_1\Delta_n,\cdots,\tau_{n-1}\Delta_n)\exp[-j(\omega_1\tau_1+\cdots+\omega_{n-1}\tau_{n-1})],$$ $w(u_1,\cdots,u_{n-1}) =$ continuous window function with bounded support, $\Delta_n = \text{bandwidth} \triangleq 1/L_n$.

39.5.3 Higher-Order Periodogram

39.5.3.1 Discrete Fourier Transform

$F_x(\omega_\lambda) = T\sum_{k=0}^{N-1} X(k)\exp(-j\omega_\lambda k)$, $\{X(k)\}$ = real-valued stationary time series with zero mean and length N, $\omega_\lambda = 2\pi\lambda/N$, $\lambda = 0,1,\cdots,N-1$

39.5.3.2 Higher-Order Periodogram

$M_n^x(\omega_{\lambda_1},\cdots,\omega_{\lambda_{n-1}}) = \frac{1}{NT}F_x(\omega_{\lambda_1})\cdots F_x(\omega_{\lambda_{n-1}})F_x^*(\omega_{\lambda_1}+\omega_{\lambda_2}+\cdots+\omega_{\lambda_{n-1}})$, $M_n^x(\cdot)$ is asymptotically unbiased.

39.5.3.3 Direct Method

1. Segment the data into K records of M samples each, i.e., $N = K\cdot M$, and subtract the average value of each segment. If the signal is deterministic, then use only one record ($N = M$). If necessary, add zeros at the end of each record to obtain a convenient length M for an FFT algorithm.
2. Assuming that $\{x^{(i)}(k)\}$, $k = 0,1,\cdots,M-1$ is the data segment $\{i\}$, generate the DFT coefficients

   $$F_x^{(i)}(\lambda) = \sum_{k=0}^{M-1} x^{(i)}(k)\exp\left\{-j\frac{2\pi}{M}k\lambda\right\},\ \lambda = 0,1,\cdots,M-1 \text{ and } i = 1,2,\cdots,K.$$
3. In general, the relationship between N_n and M is given by $M = M_n \times N_n$ where M_n is a positive odd integer; e.g., $M_n = 2J_n + 1$. In other words, M_n determines the size of higher-order spectrum smoothing over neighboring frequencies. Since M is even and M_n is odd, we compromise on the

value of N_n (closest integer). Estimate the nth-order moment spectrum, $\hat{M}_n^{(i)}(\lambda_1,\cdots,\lambda_{n-1})$, by frequency-domain averaging..

$$\hat{M}_n^{(i)}(\lambda_1,\cdots,\lambda_{n-1}) = \frac{1}{\Delta_n^{n-1}} \sum_{k_1=-J_n}^{J_n} \cdots \sum_{k_{n-1}=-J_n}^{J_n} F_x^{(i)}(\lambda_1+k_1)\cdots F_x^{(i)}(\lambda_{n-1}+k_{n-1})\cdot$$

$$F_x^{(i)*}(\lambda_1+\cdots+\lambda_{n-1}+k_1+\cdots+k_{n-1}),\quad i=1,2,\cdots,K$$

over a 2-D rectangular window of size $M_n \times M_n$. For the special case where no averaging is performed in the frequency domain, $M_n = 1(J_n = 0)$ and therefore,

$$\hat{M}_n^{(i)}(\lambda_1,\cdots,\lambda_{n-1}) = \frac{1}{\Delta_n^{n-1}} F_x^{(i)}(\lambda_1)\cdots F_x^{(i)}(\lambda_{n-1}) F_x^{(i)*}(\lambda_1+\cdots+\lambda_{n-1}),\ i=1,2,\cdots,K$$

4. Finally, the nth-order moment spectrum of the given data is obtained by averaging over the K pieces: $\hat{M}_n^x(\omega_1,\cdots,\omega_{n-1}) = \frac{1}{K}\sum_{i=1}^{K} \hat{M}_n^{(i)}(\omega_1,\cdots,\omega_{n-1})$, where $\omega_j \triangleq (2\pi\Delta_n)\lambda_j$, and $j=1,2,\cdots,n-1$.

Data: $\{X(1),\cdots,X(N)\}$, T = sampling period = 1, $\Delta_n \triangleq 1/N_n$ = required spacing between frequency samples in the higher-order spectrum domain along horizontal or vertical directions.

References

Brillinger, D. R., An introduction to polyspectra, *Ann. Math. Statist.*, Vol, 36, pp. 1351-1374, 1965.

Giannakis, G. B., and J. M. Mendel, Identification of non-minimum phase systems using higher-order statistics, *IEEE Trans. On Acoustics, Speech and Signal Processing*, Vol. 37, pp. 360-377, 1989.

Nikias, C. L., and A. P. Petropulu, *Higher-Order Spectral Analysis*, Prentice-Hall, Englewood Cliffs, NJ, 1993.

Nikias, C. L., and M. R. Raghuveer, Bispectrum estimation: a digital signal processing framework, *Proceedings IEEE*, Vol. 75, pp. 869-892, 1987.

Rao, T. S., and M. M. Gabr, *An Introduction to Bispectral Analysis and Bilinear Time Series Models*, Lecture Notes in Statistics, Vol. 24, New York, Springer-Verlag, 1984.

40

Stochastic Simulation — Random Variate Generation

40.1 Random Number Generation

40.1.1 Congruential Generators

$X_{i+1} = (aX_i + c)(\mathrm{mod}\, m)$, $i = 0,1,\cdots,n$, a = non-negative integer, c = non-negative integer, m = non-negative integer, the modulo notation (mod m) means that $X_{i+1} = aX_i + c - mk_i$ where $k_i = [(aX_i + c)/m]$, the largest positive integer in $(aX_i + c)/m$, $U_i = X_{i+1}/m$ = pseudo-random sequence in the interval $(0,1)$ X_0 = seed.

Example

If we set $a = c = X_0 = 3$, $m = 5$ implies that $X_{i+1} = (3X_i + 3)(\mathrm{mod}\, 5)$ and gives $X_i = \{3,2,4,0,3\}$ which repeats every m.

Note: Good values for generator: $a = 2^7$, $c = 1$, $m = 2^{35}$

40.1.2 Multiplicative Generator

$X_{i+1} = aX_i(\mathrm{mod}\, m)$, $i = 0,1,\cdots$, $X_0 > 0$

Note: You may use for $i > 0$, $X_{i+1} = 7^5 X_i(\mathrm{mod}\, 2^{31} - 1) = 16{,}807X_i(\mathrm{mod}\, 2^{31} - 1)$, $U_i = X_i/(2^{31} - 1)$ $i = 1,2,\cdots$

40.2 Random Variate Generation

40.2.1 Methods for Variate Generation

40.2.1.1 Inverse Transform Method

Algorithm steps

1. Generate U from a random generator that produces random variates uniformly distributed over $(0,1)$

2. $X = F_X^{-1}(U)$, $F_X(x) =$ cumulative probability distribution function, $F^{-1}(y) =$ its inverse function with $0 < y < 1$
3. Deliver X.

40.2.2.2 Acceptance-Rejection Method (single-variate case)

To carry out the method, we represent $f_X(x)$ = probability density function (p.d.f.) = $Ch(x)g(x)$ where $C \geq 1$, $h(x)$ = p.d.f., $0 < g(x) \leq 1$

Steps for Algorithm

1. Generate U from uniform distribution (0,1)
2. Generate Y from the p.d.f. $h(y)$
3. If $U \leq g(Y)$ keep Y as a variate generated from $f_x(x)$
4. Go to step 1.

40.2.1.3 Multivariate Case

Let $\underline{X} = [X_1, \cdots, X_n] = r.$ vector with p.d.f. $f_{\underline{X}}(\underline{x})$, $\underline{x} = [x_1, \cdots, x_n] \in D$, $D = \{(x_1, \cdots, x_n) : a_i \leq x_i \leq b_i$, $i = 1, \cdots n\}$ and suppose $f_X(\underline{x}) \leq M$. Generate $U_1, \cdots, U_n$ from a uniform distribution (0,1) and define $\underline{Y} = [Y_1, \cdots, Y_n]$ where $Y_i = a_i + (b_i - a_i)U_i$, $i = 1, 2, \cdots, n$. Then $P\left(Y_1 \leq x_i, i = 1, \cdots, n, \middle| U_n \leq \frac{f_X(\underline{y})}{M}\right) = F_X(\underline{x})$.

Example

Generate a random vector uniformly distributed on the surface of an n-dimensional unit sphere.

Algorithm steps

1. Generate $U_1, \cdots, U_n$ from a uniform distribution
2. $X_1 = 1 - 2U_1, \cdots, X_n = 1 - 2U_n$ and $Y^2 = \sum_{i=1}^{n} X_i^2$
3. If $Y^2 < 1$ accept $\underline{Z} = [Z_1, \cdots, Z_n]$ where $Z_i = (X_i / Y)$, $i = 1, \cdots, n$ as the desired vector
4. Go to step 1.

40.3 Generation from Continuous Distributions

40.3.1 Exponential Distribution

40.3.1.1 Exponential Distribution

$f_X(x) = \frac{1}{\beta} e^{-x/\beta}$, $0 \leq x < \infty$, $\beta > 0$ and zero otherwise. $U = \int_0^x f_X(x)dx = 1 - e^{-x/\beta}$ so that $X = -\beta \ln(1 - U)$.

Since $1 - U$ has the same distribution as U, we use $X = -\beta \ln(U)$.

Algorithm

1. Generate U from uniform distribution
2. $X = -\beta \ln(U)$
3. Accept X.

40.3.2 Gamma Distribution

40.3.2.1 Gamma Distribution

$f_X(x) = x^{\alpha-1}e^{-x/\beta} / [\beta^a\Gamma(\alpha)] \triangleq G(\alpha,\beta),\ 0 \le x < \infty,\ \alpha > 0, \beta > 0,$ and zero otherwise.

Note:

1. Let X_i, $i = 0,1,\cdots,n$ is a sequence of independent r. variables from $G(\alpha,\beta)$, then $X = \sum_{i=1}^{n} X_i$ is from $G(\alpha,\beta)$ where $\alpha = \sum_{i=1}^{n} \alpha_i$.
2. If $\alpha = m =$ integer, we obtain the variate from $G(m,p)$ distribution by summing m independent exponential random variates $\exp(\beta)$, that is

$$X = \beta\sum_{i=1}^{m}(-\ln U_i) = -\beta\ln\prod_{i=1}^{m} U_i$$

Algorithm for $G(\alpha,\beta)$ with $\alpha =$ integer

1. X = 0
2. Generate V from the distribution $e^{-x}u(x)$, $u(x)$ = unit step function
3. $X = X + V$
4. $\alpha =$ 1, $X = \beta X$, and keep X
5. $\alpha = \alpha - 1$
6. Go to step 2.

Algorithm for G(α,β) with β = 1

$$\alpha > 1,\ \lambda = (2\alpha - 1)^{1/2},\ b = \alpha - \ln 4,\ d = \alpha + 1/\lambda$$

Steps:

1. Generate U_1 and U_2 from uniform distribution (0,1)
2. $V = \lambda\ln[U_1/(1-U_2)]$
3. $X = \alpha e^V$
4. If $b + d - X \ge \ln(U_1^2U_2)$, keep X
5. Go to step 1.

40.3.3 Beta Distribution

40.3.3.1 Beta Distribution

$f_X(x) = \Gamma(\alpha+\beta)x^{\alpha-1}(1-x)^{\beta-1}/[\Gamma(\alpha)\Gamma(\beta)] \triangleq Be(\alpha,\beta),\ \alpha > 0,\ \beta > 0,\ 0 \le x \le 1.$

Algorithm

1. Generate Y_1 from $G(\alpha,1)$ (see Section 40.3.2)
2. Generate Y_2 from $G(\beta,1)$

3. $X = Y_1/(Y_1 + Y_2)$
4. Keep X.

Algorithm

1. $j = 1$
2. Generate U_j and U_{j+1} from uniform distribution (0,1)
3. $Y_1 = U_j^{1/\alpha}$
4. $Y_2 = U_{j+1}^{1/\beta}$
5. If $Y_1 + Y_2 \geq 1$, go to step 2
6. $j = j + 2$
7. Keep $X = Y_1/(Y_1 + Y_2)$.

40.3.4 Normal Distribution (Gaussian)

40.3.4.1 Normal Distribution

$f_X(x) = \frac{1}{\sigma\sqrt{2\pi}} \exp\left[-\frac{(x-\mu)^2}{2\sigma^2}\right] \triangleq N(\mu,\sigma^2),\ -\infty < x < \infty,\ \mu = \text{mean value},\ \sigma^2 = \text{variance}$, if Z is standard normal N (0,1), then $X = \mu + \sigma Z$.

Algorithm for Standard Normal N (0,1)

1. Generate U_1 and U_2 from uniform distribution (0,1)
2. Compute $Z_1 = (-2\ln U_1)^{1/2}\cos 2\pi U_2$ and $Z_2 = (-2\ln U_1)^{1/2}\sin 2\pi U_2$
3. Keep either Z_1 or Z_2

Acceptance-Rejection Method (see 40.2.1.2)

$f_X(x) = \sqrt{\frac{2}{\pi}}\, e^{-x^2/2},\ x \geq 0$ (we can assign a random sign to the r.v. generated from $f_X(x)$ and obtain an r.v. from N (0,1)). Hence, we write

$$f_X(x) = Ch(x)g(x),\ h(x) = e^{-x},\ C = \sqrt{2e/\pi},\ g(x) = \exp[-(x-1)^2/2].$$

Algorithm

1. Generate V_1 and V_2 from exponential distribution $e^{-x},\ x > 0$
2. If $V_2 \geq (V_1 - 1)^{1/2}/2$, go to step 1
3. Generate U from uniform distribution (0,1)
4. If $U \geq 0.5$, keep $Z = -V_1$
5. $Z = V_1$.

Central Limit Propert
Algorithm:

1. Generate 12 r.v. $U_1,\cdots,U_{12}$ from uniform distribution (0,1)
2. $Z = \sum_{i=1}^{12} U_1 - 6$
3. Keep Z

Additional Algorithm

1. Generate U_1 and U_2 from uniform distribution (0,1)
2. $X = \sqrt{\pi/8}\ \ln[(1+U_1)/(1-U_1)]$
3. If $U_2 \le 0.5$, keep $Z = -X$
4. Keep $Z = X$.

40.3.5 Lognormal Distribution

40.3.5.1 Lognormal Distribution

$f_Y(y) = \frac{1}{\sqrt{2\pi}\,\sigma y}\exp\left[-\frac{(\ln y-\mu)^2}{2\sigma^2}\right]$, $0 \le y < \infty$ and zero otherwise, if X is normal $N(\mu,\sigma^2)$ then $Y = e^X$ is lognormal distributed.

Algorithm

1. Generate Z from $N\,(0,1)$ (see 40.3.4.1)
2. $X = \mu + \sigma Z$
3. $Y = e^X$
4. Keep Y.

40.3.6 Cauchy Distribution

40.3.6.1 Cauchy Distribution

$f_X(x) \triangleq C(\alpha,\beta) = \frac{\beta}{\pi[\beta^2+(x-\alpha)^2]}$, $\alpha > 0,\ \beta > 0,\ -\infty < x < \infty$ $F_X(X) = \text{c.d.f.} = \frac{1}{2} + \pi^{-1}\tan^{-1}\left(\frac{X-\alpha}{\beta}\right)$

Algorithm

1. Generate U from uniform distribution
2. $X = \alpha - \beta/\tan(\pi U)$
3. Keep X

40.3.7 Weibul Distribution

40.3.7.1 Weibul Distribution

$f_X(x) \triangleq W(\alpha,\beta) = \frac{\alpha}{\beta^\alpha}x^{\alpha-1}e^{-(x/\beta)^\alpha}$, $\alpha > 0,\ \beta > 0,\ 0 \le x < \infty$ and zero otherwise.

Algorithm

1. Generate V from exponential distribution $e^{-x}u(x)$ (see 40.3.1.1)
2. $X = \beta V^{1/\alpha}$
3. Keep X

40.3.8 Chi-Square Distribution

40.3.8.1 Chi-Square Distribution

If $Z_1,\cdots,Z_k$ are variables from standard normal distribution $N\,(0,1)$ [zero mean and variance one], then

$Y = \sum_{i=1}^{k} Z_i^2$ is chi-square distributed with k degrees of freedom and is denoted by $\chi^2(k)$

Algorithm

1. Generate k variates Z's from its standard normal $N(0,1)$ random variables (see 40.3.4.1)
2. $Y = \sum_{i=1}^{k} Z_i^2$
3. Keep Y

Algorithm (k = even)

1. Generate k/2 variables $U_1,\cdots,U_{k/2}$ from uniform distribution (0,1)
2. $Y = -2\ln\left(\prod_{i=1}^{k/2} U_i\right)$
3. Keep Y

Algorithm (k = odd)

1. Generate Z from standard normal distribution $N(0,1)$, [see 40.3.4.1],
2. Generate $\frac{k}{2}-\frac{1}{2}$ U_i's from the uniform distribution
3. $Y = -2\ln\left(\prod_{i=1}^{k/2-1/2} U_i\right) + Z^2$
4. Keep Y

40.3.9 Student's t Distribution

40.3.9.1 Student's t Distribution (see Chapter 34)

Algorithm

1. Generate Z from standard normal distribution $N(0,1)$, [see 40.3.4.1]
2. Generate Y from chi-square distribution [see 40.3.8.1]
3. $X = Z/\sqrt{Y/k}$
4. Keep X (X has a student's t distribution with k degrees of freedom.

40.3.10 F Distribution

40.3.10.1 F Distribution: (see Chapter 34)

Algorithm

1. Generate Y_1. from chi-square distribution with k_1 degrees of freedom [see 40.3.8.1]
2. Generate Y_2 from chi-square distribution with k_2 degrees of freedom
3. $X = (Y_1/k_1)/(Y_2/k_2)$
4. Keep X which is F-distributed with k_1 and k_2 degrees of freedom.

Remark: If X is F-distributed with k_1 and k_2 degrees of freedom, then $1/X$ is F-distributed with k_2 and k_1 degrees of freedom.

Remark: If X is F-distributed with k_1 and k_2 degrees of freedom, then $W = (k_1X/k_2)/[1+(k_1X/k_2)]$ has a beta density with parameters $\alpha = k_1/2$ and $\beta = k_2/2$.

References

Ripley, B. D., *Stochastic Simulation*, John Wiley & Sons, New York, NY, 1987.
Rubenstein, R. Y., *Simulation and the Monte Carlo Method*, John Wiley & Sons, New York, NY, 1981.
Yakowitz, S. J., *Computational Probability and Simulation*, Addison-Wesley, Reading, MA, 1977.

41

Nonlinear Digital Filtering

41.1 Definitions

41.1.1 One- and Two-Dimensional Signals

f(t)=continuous, f(n)=discrete; f(x,y)=continuous, f(n,m)=discrete

41.1.2 Filter Length

N=2k+1, N= odd number, range of indices: $i-k, i-k+1, \ldots, i+k$

41.1.3 Data Length

$\mathbf{x}=\{x(1),\ldots,x(N_d)\}$

41.2 Mean Filter

41.2.1 Mean Filter

$$y(i)=\frac{1}{2k+1}\sum_{j=-k}^{k}x(i+j)$$

Example

$\mathbf{x}=\{4,14,18,40,10\}$. For k=1, N=2·1+1=3 and y(1)=4,

$$y(2)=\frac{1}{3}\sum_{j=-1}^{1}x(2+j)=\frac{1}{3}(4+14+18)=12,\quad y(3)=(1/3)(14+18+40)=24,\quad y(4)=68,\quad y(5)=10$$

Note: 1. Mean filtering is a linear operation 2. Mean filter attenuates noise

41.2.2 Mean Filter Algorithms

```
Input: data x={x(1),...,x(N_d)}
   moving window w_N, N=2k+1
Output: y={y(1),...,y(N)}
for i = 1 to N_d
set number at i
Sum=0
for every element x(m) of the
   image inside the window w_N
Sum=Sum+x(m)
end
y(i)=Sum/N
end
```

```
Inputs: No of rows N_R × No of columns N_C (image)
   moving window w_N=N=2k+1
Output: N_R×N_C (image)
for i=1 to N_R
for j=1 to N_C
set window w_N at (i,j)
Sum=0
for every element x(m,n) of the
   image inside w_N
Sum=Sum+x(m,n)
end
y(i,j)=Sum/N
end
end
```

41.3 Median Filter

41.3.1 Median Filter

y(i)=MED{x(i–2), x(i–1), x(i), x(i+1), x(i+2)} i=3,4,…,18 for data **x**={x(1),x(2),…,x(20)}

Example

Let **x**={1,5,3,2,10,2} and k=1 then y(1)=1,y(2)=MED{1,5,3}=MED{1,3,5}=3,
y(3)=MED{5,3,2}=MED{2,3,5}=3,
y(4)=MED{3,2,10}=MED{2,3,10}=3,y(4)=MED{3,2,10}=MED{2,3,10}=3,y(5)=MED{2,10,2}=MED{2,2,10}=2,y(6)=2. Hence, **y**={1,3,3,3,2,2}

Note: a) the filter is nonlinear, b) attenuates noise to some degree c) does not attenuate noise well in the case of noisy ramp signal d) noisy step signal is rounded at the step e) eliminates impulse noise f) if impulse noise is close to an edge it may be removed but the edge moves toward the impulse (edge jitter).

41.3.2 Median Filter Algorithms

```
Inputs: data x={x(1),...,x(N_d)}
   moving window w_N, N=2k+1
Output: data y={y(1),...,y(N_d)}
for i=1 to N_d
place window w_N at i
store the data inside the window w_N in
   x'={x'(1),...,x'(N)}
find the median of x', MED{x'}
y(i)=MED{x'}; output
end
```

```
Inputs: No of rows N_R × No of columns N_C
   moving window w_N, N=2k+1
Output: N_R×N_C (image)
for i=1 to N_R
for j=1 to N_C
place window w_N at (i,j)
store image values inside w_N
   x'={x'(1),...,x'(N)}
find median of x', MED{x'}
y(i,j)=MED{x'}; output
end
end
```

41.4 Trimmed Mean Filters

41.4.1 Motivation for Trimmed Mean Filters

The idea of trimmed filters is to reject the most probable outliers — some of the very small and the very large values.

41.4.2 (r,s)-Fold Trimmed Mean Filters

Short the samples in the window and omit r+s samples $x_{(1)},\ldots,x_{(r)}$ and $x_{(N-s+1)},x_{(N-s+2)}\cdots$

$$\mathrm{TrMean}(\mathbf{x}=\{x(1),x(2),\ldots,x(N_d)\};r,s)=\frac{1}{N-r-s}\sum_{i=r+1}^{N-s}x_{(i)}$$

41.4.2.1 (r,s)-Fold Trimmed Mean Filters Algorithm

Inputs: data $\mathbf{x}=\{x(1),\ldots,x(N_d)\}$
moving window w_N, N=2k+1, r and s integer numbers
Output: y(i)
for i=1 to N_d
place window w_N at i
store the values inside the window into
$\mathbf{x}'=\{x'(1),x'(2),\ldots,x'(N)\}$
sort $\mathbf{x}'$ to $\mathbf{x}''=\{x'_{(1)},x'_{(2)},\cdots,x'_{(N)}\}$
Sum=0
for m=r+1 to N–s
Sum=Sum+ $x'_{(m)}$
end
y(i)=Sum/(N–r–s); output
end
end
Note: $x_{(i)}\le x_{(j)}$ for all $i<j$

Inputs: No of rows N_R × No of columns N_C (image)
moving window w_N, N=2k+1, integer numbers r,s
Output: $N_R \times N_C$ (image)
for i=1 to N_R
for j=1 to N_C
place window w_N at (i,j)
store data inside window into
$\mathbf{x}=\{x(1),\ldots,x(N)\}$
sort $\mathbf{x}$ into $x'=\{x'_{(1)},x'_{(2)},\cdots,x'_{(N)}\}$
Sum=0
for m=r+s to N–s
Sum=Sum+ $x'_{(m)}$
end
y(i,j)=Sum/(N–r–s); output
end
end

Example

$\mathbf{x}$=(1,4,2,10,5,7,3,11,15), $\mathbf{x}_{()}$={1,2,3,4,5,7,10,11,15}, N=5, r=1, s=2, y(1)=1, y(2)=2, y(3)=(2+3)/2, y(4)=(3+4)/2 etc.

41.4.3 (r,s)-Fold Winsorized Filters

The r smallest values inside the window are replaced by $x_{(r+1)}$ and the s largest values inside the window are replaced by $x_{(N-s)}$

$$\mathrm{WinMean}(x=\{x(1),\ldots,x(N_d)\};r,s)=\frac{1}{N}(rx_{(r+1)}+\sum_{i=r+1}^{N-s}x_{(i)}+sx_{(N-s)}$$

Example (same as in 41.4.2.1)

y(1)=1, y(2)=2, y(3)=(1/5)(1×2+2+3+2×3)=(13/5), y(4)=(1/5)(1×3+3+4+2×4)=18/5 etc.

41.4.4 α-Trimmed Mean Filters

If r = s then $\alpha = j/N$, $0 \le j \le N/2$ is an integer, and the formula is

$$\alpha - \mathrm{TrMean}(\mathbf{x} = \{x(1),\ldots,x(N_d)\};\alpha) = \frac{1}{N-2\alpha N}\sum_{i=\alpha N+1}^{N-\alpha N} x_{(i)}$$

41.4.5 α-WinMean Filters (see 41.4.4)

$$\alpha - \mathrm{WinMean}(\mathbf{x} = \{x(1),\ldots,x(N_d)\};\alpha) = \frac{1}{N}\left(\alpha N x_{(\alpha N+1)} + \sum_{i=\alpha N+1}^{N-\alpha N} x_{(i)} + \alpha N x_{(N-\alpha N)}\right)$$

Note: a) (0,0)-fold trimmed, (0,0)-fold Winsorized, 0-trimmed, and 0-Winsorized mean filters are the same as the mean filter of the same window size; b) (k,k)-fold trimmed, (k,k)-fold Winsorized, 0.5-trimmed and 0.5-Winsorized mean filters are the same as the median filter of the same window size

Example

x=(4,14,18,40,10), N=5, $\mathbf{x}_{(\,)}$=(4,10,14,18,40) a) (2,1)-fold trimmed mean filter (14+18)/2=16, b) (1,2)-Winsorized mean filter (2×10+3×14)/5=12.4, c) 0.2-trimmed mean (14+18+40)/3=14, d) 0.2-Winsorized mean filter (2×10+14+2×18)/5=14

41.4.6 Modified Trimmed Mean Filters

A real valued constant q is first fixed and then all the samples in the range $[x_{(k+1)} - q, x_{(k+1)} + q]$ are averaged, and the results constitute the output

$$MTM(\mathbf{x} = \{x(1),\ldots,x(N_d)\}) = \frac{\sum_{i=1}^{N} a_i x(i)}{\sum_{i=1}^{N} a_i}, \quad a_i = \begin{cases} 1 & \text{if } |x(i) - x(k+1)| \le q \\ 0 & \text{otherwise} \end{cases}$$

Example

x={4,14,18,40,10}, $\mathbf{x}_{(\,)}$={4,10,14,18,40}, a) for $0 \le q < 4$ med is 14, b) for $4 \le q < 10$, (10 + 14 + 18)/3=14, c) for $10 \le q < 26$, (4 + 10 + 14 + 18)/4=11.5, d) mean (4 + 10 + 14 + 18 + 40)/5=17.2

41.4.6.1 Modified Trimmed Mean Algorithm

Inputs: data **x**={x(1),…,x(N_d)}
  moving window w_N, N=2k+1
  positive real number q
Output: **y**={y(1),…,(N_d)}
for i=1 to N_d
place window w_N at i
store the values in the window in
  x={x(1),…,x(N)}
find the median $x_{(k+1)}$ of **x**
Sum1=0
Sum2=0
for m=1 to N
if abs(x(m)–$x_{(k+1)}$)≤q then

Inputs: No of rows N_R × No of columns N_C (image)
  moving window w_N, N=2k+1
  positive real number q
Output: N_R×N_C (image)
for i=1 to N_R
for j=1 to N_C
place window w_N at (i,j)
store the values inside the window w_N in
  x={x(1),…,x(N)}
find the median $x_{(k+1)}$ of **x**
  Sum1=0
  Sum2=0
for m=1 to N

```
Sum1=Sum1+x(m)
Sum2=Sum2+1
end
y(i)=Sum1/Sum2; output
end
end
```

```
if abs(x(m)–x(k+1))≤q then
   Sum1=Sum1+x(m)
   Sum2=Sum2+1
end
y(i,j)=Sum1/Sum2; output
end
end
```

41.4.7 Double Window Modified Trimmed Mean Filters

The idea behind this filter is to suppress additive Gaussian noise with the large window and preserve the details by rejecting pixels that are far away from the median of the smaller window.

41.4.7.1 Double Window Modified Trimmed Mean Filters Algorithm

```
Inputs: data x={x(1),...,x(Nd)}
   moving window wN, N=2k+1
   moving window w1, M>N
   positive real number q
Output: y={y(1),...,y(Nd)}
for i=1 to Nd
place window wN and w1 at i
store values inside wN in
   x={x(1),...,x(N)}
find the median x(k+1) of x
sum1=0, sum2=0
store values inside w1 in
   z={x(1),...,x(M)}
for m=1 to M
if abs(x(m)–x(k+1))≤q then
   sum1=sum1+x(m)
   sum2=sum2+1
end
y(i)=sum1/sum2
end
end
```

```
Inputs: No of rows NR × No of columns NC (image)
   moving window wN, N=2k+1
   moving window w1, M>N
   positive real number q
Output: NR×NC (image)
for i=1 to NR
   for j=1 to NC
   place windows wN and w1 at (i,j)
   store the values inside wN in x={x(1),...,x(N)}
   find the median x(k+1) of x
   sum1=0, sum2=0
   store values inside w1 in
      z={z(1),...,z(M)}
   for m=1 to M
if abs(x(m)–x(k+1))≤q then
   sum1=sum1+x(m)
   sum2=sum2+1
end
y(i,j)=sum1/sum2
   end
   end
```

41.4.8 K-Nearest Neighbor Filter

The output is given by the mean of K, $1 \le K \le N$, samples whose values are closest to the value of the central value x^* inside the filter window.

Example

x={4,14,18,40,10}. Center sample is $x^*=18$ a) for K=1, 18 (itself) is the closest sample to 18 and output is 18/1=18 b) for K=2, 18 and 14 must be used and the output is (18+14)/2=16 c) for K=3, 18, 14 and 10 are closest and output is (18+14+10)/3=14.

41.4.8.1 K-Nearest Neighbor Filter Algorithm

```
Inputs: data x={x(1),...,x(Nd)}
   moving window wN, N=2k+1
   integer number K, 1≤K≤N
```

```
Inputs: No of rows NR × No of columns NC (image)
   moving window wN, N=2k+1
   integer number K, 1≤K≤N
```

Output: **y**={y(1),…,y(N_d)}
for i=1 to N_d
place window w_N at i
store values inside the window w_N
  and their differences from x^* in

$$\mathbf{z} = \{(x(1),|x(1)-x^*|),\cdots,(x(N),|x(N)-x^*|)\}$$

sort **z** with respect to $|x(i)-x^*|$ and store in

$$t = \{(x_{(1)},|x_{(1)}-x^*|),\cdots,(x_{(N)},|x_{(N)}-x^*|)\}$$

sum=0
for m=1 to K
sum=sum+$x_{(m)}$
end
y(i)=sum/K
end
end

Output: $N_R \times N_C$ (image)
for i=1 to N_R
for j=1 to N_C
  place windows w_N at (i,j)
store values inside w_N and their differences

  from x^* (pixel (i,j)) in

$$\mathbf{z} = \{(x(1),|x(1)-x^*|),\cdots,(x(N),|x(N)-x^*|)\}$$

sort **z** with respect to abs(x(i)–x^*), and store

the results in

$$t = \{(x_{(1)},|x_{(1)}-x^*|),\cdots,(x_{(N)},|x_{(N)}-x^*|)\}$$

sum=0
for m=1 to K
sum=sum+$x_{(m)}$
end
y(i,j)=sum/K
end
end

41.4.8.2 Modified K-Nearest Neighbor Filter

If we change the rule for a_i in 41.4.6 to

$$a_i = \begin{cases} 1 & \text{if } |x(i)-x^*| \le q \\ 0 & \text{otherwise} \end{cases}$$

If q is chosen to be twice the standard deviation σ of the noise, the filter is known as the *sigma filter.*

41.5 L-Filters (Order Statistic Filters)

41.5.1 Purpose of the Estimator

L-filters are running estimators making a compromise between a pure nonlinear operation (ordering) and pure operation (weighting).

41.5.2 L-Filters

$$L(x(1),\ldots,x(N);\boldsymbol{a}) = \sum_{i=1}^{N} a_i x_{(i)}, \quad \sum_{i=1}^{N} a_i = 1$$

41.5.3 L-Filters Algorithms

Inputs: data **x**={x(1),…,x(N_d)}
  moving window w_N, N=2k+1
  weight vector **a**={a_1,…,a_N}
Output: **y**={y(1),…,y(N_d)}
for i=1 to N_d

Inputs: No of rows N_R × No of columns N_C (image)
  moving window w_N, N=2k+1
  weight vector **a**={a_1,…,a_N}
Output: $N_R \times N_C$ (image)
for i=1 to N_R

```
place window w_N at i
store the data inside the window
  w_N in z={x(1),...,x(N)}
sort z and store in the vector q
  q={x_(1),...,x_(N)}
sum=0
for m=1 to N
sum=sum+a_m x_(m)
end
y(i)=sum; output
end
end
```

```
for j=1 to N_C
place window w_N at (i,j)
store values inside the window w_N in
  z={x(1),...,x(N)}
sort z and store in q={x_(1),...,x_(N)}
sum=0
for m=1 to N
sum=sum+a_m x_(m)
end
y(i,j)=sum; output
end
end
```

41.6 Weighted Median Filters

41.6.1 Purpose

With this filter we may emphasize the samples that for some reason are supposed to be more reliable, e.g., center sample x^*, and the emphasis is obtained by weighing them more heavily.

41.6.2 Repetition or Duplication Operator &

3&x=x,x,x

41.6.3 Multiset

A multiset is a collection of objects, where the repetition of objects is permitted, e.g., {3&2,1,2&4}={2,2,2,1,4,4}

41.6.4 Weighted Median Filters

WeightMed{x(1),...,x(N);**a**}=Med{a_1&x(1),...,a_N&x(N)}

41.6.5 Weighted Median Algorithm

```
Inputs: data x={x(1),...,x(N_d)}
  moving window w_N, N=2k+1
  weight vector a={a_1,...,a_N}
Output: y={y(1),...,y(N_d)}
```

$$\text{halfsum}=\left(\sum_{i=1}^{N} a_i\right)\Big/2$$

```
for i=1 to N_d
place window w_N  at i
store values inside the window and weight
in z={(x_1,a_1),...,(x_N,a_N)}
sort z with respect to x_i's and store in q
  q={(x_(1),a_(1)),...,(x_(N),a_(N))}
sum=0, m=1
repeat
sum=sum+a_(m)
m=m+1
```

```
Inputs: No of rows N_R × No of columns N_C (image)
  moving window w_N, N=2k+1
  weight vector a={a_1,...,a_N}
Output: N_R×N_C (image)
```

$$\text{halfsum}=\left(\sum_{i=1}^{N} a_i\right)\Big/2$$

```
for i=1 to N_R
 for j=1 to N_C
  place window w_N and corresponding weighs in
      z={(x_1,a_1),...,(x_N,a_N)}
  sort z with respect to x_i's and store in
  q={(x_(1),a_(1)),...,(x_(N),a_(N))}
   sum=0, m=1
   repeat
 sum=sum+a_(m)
   m=m+1
```

```
until sum≥halfsum
y(i)=x_(m-1)
end
end
```

```
until sum≥halfsum
y(i,j)=x_(m-1)
  end
  end
```

Example

x={4,14,18,40,10}, **a**={0.05,0.1,0.15,0.1,0.05}. After ordering **z**={(4,0.05),(10,0.05),(14,0.1),(18,0.15),(40,0.1)}. Halfsum=(0.05+0.05+0.1+0.15+0.1)/2=0.225. But (0.05+0.05+0.1+0.15)=0.35>0.225 and the filter output is 18.

41.7 Ranked-Order Filter

41.7.1 Purpose

These filters can be used in situations where the noise distribution is not symmetric, e.g., where there are more positive than negative impulses.

41.7.2 Ranked-Order Filter

$RO(x(1),\ldots,x(N);r)=x_{(r)},\ N=2k+1$

Note: When r=k+1 we obtain the median filter. If r<k+1 the filter introduces bias toward small values and toward large values if r>k+1.

Example

If **x**={4,14,18,40,10} and the ordered **x**-ordered={4,10,14,18,40}. Here N=5=2k+1 and k=2. The rth ranked-order filter gives the outputs: 2 for r=1, 10 for r=2, 14 for r=3, 18 for r=4, and 40 for r=5.

41.7.3 Ranked-Order Filter Algorithms

```
Inputs: data x={x(1),...,x(N_d)}
   moving window w_N, N=2k+1
   rank r
Output: y={y(1),...,y(N_d)}
for  i=1 to N_d
place window w_N  at i
store the values in the window w_N
   in z={x(1),...,x(N)}
sort z to q={x_(1),...,x_(N)}
y(i)=x_(r); output
end
end
```

```
Inputs: No. of rows N_R × No. of columns N_C (image)
   moving window w_N, N=2k+1
    rank r
Output: N_R×N_C (image)
for i=1 to N_R
for j=1 to N_C
place window w_N  at (i,j)
store values inside the window w_N in
      z={x(1),...,x(N)}
   sort z to q={x_(1),...,x_(N)}}
   y(i,j)=x_(r)
   end
   end
```

41.8 Separable Two-Dimensional Median Filters

41.8.1 Purpose

The filter has the same properties as the median filter but is much faster. It is accomplished by using two successive one-dimensional median filters of window size $M=2l+1$ ($M\times M=N$), the first along the rows and the second one along the columns of the so-obtained image.

41.8.2 Separable Two-Dimensional Median Filters

$$\text{SepMed}\begin{bmatrix} x(1,1) & x(1,2) & \cdots & x(1,M) \\ x(2,1) & x(2,2) & \cdots & x(2,M) \\ \vdots & \vdots & \vdots & \vdots \\ x(M,1) & x(M,2) & \cdots & x(M,M) \end{bmatrix} = \text{Med}\{\text{Med}\{x(1,1),\cdots,x(1,M)\},\cdots$$

$$\text{Med}\{x(M,1),\ldots,x(M,M)\}\}$$

41.8.3 Algorithm

Inputs: No. of rows N_R time No. of columns N_C; (image)
  Moving horizontal window $w_{1,M}$, $M=2l+1$
  Moving vertical window $w_{2,M}$, $M=2l+1$
Output: $N_R \times N_C$; (image)
Median filter the image by using horizontal moving window $w_{1,M}$ for each row
Store the resulting values as they are found in a column vector
Median filter the column vector using vertical moving window $w_{2,M}$

41.9 M-Filters

41.9.1 Purpose

M-estimators are generalizations of maximum likelihood estimators

41.9.2 M-Filters

$$M(x(1),\ldots,x(N_d);\ \rho) = \arg\min_{\theta\in\Theta} \sum_{i=1}^{N} \rho(x(i)-\theta)$$

41.9.3 M-Filter by the Partial Derivative ψ (with respect to θ) of ρ

The estimate $M(x(1),\ldots,x(N_d))$ satisfies the equation $\sum_{i=1}^{N} \psi(x(i)-\theta)=0$

41.9.4 ψ Function in Use

1. Skipped median, $\psi_{med(r)}(x) = \begin{cases} \sin(x) & 0 \le |x| < r \\ 0 & r \le |x| \end{cases}$

2. Andrew's sine, $\psi_{\sin(a)}(x) = \begin{cases} \sin(xa) & 0 \le |x| < \pi a \\ 0 & \pi a \le |x| \end{cases}$

3. Tukey's biweight, $\psi_{bi(r)}(x) = \begin{cases} x(r^2 - x^2)^2 & 0 \le |x| < r \\ 0 & r \le |x| \end{cases}$

4. Standard M-filters, $\psi_{st}(x) = \begin{cases} ap & x > p \\ ax & |x| \le p \\ -ap & x < -p \end{cases}$

Note: The ψ function is not strictly monotone and 41.9.3 might not have a unique solution. This can be handled in several ways:

1. Find the global minimum of $\sum_{i=1}^{N} \rho(x(i)-\theta)$
2. Select the solution of $\sum_{i=1}^{N} \psi(x(i)-\theta)=0$ nearest to the sample median
3. Use Newton's method, starting from the sample median
4. Use one-step method, starting from the sample median

Example

x={2,7,9,20,5} a) Consider standard M-filter with p=5 and a=1/5. b) Consider Tukey's beweight M-filter with r=5. If we plot the function

$$\sum_{i=1}^{N} \psi(x(i)-\theta)=\psi(2-\theta)+\psi(7-\theta)+\psi(9-\theta)+\psi(20-\theta)+\psi(5-\theta)=0 \text{ versus } \theta$$

we find that the standard filter has a zero at $\theta=7$ which is the output. However, we have various candidates for the output of the Tukey's biweight M-filter. Which one is found depends on the method used.

41.9.5 M-Filter Algorithm

Inputs: No. of rows $N_R \times$ No. of columns N_C; (image)
moving window w_N, N=2k+1
function $\psi(x)$
constant real number $c\neq 0$
number of iterations N_I
Output: $N_R \times N_C$; (image)
for i=1 to N_R
for i=1 to N_C
place window w_N at (i,j)
store values inside the window w_N in
z={x(1),x(2),…,x(N)}
find the median $\theta(0)$ of the value inside the window w_N
for k=1 to N_I
sum1=0, sum2=0
for l=1 to N
if $x(l) \neq \theta(k-1)$
sum1=sum1 $+ x(l)(\psi(x(l)-\theta(k-1)))/(x(l)-\theta(k-1))$
sum2=sum2 $+ (\psi(x(l)-\theta(k-1)))/(x(l)-\theta(k-1))$
else
sum1=sum1+cx(l)
sum2=sum2+c
end
θ(k)=sum1/sum2
end
y(i,j)=θ(k)
end
end

41.10 R-Filters

41.10.1 Purpose

R-estimators are known to be robust.

41.10.2 Wilcoxon Filter (R-Filter) with Walsh Averages

$$\mathrm{Wil}(x(1),x(2),\cdots,x(N_d)) = \mathrm{Med}\left\{\frac{x(i)+x(j)}{2} : 1 \le i \le j \le N\right\}$$

41.10.3 Wilcoxon Filter (R-Filter) with Ordered Walsh Averages

$$\mathrm{Wil}(x(1),x(2),\cdots,x(N_d)) = \mathrm{Med}\left\{\frac{x_{(i)}+x_{(j)}}{2} : 1 \le i \le j \le N\right\}$$

Example

x={4,14,18,40,10}. The averages are Med{(4+4)/2, (4+14)/2, (4+18)/2, (4+40)/2, (4+10)/2, (14+14)/2, (14+18)/2, (14+40)/2, (14+10)/2, (18+18)/2, (18+40)/2, (18+10)/2, (40+40)/2, (40+10)/2, (10+10)/2}=14

41.10.4 Wilcoxon Filter Algorithms

```
Inputs: x={x(1),...,x(N_d)}; data
   moving window w_N, N=2k+1
Output: y={y(1),...,y(N_d)}
for  i=1 to N_d
place window w_N  at i
store values inside the window in
   z={x(1),...,x(N)}
m=1
for s=1 to N
for t=s to N
Ave(m)=(x(s)+x(t))/2
m=m+1
end
end
Med{Ave(1),Ave(2),...,Ave(m–1)}
y(i)=Med
end
end
```

```
Inputs: No. of rows N_R × No. of columns N_C (image)
   moving window w_N, N=2k+1
Output: N_R×N_C (image)
for i=1 to N_R
for j=1 to N_C
place the window w_N  at (i,j)
store values inside the window in
   z={x(1),...,x(N)}
m=1
for s=1 to N
for t=s to N
Ave(m)=(x(s)+x(t))/2
   m=m+1
   end
   end
   Med{Ave(1),...,Ave(m–1)}
y(i,j)=Med
end
end
```

41.10.5 Winsorized Wilcoxon Filters

$$\mathrm{Wins-Wil}(x(1),\ldots,x(N_d);r) = \mathrm{Med}\left\{\frac{x_{(i)}+x_{(j)}}{2} : j-i<r,\ 1 \le i \le j \le N\right\},\ 1 \le r \le N$$

Example

x={4,14,18,40,10}, r=2. Hence $\mathbf{x}_{(\,)}$={4,10,14,18,40} and the average values are: (4+4)/2, (4+10)/2, (10+10)/2, (10+14)/2, (14+14)/2, (14+18)/2, (18+18)/2, (18+40)/2, (40+40)/2, or Med{4,7,10,12,14,16,18,28,40)=14.

41.10.6 Hodge–Lehmann D-Filter (related to Wilcoxon)

$$\mathrm{Ho-Leh}(x(1),\cdots,x(N_d)) = \mathrm{Med}\left\{\frac{x_{(i)} + x_{(N-i+1)}}{2} : 1 \le i \le k+1,\ N = 2k+1\right\}$$

References

Astola, J. and P. Kuosmanen, *Fundamentals of Nonlinear Digital Filtering*, CRC Press, Boca Raton, FL, 1997.

Huber P. J., *Robust Statistics*, John Wiley & Sons, New York, NY, 1981.

Pitas, I. and A. N. Venetsanopoulos, *Nonlinear Digital Filters: Principles and Applications*, Kluwer Academic Publishers, Boston, MA, 1990.

42

Wavelet Transform

Yunlong Sheng

42.1 Continuous Wavelet Transform

42.1.1 Definition

The wavelet transform of a square-integrable function *f(t)* is defined by

$$W_f(a,b) = \int_{-\infty}^{\infty} f(t)\psi_{a,b}^{*}(t)\,dt$$

where * denotes the complex conjugate, and the wavelet basis $\psi_{a,b}(t)$ is generated from a basis (mother) wavelet $\psi(t)$ by dilations and translations:

$$\psi_{a,b}(t) = \frac{1}{a}\psi\left(\frac{t-b}{a}\right)$$

where a is the scale factor, b is the translation factor, and $1/\sqrt{a}$ is the normalization factor for the normalization in terms of energy. It becomes $1/a$ for the normalization in terms of amplitude.

42.1.2 Time-Scale Joint Representation

The wavelet transform is a mapping of a 1-D time signal into a 2-D time-scale joint representation.

42.1.3 Admissible Conditions

The wavelets must be square integrable and satisfy the admissible condition:

$$\int_{-\infty}^{\infty} \psi(t)\,dt = 0 \qquad \text{or} \qquad \Psi(\omega)\big|_{\omega=0} = 0$$

where $\Psi(\omega)$ is the Fourier transform of $\psi(t)$. The admissible condition must be satisfied for the recovery of *f(t)* by the inverse wavelet transform.

42.1.4 Inverse Wavelet Transform

$$f(t) = \frac{1}{c_h} \iint_{-\infty}^{\infty} W_f(a,b) \frac{1}{\sqrt{a}} \psi\left(\frac{t-b}{a}\right) db \frac{da}{a^2}$$

where

$$c_h = \int_{-\infty}^{\infty} \frac{|\Psi(\omega)|^2}{|\omega|} d\omega < +\infty$$

and $\Psi(\omega)$ is the Fourier transform of the mother wavelet.

42.1.5 Resolution of Identity

Relation between the energy measures in the time-scale domain and the time domain:

$$\frac{1}{c_h} \int_{-\infty}^{\infty} \frac{da}{a^2} \int db < f_1, \psi_{a,b} > < \psi_{a,b}, f_2 > = < f_1, f_2 >$$

where < , > denotes the inner product. The resolution of identity leads to perfect recovery of *f(t)* by the inverse wavelet transform (42.1.4).

42.1.6 Regularity

a) It is in general required that the wavelets have the first $n + 1$ order vanishing moments:

$$M_p = \int_{-\infty}^{\infty} t^p \psi(t) dt = 0 \qquad \text{for} \qquad p = 0, 1, 2, \ldots, n$$

b) Equivalently, the derivatives of orders up to n of the Fourier transform of the wavelet are equal to zero at $\omega = 0$:

$$\Psi^{(p)}(0) = 0 \qquad \text{for} \qquad p = 0, 1, 2, \ldots, n$$

From the admissible condition (42.1.3), the zero-order moment of any wavelet must be zero.

c) Application of the wavelet transform to the Taylor series of *f(t)* yields

$$W_f(a,0) = \frac{1}{\sqrt{a}} \left[f(0) M_0 a + \frac{f'(0)}{1!} M_1 a^2 + \frac{f''(0)}{2!} M_2 a^3 + \ldots + \frac{f^{(n)}(0)}{n!} M_n a^{n+1} + O\left(a^{n+2}\right) \right]$$

Thus, the wavelet transform coefficient $W_f(a,b)$ decays as fast as a^{n+1} for a smooth signal *f(t)*.

d) Wavelet transform of regular signals: According to (42.1.3) and (42.1.6c), the wavelet transform of a constant is zero. The wavelet transform of a linear signal is zero, if the wavelet has the first-order moment vanishing: $M_1 = 0$. The wavelet transform of a quadratic signal is zero, if the wavelet has the first and second order vanishing moments: $M_1 = M_2 = 0$. The wavelet transform of a polynomial signal of degree m is zero, if the wavelet has the vanishing moments up to the order $n \geq m$.

The wavelet transform is efficient for detecting singularities and analyzing nonstationary, transient signal.

42.1.7 Localization

According to the admissible condition, the wavelet must oscillate to have a zero mean. According to the regularity the wavelet must decay as fast as t^{-n} along the time axis. As a result, the wavelet must be a "small" wave, localized in the time domain. Also, according to the regularity, $W_f(a,b)$ decays as fast as a^{n+1}, so that the wavelet transform is localized in the scale domain.

42.1.8 Scale and Resolution

The scale is related to the window size of the wavelet. A large scale means a global view, and a small scale means a detailed view. The resolution is related to the frequency of the wavelet oscillation. For some wavelets, such as the Gabor-wavelets (42.4.1), the scale and frequency may be chosen separately. For a given wavelet function, reducing the scale will reduce the window size and increase the resolution in the same time.

42.1.9 Bank of Filters

a) Bandpass filters: In the Fourier domain, the wavelet transform is

$$W_f(a,b) = \frac{\sqrt{a}}{2\pi}\int_{-\infty}^{\infty} F(\omega)\Psi^*(a\omega)\exp(j\omega b)\,d\omega$$

From the admissible condition, $\Psi(0) = 0$. We want $\Psi(\omega)$ to vanish above a certain frequency by imposing the regularity on the wavelets. Thus, $\Psi(\omega)$ is intrinsically a bandpass filter.

b) The wavelet transform is a bank of multiresolution bandpass filters.

42.1.10 Uncertainty Principle

a) Multiplying a signal by a window localizes the signal in the time domain, but blurs its frequency spectrum. Let Δt denote the width of a time window $g(t)$ and $\Delta\omega$ the bandwidth of its Fourier transform $G(\omega)$, and define:

$$\Delta t^2 = \frac{\int t^2 |g(t)|^2\,dt}{\int |g(t)|^2\,dt} \qquad \Delta\omega^2 = \frac{\int \omega^2 |G(\omega)|^2\,d\omega}{\int |G(\omega)|^2\,d\omega}$$

then,

$$\Delta t\,\Delta\omega \geq \frac{1}{2}$$

b) The Gaussian window has the minimum time-bandwidth product determined by the uncertainty principle.

42.1.11 Constant Fidelity Analysis

The wavelet multiresolution filters divide the signal into frequency subbands. The relative bandwidths (bandwidth $(\Delta\omega)_a$ divided by $(1/a)$) of the wavelets are constant:

$$(\Delta\omega)_a^2 = \frac{\int \omega^2 |\Psi(a\omega)|^2 d\omega}{\int |\Psi(a\omega)|^2 d\omega} = \frac{1}{a^2}(\Delta\omega)^2$$

42.2 Multiresolution Wavelet Decomposition

42.2.1 Discrete Wavelet Transform

The continuous wavelet transform is highly redundant. The discrete wavelet transform is evaluated with a discretely scaled and translated continuous wavelet basis. The discrete wavelet transform with the dyadic scaling factor $a = 2^i$ is effective for computer implementation.

42.2.2 Multiresolution Analysis

The multiresolution analysis allows analyzing signal in the frequency subbands. A multiresolution analysis is based on the scaling function.

42.2.3 Scaling Function

The scaling function is a continuous square integrable function, whose mean value is not equal to zero. The scaling function plays a role of an averaging function or of a lowpass filter in the multiresolution analysis.

42.2.4 Scaling Function Bases $\{\phi_i(t-k)\}$

The scaling function $\phi(t)$ is shifted by discrete translations and is dilated by dyadic scale factors:

$$\phi_{i,k}(t) = 2^{-i/2}\phi(2^{-i}t - k) \qquad \text{for } i \in Z$$

where $2^{-i/2}$ is a normalization constant. At each scale 2^i, the shifted scaling functions constitute a basis that spans a subspace V_i.

42.2.5 Wavelet Bases $\{\psi_i(t-k)\}$

Similarly, the wavelet is shifted and dilated as

$$\psi_{i,k}(t) = 2^{-i/2}\psi(2^{-i}t - k) \qquad \text{for } i \in Z$$

At each scale 2^i, the shifted wavelets constitute a basis that spans a subspace W_i.

42.2.6 Two-Scale Relation

The two-scale relation is the basic relation in the multiresolution analysis with dyadic scaling:

a) A scaling function may be decomposed as a linear combination of the scaling functions at the next higher resolution level

$$\phi(t) = \sum_k 2h(k)\phi(2t - k)$$

where the discrete sequence of coefficients $h(k)$ are useful in the wavelet decomposition as the lowpass filter.

b) Similarly, the wavelet may be decomposed as a linear combination of scaling functions at the next higher resolution level:

$$\psi(t) = \sum_k 2g(k)\phi(2t - k)$$

where the coefficients *g(k)* are useful in wavelet decomposition as the highpass filter.

42.2.7 Transfer Functions

a) The Fourier transform of the discrete lowpass filter *h(k)* is

$$H(\omega) = \sum_k h(k)\exp(-jk\omega)$$

where $j = \sqrt{-1}$, which is the z-transform of *h(k)* by the Laurent polynomial

$$H(z) = \sum_k h(k)\, z^{-k}$$

where $z = \exp(j\omega)$. $H(\omega)$ is a continuous complex valued function of 2π period.

b) The Fourier transform of the discrete highpass filter *g(k)* is

$$G(\omega) = \sum_k g(k)\exp(-jk\omega)$$

which is the z-transform of *g(k)*

$$G(z) = \sum_k g(k)\, z^{-k}$$

$G(\omega)$ is a complex-valued continuous function of 2π period.

42.2.8 Two-Scale Relations in the Frequency Domain

a) In the Fourier domain the two-scale relation becomes:

$$\Phi(\omega) = H\left(\frac{\omega}{2}\right)\Phi\left(\frac{\omega}{2}\right)$$

$$\Psi(\omega) = G\left(\frac{\omega}{2}\right)\Phi\left(\frac{\omega}{2}\right)$$

b) The two-scale relation is recursive. The recursion may be repeated infinitely, resulting in

$$\Phi(\omega) = \prod_{i=1}^{\infty} H\left(\frac{\omega}{2^i}\right)$$

$$\Psi(\omega) = G\left(\frac{\omega}{2}\right)\prod_{i=2}^{\infty} H\left(\frac{\omega}{2^i}\right)$$

42.2.9 Orthogonal Multiresolution Bases

a) Orthonormal scaling function basis: The scaling function can be orthogonal to its discrete translates at the same scale:

$$\langle \phi_k, \phi_n \rangle = \int_{-\infty}^{\infty} \phi(t-k)\phi(t-n)dt = \delta_{k,n} \qquad k,n \in Z$$

b) Orthonormal wavelet basis: The wavelet can be orthogonal to its discrete translates at the same scale:

$$\langle \psi_k, \psi_n \rangle = \int_{-\infty}^{\infty} \psi(t-k)\psi(t-n)dt = \delta_{k,n} \qquad k,n \in Z$$

Furthermore, the wavelets basis can be orthogonal to the scaling function basis at the same scale:

$$<\phi_k, \psi_n> = 2^{-i}\int_{-\infty}^{\infty} \phi(t-k)\psi(t-n)dt = 0 \qquad \text{for all } k,n \in Z$$

Thus, the subspaces are orthogonal, $V_i \perp W_i$.

42.2.10 Orthogonal Wavelet Decomposition

a) Recursive projections: Let P_i and Q_i denote the projection operators onto the orthonormal scaling function and wavelet bases at scale 2^i, respectively. The projection of $f(t)$ onto the scaling function basis at scale 2^{i-1} is then decomposed as

$$P_{i-1}f = \langle \phi_{i-1,n}, f \rangle = P_i f + Q_i f = \sum_k c_i(k)\phi_{i,k} + \sum_k d_i(k)\psi_{i,k}$$

Thus, W_i is the orthogonal complement of V_i inside V_{i-1}:

$$V_{i-1} = V_i \oplus W_i$$

where $c_i(k)$ are the approximate signal and $d_i(k)$ are the wavelet transform coefficients that represent the difference between the approximate signal and the original signal.

b) Double-shift correlation: Multiplying both sides of (42.2.10a) by $\phi_{i,n}$ and computing the inner products yields

$$c_i(k) = \langle \phi_{i,k}, P_{i-1}f \rangle = 2^{1/2}\sum_n h(n-2k)c_{i-1}(n)$$

and

$$d_i(k) = \langle \psi_{i,k}, P_{i-1,k} \rangle = 2^{1/2}\sum_n g(n-2k)c_{i-1}(n)$$

since it is easy to verify from the two-scale relation (42.2.6), the orthonormality (42.2.9) that for the bases defined in (42.2.4) and (42.2.5)

$$< \phi_{i,k}, \phi_{i-1,n} > = 2^{1/2} h(n - 2k)$$

$$< \psi_{i,k}, \phi_{i-1,n} > = 2^{1/2} g(n - 2k)$$

Note that the filters $h(n)$ and $g(n)$ are shifted by even integers. The double-shift implies the downsampling by two (decimation) of the correlation results.

c) Orthogonal wavelet series decomposition: The recursive projection (42.2.10a) continues until level $i = M$, where M is the coarsest resolution

$$f(t) = P_M L + Q_M f + Q_{M-1} f + \cdots + Q_1 f$$

and

$$f(t) = \sum_{k \varepsilon Z} 2^{M/2} c_M(k)\, \phi\left(2^{-M} t - k\right) + \sum_{i=1}^{M} \sum_{k \varepsilon Z} 2^{i/2} d_i(k)\, \psi\left(2^{-i} t - k\right)$$

d) Tree algorithm:

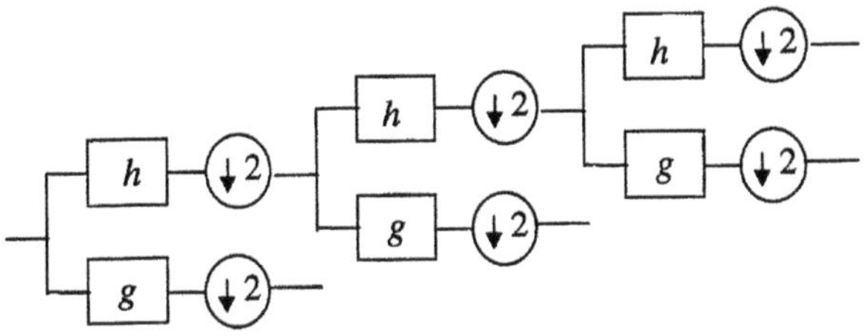

42.2.11 Recursive Reconstruction

a) Multiplying both sides of (42.2.10a) by $\phi_{i-1,n}$ and computing the inner products yields

$$c_{i-1}(n) = < P_{i-1} f, \phi_{i-1,n} > = 2^{1/2} \sum_k c_i(k) h(n - 2k) + 2^{1/2} \sum_k d_i(k) g(n - 2k)$$

On the right-hand side, the two terms are double-shift convolutions of $c_i(k)$ and $d_i(k)$ with the filters $h(n)$ and $g(n)$, respectively. $c_i(k)$ and $d_i(k)$ must be filled with zeros between each component. (upsampling or expansion) before convolving with the filters.

b) The step can be iterated until the finest scale to find $c_0(n)$. If the operators and L^* and H^*, applied to a sequence of data $\alpha(k)$, are defined as

$$\left(L^* \alpha\right)(n) = \sqrt{2} \sum_k h(n - 2k) \alpha(k)$$

$$\left(H^* \alpha\right)(n) = \sqrt{2} \sum_k g(n - 2k) \alpha(k)$$

then

$$c_0 = \sum_{i=1}^{M} \left(L^*\right)^{i-1} H^* d_i + \left(L^*\right)^M c_M$$

42.3 Filter Bank

The wavelet multiresolution filter bank permits computing the dyadic wavelet transform. The filter bank involves the Finite Impulse Response (FIR) filters. The properties of the filters may be represented equivalently in the time domain and in the frequency or z-transform domain.

The relations between the filter bank and the scaling functions and wavelets are given in the two-scale relations (42.2.6). The continuous scaling function and wavelet bases may be obtained by infinitely iterating the filter bank.

42.3.1 Two Channel Filter Bank

The two-channel filter bank, shown in Figure 42.2, is a building block of the discrete wavelet transform and of the subband coding. The lowpass $\tilde{h}(n)$ and highpass $\tilde{g}(n)$ filters are the analysis filters, which are followed by downsampling. The filters $h(n)$ and $g(n)$ are the synthesis filters for reconstruction of signal. There is upsampling before the synthesis filters.

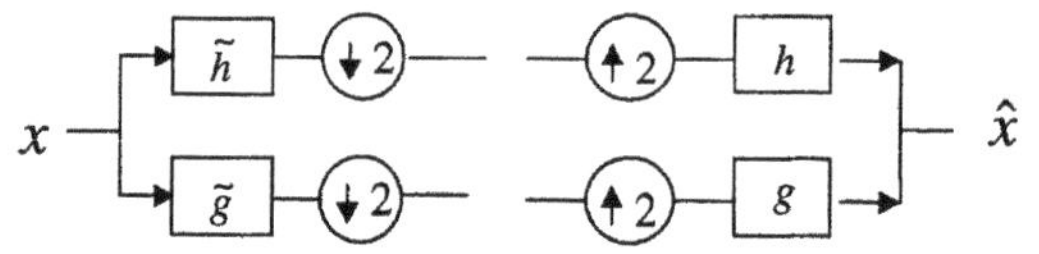

42.3.2 Downsampling

Downsampling (decimation) is to save the even-numbered components of data. Discarding the odd-numbered components leads to a loss in information. In the frequency domain this is aliasing error. *Upsampling* (expansion) is to insert zeros as the odd-numbered components into the data. Both processes are not shift-invariant and not invertible.

42.3.3 Perfect Reconstruction

a) Aliasing error introduced by the downsampling: The filters of $\tilde{H}(z)$ and $\tilde{G}(z)$, followed by downsampling and upsampling, keep only the even powers of $\tilde{H}(z)X(z)$ and of $\tilde{G}(z)X(z)$, resulting in

$$\frac{1}{2}\left[\tilde{H}(z)X(z) + \tilde{H}(-z)X(-z)\right]$$

and

$$\frac{1}{2}\left[\tilde{G}(z)X(z) + \tilde{G}(-z)X(-z)\right]$$

b) The alias terms $\tilde{H}(-z)X(-z)$ and $\tilde{G}(-z)X(-z)$ in (42.3.3.a) can be cancelled by the synthesis filters *H(z)* and *G(z)*, if the alias cancellation equation is satisfied:

$$H(z)\tilde{H}(-z^{-1}) + G(z)\tilde{G}(-z^{-1}) = 0$$

c) The building block shown in (42.3.1) would behave like an identity operation, $\hat{x}(n) = x(n)$, if the perfect reconstruction condition is satisfied:

$$H(z)\tilde{H}(z^{-1}) + G(z)\tilde{G}(z^{-1}) = 2$$

d) Modulation matrix: The conditions (42.3.3.b and c) may be summarized to

$$\tilde{M}(z^{-1})^t M(z) = 2I$$

where I is the 2×2 identity matrix and *M(z)* is the modulation matrix

$$M(z) = \begin{bmatrix} H(z) & H(-z) \\ G(z) & G(-z) \end{bmatrix}$$

42.3.4 Orthogonal Filter Bank

a) Paraunitary filter bank: When the synthesis filters are the same as the analysis filters, $H(z) = \tilde{H}(z)$ and $G(z) = \tilde{G}(z)$, and the perfect reconstruction is satisfied so that

$$M(z^{-1})^t M(z) = 2I \qquad \text{and} \qquad M(z^{-1}) M(z)^t = 2I$$

The filter bank is said to be orthogonal. The modulation matrix defined in (42.3.3.d) is a paraunitary matrix times by $\sqrt{2}$.

b) From (42.3.4.a) it follows that both *H(z)* and *G(z)* are halfband filters (see (42.3.7.c)):

$$|H(z)|^2 + |H(-z)|^2 = 2 \qquad \text{or} \qquad |H(\omega)|^2 + |H(\omega + \pi)|^2 = 2$$

$$|G(z)|^2 + |G(-z)|^2 = 2 \qquad \text{or} \qquad |G(\omega)|^2 + |G(\omega + \pi)|^2 = 2$$

c) (42.3.4.a) gives also the cross-orthogonality

$$H(z^{-1})G(z) + H(-z^{-1})G(-z) = 0$$

$$G(z^{-1})H(z) + G(-z^{-1})H(-z) = 0$$

d) Alternating flip: It is easy to verify that

$$G(z) = z^{-N} H(-z^{-1})$$

satisfies (42.3.4.c), where N is an arbitrary odd number.

e) Quadrature mirror filters (QMF): The pair of filters $H(\omega)$ and $G(\omega)$ are referred to as the quadrature mirror filters. The property of mirror filter about $\omega = \pi/2$ is shown from (42.3.4.d) as

$$|G(z)|^2 = |H(-z)|^2 \qquad \text{or} \qquad |G(\omega)|^2 = |H(\omega + \pi)|^2$$

f) From (42.3.4.e) and (42.3.4.b) it follows that $H(\omega)$ and $G(\omega)$ are complementary:

$$|H(z)|^2 + |G(z)|^2 = 2 \qquad \text{or} \qquad |H(\omega)|^2 + |G(\omega)|^2 = 2$$

$$|H(-z)|^2 + |G(-z)|^2 = 2 \qquad \text{or} \qquad |H(\omega + \pi)|^2 + |G(\omega + \pi)|^2 = 2$$

g) Since $H(\omega)$ is a lowpass filter, then

$$H(\omega)\big|_{\omega=0} = \sqrt{2} \qquad \text{and from (42.3.4.e)} \qquad G(\pi) = \sqrt{2}$$

Since $G(\omega)$ is a highpass filter

$$G(\omega)\big|_{\omega=0} = 0 \qquad \text{and from (42.3.4.e)} \qquad H(\pi) = 0$$

42.3.5 Orthogonal Filters in Time Domain

a) Double-shift orthonormality

$$2\sum h(n)\,h(n - 2k) = \delta(k)$$

$$2\sum g(n)\,g(n - 2k) = \delta(k)$$

b) Cross-filter orthogonality

$$\sum h(n)\,g(n - 2k) = 0$$

c) Even filter length: The double-shift orthonormality implies that the filters *h(n)* and *g(n)* must have even lengths.

d) Alternating flip: The highpass filter generated from the lowpass filter as

$$g(k) = (-1)^k h(N - k) \qquad k = 0,1,\ldots N$$

satisfies the cross-filter orthogonality (42.3.4.c), where N comes from the length (N+1) of the FIR filters, or is any odd number. The (42.3.5.d) is the inverse z-transform of (42.3.4.d)

42.3.6 Example

In an example of the matrix operations with 4-tips filters and the highpass filter generated by (42.3.5.d) and without considering the causality of time signal, the analysis of a signal *x(n)* is computed as:

$$
\begin{bmatrix} c(0) \\ d(0) \\ c(1) \\ d(1) \\ \vdots \\ \\ \vdots \\ \\ \\ \\ \end{bmatrix}
=
\begin{bmatrix}
h(0) & h(1) & h(2) & h(3) & & & & \\
h(3) & -h(2) & h(1) & -h(0) & & & & \\
0 & 0 & h(0) & h(1) & & \vdots & & \\
0 & 0 & h(3) & & & & & \\
 & & & & \cdots & & & \\
\vdots & & & & & & & \\
 & & & & h(0) & h(1) & h(2) & h(3) \\
 & & & & h(3) & -h(2) & h(1) & -h(0) \\
h(2) & h(3) & & & 0 & 0 & h(0) & h(1) \\
h(1) & -h(0) & & & 0 & 0 & h(3) & -h(2)
\end{bmatrix}
\begin{bmatrix} x(0) \\ x(1) \\ x(2) \\ \\ \vdots \\ \\ \\ \vdots \\ x(N-1) \\ x(N) \end{bmatrix}
$$

For iteration the output vector is permutated and only the half vector *c(n)* is again multiplied with the matrix **T**. When $N = 15$ the iterations are:

$$
\begin{bmatrix} x(0) \\ x(1) \\ x(2) \\ x(3) \\ \vdots \\ \\ \vdots \\ x(15) \end{bmatrix}
\xrightarrow{\mathbf{T}}
\begin{bmatrix} c(0) \\ d(0) \\ c(1) \\ d(1) \\ \vdots \\ \\ \vdots \\ c(7) \\ d(7) \end{bmatrix}
\xrightarrow{P}
\begin{bmatrix} c(0) \\ c(1) \\ c(2) \\ \vdots \\ \vdots \\ c(7) \\ d(0) \\ d(1) \\ \vdots \\ \vdots \\ d(7) \end{bmatrix}
\xrightarrow{\mathbf{T}}
\begin{bmatrix} c_1(0) \\ d_1(0) \\ c_1(1) \\ d_1(1) \\ c_1(2) \\ d_1(2) \\ \vdots \\ d(0) \\ d(1) \\ \vdots \\ \vdots \\ d(7) \end{bmatrix}
\xrightarrow{P}
\begin{bmatrix} c_1(0) \\ c_1(1) \\ c_1(2) \\ \vdots \\ d_1(0) \\ d_1(1) \\ \vdots \\ d_1(3) \\ d(0) \\ d(1) \\ \vdots \\ \vdots \\ d(7) \end{bmatrix}
\xrightarrow{\mathbf{T}}
\begin{bmatrix} c_2(0) \\ d_2(0) \\ c_2(1) \\ d_2(1) \\ d_1(0) \\ d_1(1) \\ d_1(2) \\ d_1(3) \\ d(0) \\ d(1) \\ \vdots \\ \vdots \\ d(7) \end{bmatrix}
\xrightarrow{P}
\begin{bmatrix} c_2(0) \\ c_2(1) \\ d_2(0) \\ d_2(1) \\ d_1(0) \\ d_1(1) \\ d_1(2) \\ d_1(3) \\ d(0) \\ d(1) \\ \vdots \\ \vdots \\ d(7) \end{bmatrix}
$$

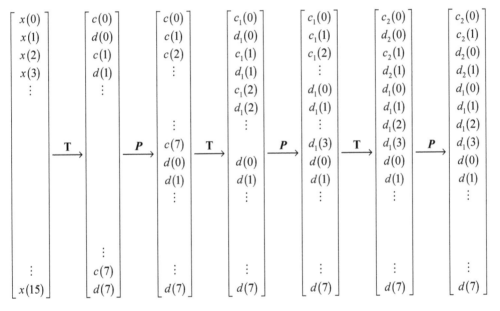

where P denotes the permutation. It is easy to verify that **T** is orthogonormal, then $\mathbf{T}^{-1} = \mathbf{T}'$, so that the synthesis filter matrix $\mathbf{T}'$ yields the perfect reconstruction.

42.3.7 Biorthogonal Filter Bank

In the two-channel filter bank (42.3.1) the synthesis filters may be different from the analysis filters; that brings more freedom in the filter design. In this case, the perfect reconstruction conditions (42.3.3.b and c) lead to the biorthogonal filter bank.

a) The choice

$$H(z) = -z^{-1}\,\tilde{G}\left(-z^{-1}\right)$$

$$G(z) = z^{-1}\,\tilde{H}\left(-z^{-1}\right)$$

satisfies the alias cancellation equation (42.3.3.b). Thus, the synthesis filters are associated to the analysis filters.

b) Product filter *P(z)*: The perfect reconstruction equation in (42.3.3.c) becomes

$$P(z) + P(-z) = 2 \qquad \text{with} \qquad P(z) = H(z)\tilde{H}(z^{-1})$$

The product *P(z)* is a half-band filter.

c) Halfband filters, defined in (42.3.7.b), have all their even powers of z equal to zero, except the zero power term (= 1). The odd powers of z in *P(z)* cancel with that in *P(–z)*. The coefficients of a half-band filter are:

$$p(2m) = \begin{cases} 1 & \text{when} \quad m = 0 \\ 0 & \text{when} \quad m \neq 0 \end{cases} \qquad \text{and} \qquad p(n) = -p(-n) \qquad \text{when} \quad n \text{ is odd}$$

in the frequency domain $P(e^{j\omega})$ is symmetric with respect to $\omega = \pi/2$.

42.3.8 Biorthogonal Filters in Time Domain

Biorthogonality: Each filter is not orthogonal to its own double integer translates. However, the filters are double-shift orthonormal to their respective duals:

$$2\sum_k \tilde{h}(k)h(k + 2n) = \delta(n)$$

$$2\sum_k \tilde{g}(k)g(k + 2n) = \delta(n)$$

and the double-shift cross-correlations between the synthesis filters and the analysis filters are zero:

$$\sum_k \tilde{h}(k)g(k + 2n) = 0$$

$$\sum_k \tilde{g}(k)h(k + 2n) = 0$$

Those equations are equivalent to the perfect reconstruction conditions (42.3.9.b and c).

42.3.9 Biorthogonal Wavelet Decomposition

Similar to the orthogonal decomposition in (42.2.10.b), but with the analysis filters $\tilde{h}(n)$ and $\tilde{g}(n)$:

$$c_i(k) = 2^{1/2}\sum_n \tilde{h}(n - 2k)\, c_{i-1}(n)$$

and

$$d_i(k) = 2^{1/2}\sum_n \tilde{g}(n - 2k)\, c_{i-1}(n)$$

42.3.10 Biorthogonal Wavelet Reconstruction

Similar to (42.2.11) but with the synthesis filters $h(n)$ and $g(n)$

$$\cdot\ c_{i-1}(n) = 2^{1/2}\sum_k c_i(k)h(n-2k) + 2^{1/2}\sum_k d_i(k)g(n-2k)$$

42.3.11 Scaling Function and Wavelets

The continuous scaling function and the wavelets can be obtained from the discrete filters $h(n)$ and $g(n)$ by infinitely iterating the filter bank: Assume that the input is the scaling function at scale 2^M itself, then $c_M(n) = \delta_{0,n}$ and $d_0 = \cdots = d_M = 0$. The scaling function may be reconstructed using the infinite recursion of the low-pass filtering, as shown in (42.2.11). The infinite recursion converges in the condition that the scaling functions are regular.

42.3.12 Regularity of the Scaling Function

The regularity of the scaling function is defined as the value of r in

$$|\Phi(\omega)| \le \left(\frac{c}{(1+|\omega|)^r}\right)$$

where c is a constant. Since $|\Phi(\omega)|$ has an exponential decay as $\omega^{-M}$where $M \le r$, then $\phi(t)$ is $(M-1)$-times differentiable and both $\phi(t)$ and $\psi(t)$ are smooth.

42.3.13 Regularity of the Filters

According to the two-scale relation (42.2.8.b) the regularity of the scaling function defined in (42.3.12) is equivalent to the regularity of the filter in the form of

$$H(\omega) = \left(\frac{1+e^{-j\omega}}{2}\right)^M F\left(e^{-j\omega}\right)$$

where $F(e^{-j\omega})$ is a polynomial in $e^{-j\omega}$, such that the infinite product

$$\left|\prod_{k=0}^{\infty} F\left(e^{-j\omega/2^k}\right)\right|$$

is convergent and bounded. At $\omega = \pi$, the derivatives up to order $M-1$ of $H(\omega)$ are equal to zero. ($H(\omega)$ has M zeros). The value of M is a measure of flatness of $H(\omega)$. Note that $H(\omega)$ has at least one zero at $\omega = \pi$ as shown in (42.3.4g).

42.4 Some Wavelets

42.4.1 Gabor-Wavelets

a) Gabor function is the basis of the windowed Fourier transform:

$$G(t) = g(t-b)\exp(j\omega' t)$$

where *g(t)* is the Gaussian window shifted to b, which controls the position of the window. In the frequency domain, the Gabor function is also a Gaussian window shifted to ω′, which is the modulation parameter, controlling the position of the window. The Gabor basis has the minimum time-bandwidth product determined by the uncertainty principle (42.1.10).

b) Gabor-Wavelet is the basis of the dilated window Fourier transform:

$$\psi(t) = \frac{1}{\sqrt{a}} g\left(\frac{t-b}{a}\right) \exp(j\omega t)$$

c) Morlet Wavelet

$$\psi(t) = \frac{1}{\sqrt{a}} g\left(\frac{t-b}{a}\right) \exp(j\omega\, t/a)$$

The function is complex-valued, which satisfies the admissible condition (42.1.3) approximately.

42.4.2 Gaussian Wavelets

The Gaussian function is perfectly local in both time and frequency and is indefinitely differentiable. Derivative of any order n of the Gaussian function can be a wavelet. In the frequency domain the Gaussian wavelets are

$$\Psi(\omega) = (j\omega)^n \exp\left(-\frac{\omega^2}{2}\right)$$

Its derivatives of orders up to n–1 are zero at $\omega = 0$.

a) The first-order derivative of Gaussian is

$$\psi(t) = -t \exp\left(-\frac{t^2}{2}\right)$$

b) Mexican-hat wavelet is the second-order derivative of Gaussian:

$$\psi(t) = (t^2 - 1) \exp\left(-\frac{t^2}{2}\right)$$

42.4.3 Haar Basis

a) The scaling function in the Haar basis is a rectangle function with a compact support of [0,1) as

$$\phi(t) = \begin{cases} 1 & \text{for } 0 \le t < 1 \\ 0 & \text{otherwise} \end{cases}$$

It is orthogonal to its own integer translates and (42.2.9a) is satisfied. In the frequency domain the scaling functions are

$$\Phi(\omega) = \left(\frac{1 - e^{-j\omega}}{j\omega}\right)$$

whose modulus has an infinite support with the decay of ω^{-1}.

b) The Haar wavelet is an antisymmetric function in [0,1)

$$\psi(x) = \begin{cases} +1 & \text{for } \; 0 \le x < 1/2 \\ -1 & \text{for } \; 1/2 \le x < 1 \\ 0 & \text{otherwise} \end{cases}$$

It is orthogonal to its own integer translates and to the scaling functions, and (42.2.9b) is satisfied. In the frequency domain, the Haar wavelets are

$$\Psi(\omega) = 2e^{-j\omega/2}\left(\frac{1 - \cos\frac{\omega}{2}}{-j\omega}\right)$$

whose modulus has an infinite support with the decay of ω^{-1}.

c) The Haar lowpass filter is given as

$$H(\omega) = \left(\frac{1 + e^{-j\omega}}{2}\right)$$

since $\Phi(\omega) = H(\omega/2)\Phi(\omega/2)$, the two-scale relation (42.2.8a) is satisfied. According to (42.2.7), the coefficients of the lowpass filter are simply $h(0) = 1$ and $h(1) = 1$.

d) The Haar's highpass filter is then:

$$G(\omega) = -\left(\frac{1 - e^{-j\omega}}{2}\right)$$

since $\Psi(\omega) = G(\omega/2)\Phi(\omega/2)$, the two-scale relation (42.2.8.a) is satisfied. According to (42.2.7), the coefficients of the highpass filter are simply $g(0) = -1$ and $g(1) = 1$.

42.4.4 Daubechies Basis

a) The Daubechies wavelets are smooth, orthogonal, and compactly supported.

b) The lowpass filter takes the form shown in (42.3.13.)

$$H(\omega) = \left(\frac{1 + e^{-j\omega}}{2}\right)^M F\left(e^{-j\omega}\right)$$

where $F(e^{-j\omega})$ is a polynomial in $e^{-j\omega}$. Thus,

$$\left|H(\omega)\right|^2 = \left(\cos^2\frac{\omega}{2}\right)^M G\left(\sin^2\frac{\omega}{2}\right)$$

c) The length (support) N of the filter $h(n)$ is chosen first. The longer the filters, the more regular the filters, and the smoother the wavelets.

d) Then, $H(\omega)$ is a polynomial in $e^{-j\omega}$ of degree N–1 and $F(e^{-j\omega})$ is a polynomial in $e^{-j\omega}$ of degree N–1–M. From the orthogonality condition (42.3.4) the polynomial $F(e^{-j\omega})$ must have a minimum degree of (M–1). Thus $M = N/2$.

e) The $H(\omega)$ has the maximum flatness of value $M = N/2$. At $\omega = \pi$, $H(\omega)$ has vanishing derivatives of orders up to M ensured by the term $(1-e^{-j\omega}/2)^M$. At $\omega = 0$, $H(\omega)$ also has vanishing derivatives of orders up to M ensured by the term $F(e^{-j\omega})$.

f) The Daubechies scaling functions and wavelets are not given in analytical form. They can be recovered by infinitely iterating the filter bank (42.3.11.). Those basis functions have no symmetry or antisymmetry. Daubechies has shown that it is impossible to obtain an orthonormal and compactly supported wavelet that is either symmetric or antisymmetric around any axis, except for the trivial Haar wavelets.

g) The lowpass filters in the Daubechies basis with the support N are

N	n	h(n)
4	0	0.482962913145
	1	0.836516303738
	2	0.224143868042
	3	−0.129409522551

N	n	h(n)
6	0	0.332670552950
	1	0.806891509311
	2	0.459877502118
	3	−0.135011020010
	4	−0.085441273882
	5	0.035226291882

N	n	h(n)
8	0	0.230377813309
	1	0.714846570553
	2	0.630880767930
	3	−0.027983769417
	4	−0.187034811719
	5	0.030841381836
	6	0.032883011667
	7	−0.010597401785

42.4.5 Splines

a) The mth order B-spline is a repeated convolution of $m+1$ rectangular functions:

$$B^m(x) = \left(B^{m-1} \otimes B^0\right)(x)$$

where $\otimes$ denotes the convolution, and

$$B^0(x) = \begin{cases} 1 & \text{for} \quad 0 \le x \le 1 \\ 0 & \quad \text{otherwise} \end{cases}$$

The mth B-spline has a support $[0,m+1]$ and is symmetric with respect to $(m+1)/2$. When $m > 0$, the spline is not orthogonal to its integer translates.

b) If the $(m\text{-}1)^{th}$ order B-spline is a scaling function, $\phi(x) = B^{m-1}(x)$, its Fourier transform is

$$\Phi(\omega) = \left(\frac{1 - e^{-j\omega}}{j\omega}\right)^m$$

c) The lowpass filter in the frequency domain is

$$H(\omega) = \left(\frac{1 + e^{-j\omega}}{2}\right)^m$$

so that the two-scale relation (42.2.8a) is satisfied. The coefficients of the lowpass filter can be obtained by a repeated convolution of m filters of the sequence (1,1):

$$(1,1) \otimes (1,1) \otimes \cdots \otimes (1,1)$$

42.4.6 Lemarie–Battle Basis

a) The Lemarie–Battle basis is obtained by imposing the orthogonality on the B-spline scaling functions. The scaling functions are given as

$$\Phi(\omega) = \frac{1}{\omega^n \sqrt{\sum_{2n}(\omega)}} \qquad \text{with} \qquad \sum_{2n}(\omega) \equiv \sum_k \frac{1}{(\omega + 2\pi k)^{2n}}$$

where $n = m+1$ and m is the order of the B-spline.

b) The lowpass filter is then

$$H(\omega) = \sqrt{\frac{\sum_{2n}(\omega)}{2^{2n}\sum_{2n}(2\omega)}}$$

so that $\Phi(\omega) = H(\omega/2)\Phi(\omega/2)$, the two-scale relation (42.2.8a) is satisfied.

c) The table gives the first 12 coefficients of the lowpass filter of the Lemarie–Battle basis with the cubic spline of $m = 3$.

n	h(n)	n	h(n)
0	0.542	6	0.012
1	0.307	7	–0.013
2	–0.035	8	0.006
3	–0.078	9	0.006
4	0.023	10	–0.003
5	–0.030	11	–0.002

42.4.7 Biorthogonal Basis

a) To satisfy the perfect reconstruction conditions (42.3.7a and b) lengths of the biorthogonal filters must be all odd or all even. The analysis filters can be: (1) Both symmetric of odd length; (2) The lowpass filter $\tilde{h}(n)$ is symmetric and the highpass filter $\tilde{g}(n)$ is antisymmetric. Both are of even length. The lowpass filter cannot be antisymmetric according to (42.3.4g)

b) The highpass filters may be obtained from the lowpass filters by the alias cancellation equation (42.3.7a). The two analysis and synthesis lowpass filters are related by the perfect reconstruction equation (42.3.7b).

c) Lazy filters:

$$\tilde{H}(z) = H(z) = 1 \qquad \text{and} \qquad \tilde{G}(z) = G(z) = z^{-1}$$

that satisfy the perfect reconstruction conditions (42.3.7a and b). The filter bank does nothing but subsampling even and odd examples of the signal.

d) The lowpass filters can be generated to satisfy (42.3.7b) by a lifting process from the lazy filters. There is an additional degree of freedom in (42.3.7b) that is useful for other properties, such as the symmetry of the filters and the dyadic filter coefficients. The regularity of the filter (see 42.3.13) may be tested, after constructing the filters.

e) The biorthogonal 9/7 filters are given by the lowpass filters as

$$\text{for analysis} \qquad \tilde{h}9 = [1 \;\; 0 \;\; -8 \;\; 16 \;\; 46 \;\; 16 \;\; -8 \;\; 0 \;\; 1]/64$$

$$\text{for synthesis} \qquad h7 = [-1 \;\; 0 \;\; 9 \;\; 16 \;\; 9 \;\; 0 \;\; -1]/16$$

which have the dyadic filter coefficients (integers divided by a power of 2), that need less time and less memory space to compute. The $\tilde{h}9$ has 2 zeros and $h7$ has 4 zeros at $\omega = \pi$ (see 42.3.13)

42.4.8 More Filters

More filter banks and filter coefficients may be found in MATLAB Wavelet Toolbox (http://www.mathworks.com) and in WaveLab (http://playfair.stanford.edu/~wavelab).

References

I. Daubechies, *Ten Lectures on Wavelets,* (*SIAM*, 1992).

Y. Sheng, Wavelet transform, in *The Transforms and Applications Handbook*, A. D. Poularikas, Ed., Chap. 10, CRC Press, Boca Raton, 1995.

G. Strang and T. Hguyen, *Wavelet and Filter Banks*, Wellesley-Cambridge, 1996.

43

Trigonometry and Hyperbolic Trigonometry

43.1 Trigonometry

43.1.1 Angle

43.1.1.1 Radian

$$180^\circ = \pi \text{ radians; } 1^\circ = \frac{\pi}{180} \text{radians; } 1 \text{ radian} = \frac{180}{\pi} \text{ degrees}$$

43.1.1.2 Right Angle

An angle of 90°

43.1.1.3 Trigonometric functions of an arbitary angle (see Figure 43.1)

$$\sin\alpha = y/r \qquad \csc\alpha = r/y$$

$$\cos\alpha = x/r \qquad \sec\alpha = r/x$$

$$\tan\alpha = y/x \qquad \cot\alpha = ctn\alpha = x/y$$

$$ex\sec\alpha = \sec\alpha - 1 \qquad \text{covers}\alpha = 1 - \sin\alpha$$

$$\text{vers}\alpha = 1 - \cos\alpha \qquad \text{hav}\alpha = \frac{1}{2}\text{vers}\alpha$$

$$\text{cis}\alpha = \cos\alpha + i\sin\alpha = e^{ia}, \alpha \text{ in radians, } i = \sqrt{-1}$$

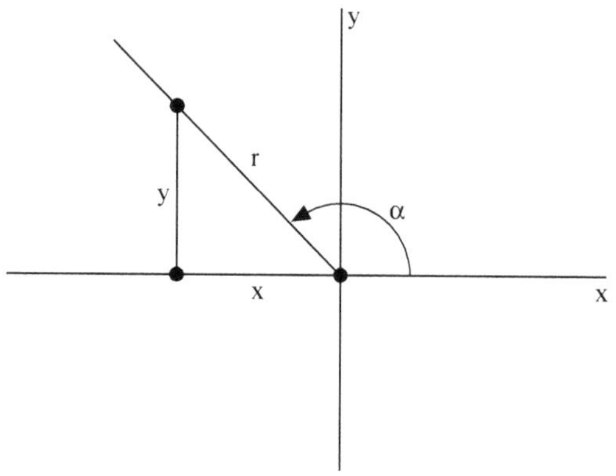

FIGURE 43.1

43.1.2 Relations of the Functions

43.1.2.1 Relations

$$\sin x = \frac{1}{\csc x} \qquad \csc x = \frac{1}{\sin x}$$

$$\cos x = \frac{1}{\sec x} \qquad \sec x = \frac{1}{\cos x}$$

$$\tan x = \frac{1}{\cot x} = \frac{\sin x}{\cos x} \qquad \sin^2 x + \cos^2 x = 1$$

$$\cot x = \frac{1}{\tan x} = \frac{\cos x}{\sin x} \qquad 1 + \tan^2 x = \sec^2 x$$

$$*\sin x = \pm\sqrt{1-\cos^2 x} \qquad 1 + \cot^2 x = \csc^2 x$$

$$*\tan x = \pm\sqrt{\sec^2 x - 1} \qquad *\cos x = \pm\sqrt{1-\sin^2 x}$$

$$*\cot x = \pm\sqrt{\csc^2 x - 1} \qquad *\sec x = \pm\sqrt{\tan^2 x + 1}$$

$$\sin x = \cos(90^\circ - x) = \sin(180^\circ - x) \qquad *\csc x = \pm\sqrt{\cot^2 x + 1}$$

$$\cos x = \sin(90^\circ - x) = -\cos(180^\circ - x)$$

$$\tan x = \cot(90^\circ - x) = -\tan(180^\circ - x)$$

$$\cot x = \tan(90^\circ - x) = -\cot(180^\circ - x)$$

$$\csc x = \cot\frac{x}{2} - \cot x$$

* The sign in front of the radical depends on the quadrant in which x falls.

43.1.3 Fundamental Identities

43.1.3.1 Fundamental Identities

Where a double sign appears in the following, the choice of sign depends upon the quadrant in which the angle terminates.

Reciprocal Relations

$$\sin\alpha = \frac{1}{\csc\alpha}, \quad \cos\alpha = \frac{1}{\sec\alpha}, \quad \tan\alpha = \frac{1}{\cot\alpha}$$

$$\csc\alpha = \frac{1}{\sin\alpha}, \quad \sec\alpha = \frac{1}{\cos\alpha}, \quad \cot\alpha = \frac{1}{\tan\alpha}$$

Product Relations

$$\sin\alpha = \tan\alpha\,\cos\alpha, \qquad \cos\alpha = \cot\alpha\,\sin\alpha$$

$$\tan\alpha = \sin\alpha\,\sec\alpha, \qquad \cot\alpha = \cos\alpha\,\csc\alpha$$

$$\sec\alpha = \csc\alpha\,\tan\alpha, \qquad \csc\alpha = \sec\alpha\,\cot\alpha$$

Quotient Relations

$$\sin\alpha = \frac{\tan\alpha}{\sec\alpha}, \qquad \cos\alpha = \frac{\cot\alpha}{\csc\alpha}, \qquad \tan\alpha = \frac{\sin\alpha}{\cos\alpha}$$

$$\csc\alpha = \frac{\sec\alpha}{\tan\alpha}, \qquad \sec\alpha = \frac{\csc\alpha}{\cot\alpha}, \qquad \cot\alpha = \frac{\cos\alpha}{\sin\alpha}$$

43.1.3.2 Pythagorian Relations

$$\sin^2\alpha + \cos^2\alpha = 1, \ 1 + \tan^2\alpha = \sec^2\alpha, \ 1 + \cot^2\alpha = \cos c^2\alpha$$

43.1.3.3 Angle-Sum and Angle-Difference Relations

$$\sin(\alpha + \beta) = \sin\alpha\cos\beta + \cos\alpha\sin\beta$$

$$\sin(\alpha - \beta) = \sin\alpha\cos\beta - \cos\alpha\sin\beta$$

$$\cos(\alpha + \beta) = \cos\alpha\cos\beta - \sin\alpha\sin\beta$$

$$\cos(\alpha - \beta) = \cos\alpha\cos\beta + \sin\alpha\sin\beta$$

$$\tan(\alpha + \beta) = \frac{\tan\alpha + \tan\beta}{1 - \tan\alpha\tan\beta}$$

$$\tan(\alpha - \beta) = \frac{\tan\alpha - \tan\beta}{1 + \tan\alpha\tan\beta}$$

$$\cot(\alpha + \beta) = \frac{\cot\beta\cot\alpha - 1}{\cot\beta + \cot\alpha}$$

$$\cot(\alpha - \beta) = \frac{\cot\beta\cot\alpha + 1}{\cot\beta - \cot\alpha}$$

$$\sin(\alpha+\beta)\sin(\alpha-\beta)=\sin^2\alpha-\sin^2\beta=\cos^2\beta-\cos^2\alpha$$

$$\cos(\alpha+\beta)\cos(\alpha-\beta=\cos^2\alpha-\sin^2\beta=\cos^2\beta-\sin^2\alpha$$

43.1.3.4 Double-Angle Relations

$$\sin 2\alpha = 2\sin\alpha\cos\alpha = \frac{2\tan\alpha}{1+\tan^2\alpha}$$

$$\cos 2\alpha = \cos^2\alpha - \sin^2\alpha = 2\cos^2\alpha - 1 = 1 - 2\sin^2\alpha = \frac{1-\tan^2\alpha}{1+\tan^2\alpha}$$

$$\tan 2\alpha = \frac{2\tan\alpha}{1-\tan^2\alpha}, \qquad \cot 2\alpha = \frac{\cot^2\alpha - 1}{2\cot\alpha}$$

43.1.3.5 Multiple-Angle Relations

$$\sin 3\alpha = 3\sin\alpha - 4\sin^3\alpha$$

$$\cos 3\alpha = 4\cos^3\alpha - 3\cos\alpha$$

$$\sin 4\alpha = 4\sin\alpha\cos\alpha - 8\sin^3\alpha\cos\alpha$$

$$\cos 4\alpha = 8\cos^4\alpha - 8\cos^2\alpha + 1$$

$$\sin 5\alpha = 5\sin\alpha - 20\sin^3\alpha + 16\sin^5\alpha$$

$$\cos 5\alpha = 16\cos^5\alpha - 20\cos^3\alpha + 5\cos\alpha$$

$$\sin 6\alpha = 32\cos^5\alpha\sin\alpha - 32\cos^3\alpha\sin\alpha + 6\cos\alpha\sin\alpha$$

$$\cos 6\alpha = 32\cos^6\alpha - 48\cos^4\alpha + 18\cos^2\alpha - 1$$

$$\sin n\alpha = 2\sin(n-1)\alpha\cos\alpha - \sin(n-2)\alpha$$

$$\cos n\alpha = 2\cos(n-1)\alpha\cos\alpha - \cos(n-2)\alpha$$

$$\tan 3\alpha = \frac{3\tan\alpha - \tan^3\alpha}{1-3\tan^2\alpha}$$

$$\tan 4\alpha = \frac{4\tan\alpha - 4\tan^3\alpha}{1-6\tan^2\alpha + \tan^4\alpha}$$

$$\tan n\alpha = \frac{\tan(n-1)\alpha + \tan\alpha}{1-\tan(n-1)\alpha\tan\alpha}$$

43.1.3.6 Function-Product Relations

$$\sin\alpha\sin\beta = \frac{1}{2}\cos(\alpha-\beta) - \frac{1}{2}\cos(\alpha+\beta)$$

$$\cos\alpha\cos\beta = \frac{1}{2}\cos(\alpha-\beta) + \frac{1}{2}\cos(\alpha+\beta)$$

$$\sin\alpha\cos\beta = \frac{1}{2}\sin(\alpha+\beta) + \frac{1}{2}\sin(\alpha-\beta)$$

$$\cos\alpha\sin\beta = \frac{1}{2}\sin(\alpha+\beta) - \frac{1}{2}\sin(\alpha-\beta)$$

43.1.3.7 Function-Sum and Function-Difference Relations

$$\sin\alpha + \sin\beta = 2\sin\frac{1}{2}(\alpha+\beta)\cos\frac{1}{2}(\alpha-\beta)$$

$$\sin\alpha - \sin\beta = 2\cos\frac{1}{2}(\alpha+\beta)\sin\frac{1}{2}(\alpha-\beta)$$

$$\cos\alpha + \cos\beta = 2\cos\frac{1}{2}(\alpha+\beta)\cos\frac{1}{2}(\alpha-\beta)$$

$$\cos\alpha - \cos\beta = -2\sin\frac{1}{2}(\alpha+\beta)\sin\frac{1}{2}(\alpha-\beta)$$

$$\tan\alpha + \tan\beta = \frac{\sin(\alpha+\beta)}{\cos\alpha\cos\beta}, \qquad \tan\alpha - \tan\beta = \frac{\sin(\alpha-\beta)}{\cos\alpha\cos\beta}$$

$$\cot\alpha + \cot\beta = \frac{\sin(\alpha+\beta)}{\sin\alpha\sin\beta}, \qquad \cot\alpha - \cot\beta = \frac{\sin(\beta-\alpha)}{\sin\alpha\sin\beta}$$

$$\frac{\sin\alpha + \sin\beta}{\sin\alpha - \sin\beta} = \frac{\tan\frac{1}{2}(\alpha+\beta)}{\tan\frac{1}{2}(\alpha-\beta)} \qquad \frac{\sin\alpha + \sin\beta}{\cos\alpha - \cos\beta} = \cot\frac{1}{2}(\beta-\alpha)$$

$$\frac{\sin\alpha + \sin\beta}{\cos\alpha + \cos\beta} = \tan\frac{1}{2}(\alpha+\beta). \qquad \frac{\sin\alpha - \sin\beta}{\cos\alpha + \cos\beta} = \tan\frac{1}{2}(\alpha-\beta)$$

43.1.3.8 Half-Angle Relations

$$\sin\frac{\alpha}{2} = \pm\sqrt{\frac{1-\cos\alpha}{2}}, \qquad \cos\frac{\alpha}{2} = \pm\sqrt{\frac{1+\cos\alpha}{2}}$$

$$\tan\frac{\alpha}{2} = \pm\sqrt{\frac{1-\cos\alpha}{1+\cos\alpha}} = \frac{1-\cos\alpha}{\sin\alpha} = \frac{\sin\alpha}{1+\cos\alpha}$$

$$\cot\frac{\alpha}{2} = \pm\sqrt{\frac{1+\cos\alpha}{1-\cos\alpha}} = \frac{1+\cos\alpha}{\sin\alpha} = \frac{\sin\alpha}{1-\cos\alpha}$$

43.1.3.9 Power Relations

$$\sin^2\alpha = \frac{1}{2}(1-\cos2\alpha), \qquad \sin^3\alpha = \frac{1}{4}(3\sin\alpha - \sin3\alpha)$$

$$\sin^4\alpha = \frac{1}{8}(3 - 4\cos2\alpha + \cos4\alpha)$$

$$\cos^2\alpha = \frac{1}{2}(1+\cos 2\alpha), \qquad \cos^3\alpha = \frac{1}{4}(3\cos\alpha + \cos 3\alpha)$$

$$\cos^4\alpha = \frac{1}{8}(3 + 4\cos 2\alpha + \cos 4\alpha)$$

$$\tan^2\alpha = \frac{1-\cos 2\alpha}{1+\cos 2\alpha}, \qquad \cot^2\alpha = \frac{1+\cos 2\alpha}{1-\cos 2\alpha}$$

43.1.3.10 Exponential Relations (a in radians)

$$e^{ia} = \cos\alpha + i\sin\alpha, \qquad i = \sqrt{-1}$$

$$\sin a = \frac{e^{ia} - e^{-ia}}{2i}, \qquad \cos a = \frac{e^{ia} + e^{-ia}}{2}$$

$$\tan a = -i\left(\frac{e^{ia} - e^{-ia}}{e^{ia} + e^{-ia}}\right) = -i\left(\frac{e^{2ia} - 1}{e^{2ia} + 1}\right)$$

43.1.3.11 Relations of Trigonometric Functions

Function	$\sin\alpha$	$\cos\alpha$	$\tan\alpha$	$\cot\alpha$	$\sec\alpha$	$\csc\alpha$
$\sin\alpha$	$\sin\alpha$	$\pm\sqrt{1-\cos^2\alpha}$	$\frac{\tan\alpha}{\pm\sqrt{1+\tan^2\alpha}}$	$\frac{1}{\pm\sqrt{1+\cot^2\alpha}}$	$\frac{\pm\sqrt{\sec^2 a - 1}}{\sec\alpha}$	$\frac{1}{\csc\alpha}$
$\cos\alpha$	$\pm\sqrt{1-\sin^2\alpha}$	$\cos\alpha$	$\frac{1}{\pm\sqrt{1+\tan^2\alpha}}$	$\frac{\cot\alpha}{\pm\sqrt{1+\cot^2\alpha}}$	$\frac{1}{\sec\alpha}$	$\frac{\pm\sqrt{\csc^2\alpha - 1}}{\csc\alpha}$
$\tan\alpha$	$\frac{\sin\alpha}{\pm\sqrt{1-\sin^2\alpha}}$	$\frac{\pm\sqrt{1-\cos^2\alpha}}{\cos\alpha}$	$\tan\alpha$	$\frac{1}{\cot\alpha}$	$\pm\sqrt{\sec^2\alpha - 1}$	$\frac{1}{\pm\sqrt{\csc^2\alpha - 1}}$
$\cot\alpha$	$\frac{\pm\sqrt{1-\sin^2\alpha}}{\sin\alpha}$	$\frac{\cos\alpha}{\pm\sqrt{1-\cos^2\alpha}}$	$\frac{1}{\tan\alpha}$	$\cot\alpha$	$\frac{1}{\pm\sqrt{\sec^2\alpha - 1}}$	$\pm\sqrt{\csc^2\alpha - 1}$
$\sec\alpha$	$\frac{1}{\pm\sqrt{1-\sin^2\alpha}}$	$\frac{1}{\cos\alpha}$	$\pm\sqrt{1+\tan^2\alpha}$	$\frac{\pm\sqrt{1+\cot^2\alpha}}{\cot\alpha}$	$\sec\alpha$	$\frac{\csc\alpha}{\pm\sqrt{\csc^2\alpha - 1}}$
$\csc\alpha$	$\frac{1}{\sin\alpha}$	$\frac{1}{\pm\sqrt{1-\cos^2\alpha}}$	$\frac{\pm\sqrt{1+\tan^2\alpha}}{\tan\alpha}$	$\pm\sqrt{1+\cot^2\alpha}$	$\frac{\sec\alpha}{\pm\sqrt{\sec^2\alpha - 1}}$	$\csc\alpha$

Note: The choice of sign depends upon the quadrant in which the angle terminates.

43.1.3.12 Identities Involving Principal Values

$$Arc\sin x + Arc\cos x = \pi/2$$

$$Arc\tan x + Arc\cot x = \pi/2$$

If $\alpha = Arc\sin x$, then

$$\sin\alpha = x, \qquad \cos\alpha = \sqrt{1-x^2}, \qquad \tan\alpha = \frac{x}{\sqrt{1-x^2}}$$

$$\csc\alpha = \frac{1}{x}, \qquad \sec\alpha = \frac{1}{\sqrt{1-x^2}}, \qquad \cot\alpha = \frac{\sqrt{1-x^2}}{x}$$

If $\alpha = Arc\cos x$, then

$$\sin\alpha = \sqrt{1-x^2}, \qquad \cos\alpha = x, \qquad \tan\alpha = \frac{\sqrt{1-x^2}}{x}$$

$$\csc\alpha = \frac{1}{\sqrt{1-x^2}}, \quad \sec\alpha = \frac{1}{x}, \qquad \cot\alpha = \frac{x}{\sqrt{1-x^2}}$$

If $\alpha = Arc\tan x$, then

$$\sin\alpha = \frac{x}{\sqrt{1+x^2}}, \qquad \cos\alpha = \frac{1}{\sqrt{1+x^2}}, \qquad \tan\alpha = x$$

$$\csc\alpha = \frac{\sqrt{1+x^2}}{x} \qquad \sec\alpha = \sqrt{1+x^2}, \qquad \cot\alpha = \frac{1}{x}$$

43.1.3.13 Plane Triangle Formulae

In the following, A, B, and C denote the angles of any plane triangle, a, b, c, the corresponding opposite sides, and $s = \frac{1}{2}(a+b+c)$.

Radius of inscribed circle:

$$r = \sqrt{\frac{(s-a)(s-b)(s-c)}{s}}$$

Radius of circumscribed circle:

$$R = \frac{a}{2\sin \mathrm{A}} = \frac{b}{2\sin B} = \frac{c}{2\sin C}$$

Law of sines:

$$\frac{a}{\sin A} = \frac{b}{\sin B} = \frac{c}{\sin C}$$

Law of cosines:

$$a^2 = b^2 + c^2 - 2bc\cos A, \qquad \cos A = \frac{b^2+c^2-a^2}{2bc}$$

$$b^2 = c^2 + a^2 - 2ca\cos B, \qquad \cos B = \frac{c^2+a^2-b^2}{2ca}$$

$$c^2 = a^2 + b^2 - 2ab\cos C, \qquad \cos C = \frac{a^2+b^2-c^2}{2ab}$$

Law of tangents:

$$\frac{b-c}{b+c}=\frac{\tan\frac{1}{2}(B-C)}{\tan\frac{1}{2}(B+C)}, \qquad \frac{c-a}{c+a}=\frac{\tan\frac{1}{2}(C-A)}{\tan\frac{1}{2}(C+A)}$$

$$\frac{a-b}{a+b}=\frac{\tan\frac{1}{2}(A-B)}{\tan\frac{1}{2}(A+B)}$$

Half-angle formulae:

$$\tan\frac{1}{2}A=\frac{r}{s-a}, \qquad \tan\frac{1}{2}B=\frac{r}{s-b}, \qquad \tan\frac{1}{2}C=\frac{r}{s-c}$$

$$\sin\frac{1}{2}A=\sqrt{\frac{(s-b)(s-c)}{bc}}, \qquad \cos\frac{1}{2}A=\sqrt{\frac{s(s-a)}{bc}}$$

$$\sin\frac{1}{2}B=\sqrt{\frac{(s-c)(s-a)}{ca}}, \qquad \cos\frac{1}{2}B=\sqrt{\frac{s(s-b)}{ca}}$$

$$\sin\frac{1}{2}C=\sqrt{\frac{(s-a)(s-b)}{ab}}, \qquad \cos\frac{1}{2}C=\sqrt{\frac{s(s-c)}{ab}}$$

Area:

$$K=\frac{1}{2}bc\sin A=\frac{1}{2}ca\sin B=\frac{1}{2}ab\sin C$$

$$K=\frac{a^2\sin B\sin C}{2\sin A}=\frac{b^2\sin C\sin A}{2\sin B}=\frac{c^2\sin A\sin B}{2\sin C}$$

$$K=\sqrt{s(s-a)(s-b)(s-c)}=rs=\frac{abc}{4R}$$

Mollweide's formulae:

$$\frac{b-c}{a}=\frac{\sin\frac{1}{2}(B-C)}{\cos\frac{1}{2}A}, \qquad \frac{c-a}{b}=\frac{\sin\frac{1}{2}(C-A)}{\cos\frac{1}{2}B}$$

$$\frac{a-b}{c}=\frac{\sin\frac{1}{2}(A-B)}{\cos\frac{1}{2}C}$$

Newton's formulae:

$$\frac{b+c}{a}=\frac{\cos\frac{1}{2}(B-C)}{\sin\frac{1}{2}A}, \qquad \frac{c+a}{b}=\frac{\cos\frac{1}{2}(C-A)}{\sin\frac{1}{2}B}$$

$$\frac{a+b}{c}=\frac{\cos\frac{1}{2}(A-B)}{\sin\frac{1}{2}C}$$

43.1.3.14 Solution of Right Triangles

a) Given acute angle A and opposite leg a.

$$B=90^\circ-A, \qquad b=a/\tan A=a\cot A, \qquad c=a/\sin A=a\csc A$$

b) Given acute angle A and adjacent leg b.

$$B=90^\circ-A, \qquad a=b\tan A, \qquad c=b/\cos A=b\sec A$$

c) Given acute angle A and hypotenuse c.

$$B=90^\circ-A, \qquad a=c\sin A, \qquad b=c\cos A$$

d) Given legs a and b.

$$c=\sqrt{a^2+b^2}, \qquad \tan A=a/b, \qquad B=90^\circ-A$$

e) Given hypotenuse c and leg a.

$$b=\sqrt{(c+a)(c-a)}, \qquad \sin A=a/c, \qquad B=90^\circ-A$$

43.1.3.15 Solution of Oblique Triangles

a) Given sides b and c and included angle A.
Nonlogarithmic solution

$$a^2=b^2+c^2-2bc\cos A, \qquad \cos B=(c^2+a^2-b^2)/2ca,$$

$$\cos C=(a^2+b^2-c^2)/2ab$$

Logarithmic solution

$$\frac{1}{2}(B+C)=90^\circ-\frac{1}{2}A, \qquad \tan\frac{1}{2}(B-C)=\frac{b-c}{b+c}\tan\frac{1}{2}(B-C).$$

$$B=\frac{1}{2}(B+C)+\frac{1}{2}(B-C), \quad C=\frac{1}{2}(B+C)-\frac{1}{2}(B-C).$$

$$a = (b \sin A)/\sin B, \qquad K = \frac{1}{2} bc \sin A$$

Check. $A + B + C = 180°$, or use Newton's formula or law of sines.

b) Given angles B and C and included side a.

$$A = 180^\circ - (B + C), \qquad b = (a \sin B)/\sin A,$$

$$c = (a \sin C)/\sin A, \qquad K = \frac{a^2 \sin B \sin C}{2 \sin A}$$

Check. $a = b\cos C + c\cos B$, or use Newton's formula or law of tangents.

c) Given sides a and c and opposite angle A.

$$\sin C = (c \sin A)/a, \qquad B = 180^\circ - (A + C),$$

$$b = (a \sin B)/\sin A, \qquad K = \frac{1}{2} ac \sin B$$

Check. $a = b\cos C + c\cos B$, or use Newton's formula or law of tangents.

Note. In this case there may be two solutions, for C may have two values: $C_1 < 90°$ and $C_2 = 180° - C_1 > 90°$. If $A + C_2 > 180°$, use only C_1.

d) Given the three sides a,b,c.

Nonlogarithmic solution

$$\cos A = (b^2 + c^2 - a^2)/2bc, \qquad \cos B = (c^2 + a^2 - b^2)/2ca,$$

$$\cos C = (a^2 + b^2 - c^2)/2ab$$

Logarithmic solution

$$s = \frac{1}{2}(a + b + c), \qquad r = \sqrt{\frac{(s-a)(s-b)(s-c)}{s}},$$

$$\tan\frac{1}{2}A = \frac{r}{s-a}, \qquad \tan\frac{1}{2}B = \frac{r}{s-b}, \qquad \tan\frac{1}{2}C = \frac{r}{s-c},$$

$$K = \sqrt{s(s-a)(s-b)(s-c)}$$

Check. $A + B + C = 180°$.

43.2 Hyperbolic Trigonometry

43.2.1 Hyperbolic Functions

43.2.1.1 Geometrical Defintions (see Figure 43.2)

Let O be the center, A the vertex, and P any point of the branch $B'AB$ of a rectangular hyperbola. Set $OM = x$, $MP = y$, $OA = a$, and

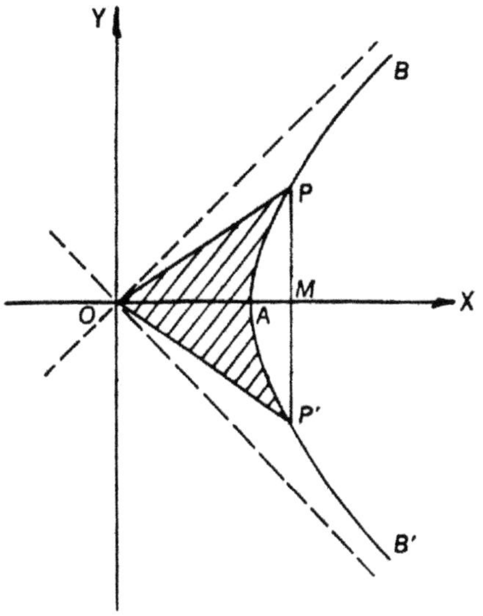

FIGURE 43.2

$$u = \frac{\text{area } O\,PAP'}{a^2}.$$

Then hyperbolic sine of u = sinh$u$ = y/a,
hyperbolic cosine of u = cosh$u$ = x/a.

43.2.1.2 Exponential Defintions

$$\text{hyperbolic sine of } u = \sinh u = \frac{1}{2}(e^{u} - e^{-u})$$

$$\text{hyperbolic cosine of } u = \cosh u = \frac{1}{2}(e^{u} + e^{-u})$$

$$\text{hyperbolic tangent of } u = \tanh u = \frac{\sinh u}{\cosh u} = \frac{e^{u} - e^{-u}}{e^{u} + e^{-u}}$$

$$\operatorname{csc}hu = \frac{1}{\sinh u}, \quad \operatorname{sec}hu = \frac{1}{\cosh u}, \quad \coth u = \frac{1}{\tanh u}$$

43.2.1.3 Fundamental Identities

$$\sinh(-u) = -\sinh u, \qquad \operatorname{csc}h(-u) = -\operatorname{csc}hu$$

$$\cosh(-u) = \cosh u, \qquad \operatorname{sec}h(-u) = \operatorname{sec}hu$$

$$\tanh(-u) = -\tanh u, \qquad \coth(-u) = -\coth u$$

$$\cosh^2 u - \sinh^2 u = 1 \qquad \tanh^2 u + \operatorname{sec}h^2 u = 1$$

$$\coth^2 u - \operatorname{csc}h^2 u = 1 \qquad \operatorname{csc}h^2 u - \operatorname{sec}h^2 u = \operatorname{csc}h^2 u \operatorname{sec}h^2 u$$

$$\sinh(u+v) = \sinh u \cosh v + \cosh u \sinh v$$

$$\sinh(u-v) = \sinh u \cosh v - \cosh u \sinh v$$

$$\cosh(u+v) = \cosh u \cosh v + \sinh u \sinh v$$

$$\cosh(u-v) = \cosh u \cosh v - \sinh u \sinh v$$

$$\tanh(u+v) = \frac{\tanh u + \tanh v}{1 + \tanh u \tanh v}$$

$$\tanh(u-v) = \frac{\tanh u - \tanh v}{1 - \tanh u \tanh v}$$

$$\sinh(u+v)\sinh(u-v) = \sinh^2 u - \sinh^2 v = \cosh^2 u - \cosh^2 v$$

$$\cosh(u+v)\cosh(u-v) = \sinh^2 u + \cosh^2 v = \cosh^2 u + \sinh^2 v$$

$$\sinh u \cosh v = \frac{1}{2}\sinh(u+v) + \frac{1}{2}\sinh(u-v)$$

$$\cosh u \sinh v = \frac{1}{2}\sinh(u+v) - \frac{1}{2}\sinh(u-v)$$

$$\cosh u \cosh v = \frac{1}{2}\cosh(u+v) + \frac{1}{2}\sinh(u-v)$$

$$\sinh u \sinh v = \frac{1}{2}\cosh(u+v) - \frac{1}{2}\cosh(u-v)$$

$$\sinh u + \sinh v = 2\sinh\frac{1}{2}(u+v)\cosh\frac{1}{2}(u-v)$$

$$\sinh u - \sinh v = 2\cosh\frac{1}{2}(u+v)\sinh\frac{1}{2}(u-v)$$

$$\cosh hu + \cosh v = 2\cosh\frac{1}{2}(u+v)\cosh\frac{1}{2}(u-v)$$

$$\cosh hu - \cosh v = 2\sinh\frac{1}{2}(u+v)\sinh\frac{1}{2}(u-v)$$

$$\sinh u = \frac{2\tanh\frac{1}{2}u}{1 - \tanh^2\frac{1}{2}u} = \frac{\tanh u}{\sqrt{1 - \tanh^2 u}}$$

$$\cosh u = \frac{1 + \tanh^2\frac{1}{2}u}{1 - \tanh^2\frac{1}{2}u} = \frac{1}{\sqrt{1 - \tanh^2 u}}$$

$$\sinh u + \cos u = \frac{1 + \tanh \frac{1}{2} u}{1 - \tanh \frac{1}{2} u}$$

$$\tanh u + \tanh v = \frac{\sinh(u+v)}{\cosh u \cosh v}$$

$$\tanh u - \tanh v = \frac{\sinh(u-v)}{\cosh u \cosh v}$$

$$\coth u + \coth v = \frac{\sinh(u+v)}{\sinh u \sinh v}$$

$$\coth u - \coth v = \frac{\sinh(v-u)}{\sinh u \sinh v}$$

$$\sinh 2u = 2 sihu \cosh u$$

$$\cosh 2u = \cosh^2 u + \sinh^2 u = 2\cosh^2 u - 1 = 1 + 2\sinh^2 u$$

$$\tanh 2u = \frac{2\tanh u}{1 + \tanh^2 u}$$

$$\sinh 3u = 3\sinh u + 4\sinh^3 u$$

$$\cosh 3u = 4\cosh^3 u - 3\cosh u$$

$$\tanh 3u = \frac{3\tanh u + \tanh^3 u}{1 + 3\tanh^2 u}$$

$$\sinh \frac{1}{2} u = \pm \sqrt{\frac{1}{2}(\cosh u - 1)}$$

$$\cosh \frac{1}{2} u = \sqrt{\frac{1}{2}(\cosh u + 1)}$$

$$\tan \frac{1}{2} u = \frac{\cosh u - 1}{\sinh u} = \frac{\sinh u}{\cosh u + 1}$$

43.2.1.4 Inverse Hyperbolic Functions*

$$\sinh^{-1} x = \log_e(x + \sqrt{x^2 + 1})$$

$\cosh^{-1} x = \log e(x \pm \sqrt{x^2 - 1}),\ x \geq 1$. The plus sign is used for the principal value.

$$\tanh^{-1} x = \frac{1}{2}\log,\left(\frac{1+x}{1-x}\right),\ x^2 < 1$$

$$\operatorname{csc}h^{-1} x = \log_e\left(\frac{1 \pm \sqrt{1+x^2}}{x}\right).$$ The plus sign is used if $x > 0$, the minus sign if $x < 0$.

$$\operatorname{sec}h^{-1} x = \log_e\left(\frac{1 \pm \sqrt{1-x^2}}{x}\right),\ 0 < x \le 1.$$ The plus sign is used for the principal values.

$$\coth^{-1} x = \frac{1}{2}\log_e\left(\frac{x+1}{x-1}\right),\ x^2 > 1$$

* $\sinh^{-1}$, $\cosh^{-1}$. etc., are sometimes replaced by arg sinh, arg cosh, etc., i.e., $\sinh^{-1}x = \text{arg}\sinh x$.

43.2.1.5 Relations with Circular Functions

$$\sinh iu = i\sin u, \qquad \sinh u = -i\sin iu$$

$$\cosh iu = \cos u, \qquad \cosh u = \cos iu$$

$$\tanh iu = i\tan u, \qquad \tanh u = -i\tan iu$$

$$\sinh(u+iv) = \sinh u\cos v + i\cosh u\sin v$$

$$\sinh(u-iv) = \sinh u\cos v - i\cosh u\sin v$$

$$\cosh(u+iv) = \cosh u\cos v + i\sinh u\sin v$$

$$\cosh(u-iv) = \cosh u\cos v - i\sinh u\sin v$$

$$\tanh(u+iv) = \frac{\sinh 2u + i\sin 2v}{\cosh 2u + \cos 2v}$$

$$\tanh(u-iv) = \frac{\sinh 2u - i\sin 2v}{\cosh 2u + \cos 2v}$$

$$\coth(u+iv) = \frac{\sinh 2u - i\sin 2v}{\cosh 2u - \cos 2v}$$

$$\coth(u-iv) = \frac{\sinh 2u + i\sin 2v}{\cosh 2u - \cos 2v}$$

$$\sinh\left(u + \frac{1}{2}\pi i\right) = i\cosh u, \qquad \cosh\left(u + \frac{1}{2}\pi i\right) = i\sinh u$$

$$\sinh(u+\pi i) = -\sinh u, \qquad \cosh(u+\pi i) = -\cosh u$$

$$\sinh(u+2\pi i) = \sinh u, \qquad \cosh(u+2\pi i) = \cosh u$$

$$e^u = \cosh u + \sinh u, \qquad e^{-u} = \cosh u - \sinh u$$

$$e^{iu} = \cos u + i\sin u, \qquad e^{-iu} = \cos u - i\sin u$$

43.2.1.6 Special Values of Hyperbolic Functions

x	0	$\frac{\pi}{2}i$	πi	$\frac{3\pi}{2}i$	∞
$\sinh x$	0	i	0	$-i$	∞
$\cosh x$	1	0	-1	0	∞
$\tanh x$	0	∞i	0	$-\infty i$	1
$\operatorname{csc}h x$	∞	$-i$	∞	i	0
$\sec h x$	1	∞	-1	∞	0
$\coth x$	∞	0	∞	0	1

44

Algebra

44.1 Factors and Expansions

$$(a \pm b)^2 = a^2 \pm 2ab + b^2$$

$$(a \pm b)^3 = a^3 \pm 3a^2b + 3ab^2 \pm b^3$$

$$(a \pm b)^4 = a^4 \pm 4a^3b + 6a^2b^2 \pm 4ab^3 + b^4$$

$$a^2 - b^2 = (a-b)(a+b)$$

$$a^2 + b^2 = (a + b\sqrt{-1})(a - b\sqrt{-1})$$

$$a^3 - b^3 = (a-b)(a^2 + ab + b^2)$$

$$a^3 + b^3 = (a+b)(a^2 - ab + b^2)$$

$$a^4 + b^4 = (a^2 + ab\sqrt{2} + b^2)(a^2 - ab\sqrt{2} + b^2)$$

$$a^n - b^n = (a-b)(a^{n-1} + a^{n-2}b + \ldots + b^{n-1})$$

$$a^n - b^n = (a+b)(a^{n-1} - a^{n-2}b + \ldots - b^{n-1}), \text{ for even values of n}$$

$$a^n + b^n = (a+b)(a^{n-1} - a^{n-2}b + \ldots + b^{n-1}), \text{ for odd values of n}$$

$$a^4 + a^2b^2 + b^4 = (a^2 + ab + b^2)(a^2 - ab + b^{2)})$$

$$(a+b+c)^2 = a^2 + b^2 + c^2 + 2ab + 2ac + 2bc$$

$$(a+b+c)^3 = a^3 + b^3 + c^3 + 3a^2(b+c) + 3b^2(a+c) + 3c^2(a+b) + 6abc$$

$$(a+b+c+d+\ldots)^2 = a^2 + b^2 + c^2 + d^2 + \ldots +$$

$$2a(b+c+d+\ldots) + 2b(c+d+\ldots) + 2c(d+\ldots) + \ldots$$

See also under Series.

44.2 Powers and Roots

$$a^x \times a^y = a^{(x+y)} \qquad a^0 = 1[\text{if } a \neq 0] \qquad (ab)^x = a^x b^x$$

$$\frac{a^x}{a^y} = a^{(x-y)}. \qquad a^{-x} = \frac{1}{a^x}. \qquad \left(\frac{a}{b}\right)^x = \frac{a^x}{b^x}.$$

$$(a^x)^y = a^{xy}. \qquad a^{1/x} = \sqrt[x]{a}. \qquad \sqrt[x]{ab} = \sqrt[x]{a}\sqrt[x]{b}$$

$$\sqrt[x]{\sqrt[y]{a}} = \sqrt[xy]{a}. \qquad a^{x/y} = \sqrt[y]{a^x}. \qquad \sqrt[x]{\frac{a}{b}} = \frac{\sqrt[x]{a}}{\sqrt[x]{b}}$$

44.3 Proportion

$$\text{If } \frac{a}{b} = \frac{c}{d}, \quad \text{then} \quad \frac{a+b}{b} = \frac{c+d}{d}, \quad \frac{a-b}{b} = \frac{c-d}{d} \quad \frac{a-b}{a+b} = \frac{c-d}{c+d}$$

44.4 Sums of Powers of Integers

44.4.1 $s_1(n) = 1 + 2 + 3 + \cdots + n = \frac{1}{2} n(n+1).$

44.4.2 $s_2(n) = 1^2 + 2^2 + 3^3 + \cdots + n^2 = \frac{1}{6} n(n+1)(2n+1).$

44.4.3 $s_3(n) = 1^3 + 2^3 + 3^3 + \cdots + n^3 = \frac{1}{4} n^2 (n+1)^2 = [s_1(n)]^2.$

44.4.4 $s_4(n) = 1^4 + 2^4 + 3^4 + \cdots + n^4 = \frac{1}{5}(3n^2 + 3n - 1) s_2(n).$

44.4.5 $s_5(n) = 1^5 + 2^5 + 3^5 + \cdots + n^5 = \frac{1}{12} n^2 (n+1)^2 (2n^2 + 2n - 1).$

44.4.6 $s_6(n) = 1^6 + 2^6 + 3^6 + \cdots + n^6 = \frac{n}{42}(n+1)(2n+1)(3n^4 + 6n^3 - 3n + 1).$

44.4.7 $s_7(n) = 1^7 + 2^7 + 3^7 + \cdots + n^7 = \frac{n^2}{24}(n+1)^2(3n^4 + 6n^3 - n^2 - 4n + 2).$

44.4.8 $s_8(n) = 1^8 + 2^8 + 3^8 + \cdots + n^8$

$$= \frac{n}{90}(n+1)(2n+1)(5n^6 + 15n^5 + 5n^4 - 15n^3 - n^2 + 9n - 3).$$

44.4.9 $s_9(n) = 1^9 + 2^9 + 3^9 + \cdots + n^9$

$$= \frac{n^2}{20}(n+1)^2(2n^6 + 6n^5 + n^4 - 8n^3 + n^2 + 6n - 3).$$

44.4.10 $s_{10}(n) = 1^{10} + 2^{10} + 3^{10} + \cdots + n^{10}$

$$= \frac{n}{66}(n+1)(2n+1)(3n^8 + 12n^7 + 8n^6 - 18n^5 - 10n^4 + 24n^3 + 2n^2 - 15n + 5).$$

44.5 Series and Sums

44.5.1 Series

44.5.1.1 Taylor Series (one variable)

$$f(x+a)=\sum_{n=0}^{N}\frac{x^n}{n!}f^{(n)}(a)+R_N,\ f^{(n)}\ \text{indicates } n\text{-times differentiation},$$

$$R_N = \text{remainder} = \frac{x^{N+1}}{(N+1)!}f^{(N+1)}(\theta a)\ \text{for}\ 0<\theta<1$$

44.5.1.2 Taylor Series (two variable)

$$f(a+x,\,b+y)=f(a,b)+xf_x(a,b)+yf_y(a,b)+\frac{1}{2!}[x^2f_{xx}(a,b,)+2xyf_{xy}(a,b)+y^2f_{yy}(a,b)]+\cdots,$$

subscripts indicate differentiation;

$$f(a+h,b+k)=f(a,b)+\left(h\frac{\partial}{\partial x}+k\frac{\partial}{\partial y}\right)f(xy)\Bigg|_{\substack{x=a\\y=b}}+\cdots+\frac{1}{n!}\left(n\frac{\partial}{\partial x}+k\frac{\partial}{\partial y}\right)^n f(x,y)\Bigg|_{\substack{x=a\\y=b}}+\cdots$$

44.5.1.3 Maclaurin

$$f(x)=f(0)+xf^{(1)}(0)+\frac{x^2}{2!}f^{(2)}(0)+\cdots+x^{n-1}\frac{f^{(n-1)}(0)}{(n-1)!}+R_N,$$

$$R_N=\frac{x^nf^{(n)}(\theta)}{n!},\quad 0<\theta<1$$

Number in exponent means differentiation of equal times.

44.5.1.4 Binomial Series

$$(x+y)^v=\sum_{n=0}^{\infty}\frac{\Gamma(v+1)}{\Gamma(v-n+1)}\frac{x^ny^{v-n}}{n!}$$

When v is a positive integer, the series terminates at $n = v$.

44.5.1.5 Arithmetic Series

$$\sum_{n=1}^{N}(a+nd)=Na+\frac{1}{2}N(N+1)d.$$

44.5.1.6 Arithmetic Power Series

$$\sum_{n=1}^{N}(a+nb)x^n=\frac{a-(a+bN)x^{n+1}}{(1-x)}+\frac{bx(1-x^n)}{(1-x)^2},\quad (x\neq 1).$$

44.5.1.7 Geometric Series

$$1+x+x^2+x^3+\cdots=\frac{1}{1-x}, \quad (|x|<1).$$

44.5.1.8 Arithmetic-Geometric Series

$$a+(a+b)x+(a+2b)x^2+(a+3b)x^3+\cdots=\frac{a}{1-x}+\frac{bx}{(1-x)^2}, \quad (|x|<1)$$

44.5.1.9 Combinational Sums

$$\bullet \sum_{k=0}^{n}\binom{x-k}{n-k}=\binom{x+1}{n}; \bullet \sum_{k=-\infty}^{m}(-1)^k\binom{x}{k}=(-1)^m\binom{x-1}{m}; \bullet \sum_{k=0}^{n}\binom{k+m}{k}=\binom{m+n+1}{n};$$

$$\bullet \sum_{k=-\infty}^{m}(-1)^k\binom{x+m}{k}=\binom{-x}{m}; \bullet \sum_{k=-\infty}^{\infty}\binom{x}{m+k}\binom{y}{n-k}=\binom{x+y}{m+n}; \bullet \sum_{k=-\infty}^{\infty}\binom{l}{m+k}\binom{x}{n+k}=$$

$$\binom{l+x}{l-m+n}; \bullet \sum_{k=-\infty}^{\infty}(-1)^k\binom{l}{m+k}\binom{x+k}{n}=(-1)^{l+m}\binom{x-m}{n-l}; \bullet \sum_{k=-\infty}^{l}(-1)^k\binom{l-k}{m}\binom{x}{k-n}=$$

$$(-1)^{l+m}\binom{x-m-l}{l-m-n}; \bullet \sum_{k=0}^{l}\binom{l-k}{m}\binom{q+k}{n}=\binom{l+q+1}{m+n+1} \quad (m\geq q)$$

44.5.2 Infinite Series

44.5.2.1 Algebraic Functions

44.5.2.1.1: $(x+y)^n=x^n+\binom{n}{1}x^{n-1}y+\binom{n}{2}x^{n-2}y^2+\cdots$

44.5.2.1.2: $(1\pm x)^n=1\pm\binom{n}{1}x+\binom{n}{2}x^2\pm\binom{n}{3}x^3+\cdots \qquad (x^2<1)$

44.5.2.1.3: $(1\pm x)^{-n}=1\mp\binom{n}{1}x+\binom{n+1}{2}x^2\mp\binom{n+2}{3}x^3+\cdots \qquad (x^2<1)$

44.5.2.1.4: $\sqrt{1+x}=1+\frac{1}{2}x-\frac{1}{8}x^2+\frac{1}{16}x^3-\frac{5}{128}x^4+\cdots \qquad (x^2<1)$

44.5.2.1.5: $(1+x)^{-1/2}=1-\frac{1}{2}x+\frac{3}{8}x^2-\frac{5}{16}x^3+\frac{35}{128}x^4+\cdots \qquad (x^2<1)$

44.5.2.1.6: $(1\pm x)^{-1}=1\mp x+x^2\mp x^3+x^4\mp x^5+\cdots \qquad (x^2<1)$

44.5.2.1.7: $(1\pm x)^{-2}=1\mp 2x+3x^2\mp 4x^3+5x^4\mp 6x^5+\cdots \qquad (x^2<1)$

44.5.2.2 Exponential Functions

44.5.2.2.1: $e=1+\frac{1}{1!}+\frac{1}{2!}+\cdots+\frac{1}{n!}+\cdots. \qquad$ (all real values of x)

44.5.2.2.2: $e^x=1+\frac{x}{1!}+\frac{x^2}{2!}+\cdots+\frac{x^n}{n!}+\cdots,$

$$= e^{a}\left[1+(x-a)+\frac{(x-a)^2}{2!}+\cdots+\frac{(x-a)^n}{n!}+\cdots\right].$$

44.5.2.2.3: $a^x = 1 + x\log_e a + \frac{(x\log_e a)^2}{2!}+\cdots+\frac{(x\log_e a)^n}{n!}+\cdots.$

44.5.2.3 Logarithmic Functions

44.5.2.3.1: $\log x = \frac{x-1}{x}+\frac{1}{2}\left(\frac{x-1}{x}\right)^2+\cdots+\frac{1}{n}\left(\frac{x-1}{x}\right)^n+\cdots, \qquad (x>1/2),$

$$= (x-1)-\frac{1}{2}(x-1)^2+\frac{1}{3}(x-1)^3-\cdots \qquad (2\ge x>0),$$

$$= 2\left[\frac{x-1}{x+1}+\frac{1}{3}\left(\frac{x-1}{x+1}\right)^3+\frac{1}{5}\left(\frac{x-1}{x+1}\right)^5+\cdots\right], \qquad (x>0).$$

$$= \log a+\frac{(x-a)}{a}-\frac{(x-a)^2}{2a^2}+\frac{(x-a)^3}{3a^2}-\cdots \qquad (0<x\le 2a).$$

44.5.2.3.2: $\log(1+x) = x-\frac{x^2}{2}+\frac{x^3}{3}-\frac{x^4}{4}+\cdots \qquad -1<x\le 1.$

44.5.2.3.3: $\log(n+1) = \log(n-1)+2\left[\frac{1}{n}+\frac{1}{3n^3}+\frac{1}{5n^5}+\cdots\right].$

44.5.2.3.4: $\log(a+x) = \log a + 2\left[\frac{x}{2a+x}+\frac{1}{3}\left(\frac{x}{2a+x}\right)^3+\frac{1}{5}\left(\frac{x}{2a+x}\right)^5+\cdots\right], \qquad (a>0, -a<x).$

44.5.2.3.5: $\log\frac{1+x}{1-x} = 2\left[x+\frac{x^3}{3}+\cdots+\frac{x^{2n-1}}{2n-1}+\cdots\right], \qquad (-1<x<1).$

44.5.2.4 Trigonometric Functions

44.5.2.4.1: $\sin x = x-\frac{x^3}{3!}+\frac{x^5}{5!}-\frac{x^7}{7!}+\cdots \qquad$ (all real values of x).

44.5.2.4.2: $\cos x = 1-\frac{x^2}{2!}+\frac{x^4}{4!}-\frac{x^6}{6!}+\cdots \qquad$ (all real values of x).

44.5.2.4.3: $\tan x = x+\frac{x^3}{3}+\frac{2x^5}{15}+\cdots+\frac{(-1)^{n-1}2^{2n}(2^{2n}-1)B_{2n}}{(2n)!}x^{2n-1}+\cdots$

($x^2<\pi^2/4$, B_n is the nth Bernoulli number).*

44.5.2.4.4: $\cot x = \frac{1}{x}-\frac{x}{3}-\frac{x^3}{45}-\frac{2x^5}{945}-\frac{x^7}{4725}-\cdots+\frac{(-1)^{n+1}2^{2n}B_{2n}}{(2n)!}x^{2n-1}+\cdots$

($x^2<\pi^2$, B_n is the nth Bernoulli number).*

44.5.2.4.5: $\sec x = 1+\frac{x^2}{2}+\frac{5}{24}x^4+\frac{61}{720}x^6+\frac{277}{8064}x^8+\cdots+\frac{(-1)^n E_{2n}}{(2n)!}x^{2n}+\cdots$

($x^2<\pi^2/4$, E_n is the n^{th}Euler number)**

44.5.2.4.6: $\csc x = \frac{1}{x}+\frac{x}{6}+\frac{7x^3}{360}+\frac{31x^5}{15120}+\cdots+\frac{(-1)^{n+1}2(2^{2n-1}-1)B_{2n}}{(2n)!}x^{2n-1}+\cdots$

($|x|<\pi$, B_n is the Bernoulli number).*

44.5.2.4.7: $\log\sin x = \log x - \frac{x^2}{6} - \frac{x^4}{180} - \frac{x^6}{2835} - \cdots \qquad (x^2 < \pi^2).$

44.5.2.4.8: $\log\cos x = -\frac{x^2}{2} - \frac{x^4}{12} - \frac{x^6}{45} - \frac{17x^6}{2520} - \cdots \qquad (x^2 < \pi^2/4).$

44.5.2.4.9: $\log\tan x = \log x + \frac{x^2}{3} + \frac{7x^4}{90} + \frac{62x^6}{2835} + \cdots \qquad (x^2 < \pi^2/4).$

44.5.2.4.10: $e^{\sin x} = 1 + x + \frac{x^2}{2!} - \frac{3x^4}{4!} - \frac{8x^5}{5!} - \frac{3x^6}{6!} + \frac{56x^7}{7!} + \cdots.$

44.5.2.4.11: $e^{\cos x} = e\left(1 - \frac{x^2}{2!} + \frac{4x^4}{4!} - \frac{31x^6}{6!} + \cdots\right).$

44.5.2.4.12: $e^{\tan x} = 1 + x + \frac{x^2}{2!} + \frac{3x^3}{3!} + \frac{9x^4}{4!} + \frac{37x^5}{5!} + \cdots \qquad (x^2 < \pi^2/4).$

44.5.2.4.13: $\sin x = \sin a + (x-a)\cos a - \frac{(x-a)^2}{2!}\sin a - \frac{(x-a)^3}{3!}\cos a + \cdots.$

* Bernoulli numbers: $\frac{te^{tx}}{e^t - 1}$ = generating functions for Bernoulli polynomials = $B_0(x) + B_1(x)t$ $+ B_2(x)\frac{t^2}{2!} + B_3(x)\frac{t^3}{3!} + \cdots$, $B_0(x) = 1$, $B_1(x) = x - \frac{1}{2}$, $B_2(x) = x^2 - x + \frac{1}{6}$, $B_3(x) = x^3 - \frac{3}{2}x^2 + \frac{x}{2}$, $B_4(x) = x^4 - 2x^3 + x^2 - \frac{1}{30}, \cdots$; B_n = Bernoulli numbers = $B_n(x)|_{x=0}$, $B_0 = 1$, $B_1 = -\frac{1}{2}$, $B_2 = \frac{1}{6}$; $B_4 = -\frac{1}{30}$, $B_6 = \frac{1}{42}$, $B_8 = -\frac{1}{30}$, $\cdots$. $B_{2n+1} = 0 (n \ge 1)$.

** Euler numbers: $\frac{2e^{xt}}{e^t + 1}$ = generating functions for Euler polynomials = $\sum_{n=0}^{\infty} E_n(x)\frac{t^n}{n!}$, $E_0(x) = 1$, $E_1(x) = (2x-1)/2$, $E_2(x) = x^2 - x$, $E_3(x) = (4x^3 - 6x^2 + 1)/4$, $E_4(x) = x^4 - 2x^3 + x$, $E_5(x) = (2x^5 - 5x^4 + 5x^2 - 1)/2$; E_n = Euler numbers = $2^n E_n(x)|_{x=\frac{1}{2}}$; $E_2 = -1$, $E_4 = 5$, $E_6 = -61$, $E_8 = 1385$, $E_{10} = -50521$, $E_{12} = 2702765$, $E_{14} = -199360981$, $E_{16} = 19391512145$

44.5.2.5 Inverse Trigonometric Functions

44.5.2.5.1: $\sin^{-1} x = x + \frac{1}{2\cdot 3}x^3 + \frac{1\cdot 3}{2\cdot 4\cdot 5}x^5 + \frac{1\cdot 3\cdot 5}{2\cdot 4\cdot 6\cdot 7}x^7 + \cdots$

$$\left(x^2 < 1, -\frac{\pi}{2} < \sin^{-1} x < \frac{\pi}{2}\right).$$

44.5.2.5.2: $\cos^{-1} x = \frac{\pi}{2} - \left(x + \frac{1}{2\cdot 3}x^3 + \frac{1.3}{2\cdot 4\cdot 5}x^5 + \frac{1.3.5}{2\cdot 4\cdot 6\cdot 7}x^7 + \cdots\right)$

$$(x^2 < 1, 0 < \cos^{-1} x < \pi).$$

44.5.2.5.3: $\tan^{-1}x = x - \frac{x^3}{3} + \frac{x^5}{5} - \frac{x^7}{7} + \cdots$ $(x^2 < 1)$.

$= \frac{\pi}{2} - \frac{1}{x} + \frac{1}{3x^3} - \frac{1}{5x^5} + \frac{1}{7x^7} - \cdots.$ $(x > 1)$.

$= -\frac{\pi}{2} - \frac{1}{x} + \frac{1}{3x^3} - \frac{1}{5x^5} + \frac{1}{7x^7} - \cdots$ $(x < -1)$.

44.5.2.5.4: $\cot^{-1}x = \frac{\pi}{2} - x + \frac{x^3}{3} - \frac{x^5}{5} + \frac{x^7}{7} - \cdots$ $(x^2 < 1)$.

44.5.2.6 Hyperbolic Functions

(For Bernoulli B_n and Euler E_n number, see Section 44.5.2.4)

44.5.2.6.1: $\sinh x = x + \frac{x^3}{3!} + \frac{x^5}{5!} + \frac{x^7}{7!} + \cdots + \frac{x^{(2n+1)}}{(2n+1)!} + \cdots$

44.5.2.6.2: $\sinh ax = \frac{2}{\pi}\sinh \pi a\left[\frac{\sin x}{a^2+1^2} - \frac{2\sin 2x}{a^2+2^2} + \frac{3\sin 3x}{a^2+3^2} + \cdots\right]$ $(|x| < \pi)$.

44.5.2.6.3: $\cosh x = 1 + \frac{x^2}{2!} + \frac{x^4}{4!} + \frac{x^6}{6!} + \cdots + \frac{x^{2n}}{(2n)!} + \cdots.$

44.5.2.6.4: $\cosh ax = \frac{2a}{\pi}\sinh \pi a\left[\frac{1}{2a^2} - \frac{\cos x}{a^2+1^2} + \frac{\cos 2x}{a^2+2^2} - \frac{\cos 3x}{a^2+3^2} + \cdots\right]$ $(|x| < \pi)$.

44.5.2.6.5: $\tanh x = x - \frac{1}{3}x^3 + \frac{2}{15}x^5 - \cdots + \frac{2^{2n}(2^{2n}-1)B_{2n}}{(2n)!}x^{2n-1} + \cdots$ $(|x| < \pi/2)$,

$= 1 - 2e^{-2x} + 2e^{-4x} - 2e^{-6x} + \ldots$ $(\mathrm{Re}\, x > 0)$,

$$= 2x\left[\frac{1}{\left(\frac{\pi}{2}\right)^2 + x^2} + \frac{1}{\left(\frac{3\pi}{2}\right)^2 + x^2} + \frac{1}{\left(\frac{5\pi}{2}\right)^2 + x^2} + \cdots\right].$$

44.5.2.6.6: $\coth x = \frac{1}{x} + \frac{x}{3} - \frac{x^3}{45} + \frac{2x^5}{945} + \cdots + \frac{2^{2n}B_{2n}}{(2n)!}x^{2n-1} + \cdots$ $(0 < |x| < \pi)$.

$= 1 + 2e^{-2x} + 2e^{-4x} + 2e^{-6x} + \cdots$ $(\mathrm{Re}\, x > 0)$,

$= \frac{1}{x} + 2x\left[\frac{1}{\pi^2 + x^2} + \frac{1}{(2\pi)^2 + x^2} + \frac{1}{(3\pi)^2 + x^2} + \cdots\right]$ $(\mathrm{Re}\, x > 0)$.

44.5.2.6.7: $\mathrm{sech}\, x = 1 - \frac{1}{2!}x^2 + \frac{5}{4!}x^4 - \frac{61}{6!}x^6 + \cdots + \frac{E_{2n}}{(2n)!}x^{2n} + \cdots$

($|x| < \pi/2$, E_n is the nth Euler number),

$= 2\left(e^{-x} - e^{-3x} + e^{-5x} - e^{-7x} + \cdots\right)$ $(\mathrm{Re}\, x > 0)$.

$$= 4\pi\left[\frac{1}{\pi^2 + 4x^2} - \frac{3}{(3\pi)^2 + 4x^2} + \frac{5}{(5\pi)^2 + 4x^2} + \cdots\right].$$

44.5.2.6.8: $\operatorname{csc}hx = \frac{1}{x} - \frac{x}{6} + \frac{7x^3}{360} + \cdots + \frac{2(2^{2n-1}-1)B_{2n}}{(2n)!}x^{2n-1} + \cdots \quad (0 < |x| < \pi).$

$= 2\left(e^{-x} + e^{-3x} + e^{-5x} + e^{-7x} + \cdots\right) \quad (\operatorname{Re} x > 0).$

$= \frac{1}{x} - \frac{2x}{\pi^2 + x^2} + \frac{2x}{(2\pi)^2 + x^2} - \frac{2x}{(3\pi)^2 + x^2} + \cdots.$

44.5.2.6.9: $\sinh nu = \sinh u\left[(2\cosh u)^{n-1} - \frac{(n-2)}{1!}(2\cosh u)^{n-3} + \frac{(n-3)(n-4)}{2!}(2\cosh u)^{n-5} - \frac{(n-4)(n-5)(n-6)}{3!}(2\cosh u)^{n-7} + \cdots\right].$

44.5.2.6.10: $\cosh nu = \frac{1}{2}\left[(2\cosh u)^{n} - \frac{n}{1!}(2\cosh u)^{n-2} + \frac{n(n-3)}{2!}(2\cosh u)^{n-4} - \frac{n(n-4)(n-5)}{3!}(2\cosh u)^{n-6} + \cdots\right].$

44.5.2.7 Inverse Hyperbolic Functions?

44.5.2.7.1: $\sinh^{-1} x = x - \frac{1}{2\cdot 3}x^3 + \frac{1\cdot 3}{2\cdot 4\cdot 5}x^5 - \frac{1\cdot 3\cdot 5}{2\cdot 4\cdot 6\cdot 7}x^7 + \cdots \quad (|x| < 1).$

$= \log(2x) + \frac{1}{2}\cdot\frac{1}{2x^2} + \frac{1\cdot 3}{2\cdot 4}\cdot\frac{1}{4x^4} + \frac{1\cdot 3\cdot 5}{2\cdot 4\cdot 6}\cdot\frac{1}{6x^6} + \cdots \quad (|x| > 1).$

44.5.2.7.2: $\cosh^{-1} x = \pm\left[\log(2x) - \frac{1}{2}\cdot\frac{1}{2x^2} - \frac{1\cdot 3}{2\cdot 4}\cdot\frac{1}{4x^4} + \cdots\right] \quad (x > 1).$

44.5.2.7.3: $\operatorname{csc}h^{-1}x = \frac{1}{x} - \frac{1}{2}\cdot\frac{1}{3x^3} - \frac{1\cdot 3}{2\cdot 4}\cdot\frac{1}{5x^5} - \frac{1\cdot 3\cdot 5}{2\cdot 4\cdot 6}\cdot\frac{1}{7x^7} + \cdots \quad (|x| > 1).$

$= \log\frac{2}{x} + \frac{1}{2}\cdot\frac{x^2}{2} - \frac{1\cdot 3}{2\cdot 4}\cdot\frac{x^4}{4} + \frac{1\cdot 3\cdot 5}{2\cdot 4\cdot 6}\cdot\frac{x^6}{6} - \cdots \quad (0 < x < 1).$

44.5.2.7.4: $\operatorname{sec}h^{-1}x = \log\frac{2}{x} - \frac{1}{2}\cdot\frac{x^2}{2} - \frac{1\cdot 3}{2\cdot 4}\cdot\frac{x^4}{4} - \frac{1\cdot 3\cdot 5}{2\cdot 4\cdot 6}\cdot\frac{x^6}{6} - \cdots \quad (0 < x < 1).$

44.5.2.7.5: $\tanh^{-1} x = x + \frac{x^3}{3} + \frac{x^5}{5} + \frac{x^7}{7} + \cdots + \frac{x^{2n+1}}{2n+1} + \cdots \quad (|x| < 1).$

44.5.2.7.6: $\coth^{-1} x = \frac{1}{x} + \frac{1}{3x^3} + \frac{1}{5x^5} + \frac{1}{7x^7} + \cdots + \frac{1}{(2n+1)x^{2n+1}} + \cdots \quad (|x| > 1).$

44.6 Partial Fractions

The technique of partial fractions allows a quotient of two polynomials to be written as a sum of simpler terms.

Given the fraction $\frac{f(x)}{x(x)}$, where both $f(x)$ and $g(x)$ are polynomials, begin by dividing $f(x)$ by $g(x)$ to produce a quotient $q(x)$ and a remainder $r(x)$, where the degree of $r(x)$ is less than the degree of $g(x)$, so that $\frac{f(x)}{g(x)} = q(x) + \frac{r(x)}{g(x)}$. Therefore, assume that the rational function has the form $\frac{r(x)}{g(x)}$, where the degree of the numerator is less than the degree of the denominator.

The techniques used depend on the factorization of $g(x)$

44.6.1 Single Linear Factor

Suppose that $g(x) = (x - a)h(x)$, where $h(a) \neq 0$. Then

$$\frac{r(x)}{g(x)} = \frac{A}{x-a} + \frac{s(x)}{h(x)},$$

where the number A is given by $r(a)/h(a)$. For example,

$$\frac{2x}{x^2-1} = \frac{1}{x-1} + \frac{1}{x+1}.$$

44.6.2 Repeated Linear Factor

Suppose that $g(x) = (x - a)^k h(x)$, where $h(a) \neq 0$. Then

$$\frac{r(x)}{g(x)} = \frac{A_1}{x-a} + \frac{A_2}{(x-a)^2} + \cdots + \frac{A_k}{(x-a)^k} + \frac{s(x)}{h(x)}.$$

where

$$A_k = \frac{r(a)}{h(a)},$$

$$A_{k-1} = \frac{d}{dx}\left(\frac{r(x)}{g(x)}\right)\bigg|_{x=a},$$

$$A_{k-2} = \frac{1}{2!}\frac{d^2}{dx^2}\left(\frac{r(x)}{g(x)}\right)\bigg|_{x=a},$$

$$A_{k-j} = \frac{1}{j!}\frac{d^j}{dx^j}\left(\frac{r(x)}{g(x)}\right)\bigg|_{x=a}.$$

44.6.3 Single Quadratic Factor

Suppose that $g(x) = (x^2 + bx + c)h(x)$, where $b^2 - 4c < 0$ (so that $x^2 + bx + c$ does not factor into real linear factors) and $h(x)$ is relatively prime to $x^2 + bx + c$. Then

$$\frac{r(x)}{g(x)} = \frac{Ax+B}{x^2+bx+c} + \frac{s(x)}{h(x)}.$$

In order to determine A and B, multiply the equation by $g(x)$ so that there are no denominators remaining, and substitute any two values for x, yielding two equations for A and B.

44.6.4 Repeated Quadratic Factor

Suppose that $g(x) = (x^2 + bx + c)^k h(x)$, where $b^2 - 4c < 0$ (so that $x^2 + bx + c$ does not factor into real linear factors) and $h(x)$ is relatively prime to $x^2 + bc + c$. Then

$$\frac{r(x)}{g(x)} = \frac{A_1x+B_1}{x^2+bx+c} + \frac{A_2x+B_2}{(x^2+bx+c)^2} + \frac{A_3x+B_3}{(x^2+bx+c)^3} + \cdots + \frac{A_kx+B_k}{(x^2+bx+c)^k} + \frac{s(x)}{h(x)}$$

In order to determine A_i and B_i, multiply the equation by $g(x)$ so that there are no denominators remaining, and substitute any $2k$ values for x, yielding $2k$ equations for A_1 and B_1.

44.7 Polynomials

44.7.1 Quadratic Poynomials

The solution of the equation $ax^2 + bx + c = 0$, where $a \neq 0$, is given by

$$x = \frac{-b \pm \sqrt{b^2 - 4ac}}{2a} \tag{2.2.1}$$

The discriminant of this equation is $b^2 - 4ac$. Suppose that a,b, and c are all real. If the discriminant is negative, then the two roots are complex numbers which are conjugate. If the discriminant is positive, then the two roots are unequal real numbers. If the discriminant is 0, then the two roots are equal.

44.7.2 Cubic Polynomials

To solve the equation $ax^3 + bx^2 + cx + d = 0$, where $a \neq 0$, begin by making the subtitution

$$y = x + \frac{b}{3a}.$$

That gives the equation

$$y^3 + 3py + q = 0,$$

where

$$p = \frac{3ac - b^2}{9a^2} \quad \text{and} \quad q = \frac{2b^3 - 9abc + 27a^2 d}{27a^3}.$$

The discriminant of this polynomial is $4p^3 + q^2$.

The solution are given by $\sqrt[3]{\alpha} - \sqrt[3]{\beta}$, $e^{\frac{2\pi i}{3}}\sqrt[3]{\alpha} - e^{\frac{4\pi i}{3}}\sqrt[3]{\beta}$ and $e^{\frac{4\pi i}{3}}\sqrt[3]{\alpha} - e^{\frac{2\pi i}{3}}\sqrt[3]{\beta}$, where

$$\alpha = \frac{-q + \sqrt{q^2 + 4p^3}}{2} \quad \text{and} \quad \beta = \frac{-q - \sqrt{q^2 + 4p^3}}{2}$$

Suppose th p and q are real numbers. If the discriminant is positive, then one root is real, and two are complex conjugates. If the discriminant is 0, then there are three real roots, of which at least two are equal. If the discriminant is negative, then there are unequal real roots.

44.7.3 Trigonometric Solution of Cubic Polynomials

In the event that the roots of the polynomial $y^3 + 3py + q = 0$ are all real, meaning that $q^2 + 4p^3 \leq 0$, then the expressions above involve complex numbers. In that case one can also express the solution in terms of trigonometric functions. Define r and θ by

$$r = \sqrt{-p^3} \quad \text{and} \quad \theta = \cos^{-1}\frac{-q}{2r}$$

Then the three roots are given by

$$2\sqrt[3]{r}\cos\frac{\theta}{3}, \qquad 2\sqrt[3]{r}\cos\frac{\theta+2\pi}{3}, \quad \text{and } 2\sqrt[3]{r}\cos\frac{\theta+4\pi}{3}.$$

44.7.4 Quartic Polynomials

To solve the equation $ax^4 + bx^3 + cx^2 + dx + e = 0$, where $a \neq 0$, start with the substitution $y = x + \frac{b}{4a}$. That gives $y^4 + py^2 + qy + r = 0$, where

$$p = \frac{8ac - 3b^2}{8a^2}, \quad q = \frac{b^3 - 4abc + 8a^2 d}{8a^3}, \text{ and } r = \frac{16ab^2c + 256a^3e - 3b^4 - 64a^2bd}{256a^4}.$$

The *cubic resolvent* of this polynomial is defined as $t^3 - pt^2 - 4rt + (4pr - q^2) = 0$. If u is a root of the cubic resolvent, then the solution of the original quartic is given by

$$y^2 \pm \sqrt{u-p}\left(y - \frac{q}{2(u-p)}\right) + \frac{u}{2} = 0.$$

44.7.5 Quintic Polynomials

Some quintic equations are solvable by radicals. If the function $f(x) = x^5 + ax + b$ (with a and b rational) is irreducible, the $f(x) = 0$ is solvable by radicals if, and only if, numbers ε, c, and e exist (with $\varepsilon = \pm 1, c \geq 0$, and $e \neq 0$) such that

$$a = \frac{5e^4(3 - 4\varepsilon c)}{c^2 + 1} \quad \text{and} \quad b = \frac{-4e^5(11\varepsilon + 2c)}{c^2 + 1}.$$

In this case, the roots are given by $x = e(\omega^j u_1 + \omega^{2j} u_2 + \omega^{3j} u_3 + \omega^{4j} u_4)$ for $j = 0, 1, 2, 3, 4$, where ω is a fifth root of unity $(\omega = \exp(2\pi i/5))$ and

$$u_1 = \left(\frac{v_1^2 v_3}{D^2}\right), \quad u_2 = \left(\frac{v_3^2 v_4}{D^2}\right), \quad u_3 = \left(\frac{v_2^2 v_1}{D^2}\right), \quad u_4 = \left(\frac{v_4^2 v_2}{D^2}\right).$$

$$v_1 = \sqrt{D} + \sqrt{D - \varepsilon\sqrt{D}}, \qquad v_2 = -\sqrt{D} - \sqrt{D + \varepsilon\sqrt{D}}.$$

$$v_3 = -\sqrt{D} + \sqrt{D + \varepsilon\sqrt{D}}, \qquad v_4 = \sqrt{D} - \sqrt{D - \varepsilon\sqrt{D}}, \text{ and } D = c^2 + 1.$$

Example

The quintic $f(x) = x^5 + 15x + 12$ has the values $\varepsilon = -1$, $c = 4/3$, and $e = 1$. Hence, the unique real root is given by

$$x = \left(\frac{-75+21\sqrt{10}}{125}\right)^{1/5} + \left(\frac{-75-21\sqrt{10}}{125}\right)^{1/5} + \left(\frac{225+72\sqrt{10}}{125}\right)^{1/5} + \left(\frac{225-72\sqrt{10}}{125}\right)^{1/5}$$

In order to determine A_1 and B_1, multiply the euqation by $g(x)$ so that there are no denominators reamining, and substitute any $2k$ values for x, yielding $2k$ equations for A_1 and B_1.

44.7.6 Polynomial Norms

The polynomial $P(x) = \sum_{j=0}^{n} a_j x^j$ has the norms:

$$\|P\|_1 = \int_0^{2\pi} |P(e^{i\theta})| \frac{d\theta}{2\pi} \qquad |P|_1 = \sum_{j=0}^{n} |a_j|$$

$$\|P\|_2 = \left(\int_0^{2\pi} |P(e^{i\theta})|^2 \frac{d\theta}{2\pi}\right)^{1/2} \qquad |P|_2 = \left(\sum_{j=1}^{n} |a_j|^2\right)^{1/2}$$

$$\|P\|_\infty = \max_{|z|=1} |P(z)| \qquad |P|_\infty = \max_j |a_j|.$$

For the double bar norms, P is considered a function on the unit circle; for the single bar norms, P is identified with its coefficients. These norms are comparable:

$$|P|_\infty \le \|P\|_1 \le |P|_2 = \|P\|_2 \le \|P\|_\infty \le |P|_1 \le n|P|_\infty.$$

45

Calculus

45.1 *Derivatives

In the following formulas u, v, w represent functions of x, while a, c, n represent fixed real numbers. All arguments in the trigonometric functions are measured in radians, and all inverse trigonometric and hyperbolic functions represent principal values.

45.1.1. $\frac{d}{dx}(a) = 0$

45.1.2. $\frac{d}{dx}(x) = 1$

45.1.3. $\frac{d}{dx}(au) = a\frac{du}{dx}$

45.1.4. $\frac{d}{dx}(u+v-w) = \frac{du}{dx} + \frac{dv}{dx} - \frac{dw}{dx}$

45.1.5. $\frac{d}{dx}(uv) = u\frac{dv}{dx} + v\frac{du}{dx}$

45.1.6. $\frac{d}{dx}(uvw) = uv\frac{dw}{dx} + vw\frac{du}{dx} + uw\frac{dv}{dx}$

* Let $y = f(x)$ and $\frac{dy}{dx} = \frac{d[f(x)]}{dx} = f'(x)$ define respectively a function and its derivative for any value x in their common domain. The differential for the function at such a value x is accordingly defined as

$$dy = d[f(x)] = \frac{dy}{dx}dx = \frac{d[f(x)]}{dx}dx = f'(x)dx$$

Each derivative formula has an associated differential formula. For example, formula 6 above has the differential formula

$$d(uvw) = uv\,dw + vw\,du + uw\,dv$$

45.1.7. $\frac{d}{dx}\left(\frac{u}{v}\right)=\frac{v\frac{du}{dx}-u\frac{dv}{dx}}{v^2}=\frac{1}{v}\frac{du}{dx}-\frac{u}{v^2}\frac{dv}{dx}$

45.1.8. $\frac{d}{dx}(u^n)=nu^{n-1}\frac{du}{dx}$

45.1.9. $\frac{d}{dx}(\sqrt{u})=\frac{1}{2\sqrt{u}}\frac{du}{dx}$

45.1.10. $\frac{d}{dx}\left(\frac{1}{u}\right)=-\frac{1}{u^2}\frac{du}{dx}$

45.1.11. $\frac{d}{dx}[f(u)]=\frac{d}{du}[f(u)]\cdot\frac{du}{dx}$

45.1.12. $\frac{d^2}{dx^2}[f(u)]=\frac{df(u)}{du}\cdot\frac{d^2u}{dx^2}+\frac{d^2f(u)}{du^2}\cdot\left(\frac{du}{dx}\right)^2$

45.1.13. $\frac{d^n}{dx^n}[uv]=\binom{n}{0}v\frac{d^nu}{dx^n}+\binom{n}{1}\frac{dv}{dx}\frac{d^{n-1}u}{dx^{n-1}}+\binom{n}{2}\frac{d^2v}{dx^2}\frac{d^{n-2}u}{dx^{n-2}}$

$+\cdots+\binom{n}{k}\frac{d^kv}{dx^k}\frac{d^{n-k}u}{dx^{n-k}}+\cdots+\binom{n}{n}u\frac{d^nv}{dx^n}$

where $\binom{n}{r}=\frac{n!}{r!(n-r)!}$ the binomial coefficient, n non-negative integer and $\binom{n}{0}=1$.

45.1.14. $\frac{du}{dx}=\frac{1}{\frac{dx}{du}}$ if $\frac{dx}{du}\neq 0$

45.1.15. $\frac{d}{dx}(\log_a u)=(\log_a e)\frac{1}{u}\frac{du}{dx}$

45.1.16. $\frac{d}{dx}(\log_e u)=\frac{1}{u}\frac{du}{dx}$

45.1.17. $\frac{d}{dx}(a^u)=a^u(\log_e a)\frac{du}{dx}$

45.1.18. $\frac{d}{dx}(e^u)=e^u\frac{du}{dx}$

45.1.19. $\frac{d}{dx}(u^v)=vu^{v-1}\frac{du}{dx}+(\log_e u)u^v\frac{dv}{dx}$

45.1.20. $\frac{d}{dx}(\sin u)=\frac{du}{dx}(\cos u)$

45.1.21. $\frac{d}{dx}(\cos u)=-\frac{du}{dx}(\sin u)$

45.1.22. $\frac{d}{dx}(\tan u)=\frac{du}{dx}(\sec^2 u)$

45.1.23. $\frac{d}{dx}(\cot u)=-\frac{du}{dx}(\csc^2 u)$

45.1.24. $\frac{d}{dx}(\sec u)=\frac{du}{dx}\sec u\cdot\tan u$

45.1.25. $\frac{d}{dx}(\csc u) = -\frac{du}{dx}\csc u \cdot \cot u$

45.1.26. $\frac{d}{dx}(\text{vers}\, u) = \frac{du}{dx}\sin u$

45.1.27. $\frac{d}{dx}(\arcsin u) = \frac{1}{\sqrt{1-u^2}} \cdot \frac{du}{dx}, \quad \left(-\frac{\pi}{2} \le \arcsin u \le \frac{\pi}{2}\right)$

45.1.28. $\frac{d}{dx}(\arccos u) = -\frac{1}{\sqrt{1-u^2}}\frac{du}{dx}, \quad (0 \le \arccos u \le \pi)$

45.1.29. $\frac{d}{dx}(\arctan u) = \frac{1}{1+u^2}\frac{du}{dx}, \quad \left(-\frac{\pi}{2} < \arctan u < \frac{\pi}{2}\right)$

45.1.30. $\frac{d}{dx}(\text{arccot}\, u) = -\frac{1}{1+u^2}\frac{du}{dx}, \quad (0 \le \text{arccot}\, u \le \pi)$

45.1.31. $\frac{d}{dx}(\text{arcsec}\, u) = \frac{1}{u\sqrt{u^2-1}}\frac{du}{dx}, \quad \left(0 \le \text{arcsec}\, u < \frac{\pi}{2}, \ -\pi \le \text{arcsec}\, u < -\frac{\pi}{2}\right)$

45.1.32. $\frac{d}{dx}(\text{arccsc}\, u) = -\frac{1}{u\sqrt{u^2-1}}\frac{du}{dx}, \quad \left(0 < \text{arccsc}\, u \le \frac{\pi}{2}, \ -\pi < \text{arccsc}\, u \le -\frac{\pi}{2}\right)$

45.1.33. $\frac{d}{dx}(\text{arcvers}\, u) = \frac{1}{\sqrt{2u-u^2}}\frac{du}{dx}, \quad (0 \le \text{arcvers}\, u \le \pi)$

45.1.34. $\frac{d}{dx}(\sinh u) = \frac{du}{dx}(\cosh u)$

45.1.35. $\frac{d}{dx}(\cosh u) = \frac{du}{dx}(\sinh u)$

45.1.36. $\frac{d}{dx}(\tanh u) = \frac{du}{dx}(\text{sech}^2\, u)$

45.1.37. $\frac{d}{dx}(\coth u) = -\frac{du}{dx}(\text{csch}^2\, u)$

45.1.38. $\frac{d}{dx}(\text{sech}\, u) = -\frac{du}{dx}(\text{sech}\, u \cdot \tanh u)$

45.1.39. $\frac{d}{dx}(\text{csch}\, u) = -\frac{du}{dx}(\text{csch}\, u \cdot \coth u)$

45.1.40. $\frac{d}{dx}(\sinh^{-1} u) = \frac{d}{dx}[\log(u + \sqrt{u^2+1})] = \frac{1}{\sqrt{u^2+1}}\frac{du}{dx}$

45.1.41. $\frac{d}{dx}(\cosh^{-1} u) = \frac{d}{dx}[\log(u + \sqrt{u^2-1})] = \frac{1}{\sqrt{u^2-1}}\frac{du}{dx}, \quad (u > 1, \ \cosh^{-1} u > 0)$

45.1.42. $\frac{d}{dx}(\tanh^{-1} u) = \frac{d}{dx}\left[\frac{1}{2}\log\frac{1+u}{1-u}\right] = \frac{1}{1-u^2}\frac{du}{dx}, \quad (u^2 < 1)$

45.1.43. $\frac{d}{dx}(\coth^{-1} u) = \frac{d}{dx}\left[\frac{1}{2}\log\frac{u+1}{u-1}\right] = \frac{1}{1-u^2}\frac{du}{dx}, \quad (u^2 > 1)$

45.1.44. $\frac{d}{dx}(\text{sech}^{-1} u) = \frac{d}{dx}\left[\log\frac{1+\sqrt{1-u^2}}{u}\right] = -\frac{1}{u\sqrt{1-u^2}}\frac{du}{dx}, \quad (0 < u < 1)$

45.1.45. $\frac{d}{dx}(\operatorname{csch}^{-1} u) = \frac{d}{dx}\left[\log\frac{1+\sqrt{1+u^2}}{u}\right] = -\frac{1}{|u|\sqrt{1+u^2}}\frac{du}{dx}$

45.1.46. $\frac{d}{dq}\int_p^q f(x)dx = f(q), \quad [p \text{ constant}]$

45.1.47. $\frac{d}{dp}\int_p^q f(x)dx = -f(p), \quad [q \text{ constant}]$

45.1.48. $\frac{d}{da}\int_p^q f(x,a)dx = \int_p^q \frac{\partial}{\partial a}[f(x,a)]dx + f(q,a)\frac{dq}{da} - f(p,a)\frac{dp}{da}$

45.2 Integration

The following is a brief discussion of some integration techniques. A more complete discussion can be found in a number of good text books. However, the purpose of this introduction is simply to discuss a few of the important techniques which may be used, in conjunction with the integral table which follows, to integrate particular functions.

No matter ow extensive the integral table, it is a fairly uncommon occurrence to find in the table the exact integral desired. Usually some form of transformation will have to be made. The simplest type of transformation, and yet the most general, is substituion. Simple forms of substituion, such as $y = ax$, are employed almost unconsciously by experienced users of integral tables. Other substitutions may require more thought. In some sections of the tables, appropriate substitutions are suggested for integrals which are similar to, but not exactly like, integrals in the table. Finding the right substitution is largely a matter of intuition and experience.

Several precautions must be observed when using substitutions:

1. Be sure to make the substitution in the *dx* term, as well as everywhere else in the integral.
2. Be sure that the function substituted is one-to-one and continuous. If this is not the case, the integral must be restricted in such a way as to make it true. See the example following.
3. With definite integrals, the limits should also be expressed in terms of the new dependent variable. With indefinite integrals, it is necessary to perform the reverse substitutions to obtain the answer in terms of the original independent variable. This may also be done for definite integrals, but it is usually easier to change the limits.

Example

$$\int \frac{x^4}{\sqrt{a^2-x^2}}dx$$

Here we make the substitution $x = |a|\sin\theta$. Then $dx = |a|\cos\theta\, d\theta$, and

$$\sqrt{a^2-x^2} = \sqrt{a^2-a^2\sin^2\theta} = |a|\sqrt{1-\sin^2\theta} = |a\cos\theta|$$

Notice the absolute value signs. It is very important to keep in mind that a square root radical always denotes the positive square root, and to assure the sign is always kept positive. Thus $\sqrt{x^2} = |x|$. Failure to observe this is a common cause of errors in integration.

Notice also that the indicated substitution is not a one-to-one function, that is, it does not have a unique inverse. Thus we must restrict the range of θ in such a way as to make the function one-to-one. Fortunately, this is easily done by solving for θ

$$\theta = \sin^{-1}\frac{x}{|a|}$$

and restricting the inverse sine to the principal values, $-\frac{\pi}{2} \le \theta \le \frac{\pi}{2}$.

Thus the integral becomes

$$\int \frac{a^4 \sin^4\theta |a| \cos\theta \, d\theta}{|a| |\cos\theta|}$$

Now, however, in the range of values chosen for θ, cosθ is always positive. Thus we may remove the absolute value signs from cosθ in the denominator. (This is one of the reasons that the principal values of the inverse trigonometric functions are defined as they are.)

Then the cosθ terms cancel, and the integral becomes

$$a^4 \int \sin^4\theta \, d\theta$$

By application of integral formulas 45.3.14.9 and 45.3.14.7, we integrate this to

$$-a^4 \frac{\sin^3\theta\cos\theta}{4} - \frac{3a^4}{8}\cos\theta\sin\theta + \frac{3a^4}{8}\theta + C$$

We now must perform the inverse substitution to get the result in terms of x. We have

$$\theta = \sin^{-1}\frac{x}{|a|}$$

$$\sin\theta = \frac{x}{|a|}$$

Then

$$\cos\theta = \pm\sqrt{1-\sin^2\theta} = \pm\sqrt{1-\frac{x^2}{a^2}} = \pm\frac{\sqrt{a^2-x^2}}{|a|}.$$

Because of the previously mentioned fact that cosθ is positive, we may omit the ± sign. The reverse substitution then produces the final answer

$$\int \frac{x^4}{\sqrt{a^2-x^2}}dx = -\frac{1}{4}x^3\sqrt{a^2-x^2} - \frac{3}{8}a^2x\sqrt{a^2-x^2} + \frac{3a^4}{8}\sin^{-1}\frac{x}{|a|} + C.$$

Any rational function of x may be integrated, if the denominator is factored into linear and irreducible quadratic factors. The function may then be broken into partial fractions, and the individual partial fractions integrated by use of the appropriate formula from the integral table. See the section on partial fractions for further information.

Many integrals may be reduced to rational functions by proper substitutions. For example,

$$z = \tan\frac{x}{2}$$

will reduce any rational function of the six trigonometric functions of x to a rational function of z. (Frequently there are other substitutions which are simpler to use, but this one will always work. See integral formula number 45.3.16.5.)

Any rational function of x and $\sqrt{ax+b}$ may be reduced to a rational function of z by making the substitution

$$z = \sqrt{ax+b}$$

Other likely substitutions will be suggested by looking at the form of the integrand.

The other main method of transforming integrals is integration by parts. This involves applying formula number 5 or 6 in the accompanying integral table. The critical factor in this method is the choice of the functions u and v. In order for the method to be successful, $v = \int dv$ and $\int v\,du$ must be easier to integrate than the original integral. Again, this choice is largely a matter of intuition and experience.

Example

$$\int x \sin x\,dx$$

Two obvious choices $u = x$, $dv = \sin x\,dx$, or $u = \sin x$, $dv = x\,dx$. Since a preliminary mental calculation indicates that $\int v\,du$ in the second choice would be more, rather than less, complicated than the original integral (it would contain x^2), we use the first choice.

$$u = x \qquad du = dx$$

$$dv = \sin x\,dx \qquad v = -\cos x$$

$$\int x\sin x\,dx = \int u\,dv = uv - \int v\,du = -x\cos x + \int \cos x\,dx$$

$$= \sin x - x\cos x$$

Of course, this result could have been obtained directly from the integral table, but it provides a simple example of the method. In more complicated examples the choice of u and v may not be so obvious, and several different choices may have to be tried. Of course, there is no guarantee that any of them will work.

Integration by parts may be applied more than once, or combined with substitution. A fairly common case is illustrated by the following example.

Example

$$\int e^x \sin x\,dx$$

Let

$$u = e^x \qquad \text{Then} \qquad du = e^x\,dx$$

$$dv = \sin x\,dx \qquad\qquad v = -\cos x$$

$$\int e^x \sin x\,dx = \int u\,dv = uv - \int v\,du = -e^x \cos x + \int e^x \cos x\,dx$$

In this latter integral,

$$\text{let} \quad u = e^x \qquad \text{Then} \qquad du = e^x\,dx$$

$$dv = \cos x\,dx \qquad\qquad v = \sin x$$

$$\int e^x \sin x\,dx = -e^x \cos x + \int e^x \cos x\,dx = -e^x \cos x + \int u\,dv$$

$$= -e^x \cos x + uv - \int v\,du$$

$$= -e^x \cos x + e^x \sin x - \int e^x \sin x\,dx$$

This looks as if a circular transformation has taken place, since we are back at the same integral we started from. However, the above equation can be solved algebraically for the required integral:

$$\int e^x \sin x\,dx = \tfrac{1}{2}(e^x \sin x - e^x \cos x)$$

In the second integration by parts, if the parts had been chosen as $u = \cos x$, $dv = e^x\,dx$. we would indeed have made a circular transformation, and returned to the starting place. In general, when doing repeated integration by parts, one should never choose the function u at any stage to be the same as the function v at the previous state, or a constant times the previous v.

The following rule is called the extended rule for integration by parts. It is the results of $n+1$ successive applications of integration by parts.

If

$$g_1(x) = \int g(x)\,dx, \qquad g_2(x) = \int g_1(x)\,dx,$$

$$g_3(x) = \int g_2(x)\,dx, \ldots, g_m(x) = \int g_{m-1}(x)\,dx, \ldots,$$

then

$$\int f(x)\cdot g(x)\,dx = f(x)\cdot g_1(x) - f'(x)\cdot g_2(x) + f''(x)\cdot g_3(x) - + \cdots$$

$$+(-1)^n f^{(n)}(x) g_{n+1}(x) + (-1)^{n+1} \int f^{(n+1)}(x) g_{n+1}(x)\,dx.$$

A useful special case of the above rule is when $f(x)$ is a polynomial of degree n. Then $f^{(n+1)}(x) = 0$, and

$$\int f(x)\cdot g(x)dx = f(x)\cdot g_1(x) - f'(x)\cdot g_2(x) + f''(x)\cdot g_3(x) - + \cdots + (-1)^n f^{(n)}(x)g_{n+1}(x) + C$$

Example

If $f(x) = x^2$, $g(x) = \sin x$

$$\int x^2 \sin x\, dx = -x^2 \cos x + 2x \sin x + 2\cos x + C$$

Another application of this formula occurs if

$$f''(x) = a\, f(x) \quad \text{and} \quad g''(x) = bg(x),$$

where a and b are unequal constants. In this case, by a process similar to that used in the above example for $\int e^x \sin x\, dx$, we get the formula

$$f(x)g(x)dx = \frac{f(x)\cdot g'(x) - f'(x)\cdot g(x)}{b-a} + C$$

This formula could have been used in the example mentioned. Here is another example.

Example

If $f(x) = e^{2x}$, $g(x) = \sin 3x$, then $a = 4$, $b = -9$, and

$$\int e^{2x} \sin 3x\, dx = \frac{3e^{2x}\cos 3x - 2e^{2x}\sin 3x}{-9-4} + C = \frac{e^{2x}}{13}(2\sin 3x - 3\cos 3x) + C$$

The following additional points should be observed when using this table.

1. A constant of integration is to be supplied with the answers for indefinite integrals.
2. Logarithmic expressions are to base e = 2.71828 …, unless otherwise specified, and are to be evaluated for the absolute value of the arguments involved therein.
3. All angles are measured in radians, and inverse trigonometric and hyperbolic functions represent principal values, unless otherwise indicated.
4. If the application of a formula produces either a zero denominator or the square root of a negative number in the result, there is always available another form of the answer which avoids this difficulty. In many of the results, the excluded values are specified, but when such are omitted it is presumed that one can tell what these should be, especially when difficulties of the type herein mentioned are obtained.
5. When inverse trigonometric functions occur in the integrals, be sure that any replacements made for them are strictly in accordance with the rules for such functions. This causes little difficulty when the argument of the inverse trigonometric function is positive, since then all angles involved are in the first quadrant. However, if the argument is negative, special care must be used. Thus if $u > 0$,

$$\sin^{-1} u = \cos^{-1}\sqrt{1-u^2} = \csc^{-1}\frac{1}{u}, \text{ etc.}$$

However, if $u < 0$,

$$\sin^{-1} u = -\cos^{-1}\sqrt{1-u^2} = -\pi - \csc^{-1}\frac{1}{u}, \text{ etc.}$$

See the section on inverse trigonometric functions for a full treatment of the allowable substitutions.

6. In integrals 259–263, the right side includes expressions of the form

$$A\tan^{-1}[B + C\tan f(x)].$$

In these formulas, the $\tan^{-1}$ does not necessarily represent the principal value. Instead of always emplying the principal branch of the inverse tangent function, one must instead use that branch of the inverse tangent function upon which $f(x)$ lies for any particular choice of x.

Example

Using Integral Formula 45.3.14.24.

$$\int_0^{4\pi} \frac{dx}{2+\sin x} = \frac{2}{\sqrt{3}}\left[\tan^{-1}\frac{2\tan\frac{x}{2}+1}{\sqrt{3}}\right]_0^{4\pi}$$

$$= \frac{2}{\sqrt{3}}\left[\tan^{-1}\frac{2\tan 2\pi+1}{\sqrt{3}} - \tan^{-1}\frac{2\tan 0+1}{\sqrt{3}}\right]$$

$$= \frac{2}{\sqrt{3}}\left[\frac{13\pi}{6} - \frac{\pi}{6}\right] = \frac{4\pi}{\sqrt{3}} = \frac{4\sqrt{3}\pi}{3}$$

Here

$$\tan^{-1}\frac{2\tan 2\pi+1}{\sqrt{3}} = \tan^{-1}\frac{1}{\sqrt{3}} = \frac{13\pi}{6},$$

since $f(x) = 2\pi$; and

$$\tan^{-1}\frac{2\tan 0+1}{\sqrt{3}} = \tan^{-1}\frac{1}{\sqrt{3}} = \frac{\pi}{6},$$

since $f(x) = 0$.

45.3 Integrals

45.3.1 Elementary Forms

45.3.1.1. $\int a\,dx = ax$

45.3.1.2. $\int a\cdot f(x)\,dx = a\int f(x)\,dx$

45.3.1.3. $\int \phi(y)\,dx = \int \frac{\phi(y)}{y'}\,dy,\quad$ where $y' = \frac{dy}{dx}$

45.3.1.4. $\int (u+v)\,dx = \int u\,dx + \int v\,dx,\quad$ where u and v are any functions of x

45.3.1.5. $\int u\,dv = u\int dv - \int v\,du = uv - \int v\,du$

45.3.1.6. $\int u\frac{dv}{dx}dx = uv - \int v\frac{du}{dx}dx$

45.3.1.7. $\int x^n\,dx = \frac{x^{n+1}}{n+1}$, except $n = -1$

45.3.1.8. $\int \frac{f'(x)dx}{f(x)} = \log f(x), \qquad (df(x) = f'(x)dx)$

45.3.1.9. $\int \frac{dx}{x} = \log x$

45.3.1.10. $\int \frac{f'(x)dx}{2\sqrt{f(x)}} = \sqrt{f(x)}, \qquad (df(x) = f'(x)dx)$

45.3.1.11. $\int e^x\,dx = e^x$

45.3.1.12. $\int e^{ax}\,dx = e^{ax}/a$

45.3.1.13. $\int b^{ax}\,dx = \frac{b^{ax}}{a\log b}, \qquad (b > 0)$

45.3.1.14. $\int \log x\,dx = x\log x - x$

45.3.1.15. $\int a^x \log a\,dx = a^x, \qquad (a > 0)$

45.3.1.16. $\int \frac{dx}{a^2 + x^2} = \frac{1}{a}\tan^{-1}\frac{x}{a}$

45.3.1.17. $$\int \frac{dx}{a^2 - x^2} = \begin{cases} \frac{1}{a}\tanh^{-1}\frac{x}{a} \\ \text{or} \\ \frac{1}{2a}\log\frac{a+x}{a-x}, \qquad (a^2 > x^2) \end{cases}$$

45.3.1.18. $$\int \frac{dx}{x^2 - a^2} = \begin{cases} -\frac{1}{a}\coth^{-1}\frac{x}{a} \\ \text{or} \\ \frac{1}{2a}\log\frac{x-a}{x+a}, \qquad (x^2 > a^2) \end{cases}$$

45.3.1.19. $$\int \frac{dx}{\sqrt{a^2 - x^2}} = \begin{cases} \sin^{-1}\frac{x}{|a|} \\ \text{or} \\ -\cos^{-1}\frac{x}{|a|}, \end{cases}$$

45.3.1.20. $\int \frac{dx}{\sqrt{x^2 \pm a^2}} = \log(x + \sqrt{x^2 \pm a^2})$

45.3.1.21. $\int \frac{dx}{x\sqrt{x^2 - a^2}} = \frac{1}{|a|}\sec^{-1}\frac{x}{a}$

45.3.1.22. $\int \frac{dx}{x\sqrt{a^2 \pm x^2}} = -\frac{1}{a}\log\left(\frac{a+\sqrt{a^2 \pm x^2}}{x}\right)$

45.3.1.23. $$\int \frac{dx}{x\sqrt{a+bx}} = \begin{cases} \frac{2}{\sqrt{-a}}\tan^{-1}\sqrt{\frac{a+bx}{-a}}, & (a<0) \\ \text{or} & \\ \frac{-2}{\sqrt{a}}\tanh^{-1}\sqrt{\frac{a+bx}{a}} & \\ \text{or} & \\ \frac{1}{\sqrt{a}}\log\frac{\sqrt{a+bx}-\sqrt{a}}{\sqrt{a+bx}+\sqrt{a}} & \end{cases}$$

45.3.2 Forms Containing $(a+bx)$

For forms containg $(a+bx)$ but not listed in the table, the substitution $u = \frac{a+bx}{x}$ may prove helpful.

45.3.2.1. $\int (a+bx)^n\,dx = \frac{(a+bx)^{n+1}}{(n+1)b}, \quad (n \neq -1)$

45.3.2.2. $\int x(a+bx)^n\,dx = \frac{1}{b^2(n+2)}(a+bx)^{n+2} - \frac{a}{b^2(n+1)}(a+bx)^{n+1}, \quad (n \neq -1, -2)$

45.3.2.3. $\int x^2(a+bx)^n\,dx = \frac{1}{b^3}\left[\frac{(a+bx)^{n+3}}{n+3} - 2a\frac{(a+bx)^{n+2}}{n+2} + a^2\frac{(a+bx)^{n+1}}{n+1}\right]$

45.3.2.4. $\int x^m(a+bx)^n\,dx = \frac{x^{m+1}(a+bx)^n}{m+n+1} + \frac{an}{m+n+1}\int x^m(a+bx)^{n-1}\,dx$

45.3.2.5. $$\int x^m(a+bx)^n\,dx = \begin{cases} \frac{1}{a(n+1)}\left[-x^{m+1}(a+bx)^{n+1} + (m+n+2)\int x^m(a+bx)^{n+1}\,dx\right] \\ \text{or} \\ \frac{1}{b(m+n+1)}\left[x^m(a+bx)^{n+1} - ma\int x^{m-1}(a+bx)^n\,dx\right] \end{cases}$$

45.3.2.6. $\int \frac{dx}{a+bx} = \frac{1}{b}\log(a+bx)$

45.3.2.7. $\int \frac{dx}{(a+bx)^2} = -\frac{1}{b(a+bx)}$

45.3.2.8. $\int \frac{dx}{(a+bx)^3} = -\frac{1}{2b(a+bx)^2}$

45.3.2.9. $$\int \frac{x\,dx}{a+bx} = \begin{cases} \frac{1}{b^2}[a+bx-a\log(a+bx)] \\ \text{or} \\ \frac{x}{b} - \frac{a}{b^2}\log(a+bx) \end{cases}$$

45.3.2.10. $\int \frac{x\,dx}{(a+bx)^2} = \frac{1}{b^2}\left[\log(a+bx) + \frac{a}{a+bx}\right]$

45.3.2.11. $\int \frac{x\,dx}{(a+bx)^3} = \frac{1}{b^2}\left[-\frac{1}{a+bx} + \frac{a}{2(a+bx)^2}\right]$

45.3.2.12. $\int \frac{x\,dx}{(a+bx)^n} = \frac{1}{b^2}\left[\frac{-1}{(n-2)(a+bx)^{n-2}} + \frac{a}{(n-1)(a+bx)^{n-1}}\right], \quad n \neq 1, 2$

45.3.2.13. $\int \frac{x^2\,dx}{a+bx} = \frac{1}{b^3}\left[\frac{1}{2}(a+bx)^2 - 2a(a+bx) + a^2\log(a+bx)\right]$

45.3.2.14. $\int \frac{x^2\,dx}{(a+bx)^2} = \frac{1}{b^3}\left[a+bx - 2a\log(a+bx) - \frac{a^2}{a+bx}\right]$

45.3.2.15. $\int \frac{x^2\,dx}{(a+bx)^3} = \frac{1}{b^3}\left[\log(a+bx) + \frac{2a}{a+bx} - \frac{a^2}{2(a+bx)^2}\right]$

45.3.2.16. $\int \frac{x^2\,dx}{(a+bx)^n} =$

$$\frac{1}{b^3}\left[\frac{-1}{(n-3)(a+bx)^{n-3}} + \frac{2a}{(n-2)(a+bx)^{n-2}} - \frac{a^2}{(n-1)(a+bx)^{n-1}}\right], \quad n \neq 1, 2, 3$$

45.3.2.17. $\int \frac{dx}{x(a+bx)} = -\frac{1}{a}\log\frac{a+bx}{x}$

45.3.2.18. $\int \frac{dx}{x(a+bx)^2} = \frac{1}{a(a+bx)} - \frac{1}{a^2}\log\frac{a+bx}{x}$

45.3.2.19. $\int \frac{dx}{x(a+bx)^3} = \frac{1}{a^3}\left[\frac{1}{2}\left(\frac{2a+bx}{a+bx}\right)^2 + \log\frac{x}{a+bx}\right]$

45.3.2.20. $\int \frac{dx}{x^2(a+bx)} = -\frac{1}{ax} + \frac{b}{a^2}\log\frac{a+bx}{x}$

45.3.2.21. $\int \frac{dx}{x^3(a+bx)} = \frac{2bx-a}{2a^2x^2} + \frac{b^2}{a^3}\log\frac{x}{a+bx}$

45.3.2.22. $\int \frac{dx}{x^2(a+bx)^2} = -\frac{a+2bx}{a^2x(a+bx)} + \frac{2b}{a^3}\log\frac{a+bx}{x}$

45.3.3 Forms Containing $c^2 \pm x^2$, $x^2 - c^2$

45.3.3.1. $\int \frac{dx}{c^2+x^2} = \frac{1}{c}\tan^{-1}\frac{x}{c}$

45.3.3.2. $\int \frac{dx}{ax^2+c} = \frac{1}{\sqrt{ac}}\tan^{-1}\left(x\sqrt{\frac{a}{c}}\right), \quad (a, c > 0)$

45.3.3.3. $\int \frac{dx}{c^2-x^2} = \frac{1}{2c}\log\frac{c+x}{c-x}, \quad (c^2 > x^2)$

45.3.3.4. $$\int \frac{dx}{ax^2+c} = \begin{cases} \frac{1}{2\sqrt{-ac}}\log\frac{x\sqrt{a}-\sqrt{-c}}{x\sqrt{a}+\sqrt{-c}}, & (a>0,\ c<0) \\ \text{or} & \\ \frac{1}{2\sqrt{-ac}}\log\frac{\sqrt{c}+x\sqrt{-a}}{\sqrt{c}-x\sqrt{-a}}, & (a<0,\ c>0) \end{cases}$$

45.3.3.5. $\int \frac{dx}{x^2 - c^2} = \frac{1}{2c}\log\frac{x-c}{x+c}, \quad (x^2 > c^2)$

45.3.4 Forms Containing $a + bx$ and $a' + b'x$

45.3.4.1. $\int \frac{dx}{(a+bx)(a'+b'x)} = \frac{1}{ab' - a'b} \cdot \log\left(\frac{a'+b'x}{a+bx}\right)$

45.3.4.2. $\int \frac{x\,dx}{(a+bx)(a'+b'x)} = \frac{1}{ab' - a'b}\left[\frac{a}{b}\log(a+bx) - \frac{a'}{b'}\log(a'+b'x)\right]$

45.3.4.3. $\int \frac{dx}{(a+bx)^2(a'+b'x)} = \frac{1}{ab' - a'b}\left(\frac{1}{a+bx} + \frac{b'}{ab'-a'b}\log\frac{a'+b'x}{a+bx}\right)$

45.3.4.4. $\int \frac{x\,dx}{(a+bx)^2(a'+b'x)} = \frac{-a}{b(ab'-a'b)(a+bx)} - \frac{a'}{(ab'-a'b)^2}\log\frac{a'+b'x}{a+bx}$

45.3.4.5. $\int \frac{x^2\,dx}{(a+bx)^2(a'+b'x)} =$

$$\frac{a^2}{b^2(ab'-a'b)(a+bx)} + \frac{1}{(ab'-a'b)^2}\left[\frac{a'^2}{b'}\log(a'+b'x) + \frac{a(ab'-2a'b)}{b^2}\log(a+bx)\right]$$

45.3.4.6. $\int \frac{dx}{(a+bx)^n(a'+b'x)^m} =$

$$\frac{1}{(m-1)(ab'-a'b)}\left[\frac{-1}{(a+bx)^{n-1}(a'+b'x)^{m-1}} - (m+n-2)b\int \frac{dx}{(a+bx)^n(a'+b'x)^{m-1}}\right]$$

45.3.4.7. $\int \frac{a+bx}{a'+b'x}dx = \frac{bx}{b'} + \frac{ab'-a'b}{b'^2}\log(a'+b'x)$

45.3.4.8.
$$\int \frac{(a+bx)^m\,dx}{(a'+b'x)^n} = \begin{cases} -\dfrac{1}{(n-1)(ab'-a'b)}\left[\dfrac{(a+bx)^{m+1}}{(a'+b'x)^{n-1}} + b(n-m-2)\displaystyle\int \dfrac{(a+bx)^m\,dx}{(a'+b'x)^{n-1}}\right] \\ \text{or} \\ -\dfrac{1}{b'(n-m-1)}\left[\dfrac{(a+bx)^m}{(a'+b'x)^{n-1}} + m(ab'-a'b)\displaystyle\int \dfrac{(a+bx)^{m-1}\,dx}{(a'+b'x)^n}\right] \\ \text{or} \\ \dfrac{-1}{(n-1)b'}\left[\dfrac{(a+bx)^m}{(a'+b'x)^{n-1}} - mb\displaystyle\int \dfrac{(a+bx)^{m-1}\,dx}{(a'+b'x)^{n-1}}\right] \end{cases}$$

45.3.5 Forms Containing $\sqrt{a+bx} = \sqrt{u}$ and $\sqrt{a'+b'x} = \sqrt{v}$ with $k = ab' - a'b$

If $k = 0$, then $v = \frac{a'}{a}u$, and other formulas should be used.

45.3.5.1. $\int \sqrt{uv}\,dx = \frac{k+2bv}{4bb'}\sqrt{uv} - \frac{k^2}{8bb'}\int \frac{dx}{\sqrt{uv}}$

45.3.5.2.
$$\int \frac{dx}{v\sqrt{u}} = \begin{cases} \dfrac{1}{\sqrt{kb'}}\log\dfrac{b'\sqrt{u} - \sqrt{kb'}}{b'\sqrt{u} + \sqrt{kb'}} \\ \text{or} \\ \dfrac{2}{\sqrt{-kb'}}\tan^{-1}\dfrac{b'\sqrt{u}}{\sqrt{-kb'}} \end{cases}$$

45.3.5.3. $$\int \frac{dx}{\sqrt{uv}} = \begin{cases} \frac{2}{\sqrt{bb'}} \tanh^{-1} \frac{\sqrt{bb'uv}}{bv}, & (bb' > 0) \\ \text{or} & \\ \frac{1}{\sqrt{bb'}} \log \frac{bv + \sqrt{bb'uv}}{bv - \sqrt{bb'uv}}, & (bb' > 0) \\ \text{or} & \\ \frac{1}{\sqrt{bb'}} \log \frac{(bv + \sqrt{bb'uv})^2}{|v|}, & (bb' > 0) \\ \text{or} & \\ \frac{2}{\sqrt{-bb'}} \tan^{-1} \frac{\sqrt{-bb'uv}}{bv}, & (bb' < 0) \end{cases}$$

45.3.5.4. $$\int \frac{x\,dx}{\sqrt{uv}} = \frac{\sqrt{uv}}{bb'} - \frac{ab' + a'b}{2bb'} \int \frac{dx}{\sqrt{uv}}$$

45.3.5.5. $$\int \frac{dx}{v\sqrt{uv}} = \frac{-2\sqrt{uv}}{kv} \quad \text{or} \quad = \frac{-2\sqrt{u}}{k\sqrt{v}}, \quad (v > 0)$$

45.3.5.6. $$\int \frac{\sqrt{v}}{\sqrt{u}} dx = \int \frac{v\,dx}{\sqrt{uv}} \quad \text{or} \quad = \frac{\sqrt{uv}}{b} - \frac{k}{2b} \int \frac{dx}{\sqrt{uv}}, \quad (v > 0)$$

45.3.5.7. $$\int v^m \sqrt{u}\,dx = \frac{1}{(2m+3)b'} \left(2v^{m+1}\sqrt{u} + k \int \frac{v^m\,dx}{\sqrt{u}} \right)$$

45.3.5.8. $$\int \frac{dx}{v^m \sqrt{u}} = -\frac{1}{(m-1)k} \left(\frac{\sqrt{u}}{v^{m-1}} + \left(m - \frac{3}{2} \right) b \int \frac{dx}{v^{m-1}\sqrt{u}} \right)$$

45.3.6 Forms Containing $(a + bx^n)$

45.3.6.1. $$\int \frac{dx}{a + bx^2} = \frac{1}{\sqrt{ab}} \tan^{-1} \frac{x\sqrt{ab}}{a}$$

45.3.6.2. $$\int \frac{dx}{a + bx^2} = \begin{cases} \frac{1}{2\sqrt{-ab}} \log \frac{a + x\sqrt{-ab}}{a - x\sqrt{-ab}}, & \\ \text{or} & \\ \frac{1}{\sqrt{-ab}} \tanh^{-1} \frac{x\sqrt{-ab}}{a}, & (ab < 0) \end{cases}$$

45.3.6.3. $$\int \frac{dx}{a^2 + b^2x^2} = \frac{1}{ab} \tan^{-1} \frac{bx}{a}$$

45.3.6.4. $$\int \frac{x\,dx}{a + bx^2} = \frac{1}{2b} \log(a + bx^2)$$

45.3.6.5. $$\int \frac{x\,dx}{a^2 + b^2x^2} = \frac{1}{2b^2} \log(a^2 + b^2x^2)$$

45.3.6.6. $$\int \frac{x^2\,dx}{a + bx^2} = \frac{x}{b} - \frac{a}{b} \int \frac{dx}{a + bx^2}$$

45.3.6.7. $$\int \frac{dx}{(a + bx^2)^2} = \frac{x}{2a(a + bx^2)} + \frac{1}{2a} \int \frac{dx}{a + bx^2}$$

45.3.6.8. $\int \frac{dx}{(x^2+a^2)^2} = \frac{1}{2a^3}\tan^{-1}\frac{x}{a} + \frac{x}{2a^2(x^2+a^2)}$

45.3.6.9. $\int \frac{dx}{a^2-b^2x^2} = \frac{1}{2ab}\log\frac{a+bx}{a-bx}$

45.3.6.10. $\int \frac{dx}{(x^2-a^2)^2} = -\frac{x}{2a^2(x^2-a^2)} + \frac{1}{4a^3}\log\frac{a+x}{a-x}$

45.3.6.11. $\int \frac{dx}{(a+bx^2)^{m+1}} = \frac{1}{2ma}\frac{x}{(a+bx^2)^m} + \frac{2m-1}{2ma}\int \frac{dx}{(a+bx^2)^m}$

45.3.6.12. $\int \frac{x\,dx}{(a+bx^2)^{m+1}} = -\frac{1}{2bm(a+bx^2)^m}$

45.3.6.13. $\int \frac{x^2\,dx}{(a+bx^2)^{m+1}} = \frac{-x}{2mb(a+bx^2)^m} + \frac{1}{2mb}\int \frac{dx}{(a+bx^2)^m}$

45.3.6.14. $\int \frac{dx}{x(a+bx^2)} = \frac{1}{2a}\log\frac{x^2}{a+bx^2}$

45.3.6.15. $\int \frac{dx}{x^2(a+bx^2)} = -\frac{1}{ax} - \frac{b}{a}\int \frac{dx}{a+bx^2}$

45.3.6.16. $\int \frac{dx}{x(a+bx^2)^{m+1}} = \frac{1}{2am(a+bx^2)^m} + \frac{1}{a}\int \frac{dx}{x(a+bx^2)^m}, \quad (m \neq 0)$

45.3.6.17. $\int \frac{dx}{x^2(a+bx^2)^{m+1}} = \frac{1}{a}\int \frac{dx}{x^2(a+bx^2)^m} - \frac{b}{a}\int \frac{dx}{(a+bx^2)^{m+1}}$

45.3.6.18. $\int \frac{dx}{a+bx^3} = \frac{k}{3a}\left[\frac{1}{2}\log\frac{(k+x)^2}{k^2-kx+x^2} + \sqrt{3}\tan^{-1}\frac{2x-k}{k\sqrt{3}}\right], \quad (bk^3 = a)$

45.3.6.19. $\int \frac{x\,dx}{a+bx^3} = \frac{1}{3bk}\left[\frac{1}{2}\log\frac{k^2-kx+x^2}{(k+x)^2} + \sqrt{3}\tan^{-1}\frac{2x-k}{k\sqrt{3}}\right], \quad (bk^3 = a)$

45.3.6.20. $\int \frac{dx}{x(a+bx^n)} = \frac{1}{an}\log\frac{x^n}{a+bx^n}$

45.3.6.21. $\int \frac{dx}{(a+bx^n)^{m+1}} = \frac{1}{a}\int \frac{dx}{(a+bx^n)^m} - \frac{b}{a}\int \frac{x^n\,dx}{(a+bx^n)^{m+1}}$

45.3.6.22. $\int \frac{x^m\,dx}{(a+bx^n)^{p+1}} = \frac{1}{b}\int \frac{x^{m-n}\,dx}{(a+bx^n)^p} - \frac{a}{b}\int \frac{x^{m-n}\,dx}{(a+bx^n)^{p+1}}$

45.3.6.23. $\int \frac{dx}{x^m(a+bx^n)^{p+1}} = \frac{1}{a}\int \frac{dx}{x^m(a+bx^n)^p} - \frac{b}{a}\int \frac{dx}{x^{m-n}(a+bx^n)^{p+1}}$

45.3.6.24. $\int x^m(a+bx^n)^p\,dx = \frac{x^{m-n+1}(a+bx^n)^{p+1}}{b(np+m+1)} - \frac{a(m-n+1)}{b(np+m+1)}\int x^{m-n}(a+bx^n)^p\,dx$

45.3.6.25. $\int x^m(a+bx^n)^p\,dx = \frac{x^{m+1}(a+bx^n)^p}{np+m+1} + \frac{anp}{np+m+1}\int x^m(a+bx^n)^{p-1}\,dx$

45.3.6.26. $\int x^{m-1}(a+bx^n)^p\,dx = \frac{1}{b(m+np)}\left[x^{m-n}(a+bx^n)^{p+1} - (m-n)a\int x^{m-n-1}(a+bx^n)^p\,dx\right]$

45.3.6.27. $\int x^{m-1}(a+bx^n)^p\,dx = \frac{1}{m+np}\left[x^m(a+bx^n)^p + npa\int x^{m-1}(a+bx^n)^{p-1}\,dx\right]$

45.3.6.28. $\int x^{m-1}(a+bx^n)^p\,dx = \frac{1}{ma}\left[x^m(a+bx^n)^{p+1} - (m+np+n)b\int x^{m+n-1}(a+bx^n)^p\,dx\right]$

45.3.6.29. $\int x^{m-1}(a+bx^n)^p\,dx = \frac{1}{an(p+1)}\left[-x^m(a+bx^n)^{p+1} + (m+np+n)\int x^{m-1}(a+bx^n)^{p+1}\,dx\right]$

45.3.7 Forms Containing $(a+bx+cx^2)$

$$X = a+bx+cx^2 \text{ and } q = 4ac-b^2$$

If $q=0$, then $X = c\left(x+\frac{b}{2c}\right)^2$, and other formulas should be used.

45.3.7.1. $\int \frac{dx}{X} = \frac{2}{\sqrt{q}}\tan^{-1}\frac{2cx+b}{\sqrt{q}}$

45.3.7.2. $\int \frac{dx}{X} = \frac{-2}{\sqrt{-q}}\tanh^{-1}\frac{2cx+b}{\sqrt{-q}}$

45.3.7.3. $\int \frac{dx}{X} = \frac{1}{\sqrt{-q}}\log\frac{2cx+b-\sqrt{-q}}{2cx+b+\sqrt{-q}}$

45.3.7.4. $\int \frac{dx}{X^2} = \frac{2cx+b}{qX} + \frac{2c}{q}\int\frac{dx}{X}$

45.3.7.5. $\int \frac{dx}{X^3} = \frac{2cx+b}{q}\left(\frac{1}{2X^2}+\frac{3c}{qX}\right) + \frac{6c^2}{q^2}\int\frac{dx}{X}$

45.3.7.6. $\int \frac{dx}{X^{n+1}} = \begin{cases} \frac{2cx+b}{nqX^n} + \frac{2(2n-1)c}{qn}\int\frac{dx}{X^n} \\ \text{or} \\ \frac{(2n)!}{(n!)^2}\left(\frac{c}{q}\right)^n\left[\frac{2cx+b}{q}\sum_{r=1}^{n}\left(\frac{q}{cX}\right)^r\left(\frac{(r-1)!r!}{(2r)!}\right)+\int\frac{dx}{X}\right] \end{cases}$

45.3.7.7. $\int \frac{x\,dx}{X} = \frac{1}{2c}\log X - \frac{b}{2c}\int\frac{dx}{X}$

45.3.7.8. $\int \frac{x\,dx}{X^2} = -\frac{bx+2a}{qX} - \frac{b}{q}\int\frac{dx}{X}$

45.3.7.9. $\int \frac{x\,dx}{X^{n+1}} = -\frac{2a+bx}{nqX^n} - \frac{b(2n-1)}{nq}\int\frac{dx}{X^n}$

45.3.7.10. $\int \frac{x^2}{X}dx = \frac{x}{c} - \frac{b}{2c^2}\log X + \frac{b^2-2ac}{2c^2}\int\frac{dx}{X}$

45.3.7.11. $\int \frac{x^2}{X^2}dx = \frac{(b^2-2ac)x+ab}{cqX} + \frac{2a}{q}\int\frac{dx}{X}$

45.3.7.12. $\int \frac{x^m\,dx}{X^{n+1}} = -\frac{x^{m-1}}{(2n-m+1)cX^n} - \frac{n-m+1}{2n-m+1}\cdot\frac{b}{c}\int\frac{x^{m-1}\,dx}{X^{n+1}} + \frac{m-1}{2n-m+1}\cdot\frac{a}{c}\int\frac{x^{m-2}\,dx}{X^{n+1}}$

45.3.7.13. $\int \frac{dx}{xX} = \frac{1}{2a}\log\frac{x^2}{X} - \frac{b}{2a}\int\frac{dx}{X}$

45.3.7.14. $\int \frac{dx}{x^2 X} = \frac{b}{2a^2}\log\frac{X}{x^2} - \frac{1}{ax} + \left(\frac{b^2}{2a^2} - \frac{c}{a}\right)\int \frac{dx}{X}$

45.3.7.15. $\int \frac{dx}{xX^n} = \frac{1}{2a(n-1)X^{n-1}} - \frac{b}{2a}\int \frac{dx}{X^n} + \frac{1}{a}\int \frac{dx}{xX^{n-1}}$

45.3.7.16. $\int \frac{dx}{x^m X^{n+1}} = -\frac{1}{(m-1)ax^{m-1}X^n} - \frac{n+m-1}{m-1}\frac{b}{a}\int \frac{dx}{x^{m-1}X^{n+1}} - \frac{2n+m-1}{m-1}\cdot\frac{c}{c}\int \frac{dx}{x^{m-2}X^{n+1}}$

45.3.8 Forms Containing $\sqrt{a+bx}$

45.3.8.1. $\int \sqrt{a+bx}\,dx = \frac{2}{3b}\sqrt{(a+bx)^3}$

45.3.8.2. $\int x\sqrt{a+bx}\,dx = -\frac{2(2a-3bx)\sqrt{(a+bx)^3}}{15b^2}$

45.3.8.3. $\int x^2\sqrt{a+bx}\,dx = \frac{2(8a^2-12abx+15b^2x^2)\sqrt{(a+bx)^3}}{105b^3}$

45.3.8.4. $\int x^m\sqrt{a+bx}\,dx = \frac{2}{b(2m+3)}\left[x^m\sqrt{(a+bx)^3} - ma\int x^{m-1}\sqrt{a+bx}\,dx\right]$

45.3.8.5. $\int \frac{\sqrt{a+bx}}{x}dx = 2\sqrt{a+bx} + a\int \frac{dx}{x\sqrt{a+bx}}$

(see No. 45.3.8.12 and No. 45.3.8.13)

45.3.8.6. $\int \frac{\sqrt{a+bx}}{x^2}dx = -\frac{\sqrt{a+bx}}{x} + \frac{b}{2}\int \frac{dx}{x\sqrt{a+bx}}$

(see No. 45.3.8.12 and No. 45.3.8.13)

45.3.8.7. $\int \frac{\sqrt{a+bx}}{x^m} = -\frac{1}{(m-1)a}\left[\frac{\sqrt{(a+bx)^3}}{x^{m-1}} + \frac{(2m-5)b}{2}\int \frac{\sqrt{a+bx}\,dx}{x^{m-1}}\right], \quad (m \neq 1)$

45.3.8.8. $\int \frac{dx}{\sqrt{a+bx}} = \frac{2\sqrt{a+bx}}{b}$

45.3.8.9. $\int \frac{x\,dx}{\sqrt{a+bx}} = -\frac{2(2a-bx)}{3b^2}\sqrt{a+bx}$

45.3.8.10. $\int \frac{x^2\,dx}{\sqrt{a+bx}} = \frac{2(8a^2-4abx+3b^2x^2)}{15b^3}\sqrt{a+bx}$

45.3.8.11. $\int \frac{x^m\,dx}{\sqrt{a+bx}} = \frac{2x^m\sqrt{a+bx}}{(2m+1)b} - \frac{2ma}{(2m+1)b}\int \frac{x^{m-1}\,dx}{\sqrt{a+bx}}$

45.3.8.12. $\int \frac{dx}{x\sqrt{a+bx}} = \frac{1}{\sqrt{a}}\log\left(\frac{\sqrt{a+bx}-\sqrt{a}}{\sqrt{a+bx}+\sqrt{a}}\right), \quad (a>0)$

45.3.8.13. $\int \frac{dx}{x\sqrt{a+bx}} = \frac{2}{\sqrt{-a}}\tan^{-1}\sqrt{\frac{a+bx}{a}}, \quad (a<0)$

45.3.8.14. $\int \frac{dx}{x^2\sqrt{a+bx}} = -\frac{\sqrt{a+bx}}{ax} - \frac{b}{2a}\int \frac{dx}{x\sqrt{a+bx}}$

45.3.8.15. $$\int \frac{dx}{x^n\sqrt{a+bx}} = \begin{cases} -\dfrac{\sqrt{a+bx}}{(n-1)ax^{n-1}} - \dfrac{(2n-3)b}{(2n-2)a}\displaystyle\int \frac{dx}{x^{n-1}\sqrt{a+bx}} \\ \text{or} \\ \dfrac{(2n-2)!}{[(n-1)!]^2}\left[\dfrac{4\sqrt{a+bx}}{b}\displaystyle\sum_{r=1}^{n}\dfrac{r!(r-1)!}{x^r(2r)!} + \left(\dfrac{-b}{4a}\right)^{n-1}\displaystyle\int \frac{dx}{x\sqrt{a+bx}}\right] \end{cases}$$

45.3.8.16. $$\int (a+bx)^{\pm\frac{n}{2}}\,dx = \frac{2(a+bx)^{\frac{2\pm n}{2}}}{b(2\pm n)}$$

45.3.8.17. $$\int x(a+bx)^{\pm\frac{n}{2}}\,dx = \frac{2}{b^2}\left[\frac{(a+bx)^{\frac{4\pm n}{2}}}{4\pm n} - \frac{a(a+bx)^{\frac{2\pm n}{2}}}{2\pm n}\right]$$

45.3.8.18. $$\int \frac{dx}{x(a+bx)^{\frac{m}{2}}} = \frac{1}{a}\int \frac{dx}{x(a+bx)^{\frac{m-2}{2}}} - \frac{b}{a}\int \frac{dx}{(a+bx)^{\frac{m}{2}}}$$

45.3.8.19. $$\int \frac{(a+bx)^{\frac{n}{2}}\,dx}{x} = b\int (a+bx)^{\frac{n-2}{2}}\,dx + a\int \frac{(a+bx)^{\frac{n-2}{2}}}{x}\,dx$$

45.3.8.20. $$\int f(x, \sqrt{a+bx})\,dx = \frac{2}{b}\int f\left(\frac{z^2-a}{b}, z\right) z\,dz \quad (z^2 = a+bx)$$

45.3.9 Forms Containing $\sqrt{x^2 \pm a^2}$

45.3.9.1. $$\int \sqrt{x^2 \pm a^2}\,dx = \tfrac{1}{2}[x\sqrt{x^2 \pm a^2} \pm a^2 \log(x + \sqrt{x^2 \pm a^2})]$$

45.3.9.2. $$\int \frac{dx}{\sqrt{x^2 \pm a^2}} = \log(x + \sqrt{x^2 \pm a^2})$$

45.3.9.3. $$\int \frac{dx}{x\sqrt{x^2 - a^2}} = \frac{1}{|a|}\sec^{-1}\frac{x}{a}$$

45.3.9.4. $$\int \frac{dx}{x\sqrt{x^2 + a^2}} = -\frac{1}{a}\log\left(\frac{a + \sqrt{x^2 + a^2}}{x}\right)$$

45.3.9.5. $$\int \frac{\sqrt{x^2 + a^2}}{x}\,dx = \sqrt{x^2 + a^2} - a\log\left(\frac{a + \sqrt{x^2 + a^2}}{x}\right)$$

45.3.9.6. $$\int \frac{\sqrt{x^2 - a^2}}{x}\,dx = \sqrt{x^2 - a^2} - |a|\sec^{-1}\frac{x}{a}$$

45.3.9.7. $$\int \frac{x\,dx}{\sqrt{x^2 \pm a^2}} = \sqrt{x^2 \pm a^2}$$

45.3.9.8. $$\int x\sqrt{x^2 \pm a^2}\,dx = \tfrac{1}{3}\sqrt{(x^2 \pm a^2)^3}$$

45.3.9.9. $$\int \sqrt{(x^2 \pm a^2)^3}\,dx = \frac{1}{4}\left[x\sqrt{(x^2 \pm a^2)^3} \pm \frac{3a^2x}{2}\sqrt{x^2 \pm a^2} + \frac{3a^4}{2}\log(x + \sqrt{x^2 \pm a^2})\right]$$

45.3.9.10. $$\int \frac{dx}{\sqrt{(x^2 \pm a^2)^3}} = \frac{\pm x}{a^2\sqrt{x^2 \pm a^2}}$$

45.3.9.11. $\displaystyle\int \frac{x\,dx}{\sqrt{(x^2 \pm a^2)^3}} = \frac{-1}{\sqrt{x^2 \pm a^2}}$

45.3.9.12. $\displaystyle\int x\sqrt{(x^2 \pm a^2)^3}\,dx = \tfrac{1}{5}\sqrt{(x^2 \pm a^2)^5}$

45.3.9.13. $\displaystyle\int x^2\sqrt{x^2 \pm a^2}\,dx = \frac{x}{4}\sqrt{(x^2 \pm a^2)^3} \mp \frac{a^2}{8}x\sqrt{x^2 \pm a^2} - \frac{a^4}{8}\log(x + \sqrt{x^2 \pm a^2})$

45.3.9.14. $\displaystyle\int x^3\sqrt{x^2 + a^2}\,dx = (\tfrac{1}{5}x^2 - \tfrac{2}{15}a^2)\sqrt{(a^2 + x^2)^3}$

45.3.9.15. $\displaystyle\int x^3\sqrt{x^2 - a^2}\,dx = \frac{1}{5}\sqrt{(x^2 - a^2)^5} + \frac{a^2}{3}\sqrt{(x^2 - a^2)^3}$

45.3.9.16. $\displaystyle\int \frac{x^2\,dx}{\sqrt{x^2 \pm a^2}} = \frac{x}{2}\sqrt{x^2 \pm a^2} \mp \frac{a^2}{2}\log(x + \sqrt{x^2 \pm a^2})$

45.3.9.17. $\displaystyle\int \frac{x^3\,dx}{\sqrt{x^2 \pm a^2}} = \frac{1}{3}\sqrt{(x^2 \pm a^2)^3} \mp a^2\sqrt{x^2 \pm a^2}$

45.3.9.18. $\displaystyle\int \frac{dx}{x^2\sqrt{x^2 \pm a^2}} = \mp\frac{\sqrt{x^2 \pm a^2}}{a^2 x}$

45.3.9.19. $\displaystyle\int \frac{dx}{x^3\sqrt{x^2 + a^2}} = -\frac{\sqrt{x^2 + a^2}}{2a^2x^2} + \frac{1}{2a^3}\log\frac{a + \sqrt{x^2 + a^2}}{x}$

45.3.9.20. $\displaystyle\int \frac{dx}{x^3\sqrt{x^2 - a^2}} = \frac{\sqrt{x^2 - a^2}}{2a^2x^2} + \frac{1}{2|a^3|}\sec^{-1}\frac{x}{a}$

45.3.9.21. $\displaystyle\int x^2\sqrt{(x^2 \pm a^2)^3}\,dx =$

$$\frac{x}{6}\sqrt{(x^2 \pm a^2)^5} \mp \frac{a^2x}{24}\sqrt{(x^2 \pm a^2)^3} - \frac{a^4x}{16}\sqrt{x^2 \pm a^2} \mp \frac{a^6}{16}\log(x + \sqrt{x^2 \pm a^2})$$

45.3.9.22. $\displaystyle\int x^3\sqrt{(x^2 \pm a^2)^3}\,dx = \frac{1}{7}\sqrt{(x^2 \pm a^2)^7} \mp \frac{a^2}{5}\sqrt{(x^2 \pm a^2)^5}$

45.3.9.23. $\displaystyle\int \frac{\sqrt{x^2 \pm a^2}\,dx}{x^2} = -\frac{\sqrt{x^2 \pm a^2}}{x} + \log(x + \sqrt{x^2 \pm a^2})$

45.3.9.24. $\displaystyle\int \frac{\sqrt{x^2 + a^2}}{x^3}dx = -\frac{\sqrt{x^2 + a^2}}{2x^2} - \frac{1}{2a}\log\frac{a + \sqrt{x^2 + a^2}}{x}$

45.3.9.25. $\displaystyle\int \frac{\sqrt{x^2 - a^2}}{x^3}dx = -\frac{\sqrt{x^2 - a^2}}{2x^2} + \frac{1}{2|a|}\sec^{-1}\frac{x}{a}$

45.3.9.26. $\displaystyle\int \frac{x^2\,dx}{\sqrt{(x^2 \pm a^2)^3}} = \frac{-x}{\sqrt{x^2 \pm a^2}} + \log(x + \sqrt{x^2 \pm a^2})$

45.3.9.27. $\displaystyle\int \frac{x^3\,dx}{\sqrt{(x^2 \pm a^2)^3}} = \sqrt{x^2 \pm a^2} \pm \frac{a^2}{\sqrt{x^2 \pm a^2}}$

45.3.9.28. $\displaystyle\int \frac{dx}{x\sqrt{(x^2 \pm a^2)^3}} = \frac{1}{a^2\sqrt{x^2 + a^2}} - \frac{1}{a^3}\log\frac{a + \sqrt{x^2 + a^2}}{x}$

45.3.9.29. $\int \frac{dx}{x\sqrt{(x^2-a^2)^3}} = -\frac{1}{a^2\sqrt{x^2-a^2}} - \frac{1}{|a^3|}\sec^{-1}\frac{x}{a}$

45.3.9.30. $\int \frac{dx}{x^2\sqrt{(x^2\pm a^2)^3}} = -\frac{1}{a^4}\left[\frac{\sqrt{x^2\pm a^2}}{x} + \frac{x}{\sqrt{x^2\pm a^2}}\right]$

45.3.9.31. $\int \frac{dx}{x^3\sqrt{(x^2\pm a^2)^3}} = -\frac{1}{2a^2x^2\sqrt{x^2+a^2}} - \frac{3}{2a^4\sqrt{x^2+a^2}} + \frac{3}{2a^5}\log\frac{a+\sqrt{x^2+a^2}}{x}$

45.3.9.32. $\int \frac{dx}{x^3\sqrt{(x^2-a^2)^3}} = \frac{1}{2a^2x^2\sqrt{x^2-a^2}} - \frac{3}{2a^4\sqrt{x^2-a^2}} - \frac{3}{2|a^5|}\sec^{-1}\frac{x}{a}$

45.3.9.33. $\int f(x, \sqrt{x^2+a^2})dx = a\int f(a\tan u, |a\sec u|)\sec^2 u\,du, \qquad (x = a\tan u)$

45.3.9.34. $\int f(x, \sqrt{x^2-a^2})dx = a\int f(a\sec u, |a\tan u|)\sec u\tan u\,du, \qquad (x = a\sec u)$

45.3.10 Forms Containing $\sqrt{a^2-x^2}$

45.3.10.1. $\int \sqrt{a^2-x^2}\,dx = \frac{1}{2}\left[x\sqrt{a^2-x^2} + a^2\sin^{-1}\frac{x}{|a|}\right]$

45.3.10.2. $\int \frac{dx}{\sqrt{a^2-x^2}} = \begin{cases} \sin^{-1}\frac{x}{|a|} \\ \text{or} \\ -\cos^{-1}\frac{x}{|a|} \end{cases}$

45.3.10.3. $\int \frac{dx}{x\sqrt{a^2-x^2}} = -\frac{1}{a}\log\left(\frac{a+\sqrt{a^2-x^2}}{x}\right)$

45.3.10.4. $\int \frac{\sqrt{a^2-x^2}}{x}dx = \sqrt{a^2-x^2} - a\log\left(\frac{a+\sqrt{a^2-x^2}}{x}\right)$

45.3.10.5. $\int \frac{x\,dx}{\sqrt{a^2-x^2}} = -\sqrt{a^2-x^2}$

45.3.10.6. $\int x\sqrt{a^2-x^2}\,dx = -\tfrac{1}{3}\sqrt{(a^2-x^2)^3}$

45.3.10.7. $\int \sqrt{(a^2-x^2)^3}\,dx = \frac{1}{4}\left[x\sqrt{(a^2-x^2)^3} + \frac{3a^2x}{2}\sqrt{a^2-x^2} + \frac{3a^4}{2}\sin^{-1}\frac{x}{|a|}\right]$

45.3.10.8. $\int \frac{dx}{\sqrt{(a^2-x^2)^3}} = \frac{x}{a^2\sqrt{a^2-x^2}}$

45.3.10.9. $\int \frac{x\,dx}{\sqrt{(a^2-x^2)^3}} = \frac{1}{\sqrt{a^2-x^2}}$

45.3.10.10. $\int x\sqrt{(a^2-x^2)^3}\,dx = -\tfrac{1}{5}\sqrt{(a^2-x^2)^5}$

45.3.10.11. $\int x^2\sqrt{a^2-x^2}\,dx=-\frac{x}{4}\sqrt{(a^2-x^2)^3}+\frac{a^2}{8}\left(x\sqrt{a^2-x^2}+a^2\sin^{-1}\frac{x}{|a|}\right)$

45.3.10.12. $\int x^3\sqrt{a^2-x^2}\,dx=(-\tfrac{1}{5}x^2-\tfrac{2}{15}a^2)\sqrt{(a^2-x^2)^3}$

45.3.10.13. $\int x^2\sqrt{(a^2-x^2)^3}\,dx=-\frac{1}{6}x\sqrt{(a^2-x^2)^5}+\frac{a^2x}{24}\sqrt{(a^2-x^2)^3}+\frac{a^4x}{16}\sqrt{a^2-x^2}+\frac{a^6}{16}\sin^{-1}\frac{x}{|a|}$

45.3.10.14. $\int x^3\sqrt{(a^2-x^2)^3}\,dx=\frac{1}{7}\sqrt{(a^2-x^2)^7}-\frac{a^2}{5}\sqrt{(a^2-x^2)^5}$

45.3.10.15. $\int\frac{x^2\,dx}{\sqrt{a^2-x^2}}=-\frac{x}{2}\sqrt{a^2-x^2}+\frac{a^2}{2}\sin^{-1}\frac{x}{|a|}$

45.3.10.16. $\int\frac{dx}{x^2\sqrt{a^2-x^2}}=-\frac{\sqrt{a^2-x^2}}{a^2x}$

45.3.10.17. $\int\frac{\sqrt{a^2-x^2}}{x^2}dx=-\frac{\sqrt{a^2-x^2}}{x}-\sin^{-1}\frac{x}{|a|}$

45.3.10.18. $\int\frac{\sqrt{a^2-x^2}}{x^3}dx=-\frac{\sqrt{a^2-x^2}}{2x^2}+\frac{1}{2a}\log\frac{a+\sqrt{a^2-x^2}}{x}$

45.3.10.19. $\int\frac{x^2\,dx}{\sqrt{(a^2-x^2)^3}}=\frac{x}{\sqrt{a^2-x^2}}-\sin^{-1}\frac{x}{|a|}$

45.3.10.20. $\int\frac{x^3\,dx}{\sqrt{a^2-x^2}}=-\frac{2}{3}(a^2-x^2)^{\frac{3}{2}}-\frac{1}{3}x^2(a^2-x^2)^{\frac{1}{2}}=-\frac{1}{3}\sqrt{a^2-x^2}\,(x^2+2a^2)$

45.3.10.21. $\int\frac{x^3\,dx}{\sqrt{(a^2-x^2)^3}}=2(a^2-x^2)^{\frac{1}{2}}+\frac{x^2}{(a^2-x^2)^{\frac{1}{2}}}=\frac{a^2}{\sqrt{a^2-x^2}}+\sqrt{a^2-x^2}$

45.3.10.22. $\int\frac{dx}{x^3\sqrt{a^2-x^2}}=-\frac{\sqrt{a^2-x^2}}{2a^2x^2}-\frac{1}{2a^3}\log\frac{a+\sqrt{a^2-x^2}}{x}$

45.3.10.23. $\int\frac{dx}{x\sqrt{(a^2-x^2)^3}}=\frac{1}{a^2\sqrt{a^2-x^2}}-\frac{1}{a^3}\log\frac{a+\sqrt{a^2-x^2}}{x}$

45.3.10.24. $\int\frac{dx}{x^2\sqrt{(a^2-x^2)^3}}=\frac{1}{a^4}\left[-\frac{\sqrt{a^2-x^2}}{x}+\frac{x}{\sqrt{a^2-x^2}}\right]$

45.3.10.25. $\int\frac{dx}{x^3\sqrt{(a^2-x^2)^3}}=-\frac{1}{2a^2x^2\sqrt{a^2-x^2}}+\frac{3}{2a^4\sqrt{a^2-x^2}}-\frac{3}{2a^5}\log\frac{a+\sqrt{a^2-x^2}}{x}$

45.3.10.26. $\int\frac{\sqrt{a^2-x^2}}{b^2+x^2}dx=\frac{\sqrt{a^2+b^2}}{|b|}\sin^{-1}\frac{x}{|a|}\frac{\sqrt{a^2+b^2}}{\sqrt{x^2+b^2}}-\sin^{-1}\frac{x}{|a|}$

45.3.10.27. $\int f(x,\sqrt{a^2-x^2})dx=a\int f(a\sin u,|a\cos u|)\cos u\,du,\qquad(x=a\sin u)$

45.3.11 Forms Containing $\sqrt{a+bx+cx^2}$

$$X = a + bx + cx^2,\ q = 4ac - b^2,\ \text{and } k = \frac{4c}{q}$$

When $q = 0$, then $X = c\left(x + \frac{b}{2c}\right)^2$ and other formulas should be used.

45.3.11.1. $\int \frac{dx}{\sqrt{X}} = \frac{1}{\sqrt{c}} \log\left(\sqrt{X} + x\sqrt{c} + \frac{b}{2\sqrt{c}}\right)$ if $c > 0$

45.3.11.2. $\int \frac{dx}{\sqrt{X}} = \frac{1}{\sqrt{c}} \sinh^{-1}\left(\frac{2cx+b}{\sqrt{4ac-b^2}}\right)$ if $c > 0$

45.3.11.3. $\int \frac{dx}{\sqrt{X}} = \frac{1}{\sqrt{-c}} \sin^{-1}\left(\frac{-2cx-b}{\sqrt{b^2-4ac}}\right)$ if $c < 0$

45.3.11.4. $\int \frac{dx}{X\sqrt{X}} = \frac{2(2cx+b)}{q\sqrt{X}}$

45.3.11.5. $\int \frac{dx}{X^2\sqrt{X}} = \frac{2(2cx+b)}{3q\sqrt{X}}\left(\frac{1}{X} + 2k\right)$

45.3.11.6. $$\int \frac{dx}{X^n\sqrt{X}} = \begin{cases} \frac{2(2cx+b)\sqrt{X}}{(2n-1)qX^n} + \frac{2k(n-1)}{2n-1}\int \frac{dx}{X^{n-1}\sqrt{X}} \\ \text{or} \\ \frac{(2cx+b)(n!)(n-1)!4^n k^{n-1}}{q[(2n)!]\sqrt{X}} \sum_{r=0}^{n-1} \frac{(2r)!}{(4kX)^r (r!)^2} \end{cases}$$

45.3.11.7. $\int \sqrt{X}\,dx = \frac{(2cx+b)\sqrt{X}}{4c} + \frac{1}{2k}\int \frac{dx}{\sqrt{X}}$

45.3.11.8. $\int X\sqrt{X}\,dx = \frac{(2cx+b)\sqrt{X}}{8c}\left(X + \frac{3}{2k}\right) + \frac{3}{8k^2}\int \frac{dx}{\sqrt{X}}$

45.3.11.9. $\int X^2\sqrt{X}\,dx = \frac{(2cx+b)\sqrt{X}}{12c}\left(X^2 + \frac{5X}{4k} + \frac{15}{8k^2}\right) + \frac{5}{16k^3}\int \frac{dx}{\sqrt{X}}$

45.3.11.10. $$\int X^n\sqrt{X}\,dx = \begin{cases} \frac{(2cx+b)X^n\sqrt{X}}{4(n+1)c} + \frac{2n+1}{2(n+1)k}\int \frac{X^n\,dx}{\sqrt{X}} \\ \text{or} \\ \frac{(2n+2)!}{[(n+1)!]^2(4k)^{n+1}}\left[\frac{k(2cx+b)\sqrt{X}}{c}\sum_{r=0}^{n-1}\frac{r!(r+1)!(4kX)^r}{(2r+2)!} + \int \frac{dx}{\sqrt{X}}\right] \end{cases}$$

45.3.11.11. $\int \frac{x\,dx}{\sqrt{X}} = \frac{\sqrt{X}}{\sqrt{c}} - \frac{b}{2c}\int \frac{dx}{\sqrt{X}}$

45.3.11.12. $\int \frac{x\,dx}{X\sqrt{X}} = -\frac{2(bx+2a)}{q\sqrt{X}}$

45.3.11.13. $\int \frac{x\,dx}{X^n\sqrt{X}} = -\frac{\sqrt{X}}{(2n-1)cX^n} - \frac{b}{2c}\int \frac{dx}{X^n\sqrt{X}}$

45.3.11.14. $\int \frac{x^2\,dx}{\sqrt{X}} = \left(\frac{x}{2c} - \frac{3b}{4c^2}\right)\sqrt{X} + \frac{3b^2-4ac}{8c^2}\int \frac{dx}{\sqrt{X}}$

45.3.11.15. $\int \frac{x^2\,dx}{X\sqrt{X}} = \frac{(2b^2-4ac)x+2ab}{cq\sqrt{X}} + \frac{1}{c}\int \frac{dx}{\sqrt{X}}$

45.3.11.16. $\int \frac{x^2\,dx}{X^n\sqrt{X}} = \frac{(2b^2-4ac)x+2ab}{(2n-1)cqX^{n-1}\sqrt{X}} + \frac{4ac+(2n-3)b^2}{(2n-1)cq}\int \frac{dx}{X^{n-1}\sqrt{X}}$

45.3.11.17. $\int \frac{x^3\,dx}{\sqrt{X}} = \left(\frac{x^2}{3c} - \frac{5bx}{12c^2} + \frac{5b^2}{8c^3} - \frac{2a}{3c^2}\right)\sqrt{X} + \left(\frac{3ab}{4c^2} - \frac{5b^3}{16c^3}\right)\int \frac{dx}{\sqrt{X}}$

45.3.11.18. $\int x\sqrt{X}\,dx = \frac{X\sqrt{X}}{3c} - \frac{b}{2c}\int \sqrt{X}\,dx$

45.3.11.19. $\int xX\sqrt{X}\,dx = \frac{X^2\sqrt{X}}{5c} - \frac{b}{2c}\int X\sqrt{X}\,dx$

45.3.11.20. $\int \frac{xX^n\,dx}{\sqrt{X}} = \frac{X^n\sqrt{X}}{(2n+1)c} - \frac{b}{2c}\int \frac{X^n\,dx}{\sqrt{X}}$

45.3.11.21. $\int x^2\sqrt{X}\,dx = \left(x - \frac{5b}{6c}\right)\frac{X\sqrt{X}}{4c} + \frac{5b^2-4ac}{16c^2}\int \sqrt{X}\,dx$

45.3.11.22. $\int \frac{dx}{x\sqrt{X}} = -\frac{1}{\sqrt{a}}\log\left(\frac{\sqrt{X}+\sqrt{a}}{x} + \frac{b}{2\sqrt{a}}\right), \quad (a>0)$

45.3.11.23. $\int \frac{dx}{x\sqrt{X}} = \frac{1}{\sqrt{-a}}\sin^{-1}\left(\frac{bx+2a}{|x|\sqrt{-q}}\right), \quad (a<0)$

45.3.11.24. $\int \frac{dx}{x\sqrt{X}} = -\frac{2\sqrt{X}}{bx}, \quad (a=0)$

45.3.11.25. $\int \frac{dx}{x^2\sqrt{X}} = -\frac{\sqrt{X}}{ax} - \frac{b}{2a}\int \frac{dx}{x\sqrt{X}}$

45.3.11.26. $\int \frac{\sqrt{X}\,dx}{x} = \sqrt{X} + \frac{b}{2}\int \frac{dx}{\sqrt{X}} + a\int \frac{dx}{x\sqrt{X}}$

45.3.11.27. $\int \frac{\sqrt{X}\,dx}{x^2} = -\frac{\sqrt{X}}{x} + \frac{b}{2a}\int \frac{dx}{x\sqrt{X}} + c\int \frac{dx}{\sqrt{X}}$

45.3.12 Forms Containing $\sqrt{2ax-x^2}$

45.3.12.1. $\int \sqrt{2ax-x^2}\,dx = \frac{1}{2}\left[(x-a)\sqrt{2ax-x^2} + a^2\sin^{-1}\frac{x-a}{|a|}\right]$

45.3.12.2. $\int \frac{dx}{\sqrt{2ax-x^2}} = \begin{cases} \cos^{-1}\frac{a-x}{|a|} \text{ or } \sin^{-1}\frac{x-a}{|a|} \\ \text{or} \\ 2\sin^{-1}\sqrt{\frac{x}{2a}} \text{ or } 2\cos^{-1}\sqrt{\frac{x}{2a}}, \quad (a>0) \end{cases}$

45.3.12.3. $\int x^n\sqrt{2ax-x^2}\,dx = -\frac{x^{n-1}(2ax-x^2)^{\frac{3}{2}}}{n+2} + \frac{(2n+1)a}{n+2}\int x^{n-1}\sqrt{2ax-x^2}\,dx, \quad (n \neq -2)$

45.3.12.4. $\int \frac{\sqrt{2ax-x^2}}{x^n}dx = \frac{(2ax-x^2)^{\frac{3}{2}}}{(3-2n)ax^n} + \frac{n-3}{(2n-3)a}\int \frac{\sqrt{2ax-x^2}}{x^{n-1}}dx, \quad \left(n \neq \frac{3}{2}\right)$

45.3.12.5. $\int \frac{x^n\, dx}{\sqrt{2ax-x^2}} = \frac{-x^{n-1}\sqrt{2ax-x^2}}{n} + \frac{a(2n-1)}{n}\int \frac{x^{n-1}}{\sqrt{2ax-x^2}}dx, \quad (n \neq 0)$

45.3.12.6. $\int \frac{dx}{x^n\sqrt{2ax-x^2}} = \frac{\sqrt{2ax-x^2}}{a(1-2n)x^n} + \frac{n-1}{(2n-1)a}\int \frac{dx}{x^{n-1}\sqrt{2ax-x^2}}, \quad \left(n \neq \frac{1}{2}\right)$

45.3.12.7. $\int \frac{dx}{(2ax-x^2)^{\frac{3}{2}}} = \frac{x-a}{a^2\sqrt{2ax-x^2}}$

45.3.12.8. $\int \frac{x\, dx}{(2ax-x^2)^{\frac{3}{2}}} = \frac{x}{a\sqrt{2ax-x^2}}$

45.3.12.9. $\int \frac{dx}{\sqrt{2ax+x^2}} = \log(x+a+\sqrt{2ax+x^2})$

45.3.13 Miscellaneous Algebraic Forms

45.3.13.1. $\int \sqrt{ax^2+c}\, dx = \begin{cases} \frac{x}{2}\sqrt{ax^2+c} + \frac{c}{2\sqrt{a}}\log(x\sqrt{a}+\sqrt{ax^2+c}), & (a>0) \\ \text{or} & \\ \frac{x}{2}\sqrt{ax^2+c} + \frac{c}{2\sqrt{-a}}\sin^{-1}\left(x\sqrt{\frac{-a}{c}}\right), & (a<0) \end{cases}$

45.3.13.2. $\int \frac{dx}{\sqrt{a+bx}\cdot\sqrt{a'+b'x}} = \begin{cases} \frac{2}{\sqrt{bb'}}\tanh^{-1}\frac{\sqrt{bb'uv}}{bv}, & (bb'>0) \\ \text{or} & \\ \frac{1}{\sqrt{bb'}}\log\frac{bv+\sqrt{bb'uv}}{bv-\sqrt{bb'uv}}, & (bb'>0) \\ \text{or} & \\ \frac{1}{\sqrt{bb'}}\log(bv+\sqrt{bb'uv})^2, & (bb'>0) \\ \text{or} & \\ \frac{2}{\sqrt{-bb'}}\tan^{-1}\frac{\sqrt{-bb'uv}}{bv}, & (bb'<0) \end{cases}$ (see integral 45.3.5.3)

45.3.13.3. $\int \sqrt{\frac{1+x}{1-x}}dx = \sin^{-1}x - \sqrt{1-x^2}$

45.3.13.4. $\int \frac{dx}{\sqrt{a \pm 2bx + cx^2}} = \frac{1}{\sqrt{c}}\log(\pm b + cx + \sqrt{c}\sqrt{a \pm 2bx + cx^2}, \quad (b^2 - ac \neq 0)$

45.3.13.5. $\int \frac{dx}{\sqrt{a \pm 2bx - cx^2}} = \frac{1}{\sqrt{c}}\sin^{-1}\frac{cx \mp b}{\sqrt{b^2+ac}}$

45.3.13.6. $\int \frac{xdx}{\sqrt{a \pm 2bx + cx^2}} =$

$$\frac{1}{c}\sqrt{a \pm 2bx + cx^2} \mp \frac{b}{\sqrt{c^3}}\log(\pm b + cx + \sqrt{c}\sqrt{a \pm 2bx + cx^2}), \quad (b^2 - ac \neq 0)$$

45.3.13.7. $\int \frac{x\,dx}{\sqrt{a \pm 2bx - cx^2}} = -\frac{1}{c}\sqrt{a \pm 2bx - cx^2} \pm \frac{b}{\sqrt{c^3}}\sin^{-1}\frac{cx \mp b}{\sqrt{b^2 + ac}}$

45.3.14 Forms Involving Trigonometric Functions

45.3.14.1. $\int \sin x\,dx = -\cos x$

45.3.14.2. $\int \cos x\,dx = \sin x$

45.3.14.3. $\int \tan x\,dx = -\log\cos x = \log\sec x$

45.3.14.4. $\int \cot x\,dx = \log\sin x = -\log\csc x$

45.3.14.5. $\int \sec x\,dx = \log(\sec x + \tan x) = \log\tan\left(\frac{\pi}{4} + \frac{x}{2}\right)$

45.3.14.6. $\int \csc x\,dx = \log(\csc x - \cot x) = \log\tan\frac{x}{2}$

45.3.14.7. $\int \sin^2 x\,dx = -\tfrac{1}{2}\cos x\sin x + \tfrac{1}{2}x = \tfrac{1}{2}x - \tfrac{1}{4}\sin 2x$

45.3.14.8. $\int \sin^3 x\,dx = -\tfrac{1}{3}\cos x(\sin^2 x + 2)$

45.3.14.9. $\int \sin^n x\,dx = -\frac{\sin^{n-1} x\cos x}{n} + \frac{n-1}{n}\int \sin^{n-2} x\,dx$

45.3.14.10. $\int \cos^2 x\,dx = \tfrac{1}{2}\sin x\cos x + \tfrac{1}{2}x = \tfrac{1}{2}x + \tfrac{1}{4}\sin 2x$

45.3.14.11. $\int \cos^3 x\,dx = \tfrac{1}{3}\sin x(\cos^2 x + 2)$

45.3.14.12. $\int \cos^n x\,dx = \frac{1}{n}\cos^{n-1} x\sin x + \frac{n-1}{n}\int \cos^{n-2} x\,dx$

45.3.14.13. $\int \sin\frac{x}{a}\,dx = -a\cos\frac{x}{a}$

45.3.14.14. $\int \cos\frac{x}{a}\,dx = a\sin\frac{x}{a}$

45.3.14.15. $\int \sin(a + bx)\,dx = -\frac{1}{b}\cos(a + bx)$

45.3.14.16. $\int \cos(a + bx)\,dx = \frac{1}{b}\sin(a + bx)$

45.3.14.17. $\int \frac{dx}{\sin x} = \begin{cases} \int \csc x\,dx = \log(\csc x - \cot x) \\ \text{or} \\ -\frac{1}{2}\log\frac{1 + \cos x}{1 - \cos x} = \log\tan\frac{x}{2} \end{cases}$

45.3.14.18. $$\int \frac{dx}{\cos x} = \begin{cases} \int \sec x\,dx = \log(\sec x + \tan x) \\ \text{or} \\ \frac{1}{2}\log\frac{1+\sin x}{1-\sin x} = \log\tan\left(\frac{\pi}{4}+\frac{x}{2}\right) \end{cases}$$

45.3.14.19. $$\int \frac{dx}{\cos^2 x} = \int \sec^2 x\,dx = \tan x$$

45.3.14.20. $$\int \frac{dx}{\cos^n x} = \frac{1}{n-1}\cdot\frac{\sin x}{\cos^{n-1} x} + \frac{n-2}{n-1}\int \frac{dx}{\cos^{n-2} x}$$

45.3.14.21. $$\int \frac{dx}{1 \pm \sin x} = \mp\tan\left(\frac{\pi}{4} \mp \frac{x}{2}\right)$$

45.3.14.22. $$\int \frac{dx}{1+\cos x} = \tan\frac{x}{2}$$

45.3.14.23. $$\int \frac{dx}{1-\cos x} = -\cot\frac{x}{2}$$

45.3.14.24. $$\int \frac{dx}{a+b\sin x} = \begin{cases} \frac{2}{\sqrt{a^2-b^2}}\tan^{-1}\frac{a\tan\frac{x}{2}+b}{\sqrt{a^2-b^2}} \\ \text{or} \\ \frac{1}{\sqrt{b^2-a^2}}\log\frac{a\tan\frac{x}{2}+b-\sqrt{b^2-a^2}}{a\tan\frac{x}{2}+b+\sqrt{b^2-a^2}} \end{cases}$$

45.3.14.25. $$\int \frac{dx}{a+b\cos x} = \begin{cases} \frac{2}{\sqrt{a^2-b^2}}\tan^{-1}\frac{\sqrt{a^2-b^2}\tan\frac{x}{2}}{a+b} \\ \text{or} \\ \frac{1}{\sqrt{b^2-a^2}}\log\left(\frac{\sqrt{b^2-a^2}\tan\frac{x}{2}+a+b}{\sqrt{b^2-a^2}\tan\frac{x}{2}-a-b}\right) \end{cases}$$

45.3.14.26. $$\int \frac{dx}{a+b\sin x+c\cos x} =$$

$$\begin{cases} \frac{1}{\sqrt{b^2+c^2-a^2}}\log\frac{b-\sqrt{b^2+c^2-a^2}+(a-c)\tan\frac{x}{2}}{b+\sqrt{b^2+c^2-a^2}+(a-c)\tan\frac{x}{2}}, & \text{if } a^2 < b^2+c^2,\ a \neq c. \\ \text{or} \\ \frac{2}{\sqrt{a^2-b^2-c^2}}\tan^{-1}\frac{b+(a-c)\tan\frac{x}{2}}{\sqrt{a^2-b^2-c^2}}, & \text{if } a^2 > b^2+c^2 \\ \text{or} \\ \frac{1}{a}\left[\frac{a-(b+c)\cos x-(b-c)\sin x}{a-(b-c)\cos x+(b+c)\sin x}\right], & \text{if } a^2 = b^2+c^2,\ a \neq c. \end{cases}$$

45.3.14.27. $\int \frac{\sin^2 x\,dx}{a+b\cos^2 x} = \frac{1}{b}\sqrt{\frac{a+b}{a}}\tan^{-1}\left(\sqrt{\frac{a}{a+b}}\tan x\right) - \frac{x}{b}, \quad [ab>0, \text{ or } |a|>|b|]$

45.3.14.28. $\int \frac{dx}{a^2\cos^2 x + b^2\sin^2 x} = \frac{1}{ab}\tan^{-1}\left(\frac{b\tan x}{a}\right)$

45.3.14.29. $\int \sqrt{1-\cos x}\,dx = \pm 2\sqrt{2}\cos\frac{x}{2},$

[use + when $(4k-2)\pi < x \le 4k\pi$, otherwise −; k an integer]

45.3.14.30. $\int \sqrt{1+\cos x}\,dx = \pm 2\sqrt{2}\sin\frac{x}{2},$

[use + when $(4k-1)\pi < x \le (4k+1)\pi$, otherwise −; k an integer]

45.3.14.31. $\int \sqrt{1+\sin x}\,dx = \pm 2\left(\sin\frac{x}{2} - \cos\frac{x}{2}\right),$

[use + if $(8k-1)\frac{\pi}{2} < x \le (8k+3)\frac{\pi}{2}$, otherwise −; k an integer]

45.3.14.32. $\int \sqrt{1-\sin x}\,dx = \pm 2\left(\sin\frac{x}{2} + \cos\frac{x}{2}\right),$

[use + if $(8k-3)\frac{\pi}{2} < x \le (8k+1)\frac{\pi}{2}$, otherwise −; k an integer]

45.3.14.33. $\int \frac{dx}{\sqrt{1-\cos x}} = \pm\sqrt{2}\log\tan\frac{x}{4},$

[use + if $4k\pi < x < (4k+2)\pi$, otherwise −; k an integer]

45.3.14.34. $\int \frac{dx}{\sqrt{1+\cos x}} = \pm\sqrt{2}\log\tan\left(\frac{x+\pi}{4}\right),$

[use + if $(4k-1)\pi < x < (4k+1)\pi$, otherwise −; k an integer]

45.3.14.35. $\int \frac{dx}{\sqrt{1-\sin x}} = \pm\sqrt{2}\log\tan\left(\frac{x}{4} - \frac{\pi}{8}\right),$

[use + if $(8k+1)\frac{\pi}{2} < x < (8k+5)\frac{\pi}{2}$, otherwise −; k an integer]

45.3.14.36. $\int \frac{dx}{\sqrt{1+\sin x}} = \pm\sqrt{2}\log\tan\left(\frac{x}{4} + \frac{\pi}{8}\right),$

[use + if $(8k-1)\frac{\pi}{2} < x < (8k+3)\frac{\pi}{2}$, otherwise −; k an integer]

45.3.14.37. $\int \sin mx \sin nx\,dx = \frac{\sin(m-n)x}{2(m-n)} - \frac{\sin(m+n)x}{2(m+n)}, \quad (m^2 \ne n^2)$

45.3.14.38. $\int x\sin^2 x\,dx = \frac{x^2}{4} - \frac{x\sin 2x}{4} - \frac{\cos 2x}{8}$

45.3.14.39. $\int x^2\sin^2 x\,dx = \frac{x^3}{6} - \left(\frac{x^2}{4} - \frac{1}{8}\right)\sin 2x - \frac{x\cos 2x}{4}$

45.3.14.40. $\int x\sin^3 x\,dx = \frac{x\cos 3x}{12} - \frac{\sin 3x}{36} - \frac{3}{4}x\cos x + \frac{3}{4}\sin x$

45.3.14.41. $\int \sin^4 x\,dx = \frac{3x}{8} - \frac{\sin 2x}{4} + \frac{\sin 4x}{32}$

45.3.14.42. $\int \cos mx \cos nx\,dx = \frac{\sin(m-n)x}{2(m-n)} + \frac{\sin(m+n)x}{2(m+n)}, \quad (m^2 \neq n^2)$

45.3.14.43. $\int x\cos^2 x\,dx = \frac{x^2}{4} + \frac{x\sin 2x}{4} + \frac{\cos 2x}{8}$

45.3.14.44. $\int x^2\cos^2 x\,dx = \frac{x^3}{6} + \left(\frac{x^2}{4} - \frac{1}{8}\right)\sin 2x + \frac{x\cos 2x}{4}$

45.3.14.45. $\int x\cos^3 x\,dx = \frac{x\sin 3x}{12} + \frac{\cos 3x}{36} + \frac{3}{4}x\sin x + \frac{3}{4}\cos x$

45.3.14.46. $\int \cos^4 x\,dx = \frac{3x}{8} + \frac{\sin 2x}{4} + \frac{\sin 4x}{32}$

45.3.14.47. $\int \frac{\sin x\,dx}{x^m} = -\frac{\sin x}{(m-1)x^{m-1}} + \frac{1}{m-1}\int \frac{\cos x\,dx}{x^{m-1}}$

45.3.14.48. $\int \frac{\cos x\,dx}{x^m} = -\frac{\cos x}{(m-1)x^{m-1}} - \frac{1}{m-1}\int \frac{\sin x\,dx}{x^{m-1}}$

45.3.14.49. $\int \tan^3 x\,dx = \tfrac{1}{2}\tan^2 x + \log\cos x$

45.3.14.50. $\int \tan^n x\,dx = \frac{\tan^{n-1} x}{n-1} - \int \tan^{n-2} x\,dx$

45.3.14.51. $\int \cot^3 x\,dx = -\tfrac{1}{2}\cot^2 x - \log\sin x$

45.3.14.52. $\int \cot^4 x\,dx = -\tfrac{1}{3}\cot^3 x + \cot x + x$

45.3.14.53. $\int \cot^n x\,dx = -\frac{\cot^{n-1} x}{n-1} - \int \cot^{n-2} x\,dx, \quad [n \neq 1]$

45.3.14.54. $\int \sin x\cos x\,dx = \tfrac{1}{2}\sin^2 x$

45.3.14.55. $\int \sin mx \cos nx\,dx = -\frac{\cos(m-n)x}{2(m-n)} - \frac{\cos(m+n)x}{2(m+n)}, \quad (m^2 \neq n^2)$

45.3.14.56. $\int \sin^2 x\cos^2 x\,dx = -\tfrac{1}{8}(\tfrac{1}{4}\sin 4x - x)$

45.3.14.57. $\int \sin x\cos^m x\,dx = -\frac{\cos^{m+1} x}{m+1}$

45.3.14.58. $\int \sin^m x\cos x\,dx = \frac{\sin^{m+1} x}{m+1}$

45.3.14.59. $\int \cos^m x\sin^n x\,dx = \frac{\cos^{m-1} x\sin^{n+1} x}{m+n} + \frac{m-1}{m+n}\int \cos^{m-2} x\sin^n x\,dx, \quad (m \neq -n)$

45.3.14.60. $\int \cos^m x\sin^n x\,dx = -\frac{\sin^{n-1} x\cos^{m+1} x}{m+n} + \frac{n-1}{m+n}\int \cos^m x\sin^{n-2} x\,dx, \quad (m \neq -n)$

45.3.14.61. $\int \frac{\cos^m x\,dx}{\sin^n x} = -\frac{\cos^{m+1} x}{(n-1)\sin^{n-1} x} - \frac{m-n+2}{n-1}\int \frac{\cos^m x\,dx}{\sin^{n-2} x}$

45.3.14.62. $$\int \frac{\cos^m x\,dx}{\sin^n x} = \frac{\cos^{m-1} x}{(m-n)\sin^{n-1} x} + \frac{m-1}{m-n}\int \frac{\cos^{m-2} x\,dx}{\sin^n x}, \quad (m \neq n)$$

45.3.14.63. $$\int \frac{\sin^m x\,dx}{\cos^n x} = -\int \frac{\cos^m\left(\frac{\pi}{2} - x\right) d\left(\frac{\pi}{2} - x\right)}{\sin^n\left(\frac{\pi}{2} - x\right)}$$

45.3.14.64. $$\int \frac{\sin x\,dx}{\cos^2 x} = \frac{1}{\cos x} = \sec x$$

45.3.14.65. $$\int \frac{\sin^2 x\,dx}{\cos x} = -\sin x + \log\tan\left(\frac{\pi}{4} + \frac{x}{2}\right)$$

45.3.14.66. $$\int \frac{\cos x\,dx}{\sin^2 x} = \frac{-1}{\sin x} = -\operatorname{cosec} x$$

45.3.14.67. $$\int \frac{dx}{\sin x \cos x} = \log\tan x$$

45.3.14.68. $$\int \frac{dx}{\sin x \cos^2 x} = \frac{1}{\cos x} + \log\tan\frac{x}{2}$$

45.3.14.69. $$\int \frac{dx}{\sin x \cos^n x} = \frac{1}{(n-1)\cos^{n-1} x} + \int \frac{dx}{\sin x \cos^{n-2} x}, \quad (n \neq 1)$$

45.3.14.70. $$\int \frac{dx}{\sin^2 x \cos x} = -\frac{1}{\sin x} + \log\tan\left(\frac{\pi}{4} + \frac{x}{2}\right)$$

45.3.14.71. $$\int \frac{dx}{\sin^2 x \cos^2 x} = -2\cot 2x$$

45.3.14.72. $$\int \frac{dx}{\sin^m x \cos^n x} = -\frac{1}{m-1} \cdot \frac{1}{\sin^{m-1} x \cdot \cos^{n-1} x} + \frac{m+n-2}{m-1}\int \frac{dx}{\sin^{m-2} x \cdot \cos^n x}$$

45.3.14.73. $$\int \frac{dx}{\sin^m x} = -\frac{1}{m-1} \cdot \frac{\cos x}{\sin^{m-1} x} + \frac{m-2}{m-1}\int \frac{dx}{\sin^{m-2} x}$$

45.3.14.74. $$\int \frac{dx}{\sin^2 x} = -\cot x$$

45.3.14.75. $$\int \tan^2 x\,dx = \tan x - x$$

45.3.14.76. $$\int \tan^n x\,dx = \frac{\tan^{n-1} x}{n-1} - \int \tan^{n-2} x\,dx, \quad (n \neq 1)$$

45.3.14.77. $$\int \cot^2 x\,dx = -\cot x - x$$

45.3.14.78. $$\int \cot^n x\,dx = -\frac{\cot^{n-1} x}{n-1} - \int \cot^{n-2} x\,dx$$

45.3.14.79. $$\int \sec^2 x\,dx = \tan x$$

45.3.14.80. $$\int \sec^n x\,dx = \int \frac{dx}{\cos^n x} = \frac{1}{n-1}\frac{\sin x}{\cos^{n-1} x} + \frac{n-2}{n-1}\int \frac{dx}{\cos^{n-2} x}$$

45.3.14.81. $$\int \csc^2 x\,dx = -\cot x$$

45.3.14.82. $\int \csc^n x\,dx = \int \frac{dx}{\sin^n x} = -\frac{1}{n-1}\frac{\cos x}{\sin^{n-1} x} + \frac{n-2}{n-1}\int \frac{dx}{\sin^{n-2} x}$

45.3.14.83. $\int x\sin x\,dx = \sin x - x\cos x$

45.3.14.84. $\int x\sin(ax)dx = \frac{1}{a^2}\sin(ax) - \frac{x}{a}\cos(ax)$

45.3.14.85. $\int x^2 \sin x\,dx = 2x\sin x - (x^2 - 2)\cos x$

45.3.14.86. $\int x^2 \sin(ax)dx = \frac{2x}{a^2}\sin(ax) - \frac{a^2x^2 - 2}{a^3}\cos(ax)$

45.3.14.87. $\int x^3 \sin x\,dx = (3x^2 - 6)\sin x - (x^3 - 6x)\cos x$

45.3.14.88. $\int x^3 \sin(ax)dx = \frac{3a^2x^2 - 6}{a^4}\sin(ax) - \frac{a^2x^3 - 6x}{a^3}\cos(ax)$

45.3.14.89. $\int x^m \sin x\,dx = -x^m \cos x + m\int x^{m-1}\cos x\,dx$

45.3.14.90. $\int x^m \sin(ax)dx = -\frac{1}{a}x^m \cos(ax) + \frac{m}{a}\int x^{m-1}\cos(ax)dx$

45.3.14.91. $\int x\cos x\,dx = \cos x + x\sin x$

45.3.14.92. $\int x\cos(ax)dx = \frac{1}{a^2}\cos(ax) + \frac{x}{a}\sin(ax)$

45.3.14.93. $\int x^2 \cos x\,dx = 2x\cos x + (x^2 - 2)\sin x$

45.3.14.94. $\int x^2 \cos(ax)dx = \frac{2x\cos(ax)}{a^2} + \frac{a^2x^2 - 2}{a^3}\sin(ax)$

45.3.14.95. $\int x^3 \cos x\,dx = (3x^2 - 6)\cos x + (x^3 - 6x)\sin x$

45.3.14.96. $\int x^3 \cos(ax)dx = \frac{(3a^2x^2 - 6)}{a^4}\cos(ax) + \frac{a^2x^3 - 6x}{a^3}\sin(ax)$

45.3.14.97. $\int x^m \cos x\,dx = x^m \sin x - m\int x^{m-1}\sin x\,dx$

45.3.14.98. $\int x^m \cos(ax)dx = \frac{1}{a}x^m \sin(ax) - \frac{m}{a}\int x^{m-1}\sin(ax)dx$

45.3.14.99. $\int \frac{\sin x}{x}dx = x - \frac{x^3}{3\cdot 3!} + \frac{x^5}{5\cdot 5!} - \frac{x^7}{7\cdot 7!} + \frac{x^9}{9\cdot 9!}\cdots$

45.3.14.100. $\int \frac{\sin(ax)}{x}dx = ax - \frac{a^3x^3}{3\cdot 3!} + \frac{a^5x^5}{5\cdot 5!} - \frac{a^7x^7}{7\cdot 7!} + \frac{a^9x^9}{9\cdot 9!} + - \cdots$

45.3.14.101. $\int \frac{\cos x}{x}dx = \log x - \frac{x^2}{2\cdot 2!} + \frac{x^4}{4\cdot 4!} - \frac{x^6}{6\cdot 6!} + \frac{x^8}{8\cdot 8!}\cdots$

45.3.14.102. $\int \frac{\cos(ax)}{x}dx = \log x - \frac{a^2x^2}{2\cdot 2!} + \frac{a^4x^4}{4\cdot 4!} - \frac{a^6x^6}{6\cdot 6!} + \frac{a^8x^8}{8\cdot 8!} - + \cdots$

45.3.15 Forms Involving Inverse Trigonometric Functions

45.3.15.1. $\int \sin^{-1} x\,dx = x\sin^{-1} x + \sqrt{1-x^2}$

45.3.15.2. $\int \cos^{-1} x\,dx = x\cos^{-1} x - \sqrt{1-x^2}$

45.3.15.3. $\int \tan^{-1} x\,dx = x\tan^{-1} x - \tfrac{1}{2}\log(1+x^2)$

45.3.15.4. $\int \cot^{-1} x\,dx = x\cot^{-1} x + \tfrac{1}{2}\log(1+x^2)$

45.3.15.5. $\int \sec^{-1} x\,dx = x\sec^{-1} x - \log(x+\sqrt{x^2-1})$

45.3.15.6. $\int \csc^{-1} x\,dx = x\csc^{-1} x + \log(x+\sqrt{x^2-1})$

45.3.15.7. $\int \text{vers}^{-1} x\,dx = (x-1)\text{vers}^{-1} x + \sqrt{2x-x^2}$

45.3.15.8. $\int \sin^{-1}\frac{x}{a}dx = x\sin^{-1}\frac{x}{a} + \sqrt{a^2-x^2}, \quad (a>0)$

45.3.15.9. $\int \cos^{-1}\frac{x}{a}dx = x\cos^{-1}\frac{x}{a} - \sqrt{a^2-x^2}, \quad (a>0)$

45.3.15.10. $\int \tan^{-1}\frac{x}{a}dx = x\tan^{-1}\frac{x}{a} - \frac{a}{2}\log(a^2+x^2)$

45.3.15.11. $\int \cot^{-1}\frac{x}{a}dx = x\cot^{-1}\frac{x}{a} + \frac{a}{2}\log(a^2+x^2)$

45.3.15.12. $\int (\sin^{-1} x)^2\,dx = x(\sin^{-1} x)^2 - 2x + 2\sqrt{1-x^2}\,(\sin^{-1} x)$

45.3.15.13. $\int (\cos^{-1} x)^2\,dx = x(\cos^{-1} x)^2 - 2x - 2\sqrt{1-x^2}\,(\cos^{-1} x)$

45.3.15.14. $\int x\sin^{-1} x\,dx = \tfrac{1}{4}[(2x^2-1)\sin^{-1} x + x\sqrt{1-x^2}]$

45.3.15.15. $\int x\sin^{-1}(ax)dx = \frac{1}{4a^2}[(2a^2x^2-1)\sin^{-1}(ax) + ax\sqrt{1-a^2x^2}]$

45.3.15.16. $\int x\cos^{-1} x\,dx = \tfrac{1}{4}[(2x^2-1)\cos^{-1} x - x\sqrt{1-x^2}]$

45.3.15.17. $\int x\cos^{-1}(ax)dx = \frac{1}{4a^2}[(2a^2x^2-1)\cos^{-1}(ax) - ax\sqrt{1-a^2x^2}]$

45.3.15.18. $\int x^n\sin^{-1} x\,dx = \frac{x^{n+1}\sin^{-1} x}{n+1} - \frac{1}{n+1}\int\frac{x^{n+1}\,dx}{\sqrt{1-x^2}}$

45.3.15.19. $\int x^n\sin^{-1}(ax)dx = \frac{x^{n+1}}{n+1}\sin^{-1}(ax) - \frac{a}{n+1}\int\frac{x^{n+1}\,dx}{\sqrt{1-a^2x^2}}, \quad (n\neq -1)$

45.3.15.20. $\int x^n\cos^{-1} x\,dx = \frac{x^{n+1}\cos^{-1} x}{n+1} + \frac{1}{n+1}\int\frac{x^{n+1}\,dx}{\sqrt{1-x^2}}$

45.3.15.21. $\int x^n \cos^{-1}(ax)\,dx = \frac{x^{n+1}}{n+1}\cos^{-1}(ax) + \frac{a}{n+1}\int \frac{x^{n+1}\,dx}{\sqrt{1-a^2x^2}}, \quad (n \neq -1)$

45.3.15.22. $\int x \tan^{-1} x\,dx = \frac{1}{2}(1+x^2)\tan^{-1} x - \frac{x}{2}$

45.3.15.23. $\int x^n \tan^{-1}(ax)\,dx = \frac{x^{n+1}}{n+1}\tan^{-1}(ax) - \frac{a}{n+1}\int \frac{x^{n+1}\,dx}{1+a^2x^2}, \quad (n \neq -1)$

45.3.15.24. $\int x \cot^{-1} x\,dx = \frac{1}{2}(1+x^2)\cot^{-1} x + \frac{x}{2}$

45.3.15.25. $\int x^n \cot^{-1} x\,dx = \frac{x^{n+1}}{n+1}\cot^{-1} x + \frac{1}{n+1}\int \frac{x^{n+1}\,dx}{1+x^2}dx$

45.3.15.26. $\int \frac{\sin^{-1} x\,dx}{x^2} = \log\left(\frac{1-\sqrt{1-x^2}}{x}\right) - \frac{\sin^{-1} x}{x}$

45.3.15.27. $\int \frac{\sin^{-1}(ax)}{x^2}dx = a\log\left(\frac{1-\sqrt{1-a^2x^2}}{x}\right) - \frac{\sin^{-1}(ax)}{x}$

45.3.15.28. $\int \frac{\cos^{-1}(ax)}{x}dx = \frac{\pi}{2}\log x - ax - \frac{1}{2\cdot 3\cdot 3}(ax)^3 - \frac{1\cdot 3}{2\cdot 4\cdot 5\cdot 5}(ax)^5 - \frac{1\cdot 3\cdot 5}{2\cdot 4\cdot 6\cdot 7\cdot 7}(ax)^7 - \cdots$

45.3.15.29. $\int \frac{\cos^{-1}(ax)\,dx}{x^2} = -\frac{1}{x}\cos^{-1}(ax) + a\log\frac{1+\sqrt{1-a^2x^2}}{x}$

45.3.15.30. $\int \frac{\tan^{-1} x\,dx}{x^2} = \log x - \frac{1}{2}\log(1+x^2) - \frac{\tan^{-1} x}{x}$

45.3.15.31. $\int \frac{\tan^{-1}(ax)\,dx}{x^2} = -\frac{1}{x}\tan^{-1}(ax) - \frac{a}{2}\log\frac{1+a^2x^2}{x^2}$

45.3.16 Forms Involving Trigonometric Substitutions

45.3.16.1. $\int f(\sin x)\,dx = 2\int f\left(\frac{2z}{1+z^2}\right)\cdot\frac{dz}{1+z^2}, \quad \left(z = \tan\frac{x}{2}\right)$

45.3.16.2. $\int f(\cos x)\,dx = 2\int f\left(\frac{1-z^2}{1+z^2}\right)\frac{dz}{1+z^2}, \quad \left(z = \tan\frac{x}{2}\right)$

45.3.16.3.* $\int f(\sin x)\,dx = \int f(u)\frac{du}{\sqrt{1-u^2}}, \quad (u = \sin x)$

45.3.16.4.* $\int f(\cos x)\,dx = -\int f(u)\frac{du}{\sqrt{1-u^2}}, \quad (u = \cos x)$

45.3.16.5.* $\int f(\sin x,\ \cos x)\,dx = \int f(u,\ \sqrt{1-u^2})\frac{du}{\sqrt{1-u^2}}, \quad (u = \sin x)$

45.3.16.6. $\int f(\sin x,\ \cos x)\,dx = 2\int f\left(\frac{2z}{1+z^2},\ \frac{1-z^2}{1+z^2}\right)\frac{dz}{1+z^2}, \quad \left(z = \tan\frac{x}{2}\right)$

* The square roots appearing in these formulas may be plus or minus, depending on the quadrant of x. Care must be used to give them the proper sign.

45.3.16.7. $\int \frac{dx}{a + b\tan x} = \frac{1}{a^2 + b^2}[ax + b\log(a\cos x + b\sin x)]$

45.3.16.8. $\int \frac{dx}{a + b\cot x} = \frac{1}{a^2 + b^2}[ax - b\log(a\sin x + b\cos x)]$

45.3.17 Logarithmic Forms

45.3.17.1. $\int \log x\, dx = x\log x - x$

45.3.17.2. $\int x\log x\, dx = \frac{x^2}{2}\log x - \frac{x^2}{4}$

45.3.17.3. $\int x^2\log x\, dx = \frac{x^3}{3}\log x - \frac{x^3}{9}$

45.3.17.4. $$\int (\log X)\, dx = \begin{cases} \left(x + \frac{b}{2c}\right)\log X - 2x + \frac{\sqrt{4ac - b^2}}{c}\tan^{-1}\frac{2cx + b}{\sqrt{4ac - b^2}}, & (b^2 - 4ac < 0) \\ \text{or} & \\ \left(x + \frac{b}{2c}\right)\log X - 2x + \frac{\sqrt{b^2 - 4ac}}{c}\tanh^{-1}\frac{2cx + b}{\sqrt{b^2 - 4ac}}, & (b^2 - 4ac > 0) \\ \text{where} & \\ X = a + bx + cx^2 & \end{cases}$$

45.3.17.5. $\int x^p\log(ax)\, dx = \frac{x^{p+1}}{p+1}\log(ax) - \frac{x^{p+1}}{(p+1)^2}, \quad (p \neq -1)$

45.3.17.6. $\int x^n\log X\, dx = \frac{x^{n+1}}{n+1}\log X - \frac{2c}{n+1}\int \frac{x^{n+2}}{X}dx - \frac{b}{n+1}\int \frac{x^{n+1}}{X}dx$ where $X = a + bx + cx^2$

45.3.17.7. $\int (\log x)^2\, dx = x(\log x)^2 - 2x\log x + 2x$

45.3.17.8. $\int (\log x)^n\, dx = x(\log x)^n - n\int (\log x)^{n-1}\, dx, \quad (n \neq -1)$

45.3.17.9. $\int \frac{(\log x)^n}{x}dx = \frac{1}{n+1}(\log x)^{n+1}$

45.3.17.10. $\int \frac{dx}{\log x} = \log(\log x) + \log x + \frac{(\log x)^2}{2\cdot 2!} + \frac{(\log x)^3}{3\cdot 3!} + \cdots$

45.3.17.11. $\int \frac{dx}{x\log x} = \log(\log x)$

45.3.17.12. $\int \frac{dx}{x(\log x)^n} = -\frac{1}{(n-1)(\log x)^{n-1}}$

45.3.17.13. $\int \frac{x^m dx}{(\log x)^n} = -\frac{x^{m+1}}{(n-1)(\log x)^{n-1}} + \frac{m+1}{n-1}\int \frac{x^m\, dx}{(\log x)^{n-1}}$

45.3.17.14. $\int x^m\log x\, dx = x^{m+1}\left[\frac{\log x}{m+1} - \frac{1}{(m+1)^2}\right]$

45.3.17.15. $\int x^m(\log x)^n\, dx = \frac{x^{m+1}(\log x)^n}{m+1} - \frac{n}{m+1}\int x^m(\log x)^{n-1}\, dx, \quad [m \neq -1]$

45.3.17.16. $\int \sin\log x\,dx = \frac{1}{2}x\sin\log x - \frac{1}{2}x\cos\log x$

45.3.17.17. $\int \cos\log x\,dx = \frac{1}{2}x\sin\log x + \frac{1}{2}x\cos\log x$

45.3.18 Exponential Forms

45.3.18.1. $\int e^x\,dx = e^x$

45.3.18.2. $\int e^{-x}\,dx = -e^{-x}$

45.3.18.3. $\int e^{ax}\,dx = \frac{e^{ax}}{a}$

45.3.18.4. $\int xe^{ax}\,dx = \frac{e^{ax}}{a^2}(ax-1)$

45.3.18.5. $$\int x^m e^{ax}\,dx = \begin{cases} \dfrac{x^m e^{ax}}{a} - \dfrac{m}{a}\displaystyle\int x^{m-1}e^{ax}\,dx \\ \text{or} \\ e^{ax}\displaystyle\sum_{r=0}^{m}(-1)^r \dfrac{m!x^{m-r}}{(m-r)!a^{r+1}} \end{cases}$$

45.3.18.6. $\int \frac{e^{ax}\,dx}{x} = \log x + \frac{ax}{1!} + \frac{a^2x^2}{2\cdot 2!} + \frac{a^3x^3}{3\cdot 3!} + \cdots$

45.3.18.7. $\int \frac{e^{ax}}{x^m}dx = -\frac{1}{m-1}\frac{e^{ax}}{x^{m-1}} + \frac{a}{m-1}\int \frac{e^{ax}}{x^{m-1}}dx$

45.3.18.8. $\int e^{ax}\log x\,dx = \frac{e^{ax}\log x}{a} - \frac{1}{a}\int \frac{e^{ax}}{x}dx$

45.3.18.9. $\int \frac{dx}{1+e^x} = x - \log(1+e^x) = \log\frac{e^x}{1+e^x}$

45.3.18.10. $\int \frac{dx}{a+be^{px}} = \frac{x}{a} - \frac{1}{ap}\log(a+be^{px})$

45.3.18.11. $\int \frac{dx}{ae^{mx}+be^{-mx}} = \frac{1}{m\sqrt{ab}}\tan^{-1}\left(e^{mx}\sqrt{\frac{a}{b}}\right), \quad (a>0,\ b>0)$

45.3.18.12. $$\int \frac{dx}{ae^{mx}-be^{-mx}} = \begin{cases} \dfrac{1}{2m\sqrt{ab}}\log\dfrac{\sqrt{a}\,e^{mx}-\sqrt{b}}{\sqrt{a}\,e^{mx}+\sqrt{b}} \\ \text{or} \\ \dfrac{-1}{m\sqrt{ab}}\tanh^{-1}\left(\sqrt{\dfrac{a}{b}}\,e^{mx}\right) \\ \text{or} \\ -\dfrac{1}{m\sqrt{ab}}\coth^{-1}\left(\sqrt{\dfrac{a}{b}}\,e^{mx}\right), \quad (a>0,\ b>0) \end{cases}$$

45.3.18.13. $\int (a^x - a^{-x})(\log a)dx = a^x + a^{-x}$

45.3.18.14. $\int e^{ax}\sin(bx)dx = \frac{e^{ax}[a\sin(bx) - b\cos(bx)]}{a^2+b^2}$

45.3.18.15. $\int e^{ax}\sin(bx)\sin(cx)dx = \frac{e^{ax}[(b-c)\sin(b-c)x + a\cos(b-c)x]}{2[a^2+(b-c)^2]}$

$-\frac{e^{ax}[(b+c)\sin(b+c)x + a\cos(b+c)x]}{2[a^2+(b-c)^2]}$

45.3.18.16. $\int e^{ax}\sin(bx)\cos(cx)dx = \frac{e^{ax}[a\sin(b-c)x - (b-c)\cos(b-c)x]}{2[a^2+(b-c)^2]}$

$+\frac{e^{ax}[a\sin(b+c)x - (b+c)\cos(b+c)x]}{2[a^2+(b+c)^2]}$

45.3.18.17. $\int e^{ax}\sin(bx)\sin(bx+c)dx = \frac{e^{ax}\cos c}{2a} - \frac{e^{ax}[a\cos(2bx+c) + 2b\sin(2bx+c)]}{2(a^2+4b^2)}$

45.3.18.18. $\int e^{ax}\sin(bx)\cos(bx+c)dx = \frac{-e^{ax}\sin c}{2a} + \frac{e^{ax}[a\sin(2bx+c) - 2b\cos(2bx+c)]}{2(a^2+4b^2)}$

45.3.18.19. $\int e^{ax}\cos(bx)dx = \frac{e^{ax}}{a^2+b^2}[a\cos(bx) + b\sin(bx)]$

45.3.18.20. $\int e^{ax}\cos(bx)\cos(cx)dx = \frac{e^{ax}[(b-c)\sin(b-c)x + a\cos(b-c)x]}{2[a^2+(b-c)^2]}$

$+\frac{e^{ax}[(b+c)\sin(b+c)x + a\cos(b+c)x]}{2[a^2+(b+c)^2]}$

45.3.18.21. $\int e^{ax}\cos(bx)\cos(bx+c)dx = \frac{e^{ax}\cos c}{2a} + \frac{e^{ax}[a\cos(2bx+c) + 2b\sin(2bx+c)]}{2(a^2+4b^2)}$

45.3.18.22. $\int e^{ax}\cos(bx)\sin(bx+c)dx = \frac{e^{ax}\sin c}{2a} + \frac{e^{ax}[a\sin(2bx+c) - 2b\cos(2bx+c)]}{2(a^2+4b^2)}$

45.3.18.23.

$$\int e^{ax}\sin^n bx\,dx = \frac{1}{a^2+n^2b^2}\left[(a\sin bx - nb\cos bx)e^{ax}\sin^{n-1}bx + n(n-1)b^2\int e^{ax}\sin^{n-2}bx\cdot dx\right]$$

45.3.18.24.

$$\int e^{ax}\cos^n bx\,dx = \frac{1}{a^2+n^2b^2}\left[(a\cos bx + nb\sin bx)e^{ax}\cos^{n-1}bx + n(n-1)b^2\int e^{ax}\cos^{n-2}bx\,dx\right]$$

45.3.18.25. $\int x^m e^x \sin x\,dx = \frac{1}{2}x^m e^x(\sin x - \cos x) - \frac{m}{2}\int x^{m-1}e^x\sin x\,dx + \frac{m}{2}\int x^{m-1}e^x\cos x\,dx$

45.3.18.26.

$$\int x^m e^{ax}\sin bx\,dx = \begin{cases} x^m e^{ax}\frac{a\sin bx - b\cos bx}{a^2+b^2} - \frac{m}{a^2+b^2}\int x^{m-1}e^{ax}(a\sin bx - b\cos bx)dx \\ \text{or} \\ e^{ax}\left[\frac{1}{\rho}x^m\sin(bx-\alpha) - \frac{m}{\rho^2}x^{m-1}\sin(bx-2\alpha)\right. \\ \left.\pm\frac{m(m-1)\cdots 1}{\rho^{m-1}}\sin\{bx-(m+1)\alpha\}\right] \\ \text{where } a + b\sqrt{-1} = \rho(\cos\alpha + \sqrt{-1}\sin\alpha) \end{cases}$$

45.3.18.27. $\int x^m e^x\cos x\,dx = \frac{1}{2}x^m e^x(\sin x + \cos x) - \frac{m}{2}\int x^{m-1}e^x\sin x\,dx - \frac{m}{2}\int x^{m-1}e^x\cos x\,dx$

45.3.18.28. $$\int x^m e^{ax}\cos bx\,dx = \begin{cases} x^m e^{ax}\dfrac{a\cos bx + b\sin bx}{a^2+b^2} - \dfrac{m}{a^2+b^2}\displaystyle\int x^{m-1}e^{ax}(a\cos bx + b\sin bx)\,dx \\ \text{or} \\ e^{ax}\left[\dfrac{1}{\rho}x^m\cos(bx-\alpha) - \dfrac{m}{\rho^2}x^{m-1}\cos(bx-2\alpha) + \cdots \right. \\ \left. \pm\dfrac{m(m-1)\cdots 1}{\rho^{m+1}}\cos(bx-(m+1)\alpha)\right] \\ \text{where } a + b\sqrt{-1} = \rho(\cos\alpha + \sqrt{-1}\sin\alpha) \end{cases}$$

45.3.18.29. $$\int e^{ax}\sin(bx)\cos(cx)\,dx = \frac{e^{ax}}{\rho}[(a\sin(bx) - b\cos(bx))\cos(cx-x) - c\sin(bx)\sin(cx-\alpha)]$$

$$\text{where } \rho = \sqrt{(a^2+b^2-c^2)^2 + 4a^2c^2}$$

$$\rho\cos\alpha = a^2+b^2-c^2$$

$$\rho\sin\alpha = 2ac$$

45.3.18.30. $$\int e^{ax}\cos^m x\sin^n x\,dx = \frac{e^{ax}\cos^{m-1}x\sin^n x\{a\cos x + (m+n)\sin x\}}{(m+n)^2+a^2}$$

$$-\frac{na}{(m+n)^2+a^2}\int e^{ax}\cos^{m-1}x\sin^{n-1}x\,dx + \frac{(m-1)(m+n)}{(m+n)^2+a^2}\int e^{ax}\cos^{m-2}x\sin^n x\,dx$$

or

$$= \frac{e^{ax}\cos^m x\sin^{n-1}x\{a\sin x - (m+n)\cos x\}}{(m+n)^2+a^2} + \frac{ma}{(m+n)^2+a^2}$$

$$\int e^{ax}\cos^{m-1}x\sin^{n-1}x\,dx + \frac{(n-1)(m+n)}{(m+n)^2+a^2}\int e^{ax}\cos^m x\sin^{n-2}x\,dx$$

or

$$= \frac{e^{ax}\cos^{m-1}x\sin^{n-1}x(a\sin x\cos x + m\sin^2 x - n\cos^2 x)}{(m+n)^2+a^2} + \frac{m(m-1)}{(m+n)^2+a^2}$$

$$\int e^{ax}\cos^{m-2}x\sin^n x\,dx + \frac{n(n-1)}{(m+n)^2+a^2}\int e^{ax}\cos^m x\sin^{n-2}x\,dx$$

or

$$= \frac{e^{ax}\cos^{m-1}x\sin^{n-1}x(a\cos x\sin x + m\sin^2 x - n\cos^2 x)}{(m+n)^2+a^2} + \frac{m(m-1)}{(m+n)^2+a^2}$$

$$\int e^{ax}\cos^{m-2}x\sin^{n-2}x\,dx + \frac{(n-m)(n+m-1)}{(m+n)^2+a^2}\int e^{ax}\cos^m x\sin^{n-2}x\,dx$$

45.3.18.31. $$\int \frac{e^{ax}}{\sin^n x}dx = -\frac{e^{ax}\{a\sin x + (n-2)\cos x\}}{(n-1)(n-2)\sin^{n-1}x} + \frac{a^2+(n-2)^2}{(n-1)(n-2)}\int\frac{e^{ax}}{\sin^{n-2}x}dx$$

45.3.18.32. $$\int \frac{e^{ax}}{\cos^n x}dx = -\frac{e^{ax}\{a\cos x - (n-2)\sin x\}}{(n-1)(n-2)\cos^{n-1}x} + \frac{a^2+(n-2)^2\}}{(n-1)(n-2)}\int\frac{e^{ax}}{\cos^{n-2}x}dx$$

45.3.18.33. $$\int e^{ax}\tan^n x\,dx = e^{ax}\frac{\tan^{n-1}x}{n-1} - \frac{a}{n-1}\int e^{ax}\tan^{n-1}x\,dx - \int e^{ax}\tan^{n-2}x\,dx$$

45.3.19 Hyperbolic Forms

45.3.19.1. $\int \sinh x\,dx = \cosh x$

45.3.19.2. $\int \cosh x\,dx = \sinh x$

45.3.19.3. $\int \tanh x\,dx = \log\cosh x$

45.3.19.4. $\int \coth x\,dx = \log\sinh x$

45.3.19.5. $\int \operatorname{sech} x\,dx = \tan^{-1}(\sinh x)$

45.3.19.6. $\int \operatorname{csch} x\,dx = \log\tanh\left(\frac{x}{2}\right)$

45.3.19.7. $\int x\sinh x\,dx = x\cosh x - \sinh x$

45.3.19.8. $\int x^n \sinh x\,dx = x^n \cosh x - n\int x^{n-1}\cosh x\,dx$

45.3.19.9. $\int x\cosh x\,dx = x\sinh x - \cosh x$

45.3.19.10. $\int x^n \cosh x\,dx = x^n \sinh x - n\int x^{n-1}\sinh x\,dx$

45.3.19.11. $\int \operatorname{sech} x\tanh x\,dx = -\operatorname{sech} x$

45.3.19.12. $\int \operatorname{csch} x\coth x\,dx = -\operatorname{csch} x$

45.3.19.13. $\int \sinh^2 x\,dx = \frac{\sinh 2x}{4} - \frac{x}{2}$

45.3.19.14.
$$\int \sinh^m x\cosh^n x\,dx = \begin{cases} \frac{1}{m+n}\sinh^{m+1} x\cosh^{n-1} x + \frac{n-1}{m+n}\int \sinh^m x\cosh^{n-2} x\,dx \\ \text{or} \\ \frac{1}{m+n}\sinh^{m-1} x\cosh^{n+1} x - \frac{m-1}{m+n}\int \sinh^{m-2} x\cosh^n x\,dx, \\ \qquad (m+n\neq 0) \end{cases}$$

45.3.19.15.
$$\int \frac{dx}{\sinh^m x\cosh^n x} = \begin{cases} -\frac{1}{(m-1)\sinh^{m-1} x\cosh^{n-1} x} - \frac{m+n-2}{m-1}\int \frac{dx}{\sinh^{m-2} x\cosh^n x}, \\ \qquad (m\neq 1) \\ \text{or} \\ \frac{1}{(n-1)\sinh^{m-1} x\cosh^{n-1} x} + \frac{m+n-2}{n-1}\int \frac{dx}{\sinh^m x\cosh^{n-2} x}, \\ \qquad (n\neq 1) \end{cases}$$

45.3.19.16. $\int \tanh^2 x\,dx = x - \tanh x$

45.3.19.17. $\int \tanh^n x\,dx = -\frac{\tanh^{n-1} x}{n-1} + \int \tanh^{n-2} x\,dx, \quad (n\neq 1)$

45.3.19.18. $\int \operatorname{sech}^2 x\,dx = \tanh x$

45.3.19.19. $\int \cosh^2 x\,dx = \frac{\sinh 2x}{4} + \frac{x}{2}$

45.3.19.20. $\int \coth^2 x\,dx = x - \coth x$

45.3.19.21. $\int \coth^n x\,dx = -\frac{\coth^{n-1} x}{n-1} + \int \coth^{n-2} x\,dx, \quad (n \neq 1)$

45.3.19.22. $\int \operatorname{csch}^2 x\,dx = -\operatorname{ctnh} x$

45.3.19.23. $\int \sinh mx \sinh nx\,dx = \frac{\sinh(m+n)x}{2(m+n)} - \frac{\sinh(m-n)x}{2(m-n)}, \quad (m^2 \neq n^2)$

45.3.19.24. $\int \cosh mx \cosh nx\,dx = \frac{\sinh(m+n)x}{2(m+n)} + \frac{\sinh(m-n)x}{2(m-n)}, \quad (m^2 \neq n^2)$

45.3.19.25. $\int \sinh mx \cosh nx\,dx = \frac{\cosh(m+n)x}{2(m+n)} + \frac{\cosh(m-n)x}{2(m-n)}, \quad (m^2 \neq n^2)$

45.3.19.26. $\int \sinh^{-1}\frac{x}{a}\,dx = x\sinh^{-1}\frac{x}{a} - \sqrt{x^2+a^2}, \quad (a > 0)$

45.3.19.27. $\int x\sinh^{-1}\frac{x}{a}\,dx = \left(\frac{x^2}{2} + \frac{a^2}{4}\right)\sinh^{-1}\frac{x}{a} - \frac{x}{4}\sqrt{x^2+a^2}, \quad (a > 0)$

45.3.19.28. $\int x^n \sinh^{-1} x\,dx = \frac{x^{n+1}}{n+1}\sinh^{-1} x - \frac{1}{n+1}\int \frac{x^{n+1}}{(1+x^2)^{\frac{1}{2}}}\,dx, \quad (n \neq -1)$

45.3.19.29. $$\int \cosh^{-1}\frac{x}{a}\,dx = \begin{cases} x\cosh^{-1}\frac{x}{a} - \sqrt{x^2-a^2}, & \left(\cosh^{-1}\frac{x}{a} > 0\right) \\ \text{or} & \\ x\cosh^{-1}\frac{x}{a} + \sqrt{x^2-a^2}, & \left(\cosh^{-1}\frac{x}{a} < 0\right), \quad (a > 0) \end{cases}$$

45.3.19.30. $\int x\cosh^{-1}\frac{x}{a}\,dx = \frac{2x^2-a^2}{4}\cosh^{-1}\frac{x}{a} - \frac{x}{4}(x^2-a^2)^{\frac{1}{2}}$

45.3.19.31. $\int x^n \cosh^{-1} x\,dx = \frac{x^{n+1}}{n+1}\cosh^{-1} x - \frac{1}{n+1}\int \frac{x^{n+1}}{(x^2-1)^{\frac{1}{2}}}\,dx, \quad (n \neq -1)$

45.3.19.32. $\int \tanh^{-1}\frac{x}{a}\,dx = x\tanh^{-1}\frac{x}{a} + \frac{a}{2}\log(a^2-x^2), \quad \left(\left|\frac{x}{a}\right| < 1\right)$

45.3.19.33. $\int \coth^{-1}\frac{x}{a}\,dx = x\coth^{-1}\frac{x}{a} + \frac{a}{2}\log(x^2-a^2), \quad \left(\left|\frac{x}{a}\right| > 1\right)$

45.3.19.34. $\int x\tanh^{-1}\frac{x}{a}\,dx = \frac{x^2-a^2}{2}\tanh^{-1}\frac{x}{a} + \frac{ax}{2}, \quad \left(\left|\frac{x}{a}\right| < 1\right)$

45.3.19.35. $\int x^n \tanh^{-1} x\,dx = \frac{x^{n+1}}{n+1}\tanh^{-1} x - \frac{1}{n+1}\int \frac{x^{n+1}}{1-x^2}\,dx, \quad (n \neq -1)$

45.3.19.36. $\int x\coth^{-1}\frac{x}{a}\,dx = \frac{x^2-a^2}{2}\coth^{-1}\frac{x}{a} + \frac{ax}{2}, \quad \left(\left|\frac{x}{a}\right| > 1\right)$

45.3.19.37. $\int x^n \coth^{-1} x\,dx = \frac{x^{n+1}}{n+1}\coth^{-1} x + \frac{1}{n+1}\int \frac{x^{n+1}}{x^2-1}dx, \quad (n \neq -1)$

45.3.19.38. $\int \operatorname{sech}^{-1} x\,dx = x\operatorname{sech}^{-1}x + \arcsin x$

45.3.19.39. $\int x\operatorname{sech}^{-1}x\,dx = \frac{x^2}{2}\operatorname{sech}^{-1}x - \frac{1}{2}(1-x^2)$

45.3.19.40. $\int x^n \operatorname{sech}^{-1}x\,dx = \frac{x^{n+1}}{n+1}\operatorname{sech}^{-1}x + \frac{1}{n+1}\int \frac{x^n}{(1-x^2)^{\frac{1}{2}}}dx, \quad (n \neq -1)$

45.3.19.41. $\int \operatorname{csch}^{-1}x\,dx = x\operatorname{csch}^{-1}x + \frac{x}{|x|}\sinh^{-1} x$

45.3.19.42. $\int x\operatorname{csch}^{-1}x\,dx = \frac{x^2}{2}\operatorname{csch}^{-1}x + \frac{1}{2}\frac{x}{|x|}\sqrt{1+x^2}$

45.3.19.43. $\int x^n \operatorname{csch}^{-1}x\,dx = \frac{x^{n+1}}{n+1}\operatorname{csch}^{-1}x + \frac{1}{n+1}\int \frac{x^n}{(x^2+1)^{\frac{1}{2}}}dx, \quad (n \neq -1)$

45.3.20 Definite Integrals

45.3.20.1. $\int_0^\infty x^{n-1}e^{-x}\,dx = \int_0^1 \left(\log\frac{1}{x}\right)^{n-1}dx = \frac{1}{n}\prod_{m=1}^{\infty}\frac{\left(1+\frac{1}{m}\right)^n}{1+\frac{n}{m}}$

$= \Gamma(n),\ n \neq 0, -1, -2, -3, \ldots$ (Gamma Function)

45.3.20.2. $\int_0^\infty t^n p^{-t}\,dt = \frac{n!}{(\log p)^{n+1}}, \quad (n = 0, 1, 2, 3, \ldots \text{and } p > 0)$

45.3.20.3. $\int_0^\infty t^{n-1}e^{-(a+1)t}\,dt = \frac{\Gamma(n)}{(a+1)^n}, \quad (n > 0,\ a > -1)$

45.3.20.4. $\int_0^1 x^m\left(\log\frac{1}{x}\right)^n dx = \frac{\Gamma(n+1)}{(m+1)^{n+1}}, \quad (m > -1,\ n > -1)$

45.3.20.5. $\Gamma(n)$ is finite if $n > 0$, $\Gamma(n+1) = n\Gamma(n)$

45.3.20.6. $\Gamma(n)\cdot\Gamma(1-n) = \frac{\pi}{\sin n\pi}$

45.3.20.7. $\Gamma(n) = (n-1)!$ if $n = \text{integer} > 0$

45.3.20.8. $\Gamma(\tfrac{1}{2}) = 2\int_0^\infty e^{-t^2}\,dt = \sqrt{\pi} = 1.7724538509\ldots = (-\tfrac{1}{2})!$

45.3.20.9. $\Gamma\left(n+\frac{1}{2}\right) = \frac{1\cdot 3\cdot 5\cdot 7\ldots(2n-1)}{2^n}\sqrt{\pi}$, where n is an integer and > 0 (see values of $\Gamma(n)$ at end of integral table)

45.3.20.10. $\int_0^1 x^{m-1}(1-x)^{n-1}\,dx = \mathrm{B}(m, n)$, (Beta function)

45.3.20.11. $\mathrm{B}(m, n) = \mathrm{B}(n, m) = \frac{\Gamma(m)\,\Gamma(n)}{\Gamma(m+n)}$, where m and n are any positive real numbers

45.3.20.12. $\int_0^1 x^{m-1}(1-x)^{n-1}\,dx = \int_0^\infty \frac{x^{m-1}\,dx}{(1+x)^{m+n}} = \frac{\Gamma(m)\,\Gamma(n)}{\Gamma(m+n)}$

45.3.20.13. $\int_a^b (x-a)^m (b-x)^n\,dx = (b-a)^{m+n+1}\frac{\Gamma(m+1)\cdot\Gamma(n+1)}{\Gamma(m+n+2)}, \quad (m>-1,\ n>-1,\ b>a)$

45.3.20.14. $\int_1^\infty \frac{dx}{x^m} = \frac{1}{m-1}, \quad [m>1]$

45.3.20.15. $\int_0^\infty \frac{dx}{(1+x)x^p} = \pi \csc p\pi, \quad [p<1]$

45.3.20.16. $\int_0^\infty \frac{dx}{(1-x)x^p} = -\pi \cot p\pi, \quad [p<1]$

45.3.20.17. $\int_0^\infty \frac{x^{p-1}\,dx}{1+x} = \frac{\pi}{\sin p\pi}$

$$= \mathrm{B}(p, 1-p) = \Gamma(p)\Gamma(1-p), \quad [0<p<1]$$

45.3.20.18. $\int_0^\infty \frac{x^{m-1}\,dx}{1+x^n} = \frac{\pi}{n \sin\frac{m\pi}{n}}, \quad [0<m<n]$

45.3.20.19. $\int_0^\infty \frac{x^a\,dx}{(m+x^b)^c} = m\frac{a+1}{b-c}\left[\frac{\Gamma\left(\frac{a+1}{b}\right)\Gamma\left(c-\frac{a+1}{b}\right)}{\Gamma(c)}\right], \quad \left(a>-1,\ b>0,\ m>0,\ c>\frac{a+1}{b}\right)$

45.3.20.20. $\int_0^\infty \frac{dx}{(1+x)\sqrt{x}} = \pi$

45.3.20.21. $\int_0^\infty \frac{a\,dx}{a^2+x^2} = \frac{\pi}{2}$, if $a>0$; 0, if $a=0$; $-\frac{\pi}{2}$, if $a<0$

45.3.20.22. $\int_0^a (a^2-x^2)^{\frac{n}{2}}\,dx = \frac{1}{2}\int_0^a (a^2-x^2)^{\frac{n}{2}}\,dx = \frac{1\cdot 3\cdot 5\ldots n}{2\cdot 4\cdot 6\ldots(n+1)}\cdot\frac{\pi}{2}\cdot a^{n+1}, \quad (n \text{ odd})$

45.3.20.23. $$\int_0^a x^m (a^2-x^2)^{\frac{n}{2}}\,dx = \begin{cases} \frac{1}{2}a^{m+n+1} B\left(\frac{m+1}{2}, \frac{n+2}{2}\right) \\ \text{or} \\ \frac{1}{2}a^{m+n+1}\frac{\Gamma\left(\frac{m+1}{2}\right)\Gamma\left(\frac{n+2}{2}\right)}{\Gamma\left(\frac{m+n+3}{2}\right)} \end{cases}$$

45.3.20.24. $$\int_0^{\pi/2} (\sin^n x)\,dx = \begin{cases} \displaystyle\int_0^{\pi/2} (\cos^n x)\,dx \\ \text{or} \\ \dfrac{1\cdot 3\cdot 5\cdot 7\ldots(n-1)}{2\cdot 4\cdot 6\cdot 8\ldots(n)}\dfrac{\pi}{2}, & (n \text{ an even integer, } n \neq 0) \\ \text{or} \\ \dfrac{2\cdot 4\cdot 6\cdot 8\ldots(n-1)}{1\cdot 3\cdot 5\cdot 7\ldots(n)}, & (n \text{ an odd integer, } n \neq 1) \\ \text{or} \\ \dfrac{\sqrt{\pi}}{2}\dfrac{\Gamma\left(\dfrac{n+1}{2}\right)}{\Gamma\left(\dfrac{n}{2}+1\right)}, & (n > -1) \end{cases}$$

45.3.20.25. $\displaystyle\int_0^\infty \frac{\sin mx\,dx}{x} = \frac{\pi}{2}$ if $m > 0$; 0, if $m = 0$; $-\dfrac{\pi}{2}$, if $m < 0$

45.3.20.26. $\displaystyle\int_0^\infty \frac{\cos x\,dx}{x} = \infty$

45.3.20.27. $\displaystyle\int_0^\infty \frac{\tan x\,dx}{x} = \frac{\pi}{2}$

45.3.20.28. $\displaystyle\int_0^\pi \sin ax \cdot \sin bx\,dx = \int_0^\pi \cos ax \cdot \cos bx\,dx = 0, \quad (a \neq b;\ a,\ b \text{ integers})$

45.3.20.29. $\displaystyle\int_0^{\pi/a} [\sin(ax)][\cos(ax)]dx = \int_0^\pi [\sin(ax)][\cos(ax)]dx = 0$

45.3.20.30. $\displaystyle\int_0^\pi [\sin(ax)][\cos(ax)]dx = \frac{2a}{a^2 - b^2}$, if $a - b$ is odd, or zero if $a - b$ is even

45.3.20.31. $\displaystyle\int_0^\infty \frac{\sin x \cos mx\,dx}{x} = 0$, if $m < -1$ or $m > 1$, $= \dfrac{\pi}{4}$, if $m = \pm 1$; $= \dfrac{\pi}{2}$, if $m^2 < 1$

45.3.20.32. $\displaystyle\int_0^\infty \frac{\sin ax \sin bx}{x^2}dx = \frac{\pi a}{2}, \quad (a \leq b)$

45.3.20.33. $\displaystyle\int_0^\pi \sin^2 mx\,dx = \int_0^\pi \cos^2 mx\,dx = \frac{\pi}{2}$

45.3.20.34. $\displaystyle\int_0^\infty \frac{\sin^2 x\,dx}{x^2} = \frac{\pi}{2}$

45.3.20.35. $\displaystyle\int \frac{\cos mx}{1+x^2}dx = \frac{\pi}{2}e^{-|m|}$

45.3.20.36. $\displaystyle\int_0^\infty \cos(x^2)dx = \int_0^\infty \sin(x^2)dx = \frac{1}{2}\sqrt{\frac{\pi}{2}}$

45.3.20.37. $\displaystyle\int_0^\infty \frac{\sin x\,dx}{\sqrt{x}} = \int_0^\infty \frac{\cos x\,dx}{\sqrt{x}} = \sqrt{\frac{\pi}{2}}$

45.3.20.38. $\displaystyle\int_0^{\pi/2} \frac{dx}{1 + a\cos x} = \frac{\cos^{-1} a}{\sqrt{1-a^2}}, \quad (a < 1)$

45.3.20.39. $\displaystyle\int_0^\infty \frac{dx}{a + b\cos x} = \frac{\pi}{\sqrt{a^2 - b^2}}, \quad (a > b \geq 0)$

45.3.20.40. $\int_0^{2\pi} \frac{dx}{1 + a\cos x} = \frac{2\pi}{\sqrt{1-a^2}}, \quad (a^2 < 1)$

45.3.20.41. $\int_0^{\infty} \frac{\cos ax - \cos bx}{x} dx = \log \frac{b}{a}$

45.3.20.42. $\int_0^{\pi/2} \frac{dx}{a^2 \sin^2 x + b^2 \cos^2 x} = \frac{\pi}{2ab}$

45.3.20.43. $\int_0^{\pi/2} \frac{dx}{(a^2 \sin^2 x + b^2 \cos^2 x)^2} = \frac{\pi(a^2 + b^2)}{4a^3 b^3}, \quad (a, b > 0)$

45.3.20.44. $\int_0^{\pi/2} \sin^{n-1} x \cos^{m-1} x \, dx = \frac{1}{2} \mathrm{B}\left(\frac{n}{2}, \frac{m}{2}\right), \quad m \text{ and } n \text{ positive integers}$

45.3.20.45. $\int_0^{\pi/2} (\sin^{2n+1} \theta) d\theta = \frac{2 \cdot 4 \cdot 6 \ldots (2n)}{1 \cdot 3 \cdot 5 \ldots (2n+1)}, \quad (n = 1, 2, 3 \ldots)$

45.3.20.46. $\int_0^{\pi/2} (\sin^{2n} \theta) d\theta = \frac{1 \cdot 3 \cdot 5 \ldots (2n-1)}{2 \cdot 4 \ldots (2n)} \left(\frac{\pi}{2}\right), \quad (n = 1, 2, 3 \ldots)$

45.3.20.47. $\int_0^{\pi/2} \sqrt{\cos\theta} \, d\theta = \frac{(2\pi)^{\frac{3}{2}}}{[\Gamma(\frac{1}{4})]^2}$

45.3.20.48. $\int_0^{\pi/2} (\tan^h \theta) d\theta = \frac{\pi}{2\cos\left(\frac{h\pi}{2}\right)}, \quad (0 < h < 1)$

45.3.20.49. $\int_0^{\infty} \frac{\tan^{-1}(ax) - \tan^{-1}(bx)}{x} dx = \frac{\pi}{2} \log \frac{a}{b}, \quad (a, b > 0)$

45.3.20.50. The area enclosed by a curve defined through the equation $x^{\frac{b}{c}} + y^{\frac{b}{c}} = a^{\frac{b}{c}}$ where $a > 0$, c a positive odd integer and b a positive even integer is given by $\frac{\left[\Gamma\left(\frac{c}{b}\right)\right]^2}{\Gamma\left(\frac{2c}{b}\right)} \left(\frac{2ca^2}{b}\right)$

45.3.20.51. $I = \iiint\limits_R x^{h-1} y^{m-1} z^{n-1} \, dv$, where R denotes the region of space bounded by the co-ordinate planes and that portion of the surface $\left(\frac{x}{a}\right)^p + \left(\frac{y}{b}\right)^q + \left(\frac{z}{c}\right)^k = 1$, which lies in the first octant and where $h, m, n, p, q, k, a, b, c,$ denote positive real numbers is given by

$$\int_0^a x^{h-1} dx \int_0^{b\left[1-\left(\frac{x}{a}\right)^p\right]^{\frac{1}{q}}} y^m dy \int_0^{c\left[1-\left(\frac{x}{a}\right)^p-\left(\frac{y}{b}\right)^q\right]^{\frac{1}{k}}} z^{n-1} dz = \frac{a^h b^m c^n}{pqk} \frac{\Gamma\left(\frac{h}{p}\right)\Gamma\left(\frac{m}{q}\right)\Gamma\left(\frac{n}{k}\right)}{\Gamma\left(\frac{h}{p}+\frac{m}{q}+\frac{n}{k}+1\right)}$$

45.3.20.52. $\int_0^{\infty} e^{-ax} dx = \frac{1}{a}, \quad (a > 0)$

45.3.20.53. $\int_0^{\infty} \frac{e^{-ax} - e^{-bx}}{x} dx = \log \frac{b}{a}, \quad (a, b > 0)$

45.3.20.54. $\int_0^{\infty} x^n e^{-ax}\,dx = \frac{\Gamma(n+1)}{a^{n+1}}, \quad (n > -1,\ a > 0)$

$= \frac{n!}{a^{n+1}}, \quad (n \text{ pos. integ.},\ a > 0)$

45.3.20.55. $\int_0^{\infty} e^{-a^2x^2}\,dx = \frac{1}{2a}\sqrt{\pi} = \frac{1}{2a}\Gamma\left(\frac{1}{2}\right), \quad (a > 0)$

45.3.20.56. $\int_0^{\infty} xe^{-x^2}\,dx = \frac{1}{2}$

45.3.20.57. $\int_0^{\infty} x^2e^{-x^2}\,dx = \frac{\sqrt{\pi}}{4}$

45.3.20.58. $\int_0^{\infty} x^{2n}e^{-ax^2}\,dx = \frac{1\cdot 3\cdot 5\ldots(2n-1)}{2^{n+1}a^n}\sqrt{\frac{\pi}{a}}$

45.3.20.59. $\int_0^{1} x^m e^{-ax}\,dx = \frac{m!}{a^{m+1}}\left[1 - e^{-a}\sum_{r=0}^{m}\frac{a^r}{r!}\right]$

45.3.20.60. $\int_0^{\infty} e^{\left(-x^2 - \frac{a^2}{x^2}\right)}\,dx = \frac{e^{-2a}\sqrt{\pi}}{2}, \quad (a \geq 0)$

45.3.20.61. $\int_0^{\infty} e^{-nx}\sqrt{x}\,dx = \frac{1}{2n}\sqrt{\frac{\pi}{n}}$

45.3.20.62. $\int_0^{\infty} \frac{e^{-nx}}{\sqrt{x}}\,dx = \sqrt{\frac{\pi}{n}}$

45.3.20.63. $\int_0^{\infty} e^{-ax}\cos mx\,dx = \frac{a}{a^2+m^2}, \quad (a > 0)$

45.3.20.64. $\int_0^{\infty} e^{-ax}\sin mx\,dx = \frac{m}{a^2+m^2}, \quad (a > 0)$

45.3.20.65. $\int_0^{\infty} xe^{-ax}[\sin(bx)]dx = \frac{2ab}{(a^2+b^2)^2}, \quad (a > 0)$

45.3.20.66. $\int_0^{\infty} xe^{-ax}[\cos(bx)]dx = \frac{a^2-b^2}{(a^2+b^2)^2}, \quad (a > 0)$

45.3.20.67. $\int_0^{\infty} x^n e^{-ax}[\sin(bx)]dx = \frac{n![(a-ib)^{n+1} - (a+ib)^{n+1}]}{2(a^2+b^2)^{n+1}}, \quad (i^2 = -1,\ a \geq 0)$

45.3.20.68. $\int_0^{\infty} x^n e^{-ax}[\cos(bx)]dx = \frac{n![(a-ib)^{n+1} + (a+ib)^{n+1}]}{2(a^2+b^2)^{n+1}}, \quad (i^2 = -1,\ a \geq 0)$

45.3.20.69. $\int_0^{\infty} \frac{e^{-ax}\sin x}{x}\,dx = \cot^{-1} a, \quad (a \geq 0)$

45.3.20.70. $\int_0^{\infty} e^{-a^2x^2}\cos bx\,dx = \frac{\sqrt{\pi}}{2a}e^{\frac{-b^2}{4a^2}}, \quad (ab \neq 0)$

45.3.20.71. $\int_0^{\infty} e^{-t\cos\phi}t^{b-1}[\sin(t\sin\phi)]dt = [\Gamma(b)]\sin(b\phi), \quad \left(b > 0,\ -\frac{\pi}{2} < \phi < \frac{\pi}{2}\right)$

45.3.20.72. $\int_0^{\infty} e^{-t\cos\phi}t^{b-1}[\cos(t\sin\phi)]dt = [\Gamma(b)]\cos(b\phi), \quad \left(b > 0,\ -\frac{\pi}{2} < \phi < \frac{\pi}{2}\right)$

45.3.20.73. $\int_0^\infty t^{b-1}\cos t\,dt = [\Gamma(b)]\cos\left(\frac{b\pi}{2}\right), \quad (0<b<1)$

45.3.20.74. $\int_0^\infty t^{b-1}(\sin t)\,dt = [\Gamma(b)]\sin\left(\frac{b\pi}{2}\right), \quad (0<b<1)$

45.3.20.75. $\int_0^1 (\log x)^n\,dx = (-1)^n \cdot n!$

45.3.20.76. $\int_0^1 \left(\log\frac{1}{x}\right)^{\frac{1}{2}} dx = \frac{\sqrt{\pi}}{2}$

45.3.20.77. $\int_0^1 \left(\log\frac{1}{x}\right)^{-\frac{1}{2}} dx = \sqrt{\pi}$

45.3.20.78. $\int_0^1 \left(\log\frac{1}{x}\right)^{n} dx = n!$

45.3.20.79. $\int_0^1 x\log(1-x)\,dx = -\frac{3}{4}$

45.3.20.80. $\int_0^1 x\log(1+x)\,dx = \frac{1}{4}$

45.3.20.81. $\int_0^1 \frac{\log x}{1+x}\,dx = -\frac{\pi^2}{12}$

45.3.20.82. $\int_0^1 \frac{\log x}{1-x}\,dx = -\frac{\pi^2}{6}$

45.3.20.83. $\int_0^1 \frac{\log x}{1-x^2}\,dx = -\frac{\pi^2}{8}$

45.3.20.84. $\int_0^1 \log\left(\frac{1+x}{1-x}\right)\cdot\frac{dx}{x} = \frac{\pi^2}{4}$

45.3.20.85. $\int_0^1 \frac{\log x\,dx}{\sqrt{1-x^2}} = -\frac{\pi}{2}\log 2$

45.3.20.86. $\int_0^1 x^m\left[\log\left(\frac{1}{x}\right)\right]^n dx = \frac{\Gamma(n+1)}{(m+1)^{n+1}}, \text{ if } m+1>0, n+1>0$

45.3.20.87. $\int_0^1 \frac{(x^p - x^q)\,dx}{\log x} = \log\left(\frac{p+1}{q+1}\right), \quad (p+1>0, q+1>0)$

45.3.20.88. $\int_0^1 \frac{dx}{\sqrt{\log\left(\frac{1}{x}\right)}} = \sqrt{\pi}$

45.3.20.89. $\int_0^\infty \log\left(\frac{e^x+1}{e^x-1}\right)dx = \frac{\pi^2}{4}$

45.3.20.90. $\int_0^{\pi/2} \log\sin x\,dx = \int_0^{\pi/2} \log\cos x\,dx = -\frac{\pi}{2}\log 2$

45.3.20.91. $\int_0^{\pi/2} \log\sec x\,dx = \int_0^{\pi/2} \log\csc x\,dx = \frac{\pi}{2}\log 2$

45.3.20.92. $\int_0^{\pi} x \log \sin x \, dx = -\frac{\pi^2}{2} \log 2$

45.3.20.93. $\int_0^{\pi/2} \sin x \log \sin x \, dx = \log 2 - 1$

45.3.20.94. $\int_0^{\pi/2} \log \tan x \, dx = 0$

45.3.20.95. $\int_0^{\pi} \log(a \pm b \cos x) \, dx = \pi \log\left(\frac{a + \sqrt{a^2 - b^2}}{2}\right), \quad (a \geqq b)$

45.3.20.96. $\int_0^{\infty} \frac{dx}{\cosh ax} = \frac{\pi}{2a}$

45.3.20.97. $\int_0^{\infty} \frac{x \, dx}{\sinh ax} = \frac{\pi^2}{4a^2}$

45.3.20.98. $\int_0^{\infty} e^{-ax} \cosh bx \, dx = \frac{a}{a^2 - b^2}, \quad (0 \leq |b| < a)$

45.3.20.99. $\int_0^{\infty} e^{-ax} \sinh bx \, dx = \frac{b}{a^2 - b^2}, \quad (0 \leq |b| < a)$

45.3.20.100. $\int_{+\infty}^{1} \frac{e^{-xu}}{u} du = \gamma + \log x - x + \frac{x^2}{2 \cdot 2!} - \frac{x^3}{3 \cdot 3!} + \frac{x^4}{4 \cdot 4!} - \ldots,$

where $\gamma = \lim_{z \to \infty}\left(1 + \frac{1}{2} + \frac{1}{3} + \cdots + \frac{1}{z} - \log z\right)$

$= 0.5772157\ldots, \quad (0 < x < \infty)$

45.3.20.101. $\int_0^{\pi/2} \frac{dx}{\sqrt{1 - k^2 \sin^2 x}} = \frac{\pi}{2}\left[1 + \left(\frac{1}{2}\right)^2 k^2 + \left(\frac{1 \cdot 3}{2 \cdot 4}\right)^2 k^4 + \left(\frac{1 \cdot 3 \cdot 5}{2 \cdot 4 \cdot 6}\right)^2 k^6 + \cdots\right], \text{ if } k^2 < 1$

45.3.20.102. $\int_0^{\pi/2} \sqrt{1 - k^2 \sin^2 x} \, dx = \frac{\pi}{2}\left[1 - \left(\frac{1}{2}\right)^2 k^2 - \left(\frac{1 \cdot 3}{2 \cdot 4}\right)^2 \frac{k^4}{3} - \left(\frac{1 \cdot 3 \cdot 5}{2 \cdot 4 \cdot 6}\right)^2 \frac{k^6}{5} + \cdots\right], \text{ if } k^2 < 1$

45.3.20.103. $\int_0^{\infty} e^{-x} \log x \, dx = -\gamma = 0.5772157\ldots$

45.3.20.104. $\int_0^{\infty} \left(\frac{1}{1 - e^{-x}} - \frac{1}{x}\right) e^{-x} \, dx = \gamma = 0.5772157\ldots$ [Euler's Constant]

45.3.20.105. $\int_0^{\infty} \frac{1}{x}\left(\frac{1}{1 + x} - e^{-x}\right) dx = \gamma = 0.5772157\ldots$

Index

K

L

M

N

O

P

R

S

T

U

V

W

Z